THE MECHANICS OF CONSTITUTIVE MODELING

THE MECHANICS OF CONSTITUTIVE MODELING

Niels Saabye Ottosen

Matti Ristinmaa

Division of Solid Mechanics,
Lund University, Sweden

2005

ELSEVIER

AMSTERDAM • BOSTON • HEIDELBERG • LONDON • NEW YORK • OXFORD
PARIS • SAN DIEGO • SINGAPORE • SYDNEY • TOKYO

ELSEVIER B.V.
Radarweg 29
P.O. Box 211, 1000
AE Amsterdam
The Netherlands

ELSEVIER Inc.
525 B Street, Suite 1900
San Diego, CA 92101-4495
USA

ELSEVIER Ltd
The Boulevard, Langford
Kidlington, Oxford OX5 1GB
UK

ELSEVIER Ltd
84 Theobalds Road
London WC1X 8RR
UK

First edition 2005

ISBN: 0-008-044606-X

Printed in Great Britain.

CONTENTS

PREFACE

This book is the fruit of several years of courses for graduate students, Ph.D. students and researchers. The intention of the book is to provide a thorough exposition of elasto-plasticity, viscoplasticity, creep and viscoelasticity regarding both their constitutive modeling and their computational treatment.

Any text is subject to limitations, and we have chosen to assume the existence of small displacement gradients. Since research activities within constitutive mechanics are highly intensive, a huge number of constitutive models and corresponding numerical strategies can be found in the literature. It is not our intention to try to cover as many of these models as possible. Instead, the exposition is geared to a presentation of various important concepts. Even so, a number of specific models must still be addressed, however, each of these formulations will involve fundamental features not covered by the other models. For every topic considered, the exposition is self-contained, with an extensive list of references where the reader can find further details and ramifications.

For constitutive modeling, our viewpoint will be purely phenomenological with the focus on the macroscopical material response. Although major emphasis will be placed on issues within plasticity theory, it will turn out that advantage can be taken of a number of these concepts when dealing with viscoplasticity and creep.

Approximately, the first half of the book deals with quite classical concepts within constitutive modeling. Experimental evidence for various phenomena are used as a basis for construction of constitutive theories. Against this background, the second half of the book treats the thermodynamic framework for constitutive modeling and the computational treatment of plasticity, viscoplasticity, creep and viscoelasticity.

The book starts with topics of a relatively elementary nature; gradually, the topics and style become more abstract and the later part of the book deals with issues of current research interest. Whereas many other books focus on a specific group of materials, for instance, steel or concrete, we try to be as general as possible and, thus, the exposition should be of interest for a broad engineering audience.

For graduate students, we have successfully created courses comprising most of the first half of the book together with some of the computational topics. The second half of the book has turned out to be useful for Ph.D. courses and other research programs.

In the running battle against misprints, we have had notable help from our colleague tech. Dr. Mathias Wallin and our Ph.D. students, tech. Lic. Magnus Fredriksson, M. Sc. Anders Harrysson, M. Sc. Magnus Harrysson, M. Sc. Håkan Hallberg and tech. Lic. Paul Håkansson; a special tribute to Mag. art. Kerstin Saabye Wullt for her judicious linguistic support and for her enthusiasm.

Lund University, Sweden, March 2005
Niels Saabye Ottosen and Matti Ristinmaa

1 NOTATIONS AND CARTESIAN TENSORS

In this chapter we will present the set of notations that will be used and we will find it convenient to use both *matrix* and *tensor notation* depending on the particular application. As *index notation* is an integrated part of tensor algebra, the advantage of this notation will be illustrated and we will then provide an elementary discussion of the concept of tensors and why they appear in a naturally manner when formulating physical relations.

1.1 Matrix notation

In general, a matrix consists of a collection of certain quantities which are termed the *components* of the matrix. The components are ordered in rows and columns and if the number of rows or columns is equal to one, the matrix is *one-dimensional*; otherwise it is *two-dimensional*. A treatment of matrix algebra can be found in many textbooks, for instance Ayres (1962), Eves (1980) or Strang (1980). The intention here is not to provide a resume of matrix algebra, but simply to present sufficient information of the notation used.

A column matrix is denoted by a bold-face, usually lower-case letter, for instance

$$\boldsymbol{a} = \begin{bmatrix} a_1 \\ a_2 \\ a_3 \end{bmatrix} \quad ; \quad \boldsymbol{b} = \begin{bmatrix} b_1 \\ b_2 \\ b_3 \end{bmatrix} \quad ; \quad \boldsymbol{c} = \begin{bmatrix} c_1 \\ c_2 \\ c_3 \\ c_4 \end{bmatrix} \tag{1.1}$$

where c_1, c_2, c_3 and c_4 are the components of the matrix $\boldsymbol{c}$. The *dimension* of a matrix is given by the number of rows and columns, i.e. the *column matrix* $\boldsymbol{c}$ of (1.1) has the dimension 4x1. The *transpose* $\boldsymbol{a}^T$ of $\boldsymbol{a}$ is given by the *row matrix*

$$\boldsymbol{a}^T = [a_1 \;\; a_2 \;\; a_3]$$

The *length* of $\boldsymbol{a}$ or $\boldsymbol{a}^T$ is denoted by $|\boldsymbol{a}|$ and we have

$$|\boldsymbol{a}| = (a_1^2 + a_2^2 + a_3^2)^{1/2}$$

The *scalar product* of two column matrices $\boldsymbol{a}$ and $\boldsymbol{b}$ having the same dimensions is defined according to

$$\boldsymbol{a}^T\boldsymbol{b} = \boldsymbol{b}^T\boldsymbol{a} = [b_1 \quad b_2 \quad b_3] \begin{bmatrix} a_1 \\ a_2 \\ a_3 \end{bmatrix} = b_1a_1 + b_2a_2 + b_3a_3 \tag{1.2}$$

where $\boldsymbol{a}$ and $\boldsymbol{b}$ in the present case are given by (1.1). Therefore, the length $|\boldsymbol{a}|$ of $\boldsymbol{a}$ can be written as

$$|\boldsymbol{a}| = (\boldsymbol{a}^T\boldsymbol{a})^{1/2}$$

A two-dimensional matrix is denoted by a bold-face, usually an upper-case letter, for instance

$$\boldsymbol{B} = \begin{bmatrix} B_{11} & B_{12} & B_{13} \\ B_{21} & B_{22} & B_{23} \\ B_{31} & B_{32} & B_{33} \end{bmatrix} \quad ; \quad \boldsymbol{C} = \begin{bmatrix} C_{11} & C_{12} \\ C_{21} & C_{22} \\ C_{31} & C_{32} \end{bmatrix} \tag{1.3}$$

where $\boldsymbol{B}$ is termed a *square matrix* since the number of rows and columns is equal. The transpose $\boldsymbol{B}^T$ of $\boldsymbol{B}$ is obtained by interchanging rows and columns in $\boldsymbol{B}$, i.e.

$$\boldsymbol{B}^T = \begin{bmatrix} B_{11} & B_{21} & B_{31} \\ B_{12} & B_{22} & B_{32} \\ B_{13} & B_{23} & B_{33} \end{bmatrix}$$

and the matrix $\boldsymbol{B}$ is *symmetric* if $\boldsymbol{B} = \boldsymbol{B}^T$. The unit matrix $\boldsymbol{I}$ is defined

$$\boldsymbol{I} = \begin{bmatrix} 1 & 0 & 0 \\ 0 & 1 & 0 \\ 0 & 0 & 1 \end{bmatrix} \tag{1.4}$$

A *zero matrix* is defined as a matrix where all components are zero. Examples are

$$\boldsymbol{0} = \begin{bmatrix} 0 & 0 & 0 \\ 0 & 0 & 0 \end{bmatrix} \quad ; \quad \boldsymbol{0} = \begin{bmatrix} 0 \\ 0 \end{bmatrix}$$

We note that the *inverse* $\boldsymbol{B}^{-1}$ of a square matrix $\boldsymbol{B}$ is defined by

$$\boldsymbol{B}^{-1}\boldsymbol{B} = \boldsymbol{B}\boldsymbol{B}^{-1} = \boldsymbol{I} \tag{1.5}$$

and that $\boldsymbol{B}^{-1}$ exists if the *determinant* $\det \boldsymbol{B}$ of $\boldsymbol{B}$ is different from zero. If $\det \boldsymbol{B} \neq 0$, then $\boldsymbol{B}$ is *nonsingular*; otherwise it is *singular*. For matrices having the correct dimension the *matrix product* $\boldsymbol{AB}$ exists and we recall that

$$(\boldsymbol{AB})^T = \boldsymbol{B}^T\boldsymbol{A}^T \quad ; \quad (\boldsymbol{AB})^{-1} = \boldsymbol{B}^{-1}\boldsymbol{A}^{-1}$$

and for two square matrices we have

$$\det(\boldsymbol{AB}) = \det \boldsymbol{A} \ \det \boldsymbol{B} \tag{1.6}$$

For a square matrix $\boldsymbol{A}$, consider the quantity $\boldsymbol{x}^T\boldsymbol{A}\boldsymbol{x}$, which is a number; this quantity is called a *quadratic form.* If

$$\boxed{\boldsymbol{x}^T\boldsymbol{A}\boldsymbol{x} > 0 \qquad \textit{for all } \boldsymbol{x} \neq 0} \tag{1.7}$$

then the matrix $\boldsymbol{A}$ is said to be *positive definite*. It is recalled that

$$\boxed{\textit{If } \boldsymbol{A} \textit{ is positive definite then } \det \boldsymbol{A} \neq 0}$$

We also mention that a matrix $\boldsymbol{A}$ is called *positive semi-definite* if

$$\boxed{\boldsymbol{x}^T\boldsymbol{A}\boldsymbol{x} \geq 0 \qquad \textit{for all } \boldsymbol{x} \neq 0}$$

1.2 Cartesian coordinate system

Whenever a coordinate system is employed in the following, we will for simplicity only make use of the standard *orthogonal, rectangular and right-handed coordinate system* shown in Fig. 1.1. The word rectangular signifies that the coordinate axes are straight orthogonal lines. For reasons that will be unfolded in a moment we label the coordinate axes by x_1, x_2 and x_3 instead of the usual notation of x, y and z.

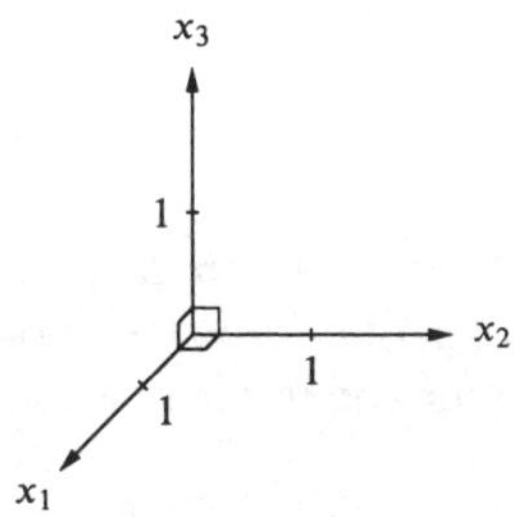

Figure 1.1: Cartesian coordinate system.

In order to maintain the standard definition of distance between two points in this coordinate system, the unit length along all the coordinate axes is equal to the unit length scale. Such a coordinate system is termed a *Cartesian coordinate system* in recognition of the French philosopher and mathematician Descartes (1596-1650), whose Latin name is Cartesius and who introduced the concept of a coordinate system. It is obvious that a certain set of coordinates, i.e. a certain set of x_1-, x_2- and x_3-values defines uniquely the position of a point in the coordinate system.

1.3 Index notation

Index notation is often used in tensor algebra and it is therefore often termed as *tensor notation*. Index notation implies that complicated expressions can be written in a very compact fashion that emphasizes the physical content of these expressions and greatly facilitates mathematical manipulations.

The coordinate axes x_1, x_2 and x_3 in Fig. 1.1 can be written more briefly as x_i, where the index i takes the values $i = 1, 2$ and 3. The column matrix $\boldsymbol{a}$ given by (1.1) can then be written as $[a_i]$ where the brackets [ ] around a_i emphasize that we in the present case interpret the quantity a_i as a matrix. Therefore

$$\boldsymbol{a} = [a_i] = \begin{bmatrix} a_1 \\ a_2 \\ a_3 \end{bmatrix}$$

where, again, the index i takes the values 1, 2 and 3. In what follows, Latin indices, unless otherwise specified, assume the values 1, 2 and 3; on the other hand, Greek indices will extend over a range to be specified in each case. If reference is made to a_i we refer to the entire quantity given by a_i whereas a specific component of a_i like the one given by, for instance, $i = 2$ is referred to as a_2.

An important convention in index notation is the so-called *summation convention*, which states that if an index is repeated twice then a summation over this index is implied. As an example, the product $b_i a_i$, where the index i is repeated twice, means

$$b_i a_i = b_1 a_1 + b_2 a_2 + b_3 a_3$$

and a comparison with (1.2) shows that $\boldsymbol{b}^T \boldsymbol{a} = b_i a_i$. It is also a convention in index notation that an index cannot be repeated more than twice. If it is repeated twice, it is called a *dummy index* and if it is not repeated, it is called a *free index*, i.e

$$\textit{index}= \begin{cases} \textit{free if it appears once} \\ \textit{dummy if it appears twice} \end{cases}$$

An index can only be free or dummy

It is obvious that the specific letter used for a dummy index is immaterial and we have, for instance, $b_i a_i = b_k a_k$. However, for a free index the specific letter used is of extreme importance. It should also be noted that whereas the position of a quantity in a matrix expression is of significance - we have for example $\boldsymbol{b}^T \boldsymbol{a} \neq \boldsymbol{a}\boldsymbol{b}^T$ - this is not the case in index notation where, for instance, $b_i a_i = a_i b_i = \boldsymbol{a}^T \boldsymbol{b} = \boldsymbol{b}^T \boldsymbol{a}$.

It is also possible to work with quantities having two indices and it is evident that the matrix $\boldsymbol{B}$ given by (1.3) can be written as

$$\boldsymbol{B} = [B_{ij}] = \begin{bmatrix} B_{11} & B_{12} & B_{13} \\ B_{21} & B_{22} & B_{23} \\ B_{31} & B_{32} & B_{33} \end{bmatrix}$$

where the brackets [ ] around B_{ij} again emphasize that in the present case we interpret the quantity B_{ij} as a matrix.

Using the summation convention, it follows that the inhomogeneous equation system $\boldsymbol{Bx} = \boldsymbol{a}$ can be written as $B_{ij}x_j = a_i$ and that

$$B_{ii} = B_{11} + B_{22} + B_{33}$$

From the rules defined, it follows that each term in an expression must possess the same number of free indices, i.e. whereas $B_{ij}x_j = a_i$ is a valid expression, the formulations $B_{ij}x_j = C$ and $B_{ij}x_j = A_{ij}$ are invalid. The operation, where two free indices are made equal to each other, so that a dummy index arises, is called *contraction*. As an example, contraction of A_{ij} gives A_{ii}.

The *Kronecker delta* δ_{ij} plays an essential role in index notation and tensor algebra and it is defined as

$$\boxed{\delta_{ij} = \begin{cases} 1 & \text{if } i = j \\ 0 & \text{if } i \neq j \end{cases}} \tag{1.8}$$

i.e. it is equal to the unit matrix $\boldsymbol{I}$ given by (1.4). Using the summation convention it follows that

$$B_{ij}\delta_{jk} = B_{ik} \tag{1.9}$$

This result follows from the fact that δ_{ij} only contributes with the value of unity provided that j and k take the same value. Alternatively, the trivial use of the summation convention yields

$$B_{ij}\delta_{jk} = B_{i1}\delta_{1k} + B_{i2}\delta_{2k} + B_{i3}\delta_{3k}$$

and an evaluation of this relation for each i- and k-value results in expression (1.9). Another example of the use of Kronecker's delta arises from the matrix equation $\boldsymbol{AB} = \boldsymbol{I}$. In index notation this is written as

$$A_{ik}B_{kj} = \delta_{ij}$$

which shows that A_{ik} is the inverse of B_{ik}, cf. (1.5). A final important illustration of the use of Kronecker's delta is the expression

$$\frac{\partial a_i}{\partial a_j} = \delta_{ij}$$

This identity follows from the fact that $\partial a_i/\partial a_j$ is zero if $i \neq j$ and unity if i and j are equal.

In accordance with the matrix notation, it follows that the quantity M_{ij} is *symmetric* if

$$M_{ij} = M_{ji}$$

Moreover, a quantity N_{ij} is termed *anti-symmetric* - or *skew-symmetric* - if

$$N_{ij} = -N_{ji}$$

holds. This implies that all diagonal terms in N_{ij} are equal to zero.

Suppose we have an arbitrary quantity P_{ij}. It is always possible to write P_{ij} according to

$$\boxed{P_{ij} = P^s_{ij} + P^a_{ij}} \tag{1.10}$$

where the symmetric part P^s_{ij} of P_{ij} is defined by

$$\boxed{P^s_{ij} = \frac{1}{2}(P_{ij} + P_{ji})} \tag{1.11}$$

and the anti-symmetric part P^a_{ij} of P_{ij} is defined by

$$\boxed{P^a_{ij} = \frac{1}{2}(P_{ij} - P_{ji})} \tag{1.12}$$

It is easy to prove that this decomposition of P_{ij} is unique. To show this, assume that

$$P_{ij} = \tilde{P}^s_{ij} + \tilde{P}^a_{ij} \tag{1.13}$$

where $\tilde{P}^s_{ij}$ is symmetric and $\tilde{P}^a_{ij}$ is anti-symmetric and possibly different from P^s_{ij} and P^a_{ij} respectively. Interchanging i and j in (1.13) and using that $\tilde{P}^s_{ij}$ is symmetric and $\tilde{P}^a_{ij}$ is anti-symmetric, we obtain

$$P_{ji} = \tilde{P}^s_{ij} - \tilde{P}^a_{ij} \tag{1.14}$$

Addition and subtraction of (1.13) and (1.14) give

$$\tilde{P}^s_{ij} = \frac{1}{2}(P_{ij} + P_{ji})\,; \qquad \tilde{P}^a_{ij} = \frac{1}{2}(P_{ij} - P_{ji})$$

A comparison with (1.11) and (1.12) shows that $\tilde{P}^s_{ij} = P^s_{ij}$ and $\tilde{P}^a_{ij} = P^a_{ij}$, i.e. the decomposition (1.10) of P_{ij} into a symmetric and a anti-symmetric part is unique.

A problem often encountered is the multiplication of a symmetric quantity A^s_{ij} with a quantity B_{ij} not necessarily symmetric. It turns out that

$$\boxed{A^s_{ij}B_{ij} = A^s_{ij}B^s_{ij}}$$

This result follows if we can prove that $A^s_{ij}B^a_{ij} = 0$. To show this we have

$$\begin{aligned} A^s_{ij}B^a_{ij} &= A^s_{ji}B^a_{ji} && \text{(redefinition of dummy indices)} \\ &= A^s_{ij}B^a_{ji} && \text{(symmetry of } A^s_{ij}) \\ &= -A^s_{ij}B^a_{ij} && \text{(anti-symmetry of } B^a_{ij}) \end{aligned}$$

which implies that $A^s_{ij}B^a_{ij} = 0$.

A so-called *comma convention* is also used in index notation. It states that whenever a quantity is differentiated with respect to the coordinates x_i, we use a comma to indicate this differentiation. Examples are

$$\boxed{\frac{\partial f}{\partial x_i} = f_{,i}\,; \qquad \frac{\partial a_i}{\partial x_j} = a_{i,j}}$$

We finally observe that in matrix notation we are restricted to working with one- and two-dimensional arrays. This is not the case in index notation where, for instance, the quantity e_{ijk} exists and comprises $3 \times 3 \times 3 = 27$ components. Likewise, the quantity D_{ijkl} exists and comprises $3 \times 3 \times 3 \times 3 = 81$ components.

1.4 Change of coordinate system

Before we present the concept of a tensor, we may first note that the essential issue of a tensor is that it behaves in a certain manner when a coordinate change is performed. Let us therefore first discuss coordinate changes between Cartesian coordinate systems.

Such a coordinate change can only occur in form of a *translation* and/or a *rotation* of the original coordinate system. Letting the old coordinates be denoted by x_i and the new coordinates by x'_i, Fig. 1.2 illustrates possible coordinate changes.

Consider first a translation of the coordinate system, Fig. 1.2a). It is obvious that we have

$$x_i = x'_i + c_i$$

i.e.

$$x'_i = x_i - c_i \qquad \text{or} \qquad \boldsymbol{x}' = \boldsymbol{x} - \boldsymbol{c} \tag{1.15}$$

where c_i is constant and contains information of the translation of the old origin to the new origin. To be specific, c_i is the coordinates of the new origin measured

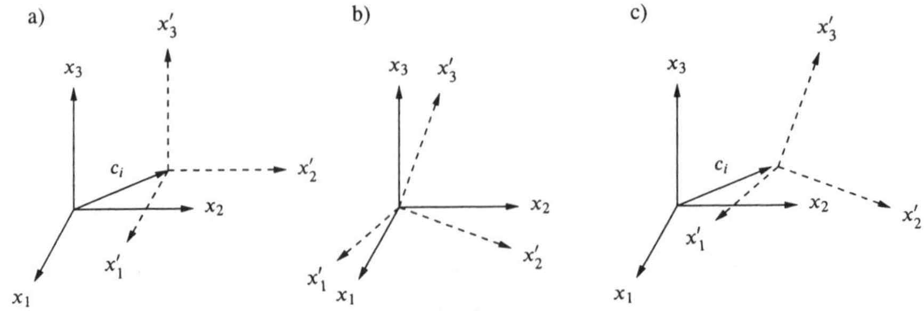

Figure 1.2: Coordinate changes; a) translation; b) rotation; c) translation and rotation.

in the old x_i-coordinate system. That is, if the coordinates x_i to a given fixed point are known in the old coordinate system, (1.15) provides the corresponding coordinates to that point in the new coordinate system.

Consider next a rotation of the coordinate system, Fig. 1.2b). In this case we expect x'_1 to be a linear function of x_1, x_2 and x_3, i.e.

$$x'_1 = A_{11}x_1 + A_{12}x_2 + A_{13}x_3$$

where A_{11}, A_{12} and A_{13} are certain constants. Likewise, for x'_2 and x'_3 we expect that

$$x'_2 = A_{21}x_1 + A_{22}x_2 + A_{23}x_3$$
$$x'_3 = A_{31}x_1 + A_{32}x_2 + A_{33}x_3$$

Using index and matrix notation these expressions can be combined into

$$x'_i = A_{ij}x_j \qquad \text{or} \qquad \boldsymbol{x}' = \boldsymbol{A}\boldsymbol{x} \tag{1.16}$$

where the *transformation matrix* $\boldsymbol{A}$ for a given rotation is constant, i.e. independent of the coordinates.

From (1.15) and (1.16) it follows that the most general coordinate transformation, which comprises a translation and a rotation, can be written as

$$\boxed{x'_i = A_{ij}(x_j - c_j) \qquad or \qquad \boldsymbol{x}' = \boldsymbol{A}(\boldsymbol{x} - \boldsymbol{c})} \tag{1.17}$$

If $A_{ij} = \delta_{ij}$, this expression corresponds to a translation only and if $c_i = 0$ we only have a rotation of the coordinate system; finally, if $x_j = c_j$ then $x'_i = 0$ as expected. Since A_{ij} and c_i are constant quantities, it follows from (1.17) that

$$dx'_i = A_{ij}dx_j \tag{1.18}$$

The transformation matrix A_{ij} turns out to have a remarkable property. To see this, consider a point having the coordinates x_j and a neighboring point having the coordinates $x_j + dx_j$. According to Pythagoras' theorem, the distance ds between these two points is given by

$$ds^2 = (x_j + dx_j - x_j)(x_j + dx_j - x_j)$$

i.e.

$$ds^2 = dx_j dx_j = dx_j dx_k \delta_{kj} \tag{1.19}$$

The distance between two fixed points must be the same irrespective of the Cartesian coordinate system used. Therefore, in another coordinate system we have

$$ds^2 = dx'_i dx'_i = A_{ij} dx_j A_{ik} dx_k \tag{1.20}$$

where (1.18) has been used. Subtraction of (1.19) and (1.20) results in

$$(\delta_{jk} - A_{ij} A_{ik}) dx_j dx_k = 0$$

As this expression holds for arbitrary dx_j-values, we conclude that $A_{ij} A_{ik} = \delta_{jk}$, i.e.

$$\boxed{A_{ki} A_{kj} = \delta_{ij}} \tag{1.21}$$

In matrix notation, (1.21) reads

$$\boldsymbol{A}^T \boldsymbol{A} = \boldsymbol{I} \tag{1.22}$$

From (1.22) and (1.5), we conclude that the transformation matrix $\boldsymbol{A}$ is *orthogonal*, i.e.

$$\boxed{\boldsymbol{A}^T = \boldsymbol{A}^{-1}} \tag{1.23}$$

From (1.23) follows that

$$\boldsymbol{A}\boldsymbol{A}^T = \boldsymbol{I} \tag{1.24}$$

which in index notation takes the form

$$\boxed{A_{ik} A_{jk} = \delta_{ij}} \tag{1.25}$$

The similarity in structure when compared with (1.21) should be noticed. However, we notice the different positions of the dummy index in (1.21) and (1.25) and we emphasize that the transformation matrix A_{ij} is, in general, unsymmetric.

Since $\det \boldsymbol{A} = \det \boldsymbol{A}^T$, we find from (1.24) and (1.6) that

$$\det(\boldsymbol{A}\boldsymbol{A}^T) = \det \boldsymbol{A} \cdot \det \boldsymbol{A}^T = (\det \boldsymbol{A})^2 = \det \boldsymbol{I} = 1$$

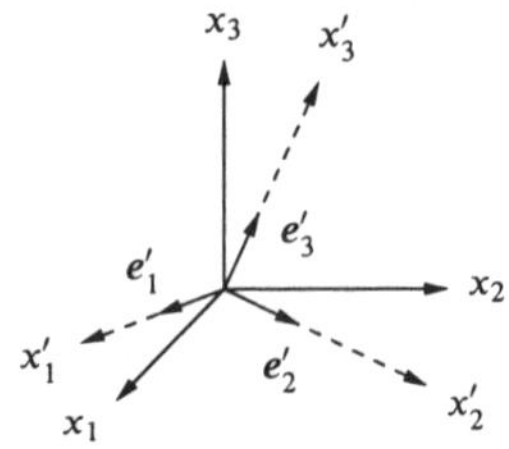

Figure 1.3: Rotation of coordinate system.

i.e. $\det \boldsymbol{A} = \pm 1$. For no rotation of the coordinate system, we have $\boldsymbol{A} = \boldsymbol{I}$, i.e. $\det \boldsymbol{A} = 1$ and of continuity reasons, we conclude that

$$\boxed{\det \boldsymbol{A} = 1} \tag{1.26}$$

holds for all right-handed coordinate systems.

Let us see, how in practice we may determine the components of the transformation matrix. For this purpose, consider the rotation of the coordinate system as shown in Fig. 1.3. Let $\boldsymbol{e}'_1$, $\boldsymbol{e}'_2$ and $\boldsymbol{e}'_3$ denote unit vectors along the x'_1-, x'_2- and x'_3-axis respectively. The components of these unit vectors in the old x_i-system are given by

$$\boldsymbol{e}'_1 = \begin{bmatrix} e_{11} \\ e_{12} \\ e_{13} \end{bmatrix} \quad ; \quad \boldsymbol{e}'_2 = \begin{bmatrix} e_{21} \\ e_{22} \\ e_{23} \end{bmatrix} \quad ; \quad \boldsymbol{e}'_3 = \begin{bmatrix} e_{31} \\ e_{32} \\ e_{33} \end{bmatrix}$$

where the first index refers to the axis in the x'_i-system and the second index to the component measured in the x_i-system. According to (1.17), we have

$$\boldsymbol{x} = \boldsymbol{A}^T \boldsymbol{x}' \tag{1.27}$$

It turns out that

$$\boxed{\boldsymbol{A}^T = \left[\ \boldsymbol{e}'_1, \boldsymbol{e}'_2, \boldsymbol{e}'_3\ \right] = \begin{bmatrix} e_{11} & e_{21} & e_{31} \\ e_{12} & e_{22} & e_{32} \\ e_{13} & e_{23} & e_{33} \end{bmatrix}} \tag{1.28}$$

To check this expression for $\boldsymbol{A}^T$, let us recall our definitions of $\boldsymbol{e}'_1$, $\boldsymbol{e}'_2$ and $\boldsymbol{e}'_3$. Then, setting $\boldsymbol{x}' = [1\ \ 0\ \ 0]^T$ in (1.27) gives with (1.28) that $\boldsymbol{x} = \boldsymbol{e}'_1$, whereas $\boldsymbol{x}' = [0\ \ 1\ \ 0]^T$ gives $\boldsymbol{x} = \boldsymbol{e}'_2$ and $\boldsymbol{x}' = [0\ \ 0\ \ 1]^T$ provides $\boldsymbol{x} = \boldsymbol{e}'_3$. These results are in accordance with our definitions of $\boldsymbol{e}'_1$, $\boldsymbol{e}'_2$ and $\boldsymbol{e}'_3$ and therefore prove the correctness of (1.28).

Considering $\boldsymbol{e}'_1$ we observe that the component e_{11} is $\cos\theta_{11}$, where θ_{11} is the angle between the x'_1-axis and the x_1-axis. Likewise, component $e_{12} = \cos\theta_{12}$ where θ_{12} is the angle between the x'_1-axis and the x_2-axis. Finally, component $e_{13} = \cos\theta_{13}$ where θ_{13} is the angle between the x'_1-axis and the x_3-axis. Similar interpretations hold for the components of $\boldsymbol{e}'_2$ and $\boldsymbol{e}'_3$.

1.5 Cartesian tensors

Now we will present an elementary discussion of the concept of tensors and why they appear naturally when formulating physical relations. As we only use Cartesian coordinate systems, no difference exists between so-called covariant and contravariant tensors and therefore, by a tensor we always mean a *Cartesian tensor*. This simplifies the concept of a tensor significantly and for our purpose, there is no need to work with general tensors. For a discussion of general tensor analysis the reader is referred, for instance, to Malvern (1969), Segel (1987), Sokolnikoff (1951) or Spain (1965).

As previously mentioned, the essential issue of a tensor is that it behaves in a certain manner when a change of coordinate system is performed. We shall now establish this relation.

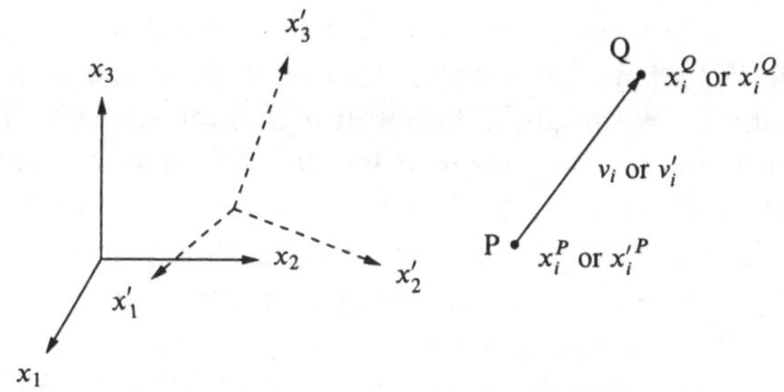

Figure 1.4: Vector from P to Q

We define a *vector* in the usual manner as a quantity having a length and a direction. In Fig. 1.4, the two fixed points P and Q have the coordinates x_i^P and x_i^Q in the old coordinate system and the coordinates $x_i'^P$ and $x_i'^Q$ in the new x_i'-coordinate system. The components of the vector v_i from P to Q in the old x_i-system are then given by

$$v_i = x_i^Q - x_i^P \tag{1.29}$$

where v_1, v_2 and v_3 are the components of the vector in the x_1-, x_2- and x_3-direction respectively. Likewise, the components of the vector v_i' from P to Q measured in the new x_i'-system are given by

$$v_i' = x_i'^Q - x_i'^P \tag{1.30}$$

where v_1', v_2' and v_3' are the components of the vector in the x_1'-, x_2'- and x_3'-direction respectively. From (1.17) we have

$$x_i'^Q = A_{ij}x_j^Q - c_i\ ; \qquad x_i'^P = A_{ij}x_j^P - c_i$$

Insertion into (1.30) and recognition of (1.29) result in

$$\boxed{v'_i = A_{ij}v_j \qquad or \qquad \boldsymbol{v}' = \boldsymbol{A}\boldsymbol{v}} \tag{1.31}$$

We have now established the important relation that shows how the components of a vector changes if a coordinate transformation is made. Here we have derived (1.31) from the usual definition of a vector, but we will now define a *quantity* v_i *as a vector if it transforms according to* (1.31). A vector is also called a *first-order tensor*, where first order refers to the fact that v_i only possesses one index. Now we have an indication of the statement expressed previously that tensors are quantities which behave in a certain manner when a coordinate change is performed.

It is of extreme importance that whereas any quantity containing three pieces of information can be written in the index form b_i, this does not make b_i a vector i.e. a first-order tensor as b_i will not, in general, transform according to (1.31). As an example, assume that $\boldsymbol{a}$ is a vector and consider the quantity $b_i = (|\boldsymbol{a}|, \theta_1, \theta_2)$ where $|\boldsymbol{a}|$ = the length of $\boldsymbol{a}$, θ_1 = the angle between $\boldsymbol{a}$ and the x_1-axis and θ_2 = the angle between $\boldsymbol{a}$ and the x_2-axis. In this case, b_i is certainly not a vector, since each of the components of b_i maintains its value irrespectively of the coordinate system, i.e. b_i does not fulfill the transformation rule (1.31). It is now apparent why we have chosen to use the name column matrix for **a** given by (1.1). Even though a vector a_i can be written in the same manner, the column matrix $\boldsymbol{a}$ is not necessary a vector.

Using (1.21) it is straightforward to invert (1.31) because multiplication by A_{ik} gives $v_k = A_{ik}v'_i$, i.e.

$$v_i = A_{ji}v'_j \qquad \text{or} \qquad \boldsymbol{v} = \boldsymbol{A}^T\boldsymbol{v}' \tag{1.32}$$

As indicated below, it is easy to show formally that velocity and acceleration vectors indeed are vectors.

Consider a specific particle of a body. This particle is described by its coordinates, which are functions of time, i.e. $x_i = x_i(t)$ where t is the time. The velocity components v_i are then defined by

$$v_i = \dot{x}_i \tag{1.33}$$

where a dot denotes the derivative with respect to time and v_1, v_2 and v_3 are the components of v_i in the x_1-, x_2- and x_3-direction respectively. Likewise, in a new coordinate system the velocity v'_i is defined by

$$v'_i = \dot{x}'_i \tag{1.34}$$

Differentiating (1.17) with respect to time and noting that A_{ij} and c_i are constants it follows directly from the definitions (1.33) and (1.34) that v_i transforms according to (1.31); i.e. the velocity v_i is indeed a vector. Differentiating (1.31) with respect to time and assuming that v_i is the velocity vector it appears that also the acceleration vector is, in fact, a vector.

As a force vector is defined as a quantity having a length and direction it follows in complete analogy with (1.29) and (1.30), which lead to (1.31), that a force vector is, in fact, a vector.

We have already touched upon quantities containing one piece of information and which take the same value irrespectively of the coordinate system. Such a quantity b is called a *scalar*, an *invariant* or a *zero-order* tensor and it transforms according to

$$\boxed{b' = b}$$

i.e. it takes the same value in the old coordinate system x_i and in the new coordinate system x_i'. A specific example of a scalar is the distance between two fixed points in the space as used in (1.19) and (1.20).

We have dwelt on the fact that tensors are quantities which transform in a particular manner when coordinate changes are made. It is now timely to ask why tensors are of relevance for our present purpose. The reason for this is of extraordinary importance, because it turns out that the relations of physics are conveniently expressed in terms of tensors. To illustrate this important aspect we write Newton's second law for a particle in the old coordinate system x_i according to

$$F_i = ma_i \tag{1.35}$$

where F_i is the force vector, m is the mass and a_i the acceleration vector. The vectors F_i and a_i are interpreted in the usual way that, for instance, F_2 is the component of F_i in the x_2-direction. When writing (1.35), we did not specify our coordinate system in any manner so in another coordinate system x_i', we expect that Newton's second law takes the form $F_i' = m'a_i'$, i.e.

$$F_i' = ma_i' \tag{1.36}$$

where it has been assumed that the mass m is an invariant, i.e. independent of the coordinate system. As F_i and a_i are vectors, they transform according to (1.31) i.e. we have

$$F_j' = A_{ji}F_i \quad ; \qquad a_j' = A_{ji}a_i$$

Multiply (1.35) by A_{ji} and use the expressions above to obtain

$$F_j' = ma_j'$$

which is precisely the form stipulated in (1.36). It appears that irrespectively of the coordinate system, we write Newton's law in the same form, either (1.35) or (1.36), and this is possible only because F_i and a_i, in fact, are vectors, i.e. first-order tensors and because the mass m is an invariant, i.e. a zero-order tensor. Therefore, the occurrence of vectors and scalars in physical relations is a result of the fact that we expect physical laws to be independent of the particular coordinate system we choose to work with.

Above we illustrated that if a quantity like b_i appears in a physical relation, we expect it to be a vector. Let us pursue the argument above and assume that we have a physical relation which in the x_i-coordinate system states that

$$b_i = B_{ij}c_j \tag{1.37}$$

where b_i and c_i are assumed to be vectors and $\boldsymbol{B}_{ij}$ some quantity. When writing (1.37) we did not specify our coordinate system in any manner, so we expect that in another coordinate system x_i' the same physical relation is expressed through

$$b_i' = B_{ij}'c_j' \tag{1.38}$$

or

$$b_k' = B_{kl}'c_l' \tag{1.39}$$

Multiply (1.37) by A_{ki} and use (1.31) to obtain

$$b_k' = A_{ki}B_{ij}c_j$$

Transformation of c_j according to (1.32) yields

$$b_k' = A_{ki}B_{ij}A_{lj}c_l' \tag{1.40}$$

Subtraction of (1.39) and (1.40) provides

$$(B_{kl}' - A_{ki}B_{ij}A_{lj})c_l' = 0$$

This expression should hold for arbitrary c_l'-values and $\boldsymbol{B}_{ij}$ must therefore transform according to

$$\boxed{B_{kl}' = A_{ki}B_{ij}A_{lj} \qquad or \qquad \boldsymbol{B}' = \boldsymbol{A}\boldsymbol{B}\boldsymbol{A}^T} \tag{1.41}$$

We have found that if it is allowable to write a physical relation as (1.37) in one coordinate system and as (1.38) in another coordinate system - and we certainly expect this to be acceptable - then the quantity $\boldsymbol{B}_{ij}$ must transform according to (1.41). A quantity $\boldsymbol{B}_{ij}$, which transforms according to (1.41), is defined to be a *second-order tensor*. It is obvious that whereas any square matrix containing 3x3 components can be written in index notation as $\boldsymbol{B}_{ij}$, this does not make $\boldsymbol{B}_{ij}$ a second-order tensor. Only those $\boldsymbol{B}_{ij}$-quantities, which transform according to (1.41), are second-order tensors.

We started with (1.37) where b_i and c_i were assumed to be vectors and $\boldsymbol{B}_{ij}$ some quantity. We then concluded that $\boldsymbol{B}_{ij}$ must be a second-order tensor, which transforms according to (1.41). This conclusion is an example of the use of the so-called *quotient theorem*.

Using (1.21) it is easy to invert (1.41). Multiplication of (1.41) by A_{km} gives

$$A_{km}B'_{kl} = B_{mj}A_{lj}$$

and multiplication by A_{ln} yields

$$B_{mn} = A_{km}B'_{kl}A_{ln}$$

which can be written as

$$B_{kl} = A_{ik}B'_{ij}A_{jl} \qquad \text{or} \qquad \boldsymbol{B} = \boldsymbol{A}^T\boldsymbol{B}'\boldsymbol{A} \tag{1.42}$$

Let us finally consider the following physical relation expressed in the x_i-coordinate system by

$$B_{ij} = D_{ijkl}M_{kl}$$

where B_{ij} and M_{kl} are assumed to be second-order tensors. In the x'_i-coordinate system we expect the relation

$$B'_{ij} = D'_{ijkl}M'_{kl}$$

If this is true, then by arguments like before, it is easily shown that the quantity D_{ijkl} must transform according to

$$\boxed{D'_{ijkl} = A_{im}A_{jn}D_{mnpq}A_{kp}A_{lq}} \tag{1.43}$$

Such a quantity is defined as a *fourth-order tensor*. Using (1.21) it follows in a straightforward manner that

$$D_{ijkl} = A_{mi}A_{nj}D'_{mnpq}A_{pk}A_{ql}$$

At this point it should be evident why tensors appear naturally when formulating physical relations. Thus it is not surprising that, in the following chapters, we will encounter a variety of tensors which characterize different physical phenomena. Indeed, if we start with tensors in a certain expression and then, after certain manipulations are left with quantities that are not tensors, we have a clear warning that something dubious has probably sneaked into our considerations. Although it is not a necessity to use a tensor formulation when dealing with physical phenomena, we have seen strong indications why it is very convenient to do so. Matrix formulations are often used instead of tensors, the main reason being that matrices are convenient when it comes to numerical computations. Often, tensors are used to derive the general relations governing the specific problem investigated and hereafter a corresponding matrix formulation is obtained from the tensor formulation in order to facilitate later numerical calculations. Later we will see applications of that approach.

To represent tensors both a component representation, i.e. index notation, as well as so-called *index free notation* can be used, cf. Gurtin (1981). It turns out that the index representation of tensors as presented above is very attractive when manipulating various expressions. The index free notation of tensors - also called *direct notation* - makes no reference to the base vectors of the coordinate system in question and thereby it makes no reference to components. The index free notation is therefore advantageous when representing general concepts. As an example, consider a tensor, which in index notation is written as A_{ij}. We may also adopt the matrix notation for the same quantity and we then denote it by $\boldsymbol{A}$. If an index free formulation is adopted for the tensor in question, the standard notation would also be $\boldsymbol{A}$. It appears that the same notation $\boldsymbol{A}$ is used in the literature both for the matrix notation and for the free index notation of tensors. Since matrices are not necessarily tensors, this implies a certain ambiguity in the interpretation. To avoid this problem, we will, in the present text, avoid the use of index free notation and all boldface letters, like $\boldsymbol{A}$, shall be viewed as matrices.

1.6 Example of tensors - Isotropic tensors

We have previously defined an invariant as a quantity that takes the same value in all coordinate systems. It is of considerable interest that if a quantity is known to be a tensor, it is then possible to establish various invariants from this tensor. As an example, assume that B_{ij} is a second-order tensor. As B_{ij} then obeys the transformation rule (1.41), we obtain by contraction and use of (1.21) that

$$B'_{kk} = A_{ki}B_{ij}A_{kj} = \delta_{ij}B_{ij} = B_{ii}$$

i.e. the quantity B_{ii} is an invariant. We shall later see that other invariants may be obtained from a second-order tensor.

We have previously introduced Kronecker's δ_{ij}, cf. (1.8), and we may ask whether δ_{ij} is a second-order tensor. When defining δ_{ij} no reference was made to any coordinate system, i.e. we must have that $\delta'_{ij} = \delta_{ij}$. With (1.25), it then follows that

$$\delta'_{ij} = \delta_{ij} = A_{ik}A_{jk} = A_{ik}\delta_{kl}A_{jl}$$

It appears that (1.41) is fulfilled, which means that δ_{ij} is a second-order tensor. In fact, it is called an *isotropic tensor*, because it takes the same value irrespectively of the coordinate system.

Let us now derive the most general isotropic second-order tensor. From (1.41) and imposing that B_{ij} is isotropic, we have

$$B_{kl} = A_{ki}B_{ij}A_{lj} \tag{1.44}$$

Multiplication by A_{km} and using that $A_{km}A_{ki} = \delta_{mi}$ gives

$$A_{km}B_{kl} = B_{mj}A_{lj} \qquad \text{i.e} \qquad A_{kj}\delta_{jm}B_{kl} = B_{mj}A_{kj}\delta_{kl}$$

We then obtain

$$(\delta_{mj}B_{kl} - B_{mj}\delta_{kl})A_{kj} = 0$$

Since the transformation matrix A_{kj} is arbitrary, it follows that the expression within the parenthesis must be equal to zero, i.e. $\delta_{mj}B_{kl} - B_{mj}\delta_{kl} = 0$. To evaluate this expression, choose $j = m = 1$; it then follows that $B_{kl} - B_{11}\delta_{kl} = 0$. Likewise, for $j = m = 2$ we obtain $B_{kl} - B_{22}\delta_{kl} = 0$, whereas $j = m = 3$ gives $B_{kl} - B_{33}\delta_{kl} = 0$. We then conclude that

The most general isotropic second-order tensor is given by $k\delta_{ij}$ where k is an arbitrary invariant (1.45)

We will next demonstrate that differentiation, multiplication, addition and subtraction of tensors lead to quantities that also are tensors.

Suppose that f is an invariant, then in applications we will often encounter the quantity $\partial f/\partial x_i = f_{,i}$; this quantity is called the *gradient* of f and we shall prove that it is a vector, i.e. a first-order tensor. From (1.17) and since A_{ij} and c_i are constant quantities, we obtain

$$\frac{\partial x'_i}{\partial x_k} = A_{ik} \tag{1.46}$$

We then obtain

$$\frac{\partial f}{\partial x_i} = \frac{\partial f}{\partial x'_j}\frac{\partial x'_j}{\partial x_i} = A_{ji}\frac{\partial f}{\partial x'_j}$$

and a comparison with (1.32) proves that $\partial f/\partial x_i = f_{,i}$ is a vector, i.e. a first-order tensor.

Likewise, considering the quantity $\partial f/\partial B_{ij}$ where f is an invariant and B_{ij} is a second-order tensor, we shall prove that $\partial f/\partial B_{ij}$ is a second-order tensor. We have

$$\frac{\partial f}{\partial B_{ij}} = \frac{\partial f}{\partial B'_{kl}}\frac{\partial B'_{kl}}{\partial B_{ij}}$$

As B_{ij} is a second-order tensor, (1.41) implies that

$$\frac{\partial B'_{kl}}{\partial B_{ij}} = A_{km}\frac{\partial B_{mn}}{\partial B_{ij}}A_{ln} = A_{km}\delta_{mi}\delta_{nj}A_{ln} = A_{ki}A_{lj}$$

i.e.

$$\frac{\partial f}{\partial B_{ij}} = A_{ki}\frac{\partial f}{\partial B'_{kl}}A_{lj}$$

A comparison with (1.41) proves that $\partial f/\partial B_{ij}$ is a second- order tensor.

Consider now the quantity $\partial v_i/\partial x_j = v_{i,j}$ where v_i is a vector. According to (1.32) we have

$$\frac{\partial v_i}{\partial x_j} = A_{ki}\frac{\partial v'_k}{\partial x_j} = A_{ki}\frac{\partial v'_k}{\partial x'_l}\frac{\partial x'_l}{\partial x_j}$$

and use of (1.46) implies that

$$\frac{\partial v_i}{\partial x_j} = A_{ki}\frac{\partial v'_k}{\partial x'_l}A_{lj} \tag{1.47}$$

A comparison with (1.41) proves that $\partial v_i/\partial x_j = v_{i,j}$ is a second-order tensor.

For the often encountered quantity $\partial B_{ij}/\partial x_j$ where B_{ij} is a second-order tensor, we find according to (1.42) that

$$\frac{\partial B_{ij}}{\partial x_j} = A_{ki}\frac{\partial B'_{kl}}{\partial x_j}A_{lj} = A_{ki}\frac{\partial B'_{kl}}{\partial x'_m}\frac{\partial x'_m}{\partial x_j}A_{lj}$$

and use of (1.46) implies that

$$\frac{\partial B_{ij}}{\partial x_j} = A_{ki}\frac{\partial B'_{kl}}{\partial x'_m}A_{mj}A_{lj}$$

From expression (1.25) we conclude that

$$\frac{\partial B_{ij}}{\partial x_j} = A_{ki}\frac{\partial B'_{kl}}{\partial x'_l}$$

and a comparison with (1.32) reveals that $\partial B_{ij}/\partial x_j = B_{ij,j}$ is a vector, i.e. a first-order tensor.

Finally, consider the quantity $a_i b_j$ where a_i and b_j are assumed to be vectors. As a_i and b_j are vectors, it follows from (1.32) that

$$a_i b_j = A_{ki}a'_k b'_l A_{lj} \tag{1.48}$$

which, according to (1.42), shows that $a_i b_j$ is a second-order tensor. Occasionally, the product $a_i b_j$ is called a *dyad* and the sum of dyads is termed a *dyadic*. Analogous with the derivation of (1.48), it follows directly that products of tensors result in the creation of new tensors. As an example, the product $a_i B_{ij}$ is a vector, if a_i and B_{ij} is a vector and a second-order tensor respectively. Moreover, it follows directly that the sum or difference of two tensors results in a new tensor.

Because of the transformation properties, a tensor is known completely in all coordinate systems if it is known in one of them. In particular, if all the components vanish in one system, they vanish in all. This seemingly trivial statement is helpful in various mathematical proofs. As an example, it will be shown later that static equilibrium requires

$$\sigma_{ij,j} + b_i = 0$$

where σ_{ij} is the stress tensor and b_i the body forces per unit volume. Defining the quantity D_i as $D_i = \sigma_{ij,j} + b_i$, we recognize that D_i is a first-order tensor, i.e. a vector; equilibrium is then expressed by $D_i = 0$. Therefore, if the vector $D_i = 0$ in one coordinate system it is zero in all coordinate systems. That is, if the body is in equilibrium in one coordinate system, it is unnecessary to reinvestigate equilibrium in any other coordinate system.

After these preliminary remarks about notations and tensors and as the reader should now be familiar with index notation, it is timely to proceed with what is our main interest: *the behavior of materials*. The first thing of relevance is the ability to describe the deformation of the material in a proper manner and this is the subject of the next chapter.

2 STRAIN TENSOR

As we are concerned with the behavior of deformable bodies, it is essential to establish a quantity that only describes the deformation of the body, i.e. it should not be influenced by any rigid-body motions. Such a quantity, the *strain tensor*, will be derived in the present chapter.

We will present a detailed derivation of a number of properties of the strain tensor not only because of the importance of these properties, but also because it turns out that many of the properties can be transferred directly to the stress tensor that is treated in the next chapter. For further studies of these topics, we may refer to Fung (1965), Malvern (1969), Sokolnikoff (1946) and Spencer (1980).

2.1 Introduction

In the *reference configuration* before any deformation, a material point, i.e. a *particle*, has the position vector x_i in the fixed coordinate system. After deformation and in the same coordinate system, this material point has the position vector x_i^* given by

$$\boxed{x_i^* = x_i + u_i} \tag{2.1}$$

where $u_i = u_i(x_i, t)$ is the *displacement vector* and t denotes the time.

Referring to Fig. 2.1, we consider in the reference configuration the material points P and Q which are located infinitely close to each other; due to the deformation these positions change to the positions P* and Q* respectively. According to (2.1), point P given by the x_i-vector moves to the point P* given by the vector $x_i^* = x_i + u_i(x_i, t)$. Likewise, point Q given by the vector $x_i + dx_i$ moves to the point Q* given by the vector $x_i^* + dx_i^* = x_i + dx_i + u_i(x_i + dx_i, t)$. These geometric issues are illustrated in Fig. 2.1. From (2.1) it follows directly that

$$dx_i^* = dx_i + du_i = dx_i + u_{i,j}dx_j \,; \qquad du_i = u_{i,j}dx_j \tag{2.2}$$

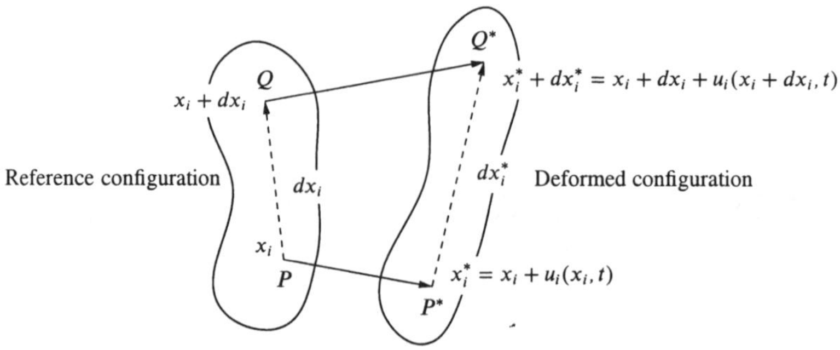

Figure 2.1: Displacements of neighboring material points P and Q

where the *displacement gradient* $u_{i,j}$ is defined by

$$u_{i,j} = \frac{\partial u_i}{\partial x_j} \tag{2.3}$$

Equation (2.2) can also be written as

$$dx_i^* = (\delta_{ij} + u_{i,j})dx_j \tag{2.4}$$

Thus, due to the deformation the vector dx_i changes to the vector dx_i^* according to (2.4) and illustrated in Fig. 2.1.

Referring to this figure, let ds denote the length of the vector $\overline{PQ}$ and ds^* the length of the vector $\overline{P^*Q^*}$, i.e. $ds = |\overline{PQ}|$ and $ds^* = |\overline{P^*Q^*}|$. We then obtain

$$ds^2 = dx_j dx_j \; ; \qquad ds^{*^2} = dx_k^* dx_k^*$$

Using (2.4), we get

$$ds^{*^2} = (\delta_{kj} + u_{k,j})(\delta_{ki} + u_{k,i})dx_j dx_i$$

and as $dx_j dx_j = \delta_{ij} dx_i dx_j$, it appears that

$$ds^{*^2} - ds^2 = (u_{i,j} + u_{j,i} + u_{k,i}u_{k,j})dx_j dx_i$$

This equation can be written as

$$\boxed{ds^{*^2} - ds^2 = 2dx_i E_{ij} dx_j} \tag{2.5}$$

where the *strain tensor* E_{ij} is defined by

$$\boxed{E_{ij} = \tfrac{1}{2}(u_{i,j} + u_{j,i} + u_{k,i}u_{k,j})} \tag{2.6}$$

It appears that E_{ij} is symmetric, i.e. $E_{ij} = E_{ji}$. This strain tensor was introduced by Green and St. Venant and it is often called *Green's strain tensor*.

Here we have described the displacement vector u_i as function of its position x_i before any deformations, i.e. $u_i = u_i(x_i, t)$ and such an approach is called a *Lagrangian description*. For that reason E_{ij} is often called *Lagrange's strain tensor* (occasionally, in the literature it is called the *Green-Lagrange strain tensor*; in fact, it was introduced by Green in 1841 and by St.-Venant in 1844). The alternative approach is the *Eulerian description*, often employed in fluid mechanics, where the displacement vector u_i is given as function of the current coordinates x_i^* i.e. $u_i = u_i(x_i^*, t)$.

That E_{ij} is, indeed, a second-order tensor follows from the fact that u_i is a vector and $u_{i,j}$ therefore is a second-order tensor, cf. (1.47).

2.2 Small strain tensor

In the following, we will only consider situations where the displacement gradients $u_{i,j}$ are small, i.e. each component is small when compared to unity

$$\boxed{|u_{i,j}| \ll 1}$$

In that case, the quadratic term in (2.6) can be ignored and the Lagrange strain E_{ij} can be approximated by the *small strain tensor* ε_{ij} defined by

$$\boxed{\varepsilon_{ij} = \tfrac{1}{2}(u_{i,j} + u_{j,i})} \tag{2.7}$$

which is also symmetric, i.e.

$$\boxed{\varepsilon_{ij} = \varepsilon_{ji}}$$

It is obvious that ε_{ij} is a second-order tensor.

2.3 Rigid-body motions

Our aim was to establish a quantity, the strain tensor, that is independent of rigid-body motions. Let us now prove that ε_{ij} possesses this property.

Any rigid-body motion is characterized by the fact that during motion, the vector $\overline{PQ}$ of two neighboring material points, cf. Fig. 2.1, changes into the vector $\overline{P^*Q^*}$ in such a way that its length remains constant. As $ds = |\overline{PQ}|$ and $ds^* = |\overline{P^*Q^*}|$, we can then write that during any rigid-body motion, we have

$$ds^{*2} - ds^2 = 0$$

Making use of (2.5) and (2.6) and noting that dx_i is arbitrary, we conclude that

$$2E_{ij} = u_{i,j} + u_{j,i} + u_{k,i}u_{k,j} = 0$$

We observe that Green-Lagrange's strain tensor is unaffected by rigid-body motions and within our approximation of small displacement gradients we have $E_{ij} = \varepsilon_{ij}$, i.e. rigid-body motions do not influence the small strain tensor, which proves the desired property of this strain tensor.

2.4 Physical significance of the strain tensor

We shall now evaluate the physical significance of the strain tensor ε_{ij} and its components. Within our assumption of small displacement gradients, we have $E_{ij} = \varepsilon_{ij}$, i.e. (2.5) reads

$$\frac{ds^{*2} - ds^2}{ds^2} = 2\frac{dx_i}{ds}\varepsilon_{ij}\frac{dx_j}{ds} \tag{2.8}$$

where it is recalled from Fig. 2.1 that ds is the length of the vector dx_i between the two neighboring particles P and Q before any deformation takes place and ds^* is the distance between these two particles after the deformation. Therefore

$$n_i = \frac{dx_i}{ds}$$

is a unit vector in the direction of dx_i. From this expression and (2.8) follow that

$$\frac{ds^{*2} - ds^2}{2ds^2} = n_i \varepsilon_{ij} n_j \tag{2.9}$$

As the displacement gradients are small the components of ε_{ij} are also small and this implies that the left-hand side of (2.9) is small. Consequently ds^* is close to ds and we then obtain

$$\frac{ds^{*2} - ds^2}{2ds^2} = \frac{(ds^* + ds)(ds^* - ds)}{2ds^2} \approx \frac{2ds(ds^* - ds)}{2ds^2} = \frac{ds^* - ds}{ds}$$

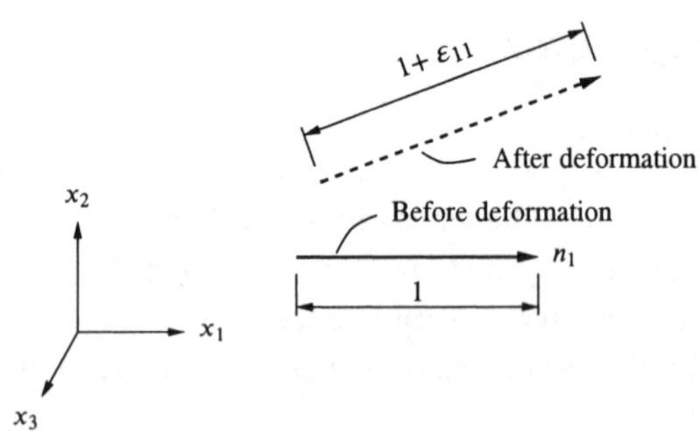

Figure 2.2: Illustration of strain component ε_{11}

We define the *relative elongation* or the *normal strain* ε of the vector $\overline{PQ}$ deforming into the $\overline{P^*Q^*}$-vector, cf. Fig. 2.1, by

$$\boxed{\varepsilon = \frac{ds^* - ds}{ds}} \tag{2.10}$$

in accordance with the elementary definition of normal strain. A combination of (2.9) - (2.10) yields

$$\boxed{\varepsilon = n_i \varepsilon_{ij} n_j \qquad or \qquad \varepsilon = \boldsymbol{n}^T \boldsymbol{\varepsilon}\, \boldsymbol{n}} \tag{2.11}$$

As an example, choose the direction n_i so that $n_i = (1, 0, 0)$, then we obtain $\varepsilon = \varepsilon_{11}$ as illustrated in Fig. 2.2. Likewise, for $n_i = (0, 1, 0)$, we obtain $\varepsilon = \varepsilon_{22}$ whereas $n_i = (0, 0, 1)$ yields $\varepsilon = \varepsilon_{33}$. Therefore, we have achieved a physical interpretation of all the diagonal terms of the strain tensor and it appears from (2.11) that the normal strain, i.e. the relative elongation, in an arbitrary direction given by the unit vector n_i, is known once the strain tensor is known.

To obtain a physical interpretation of the off-diagonal terms in the strain tensor, consider two directions $dx_i^{(1)}$ and $dx_i^{(2)}$ in the reference configuration before any deformations take place. These two directions are taken to be orthogonal, i.e.

$$dx_i^{(1)}\, dx_i^{(2)} = 0 \tag{2.12}$$

In accordance with Fig. 2.3, the lengths of $dx_i^{(1)}$ and $dx_i^{(2)}$ are given by $ds^{(1)}$

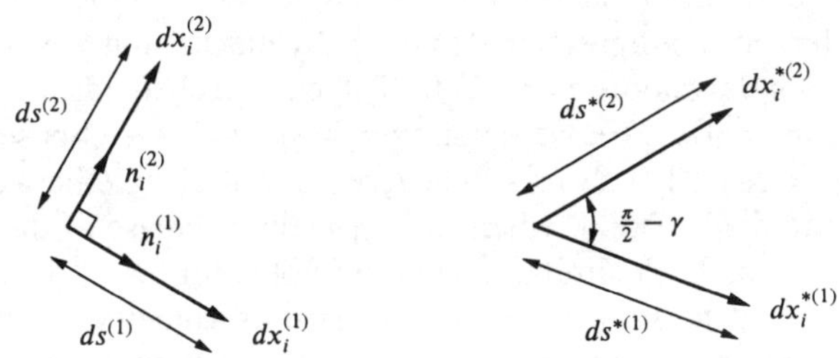

Figure 2.3: Change of orthogonal angle in reference configuration due to the deformation.

and $ds^{(2)}$ respectively, i.e. we have the following two orthogonal unit vectors

$$n_i^{(1)} = \frac{dx_i^{(1)}}{ds^{(1)}} \quad ; \qquad n_i^{(2)} = \frac{dx_i^{(2)}}{ds^{(2)}} \tag{2.13}$$

Due to the deformation, the vector $dx_i^{(1)}$ changes to $dx_i^{*(1)}$ with length $ds^{*(1)}$ whereas the vector $dx_i^{(2)}$ changes to $dx_i^{*(2)}$ with length $ds^{*(2)}$, cf. Fig. 2.3. The angle $90^o - \gamma$ between $dx^{*(1)}$ and $dx^{*(2)}$ is then given by

$$\cos(90^o - \gamma) = \frac{dx_i^{*(1)}}{ds^{*(1)}} \frac{dx_i^{*(2)}}{ds^{*(2)}} \tag{2.14}$$

From (2.4), we have

$$dx_i^{*(1)} = (\delta_{ij} + u_{i,j}) dx_j^{(1)} \; ; \qquad dx_i^{*(2)} = (\delta_{ik} + u_{i,k}) dx_k^{(2)}$$

Insertion into (2.14) yields

$$\sin\gamma = (\delta_{jk} + u_{k,j} + u_{j,k} + u_{i,j}u_{i,k})\frac{dx_j^{(1)}}{ds^{*(1)}}\frac{dx_k^{(2)}}{ds^{*(2)}} \tag{2.15}$$

Due to the small strain approximation, we can ignore the quadratic term and set $ds^{*(1)} \approx ds^{(1)}$ and $ds^{*(2)} \approx ds^{(2)}$. Consequently, (2.15) reduces with (2.7) to

$$\sin\gamma = \frac{dx_k^{(1)}}{ds^{(1)}}\frac{dx_k^{(2)}}{ds^{(2)}} + 2\varepsilon_{jk}\frac{dx_j^{(1)}}{ds^{(1)}}\frac{dx_k^{(2)}}{ds^{(2)}}$$

As we assume small strains we have $\sin\gamma \approx \gamma$. With (2.12) and (2.13) we then obtain

$$\gamma = 2\varepsilon_{jk}\, n_j^{(1)}\, n_k^{(2)}$$

To emphasize that the vectors $n_i^{(1)}$ and $n_i^{(2)}$ are orthogonal, we write $m_i = n_i^{(1)}$ and $n_i = n_i^{(2)}$ and the expression above takes the more convenient form

$$\gamma = 2m_i\varepsilon_{ij}n_j \tag{2.16}$$

Hence, due to the deformation the right angle between the unit vectors n_i and m_i in the reference configuration decreases by the amount γ given by (2.16).

As an example, choose $n_i = (1, 0, 0)$ then $\varepsilon_{ij}n_j$ becomes $\varepsilon_{ij}n_j = \varepsilon_{i1}$. If we then choose $m_i = (0, 1, 0)$, we obtain $\gamma = 2\varepsilon_{21}$ and if we choose $m_i = (0, 0, 1)$, we obtain $\gamma = 2\varepsilon_{31}$. I.e. $2\varepsilon_{21}$ is the decrease of the angle between the x_2- and x_1-axes due to deformation, whereas $2\varepsilon_{31}$ is the decrease of the angle between the x_3- and x_1-axes. A similar evaluation holds for $2\varepsilon_{23}$. These off-diagonal terms of the strain tensor are called *shear strains* as they describe the shearing, i.e. the distortion of the material. With obvious notation we can then write

$$\boxed{\gamma_{nm} = 2\varepsilon_{nm}}$$

where

$$\boxed{\varepsilon_{nm} = n_i\varepsilon_{ij}m_j = m_i\varepsilon_{ij}n_j \quad or \quad \varepsilon_{nm} = \mathbf{n}^T\boldsymbol{\varepsilon}\,\mathbf{m} = \mathbf{m}^T\boldsymbol{\varepsilon}\,\mathbf{n}} \tag{2.17}$$

In this expression n_i and m_i are arbitrary unit vectors, which are orthogonal in the reference configuration. The angle decrease γ_{nm} between n_i and m_i caused by the deformation is termed the *engineering shear strain* to be distinguished from the tensorial shear strain ε_{nm}. The shearing between two directions parallel with the x_1- and x_2-axes is illustrated in Fig. 2.4.

It appears that the strain tensor contains information by which relative elongation in arbitrary directions and angle changes between arbitrary orthogonal directions can be determined. Consequently, the strain tensor describes the deformation completely and, in addition, we have achieved a direct physical interpretation of all the components of this tensor. These results were already obtained by Cauchy in 1822.

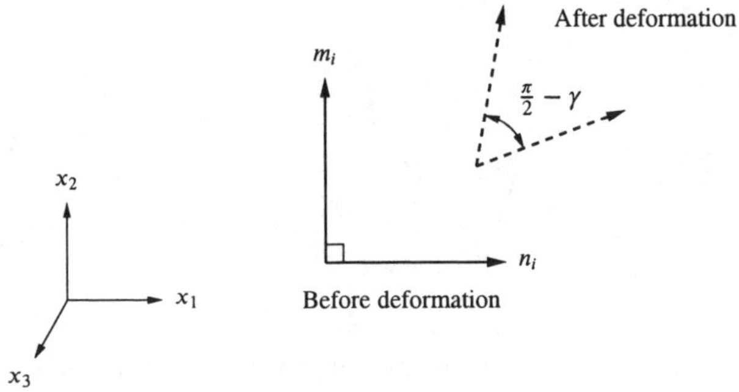

Figure 2.4: Illustration of shear component $\varepsilon_{12} = \gamma_{12}/2$

2.5 Change of coordinate system

The implications of coordinate system changes are important in many connections and we have already discussed this aspect in detail in Chapter 1.

Let us consider the change from the old x_i-coordinate system to the new x'_i-coordinate system. From (1.17) we have

$$x'_i = A_{ij}(x_j - c_i) \qquad \text{or} \qquad \boldsymbol{x}' = \boldsymbol{A}(\boldsymbol{x} - \boldsymbol{c}) \tag{2.18}$$

where A_{ij} is the transformation matrix. Suppose that we know the components of ε_{ij} in the x_i-system and suppose that we want to determine the components of ε'_{ij} in the x'_i-system. We have already proved that ε_{ij} is a second-order tensor, i.e. it follows directly from (1.41) that

$$\boxed{\varepsilon'_{ij} = A_{ik}\varepsilon_{kl}A_{jl} \qquad or \qquad \boldsymbol{\varepsilon}' = \boldsymbol{A}\boldsymbol{\varepsilon}\boldsymbol{A}^T} \tag{2.19}$$

The inverse relations follow from (1.42), i.e.

$$\varepsilon_{ij} = A_{ki}\varepsilon'_{kl}A_{lj} \qquad \text{or} \qquad \boldsymbol{\varepsilon} = \boldsymbol{A}^T\boldsymbol{\varepsilon}'\boldsymbol{A} \tag{2.20}$$

2.6 Principal strains and principal directions - Invariants

We have previously obtained a physical interpretation of the strain tensor components. However, it turns out that for a special choice of coordinate system, the strain tensor takes a particularly simple form.

For this purpose, consider a direction in the reference configuration given by the unit vector $\boldsymbol{n}$. We then define the vector $\boldsymbol{q}$ by

$$\boldsymbol{q} = \boldsymbol{\varepsilon}\boldsymbol{n} \tag{2.21}$$

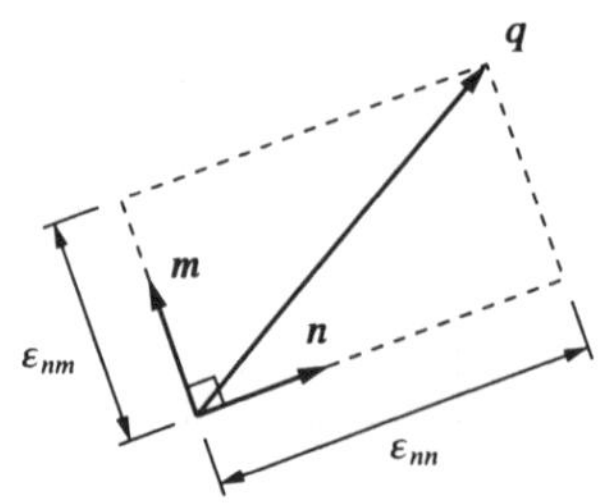

Figure 2.5: The vector $\boldsymbol{q} = \boldsymbol{\varepsilon n}$ and its components after direction $\boldsymbol{n}$ and $\boldsymbol{m}$

Referring to Fig. 2.5, the unit vector $\boldsymbol{m}$ is orthogonal to $\boldsymbol{n}$. Following Fig. 2.5 and in accordance with (2.11) and (2.21), the component of $\boldsymbol{q}$ in the direction of $\boldsymbol{n}$ is given by

$$\varepsilon_{nn} = \boldsymbol{n}^T \boldsymbol{q} \tag{2.22}$$

where ε_{nn} is the normal strain in the direction $\boldsymbol{n}$. Likewise from (2.17) and (2.21), the component of $\boldsymbol{q}$ in the direction of $\boldsymbol{m}$ is given by

$$\varepsilon_{nm} = \boldsymbol{m}^T \boldsymbol{q}$$

where ε_{nm} is the shear strain between the directions $\boldsymbol{n}$ and $\boldsymbol{m}$.

We now look for the situation where the direction $\boldsymbol{n}$ is chosen so that $\boldsymbol{q}$ is collinear with $\boldsymbol{n}$, i.e. the shear strain $\varepsilon_{nm} = 0$. To achieve this situation, we must have

$$\boldsymbol{q} = \lambda \boldsymbol{n} \tag{2.23}$$

where λ is an unknown parameter and from (2.22) we conclude that $\varepsilon_{nn} = \lambda$. Use of (2.21) in (2.23) yields the following requirement

$$\boxed{(\boldsymbol{\varepsilon} - \lambda \boldsymbol{I})\boldsymbol{n} = \boldsymbol{0} \qquad or \qquad (\varepsilon_{ij} - \lambda\delta_{ij})n_j = 0} \tag{2.24}$$

where $\boldsymbol{0}$ is defined as $\boldsymbol{0}^T = [0\ 0\ 0]$.

Expression (2.24) is an example of the well-known *eigenvalue problem.* It consists of a quadratic set of homogeneous equations and if a nontrivial solution $\boldsymbol{n}$ is to exist, we must require

$$\det(\boldsymbol{\varepsilon} - \lambda \boldsymbol{I}) = 0 \tag{2.25}$$

As $\boldsymbol{\varepsilon} - \lambda \boldsymbol{I}$ is a 3 x 3 matrix, the expression above provides a cubic equation for the determination of λ - the so-called *characteristic equation.* That is, (2.25) is fulfilled by three values of λ - the *eigenvalues* λ_1, λ_2 and λ_3. When λ_1, λ_2 and λ_3 have been determined, then substitution of λ_1 in (2.24) provides the solution $\boldsymbol{n}_1$, substitution of λ_2 provides the solution $\boldsymbol{n}_2$ and substitution of λ_3 yields the solution $\boldsymbol{n}_3$. The solutions $\boldsymbol{n}_1$, $\boldsymbol{n}_2$ and $\boldsymbol{n}_3$ are the *eigenvectors*. In accordance with the theory of homogeneous equations the lengths of the eigenvectors will be undetermined whereas the direction will be known. Accordingly, it is always possible to choose a solution so that $\boldsymbol{n}$ becomes a unit vector and this situation will be assumed in the following. In the present context, the λ-values are most frequently called the *principal strains*, whereas the $\boldsymbol{n}$-vectors are called the *principal strain directions*.

The importance of the λ-values comes from the fact that they are *invariants*, i.e. they take the same values irrespective of the coordinate system. From a physical point of view, this is rather obvious as the magnitude of a principal strain λ was found above to be given by the relative elongation ε_{nn} in the fixed direction $\boldsymbol{n}$ and this relative elongation must be independent of the coordinate system chosen. To prove this formally, assume that we change the coordinate system from the old x_i-system to the new x'_i-system in accordance with (2.18). Following (2.24), the principal directions and principal strains in the new coordinate system are determined by

$$\boldsymbol{\varepsilon}'\boldsymbol{n}' = \lambda'\boldsymbol{n}' \tag{2.26}$$

where λ' denotes the principal strain in the new coordinate system. Since $\boldsymbol{n}$ is a vector, we have from (1.31) that

$$\boldsymbol{n}' = \boldsymbol{A}\boldsymbol{n}$$

Use of this expression and (2.19) in (2.26) yields

$$\boldsymbol{A}\boldsymbol{\varepsilon}\boldsymbol{A}^T\boldsymbol{A}\boldsymbol{n} = \lambda'\boldsymbol{A}\boldsymbol{n}$$

Premultiplication by $\boldsymbol{A}^T$ and using that $\boldsymbol{A}^T\boldsymbol{A} = \boldsymbol{I}$, we find

$$\boldsymbol{\varepsilon}\boldsymbol{n} = \lambda'\boldsymbol{n}$$

and a comparison with (2.24) proves that $\lambda = \lambda'$ implying that the λ-values are invariants, i.e. independent of the coordinate system. However, since the components of the eigenvector $\boldsymbol{n}'$ are now measured in the new x'_i-coordinate system, these components differ from the components of the eigenvector $\boldsymbol{n}$.

Evaluation of the cubic equation (2.25) – the characteristic equation – gives after some algebra

$$\boxed{-\lambda^3 + \theta_1\lambda^2 - \theta_2\lambda + \theta_3 = 0} \tag{2.27}$$

where θ_1, θ_2 and θ_3 are defined by

$$\begin{aligned}
\theta_1 &= \varepsilon_{11} + \varepsilon_{22} + \varepsilon_{33} = \varepsilon_{ii} \\
\theta_2 &= \varepsilon_{11}\varepsilon_{22} + \varepsilon_{22}\varepsilon_{33} + \varepsilon_{11}\varepsilon_{33} - \varepsilon_{23}^2 - \varepsilon_{12}^2 - \varepsilon_{13}^2 = \tfrac{1}{2}\theta_1^2 - \tfrac{1}{2}\varepsilon_{ij}\varepsilon_{ji} \\
\theta_3 &= \varepsilon_{11}\varepsilon_{22}\varepsilon_{33} - \varepsilon_{11}\varepsilon_{23}^2 - \varepsilon_{22}\varepsilon_{13}^2 - \varepsilon_{33}\varepsilon_{12}^2 + 2\varepsilon_{12}\varepsilon_{13}\varepsilon_{23} = \det(\varepsilon_{ij})
\end{aligned} \tag{2.28}$$

As the λ-values are invariants determined by the values of θ_1, θ_2 and θ_3 it is obvious that also the θ_1-, θ_2- and θ_3-values are invariants. They are called the *Cauchy-strain invariants* and any combination of these invariants is also an invariant.

An important issue is that the eigenvectors are orthogonal and that the eigenvalues are real; this is a consequence of the matrix $\boldsymbol{\varepsilon}$ being real and symmetric and it is a well-known result in mathematics. However, we will take the opportunity to prove it here.

To prove that the eigenvectors are orthogonal, assume that we have determined the two eigenvalues λ_1 and λ_2 and the corresponding two eigenvectors $\boldsymbol{n}_1$ and $\boldsymbol{n}_2$. We then have

$$\boldsymbol{\varepsilon n}_1 = \lambda_1 \boldsymbol{n}_1$$

$$\boldsymbol{\varepsilon n}_2 = \lambda_2 \boldsymbol{n}_2 \tag{2.29}$$

Transpose the first equation, utilize that $\boldsymbol{\varepsilon}$ is symmetric and postmultiply it by $\boldsymbol{n}_2$ to obtain

$$\boldsymbol{n}_1^T \boldsymbol{\varepsilon n}_2 = \lambda_1 \boldsymbol{n}_1^T \boldsymbol{n}_2 \tag{2.30}$$

Premultiply (2.29) by $\boldsymbol{n}_1^T$ to obtain

$$\boldsymbol{n}_1^T \boldsymbol{\varepsilon n}_2 = \lambda_2 \boldsymbol{n}_1^T \boldsymbol{n}_2 \tag{2.31}$$

Subtraction of (2.30) and (2.31) yields

$$(\lambda_1 - \lambda_2)\boldsymbol{n}_1^T \boldsymbol{n}_2 = 0$$

If we assume that $\lambda_1 \neq \lambda_2$ then it follows that $\boldsymbol{n}_1$ and $\boldsymbol{n}_2$ must be orthogonal. Similar arguments hold between $\boldsymbol{n}_1$ and $\boldsymbol{n}_3$ and between $\boldsymbol{n}_2$ and $\boldsymbol{n}_3$, i.e. we obtain the following fundamental property

$$\boxed{\boldsymbol{n}_1^T \boldsymbol{n}_2 = \boldsymbol{n}_1^T \boldsymbol{n}_3 = \boldsymbol{n}_2^T \boldsymbol{n}_3 = 0 \quad \textit{orthogonality of eigenvectors}} \tag{2.32}$$

When proving this orthogonality, it was assumed that the principal strains were unequal. What happens if some of them are equal? Suppose that in a certain coordinate system, we have the following strain tensor

$$\boldsymbol{\varepsilon} = \begin{bmatrix} a & 0 & 0 \\ 0 & a & 0 \\ 0 & 0 & a \end{bmatrix} = a\boldsymbol{I}$$

It is obvious that in this coordinate system the principal strains are all equal and given by the quantity a. Suppose now that the coordinate system is changed from the present x_i-system to the new x'_i-system in accordance with (2.18). In this new x'_i-system, the strain tensor transforms into the one given by (2.19), i.e.

$$\boldsymbol{\varepsilon}' = \boldsymbol{A\varepsilon A}^T = a\boldsymbol{AIA}^T = a\boldsymbol{AA}^T = a\boldsymbol{I} = \boldsymbol{\varepsilon}$$

Consequently, we have proved that if all three principal strains are equal, then any coordinate system corresponds to the principal directions.

Suppose now that in a certain coordinate system, we have the following strain tensor

$$\boldsymbol{\varepsilon} = \begin{bmatrix} a & 0 & 0 \\ 0 & b & 0 \\ 0 & 0 & b \end{bmatrix} = b\boldsymbol{I} + \begin{bmatrix} a-b & 0 & 0 \\ 0 & 0 & 0 \\ 0 & 0 & 0 \end{bmatrix}$$

i.e. two of the principal strains are equal. Suppose furthermore that we rotate the coordinate system according to (2.18). However, we will make the special choice that this rotation consists of a rotation about the x_1-axis. This implies that $\boldsymbol{e}_1'^T = [1\ 0\ 0]$, cf. Fig. 1.3. According to (2.19) and (1.28), the strain tensor in the new x_i'-system becomes

$$\boldsymbol{\varepsilon}' = b\boldsymbol{A}\boldsymbol{I}\boldsymbol{A}^T + \boldsymbol{A} \begin{bmatrix} a-b & 0 & 0 \\ 0 & 0 & 0 \\ 0 & 0 & 0 \end{bmatrix} \boldsymbol{A}^T$$

$$= b\boldsymbol{I} + \begin{bmatrix} \boldsymbol{e}_1'^T \\ \boldsymbol{e}_2'^T \\ \boldsymbol{e}_3'^T \end{bmatrix} \begin{bmatrix} a-b & 0 & 0 \\ 0 & 0 & 0 \\ 0 & 0 & 0 \end{bmatrix} [\, \boldsymbol{e}_1'\ \boldsymbol{e}_2'\ \boldsymbol{e}_3' \,]$$

As we only consider a rotation about the x_1-axis, i.e. $\boldsymbol{e}_1'^T = [1\ 0\ 0]$, we obtain

$$\boldsymbol{\varepsilon}' = b\boldsymbol{I} + \begin{bmatrix} a-b & 0 & 0 \\ 0 & 0 & 0 \\ 0 & 0 & 0 \end{bmatrix} [\, \boldsymbol{e}_1'\ \boldsymbol{e}_2'\ \boldsymbol{e}_3' \,]$$

i.e.

$$\boldsymbol{\varepsilon}' = b\boldsymbol{I} + \begin{bmatrix} a-b & 0 & 0 \\ 0 & 0 & 0 \\ 0 & 0 & 0 \end{bmatrix} = \boldsymbol{\varepsilon}$$

Consequently, we have proved that if two of the principal strains are equal, then any coordinate system obtained by rotation about that axis, which corresponds to the principal strain different from the other principal strains, corresponds to the principal directions.

In conclusion, we find that it is always allowable to take the principal directions as orthogonal directions in accordance with (2.32).

Remembering the physical interpretation of λ, cf. the discussion of (2.23), it is evident that the λ-values must be real. However, a formal proof is readily achieved. For the eigenvalue λ and the corresponding eigenvector $\boldsymbol{n}$, $\boldsymbol{\varepsilon n} = \lambda \boldsymbol{n}$ holds. Take the complex conjugate of this equation to obtain

$$\boldsymbol{\varepsilon n}^* = \lambda^* \boldsymbol{n}^* \tag{2.33}$$

where an asterisk * for the time being denotes the complex conjugate and where it has been used that $\boldsymbol{\varepsilon}$ is real and that $(\lambda \boldsymbol{n})^* = \lambda^* \boldsymbol{n}^*$. Premultiplying $\boldsymbol{\varepsilon n} = \lambda \boldsymbol{n}$ by $\boldsymbol{n}^{*T}$ gives

$$\boldsymbol{n}^{*T} \boldsymbol{\varepsilon n} = \lambda \boldsymbol{n}^{*T} \boldsymbol{n} \tag{2.34}$$

whereas transposing (2.33), utilizing the symmetry of $\boldsymbol{\varepsilon}$ and postmultiplying by $\boldsymbol{n}$ provides

$$\boldsymbol{n}^{*T} \boldsymbol{\varepsilon n} = \lambda^* \boldsymbol{n}^{*T} \boldsymbol{n} \tag{2.35}$$

Then, finally, subtraction of (2.34) and (2.35) yields

$$(\lambda - \lambda^*) \boldsymbol{n}^{*T} \boldsymbol{n} = 0$$

However, $\boldsymbol{n}^{*T} \boldsymbol{n}$ is certainly different from zero implying that $\lambda = \lambda^*$ and it has then been proved that the eigenvalues are real. It follows immediately that also the eigenvectors are real, i.e.

The eigenvalues and eigenvectors are real

We are now in a position to illustrate a significant feature related to the eigenvalues and eigenvectors. As $\boldsymbol{n}_1$, $\boldsymbol{n}_2$ and $\boldsymbol{n}_3$ are orthogonal, we can change our coordinate system from the x_i-system to a x'_i-system collinear with the $\boldsymbol{n}_1$-, $\boldsymbol{n}_2$– and $\boldsymbol{n}_3$- directions. Following (2.18) and (1.28), we then have

$$\boldsymbol{x}' = \boldsymbol{A}\boldsymbol{x} - \boldsymbol{c} \qquad \text{where} \qquad \boldsymbol{A}^T = \begin{bmatrix} \boldsymbol{n}_1 & \boldsymbol{n}_2 & \boldsymbol{n}_3 \end{bmatrix}$$

In this new x'_i-system the strain tensor becomes, cf. (2.19)

$$\boldsymbol{\varepsilon}' = \begin{bmatrix} \boldsymbol{n}_1^T \\ \boldsymbol{n}_2^T \\ \boldsymbol{n}_3^T \end{bmatrix} \boldsymbol{\varepsilon} \begin{bmatrix} \boldsymbol{n}_1 & \boldsymbol{n}_2 & \boldsymbol{n}_3 \end{bmatrix}$$

$$= \begin{bmatrix} \boldsymbol{n}_1^T \\ \boldsymbol{n}_2^T \\ \boldsymbol{n}_3^T \end{bmatrix} \begin{bmatrix} \boldsymbol{\varepsilon n}_1 & \boldsymbol{\varepsilon n}_2 & \boldsymbol{\varepsilon n}_3 \end{bmatrix}$$

Using that $\boldsymbol{\varepsilon n}_1 = \lambda_1 \boldsymbol{n}_1$ and the similar relations, cf. (2.24), we obtain

$$\boldsymbol{\varepsilon}' = \begin{bmatrix} \boldsymbol{n}_1^T \\ \boldsymbol{n}_2^T \\ \boldsymbol{n}_3^T \end{bmatrix} \begin{bmatrix} \lambda_1 \boldsymbol{n}_1 & \lambda_2 \boldsymbol{n}_2 & \lambda_3 \boldsymbol{n}_3 \end{bmatrix}$$

$$= \begin{bmatrix} \lambda_1 \boldsymbol{n}_1^T \boldsymbol{n}_1 & \lambda_2 \boldsymbol{n}_1^T \boldsymbol{n}_2 & \lambda_3 \boldsymbol{n}_1^T \boldsymbol{n}_3 \\ \lambda_1 \boldsymbol{n}_2^T \boldsymbol{n}_1 & \lambda_2 \boldsymbol{n}_2^T \boldsymbol{n}_2 & \lambda_3 \boldsymbol{n}_2^T \boldsymbol{n}_3 \\ \lambda_1 \boldsymbol{n}_3^T \boldsymbol{n}_1 & \lambda_2 \boldsymbol{n}_3^T \boldsymbol{n}_2 & \lambda_3 \boldsymbol{n}_3^T \boldsymbol{n}_3 \end{bmatrix}$$

However, as the $\boldsymbol{n}$-vectors are unit vectors orthogonal to each other we finally obtain

$$\boldsymbol{\varepsilon}' = \begin{bmatrix} \lambda_1 & 0 & 0 \\ 0 & \lambda_2 & 0 \\ 0 & 0 & \lambda_3 \end{bmatrix} \tag{2.36}$$

Accordingly, we have obtained the important result that if the coordinate system is chosen collinearly with the principal directions $\boldsymbol{n}_1$, $\boldsymbol{n}_2$ and $\boldsymbol{n}_3$, then the strain tensor becomes diagonal and the normal strains become equal to λ_1, λ_2 and λ_3. This result is in accordance with the physical conditions, which were specified in the beginning when the eigenvalue problem was formulated. This important result also illustrates why the eigenvalues are called the principal strains and the eigenvectors the principal directions. The principal strains are often denoted by ε_1, ε_2 and ε_3, i.e. $\varepsilon_1 = \lambda_1$, $\varepsilon_2 = \lambda_2$ and $\varepsilon_3 = \lambda_3$.

The above result can be summarized by stating that if the coordinate system is collinear with the principal directions we have in accordance with (2.19) and (2.36) that

$$\boxed{\boldsymbol{\varepsilon}' = \boldsymbol{A}\boldsymbol{\varepsilon}\boldsymbol{A}^T = \begin{bmatrix} \varepsilon_1 & 0 & 0 \\ 0 & \varepsilon_2 & 0 \\ 0 & 0 & \varepsilon_3 \end{bmatrix} \quad \textit{for} \quad \boldsymbol{A}^T = \begin{bmatrix} \boldsymbol{n}_1 & \boldsymbol{n}_2 & \boldsymbol{n}_3 \end{bmatrix}} \tag{2.37}$$

2.7 Extremum values of the normal strain

The normal strain ε in any direction n_i is determined by (2.11), i.e.

$$\varepsilon = n_i \varepsilon_{ij} n_j$$

For different directions of n_i, different ε-values are achieved. It will now be proved that the normal strain ε takes stationary values, i.e. maximum or minimum values, when the direction n_i is in the direction of one of the principal axes.

To find the stationary values of ε, the n_i-vector is varied. However, the n_i-components cannot be varied arbitrarily, as we have the constraint

$$n_i n_i - 1 = 0$$

Accordingly, we employ the *method of Lagrange* and find stationary values of the function

$$\psi = n_i \varepsilon_{ij} n_j - \alpha(n_i n_i - 1) \tag{2.38}$$

where now the n_i-components and α are independent quantities, α being a Lagrangian multiplier. From (2.38), where $\psi = \psi(n_i, \alpha)$ we obtain

$$\frac{\partial \psi}{\partial n_k} = \varepsilon_{kj} n_j + \varepsilon_{ki} n_i - \alpha(n_k + n_k) = 0 \tag{2.39}$$

and

$$\frac{\partial \psi}{\partial \alpha} = n_i n_i - 1 = 0 \tag{2.40}$$

Equation (2.39) can be written as

$$\varepsilon_{kj} n_j - \alpha n_k = 0 \qquad \text{or} \qquad (\varepsilon_{ij} - \alpha \delta_{ij}) n_j = 0 \tag{2.41}$$

Therefore, stationary values for the normal strain ε are obtained by solution of the homogeneous equation system (2.41) subject to the condition (2.40). We immediately observe that this is exactly the same eigenvalue problem as stated by (2.24) proving that stationary values, i.e. maximum and minimum values, of the normal strain ε occur in the principal directions.

2.8 Cayley-Hamilton's theorem

We will now prove an interesting relation for the strain tensor (occasionally also called the strain matrix).

Considering the eigenvalue problem (2.24), we premultiply this equation by $\boldsymbol{\varepsilon}$, i.e.

$$\boldsymbol{\varepsilon}^2 \boldsymbol{n} = \lambda \boldsymbol{\varepsilon} \boldsymbol{n} = \lambda^2 \boldsymbol{n}$$

where the notation

$$\boldsymbol{\varepsilon}^2 = \boldsymbol{\varepsilon}\boldsymbol{\varepsilon} \tag{2.42}$$

has been used. Proceeding, we obtain the general result

$$\boldsymbol{\varepsilon}^\alpha \boldsymbol{n} = \lambda^\alpha \boldsymbol{n}\,; \qquad \alpha = 0, \pm 1, \pm 2 \ldots \tag{2.43}$$

where α is any integer (positive, negative or zero). If α is negative, say α = -2 then, in accordance with (2.42), we define

$$\boldsymbol{\varepsilon}^{-2} = \boldsymbol{\varepsilon}^{-1}\boldsymbol{\varepsilon}^{-1}$$

Hence, (2.43) holds even for negative values of the integer α provided that $\boldsymbol{\varepsilon}^{-1}$ exists i.e. provided that det $\boldsymbol{\varepsilon} \neq 0$. Moreover, in accordance with the usual definition that $x^0 = 1$ we make the following definition

$$\boldsymbol{\varepsilon}^0 = \boldsymbol{I}$$

From this definition follows that (2.43) holds even when $\alpha = 0$.

Equation (2.43) shows that if $\boldsymbol{\varepsilon}$ has the eigenvalue λ and eigenvector $\boldsymbol{n}$, then $\boldsymbol{\varepsilon}^\alpha$ will have the same eigenvector and the eigenvalue λ^α. Now, multiply the characteristic equation for λ, as given by (2.27), by $\boldsymbol{n}$ to obtain

$$-\lambda^3 \boldsymbol{n} + \theta_1 \lambda^2 \boldsymbol{n} - \theta_2 \lambda \boldsymbol{n} + \theta_3 \boldsymbol{n} = \boldsymbol{0} \tag{2.44}$$

where $\mathbf{0}$ is given by $\mathbf{0}^T = [0\ 0\ 0]$. Use of (2.43) in (2.44) gives

$$(-\boldsymbol{\varepsilon}^3 + \theta_1 \boldsymbol{\varepsilon}^2 - \theta_2 \boldsymbol{\varepsilon} + \theta_3 \boldsymbol{I})\boldsymbol{n} = \mathbf{0}$$

We know that this equation is fulfilled for $\boldsymbol{n}$ given by any of the three eigenvectors, i.e. these three matrix equations can be combined into the following format

$$(-\boldsymbol{\varepsilon}^3 + \theta_1 \boldsymbol{\varepsilon}^2 - \theta_2 \boldsymbol{\varepsilon} + \theta_3 \boldsymbol{I}) \begin{bmatrix} \boldsymbol{n}_1 & \boldsymbol{n}_2 & \boldsymbol{n}_3 \end{bmatrix} = \mathbf{0} \tag{2.45}$$

where $\mathbf{0}$ now denotes the 3 x 3 null matrix. As the unit vectors $\boldsymbol{n}_1$, $\boldsymbol{n}_2$ and $\boldsymbol{n}_3$ are orthogonal, we have according to (1.28) that

$$\begin{bmatrix} \boldsymbol{n}_1 & \boldsymbol{n}_2 & \boldsymbol{n}_3 \end{bmatrix} = \boldsymbol{A}^T$$

where $\boldsymbol{A}$ is some transformation matrix. Expression (2.45) therefore takes the form

$$(-\boldsymbol{\varepsilon}^3 + \theta_1 \boldsymbol{\varepsilon}^2 - \theta_2 \boldsymbol{\varepsilon} + \theta_3 \boldsymbol{I})\boldsymbol{A}^T = \mathbf{0}$$

Postmultiplication by $\boldsymbol{A}$ and noting that $\boldsymbol{A}^T \boldsymbol{A} = \boldsymbol{I}$ give

$$\boxed{-\boldsymbol{\varepsilon}^3 + \theta_1 \boldsymbol{\varepsilon}^2 - \theta_2 \boldsymbol{\varepsilon} + \theta_3 \boldsymbol{I} = \mathbf{0}} \tag{2.46}$$

This equation is similar to the characteristic equation for λ, cf. (2.27) and the result is thus often stated by saying that

The strain matrix satisfies its own characteristic equation

This important result is the *Cayley-Hamilton theorem.* Note that (2.46) is a matrix equation.

A significant implication of (2.46) is that an expression involving the term $\boldsymbol{\varepsilon}^3$ can always be simplified so that it only involves terms of $\boldsymbol{\varepsilon}^2$, $\boldsymbol{\varepsilon}$ and $\boldsymbol{I}$. More generally, if we multiply (2.46) by $\boldsymbol{\varepsilon}^\alpha$, where α is any integer (positive, negative or zero), we obtain

$$\boxed{\boldsymbol{\varepsilon}^{3+\alpha} = \theta_1 \boldsymbol{\varepsilon}^{2+\alpha} - \theta_2 \boldsymbol{\varepsilon}^{1+\alpha} + \theta_3 \boldsymbol{\varepsilon}^\alpha}$$

If $\alpha \geq 0$ this means that any $\boldsymbol{\varepsilon}^{3+\alpha}$-term can be replaced by lower order powers of $\boldsymbol{\varepsilon}$. If $\alpha \leq 0$ (which presumes that $\boldsymbol{\varepsilon}^{-1}$ exists), then any $\boldsymbol{\varepsilon}^\alpha$-term can be replaced by higher order powers of $\boldsymbol{\varepsilon}$. Such manipulations are of importance in the so-called *representation theorems,* to be discussed later in Chapter 6.

2.9 Deviatoric strains

Instead of the full strain tensor, it is often convenient to operate with the so-called *deviatoric strain tensor* e_{ij} defined by

$$\boxed{e_{ij} = \varepsilon_{ij} - \tfrac{1}{3}\varepsilon_{kk}\delta_{ij}} \tag{2.47}$$

where $\frac{1}{3}\,\varepsilon_{kk}\,\delta_{ij}$ is the *volumetric* or *spherical strain tensor*, which only involves diagonal terms. As both ε_{ij} and δ_{ij} are second-order tensors, it follows directly that so is e_{ij}. Therefore, by analogy with (2.19) and (2.20) we have

$$\boxed{e'_{ij} = A_{ik}e_{kl}A_{jl} \qquad or \qquad \boldsymbol{e}' = \boldsymbol{A}\boldsymbol{e}\boldsymbol{A}^T}$$

and

$$e_{ij} = A_{ki}e'_{kl}A_{lj} \qquad \text{or} \qquad \boldsymbol{e} = \boldsymbol{A}^T\boldsymbol{e}'\boldsymbol{A}$$

Moreover, we observe from definition (2.47) that

$$\boxed{e_{ii} = 0} \tag{2.48}$$

In a principal coordinate system the principal strains become ε_1, ε_2 and ε_3. Referring to (2.10), this means that the relative volume change due to the deformation becomes

$$\frac{dV^* - dV}{dV} = \frac{(1+\varepsilon_1)dx_1(1+\varepsilon_2)dx_2(1+\varepsilon_3)dx_3 - dx_1dx_2dx_3}{dx_1dx_2dx_3}$$

where dV is the infinitesimal volume before deformation, which owing to the deformation changes to dV^*. In accordance with our assumption of small strains, we ignore higher order strain terms and the expression above becomes

$$\boxed{\frac{dV^* - dV}{dV} = \varepsilon_1 + \varepsilon_2 + \varepsilon_3 = \varepsilon_{kk}} \tag{2.49}$$

We conclude that ε_{kk} is equal to the relative volume change, i.e. an incompressible material is characterized by ε_{kk}=0. Moreover, it may be recalled that ε_{kk} is an invariant.

Referring to (2.47), it appears that the off-diagonal terms of e_{ij} and ε_{ij} are identical. Consequently, it can be concluded that the volumetric strain tensor only influences the volumetric changes whereas the deviatoric strain tensor only influences the shearing (distortion) of the material.

Returning to the eigenvalue problem (2.24), we may eliminate ε_{ij} by means of (2.47) to obtain.

$$\left[\boldsymbol{e} - (\lambda - \frac{\varepsilon_{kk}}{3})\boldsymbol{I}\right]\boldsymbol{n} = \boldsymbol{0}$$

It is concluded that the eigenvalues of e_{ij} are given by λ - $\varepsilon_{kk}/3$ whereas the eigenvectors, i.e. the principal directions, are identical for the deviatoric strain tensor and the strain tensor. The fact that the principal directions of e_{ij} and ε_{ij} are identical follows also directly from the observation that they have identical off-diagonal terms, i.e. when ε_{ij} is diagonal, so is e_{ij}.

2.10 Important strain invariants

We have seen quite a number of different invariants and it might be convenient to summarize these invariants and make use of the opportunity to introduce additional invariants which later turn out to be of importance.

The Cauchy invariants are given by (2.28)

$$\theta_1 = \varepsilon_{ii} \;\; ; \quad \theta_2 = \frac{1}{2}\theta_1^2 - \frac{1}{2}\varepsilon_{ij}\varepsilon_{ji} \;\; ; \quad \theta_3 = \det(\varepsilon_{ij}) = \varepsilon_1\varepsilon_2\varepsilon_3 \tag{2.50}$$

In general, to prove that a quantity is an invariant, we must demonstrate that it takes the same value in all coordinate systems. As a prototype of such an evaluation we consider

$$\varepsilon'_{ij}\varepsilon'_{ij} = A_{ik}\varepsilon_{kl}A_{jl}A_{is}\varepsilon_{st}A_{jt} = \delta_{ks}\varepsilon_{kl}\varepsilon_{st}\delta_{lt} = \varepsilon_{sl}\varepsilon_{sl}$$

where advantage is taken of the transformation rule (2.19) as well as of (1.21). This demonstrates that the quantity $\varepsilon_{ij}\varepsilon_{ij}$ is an invariant. Likewise, it is easily shown that ε_{ii} and $\varepsilon_{ij}\varepsilon_{jk}\varepsilon_{ki}$ are invariants. We can therefore list the following so-called *generic invariants* , where the term 'generic' reflects the systematic manner of their definition

$$\begin{aligned} \tilde{I}_1 &= \varepsilon_{ii} = \varepsilon_1 + \varepsilon_2 + \varepsilon_3 \\ \tilde{I}_2 &= \tfrac{1}{2}\varepsilon_{ij}\varepsilon_{ji} = \tfrac{1}{2}(\varepsilon_1^2 + \varepsilon_2^2 + \varepsilon_3^2) \\ \tilde{I}_3 &= \tfrac{1}{3}\varepsilon_{ij}\varepsilon_{jk}\varepsilon_{ki} = \tfrac{1}{3}(\varepsilon_1^3 + \varepsilon_2^3 + \varepsilon_3^3) \end{aligned} \tag{2.51}$$

Occasionally, it is convenient to express these invariants in matrix notation and for that purpose, we define the *trace* of a 3 x 3 square matrix $\boldsymbol{B}$ by

$$tr\ \boldsymbol{B} = B_{ii}$$

i.e.

$$\tilde{I}_1 = tr\ \varepsilon$$

Define the quantity B_{ij} by

$$B_{ij} = \varepsilon_{ik}\varepsilon_{kj} \qquad \text{or} \qquad \boldsymbol{B} = \varepsilon\varepsilon = \varepsilon^2$$

i.e. $tr\ \boldsymbol{B} = B_{ii} = \varepsilon_{ik}\ \varepsilon_{ki}$ and we therefore obtain

$$\tilde{I}_2 = \frac{1}{2}\ tr\ (\varepsilon^2) \qquad \text{and likewise} \qquad \tilde{I}_3 = \frac{1}{3}\ tr\ (\varepsilon^3)$$

It turns out that it is possible to obtain a unique relation between the Cauchy-invariants θ_1, θ_2 and θ_3 and the generic invariants $\tilde{I}_1$, $\tilde{I}_2$ and $\tilde{I}_3$. We have

$$\tilde{I}_1 = \theta_1 \;\; ; \quad \tilde{I}_2 = \frac{1}{2}\theta_1^2 - \theta_2 \;\; ; \quad \tilde{I}_3 = \frac{1}{3}\theta_1^3 - \theta_1\theta_2 + \theta_3 \tag{2.52}$$

The first two of these solutions follows directly from (2.50a), (2.51a) and (2.50b), (2.51b) respectively. Equation (2.52c) is easily established, for instance by taking the trace of matrix expression (2.46) given by Cayley-Hamilton's theorem and then using (2.52a) and (2.52b).

The inverse relations of (2.52) provide the following expressions

$$\theta_1 = \tilde{I}_1 \quad ; \quad \theta_2 = \frac{1}{2}\tilde{I}_1^2 - \tilde{I}_2 \quad ; \quad \theta_3 = \tilde{I}_3 + \frac{1}{6}\tilde{I}_1^3 - \tilde{I}_1\tilde{I}_2$$

It appears that a unique relation exists between θ_1, θ_2, θ_3 and $\tilde{I}_1$, $\tilde{I}_2$, $\tilde{I}_3$.

Now, let us turn to the *generic invariants* of the deviatoric strain tensor defined by analogy with (2.51). We have

$$\boxed{\begin{aligned} \tilde{J}_1 &= e_{ii} = tr\ \boldsymbol{e} = e_1 + e_2 + e_3 = 0 \\ \tilde{J}_2 &= \tfrac{1}{2}e_{ij}e_{ji} = \tfrac{1}{2}tr\ (\boldsymbol{e}^2) = \tfrac{1}{2}(e_1^2 + e_2^2 + e_3^2) \\ \tilde{J}_3 &= \tfrac{1}{3}e_{ij}e_{jk}e_{ki} = \tfrac{1}{3}tr\ (\boldsymbol{e}^3) = \tfrac{1}{3}(e_1^3 + e_2^3 + e_3^3) = e_1\ e_2\ e_3 \end{aligned}} \qquad (2.53)$$

To prove the last relation that $\tilde{J}_3 = e_1e_2e_3$, we first observe that

$$(e_2 + e_3)^3 = e_2^3 + e_3^3 + 3e_2e_3(e_2 + e_3)$$

and since $e_2 + e_3 = -e_1$, we obtain

$$-e_1^3 = e_2^3 + e_3^3 - 3e_1e_2e_3$$

From the definition of $\tilde{J}_3 = \frac{1}{3}(e_1^3 + e_2^3 + e_3^3)$, it then follows that

$$\tilde{J}_3 = e_1e_2e_3$$

which was to be proved.

Moreover, using the definition of the deviatoric strain tensor as given by (2.47) in (2.53), we obtain

$$\tilde{J}_2 = \tilde{I}_2 - \frac{1}{6}\tilde{I}_1^2 \quad ; \quad \tilde{J}_3 = \tilde{I}_3 - \frac{2}{3}\tilde{I}_1\tilde{I}_2 + \frac{2}{27}\tilde{I}_1^3 \qquad (2.54)$$

and the inverse relations become

$$\tilde{I}_2 = \tilde{J}_2 + \frac{1}{6}\tilde{I}_1^2 \quad ; \quad \tilde{I}_3 = \tilde{J}_3 + \frac{2}{3}\tilde{I}_1\tilde{J}_2 + \frac{1}{27}\tilde{I}_1^3 \qquad (2.55)$$

Therefore, instead of using the set of invariants $\tilde{I}_1$, $\tilde{I}_2$, $\tilde{I}_3$ we may equally well use the set $\tilde{I}_1$, $\tilde{J}_2$, $\tilde{J}_3$.

An *octahedral plane* is defined as a plane where the normal to that plane makes equal angles to the three principal strain directions. Eight such planes exist and one example is shown in Fig. 2.6 where the axes 1,2 and 3 refer to

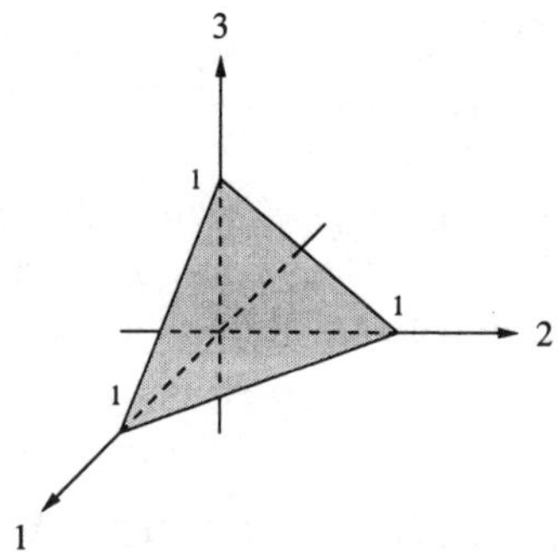

Figure 2.6: Example of octahedral plane.

the principal strain directions. For the normal to the octahedral plane shown in Fig. 2.6, we have

$$\boldsymbol{n} = \frac{1}{\sqrt{3}} \begin{bmatrix} 1 \\ 1 \\ 1 \end{bmatrix}$$

In the coordinate system collinear with the principal strain directions, the strain tensor takes the form

$$\boldsymbol{\varepsilon} = \begin{bmatrix} \varepsilon_1 & 0 & 0 \\ 0 & \varepsilon_2 & 0 \\ 0 & 0 & \varepsilon_3 \end{bmatrix}$$

The vector $\boldsymbol{q}$ is defined by $\boldsymbol{q} = \boldsymbol{\varepsilon n}$ cf. (2.21). It then follows from Fig. 2.5 that the normal strain ε_0 and tensorial shear strain $\gamma_0/2$ on the octahedral plane are given by

$$\varepsilon_o = \boldsymbol{n}^T \boldsymbol{q} \, ; \qquad \frac{\gamma_0}{2} = \sqrt{\boldsymbol{q}^T \boldsymbol{q} - \varepsilon_0{}^2}$$

where ε_0 is called the *octahedral normal strain* and γ_0 is called the *octahedral shear strain*. It follows that

$$\varepsilon_o = \frac{1}{3}\tilde{I}_1 \, ; \qquad \frac{\gamma_0}{2} = \sqrt{\frac{1}{3}(\varepsilon_1^2 + \varepsilon_2^2 + \varepsilon_3^2) - \frac{1}{9}\tilde{I}_1^2}$$

According to (2.47), we have

$$\varepsilon_1 = e_1 + \frac{1}{3}\tilde{I}_1 \, ; \qquad \varepsilon_2 = e_2 + \frac{1}{3}\tilde{I}_1 \, ; \qquad \varepsilon_3 = e_3 + \frac{1}{3}\tilde{I}_1$$

i.e.

$$\frac{\gamma_0}{2} = \sqrt{\frac{1}{3}\left[e_1^2 + e_2^2 + e_3^2 + \frac{1}{3}\tilde{I}_1^2 + \frac{2}{3}(e_1 + e_2 + e_3)\tilde{I}_1\right] - \frac{1}{9}\tilde{I}_1^2}$$

Due to (2.48) and (2.53b), we conclude that

$$\boxed{\varepsilon_0 = \frac{1}{3}\tilde{I}_1\,; \qquad \gamma_0 = 2\sqrt{\frac{2}{3}\tilde{J}_2}} \tag{2.56}$$

It is easily shown that these relations hold not only for the octahedral plane shown in Fig. 2.6, but also for all the other octahedral planes. Finally, it is emphasized that γ_0 is the engineering shear strain as already suggested by the notation.

2.11 Change of coordinate system - Mohr's circle

We have previously discussed the consequences of choosing a different coordinate system. Let us now consider the special case where the coordinate system is rotated about the x_3-axis as shown in Fig. 2.7.

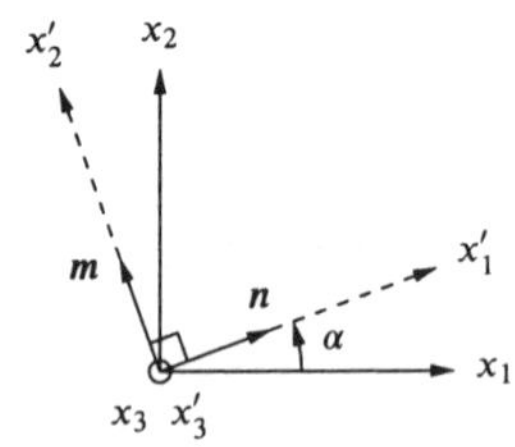

Figure 2.7: Rotation of coordinate system about the x_3-axis.

Thus, we change the coordinate system from the x_i-system to the x_i'-system, where the x_3- and x_3'-axes are identical and we shall investigate the strain tensor in this new coordinate-system. In the old x_i-coordinate system the unit vector $\boldsymbol{n}$ along the x_1'-axis and the unit vector $\boldsymbol{m}$ along the x_2'-axis have the components

$$[n_i] = \begin{bmatrix} \cos\alpha \\ \sin\alpha \\ 0 \end{bmatrix} \;; \qquad [m_i] = \begin{bmatrix} -\sin\alpha \\ \cos\alpha \\ 0 \end{bmatrix}$$

where α is positive, when going in the counter-clockwise direction of the x_1x_2-plane, cf. Fig. 2.7. We then obtain

$$[\varepsilon_{ij}][n_j] = \begin{bmatrix} \varepsilon_{11} & \varepsilon_{12} & \varepsilon_{13} \\ \varepsilon_{12} & \varepsilon_{22} & \varepsilon_{23} \\ \varepsilon_{13} & \varepsilon_{23} & \varepsilon_{33} \end{bmatrix} \begin{bmatrix} \cos\alpha \\ \sin\alpha \\ 0 \end{bmatrix} = \begin{bmatrix} \varepsilon_{11}\cos\alpha + \varepsilon_{12}\sin\alpha \\ \varepsilon_{12}\cos\alpha + \varepsilon_{22}\sin\alpha \\ \varepsilon_{13}\cos\alpha + \varepsilon_{23}\sin\alpha \end{bmatrix}$$

With (2.11), the normal strain ε_{nn} in the direction of $\boldsymbol{n}$ becomes

$$\varepsilon_{nn} = n_i\varepsilon_{ij}n_j = \varepsilon_{11}\cos^2\alpha + \varepsilon_{22}\sin^2\alpha + \varepsilon_{12}\sin 2\alpha \tag{2.57}$$

Similarly, using (2.17), the shear strain ε_{nm} between the orthogonal axes $\boldsymbol{n}$ and $\boldsymbol{m}$ becomes

$$\varepsilon_{nm} = m_i \varepsilon_{ij} n_j = -\frac{1}{2}(\varepsilon_{11} - \varepsilon_{22}) \sin 2\alpha + \varepsilon_{12} \cos 2\alpha \tag{2.58}$$

It is obvious that when replacing α by $\alpha + \pi/2$ in (2.57), we obtain an expression for the normal strain ε_{mm} in the $\boldsymbol{m}$-direction, as this replacement corresponds to a rotation of 90° of the nm-coordinate system, i.e.

$$\varepsilon_{mm} = \varepsilon_{11} \sin^2\alpha + \varepsilon_{22} \cos^2\alpha - \varepsilon_{12} \sin 2\alpha \tag{2.59}$$

Thus, instead of the components ε_{11}, ε_{22}, ε_{12} in the old coordinate system, we have determined the components ε_{nn}, ε_{mm}, ε_{nm} in the new coordinate system by means of (2.57) - (2.59).

Occasionally, it is of interest to be able to determine ε_{11}, ε_{22}, ε_{12} provided that ε_{nn}, ε_{mm}, ε_{nm} are known. Let us assume that the nmx_3-axes comprise the original coordinate system and let us rotate this coordinate system the angle $-\alpha$ about the x_3-axis. We can then obtain the required result directly from (2.57) - (2.59) by replacing α with $-\alpha$ and ε_{11}, ε_{22}, ε_{12} with ε_{nn}, ε_{mm}, ε_{nm} respectively. This leads to

$$\boxed{\begin{aligned} \varepsilon_{11} &= \varepsilon_{nn} \cos^2\alpha + \varepsilon_{mm} \sin^2\alpha - \varepsilon_{nm} \sin 2\alpha \\ \varepsilon_{22} &= \varepsilon_{nn} \sin^2\alpha + \varepsilon_{mm} \cos^2\alpha + \varepsilon_{nm} \sin 2\alpha \\ \varepsilon_{12} &= \tfrac{1}{2}(\varepsilon_{nn} - \varepsilon_{mm}) \sin 2\alpha + \varepsilon_{nm} \cos 2\alpha \end{aligned}} \tag{2.60}$$

where the angle α still is measured positive in the counter-clockwise direction of the x_1x_2-plane, cf. Fig. 2.7.

It appears from (2.57) - (2.59) that $\varepsilon_{nn}(\alpha+\pi) = \varepsilon_{nn}(\alpha)$, $\varepsilon_{mm}(\alpha+\pi) = \varepsilon_{mm}(\alpha)$ and $\varepsilon_{nm}(\alpha+\pi) = \varepsilon_{nm}(\alpha)$, i.e. the strain components vary with a period of 180°. This property complies with the physical interpretation of the normal and the shear strain. Thus it appears that it is sufficiently general only to consider α-values in the range

$$\boxed{0 \leq \alpha < \pi} \tag{2.61}$$

It is of interest that (2.58) implies that a particular α-value exists, $\alpha = \psi$, for which the shear strain ε_{nm} becomes zero. This angle ψ is determined by

$$\boxed{\tan 2\psi = \frac{2\varepsilon_{12}}{\varepsilon_{11} - \varepsilon_{22}}} \tag{2.62}$$

which even applies when $\varepsilon_{11} = \varepsilon_{22}$, since (2.58) for $\varepsilon_{11} = \varepsilon_{22}$ provides $\cos 2\psi = 0$, i.e. the same solution as (2.62).

Therefore, when $\alpha = \psi$, we find that $\varepsilon_{nm} = 0$; however the shear strains ε_{nx_3} and ε_{mx_3} will, in general, be different from zero. Consequently, only if the x_3-

direction is a principal direction, will the directions $\boldsymbol{n}$ and $\boldsymbol{m}$ defined by $\alpha = \psi$ also be principal directions. This situation will be assumed in the following.

Consequently, when $\alpha = \psi$, where ψ is determined by (2.62), we have determined the position of the new coordinate system, so that it corresponds to the principal directions. In accordance with our previous discussion of principal strains and directions, the principal directions determined by ψ satisfying (2.62) are perpendicular to each other. In the present situation this implies that if the new coordinate system, collinear with the principal directions, is rotated by $\pm 90^o$, $\pm 180^o$ etc., then these new positions of the coordinate system will also be principal directions. Therefore, when we have determined the angle ψ, then any angle $\psi \pm n\pi/2$, where $n = 1, 2, 3$ .... will also be a valid ψ-value for which the shear strain $\varepsilon_{nm} = 0$.

With this discussion in mind, it is advantageous to accept some convention, when selecting a ψ-value which satisfies (2.62). In accordance with the convention (2.61) and as ψ is just a special value of α, for which no shear strains exist, it is advantageous to have the same range for ψ as for α. Therefore, we accept the following range for ψ

$$\boxed{0 \leq \psi < \pi} \tag{2.63}$$

However, as the tan-function is periodic with a period of π, this implies that when solving (2.62) subjected to (2.63), two ψ-solutions will result. Both ψ-solutions are acceptable and we will later – after having introduced Mohr's circle of strain – present a procedure, by which it is possible to select one uniquely defined ψ-value.

It turns out that an elegant geometrical interpretation can be made of (2.57) and (2.58). Define first the quantities a and b by

$$a = \frac{1}{2}(\varepsilon_{11} + \varepsilon_{22}), \qquad b = \frac{1}{2}(\varepsilon_{11} - \varepsilon_{22}) \tag{2.64}$$

Observing that $\cos^2\alpha = (1{+}\cos 2\alpha)/2$ and $\sin^2\alpha{=}(1{-}\cos 2\alpha)/2$, (2.57) and (2.58) can be written as

$$\begin{aligned} \varepsilon_{nn} - a &= b\cos 2\alpha + \varepsilon_{12}\sin 2\alpha \\ \varepsilon_{nm} &= -b\sin 2\alpha + \varepsilon_{12}\cos 2\alpha \end{aligned} \tag{2.65}$$

By squaring each of these equations and adding, we obtain

$$(\varepsilon_{nn} - a)^2 + \varepsilon_{nm}^2 = R^2 \tag{2.66}$$

where the quantity R is defined by

$$R = \sqrt{b^2 + \varepsilon_{12}^2} = \sqrt{\left(\frac{\varepsilon_{11} - \varepsilon_{22}}{2}\right)^2 + \varepsilon_{12}^2} \tag{2.67}$$

Since we are considering a given strain state, the quantity R is constant. Returning to expression (2.66), we observe that it is exactly the equation for a circle in

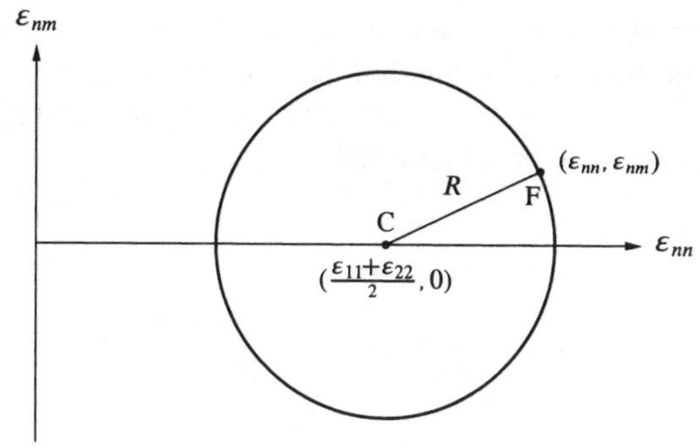

Figure 2.8: Mohr's circle of strain.

a ε_{nn}, ε_{nm}-coordinate system with the center of the circle at $(a, 0)$ and a radius R. Since the point $(\varepsilon_{nn}, \varepsilon_{nm})$ is located on this circle, we have obtained *Mohr's circle of strain* (Mohr, 1882) as shown in Fig. 2.8.

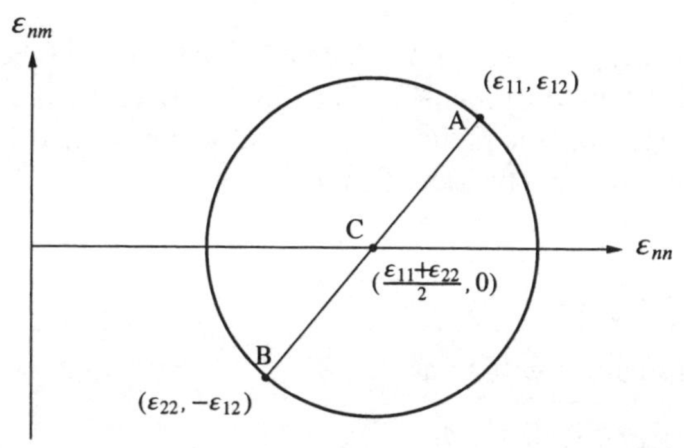

Figure 2.9: Graphical construction of Mohr's circle of strain.

The construction above of Mohr's circle involves the analytical determination of the center and of the radius. For practical purposes, however, a direct graphical construction is to be preferred. To this end, we note from (2.57) and (2.58) that

$$\varepsilon_{nn}(\alpha = 0) = \varepsilon_{11} \; ; \qquad \varepsilon_{nm}(\alpha = 0) = \varepsilon_{12} \qquad \text{point A}$$

and

$$\varepsilon_{nn}(\alpha = \frac{\pi}{2}) = \varepsilon_{22} \; ; \qquad \varepsilon_{nm}(\alpha = \frac{\pi}{2}) = -\varepsilon_{12} \qquad \text{point B}$$

These $(\varepsilon_{nn}, \varepsilon_{nm})$-values also correspond to points on Mohr's circle and they are referred to as point A and B respectively, in Fig. 2.9. If a line is drawn between point A and B, the midpoint of this line has the coordinates $(\frac{1}{2}(\varepsilon_{11} + \varepsilon_{22}), 0)$, i.e. this midpoint is precisely the center of Mohr's circle. Having identified this center, Mohr's circle can be drawn directly.

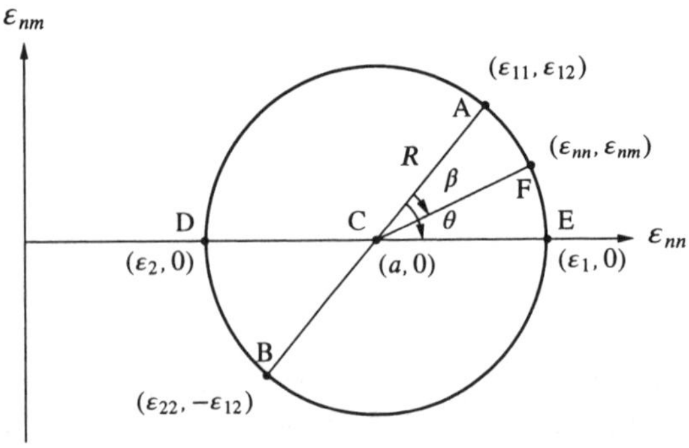

Figure 2.10: Identification of different angles $(\varepsilon_1 \geq \varepsilon_2)$.

The strain point $(\varepsilon_{nn}(\alpha), \varepsilon_{nm}(\alpha))$ is located somewhere on the circle and according to Fig. 2.10, its position F can be characterized by the angle β that is measured clockwise from the radius CA; also the angle θ is shown in this figure. From Fig. 2.10, point F can be identified as

$$\begin{aligned} \varepsilon_{nn} &= a + R\cos(\theta - \beta) \\ \varepsilon_{nm} &= R\sin(\theta - \beta) \end{aligned}$$

These expressions can be written as

$$\begin{aligned} \varepsilon_{nn} &= a + R\cos\theta\cos\beta + R\sin\theta\sin\beta \\ \varepsilon_{nm} &= R\sin\theta\cos\beta - R\cos\theta\sin\beta \end{aligned} \tag{2.68}$$

From Fig. 2.10, it also follows that $\frac{1}{2}(\varepsilon_{11} - \varepsilon_{22}) = R\cos\theta$ which with (2.64) results in $b = R\cos\theta$; we also have $\varepsilon_{12} = R\sin\theta$. Then (2.68) becomes

$$\begin{aligned} \varepsilon_{nn} &= a + b\cos\beta + \varepsilon_{12}\sin\beta \\ \varepsilon_{nm} &= \varepsilon_{12}\cos\beta - b\sin\beta \end{aligned}$$

and a comparison with (2.65) proves that $\beta = 2\alpha$.

We then obtain *Mohr's circle of strain* in the following fashion. For given strain components $\varepsilon_{11}, \varepsilon_{22}$ and ε_{12} plot the points A $(\varepsilon_{11}, \varepsilon_{12})$ and B $(\varepsilon_{22}, -\varepsilon_{12})$ in the $\varepsilon_{nn}, \varepsilon_{nm}$-coordinate system shown in Fig. 2.11. Draw a line between these two points. The intersection of this line with the ε_{nn}-axis defines the center C

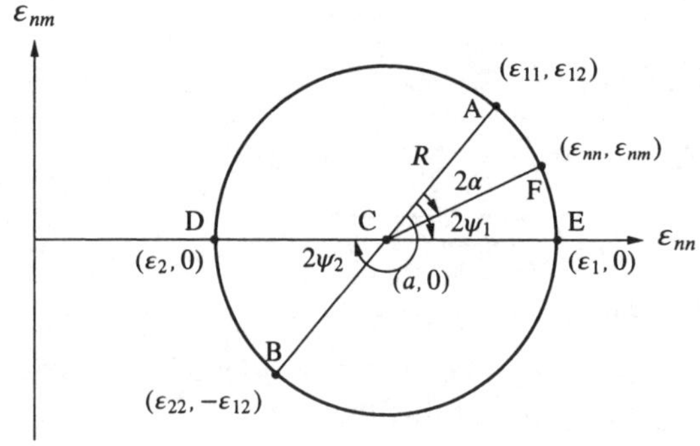

Figure 2.11: Identification of different angles ($\varepsilon_1 \geq \varepsilon_2$).

of a circle that contains the points A and B. Draw this circle. For an arbitrary angle α, cf. Fig. 2.7, the corresponding values of ε_{nn} and ε_{nm} are then given by point F in Fig. 2.11. Note that

The angle 2α in Mohr's circle in Fig. 2.11 is measured positive in the clockwise direction

and compare with Fig. 2.7 where α is measured positive in the counter-clockwise direction.

Note also that the graphical interpretation given by Mohr's circle only holds when both the $xy-$ and $\varepsilon_{nn}, \varepsilon_{nm}$-coordinate systems are right-handed coordinate systems; otherwise Mohr's circle of strain has to be modified.

The two points where the circle intersects the ε_{nn}-axis correspond to α-values where the shear strain $\varepsilon_{nm} = 0$; these two points therefore correspond to the principal strains and if we adopt the convention that

$$\boxed{\varepsilon_1 > \varepsilon_2} \tag{2.69}$$

the principal strains are given by the points E and D in Fig. 2.11. If we furthermore make the convention that the α-value corresponding to ε_1 is called ψ_1 and the α-value corresponding to ε_2 is called ψ_2, we then have the situation illustrated in Fig. 2.11. Due to (2.69) it appears for Fig. 2.11 that

$$\boxed{0 \leq \psi_1 < \pi} \tag{2.70}$$

To identify this angle, (2.62) could be used, but owing to the periodicity of the function $\tan 2\psi$, it is more convenient to observe from Fig. 2.11 that

$$\begin{aligned} \cos 2\psi_1 &= \frac{\varepsilon_{11} - \varepsilon_{22}}{2R} \\ \sin 2\psi_1 &= \frac{\varepsilon_{12}}{R} \\ \textit{where} \quad R &= \sqrt{\left(\frac{\varepsilon_{11} - \varepsilon_{22}}{2}\right)^2 + \varepsilon_{12}^2} \end{aligned}$$

Inserting the values for ε_{11}, ε_{22} and ε_{12} we can identify the signs for $\sin 2\psi_1$ and $\cos 2\psi_1$ and thereby the quadrant that the angle $2\psi_1$ is located in and thereby easily determine ψ_1 so that (2.70) is fulfilled.

Note the convention that elongation is considered positive and that $\varepsilon_1 \geq \varepsilon_2$. It is then emphasised that the angle $2\psi_1$ is the angle to the largest principal strain ε_1.

It is also concluded from Fig. 2.11 that the principal strains are given by

$$\left.\begin{matrix} \varepsilon_1 \\ \varepsilon_2 \end{matrix}\right\} = \frac{1}{2}(\varepsilon_{11} - \varepsilon_{22}) \pm R \tag{2.71}$$

We finally observe from Fig. 2.11 that the extremum values for the shear strain are determined by the radius R, i.e. (2.67) gives with the convention $\varepsilon_1 \geq \varepsilon_2$

$$\varepsilon_{nm,max} = \frac{1}{2}(\varepsilon_1 - \varepsilon_2)\ ; \qquad \varepsilon_{nm,min} = -\frac{1}{2}(\varepsilon_1 - \varepsilon_2)$$

2.12 Special states of strain

Several special states of strain, which are often encountered in practice, will now be discussed.

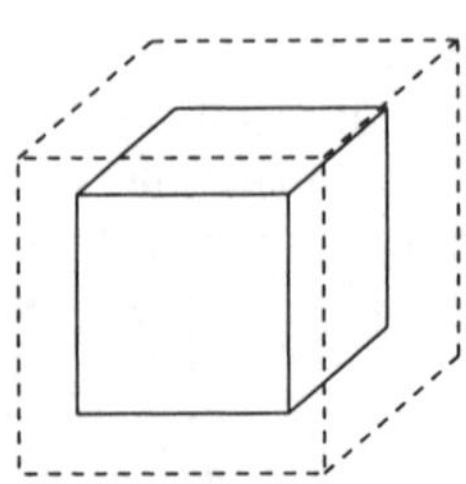

Figure 2.12: Uniform dilatation.

A state of *uniform dilatation* occurs, if the strain tensor is given by

$$\varepsilon_{ij} = b\,\delta_{ij}$$

where b is an arbitrary scalar. It appears from (2.47) that the deviatoric strain tensor e_{ij} becomes $e_{ij} = 0$ and according to the discussion of (2.49), the strain state corresponds to a uniform dilatation, i.e. a volume change, where the extension - or contraction - in any direction is the same and equal to the parameter b, cf. Fig. 2.12.

Uniaxial strain occurs if the displacement vector u_i is given by

$$[u_i] = \begin{bmatrix} u_1(x_1, t) \\ 0 \\ 0 \end{bmatrix}$$

which implies that $\varepsilon_{11} = \partial u_1/\partial x_1$ and all other strain components being zero, cf. Fig.2.13.

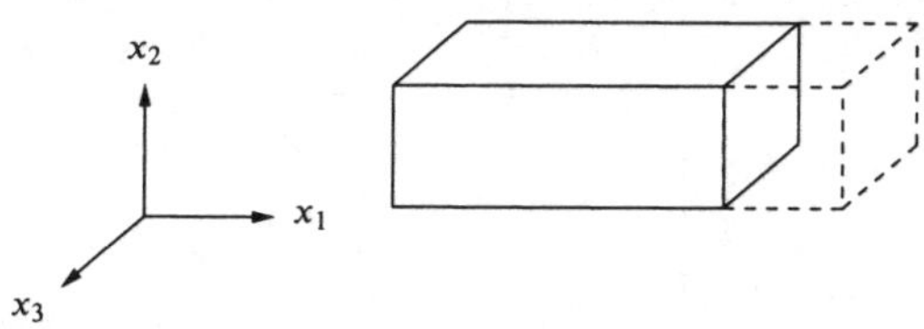

Figure 2.13: Uniaxial strain.

Plane strain or *plane deformation* occurs if the displacement vector u_i is given by

$$[u_i] = \begin{bmatrix} u_1(x_1, x_2, t) \\ u_2(x_1, x_2, t) \\ 0 \end{bmatrix}$$

which implies

$$[\varepsilon_{ij}] = \begin{bmatrix} \varepsilon_{11} & \varepsilon_{12} & 0 \\ \varepsilon_{21} & \varepsilon_{22} & 0 \\ 0 & 0 & 0 \end{bmatrix} \tag{2.72}$$

This strain state occurs often in practice when a long prismatic or cylindrical body is loaded by forces which are perpendicular to the longitudinal elements and which do not vary along the length. In this case, it can be assumed that all cross sections are in the same state and if, moreover, the body is restricted from moving in the length direction, a state of plane strain exists. An example is an internally pressurized tube with end sections confined between smooth and rigid walls, Fig. 2.14.

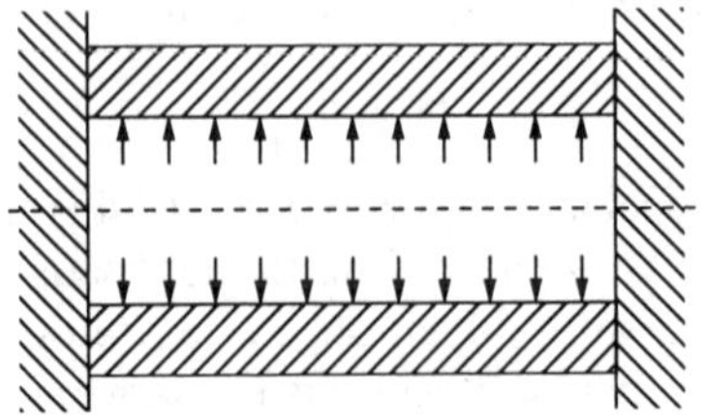

Figure 2.14: Example of plane strain. Pressurized tube with end sections confined between smooth and rigid walls.

So-called *generalized plane strain* or *generalized plane deformation* occurs if

$$[u_i] = \begin{bmatrix} u_1(x_1, x_2, t) \\ u_2(x_1, x_2, t) \\ u_3(x_1, x_2, t) \end{bmatrix}$$

which leads to

$$[\varepsilon_{ij}] = \begin{bmatrix} \varepsilon_{11} & \varepsilon_{12} & \varepsilon_{13} \\ \varepsilon_{21} & \varepsilon_{22} & \varepsilon_{23} \\ \varepsilon_{31} & \varepsilon_{32} & 0 \end{bmatrix}$$

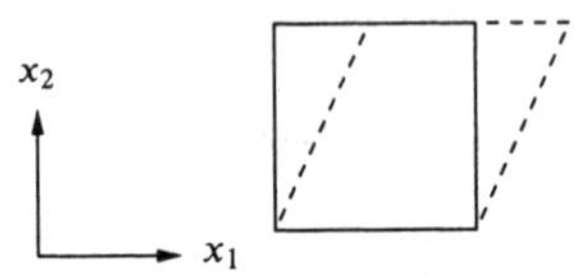

Figure 2.15: Simple shear.

Finally, a state of *simple shear* exists if

$$[\varepsilon_{ij}] = \begin{bmatrix} 0 & \varepsilon_{12} & 0 \\ \varepsilon_{21} & 0 & 0 \\ 0 & 0 & 0 \end{bmatrix}$$

corresponding to $u_1 = u_1(x_2, t)$ and $u_2 = u_3 = 0$, as illustrated in Fig. 2.15. It appears that for simple shear, we have $\varepsilon_{ii} = 0$, i.e. no volume change and it is easily shown that the principal strains become $\varepsilon_1 = \varepsilon_{12}$, $\varepsilon_2 = -\varepsilon_{12}$ and $\varepsilon_3 = 0$.

3 STRESS TENSOR

Having obtained a description of the deformation of the body, the next important point is to be able to describe the loading on the body at an arbitrary point. It turns out that this description is provided by the so-called *stress tensor*. Just like the strain tensor, the stress tensor turns out to be a symmetric second-order tensor and we will therefore take advantage of a number of properties that were proven in Chapter 2 for the strain tensor. For further studies of the stress tensor, the reader may consult, for instance, Fung (1965), Malvern (1969), Sokolnikoff (1946) or Spencer (1980).

3.1 Introduction

The body is supposed to be continuous and two kinds of forces are assumed: *body forces* (i.e. force per unit volume) and *surface forces* (i.e. force per unit area).

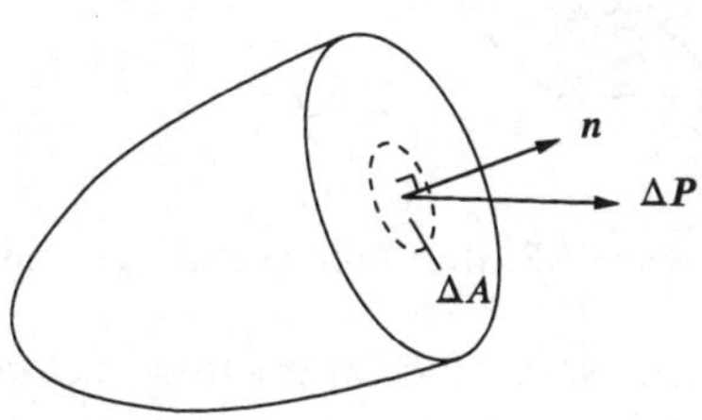

Figure 3.1: Force $\Delta \boldsymbol{P}$ on area ΔA with outer unit normal vector $\boldsymbol{n}$.

Consider a surface of the body as shown in Fig. 3.1. This surface can be an external surface or an internal surface obtained by a section of the body. The vector $\boldsymbol{n}$ is a unit vector normal to the surface and directed out of the body. The incremental force vector $\Delta \boldsymbol{P}$ acts on the incremental surface area ΔA. When ΔA approaches zero, it is assumed that the ratio $\Delta \boldsymbol{P}/\Delta A$ approaches a value $\boldsymbol{t}$, i.e.

$$t = \left(\frac{\Delta P}{\Delta A}\right)_{\Delta A \to 0} \quad ; \qquad t = \begin{bmatrix} t_1 \\ t_2 \\ t_3 \end{bmatrix}$$

The vector $\boldsymbol{t}$, with components t_1, t_2 and t_3 in the x_1-, x_2- and x_3-directions respectively, is termed the *traction vector* and has the unit [N/m^2].

In principle, the surface forces around a point may also give rise to a moment about that point even when the ΔA-area approaches zero. Likewise, the body forces might result in a moment about a point even when the volume approaches zero. In these situations, so-called *couple stresses* will be present, but in classical continuum mechanics, these possible couple stresses are ignored; consideration of couple stresses leads to the so-called *polar continuum mechanics*, cf. Fung (1965), Malvern (1969), and Jaunzemis (1967) for introductory remarks and Cosserat and Cosserat (1909), Mindlin and Tiersten (1962), Mindlin (1964), Koiter (1964) and Eringen (1999) for more details.

The traction vector $\boldsymbol{t}$ defined above is related to a surface with the outer unit normal vector $\boldsymbol{n}$. It is obvious that the traction vector will, in general, be different when other sections through the same point are considered. What we are looking for is a quantity - the *stress tensor* - which for a particular point contains all the information necessary to determine the traction vector for arbitrary sections through that point.

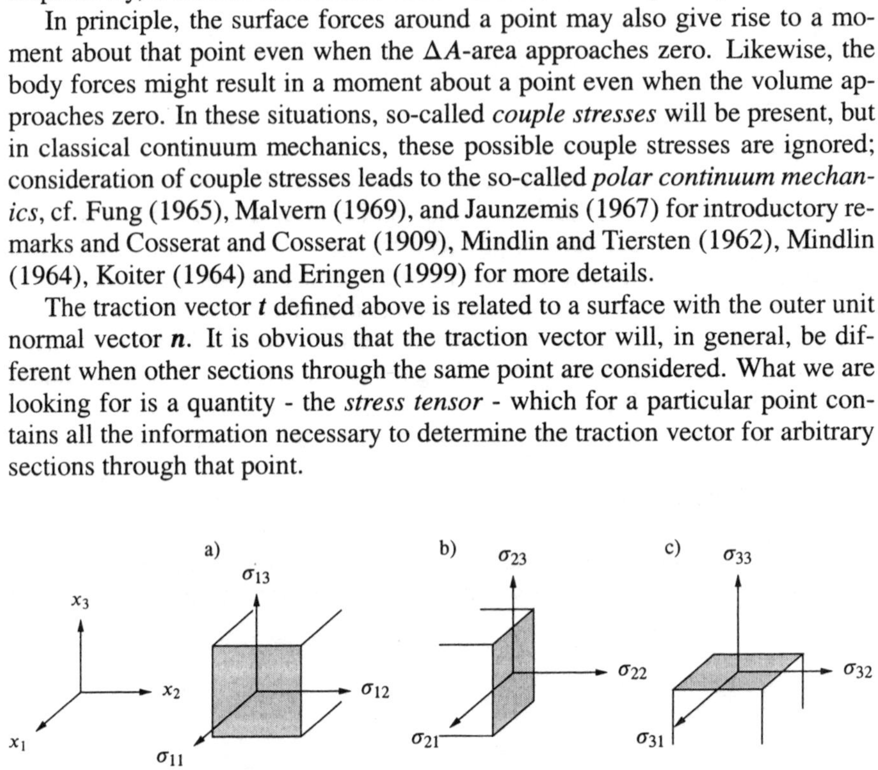

Figure 3.2: Illustration of stress components.

Let us first consider some special traction vectors, namely those obtained when sections perpendicular to the coordinate axes are considered. Assume that the outer normal vector $\boldsymbol{n}$ (see Fig. 3.1) is taken in the direction of the x_1-axis. The corresponding traction vector is denoted by $\boldsymbol{t}_1$ and we can resolve this vector into its components along the coordinate axes, i.e.

$$\boldsymbol{t}_1^T = \begin{bmatrix} \sigma_{11} & \sigma_{12} & \sigma_{13} \end{bmatrix} \tag{3.1}$$

where σ_{11}, σ_{12} and σ_{13} denote the components of $\boldsymbol{t}_1$ in the x_1-, x_2- and x_3-directions respectively. These components are illustrated in Fig. 3.2a.

Likewise, if the outer normal unit vector $\boldsymbol{n}$ is taken in the direction of the x_2-axis, we denote the corresponding traction vector by $\boldsymbol{t}_2$, i.e.

$$\boldsymbol{t}_2^T = \begin{bmatrix} \sigma_{21} & \sigma_{22} & \sigma_{23} \end{bmatrix} \tag{3.2}$$

where σ_{21}, σ_{22} and σ_{23} denote the components of $\boldsymbol{t}_2$ in the x_1-, x_2- and x_3-directions respectively, cf. Fig. 3.2b. Finally, if the outer normal unit vector $\boldsymbol{n}$ is taken in the direction of the x_3-axis, we denote the corresponding traction vector by $\boldsymbol{t}_3$, i.e.

$$\boldsymbol{t}_3^T = \begin{bmatrix} \sigma_{31} & \sigma_{32} & \sigma_{33} \end{bmatrix} \tag{3.3}$$

where σ_{31}, σ_{32} and σ_{33} denote the components of $\boldsymbol{t}_3$ in the x_1-, x_2- and x_3-directions respectively, cf. Fig. 3.2c.

The components given by (3.1)-(3.3) are termed the *stress components* and σ_{11}, σ_{22}, σ_{33} are called *normal stresses*, whereas σ_{12}, σ_{13}, σ_{21}, σ_{23}, σ_{31} and σ_{32} are referred to as *shear stresses*. We observe the consistent notation of the stress components where, for instance, σ_{23} is the x_3-component of the traction vector for a surface with the outer unit vector in the x_2-direction. Likewise, σ_{12} is the x_2-component of the traction vector for a surface with the outer unit vector in the x_1-direction.

Using the special traction vectors considered above, we define the quantity σ_{ij} by

$$\boxed{[\sigma_{ij}] = \begin{bmatrix} \boldsymbol{t}_1^T \\ \boldsymbol{t}_2^T \\ \boldsymbol{t}_3^T \end{bmatrix} = \begin{bmatrix} \sigma_{11} & \sigma_{12} & \sigma_{13} \\ \sigma_{21} & \sigma_{22} & \sigma_{23} \\ \sigma_{31} & \sigma_{32} & \sigma_{33} \end{bmatrix}} \tag{3.4}$$

We shall later prove that σ_{ij} is a second-order tensor and σ_{ij} is therefore called the *stress tensor*.

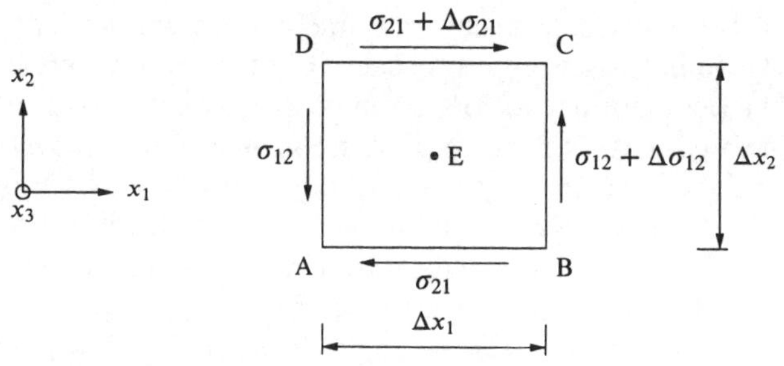

Figure 3.3: Moment about an axis through the center E and parallel to the x_3-axis.

Let us first prove that σ_{ij} is symmetric. From the body we cut a small parallelepiped with planes parallel to the coordinate planes. We then consider the moment equilibrium about an axis through the center E of this parallelepiped and parallel to the x_3-axis, cf. Fig. 3.3. It appears that body forces do not provide a moment about this axis. It is also obvious that only forces acting on planes parallel to the moment axis can contribute to the moment equilibrium. On these planes, only shear stresses normal to the moment axis can give rise to the moments; see Fig. 3.3.

Referring to this figure, the positive direction of the shear stresses along BC and DC is in accordance with the previous interpretation of the stress components, cf. Fig. 3.2. The positive direction of the shear stresses along AB and AD follows from the *law of action and reaction*. Taking moments as positive in the counter-clockwise direction, moment equilibrium about point E yields

$$(\sigma_{12} + \Delta\sigma_{12})\Delta x_2 \Delta x_3 \frac{1}{2}\Delta x_1 - (\sigma_{21} + \Delta\sigma_{21})\Delta x_1 \Delta x_3 \frac{1}{2}\Delta x_2$$
$$+ \sigma_{12}\Delta x_2 \Delta x_3 \frac{1}{2}\Delta x_1 - \sigma_{21}\Delta x_1 \Delta x_3 \frac{1}{2}\Delta x_2 = 0$$

i.e.

$$2\sigma_{12} - 2\sigma_{21} + \Delta\sigma_{12} - \Delta\sigma_{21} = 0$$

Letting Δx_1, Δx_2 and Δx_3 approach zero, both $\Delta\sigma_{12}$ and $\Delta\sigma_{21}$ also approach zero; that is, moment equilibrium requires that $\sigma_{12} = \sigma_{21}$. Likewise, considering moment equilibrium about axes parallel to the x_1- and x_2-axes implies that $\sigma_{23} = \sigma_{32}$ and $\sigma_{13} = \sigma_{31}$ respectively. In conclusion, we have proved that σ_{ij} is symmetric, i.e.

$$\boxed{\sigma_{ij} = \sigma_{ji} \qquad or \qquad \boldsymbol{\sigma} = \boldsymbol{\sigma}^T}$$

Our aim was to establish a quantity which contains all the information necessary to determine the traction vector $\boldsymbol{t}$ for arbitrary sections through the point in question. We shall now prove that the stress tensor σ_{ij} contains this information.

Consider the small tetrahedron shown in Fig. 3.4a). At the surface ABC with the outer unit normal vector $\boldsymbol{n}$, we have the traction vector $\boldsymbol{t}$. On the planes parallel to the coordinate planes, the traction vectors are $-\boldsymbol{t}_1$, $-\boldsymbol{t}_2$ and $-\boldsymbol{t}_3$, cf. (3.1)-(3.3); (minus signs appear because of the law of action and reaction and because the outer normal vectors are in the negative direction of the coordinate axes). The area ABC is denoted by ΔA, the area AOC by ΔA_1, the area AOB by ΔA_2 and the area BOC by ΔA_3. In Fig. 3.4b) the line CP is orthogonal to the line AB. As $\boldsymbol{n}$ is perpendicular to the surface ABC, it is also perpendicular to the lines CP and AB. The vector $\boldsymbol{n}$ is therefore located in the plane OCP. The components of the unit vector n_i are given by $n_i = (n_1, n_2, n_3)$ and by definition we have $n_2 = \cos\theta$ where θ is the angle shown in Fig. 3.4b). From Fig. 3.4b) follows that

$$\Delta A_2 = \frac{1}{2}|OP| \cdot |AB| ; \qquad |OP| = |CP|\cos\theta = |CP|n_2 \qquad \text{i.e.}$$
$$\Delta A_2 = \frac{1}{2}|CP| \cdot |AB|n_2 = n_2 \Delta A$$

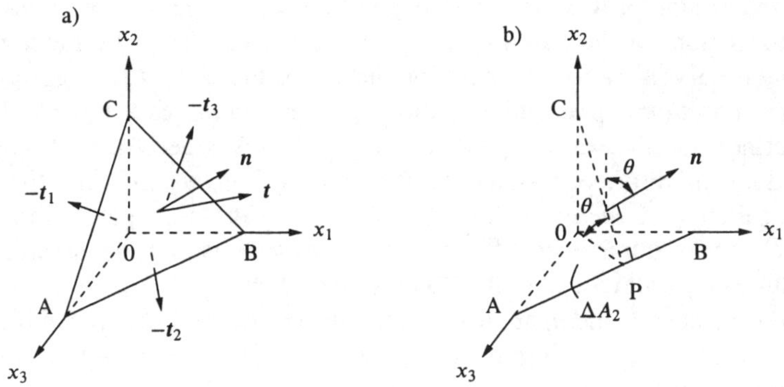

Figure 3.4: a) Traction vectors on a tetrahedron: $\boldsymbol{t}$ acts on ABC, -$\boldsymbol{t}_1$ on AOC, -$\boldsymbol{t}_2$ on AOB and -$\boldsymbol{t}_3$ on BOC; b) determination of ΔA by geometrical arguments. Vector $\boldsymbol{n}$ is located in the plane OCP.

By analogous arguments we find that

$$\Delta A_1 = n_1 \Delta A \; ; \qquad \Delta A_2 = n_2 \Delta A \; ; \qquad \Delta A_3 = n_3 \Delta A \tag{3.5}$$

The condition of force equilibrium of the tetrahedron of Fig. 3.4a) requires that

$$\boldsymbol{t}\Delta A - \boldsymbol{t}_1 \Delta A_1 - \boldsymbol{t}_2 \Delta A_2 - \boldsymbol{t}_3 \Delta A_3 + \boldsymbol{b}\Delta V = \boldsymbol{0} \tag{3.6}$$

where $\boldsymbol{b}$ is the body force per unit volume and ΔV is the volume of the small tetrahedron. The body force $\boldsymbol{b}$ has the components

$$\boxed{\boldsymbol{b}^T = [b_1 \;\; b_2 \;\; b_3]}$$

Use of (3.5) in (3.6) gives

$$\boldsymbol{t} - \boldsymbol{t}_1 n_1 - \boldsymbol{t}_2 n_2 - \boldsymbol{t}_3 n_3 + \boldsymbol{b}\frac{\Delta V}{\Delta A} = \boldsymbol{0}$$

Letting the size of the tetrahedron shrink towards zero, we have $\Delta V / \Delta A \to 0$ (volume has the unit m^3 and area has the unit m^2) and we then obtain

$$\boldsymbol{t} = \boldsymbol{t}_1 n_1 + \boldsymbol{t}_2 n_2 + \boldsymbol{t}_3 n_3$$

which may be written as

$$\boldsymbol{t} = [\boldsymbol{t}_1 \;\; \boldsymbol{t}_2 \;\; \boldsymbol{t}_3] \begin{bmatrix} n_1 \\ n_2 \\ n_3 \end{bmatrix} = \boldsymbol{\sigma}^T \boldsymbol{n}$$

where (3.4) was used. Due to the symmetry of $\boldsymbol{\sigma}$ we arrive at

$$\boxed{t_i = \sigma_{ij} n_j \qquad or \qquad \boldsymbol{t} = \boldsymbol{\sigma}\boldsymbol{n}} \tag{3.7}$$

This expression proves that knowledge of the stress tensor $\boldsymbol{\sigma}$ provides sufficient information for the traction vector $\boldsymbol{t}$ to be derived for any direction $\boldsymbol{n}$. It should be observed that on the exterior surface of the body, (3.7) represents a *boundary condition* expressing a relation between the forces acting on the external surface and the stress tensor. Equation (3.7) was derived by Cauchy in 1822 and it is therefore occasionally referred to as *Cauchy's formula*; the stress tensor is called the *Cauchy stress tensor*. When considering large deformations, it turns out that a number of different stress tensors exist, but for small strains and rotations they all reduce to the Cauchy stress tensor.

Moreover, since t_i and n_i are first-order tensors (vectors), it follows from the quotient theorem (cf. the argument that led to (1.41)) that σ_{ij} - indeed - is a second-order tensor.

3.2 Change of coordinate system

If we instead of the x_i-coordinate system change to a x'_i-coordinate system, we have as usual that

$$x'_i = A_{ij}(x_j - c_j) \qquad \text{or} \qquad \boldsymbol{x}' = \boldsymbol{A}(\boldsymbol{x} - \boldsymbol{c})$$

where A_{ij} is the transformation matrix and where $\boldsymbol{A}^T\boldsymbol{A} = \boldsymbol{I}$, cf. (1.17) and (1.22).

Since σ_{ij} is known to be a second-order tensor, we can directly from (1.41) and (1.42) write the following relations between the components σ_{ij} in the x_i-system and the components σ'_{ij} in the x'_i-system.

$$\boxed{\sigma'_{ij} = A_{ik}\sigma_{kl}A_{jl} \qquad or \qquad \boldsymbol{\sigma}' = \boldsymbol{A}\boldsymbol{\sigma}\boldsymbol{A}^T} \tag{3.8}$$

and

$$\sigma_{ij} = A_{ki}\sigma'_{kl}A_{lj} \qquad \text{or} \qquad \boldsymbol{\sigma} = \boldsymbol{A}^T\boldsymbol{\sigma}'\boldsymbol{A} \tag{3.9}$$

3.3 Principal stresses and principal directions - Invariants

The traction vector $\boldsymbol{t}$ on a surface with the outer normal unit vector $\boldsymbol{n}$ is given by (3.7). The traction vector $\boldsymbol{t}$ can be resolved into a component parallel to $\boldsymbol{n}$ and a component perpendicular to $\boldsymbol{n}$. The component parallel to $\boldsymbol{n}$ is called the normal stress in direction $\boldsymbol{n}$ and denoted by σ_n. From (3.7) we obtain

$$\boxed{\sigma_n = n_i t_i = n_i\sigma_{ij}n_j \qquad or \qquad \sigma_n = \boldsymbol{n}^T\boldsymbol{t} = \boldsymbol{n}^T\boldsymbol{\sigma}\boldsymbol{n}} \tag{3.10}$$

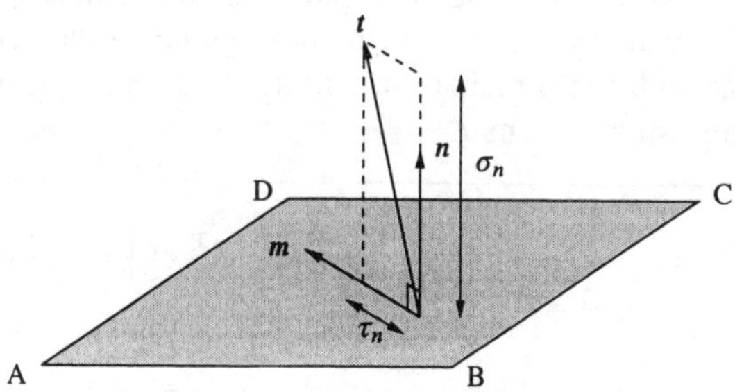

Figure 3.5: Illustration of normal stress σ_n and shear stress τ_n.

The component of $\boldsymbol{t}$ perpendicular to $\boldsymbol{n}$ is called the shear stress and is denoted by τ_n. Both σ_n and τ_n are illustrated in Fig. 3.5, where the unit vector $\boldsymbol{m}$ is perpendicular to $\boldsymbol{n}$ and located in the plane ABCD. It readily appears that

$$\boxed{\tau_n = m_i t_i = m_i \sigma_{ij} n_j \qquad or \qquad \tau_n = \boldsymbol{m}^T \boldsymbol{t} = \boldsymbol{m}^T \boldsymbol{\sigma n}} \tag{3.11}$$

Alternatively, we may write

$$\tau_n^2 = t_i t_i - \sigma_n^2$$

With these preliminary results, we may obtain a physical interpretation of the important eigenvalue problem of the stress tensor and with the solution of the eigenvalue problem, we arrive at the stress invariants. Moreover, it turns out that for a special choice of coordinate system, the stress tensor takes a particularly simple form.

Returning to Fig. 3.5, we look for a situation where the traction vector $\boldsymbol{t}$ is collinear with the unit vector $\boldsymbol{n}$. From Fig. 3.5, the direction $\boldsymbol{n}$ should be chosen so that

$$t_i = \lambda n_i \tag{3.12}$$

where λ is some factor and (3.10) implies that $\lambda = \sigma_n$. Since n_i and m_i are orthogonal, (3.11) gives in the present situation that the shear stress $\tau_n = 0$. Insertion of (3.7) into (3.12) yields

$$\boxed{(\sigma_{ij} - \lambda \delta_{ij}) n_j = 0 \qquad or \qquad (\boldsymbol{\sigma} - \lambda \boldsymbol{I})\boldsymbol{n} = \boldsymbol{0}} \tag{3.13}$$

This constitutes an eigenvalue problem and a comparison with (2.24) shows a complete analogy. Therefore all the conclusions that were derived for the strain tensor apply also for the stress tensor. That is, the characteristic equation

$$\det(\boldsymbol{\sigma} - \lambda \boldsymbol{I}) = 0$$

determines the three principal stresses $\sigma_1 = \lambda_1$, $\sigma_2 = \lambda_2$ and $\sigma_3 = \lambda_3$ and for each λ-value, (3.13) provides the corresponding principal direction $\boldsymbol{n}$. The principal stresses and directions are real; the principal stresses are invariants and the principal directions may always be taken to be orthogonal. If the coordinate system is taken collinear with the principal directions $\boldsymbol{n}_1$, $\boldsymbol{n}_2$ and $\boldsymbol{n}_3$, the stress tensor takes the following simple form

$$\boxed{\boldsymbol{\sigma}' = \boldsymbol{A}\boldsymbol{\sigma}\boldsymbol{A}^T = \begin{bmatrix} \sigma_1 & 0 & 0 \\ 0 & \sigma_2 & 0 \\ 0 & 0 & \sigma_3 \end{bmatrix} \quad \textit{where} \quad \boldsymbol{A}^T = \begin{bmatrix} \boldsymbol{n}_1 \; \boldsymbol{n}_2 \; \boldsymbol{n}_3 \end{bmatrix}}$$

cf. (2.37)

Also the stress tensor satisfies the Cayley-Hamilton theorem. Moreover, the coefficients in the characteristic equation are the Cauchy-stress invariants, but of more importance are the following generic stress invariants

$$\boxed{I_1 = \sigma_{ii} \; ; \qquad I_2 = \tfrac{1}{2}\sigma_{ij}\sigma_{ji} \; ; \qquad I_3 = \tfrac{1}{3}\sigma_{ij}\sigma_{jk}\sigma_{ki}} \tag{3.14}$$

where the term 'generic' refers to the systematic definition of these invariants (we may refer to (2.51) for a comparison with the corresponding strain invariants).

3.4 Stress deviator tensor

Similarly to the exposition of the strain tensor, we define the *stress deviator tensor* by

$$\boxed{s_{ij} = \sigma_{ij} - \tfrac{1}{3}\sigma_{kk}\delta_{ij}} \tag{3.15}$$

where $\sigma_{kk}/3$ is called the *hydrostatic stress*. The σ_{ij}- and s_{ij}-tensors have identical off-diagonal elements and thus they have identical principal directions.

The generic invariants of the stress deviator tensor are given by

$$\boxed{J_1 = s_{ii} = 0 \; ; \qquad J_2 = \tfrac{1}{2}s_{ij}s_{ji} \; ; \qquad J_3 = \tfrac{1}{3}s_{ij}s_{jk}s_{ki}} \tag{3.16}$$

Similar to (2.53c), we have

$$J_3 = s_1 s_2 s_3 \tag{3.17}$$

Moreover, similar to (2.54) and (2.55) we find the following relations

$$J_2 = I_2 - \frac{1}{6}I_1^2 \; ; \qquad J_3 = I_3 - \frac{2}{3}I_1 I_2 + \frac{2}{27}I_1^3 \tag{3.18}$$

and

$$I_2 = J_2 + \frac{1}{6}I_1^2 \; ; \qquad I_3 = J_3 + \frac{2}{3}I_1 I_2 - \frac{2}{27}I_1^3 \tag{3.19}$$

Therefore, instead of using the set of invariants I_1, I_2 and I_3 we may equally well use the set I_1, J_2 and J_3.

Finally, and in analogy with (2.56), we have the *octahedral normal stress* σ_0 and *octahedral shear stress* τ_0 defined by

$$\boxed{\sigma_0 = \frac{1}{3}I_1\;; \qquad \tau_0 = \sqrt{\frac{2}{3}J_2}} \tag{3.20}$$

where σ_0 and τ_0 are the normal stress shear stress respectively, that act on an octahedral plane. Here, the normal to an octahedral plane makes equal angles to the principal stress directions; when comparing (3.20) with (2.56) note the difference between engineering shear strain and tensorial shear strain.

3.5 Change of coordinate system - Mohr's circle

Let us consider a coordinate system obtained by rotating the old coordinate system by an angle α in the counter-clockwise direction around the x_3-axis, cf. Fig. 3.6a).

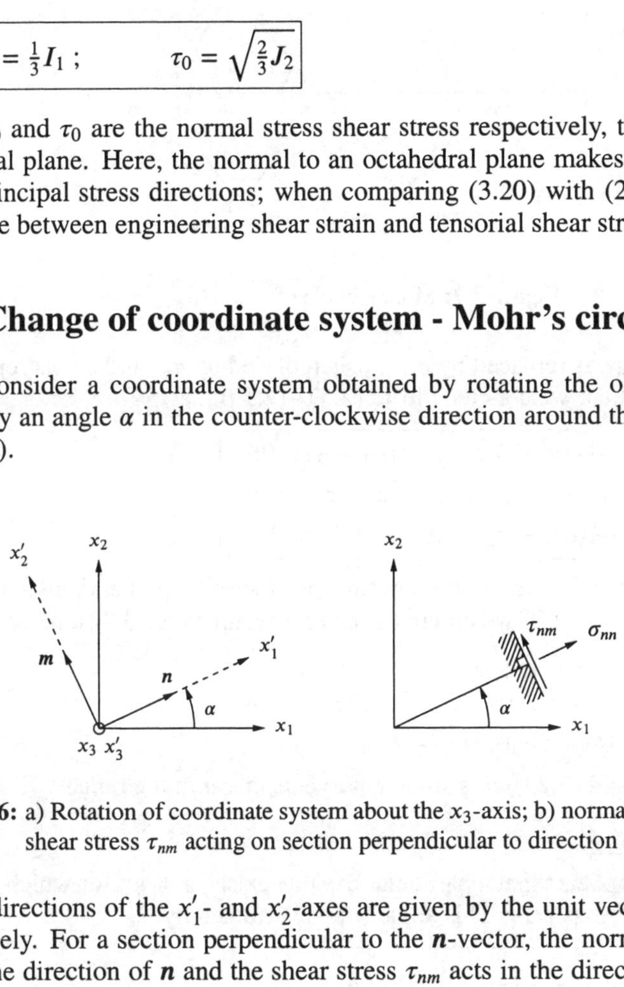

Figure 3.6: a) Rotation of coordinate system about the x_3-axis; b) normal stress σ_{nn} and shear stress τ_{nm} acting on section perpendicular to direction $\boldsymbol{n}$.

The directions of the x_1'- and x_2'-axes are given by the unit vectors $\boldsymbol{n}$ and $\boldsymbol{m}$ respectively. For a section perpendicular to the $\boldsymbol{n}$-vector, the normal stress σ_{nn} acts in the direction of $\boldsymbol{n}$ and the shear stress τ_{nm} acts in the direction of $\boldsymbol{m}$, cf. Fig. 3.6b). According to (3.10) and (3.11) we have

$$\sigma_{nn} = n_i\sigma_{ij}n_j\;; \qquad \tau_{nm} = m_i\sigma_{ij}n_j \tag{3.21}$$

We emphasize that σ_{nn} and τ_{nm} are measured positive in the direction of $\boldsymbol{n}$ and $\boldsymbol{m}$ respectively.

Comparing (3.21) with (2.11) and (2.17), we observe that the expressions for σ_{nn} and τ_{nm} are identical with the expressions for ε_{nn} and ε_{nm}. That is, all that has been derived for Mohr's circle of strain applies directly for Mohr's circle of

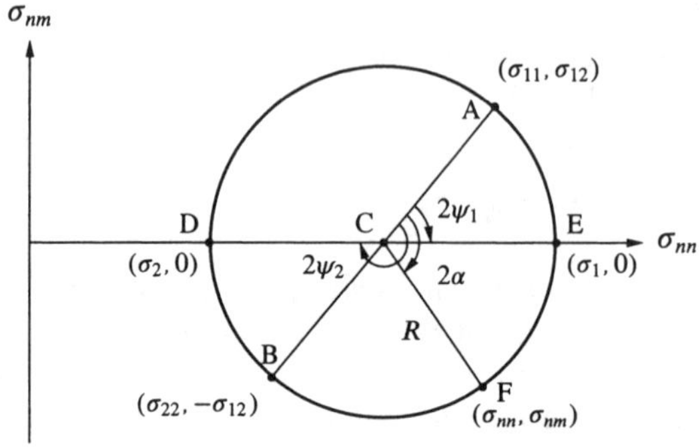

Figure 3.7: Mohr's circle of stress ($\sigma_1 \geq \sigma_2$).

stress, when ε_{ij} is replaced by σ_{ij}, ε_{nn} is replaced by σ_{nn} and ε_{nm} is replaced by τ_{nm}. In particular, we conclude from (2.57)-(2.59) that

$$\boxed{\begin{aligned} \sigma_{nn} &= \sigma_{11}\cos^2\alpha + \sigma_{22}\sin^2\alpha + \sigma_{12}\sin 2\alpha \\ \sigma_{mm} &= \sigma_{11}\sin^2\alpha + \sigma_{22}\cos^2\alpha - \sigma_{12}\sin 2\alpha \\ \tau_{nm} &= -\tfrac{1}{2}(\sigma_{11} - \sigma_{22})\sin 2\alpha + \sigma_{12}\cos 2\alpha \end{aligned}} \tag{3.22}$$

where σ_{mm} is the normal stress acting on a section perpendicular to the $\boldsymbol{m}$-direction. From (2.60) we obtain the inverse relations to (3.22), namely

$$\begin{aligned} \sigma_{11} &= \sigma_{nn}\cos^2\alpha + \sigma_{mm}\sin^2\alpha - \tau_{nm}\sin 2\alpha \\ \sigma_{22} &= \sigma_{nn}\sin^2\alpha + \sigma_{mm}\cos^2\alpha + \tau_{nm}\sin 2\alpha \\ \sigma_{12} &= \tfrac{1}{2}(\sigma_{nn} - \sigma_{mm})\sin 2\alpha + \tau_{nm}\cos 2\alpha \end{aligned} \tag{3.23}$$

In (3.22) and (3.23), it is sufficient to consider α in the range

$$\boxed{0 \leq \alpha < \pi}$$

From (3.22) appears that a particular α-value exists, $\alpha = \psi$, for which the shear stress τ_{nm} becomes zero. This angle ψ is determined by

$$\boxed{\tan 2\psi = \frac{2\tau_{nm}}{\sigma_{nn} - \sigma_{mm}}\,; \qquad 0 \leq \psi < \pi}$$

If the x_3'-direction, cf. Fig. 3.6, is a principal stress direction, it follows that the directions $\boldsymbol{n}$ and $\boldsymbol{m}$ defined by $\alpha = \psi$ are also principal directions. This situation will be assumed below.

Similar to (2.71), the principal stresses in the x_1x_2-plane are given by

$$\boxed{\left.\begin{matrix}\sigma_1 \\ \sigma_2\end{matrix}\right\} = \tfrac{1}{2}(\sigma_{11} + \sigma_{22}) \pm \sqrt{[\tfrac{1}{2}(\sigma_{11} - \sigma_{22})]^2 + \sigma_{12}^2}\,; \qquad (\sigma_1 \geq \sigma_2)}$$

Moreover, *Mohr's circle of stress* (Mohr, 1882) takes the form shown in Fig. 3.7. Similar to the comments related to Fig. 2.11, we conclude that the value of the angle ψ_1, apparent from Fig. 3.7, is the angle α towards the principal direction having the largest principal stress.

Finally, from Fig. 3.7 we conclude that

$$\boxed{\tau_{nm,max} = \frac{1}{2}(\sigma_1 - \sigma_2)\,; \qquad \tau_{nm,min} = -\frac{1}{2}(\sigma_1 - \sigma_2)} \tag{3.24}$$

where $\sigma_1 \geq \sigma_2$.

3.6 Special states of stress

Several special states of stress, which are often encountered in practice, will now be discussed.

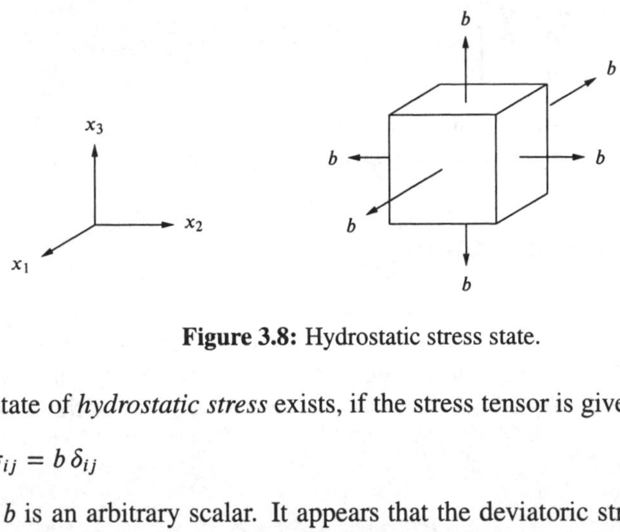

Figure 3.8: Hydrostatic stress state.

A state of *hydrostatic stress* exists, if the stress tensor is given by

$$\sigma_{ij} = b\,\delta_{ij}$$

where b is an arbitrary scalar. It appears that the deviatoric stress tensor s_{ij} is zero and that the loading consists of equal normal stresses having the amount b, cf. Fig. 3.8.

Uniaxial stress occurs if the stress tensor is given by

$$[\sigma_{ij}] = \begin{bmatrix} \sigma_{11} & 0 & 0 \\ 0 & 0 & 0 \\ 0 & 0 & 0 \end{bmatrix}$$

Plane stress exists if the stress tensor is given by

$$[\sigma_{ij}] = \begin{bmatrix} \sigma_{11} & \sigma_{12} & 0 \\ \sigma_{21} & \sigma_{22} & 0 \\ 0 & 0 & 0 \end{bmatrix} \tag{3.25}$$

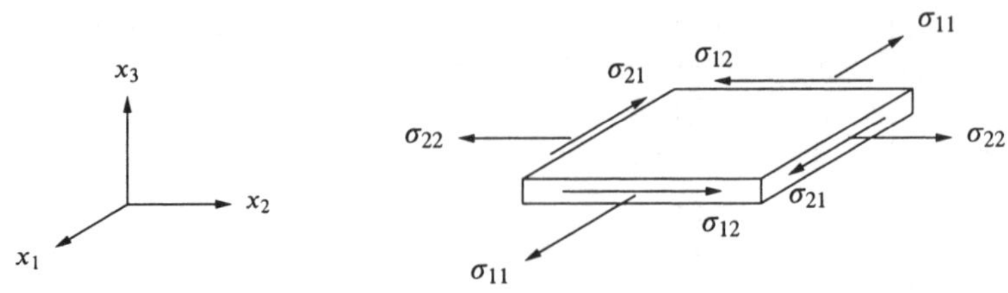

Figure 3.9: Plane stress.

and a disc loaded by *in-plane* stresses comprises an illustration of this type of loading, cf. Fig. 3.9.

Finally, a state of *pure shear* exists, if

$$[\sigma_{ij}] = \begin{bmatrix} 0 & \sigma_{12} & 0 \\ \sigma_{21} & 0 & 0 \\ 0 & 0 & 0 \end{bmatrix}$$

which holds for pure torsion of a cylindrical specimen. It is easily shown that the principal stresses become $\sigma_1 = \sigma_{12}$, $\sigma_2 = -\sigma_{12}$ and $\sigma_3 = 0$.

3.7 Equations of motion

We have previously used the equilibrium condition on an infinitesimal tetrahedron (Fig. 3.4) to obtain the connection between the traction vector and the stress tensor, (3.7), which can be considered as a static boundary condition. Let us now formulate the equations of motion for an arbitrary part of the body. The

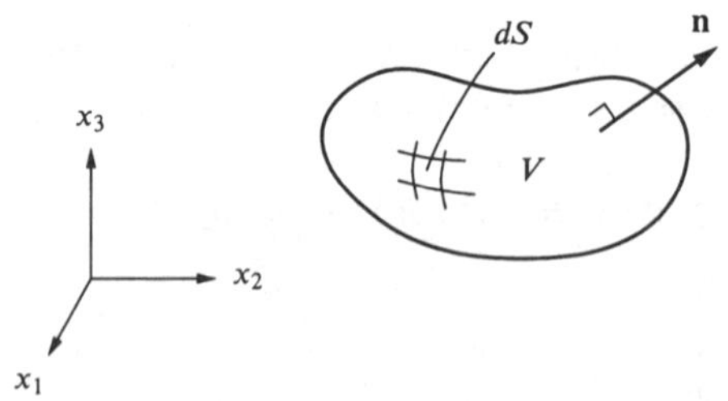

Figure 3.10: Volume V with surface boundary S and outer normal unit vector $\boldsymbol{n}$.

arbitrary part of the body has the volume V and the outer surface S with the outer normal unit vector $\boldsymbol{n}$, as shown in Fig. 3.10. The forces acting on this arbitrary body are given by the traction vector t_i along the boundary surface S

and the body force b_i per unit volume in the region V. The acceleration vector is denoted by $\ddot{u}_i$ where u_i is the displacement vector and a dot denotes the time derivative. *Newton's second law* states that

$$\int_S t_i dS + \int_V b_i dV = \int_V \rho \ddot{u}_i dV \tag{3.26}$$

where ρ is the mass density.

Before reformulating this equation, we recall the *divergence theorem of Gauss*, which states that for an arbitrary vector $\boldsymbol{q}$, the following relation holds

$$\int_V \text{div}\boldsymbol{q}\, dV = \int_S \boldsymbol{q}^T \boldsymbol{n}\, dS$$

We have per definition that

$$\text{div}\boldsymbol{q} = \frac{\partial q_1}{\partial x_1} + \frac{\partial q_2}{\partial x_2} + \frac{\partial q_3}{\partial x_3} = q_{i,i}\,; \qquad \boldsymbol{q}^T \boldsymbol{n} = q_i n_i$$

i.e. the divergence theorem can be written as

$$\boxed{\int_V q_{i,i} dV = \int_S q_i n_i dS} \tag{3.27}$$

If we choose q_i as the quantity c_{1i}, a relation analogous with (3.27) is obtained. Likewise, similar relations can be obtained by choosing q_i as c_{2i} and q_i as c_{3i}. Collecting all these results, we obtain for an arbitrary quantity c_{ij}, that (3.27) generalizes to

$$\boxed{\int_V c_{ij,j} dV = \int_S c_{ij} n_j dS} \tag{3.28}$$

With (3.7), we may then reformulate (3.26) by means of (3.28) to obtain

$$\int_V (\sigma_{ij,j} + b_i - \rho \ddot{u}_i) dV = 0$$

As this equation holds for arbitrary regions V of the body, we conclude that

$$\boxed{\sigma_{ij,j} + b_i = \rho \ddot{u}_i} \tag{3.29}$$

These equations express the *equations of motion* for the body and they are of course of fundamental importance in a number of applications; they were derived by Cauchy already in 1822.

3.8 Weak formulation - Principle of virtual work

From the equations of motion (3.29), i.e. the balance equations, we shall now derive one of the most powerful principles in mechanics namely the corresponding so-called *weak formulation.* The establishment of this weak form can be applied to any balance equation and within solid mechanics it is most often referred to as the celebrated *principle of virtual work.*

We multiply the equations of motion (3.29) by an arbitrary vector v_i - the *weight vector* - and integrate over the volume to obtain

$$\int_V v_i(\sigma_{ij,j} + b_i - \rho\ddot{u}_i)dV = 0$$

This equation may be written as

$$\int_V \left[(\sigma_{ij}v_i)_{,j} - \sigma_{ij}v_{i,j}\right] dV + \int_V (v_i b_i - \rho v_i \ddot{u}_i)dV = 0 \tag{3.30}$$

From the divergence theorem (3.27) and (3.7) we have

$$\int_V (\sigma_{ij}v_i)_{,j}dV = \int_S \sigma_{ij}v_i n_j dS = \int_S v_i t_i dS$$

Use of this expression in (3.30) gives

$$\int_V \rho v_i \ddot{u}_i dV + \int_V v_{i,j}\sigma_{ij}dV = \int_S v_i t_i dS + \int_V v_i b_i dV \tag{3.31}$$

This is the result aimed at, but we may occasionally use a slight reformulation. We recall that v_i is an arbitrary vector which, in general, has nothing to do with the displacement vector u_i. However, we may determine a quantity ε^v_{ij} defined by

$$\boxed{\varepsilon^v_{ij} = \frac{1}{2}(v_{i,j} + v_{j,i})} \tag{3.32}$$

i.e. the tensor ε^v_{ij} is related to the arbitrary vector v_i in the same manner as the strain tensor ε_{ij} is related to the displacement vector u_i; therefore ε^v_{ij} is the 'strain' associated with v_i. Moreover, due to the symmetry of σ_{ij} we have

$$v_{i,j}\sigma_{ij} = \frac{1}{2}(v_{i,j}\sigma_{ij} + v_{j,i}\sigma_{ji}) = \frac{1}{2}(v_{i,j}\sigma_{ij} + v_{j,i}\sigma_{ij}) = \varepsilon^v_{ij}\sigma_{ij}$$

With this result, we may write (3.31) as

Weak form of equations of motion = principle of virtual work

$$\int_V \rho v_i \ddot{u}_i dV + \int_V \varepsilon^v_{ij}\sigma_{ij}dV = \int_S v_i t_i dS + \int_V v_i b_i dV \tag{3.33}$$

This is the *weak form* of the equations of motion. However, we may think of the weight vector v_i as being some fictitious displacement - a so-called *virtual displacement*. In that case the expression on the right-hand side of (3.31) or (3.33) may be interpreted as the external work done during the 'virtual' displacement v_i suggesting the terminology of (3.31) and (3.33) being the *principle of virtual work*. We emphasize that v_i is an arbitrary vector that has nothing to do with the displacement vector u_i. Naturally, we may choose $v_i = u_i$, but in general this is not the case.

Moreover, we stress that (3.31) and (3.33) follow from the equations of motion alone and they hold therefore for any kind of material behavior. As the material response is described by certain relations between stresses and strains - so-called *constitutive relations* - we conclude that (3.31) and (3.33) hold for any kind of constitutive relation.

We have already used the terminology that (3.33) is the 'weak' form of the equations of motion. On the other hand, the differential equations of motion given by (3.29) is often called the *strong form* of the equations of motion. The reason for this terminology is whereas the strong form (3.29) contains derivatives of the stress tensor, the weak form does not and this is utilized in the numerical solution technique provided by the *finite element* formulation, cf. for instance Hughes (1987) and Ottosen and Petersson (1992). The weak form - i.e. the principle of virtual work - forms the basis not only for the finite element method, but also for other numerical solution techniques and it is also central for the establishment of a number of important theorems in solid mechanics. Thus, the weak form is one of the most important principles within solid mechanics.

In fact, the approach by which the weak form was derived above can be applied to arbitrary balance differential equations; therefore it provides a very powerful means for the establishment of various approximate solution techniques within different fields of physics.

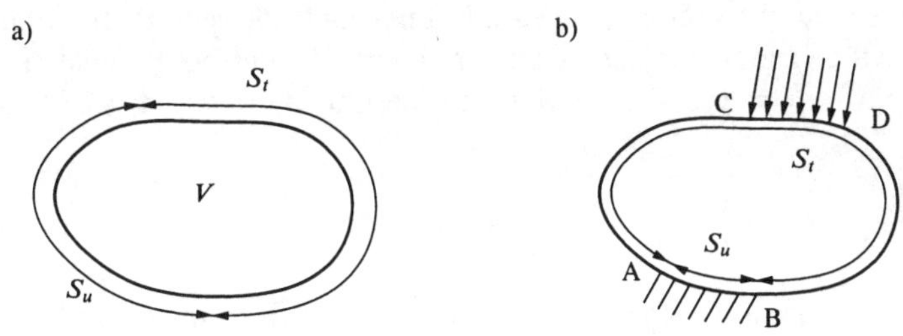

Figure 3.11: Boundary conditions; a) general situation, b) specific situation.

To solve the *field equations* within solid mechanics, i.e. the kinematic relations involving the strains and the displacements, the equations of motion and the constitutive relations relating stresses and strains, *boundary conditions* are required. They are given in the following form

$$\boxed{\begin{array}{ll} \boldsymbol{u} & \textit{is given along } \; S_u \\ \boldsymbol{t} & \textit{is given along } \; S_t \end{array}} \tag{3.34}$$

i.e. on the part S_u of the boundary, the displacements are known and prescribed whereas on the part S_t of the boundary, the traction vector is known and prescribed. The total boundary S consists of S_u and S_t. Figure 3.11a) illustrates the situation in general and in Fig. 3.11b) a specific example is shown. Here the support AB is the part S_u whereas ACDB is the part S_t (here the traction vector is zero along AC and DB and the traction vector is non-zero and known along CD).

The boundary condition where $\boldsymbol{u}$ is given along S_u is called a *kinematic boundary condition* and as the displacement vector $\boldsymbol{u}$ is the fundamental quantity to be determined, one also refers to this boundary condition as being an *essential boundary condition*. On the other hand, the boundary condition where $\boldsymbol{t}$ is given along S_t is called a *static boundary condition* and as the traction vector $\boldsymbol{t}$ emerges as a logical consequence in the boundary term in the weak form (3.33), it is also called a *natural boundary condition*.

We obtained the weak form of the equations of motion from their strong form (3.29), so the strong form implies the weak form. It is of significant interest, however, that if we accept the weak form and observe that the weight vector v_i is arbitrary, it then turns out that the weak form implies the strong form. The two forms are therefore equivalent expressions for the equations of motion, i.e.

$$\boxed{\begin{array}{c} \textit{Strong form of equations of motion} \\ \Longleftrightarrow \\ \textit{Weak form of equations of motion} \end{array}} \tag{3.35}$$

To prove this statement we assume that the weak form (3.33) holds. Choose the arbitrary weight vector v_i as an arbitrary constant vector; this implies that we choose the virtual displacements v_i as a virtual rigid-body translation. Since we then have $\varepsilon_{ij}^v = 0$ and as v_i does not depend on position, (3.33) takes the form

$$v_i[\int_V \rho \ddot{u}_i dV - \int_S t_i dS - \int_V b_i dV] = 0$$

and as this expression holds for arbitrary vectors v_i, it follows that

$$\int_V \rho \ddot{u}_i dV = \int_S t_i dS + \int_V b_i dV$$

But this expression is precisely Newton's second law, cf. (3.26), which leads to the strong form of the equations of motion given by (3.29). In the first place, we started out with the strong form and then derived the weak form. Now we have shown that if we start out with the weak form we can obtain the strong form. This proves the correctness of statement (3.35). For other implications of the weak form, i.e. the principle of virtual work, the reader is referred to the detailed discussion provided by Maugin (1980).

Finally, we mention that some authors present the principle of virtual work in a form where the virtual displacements must fulfill $v_i = 0$ along S_u, cf. (3.34), and with v_i otherwise being arbitrary. Certainly, this format is allowable, but it implies an unnecessary restriction on the general formulation of the principle of virtual work. Here, we therefore adopt the formulation of this important principle in its most general form.

4 HYPER-ELASTICITY

In the previous chapters we established the concepts of stresses and strains. No reference was made to the material as such, and we emphasize that within the assumption of small strains and small rotations, the results hold for any material which may be treated as a continuum. It is obvious, however, that stresses and strains must be related in some way or another and the specific manner of this relation is controlled by the specific material in question. The expression between stresses and strains is called the *constitutive relation* and a variety of such relations has been established. Examples are *elasticity, plasticity, viscoelasticity, viscoplasticity* and *creep*. In the present chapter, we will consider the simplest constitutive theory, namely *hyper-elasticity*.

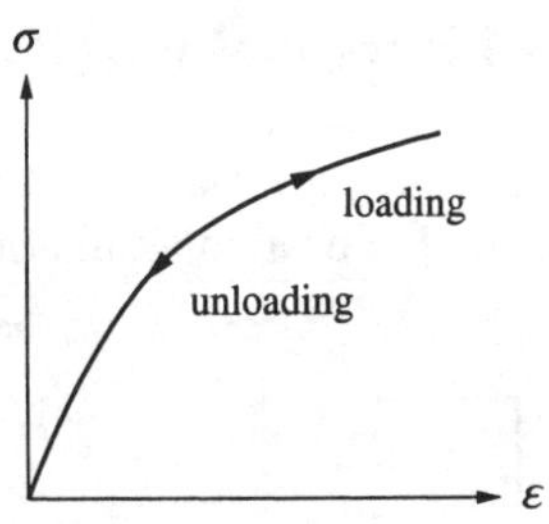

Figure 4.1: Nonlinear elasticity for uniaxial loading.

Let us first define what is meant by elasticity:

Elastic response is independent of the load history

Alternatively, one says that the material response is *path independent* and it follows that the response for loading or unloading follows the same path as illustrated in Fig. 4.1. After removal of the loading, the material therefore returns to its original configuration. We emphasize that elastic response, in general, is nonlinear, as also illustrated in Fig. 4.1.

In accordance with these statements, the stresses are uniquely given by the strains, i.e. we have the constitutive relation

$$\boxed{\sigma_{ij} = \sigma_{ij}(\varepsilon_{kl})} \tag{4.1}$$

If we say nothing more than that, we have the most general form of elasticity, namely so-called *Cauchy-elasticity* (in honor of Cauchy, 1789-1857, who in 1822 formulated the constitutive law for isotropic linear elastic materials). This type of elasticity will be treated in the next chapter. However, it turns out that a slightly restricted form of elasticity can be established by considerations to the strain energy. This restricted form of elasticity is called *hyper-elasticity* and it turns out to be very simple to derive the most general form of nonlinear hyper-elasticity for isotropic materials.

In general, we observe that if the constitutive relation depends on position, then the material is termed *inhomogeneous*; otherwise, it is termed *homogeneous*.

In addition to a treatment of hyper-elasticity, in the present chapter we will introduce and discuss a number of questions that are of general importance and applicable to a number of constitutive theories like, for instance, plasticity theory. Examples are matrix formulation of constitutive laws, discussion of symmetry properties of the constitutive matrix and topics within anisotropy. Thus, the present chapter is rather extensive, but in the following chapters we will take advantage of the concepts introduced here.

4.1 Strain energy and hyper-elasticity

Let us first introduce the concept of *strain energy* W per unit volume of the body, i.e. W has the unit [Nm/m^3]. For a uniaxial stress state, the incremental strain energy is defined by

$$dW = \sigma d\varepsilon \quad \text{i.e.} \quad W = \int_0^{\varepsilon} \sigma(\varepsilon^*) d\varepsilon^* \tag{4.2}$$

where advantage was taken of (4.1) and where ε^* is an integration variable whereas ε denotes the current strain. Equation (4.2) is illustrated in Fig. 4.2.

Adopting this approach to the general situation we obtain

$$\boxed{dW = \sigma_{ij} d\varepsilon_{ij} \quad \textit{i.e.} \quad W = \int_0^{\varepsilon_{ij}} \sigma_{ij}(\varepsilon_{kl}^*) d\varepsilon_{ij}^*} \tag{4.3}$$

where ε_{ij} denote the current strains whereas ε_{ij}^* denotes the integration variables.

Even though the current stresses σ_{ij} only depend on the current strains ε_{ij}, cf. (4.1), we will in general have that the strain energy W as determined by

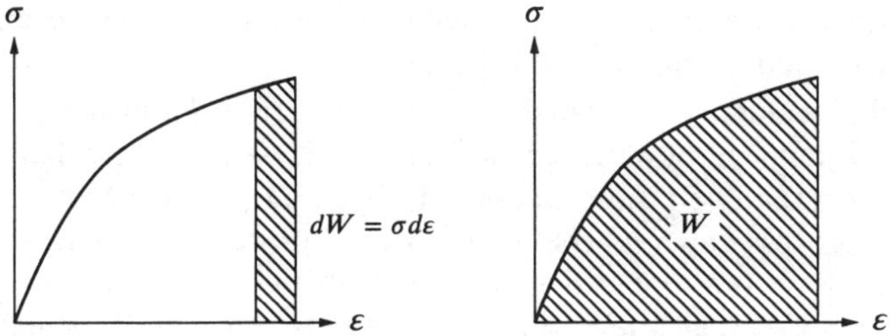

Figure 4.2: Incremental strain energy dW and strain energy W for uniaxial loading.

(4.3b) depends both on the current strains ε_{ij} as well as on the manner in which these strains were achieved, i.e.

$$W = W(\varepsilon_{ij}\,, \text{load history}) \tag{4.4}$$

This is just to say that W as determined by (4.3b) depends not only on the current strains, but also on the integration path, where the integration path represents the load history.

We will now make the assumption that W is independent on the integration path and (4.4) then reduces to

$$\boxed{W = W(\varepsilon_{ij})} \tag{4.5}$$

From this expression follows that

$$dW = \frac{\partial W}{\partial \varepsilon_{ij}} d\varepsilon_{ij} \tag{4.6}$$

Subtraction of (4.6) from (4.3a) gives

$$(\sigma_{ij} - \frac{\partial W}{\partial \varepsilon_{ij}}) d\varepsilon_{ij} = 0 \tag{4.7}$$

In general, the incremental strains $d\varepsilon_{ij}$ can be chosen arbitrarily and independently of each other and we therefore conclude from (4.7) that

$$\boxed{\sigma_{ij} = \frac{\partial W}{\partial \varepsilon_{ij}}} \tag{4.8}$$

There is one exception where the incremental strains $d\varepsilon_{ij}$ cannot be chosen arbitrarily, namely the case of incompressible response and we will return to this special situation in Section 4.13. Since σ_{ij} is obtained from W by a differentiation, one uses the phrase that W serves as a *potential function* for the stresses.

We observe that (4.5) and (4.8) imply (4.1) and a material that obeys the constitutive relation (4.5) and thereby (4.8) is called a *hyper-elastic material*;

'hyper' meaning 'to a higher degree'. Occasionally, the term *Green-elasticity* is used since this formulation was adopted by Green in 1839 and even today most work on elasticity is based on this format.

Another feature often related to elasticity is *reversibility* also from a thermodynamical point of view. Later, in Chapter 21 we will show that two thermodynamical reversible processes result in hyper-elasticity, namely, a reversible, adiabatic process and a reversible, isothermal process. Therefore, the term hyper - meaning 'to a higher degree' - is attributed to this kind of elasticity, as hyper-elasticity implies reversibility not only between stresses and strains, but also reversibility in the thermodynamical sense. In the next chapter we will encounter elasticity models (Cauchy-elasticity), which only implies reversibility between stresses and strains.

Let us return to (4.3b) and (4.8) and the issue of the strain energy W being independent of the integration path i.e. independent of the load history. As an illustration consider the quantity Q given by

$$Q = \int_A^B (L\ dx + M\ dy)$$

which means that Q is obtained as an integration along some curve in the xy-plane from point A to point B; moreover, $L = L(x, y)$ and $M = M(x, y)$. From standard mathematics, it is well known that Q only depends on the end points A and B and not on the path between A and B if $L\ dx + M\ dy$ is a *perfect differential.* The necessary and sufficient condition for $L\ dx + M\ dy$ being a perfect differential is

$$\frac{\partial L}{\partial y} = \frac{\partial M}{\partial x}$$

Generalizing these concepts to (4.3b), we see that W is independent on the integration path if

$$\boxed{\frac{\partial \sigma_{ij}}{\partial \varepsilon_{kl}} = \frac{\partial \sigma_{kl}}{\partial \varepsilon_{ij}}} \tag{4.9}$$

and use of (4.8) demonstrates this condition to be fulfilled – as expected.

Using the transformation rule for the second-order tensors σ_{ij} and $d\varepsilon_{ij}$, it is readily shown that $\sigma_{ij} d\varepsilon_{ij}$ is an invariant. It therefore follows from (4.3a) that dW is an invariant, i.e. we have

$$\boxed{\textit{Strain energy } W \textit{ is an invariant}} \tag{4.10}$$

4.2 Complementary energy and hyper-elasticity

Having established the strain energy W and the fundamental relation (4.8), we will now perform an interesting reformulation. Define the function C - the *complementary energy* per unit volume - by

$$\boxed{C = \sigma_{ij}\varepsilon_{ij} - W(\varepsilon_{ij})} \tag{4.11}$$

It is obvious that C only depends on the current state and not on the manner in which this state was established. By differention we obtain

$$dC = d\sigma_{ij}\varepsilon_{ij} + \sigma_{ij}d\varepsilon_{ij} - \frac{\partial W}{\partial \varepsilon_{ij}}d\varepsilon_{ij}$$

which together with (4.8) gives

$$dC = \varepsilon_{ij}d\sigma_{ij} \tag{4.12}$$

Instead of (4.1), we assume that the inverse relation exists i.e.

$$\boxed{\varepsilon_{ij} = \varepsilon_{ij}(\sigma_{kl})} \tag{4.13}$$

and we obtain

$$\boxed{C(\sigma_{ij}) = \int_0^{\sigma_{ij}} \varepsilon_{kl}(\sigma^*_{mn})d\sigma^*_{kl}} \tag{4.14}$$

where σ_{ij} is the current stress state whereas σ^*_{kl} denotes the integration variable.

We mentioned that the complementary energy C only depends on the current state and not on the history. Moreover, we found from (4.14) that $C = C(\sigma_{ij})$. This may seem a little surprising since ε_{ij} enters (4.11). To convince ourselves that $C = C(\sigma_{ij})$, we assume that $C = C(\sigma_{ij}, \varepsilon_{ij})$ and obtain

$$dC = \frac{\partial C}{\partial \sigma_{ij}}d\sigma_{ij} + \frac{\partial C}{\partial \varepsilon_{ij}}d\varepsilon_{ij}$$

and a comparison with (4.12) indicates that $\partial C/\partial \varepsilon_{ij} = 0$ i.e. $C = C(\sigma_{ij})$. We therefore have

$$dC = \frac{\partial C}{\partial \sigma_{ij}}d\sigma_{ij} \tag{4.15}$$

It appears that by the format (4.11) we have shifted the old variable ε_{ij} in $W = W(\varepsilon_{ij})$ into a new variable σ_{ij} in $C = C(\sigma_{ij})$ without knowing the explicit relation between ε_{ij} and σ_{ij}. The format (4.11) is an example of the use of the *Legendre transformation* that is frequently used in mechanics and – in particular – in thermodynamics; in Chapter 21, we will encounter a number of applications of the Legendre transformation. Subtraction of (4.15) from (4.12) yields

$$(\varepsilon_{ij} - \frac{\partial C}{\partial \sigma_{ij}})d\sigma_{ij} = 0$$

Since this relation holds for arbitrary stress states, it follow that

$$\boxed{\varepsilon_{ij} = \frac{\partial C}{\partial \sigma_{ij}}; \qquad C = C(\sigma_{ij})} \tag{4.16}$$

i.e. the complementary energy C serves as a potential function for ε_{ij}. In the uniaxial case, an illustration of C given by (4.14) is shown in Fig. 4.3.

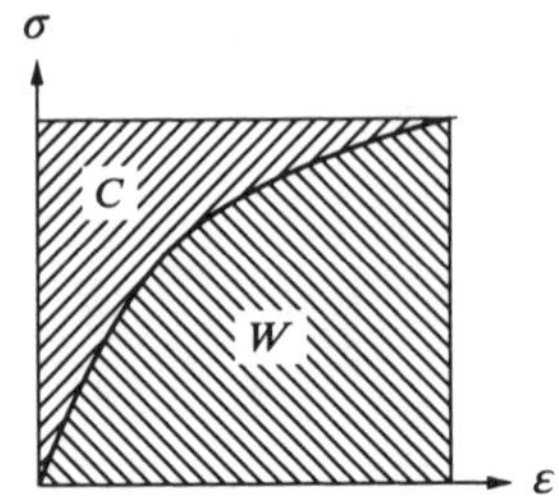

Figure 4.3: Complementary energy C and strain energy W for uniaxial loading.

By arguments similar to those adopted when we evaluated the strain energy W, cf. (4.10), it follows that

$$\boxed{\textit{Complementary energy } C \textit{ is an invariant}} \tag{4.17}$$

Moreover, from (4.16) appears that

$$\boxed{\frac{\partial \varepsilon_{ij}}{\partial \sigma_{kl}} = \frac{\partial \varepsilon_{kl}}{\partial \sigma_{ij}}} \tag{4.18}$$

4.3 Linear hyper-elasticity - Anisotropy

A material is *anisotropic*, if it behaves differently when loaded in the same manner in different directions. As an illustration, consider the piece of wood shown in Fig. 4.4. In Fig. 4.4a), we have uniaxial tension along the x_1−axis and we may express the relation between σ_{11} and ε_{11} as $\sigma_{11} = E_a\varepsilon_{11}$ where E_a is some experimentally determined stiffness parameter. Likewise, in Fig. 4.4b) we also have uniaxial tension along the x_1−axis and the relation between σ_{11} and ε_{11} can now be written $\sigma_{11} = E_b\varepsilon_{11}$ where E_b again is some experimentally determined stiffness parameter. Comparison of Figs. 4.4a) and 4.4b) clearly indicates that $E_a \neq E_b$. We are thereby led to the following general conclusion:

Material anisotropy means that the constitutive relation takes different forms depending on the Cartesian coordinate system we use

(4.19)

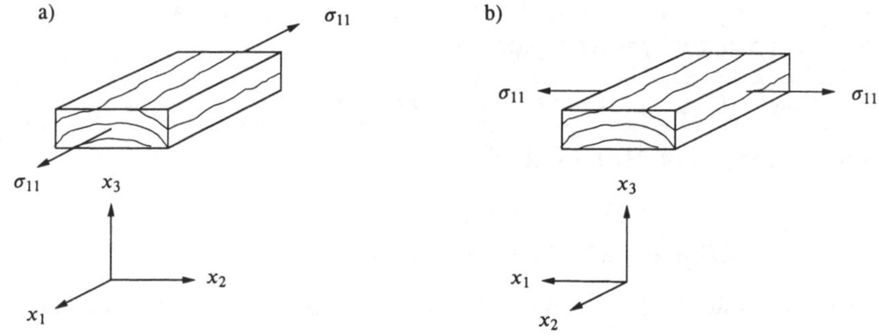

Figure 4.4: Example of anisotropy. Piece of wood loaded by the same uniaxial tension in different directions.

Let us assume that the constitutive law between stresses and strains is linear and let us investigate the general properties of this relation for a hyper-elastic material. The most general linear relation must be of the form

$$\boxed{\sigma_{ij} = D_{ijkl}\varepsilon_{kl} \; ; \qquad D_{ijkl} = D_{ijkl}(x_i)} \tag{4.20}$$

where D_{ijkl} is the *elastic stiffness tensor*. That D_{ijkl} indeed is a tensor of fourth order follows from the quotient theorem and the fact that σ_{ij} and ε_{ij} are second-order tensors, cf. the derivation of (1.41) from (1.37). The formulation (4.20) covers both anisotropic and isotropic elastic materials and in another coordinate system x'_i, we have $\sigma'_{ij} = D'_{ijkl}\varepsilon'_{kl}$ where D'_{ijkl} is related to D_{ijkl} via (1.43). For anisotropic materials, we have $D_{ijkl} \neq D'_{ijkl}$ in accordance with (4.19). Due to the linearity, D_{ijkl} is independent of the amount of loading. However, D_{ijkl} is allowed to depend on the position x_i in which case we have an inhomogeneous material; otherwise it is homogeneous. The constitutive relation (4.20) is referred to as *Hooke's law* or *Hooke's generalized law* since it generalizes the uniaxial form $\sigma = E\varepsilon$ suggested by Hooke in 1676.

Writing all terms in (4.20) explicitly and using the symmetry of ε_{ij}, we obtain

$$\begin{aligned}\sigma_{ij} = \; & D_{ij11}\varepsilon_{11} + D_{ij22}\varepsilon_{22} + D_{ij33}\varepsilon_{33} + \\ & (D_{ij12} + D_{ij21})\varepsilon_{12} + (D_{ij13} + D_{ij31})\varepsilon_{13} + (D_{ij23} + D_{ij32})\varepsilon_{23}\end{aligned}$$

It appears that it is no restriction to rename the term $D_{ij12} + D_{ij21}$ by $2D_{ij12}$. If the new term D_{ij12} now has the symmetry property $D_{ij12} = D_{ij21}$ then we would again obtain the same result as (4.21). Likewise, we rename the term $D_{ij13} + D_{ij31}$ by $2D_{ij13}$ and the term $D_{ij23} + D_{ij32}$ by $2D_{ij23}$ where now $D_{ij13} = D_{ij31}$ as well as $D_{ij23} = D_{ij32}$. By this operation, we have achieved that $D_{ijkl} = D_{ijlk}$.

Moreover, since the stress tensor σ_{ij} is symmetric, it follows directly from (4.20) that $D_{ijkl} = D_{jikl}$. Thus, from the symmetries of σ_{ij} and ε_{ij} we generally have the so-called *minor symmetry* properties

$$\boxed{D_{ijkl} = D_{jikl} \quad ; \quad D_{ijkl} = D_{ijlk} \quad \textit{minor symmetry}} \tag{4.21}$$

With the property (4.21b), (4.21) may be written

$$\begin{aligned}\sigma_{ij} &= D_{ij11}\varepsilon_{11} + D_{ij22}\varepsilon_{22} + D_{ij33}\varepsilon_{33} \\ &\quad +2D_{ij12}\varepsilon_{12} + 2D_{ij13}\varepsilon_{13} + 2D_{ij23}\varepsilon_{23}\end{aligned} \tag{4.22}$$

Up until now, we have just based our considerations on (4.20) and not used that the material is assumed to be hyper-elastic. However, hyper-elasticity implies (4.9) and if (4.20) is inserted into this relation, we immediately conclude the following additional so-called *major symmetry* property

$$\boxed{D_{ijkl} = D_{klij} \qquad \textit{for hyper-elasticity; major symmetry}} \tag{4.23}$$

We saw previously that the strain energy W plays a central role in hyper-elasticity, so it is tempting to derive W for the present material model. For this purpose, we obtain with Hooke's law (4.20) and the symmetry property (4.23) that

$$\begin{aligned} d(\frac{1}{2}\varepsilon_{ij}D_{ijkl}\varepsilon_{kl}) &= \frac{1}{2}d\varepsilon_{ij}D_{ijkl}\varepsilon_{kl} + \frac{1}{2}\varepsilon_{ij}D_{ijkl}d\varepsilon_{kl} \\ &= \frac{1}{2}d\varepsilon_{ij}\sigma_{ij} + \frac{1}{2}\sigma_{kl}d\varepsilon_{kl} \\ &= \sigma_{ij}d\varepsilon_{ij} \end{aligned}$$

From (4.3) then follows that

$$\boxed{W = \frac{1}{2}\varepsilon_{ij}D_{ijkl}\varepsilon_{kl} \qquad or \qquad W = \frac{1}{2}\sigma_{ij}\varepsilon_{ij}} \tag{4.24}$$

Since ε_{ij} is a second-order tensor and D_{ijkl} is a fourth-order tensor, it follows that the quantity $\varepsilon_{ij}D_{ijkl}\varepsilon_{kl}$ is an invariant. It then follows from (4.24) that the strain energy W is an invariant and this conclusion is in accordance with (4.10). Moreover, adopting (4.24a) together with the general expression (4.8), we recover Hooke's generalized law (4.20).

For uniaxial loading where $\sigma = E\varepsilon$ holds, we have $W = \frac{1}{2}E\varepsilon^2$ which is positive since we certainly expect *Young's modulus* E to be positive. It seems natural also to expect that the strain energy is positive for the general case, i.e.

$$\boxed{W > 0} \tag{4.25}$$

Referring to (4.24a) and borrowing from the terminology of matrices, it appears that the linear elastic stiffness tensor D_{ijkl} is *positive definite* (i.e. for any $\varepsilon_{ij} \neq 0$, we have $\varepsilon_{ij}D_{ijkl}\varepsilon_{kl} > 0$, cf. (1.7)). This property turns out to be of importance in order to ensure that a linear elastic boundary value problem possesses a unique solution (see Chapter 24).

As the entire concept of linear elasticity relies on the concept of a one-to-one relation between stresses and strains, it is possible to invert the constitutive relation (4.20) to obtain

$$\boxed{\varepsilon_{ij} = C_{ijkl}\sigma_{kl}} \tag{4.26}$$

where C_{ijkl} is termed the *elastic flexibility tensor* and C_{ijkl} is independent of the amount of loading. In the next section, we shall provide a formal proof that this inversion is mathematically possible. Based on the symmetry of ε_{ij} and σ_{ij}, we obtain similarly to (4.21) that

$$\boxed{C_{ijkl} = C_{jikl} \quad ; \quad C_{ijkl} = C_{ijlk} \quad \textit{minor symmetry}} \tag{4.27}$$

We now require that the constitutive law (4.26), i.e. Hooke's generalized law, should comply with the concepts of hyper-elasticity. Insertion of (4.26) into (4.18) then reveals that

$$\boxed{C_{ijkl} = C_{klij} \qquad \textit{for hyper-elasticity; major symmetry}} \tag{4.28}$$

in accordance with (4.23).

The complementary energy C is defined by (4.11) which with (4.24b) proves that for linear hyper-elastic materials, we have

$$\boxed{C = W = \frac{1}{2}\sigma_{ij}\varepsilon_{ij} > 0} \tag{4.29}$$

Recalling the interpretations of C and W, cf. Fig. 4.3, and the assumed linearity between stresses and strains, this relation is certainly not surprising. Insertion of Hooke's law in the form of (4.26) into (4.29) gives

$$\boxed{C = \frac{1}{2}\sigma_{ij}C_{ijkl}\sigma_{kl} > 0} \tag{4.30}$$

Since σ_{ij} is a second-order tensor and C_{ijkl} a fourth-order tensor, it follows that the quantity $\sigma_{ij}C_{ijkl}\sigma_{kl}$ is an invariant. It then follows from (4.30) that C is an invariant and this conclusion is in accordance with (4.17). We also conclude from (4.30) that C_{ijkl} is positive definite. Moreover, (4.14) shows that $dC = \varepsilon_{ij}d\sigma_{ij}$.

Insertion of the constitutive relation (4.26) into (4.20) yields

$$\sigma_{ij} = D_{ijmn}C_{mnkl}\sigma_{kl}$$

Direct inspection shows that the expression $\sigma_{ij} = \frac{1}{2}(\delta_{ik}\delta_{jl} + \delta_{il}\delta_{jk})\sigma_{kl}$ holds and we also observe that both the left-hand and right-hand side of this relation fulfill the symmetry in i and j. Therefore, we obtain

$$[\frac{1}{2}(\delta_{ik}\delta_{jl} + \delta_{il}\delta_{jk}) - D_{ijmn}C_{mnkl}]\sigma_{kl} = 0$$

and as this expression holds for arbitrary stresses σ_{kl}, we conclude that

$$\boxed{D_{ijmn}C_{mnkl} = \frac{1}{2}(\delta_{ik}\delta_{jl} + \delta_{il}\delta_{jk})} \tag{4.31}$$

In fact, (4.31) shows that the stiffness tensor is the inverse of the flexibility tensor. If we instead insert the constitutive relation (4.20) into (4.26) we obtain in a similar manner

$$\boxed{C_{ijmn}D_{mnkl} = \frac{1}{2}(\delta_{ik}\delta_{jl} + \delta_{il}\delta_{jk})} \tag{4.32}$$

4.4 Linear elasticity – Matrix formulation

As discussed in Chapter 1, tensor formulation is usually most convenient when deriving a theory whereas matrix notation is of advantage when it comes to numerical applications. As our objective is to be able to solve general boundary value problems, which in practice implies the use of numerical solution strategies in terms of the finite element method, we need to reformulate the tensor equations of linear elasticity into their corresponding matrix form.

Let us assume that Hooke's law is written in the form

$$\sigma_{ij} = D_{ijkl}\varepsilon_{kl} \tag{4.33}$$

With the summation convention and the symmetry property $D_{ijkl} = D_{ijlk}$, (4.33) can be written as

$$[\sigma_{ij}] = \begin{bmatrix} D_{ij11} & D_{ij22} & D_{ij33} & D_{ij12} & D_{ij13} & D_{ij23} \end{bmatrix} \begin{bmatrix} \varepsilon_{11} \\ \varepsilon_{22} \\ \varepsilon_{33} \\ 2\varepsilon_{12} \\ 2\varepsilon_{13} \\ 2\varepsilon_{23} \end{bmatrix} \tag{4.34}$$

in accordance with (4.22). Define the following matrices

$$\boxed{\boldsymbol{\sigma} = \begin{bmatrix} \sigma_{11} \\ \sigma_{22} \\ \sigma_{33} \\ \sigma_{12} \\ \sigma_{13} \\ \sigma_{23} \end{bmatrix}; \quad \boldsymbol{\varepsilon} = \begin{bmatrix} \varepsilon_{11} \\ \varepsilon_{22} \\ \varepsilon_{33} \\ 2\varepsilon_{12} \\ 2\varepsilon_{13} \\ 2\varepsilon_{23} \end{bmatrix}} \tag{4.35}$$

and

$$\boldsymbol{D} = \begin{bmatrix} D_{1111} & D_{1122} & D_{1133} & D_{1112} & D_{1113} & D_{1123} \\ D_{2211} & D_{2222} & D_{2233} & D_{2212} & D_{2213} & D_{2223} \\ D_{3311} & D_{3322} & D_{3333} & D_{3312} & D_{3313} & D_{3323} \\ D_{1211} & D_{1222} & D_{1233} & D_{1212} & D_{1213} & D_{1223} \\ D_{1311} & D_{1322} & D_{1333} & D_{1312} & D_{1313} & D_{1323} \\ D_{2311} & D_{2322} & D_{2333} & D_{2312} & D_{2313} & D_{2323} \end{bmatrix} \tag{4.36}$$

then (4.34) can be written as

$$\boxed{\boldsymbol{\sigma} = \boldsymbol{D}\boldsymbol{\varepsilon}} \tag{4.37}$$

It may be slightly confusing that we use the notation $\boldsymbol{\sigma}$ both for the column matrix defined by (4.35) as well as for the stress tensor. Likewise, we use $\boldsymbol{\varepsilon}$ to denote the column matrix defined by (4.35) as well as the strain tensor. Unfortunately, such double designation cannot be avoided, but remembering that $\boldsymbol{\sigma}$ and $\boldsymbol{\varepsilon}$ defined by (4.35) most often will appear in connection with the matrix **D** defined by (4.36), uncertainty of what is meant by $\boldsymbol{\sigma}$ and $\boldsymbol{\varepsilon}$ will hardly emerge in practice.

It is of interest that the engineering shear strains $2\varepsilon_{12}, 2\varepsilon_{13}, 2\varepsilon_{23}$ occur in $\boldsymbol{\varepsilon}$ defined by (4.35). In principle, we could equally well have placed the digit 2 at appropriate locations in the **D**-matrix. By not doing so we obtain the significant advantage that owing to the major symmetry property $D_{ijkl} = D_{klij}$ that holds for hyper-elasticity, **D** as defined by (4.36) becomes symmetric, i.e.

$$\boxed{\boldsymbol{D} = \boldsymbol{D}^T \qquad when\ D_{ijkl} = D_{klij}} \tag{4.38}$$

A further advantage is obtained by the definition given in (4.35) since it appears that

$$\boxed{\sigma_{ij}\varepsilon_{ji} = \boldsymbol{\sigma}^T\boldsymbol{\varepsilon} = \boldsymbol{\varepsilon}^T\boldsymbol{\sigma}} \tag{4.39}$$

where $\sigma_{ij}\varepsilon_{ji}$ clearly is an invariant.

From $\sigma_{ij}\varepsilon_{ji} = \boldsymbol{\varepsilon}^T\boldsymbol{\sigma}$ we conclude from (4.24b), (4.25) and (4.37) that the strain energy W is given by

$$\boxed{W = \frac{1}{2}\boldsymbol{\varepsilon}^T\boldsymbol{D}\boldsymbol{\varepsilon} > 0} \tag{4.40}$$

i.e. $\boldsymbol{D}$ is positive definite. It therefore possesses an inverse defined by

$$\boxed{\boldsymbol{C} = \boldsymbol{D}^{-1}}$$

i.e. (4.37) leads to

$$\boxed{\boldsymbol{\varepsilon} = \boldsymbol{C}\boldsymbol{\sigma}} \tag{4.41}$$

The matrix $\boldsymbol{D}$ is called the *linear elastic stiffness matrix* whereas $\boldsymbol{C}$ is called the *linear elastic flexibility matrix* . From (4.38), it is obvious that

$$\boxed{\boldsymbol{C} = \boldsymbol{C}^T \qquad \textit{for hyper-elasticity}}$$

It is evident that (4.41) in tensor notation takes the form stipulated by (4.26) and we have therefore proved formally that the inversion of (4.20) that leads to (4.26) is mathematically possible. Finally, we obtain from $\sigma_{ij}\varepsilon_{ji} = \boldsymbol{\sigma}^T\boldsymbol{\varepsilon}$, (4.29) and (4.41) that the complementary energy C is given by

$$\boxed{C = \frac{1}{2}\boldsymbol{\sigma}^T\boldsymbol{C}\boldsymbol{\sigma}}$$

Returning to the definition of the column matrices $\boldsymbol{\sigma}$ and $\boldsymbol{\varepsilon}$ defined by (4.35), we mentioned previously that other definitions are possible. However, the present definitions are the classical ones and as mentioned they possess certain advantages; they were introduced by Voigt (1928). We may mention that if $\boldsymbol{\sigma}$ and $\boldsymbol{\varepsilon}$ are defined as $\boldsymbol{\sigma}^T = [\sigma_{11}\ \sigma_{22}\ \sigma_{33}\sqrt{2}\sigma_{12}\ \sqrt{2}\sigma_{13}\sqrt{2}\sigma_{23}]$ and $\boldsymbol{\varepsilon}^T = [\varepsilon_{11}\ \varepsilon_{22}\ \varepsilon_{33}\ \sqrt{2}\varepsilon_{12}\ \sqrt{2}\varepsilon_{13}\ \sqrt{2}\varepsilon_{23}]$ then the corresponding stiffness matrix also becomes symmetric. Moreover, $\boldsymbol{\sigma}^T\boldsymbol{\sigma} = \sigma_{ij}\sigma_{ji}$ and $\boldsymbol{\varepsilon}^T\boldsymbol{\varepsilon} = \varepsilon_{ij}\varepsilon_{ji}$ as well as $\boldsymbol{\sigma}\boldsymbol{\varepsilon} = \sigma_{ij}\varepsilon_{ji}$. This approach was adopted by Argyris (1965a) and by Horgan (1973) and its further ramifications is discussed by Pedersen (1995) as well as Cowin and Mehrabadi (1995).

4.5 Change of coordinate system when using matrix format

Suppose that in the x_i−coordinate system, we have determined the stress tensor σ_{ij}, the strain tensor ε_{ij} and the constitutive tensor D_{ijkl}. In another coordinate system given by x'_i, the stress tensor σ'_{ij}, the strain tensor ε'_{ij} and the constitutive tensor D'_{ijkl} are given by the corresponding transformation formulas that are characteristic for tensors, cf. (3.8), (1.43) and (2.19). However, if we work in the matrix form $\boldsymbol{\sigma} = \boldsymbol{D}\boldsymbol{\varepsilon}$, cf. (4.37), then neither $\boldsymbol{\sigma}$ nor $\boldsymbol{\varepsilon}$ and nor $\boldsymbol{D}$ are tensors so we need to establish rules for how these quantities transform.

Let us first establish the transformation rules for the column matrices $\boldsymbol{\sigma}$ and $\boldsymbol{\varepsilon}$. From the tensor transformation rule (3.8) we have

$$\sigma'_{ij} = A_{ik}\sigma_{kl}A_{jl} = L_{ijkl}\sigma_{kl} \qquad \text{where} \qquad L_{ijkl} = A_{ik}A_{jl}$$

from which we can establish the following transformation rule in matrix format

$$\boxed{\boldsymbol{\sigma}' = \boldsymbol{L}\boldsymbol{\sigma}} \tag{4.42}$$

where $\boldsymbol{L}$ is a 6x6 transformation matrix with components that can be determined in a trivial, but tedious manner. We noted in (4.39) that $\boldsymbol{\sigma}^T\boldsymbol{\varepsilon}$ is an invariant. Therefore, we have with (4.42) that

$$\boldsymbol{\sigma}^T\boldsymbol{\varepsilon} = \boldsymbol{\sigma}'^T\boldsymbol{\varepsilon}' \quad \text{i.e.} \quad \boldsymbol{\sigma}^T(\boldsymbol{\varepsilon} - \boldsymbol{L}^T\boldsymbol{\varepsilon}') = 0$$

and as this expression holds for arbitrary values of $\boldsymbol{\sigma}^T$, we conclude that

$$\boxed{\boldsymbol{\varepsilon} = \boldsymbol{L}^T\boldsymbol{\varepsilon}'} \tag{4.43}$$

It is important to observe that the transformation matrix $\boldsymbol{L}$ is not orthogonal, i.e. $\boldsymbol{L}^T \neq \boldsymbol{L}^{-1}$. To prove this, consider the quantity $\boldsymbol{\sigma}^T\boldsymbol{\sigma}$ which is certainly not an invariant (note that $\boldsymbol{\sigma}^T\boldsymbol{\sigma} \neq \sigma_{ij}\sigma_{ij}$). With (4.42), we obtain $\boldsymbol{\sigma}'^T\boldsymbol{\sigma}' = \boldsymbol{\sigma}^T\boldsymbol{L}^T\boldsymbol{L}\boldsymbol{\sigma}$, i.e. if $\boldsymbol{L}^T\boldsymbol{L} = \boldsymbol{I}$ then $\boldsymbol{\sigma}'^T\boldsymbol{\sigma}' = \boldsymbol{\sigma}^T\boldsymbol{\sigma}$ in contradiction with the fact that $\boldsymbol{\sigma}^T\boldsymbol{\sigma}$ is not an invariant.

Having established the transformation rules for $\boldsymbol{\sigma}$ and $\boldsymbol{\varepsilon}$, we next turn our attention to the constitutive matrix $\boldsymbol{D}$. In the x_i–coordinate system, we have

$$\boldsymbol{\sigma} = \boldsymbol{D}\boldsymbol{\varepsilon} \tag{4.44}$$

and in the x_i'–system, the constitutive relation reads

$$\boldsymbol{\sigma}' = \boldsymbol{D}'\boldsymbol{\varepsilon}' \tag{4.45}$$

Insertion of (4.42) and (4.43) into (4.44) yields

$$\boldsymbol{L}^{-1}\boldsymbol{\sigma}' = \boldsymbol{D}\boldsymbol{L}^T\boldsymbol{\varepsilon}' \quad \text{i.e.} \quad \boldsymbol{\sigma}' = \boldsymbol{L}\boldsymbol{D}\boldsymbol{L}^T\boldsymbol{\varepsilon}'$$

and a comparison with (4.45) reveals that

$$\boxed{\boldsymbol{D}' = \boldsymbol{L}\boldsymbol{D}\boldsymbol{L}^T} \tag{4.46}$$

Evidently, this transformation rule is not restricted to linear hyper-elasticity and it holds for any constitutive relation that is given in a format similar to (4.44).

4.6 Anisotropy in linear hyper-elasticity

We mentioned previously, cf. (4.19), that anisotropy, in general, means that the constitutive relation takes different forms depending on the Cartesian coordinate system we use. Anisotropy appears when the material behaves differently when loaded in the same manner in different directions; a piece of wood may be taken as an example, cf. Fig. 4.4.

On the other hand, if the material behaves identically when loaded in the same manner in all directions, the material is said to be *isotropic*. We are then led to the following conclusion:

Material isotropy means that the constitutive relation remains the same irrespective of the Cartesian coordinate system we use

(4.47)

Isotropy means that the material has no properties that depend on the direction and a piece of steel may be taken as an example of an isotropic material. In the x_i-coordinate system, we have $\sigma_{ij} = D_{ijkl}\varepsilon_{kl}$ and in another coordinate system x'_i we have $\sigma'_{ij} = D'_{ijkl}\varepsilon'_{kl}$ where $D_{ijkl} = D'_{ijkl}$ according to (4.47) holds for isotropic materials. In agreement with the discussion in Section 1.6, the elastic stiffness tensor D_{ijkl} for isotropic materials is then an isotropic fourth-order tensor and we shall later derive its explicit format.

In (4.37), we expressed Hooke's law in matrix form as

$$\boldsymbol{\sigma} = \boldsymbol{D}\boldsymbol{\varepsilon} \tag{4.48}$$

where the assumption of hyper-elasticity implies that $\boldsymbol{D} = \boldsymbol{D}^T$. It is convenient to redefine the components of $\boldsymbol{D}$ given by (4.36) according to

$$\boldsymbol{D} = \begin{bmatrix} D_{11} & D_{12} & D_{13} & D_{14} & D_{15} & D_{16} \\ D_{21} & D_{22} & D_{23} & D_{24} & D_{25} & D_{26} \\ D_{31} & D_{32} & D_{33} & D_{34} & D_{35} & D_{36} \\ D_{41} & D_{42} & D_{43} & D_{44} & D_{45} & D_{46} \\ D_{51} & D_{52} & D_{53} & D_{54} & D_{55} & D_{56} \\ D_{61} & D_{63} & D_{63} & D_{64} & D_{65} & D_{66} \end{bmatrix} \tag{4.49}$$

It appears that for a completely anisotropic material, the stiffness matrix $\boldsymbol{D}$ is fully populated. Due to the symmetry of $\boldsymbol{D}$, however, it comprises 21 independent components. We will now evaluate different forms of $\boldsymbol{D}$ for different forms of anisotropy where some kind of material symmetry exists.

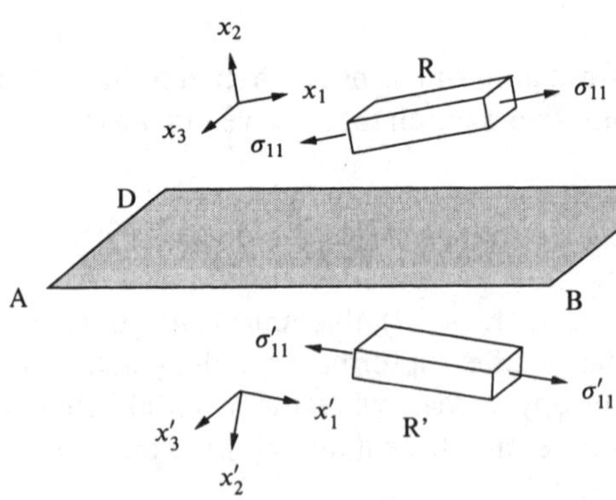

Figure 4.5: Region R' is a reflection of region R; coordinate system x'_i is a reflection of coordinate system x_i.

To illustrate such a symmetry property, consider two regions R and R' of a homogeneous material. According to Fig. 4.5, these regions are reflections of each other about the plane ABCD. Moreover, the coordinate system x'_i is a reflection (i.e. mirror image) of coordinate system x_i about this plane. Suppose that region R is subjected to a stress state where $\sigma_{11} \neq 0$ and all other stress components are equal to zero, cf. Fig. 4.5; via Hooke's law $\boldsymbol{\varepsilon} = \boldsymbol{C}\boldsymbol{\sigma}$, this stress state results in a strain state ε_{ij}. Likewise, let region R' be subjected to the stress state $\sigma'_{11} \neq 0$ and all other stress components equal to zero; this gives rise to the strain state ε'_{ij}. Take the stress component $\sigma_{11} = \sigma'_{11}$; assume that this implies that all strain components $\varepsilon_{ij} = \varepsilon'_{ij}$. To generalize this situation, consider an arbitrary stress state and take the stress components $\sigma_{ij} = \sigma'_{ij}$. Assume that this implies that all strain components $\varepsilon_{ij} = \varepsilon'_{ij}$; in that case the material is said to possess a *plane of elastic symmetry* and, in the present case, this symmetry plane is plane ABCD.

From the discussion above, we arrive at the following definition

> *If the constitutive relation takes the same form for every pair of Cartesian coordinate systems that are mirror images (reflections) of each other in a certain plane, this plane is a plane of elastic symmetry* (4.50)

cf. Malvern (1969) and Love (1944). As an example, we may assume that the x_1x_2–plane is a plane of elastic symmetry, cf. Fig. 4.6a). In this coordinate system, Hooke's law is given by (4.48). Figure 4.6b) illustrates that the x'_i–coordinate system is a mirror image (reflection) of the x_i–system given in Fig. 4.6a), i.e. $x'_1 = x_1$, $x'_2 = x_2$ and $x'_3 = -x_3$. In the x'_i–coordinate system, Hooke's law reads

$$\boldsymbol{\sigma}' = \boldsymbol{D}\boldsymbol{\varepsilon}' \tag{4.51}$$

where it was used that the constitutive matrix is the same in the two coordinate systems since the x_1x_2–plane is assumed to be a plane of elastic symmetry.

Having motivated and discussed the implications of elastic symmetry, we now consider the situation where the stress states $\boldsymbol{\sigma}$ and $\boldsymbol{\sigma}'$ are the same, but

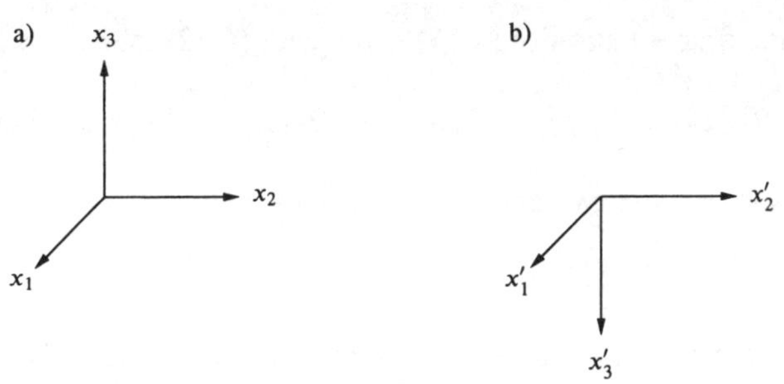

Figure 4.6: Coordinate change when x_1x_2–plane is a plane of elastic symmetry.

measured in different coordinate systems. This implies that also the strain states $\boldsymbol{\varepsilon}$ and $\boldsymbol{\varepsilon}'$ are the same, but measured in different coordinate systems. According to (1.27) and (1.28), the coordinate change in Fig. 4.6 implies

$$\boldsymbol{x} = \boldsymbol{A}^T \boldsymbol{x}' \qquad \text{where} \qquad \boldsymbol{A}^T = \begin{bmatrix} 1 & 0 & 0 \\ 0 & 1 & 0 \\ 0 & 0 & -1 \end{bmatrix}$$

From the transformation rules (3.8) and (2.19) we have $\boldsymbol{\sigma}' = \boldsymbol{A}\boldsymbol{\sigma}\boldsymbol{A}^T$ and $\boldsymbol{\varepsilon}' = \boldsymbol{A}\boldsymbol{\varepsilon}\boldsymbol{A}^T$ respectively, and this leads to

$$\begin{aligned} &\sigma'_{13} = -\sigma_{13}\ ;\ \sigma'_{23} = -\sigma_{23} \qquad \text{otherwise} \qquad \sigma'_{ij} = \sigma_{ij} \\ &\varepsilon'_{13} = -\varepsilon_{13}\ ;\ \varepsilon'_{23} = -\varepsilon_{23} \qquad \text{otherwise} \qquad \varepsilon'_{ij} = \varepsilon_{ij} \end{aligned} \tag{4.52}$$

It seems appropriate to point out that when we in Chapter 1 discussed a change of coordinate system, a translation and a rotation of the coordinate system was considered. This is the most general change of a coordinate system by which a right-handed (left-handed) coordinate system is preserved as a right-handed (left-handed) coordinate system. However, if a reflection of the coordinate system is involved, as shown in Fig. 4.6, the right-handed coordinate system x_i changes into the left-handed coordinate system x'_i. Whereas det$\boldsymbol{A}$=1 holds for all changes of right-handed coordinate systems, cf. (1.26), we now have det$\boldsymbol{A}$=-1, cf. the transformation matrix above. Apart from that, it is evident that we still have the transformation rule $\boldsymbol{x}' = \boldsymbol{A}(\boldsymbol{x} - \boldsymbol{c})$, cf. (1.17), as well as the previously established tensor transformation rules that still hold even when reflections of the coordinate system are involved. In the literature, a transformation where det$\boldsymbol{A}$=1 and $\boldsymbol{A}^T\boldsymbol{A} = \boldsymbol{I}$ is called a *proper orthogonal transformation* whereas det$\boldsymbol{A}$=-1 and $\boldsymbol{A}^T\boldsymbol{A} = \boldsymbol{I}$ is called an *improper orthogonal transformation.*

The first equation of (4.48) provides

$$\sigma_{11} = D_{11}\varepsilon_{11} + D_{12}\varepsilon_{22} + D_{13}\varepsilon_{33} + 2D_{14}\varepsilon_{12} + 2D_{15}\varepsilon_{13} + 2D_{16}\varepsilon_{23}$$

Likewise, the first equation of (4.51) gives with (4.52) that

$$\sigma_{11} = D_{11}\varepsilon_{11} + D_{12}\varepsilon_{22} + D_{13}\varepsilon_{33} + 2D_{14}\varepsilon_{12} - 2D_{15}\varepsilon_{13} - 2D_{16}\varepsilon_{23}$$

A comparison of these two expressions reveals that

$$D_{15} = D_{16} = 0$$

Proceeding in the same manner for the remaining equations of (4.48) and (4.51) results in

$$D_{25} = D_{26} = D_{35} = D_{36} = D_{45} = D_{46} = 0$$

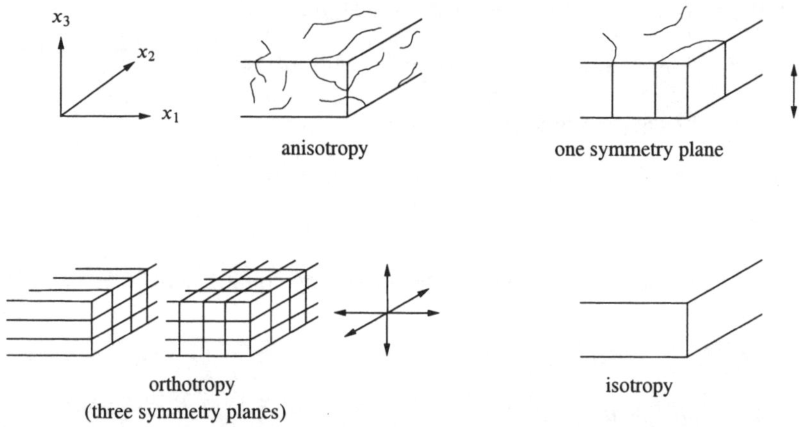

Figure 4.7: Illustration of increasing degree of symmetry.

We have therefore proved that if the x_1x_2–plane is a plane of elastic symmetry, then the constitutive matrix $\boldsymbol{D}$ takes the form

$$\boldsymbol{D} = \begin{bmatrix} D_{11} & D_{12} & D_{13} & D_{14} & 0 & 0 \\ D_{21} & D_{22} & D_{23} & D_{24} & 0 & 0 \\ D_{31} & D_{32} & D_{33} & D_{34} & 0 & 0 \\ D_{41} & D_{42} & D_{43} & D_{44} & 0 & 0 \\ 0 & 0 & 0 & 0 & D_{55} & D_{56} \\ 0 & 0 & 0 & 0 & D_{65} & D_{66} \end{bmatrix} \quad \textit{one symmetry plane} \tag{4.53}$$

Therefore, recalling the symmetry of $\boldsymbol{D}$, it appears that one plane of symmetry reduces the 21 independent components of $\boldsymbol{D}$ to 13 independent components. The case of the x_1x_2–plane being a plane of elastic symmetry is illustrated in Fig. 4.7.

If not only the x_1x_2-plane is a symmetry plane, but also the x_1x_3-plane is a symmetry plane, then we obtain

$$\boldsymbol{x} = \boldsymbol{A}^T\boldsymbol{x}' \qquad \text{where} \qquad \boldsymbol{A}^T = \begin{bmatrix} 1 & 0 & 0 \\ 0 & -1 & 0 \\ 0 & 0 & 1 \end{bmatrix}$$

and with $\boldsymbol{\sigma}' = \boldsymbol{A}\boldsymbol{\sigma}\boldsymbol{A}^T$ and $\boldsymbol{\varepsilon}' = \boldsymbol{A}\boldsymbol{\varepsilon}\boldsymbol{A}^T$, this implies that

$$\begin{aligned} \sigma'_{12} &= -\sigma_{12} \ ; \ \sigma'_{32} = -\sigma_{32} && \text{otherwise} && \sigma'_{ij} = \sigma_{ij} \\ \varepsilon'_{12} &= -\varepsilon_{12} \ ; \ \varepsilon'_{32} = -\varepsilon_{32} && \text{otherwise} && \varepsilon'_{ij} = \varepsilon_{ij} \end{aligned} \tag{4.54}$$

From $\boldsymbol{\sigma} = \boldsymbol{D}\boldsymbol{\varepsilon}$ and the first equation of (4.53) we obtain

$$\sigma_{11} = D_{11}\varepsilon_{11} + D_{12}\varepsilon_{22} + D_{13}\varepsilon_{33} + 2D_{14}\varepsilon_{12}$$

Likewise, the first equation of (4.51) gives with (4.54)

$$\sigma_{11} = D_{11}\varepsilon_{11} + D_{12}\varepsilon_{22} + D_{13}\varepsilon_{33} - 2D_{14}\varepsilon_{12}$$

A comparison of these two expressions reveals that

$$D_{14} = 0$$

Proceeding in the same manner for the remaining equations of (4.53) and (4.51) results in

$$D_{24} = D_{34} = D_{56} = 0$$

i.e. (4.53) reduces to

$$\boldsymbol{D} = \begin{bmatrix} D_{11} & D_{12} & D_{13} & 0 & 0 & 0 \\ D_{21} & D_{22} & D_{23} & 0 & 0 & 0 \\ D_{31} & D_{32} & D_{33} & 0 & 0 & 0 \\ 0 & 0 & 0 & D_{44} & 0 & 0 \\ 0 & 0 & 0 & 0 & D_{55} & 0 \\ 0 & 0 & 0 & 0 & 0 & D_{66} \end{bmatrix} \quad \textit{orthotropy} \tag{4.55}$$

If finally, not only the x_1x_2- and x_1x_3-planes are planes of symmetry, but also the x_2x_3-plane is a plane of symmetry, we obtain

$$\boldsymbol{x} = \boldsymbol{A}^T\boldsymbol{x}' \qquad \text{where} \qquad \boldsymbol{A}^T = \begin{bmatrix} -1 & 0 & 0 \\ 0 & 1 & 0 \\ 0 & 0 & 1 \end{bmatrix}$$

i.e.

$$\begin{aligned} &\sigma'_{12} = -\sigma_{12}\ ;\ \sigma'_{13} = -\sigma_{13} \qquad \text{otherwise} \qquad \sigma'_{ij} = \sigma_{ij} \\ &\varepsilon'_{12} = -\varepsilon_{12}\ ;\ \varepsilon'_{13} = -\varepsilon_{13} \qquad \text{otherwise} \qquad \varepsilon'_{ij} = \varepsilon_{ij} \end{aligned} \tag{4.56}$$

However, a comparison of (4.56) with (4.52) and (4.54) shows that no new information can be achieved and as a material possessing three orthogonal planes of elastic symmetry is called *orthotropic* we conclude

$$\begin{aligned} \textit{Orthotropy} &= \textit{three orthogonal symmetry planes} \\ &= \textit{two orthogonal symmetry planes} \end{aligned} \tag{4.57}$$

From (4.55), we then conclude that an orthotropic material possesses nine independent elastic parameters. Several important engineering materials are orthotropic and examples are wood, paper, rolled steel and rolled aluminum and an illustration is given in Fig. 4.7. It is emphasized that (4.55) holds when the coordinate axes are chosen as parallel to the *material axes of orthotropy*. If this is not the case, the coordinate system must be rotated accordingly and the stress and strain components in the new aligned coordinate system must be determined using (3.8) and (2.19) before use can be made of (4.55).

From a physical point of view, the form (4.55) of the orthotropic $\boldsymbol{D}$-matrix is certainly not surprising since it implies that normal strains only result in normal stresses and that a shear strain only affects the corresponding shear stress. In view of the orthotropic symmetry properties this is exactly the response we would expect and, in fact, (4.55) could be written down directly just utilizing these physical evident properties. The reason that we have here adopted a more formal route to derive (4.55) is that when it comes to yield criteria for orthotropic materials, the direct physical approach is difficult to apply and the benefits of the formal approach introduced already now are then considerable, cf. Section 8.13. In Section 6.5, we will return to orthotropic elasticity, but adopt another and very elegant approach.

Let us finally invert (4.55) and obtain Hooke's law in the format

$$\begin{bmatrix} \varepsilon_{11} \\ \varepsilon_{22} \\ \varepsilon_{33} \\ 2\varepsilon_{12} \\ 2\varepsilon_{13} \\ 2\varepsilon_{23} \end{bmatrix} = \begin{bmatrix} \frac{1}{E_1} & -\frac{\nu_{21}}{E_2} & -\frac{\nu_{31}}{E_3} & 0 & 0 & \\ -\frac{\nu_{12}}{E_1} & \frac{1}{E_2} & -\frac{\nu_{32}}{E_3} & 0 & 0 & \\ \frac{\nu_{13}}{E_1} & -\frac{\nu_{23}}{E_2} & \frac{1}{E_3} & 0 & 0 & \\ 0 & 0 & 0 & \frac{1}{G_{12}} & 0 & 0 \\ 0 & 0 & 0 & 0 & \frac{1}{G_{13}} & 0 \\ 0 & 0 & 0 & 0 & 0 & \frac{1}{G_{23}} \end{bmatrix} \begin{bmatrix} \sigma_{11} \\ \sigma_{22} \\ \sigma_{33} \\ \sigma_{12} \\ \sigma_{13} \\ \sigma_{23} \end{bmatrix} \qquad (4.58)$$

Since the flexibility matrix is symmetric, we have

$$\frac{\nu_{21}}{E_2} = \frac{\nu_{12}}{E_1}; \quad \frac{\nu_{31}}{E_3} = \frac{\nu_{13}}{E_1}; \quad \frac{\nu_{32}}{E_3} = \frac{\nu_{23}}{E_2}$$

The notation for Poisson's ratios needs an explanation. Suppose that we have uniaxial loading in the x_2-direction; all stress components except σ_{22} are zero. From (4.58) follows that $\varepsilon_{22} = \sigma_{22}/E_2$ whereas $\varepsilon_{11} = -\nu_{21}\varepsilon_{22}$ and $\varepsilon_{33} = -\nu_{23}\varepsilon_{22}$. Therefore, ν_{21} is Poisson's ratio associated with loading in the x_2-direction and strain in the x_1-direction and a similar interpretation holds for ν_{23}.

Apart from orthotropy, other symmetry properties like various crystal symmetries may occasionally be of importance and reference is given to, for instance, Love (1944), Green and Adkins (1960), Lekhnitskii (1981) and Cowin and Mehrabadi (1995) for further information.

We have already mentioned, cf. (4.47), that isotropy means that the constitutive relation is the same irrespective of the Cartesian coordinate system we use. This implies that every plane is a plane of elastic symmetry and, referring for instance to Sokolnikoff (1946) or Malvern (1969), the constitutive matrix then takes the form

$$D = \frac{E}{(1+\nu)(1-2\nu)}\begin{bmatrix} 1-\nu & \nu & \nu & 0 & 0 & 0 \\ \nu & 1-\nu & \nu & 0 & 0 & 0 \\ \nu & \nu & 1-\nu & 0 & 0 & 0 \\ 0 & 0 & 0 & \frac{1}{2}(1-2\nu) & 0 & 0 \\ 0 & 0 & 0 & 0 & \frac{1}{2}(1-2\nu) & 0 \\ 0 & 0 & 0 & 0 & 0 & \frac{1}{2}(1-2\nu) \end{bmatrix} \qquad (4.59)$$

We now have only two independent material parameters and they are *Young's modulus E* and *Poisson's ratio* ν. We shall deal with isotropic linear elasticity in more details in Section 4.9 and we shall then derive (4.59) by another approach. For isotropic materials, it is straightforward to invert $\boldsymbol{D}$ to obtain

$$C = \frac{1}{E}\begin{bmatrix} 1 & -\nu & -\nu & 0 & 0 & 0 \\ -\nu & 1 & -\nu & 0 & 0 & 0 \\ -\nu & -\nu & 1 & 0 & 0 & 0 \\ 0 & 0 & 0 & 2(1+\nu) & 0 & 0 \\ 0 & 0 & 0 & 0 & 2(1+\nu) & 0 \\ 0 & 0 & 0 & 0 & 0 & 2(1+\nu) \end{bmatrix} \qquad (4.60)$$

We finally mention that whereas each stress component for general anisotropic elasticity couples with all strain components, cf. (4.49), isotropy implies that the normal stresses are given by the normal strains and each shear stress is determined entirely by the corresponding shear strain, cf. (4.59).

4.7 Initial strains - Thermoelasticity

Let us assume that there exists a given strain state ε^o_{ij} for which the strain energy is zero. In analogy with (4.24), this may be achieved by assuming that

$$\boxed{W = \frac{1}{2}(\varepsilon_{ij} - \varepsilon^o_{ij})D_{ijkl}(\varepsilon_{kl} - \varepsilon^o_{kl})} \qquad (4.61)$$

From (4.8) then follows that

$$\boxed{\sigma_{ij} = D_{ijkl}(\varepsilon_{kl} - \varepsilon^o_{kl})} \qquad (4.62)$$

It appears that by the choice (4.61), we have obtained an extension of Hooke's law, which implies that $\sigma_{ij} = 0$ when $\varepsilon_{ij} = \varepsilon^o_{ij}$. The strains ε^o_{ij} are called *initial strains*. In obvious matrix notation (4.62) reads

$$\sigma = D(\varepsilon - \varepsilon^o)$$

An important example of initial strains is *thermal strains*, i.e. strains caused by thermal expansion of the material. When thermal strains are considered in Hooke's law (4.62), one speaks of *thermoelasticity*.

To determine the thermal strains, consider a specimen of isotropic and homogeneous material which is free to expand or contract as a result of a uniform change in temperature. As the material is free to expand or contract, $\sigma_{ij} = 0$ holds and the total strains ε_{ij} and the thermal strains ε^o_{ij} are equal. We have

$$\varepsilon^o_{ij} = \alpha \Delta T \delta_{ij} \tag{4.63}$$

where α is the *thermal expansion coefficient* and ΔT is the change of temperature from some reference temperature where no thermal strains exist. Note that in accordance with the assumption of isotropy, the thermal normal strains are equal and no thermal shear strains exist.

For a anisotropic material, (4.63) is replaced by

$$\varepsilon^o_{ij} = \alpha_{ij} \Delta T \tag{4.64}$$

where α_{ij} is some symmetric second-order tensor that characterizes the thermal expansion properties of the material. Consider orthotropy, for instance, we have

$$[\alpha_{ij}] = \begin{bmatrix} \alpha_{11} & 0 & 0 \\ 0 & \alpha_{22} & 0 \\ 0 & 0 & \alpha_{33} \end{bmatrix}$$

Finally, we note that (4.62) may be written as

$$\boxed{\varepsilon_{ij} = \varepsilon^e_{ij} + \varepsilon^o_{ij}}$$

where

$$\boxed{\varepsilon^e_{ij} = C_{ijkl}\sigma_{kl} \quad ; \qquad \varepsilon^o_{ij} = \alpha_{ij}\Delta T}$$

I.e. the *total strains* ε_{ij} consist of the sum of the *elastic strains* ε^e_{ij} determined by Hooke's law and the thermal strains ε^o_{ij}. The formulation (4.62) was introduced by Duhamel in 1837.

4.8 Most general isotropic hyper-elasticity

In the previous sections, we discussed various aspects of linear hyper-elasticity, its matrix formulation and different forms of anisotropy. We will now derive the most general form of hyper-elasticity for isotropic materials and we shall see that this opens for modeling of nonlinear elastic behavior.

The strain energy depends on the current strains ε_{ij} through $W = W(\varepsilon_{ij})$ and according to (4.10), the strain energy is an invariant. The strain tensor ε_{ij} can be expressed by the principal strains ε_1, ε_2 and ε_3 and the corresponding principal strain directions. Isotropy means that the material has no directional properties and this implies that we may write the strain energy as

$W = W(\varepsilon_1, \varepsilon_2, \varepsilon_3)$. As the principal strains are given uniquely by the strain invariants, we may equally well write the strain energy W as

$$\boxed{W = W(\tilde{I}_1, \tilde{I}_2, \tilde{I}_3)} \tag{4.65}$$

where the generic strain invariants are given (2.51), i.e.

$$\tilde{I}_1 = \varepsilon_{kk}\ ; \quad \tilde{I}_2 = \frac{1}{2}\varepsilon_{kl}\varepsilon_{lk}\ ; \quad \tilde{I}_3 = \frac{1}{3}\varepsilon_{kl}\varepsilon_{lm}\varepsilon_{mk} \tag{4.66}$$

Formulation (4.65) is evidently in accordance with (4.10), stating that W is an invariant. The choice of the set of invariants given by (4.66) is particularly convenient, since we have the following neat relations

$$\frac{\partial \tilde{I}_1}{\partial \varepsilon_{ij}} = \delta_{ij}\ ; \quad \frac{\partial \tilde{I}_2}{\partial \varepsilon_{ij}} = \varepsilon_{ij}\ ; \quad \frac{\partial \tilde{I}_3}{\partial \varepsilon_{ij}} = \varepsilon_{ik}\varepsilon_{kj} \tag{4.67}$$

We are now in a position to derive the most general constitutive law for isotropic hyper-elastic materials. From (4.8) and (4.65) we obtain

$$\sigma_{ij} = \frac{\partial W}{\partial \tilde{I}_1}\frac{\partial \tilde{I}_1}{\partial \varepsilon_{ij}} + \frac{\partial W}{\partial \tilde{I}_2}\frac{\partial \tilde{I}_2}{\partial \varepsilon_{ij}} + \frac{\partial W}{\partial \tilde{I}_3}\frac{\partial \tilde{I}_3}{\partial \varepsilon_{ij}} \tag{4.68}$$

With the notation

$$\boxed{\phi_1 = \frac{\partial W}{\partial \tilde{I}_1}\ ; \quad \phi_2 = \frac{\partial W}{\partial \tilde{I}_2}\ ; \quad \phi_3 = \frac{\partial W}{\partial \tilde{I}_3}} \tag{4.69}$$

(4.68) reduces with (4.67) to

$$\boxed{\sigma_{ij} = \phi_1\delta_{ij} + \phi_2\varepsilon_{ij} + \phi_3\varepsilon_{ik}\varepsilon_{kj}} \tag{4.70}$$

Instead of the index notation, we may write $\sigma_{ij} = \boldsymbol{\sigma}$ and $\varepsilon_{ij} = \boldsymbol{\varepsilon}$, i.e. (4.70) can be written as

$$\boxed{\boldsymbol{\sigma} = \phi_1\boldsymbol{I} + \phi_2\boldsymbol{\varepsilon} + \phi_3\boldsymbol{\varepsilon}^2} \tag{4.71}$$

From the definition of the parameters ϕ_1, ϕ_2 and ϕ_3 it follows directly that they may depend on the strain invariants. However, these parameters are not independent, since (4.69) results in the following constraints

$$\boxed{\frac{\partial \phi_1}{\partial \tilde{I}_2} = \frac{\partial \phi_2}{\partial \tilde{I}_1}\ ; \quad \frac{\partial \phi_1}{\partial \tilde{I}_3} = \frac{\partial \phi_3}{\partial \tilde{I}_1}\ ; \quad \frac{\partial \phi_2}{\partial \tilde{I}_3} = \frac{\partial \phi_3}{\partial \tilde{I}_2}} \tag{4.72}$$

It is of interest that (4.70) or (4.71) shows that the stress tensor and the strain tensor have identical principal directions. To see this, assume that the coordinate system is chosen collinearly with the principal directions of the strain tensor, i.e. $\boldsymbol{\varepsilon}$ and $\boldsymbol{\varepsilon}^2$ become diagonal matrices. It follows immediately from (4.71) that this implies that also $\boldsymbol{\sigma}$ becomes diagonal proving that $\boldsymbol{\varepsilon}$ and $\boldsymbol{\sigma}$ have identical principal directions.

The most general form of isotropic hyper-elasticity is given by (4.70) and it appears that the presence of the quadratic term $\varepsilon_{ik}\varepsilon_{kj}$ implies a nonlinear relation between stresses and strains. It is of importance that even though we restrict ourselves to small strain theory, we cannot ignore the quadratic term $\varepsilon_{ik}\varepsilon_{kj}$ when compared to the linear term ε_{ij}, since this depends entirely on the magnitude of the parameters ϕ_2 and ϕ_3. Moreover, even if we choose $\phi_3 = 0$, i.e. the strain energy W does not depend on the third strain invariant $\tilde{I}_3$, cf. (4.69), we may obtain a nonlinear formulation if ϕ_1 and ϕ_2 are not constants. The implications of this approach is discussed in Section 4.10.

Another point of interest is that the presence of the three parameters ϕ_1, ϕ_2 and ϕ_3 means that the most general form of isotropic nonlinear hyper-elasticity involves three material parameters.

In the discussion above, we used the strain energy W as the vehicle for our derivation. However, we could equally well use the complementary energy C as the basis for our discussions and we shall now see the implications of this approach.

In general, the strain energy is given by $W = W(\varepsilon_{ij})$ and for isotropic materials this reduces to $W = W(\tilde{I}_1, \tilde{I}_2, \tilde{I}_3)$, cf. (4.65). In general, the complementary energy is given by $C = C(\sigma_{ij})$ and similarly to the result above, we have for isotropic materials

$$\boxed{C = C(I_1, I_2, I_3)} \tag{4.73}$$

where the generic stress invariants are given by (3.14), i.e.

$$I_1 = \sigma_{kk}\ ; \quad I_2 = \frac{1}{2}\sigma_{kl}\sigma_{lk}\ ; \quad I_3 = \frac{1}{3}\sigma_{kl}\sigma_{lm}\sigma_{mk}$$

We have the following relations

$$\frac{\partial I_1}{\partial \sigma_{ij}} = \delta_{ij}\ ; \quad \frac{\partial I_2}{\partial \sigma_{ij}} = \sigma_{ij}\ ; \quad \frac{\partial I_3}{\partial \sigma_{ij}} = \sigma_{ik}\sigma_{kj} \tag{4.74}$$

From (4.16) and (4.73) follow that

$$\varepsilon_{ij} = \frac{\partial C}{\partial I_1}\frac{\partial I_1}{\partial \sigma_{ij}} + \frac{\partial C}{\partial I_2}\frac{\partial I_2}{\partial \sigma_{ij}} + \frac{\partial C}{\partial I_3}\frac{\partial I_3}{\partial \sigma_{ij}} \tag{4.75}$$

With the notation

$$\boxed{\psi_1 = \frac{\partial C}{\partial I_1}\ ; \quad \psi_2 = \frac{\partial C}{\partial I_2}\ ; \quad \psi_3 = \frac{\partial C}{\partial I_3}} \tag{4.76}$$

(4.75) reduces with (4.74) to

$$\boxed{\varepsilon_{ij} = \psi_1\delta_{ij} + \psi_2\sigma_{ij} + \psi_3\sigma_{ik}\sigma_{kj}} \tag{4.77}$$

which in matrix notation reads

$$\boxed{\boldsymbol{\varepsilon} = \psi_1 \boldsymbol{I} + \psi_2 \boldsymbol{\sigma} + \psi_3 \boldsymbol{\sigma}^2} \qquad (4.78)$$

These results may be compared with the forms given by (4.70) and (4.71).

From the definition of the parameters ψ_1, ψ_2 and ψ_3 it appears that, in general, they depend on the stress invariants. However, these parameters are not independent, since (4.76) results in the following constraints

$$\boxed{\frac{\partial \psi_1}{\partial I_2} = \frac{\partial \psi_2}{\partial I_1}\,; \quad \frac{\partial \psi_1}{\partial I_3} = \frac{\partial \psi_3}{\partial I_1}\,; \quad \frac{\partial \psi_2}{\partial I_3} = \frac{\partial \psi_3}{\partial I_2}} \qquad (4.79)$$

Formulation (4.70) or (4.77) corresponds to isotropic hyper-elasticity in its most general form and it evidently includes nonlinear elasticity. A detailed discussion of the various types of nonlinearity that can be modeled by this formulation is given by Evans and Pister (1966).

4.9 Isotropic linear elasticity

We have already touched upon linear isotropic hyper-elasticity, cf. (4.59), but we shall now treat it in a more consistent manner.

The general format of isotropic hyper-elasticity is given by (4.70) where ϕ_1, ϕ_2 and ϕ_3 are defined by (4.69). To obtain a linear relation between stresses and strains, we must have $\phi_3 = 0$. Moreover, let us assume that

$$\phi_1 = \lambda \tilde{I}_1\,; \quad \phi_2 = 2\mu\,; \quad \phi_3 = 0 \qquad (4.80)$$

where λ and μ are constants – the so-called *Lamé parameters* introduced by Lamé in 1852. It appears that this choice fulfills the constraints given by (4.72). We also observe that $\phi_3 = 0$ implies that the strain energy W does not depend on $\tilde{I}_3$, cf. (4.69). I.e. (4.69) and (4.80) lead to the following expression for the strain energy

$$W = \frac{1}{2}\lambda \tilde{I}_1^2 + 2\mu \tilde{I}_2$$

With (4.70) and (4.80) and as $\tilde{I}_1 = \varepsilon_{kk}$ we find that

$$\sigma_{ij} = \lambda\, \varepsilon_{kk}\delta_{ij} + 2\mu\varepsilon_{ij} \qquad (4.81)$$

Since λ and μ are material constants, we observe that (4.81) expresses a linearity between stresses and strains, i.e. we have recovered Hooke's generalized law for isotropic materials. We mentioned previously in relation to (4.59) that linear isotropic elasticity is characterized by two material parameters and the present derivation provides a formal proof for that. These two parameters may be expressed in terms of the more familiar material parameters E = Young's modulus and ν = Poisson's ratio through

$$\lambda = \frac{\nu E}{(1+\nu)(1-2\nu)}\,; \qquad \mu = \frac{E}{2(1+\nu)} \tag{4.82}$$

Moreover, the *shear modulus* G, where $G = \mu$, and the *bulk modulus* K are defined by

$$\boxed{G = \frac{E}{2(1+\nu)}\,; \qquad K = \frac{E}{3(1-2\nu)}} \tag{4.83}$$

With (4.82) and (4.83), Hooke's law (4.81) takes the form

$$\boxed{\sigma_{ij} = 2G\left[\varepsilon_{ij} + \frac{\nu}{1-2\nu}\varepsilon_{kk}\delta_{ij}\right]} \tag{4.84}$$

An interesting reformulation of Hooke's law can be obtained as follows. Contraction of (4.84) yields

$$\boxed{\sigma_{kk} = 3K\varepsilon_{kk}} \tag{4.85}$$

From the definition of deviatoric stresses $s_{ij} = \sigma_{ij} - \sigma_{kk}\delta_{ij}/3$ we obtain from (4.84) and (4.85) that

$$s_{ij} = 2G\left(\varepsilon_{ij} - \frac{1}{3}\varepsilon_{kk}\delta_{ij}\right)$$

which, by means of the definition of the deviatoric strains e_{ij}, takes the form

$$\boxed{s_{ij} = 2G\, e_{ij}} \tag{4.86}$$

We recall that the off-diagonal terms of the deviatoric stress and strain tensors are equal to the shear stresses and shear strains respectively. From (4.85) and (4.86) follow that

Hooke's linear isotropic law implies decoupling between volumetric and deviatoric response	(4.87)

Let us now make a further reformulation of (4.84). We have that $\varepsilon_{kk} = \delta_{kl}\varepsilon_{kl}$. Moreover, direct inspection shows that $\varepsilon_{ij} = \frac{1}{2}(\delta_{ik}\delta_{jl} + \delta_{il}\delta_{jk})\varepsilon_{kl}$ where both the left-hand and right-hand side of this expression are symmetric in i and j. With these two reformulations, (4.84) takes the form

$$\boxed{\sigma_{ij} = D_{ijkl}\varepsilon_{kl}} \tag{4.88}$$

where

$$\boxed{D_{ijkl} = 2G\left[\frac{1}{2}(\delta_{ik}\delta_{jl} + \delta_{il}\delta_{jk}) + \frac{\nu}{1-2\nu}\delta_{ij}\delta_{kl}\right]} \tag{4.89}$$

and D_{ijkl} is the *elastic isotropic stiffness tensor*. We have then recovered the formulation (4.20) with an explicit expression for D_{ijkl} valid for an isotropic material. Expression (4.89) clearly fulfills the general symmetry properties given by (4.21) and (4.23). Writing all components in (4.84) - or (4.88) - and changing to a matrix formulation, we recover the stiffness matrix $\boldsymbol{D}$ given by (4.59).

Insertion of (4.89) into the transformation formula (1.43) provides

$$\boxed{D'_{ijkl} = D_{ijkl}} \tag{4.90}$$

Since the isotropic elastic stiffness tensor takes the same form in all coordinate systems, it is an *isotropic fourth-order tensor*. As a comparison, we have previously shown that Kronecker's delta δ_{ij} is an isotropic second-order tensor, cf. (1.45). That D_{ijkl} is an isotropic tensor is certainly not surprising since we are considering an isotropic material, which means that we expect Hooke's law to be the same irrespective of which Cartesian coordinate system we choose.

To further scrutinize different expressions of Hooke's law, we will now make an inversion so that we obtain the strains as function of the stresses. This inversion is easily accomplished when noting Hooke's law in the format given by (4.85) and (4.86). Finding ε_{kk} and e_{ij} from these expressions and using the definition $\varepsilon_{ij} = e_{ij} + \delta_{ij}\varepsilon_{kk}/3$, we arrive at

$$\boxed{\varepsilon_{ij} = \frac{1}{2G}\left[\sigma_{ij} - \frac{\nu}{1+\nu}\sigma_{kk}\delta_{ij}\right]} \tag{4.91}$$

Since $\sigma_{ij} = \frac{1}{2}(\delta_{ik}\delta_{jl} + \delta_{il}\delta_{jk})\sigma_{kl}$ and $\sigma_{kk} = \delta_{kl}\sigma_{kl}$, we obtain

$$\boxed{\varepsilon_{ij} = C_{ijkl}\sigma_{kl}} \tag{4.92}$$

where

$$\boxed{C_{ijkl} = \frac{1}{2G}\left[\frac{1}{2}(\delta_{ik}\delta_{jl} + \delta_{il}\delta_{jk}) - \frac{\nu}{1+\nu}\delta_{ij}\delta_{kl}\right]} \tag{4.93}$$

and C_{ijkl} is termed the *elastic isotropic flexibility tensor*. We have then recovered the formulation (4.26) with an explicit expression for C_{ijkl} valid for an isotropic material. Expression (4.93) clearly fulfills the general symmetry properties given by (4.27) and (4.28).

Insertion of (4.93) into the transformation formula similar to (1.43) shows that

$$\boxed{C'_{ijkl} = C_{ijkl}} \tag{4.94}$$

i.e. the flexibility tensor is an isotropic tensor - as expected. It is straightforward to convince oneself that the explicit expressions given by (4.89) and (4.93) fulfill the general inversion properties stated by (4.31) and (4.32). Writing all components in (4.91) - or (4.92) - and changing to a matrix formulation, we recover the flexibility matrix $\boldsymbol{C}$ given by (4.60).

Due to (4.24) and (4.25), the stiffness tensor D_{ijkl} must be positive definite. Let us now see, what kind of constraints this requirement sets on the two material parameters E and ν. From (4.24) and (4.25) follow that

$$W = \frac{1}{2}\varepsilon_{ij}\sigma_{ij} > 0$$

Since $\varepsilon_{ij} = e_{ij} + \frac{1}{3}\delta_{ij}\varepsilon_{kk}$ and $\sigma_{ij} = s_{ij} + \frac{1}{3}\delta_{ij}\sigma_{kk}$, we then find

$$W = \frac{1}{2}\left(e_{ij}s_{ij} + \frac{1}{3}\varepsilon_{kk}\sigma_{mm}\right) > 0$$

where advantage was taken of the fact that $e_{ii} = s_{ii} = 0$. Use of (4.85) and (4.86) then provide

$$W = \frac{1}{2}\left(2Ge_{ij}e_{ij} + K\varepsilon_{kk}\varepsilon_{mm}\right) > 0 \tag{4.95}$$

Both $e_{ij}e_{ij}$ and $\varepsilon_{kk}\varepsilon_{mm}$ are non-negative numbers. Moreover, e_{ij} and ε_{kk} can be chosen arbitrarily and independently of each other. Thus, we are led to the requirement $G > 0$ and $K > 0$. With reference to (4.83) and as we certainly must require that $E > 0$, we conclude that the elastic parameters must fulfill the constraints

$$\boxed{E > 0\,; \quad -1 < \nu < \frac{1}{2}} \tag{4.96}$$

in order to fulfill (4.95).

4.10 Nonlinear isotropic Hooke formulation

In order to provide a firm basis, we have given quite an extensive description of linear material behavior. However, our objective is to investigate nonlinear material behavior, so it is timely to change our focus of interest.

Our first objective is to derive a nonlinear elasticity theory that can be expressed in a form similar to Hooke's law

$$\sigma_{ij} = D_{ijkl}\varepsilon_{kl} \tag{4.97}$$

but where D_{ijkl} now depends on the amount of loading. To derive such a model, we shall restrict ourselves to isotropic elasticity.

The most general isotropic nonlinear elasticity may be derived from the complementary energy C. According to (4.73), C depends on the stress invariants I_1, I_2 and I_3, but we could equally well use the invariants given by I_1, J_2 and

J_3, cf. (3.18) and (3.19). For our present purpose, this latter set of invariants turns out out to give a more convenient description. We therefore have

$$C = C(I_1, J_2, J_3)$$

where

$$I_1 = \sigma_{kk} \quad ; \quad J_2 = \frac{1}{2}s_{kl}s_{lk} \quad ; \quad J_3 = \frac{1}{3}s_{kl}s_{lm}s_{mk}$$

From (4.16) we then obtain with the chain rule

$$\varepsilon_{ij} = \frac{\partial C}{\partial I_1}\delta_{ij} + \frac{\partial C}{\partial J_2}s_{ij} + \frac{\partial C}{\partial J_3}(s_{ik}s_{kj} - \frac{2}{3}J_2\delta_{ij}) \tag{4.98}$$

Since our objective is to derive a nonlinear elasticity theory that can be expressed in the Hooke format (4.97), we must in (4.98) require that the term containing the quadratic quantity $s_{ik}s_{kj}$ must disappear, i.e. $\partial C/\partial J_3 = 0$. This leads to

$$C = C(I_1, J_2)$$

With this restriction, (4.98) reduces to

$$\varepsilon_{ij} = \frac{\partial C}{\partial I_1}\delta_{ij} + \frac{\partial C}{\partial J_2}s_{ij}$$

Contraction of this equation and use of the definition of the deviatoric strains e_{ij} lead to

$$\varepsilon_{kk} = 3\frac{\partial C}{\partial I_1} \quad ; \quad e_{ij} = \frac{\partial C}{\partial J_2}s_{ij} \tag{4.99}$$

Let us make the following choices

$$\frac{\partial C}{\partial I_1} = \frac{\sigma_{kk}}{9K} \quad ; \quad \frac{\partial C}{\partial J_2} = \frac{1}{2G} \tag{4.100}$$

where the material functions K and G, in general, depend on the stress invariants I_1 and J_2, i.e.

$$\boxed{K = K(I_1, J_2) \quad ; \quad G = G(I_1, J_2)} \tag{4.101}$$

With (4.100), (4.99) reduces to

$$\boxed{\sigma_{kk} = 3K\varepsilon_{kk} \quad ; \quad s_{ij} = 2Ge_{ij}} \tag{4.102}$$

and we have therefore obtained a formulation that is completely similar to linear elasticity, cf. (4.85) and (4.86). In analogy with linear elasticity, (4.102) may be written in the format of (4.97) where D_{ijkl} is given by (4.89) and where G and ν (or E and ν) now depend on I_1 and J_2, cf. (4.101) and (4.83).

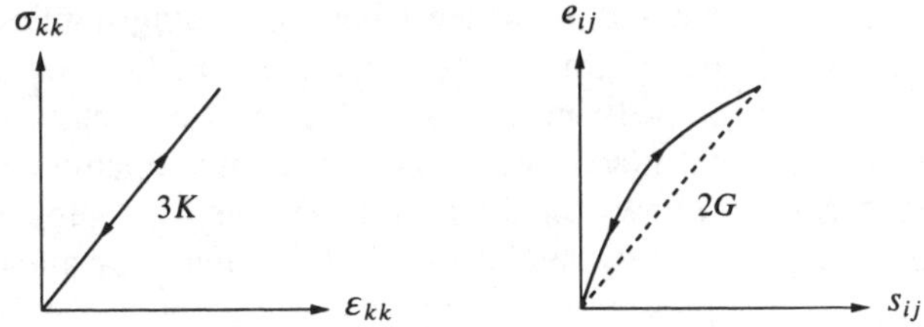

Figure 4.8: Illustration of response predicted by (4.102) and (4.105) - metals and steel.

By the above approach, we have created a nonlinear Hooke formulation with quite remarkable properties, as will be illustrated shortly.

The bulk modulus K and shear modulus G given by (4.101) cannot take arbitrary expressions of I_1 and J_2. Since $\partial(\partial C/\partial I_1)/\partial J_2 = \partial(\partial C/\partial J_2)/\partial I_1$, we obtain from (4.100) the following constraint

$$\boxed{\frac{\sigma_{kk}}{3}\frac{\partial}{\partial J_2}(\frac{1}{3K}) = \frac{\partial}{\partial I_1}(\frac{1}{2G})} \tag{4.103}$$

Let us now investigate whether the theory derived can be used to model the nonlinear behavior of important engineering materials. The general experimental evidence for nonlinear time-independent behavior of metals and steel can be summarized as follows:

For metals and steel, the volumetric response is linear elastic and all nonlinearity is related to the deviatoric response (4.104)

This implies that the volumetric and deviatoric responses are uncoupled. These features can be captured in a very simple manner by the nonlinear Hooke formulation presented above. For this purpose and referring to (4.101), we make the following choice

$$K = \text{constant} \quad ; \quad G = G(J_2) \tag{4.105}$$

which evidently fulfills the constraint (4.103). Relations (4.102) in combination with (4.105) fulfill precisely the general requirements given by (4.104) and we have then obtained the so-called *deformation plasticity theory* proposed by Hencky (1924) and used extensively in older literature. The principal response predicted by (4.102) and (4.105) is illustrated in Fig. 4.8. It appears that the volumetric and deviatoric response is uncoupled just like in linear elasticity theory, cf. (4.87). The deformation plasticity theory states a relation between the current stresses and the current strains independently of the load history. Therefore, in contrast to experimental evidence where unloading implies the occurrence of permanent (plastic) strains, the present model predicts that loading and unloading follow the same path and this is the major disadvantage of deformation plasticity. In 'real' plasticity theory, the constitutive relations are given in an incremental form that allows for different behaviors in loading and unloading. However, in Section 9.7 we shall show that the 'real' so-called von Mises plasticity theory and the present deformation plasticity theory are identical for proportional increasing loading.

The nonlinear time-independent response of soil, rocks and concrete is more complicated and the general experimental evidence may be summarized as follows

For concrete, soil and rocks, the volumetric and deviatoric response is coupled and both the volumetric and deviatoric response are nonlinear (4.106)

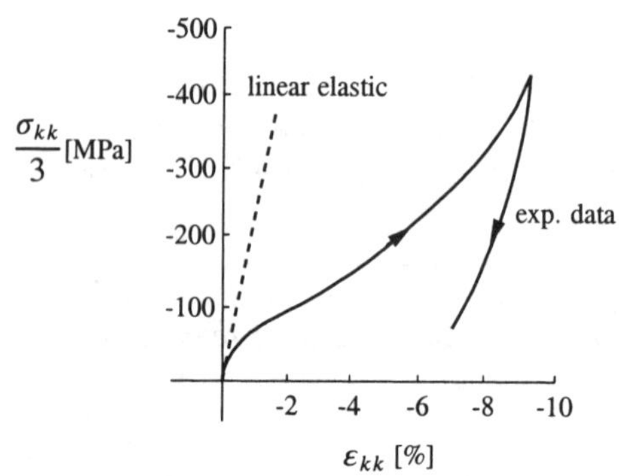

Figure 4.9: Hydrostatic compression of concrete ($\sigma_1 = \sigma_2 = \sigma_3 < 0$). Experimental data of Green and Swanson (1973); uniaxial compressive strength=48.5 MPa.

As examples of such behavior, Fig. 4.9 illustrated the nonlinear response of concrete for purely hydrostatic compression and Fig. 4.10 illustrates the nonlinear response of concrete for uniaxial compressive loading. Especially the volumetric response in Fig. 4.10b) is of interest, as it shows that the volume starts to increase when the failure state, i.e. the peak stress, is approached.

An example of the coupled response of concrete, soil and rocks is shown in Fig. 4.10. For constant hydrostatic stress $I_1 = \sigma_{kk}$ = constant, increase of the deviatoric stresses not only leads to changes of the deviatoric strains, as given by (4.102b), but also to a change of the volumetric strain ε_{ii}, cf. Fig. 4.11. For

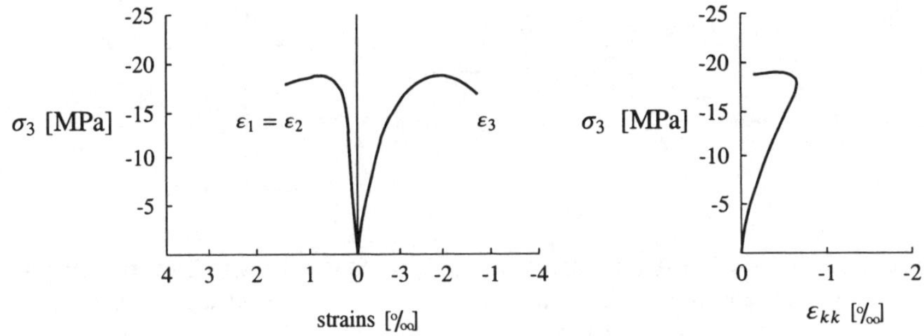

Figure 4.10: Uniaxial compression of concrete ($\sigma_1 = \sigma_2 = 0, \sigma_3 < 0$). Experimental data of Kupfer (1973); uniaxial compressive strength=18.7 MPa. a) stress-strain curves; b) development of volumetric strain ε_{ii}: first the volume decreases and then it increases.

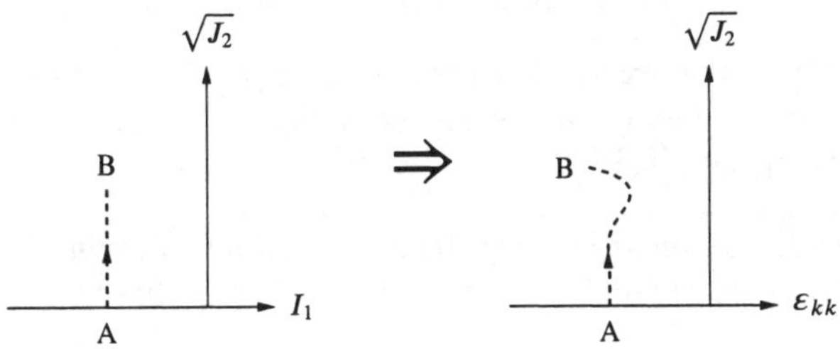

Figure 4.11: Coupled deviatoric and volumetric response characteristic for concrete, soil and rocks. Constant hydrostatic stress and increasing deviatoric stresses result in a change of volumetric strain.

σ_{kk} = constant, we obtain with (4.102a) and (4.101) that

$$\underbrace{\sigma_{kk}}_{\text{const.}} = 3K(\underbrace{I_1}_{\text{const.}}, \overbrace{J_2}^{\text{varies}})\,\varepsilon_{kk}$$

It is evident that fulfillment of this relation requires a change of ε_{kk} even though σ_{kk} is constant. This means that our nonlinear elasticity model allows for the coupled response characteristic for concrete, soil and rocks.

Another example of coupled response is illustrated in Fig. 4.12. For constant deviatoric stresses (s_{ij} = constant i.e. J_2 = constant), change of the hydrostatic stress I_1 not only results in a change of the volumetric strain ε_{ii} according to (4.102a), but also in a change of the deviatoric strains. For s_{ij} = constant,

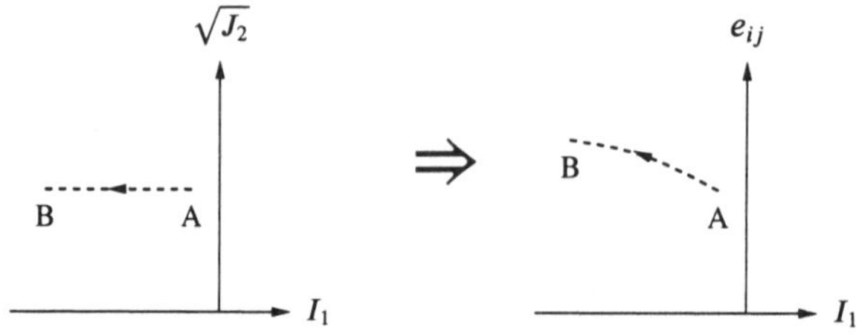

Figure 4.12: Coupled deviatoric and volumetric response characteristic for concrete, soil and rocks. Constant deviatoric stresses and change of hydrostatic stress result in a change of deviatoric strains.

(4.101) and (4.102b) yield

$$\underbrace{s_{ij}}_{\text{const.}} = 2G(\overbrace{I_1}^{\text{varies}}, \underbrace{J_2}_{\text{const.}})\, e_{ij}$$

and fulfillment of this expression requires a change of e_{ij} even though s_{ij} is constant, i.e. this expression allows the complicated response just described.

In conclusion, we obtain

A simple nonlinear Hooke formulation allows coupling effects between volumetric and deviatoric responses

and this coupling effect is characteristic for soils, rock and concrete.

It is therefore no surprise that a number of specific models have been proposed in the literature, where the moduli K and G have been calibrated to experimental data for concrete, soil and rocks. For obvious reasons, these models are often called *variable moduli models* and for a more detailed discussion, we may refer to Chen and Saleeb (1982) and Desai and Siriwardane (1984). We shall return to such models in Section 5.4, where it turns out that within the concept of Cauchy-elasticity, it is allowable to obtain a model similar to (4.102). However, in contrast to (4.101), Cauchy-elasticity allows that K and G even may depend on the third stress invariant J_3 and this makes for a much more accurate calibration of K and G to experimental data for concrete, soil and rocks. Finally, we recall the major drawback of nonlinear elasticity, namely that unloading follows the same path as loading. Remedies to remove this drawback will also be discussed in Section 5.4.

4.11 Plane strain

In plane strain, the only non-zero strains are ε_{11}, ε_{22} and ε_{12}, cf. (2.72). Using these conditions in (4.37) and assuming isotropy, we obtain with (4.59) that

$$\boxed{\boldsymbol{\sigma} = \boldsymbol{D}\boldsymbol{\varepsilon}} \tag{4.107}$$

where

$$\boldsymbol{\sigma} = \begin{bmatrix} \sigma_{11} \\ \sigma_{22} \\ \sigma_{12} \end{bmatrix} ; \quad \boldsymbol{\varepsilon} = \begin{bmatrix} \varepsilon_{11} \\ \varepsilon_{22} \\ 2\varepsilon_{12} \end{bmatrix} \tag{4.108}$$

and

$$\boxed{\boldsymbol{D} = \frac{E}{(1+\nu)(1-2\nu)} \begin{bmatrix} 1-\nu & \nu & 0 \\ \nu & 1-\nu & 0 \\ 0 & 0 & \frac{1}{2}(1-2\nu) \end{bmatrix}} \tag{4.109}$$

With these definitions, (4.107) describes the *in-plane* conditions. Moreover, from (4.37) and (4.59) the *out-of-plane* stresses are given by

$$\sigma_{33} = \frac{E\nu}{(1+\nu)(1-2\nu)}(\varepsilon_{11} + \varepsilon_{22}) ; \quad \sigma_{13} = \sigma_{23} = 0$$

Alternatively, we may use (4.88) and (4.89) to obtain the following in-plane description

$$\boxed{\sigma_{\alpha\beta} = D_{\alpha\beta\gamma\lambda}\varepsilon_{\gamma\lambda}}$$

where Greek subscripts take the range 1, 2 and where

$$\boxed{D_{\alpha\beta\gamma\lambda} = 2G[\frac{1}{2}(\delta_{\alpha\gamma}\delta_{\beta\lambda} + \delta_{\alpha\lambda}\delta_{\beta\gamma}) + \frac{\nu}{1-2\nu}\delta_{\alpha\beta}\delta_{\gamma\lambda}]} \tag{4.110}$$

4.12 Plane stress

In plane stress, the only non-zero stresses are σ_{11}, σ_{22} and σ_{12}, cf. (3.25). Using these conditions in (4.41) and assuming isotropy, we obtain with (4.60) the following in-plane description

$$\begin{bmatrix} \varepsilon_{11} \\ \varepsilon_{22} \\ 2\varepsilon_{12} \end{bmatrix} = \frac{1}{E} \begin{bmatrix} 1 & -\nu & 0 \\ -\nu & 1 & 0 \\ 0 & 0 & 2(1+\nu) \end{bmatrix} \begin{bmatrix} \sigma_{11} \\ \sigma_{22} \\ \sigma_{12} \end{bmatrix} \tag{4.111}$$

as well as

$$\varepsilon_{33} = -\frac{\nu}{E}(\sigma_{11} + \sigma_{22}) ; \quad \varepsilon_{13} = \varepsilon_{23} = 0$$

Inversion of (4.111) gives

$$\boxed{\boldsymbol{\sigma} = \boldsymbol{D}^*\boldsymbol{\varepsilon}}$$

where $\boldsymbol{\sigma}$ and $\boldsymbol{\varepsilon}$ are defined by (4.108) and $\boldsymbol{D}^*$ is given by

$$\boxed{\boldsymbol{D}^* = \frac{E}{1-\nu^2}\begin{bmatrix} 1 & \nu & 0 \\ \nu & 1 & 0 \\ 0 & 0 & \frac{1}{2}(1-\nu) \end{bmatrix}} \tag{4.112}$$

Note that $\boldsymbol{D}^*$ differs from $\boldsymbol{D}$ given by (4.109).

Using tensor notation, (4.92) and (4.93) gives for the in-plane behavior

$$\varepsilon_{\alpha\beta} = C_{\alpha\beta\gamma\lambda}\sigma_{\gamma\lambda} \tag{4.113}$$

where

$$C_{\alpha\beta\gamma\lambda} = \frac{1}{2G}[\frac{1}{2}(\delta_{\alpha\gamma}\delta_{\beta\lambda} + \delta_{\alpha\lambda}\delta_{\beta\gamma}) - \frac{\nu}{1+\nu}\delta_{\alpha\beta}\delta\gamma\lambda]$$

Define the tensor $D^*_{\alpha\beta\gamma\lambda}$ by

$$\boxed{D^*_{\alpha\beta\gamma\lambda} = 2G[\frac{1}{2}(\delta_{\alpha\gamma}\delta_{\beta\lambda} + \delta_{\alpha\lambda}\delta_{\beta\gamma}) + \frac{\nu}{1-\nu}\delta_{\alpha\beta}\delta_{\gamma\lambda}]} \tag{4.114}$$

It appears that

$$D^*_{\alpha\beta\gamma\lambda}C_{\gamma\lambda\theta\kappa} = \frac{1}{2}(\delta_{\alpha\theta}\delta_{\beta\kappa} + \delta_{\alpha\kappa}\delta_{\beta\theta}) \tag{4.115}$$

Note the difference in the term containing Poisson's ratio when comparing (4.110) and (4.114). With (4.115), multiplication of (4.113) by $D^*_{\kappa\theta\alpha\beta}$ provides

$$\boxed{\sigma_{\alpha\beta} = D^*_{\alpha\beta\gamma\lambda}\varepsilon_{\gamma\lambda}}$$

4.13 Incompressible linear hyper-elasticity

When deriving the fundamental relation (4.8), i.e. $\sigma_{ij} = \partial W/\partial\varepsilon_{ij}$, we argued that $d\varepsilon_{ij}$ could be chosen arbitrarily in (4.7). However, this argument fails if some restrictions exist on $d\varepsilon_{ij}$ and this is the case if the material behaves *incompressibly*, i.e. $\varepsilon_{ii} = 0$.

To circumvent this obstacle, we may instead use the formulation (4.16), i.e.

$$\varepsilon_{ij} = \frac{\partial C}{\partial\sigma_{ij}} \tag{4.116}$$

When deriving this relation, we used that the incremental stresses $d\sigma_{ij}$ can be chosen arbitrarily, i.e. (4.116) holds in general. In the following let us restrict ourselves to isotropic elasticity, i.e. the complementary energy C depends only on the invariants of the stress tensor. It turns out to be convenient to choose the following invariants

$$C = C(I_1, J_2, J_3)$$

From (4.116) we then obtain in accordance with (4.98)

$$\varepsilon_{ij} = \frac{\partial C}{\partial I_1}\delta_{ij} + \frac{\partial C}{\partial J_2}s_{ij} + \frac{\partial C}{\partial J_3}(s_{ik}s_{kj} - \frac{2}{3}J_2\delta_{ij}) \tag{4.117}$$

Let us furthermore restrict ourselves to linear elasticity. Then (4.117) shows that we must have $\partial C/\partial J_3 = 0$, i.e.

$$C = C(I_1, J_2)$$

i.e.

$$\varepsilon_{ij} = \frac{\partial C}{\partial I_1}\delta_{ij} + \frac{\partial C}{\partial J_2}s_{ij} \tag{4.118}$$

The incompressibility condition $\varepsilon_{ii} = 0$ then implies that $\partial C/\partial I_1 = 0$, i.e. $C = C(J_2)$. Moreover, since (4.118) should provide a linear relation between stresses and strains, we choose

$$C = \frac{J_2}{2G}$$

where G is a constant material parameter. With this expression, (4.118) becomes

$$\boxed{\varepsilon_{ij} = e_{ij} = \frac{s_{ij}}{2G}; \quad \varepsilon_{ii} = 0} \tag{4.119}$$

and we immediately identify the material parameter G to be the shear modulus, cf. (4.86).

Expression (4.119) states the constitutive law for an isotropic, incompressible and linear elastic solid. It is of major importance that, for any given strain state, it is only possible to identify the deviatoric stresses s_{ij} and not the full stress tensor σ_{ij} since the hydrostatic stress σ_{kk} is not determined by the constitutive relation. On the other hand, a change of the hydrostatic stress does not change the strain state.

As an example, we may imagine an incompressible material subject to a given stress state and a known strain state. On top of this stress state, we superpose a large hydrostatic stress by submerging the material onto the bottom of the ocean. However, this will not change the strain state. We therefore conclude that

> *The constitutive law for a linear elastic, isotropic and incompressible material provides no information of the hydrostatic stress*

Since the hydrostatic stress is not determined by the constitutive relation, it is entirely determined by the static conditions of the boundary value problem in question.

Finally, a comparison of (4.119) with (4.85) and (4.86) shows that incompressibility occurs when the bulk modulus $K \to \infty$, i.e. $\nu \to 1/2$, and rubber provides an example of an incompressible material.

5 CAUCHY-ELASTICITY

We will now deal with a type of elasticity that is more general than hyper-elasticity. It is recalled that the essential property of elasticity is that there exists a one-to-one relation between strains and stresses, i.e.

$$\varepsilon_{ij} = \varepsilon_{ij}(\sigma_{kl}) \tag{5.1}$$

This expression is completely similar to (4.13), but now we do not assume that the strain energy or complementary energy is independent of the load history. As mentioned previously, the corresponding type of elasticity is called *Cauchy-elasticity* and we shall here derive the most general format of isotropic Cauchy-elasticity. Before that can be done, some introductory concepts have to be presented.

5.1 Response function, principle of coordinate invariance and isotropic tensor function

It turns out to be more instructive to write (5.1) as

$$\boxed{\varepsilon_{ij} = g_{ij}(\sigma_{kl})} \tag{5.2}$$

Evidently, $\varepsilon_{ij} = g_{ij}$, but the formulation (5.2) emphasizes that g_{ij} is a function that depends on the stress tensor σ_{kl}. The tensor function g_{ij} is called the *response function* and it determines the operations that are to be performed on the stress tensor σ_{kl} in order to determine the strain tensor ε_{ij}.

In another coordinate system x'_i, the constitutive law (5.2) states that

$$\varepsilon'_{ij} = g^*_{ij}(\sigma'_{kl}) \tag{5.3}$$

where g^* is the response function in the x'_i-coordinate system. In general, the operations that the response function g^*_{ij} performs on σ'_{kl} are different from the operations that the response function g_{ij} performs on σ_{kl}.

As an illustration, assume that the constitutive law (5.2) is linear, i.e.

$$\varepsilon_{ij} = C_{ijkl}\sigma_{kl} \tag{5.4}$$

This is just Hooke's law where C_{ijkl} is the flexibility tensor, cf. (4.26). We may write (5.4) as

$$\varepsilon_{ij} = g_{ij}(\sigma_{kl}) \qquad \text{where} \qquad g_{ij}(\sigma_{kl}) = C_{ijkl}\sigma_{kl} \tag{5.5}$$

In another coordinate system, we have

$$\varepsilon'_{ij} = C'_{ijkl}\sigma'_{kl}$$

or

$$\varepsilon'_{ij} = g^*_{ij}(\sigma'_{kl}) \qquad \text{where} \qquad g^*_{ij}(\sigma'_{kl}) = C'_{ijkl}\sigma'_{kl} \tag{5.6}$$

Referring to the discussion in Sections 4.3 and 4.6, we have for anisotropic materials that $C'_{ijkl} \neq C_{ijkl}$. In fact, the two flexibility tensors are related through the transformation rule similar to (1.43), i.e.

$$C'_{ijkl} = A_{im}A_{jn}C_{mnpq}A_{kp}A_{lq} \tag{5.7}$$

and it is only for isotropic materials that $C'_{ijkl} = C_{ijkl}$, cf. (4.94). To further illustrate the concept of a response function, we observe that for anisotropic materials where $C'_{ijkl} \neq C_{ijkl}$ holds, the response function g^*_{ij} given by (5.6) operates on σ'_{kl} in a different manner than the response function g_{ij} given by (5.5) operates on σ_{kl}. It seems tempting to use the notation $\varepsilon'_{ij} = g'_{ij}(\sigma_{kl})$ in the x'_i-coordinate system whereas we prefer to write $\varepsilon'_{ij} = g^*_{ij}(\sigma_{kl})$. The reason is that the quantity g_{ij} is a tensor function, i.e. it defines the operations to be performed on the argument σ_{kl}. The notation g^*_{ij} therefore underlines that in the x'_i-coordinate system, the response function takes the form g^*_{ij}.

Consider a material point at which in the x_i-coordinate system we have the stress components σ_{ij} and the strain components ε_{ij}. In another x'_i-coordinate system, the same stress state and the same strain state are now given by the stress components σ'_{ij} and the strain components ε'_{ij}. Since σ_{ij} and ε_{ij} correspond to σ'_{ij} and ε'_{ij} respectively, and the only difference is that they are measured in different coordinate systems, we have according to (2.19) and (3.8) that

$$\varepsilon'_{ij} = A_{ik}\varepsilon_{kl}A_{jl}\ ; \qquad \sigma'_{kl} = A_{kp}\sigma_{pq}A_{lq}$$

Insertion of these expressions into (5.3) gives

$$A_{ik}\varepsilon_{kl}A_{jl} = g^*_{ij}(A_{kp}\sigma_{pq}A_{lq})$$

and use of (5.2) on the left-hand side results in

$$A_{ik}g_{kl}(\sigma_{mn})A_{jl} = g^*_{ij}(A_{kp}\sigma_{pq}A_{lq}) \qquad \text{coordinate invariance} \tag{5.8}$$

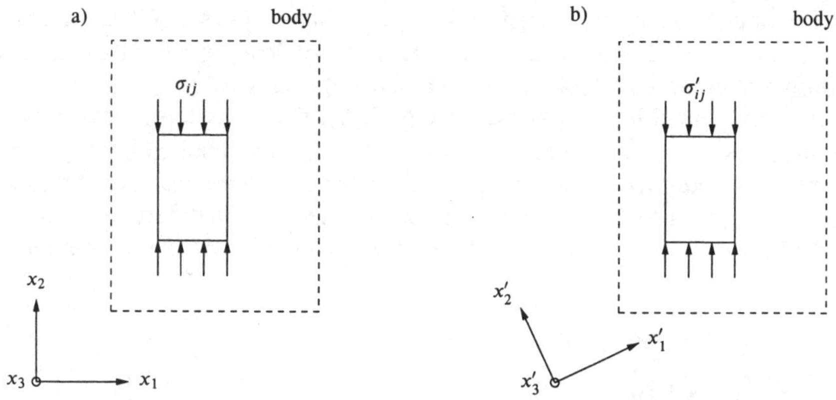

Figure 5.1: Body from which a homogeneously loaded piece of material is considered in the: a) x_i-coordinate system, b) x'_i-coordinate system. The components σ_{ij} and σ'_{ij} are different, but they refer to the same loading measured in different coordinate systems.

This important result is often referred to as being a result of the so-called *coordinate invariance* principle. Using matrix notation, (5.8) evidently reads

$$\boldsymbol{A}\boldsymbol{g}(\boldsymbol{\sigma})\boldsymbol{A}^T = \boldsymbol{g}^*(\boldsymbol{A}\boldsymbol{\sigma}\boldsymbol{A}^T) \qquad \textit{coordinate invariance} \tag{5.9}$$

The principle of coordinate invariance can be stated as

Coordinate invariance principle:
The material response is independent of the coordinate system we choose (5.10)

cf. Truesdell and Toupin (1960).

The principle of coordinate invariance is trivially fulfilled when the constitutive law is written as a tensor relation and the principle of coordinate invariance as expressed by (5.9) follows trivially from tensor algebra. Indeed, as discussed in Chapter 1 the advantage of writing a constitutive relation in tensor format is that it ensures that if this relation holds in one coordinate system, it holds in all coordinate systems. However, the advantage of the format (5.9) is that it involves the, as yet, unknown response function and we shall see later that this format allow us to identify the response function for isotropic materials.

Leaving tensor algebra aside, it may be instructive to illustrate the principle of coordinate invariance in a simple fashion. For that purpose, we consider a piece of a material and assume that this piece is loaded in a homogeneous manner. In Fig. 5.1a) the loading is given by the stress components σ_{ij} measured in the x_i-coordinate system whereas in Fig. 5.1b) the loading is given by the stress components σ'_{ij} measured in the x'_i-coordinate system. We assume that the stress components σ_{ij} and σ'_{ij} refer to the same loading, but just measured in different coordinate systems. In Fig. 5.1a), the stresses σ_{ij} gives rise to the strains determined by ε_{ij} whereas in Fig. 5.1b), the stresses σ'_{ij} results in the strains given by ε'_{ij}. Since σ_{ij} and σ'_{ij} refer to the same loading, but measured in different coordinate systems, it is reasonable to assume that the strains ε_{ij} and ε'_{ij} must refer to the same deformation, but measured in different coordinate systems. This is the physical content of the principle of coordinate invariance.

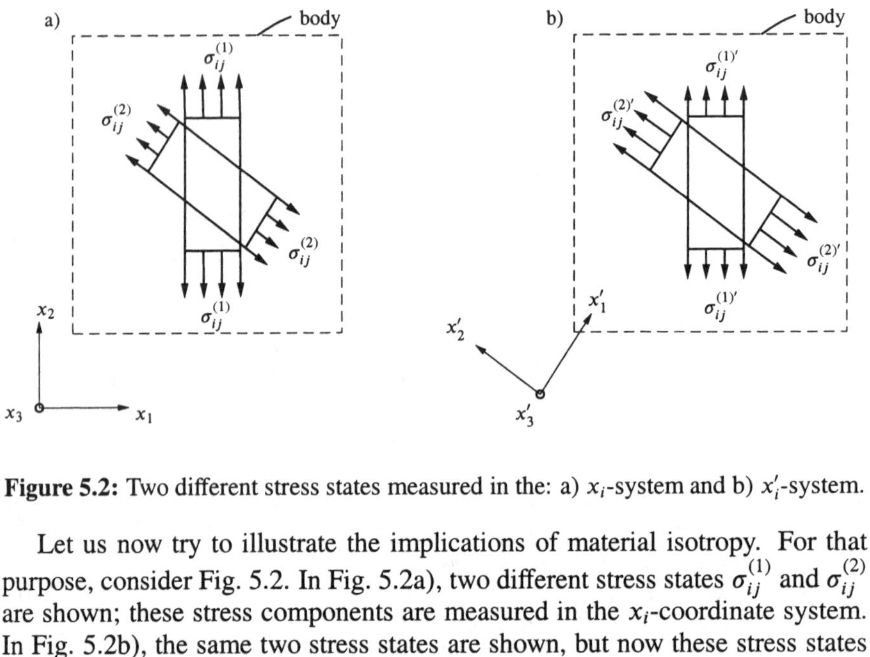

Figure 5.2: Two different stress states measured in the: a) x_i-system and b) x'_i-system.

Let us now try to illustrate the implications of material isotropy. For that purpose, consider Fig. 5.2. In Fig. 5.2a), two different stress states $\sigma_{ij}^{(1)}$ and $\sigma_{ij}^{(2)}$ are shown; these stress components are measured in the x_i-coordinate system. In Fig. 5.2b), the same two stress states are shown, but now these stress states are given by $\sigma_{ij}^{(1)'}$ and $\sigma_{ij}^{(2)'}$ measured in the x'_i-coordinate system. Therefore, $\sigma_{ij}^{(1)}$ and $\sigma_{ij}^{(1)'}$ refer to the same stress state, but merely measured in different coordinate systems. Likewise, $\sigma_{ij}^{(2)}$ and $\sigma_{ij}^{(2)'}$ refer to the same stress state, but merely measured in different coordinate systems. It follows that

$$\boldsymbol{\sigma}^{(1)'} = \boldsymbol{A}\boldsymbol{\sigma}^{(1)}\boldsymbol{A}^T \; ; \qquad \boldsymbol{\sigma}^{(2)'} = \boldsymbol{A}\boldsymbol{\sigma}^{(2)}\boldsymbol{A}^T$$

According to (5.2), we have

$$\varepsilon_{ij}^{(1)} = g_{ij}(\sigma_{ij}^{(1)}) \; ; \qquad \varepsilon_{ij}^{(2)'} = g_{ij}^{*}(\sigma_{ij}^{(2)'}) \tag{5.11}$$

It is always possible to choose the two stress states such that

$$\sigma_{ij}^{(1)} = \sigma_{ij}^{(2)'}$$

i.e. the stress components $\sigma_{ij}^{(1)}$ are equal to the stress components $\sigma_{ij}^{(2)'}$; this situation is illustrated in Fig. 5.2. Insertion into (5.11) gives

$$\varepsilon_{ij}^{(1)} = g_{ij}(\sigma_{kl}^{(1)})\ ; \qquad \varepsilon_{ij}^{(2)'} = g_{ij}^{*}(\sigma_{kl}^{(1)}) \tag{5.12}$$

We now assume that the material is isotropic. Since the material therefore has no directional properties and as $\sigma_{ij}^{(1)} = \sigma_{ij}^{(2)'}$, we certainly expect that

$$\varepsilon_{ij}^{(1)} = \varepsilon_{ij}^{(2)'}$$

and a comparison with (5.12) shows that $g_{ij}(\sigma_{kl}^{(1)}) = g_{ij}^{*}(\sigma_{kl}^{(1)})$ holds for isotropic materials, i.e. for isotropy, g_{ij} and g_{ij}^{*} express the same function. We have previously defined what is meant by anisotropy and isotropy, cf. (4.19) and (4.47). With the discussion above, we may express these definitions in the following more precise form

For isotropic materials, the response function is the same in all Cartesian coordinate systems, i.e. $g_{ij}(\sigma_{kl}) = g_{ij}^{*}(\sigma_{kl})$
For anisotropic materials, the response function depends on the Cartesian coordinate system, i.e. $g_{ij}(\sigma_{kl}) \neq g_{ij}^{*}(\sigma_{kl})$ (5.13)

In these comparisons of the response functions, it is important that the same argument enters in the functions g_{ij} and g_{ij}^{*}. Thus, isotropy could equally well be expressed as $g_{ij}(\sigma'_{kl}) = g_{ij}^{*}(\sigma'_{kl})$; the important point is that for isotropy, g_{ij} and g_{ij}^{*} define the same function.

To illustrate (5.13), we consider linear elasticity given by (5.5) and (5.6). For isotropic materials where $C_{ijkl} = C'_{ijkl}$ holds, the response function g_{ij} given by (5.5) operates on σ_{kl} in the same manner as the response function g_{ij}^{*} given by (5.6) operates on σ'_{kl}. Therefore, g_{ij} and g_{ij}^{*} define the same function and we have $g_{ij}(\sigma_{kl}) = g_{ij}^{*}(\sigma_{kl})$ in accordance with (5.13). On the other hand, for anisotropic materials where $C_{ijkl} \neq C'_{ijkl}$ holds, the response function g_{ij} given by (5.5) operates on σ_{kl} in a different manner than the response function g_{ij}^{*} given by (5.6) operates on σ'_{kl}. For anisotropic materials, g_{ij} and g_{ij}^{*} are therefore different functions in accordance with (5.13).

The response function g_{ij} refers to the x_i-coordinate system whereas the response function g_{ij}^{*} refers to the x'_i-coordinate system and for isotropic materials, these response functions define the same function. This may be expressed by saying that g_{ij} and g_{ij}^{*} possess the same form. In the literature, one therefore often uses the phrase that the response function for isotropic materials is *form-invariant* for arbitrary coordinate transformations, cf. Eringen (1975a) p.139. It is also referred to as the *principle of material invariance*, since the response function is invariant, i.e. has the same form, for arbitrary coordinate transformations, cf. Eringen (1975a).

With this discussion, let us now combine the coordinate invariance principle with the assumption of isotropic material behavior. Following (5.13), the assumption of isotropic material behavior implies that g_{ij} and g^*_{ij} expresses the same function, i.e.

$$g^*(A\sigma A^T) = g(A\sigma A^T)$$

Use of this expression in (5.9) provides

$$\boxed{Ag(\sigma)A^T = g(A\sigma A^T) \qquad \textit{coordinate invariance + isotropy}} \tag{5.14}$$

It is evident that an arbitrary response function g_{ij} cannot be expected to fulfill this relation; indeed, we will in a moment identify the most general response function that fulfills (5.14)

The response function g_{ij} is a tensor function and in case of isotropic materials, we have $g_{ij}(\sigma_{kl}) = g^*_{ij}(\sigma_{kl})$. In that case, we say that g_{ij} is an *isotropic tensor function* (of second-order).

An isotropic tensor function is not the same as an isotropic tensor. According to Section 1.6, the components of an isotropic tensor are the same in all Cartesian coordinate systems; examples are Kronecker's delta δ_{ij}, cf. (1.45), the isotropic stiffness tensor D_{ijkl} which obeys the transformation rule $D'_{ijkl} = D_{ijkl}$, cf. (4.90), and the isotropic flexibility tensor C_{ijkl} which obeys the transformation rule $C'_{ijkl} = C_{ijkl}$, cf. (4.94). On the other hand, for an isotropic tensor function $g_{ij}(\sigma_{kl})$ we have

$$\boxed{\begin{array}{c}\textit{The tensor function } g_{ij} \textit{ is isotropic if}\\ g_{ij}(\sigma_{kl}) = g^*_{ij}(\sigma_{kl})\end{array}} \tag{5.15}$$

The important point in (5.15) is that the same argument enters g_{ij} and g^*_{ij}. Since the argument is the same, we could equally well have defined an isotropic tensor function as $g_{ij}(\sigma'_{kl}) = g^*_{kl}(\sigma'_{kl})$; the essential issue is that g_{ij} and g^*_{ij} define the same function.

While a tensor is isotropic if its components are unchanged by a coordinate transformation, a tensor function is isotropic if it expresses the same function in different coordinate systems. For isotropic materials, the response functions are isotropic.

5.2 Most general isotropic Cauchy-elasticity

After these general remarks, we are now in a position to derive the most general form of isotropic Cauchy-elasticity.

Our starting point is (5.14), which follows from the coordinate invariance principle and the assumption of material isotropy, and the one-to-one relation between strains and stresses given by (5.2). It turns out that this information alone allows us to conclude that the most general form of isotropic Cauchy-elasticity is given by

$$\boxed{\boldsymbol{\varepsilon} = \alpha_1 \boldsymbol{I} + \alpha_2 \boldsymbol{\sigma} + \alpha_3 \boldsymbol{\sigma}^2} \tag{5.16}$$

where the functions α_1,α_2 and α_3, in general, depend on the stress invariants, i.e.

$$\boxed{\alpha_1 = \alpha_1(I_1, I_2, I_3) \ ; \quad \alpha_2 = \alpha_2(I_1, I_2, I_3) \ ; \quad \alpha_3 = \alpha_3(I_1, I_2, I_3)} \tag{5.17}$$

The proof of this remarkable result is given in the next section.

It is of considerable interest, that (5.16) is exactly of the same form as that derived for hyper-elasticity, cf. (4.78). However, there is one significant difference, namely that whereas the coefficients α_1,α_2 and α_3 of (5.16) can take any form, the corresponding coefficients ψ_1,ψ_2 and ψ_3 of (4.78) are related through the constraints (4.79). We have already in Section 4.10 discussed that these constraints have significant consequences when modeling different nonlinear material behaviors and we therefore conclude that Cauchy-elasticity offers significant advantages as compared with hyper-elasticity. We will return to this topic in Section 5.4.

Since Cauchy-elasticity is a more general form of elasticity than hyper-elasticity, it may be of interest to investigate under what circumstances Cauchy-elasticity reduces to hyper-elasticity.

Let us first rewrite (5.16) as

$$\boxed{\varepsilon_{ij} = \alpha_1 \delta_{ij} + \alpha_2 \sigma_{ij} + \alpha_3 \sigma_{ik} \sigma_{kj}} \tag{5.18}$$

Define the quantity dF by

$$dF = \varepsilon_{ij} d\sigma_{ij} \tag{5.19}$$

where ε_{ij} is given by (5.18). If we integrate up until the current stress state, F will depend not only on the current stress state, but also on the integration path, i.e. the load history. Let us investigate the conditions for which F is independent of the integration path. This requires that dF must be a perfect differential and thereby

$$\frac{\partial \varepsilon_{ij}}{\partial \sigma_{kl}} = \frac{\partial \varepsilon_{kl}}{\partial \sigma_{ij}} \tag{5.20}$$

cf. the discussion leading to (4.9). We observe that α_1,α_2 and α_3 depend on the stress invariants and we recall that

$$\frac{\partial I_1}{\partial \sigma_{kl}} = \delta_{kl} \ ; \qquad \frac{\partial I_2}{\partial \sigma_{kl}} = \sigma_{kl} \ ; \qquad \frac{\partial I_3}{\partial \sigma_{kl}} = \sigma_{km} \sigma_{ml}$$

A little algebra will then show that (5.20) is fulfilled if and only if

$$\frac{\partial \alpha_1}{\partial I_2} = \frac{\partial \alpha_2}{\partial I_1} \ ; \qquad \frac{\partial \alpha_1}{\partial I_3} = \frac{\partial \alpha_3}{\partial I_1} \ ; \qquad \frac{\partial \alpha_2}{\partial I_3} = \frac{\partial \alpha_3}{\partial I_2}$$

i.e. α_1,α_2 and α_3 must satisfy the same constraints as the functions ψ_1,ψ_2 and ψ_3 for a hyper-elastic material, cf. (4.79). When (5.20) is fulfilled, we immediately conclude that F as given by (5.19) becomes the complementary energy C, cf. (4.14) and (4.18) and we have then proved formally that Cauchy-elasticity contains hyper-elasticity as a special case.

Returning to (5.16), it is evident that this expression is in accordance with (2). Moreover, it is easy to see that it also fulfills (5.14). To show this, we write (5.16) as

$$\boldsymbol{\varepsilon} = \boldsymbol{g}(\boldsymbol{\sigma}) \qquad \text{where} \qquad \boldsymbol{g}(\boldsymbol{\sigma}) = \alpha_1 \boldsymbol{I} + \alpha_2 \boldsymbol{\sigma} + \alpha_3 \boldsymbol{\sigma\sigma} \tag{5.21}$$

Replacing $\boldsymbol{\sigma}$ by $\boldsymbol{A\sigma A}^T$ and noting that α_1,α_2 and α_3 only depend on the stress invariants and therefore are unaffected by this operation, we obtain

$$\boldsymbol{g}(\boldsymbol{A\sigma A}^T) = \alpha_1 \boldsymbol{I} + \alpha_2 \boldsymbol{A\sigma A}^T + \alpha_3 \boldsymbol{A\sigma A}^T \boldsymbol{A\sigma A}^T$$

Since $\boldsymbol{AA}^T = \boldsymbol{I}$, it follows that

$$\boldsymbol{g}(\boldsymbol{A\sigma A}^T) = \boldsymbol{A}(\alpha_1 \boldsymbol{I} + \alpha_2 \boldsymbol{\sigma} + \alpha_3 \boldsymbol{\sigma\sigma})\boldsymbol{A}^T$$

and use of (5.21) yields

$$\boldsymbol{g}(\boldsymbol{A\sigma A}^T) = \boldsymbol{A}\boldsymbol{g}(\boldsymbol{\sigma})\boldsymbol{A}^T$$

in accordance with (5.14). We have then shown that (5.16) is an allowable form and in the next section we will prove that (5.16) is, in fact, the most general form.

Instead of the constitutive law (5.2), we may assume that

$$\sigma_{ij} = g_{ij}(\varepsilon_{kl})$$

In complete analogy with the discussion leading to (5.16), it follows that the most general isotropic Cauchy-elasticity may also be expressed as

$$\boxed{\boldsymbol{\sigma} = \beta_1 \boldsymbol{I} + \beta_2 \boldsymbol{\varepsilon} + \beta_3 \boldsymbol{\varepsilon}^2}$$

where β_1,β_2 and β_3 depend on the strain invariants; otherwise they are arbitrary.

5.3 Proof of most general form of isotropic Cauchy-elasticity

We claimed that (5.16) presents the most general form of isotropic Cauchy-elasticity. Moreover, we saw that it was easy to show that (5.16) fulfills the coordinate invariance principle and assumption of isotropy as given by (5.14). To prove that (5.16), in fact, is the most general form is much more complex and we will next present proof of this statement.

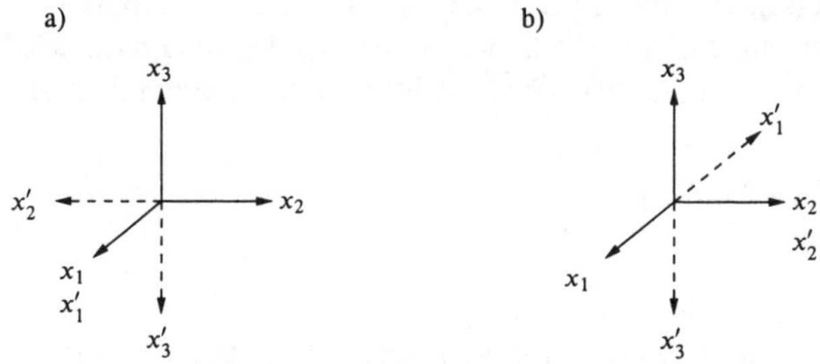

Figure 5.3: Different coordinate changes; a) 180°-rotation about the x_1-axis, b) 180°-rotation about the x_2-axis.

We will first prove that the assumption of isotropy implies that ε_{ij} and σ_{ij} have identical principal directions. Indeed this result is not surprising, since it says that an isotropic material loaded only by normal stresses will deform without appearance of shear strains. However, we will now provide formal proof of this property.

From (5.2), (5.3) and the assumption of isotropy (5.13) follow that

$$\boldsymbol{\varepsilon} = \boldsymbol{g}(\boldsymbol{\sigma}) \quad ; \quad \boldsymbol{\varepsilon}' = \boldsymbol{g}(\boldsymbol{\sigma}') \tag{5.22}$$

where $\boldsymbol{\sigma}$ and $\boldsymbol{\varepsilon}$ correspond to $\boldsymbol{\sigma}'$ and $\boldsymbol{\varepsilon}'$ respectively, but measured in different coordinate systems. Moreover, from (2.19) and (3.8) we have

$$\boldsymbol{\varepsilon}' = \boldsymbol{A}\boldsymbol{\varepsilon}\boldsymbol{A}^T \quad ; \quad \boldsymbol{\sigma}' = \boldsymbol{A}\boldsymbol{\sigma}\boldsymbol{A}^T \tag{5.23}$$

Now, choose the x_i-coordinate system collinear with the principal directions of $\boldsymbol{\sigma}$, i.e. $\boldsymbol{\sigma}$ becomes diagonal whereas nothing is known about $\boldsymbol{\varepsilon}$. We therefore have

$$\boldsymbol{\sigma} = \begin{bmatrix} \sigma_1 & 0 & 0 \\ 0 & \sigma_2 & 0 \\ 0 & 0 & \sigma_3 \end{bmatrix} ; \quad \boldsymbol{\varepsilon} = \begin{bmatrix} \varepsilon_{11} & \varepsilon_{12} & \varepsilon_{13} \\ \varepsilon_{21} & \varepsilon_{22} & \varepsilon_{23} \\ \varepsilon_{31} & \varepsilon_{32} & \varepsilon_{33} \end{bmatrix} \tag{5.24}$$

Let us assume that the x_i'-coordinate system is obtained as a 180°- rotation about the x_1-axis, cf. Fig. 5.3a). Referring to (1.28), the transformation matrix is then given by

$$\boldsymbol{A}^T = \begin{bmatrix} 1 & 0 & 0 \\ 0 & -1 & 0 \\ 0 & 0 & -1 \end{bmatrix} = \boldsymbol{A} \tag{5.25}$$

Insertion of (5.24) and (5.25) in (5.23) yields

$$\boldsymbol{\varepsilon}' = \begin{bmatrix} \varepsilon_{11} & -\varepsilon_{12} & -\varepsilon_{13} \\ -\varepsilon_{21} & \varepsilon_{22} & \varepsilon_{23} \\ -\varepsilon_{31} & \varepsilon_{32} & \varepsilon_{33} \end{bmatrix} ; \quad \boldsymbol{\sigma}' = \begin{bmatrix} \sigma_1 & 0 & 0 \\ 0 & \sigma_2 & 0 \\ 0 & 0 & \sigma_3 \end{bmatrix} \tag{5.26}$$

Consequently, we have $\boldsymbol{\sigma}' = \boldsymbol{\sigma}$ and it then follows from (5.22) that we must have $\boldsymbol{\varepsilon}' = \boldsymbol{\varepsilon}$. A comparison of (5.24) and (5.26) then reveals that $\varepsilon_{12} = \varepsilon_{13} = 0$.

Assume next that the x_i'-coordinate system is obtained as a 180°-rotation about the x_2-axis, cf. Fig. 5.3b). The transformation matrix is then given by

$$\boldsymbol{A}^T = \begin{bmatrix} -1 & 0 & 0 \\ 0 & 1 & 0 \\ 0 & 0 & -1 \end{bmatrix} = \boldsymbol{A}$$

and using the same argumentation as above, it follows that $\varepsilon_{23} = 0$

Consequently, it has been shown that if $\boldsymbol{\sigma}$ is diagonal, so is $\boldsymbol{\varepsilon}$. I.e. the assumption of isotropy implies that the principal directions for $\boldsymbol{\sigma}$ and $\boldsymbol{\varepsilon}$ coincide.

Therefore, if we choose the coordinate system collinearly with the principal directions of $\boldsymbol{\sigma}$, only principal strains are activated and (5.22a) implies

$$\varepsilon_1 = h_1(\sigma_1, \sigma_2, \sigma_3) \ ; \quad \varepsilon_2 = h_2(\sigma_1, \sigma_2, \sigma_3) \ ; \quad \varepsilon_3 = h_3(\sigma_1, \sigma_2, \sigma_3) \qquad (5.27)$$

where h_1,h_2 and h_3 are some functions. To summarize, we then have $\varepsilon_i = h_i(\sigma_k)$.

Consider now two different states: $\varepsilon_i^o = h_i(\sigma_k^o)$ which is assumed to be known and $\varepsilon_i = h_i(\sigma_k)$ which will be evaluated. Since h_i is a smooth function, a Taylor expansion of ε_i about the state σ_i^o gives

$$\begin{aligned} \varepsilon_i = h_i^o + \left(\frac{\partial h_i}{\partial \sigma_k}\right)^o (\sigma_k - \sigma_k^o) + \frac{1}{2}\left(\frac{\partial^2 h_i}{\partial \sigma_k \partial \sigma_l}\right)^o (\sigma_k - \sigma_k^o)(\sigma_l - \sigma_l^o) \\ + \frac{1}{6}\left(\frac{\partial^3 h_i}{\partial \sigma_k \partial \sigma_l \partial \sigma_p}\right)^o (\sigma_k - \sigma_k^o)(\sigma_l - \sigma_l^o)(\sigma_p - \sigma_p^o) + \ldots \end{aligned}$$

The superscript o is used to indicate that the quantities are evaluated at state σ_k^o. To simplify the following derivation let us now consider $i = 1$. Collecting terms, it then follows that the above expression can be written as

$$\begin{aligned} \varepsilon_1 =& [h_1^o + \left(\frac{\partial h_1}{\partial \sigma_\alpha}\right)^o (\sigma_\alpha - \sigma_\alpha^o) - \left(\frac{\partial h_1}{\partial \sigma_1}\right)^o \sigma_1^o + \ldots] \\ &+ [\left(\frac{\partial h_1}{\partial \sigma_1}\right)^o + \left(\frac{\partial^2 h_1}{\partial \sigma_\alpha \partial \sigma_1}\right)^o (\sigma_\alpha - \sigma_\alpha^o) + \ldots]\sigma_1 \\ &+ [\frac{1}{2}\left(\frac{\partial^2 h_1}{\partial \sigma_1 \partial \sigma_1}\right)^o + \frac{1}{2}\left(\frac{\partial^3 h_i}{\partial \sigma_\alpha \partial \sigma_1 \partial \sigma_1}\right)^o (\sigma_\alpha - \sigma_\alpha^o) + \ldots]\sigma_1^2 \\ &+ [\frac{1}{6}\left(\frac{\partial^3 h_i}{\partial \sigma_1 \partial \sigma_1}\right)^o + \frac{1}{6}\left(\frac{\partial^4 h_i}{\partial \sigma_\alpha \partial \sigma_1 \partial \sigma_1 \partial \sigma_1}\right)^o (\sigma_\alpha - \sigma_\alpha^o) + \ldots]\sigma_1^3 + \ldots \end{aligned}$$

where α takes the values 2 and 3. Introducing a more compact notation we find that

$$\varepsilon_1 = b_1 + b_2\sigma_1 + b_3\sigma_1^2 + b_4\sigma_1^3 + \cdots \qquad (5.28)$$

where b_1, b_2,... depend on σ_2 and σ_3.

The eigenvalue problem for the stress tensor is similar to the eigenvalue problem for the strain tensor. In complete conformity with (2.27), we therefore have for the principal stress σ_1 that

$$-\sigma_1^3 + \theta_1\sigma_1^2 - \theta_2\sigma_1 + \theta_3 = 0 \tag{5.29}$$

where θ_1,θ_2 and θ_3 are the Cauchy-stress invariants defined similar to (2.28). From (5.29) follows that

$$\sigma_1^{3+\alpha} = \theta_1\sigma_1^{2+\alpha} - \theta_2\sigma_1^{1+\alpha} + \theta_3\sigma_1^{\alpha} \tag{5.30}$$

where α is any non-negative integer. Repeated use of (5.30) in (5.28) means that all higher-order terms in (5.28) can be eliminated and we are then left with

$$\varepsilon_1 = p_1 + q_1\sigma_1 + r_1\sigma_1^2 \tag{5.31}$$

where p_1, q_1 and r_1 are some quantities that may depend on the stress invariants. Likewise, making a Taylor-expansion of the function h_2 and h_3 in (5.27) about σ_k^o, we obtain

$$\begin{aligned}\varepsilon_2 &= p_2 + q_2\sigma_2 + r_2\sigma_2^2\\ \varepsilon_3 &= p_3 + q_3\sigma_3 + r_3\sigma_3^2\end{aligned} \tag{5.32}$$

where p_2, q_2, r_2 as well as p_3, q_3 and r_3 are some quantities that may depend on the stress invariants.

From (5.31) and (5.32), we may write $\boldsymbol{\varepsilon}$ as

$$\boldsymbol{\varepsilon} = \boldsymbol{P} + \boldsymbol{Q}\boldsymbol{\sigma} + \boldsymbol{R}\boldsymbol{\sigma}^2 \tag{5.33}$$

where

$$\boldsymbol{P} = \begin{bmatrix} p_1 & 0 & 0\\ 0 & p_2 & 0\\ 0 & 0 & p_3 \end{bmatrix};\ \boldsymbol{Q} = \begin{bmatrix} q_1 & 0 & 0\\ 0 & q_2 & 0\\ 0 & 0 & q_3 \end{bmatrix};\ \boldsymbol{R} = \begin{bmatrix} r_1 & 0 & 0\\ 0 & r_2 & 0\\ 0 & 0 & r_3 \end{bmatrix}$$

The only thing we know about $\boldsymbol{P}$,$\boldsymbol{Q}$ and $\boldsymbol{R}$ is that they are diagonal matrices which may depend on the stress invariants. We will now prove that they are, in fact, second-order tensors; indeed, it will turn out that they are isotropic second-order tensors

To show this, we may write (5.33) as

$$\boldsymbol{\varepsilon} = \boldsymbol{g}(\boldsymbol{\sigma}) \qquad \text{where} \qquad \boldsymbol{g}(\boldsymbol{\sigma}) = \boldsymbol{P} + \boldsymbol{Q}\boldsymbol{\sigma} + \boldsymbol{R}\boldsymbol{\sigma}\boldsymbol{\sigma} \tag{5.34}$$

The coordinate invariance principle and the assumption of isotropy led to (5.14), which may be written as

$$\boldsymbol{g}(\boldsymbol{\sigma}) = \boldsymbol{A}^T\boldsymbol{g}(\boldsymbol{A}\boldsymbol{\sigma}\boldsymbol{A}^T)\boldsymbol{A} \tag{5.35}$$

Since $\boldsymbol{\sigma}' = \boldsymbol{A}\boldsymbol{\sigma}\boldsymbol{A}^T$ and as $\boldsymbol{P}$, $\boldsymbol{Q}$ and $\boldsymbol{R}$ only depend on the stress invariants, these matrices take the same value irrespective of whether they are evaluated for the stress $\boldsymbol{\sigma}$ or for the stress $\boldsymbol{\sigma}' = \boldsymbol{A}\boldsymbol{\sigma}\boldsymbol{A}^T$. With $\boldsymbol{g}(\boldsymbol{\sigma})$ given by (5.34) we therefore obtain from (5.35)

$$\boldsymbol{P} + \boldsymbol{Q}\boldsymbol{\sigma} + \boldsymbol{R}\boldsymbol{\sigma}\boldsymbol{\sigma} = \boldsymbol{A}^T(\boldsymbol{P} + \boldsymbol{Q}\boldsymbol{A}\boldsymbol{\sigma}\boldsymbol{A}^T + \boldsymbol{R}\boldsymbol{A}\boldsymbol{\sigma}\boldsymbol{A}^T\boldsymbol{A}\boldsymbol{\sigma}\boldsymbol{A}^T)\boldsymbol{A}$$

Since $\boldsymbol{A}^T\boldsymbol{A} = \boldsymbol{I}$, this can be written as

$$\overline{\boldsymbol{P}} + \overline{\boldsymbol{Q}}\boldsymbol{\sigma} + \overline{\boldsymbol{R}}\boldsymbol{\sigma}^2 = \boldsymbol{0} \tag{5.36}$$

where

$$\overline{\boldsymbol{P}} = \boldsymbol{P} - \boldsymbol{A}^T\boldsymbol{P}\boldsymbol{A}\ ; \quad \overline{\boldsymbol{Q}} = \boldsymbol{Q} - \boldsymbol{A}^T\boldsymbol{Q}\boldsymbol{A}\ ; \quad \overline{\boldsymbol{R}} = \boldsymbol{R} - \boldsymbol{A}^T\boldsymbol{R}\boldsymbol{A} \tag{5.37}$$

The coordinate system was taken to be collinear with the principal stress directions implying that $\boldsymbol{\sigma}$ and $\boldsymbol{\sigma}^2$ are diagonal matrices in (5.36). For a given principal stress state σ_1,σ_2 and σ_3, the matrices $\boldsymbol{\sigma}$,$\boldsymbol{\sigma}^2$,$\boldsymbol{P}$,$\boldsymbol{Q}$ and $\boldsymbol{R}$ are then given and fixed quantities. However, we may in (5.37) choose the transformation matrix $\boldsymbol{A}$ in an arbitrary fashion. To fulfill (5.36) under these circumstances, we are led to the requirement

$$\boldsymbol{P} = \boldsymbol{A}^T\boldsymbol{P}\boldsymbol{A}\ ; \quad \boldsymbol{Q} = \boldsymbol{A}^T\boldsymbol{Q}\boldsymbol{A}\ ; \quad \boldsymbol{R} = \boldsymbol{A}^T\boldsymbol{R}\boldsymbol{A}$$

We conclude that not only are $\boldsymbol{P}$,$\boldsymbol{Q}$ and $\boldsymbol{R}$ second-order tensors, but they are also isotropic second-order tensors, cf. (1.44). According to (1.45), the only isotropic second-order tensor is $k\delta_{ij}$ where the factor k may depend on some invariants. We are then led to

$$\boldsymbol{P} = \alpha_1\boldsymbol{I}\ ; \quad \boldsymbol{Q} = \alpha_2\boldsymbol{I}\ ; \quad \boldsymbol{R} = \alpha_3\boldsymbol{I}$$

where α_1,α_2 and α_3 may depend on the stress invariants. With these results, (5.33) takes the form

$$\boldsymbol{\varepsilon} = \alpha_1\boldsymbol{I} + \alpha_2\boldsymbol{\sigma} + \alpha_3\boldsymbol{\sigma}^2 \tag{5.38}$$

This result was derived on condition that the coordinate system is collinear with the principal stress directions. However, as $\boldsymbol{\varepsilon}$, $\boldsymbol{I}$ and $\boldsymbol{\sigma}$ are second-order tensors, (5.38) holds for arbitrary coordinate systems and we have then retrieved the format already stipulated in (5.16).

The result (5.38) derived from (5.2), the coordinate invariance principle and the assumption of isotropy is an example of a so-called *representation theorem* and we shall return to this topic in the next chapter; the result (5.38) was first derived by Prager (1945) (for incompressible materials) and Reiner (1945).

5.4 Nonlinear isotropic Hooke formulation

Let us return to the general format given by (5.16) i.e.

$$\varepsilon_{ij} = \alpha_1 \delta_{ij} + \alpha_2 \sigma_{ij} + \alpha_3 \sigma_{ik} \sigma_{kj} \tag{5.39}$$

where α_1,α_2 and α_3 may depend on the stress invariants. Instead of the stress invariants indicated in (5.17), we may just as well choose

$$\alpha_1 = \alpha_1(I_1, J_2, J_3) \ ; \quad \alpha_2 = \alpha_2(I_1, J_2, J_3) \ ; \quad \alpha_3 = \alpha_3(I_1, J_2, J_3)$$

Similar to Section 4.10, our objective is to derive a nonlinear Hooke formulation, which implies that the quadratic term $\sigma_{ik}\sigma_{kj}$ must disappear. We therefore choose

$$\alpha_3 = 0$$

With this result, (5.39) may be written

$$\varepsilon_{kk} = 3\alpha_1 + \alpha_2 \sigma_{kk} ; \qquad e_{ij} = \alpha_2 s_{ij} \tag{5.40}$$

Let us choose

$$\alpha_1 = (\frac{1}{9K} - \frac{1}{6G})\sigma_{kk} ; \qquad \alpha_2 = \frac{1}{2G}$$

then (5.40) becomes

$$\boxed{\sigma_{kk} = 3K\varepsilon_{kk} ; \qquad s_{ij} = 2Ge_{ij}} \tag{5.41}$$

and it is evident that it is allowable to let K and G be arbitrary functions of the stress invariants, i.e.

$$\boxed{K = K(I_1, J_2, J_3); \qquad G = G(I_1, J_2, J_3)} \tag{5.42}$$

A comparison of (5.41) with (4.85) and (4.86) shows that K and G are the bulk and shear modulus respectively.

It may be of interest to compare (5.41) and (5.42) with the corresponding nonlinear Hooke formulation for hyper-elasticity. In the latter case, this formulation is given by (4.101)-(4.103). It is of considerable importance that the Cauchy-formulation is free from the constraint given by (4.103). Not only that, but K and G are now allowed also to depend on the third stress invariant J_3 and as we will see in Chapter 8 this is of major importance when modeling the behavior of materials like concrete, soil and rocks.

Indeed, (5.41) and (5.42) have been successfully applied to model such materials and, as examples, we may refer to Ottosen (1979) and Kotsovos (1980) for concrete and to DiMaggio and Sandler (1971) for soil (using hyper-elasticity); moreover, Chen and Saleeb (1982) present a comprehensive review

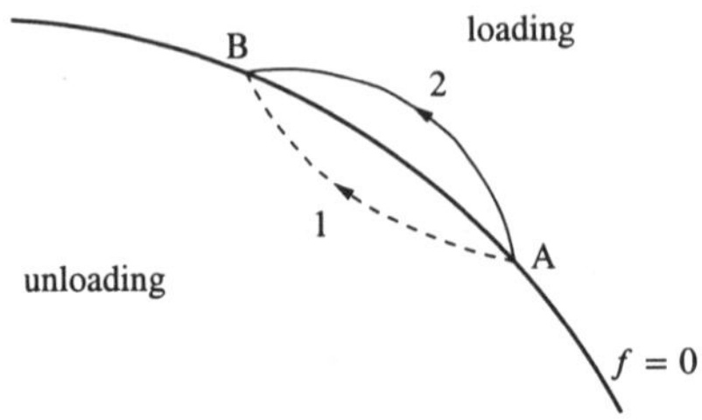

Figure 5.4: Stress space with surface $f = 0$ that divides the stress space into a space where unloading occurs and a space where loading occurs.

of such concrete and soil models. Recalling relation (4.91), we may write (5.41) in the alternative manner

$$\varepsilon_{ij} = -\frac{\nu}{E}\sigma_{kk}\delta_{ij} + \frac{1+\nu}{E}\sigma_{ij} \tag{5.43}$$

where Young's modulus E and Poisson's ratio ν now may depend on all the stress invariants, i.e.

$$E = E(I_1, J_2, J_3) \quad ; \qquad \nu = \nu(I_1, J_2, J_3)$$

Whereas a nonlinear Hooke formulation may provide close predictions to a variety of materials during loading, the major drawback is that unloading follows the same path as loading. This is certainly not a realistic prediction and various attempts have therefore been proposed in order to improve the response during unloading. However, as shown below there is only one manner in which the predictions during unloading may be improved in a consistent manner, Ottosen (1980)

For this purpose, assume that we introduce a criterion that tells us whether loading or unloading occurs. In Fig. 5.4 that illustrates the stress space, this criterion is expressed by the surface $f = 0$. We shall assume that Hooke's law also applies during unloading and that Young's modulus and Poisson's ratio are fixed material parameters during unloading.

Following Handelman *et al.* (1947), the essential property that we want to satisfy is that an infinitely small change of the applied loading only will result in an infinitely small change of the response. This requirement may be called the *continuity requirement*.

Consider path 1 in the unloading space of Fig. 5.4; the elastic parameters are fixed during unloading. Let path 1 approach the surface $f = 0$ so that path 1 in the limit follows the surface $f = 0$. For continuity reasons, we conclude that the elastic parameters are unchanged also when the path follows the surface $f = 0$.

Consider again path 1 in the unloading space and let E_u and ν_u be Young's modulus and Poisson's ratio respectively, during unloading. When moving from Point A to B along path 1, cf. Fig. 5.4, and recalling that E_u and ν_u are fixed quantities during unloading, we have according to (5.43) that the strain change during unloading $\Delta\varepsilon_{ij}^u$ is given by

$$\Delta\varepsilon_{ij}^u = -\frac{\nu_u}{E_u}\Delta\sigma_{kk}\delta_{ij} + \frac{1+\nu_u}{E_u}\Delta\sigma_{ij} \tag{5.44}$$

Let E_l and ν_l be Young's modulus and Poisson's ratio respectively, during loading. Moreover, let path 2 in the loading space, cf. Fig. 5.4, be infinitely close to the surface $f = 0$. Since the elastic parameters are unchanged when moving along $f = 0$, also E_l and ν_l are unchanged for this type of loading. According to (5.43), when moving from point A to B along path 2, the strain change during loading $\Delta\varepsilon_{ij}^l$ is then given by

$$\Delta\varepsilon_{ij}^l = -\frac{\nu_l}{E_l}\Delta\sigma_{kk}\delta_{ij} + \frac{1+\nu_l}{E_l}\Delta\sigma_{ij} \tag{5.45}$$

We have assumed that both path 1 and 2 are infinitely close to the surface $f = 0$. For continuity reasons, we must require that $\Delta\varepsilon_{ij}^u = \Delta\varepsilon_{ij}^l$, i.e. (5.44) and (5.45) provide

$$(-\frac{\nu_u}{E_u} + \frac{\nu_l}{E_l})\Delta\sigma_{kk}\delta_{ij} + (\frac{1+\nu_u}{E_u} - \frac{1+\nu_l}{E_l})\Delta\sigma_{ij} = 0$$

and this relation can only be satisfied if

$$\boxed{E_u = E_l \quad ; \qquad \nu_u = \nu_l} \tag{5.46}$$

That is, if unloading occurs, E_u and ν_u take those values of E_l and ν_l respectively, that were relevant immediately before unloading. We recall that during unloading, E_u and ν_u are fixed quantities.

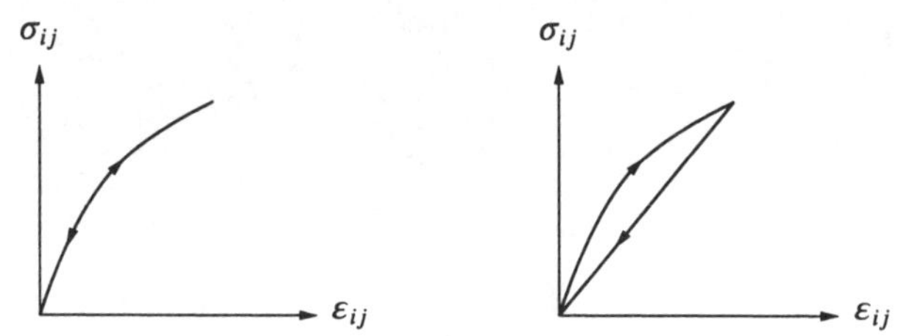

Figure 5.5: a) Nonlinear elastic response; b) elastic-fracturing response, secant approach.

If no loading/unloading criterion is introduced, our nonlinear elastic model of (5.43) predicts the same response in unloading as in loading, cf. Fig. 5.5a).

However, if a loading/unloading criterion is introduced, the only manner in which a consistent approach can be obtained that fulfills the continuity requirement, is to adopt the procedure indicated in (5.46) and illustrated in Fig. 5.5b). The material response shown in Fig. 5.5b) is often referred to as an *elastic fracturing material* after Dougill (1976). The response shown in this figure may also be termed as a *secant approach* since the linear unloading response is given by the secant values of E and ν just before unloading.

Whereas the unloading response in Fig. 5.5b) is much more realistic than that in Fig. 5.5a), it is still a crude approximation to the real unloading response of most materials. We shall later see that the plasticity theory makes for realistic predictions not only of the loading response, but also of the unloading response.

6 REPRESENTATION THEOREMS

Establishment of constitutive relations for nonlinear material behavior is in general not trivial and a number of methods exists. Often, the only initial information available is the knowledge that a certain quantity depends on some other quantities, but the explicit constitutive relation is unknown. Evidently, knowledge of experimental results provides important information, but experimental data are often of a form that only enables one to determine the explicit constitutive relation for particular load paths - uniaxial loading, for instance. The question of determination of the explicit constitutive relation for arbitrary load paths then remains. In recent years, increasing use has been made of certain powerful theorems, which enables one to determine the most general explicit form of constitutive relations. These theorems are based on so-called *representation theorems*, which determine the most general forms of various scalar and tensor functions that satisfy both the coordinate invariance principle as well as the material symmetry in question.

To illustrate this approach, we may again consider Cauchy-elasticity where it is known that a one-to-one relation exists between strains and stresses, i.e.

$$\boldsymbol{\varepsilon} = \boldsymbol{g}(\boldsymbol{\sigma}) \tag{6.1}$$

where $\boldsymbol{g}$ is the response function. The material is assumed to be isotropic and we want to investigate whether it is possible to obtain a more explicit form for (6.1). Obviously, by performing a number of experiments, it is possible to derive some information of a more explicit form for (6.1). However, as shown in the previous chapter, just the coordinate invariance principle and the assumption of isotropy alone allow us to conclude the most general explicit form of (6.1). The result was given by (5.16) and below we will recall the essential steps as they provide an example of a result given in terms of a representation theorem.

In another coordinate system, the constitutive relation (6.1) reads

$$\boldsymbol{\varepsilon}' = \boldsymbol{g}^*(\boldsymbol{\sigma}') \tag{6.2}$$

Suppose that $\boldsymbol{\varepsilon}$ and $\boldsymbol{\sigma}$ correspond to $\boldsymbol{\varepsilon}'$ and $\boldsymbol{\sigma}'$ respectively, but just measured in another coordinate system. Then we have

$$\boldsymbol{\varepsilon}' = \boldsymbol{A}\boldsymbol{\varepsilon}\boldsymbol{A}^T \quad ; \qquad \boldsymbol{\sigma}' = \boldsymbol{A}\boldsymbol{\sigma}\boldsymbol{A}^T \tag{6.3}$$

The coordinate invariance principle, (5.10), states that the material response is independent of the coordinate system. Hence, the predictions provided by (6.1) and (6.2) must be identical. With (6.3a) in (6.2), we therefore have

$$\boldsymbol{A}\boldsymbol{\varepsilon}\boldsymbol{A}^T = \boldsymbol{g}^*(\boldsymbol{\sigma}')$$

and use of (6.1) leads to

$$\boldsymbol{A}\boldsymbol{g}(\boldsymbol{\sigma})\boldsymbol{A}^T = \boldsymbol{g}^*(\boldsymbol{\sigma}') \qquad \text{coordinate invariance} \tag{6.4}$$

Referring to (5.13), the assumption of isotropy means that the response function is the same in all coordinate systems, i.e.

$$\boldsymbol{g}^*(\boldsymbol{\sigma}') = \boldsymbol{g}(\boldsymbol{\sigma}') \qquad \text{isotropy} \tag{6.5}$$

Insertion of (6.5) and (6.3b) in (6.4) then leads to

$$\boldsymbol{A}\boldsymbol{g}(\boldsymbol{\sigma})\boldsymbol{A}^T = \boldsymbol{g}(\boldsymbol{A}\boldsymbol{\sigma}\boldsymbol{A}^T) \qquad \text{coordinate invariance + isotropy}$$

which means that the response function $\boldsymbol{g}$ is an *isotopic second-order tensor function*. In Chapter 5, we showed that this requirement implies that the most general format of (6.1) is provided by (5.16), i.e.

$$\boldsymbol{\varepsilon} = \alpha_1 \boldsymbol{I} + \alpha_2 \boldsymbol{\sigma} + \alpha_3 \boldsymbol{\sigma}^2 \tag{6.6}$$

and this representation theorem is evidently of a very powerful nature.

Representation theorems are available for a number of relations of which (6.6) only represents a very simple example. In general, representation theorems turn out to be rather difficult to prove, but since they are often of great help when establishing constitutive theories, we will present below some useful results. In the literature, there has been a long standing debate as to what should be considered as correct representation theorems. A concise theoretical formulation was established by Rivlin and Ericksen (1955) and Truesdell (1955a) and (6.6) is in the literature often referred to as the *Rivlin-Ericksen representation theorem*. Later, important contributions were provided by, for instance, Wang (1970), Smith (1971), Spencer (1971) and Boehler (1977). The review articles of Murakami and Sawczuk (1981) and Zheng (1994) contain comprehensive discussions of various results and their historical development.

In the following, we shall consider scalar functions as well as isotropic tensor functions of second order that depend on certain quantities and we will present the corresponding representation theorems. To keep the formulation as general as possible let

$\boldsymbol{M}, \boldsymbol{N}, \boldsymbol{P}, \boldsymbol{S} =$	*symmetric second-order tensors*
$H_\alpha =$	*scalar quantities* ($\alpha = 1, 2, \ldots$)

We observe that in another coordinate system $\boldsymbol{x}'$, the scalars H_α are unchanged whereas $\boldsymbol{M}'$, $\boldsymbol{N}'$, $\boldsymbol{P}'$ and $\boldsymbol{S}'$ in accordance with (1.41) become

$$\boldsymbol{M}' = \boldsymbol{A}\boldsymbol{M}\boldsymbol{A}^T\,; \quad \boldsymbol{N}' = \boldsymbol{A}\boldsymbol{N}\boldsymbol{A}^T\,; \quad \boldsymbol{P}' = \boldsymbol{A}\boldsymbol{P}\boldsymbol{A}^T\,; \quad \boldsymbol{S}' = \boldsymbol{A}\boldsymbol{S}\boldsymbol{A}^T \tag{6.7}$$

Representation theorems that even involve, for instance, vectors and anti-symmetric second-order tensors are also available and the reader may consult the literature mentioned above on this subject.

6.1 Scalar functions

Consider some quantity which is assumed to be a scalar, i.e. an invariant. Let us further assume that this quantity in some coordinate system is given by

$$\boxed{g = g(\boldsymbol{N}, \boldsymbol{P}, \boldsymbol{S}, H_\alpha)} \tag{6.8}$$

where g expresses some property of the material and it is therefore a response function. Since the quantity g is a scalar, this implies that it takes the same value in all coordinate systems; it is an invariant. In the x_i-coordinate system, we have $g = g(\boldsymbol{N}, \boldsymbol{P}, \boldsymbol{S}, H_\alpha)$ whereas in another coordinate system x'_i, we have $g^* = g^*(\boldsymbol{N}', \boldsymbol{P}', \boldsymbol{S}', H'_\alpha)$. In accordance with the discussion following (5.7), the response function is denoted g in the x_i-coordinate system and g^* in the x'_i-coordinate system. Since the quantity takes the same value in all coordinate systems, we have

$$g(\boldsymbol{N}, \boldsymbol{P}, \boldsymbol{S}, H_\alpha) = g^*(\boldsymbol{N}', \boldsymbol{P}', \boldsymbol{S}', H'_\alpha) \qquad \text{coordinate invariance} \tag{6.9}$$

Assume that the material is isotropic. Isotropy means that the response function is the same in all coordinate systems, cf. (5.13). This implies that

$$g^*(\boldsymbol{N}', \boldsymbol{P}', \boldsymbol{S}', H'_\alpha) = g(\boldsymbol{N}', \boldsymbol{P}', \boldsymbol{S}', H'_\alpha) \qquad \text{isotropy}$$

Insertion of this expression in (6.9) and use of (6.7) then leads to

$$\boxed{g(\boldsymbol{N}, \boldsymbol{P}, \boldsymbol{S}, H_\alpha) = g(\boldsymbol{A}\boldsymbol{N}\boldsymbol{A}^T, \boldsymbol{A}\boldsymbol{P}\boldsymbol{A}^T, \boldsymbol{A}\boldsymbol{S}\boldsymbol{A}^T, H_\alpha) \quad \textit{coord. inv.+iso.}} \tag{6.10}$$

A scalar function which fulfills this expression is called an *isotropic scalar tensor function.* It is evident that if $\boldsymbol{N}$, $\boldsymbol{P}$ and $\boldsymbol{S}$ enter the function g in terms of invariants, then (6.10) is fulfilled. Moreover, referring to Zheng (1994), for instance, it turns out that the most general form of g, that fulfills requirement (6.10), is given by

$$\boxed{\begin{aligned} g = g(&I_{1N}, I_{2N}, I_{3N}, I_{1P}, I_{2P}, I_{3P}, I_{1S}, I_{2S}, I_{3S}, \\ &J^o_1, J^o_2, J^o_3, J^o_4, J^o_5, J^o_6, J^o_7, J^o_8, J^o_9, J^o_{10}, J^o_{11}, J^o_{12}, J^o_{13}, H_\alpha) \end{aligned}} \tag{6.11}$$

where the three invariants of $\boldsymbol{N}$, as usual, are given by

$$I_{1N} = tr\, \boldsymbol{N}\ ; \quad I_{2N} = \frac{1}{2} tr\, (\boldsymbol{N}^2)\ ; \quad I_{3N} = \frac{1}{3} tr\, (\boldsymbol{N}^3) \tag{6.12}$$

the three invariants of $\boldsymbol{P}$ are given by

$$I_{1P} = tr\, \boldsymbol{P}\ ; \quad I_{2P} = \frac{1}{2} tr\, (\boldsymbol{P}^2)\ ; \quad I_{3P} = \frac{1}{3} tr\, (\boldsymbol{P}^3) \tag{6.13}$$

and, likewise, the three invariants of $\boldsymbol{S}$ are given by

$$I_{1S} = tr\,\boldsymbol{S}\ ;\quad I_{2S} = \frac{1}{2} tr\,(\boldsymbol{S}^2)\ ;\quad I_{3S} = \frac{1}{3} tr\,(\boldsymbol{S}^3) \tag{6.14}$$

However, the so-called *joint invariants* are given by

$$\boxed{\begin{array}{ll} J_1^o = tr\,(\boldsymbol{NP})\ ; & J_2^o = tr\,(\boldsymbol{NP}^2) \\ J_3^o = tr\,(\boldsymbol{N}^2\boldsymbol{P})\ ; & J_4^o = tr\,(\boldsymbol{N}^2\boldsymbol{P}^2) \\ J_5^o = tr\,(\boldsymbol{NS})\ ; & J_6^o = tr\,(\boldsymbol{NS}^2) \\ J_7^o = tr\,(\boldsymbol{N}^2\boldsymbol{S})\ ; & J_8^o = tr\,(\boldsymbol{N}^2\boldsymbol{S}^2) \\ J_9^o = tr\,(\boldsymbol{PS})\ ; & J_{10}^o = tr\,(\boldsymbol{PS}^2) \\ J_{11}^o = tr\,(\boldsymbol{P}^2\boldsymbol{S})\ ; & J_{12}^o = tr\,(\boldsymbol{P}^2\boldsymbol{S}^2) \\ J_{13}^o = tr\,(\boldsymbol{NPS}) & \end{array}} \tag{6.15}$$

In index notation (6.12) - (6.15) read

$$\begin{aligned} I_{1N} &= N_{kk}\ ;\quad I_{2N} = \frac{1}{2} N_{kl}N_{lk}\ ;\quad I_{3N} = \frac{1}{3} N_{kl}N_{lm}N_{mk} \\ I_{1P} &= P_{kk}\ ;\quad I_{2P} = \frac{1}{2} P_{kl}P_{lk}\ ;\quad I_{3P} = \frac{1}{3} P_{kl}P_{lm}P_{mk} \\ I_{1S} &= S_{kk}\ ;\quad I_{2S} = \frac{1}{2} S_{kl}S_{lk}\ ;\quad I_{3S} = \frac{1}{3} S_{kl}S_{lm}S_{mk} \end{aligned}$$

and

$$\begin{array}{ll} J_1^o = N_{kl}P_{lk}\ ; & J_2^o = N_{kl}P_{lm}P_{mk} \\ J_3^o = N_{kl}N_{lm}P_{mk}\ ; & J_4^o = N_{kl}N_{lm}P_{mn}P_{nk} \\ J_5^o = N_{kl}S_{lk}\ ; & J_6^o = N_{kl}S_{lm}S_{mk} \\ J_7^o = N_{kl}N_{lm}S_{mk}\ ; & J_8^o = N_{kl}N_{lm}S_{mn}S_{nk} \\ J_9^o = P_{kl}S_{lk}\ ; & J_{10}^o = P_{kl}S_{lm}S_{mk} \\ J_{11}^o = P_{kl}P_{lm}S_{mk}\ ; & J_{12}^o = P_{kl}P_{lm}S_{mn}S_{nk} \\ J_{13}^o = N_{kl}P_{lm}S_{mk} & \end{array}$$

It follows that a total of 22 invariants appears in (6.11). As these invariants are independent of each other and all other invariants of $\boldsymbol{N}$, $\boldsymbol{P}$ and $\boldsymbol{S}$ can be expressed in terms of the 22 invariants defined, these invariants provide a so-called *minimal function basis*. Occasionally, one uses the phrase that these 22 invariants comprise an *irreducible set of invariants*. We also observe that as the number of variables in the g function increases, the number of irreducible invariants increases dramatically: the function $g = g(\boldsymbol{N})$ implies three invariants, $g = g(\boldsymbol{N}, \boldsymbol{P})$ implies ten invariants whereas $g = g(\boldsymbol{N}, \boldsymbol{P}, \boldsymbol{S})$ implies 22 invariants.

As an example of the use of (6.8) and (6.11) assume that the strain energy for an isotropic material is given in the form $W = W(\varepsilon_{ij})$. A comparison with (6.8) shows that $N_{ij} = \varepsilon_{ij}$ and $P_{ij} = S_{ij} = H_\alpha = 0$. In this case (6.11) provides

$$W = W(I_{1\varepsilon}, I_{2\varepsilon}, I_{3\varepsilon})$$

and noting that $I_{1\varepsilon} = \tilde{I}_1$, $I_{2\varepsilon} = \tilde{I}_2$ and $I_{3\varepsilon} = \tilde{I}_3$, cf. (6.12) and (2.51), we have obtained a form that precisely corresponds to (4.65).

Let us next determine the expression $\partial g/\partial N_{ij}$ and let us for simplicity assume that $g = g(\boldsymbol{N}, \boldsymbol{P})$. From (6.11) we find that

$$\begin{aligned}\frac{\partial g}{\partial N_{ij}} &= \frac{\partial g}{\partial I_{1N}}\frac{\partial I_{1N}}{\partial N_{ij}} + \frac{\partial g}{\partial I_{2N}}\frac{\partial I_{2N}}{\partial N_{ij}} + \frac{\partial g}{\partial I_{3N}}\frac{\partial I_{3N}}{\partial N_{ij}} \\ &\quad + \frac{\partial g}{\partial J_1^o}\frac{\partial J_1^o}{\partial N_{ij}} + \frac{\partial g}{\partial J_2^o}\frac{\partial J_2^o}{\partial N_{ij}} + \frac{\partial g}{\partial J_3^o}\frac{\partial J_3^o}{\partial N_{ij}} + \frac{\partial g}{\partial J_4^o}\frac{\partial J_4^o}{\partial N_{ij}}\end{aligned}$$

i.e.

$$\boxed{\begin{aligned}&g = g(\boldsymbol{N}, \boldsymbol{P}) \qquad \textit{implies} \\ &\frac{\partial g}{\partial N_{ij}} = \phi_1\delta_{ij} + \phi_2 N_{ij} + \phi_3 N_{ik}N_{kj} + \phi_4 P_{ij} + \phi_5 P_{ik}P_{kj} \\ &\qquad + \phi_6(N_{ik}P_{kj} + P_{ik}N_{kj}) + \phi_7(N_{ik}P_{kl}P_{lj} + P_{ik}P_{kl}N_{lj})\end{aligned}} \tag{6.16}$$

where $\phi_1 \ldots \phi_7$ are given by $\partial g/\partial I_{1N} \ldots \partial g/\partial J_4^o$ and where we therefore have the constraints

$$\frac{\partial \phi_i}{\partial I_j} = \frac{\partial \phi_j}{\partial I_i} \tag{6.17}$$

In this expression i goes from 1 to 7 and the notation $I_1 \ldots I_7 = I_{1N} \ldots J_4^o$ has been used for convenience. The ϕ-quantities in (6.16) depend, in general, on ten of the invariants defined by (6.12)-(6.15).

As an example, consider hyper-elasticity and let g be chosen as the strain energy $W(\varepsilon_{ij})$ i.e. $N_{ij} = \varepsilon_{ij}$ and $P_{ij} = S_{ij} = H_\alpha = 0$. In this case (6.16) reduces to the previous expressions given by (4.8) and (4.71) where we notice that the constraints (6.17) are similar to (4.72).

6.2 Second-order tensor functions

Let us assume that the quantity $\boldsymbol{M}$ in some coordinate system depends on $\boldsymbol{N}$, $\boldsymbol{P}$ and H_α through the tensor function $\boldsymbol{f}$ of second order, i.e.

$$\boxed{\boldsymbol{M} = \boldsymbol{f}(\boldsymbol{N}, \boldsymbol{P}, H_\alpha)} \tag{6.18}$$

We assume that (6.18) expresses some constitutive relation, i.e. $\boldsymbol{f}$ is a response function. In another coordinate system $\boldsymbol{x}'$, this relation reads

$$\boldsymbol{M}' = \boldsymbol{f}^*(\boldsymbol{N}', \boldsymbol{P}', H_\alpha) \tag{6.19}$$

The coordinate invariance principle states that (6.18) and (6.19) must describe the same physical property that is just measured in different coordinate systems. Therefore, use of (6.7) on the left-hand side of (6.19) gives

$$\boldsymbol{A}\boldsymbol{M}\boldsymbol{A}^T = \boldsymbol{f}^*(\boldsymbol{N}', \boldsymbol{P}', H_\alpha)$$

which with (6.18) takes the form

$$\boldsymbol{A}\boldsymbol{f}(\boldsymbol{N}, \boldsymbol{P}, H_\alpha)\boldsymbol{A}^T = \boldsymbol{f}^*(\boldsymbol{N}', \boldsymbol{P}', H_\alpha) \qquad \text{coordinate invariance} \tag{6.20}$$

Assuming the material to be isotropic, this means according to (5.13) that the response function is the same in all coordinate systems, i.e.

$$\boldsymbol{f}^*(\boldsymbol{N}', \boldsymbol{P}', H_\alpha) = \boldsymbol{f}(\boldsymbol{N}', \boldsymbol{P}', H_\alpha) \qquad \text{isotropy}$$

Insertion of this expression in (6.20) and observation of (6.7) then imply

$$\boxed{\boldsymbol{A}\boldsymbol{f}(\boldsymbol{N}, \boldsymbol{P}, H_\alpha)\boldsymbol{A}^T = \boldsymbol{f}(\boldsymbol{A}\boldsymbol{N}\boldsymbol{A}^T, \boldsymbol{A}\boldsymbol{P}\boldsymbol{A}^T, H_\alpha) \quad \textit{coord. inv.+iso.}} \tag{6.21}$$

A second-order tensor function which fulfills this expression is called an *isotropic second-order tensor function*.

In order to fulfill (6.21), it turns out, cf. for instance, Zheng (1994), that the most general form of (6.18) is provided by

$$\boxed{\boldsymbol{M} = \sum_{i=1}^{8} \alpha_i \boldsymbol{G}_i} \tag{6.22}$$

where α_i are scalar functions of H_α as well as of the ten invariants formed by $\boldsymbol{N}$ and $\boldsymbol{P}$, i.e. the ten invariants defined in (6.12)-(6.15) and where the $\boldsymbol{G}_i$-tensors are given by

$$\begin{aligned} &\boldsymbol{G}_1 = \boldsymbol{I}\ ;\ \ \boldsymbol{G}_2 = \boldsymbol{N}\ ;\ \ \boldsymbol{G}_3 = \boldsymbol{N}^2\ ;\ \ \boldsymbol{G}_4 = \boldsymbol{P}\ ;\ \ \boldsymbol{G}_5 = \boldsymbol{P}^2 \\ &\boldsymbol{G}_6 = \boldsymbol{N}\boldsymbol{P} + \boldsymbol{P}\boldsymbol{N}\ ;\ \ \boldsymbol{G}_7 = \boldsymbol{N}^2\boldsymbol{P} + \boldsymbol{P}\boldsymbol{N}^2\ ; \\ &\boldsymbol{G}_8 = \boldsymbol{N}\boldsymbol{P}^2 + \boldsymbol{P}^2\boldsymbol{N} \end{aligned} \tag{6.23}$$

The $\boldsymbol{G}_i$-matrices (tensors) are often termed *tensor generators* since $\boldsymbol{M}$ according to (6.22) is generated by the $\boldsymbol{G}-$matrices. Using (6.23), (6.22) can be written as

$$\boxed{\begin{aligned}\boldsymbol{M} &= \alpha_1\boldsymbol{I} + \alpha_2\boldsymbol{N} + \alpha_3\boldsymbol{N}^2 + \alpha_4\boldsymbol{P} + \alpha_5\boldsymbol{P}^2 + \alpha_6(\boldsymbol{NP} + \boldsymbol{PN})\\ &+ \alpha_7(\boldsymbol{N}^2\boldsymbol{P} + \boldsymbol{PN}^2) + \alpha_8(\boldsymbol{NP}^2 + \boldsymbol{P}^2\boldsymbol{N})\end{aligned}} \tag{6.24}$$

According to (6.18) $\boldsymbol{f} = \boldsymbol{M}$ and it is easy to check that with $\boldsymbol{f}$ given by (6.24), (6.21) is fulfilled; however, to show that solution (6.24) is, in fact, the most general solution is much more involved and we refer to the previously mentioned literature for available proofs.

As an example consider Cauchy-elasticity and let $\boldsymbol{M}=\boldsymbol{\varepsilon}$, $\boldsymbol{N}=\boldsymbol{\sigma}$ and $\boldsymbol{P}=\boldsymbol{H}_\alpha=0$. In this case (6.24) reduces to our well-known form given by (6.6). We recall that the scalars $\alpha_1 \ldots \alpha_8$ in (6.24) are completely arbitrary.

It is of interest to observe that the formats (6.16) and (6.24) are almost identical, except that the term corresponding to α_7 does not appear in (6.16). Moreover, the α_i-functions in (6.24) are completely arbitrary functions dependent on the ten invariants in (6.12)-(6.15), whereas the ϕ_i-functions in (6.16) are restricted by the constraints (6.17). This is completely analogous with the difference between Cauchy- and hyper-elasticity.

It is emphasized, that the results above provide very powerful tools when constructing constitutive models and that modern constitutive theories make increasing use of these results. Quite often this approach is referred to as the *tensor function approach.* In the next chapter, we shall see how these results can be used to establish the most general incremental time-independent constitutive relation. Before this is done, we shall demonstrate some further examples of the usefulness of the representation theorems (6.11) and (6.24).

6.3 Thermoelasticity

As a first example, assume that the strains depend on the stresses as well as the temperature difference ΔT measured from some reference state where no thermal expansion exists, i.e. we assume that

$$\varepsilon_{ij} = \varepsilon_{ij}(\sigma_{kl}, \Delta T)$$

A comparison with (6.18) shows that $M_{ij} = \varepsilon_{ij}$, $N_{ij} = \sigma_{ij}$, $P_{ij} = 0$ and $H_\alpha = \Delta T$, i.e. (6.24) provides

$$\varepsilon_{ij} = \alpha_1\delta_{ij} + \alpha_2\sigma_{ij} + \alpha_3\sigma_{ik}\sigma_{kj} \tag{6.25}$$

where α_1, α_2 and α_3 may depend on the stress invariants σ_{kk}, $\sigma_{kl}\sigma_{lk}$, $\sigma_{kl}\sigma_{lm}\sigma_{mk}$ as well as on ΔT. Let us choose

$$\alpha_1 = -\frac{\nu}{E}\sigma_{kk} + \alpha\Delta T\,; \qquad \alpha_2 = \frac{1+\nu}{E}\,; \qquad \alpha_3 = 0$$

then (6.25) reduces to

$$\boxed{\varepsilon_{ij} = \varepsilon_{ij}^e + \varepsilon_{ij}^o}$$

where

$$\boxed{\varepsilon_{ij}^e = \frac{1+\nu}{E}\sigma_{ij} - \frac{\nu}{E}\sigma_{kk}\delta_{ij}\,; \qquad \varepsilon_{ij}^o = \alpha\Delta T\delta_{ij}}$$

A comparison with (4.91) shows that ε_{ij}^e are the elastic strains determined by the usual Hooke formulation and a comparison with (4.63) shows that ε_{ij}^o represents the thermal strains and we have then recovered the constitutive law for isotropic thermoelasticity.

The thermoelasticity presented above is uncoupled in the sense that the stress state does not influence the thermal strains. However, for some materials, for instance concrete at high temperatures, it turns out that the thermal strains depend on the stress state. Also in that complex case, the representation theorem (6.24) may be used in an elegant manner to obtain a suitable constitutive law, as demonstrated by Sawzcuk (1984) and Thelandersson (1987).

6.4 Viscoelasticity

Let us next turn our attention to materials where the response depends not only on the loading itself, but also on the duration of the loading. In that case one speaks of *time-dependent behavior* or *creep* and we shall return to this subject in Chapter 14.

Quite a number of creep models are constructed on the basis of certain combinations of linear springs and dashpots. These components are illustrated in Fig. 6.1 and the uniaxial constitutive relations are

$$\sigma^e = E\varepsilon^e\,; \qquad \sigma^v = \eta\dot{\varepsilon}^v \tag{6.26}$$

where superscripts e and v refer to elastic and *viscous* behavior respectively. Moreover, a dot denotes the time derivative, i.e. $\dot{\varepsilon}^v = d\varepsilon^v/dt$ and η is the *viscosity coefficient* [Pa·s]. The time derivative is also referred to as the *rate*. We note that the dashpot responds as a rigid member for a sudden application of the load.

As an example, consider the so-called *Maxwell model* illustrated in Fig. 6.2a). Based on (6.26) and

$$\dot{\varepsilon} = \dot{\varepsilon}^e + \dot{\varepsilon}^v\,; \qquad \sigma = \sigma^e = \sigma^v \tag{6.27}$$

it is readily shown that the constitutive relation for a Maxwell model reads

$$\dot{\varepsilon} = \frac{\dot{\sigma}}{E} + \frac{\sigma}{\eta} \tag{6.28}$$

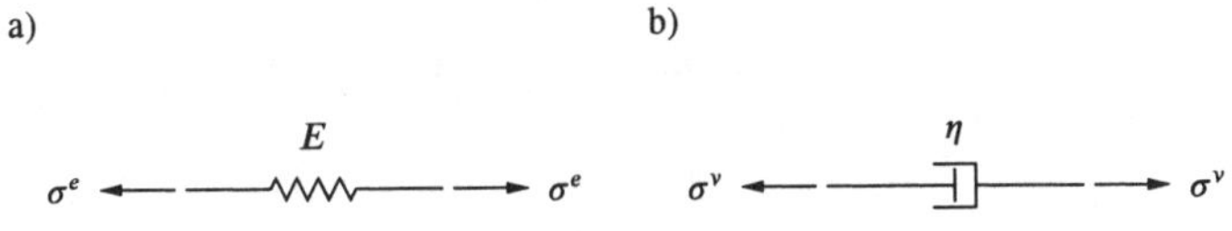

Figure 6.1: a) linear spring; b) dashpot.

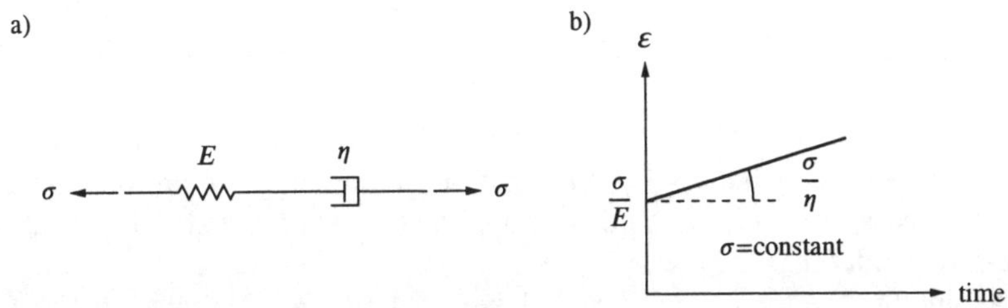

Figure 6.2: Maxwell model; a) configuration; b) response for sudden applied constant load σ.

This constitutive relation is of interest for two reasons.

First, the constitutive relation is formulated in a rate form. This means that the current strain must be obtained by an integration over the stress history; i.e. the current strain is not given as a function of the current stress, but rather as a function of the stress history. To show this explicitly, multiplication of (6.28) by dt and integration from the unloaded state up to the current state gives

$$\varepsilon = \frac{\sigma}{E} + \frac{1}{\eta}\int_0^t \sigma dt \qquad \text{where} \quad \sigma = \sigma(t) \tag{6.29}$$

where t is the current time and it is assumed that the stress is given as a function of time, i.e. $\sigma = \sigma(t)$. As an example, suppose that the stress history is a sudden application of a constant stress. In that case (6.29) gives

$$\varepsilon = \frac{\sigma}{E} + \frac{\sigma t}{\eta} \tag{6.30}$$

as illustrated in Fig. 6.2b). It appears that the current strain not only depend on the stress, but also on the load duration and, evidently, this is why one speaks of time-dependent behavior or creep.

The second point of interest is the question why (6.28) models a time-dependent behavior. This hinges on the fact that the quantity dt enters the constitutive relation in an inhomogeneous fashion. As an example of a homogeneous equation in dt, suppose that we are faced with the constitutive relation

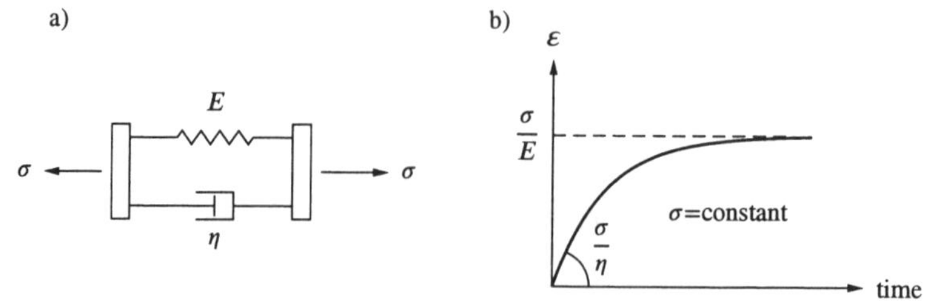

Figure 6.3: Kelvin model; a) configuration; b) response for sudden applied constant load σ.

$\dot{\varepsilon} = \dot{\sigma}/E + k\sigma\dot{\sigma}$, where k is some material parameter. Multiplication by dt gives $d\varepsilon = d\sigma/E + k\sigma d\sigma$ and it is apparent that the constitutive relation is time-independent.

As another example, the so-called *Kelvin model* is illustrated in Fig 6.3a). Based on (6.26) and

$$\sigma = \sigma^e + \sigma^v \; ; \qquad \varepsilon = \varepsilon^e = \varepsilon^v \tag{6.31}$$

the constitutive relation is easily shown to have the following form

$$\sigma = E\,\varepsilon + \eta\dot{\varepsilon} \tag{6.32}$$

The response for a sudden applied constant load is easily shown to be $\varepsilon = \frac{\sigma}{E}(1 - e^{-E/\eta\, t})$ and it is illustrated in Fig. 6.3b). We observe that if the stress is doubled then, according to (6.28) and (6.32), the strain is also doubled. Due to this linearity, the creep models considered here are examples of so-called *viscoelasticity*.

We will now see how representation theorems allow us to obtain generalized forms of (6.28) and (6.32) applicable to general stress states.

Consider first the Maxwell model (6.28). Generalizing (6.27), we have

$$\dot{\varepsilon}_{ij} = \dot{\varepsilon}^e_{ij} + \dot{\varepsilon}^v_{ij} \; ; \qquad \sigma_{ij} = \sigma^e_{ij} = \sigma^v_{ij} \tag{6.33}$$

i.e. the total strain rate is split into an elastic strain rate and a viscous strain rate. In accordance with (6.26), we take

$$\varepsilon^e_{ij} = \varepsilon^e_{ij}(\sigma_{kl}) \; ; \qquad \dot{\varepsilon}^v_{ij} = \dot{\varepsilon}^v_{ij}(\sigma_{kl})$$

Starting with the elastic part ε^e_{ij}, we choose in (6.18) $M_{ij} = \varepsilon^e_{ij}$, $N_{ij} = \sigma_{ij}$, $P_{ij} = H_\alpha = 0$ and (6.24) then provides

$$\varepsilon^e_{ij} = \alpha_1\delta_{ij} + \alpha_2\sigma_{ij} + \alpha_3\sigma_{ik}\sigma_{kj} \tag{6.34}$$

where α_1, α_2 and α_3 may depend on the stress invariants. Since a linear theory is aimed at, we must choose $\alpha_3 = 0$. Moreover, led by (4.91), we choose α_1 and α_2 to be in accordance with that expression, i.e.

$$\alpha_1 = -\frac{\nu}{2G(1+\nu)}\sigma_{kk}\,; \qquad \alpha_2 = \frac{1}{2G}\,; \qquad \alpha_3 = 0$$

It then follows that (6.34) becomes

$$\varepsilon^e_{ij} = -\frac{\nu}{2G(1+\nu)}\sigma_{kk}\delta_{ij} + \frac{1}{2G}\sigma_{ij} \tag{6.35}$$

Turning to the viscoelastic part $\dot{\varepsilon}^v_{ij} = \dot{\varepsilon}^v_{ij}(\sigma_{kl})$ similarly to (6.34), we obtain

$$\dot{\varepsilon}^v_{ij} = \beta_1\delta_{ij} + \beta_2\sigma_{ij} + \beta_3\sigma_{ik}\sigma_{kj} \tag{6.36}$$

where β_1, β_2 and β_3 may depend on the stress invariants. Again we choose $\beta_3 = 0$ in order to obtain a linear relation. Moreover, we choose

$$\beta_1 = -\frac{\sigma_{kk}}{\eta_1}\,; \qquad \beta_2 = \frac{1}{\eta_2}\,; \qquad \beta_3 = 0$$

where η_1 and η_2 are viscosity coefficients. Expression (6.36) then takes the form

$$\dot{\varepsilon}^v_{ij} = -\frac{\sigma_{kk}}{\eta_1}\delta_{ij} + \frac{\sigma_{ij}}{\eta_2} \tag{6.37}$$

It appears the $\dot{\varepsilon}^v_{kk} = (-\frac{3}{\eta_1} + \frac{1}{\eta_2})\sigma_{kk}$. It is an experimental experience that the viscous strain is often incompressible and this is obtained by choosing $\eta_1 = 3\eta_2$. In the general case, however, combination of (6.33a), (6.35) and (6.37) gives

$$\boxed{\dot{\varepsilon}_{ij} = \frac{1}{2G}(\dot{\sigma}_{ij} - \frac{\nu}{1+\nu}\dot{\sigma}_{kk}\delta_{ij}) - \frac{\sigma_{kk}}{\eta_1}\delta_{ij} + \frac{\sigma_{ij}}{\eta_2} \qquad \textit{Maxwell}} \tag{6.38}$$

The generalization of the uniaxial Maxwell model (6.28) is then given by (6.38) where we put $\eta_1 = 3\eta_2$ if, as often is the case, the viscous strain is incompressible.

Let us next obtain a generalization of the uniaxial Kelvin model (6.32). Here we start by a generalization of (6.31) which then reads

$$\sigma_{ij} = \sigma^e_{ij} + \sigma^v_{ij}\,; \qquad \varepsilon_{ij} = \varepsilon^e_{ij} = \varepsilon^v_{ij} \tag{6.39}$$

i.e. the stress tensor is split into an elastic part and a viscous part. In accordance with (6.26), we take

$$\sigma^e_{ij} = \sigma^e_{ij}(\varepsilon_{kl})\,; \qquad \sigma^v_{ij} = \sigma^v_{ij}(\dot{\varepsilon}_{kl})$$

Starting with σ^e_{ij}, we choose in (6.18) $M_{ij} = \sigma^e_{ij}$, $N_{ij} = \varepsilon_{ij}$ and $P_{ij} = H_\alpha = 0$, then the representation theorem (6.24) gives

$$\sigma^e_{ij} = \alpha_1\delta_{ij} + \alpha_2\varepsilon_{ij} + \alpha_3\varepsilon_{ik}\varepsilon_{kj} \tag{6.40}$$

where α_1, α_2 and α_3 may depend on the strain invariants. Led by (4.84), we choose

$$\alpha_1 = 2G\frac{\nu}{1-2\nu}\varepsilon_{kk}\ ; \qquad \alpha_2 = 2G\ ; \qquad \alpha_3 = 0$$

i.e. we obtain

$$\sigma^e_{ij} = 2G\frac{\nu}{1-2\nu}\varepsilon_{kk}\delta_{ij} + 2G\varepsilon_{ij} \tag{6.41}$$

Turning to the viscous part of the stress tensor $\sigma^v_{ij} = \sigma^v_{ij}(\dot{\varepsilon}_{kl})$ similarly to (6.40), we obtain

$$\sigma^v_{ij} = \beta_1\delta_{ij} + \beta_2\dot{\varepsilon}_{ij} + \beta_3\dot{\varepsilon}_{ik}\dot{\varepsilon}_{kj} \tag{6.42}$$

where β_1, β_2 and β_3 may depend on the invariants of the strain rate. To obtain a linear theory $\beta_3 = 0$ is chosen. Moreover, we choose

$$\beta_1 = -\eta_1\dot{\varepsilon}_{kk}\ ; \qquad \beta_2 = \eta_2\ ; \qquad \beta_3 = 0$$

where η_1 and η_2 are some viscosity coefficients, i.e. (6.42) reduces to

$$\sigma^v_{ij} = -\eta_1\dot{\varepsilon}_{kk}\delta_{ij} + \eta_2\dot{\varepsilon}_{ij} \tag{6.43}$$

Often it is observed experimentally that only the deviatoric part of σ^v_{ij} is of importance and this can be achieved by choosing $\eta_1 = \frac{1}{3}\eta_2$. Combination of (6.39a), (6.41) and (6.43) gives

$$\boxed{\sigma_{ij} = 2G(\varepsilon_{ij} + \frac{\nu}{1-2\nu}\varepsilon_{kk}\delta_{ij}) - \eta_1\dot{\varepsilon}_{kk}\delta_{ij} + \eta_2\dot{\varepsilon}_{ij} \qquad \textit{Kelvin}}$$

which is then the generalization of the Kelvin model (6.32). If $\eta_1 = \frac{1}{3}\eta_2$ the two last terms reduce to $\eta_2\dot{e}_{ij}$, i.e. only the deviatoric part displays a time-dependent behavior.

In the derivations above, the formulation was based on (6.33) for the Maxwell model and (6.39) for the Kelvin model and the advantages of this approach will become more evident when we focus on more complex viscoelastic models in Chapter 14. At the present stage, an alternative and perhaps more direct manner to derive the Maxwell model would be to choose $M_{ij} = \dot{\varepsilon}_{ij}$, $N_{ij} = \sigma_{ij}$, $P_{ij} = \dot{\sigma}_{ij}$ and $H_\alpha = 0$ in (6.18). Moreover, by choosing $\alpha_3 = \alpha_5 = \alpha_6 = \alpha_7 = \alpha_8 = 0$ and

$$\alpha_1 = -\frac{\sigma_{kk}}{\eta_1} - \frac{\nu}{2G(1+\nu)}\dot{\sigma}_{kk}\ ; \qquad \alpha_2 = \frac{1}{\eta_2}\ ; \qquad \alpha_3 = \frac{1}{2G}$$

in (6.24), we will retrieve the generalized Maxwell model, cf. (6.38). A similar approach could be adopted for establishment of the Kelvin model. However, while these approaches are prosperous within the present context they are not applicable within the general framework of viscoelasticity discussed in Chapter 14.

The models of Maxwell and Kelvin represent the two simplest models of viscoelasticity and more elaborate models can be obtained by various combinations of springs and dashpots, e.g. by connecting a Maxwell and a Kelvin model in series and thereby obtaining the so-called *Burgers model.* These more advanced models of viscoelasticity can also be derived using representation theorems for several variables. In Chapter 14, we will present a more detailed discussion of viscoelasticity.

6.5 Orthotropic linear elasticity

In Section 4.6, we derived the stiffness matrix $\boldsymbol{D}$ present in Hooke's law $\boldsymbol{\sigma} = \boldsymbol{D\varepsilon}$ when the material is orthotropic; the result was given by (4.55). It is recalled that orthotropy means that the material possesses three symmetry planes and the result (4.55) holds when the coordinate planes are chosen to coincide with these symmetry planes. If another coordinate system is used, the stiffness matrix $\boldsymbol{D}$ has to be transformed in accordance with (4.46).

We will now derive linear elasticity for orthotropic materials by a completely different method, which makes use of representation theorems and which makes for an entirely different approach to deal with various kinds of anisotropy.

According to (4.50), the constitutive relation takes the same form in coordinate systems that are mirror images (reflections) of each other in a symmetry plane. One may also say that, for reflections about a symmetry plane, the constitutive relation is isotropic. For these reflections, the relation between the old coordinate system $\boldsymbol{x}$ and the new coordinate system $\boldsymbol{x}'$ is given by

$$\boldsymbol{x} = \boldsymbol{A}^T \boldsymbol{x}'$$

where the transformation matrix $\boldsymbol{A}$ is orthogonal. Choosing the coordinate planes to coincide with the three symmetry planes we found in Section 4.6 that the reflections are determined by

Symmetry group S:

$$\boldsymbol{A}^T = \begin{bmatrix} 1 & 0 & 0 \\ 0 & 1 & 0 \\ 0 & 0 & -1 \end{bmatrix}; \quad \boldsymbol{A}^T = \begin{bmatrix} -1 & 0 & 0 \\ 0 & 1 & 0 \\ 0 & 0 & 1 \end{bmatrix}; \quad \boldsymbol{A}^T = \begin{bmatrix} 1 & 0 & 0 \\ 0 & -1 & 0 \\ 0 & 0 & 1 \end{bmatrix} \tag{6.44}$$

for reflections about the x_1x_2-plane, x_2x_3-plane and x_3x_1-plane respectively. For these reflections, the constitutive relation is isotropic. The reflections, i.e. the symmetry group defined by (6.44) is called the symmetry group S and it characterizes the symmetries related to orthotropy.

For orthotropy, assume that we can identify three second-order tensors $M_{ij}^{(1)}$, $M_{ij}^{(2)}$ and $M_{ij}^{(3)}$ that characterize the three symmetry planes; these $\boldsymbol{M}^{(\alpha)}$-tensors are called *structural tensors*. Since they are second-order tensors, the components $\boldsymbol{M}^{(\alpha)'}$ in a new coordinate system are related to the components $\boldsymbol{M}^{(\alpha)}$ in the old coordinate system through

$$\boldsymbol{M}^{(\alpha)'} = \boldsymbol{A}\boldsymbol{M}^{(\alpha)}\boldsymbol{A}^T$$

The structural tensors are chosen such that they are unchanged for the symmetry in question. In the present case, the symmetry group S is defined in (6.44), i.e. we have

$$\boldsymbol{M}^{(\alpha)} = \boldsymbol{A}\boldsymbol{M}^{(\alpha)}\boldsymbol{A}^T \quad \text{for} \quad \boldsymbol{A} \in S \tag{6.45}$$

Consider the strain energy W and assume that W depends on the strains and on the three structural tensors, i.e.

$$W = W(\boldsymbol{\varepsilon}, \boldsymbol{M}^{(1)}, \boldsymbol{M}^{(2)}, \boldsymbol{M}^{(3)}) \tag{6.46}$$

Assume that the strain energy is an isotropic scalar tensor function. According to (6.10) we then have

$$\begin{aligned} &W(\boldsymbol{\varepsilon}, \boldsymbol{M}^{(1)}, \boldsymbol{M}^{(2)}, \boldsymbol{M}^{(3)}) \\ &= W(\boldsymbol{A}\boldsymbol{\varepsilon}\boldsymbol{A}^T, \boldsymbol{A}\boldsymbol{M}^{(1)}\boldsymbol{A}^T, \boldsymbol{A}\boldsymbol{M}^{(2)}\boldsymbol{A}^T, \boldsymbol{A}\boldsymbol{M}^{(3)}\boldsymbol{A}^T) \end{aligned} \tag{6.47}$$

for arbitrary $\boldsymbol{A}$ matrices. However, when the transformation matrix $\boldsymbol{A}$ is chosen in accordance with the symmetry group S given by (6.44), (6.45) holds and (6.47) then reduces to

$$\begin{aligned} &W(\boldsymbol{\varepsilon}, \boldsymbol{M}^{(1)}, \boldsymbol{M}^{(2)}, \boldsymbol{M}^{(3)}) \\ &= W(\boldsymbol{A}\boldsymbol{\varepsilon}\boldsymbol{A}^T, \boldsymbol{M}^{(1)}, \boldsymbol{M}^{(2)}, \boldsymbol{M}^{(3)}) \end{aligned} \quad \text{for} \quad \boldsymbol{A} \in S \tag{6.48}$$

Suppose that the material is isotropic. Then the strain energy is given by $W = W(\boldsymbol{\varepsilon})$ and for an arbitrary transformation matrix $\boldsymbol{A}$ we have $W(\boldsymbol{\varepsilon}) = W(\boldsymbol{A}\boldsymbol{\varepsilon}\boldsymbol{A}^T)$. Therefore, expression (6.48) shows that for the reflections defined by (6.44), the relation for the strain energy is as if the material were isotropic and this is exactly what is meant by orthotropy. Therefore, with the format (6.46) and the structural tensors fulfilling property (6.45) we have achieved an intriguing new route to treat orthotropy as well as other kinds of material symmetries. Since (6.47) holds, we can use the representation theorem (6.11) on the format (6.46). Before this is performed, we have to identify the specific structural $\boldsymbol{M}^{(\alpha)}$-tensors.

Orthotropy means that three orthogonal symmetry planes exist. We could equally well speak of three orthogonal directions defined by the orthogonal unit vectors $\boldsymbol{v}^{(1)}$, $\boldsymbol{v}^{(2)}$ and $\boldsymbol{v}^{(3)}$, i.e.

$$\begin{aligned} &\boldsymbol{v}^{(1)T}\boldsymbol{v}^{(1)} = 1; \quad &&\boldsymbol{v}^{(2)T}\boldsymbol{v}^{(2)} = 1; \quad &&\boldsymbol{v}^{(3)T}\boldsymbol{v}^{(3)} = 1 \\ &\boldsymbol{v}^{(1)T}\boldsymbol{v}^{(2)} = 0; \quad &&\boldsymbol{v}^{(2)T}\boldsymbol{v}^{(3)} = 0; \quad &&\boldsymbol{v}^{(3)T}\boldsymbol{v}^{(1)} = 0 \end{aligned}$$

where the $\boldsymbol{v}^{(\alpha)}$-vectors are termed *material directions*. Suppose that we have identified a material direction, then its opposite direction also qualifies as a material direction. If the material directions $\boldsymbol{v}^{(\alpha)}$ are to describe the material, they must therefore appear in the form of equal powers of $\boldsymbol{v}^{(\alpha)}$. This is achieved by defining the structural tensors according to

$$\boxed{\boldsymbol{M}^{(1)} = \boldsymbol{v}^{(1)}\boldsymbol{v}^{(1)T}\,; \qquad \boldsymbol{M}^{(2)} = \boldsymbol{v}^{(2)}\boldsymbol{v}^{(2)T}\,; \qquad \boldsymbol{M}^{(3)} = \boldsymbol{v}^{(3)}\boldsymbol{v}^{(3)T}} \tag{6.49}$$

It follows directly that

$$\begin{aligned} &\boldsymbol{M}^{(1)}\boldsymbol{M}^{(1)} = \boldsymbol{M}^{(1)}\,; \quad &&\boldsymbol{M}^{(2)}\boldsymbol{M}^{(2)} = \boldsymbol{M}^{(2)}\,; \quad &&\boldsymbol{M}^{(3)}\boldsymbol{M}^{(3)} = \boldsymbol{M}^{(3)} \\ &\boldsymbol{M}^{(1)}\boldsymbol{M}^{(2)} = 0\,; &&\boldsymbol{M}^{(2)}\boldsymbol{M}^{(3)} = 0\,; &&\boldsymbol{M}^{(3)}\boldsymbol{M}^{(1)} = 0 \end{aligned} \tag{6.50}$$

Suppose that the coordinate system is chosen collinearly with the material directions. Then we have $\boldsymbol{v}^{(1)T} = [1\,0\,0]$, $\boldsymbol{v}^{(2)T} = [0\,1\,0]$ and $\boldsymbol{v}^{(3)T} = [0\,0\,1]$ and (6.49) becomes

$$\boldsymbol{M}^{(1)} = \begin{bmatrix} 1 & 0 & 0 \\ 0 & 0 & 0 \\ 0 & 0 & 0 \end{bmatrix}; \quad \boldsymbol{M}^{(2)} = \begin{bmatrix} 0 & 0 & 0 \\ 0 & 1 & 0 \\ 0 & 0 & 0 \end{bmatrix}; \quad \boldsymbol{M}^{(3)} = \begin{bmatrix} 0 & 0 & 0 \\ 0 & 0 & 0 \\ 0 & 0 & 1 \end{bmatrix} \tag{6.51}$$

and it is trivial to check that these structural tensors fulfill requirement (6.45).

For the specific choice of coordinate system which leads to (6.51), we obtain

$$\boldsymbol{I} = \boldsymbol{M}^{(1)} + \boldsymbol{M}^{(2)} + \boldsymbol{M}^{(3)} \tag{6.52}$$

However, since the $\boldsymbol{M}^{(\alpha)}$-quantities are second-order tensors, cf. (6.49), it follows that (6.52) holds for arbitrary coordinate systems. Relation (6.52) shows that the three structural tensors depend on each other; if two structural tensors are known, the last can be derived from (6.52). This is another way of expressing that orthotropy can be defined as the existence of three orthogonal symmetry planes or as the existence of two orthogonal symmetry planes, cf. (4.57).

Similar to (6.52), we will now derive another useful relation. For the strain tensor $\boldsymbol{\varepsilon}$ we evidently have $\boldsymbol{\varepsilon}\boldsymbol{I} = \boldsymbol{\varepsilon}$ which together with (6.52) becomes

$$\boldsymbol{\varepsilon}\boldsymbol{M}^{(1)} + \boldsymbol{\varepsilon}\boldsymbol{M}^{(2)} + \boldsymbol{\varepsilon}\boldsymbol{M}^{(3)} = \boldsymbol{\varepsilon}$$

Likewise, we have that $\boldsymbol{I}\boldsymbol{\varepsilon} = \boldsymbol{\varepsilon}$, i.e.

$$\boldsymbol{M}^{(1)}\boldsymbol{\varepsilon} + \boldsymbol{M}^{(2)}\boldsymbol{\varepsilon} + \boldsymbol{M}^{(3)}\boldsymbol{\varepsilon} = \boldsymbol{\varepsilon}$$

Adding these two expressions gives

$$\boldsymbol{\varepsilon} = \frac{1}{2}(\boldsymbol{\varepsilon}\boldsymbol{M}^{(1)} + \boldsymbol{M}^{(1)}\boldsymbol{\varepsilon}) + \frac{1}{2}(\boldsymbol{\varepsilon}\boldsymbol{M}^{(2)} + \boldsymbol{M}^{(2)}\boldsymbol{\varepsilon}) + \frac{1}{2}(\boldsymbol{\varepsilon}\boldsymbol{M}^{(3)} + \boldsymbol{M}^{(3)}\boldsymbol{\varepsilon}) \tag{6.53}$$

Here, we have considered the strain tensor $\boldsymbol{\varepsilon}$, but it is evident that a relation similar to (6.53) holds for an arbitrary symmetric second-order tensor.

With this discussion, we will now use a representation theorem to determine the format of the strain energy. Since the three structural tensors depend on each other through (6.52), we choose to work with $\boldsymbol{M}^{(1)}$ and $\boldsymbol{M}^{(2)}$, only. The strain energy is then taken as

$$W = W(\boldsymbol{\varepsilon}, \boldsymbol{M}^{(1)}, \boldsymbol{M}^{(2)})$$

Following (6.11), the strain energy is taken as function of the invariants defined by (6.12)-(6.15) where $\boldsymbol{N} = \boldsymbol{\varepsilon}$, $\boldsymbol{P} = \boldsymbol{M}^{(1)}$, $\boldsymbol{S} = \boldsymbol{M}^{(2)}$ and $H_\alpha = 0$. Due to (6.50) and (6.49), we have, for instance, $I_{1M^{(1)}} = 1$, $I_{2M^{(1)}} = 1/2$, $I_{3M^{(1)}} = 1/3$, $J_4^0 = tr(\boldsymbol{\varepsilon}^2 \boldsymbol{M}^{(1)2}) = tr(\boldsymbol{\varepsilon}^2 \boldsymbol{M}^{(1)}) = J_3^0$ and $J_9^0 = tr(\boldsymbol{M}^{(1)} \boldsymbol{M}^{(2)}) = 0$. We therefore obtain

$$\begin{aligned} W =& W(tr\boldsymbol{\varepsilon}, \frac{1}{2}tr(\boldsymbol{\varepsilon}^2), \frac{1}{3}tr(\boldsymbol{\varepsilon}^3), \\ & tr(\boldsymbol{\varepsilon}\boldsymbol{M}^{(1)}), tr(\boldsymbol{\varepsilon}^2\boldsymbol{M}^{(1)}), tr(\boldsymbol{\varepsilon}\boldsymbol{M}^{(2)}), tr(\boldsymbol{\varepsilon}^2\boldsymbol{M}^{(2)})) \end{aligned} \tag{6.54}$$

To obtain an expression, where all the three structural tensors enter in a symmetric fashion, we make use of (6.53) to obtain

$$tr\boldsymbol{\varepsilon} = tr(\boldsymbol{\varepsilon}\boldsymbol{M}^{(1)}) + tr(\boldsymbol{\varepsilon}\boldsymbol{M}^{(2)}) + tr(\boldsymbol{\varepsilon}\boldsymbol{M}^{(3)}) \tag{6.55}$$

Replacing $\boldsymbol{\varepsilon}$ with $\boldsymbol{\varepsilon}^2$, we obtain in a similar way

$$tr(\boldsymbol{\varepsilon}^2) = tr(\boldsymbol{\varepsilon}^2\boldsymbol{M}^{(1)}) + tr(\boldsymbol{\varepsilon}^2\boldsymbol{M}^{(2)}) + tr(\boldsymbol{\varepsilon}^2\boldsymbol{M}^{(3)})$$

Insertion of this expression and (6.55) into (6.54) provides

$$W = W(I_1, I_2, I_3, I_4, I_5, I_6, I_7) \tag{6.56}$$

where

$$\begin{aligned} &I_1 = tr(\boldsymbol{\varepsilon}\boldsymbol{M}^{(1)}); \quad I_2 = tr(\boldsymbol{\varepsilon}\boldsymbol{M}^{(2)}); \quad I_3 = tr(\boldsymbol{\varepsilon}\boldsymbol{M}^{(3)}) \\ &I_4 = tr(\boldsymbol{\varepsilon}^2\boldsymbol{M}^{(1)}); \quad I_5 = tr(\boldsymbol{\varepsilon}^2\boldsymbol{M}^{(2)}); \quad I_6 = tr(\boldsymbol{\varepsilon}^2\boldsymbol{M}^{(3)}); \quad I_7 = \tfrac{1}{3}tr(\boldsymbol{\varepsilon}^3) \end{aligned} \tag{6.57}$$

In hyper-elasticity where a strain energy exists, we have $\sigma_{ij} = \partial W/\partial \varepsilon_{ij}$, which with (6.56) and (6.57) lead to

$$\begin{aligned} \boldsymbol{\sigma} =& \frac{\partial W}{\partial I_1}\boldsymbol{M}^{(1)} + \frac{\partial W}{\partial I_2}\boldsymbol{M}^{(2)} + \frac{\partial W}{\partial I_3}\boldsymbol{M}^{(3)} + \frac{\partial W}{\partial I_4}(\boldsymbol{\varepsilon}\boldsymbol{M}^{(1)} + \boldsymbol{M}^{(1)}\boldsymbol{\varepsilon}) \\ &+ \frac{\partial W}{\partial I_5}(\boldsymbol{\varepsilon}\boldsymbol{M}^{(2)} + \boldsymbol{M}^{(2)}\boldsymbol{\varepsilon}) + \frac{\partial W}{\partial I_6}(\boldsymbol{\varepsilon}\boldsymbol{M}^{(3)} + \boldsymbol{M}^{(3)}\boldsymbol{\varepsilon}) + \frac{\partial W}{\partial I_7}\boldsymbol{\varepsilon}^2 \end{aligned} \tag{6.58}$$

in accordance with Boehler (1979). Since we want to determine the constitutive relation for linear elasticity, we choose

$$\begin{aligned}
\frac{\partial W}{\partial I_1} &= \alpha_1 I_1 + \beta_1 I_2 + \beta_2 I_3 \\
\frac{\partial W}{\partial I_2} &= \alpha_2 I_1 + \alpha_3 I_2 + \beta_3 I_3 \\
\frac{\partial W}{\partial I_3} &= \alpha_4 I_1 + \alpha_5 I_2 + \alpha_6 I_3 \\
\frac{\partial W}{\partial I_4} &= \alpha_7\,; \quad \frac{\partial W}{\partial I_5} = \alpha_8\,; \quad \frac{\partial W}{\partial I_6} = \alpha_9\,; \quad \frac{\partial W}{\partial I_7} = 0
\end{aligned} \tag{6.59}$$

where $\alpha_1 \cdots \alpha_9$ and β_1, β_2, β_3 are constants and where the reason for the specific notation will become evident in a moment. We have

$$\frac{\partial}{\partial I_2}(\frac{\partial W}{\partial I_1}) = \frac{\partial}{\partial I_1}(\frac{\partial W}{\partial I_2});\quad \frac{\partial}{\partial I_3}(\frac{\partial W}{\partial I_1}) = \frac{\partial}{\partial I_1}(\frac{\partial W}{\partial I_3});\quad \frac{\partial}{\partial I_3}(\frac{\partial W}{\partial I_2}) = \frac{\partial}{\partial I_2}(\frac{\partial W}{\partial I_3})$$

which leads to

$$\beta_1 = \alpha_2\,; \quad \beta_2 = \alpha_4\,; \quad \beta_3 = \alpha_5$$

Insertion of these expressions and (6.59) into (6.58) then provides the result sought

$$\boxed{\begin{aligned}
\boldsymbol{\sigma} = {} & [\alpha_1 tr(\boldsymbol{\varepsilon M}^{(1)}) + \alpha_2 tr(\boldsymbol{\varepsilon M}^{(2)}) + \alpha_4 tr(\boldsymbol{\varepsilon M}^{(3)})]\boldsymbol{M}^{(1)} \\
& + [\alpha_2 tr(\boldsymbol{\varepsilon M}^{(1)}) + \alpha_3 tr(\boldsymbol{\varepsilon M}^{(2)}) + \alpha_5 tr(\boldsymbol{\varepsilon M}^{(3)})]\boldsymbol{M}^{(2)} \\
& + [\alpha_4 tr(\boldsymbol{\varepsilon M}^{(1)}) + \alpha_5 tr(\boldsymbol{\varepsilon M}^{(2)}) + \alpha_6 tr(\boldsymbol{\varepsilon M}^{(3)})]\boldsymbol{M}^{(3)} \\
& + \alpha_7(\boldsymbol{\varepsilon M}^{(1)} + \boldsymbol{M}^{(1)}\boldsymbol{\varepsilon}) + \alpha_8(\boldsymbol{\varepsilon M}^{(2)} + \boldsymbol{M}^{(2)}\boldsymbol{\varepsilon}) \\
& + \alpha_9(\boldsymbol{\varepsilon M}^{(3)} + \boldsymbol{M}^{(3)}\boldsymbol{\varepsilon})
\end{aligned}} \tag{6.60}$$

in accordance with Boehler (1979). It appears that this expression contains nine material parameters ($\alpha_1 \cdots \alpha_9$) in agreement with (4.55). Indeed, if the coordinate system is chosen collinearly with the material directions then (6.51) holds and it is trivial from (6.60) to recover exactly the format given by (4.55).

One advantage of the format (6.60) is that it holds in all coordinate systems whereas (4.55) requires the coordinate system to be collinear with the material directions. Another advantage is that (6.60) is a tensorial expression and that, by use of the concept of structural tensors we have opened for a treatment of anisotropy in a very elegant fashion.

Smith and Rivlin (1957) introduced the concept of an *anisotropic tensor*, which has the property that it is unchanged by coordinate changes belonging to the symmetry group characterizing the material in question; an example is given by (6.45). This idea was further developed by Ericksen (1960)

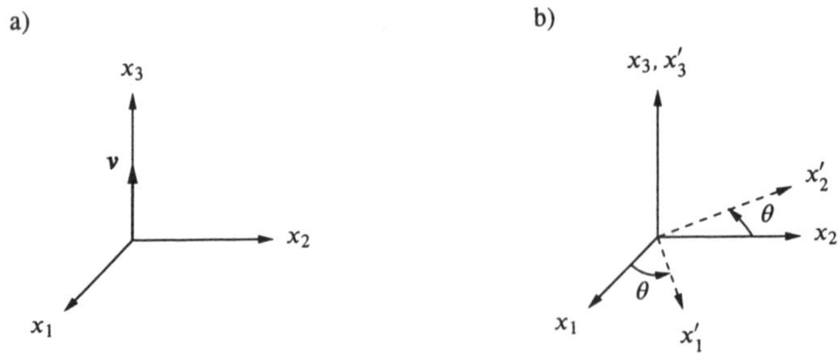

Figure 6.4: Transverse isotropy; a) material direction $\boldsymbol{v}$ in the x_3-direction, b) rotation of the coordinate system about $\boldsymbol{v}$.

for transverse isotropic fluids where the quantity $\boldsymbol{M} = \boldsymbol{v}\boldsymbol{v}^T$ was introduced and where $\boldsymbol{v}$ is the privileged direction of the medium; transversely isotropic solids were treated by Boehler and Sawczuk (1976). A systematic approach was established by Boehler (1978, 1979) and a comprehensive review is given by Boehler (1987a,b,c) where also the terminology of $\boldsymbol{M}$ being a *structural tensor* is adopted.

6.6 Transverse isotropic linear elasticity

We will now adopt the concept of structural tensors to establish linear elasticity for *transverse isotropy*. In a transversely isotropic material there exists a material direction defined by the unit vector $\boldsymbol{v}$ such that the constitutive relation is unchanged for arbitrary rotations of the coordinate system about that axis; transverse isotropy is occasionally termed as a material possessing an *axis of elastic symmetry*.

Choose the coordinate system such that the x_3-axis is in the $\boldsymbol{v}$-direction, cf. Fig. 6.4. Following (1.28) we obtain

$$\text{Symmetry group S:} \qquad \boldsymbol{A}^T = \begin{bmatrix} \cos\theta & -\sin\theta & 0 \\ \sin\theta & \cos\theta & 0 \\ 0 & 0 & 1 \end{bmatrix} \tag{6.61}$$

The structural tensor $\boldsymbol{M}$ is taken as

$$\boxed{\boldsymbol{M} = \boldsymbol{v}\boldsymbol{v}^T}$$

and for the coordinate system shown in Fig. 6.4a) we get $\boldsymbol{v}^T = [0\,0\,1]$, i.e.

$$\boldsymbol{M} = \begin{bmatrix} 0 & 0 & 0 \\ 0 & 0 & 0 \\ 0 & 0 & 1 \end{bmatrix} \tag{6.62}$$

It is trivial to check that for the coordinate rotations shown in Fig. 6.4b) we obtain

$$\boldsymbol{M} = \boldsymbol{A}\boldsymbol{M}\boldsymbol{A}^T \quad \text{for} \quad \boldsymbol{A} \in S$$

The strain energy is taken as

$$W = W(\boldsymbol{\varepsilon}, \boldsymbol{M}) \tag{6.63}$$

and assuming that W is an isotropic scalar tensor function, we have in accordance with (6.10)

$$W(\boldsymbol{\varepsilon}, \boldsymbol{M}) = W(\boldsymbol{A}\boldsymbol{\varepsilon}\boldsymbol{A}^T, \boldsymbol{A}\boldsymbol{M}\boldsymbol{A}^T)$$

In the particular, when the transformation matrix $\boldsymbol{A}$ belongs to the symmetry group S defined by (6.61), we obtain

$$W(\boldsymbol{\varepsilon}, \boldsymbol{M}) = W(\boldsymbol{A}\boldsymbol{\varepsilon}\boldsymbol{A}^T, \boldsymbol{M}) \quad \text{for} \quad \boldsymbol{A} \in S$$

i.e. the expression for the strain energy changes as if the material is isotropic; this is precisely what transverse isotropy means.

For expression (6.63), we now use the representation theorem (6.11) with $\boldsymbol{N} = \boldsymbol{\varepsilon}$, $\boldsymbol{P} = \boldsymbol{M}$ and $\boldsymbol{S} = \boldsymbol{H}_\alpha = 0$. Observing that, for instance, $\boldsymbol{M} = \boldsymbol{M}^2$ we obtain

$$W = W(I_1, I_2, I_3, I_4, I_5)$$

where

$$I_1 = tr\boldsymbol{\varepsilon}; \quad I_2 = \tfrac{1}{2}tr(\boldsymbol{\varepsilon}^2); \quad I_3 = \tfrac{1}{3}tr(\boldsymbol{\varepsilon}^3)$$
$$I_4 = tr(\boldsymbol{\varepsilon}\boldsymbol{M}); \quad I_5 = tr(\boldsymbol{\varepsilon}^2\boldsymbol{M})$$

For hyper-elasticity, we have $\sigma_{ij} = \partial W/\partial \varepsilon_{ij}$ and we then get

$$\boldsymbol{\sigma} = \frac{\partial W}{\partial I_1}\boldsymbol{I} + \frac{\partial W}{\partial I_2}\boldsymbol{\varepsilon} + \frac{\partial W}{\partial I_3}\boldsymbol{\varepsilon}^2 + \frac{\partial W}{\partial I_4}\boldsymbol{M} + \frac{\partial W}{\partial I_5}(\boldsymbol{\varepsilon}\boldsymbol{M} + \boldsymbol{M}\boldsymbol{\varepsilon}) \tag{6.64}$$

in accordance with Boehler (1987b). Since linear elasticity is sought, we choose

$$\begin{aligned}
\frac{\partial W}{\partial I_1} &= \alpha_1 I_1 + \beta I_4 \\
\frac{\partial W}{\partial I_2} &= \alpha_2 \\
\frac{\partial W}{\partial I_3} &= 0 \\
\frac{\partial W}{\partial I_4} &= \alpha_3 I_1 + \alpha_4 I_4 \\
\frac{\partial W}{\partial I_5} &= \alpha_5
\end{aligned} \tag{6.65}$$

where $\alpha_1 \cdots \alpha_5$ and β are constants and where the reason for the specific notation will become evident in a moment. We have

$$\frac{\partial}{\partial I_4}(\frac{\partial W}{\partial I_1}) = \frac{\partial}{\partial I_1}(\frac{\partial W}{\partial I_4}) \quad \Rightarrow \quad \beta = \alpha_3$$

Insertion of this result as well as (6.65) in (6.64) yields

$$\boxed{\begin{aligned}\boldsymbol{\sigma} = &[\alpha_1 tr\boldsymbol{\varepsilon} + \alpha_3 tr(\boldsymbol{\varepsilon M})]\,\boldsymbol{I} + \alpha_2\boldsymbol{\varepsilon} \\ &+ [\alpha_3 tr\boldsymbol{\varepsilon} + \alpha_4 tr(\boldsymbol{\varepsilon M})]\,\boldsymbol{M} + \alpha_5(\boldsymbol{\varepsilon M} + \boldsymbol{M\varepsilon})\end{aligned}} \tag{6.66}$$

in accordance with Boehler (1975). It appears that transverse isotropy involves five independent material parameters; moreover, (6.66) holds in an arbitrary coordinate system.

If the coordinate system is chosen such that the material direction $\boldsymbol{v}$ is in the direction of the x_3-axis, cf. Fig. 6.4, the structural tensor is given by (6.62). Then it is easily shown that (6.66) takes the format

$$\begin{bmatrix}\sigma_{11}\\ \sigma_{22}\\ \sigma_{33}\\ \sigma_{12}\\ \sigma_{13}\\ \sigma_{23}\end{bmatrix} = \begin{bmatrix} \alpha_1+\alpha_2 & \alpha_1 & \alpha_1+\alpha_3 & 0 & 0 & 0\\ \alpha_1 & \alpha_1+\alpha_2 & \alpha_1+\alpha_3 & 0 & 0 & 0\\ \alpha_1+\alpha_3 & \alpha_1+\alpha_3 & \begin{matrix}\alpha_1+\alpha_2+\alpha_4\\ +2(\alpha_3+\alpha_5)\end{matrix} & 0 & 0 & 0\\ 0 & 0 & 0 & \frac{1}{2}\alpha_2 & 0 & 0\\ 0 & 0 & 0 & 0 & \frac{1}{2}(\alpha_2+\alpha_5) & 0\\ 0 & 0 & 0 & 0 & 0 & \frac{1}{2}(\alpha_2\alpha_5)\end{bmatrix} \begin{bmatrix}\varepsilon_{11}\\ \varepsilon_{22}\\ \varepsilon_{33}\\ 2\varepsilon_{12}\\ 2\varepsilon_{13}\\ 2\varepsilon_{23}\end{bmatrix}$$

By comparison with (4.59) and Fig. 6.4, it appears that the material is isotropic in the x_1x_2-plane.

7 HYPO - ELASTICITY

Let us postulate a constitutive relation between the time-derivative of the stress tensor and the time-derivative of the strain tensor in the form

$$\dot{\sigma}_{ij} = f_{ij}(\sigma_{kl}, \dot{\varepsilon}_{mn}) \tag{7.1}$$

Here a dot denotes the time-derivative, i.e. $\dot{\sigma}_{ij} = d\sigma_{ij}/dt$, where t is the time. From (7.1) appears that the time-derivative of the stress tensor is assumed to depend not only on the time-derivative of the strain tensor, but also of the current stress tensor. Thus, (7.1) provides a very general constitutive relation.

Since (7.1) is given in rate form, an integration along the load path is necessary in order to obtain the current stresses or strains. The constitutive relation (7.1) therefore expresses a material, where the response depends on the history. In the remaining part of the book, we will be concerned with such kinds of constitutive relations.

Expression (7.1) is clearly of the form investigated in (6.18) and replacing $\boldsymbol{M}$ by $\dot{\sigma}_{ij}$, $\boldsymbol{N}$ by σ_{ij} and $\boldsymbol{P}$ by $\dot{\varepsilon}_{ij}$ and using the result of (6.24), it appears that (7.1) is equivalent with

$$\begin{aligned}\dot{\sigma}_{ij} = \; & \alpha_1\delta_{ij} + \alpha_2\sigma_{ij} + \alpha_3\sigma_{ik}\sigma_{kj} + \alpha_4\dot{\varepsilon}_{ij} + \alpha_5\dot{\varepsilon}_{ik}\dot{\varepsilon}_{kj} \\ & + \alpha_6(\sigma_{ik}\dot{\varepsilon}_{kj} + \dot{\varepsilon}_{ik}\sigma_{kj}) + \alpha_7(\sigma_{ik}\sigma_{kl}\dot{\varepsilon}_{lj} + \dot{\varepsilon}_{ik}\sigma_{kl}\sigma_{lj}) \\ & + \alpha_8(\sigma_{ik}\dot{\varepsilon}_{kl}\dot{\varepsilon}_{lj} + \dot{\varepsilon}_{ik}\dot{\varepsilon}_{kl}\sigma_{lj})\end{aligned} \tag{7.2}$$

where $\alpha_1 \ldots \alpha_8$ are scalar functions of the ten invariants defined in (6.12) - (6.15), i.e.

$$\begin{aligned}\alpha_i = \alpha_i[& tr\sigma,\ tr(\sigma^2),\ tr(\sigma^3),\ tr\dot{\varepsilon},\ tr(\dot{\varepsilon}^2),\ tr(\dot{\varepsilon}^3),\ tr(\sigma\dot{\varepsilon}), \\ & tr(\sigma\dot{\varepsilon}^2),\ tr(\sigma^2\dot{\varepsilon}),\ tr(\sigma^2\dot{\varepsilon}^2)]\end{aligned} \tag{7.3}$$

7.1 Time-independent response

We now require that (7.1) corresponds to a time-independent constitutive relation, i.e. the same response is obtained irrespective of the loading rate. This can only be accomplished if (7.2) is a homogeneous function of time, i.e. the only way time can appear in (7.2) is through the factor $1/dt$ present in all terms of (7.2), cf. the discussion following (6.30). This, in turn, implies that (7.1) can be written as

$$\boxed{\dot{\sigma}_{ij} = D_{ijkl}(\sigma_{mn})\dot{\varepsilon}_{kl} \quad \text{or} \quad d\sigma_{ij} = D_{ijkl}(\sigma_{mn})d\varepsilon_{kl}} \tag{7.4}$$

where the latter form clearly shows that time does not influence the response of the material.

To achieve the form given by (7.4), each term on the right-hand side of (7.2) must only contain the time-derivative of the strain tensor to the power of one. We immediately conclude that

$$\alpha_5 = \alpha_8 = 0 \tag{7.5}$$

Let us next consider the remaining terms in (7.2). Starting with the term containing α_1 that must contain the time-derivative of the strain tensor to the power of one, we conclude from (7.3) that the most general form is obtained by writing α_1 as

$$\begin{aligned}\alpha_1 = \; & \beta_1[tr\boldsymbol{\sigma},\, tr(\boldsymbol{\sigma}^2),\, tr(\boldsymbol{\sigma}^3)]tr\dot{\boldsymbol{\varepsilon}} + \beta_4[tr\boldsymbol{\sigma},\, tr(\boldsymbol{\sigma}^2),\, tr(\boldsymbol{\sigma}^3)]tr(\boldsymbol{\sigma}\dot{\boldsymbol{\varepsilon}}) \\ & + \beta_8[tr\boldsymbol{\sigma},\, tr(\boldsymbol{\sigma}^2),\, tr(\boldsymbol{\sigma}^3)]tr(\boldsymbol{\sigma}^2\dot{\boldsymbol{\varepsilon}})\end{aligned}$$

where the reason for the notation β_1, β_4 and β_8 will appear later on. As $tr\boldsymbol{\sigma}$, $tr\boldsymbol{\sigma}^2$ and $tr\boldsymbol{\sigma}^3$ are equivalent with the three invariants I_1, I_2 and I_3 of the stress tensor, we can write the above relation for α_1 as

$$\alpha_1 = \beta_1\dot{\varepsilon}_{kk} + \beta_4\sigma_{mn}\dot{\varepsilon}_{nm} + \beta_8\sigma_{lm}\sigma_{mn}\dot{\varepsilon}_{nl} \tag{7.6}$$

where β_1, β_4 and β_8 are functions of the three invariants of the stress tensor. Likewise, for the terms in (7.2), that involve α_2 and α_3, we conclude that

$$\alpha_2 = \beta_3\dot{\varepsilon}_{kk} + \beta_7\sigma_{mn}\dot{\varepsilon}_{nm} + \beta_{11}\sigma_{lm}\sigma_{mn}\dot{\varepsilon}_{nl} \tag{7.7}$$

and

$$\alpha_3 = \beta_6\dot{\varepsilon}_{kk} + \beta_{10}\sigma_{mn}\dot{\varepsilon}_{nm} + \beta_{12}\sigma_{lm}\sigma_{mn}\dot{\varepsilon}_{nl} \tag{7.8}$$

where the $\beta-$parameters may depend on the stress invariants. The terms in (7.2) that involve α_4, α_6 and α_7 already contain the time-derivative of the strain tensor to the power of one. We shift the notation and write

$$\alpha_4 = \beta_2 \qquad \alpha_6 = \beta_5 \qquad \alpha_7 = \beta_9 \tag{7.9}$$

where also β_2, β_5 and β_9 may only depend on the three invariants of the stress tensor.

With (7.5)-(7.9) in (7.2), we then achieve the following most general time-independent incremental constitutive relation

$$\boxed{\begin{aligned}\dot{\sigma}_{ij} &= \beta_1 \dot{\varepsilon}_{kk}\delta_{ij} + \beta_2\dot{\varepsilon}_{ij} + \beta_3\dot{\varepsilon}_{kk}\sigma_{ij} + \beta_4\sigma_{mn}\dot{\varepsilon}_{mn}\delta_{ij} \\ &+\beta_5(\sigma_{ik}\dot{\varepsilon}_{kj} + \dot{\varepsilon}_{ik}\sigma_{kj}) + \beta_6\dot{\varepsilon}_{mm}\sigma_{ik}\sigma_{kj} + \beta_7\sigma_{mn}\dot{\varepsilon}_{nm}\sigma_{ij} \\ &+\beta_8\sigma_{lm}\sigma_{mn}\dot{\varepsilon}_{nl}\delta_{ij} + \beta_9(\sigma_{ik}\sigma_{kl}\dot{\varepsilon}_{lj} + \dot{\varepsilon}_{ik}\sigma_{kl}\sigma_{lj}) \\ &+\beta_{10}\sigma_{mn}\dot{\varepsilon}_{nm}\sigma_{ik}\sigma_{kj} + \beta_{11}\sigma_{lm}\sigma_{mn}\dot{\varepsilon}_{nl}\sigma_{ij} \\ &+\beta_{12}\sigma_{lm}\sigma_{mn}\dot{\varepsilon}_{nl}\sigma_{ik}\sigma_{kj}\end{aligned}} \tag{7.10}$$

where $\beta_1 \ldots \beta_{12}$ may depend on the three stress invariants. In a symbolic form, this expression may be written as (7.4).

The constitutive relation given by (7.4) or (7.10) is the most general so-called *hypo-elastic model.* As this relation is given in an incremental fashion, the corresponding material response will, in general, be *path-dependent*, i.e. the stresses corresponding to a specific strain state will generally depend on the strain path which led to that strain state. Hypo-elasticity was introduced by Truesdell (1955a,b).

It is evident that a material obeying the constitutive relation (7.4) will behave in a nonlinear fashion. However, (7.4) describes a material that is incrementally linear, since a doubling of the strain increment $d\varepsilon_{kl}$ will double the stress increment $d\sigma_{ij}$.

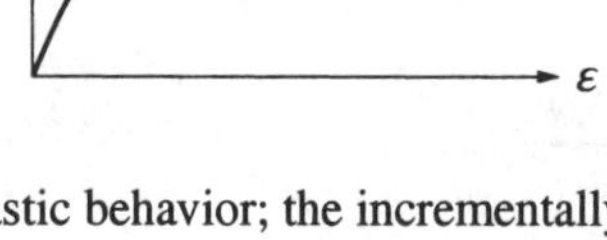

Figure 7.1: Hypo-elastic behavior; the incrementally reversible behavior is seen.

In (7.4), $D_{ijkl}(\sigma_{mn})$ is considered to be a continuous function of the stresses and this is a characteristic feature of hypo-elastic behavior. The word 'elasticity' is used (Prager (1961), pp. 142-145), since just like infinitely small strain changes $d\varepsilon_{kl}$ result in infinitely small stress changes $d\sigma_{ij}$, so do infinitely small strain changes $-d\varepsilon_{kl}$ result in infinitely small stress changes $-d\sigma_{ij}$. Thus, for infinitely small changes, the behavior is reversible, as it is for ordinary elastic behavior. In general however, the response for finite changes is path dependent, since an integration is required to obtain the total stresses from a given strain history, thus explaining the term 'hypo' meaning 'to a lesser degree', cf. Fig. 7.1. Since the response is path-dependent, it follows that the behavior in unloading will be different from that of loading (apart from incremental loading and unloading).

Later, in Chapter 10, we shall see that for general elasto-plastic behavior, an equation similar to (7.4) applies, where D_{ijkl} now corresponds to the so-called elasto-plastic stiffness tensor. However, in this case D_{ijkl} is not a continuous function of the stresses, as it changes in an abrupt manner depending on whether plastic loading or elastic unloading occurs, hereby causing the irreversibility also for infinitely small changes that is intimately connected with plastic behavior.

If $\beta_3 = \beta_4 = \ldots = \beta_{12} = 0$, then (7.10) reduces to

$$\dot{\sigma}_{ij} = \beta_1 \dot{\varepsilon}_{kk} \delta_{ij} + \beta_2 \dot{\varepsilon}_{ij}$$

which is called a *hypo-elastic material of grade zero*, as the stress tensor only enters to a power of zero on the right-hand side. Letting $\beta_1 = \lambda$ and $\beta_2 = 2\mu$, where λ and μ are the Lamé constants, we recover the usual Hooke's law in an incremental form, cf. (4.81). Likewise, a material is said to be *hypo-elastic of grade one*, if only powers of σ_{ij} up to one are allowed. A comparison with (7.10) shows that a hypo-elastic material of grade one includes the first five terms, whereas $\beta_6 = \ldots = \beta_{12} = 0$ must apply. It appears that the most general form of (7.10) corresponds to a hypo-elastic material of grade four.

The general conditions for which (7.10) represents elastic behavior, i.e. when the relation can be integrated to allow the total current stresses to be related to the total current strains independent of the load history, have been investigated by Bernstein (1960) and Coon and Evans (1971). In this case $d\sigma_{ij} = D_{ijkl}(\sigma_{st}) d\varepsilon_{kl}$ must be a perfect differential, and the integrability conditions therefore become (similar to (4.3a) and (4.9))

$$\frac{\partial D_{ijkl}}{\partial \varepsilon_{mn}} = \frac{\partial D_{ijmn}}{\partial \varepsilon_{kl}}$$

or

$$\frac{\partial D_{ijkl}}{\partial \sigma_{pq}} \frac{\partial \sigma_{pq}}{\partial \varepsilon_{mn}} = \frac{\partial D_{ijmn}}{\partial \sigma_{pq}} \frac{\partial \sigma_{pq}}{\partial \varepsilon_{kl}}$$

where D_{ijkl} can be established from (7.10). The consequences of this restriction are discussed in detail in the references mentioned above.

Finally, let us evaluate the concept of *failure conditions*. For uniaxial loading, we then assume that the stress-strain curve exhibits a peak stress (i.e. a 'failure' stress) as shown in Fig. 7.2. At the peak stress, it is possible to change the strain without changing the stress, i.e. $d\varepsilon \neq 0$ and $d\sigma = 0$.

Following Coon and Evans (1972), let us generalize this observation to the constitutive relation (7.4). Using the technique discussed in Section 4.4, the matrix formulation of (7.4) becomes

$$d\boldsymbol{\sigma} = \boldsymbol{D} d\boldsymbol{\varepsilon}$$

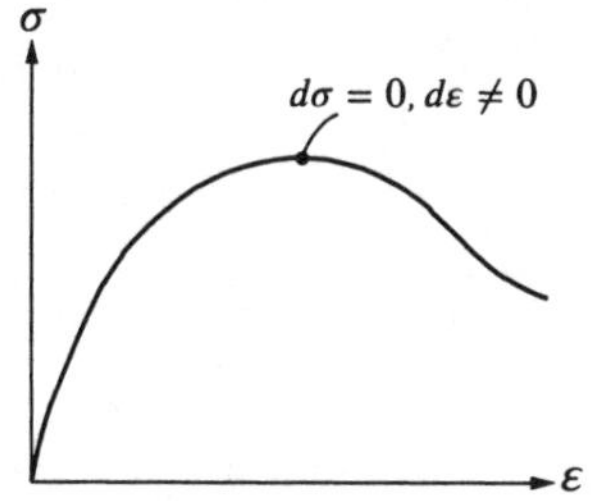

Figure 7.2: Uniaxial loading; condition at peak stress.

If this constitutive relation implies that peak stresses - i.e. 'failure' stresses - exist, then at peak we must have that $d\boldsymbol{\varepsilon} \neq \mathbf{0}$ implies $d\boldsymbol{\sigma} = \mathbf{0}$. The condition for failure therefore becomes that the homogeneous equation system

$$\boldsymbol{D}d\boldsymbol{\varepsilon} = \mathbf{0}$$

must possess a non-trivial solution $d\boldsymbol{\varepsilon} \neq \mathbf{0}$. This is only possible if

$$\det \boldsymbol{D} = 0 \tag{7.11}$$

i.e. the constitutive relation (7.4) inherently contains a condition for failure, if such peak stresses exist, and this condition is given by (7.11). The relevance of this approach for establishing the failure conditions for concrete using hypo-elasticity was discussed by Coon and Evans (1972).

Specific hypo-elastic models suggested for concrete and soil are evaluated by Chen and Saleeb (1982) and Desai and Siriwardane (1984).

8 FAILURE AND INITIAL YIELD CRITERIA

In the previous chapters, we discussed constitutive relations with an increasing degree of complexity. We started out with linear and nonlinear elasticity where a one-to-one relation exists between the current stresses and current strains. We then moved to materials where the constitutive relation is given in an incremental fashion, i.e. the current stress or strain state can only be obtained by an integration of the load history. This means that the response becomes history dependent.

For a time-independent material response, the most important class of incremental constitutive relations that are able to distinguish between loading and unloading in a realistic manner is provided by the *plasticity theory*; in the following chapters, we will discuss various aspects of this theory.

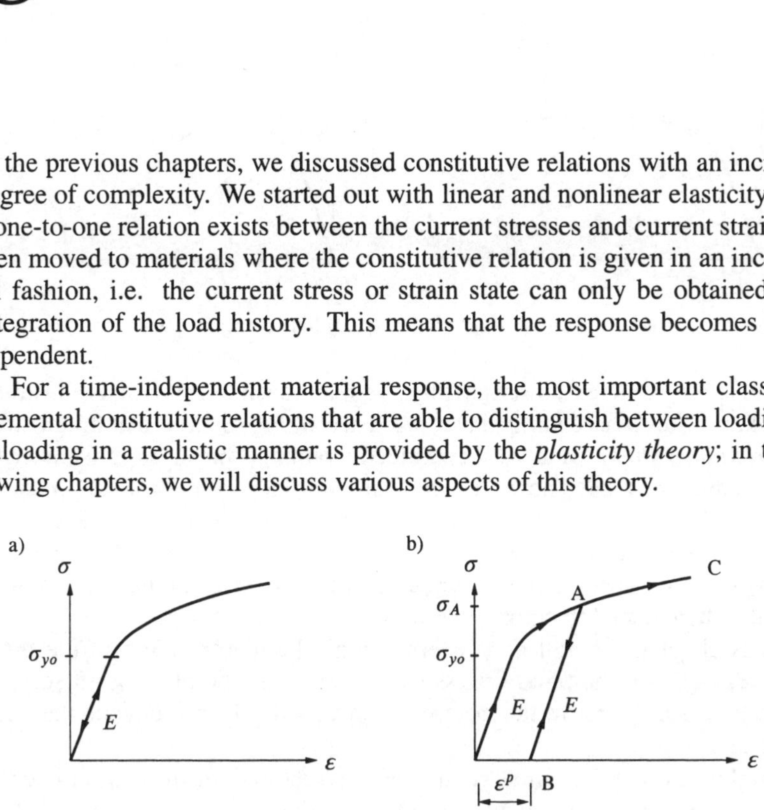

Figure 8.1: a) Loading below the initial yield stress σ_{yo}; b) loading above the initial yield stress.

As a beginning to the plasticity theory, in this chapter we will deal with criteria which tell us whether plastic deformations - i.e. yielding of the material - or failure occurs. For this purpose consider the uniaxial stress-strain curve shown in Fig. 8.1. If the stress is below the *initial yield stress* σ_{yo}, the material is assumed to behave linear elastic with a stiffness given by Young's modulus E, cf. Fig. 8.1a). If the material is loaded to the stress σ_A, cf. Fig. 8.1b), *yielding* occurs and at unloading to point B we are left with the *plastic strain* ε^p. The unloading from A to B is assumed to occur elastically with the stiffness E. Reloading from point B first follows the linear path BA and at point A yielding is again activated and the path AC is then traced as if the unloading at point A had never occurred. It appears that for unloading from point A and subsequent reloading, the stress σ_A that is needed to activate further plastic deformations has increased when compared with the initial yield stress σ_{yo}. We therefore have a *hardening* effect.

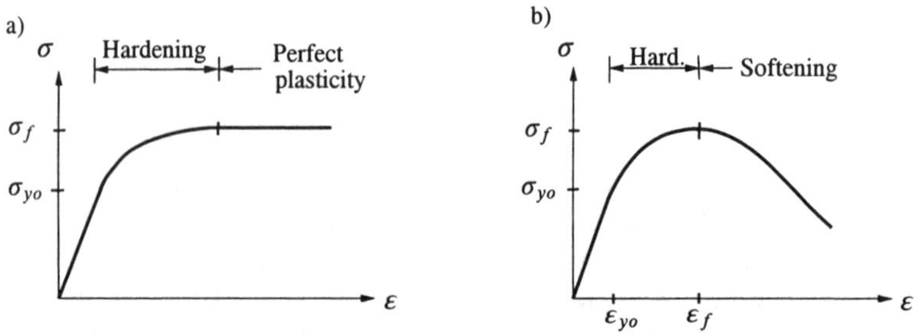

Figure 8.2: a) Hardening and perfect plasticity; b) hardening and softening plasticity.

If the strain is increased sufficiently, we may reach the situations illustrated in Fig. 8.2. After a hardening phase, we reach in Fig. 8.2a) a maximum stress σ_f - the *failure stress* - and with continued straining, the stress remains at the value σ_f; that is we have now reached a situation of *perfect* or *ideal plasticity*. In Fig. 8.2b), on the other hand, after having obtained the failure stress σ_f, the stress decreases with continued deformation. This so-called *softening* behavior is typical for materials like concrete, soil and rocks and other cementitious materials when loaded in compression.

In this chapter, we will be concerned with the identification of the initial yield stress σ_{yo} and the failure stress σ_f. Later on, in the plasticity theory, we will establish rules for how the material behaves when loaded beyond the initial yield stress.

It is obvious that the initial yield stress and the failure stress are important engineering quantities. Whereas their identification is trivial for uniaxial stress states, this is not the case for general stress states. In general, the stress state is defined by the stress tensor which comprises six independent stress components; the task is therefore to determine critical combinations of these components that result in initial yielding or failure of the material. We will see that for isotropic materials, it is possible to obtain a large amount of information on the general form of such criteria without knowing the specific material.

It is worthwhile to scrutinize the term 'failure stress' a little further. As apparent from Fig. 8.2, this term is slightly misleading since the material does not necessarily lose its load-carrying capacity when the 'failure' stress σ_f is

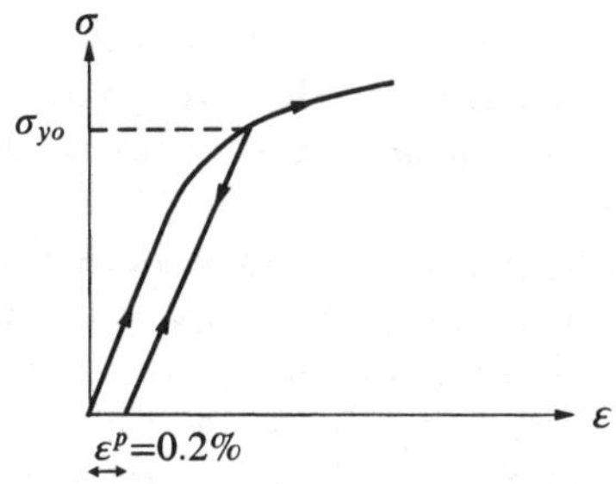

Figure 8.3: Definition of σ_{yo} based on 0.2% off-set strain

reached. Rather, the stress σ_f refers to that peak stress a homogeneously loaded specimen can carry and therefore the term *ultimate stress* seems more appropriate. However, by tradition we will use the word failure stress.

While the identification of the failure stress is unambiguous, this is not the case for the initial yield stress σ_{yo}. The reason is that most materials exhibit a smooth transition from the elastic region to the elastic-plastic region with no distinct point where yielding is initiated; examples are shown in Fig. 8.2. The identification of the initial yield stress σ_{yo} therefore becomes a matter of convention. In handbooks, the initial yield stress σ_{yo} for metals and steel is most often identified as the so-called $\sigma_{0.2}$-*stress*, i.e. the stress at which the remaining plastic strain after unloading equals 0.2%. This definition is illustrated in Fig. 8.3 and the remaining plastic strain is often called the *off-set strain*. Since a plastic strain of 0.2% is not insignificant, this definition of σ_{yo} is rather crude and in most scientific experimental investigations a much smaller off-set strain is used.

For general stress states, the conditions for failure or initial yielding are called *failure* or *initial yield criteria* respectively. Since they can be treated in a unified manner, we will often simply use the word criterion. We will see that stress invariants play an extremely important role in failure and yield criteria and we will therefore make extensive use of the results obtained in Chapter 3.

In general, the material is anisotropic, i.e. for a given loading the orientation of the material influences its response. We seek a criterion, i.e. a function, which takes the value of zero when the conditions for initial yielding or failure are fulfilled.

Consider a specimen of a homogeneous material loaded by a homogeneous stress state. Considering proportional loading, we will assume that the yield or failure criterion is independent of the loading rate. Under these conditions, the initial yield or failure criterion can only depend on the stress tensor σ_{ij}, i.e.

$$\boxed{F(\sigma_{ij}) = 0} \tag{8.1}$$

When this condition is fulfilled, initial yielding and failure occur in the material. We know from Sections 6.5 and 6.6 that if anisotropic materials are considered, say orthotropy, then the yield or failure criterion, in addition to the stresses σ_{ij}, also depends on some structural tensors. However, at the present stage it is not of importance to include such structural tensors among the variables entering the yield or failure criterion.

By convention the function F is normalized in such a manner that if the stress state is below the yield or failure limit then $F(\sigma_{ij}) < 0$. This implies that if the stress state is above the yield or failure limit then $F(\sigma_{ij}) > 0$. The conditions that $F(\sigma_{ij}) < 0$, $F(\sigma_{ij}) = 0$ and $F(\sigma_{ij}) > 0$ hold when the stress state is below, equal to and above the yield or failure limit respectively, were established in an arbitrary x_i-coordinate system. To make sense they must therefore also hold when we adopt another x'_i-coordinate system. This implies that the value of F is an invariant, i.e.

The yield or failure criterion F is an invariant (8.2)

For anisotropic materials, various expressions of the criterion (8.1) are discussed in Section 8.13. However, we will first consider the important case of isotropic materials.

The stress tensor σ_{ij} can also be expressed by the principal stresses σ_1, σ_2 and σ_3 and the corresponding principal stress directions. As an isotropic material has no directional properties, it is expected that we can write $F = F(\sigma_1, \sigma_2, \sigma_3)$. As the principal stresses are given uniquely in terms of the three stress invariants, we may equally well write F as

$$F(I_1, I_2, I_3) = 0 \tag{8.3}$$

where the 'generic' stress invariants I_1, I_2 and I_3 are defined by

$$I_1 = \sigma_{ii} \; ; \quad I_2 = \frac{1}{2}\sigma_{ij}\sigma_{ji} \; ; \quad I_3 = \frac{1}{3}\sigma_{ij}\sigma_{jk}\sigma_{ki}$$

cf. (3.14). We will now show this result in a more formal manner.

In the x_i-coordinate system, we have $F(\sigma_{ij})$ and if we instead adopt the x'_i-coordinate system, we have $F^*(\sigma'_{ij})$. Since the criterion is an invariant, we conclude that

$$F(\sigma_{ij}) = F^*(\sigma'_{ij}) \qquad \text{coordinate invariance} \tag{8.4}$$

This condition is just a result of F being an invariant, i.e. a zero-order tensor. The function F is a response function and in accordance with the discussion following (5.7), the response function is denoted F in the x_i-coordinate system and F^* in the x'_i-coordinate system.

Isotropy means that the response function is the same in all coordinate systems, cf. (5.13). This implies

$$F(\sigma'_{ij}) = F^*(\sigma'_{ij}) \qquad \text{isotropy} \tag{8.5}$$

Insertion of (8.5) in (8.4) and noting that $\sigma'_{ij} = A_{ik}\sigma_{kl}A_{jl}$, cf. (3.8), we obtain

$$F(\sigma_{ij}) = F(A_{ik}\sigma_{kl}A_{jl}) \qquad \text{coordinate invariance + isotropy} \tag{8.6}$$

and this result shows that F is an isotropic scalar tensor function. Referring to (6.10) and (6.11), the result (8.3) then follows directly.

As already touched upon, instead of (8.3) we may write alternatively

$$\boxed{F(\sigma_1, \sigma_2, \sigma_3) = 0} \tag{8.7}$$

where σ_1, σ_2 and σ_3 are the principal stresses. The principal stresses are invariants, but it is observed that σ_1 is the principal stress in the first principal direction, σ_2 is the principal stress in the second principal direction and σ_3 is the principal stress in the third principal direction. However, since the yield or failure criterion (8.7) only depends on (true) invariants having no directional preferences, the ordering of σ_1, σ_2 and σ_3 i (8.7) is indifferent. That is, (8.7) should be interpreted as a function of principal stresses without any reference to specific principal axes. To emphasize this, we may write (8.7) as

$$F(\sigma_1, \sigma_2, \sigma_3) = F(\sigma_2, \sigma_1, \sigma_3) = F(\sigma_1, \sigma_3, \sigma_2)$$
$$= F(\sigma_3, \sigma_1, \sigma_2) = F(\sigma_3, \sigma_2, \sigma_1) = F(\sigma_2, \sigma_3, \sigma_1) = 0$$

However, if we adopt some convention, for instance that $\sigma_1 \geq \sigma_2 \geq \sigma_3$, the formulation (8.7) suffices.

Determination of the principal stresses requires the solution of an eigenvalue problem and this obstacle is avoided by expressing the criterion in terms of the stress invariants. However, instead of the invariants used in the format (8.3), it turns out to be more convenient to use another set of invariants and write the yield or failure criterion as $F(I_1, J_2, J_3) = 0$ or, even more advantageously, as

$$\boxed{F(I_1, J_2, \cos 3\theta) = 0} \tag{8.8}$$

where

$$J_2 = \frac{1}{2}s_{ij}s_{ji}\ ; \quad \cos 3\theta = \frac{3\sqrt{3}}{2}\frac{J_3}{J_2^{3/2}}\ ; \quad J_3 = \frac{1}{3}s_{ij}s_{jk}s_{ki} \tag{8.9}$$

where s_{ij} is the deviatoric stress tensor, cf. (3.15). The invariant $cos3\theta$ has not been discussed previously, but in a moment we will provide such a discussion. Generally speaking, old criteria use the format (8.7), whereas modern criteria take advantage of the formulation provided by (8.8). One advantage of the format (8.8) is that it separates the influence of the hydrostatic stress I_1 from the influence of the deviatoric stresses expressed by J_2 and $\cos 3\theta$. Moreover, the invariants I_1, J_2 and $\cos 3\theta$ may be given an illuminating geometrical interpretation as shown in a moment.

Identification of failure and initial yield criteria is one of the classical topics in constitutive mechanics and the literature on this subject is therefore very extensive. The intention of this chapter is not to provide an overview of all the different criteria proposed, but rather to present some main contributions. In addition to being of practical interest, each of the criteria presented therefore involves features not considered by the other criteria dealt with. Thus, the exposition in this chapter aims at a presentation of the mainstream within failure and initial criteria. For further general information, the reader is referred to Chen (1982), Chen and Saleeb (1982), Chen and Han (1988), Desai and Siriwardane (1984), Jeager and Cook (1976), Hill (1950), Mendelsohn (1968), Nadai (1950) and Paul (1968); a very comprehensive review is provided by Yu (2002).

8.1 Haigh-Westergaard coordinate system - Geometrical interpretation of stress invariants

It is evident that (8.7) may be interpreted as a surface in the Cartesian coordinate system with axes σ_1, σ_2 and σ_3 - the so-called *Haigh-Westergaard coordinate system*, cf. Haigh (1920) and Westergaard (1920). Moreover, with this interpretation it will turn out that it is possible to identify certain geometrical quantities related to the stress invariants I_1, J_2 and $\cos 3\theta$.

For this purpose, consider an arbitrary point P with coordinates $(\sigma_1, \sigma_2, \sigma_3)$ in the Haigh-Westergaard coordinate system, cf. Fig. 8.4a). In this stress space, we may identify the unit vector n_i along the space diagonal. This vector is given by

$$n_i = \frac{1}{\sqrt{3}}(1, 1, 1) \tag{8.10}$$

If the stress point is located along the space diagonal, all principal stresses are equal and the space diagonal is therefore called the *hydrostatic axis*.

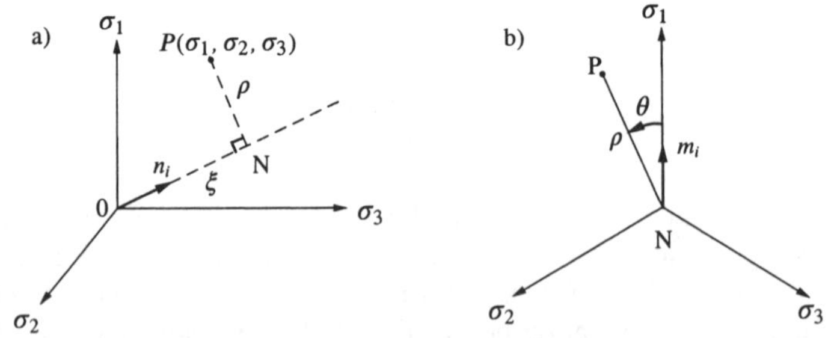

Figure 8.4: a) Haigh-Westergaard coordinate system; b) deviatoric plane perpendicular to the hydrostatic axis and containing line NP.

For any stress point P we may locate a plane which is perpendicular to the hydrostatic axis and which contains the point P. This plane is called the *deviatoric plane* and it contains the line PN in Fig. 8.4a). When viewed in the direction of the hydrostatic axis, the projections of the σ_1-, σ_2- and σ_3-axes on the deviatoric plane are shown in Fig. 8.4b). The particular deviatoric plane that contains the origin O of the stress space is occasionally called the *π-plane*.

The position of the arbitrary stress point P is given by the Cartesian coordinates $(\sigma_1, \sigma_2, \sigma_3)$. However, instead of these coordinates, we may equally well use the coordinates (ξ, ρ, θ) illustrated in Fig. 8.4. The coordinate ξ is then the distance from the origin O to the point N and ξ is a positive or negative quantity, if the vector $\overline{ON}$ has the same or opposite direction as the unit vector n_i respectively. The coordinate ρ denotes the distance $|\overline{NP}|$ in the deviatoric plane of the point P to the hydrostatic axis. Finally, θ is the angle in the deviatoric plane between the projection of the σ_1-axis on the deviatoric plane and the line NP. It appears that ρ and θ are the polar coordinates of point P in the deviatoric plane.

With this qualitative description, we will now derive explicit expressions for the coordinates (ξ, ρ, θ). From Fig. 8.4a) and (8.10), the coordinate ξ is given by

$$\xi = \boldsymbol{n}^T\overline{OP} = \frac{1}{\sqrt{3}}\begin{bmatrix} 1 & 1 & 1 \end{bmatrix}\begin{bmatrix} \sigma_1 \\ \sigma_2 \\ \sigma_3 \end{bmatrix}$$

i.e.

$$\boxed{\xi = \frac{I_1}{\sqrt{3}}} \tag{8.11}$$

It follows that the vector $\overline{ON} = \xi\mathbf{n}$ is given by

$$\overline{ON} = \frac{I_1}{3}\begin{bmatrix} 1 \\ 1 \\ 1 \end{bmatrix}$$

The vector $\overline{NP}$ in the deviatoric plane then becomes

$$\overline{NP} = \overline{OP} - \overline{ON} = \begin{bmatrix} \sigma_1 \\ \sigma_2 \\ \sigma_3 \end{bmatrix} - \frac{I_1}{3}\begin{bmatrix} 1 \\ 1 \\ 1 \end{bmatrix} = \begin{bmatrix} s_1 \\ s_2 \\ s_3 \end{bmatrix} \tag{8.12}$$

where s_1, s_2 and s_3 are the principal deviatoric stresses. We recall that $\overline{NP}$ is located in the deviatoric plane and since $\overline{NP}$ is given entirely in terms of the deviatoric stresses, this suggests the notation of the deviatoric plane. The length $\rho = \left|\overline{NP}\right|$ of the vector $\overline{NP}$ is given by $\rho^2 = \overline{NP}^T\overline{NP} = s_1^2 + s_2^2 + s_3^2$ i.e.

$$\boxed{\rho = \sqrt{2J_2}} \tag{8.13}$$

where the invariant J_2 is defined by (8.9). It should be observed that, by definition, both ρ and J_2 are non-negative quantities.

With (8.11) and (8.13), we have seen that the coordinates ξ and ρ can be expressed in terms of stress invariants. To obtain an expression for the angle θ, cf. Fig. 8.4b), some further manipulations are necessary.

Referring to Fig. 8.4b), the unit vector m_i located in the deviatoric plane and directed along the projection of the σ_1−axis on the deviatoric plane must have the form

$$m_i = (a, -b, -b)$$

where $a > 0$ and $b > 0$. Since m_i is orthogonal to the hydrostatic axis we have $m_i n_i = 0$, cf. Fig. 8.4, and this leads to $b = a/2$. Moreover, as m_i is a unit vector, we conclude that

$$m_i = \frac{1}{\sqrt{6}}(2, -1, -1) \tag{8.14}$$

The angle θ is measured from the m_i-vector in the counter-clockwise direction towards the vector $\overline{NP}$, i.e. we obtain with (8.12) and (8.14)

$$\rho \cos\theta = \boldsymbol{m}^T \overline{\boldsymbol{NP}} = \frac{1}{\sqrt{6}} \begin{bmatrix} 2 & -1 & -1 \end{bmatrix} \begin{bmatrix} s_1 \\ s_2 \\ s_3 \end{bmatrix} = \frac{2s_1 - s_2 - s_3}{\sqrt{6}}$$

With ρ given by (8.13) and since $s_2 + s_3 = -s_1$ we obtain

$$\cos\theta = \frac{\sqrt{3}}{2}\frac{s_1}{\sqrt{J_2}} \tag{8.15}$$

Use of the trigonometric identity $\cos 3\theta = 4\cos^3\theta - 3\cos\theta$ then results in

$$\cos 3\theta = \frac{3\sqrt{3}}{2J_2^{3/2}} s_1(s_1^2 - J_2) \tag{8.16}$$

To obtain a more convenient expression for $\cos 3\theta$, we shall perform some algebraic manipulations. From the definition of J_2, cf. (8.9), and since $s_1 = -(s_2 + s_3)$ we find

$$s_1^2 - J_2 = \frac{1}{2}(s_1^2 - s_2^2 - s_3^2) = \frac{1}{2}[(s_2 + s_3)^2 - s_2^2 - s_3^2] = s_2 s_3 \tag{8.17}$$

We next note from (3.17) that the invariant J_3 also can be written as

$$J_3 = s_1 s_2 s_3$$

Finally, use of this expression and (8.17) in (8.16) provides the result

$$\boxed{\cos 3\theta = \frac{3\sqrt{3}}{2}\frac{J_3}{J_2^{3/2}}} \tag{8.18}$$

i.e. we have established the relation already stipulated in (8.9) and we have expressed the angle θ in terms of stress invariants. The angle θ is often called the *Lode angle* after Lode (1926). Clearly, the angle θ is also given by (8.15), but the advantage of (8.18) is that here θ is expressed in terms of the stress invariants and not the principal stresses, *per se*. This implies that the eigenvalue problem does not have to be solved as the stress invariants are obtained directly from the stress tensor.

Let us return to the formulation (8.8), i.e.

$$\boxed{F(I_1, J_2, \cos 3\theta) = 0} \tag{8.19}$$

It appears that we have established a very convenient formulation where all the stress invariants can be interpreted geometrically. Moreover, formulation (8.19) separates the influence of the hydrostatic stress I_1 from the influence of the deviatoric stresses as expressed by J_2 and $\cos 3\theta$. Whereas the invariant J_2 tells us about the influence of the magnitude of the deviatoric stresses, cf. (8.13), the invariant $\cos 3\theta$ informs us about the influence of the direction of the deviatoric stresses. In addition, the presence of the term $\cos 3\theta$ enables us to derive a number of symmetry properties of the failure or initial yield criterion, as shown next.

8.2 Symmetry properties of the failure or initial yield curve in the deviatoric plane

It is evident that the failure or initial yield surface intersects the deviatoric plane in a certain curve, cf. Fig. 8.4. It turns out that due to the presence of the term $\cos 3\theta$ in criterion (8.19), we are able to derive a number of general symmetry properties.

Referring to the general criterion (8.19), the trace of this surface with an arbitrary deviatoric plane is obtained for I_1 = constant. As the cos-function is periodic with a period of 360°, we conclude that the failure or yield curve in

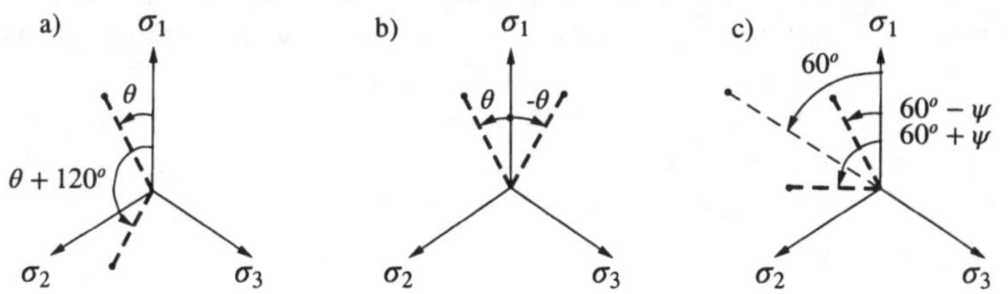

Figure 8.5: General symmetry properties of the failure or initial yield curve in the deviatoric plane; a) 120° period; b) symmetry about $\theta = 0°$; c) symmetry about $\theta = 60°$.

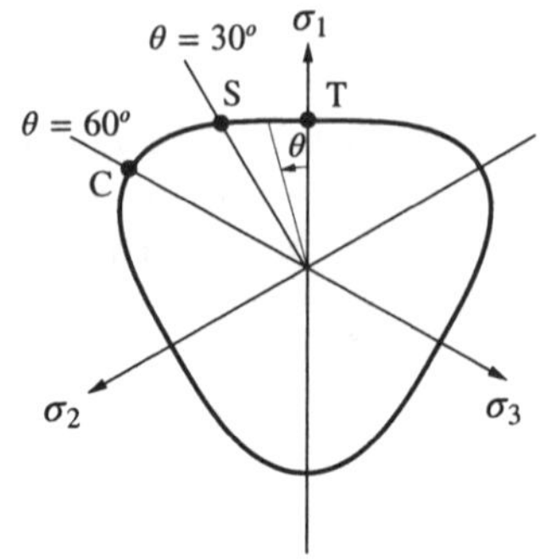

Figure 8.6: Possible shape of failure or initial yield curve in the deviatoric plane. T=tensile meridian, C=compressive meridian, S=shear meridian.

the deviatoric plane is periodic with a period of 120°. That is, the distance ρ, cf. Fig. 8.4b), is the same for θ and for $\theta + 120^o$ as well as for $\theta + 240^o$; this symmetry property is illustrated in Fig. 8.5a). Moreover, as $\cos 3\theta = \cos(-3\theta)$ we find that the curve in the deviatoric plane is symmetric about $\theta = 0^o$; this symmetry property is illustrated in Fig. 8.5b). Due to the periodicity of 120°, the curve is also symmetric about $\theta = 120^o$ and $\theta = 240^o$. Finally, setting $\theta = 60^o + \psi$, i.e. $\psi = 0^o$ corresponds to $\theta = 60^o$, we obtain $\cos 3\theta = -\cos 3\psi$ and setting $\theta = 60^o - \psi$ yields $\cos 3\theta = -\cos 3\psi$. Accordingly, we have the same distance ρ for $\theta = 60^o + \psi$ and $\theta = 60^o - \psi$; it is concluded that the curve in the deviatoric plane is symmetric about $\theta = 60^o$ and thereby also symmetric about $\theta = 180^o$ and $\theta = 300^o$. This symmetry property is illustrated in Fig. 8.5c). In conclusion, the symmetry properties shown in Fig. 8.5 imply that the curve in the deviatoric plane is completely characterized by its form for $0^o \leq \theta \leq 60^o$ and that this shape is repeated in the remaining sectors of the deviatoric plane. We observe that this far-reaching conclusion is a consequence of the material being isotropic.

A possible shape of the failure or initial yield curve in the deviatoric plane fulfilling the above-mentioned 60°-symmetry property is illustrated in Fig. 8.6. Here, the curve is shown as a *convex* curve, a property that does not follow from the mathematical analysis above, but which is strongly confirmed by experimental evidence, irrespective of the material in question.

We found above that the shape of the curve in the deviatoric plane is characterized by its form for $0^o \leq \theta \leq 60^o$ and it may be of interest to identify the corresponding stress range. For this purpose, let us arrange the principal stresses according to

$$\boxed{\sigma_1 \geq \sigma_2 \geq \sigma_3} \tag{8.20}$$

where tension is considered as a positive quantity. This allows us to write

$$\sigma_2 = (1-\alpha)\sigma_1 + \alpha\sigma_3 \; ; \qquad 0 \leq \alpha \leq 1$$

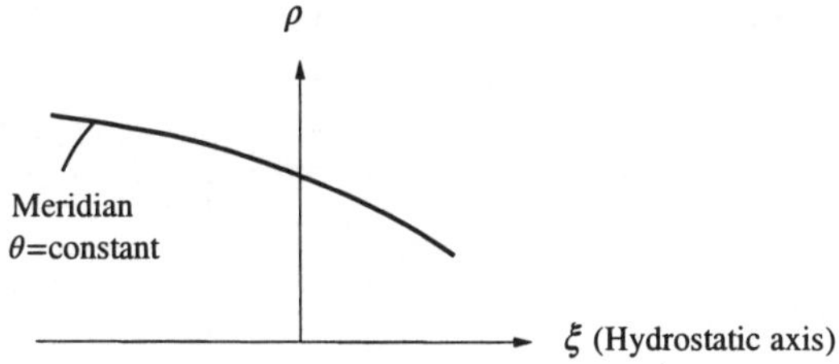

Figure 8.7: Meridian plane obtained by the intersection of the failure or initial yield surface with a plane containing the hydrostatic axis.

i.e.

$$s_1 = \tfrac{1}{3}(1+\alpha)(\sigma_1 - \sigma_3) \ ; \quad s_2 = \tfrac{1}{3}(1-2\alpha)(\sigma_1 - \sigma_3)$$

$$s_3 = -\tfrac{1}{3}(2-\alpha)(\sigma_1 - \sigma_3)$$

With these expressions, (8.15) takes the form

$$\cos\theta = \frac{1+\alpha}{2\sqrt{\alpha^2 - \alpha + 1}} \tag{8.21}$$

For α in the range $0 \leq \alpha \leq 1$, it follows that $0 \leq \theta \leq 60^o$, i.e. with the ordering of the principal stresses given by (8.20), all stress states are covered by the range $0 \leq \theta \leq 60^o$.

The *meridians* of the failure or initial yield surface are the curves where θ =constant applies, i.e. they are obtained by the intersection of the failure or initial yield surface with a plane containing the hydrostatic axis. Meridians may conveniently be depicted in a ξ, ρ-coordinate system or in a $I_1, \sqrt{J_2}$-coordinate system, the so-called *meridian plane*, cf. Fig. 8.7. Three meridians are of particular interest.

If $\sigma_1 > \sigma_2 = \sigma_3$ applies then $\alpha = 1$ and it follows from (8.21) that $\theta = 0^o$. This meridian is termed the *tensile meridian*, as the stress state $\sigma_1 > \sigma_2 = \sigma_3$ corresponds to a hydrostatic stress state superposed by a tensile stress in the σ_1-direction. We have

$$\boxed{\sigma_1 > \sigma_2 = \sigma_3 \quad i.e.\ \theta = 0^o \quad \Rightarrow \quad tensile\ meridian}$$

Uniaxial tensile stress states are located on the tensile meridian, cf. Fig. 8.8a), and so are biaxial compressive stress states when the two compressive principal stresses are equal.

If $\sigma_1 = \sigma_2 > \sigma_3$ holds then $\alpha = 0$ and (8.21) shows that $\theta = 60^o$. This meridian is termed the *compressive meridian*, as the stress state $\sigma_1 = \sigma_2 > \sigma_3$ corresponds to a hydrostatic stress state superposed by a compressive stress in the σ_3-direction. Consequently

$$\boxed{\sigma_1 = \sigma_2 > \sigma_3 \quad i.e.\ \theta = 60^o \quad \Rightarrow \quad compressive\ meridian} \tag{8.22}$$

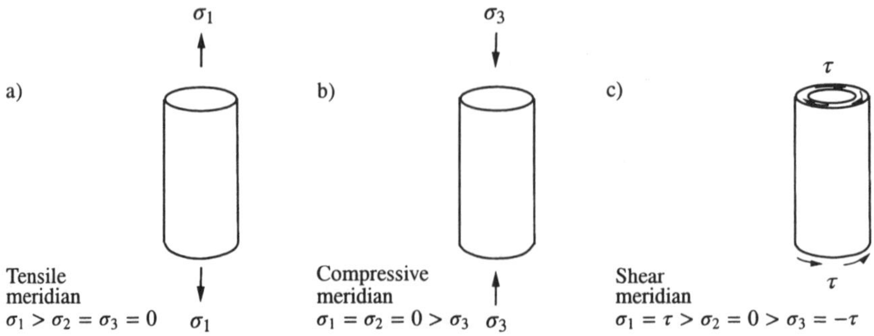

Figure 8.8: Simple examples of stress states located on different meridians; a) uniaxial tension; b) uniaxial compression; c) pure shear.

Uniaxial compressive stress conditions are therefore located on the compressive meridian, cf. Fig. 8.8b).

Finally, if $\sigma_1 > \sigma_2 = (\sigma_1 + \sigma_3)/2 > \sigma_3$ then $\alpha = 1/2$ and it follows from (8.21) that $\theta = 30^o$. This meridian is termed the *shear meridian*, as the stress state $\sigma_1 > \sigma_2 = (\sigma_1 + \sigma_3)/2 > \sigma_3$ corresponds to a hydrostatic stress state superposed by a positive stress, τ, in the σ_1-direction and a negative stress, $-\tau$, in the σ_3-direction. That is

$$\boxed{\sigma_1 > \sigma_2 = \frac{\sigma_1 + \sigma_3}{2} > \sigma_3 \quad i.e. \;\; \theta = 30^o \quad \Rightarrow \quad \textit{shear meridian}}$$

A stress state corresponding to pure shear is therefore located on the shear meridian, cf. Fig. 8.8c).

The points where the tensile, compressive and shear meridians intersect the deviatoric plane are illustrated in Fig. 8.6. (Points T, C and S).

To identify points on certain meridians for multiaxial stress states, the *von Kármán pressure cell* is often used, especially for soil and cementitious materials like concrete and rocks. This type of pressure cell is named after von Kármán in recognition of his triaxial tests on marble and sandstone using this kind of equipment, cf. von Kármán (1911). A cylindrical specimen is inserted into a pressure chamber, cf. Fig. 8.9. The oil inside the pressure chamber is pressurized providing stresses on the lateral surface of the specimen and, via a piston, an ordinary testing machine supplies a pressure to the end surfaces of the specimen.

It is evident that the von Kármán pressure cell enables one to test materials along two meridians only: the tensile meridian for which $\sigma_1 > \sigma_2 = \sigma_3$ and the compressive meridian where $\sigma_1 = \sigma_2 > \sigma_3$, cf. Fig. 8.10 (recall that tension is considered as a positive quantity).

Previously, we deduced the general symmetry properties of the curve in the deviatoric plane, cf. Figs. 8.5 and 8.6, and we will now establish some additional symmetry properties which hold under some specialized conditions.

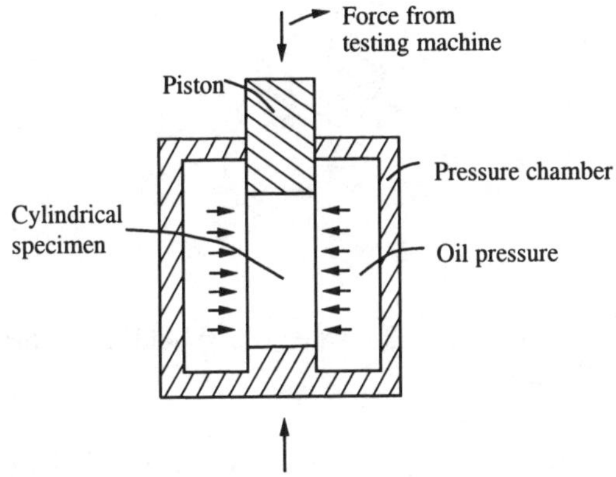

Figure 8.9: Principal sketch of von Kármán pressure cell enabling one to test cylindrical specimens under multiaxial stress states.

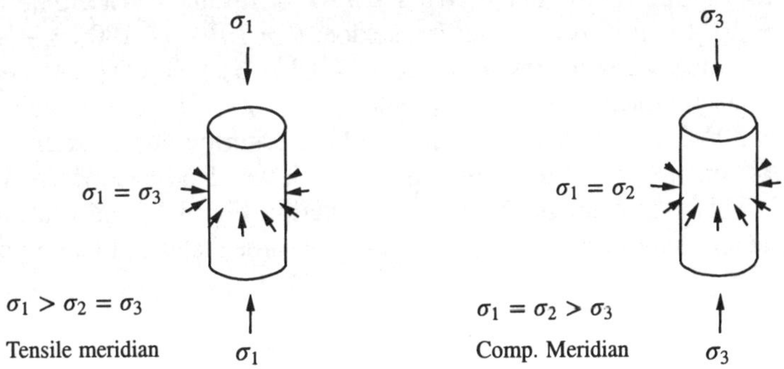

Figure 8.10: Stress states that may be obtained in a von Kármán pressure cell.

Let us assume that the occurrence of failure or initial yield is independent of the hydrostatic stress I_1, i.e. (8.19) reduces to

$$F(J_2, \cos 3\theta) = 0 \tag{8.23}$$

This implies that the corresponding surface in the principal stress space consists of a cylindrical surface with the meridians parallel with the hydrostatic axis. Consequently irrespective of the deviatoric plane considered the same trace of the failure or initial yield surface is obtained. For metals and steel, yielding turns out to be independent of the hydrostatic stress, i.e. (8.23) is a valid assumption.

Let us further assume that criterion (8.23) is fulfilled both for the stress state σ_{ij} and the stress state $-\sigma_{ij}$. As an example, for metals and steel the initial yield stress is the same for uniaxial tension and uniaxial compression. Let us now investigate the consequences of the two assumptions mentioned above. For the stress state σ_{ij}, we may determine J_2, J_3 and thereby $\cos 3\theta$, cf. (8.18).

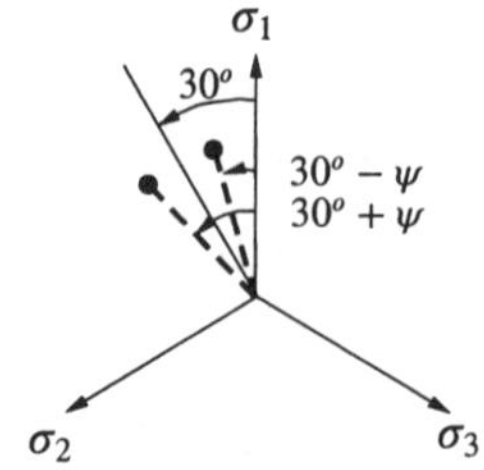

Figure 8.11: Symmetry about $\theta = 30^o$ when (8.23) holds and when σ_{ij} and $-\sigma_{ij}$ both fulfill the criterion.

Likewise, for the stress state $\sigma^*_{ij} = -\sigma_{ij}$, we obtain $J^*_2 = J_2, J^*_3 = -J_3$, i.e. $\cos 3\theta^* = -\cos 3\theta$. This means that for the same J_2–value, criterion (8.23) is fulfilled both for $\cos 3\theta$ and for $-\cos 3\theta$; i.e. both θ as well as $\theta \pm 180^o$ fulfill the criterion. Consider now $\theta = 30^o + \psi$, i.e. $\psi = 0^o$ corresponds to $\theta = 30^o$, cf. Fig. 8.11. We found above that both θ and $\theta - 180^o$ fulfill the criterion; therefore, when $\theta = 30^o + \psi$ fulfills the criterion, so does $\theta = 30^o + \psi - 180^o = -150^o + \psi$. The latter value leads to $\cos 3\theta = \cos(-450^o + 3\psi) = \cos(-90^o + 3\psi) = \cos(90^o - 3\psi)$, which may be interpreted as $\theta = 30^o - \psi$. It is concluded that both $\theta = 30^o + \psi$ and $\theta = 30^o - \psi$ fulfill the criterion and in addition to the general symmetry properties shown in Fig. 8.5, we also have symmetry about $\theta = 30^o$. This symmetry property is illustrated in Fig. 8.11 and it implies that the tensile and compressive meridians have the same distance to the hydrostatic axis.

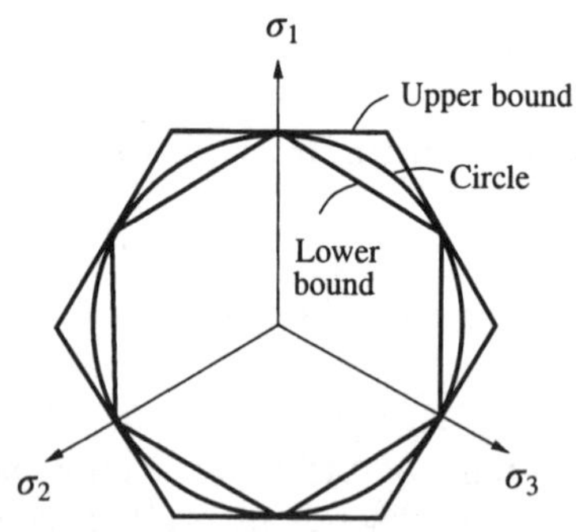

Figure 8.12: Upper and lower bounds for curve in the deviatoric plane when (8.23) holds and when both σ_{ij} and $-\sigma_{ij}$ fulfill the criterion.

The general symmetry properties imply that the curve in the sector $0^o \leq \theta \leq 60^o$ is repeated in the remaining sectors of the deviatoric plane, cf. Fig. 8.6. With the assumptions described above, we also have a symmetry line about $\theta = 30^o$. If we, in addition, assume that the trace in the deviatoric plane is convex – an assumption that is strongly supported by experimental evidence for all materials – we are led to the upper and lower bounds for the trace in the deviatoric plane as shown in Fig. 8.12.

Let us recall the assumptions that led to these upper and lower bounds. As discussed, these assumptions are in close agreement with the initial yield properties for metals and steel and it is therefore convenient to make the following summation

Initial yield of metals and steel is characterized in that: * *the hydrostatic stress has no influence* * *if* σ_{ij} *results in yielding so does* $-\sigma_{ij}$ * *the trace in the deviatoric plane is convex*	(8.24)

Recalling that (8.23) implies that the surface in the stress space is a cylindrical surface with the meridians parallel with the hydrostatic axes and observing the upper and lower bounds in the deviatoric plane illustrated in Fig. 8.12, we conclude that there are very narrow bounds within which a valid initial yield criterion for metals and steel can be chosen. Indeed, we shall later see that the circle of Fig. 8.12 corresponds to the *von Mises yield criterion* whereas the lower bound of Fig. 8.12 corresponds to the *Tresca yield criterion*.

Having discussed issues that are of relevance for metals and steel, it may be of interest to evaluate the general experimental evidence for another large group of materials, namely concrete, soil and rocks. These materials are characterized by smooth stress-strain curves exhibiting no well defined initial yield stress. Moreover, the analysis of constructions involving these materials is often focused on the determination of the ultimate load capacity and whereas the ratio $\varepsilon_f/\varepsilon_{yo}$, cf. Fig. 8.2b, is large for metals and steel, it is much smaller for concrete, soil and rocks. For these reasons, the failure criterion is of primary importance for concrete, soil and rocks. Quite generally, the experimental evidence for these materials may be summarized as follows

Failure of concrete, soil and rocks is characterized in that: * *the hydrostatic stress has a strong influence* * *inclusion of the term* $\cos 3\theta$ *is of importance* * *the failure surface is convex*	(8.25)

It follows that we expect the failure curve in the deviatoric plane to take the form sketched in Fig. 8.6. We finally observe that experimental observations for cast iron fall somewhere between the characteristics defined by (8.24) and (8.25).

Since elasticity, per definition, only occurs within the initial yield surface, this surface is independent of the previous load history history. On the other hand, to reach the failure surface significant inelastic strains are developed and, in principle, the failure surface is therefore expected to depend on the load history. However, experimental evidence shows for concrete (cf. Chinn and Zimmerman (1965), Schickert and Winkler (1977)), soil (cf. Scott (1963)) and

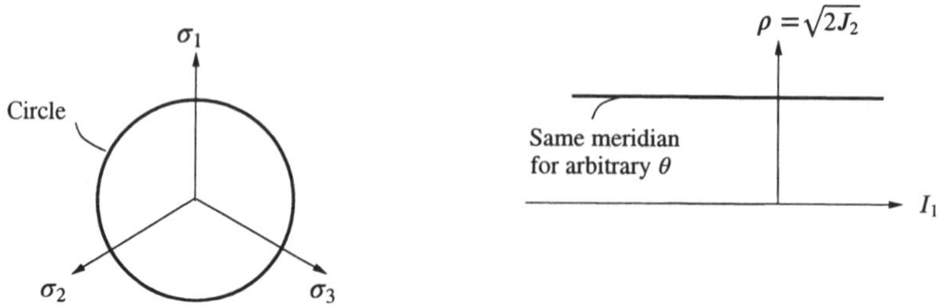

Figure 8.13: von Mises criterion (8.26); a) deviatoric plane; b) meridian plane.

rocks (cf. Swanson and Brown (1971)) that whether the loading is proportional or non-proportional only influences the failure surface to a very modest degree.

8.3 von Mises criterion

For initial yielding of metals and steel, the general experimental evidence is summarized in (8.24). As the hydrostatic stress I_1 has no influence on the yielding, the general criterion (8.19) reduces to (8.23). The simplest assumption is then to ignore the influence of the complicated term $\cos 3\theta$, which leads to $F(J_2) = 0$, i.e. it is assumed that J_2 takes a constant value at yielding, i.e.

$$\sqrt{J_2} - c = 0$$

where c is a constant. This relation may be written in various manners, but the most convenient expression is obtained by

$$\boxed{\sqrt{3J_2} - \sigma_{yo} = 0} \qquad (8.26)$$

where, for convenience, the factor 3 in front of J_2 is inserted since $\sqrt{3J_2}$ for uniaxial tension reduces to $\sqrt{3J_2} = \sigma$. According to the criterion, $\sqrt{3J_2}$ takes a constant value for initial yielding and this constant value then becomes σ_{yo}, i.e. the initial yield stress in tension, cf. Fig. 8.1.

Criterion (8.26) is independent of the hydrostatic stress I_1, i.e. it represents a cylindrical surface in the principal stress space with the meridians being parallel with the hydrostatic axis. This means that only the deviatoric stresses influence the criterion. Moreover, it is evident that (8.26) in the deviatoric plane represents a circle, i.e. all meridians are located at the same distance to the hydrostatic axis. These properties are illustrated in Fig. 8.13 and it appears that the circle in the deviatoric plane falls between the lower and upper bounds shown in Fig. 8.12. With these properties, the appearance of the yield surface in the principal stress space takes the form shown in Fig. 8.14.

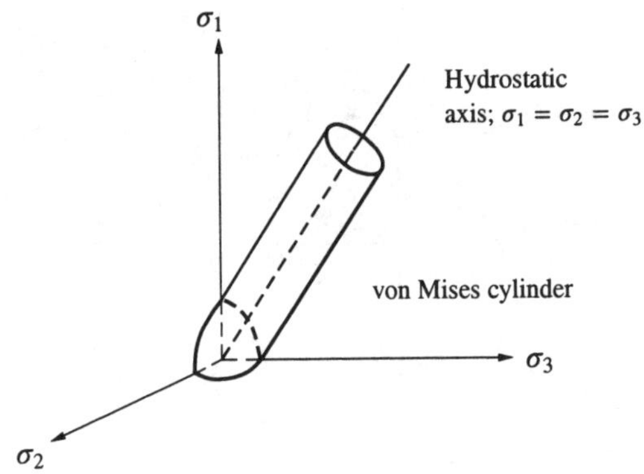

Figure 8.14: von Mises surface in the principal stress space.

The criterion (8.26) was suggested by von Mises (1913) and it is therefore called the *von Mises criterion*; it was anticipated, to some extent, by the proposal of Huber (1904) and the criterion is thus occasionally called the *Huber-von Mises criterion.* Hencky (1924) suggested an interesting physical interpretation of the criterion. Inside the initial yield surface given by (8.26), the material behaves linear elastic. According to (4.95), the strain energy W of a linear elastic and isotropic material can be written as

$$W = W_d + W_v$$

where

$$W_d = Ge_{ij}e_{ij} \; ; \quad W_v = \frac{1}{2}K\varepsilon_{kk}\varepsilon_{mm}$$

It appears that W_d represents the *deviatoric strain energy* whereas W_v is the *volumetric strain energy*. Moreover, as e_{ij} and ε_{kk} are decoupled W_d and W_v are also decoupled. With Hooke's law (4.86) for the deviatoric response, we obtain

$$W_d = \frac{1}{4G}s_{ij}s_{ij} = \frac{1}{2G}J_2$$

i.e. the von Mises yield criterion may be interpreted by saying that initial yield occurs when the deviatoric strain energy achieves a certain value.

Referring to (3.20) the octahedral shear stress τ_o is given by $\tau_o = \sqrt{\frac{2}{3}J_2}$. Another physical interpretation of the von Mises criterion is therefore to claim that the yielding occurs when the octahedral shear stress τ_o that acts on the octahedral plane exceeds a certain value. Expressed in the principal stresses, criterion (8.26) takes the form

$$\boxed{\sqrt{\frac{1}{2}[(\sigma_1 - \sigma_2)^2 + (\sigma_1 - \sigma_3)^2 + (\sigma_2 - \sigma_3)^2]} - \sigma_{yo} = 0} \qquad (8.27)$$

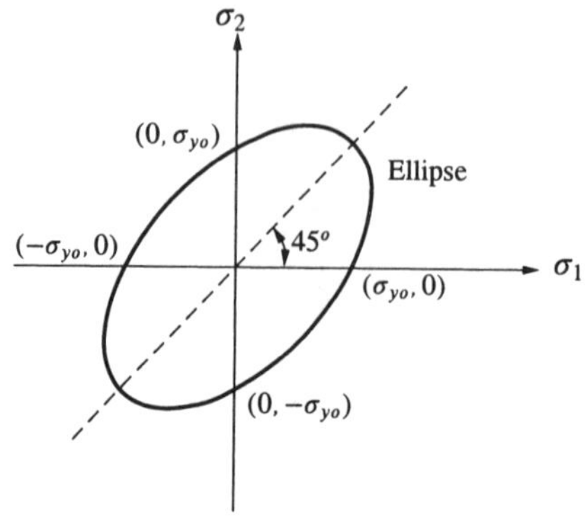

Figure 8.15: von Mises ellipse in the $\sigma_1\sigma_2$−plane.

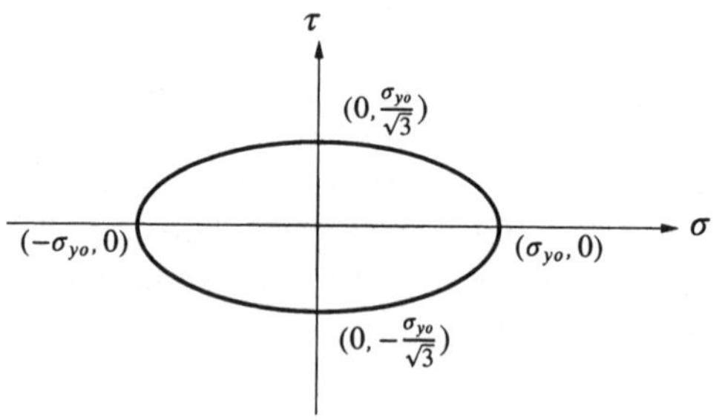

Figure 8.16: von Mises ellipse in the $\sigma\tau$−plane.

For plane stress conditions where $\sigma_3 = 0$ holds, it follows that

$$\sqrt{\sigma_1^2 + \sigma_2^2 - \sigma_1\sigma_2} - \sigma_{yo} = 0 \tag{8.28}$$

which represents an ellipse in the $\sigma_1\sigma_2$-plane as shown in Fig. 8.15. Another stress state of interest is obtained by simultaneous uniaxial stressing and torsion of, for instance, a thin-walled tube. This stress state is given by

$$[\sigma_{ij}] = \begin{bmatrix} \sigma & \tau & 0 \\ \tau & 0 & 0 \\ 0 & 0 & 0 \end{bmatrix} \quad \text{i.e.} \quad [s_{ij}] = \begin{bmatrix} \frac{2}{3}\sigma & \tau & 0 \\ \tau & -\frac{1}{3}\sigma & 0 \\ 0 & 0 & -\frac{1}{3}\sigma \end{bmatrix}$$

We then obtain from (8.26) that

$$\sqrt{\sigma^2 + 3\tau^2} - \sigma_{yo} = 0 \tag{8.29}$$

which represents an ellipse in the $\sigma\tau$−plane, cf. Fig. 8.16. It may be of interest to determine the initial yield shear stress τ_{yo} when $\sigma = 0$ and from (8.29) we find that

$$\tau_{yo} = \frac{\sigma_{yo}}{\sqrt{3}} \tag{8.30}$$

We have scrutinized the principal ramifications of the von Mises criterion in great detail and in Section 8.8 we will compare its predictions with experimental data for metals and steel.

8.4 Drucker-Prager criterion

We now shift our focus of interest to materials like concrete, soil and rocks. The general failure properties for such materials are summarized in (8.25) and we note that the hydrostatic stress I_1 is of paramount importance. All terms in the general criterion (8.19) must therefore be considered. Even though the term $\cos 3\theta$ is of great importance, this term complicates the criterion considerably; thus we may, as an approximation, simply ignore its influence. We are thereby left with

$$F(I_1, J_2) = 0 \tag{8.31}$$

This formulation was suggested by Schleicher (1926) and (8.31) is often referred to as an *octahedral criterion* since the octahedral normal stress σ_o and the octahedral shear stress τ_o are related to I_1 and J_2 via (3.20), i.e.

$$\sigma_o = \frac{1}{3} I_1 \; ; \quad \tau_o = \sqrt{\frac{2}{3} J_2} \tag{8.32}$$

The simplest possible explicit form of (8.31) is a linear relation between I_1 and $\sqrt{J_2}$, i.e.

$$\boxed{\sqrt{3J_2} + \alpha I_1 - \beta = 0} \tag{8.33}$$

where α and β are material parameters. Moreover, α is dimensionless whereas β has the dimension of stress. The reason for the factor 3 in front of J_2 is that for $\alpha = 0$, (8.33) then reduces exactly to the von Mises criterion (8.26). Criterion (8.33) was suggested by Drucker and Prager (1952) and it is therefore called the *Drucker-Prager criterion.*

Both the octahedral normal stress σ_o and the octahedral shear stress τ_o act on the octahedral plane. Thus, a physical interpretation of the Drucker-Prager criterion is to claim that failure (or yielding) occurs when the octahedral shear stress τ_o exceeds a certain value that depends on the octahedral normal stress.

The deviatoric plane is defined by I_1 = constant, i.e. (8.33) implies that $\sqrt{3J_2}$ is constant in the deviatoric plane. Therefore, the trace in the deviatoric

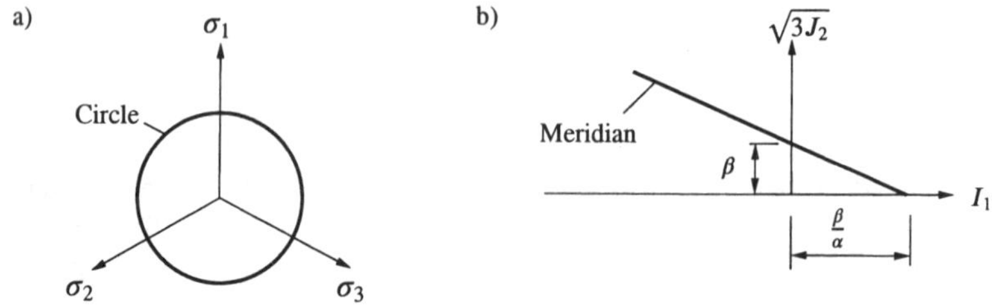

Figure 8.17: Drucker-Prager criterion; a) deviatoric plane; b) meridian plane.

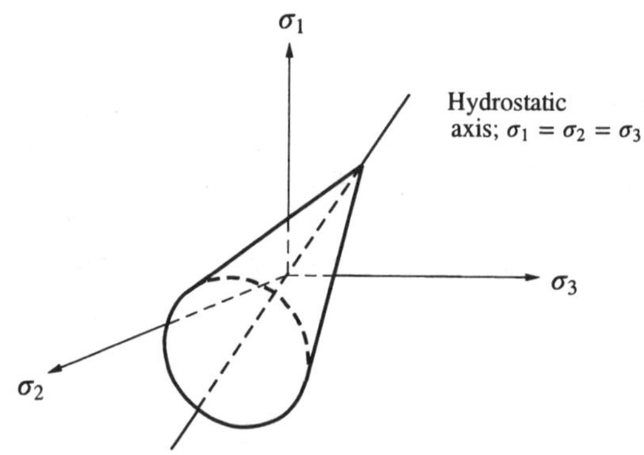

Figure 8.18: Drucker-Prager surface in the principal stress space.

plane is described by a circle. It follows that we have the same meridian irrespective of the θ-angle and that this meridian makes a certain slope with the hydrostatic axis. These features are shown in Fig. 8.17.

Since $\sqrt{3J_2}$ is a non-negative quantity, we conclude from (8.33) with $I_1 = 0$ that β is a positive material parameter. Moreover, as illustrated in Fig. 8.17b) the Drucker-Prager surface intersects the I_1-axis for $I_1 = \beta/\alpha$. Considering failure conditions, it is evident that a material like rock or concrete will break for a sufficiently large hydrostatic tension. Therefore, the dimensionless parameter α must be a non-negative quantity as illustrated in Fig. 8.17b).

In accordance with the properties given by Fig. 8.17, the appearance of the Drucker-Prager criterion in the principal stress space takes the form of a circular cone, as shown in Fig. 8.18.

For plane stress conditions, i.e. $\sigma_3 = 0$, (8.33) reduces to

$$\sqrt{\sigma_1^2 + \sigma_2^2 - \sigma_1\sigma_2} + \alpha(\sigma_1 + \sigma_2) - \beta = 0 \tag{8.34}$$

which obviously reduces to the von Mises expression (8.28) for $\alpha = 0$. As shown in Fig. 8.19, (8.34) represents an off-center ellipse in the $\sigma_1\sigma_2$–plane. The *uniaxial tensile strength* σ_t, *uniaxial compressive strength* σ_c, *biaxial tensile*

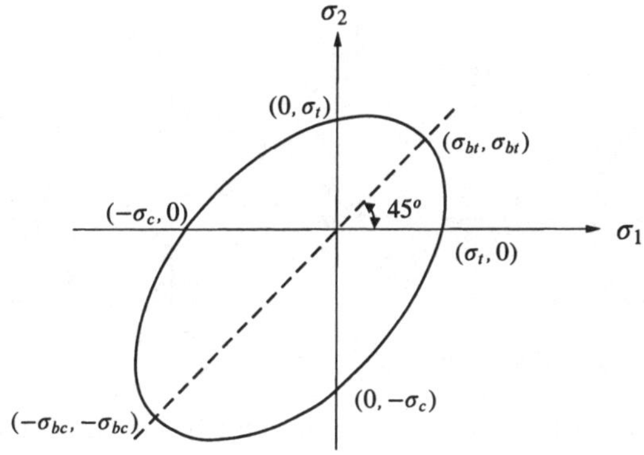

Figure 8.19: Drucker-Prager off-center ellipse in the $\sigma_1\sigma_2$–plane.

strength σ_{bt} and the *biaxial compressive strength* σ_{bc} are illustrated in Fig. 8.19 and from (8.34) we derive that

$$\sigma_t = \frac{\beta}{1+\alpha} \; ; \quad \sigma_c = \frac{\beta}{1-\alpha}$$

and

$$\sigma_{bt} = \frac{\beta}{1+2\alpha} \; ; \quad \sigma_{bc} = \frac{\beta}{1-2\alpha} \tag{8.35}$$

These relations may be used to identify the material paramcters α and β.

We will later, in Section 8.10, compare the predictions of the Drucker-Prager criterion with experimental data, but we may already at this state emphasize that due to the elimination of the $\cos 3\theta$-term, the Drucker-Prager criterion should be used with caution. In practice, it can only be used with sufficient accuracy when α is small, i.e. when the influence of the hydrostatic stress I_1 is moderate. Cast iron may be representative of such a material.

8.5 Coulomb criterion

We will again consider failure characteristics for concrete, soil and rocks, but instead of the formulation (8.19), we will adopt the description given by (8.7), i.e.

$$F(\sigma_1, \sigma_2, \sigma_3) = 0 \tag{8.36}$$

with the convention that

$$\boxed{\sigma_1 \geq \sigma_2 \geq \sigma_3}$$

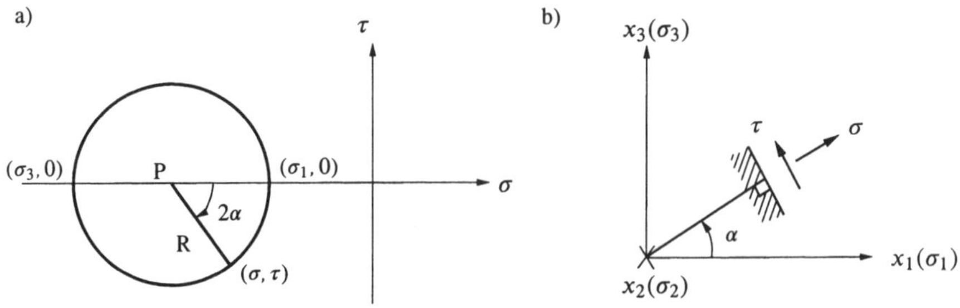

Figure 8.20: a) Mohr's circle of stress; b) corresponding interpretation of σ and τ.

In general, (8.36) is quite complicated and in order to simplify the expression, it is tempting to assume that the intermediate principal stress σ_2 is of minor importance, i.e. we assume that

$$F(\sigma_1, \sigma_3) = 0$$

The most simple expression of this form is then provided by a linear relation between σ_1 and σ_3, i.e.

$$k\sigma_1 - \sigma_3 - m = 0 \tag{8.37}$$

where k and m are material parameters. Requiring that this expression should predict the uniaxial compressive strength value σ_c, the stress state $(\sigma_1, \sigma_2, \sigma_3) = (0, 0, -\sigma_c)$ should fulfill (8.37) and we find

$$\boxed{k\sigma_1 - \sigma_3 - \sigma_c = 0} \tag{8.38}$$

This so-called *Coulomb criterion* was suggested by Coulomb (1776) and is the oldest criterion ever proposed.

To comply with the present exposition, we have here chosen to derive the Coulomb criterion from (8.36), but traditionally the Coulomb criterion is established in a different manner. As this traditional establishment makes for some interesting interpretations, we will now form a bridge between the two viewpoints.

Following Section 3.5, Mohr's circle of stress is shown in Fig. 8.20a). In Fig. 8.20b), the x_1, x_2, x_3–coordinate system is collinear with the principal directions of σ_1, σ_2 and σ_3 and the interpretation of the stress point (σ, τ) in Fig. 8.20a) is shown in Fig. 8.20b). That is, the normal stress σ and the shear stress τ act on the section having a normal which makes the angle α with the σ_1–direction. From Fig. 8.20a), the center position P and the radius R of Mohr's circle are given by

$$P = \frac{1}{2}(\sigma_1 + \sigma_3) \; ; \quad R = \frac{1}{2}(\sigma_1 - \sigma_3) \tag{8.39}$$

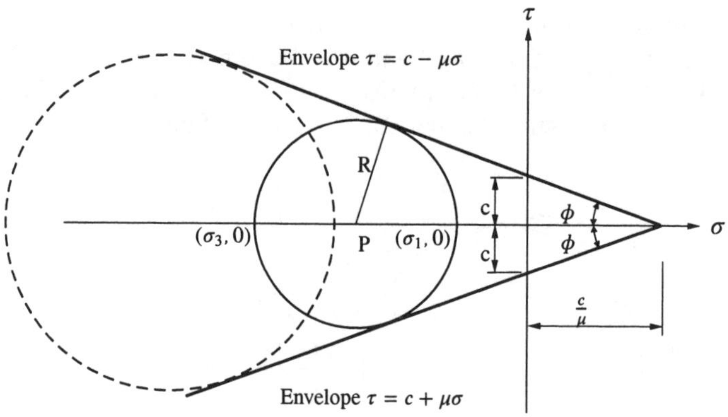

Figure 8.21: Coulomb criterion in Mohr diagram.

It is assumed that the stress state fulfills Coulomb's criterion. Therefore, insertion of σ_3 as determined by (8.38) into (8.39) yields

$$P = \frac{1}{2}[(k+1)\sigma_1 - \sigma_c] \; ; \quad R = \frac{1}{2}[\sigma_c - (k-1)\sigma_1]$$

and elimination of σ_1 provides

$$R = \frac{\sigma_c}{k+1} - \frac{k-1}{k+1}P$$

Thus, the radius R varies linearly with the center position P. Consequently, and as shown in Fig. 8.21, all Mohr's circles of stress that fulfill the Coulomb criterion have two symmetrically positioned straight lines as their envelopes. These straight lines can be written as

$$\boxed{|\tau| = c - \mu\sigma} \tag{8.40}$$

where c and μ are non-negative material parameters. It appears that (8.40) provides an alternative formulation of the Coulomb criterion.

Referring to Fig. 8.21 and (8.40), we see that $|\tau| = c$ is the shear strength when the normal stress $\sigma = 0$, i.e. c is the *cohesion* of the material. If σ is negative, i.e. σ corresponds to a pressure, it follows that the shear strength $|\tau|$ is increased and μ is therefore called the *friction coefficient* of the material. Conseqently, we have obtained a direct physical interpretation of the Coulomb criterion and, most frequently, this criterion is postulated directly in the form given by (8.40).

With this discussion, it is no surprise that the linear expression (8.40) may be generalized to achieve

$$\boxed{|\tau| = h(\sigma)} \tag{8.41}$$

where $h(\sigma)$ denotes an arbitrary function of σ. This so-called *Mohr criterion* was suggested by Mohr (1900) and it is illustrated in Fig. 8.22. Just like the Coulomb criterion, the Mohr criterion (8.41) serves as the envelope of all Mohr's circles of stress when the material is loaded to failure.

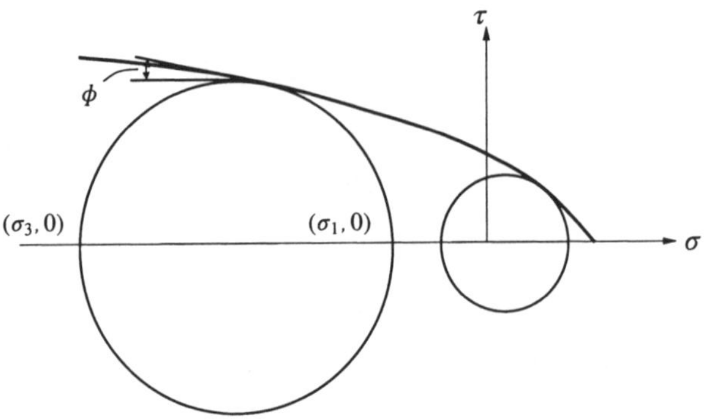

Figure 8.22: Mohr criterion; illustration of current friction angle ϕ.

Returning to the Coulomb criterion, it is of interest to compare the material parameters k and σ_c in (8.38) with the material parameters c and μ in (8.40). First, we observe from Fig. 8.21 that

$$\boxed{\tan\phi = \mu}$$

where ϕ is termed the *friction angle*. Let us next consider a hydrostatic stress state $(\sigma_1, \sigma_2, \sigma_3) = (\sigma, \sigma, \sigma)$. It follows from (8.38) that $\sigma = \sigma_c/(k-1)$ and from Fig. 8.21, we have $\sigma = c/\mu$. This provides

$$\frac{c}{\mu} = \frac{\sigma_c}{k-1} \tag{8.42}$$

Observing that P for the situation displayed in Fig. 8.21 is a negative quantity, cf. (8.39), we obtain from Fig. 8.21, (8.42) and (8.39) that

$$\sin\phi = \frac{R}{\dfrac{c}{\mu} - P} = \frac{\frac{1}{2}(\sigma_1 - \sigma_3)}{\dfrac{\sigma_c}{k-1} - \dfrac{1}{2}(\sigma_1 + \sigma_3)}$$

i.e.

$$\frac{1+\sin\phi}{1-\sin\phi}\sigma_1 - \sigma_3 - \frac{2\sigma_c}{k-1}\frac{\sin\phi}{1-\sin\phi} = 0$$

A comparison with (8.38) reveals that

$$\boxed{k = \frac{1+\sin\phi}{1-\sin\phi}} \quad i.e. \;\; k \geq 1$$

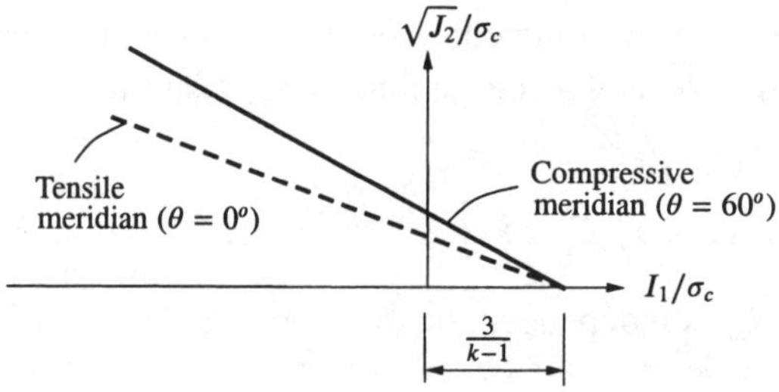

Figure 8.23: Coulomb criterion in meridian plane.

i.e.

$$\sin\phi = \frac{k-1}{k+1} \tag{8.43}$$

As $\tan\phi = \sin\phi/\sqrt{1-\sin^2\phi}$, this implies

$$\boxed{\mu = \tan\phi = \frac{k-1}{2\sqrt{k}}} \tag{8.44}$$

Moreover, this expression and (8.42) yield

$$\boxed{c = \frac{\sigma_c}{2\sqrt{k}}} \tag{8.45}$$

With this discussion of the relations between the various material parameters, we return to the Coulomb criterion given in the form of (8.38). Suppose that the stress state $(\sigma_1, \sigma_2, \sigma_3)$ fulfills the criterion. Let us on this stress state superpose a hydrostatic stress state given by the quantity p; this results in the stress state $(\sigma_1 + p, \sigma_2 + p, \sigma_3 + p)$. It is evident that this new stress state fulfills criterion (8.38) only if $p(k-1) = 0$, i.e. if $k = 1$. It is concluded that criterion (8.38) depends on the hydrostatic stress state if $k \neq 1$, i.e.

The Coulomb criterion (8.38) depends on the hydrostatic stress if $k \neq 1$	(8.46)

Criterion (8.38) defines a plane in the principal stress space, i.e. the meridians take the form of straight lines. Moreover, the trace in the deviatoric plane ($0^o \leq \theta \leq 60^o$ corresponding to $\sigma_1 \geq \sigma_2 \geq \sigma_3$) is also a straight line. Let us now scrutinize the compressive and tensile meridians.

Along the compressive meridian ($\sigma_1 = \sigma_2 > \sigma_3,\ \theta = 60^o$), we obtain

$$I_{1c} = 2\sigma_1 + \sigma_3 \quad ; \quad \sqrt{J_{2c}} = \frac{1}{\sqrt{3}}(\sigma_1 - \sigma_3) \tag{8.47}$$

where subscript c refers to the compressive meridian. Expressing σ_1 and σ_3 in terms of I_{1c} and $\sqrt{J_{2c}}$ and insertion into (8.38) result in

$$\sqrt{J_{2c}} + \frac{k-1}{\sqrt{3}(k+2)} I_{1c} - \frac{\sqrt{3}\sigma_c}{k+2} = 0 \tag{8.48}$$

As expected $\sqrt{J_{2c}}$ is independent on the hydrostatic stress when k=1.

Along the tensile meridian ($\sigma_1 > \sigma_2 = \sigma_3,\ \theta = 0^o$), we have

$$I_{1t} = \sigma_1 + 2\sigma_3 \ ; \ \sqrt{J_{2t}} = \frac{1}{\sqrt{3}}(\sigma_1 - \sigma_3) \tag{8.49}$$

where subscript t refers to the tensile meridian. Expressing σ_1 and σ_3 in terms of I_{1t} and $\sqrt{J_{2t}}$ and insertion into (8.38) provide

$$\sqrt{J_{2t}} + \frac{k-1}{\sqrt{3}(2k+1)} I_{1t} - \frac{\sqrt{3}\sigma_c}{2k+1} = 0 \tag{8.50}$$

Expressions (8.48) and (8.50) are illustrated in Fig. 8.23

To determine the trace in the deviatoric plane, we recall that here $I_1 = I_{1t} = I_{1c}$, i.e. elimination of this quantity from (8.48) and (8.50) yields

$$\frac{\rho_c}{\rho_t} = \frac{\sqrt{2J_c}}{\sqrt{2J_t}} = \frac{2k+1}{k+2} \tag{8.51}$$

where $\rho = \sqrt{2J_2}$, cf. (8.13). Recalling that the trace of the Coulomb criterion in the deviatoric plane is a straight line when $\sigma_1 \geq \sigma_2 \geq \sigma_3$, i.e. $0^o \leq \theta \leq 60^o$, we obtain the result shown in principle in Fig. 8.24.

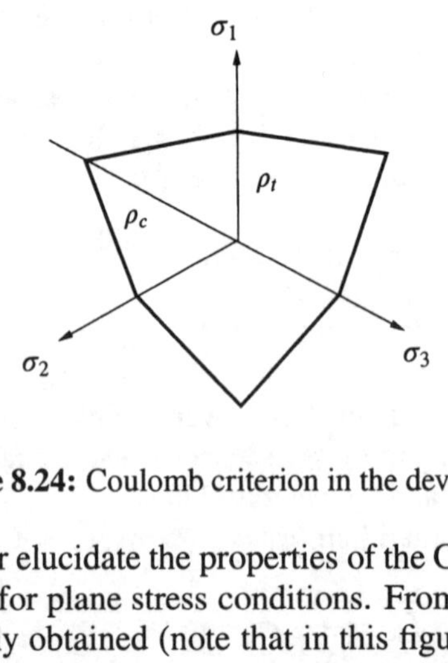

Figure 8.24: Coulomb criterion in the deviatoric plane.

In order to further elucidate the properties of the Coulomb criterion, we consider its predictions for plane stress conditions. From (8.38), the results shown in Fig. 8.25 are easily obtained (note that in this figure the usual convention of

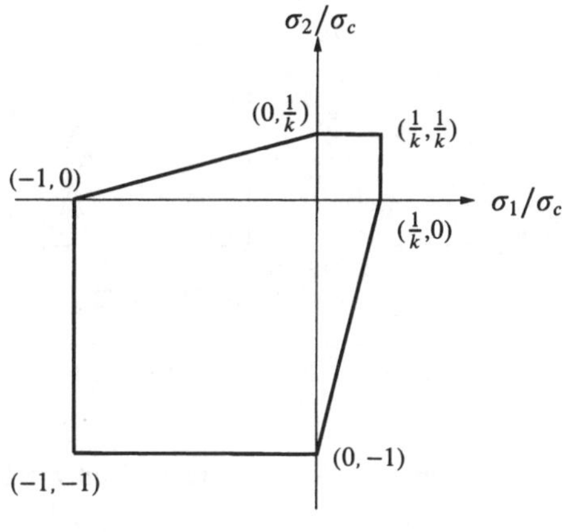

Figure 8.25: Coulomb criterion for plane stress conditions.

$\sigma_1 \geq \sigma_2 \geq \sigma_3$ has been abandoned). It appears that the predicted uniaxial tensile strength becomes $\sigma_t = \sigma_c/k$.

Due to its simplicity, the Coulomb criterion is widely used in analytical applications, cf. for instance Chen (1975) for soil applications and Nielsen (1984) for concrete applications. In numerical applications, however, its use is impeded by the corners of the surface, cf. Fig. 8.24. By calibration of the parameter k, the criterion can be used to model a large variety of material, but, as we will see later, the ignorance of the influence of the intermediate principal stress σ_2 implies that the criterion, in general, will underestimate the experimentally determined failure stresses.

8.6 Mohr's failure mode criterion

We have discussed the assumption of a Coulomb criterion (8.38) or (8.40) in some detail. In this discussion, the Coulomb criterion is simply a criterion that provides information on the magnitude of the failure stresses. Another feature often related to the Coulomb criterion is a *failure mode criterion.* Often in the literature, the failure mode criterion is presented as if it is an integrated part of the Coulomb criterion. We emphasize that this is not the case, as the failure mode criterion is an additional postulate which, in principle, has nothing to do with the failure criterion itself.

The failure mode criterion dates back to Mohr (1900) and it is based on Fig. 8.20b) and the discussion following (8.40). From Figs. 8.20 and 8.21, we may have the situation shown in Fig. 8.26.

Consider the stress state (σ, τ) that satisfies the Coulomb criterion (8.40) in accordance with Fig. 8.26a). From Mohr's circle of stress, the interpretation of σ and τ is displayed in Fig. 8.26b) where the $x_1x_2x_3$-coordinate system is

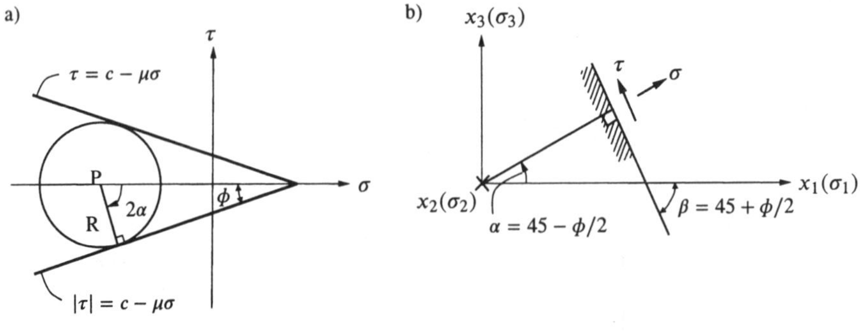

Figure 8.26: a) Coulomb criterion and Mohr's circle; b) interpretation of σ and τ.

collinear with the principal directions of σ_1, σ_2 and σ_3. From Fig. 8.26a) follows that $2\alpha + 90^o + \phi = 180^o$, i.e.

$$\alpha = 45^o - \frac{\phi}{2}$$

It seems tempting to assume that the plane illustrated in Fig. 8.26b) on which the failure stresses σ and τ act is also a *failure plane* where failure takes place in the form of sliding. This plane is also called a *slip plane* since the failure mode is postulated to be a movement along the plane. The angle β which the failure plane makes with the largest principal stress direction (σ_1) becomes $\beta = 45^o + \phi/2$ as shown in Fig. 8.26b).

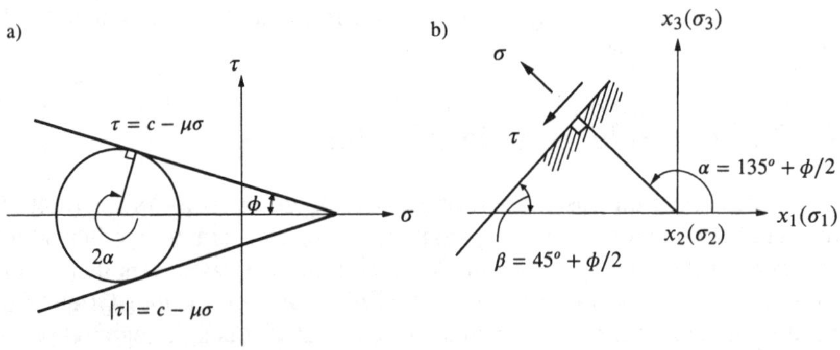

Figure 8.27: a) Coulomb criterion and Mohr's circle; b) interpretation of σ and τ.

Consider next the situation where the stress state (σ, τ) that satisfies the Coulomb criterion is located as displayed in Fig. 8.27a); the corresponding interpretation of σ and τ is shown in Fig. 8.27b). From Fig. 8.27a) follows that $360^o - 2\alpha + 90^o + \phi = 180^o$, i.e.

$$\alpha = 135^o + \frac{\phi}{2}$$

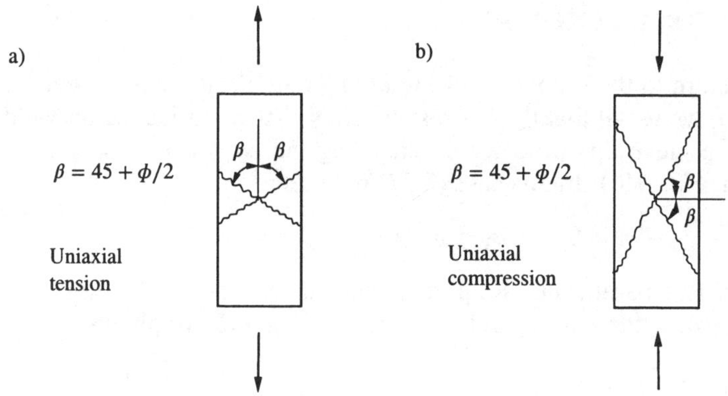

Figure 8.28: Illustration of Mohr's failure mode criterion; a) uniaxial tension; b) uniaxial compression.

Again it is assumed that the plane in Fig. 8.27b) on which the failure stresses σ and τ act is a failure plane. The angle β which this failure plane makes with the largest principal stress direction (σ_1) becomes $\beta = 45^o + \phi/2$ as illustrated in Fig. 8.27b). It is observed that the two failure planes shown in Figs. 8.26b) and 8.27b) both contain the direction of the intermediate principal stress direction (σ_2).

From the discussion above we conclude that

Mohr's failure mode criterion postulates that two failure planes exist. These planes contain the direction of the intermediate principal stress direction and they both make the angle $\beta = 45^o + \phi/2$ with the largest principal stress direction	(8.52)

It is emphasized that the angle $\beta = 45^o + \phi/2$ is the angle to the largest principal stress direction (σ_1) and that $\sigma_1 \geq \sigma_2 \geq \sigma_3$, where tension is considered as a positive quantity. From the discussion above follows directly that if a Mohr criterion is used and if ϕ denotes the current friction angle, cf. Fig. 8.22, then conclusion (8.52) also holds. Conclusion (8.52) is illustrated in Fig. 8.28.

It turns out that Mohr's failure mode criterion is often in fair agreement with experimental results for a variety of materials. However, it was emphasized previously that this ingenious failure mode criterion is a postulate and it is of considerable interest that later, in Chapter 24, we will prove that it follows strictly if so-called associated plasticity is adopted whereas so-called non-associated plasticity gives rise to other failure angles.

8.7 Tresca criterion

Let us return to the properties of initial yielding of metals and steel. Referring to (8.24), we recall that the hydrostatic stress has no influence on yielding and we may achieve this property by choosing the parameter $k = 1$ in Coulomb criterion, cf. (8.46). In this case (8.37) reduces to

$$\sigma_1 - \sigma_3 - m = 0 \; ; \quad \sigma_1 \geq \sigma_2 \geq \sigma_3 \tag{8.53}$$

where m is a parameter. Requiring that this expression for uniaxial tension should provide the initial yield stress σ_{yo}, cf. Fig. 8.1, we obtain

$$\boxed{\sigma_1 - \sigma_3 - \sigma_{yo} = 0} \tag{8.54}$$

This so-called *Tresca criterion* was suggested by Tresca (1864) in relation to his work on plastic response of metals. Referring to (3.24), it appears that Tresca's criterion states that the material yields when the maximum shear stress achieves a certain value.

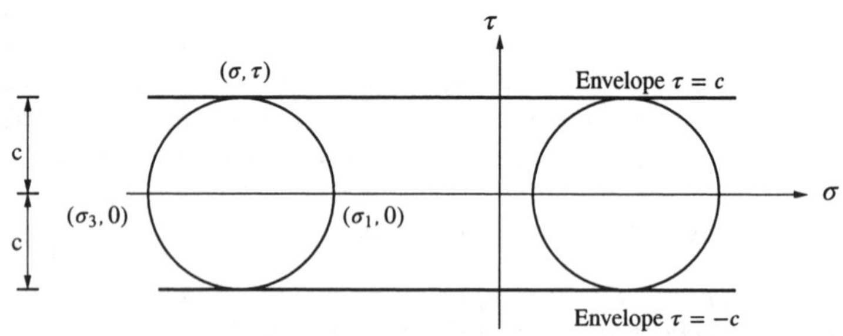

Figure 8.29: Tresca criterion in Mohr diagram.

Since $k = 1$, we conclude from (8.43) and (8.45) that the friction angle ϕ and the cohesion c are given by

$$\boxed{\phi = 0 \; ; \quad c = \frac{\sigma_{yo}}{2}}$$

i.e. in terms of the formulation (8.40), (8.53) is equivalent with

$$\boxed{|\,\tau| = c}$$

Therefore, in analogy with Fig. 8.21, we achieve the interpretation of this expression as shown in Fig. 8.29

To investigate the Tresca criterion (8.54) in the principal stress space, we obtain from (8.51) and Fig. 8.24 with $k = 1$ that $\rho_t = \rho_c$, i.e. we obtain the trace in the deviatoric plane as shown in Fig. 8.30a). It appears that Tresca's criterion corresponds to the lower bound displayed in Fig. 8.12. Moreover,

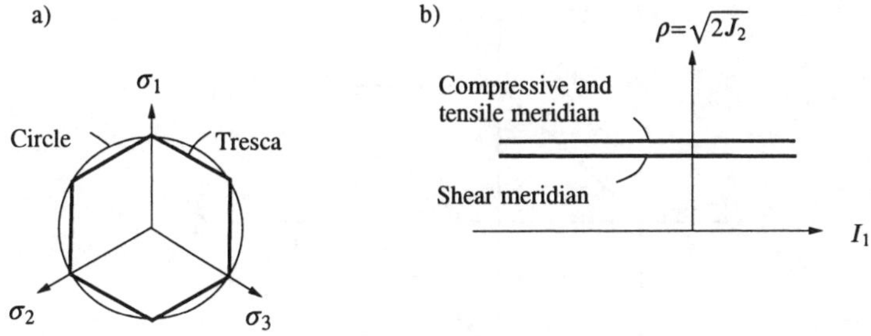

Figure 8.30: Tresca criterion; a) deviatoric plane; b) meridian plane

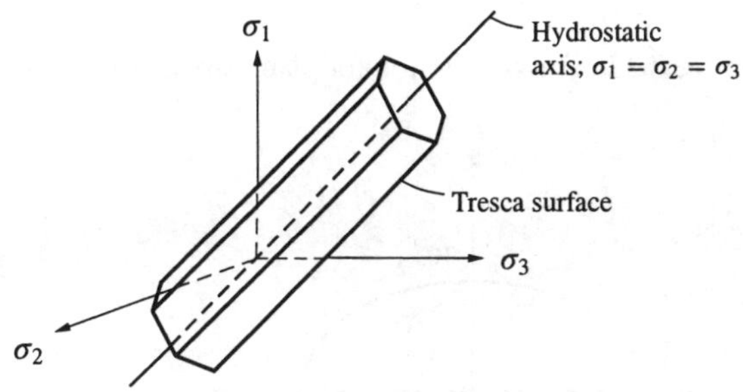

Figure 8.31: Tresca surface in the principal stress space.

from Fig. 8.23 with $k = 1$ the appearance in the meridian plane takes the form shown in Fig. 8.30b.

With these properties the illustration of the yield surface in the principal stress space is given in Fig. 8.31.

For plane stress conditions, (8.54) or Fig. 8.25 with $k = 1$ gives the result illustrated in Fig. 8.32 (note that in this figure the usual convention of $\sigma_1 \geq \sigma_2 \geq \sigma_3$ has been abandoned). This figure may be compared with the corresponding result for the von Mises criterion, cf. Fig. 8.15. Moreover, for simultaneous uniaxial stressing and torsion of, for instance, a thin-walled tube, we have

$$[\sigma_{ij}] = \begin{bmatrix} \sigma & \tau & 0 \\ \tau & 0 & 0 \\ 0 & 0 & 0 \end{bmatrix}$$

Let σ'_{ij} denote the stress components when the coordinate axes are collinear with the principal stress directions, i.e.

$$[\sigma'_{ij}] = \begin{bmatrix} \frac{\sigma}{2} + \frac{1}{2}\sqrt{\sigma^2 + 4\tau^2} & 0 & 0 \\ 0 & 0 & 0 \\ 0 & 0 & \frac{\sigma}{2} - \frac{1}{2}\sqrt{\sigma^2 + 4\tau^2} \end{bmatrix}$$

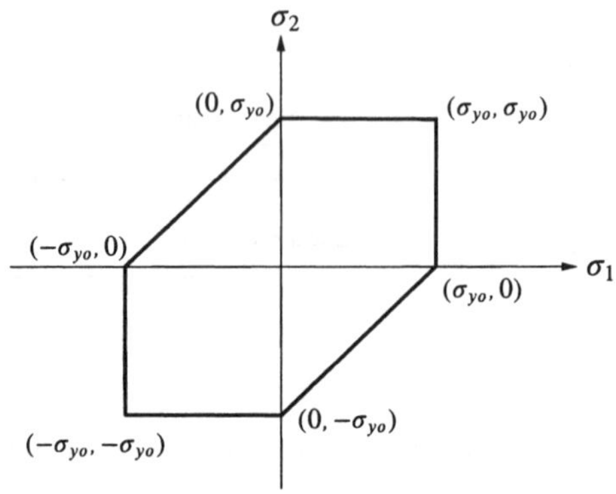

Figure 8.32: Tresca criterion for plane stress conditions.

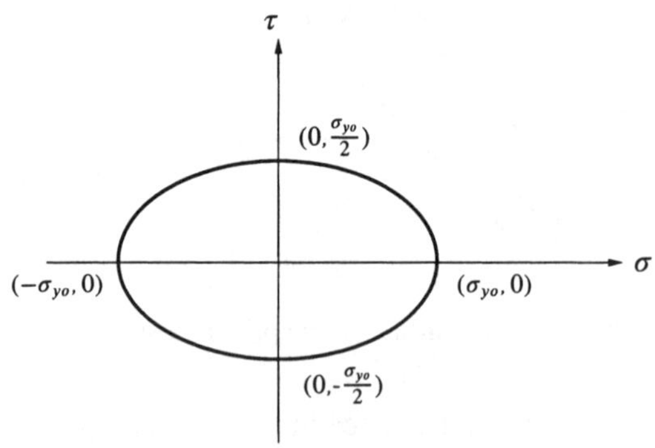

Figure 8.33: Tresca ellipse in the $\sigma\tau$-plane.

From this expression and (8.54) we obtain

$$\sqrt{\sigma^2 + 4\tau^2} - \sigma_{yo} = 0 \tag{8.55}$$

which represents an ellipse in the $\sigma\tau$–plane, cf. Fig. 8.33. Expression (8.55) may be compared with the corresponding result (8.29) using the von Mises criterion. From (8.55) with $\sigma = 0$, the initial yield shear stress becomes

$$\tau_{yo} = \frac{\sigma_{yo}}{2} \tag{8.56}$$

As this shear stress is located along the shear meridian and as the maximum deviation between the von Mises criterion and the Tresca criterion occurs along this meridian, cf. Fig. 8.30a), a comparison between (8.30) and (8.56) reveals that any Tresca yield stress, at most, is 13.4% lower than the corresponding von Mises yield stress.

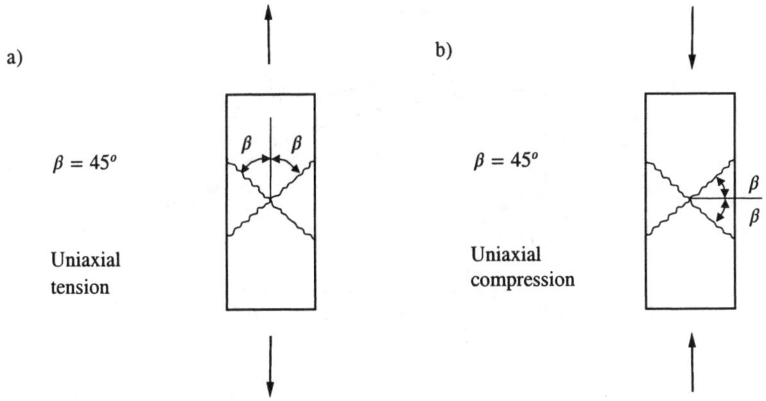

Figure 8.34: Mohr's failure mode criterion for Tresca yielding; a) uniaxial tension; b) uniaxial compression.

Since the friction angle $\phi = 0$ for Tresca's criterion, Mohr's failure mode criterion (8.52) states that any failure plane (slip plane) makes 45° with the maximum principal stress direction. This is illustrated in Fig. 8.34 and this prediction is in close agreement with experimental results for metals and steel.

8.8 Experimental results for metals and steel – von Mises versus Tresca

We have discussed the principal properties of different criteria in great detail, so it is timely to compare their predictions with experimental results. For this purpose, we will first concentrate on initial yielding of metals and steel. Later, in Sections 8.10 and 8.12, we will focus on failure stresses for concrete and, taken together, these comparisons will provide important information on how different materials behave. As a result, they will enable the reader to get a grasp of the accuracy that may be obtained using different criteria.

For initial yielding of metals and steel, we have already summarized the general experimental evidence in (8.24). Moreover, in relation to Figs. 8.12, 8.13 and 8.30 it can then be argued that Tresca's criterion must provide a lower bound whereas the von Mises criterion is located between the lower and upper bound. We also found that any Tresca yield stress, at most, is 13.4% lower than the corresponding von Mises yield stress. Let us now investigate whether these conclusions are in accordance with experimental data.

It was claimed that initial yielding of metals and steel is independent of the hydrostatic stress I_1. According to the extensive test series of Bridgman (1952), this assumption is closely fulfilled when $|I_1| \leq$ about $4\sigma_{yo}$, i.e. for all cases of

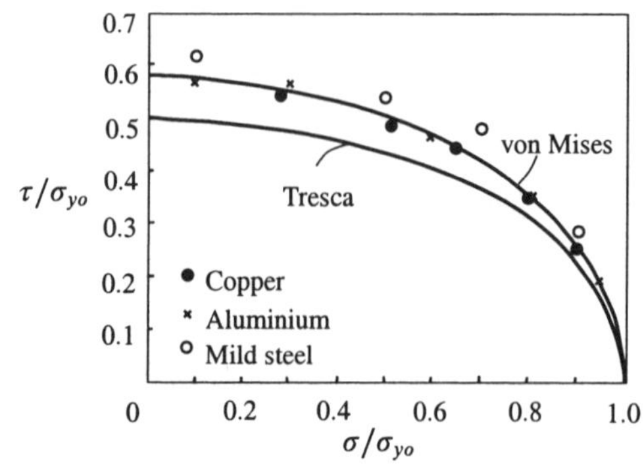

Figure 8.35: Experimental results of Taylor and Quinney (1931).

practical interest. The next issue mentioned in the summary (8.24) is that if the stress state σ_{ij} results in initial yielding so does the stress state $-\sigma_{ij}$. Also this assumption is closely fulfilled and as an example, the initial yield stress is the same for uniaxial tension and uniaxial compression. The last issue mentioned in (8.24) is the convexity of the yield surface and we will see that this assumption is also closely fulfilled.

The classical results of Taylor and Quinney (1931) shown in Fig. 8.35 were obtained by subjecting thin-walled tubes to combined tension and torsion. The figure also shows the ellipses of von Mises and Tresca in accordance with (8.29) and (8.55) and it appears that the von Mises criterion fits the experimental data considerably better than the Tresca criterion.

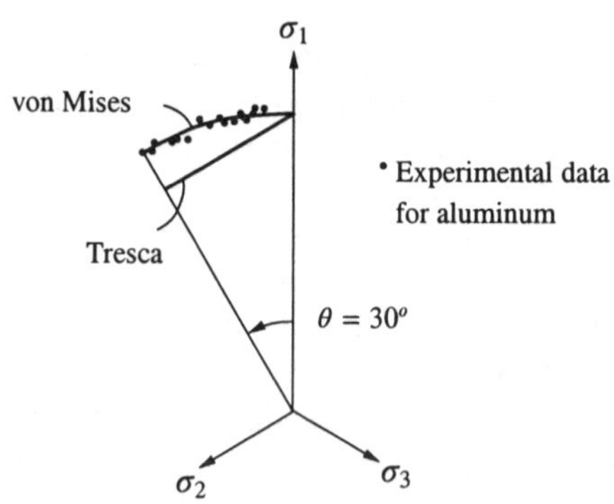

Figure 8.36: Deviatoric plane; experimental data of Lianis and Ford (1957).

The same conclusion may be drawn from the experimental results of Lianis and Ford (1957). They tested commercially pure aluminum specially treated so that a well defined yield stress is obtained. They used a specially designed notched specimen whereby arbitrary uniform states of combined stresses can be produced; the results are illustrated in the deviatoric plane in Fig. 8.36 together with the predictions of von Mises and Tresca. This figure also demonstrates the convexity of the yield surface.

It is concluded that the general experimental evidence summarized in (8.24) is well-founded and that the von Mises criterion fits the experimental data very closely and it should therefore, in general, be preferred as compared with the Tresca criterion.

8.9 Rankine criterion and modified Coulomb criterion

For brittle materials like concrete and rocks, the Coulomb criterion (8.38) is often used and the parameter k may be fitted to obtain a close agreement with experimental failure stresses for compressive stresses. As an example, in the next section we will see that $k \approx 4$ provides a good approximation for concrete. However, from Fig. 8.25 the Coulomb criterion is seen to predict a uniaxial tensile strength σ_t equal to σ_c/k, i.e. $\sigma_t \approx 0.25\,\sigma_c$ for concrete. This value is far too large for concrete where typical values for σ_t amounts to 5–12 % of the uniaxial compressive strength σ_c.

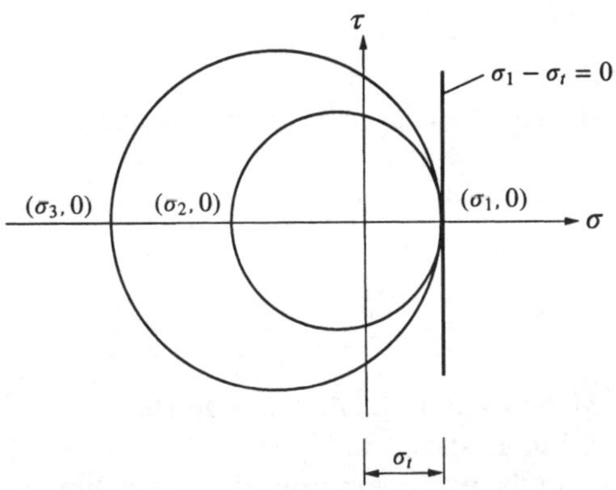

Figure 8.37: Rankine criterion viewed as a Coulomb criterion with the friction angle $\phi = 90^o$.

To remedy this deficiency one may assume a failure criterion in the form

$$\boxed{\sigma_1 - \sigma_t = 0\,; \qquad \sigma_1 \geq \sigma_2 \geq \sigma_3} \tag{8.57}$$

This so-called *Rankine criterion* was proposed by Rankine (1858) and, for obvious reasons, it is occasionally referred to as the *maximum principal stress criterion*. In a Mohr diagram, (8.57) takes the form shown in Fig. 8.37 and it is evident that (8.57) may be viewed as the envelope of all Mohr's stress circles

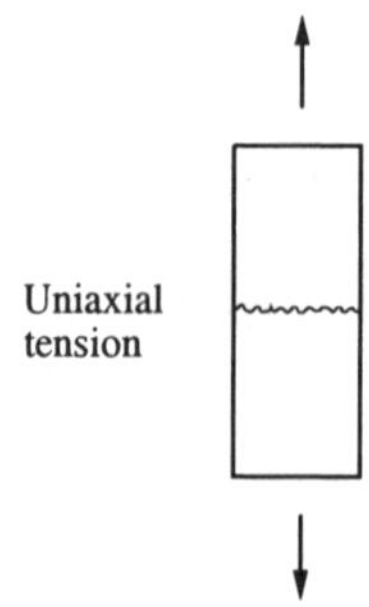

Figure 8.38: Mohr's failure mode criterion for Rankine failure.

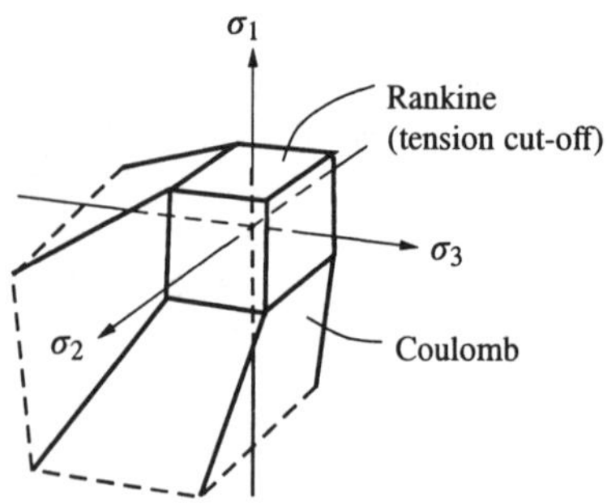

Figure 8.39: Modified Coulomb criterion in the principal stress space.

for which the stress state fulfills (8.57). That is, (8.57) may be viewed as a Coulomb criterion where the friction angle ϕ is given by

$$\boxed{\phi = 90^o} \tag{8.58}$$

cf. Fig. 8.21.

With (8.58) and Mohr's failure mode criterion (8.52) we obtain the failure plane for uniaxial tension as shown in Fig. 8.38; only one failure plane exists and it is perpendicular to the maximum principal stress direction. This result is in close agreement with experimental results for concrete and rocks where the failure manifests itself as a crack perpendicular to the maximum principal stress direction.

Let us return to the Coulomb criterion and its prediction given by Fig. 8.25. To remedy the too high uniaxial tensile strength predicted by the Coulomb criterion, we may use a combined failure criterion which states that failure is obtained, if

$$\boxed{k\sigma_1 - \sigma_3 - \sigma_c = 0 \quad \text{or} \quad \sigma_1 - \sigma_t = 0} \tag{8.59}$$

is fulfilled. This is the *modified Coulomb criterion* which due to its simplicity often is used in analytical calculations, cf. Nielsen (1984).

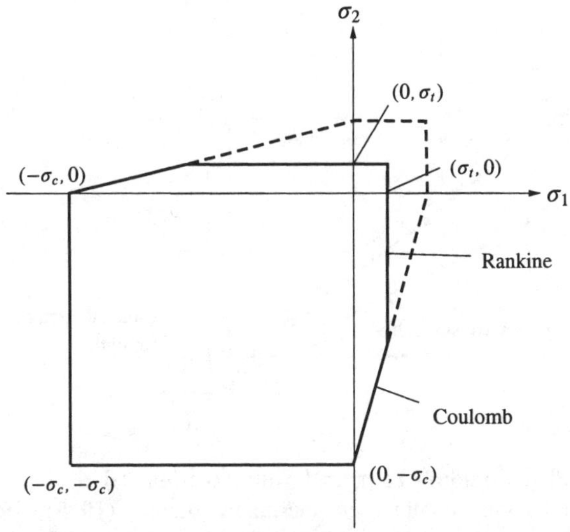

Figure 8.40: Modified Coulomb criterion for plane stress conditions.

Often the Rankine criterion in (8.57) is called a *tension cut-off criterion* and the appearance of the modified Coulomb criterion in the principal stress space is shown in Fig. 8.39. For biaxial stress states the modified Coulomb criterion is illustrated in Fig. 8.40.

8.10 Experimental results for concrete versus the modified Coulomb criterion

The predictions of the modified Coulomb criterion (8.59) will now be compared with experimental failure results for concrete. In this comparison we will adopt

$$k = 4\,; \qquad \text{i.e.} \qquad \phi = 36.9^o$$

where the friction angle ϕ was determined using (8.44).

Figure 8.41 shows the comparison along the compressive and tensile meridians for stresses ranging from tensile to very large triaxial compressive stresses. It appears that the modified Coulomb criterion provides a fair estimate.

For biaxial stress conditions, the modified Coulomb criterion is compared with the experimental results of Kupfer *et al.* (1969) in Fig. 8.42. Just like in Fig. 8.41, the modified Coulomb criterion underestimates the experimental results. It seems like the modified Coulomb criterion for stresses of most practical interest provides predictions which deviate up to 30% from the correct ones.

It appears that Figs. 8.41 and 8.42 confirm the general experimental evidence for concrete, soil and rocks already summarized in (8.25). There it was stated

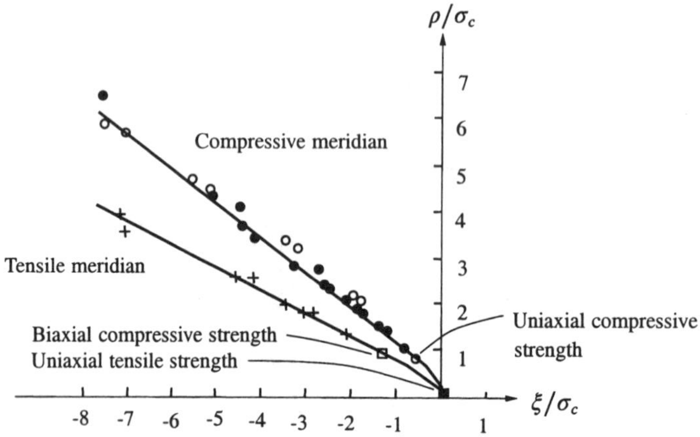

Figure 8.41: Modified Coulomb compared with experimental data in the meridian plane. Along compressive meridian: Balmer (1949)∘, Richart *et al.* (1928)•; along tensile meridian: Richart *et al.* (1928)+, Kupfer *et al.* (1969)□. Moreover, $\sigma_t = 0.08\sigma_c$ is assumed.

that inclusion of the term $\cos 3\theta$ in the failure criterion is of importance. This conclusion is evident from Fig. 8.41 and to substantiate this observation, we may adopt the Drucker-Prager criterion, which lacks the influence of the angle θ, cf. (8.33). Assume that the parameters α and β in (8.33) are calibrated along the compressive meridian. Along this meridian (8.47) holds and insertion into (8.33) yields

$$\frac{1+2\alpha}{1-\alpha}\sigma_1 - \sigma_3 - \frac{\beta}{1-\alpha} = 0$$

If the Coulomb criterion and the Drucker-Prager criterion are calibrated so that they coincide along the compression meridian, a comparison with (8.59) reveals that

$$k = \frac{1+2\alpha}{1-\alpha}; \qquad \sigma_c = \frac{\beta}{1-\alpha}$$

i.e.

$$\alpha = \frac{k-1}{2+k}; \qquad \beta = \frac{3\sigma_c}{2+k} \tag{8.60}$$

The biaxial compressive strength σ_{bc} is located along the tensile meridian, cf. Fig. 8.41, and σ_{bc} as predicted by the Drucker-Prager criterion is given by (8.35) which together with (8.60) results in

$$\sigma_{bc} = \frac{3\sigma_c}{4-k} \tag{8.61}$$

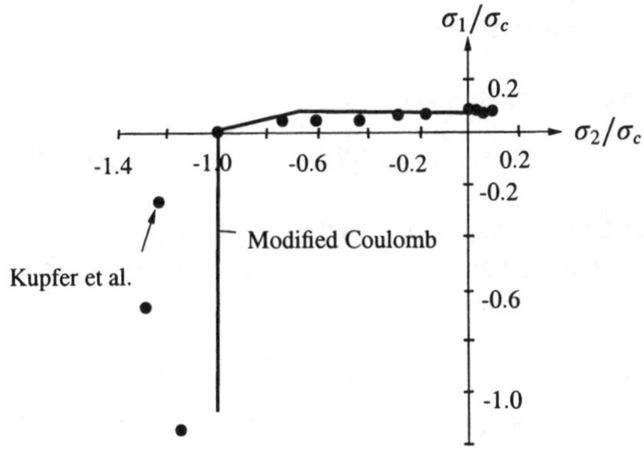

Figure 8.42: Modified Coulomb criterion compared with biaxial test results of Kupfer *et al.* (1969); $\sigma_t = 0.08\sigma_c$ is used.

For concrete $k \approx 4$ and (8.61) shows that if $k \to 4$ then σ_{bc} as predicted by the Drucker-Prager criterion approaches infinity! This surprising result was obtained even though the Drucker-Prager criterion was calibrated to fit the experimental data along the compression meridian closely. Figure 8.42 shows that in reality $\sigma_{bc} \approx 1.2\sigma_c$ and we conclude that the Drucker-Prager criterion should be used with caution. As already stated in Section 8.4, it may only be used with sufficient accuracy when the parameter α is small, i.e. when the influence of the hydrostatic stress I_1 is moderate. Cast iron may be representative of such a material.

8.11 4-parameter criterion

Concerning failure of concrete and rocks, we have discussed the (modified) Coulomb criterion and the Drucker-Prager criterion. The Coulomb criterion lacks the influence of the intermediate principal stress σ_2 and the surface in the principal stress space possesses sharp corners. On the other hand, the Drucker-Prager criterion lacks the influence of the $\cos 3\theta$-term. We will now discuss a criterion which avoids these deficiencies at the expense of a more complicated formulation.

The so-called *4-parameter criterion* was proposed by Ottosen (1977) and it reads

$$\boxed{A\frac{J_2}{\sigma_c^2} + \lambda\frac{\sqrt{J_2}}{\sigma_c} + B\frac{I_1}{\sigma_c} - 1 = 0} \qquad (8.62)$$

where the function $\lambda = \lambda(\cos 3\theta)$ is defined by

$$\lambda = \begin{cases} K_1 \cos[\frac{1}{3} \arccos(K_2 \cos 3\theta)] & \text{if} \quad \cos 3\theta \geq 0 \\ K_1 \cos[\frac{\pi}{3} - \frac{1}{3} \arccos(-K_2 \cos 3\theta)] & \text{if} \quad \cos 3\theta \leq 0 \end{cases} \tag{8.63}$$

Moreover, the four dimensionless parameters A, B, K_1 and K_2 are quantities to be determined from experimental evidence. We will require that A, B and K_1 are non-negative quantities and it appears from (8.63) that $0 \leq K_2 \leq 1$, i.e.

$$A \geq 0\,; \quad B \geq 0\,; \quad K_1 \geq 0\,; \quad 0 \leq K_2 \leq 1 \tag{8.64}$$

For reasons that will become apparent later on, K_1 is called a *size factor* whereas K_2 is a *shape factor*. Moreover, it appears from (8.63) and (8.64) that $0 \leq \lambda \leq K_1$.

It follows directly, that (8.62) reduces to the Drucker-Prager criterion (8.33) if $A = K_2 = 0$ and if, in addition, also $B = 0$ we obtain the von Mises criterion (8.26).

The failure surface given by (8.62) intersects the hydrostatic axis only at one point where I_1 takes the positive value given by

$$I_1 = \frac{\sigma_c}{B} \tag{8.65}$$

Moreover, (8.62) shows that the meridians are smooth, convex and curved and to obtain further insight we may solve (8.62) to obtain

$$\frac{\sqrt{J_2}}{\sigma_c} = \frac{1}{2A}\left[-\lambda + \sqrt{\lambda^2 - 4A(B\frac{I_1}{\sigma_c} - 1)}\right] \tag{8.66}$$

As $BI_1/\sigma_c - 1 \leq 0$, cf. (8.65), and as $A \geq 0$ the discriminant of (8.66) is always non-negative. The trace of the failure surface in the deviatoric plane is given by (8.66) for I_1=constant.

If $K_2 = 0$ then λ as given by (8.63) becomes a constant quantity independent of the angle θ. In this case, (8.66) shows that the trace in the deviatoric plane becomes a circle. On the other hand, if $0 < K_2 \leq 1$ then λ depends on the angle θ, i.e. (8.66) shows that also the trace in the deviatoric plane depends on the angle θ. This suggests the terminology of K_2 being a shape factor and it follows directly from (8.63) and (8.66) that K_1 influences the size of the trace in the deviatoric plane.

It is not difficult to prove that if the function $r = 1/\lambda(cos3\theta)$ describes a smooth and convex curve in the polar coordinate system (r, θ), then the trace of the failure surface in the deviatoric plane as given by (8.66) is also a smooth and convex curve; the details of this proof are given by Ottosen (1989).

It is therefore required that the function $r = 1/\lambda(cos3\theta)$ describes a smooth and convex curve in the polar coordinate system (r, θ). The smooth convex contour lines of a deflected membrane loaded by a lateral pressure and supported

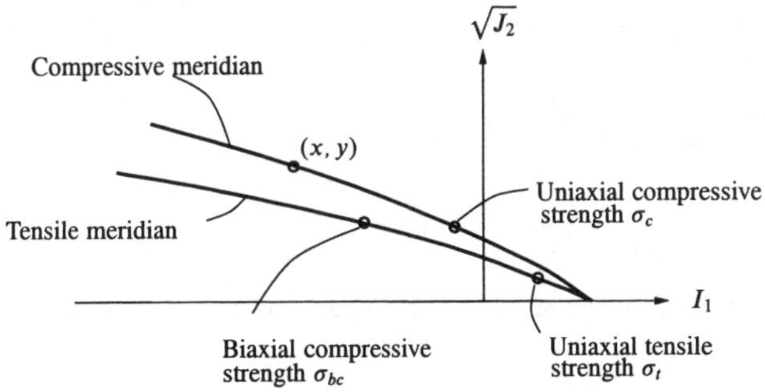

Figure 8.43: Sketch of the four failure states used to calibrate the 4-parameter criterion.

along the edges of an equilateral triangle fulfill these requirements and expression (8.63) was, in fact, derived on this basis. The details of this derivation were given by Ottosen (1975) and they can also be found in Chen (1982).

For the criterion given by (8.62) and (8.63), we have therefore shown that the trace of the failure surface in the deviatoric plane is smooth and convex and thus, it generally takes the form illustrated in Fig. 8.6. We also recall that the meridians are smooth, curved and convex.

To calibrate the four parameters A, B, K_1 and K_2 appearing in (8.62) and (8.63), knowledge of four arbitrary failure stress states is necessary. In practice, the following failure stress states are conveniently used:

1) uniaxial compressive strength σ_c
2) biaxial compressive strength σ_{bc}
3) uniaxial tensile strength σ_t (8.67)
4) an arbitrary failure state $(I_1, \sqrt{J_2}) = (x, y)$ along the compressive meridian

where we recall that σ_c, σ_{bc} and σ_t are all positive quantities. It appears that failure states 1) and 4) are located on the compressive meridian whereas failure states 2) and 3) are located on the tensile meridian, cf. Fig. 8.43.

Naturally, a numerical approach may be used to determine the four parameters A, B, K_1 and K_2 from the failure states given by (8.67). However, an explicit analytical approach is clearly preferable and in the following we will present such an approach.

Let us first determine the expressions λ_t and λ_c for λ along the tensile and compressive meridian respectively. It follows from (8.63) that

$$\begin{aligned} \lambda_t &= \lambda(\theta = 0^o) = K_1 \cos(\tfrac{1}{3} \arccos K_2) \\ \lambda_c &= \lambda(\theta = 60^o) = K_1 \cos(\tfrac{\pi}{3} - \tfrac{1}{3} \arccos K_2) \end{aligned} \qquad (8.68)$$

From the failure states 2) and 3) of (8.67) - both located on the tensile meridian

– use of (8.49) and (8.62) yields

$$A\frac{\sigma_{bc}^2}{3\sigma_c^2}+\lambda_t\frac{\sigma_{bc}}{\sigma_c\sqrt{3}}-B\frac{2\sigma_{bc}}{\sigma_c}-1=0$$
$$A\frac{\sigma_t^2}{3\sigma_c^2}+\lambda_t\frac{\sigma_t}{\sigma_c\sqrt{3}}+B\frac{\sigma_t}{\sigma_c}-1=0 \tag{8.69}$$

Likewise, from the failure states 1) and 4) located on the compressive meridian, we find from (8.47) and (8.62)

$$\frac{A}{3}+\lambda_c\frac{1}{\sqrt{3}}-B-1=0$$
$$A\frac{y^2}{\sigma_c^2}+\lambda_c\frac{y}{\sigma_c}+B\frac{x}{\sigma_c}-1=0 \tag{8.70}$$

Elimination of λ_t from (8.69) and elimination of λ_c from (8.70) result in the following equations

$$A-\frac{9\sigma_c}{\sigma_{bc}-\sigma_t}B=-\frac{3\sigma_c^2}{\sigma_{bc}\sigma_t}$$
$$-\frac{y}{\sigma_c}A-\kappa B=\sqrt{3} \tag{8.71}$$

where the dimensionless parameter κ is defined by

$$\boxed{\kappa=\frac{x+y\sqrt{3}}{y-\dfrac{\sigma_c}{\sqrt{3}}}} \tag{8.72}$$

It appears that κ is a known quantity. Elimination of A from (8.71a) and (8.71b) gives

$$\boxed{B=\frac{\dfrac{3\sigma_c y}{\sigma_{bc}\sigma_t}-\sqrt{3}}{\kappa+\dfrac{9y}{\sigma_{bc}-\sigma_t}}} \tag{8.73}$$

Having determined the parameter B, A can be determined from (8.71b), i.e.

$$\boxed{A=-\frac{\sigma_c}{y}(\kappa B+\sqrt{3})}$$

Having determined the parameters A and B, (8.69a) and (8.70a) provide

$$\boxed{\begin{aligned}\lambda_t&=\sqrt{3}\left[\frac{\sigma_c}{\sigma_{bc}}+2B-\frac{\sigma_{bc}}{3\sigma_c}A\right]\\ \lambda_c&=\sqrt{3}\left[1+B-\frac{A}{3}\right]\end{aligned}} \tag{8.74}$$

i.e. λ_t and λ_c are now known quantities that, according to (8.68), depend on K_1 and K_2 and thus, the next issue is to determine these parameters.

Define the quantity α by

$$\alpha = \frac{1}{3}\arccos K_2 \tag{8.75}$$

From (8.68) and the relation $\cos(\frac{\pi}{3} - \alpha) = \cos\frac{\pi}{3}\cos\alpha + \sin\frac{\pi}{3}\sin\alpha$, we then obtain

$$\frac{\lambda_t}{K_1} = \cos\alpha\,; \qquad \frac{\lambda_c}{K_1} = \frac{1}{2}\cos\alpha + \frac{\sqrt{3}}{2}\sin\alpha \tag{8.76}$$

Since $\sin\alpha = \pm\sqrt{1-\cos^2\alpha}$, use of (8.76a) in (8.76b) gives

$$\boxed{K_1 = \frac{2}{\sqrt{3}}\sqrt{\lambda_t^2 + \lambda_c^2 - \lambda_t\lambda_c}} \tag{8.77}$$

where it was used that K_1 is non-negative, cf. (8.64). Since λ_t and λ_c are known also the parameter K_1 can now be determined. From (8.75) follows that $\cos 3\alpha = K_2$ and with the identity $\cos 3\alpha = 4\cos^3\alpha - 3\cos\alpha$ and (8.76a), we conclude that

$$\boxed{K_2 = 4\left(\frac{\lambda_t}{K_1}\right)^3 - 3\frac{\lambda_t}{K_1}} \tag{8.78}$$

From the failure states (8.67), we are therefore able to determine the four parameters A, B, K_1 and K_2 by means of explicit analytical expressions.

As a simple example of the use of these formulas and considering (8.62) and (8.63) as an initial yield criterion, we assume in accordance with experimental evidence for metals and steel that $\sigma_c = \sigma_t = \sigma_{bc} = \sigma_{yo}$. Then (8.73) shows that $B = 0$, i.e. there is no influence of the hydrostatic stress. Expression (8.74) then implies that $\lambda_t = \lambda_c = \sqrt{3}(1 - A/3)$ and we would expect that there is no influence of the angle θ. This is easily verified since (8.77) with $\lambda_t = \lambda_c$ yields $K_1 = 2\lambda_t/\sqrt{3}$, i.e. (8.78) shows that $K_2 = 0$, as expected. Finally, use of $\lambda = \lambda_t = \lambda_c = \sqrt{3}(1 - A/3)$ and $B = 0$ in (8.66) results in

$$\sqrt{J_2} = \frac{\sigma_c}{\sqrt{3}}$$

and we have then recaptured the von Mises criterion, cf. (8.26). In the next section, we shall see some more advanced applications of the calibration formulas derived above.

8.12 Experimental results for concrete versus the 4-parameter criterion

The 4-parameter criterion (8.62) and (8.63) is a very general criterion that can provide close predictions for a variety of materials. In this section we will compare its predictions with experimental failure stresses for concrete.

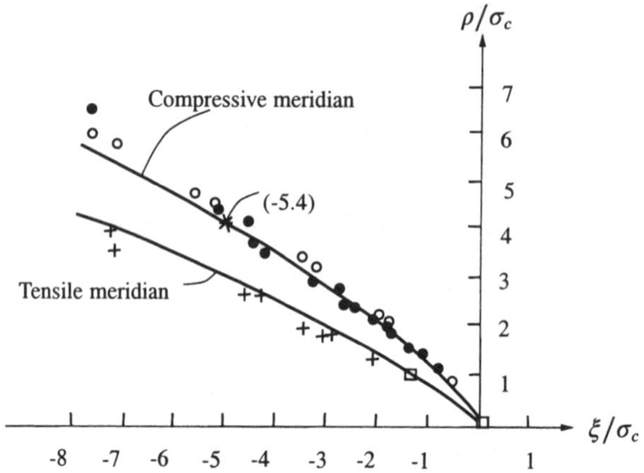

Figure 8.44: 4-parameter criterion compared with experimental data in the meridian plane. Along compressive meridian: Balmer (1949)∘, Richart *et al.* (1928)•; along tensile meridian: Richart *et al.* (1928)+, Kupfer *et al.* (1969)□. Moreover, $\sigma_t = 0.10\sigma_c$ is assumed.

To specify the failure states (8.67), we choose

$$\begin{aligned} &1) \text{ uniaxial compressive strength } \sigma_c \\ &2) \text{ biaxial compressive strength } \sigma_{bc} = 1.16\,\sigma_c \\ &3) \text{ uniaxial tensile strength } \sigma_t = 0.08\,\sigma_c,\ 0.10\,\sigma_c \text{ and } 0.12\,\sigma_c \\ &4) \text{ failure state on the compressive meridian} \\ &\quad (I_1, \sqrt{J_2}) = (x, y) = (\xi\sqrt{3}, \rho/\sqrt{2}) = (-5\sqrt{3}\sigma_c, 4\sigma_c/\sqrt{2}) \end{aligned} \tag{8.79}$$

Failure state 2) corresponds to the experimental data of Kupfer *et al.* (1969), cf. Fig. 8.42, and the uniaxial tensile strengths given by 3) are typical for concrete. Finally, failure state 4), where relations (8.11) and (8.13) are recalled, is supported by the experimental results shown in Fig. 8.41. With (8.79) and the formulas (8.72) - (8.74), (8.77) and (8.78), the resulting parameter values as well as information of λ_t and λ_c are given in Tables 8.1 and 8.2.

Although the parameters A, B, K_1 and K_2 show considerable dependence on the σ_t/σ_c-ratio, the failure stresses, when only compressive stresses occur, are influenced to a minor extent. Using $\sigma_t/\sigma_c = 0.10$ as a reference, the difference for compressive stresses amounts to less than 2.5 %.

σ_t/σ_c	A	B	K_1	K_2
0.08	1.8076	4.0962	14.4863	0.9914
0.10	1.2759	3.1962	11.7365	0.9801
0.12	0.9218	2.5969	9.9110	0.9647

Table 8.1: Parameter values and their dependence on the σ_t/σ_c– ratio.

σ_t/σ_c	λ_t	λ_c	λ_c/λ_t
0.08	14.4725	7.7834	0.5378
0.10	11.7109	6.5315	0.5577
0.12	9.8720	5.6979	0.5772

Table 8.2: λ_t- and λ_c-values and their dependence on the σ_t/σ_c–ratio.

With these parameter values, the predictions of the 4-parameter criterion are compared with the same experimental data as used in Figs. 8.41 and 8.42. Figure 8.44 shows results along the compressive and tensile meridians and for clarity, the calibration point 4) where $(\xi/\sigma_c, \rho/\sigma_c)$ = (-5, 4) is indicated. It may be noted that failure stresses for which $\xi < -5\sigma_c$ are very seldom found in practice and that other experimental data indicates smaller ρ-values in this stress range along the compressive meridian than those shown in Fig. 8.44, cf. Ottosen (1977). For biaxial stress states, a comparison of the predictions and the experimental results of Kupfer *et al.* (1969) is shown in Fig. 8.45. On the whole, Figs. 8.44 and 8.45 show a satisfactory agreement.

The experimental triaxial test results illustrated in Fig. 8.44 are quite old (Richart *et al.* (1928), Balmer (1949)) and especially the failure stresses of Balmer (1949) are larger than other triaxial test data. Thus, it may be of interest to compare the predictions of the 4-parameter criterion with the very accurate triaxial and biaxial results obtained by Schickert and Winkler (1977).

To calibrate the parameters to these data, we now choose the following failure states

$$\begin{aligned}
&1) \text{ uniaxial compressive strength } \sigma_c \\
&2) \text{ biaxial compressive strength } \sigma_{bc} = 1.21\,\sigma_c \\
&3) \text{ uniaxial tensile strength } \sigma_t = 0.10\,\sigma_c \\
&4) \text{ failure state on the compressive meridian } (I_1, \sqrt{J_2}) \\
&\quad = (x, y) = (\xi\sqrt{3}, \rho/\sqrt{2}) = (-5\sqrt{3}\sigma_c, 3.28\sigma_c/\sqrt{2})
\end{aligned} \tag{8.80}$$

In accordance with the previous discussion, the failure state 4) corresponds to

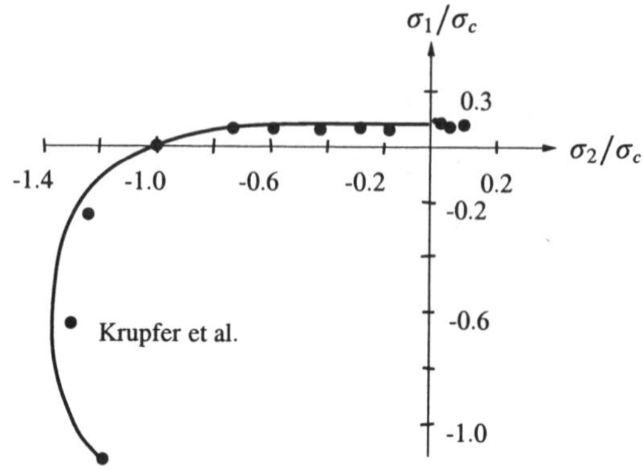

Figure 8.45: 4-parameter criterion compared with biaxial test results of Kupfer *et al.* (1969). Moreover, $\sigma_t = 0.08\,\sigma_c$ is assumed.

A	B	K_1	K_2	λ_c	λ_t	λ_c/λ_t
3.2244	3.4555	11.1538	0.9962	5.8553	11.1491	0.5252

Table 8.3: Parameter values using experimental data of Schickert and Winkler (1977); $\sigma_t/\sigma_c = 0.10$.

somewhat lower failure stresses, cf. (8.79). With (8.80), the parameter values given in Table 8.3 are obtained.

The triaxial results of Schickert and Winkler are of special interest as they were obtained not only along the compressive and tensile meridian, but also along the shear meridian. Moreover, both proportional and non-proportional load paths were used for the triaxial tests and the corresponding effect on the failure stresses may therefore be evaluated. The results of Schickert and Winkler (1977) were presented in terms of octahedral stresses and we recall from (8.32) that $\sigma_o = I_1/3$ and $\tau_o = \sqrt{2J_2/3}$. For triaxial loading, the following four different load paths were used:

* Path 1 : hydrostatic loading until a prescribed σ_o- value. This σ_o-value is now held constant and τ_o is then increased along the compressive meridian.
* Path 2 : hydrostatic loading until a prescribed σ_o- value. This σ_o-value is now held constant and τ_o is then increased along the shear meridian.

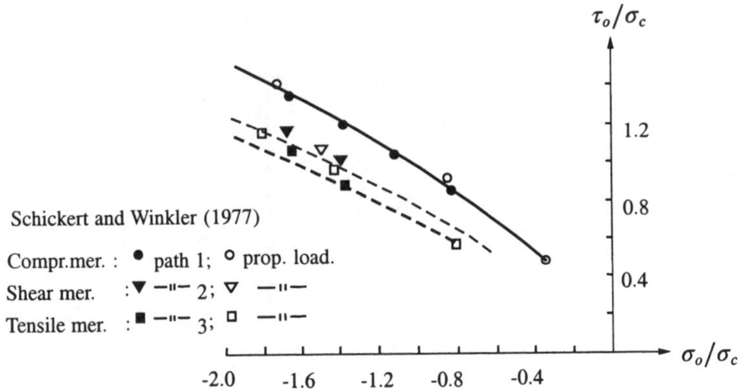

Figure 8.46: 4-parameter criterion compared with experimental data of Schickert and Winkler (1977) in the meridian plane.

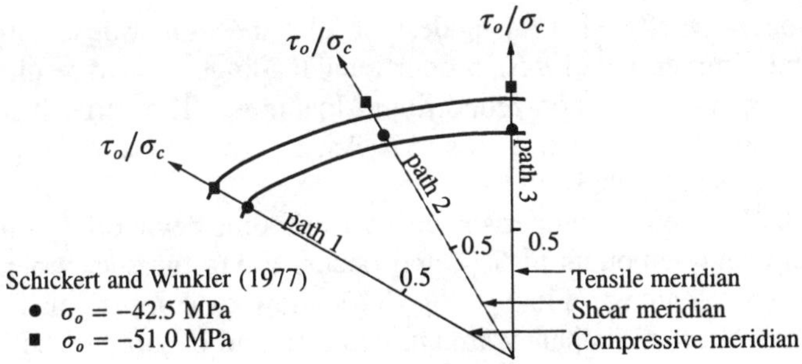

Figure 8.47: 4-parameter criterion compared with experimental data of Schickert and Winkler (1977) in two different deviatoric planes.

* Path 3 : hydrostatic loading until a prescribed σ_o- value. This σ_o-value is now held constant and τ_o is then increased along the tensile meridian.

* Proportional loading along the compressive, shear and tensile meridian.

Figures 8.46 - 8.48 show a comparison of predictions with experimental data. Every experimental point is a mean value based upon three to six tests. The uniaxial compressive strength of the tested concrete is $\sigma_c = 30.6$ MPa.

Figure 8.46 shows the compressive, shear and tensile meridians. The agreement along the compressive meridian is close as the (x,y)-value previously discussed was chosen so that deviations were minimized along that meridian. The largest discrepancy amounts to 7 % and occurs along the tensile meridian for large hydrostatic loading.

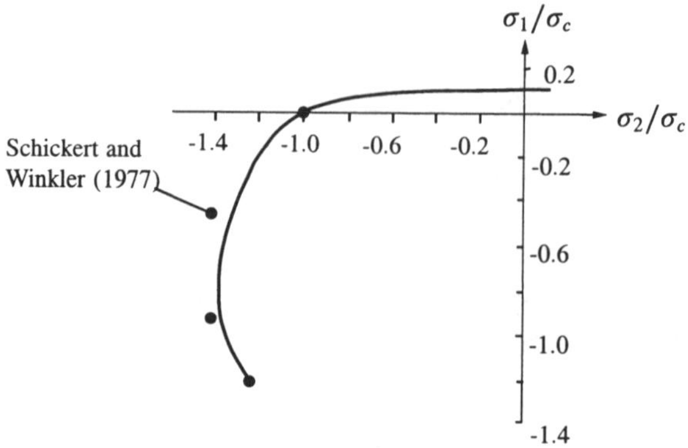

Figure 8.48: 4-parameter criterion compared with biaxial test results of Schickert and Winkler (1977).

Even though the effect is very modest, cf. also the similar observation made by Chinn and Zimmerman (1965), proportional loadings seem to result in higher strengths as compared to non-proportional loadings. The same insignificant tendency is observed for triaxial tests of rocks, cf. Swanson and Brown (1971) and for soil, cf. Scott (1963).

Figure 8.47 shows a comparison in two different deviatoric planes. The largest discrepancy amounts to 5 % and occurs along the shear meridian for large hydrostatic loading. Finally, Fig. 8.48 shows a comparison for biaxial loading. The largest discrepancy amounts to 8 % and occurs for the load path $\sigma_1/\sigma_2 = 1/3$. Based on this extensive comparison with a range of different experimental data, it seems fair to state that it is possible to make close predictions of the failure conditions for general stress conditions. This agreement is in accordance with the general consensus expressed by ASCE (1982), Chen (1982) and Eibl *et al.* (1983) and we may note that the 4-parameter criterion is included in the model code of CEB (Comité Euro-International du Béton) as well as in the Euro-code.

It is observed that other advanced failure criteria for concrete are available that also provide accurate predictions, for instance, the proposals of Hsieh *et al.* (1982), Willam and Warnke (1974) and Podgorski (1985). A detailed description of the first two of these criteria may also be found in Chen (1982). We also refer to the very comprehensive experimental program reported by Gerstle *et al.* (1980)

We finally recall that qualitatively the same features are observed for the failure conditions of concrete, rocks and soil. In that respect, we refer to the failure condition of Krenk (1996) and the proposal of Lade (1977) that is often adopted within soil mechanics.

8.13 Anisotropic criteria

We have presented a rather detailed discussion of concepts and specific criteria for materials that are isotropic. However, many engineering materials are not isotropic and, as examples, we might mention that ice, rolled steel and aluminum exhibit orthotropic properties whereas wood, paper and stratified rocks are strongly orthotropic, cf. the discussion of anisotropy given in Sections 4.3 and 4.6. We will therefore conclude this chapter by a discussion of criteria that account for anisotropic effects. It is emphasized that the intention of this chapter goes beyond a mere presentation of different criteria. Rather, it is to present a mainstream where each of the criteria discussed involves features not covered by the other criteria.

According to (8.2) the yield function is an invariant. For isotropic and anisotropic elasticity, the elastic stiffness tensor D_{ijkl} characterizes the stiffness properties of the material. We may similarly assume that initial yielding for an isotropic or anisotropic material is characterized by the fourth-order tensor P_{ijkl}. The quantity $\sigma_{ij}P_{ijkl}\sigma_{kl}$ is then an invariant and we may then postulate the following criterion

$$\sigma_{ij}P_{ijkl}\sigma_{kl} - 1 = 0 \tag{8.81}$$

If, furthermore, initial yielding is independent of the hydrostatic stress, a valid form of the yield criterion is

$$s_{ij}P_{ijkl}s_{kl} - 1 = 0 \tag{8.82}$$

Depending on the choice of P_{ijkl}, this form can express yielding of various anisotropic materials. As an example, in the extreme case where isotropy is involved, we choose P_{ijkl} as the following isotropic fourth-order tensor

$$P_{ijkl} = \frac{3}{4\sigma_{yo}^2}(\delta_{ik}\delta_{jl} + \delta_{il}\delta_{jk})$$

(8.82) then becomes

$$\frac{3}{2\sigma_{yo}^2}s_{ij}s_{ij} - 1 = 0 \tag{8.83}$$

which is exactly the von Mises criterion, cf. (8.26). With reference to (4.89), the most general isotropic fourth-order tensor P_{ijkl} also includes the term $\delta_{ij}\delta_{kl}$, but since $s_{ii} = 0$ the effect of this term in P_{ijkl} drops out. Written explicitly, (8.83) becomes

$$\frac{1}{2\sigma_{yo}^2}[3s_{11}^2 + 3s_{22}^2 + 3s_{33}^2 + 6s_{12}^2 + 6s_{13}^2 + 6s_{23}^2] - 1 = 0 \tag{8.84}$$

Since $s_{ii} = 0$, we have $3s_{11}^2 = 2s_{11}^2 + s_{11}^2 = 2s_{11}^2 - s_{11}(s_{22} + s_{33})$ as well as $3s_{22}^2 = 2s_{22}^2 - s_{22}(s_{11} + s_{33})$ and $3s_{33}^2 = 2s_{33}^2 - s_{33}(s_{11} + s_{22})$ and (8.84) then takes the following alternative form

$$\frac{1}{2\sigma_{yo}^2}[(s_{11} - s_{22})^2 + (s_{11} - s_{33})^2 + (s_{22} - s_{33})^2 + 6s_{12}^2 + 6s_{13}^2 + 6s_{23}^2] - 1 = 0 \tag{8.85}$$

Let us return to the format (8.81). Changing to a matrix format, it reads

$$\boxed{\boldsymbol{\sigma}^T \boldsymbol{P} \boldsymbol{\sigma} - 1 = 0} \tag{8.86}$$

where $\boldsymbol{\sigma}$ is the column matrix defined by (4.35) and $\boldsymbol{P}$ is a 6×6 matrix. This matrix format is simpler to work with than its tensorial counterpart. However, the price we pay is that (8.86) now refers to a specific coordinate system and proper transformations need to be performed if another coordinate system is chosen. Since the stress components only appear in terms of quadratic expressions, assumption (8.86) implies that if the stress state σ_{ij} results in initial yielding, so does the reversed stress state $-\sigma_{ij}$; for instance, tension and compression in the same direction exhibit the same yield stress. This is a valid assumption for metals and steel.

Suppose that $\boldsymbol{P}$ consists of a symmetric part $\boldsymbol{P}^s$ and an anti-symmetric part $\boldsymbol{P}^a$, i.e. $\boldsymbol{P} = \boldsymbol{P}^s + \boldsymbol{P}^a$, where, according to the definitions of symmetry and anti-symmetry, we have $(\boldsymbol{P}^s)^T = \boldsymbol{P}^s$ and $(\boldsymbol{P}^a)^T = -\boldsymbol{P}^a$. Consider the quantity $b = \boldsymbol{\sigma}^T \boldsymbol{P}^a \boldsymbol{\sigma}$. Since b is a number, we have $b = \boldsymbol{\sigma}^T \boldsymbol{P}^a \boldsymbol{\sigma} = (\boldsymbol{\sigma}^T \boldsymbol{P}^a \boldsymbol{\sigma})^T = \boldsymbol{\sigma}^T (\boldsymbol{P}^a)^T \boldsymbol{\sigma} = -\boldsymbol{\sigma}^T \boldsymbol{P}^a \boldsymbol{\sigma} = -b$ and it is concluded that $b = \boldsymbol{\sigma}^T \boldsymbol{P}^a \boldsymbol{\sigma} = 0$. Without loss of generality, we can therefore take the matrix $\boldsymbol{P}$ in (8.86) to be symmetric, i.e.

$$\boldsymbol{P} = \boldsymbol{P}^T$$

It is concluded that $\boldsymbol{P}$ in the most general case of anisotropy comprises 21 independent components, i.e. 21 material parameters. This is similar to the case when the elastic stiffness $\boldsymbol{D}$ is considered, cf. (4.49).

The anisotropic criterion (8.86) with 21 material parameters for a fully anisotropic material was proposed by von Mises (1928). It seems to be overlooked in the literature that this criterion apparently was the first criterion for anisotropic materials and in a moment we will see that it contains as a special case, a criterion that is often used in practice.

Using the symmetry property of $\boldsymbol{P}$, it can be written as

$$\boldsymbol{P} = \begin{bmatrix} A & -F & -G & P_{14} & P_{15} & P_{16} \\ -F & B & -H & P_{24} & P_{25} & P_{26} \\ -G & -H & C & P_{34} & P_{35} & P_{36} \\ P_{14} & P_{24} & P_{34} & 2L & P_{45} & P_{46} \\ P_{15} & P_{25} & P_{35} & P_{45} & 2M & P_{56} \\ P_{16} & P_{26} & P_{36} & P_{46} & P_{56} & 2N \end{bmatrix} \tag{8.87}$$

where the special notation for the components will turn out to be convenient for later purposes.

Let us split the stress components into a deviatoric part and a hydrostatic part. In matrix notation, this can be written as

$$\boldsymbol{\sigma} = \boldsymbol{s} + \boldsymbol{e} \tag{8.88}$$

where

$$\boldsymbol{\sigma} = \begin{bmatrix} \sigma_{11} \\ \sigma_{22} \\ \sigma_{33} \\ \sigma_{12} \\ \sigma_{13} \\ \sigma_{23} \end{bmatrix} ; \quad \boldsymbol{s} = \begin{bmatrix} s_{11} \\ s_{22} \\ s_{33} \\ s_{12} \\ s_{13} \\ s_{23} \end{bmatrix} ; \quad \boldsymbol{e} = \frac{1}{3} I_1 \begin{bmatrix} 1 \\ 1 \\ 1 \\ 0 \\ 0 \\ 0 \end{bmatrix} \tag{8.89}$$

Insertion of (8.88) into (8.86) gives

$$\boldsymbol{s}^T \boldsymbol{P} \boldsymbol{s} + (2\boldsymbol{s}^T + \boldsymbol{e}^T)\boldsymbol{P}\boldsymbol{e} - 1 = 0 \tag{8.90}$$

Let us assume that initial yielding of the anisotropic material only depends on the deviatoric stresses; this will be the case, for instance, when considering anisotropic metals and steel. Since the deviatoric part $\boldsymbol{s}$ and the hydrostatic part $\boldsymbol{e}$ can be chosen independent of each other, (8.90) shows that we must require

$$\boldsymbol{P}\boldsymbol{e} = \boldsymbol{0} \tag{8.91}$$

With $\boldsymbol{P}$ and $\boldsymbol{e}$ given by (8.87) and (8.89) respectively, it appears that (8.91) provides 6 equations and the number of independent material parameters has been reduced from 21 to 15 just by requiring yielding to depend on the deviatoric stresses alone. Moreover with (8.91), (8.90) reduces to

$$\boxed{\boldsymbol{s}^T \boldsymbol{P} \boldsymbol{s} - 1 = 0} \tag{8.92}$$

Let us next assume that the material behaves orthotropically with respect to yielding. Previously, we discussed the stiffness matrix $\boldsymbol{D}$ when the material behaves orthotropically, cf. the discussion leading to (4.55). There, orthotropy implies that there exist three planes – so-called elastic symmetry planes – so that $\boldsymbol{D}$ is unchanged for two coordinate systems that are mirror images of each other in these symmetry planes, cf. (4.50). The result of orthotropy, i.e. three elastic symmetry planes, then implies that the matrix $\boldsymbol{D}$ is given by (4.55) where nine independent material parameters exist. In exactly the same manner and considering now orthotropic yield properties, this means that the matrix $\boldsymbol{P}$ is unchanged for two coordinate systems that are mirror images of each other in the three symmetry planes that exist with respect to yielding. While $\boldsymbol{P}$ in the general case is given by (8.87), we conclude that for orthotropy with respect to yielding, and with the symmetry planes being parallel with the coordinate planes, we require $s'^T P s' = s^T P s$ where s' is related to s by expressions similar to (4.52), (4.54) and (4.56). Trivial calculations show that

$$P = \begin{bmatrix} A & -F & -G & 0 & 0 & 0 \\ -F & B & -H & 0 & 0 & 0 \\ -G & -H & C & 0 & 0 & 0 \\ 0 & 0 & 0 & 2L & 0 & 0 \\ 0 & 0 & 0 & 0 & 2M & 0 \\ 0 & 0 & 0 & 0 & 0 & 2N \end{bmatrix} \tag{8.93}$$

i.e. nine independent materials in complete similarity with (4.55). In order to be as general as possible, we have here distinguished between elastic symmetry planes and the symmetry planes with respect to yielding, but in practice these two sets of planes are expected to coincide.

Since only deviatoric stresses were assumed to influence initial yielding, we have restriction (8.91) which with (8.93) leads to $A = F + G$, $B = F + H$ and $C = G + H$. Thus, we have only six independent material parameters and (8.93) then reduces to

$$\boxed{P = \begin{bmatrix} F+G & -F & -G & 0 & 0 & 0 \\ -F & F+H & -H & 0 & 0 & 0 \\ -G & -H & G+H & 0 & 0 & 0 \\ 0 & 0 & 0 & 2L & 0 & 0 \\ 0 & 0 & 0 & 0 & 2M & 0 \\ 0 & 0 & 0 & 0 & 0 & 2N \end{bmatrix}} \tag{8.94}$$

where F, G, H, L, M and N are material parameters that characterize the orthotropy of the material. Then (8.92) takes the form

$$\boxed{\begin{aligned} F(s_{11} - s_{22})^2 + G(s_{11} - s_{33})^2 + H(s_{22} - s_{33})^2 \\ + 2Ls_{12}^2 + 2Ms_{13}^2 + 2Ns_{23}^2 - 1 = 0 \end{aligned}} \tag{8.95}$$

This is *Hill's orthotropic yield criterion* proposed by Hill (1948a, 1950); a comparison with (8.85) shows that it comprises a generalization of the classical von Mises criterion for anisotropic materials.

To identify the material parameters F, G, H, L, M and N consider the orthotropic material, say wood or rolled steel, illustrated in Fig. 8.49. Here the coordinate axes are aligned with the material axes of orthotropy. Consider a uniaxial stress in the x_1-direction, i.e. $s_{11} = 2\sigma_{11}/3$, $s_{22} = s_{33} = -\sigma_{11}/3$ and $s_{12} = s_{13} = s_{23} = 0$; then (8.95) reduces to

$$(F + G)\sigma_{11}^2 - 1 = 0$$

Let the initial yield stress in the x_1-direction be denoted by σ_{yo}^{11}, then the equation above gives $F + G = 1/(\sigma_{yo}^{11})^2$. Performing the same exercise in the x_2- and

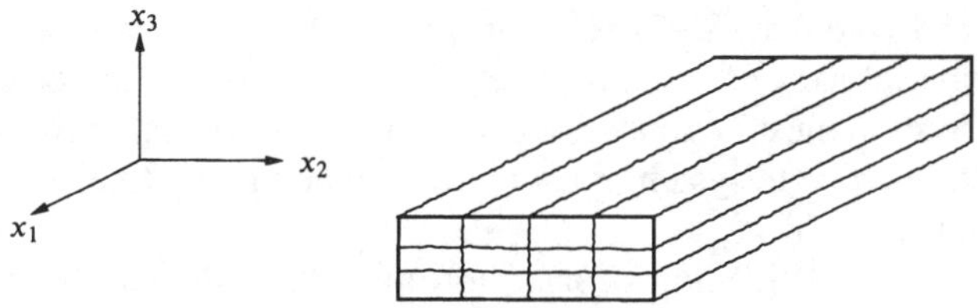

Figure 8.49: Orthotropic material, for instance wood or rolled steel, with coordinate system aligned with the material axes of orthotropy.

x_3-direction, we obtain

$$F + G = \frac{1}{(\sigma_{yo}^{11})^2}\,; \qquad F + H = \frac{1}{(\sigma_{yo}^{22})^2}\,; \qquad G + H = \frac{1}{(\sigma_{yo}^{33})^2}$$

where σ_{yo}^{22} and σ_{yo}^{33} are the initial yield stresses in the x_2- and x_3-direction respectively. From these expressions, we conclude that the parameters F, G and H are determined by

$$\boxed{\begin{aligned} F &= \tfrac{1}{2}\left[\frac{1}{(\sigma_{yo}^{11})^2} + \frac{1}{(\sigma_{yo}^{22})^2} - \frac{1}{(\sigma_{yo}^{33})^2}\right] \\ G &= \tfrac{1}{2}\left[\frac{1}{(\sigma_{yo}^{11})^2} + \frac{1}{(\sigma_{yo}^{33})^2} - \frac{1}{(\sigma_{yo}^{22})^2}\right] \\ H &= \tfrac{1}{2}\left[\frac{1}{(\sigma_{yo}^{22})^2} + \frac{1}{(\sigma_{yo}^{33})^2} - \frac{1}{(\sigma_{yo}^{11})^2}\right] \end{aligned}} \qquad (8.96)$$

Let τ_{yo}^{12} denote the initial yield shear stress when the orthotropic material shown in Fig. 8.49 is subjected to the shear stress $\sigma_{12} = s_{12}$; similar notations hold for the initial yield shear stresses τ_{yo}^{13} and τ_{yo}^{23}. Equation (8.95) then leads to

$$\boxed{L = \frac{1}{2(\tau_{yo}^{12})^2}\,; \qquad M = \frac{1}{2(\tau_{yo}^{13})^2}\,; \qquad N = \frac{1}{2(\tau_{yo}^{23})^2}} \qquad (8.97)$$

For an isotropic material, (8.96) implies that $F = G = H = 1/(2\sigma_{yo}^2)$ and (8.97) gives $L = M = N = 1/(2\tau_{yo}^2)$. If, furthermore, $\tau_{yo} = \sigma_{yo}/\sqrt{3}$ then

$$\left.\begin{aligned} F = G = H &= \frac{1}{2\sigma_{yo}^2} \\ L = M = N &= \frac{3}{2\sigma_{yo}^2} \end{aligned}\right\} \quad \Rightarrow \quad \text{von Mises criterion} \qquad (8.98)$$

and it appears that (8.95) reduces to the von Mises criterion (8.85).

It is emphasized that (8.95) holds only when the coordinate axes are aligned with the material axes of orthotropy, cf. Fig. 8.49. If this is not the case, the coordinate system must be rotated accordingly and the stress components in the new aligned coordinate system must be determined using (3.8) or (3.9) before use can be made of (8.95).

Hill's criterion (8.95) holds for orthotropic materials, but the degree of orthotropy cannot be arbitrarily large. To identify these restrictions in a rational fashion, we eliminate s_{33} in (8.95) using $s_{33} = -(s_{11} + s_{22})$ to obtain

$$\begin{aligned}(F+4G+H)s_{11}^2 + (F+G+4H)s_{22}^2 + 2(-F+2G+2H)s_{11}s_{22} \\ + 2Ls_{12}^2 + 2Ms_{13}^2 + 2Ns_{23}^2 - 1 = 0\end{aligned} \tag{8.99}$$

Here the quantities s_{11}, s_{22}, s_{12}, s_{13} and s_{23} may take arbitrary values independent of each other. If an arbitrary set of s_{11}, s_{22}, s_{12}, s_{13} and s_{23} is increased by a sufficient amount, we must require that it is possible to fulfill (8.99); otherwise there exist arbitrary large deviatoric stress states that do not result in yielding. This is just to say that (8.99) must correspond to a closed surface in the s_{11}, s_{22}, s_{12}, s_{13} and s_{23}-space. Since L, M and N are positive quantities, cf. (8.97), we conclude that increase of the shear stresses will always eventually result in yielding. That is, the quantities that may give rise to restrictions on the degree of orthotropy are related to the first three terms of (8.99). These terms may be written in the following form

$$\begin{aligned}&(F+4G+H)s_{11}^2 + (F+G+4H)s_{22}^2 + 2(-F+2G+2H)s_{11}s_{22} \\ &= \begin{bmatrix} s_{11} & s_{22} \end{bmatrix} \begin{bmatrix} F+4G+H & -F+2G+2H \\ -F+2G+2H & F+G+4H \end{bmatrix} \begin{bmatrix} s_{11} \\ s_{22} \end{bmatrix}\end{aligned} \tag{8.100}$$

To avoid the situation that arbitrarily large values of s_{11} and s_{22} will not result in yielding, we must require that the quadratic matrix in (8.100) is positive definite. Thus, the eigenvalues of the matrix must be positive and evaluation of these eigenvalues results in

$$2F + 5(G+H) > 0\,; \qquad F(G+H) + GH > 0$$

From (8.96) appears that the first of these inequalities is always satisfied whereas the second inequality leads to the constraint

$$\boxed{\frac{4}{(\sigma_{yo}^{11})^2(\sigma_{yo}^{22})^2} > \left[\frac{1}{(\sigma_{yo}^{33})^2} - \left(\frac{1}{(\sigma_{yo}^{11})^2} + \frac{1}{(\sigma_{yo}^{22})^2}\right)\right]^2} \tag{8.101}$$

which restricts the applicability range for Hill's criterion.

It may be somewhat surprising that Hill's criterion implies restriction (8.101). However, the key point is that we have assumed the quadratic expression (8.92)

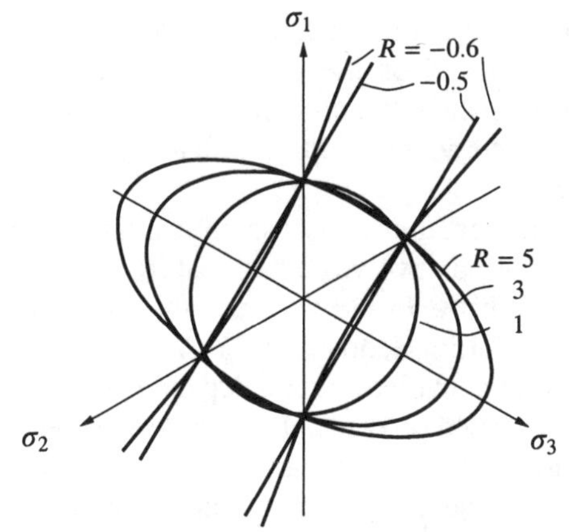

Figure 8.50: Yield curve in deviatoric plane for Hill's criterion; $\sigma_{yo}^{22} = \sigma_{yo}^{11}$, $(\sigma_{yo}^{33})^2 = (\sigma_{yo}^{11})^2\frac{1}{2}(1 + R)$ i.e. $R = 1$ corresponds to the von Mises criterion.

and this expression only allows the yield surface to be a closed surface in the deviatoric stress space when (8.101) is fulfilled. To illustrate this aspect, consider a situation where the material axes of orthotropy coincide with the principal stress axes. Then (8.95) - with $s_{12} = s_{13} = s_{23} = 0$ - describes the yield curve in the deviatoric plane. Assume for simplicity that $\sigma_{yo}^{22} = \sigma_{yo}^{11}$. In that case and following Hosford and Backofen (1964), we introduce the dimensionless parameter R to express σ_{yo}^{33} via $(\sigma_{yo}^{33})^2 = \frac{1}{2}(\sigma_{yo}^{11})^2(1 + R)$; for $R = 1$, we recover the isotropic case, i.e. the von Mises criterion. Restriction (8.101) then leads to $R > -\frac{1}{2}$, i.e. $\sigma_{yo}^{33} > \frac{1}{2}\sigma_{yo}^{11}$. The yield curves in the deviatoric plane for various R-values are shown in Fig. 8.50 and it appears that when R approaches the limit $R = -\frac{1}{2}$, the trace changes from a closed curve to two parallel lines.

Apart from the classical Hill criterion treated here, we may refer to Hill (1993) for a generalized version; this generalization is widely used and as an example, Tryding (1994) made a calibration applicable to paper.

With this detailed discussion, it is easy to generalize the ideas. In (8.81), P_{ijkl} was assumed to characterize the anisotropy of the material. We may even include the effect of some second-order tensor q_{ij} that also characterizes the anisotropy. Both σ_{ij}, P_{ijkl} and q_{ij} must appear in the form of different combinations of invariants and an evident possibility is

$$\sigma_{ij}P_{ijkl}\sigma_{kl} + \sigma_{ij}q_{ij} - 1 = 0$$

which in matrix notation reads

$$\boxed{\boldsymbol{\sigma}^T \boldsymbol{P}\boldsymbol{\sigma} + \boldsymbol{\sigma}^T \boldsymbol{q} - 1 = 0}$$

where $\boldsymbol{P}$ is a 6x6 matrix and $\boldsymbol{q}$ is a 6x1 column matrix. This is the *Tsai-Wu anisotropic yield criterion* suggested by Tsai and Wu (1971). In this format, the stress components enter both in quadratic terms and in linear terms, i.e. if the stress state σ_{ij} is located on the yield surface, the reversed stress state $-\sigma_{ij}$ is not; for instance, tension and compression in the same direction exhibit different yield stresses. Moreover, it appears that the effect of the hydrostatic stress is accounted for.

If orthotropy is considered, then a discussion similar to that following (4.52) shows that the $\boldsymbol{q}$-matrix cannot involve components that affect the shear stresses, i.e. $\boldsymbol{q}^T = [q_1\ q_2\ q_3\ 0\ 0\ 0]$. If we, in addition, take the orthotropic $\boldsymbol{P}$-matrix given by (8.94), the Tsai-Wu criterion reduces to the *Hoffman* criterion proposed by Hoffman (1967) which then involves nine independent parameters.

Here, we have concentrated on some classic criteria where the stress components enter via quadratic terms (and linear terms). However, in order to obtain closer fits to experimental data, recent research seems to favor criteria where other powers of the stress components are used. The reader may consult, for instance, Barlat *et al.* (1991) and Karafillis and Boyce (1993) for further information.

We might also mention that calibration of Hoffman's criterion to columnar-grained ice that is orthotropic and where initial yielding depends on the hydrostatic stress is discussed by Chen and Han (1988) and that a further general discussion of anisotropic yield criteria is presented by Rathkjen (1986).

The discussion above has followed a quite classical route and, instead, it may be of interest to pursue the concept provided by structural tensors as discussed in Sections 6.5 and 6.6.

As an illustration, consider orthotropy where we have three orthogonal material directions each identified by the orthogonal unit vectors $\boldsymbol{v}^{(1)}$, $\boldsymbol{v}^{(2)}$ and $\boldsymbol{v}^{(3)}$. In analogy with (6.49) we now define the structural tensors according to

$$\boxed{\boldsymbol{M}^{(1)} = \boldsymbol{v}^{(1)}\boldsymbol{v}^{(1)T}\,; \qquad \boldsymbol{M}^{(2)} = \boldsymbol{v}^{(2)}\boldsymbol{v}^{(2)T}\,; \qquad \boldsymbol{M}^{(3)} = \boldsymbol{v}^{(3)}\boldsymbol{v}^{(3)T}} \tag{8.102}$$

and from (6.52), we note that

$$\boldsymbol{I} = \boldsymbol{M}^{(1)} + \boldsymbol{M}^{(2)} + \boldsymbol{M}^{(3)} \tag{8.103}$$

Suppose that the coordinate system is chosen collinearly with the material directions. Then we have $\boldsymbol{v}^{(1)T} = [1\,0\,0]$, $\boldsymbol{v}^{(2)T} = [0\,1\,0]$ and $\boldsymbol{v}^{(3)T} = [0\,0\,1]$ and (8.102) becomes

$$\boldsymbol{M}^{(1)} = \begin{bmatrix} 1 & 0 & 0 \\ 0 & 0 & 0 \\ 0 & 0 & 0 \end{bmatrix}; \quad \boldsymbol{M}^{(2)} = \begin{bmatrix} 0 & 0 & 0 \\ 0 & 1 & 0 \\ 0 & 0 & 0 \end{bmatrix}; \quad \boldsymbol{M}^{(3)} = \begin{bmatrix} 0 & 0 & 0 \\ 0 & 0 & 0 \\ 0 & 0 & 1 \end{bmatrix} \tag{8.104}$$

The yield criterion is now taken as a function of the stresses and the structural tensors. However, since these structural tensors depend on each other through (8.103), we choose to work with $\boldsymbol{M}^{(1)}$ and $\boldsymbol{M}^{(2)}$, only, i.e. $F = F(\boldsymbol{\sigma}, \boldsymbol{M}^{(1)}, \boldsymbol{M}^{(2)})$.

Instead of the coordinate system $\boldsymbol{x}$, we may consider another coordinate system $\boldsymbol{x}'$ where $\boldsymbol{x} = \boldsymbol{A}^T\boldsymbol{x}'$ and $\boldsymbol{A}$ is the orthogonal transformation matrix. If it is assumed that the yield function F is an isotropic scalar tensor function then

$$F(\boldsymbol{\sigma}, \boldsymbol{M}^{(1)}, \boldsymbol{M}^{(2)}) = F(\boldsymbol{A}\boldsymbol{\sigma}\boldsymbol{A}^T, \boldsymbol{A}\boldsymbol{M}^{(1)}\boldsymbol{A}^T, \boldsymbol{A}\boldsymbol{M}^{(2)}\boldsymbol{A}^T) \qquad (8.105)$$

for arbitrary $\boldsymbol{A}$ matrices, cf. (6.10). Let S denote the symmetry group in question, i.e. reflections of the coordinate system in the three symmetry planes. For these reflections the structural tensors are unchanged, i.e.

$$\boldsymbol{M}^{(\alpha)} = \boldsymbol{A}\boldsymbol{M}^{(\alpha)}\boldsymbol{A}^T \quad \text{for} \quad \boldsymbol{A} \in S$$

From (8.105), we therefore get

$$F(\boldsymbol{\sigma}, \boldsymbol{M}^{(1)}, \boldsymbol{M}^{(2)}) = F(\boldsymbol{A}\boldsymbol{\sigma}\boldsymbol{A}^T, \boldsymbol{M}^{(1)}, \boldsymbol{M}^{(2)}) \quad \text{for} \quad \boldsymbol{A} \in S$$

That is, for the symmetry group in question, the relation for the yield criterion is as if the material were isotropic, cf (8.6), and this is exactly what is meant by orthotropy.

The representation theorem (6.11) is now applied to $F = F(\boldsymbol{\sigma}, \boldsymbol{M}^{(1)}, \boldsymbol{M}^{(2)})$. The result is completely similar to the discussion in Section 6.5 and in analogy with (6.56) and (6.57) we obtain

$$F = F(I_1, I_2, I_3, I_4, I_5, I_6, I_7) \qquad (8.106)$$

where

$$I_1 = tr(\boldsymbol{\sigma}\boldsymbol{M}^{(1)}); \quad I_2 = tr(\boldsymbol{\sigma}\boldsymbol{M}^{(2)}); \quad I_3 = tr(\boldsymbol{\sigma}\boldsymbol{M}^{(3)})$$
$$I_4 = tr(\boldsymbol{\sigma}^2\boldsymbol{M}^{(1)}); \quad I_5 = tr(\boldsymbol{\sigma}^2\boldsymbol{M}^{(2)}); \quad I_6 = tr(\boldsymbol{\sigma}^2\boldsymbol{M}^{(3)}); \quad I_7 = \tfrac{1}{3}tr(\boldsymbol{\sigma}^3)$$

Suppose that we want to derive Hill's criterion; in this criterion, the stress components occur in a quadratic form. In (8.106) we therefore ignore the invariant I_7 and obtain

$$\begin{aligned}\alpha_1(I_1 - I_2)^2 + \alpha_2(I_1 - I_3)^2 + \alpha_3(I_2 - I_3)^2 \\ + \alpha_4(I_4 - I_1^2) + \alpha_5(I_5 - I_2^2) + \alpha_6(I_6 - I_3^2) \\ + \beta_1 I_1 I_2 + \beta_2 I_1 I_3 + \beta_3 I_2 I_3 - 1 = 0\end{aligned} \qquad (8.107)$$

where $\alpha_1 \cdots \alpha_6$ and β_1, β_2 and β_3 are constants and where the reason for the specific notation will become evident in a moment. It is observed that (8.107) is the most general quadratic expression that exists and it is an explicit version of the anisotropic von Mises criterion (1928) applicable to orthotropic materials, cf. (8.81) and (8.86).

Since $\sigma_{ij} = s_{ij} + \frac{1}{3}\delta_{ij}\sigma_{kk}$, it follows with evident notation that

$$\begin{aligned} tr(\boldsymbol{\sigma}\boldsymbol{M}^{(\alpha)}) &= tr(\boldsymbol{s}\boldsymbol{M}^{(\alpha)}) + \tfrac{1}{3}\sigma_{kk} \\ tr(\boldsymbol{\sigma}^2\boldsymbol{M}^{(\alpha)}) &= tr(\boldsymbol{s}^2\boldsymbol{M}^{(\alpha)}) + \tfrac{2}{3}\sigma_{kk}tr(\boldsymbol{s}\boldsymbol{M}^{(\alpha)}) + \tfrac{1}{9}\sigma_{kk}\sigma_{ss}\end{aligned} \qquad (8.108)$$

Moreover, define the invariants

$$\boxed{\begin{aligned} &I_{d1}=tr(sM^{(1)}); \quad I_{d2}=tr(sM^{(2)}); \quad I_{d3}=tr(sM^{(3)})\\ &I_{d4}=tr(s^2M^{(1)}); \quad I_{d5}=tr(s^2M^{(2)}); \quad I_{d6}=tr(s^2M^{(3)}) \end{aligned}} \tag{8.109}$$

Use of (8.108) and (8.109) in (8.107) results in

$$\begin{aligned} &\alpha_1(I_{d1}-I_{d2})^2+\alpha_2(I_{d1}-I_{d3})^2+\alpha_3(I_{d2}-I_{d3})^2\\ &+\alpha_4(I_{d4}-I_{d1}^2)+\alpha_5(I_{d5}-I_{d2}^2)+\alpha_6(I_{d6}-I_{d3}^2)\\ &+\beta_1 I_{d1}I_{d2}+\beta_2 I_{d1}I_{d3}+\beta_3 I_{d2}I_{d3}\\ &+\frac{1}{9}\sigma_{kk}\sigma_{ss}(\beta_1+\beta_2+\beta_3)\\ &+\frac{1}{3}\sigma_{kk}\beta_1(I_{d1}+I_{d2})+\frac{1}{3}\sigma_{kk}\beta_2(I_{d1}+I_{d3})\\ &+\frac{1}{3}\sigma_{kk}\beta_3(I_{d2}+I_{d3})-1=0 \end{aligned} \tag{8.110}$$

We want a criterion which is independent of the hydrostatic stress σ_{kk}. Thus, the parameters β_1, β_2 and β_3 are chosen such that the last four terms in front of the term -1 which involve the hydrastatic stress σ_{kk} disappear. Therefore

$$\beta_1=\beta_2=\beta_3=0 \tag{8.111}$$

Redefine the $\alpha_1 \cdots \alpha_6$ parameters according to

$$\begin{aligned} &\alpha_1=F; && \alpha_2=G; && \alpha_3=H\\ &\alpha_4=L+M-N; && \alpha_5=N-M+L; && \alpha_6=N+M-L \end{aligned} \tag{8.112}$$

Finally, inserting (8.111) and (8.112) into (8.110) gives the result sought

$$\boxed{\begin{aligned} &\text{Hill's criterion expressed in terms of structural tensors}\\ &F(I_{d1}-I_{d2})^2+G(I_{d1}-I_{d3})^2+H(I_{d2}-I_{d3})^2\\ &+(L+M-N)(I_{d4}-I_{d1}^2)+(N-M+L)(I_{d5}-I_{d2}^2)\\ &+(N+M-L)(I_{d6}-I_{d3}^2)-1=0 \end{aligned}}$$

which corresponds to the result of Dafalias and Rashid (1989). Suppose that the coordinate axes are taken collinearly with the material directions; in that case the structural tensors are given by (8.104) and the expression above then reduces to the classical formulation given by (8.95).

9 INTRODUCTION TO PLASTICITY THEORY

The present chapter serves as a prelude to the next chapter where the general *plasticity theory* will be presented in a rather abstract and formal manner. Before we get to that point, it is necessary to become familiar with the various ingredients in the plasticity theory and this is the subject of the present chapter.

Plasticity theory is concerned with time-independent behavior that is nonlinear and where strains exist when the material is unloaded; these residual strains are the plastic strains. This is in contrast to nonlinear elasticity, where the body recovers its original configuration when unloaded. In Chapter 8, we discussed various initial yield criteria, i.e. conditions for which plastic effects are initiated. When the stress state exceeds the initial yield criterion, plastic strains will develop and this topic is one of the major issues that we will now address.

The basic behavior of an elasto-plastic material is summarized in Fig. 9.1. The behavior is linear elastic with stiffness E until the *initial yield stress* σ_{yo} is reached; after that *plastic strains* develop. Unloading from point A, see Fig. 9.1, occurs elastically with the stiffness E so that at complete unloading to point B, the residual strain amounts to the plastic strain ε^p developed at point A. There-

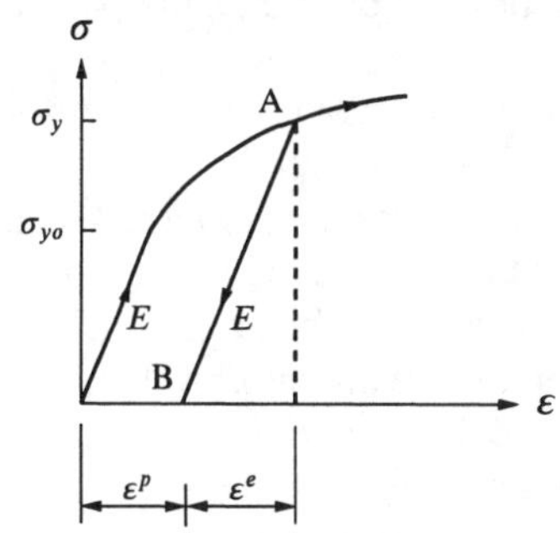

Figure 9.1: Basic response of elasto-plastic material.

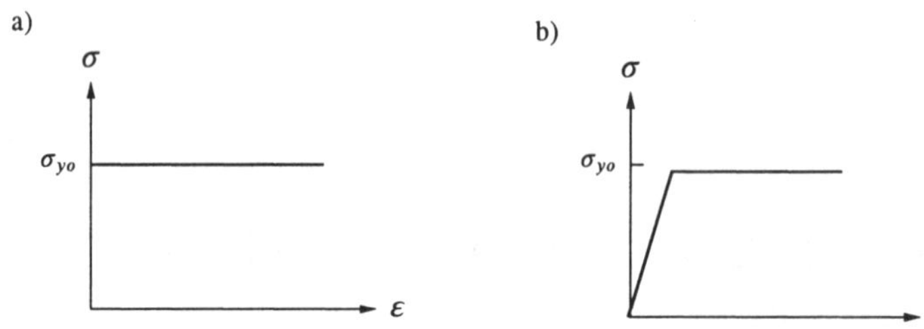

Figure 9.2: a) Stiff-ideal plastic behavior; b) elastic-ideal plastic behavior.

fore, at point A, the *total strain* ε consists of the sum of the *elastic* and *plastic* strains, i.e.

$$\boxed{\varepsilon = \varepsilon^e + \varepsilon^p} \tag{9.1}$$

If we reload again from point B, cf. Fig. 9.1, the material responds elastically until the stress reaches the value σ_y at point A. The value σ_y is therefore the *current yield stress* which, in general, differs from the initial yield stress σ_{yo}. On loading beyond point A the material behaves as if the previous unloading from point A had never occurred. Moreover, the response shown in Fig. 9.1 is assumed to be independent of time; this implies that we obtain the same response irrespective of the loading rate.

The behavior sketched in Fig. 9.1 is our model for the real material behavior, but it turns out that this model behavior closely agrees with the real behavior of elasto-plastic materials.

To characterize plastic behavior, a number of idealized responses have been identified. For the simplest response shown in Fig. 9.2a), the behavior is termed *stiff-ideal plastic* since no deformation occurs before the yield point has been reached and since the yield stress is unaffected by the amount of plastic strains. With obvious notation, the behavior shown in Fig. 9.2b) is termed *elastic-ideal plastic* behavior. Instead of ideal plasticity, the phrase *perfect plasticity* is often used.

In Fig. 9.3a) *hardening plasticity* is displayed; formally, the phrase elastic-hardening plasticity should be used, but the word elastic is ignored since it is evident that we will consider the elastic response. The hardening response shown in Fig. 9.3a) means that the current yield stress σ_y increases with increasing plastic strain, cf. Fig. 9.1, and this behavior is characteristic for alloyed steel and aluminum; moreover, aluminum lacks a sharply defined initial yield stress. In Fig. 9.3b), combined ideal and hardening plasticity is shown and this behavior is characteristic for mild steel.

Finally, Fig. 9.4 shows the development of hardening plasticity followed by *softening plasticity*; this response is typical for concrete, soil and rocks loaded

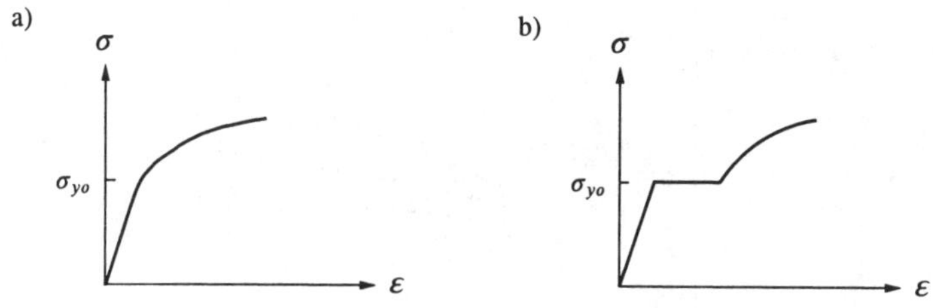

Figure 9.3: a) Hardening plasticity characteristic for alloyed steel and aluminum; b) combined ideal and hardening plasticity characteristic for mild steel.

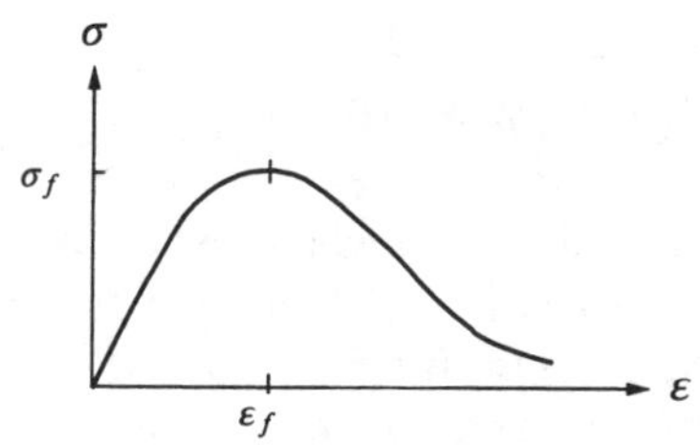

Figure 9.4: Hardening plasticity followed by softening plasticity; characteristic for rocks and concrete in compression.

in compression. Naturally, to achieve the softening branch it is required that the testing machine be operated by means of prescribed displacements and we shall return to this subject in relation to Fig. 9.27. In Fig. 9.4, σ_f denotes the failure stress and ε_f the corresponding strain.

For uniaxial loading, it is straightforward to establish various plasticity models. In essence, it merely consists of adjusting a mathematical curve so that it fits the experimental stress-strain data during loading and adopting a linear elastic response during unloading. We will first mention some well-known proposals that can be used to fit the experimental stress-strain data during uniaxial loading. However, for three-dimensional stress conditions, it is not possible to adopt this simple curve fitting technique; there are simply too many variables and too many different load cases. Thus, in that case, there exists a need for establishment of a general framework for plasticity formulations which will be discussed in detail in this and the following chapter. Before that we will consider some simple curve fitting techniques for uniaxial loading.

When no sharply defined initial yield stress exists, the uniaxial hardening stress-strain curve may be approximated by the Ramberg and Osgood (1943)

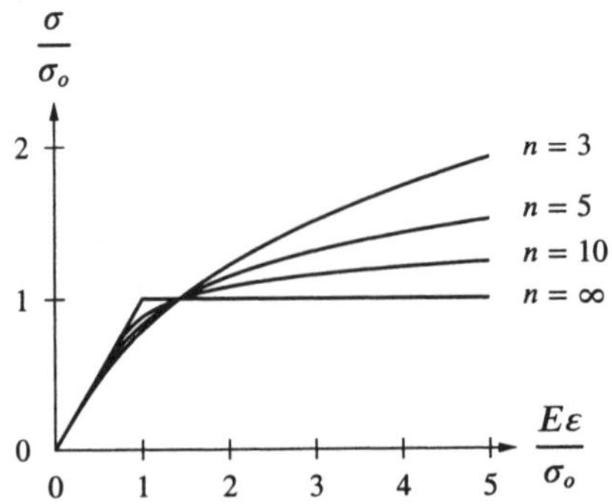

Figure 9.5: Ramberg-Osgood curve for $\alpha = 3/7$.

formula

$$\varepsilon = \frac{\sigma}{E}[1 + \alpha(\frac{\sigma}{\sigma_o})^{n-1}] \; ; \qquad n > 1 \tag{9.2}$$

where α and n are dimensionless parameters whereas σ_o is a parameter with the dimension of stress. Expression (9.2) implies that plastic strains develop right from the onset of the loading. It appears that if $\sigma = \sigma_o$ then (9.2) predicts $\varepsilon = \sigma_o(1 + \alpha)/E$, i.e. the parameter σ_o may be interpreted as the stress value at which the curve given by (9.2) intersects the straight line given by $\sigma = \frac{E}{1+\alpha}\varepsilon$. The value $E/(1 + \alpha)$ may therefore be viewed as the secant modulus when the stress is σ_o. Most often the parameter α is chosen as $\alpha = 3/7$ implying that σ_o becomes the stress at which the secant modulus $E/(1 + \alpha)$ is $7E/10$. For this $\alpha-$value, the appearance of (9.2) is shown in Fig. 9.5. From this figure, it appears that (9.2) for $n \to \infty$ corresponds to ideal plasticity.

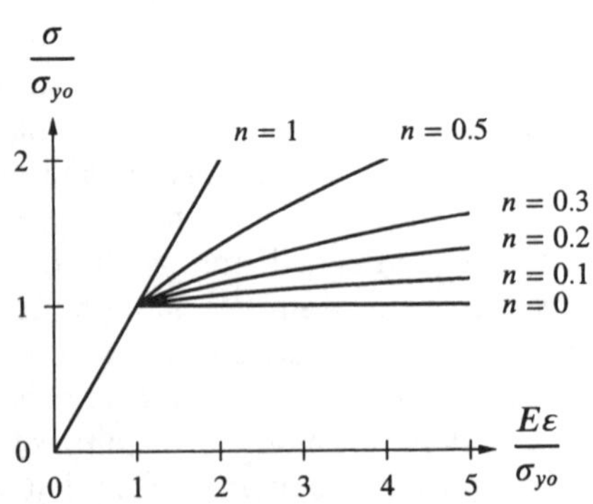

Figure 9.6: Ludwik curve.

If a sharply defined initial yield stress is required, the hardening stress-strain curve may be approximated by the following expression proposed by Ludwik (1909)

$$\sigma = \begin{cases} E\varepsilon & \text{when } \varepsilon \leq \frac{\sigma_{yo}}{E} \\ \sigma_{yo}(\frac{E\varepsilon}{\sigma_{yo}})^n & \text{when } \varepsilon \geq \frac{\sigma_{yo}}{E} \quad (0 \leq n \leq 1) \end{cases}$$

This expression is shown in Fig. 9.6 and it appears that the slope $d\sigma/d\varepsilon$ changes discontinuously at $\sigma = \sigma_{yo}$ (except when $n = 1$); moreover, ideal plasticity is recovered for $n = 0$.

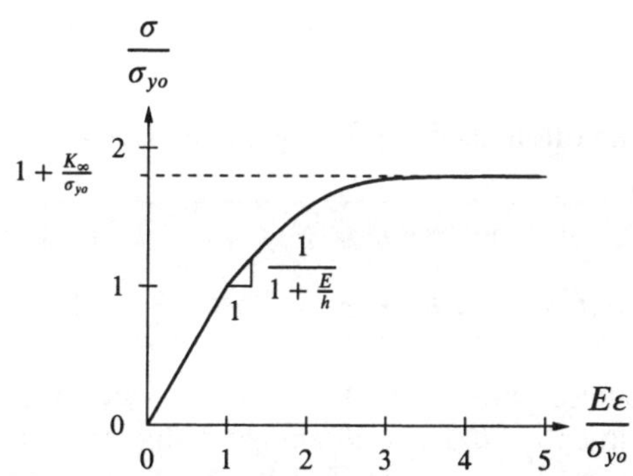

Figure 9.7: Exponential law curve when $E/\sigma_{yo} = 500$, $K_\infty/\sigma_{yo} = 0.8$ and $E/h = 1$.

Another expression that exhibits a sharply defined initial yield stress is given by the following *exponential law*

$$\sigma = \begin{cases} E\varepsilon & \text{when } \varepsilon \leq \frac{\sigma_{yo}}{E} \\ \sigma_{yo} + K_\infty(1 - e^{-\frac{h}{K_\infty}\varepsilon^p}) & \text{when } \varepsilon \geq \frac{\sigma_{yo}}{E} \end{cases} \tag{9.3}$$

where K_∞ and h are parameters with the dimension of stress. Using (9.1), we obtain $\varepsilon^p = \varepsilon - \frac{\sigma}{E}$ which means that (9.3) alternatively may be written as

$$\frac{\sigma}{\sigma_{yo}} = 1 + \frac{K_\infty}{\sigma_{yo}}[1 - e^{-\frac{\sigma_{yo}}{K_\infty}\frac{h}{E}(\frac{E\varepsilon}{\sigma_{yo}} - \frac{\sigma}{\sigma_{yo}})}]$$

Using E/σ_{yo}=500, which is typical for steel and choosing K_∞/σ_{yo}=0.8 and E/h=1, we obtain the curve shown in Fig. 9.7; the interpretation of the parameters K_∞ and h also appears from this figure.

To achieve an approximation for a hardening stress-strain curve that exhibits a sharply defined initial yield stress as well as a continuous variation of the slope

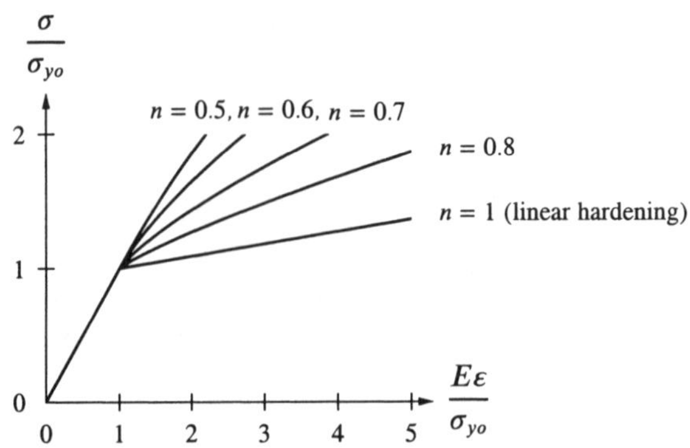

Figure 9.8: Power law curve when $E/\sigma_{yo} = 500$ and $k = 50$.

$d\sigma/d\varepsilon$ also at $\sigma = \sigma_{yo}$, an often used expression is the following *power law*

$$\boxed{\sigma = \begin{cases} E\varepsilon & when \quad \varepsilon \leq \frac{\sigma_{yo}}{E} \\ \sigma_{yo} + k\sigma_{yo}(\varepsilon^p)^n & when \quad \varepsilon \geq \frac{\sigma_{yo}}{E} \quad (0 < n \leq 1) \end{cases}} \tag{9.4}$$

where k is a dimensionless parameter. For $k = 0$, ideal plasticity emerges whereas $n = 1$ implies so-called *linear hardening* (in that case, the slope $d\sigma/d\varepsilon$ varies discontinuously at $\sigma = \sigma_{yo}$). Since $\varepsilon^p = \varepsilon - \sigma/E$, (9.4) may alternatively be written as

$$\frac{E\varepsilon}{\sigma_{yo}} = \frac{\sigma}{\sigma_{yo}} + \frac{E}{\sigma_{yo}}\frac{1}{k^{1/n}}\left(\frac{\sigma}{\sigma_{yo}} - 1\right)^{1/n} \quad \text{when} \quad \frac{E\varepsilon}{\sigma_{yo}} \geq 1$$

Using again $E/\sigma_{yo} = 500$ and choosing the parameter $k = 50$, we obtain the curves shown in Fig. 9.8.

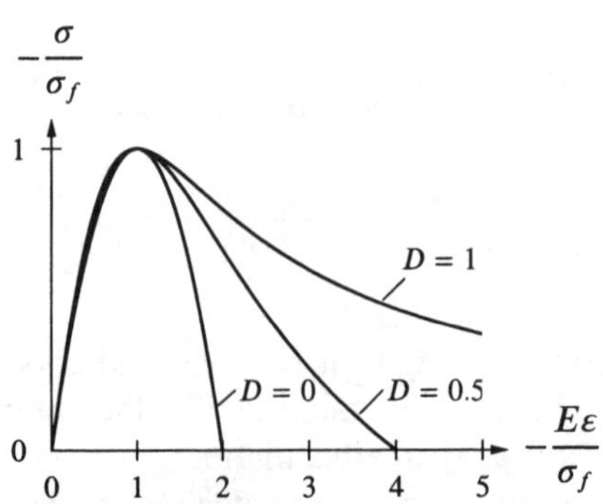

Figure 9.9: Sargin curve for $A = 2$.

To approximate the hardening/softening curve of Fig. 9.4, an expression proposed by Sargin (1971) is conveniently used. Since Fig. 9.4 is typical for concrete and rocks in compression, we shall write

$$\boxed{-\frac{\sigma}{\sigma_f} = \frac{-A\frac{\varepsilon}{\varepsilon_f} + (D-1)(\frac{\varepsilon}{\varepsilon_f})^2}{1-(A-2)\frac{\varepsilon}{\varepsilon_f} + D(\frac{\varepsilon}{\varepsilon_f})^2}} \tag{9.5}$$

Here σ_f and ε_f are the failure values of the stress and strain respectively, cf. Fig. 9.4; as expected, (9.5) then provides $\sigma = -\sigma_f$ when $\varepsilon = -\varepsilon_f$. Moreover, the parameter A is defined by $A = E/E_f$ in which $E_f = \sigma_f/\varepsilon_f$, i.e. E_f is the secant modulus at failure. Finally, the dimensionless parameter D is a parameter that mainly influences the descending curve in the post-failure region. To achieve that (9.5) reflects: 1) an increasing function without inflexion points before failure; 2) a decreasing function with at most one inflexion point after failure; and 3) a residual strength equal to zero at sufficiently large strain, it turns out that we must require

$$A > \tfrac{4}{3}$$

$$(1-\tfrac{A}{2})^2 \le D \le 1 + A(A-2) \quad \text{when} \quad A \le 2$$

$$0 \le D \le 1 \quad \text{when} \quad A \ge 2$$

Sargin's expression is illustrated in Fig. 9.9 for $A = 2$. It appears that different softening behaviors can be simulated by means of the parameter D and that this only insignificantly affects the behavior before failure. Finally, we refer to Popovicz (1970) for other uniaxial stress-strain expressions relevant for concrete and rocks.

Whereas it is straightforward to posit different proposals for the uniaxial stress-strain curve, the elasto-plastic response for general three-dimensional loading is much more complex. In order to address this problem, we will first discuss various issues related to the yield surface.

9.1 Change of yield surface due to loading - Hardening rules

Since the yield stress most often varies with the plastic deformation and since the yield surface is the generalization of the yield stress to general stress states, it is evident that the yield surface will change with the plastic loading. This change of yield surface is called the *hardening rule*, i.e.

Hardening rule = rule for how the yield surface changes with the plastic loading

Since the yield surface is fundamental to the plasticity theory, we will first discuss this issue.

In general, we describe the *initial yield surface* by

$$\boxed{F(\sigma_{ij}) = 0\,; \quad \textit{initial yield surface}}$$

which for isotropic materials can also be written in the form given by (8.19). We know from Sections 6.5, 6.6 and 8.13 that if anisotropic materials are considered, say orthotropy, then structural tensors should also be included in the expression for the yield criterion. However, for the following discussion this aspect is not of importance; thus, we will not include structural tensors in the yield criterion even when anisotropy is present.

Since the yield surface in general varies with the development of plastic strains, we may express the *current yield surface* by

$$f(\sigma_{ij}, K_1, K_2, \ldots) = 0 \tag{9.6}$$

where we have introduced the so-called *hardening parameters* K_1, K_2,... that characterize the manner in which the current yield surface changes its size, shape and position with plastic loading. Before any plasticity is initiated, we know per definition that $K_\alpha = 0$. As yet, the number of hardening parameters is unknown, and, as indicated, we may have one, two or more hardening parameters. Moreover, at this point we do not know the type of the hardening parameters, which may be scalars or higher-order tensors. Therefore, we may collect all these hardening parameters into the notation K_α and use the following definition

$$\boxed{\begin{aligned} &K_\alpha = \textit{hardening parameters}\ (\alpha = 1, 2, \ldots) \\ &K_\alpha = 0 \qquad \textit{initially} \end{aligned}} \tag{9.7}$$

i.e. (9.6) can be written as

$$\boxed{f(\sigma_{ij}, K_\alpha) = 0\,; \quad \textit{current yield surface}} \tag{9.8}$$

Since $K_\alpha = 0$ holds initially, it follows that

$$f(\sigma_{ij}, 0) = F(\sigma_{ij}) \tag{9.9}$$

i.e. when the hardening parameters are zero, the current yield surface coincides with the initial yield surface. Through the hardening parameters, (9.8) describes how the size, shape and position of the current yield surface vary with plastic loading and the explicit manner in which this occurs is given by the hardening rule, i.e.

Choice of hardening parameters = choice of hardening rule

The hardening parameters K_α vary with the plastic loading. To model this, we assume that there exist some *internal variables* that characterize the condition, i.e. the state of the elasto-plastic material. As internal variables we may, for instance, use the plastic strains ε_{ij}^p or some combinations of this tensor; we will revert to this choice of the internal variables later and discuss it in detail, see for instance Section 9.6. The important point is that the internal variables are used to memorize the plastic loading history. As the internal variables characterize the state of the material, they are often termed *state variables*. In principle, the only variables that we can directly observe and measure are the total strains and the temperature. As we cannot directly observe or directly measure the internal variables, they are occasionally termed *hidden* variables in the literature. We shall follow the trend in recent literature and exclusively use the word 'internal variables' since the terminology of 'hidden' variables may act as a psychological block to acceptance. Summarizing these introductory remarks, we have

$$\boxed{\textit{internal variables} = \textit{state variables}} \tag{9.10}$$

In analogy with the notation above, we shall let κ_α denote the internal variables, i.e.

$$\boxed{\begin{array}{l} \kappa_\alpha = \textit{internal variables}\ (\alpha = 1, 2 \ldots) \\ \kappa_\alpha = 0 \qquad \textit{initially} \end{array}} \tag{9.11}$$

Since the internal variables memorize the plastic loading history, they are, per definition, zero before any plasticity is initiated. As before, the above notation means that we may have one, two or more internal variables and at the present time we do not know whether κ_α are scalars or higher-order tensors. Since the internal variables κ_α characterize the elasto-plastic material, we can assume that

$$\boxed{K_\alpha = K_\alpha(\kappa_\beta)} \tag{9.12}$$

i.e. the hardening parameters K_α depend on the internal variables κ_α. It seems natural to assume that the number of hardening parameters equals the number of internal variables; otherwise (9.12) will not provide a unique relation between the set of hardening parameters and the set of internal variables, From (9.12) follows that

$$\dot{K}_\alpha = \frac{\partial K_\alpha}{\partial \kappa_\beta} \dot{\kappa}_\beta \tag{9.13}$$

where, as previously, a dot denotes the rate, i.e. the change. Here, the summation convention is also adopted for Greek letters. As the internal variables κ_α characterize the state of the elasto-plastic material, they can only change during plastic loading, i.e. $\dot{\kappa}_\alpha$ must be zero for an elastic response and, in view of (9.13), we conclude

$$\boxed{\dot{\kappa}_\alpha = \dot{K}_\alpha = 0\,; \quad \textit{for elastic behavior}} \tag{9.14}$$

With this general discussion, let us return to the hardening rule. Starting with the simplest case of ideal plasticity, as illustrated in Fig. 9.2b), the yield

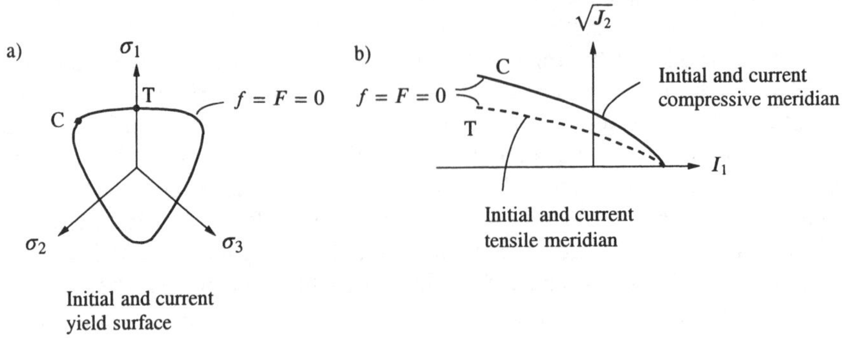

Figure 9.10: Ideal plasticity where the yield surface remains fixed; a) deviatoric plane; b) meridian plane.

surface is unaffected by the plastic deformations, i.e. it remains fixed in the stress space. This situation is illustrated in Fig. 9.10, where C refers to the compression meridian and T to the tensile meridian, cf. Fig. 8.6. In that case, no hardening parameters exist, i.e. (9.8) reduces with (9.9) to

$$\boxed{f(\sigma_{ij}, K_\alpha) = F(\sigma_{ij}) = 0\,; \quad \textit{ideal plasticity}} \tag{9.15}$$

i.e. the current yield surface coincides with the initial yield surface. We conclude that

For ideal plasticity, the yield surface remains fixed in the stress space

Let us next assume that the shape and position of the yield surface remain fixed whereas the size of the yield surface changes. This situation is called *isotropic hardening* and is usually attributed to Hill (1950). As an example, we may consider the von Mises criterion where the initial yield surface is given by

$$F(\sigma_{ij}) = \sqrt{3J_2} - \sigma_{yo} = 0 \tag{9.16}$$

cf. (8.26). We may accomplish isotropic hardening by writing

$$f(\sigma_{ij}, K_\alpha) = \sqrt{3J_2} - \sigma_{yo} - K = 0 \tag{9.17}$$

where, for convenience, we have assumed that only one hardening parameter, K, controls the change of size of the yield surface. In turn, this implies the existence of only one internal parameter κ. The function $K(\kappa)$ describes how the size of the yield surface changes with the development of plastic strains and in accordance with (9.7) and (9.11) we have $K(0) = 0$ so that (9.17) reduces

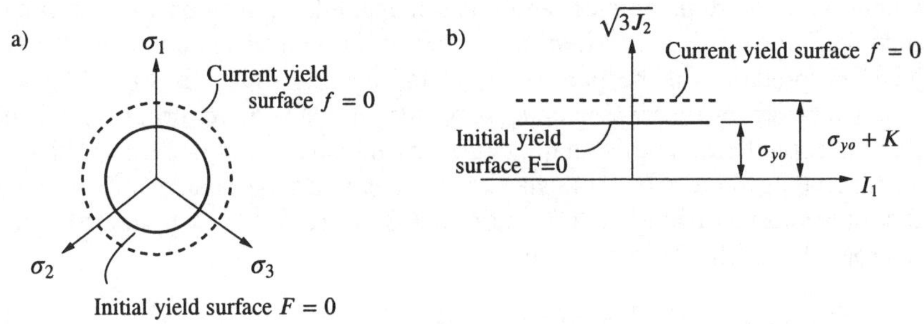

Figure 9.11: Isotropic hardening of the von Mises criterion; a) deviatoric plane; b) meridian plane.

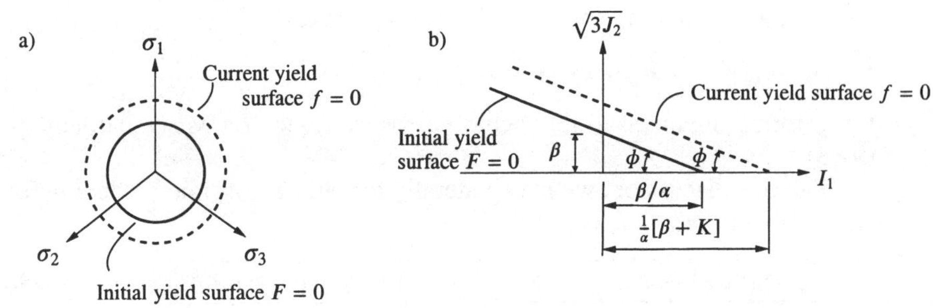

Figure 9.12: Isotropic hardening of the Drucker-Prager criterion; a) deviatoric plane; b) meridian plane.

to (9.16) before the development of plastic strains. Instead of the formulation (9.17), we may write

$$f(\sigma_{ij}, K_\alpha) = F(\sigma_{ij}) - K = 0 \tag{9.18}$$

Isotropic hardening of the von Mises criterion is shown in Fig. 9.11. In this figure, the yield surface expands with increasing plastic deformation and this increase of the current yield stress evidently corresponds to the case of hardening plasticity illustrated in Figs. 9.1 and 9.3a). Mathematically, this is obtained by letting the function $K(\kappa)$ in (9.18) increase with increasing plastic deformation. It is of interest that if, at some stage, we let the function $K(\kappa)$ decrease with increasing plastic deformation then the von Mises surface shrinks in size and this decrease of the current yield stress corresponds to softening plasticity as illustrated in Fig. 9.4. By tradition, the terminology here is somewhat vague since we have achieved softening plasticity by means of the isotropic 'hardening' concept.

As the next example of isotropic hardening, consider the Drucker-Prager criterion. Referring to (8.33), the initial yield surface is here given by

$$F(\sigma_{ij}) = \sqrt{3J_2} + \alpha I_1 - \beta = 0$$

where α and β are parameters and α is dimensionless. We observe that if $\alpha = 0$ then the Drucker-Prager criterion reduces to the von Mises criterion of (9.16). The interpretation of the parameters α and β is illustrated in Fig. 9.12b). To obtain an isotropic hardening concept for the Drucker-Prager criterion, we recall that isotropic hardening is characterized by the shape and position of the yield surface being fixed while the size of the yield surface changes. Referring to the interpretation of the parameters α and β in Fig. 9.12b) we therefore obtain isotropic hardening by the formulation

$$f(\sigma_{ij}, K_\alpha) = \sqrt{3J_2} + \alpha I_1 - \beta - K = 0 \tag{9.19}$$

where the single hardening parameter, K, only depends on one internal variable κ, i.e. $K = K(\kappa)$. This isotropic hardening formulation is illustrated in Fig. 9.12 and we observe that it is possible to write (9.19) as

$$f(\sigma_{ij}, K_\alpha) = F(\sigma_{ij}) - K = 0$$

i.e. a format identical to that achieved for isotropic von Mises hardening, cf. (9.18).

With this discussion, we may generally formulate isotropic hardening for an arbitrary yield function as

$$\boxed{f(\sigma_{ij}, K_\alpha) = F(\sigma_{ij}) - K = 0\,; \qquad \textit{isotropic hardening}} \tag{9.20}$$

which may be expressed as

For isotropic hardening, the position and shape of the yield surface remain fixed whereas the size of the yield surface changes with plastic deformation

Returning to isotropic hardening of the von Mises criterion, it is obvious that we may write (9.17) as

$$\sqrt{3J_2} - \sigma_y(\kappa) = 0\,; \qquad \sigma_y(\kappa) = \sigma_{yo} + K(\kappa) \tag{9.21}$$

where σ_y is the current yield stress. For uniaxial loading, (9.21) reduces to $|\sigma| = \sigma_y$. As illustrated in Fig. 9.13a), this implies that if we reverse the loading from point A where $\sigma = \sigma_y$, the isotropic hardening model will predict elastic unloading until we reach point B where $\sigma = -\sigma_y$. As a result, even after plastic strains have developed, the isotropic hardening model of von Mises predicts the same yield stress in tension and in compression.

This prediction does not agree well with experimental results for metals and steel. Referring to Fig. 9.13b), experimental results show that point B, where plastic effects are again encountered, occurs much earlier than that predicted by the isotropic hardening model. This phenomenon was first observed by Bauschinger (1886) and is therefore called the *Bauschinger effect*. Let us now see how we can approximate this effect.

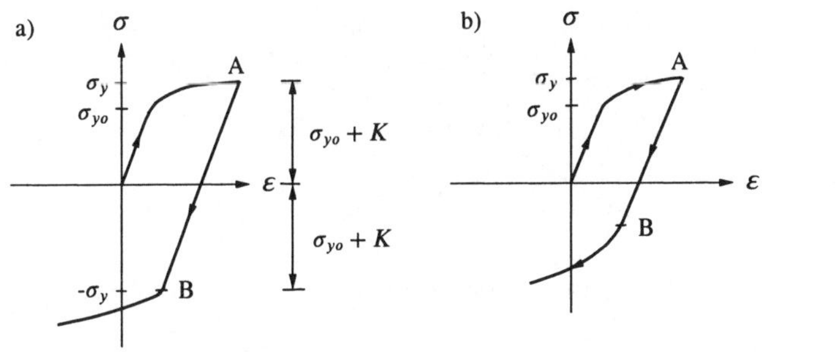

Figure 9.13: a) isotropic hardening; b) Bauschinger effect.

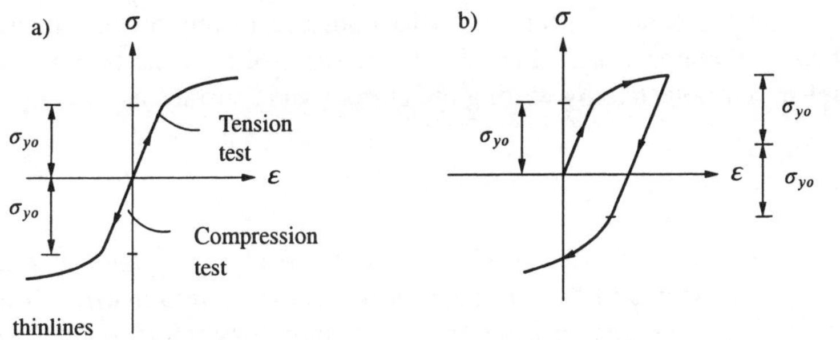

Figure 9.14: Metals and steel; a) tension and compression test; b) kinematic hardening model for uniaxial loading.

For metals and steel and in accordance with the von Mises model, the initial yield stress is the same whether we load in tension or in compression, cf. Fig. 9.14a). That means, the difference between these two yield points is $2\sigma_{yo}$. In an effort to approximate the Bauschinger effect, we assume that the difference between the two yield points is maintained at the value of $2\sigma_{yo}$ even after plastic deformations have occurred. This assumption is illustrated in Fig. 9.14b) and it appears to be a reasonable approximation of the real material behavior shown in Fig. 9.13b). This approximation to hardening is called *kinematic hardening* and it was introduced by Melan (1938) and later by Prager (1955).

Let us see how we can formulate the kinematic hardening assumption within a von Mises concept. Let us first rewrite the initial von Mises criterion (9.16) by using the definition (3.16) of the invariant J_2. This provides

$$F(\sigma_{ij}) = (\frac{3}{2}s_{ij}s_{ij})^{1/2} - \sigma_{yo} = 0 \tag{9.22}$$

In accordance with Fig. 9.14, we can assume that the size and form of the yield surface are unchanged during plastic loading. We are therefore left with the

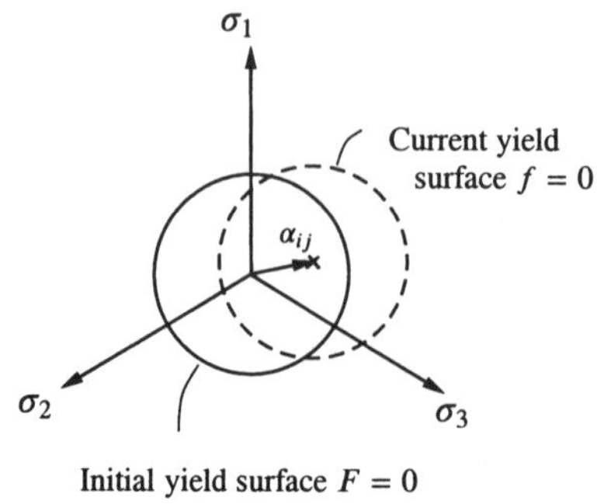

Figure 9.15: Deviatoric plane; kinematic hardening of the von Mises criterion.

possibility that the position of the yield surface changes as a result of the plastic loading. Let the center of the von Mises surface in the deviatoric plane be described by the tensor α_{ij}, where the initial value of α_{ij} is zero. Then we can accomplish our objective by writing the current yield surface as

$$f(\sigma_{ij}, K_\alpha) = \left[\frac{3}{2}(s_{ij} - \alpha_{ij})(s_{ij} - \alpha_{ij})\right]^{1/2} - \sigma_{yo} = 0 \qquad (9.23)$$

which clearly reduces to (9.22) when $\alpha_{ij} = 0$. We see that there is only one hardening parameter and that it takes the form of the tensor α_{ij}. Often, the parameter α_{ij} is called the *back-stress*. This terminology refers to the deviatoric stresses being referred 'back' to the center α_{ij}. The kinematic hardening model of a von Mises material is illustrated in Fig. 9.15 which shows that the only thing that happens to the initial yield surface is that it moves as a rigid body in the stress space due to the plastic deformations.

Generalizing these ideas to arbitrary yield functions, we see that kinematic hardening is modelled by

$$\boxed{f(\sigma_{ij}, K_\alpha) = F(\sigma_{ij} - \alpha_{ij}) = 0\,; \qquad \textit{kinematic hardening}} \qquad (9.24)$$

where we have one hardening parameter in terms of the tensor α_{ij} which describes the position of the current yield surface. Equation (9.24) may be expressed as

For kinematic hardening, the size and shape of the yield surface remain fixed whereas the position of yield surface changes with plastic deformation

Most materials consist of different constituents; concrete is a mixture of aggregate in a matrix of cement paste whereas metals and steel consist of polycrystals. When loading such a material into the plastic regime, the differences in stiffness and yield properties of the constituents imply that they experience

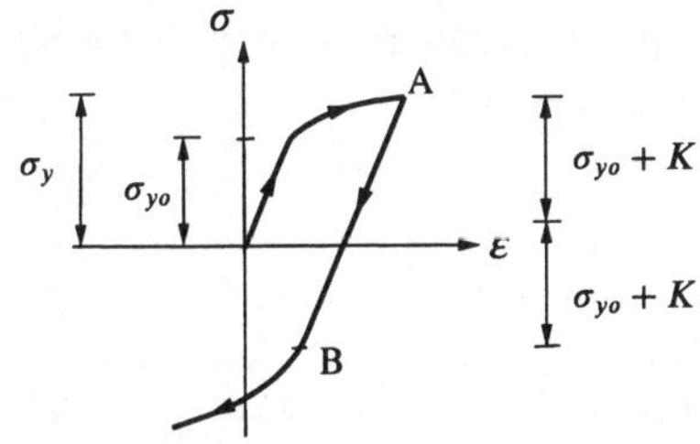

Figure 9.16: Uniaxial loading; mixed hardening of a von Mises material.

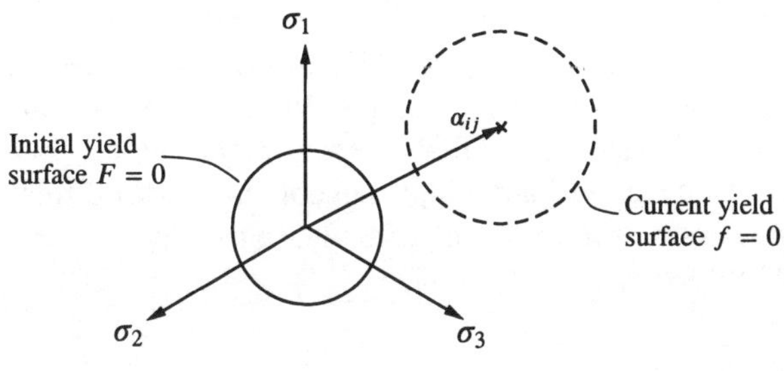

Figure 9.17: Deviatoric plane; mixed hardening of a von Mises material.

different plastic straining. When reversing the loading, this 'mismatch' of the constituents means that some constituents again enter the plastic regime before others and this manifests itself on the macrolevel as the Bauschinger effect and thereby kinematic hardening.

The two classic hardening rules: isotropic and kinematic hardening, may be combined into what is called *mixed hardening*. This concept was introduced by Hodge (1957). For the von Mises criterion, combination of the isotropic hardening given by (9.17) and the kinematic hardening expressed by (9.23) lead to

$$f(\sigma_{ij}, K_\alpha) = \left[\frac{3}{2}(s_{ij} - \alpha_{ij})(s_{ij} - \alpha_{ij})\right]^{1/2} - \sigma_{yo} - K = 0 \tag{9.25}$$

where the set of hardening parameters K_α consists of α_{ij} and K, i.e. $K_\alpha = \{\alpha_{ij}, K\}$. It appears that (9.25) for $K = 0$ reduces to kinematic hardening given by (9.23) whereas (9.25) for $\alpha_{ij} = 0$ reduces to isotropic hardening given by (9.17).

With reference to Figs. 9.13a) and 9.14b) the response predicted by mixed hardening in the case of uniaxial loading is illustrated in Fig. 9.16. Moreover, Fig. 9.17 shows the evolution of the yield surface in the deviatoric plane. Generalizing the concept of mixed hardening to an arbitrary yield function, we obtain

$$\boxed{f(\sigma_{ij}, K_\alpha) = F(\sigma_{ij} - \alpha_{ij}) - K = 0\,; \quad \textit{mixed hardening}} \tag{9.26}$$

where the set of hardening parameters K_α both consists of the back-stress α_{ij} and the hardening parameter K, i.e.

$$K_\alpha = \{\alpha_{ij}, K\}$$

Equation (9.26) may be expressed as

For mixed hardening, i.e. a combination of isotropic and kinematic hardening, the shape of the yield surface remains fixed whereas the size and position of the yield surface change with plastic deformation	(9.27)

In principle, it is possible to allow for a change not only in the size and position of the yield surface, but even of its shape. In that case, one speaks of *distortional hardening* - occasionally called *anisotropic hardening* - but for simplicity this more advanced mixed hardening concept shall not be treated here. The interested reader may, for instance, consult Baltov and Sawczuk (1965) and Axelsson (1979).

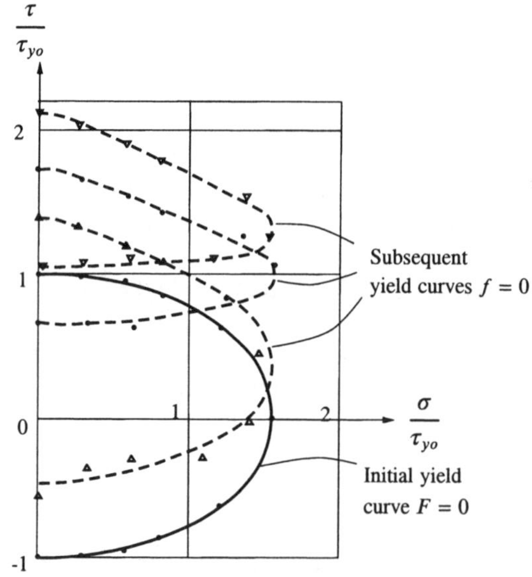

Figure 9.18: Aluminum alloy tested in combined torsion and tension by Ivey (1961); τ_{yo} is the initial yield stress in pure shear.

With this discussion it may be of interest to see what experimental results for metals and steel indicate. In Fig. 9.18, an aluminum alloy was tested in combined torsion and tension by Ivey (1961). In accordance with the von Mises criterion for such stress states, cf. (8.29), the initial yield curve can be described by an ellipse. The three subsequent yield curves show a pronounced

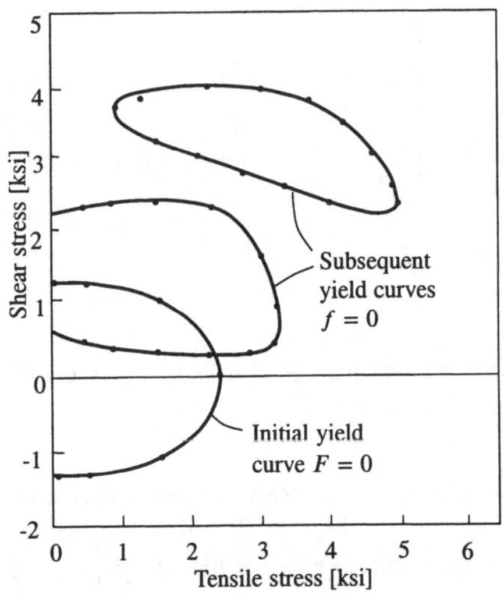

Figure 9.19: Pure aluminum tested in combined torsion and tension by Phillips and Tang (1972).

kinematic hardening as well as effects of isotropic hardening (or rather isotropic 'softening'). However, even an distortion of the form of the yield curve can be observed.

Figure 9.19 shows the results of Phillips and Tang (1972) for pure aluminum also tested in combined torsion and tension, but for a different load path than that used in Fig. 9.18. Again the kinematic hardening effect is pronounced and some isotropic hardening (softening) as well as distortion of the yield curve may be observed. For further experimental evidence and a historical account, the reader may consult Michno and Findley (1976).

In practice, it is seldom that models are used that go beyond combinations of isotropic and kinematic hardening. With reference to Figs. 9.18 and 9.19, this may seem to be a rather crude approximation. However, one should be aware of the fact that experimentally determined yield curves are highly sensitive to how the yield limit is defined. A detailed discussion of this aspect is given by Axelsson (1979). The yield limit is usually defined as some kind of deviation from linear response, but the threshold value used to identify this deviation influences the yield curve obtained most significantly. This means that for the same observed stress-strain curves, the position of the yield curve depends significantly on the experimental procedure. In practice, one is more interested in an accurate prediction of the stress-strain curves than in the determination of the yield curve as such. This implies that combined isotropic and kinematic hardening in most cases allows a prediction that is of sufficient engineering accuracy and, in practice, either isotropic or kinematic hardening is even adopted

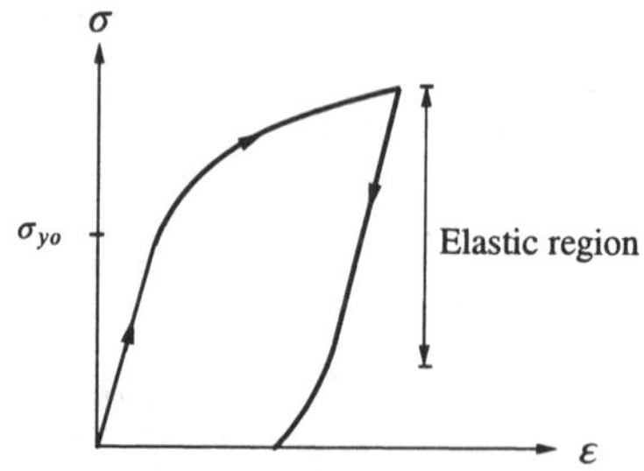

Figure 9.20: Due to the effect of kinematic hardening, plastic strains may develop during the unloading phase

and considered to be sufficiently accurate for not too complicated load paths. In general, isotropic hardening is adopted if the loading mainly increases whereas kinematic hardening is used when reversed loadings, i.e. cyclic loadings, are of interest.

Another point of interest is that both Figs. 9.18 and 9.19 show that the origin of the stress space may be located outside the current yield surface. When the material is completely unloaded, this implies that plastic strains develop during the unloading phase. This behavior, which at first sight may seem surprising, is illustrated in Fig. 9.20.

9.2 Development of plastic strains - Introductory remarks

Previously, we have discussed the yield surface and its change during plastic deformation, i.e. the hardening rule, in great detail. Let us now turn to the important issue of determining the development of the plastic strains.

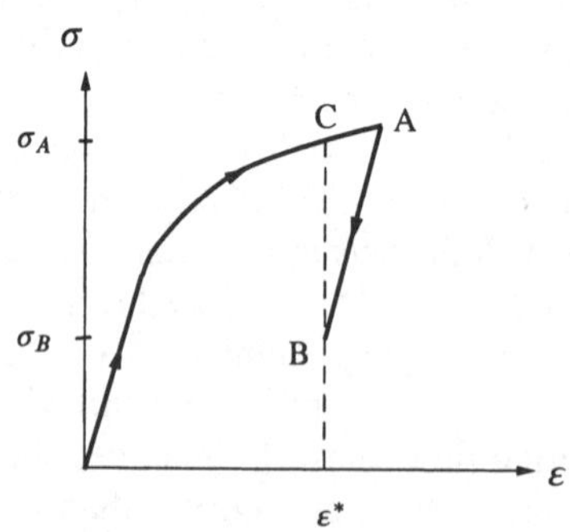

Figure 9.21: For a given strain ε^*, the corresponding stress is unknown unless we know the load history.

Consider the uniaxial loading in Fig. 9.21 where we unload to point B where the strain is ε^*. It is obvious that if only the strain value ε^* is known, we do not know whether the corresponding stress is σ_B or σ_C. We conclude that in plasticity, no unique relation exists between the stress state σ_{ij} and the strain state ε_{ij}. Therefore, the constitutive relation for elasto-plasticity must be of an incremental nature. This means that for a given strain state the corresponding stress state is obtained by an integration of the incremental constitutive relations and the result of this integration will depend on the integration path, i.e. the load history. This load history dependence is illustrated in Fig. 9.21.

Referring to (9.1), the total strains ε_{ij} are assumed to consist of the elastic and plastic strains, i.e.

$$\boxed{\varepsilon_{ij} = \varepsilon_{ij}^e + \varepsilon_{ij}^p} \tag{9.28}$$

The elastic strains are determined by Hooke's law (4.20) or (4.26) i.e.

$$\boxed{\sigma_{ij} = D_{ijkl}\varepsilon_{kl}^e \quad \text{or} \quad \varepsilon_{ij}^e = C_{ijkl}\sigma_{kl}}$$

Therefore, the incremental constitutive relation must be related to the increment $\dot{\varepsilon}_{ij}^p$ of the plastic strains. We recall that a dot denotes the time derivative, i.e. $\dot{\varepsilon}_{ij}^p = d\varepsilon_{ij}^p/dt$ and since we want to establish a plasticity theory that is independent of the loading rate, i.e. time does not influence the response, we merely use the format $\dot{\varepsilon}_{ij}^p$ instead of $d\varepsilon_{ij}^p$ to simplify the notation. This discussion is similar to the one given in Chapter 7.

In order to motivate the constitutive relationship for the plastic strains it turns out to be instructive to follow the historical development of the plasticity theory adopting a broad conceptual viewpoint.

Figure 9.22: Movement of a plate in a Newton fluid.

In the 19th century, the scientific community was concerned with the simplest possible plasticity model, namely that of rigid–ideal plasticity, cf. Fig. 9.2a). With the word 'yielding' of materials, it seems tempting to associate this behavior with the response of viscous fluids. In Fig. 9.22, a plate is moved in a viscous fluid close to a rigid wall. Assuming the viscous fluid to be a linear Newton fluid, the shear stress τ in the fluid is then given by

$$\dot{\gamma} = \frac{\tau}{\mu}$$

where μ is the viscosity of the fluid, i.e.μ has the dimension $[Pa \cdot s]$ and γ is the engineering shear strain. We may also refer to the behavior of the dashpot shown in Fig. 6.1b) which obeys a similar constitutive relation, cf. (6.26). Referring to Fig. 9.22, we have $\gamma = 2\varepsilon_{12}$ and $\tau = s_{12}$ i.e. the relationship above may be written as $\dot{\varepsilon}_{12} = s_{12}/(2\mu)$. Generalizing this relation we obtain

$$\dot{\varepsilon}_{ij} = \frac{s_{ij}}{2\mu} \tag{9.29}$$

With this discussion, and considering rigid-ideal plastic materials where the elastic strains are zero, it is tempting to assume that the constitutive relation for such materials is given by

$$\dot{\varepsilon}_{ij} = \dot{\beta} s_{ij}\,; \qquad \dot{\beta} \geq 0 \tag{9.30}$$

Whereas (9.29) depends on time, we have by introduction of the quantity $\dot{\beta} = d\beta/dt$ ensured that (9.30) is independent of time, i.e. (9.30) is an expression that is homogeneous in dt. Since μ in (9.29) is positive, we introduce the constraint $\dot{\beta} \geq 0$ in (9.30) where $\dot{\beta} = 0$ implies that no strains develop; apart from that, the quantity $\dot{\beta}$ is at this stage unknown. It is of interest that the formulation (9.30) is precisely the formulation proposed by Saint-Venant (1870) for plane strain and by Lévy (1870) for general conditions. Later, von Mises (1913) also arrived at the same constitutive law which often is called the *Lévy-von Mises equations*.

Later on, interest shifted towards the behavior of elastic-ideal plastic materials and with (9.28) and (9.30) it seems natural to assume that

$$\dot{\varepsilon}^p_{ij} = \dot{\beta} s_{ij}\,; \qquad \dot{\beta} \geq 0 \tag{9.31}$$

and this format was suggested by Prandtl (1924) for two dimensions and by Reuss (1930) for three dimensions; expression (9.31) is therefore called the *Prandtl-Reuss equations*. We observe that $\dot{\beta} = 0$ implies that plastic strains do not develop. In general, a constitutive relation for $\dot{\varepsilon}^p_{ij}$ is called a *flow rule* and (9.31) is an example of a flow rule. It is of considerable interest that (9.31) implies that the plastic volumetric strain is $\dot{\varepsilon}^p_{ii} = 0$, i.e. plastic incompressibility and this property is in very close agreement with the behavior of metals and steels. For such materials, the experimental observation that the plastic response only depends on the deviatoric part of the stresses is also reflected by the flow rule (9.31).

In solid mechanics, as well as in other branches of mechanics, many problems may be formulated by means of a potential function. This means that one quantity is obtained by differentiation of a scalar function, the *potential function*.

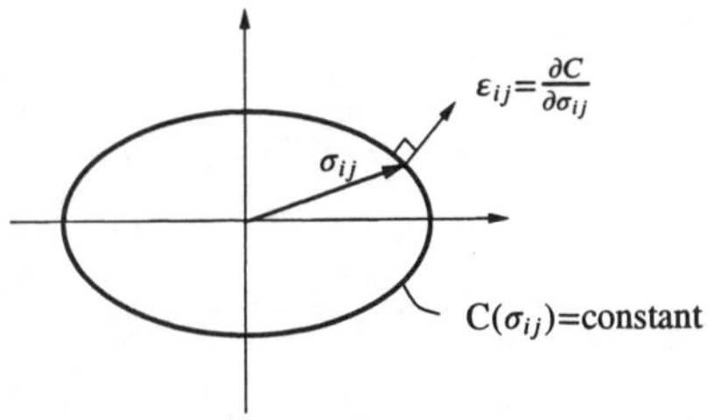

Figure 9.23: Linear elasticity. Normality of strain tensor ε_{ij} to the surface in the stress space described by $C(\sigma_{ij})$ = constant.

Well known examples are Airy's stress function for two-dimensional elasticity problems and Prandtl's stress function for torsion of non-circular elastic shafts.

In Chapter 4, we also encountered such a potential function, namely the complementary energy $C(\sigma_{ij})$ from which according to (4.16) we obtain the strains by a differentiation, i.e.

$$\varepsilon_{ij} = \frac{\partial C}{\partial \sigma_{ij}} \tag{9.32}$$

and this relation is characteristic for hyper-elasticity. For linear elasticity, the complementary energy C is given by (4.30) i.e.

$$C = \frac{1}{2}\sigma_{ij}\varepsilon_{ij} = \frac{1}{2}\sigma_{ij}C_{ijkl}\sigma_{kl} > 0 \tag{9.33}$$

which proves that the flexibility tensor C_{ijkl} is positive definite. From (9.32) and (9.33) we conclude that

$$\sigma_{ij}\varepsilon_{ij} = \sigma_{ij}\frac{\partial C}{\partial \sigma_{ij}} > 0 \tag{9.34}$$

If we consider the expression $C(\sigma_{ij})$ = constant, then this expression describes a surface in the stress space as illustrated in Fig. 9.23. According to (9.32), the strain tensor is orthogonal to this surface and following (9.34) the scalar product $\sigma_{ij}\varepsilon_{ij}$ is positive, i.e. ε_{ij} is directed outwards, as shown in Fig. 9.23. We have observed that the strain tensor ε_{ij} is normal to the surface $C(\sigma_{ij})$ = constant. Let us next prove that C is *convex*. For a one-dimensional function $g(x)$, convexity requires that $d^2g/dx^2 > 0$, cf. Fig. 9.24. For the multi-dimensional function $C(\sigma_{ij})$, the requirement of convexity is that the quantity $\partial^2 C/\partial\sigma_{ij}\partial\sigma_{kl}$ is positive definite, cf. for instance the Appendix. From Hooke's law $\varepsilon_{ij} = C_{ijkl}\sigma_{kl}$ and (9.32) we obtain

$$\frac{\partial \varepsilon_{ij}}{\partial \sigma_{kl}} = C_{ijkl} = \frac{\partial^2 C}{\partial \sigma_{kl}\partial \sigma_{ij}}$$

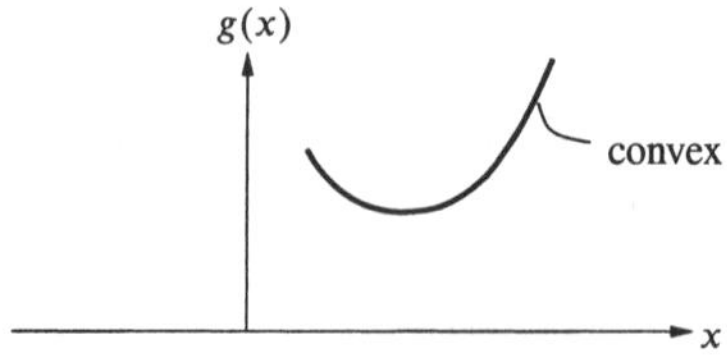

Figure 9.24: Convex function in one dimension.

Since C_{ijkl} is positive definite so is $\partial^2 C/\partial\sigma_{ij}\partial\sigma_{kl}$, i.e. C is convex.

We have discussed hyper-elasticity in some detail and we have shown that the concepts of a potential function, normality and convexity apply for this group of materials. We shall now see that similar concepts hold for the plasticity theory.

Previously, we established the Prandtl-Reuss equations given by (9.31) and we have argued for the natural wish to try to formulate problems in terms of potential functions. Around 1930, the von Mises criterion was well established and since this criterion provides a scalar function in terms of the yield function, it seems natural to investigate whether it can be used as a potential function. According to (9.21) we have

$$f(\sigma_{ij}, K) = \sqrt{3J_2} - \sigma_y(\kappa) \; ; \qquad f(\sigma_{ij}, K) = 0$$

where $\sigma_y(\kappa) = \sigma_{yo} + K(\kappa)$. Differentiation gives

$$\frac{\partial f}{\partial \sigma_{ij}} = \frac{3s_{ij}}{2\sqrt{3J_2}} = \frac{3}{2}\frac{s_{ij}}{\sigma_y}$$

It is of interest that this expression and (9.31) may be combined into

$$\boxed{\dot{\varepsilon}_{ij}^p = \dot{\lambda}\frac{\partial f}{\partial \sigma_{ij}}; \qquad \dot{\lambda} \geq 0} \tag{9.35}$$

where $\dot{\lambda} = 2\sigma_y\dot{\beta}/3$, i.e. the incremental plastic strains can be derived by using the von Mises function as a potential function.

For stress states inside the yield surface, we have $f < 0$, i.e. $\partial f/\partial\sigma_{ij}$ is normal to the yield surface $f = 0$ and directed outwards. Referring to Fig. 9.25, we have then established the important property of $\dot{\varepsilon}_{ij}^p$ being normal to the yield surface. Moreover, the von Mises surface is convex and as discussed in Chapter 8, experimental evidence shows that yield surfaces are, in general, convex.

The above discussion has led us to the important concept of *normality* expressed by the flow rule given by (9.35) in which the yield surface serves as a *potential function* for the determination of the incremental plastic strains. This flow rule is therefore called the *associated* flow rule since the yield criterion is

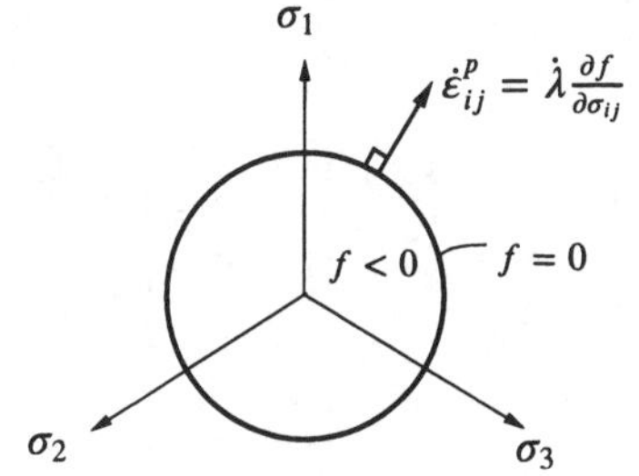

Figure 9.25: Normality of the incremental plastic strains to the von Mises surface in the deviatoric plane.

associated with - i.e. taken as - the potential function. Moreover, the *convexity* of the yield surface has also been discussed.

The flow rule (9.35) is simple and elegant and it was derived using the von Mises criterion as potential function. In principle however, it can be used in combination with any yield function and it was, in fact, proposed by von Mises (1928). We observe that only the direction of the incremental plastic strains is given by (9.35) and also to determine their magnitude we need to determine the so-called *plastic multiplier* $\dot{\lambda}$. This topic will be discussed later on.

Before turning to this point, it may be argued that the background for the establishment of associated flow rule (9.35) seems rather vague. However, since the use of this rule is elegant and provides close agreement with experimental results for metals and steel there have – over the years – been many attempts to strengthen the background for this flow rule. One of the most important attempts was proposed by Drucker (1951) and for that reason we will scrutinize his suggestion.

9.3 Drucker's postulate and its consequences

We have already touched upon what is meant by hardening and softening plasticity, and for uniaxial loading these phenomena are illustrated in Figs. 9.3 and 9.4. To obtain definitions applicable to general stress states, we will adopt the proposal of Drucker (1951, 1964) and it will turn out that this postulate leads to the associated flow rule as well as to the convexity of the yield surface.

Drucker's postulate makes use of a *stress cycle* and to illustrate this concept, we consider uniaxial loading of a hardening material, cf. Fig. 9.26a). First the material has been loaded to point B and then unloaded elastically to point A. The state indicated by point A with the stress σ^* is now considered as the existing state of the material. We now imagine that an additional load is first applied to the material; this brings us to point B with the stress σ. The additional load is now increased by the infinitesimal amount $d\sigma$ and this brings us to point C with

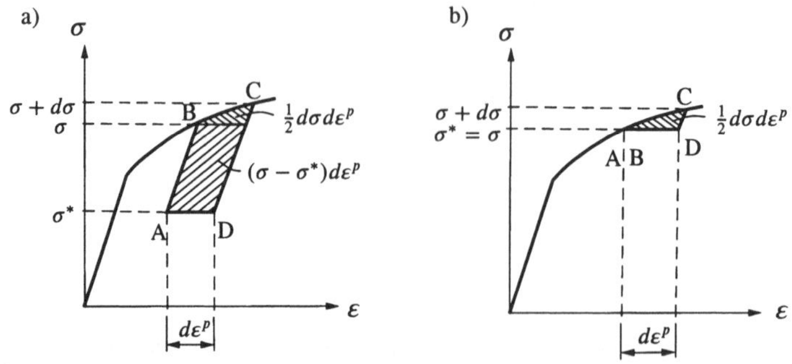

Figure 9.26: Hardening plasticity in uniaxial tension; a) stress cycle when the starting point A is below the current yield stress; b) stress cycle when the starting point A coincides with the current yield stress.

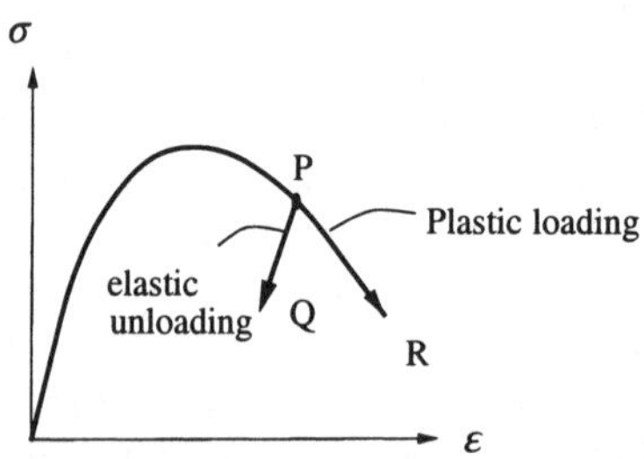

Figure 9.27: Softening material in uniaxial tension.Illustration of elastic unloading and plastic loading.

the stress $\sigma + d\sigma$. Then the entire additional load is removed and the material therefore unloads elastically to point D with the stress σ^* equal to the stress at point A. It appears that the additional load has carried the material through a complete *stress cycle*. This additional load which carries the material through a complete stress cycle is occasionally called an *external agency*.

As a special example of a stress cycle, we may imagine that points A and B coincide, cf. Fig. 9.26b). Also in this case is it possible for the material to go through a complete stress cycle by first applying a small additional load and then removing this load.

We have seen that it is always possible to let a hardening material go through a complete stress cycle and it is of interest to investigate whether the same is possible for a softening material. Before doing so we consider a softening material that has been loaded to point P in Fig. 9.27. It is of importance that a stress larger than the stress at point P can never be achieved. At point P, the material can only respond in two manners: either it unloads elastically towards point Q by decreasing the stress or strain or it responds plastically from point

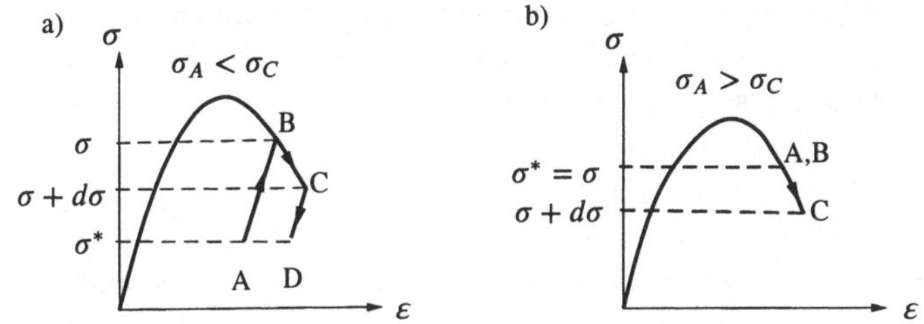

Figure 9.28: Softening material in uniaxial tension; a) stress cycle is possible; b) stress cycle is not possible.

P towards point R. However, such plastic loading can only occur if we force the total strain to increase. In particular, we cannot reach point R from point P either by prescribing an increasing or decreasing stress; increasing the stress is impossible and prescribing a decreasing stress would imply elastic unloading along the path PQ. We therefore conclude that

Elasto-plastic loading of a softening material can only be accomplished by strain control	(9.36)

This is in direct contrast to a hardening material where plastic loading can be accomplished by either prescribing an increasing stress or an increasing strain and we shall later see that this difference has important consequences for the establishment of general loading and unloading criteria, cf. the discussion relating to (10.38). Moreover, conclusion (9.36) has implications when testing softening materials in a testing machine. The only possibility for an experimental determination of the softening branch of the stress-strain curve is by prescribing an increasing strain, i.e. prescribing an increasing displacement on the specimen. This means that the testing machine must be operated in a *displacement controlled* manner.

With this discussion in mind, we consider a softening material and the situation shown in Fig. 9.28a) where $d\sigma$ is negative. Moreover, $\sigma_A < \sigma_C$ and by strain controlling the loading, it is possible for the additional loading - the external agency - to carry the material through a complete stress cycle ABCD. However, if point A and B coincide, cf. Fig. 9.28b), then by increasing the strain we reach point C where $\sigma_C < \sigma_A = \sigma_B$ and we have lost the possibility of reaching the stress given by $\sigma_A = \sigma_B$ irrespective of what load path we imagine. Consequently, a softening material in the state given by Fig. 9.28b) is not able to go through a stress cycle.

We have seen that the possibility for the material to go through a complete stress cycle is a property that distinguishes hardening plasticity from softening plasticity. Let us therefore return to the stress cycle for a hardening material shown in Fig. 9.26 and evaluate the work that is required by the additional load, i.e. by the external agency, to force the material to go through a stress cycle. Letting $W_{ex.ag.}$ denote the work per unit volume performed by the external agency during a complete stress cycle, we have

$$W_{ex.ag.} = \int_{ABCD} (\sigma - \sigma^*) d\varepsilon$$

The strain increment $d\varepsilon$ consists of its elastic and plastic components, i.e. $d\varepsilon = d\varepsilon^e + d\varepsilon^p$ and as plastic strains only develop during load path BC, we obtain

$$W_{ex.ag.} = \int_{ABCD} (\sigma - \sigma^*) d\varepsilon^e + \int_{BC} (\sigma - \sigma^*) d\varepsilon^p$$

As $d\varepsilon^e = d\sigma / E$, where Young's modulus E is assumed to be constant, and as σ^* is a constant quantity, we obtain

$$W_{ex.ag.} = \frac{1}{E} \int_{ABCD} \sigma d\sigma - \frac{\sigma^*}{E} \int_{ABCD} d\sigma + \int_{BC} (\sigma - \sigma^*) d\varepsilon^p$$

The path ABCD corresponds to a complete stress cycle and the contribution from the first two terms is therefore equal to zero. Denoting the plastic strain increment from B to C by $d\varepsilon^p$ ($d\varepsilon^p > 0$) and using the trapezoidal rule to the last term we obtain

$$W_{ex.ag.} = \frac{1}{2} \left[(\sigma - \sigma^*) + (\sigma + d\sigma - \sigma^*) \right] d\varepsilon^p = (\sigma - \sigma^*) d\varepsilon^p + \frac{1}{2} \, d\sigma d\varepsilon^p$$

The quantities appearing in this expression are illustrated in Fig. 9.26, where $\sigma > \sigma^*$ corresponds to Fig. 9.26a) whereas $\sigma = \sigma^*$ corresponds to Fig. 9.26b). It is obvious that we have $W_{ex.ag.} > 0$, i.e. the external agency must perform a positive work in order to force the hardening material to go through a stress cycle.

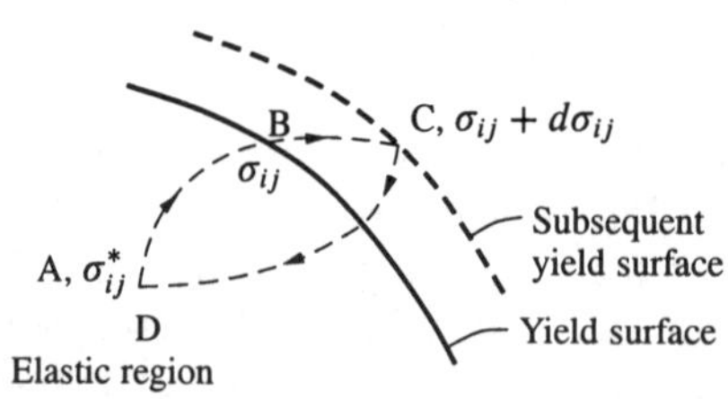

Figure 9.29: Stress cycle for hardening material produced by external agency.

To extend these observations to general stress states, we adopt the postulate of Drucker (1951, 1964). Again the external agency slowly applies an additional set of stresses to the already stressed material, thereby creating plastic strains and then slowly removing the added set. For hardening plasticity, the work done by the external agency during this stress cycle is postulated to be positive. The situation is sketched in Fig. 9.29, where σ^*_{ij} is the existing stress state at point A located on or inside the current yield surface. Additional stresses are then applied bringing the stress state to the yield surface at point B having the stress state σ_{ij}. Only changes in the elastic strains have occurred so far. Now, due to the external agency, a small stress increment $d\sigma_{ij}$ takes the stress state to point C on the subsequent yield surface thereby creating incremental plastic strains as well as incremental elastic strains. The external agency then releases the applied additional stresses thereby bringing the stress state back to the original stress state σ^*_{ij} along some stress path CD not necessarily coinciding with path ABC. According to the postulate of Drucker, the work done by the external agency during the stress cycle defined must be positive for hardening plasticity, i.e.

$$W_{ex-ag.} = \int_{ABCD} (\sigma_{ij} - \sigma^*_{ij}) d\varepsilon_{ij} > 0$$

where $d\varepsilon_{ij} = d\varepsilon^e_{ij} + d\varepsilon^p_{ij}$. Observing that plastic strains only occur during load path BC, and as σ^*_{ij} is a constant tensor, we obtain

$$W_{ex.ag.} = \int_{ABCD} \sigma_{ij} d\varepsilon^e_{ij} - \sigma^*_{ij} \int_{ABCD} d\varepsilon^e_{ij} + \int_{BC} (\sigma_{ij} - \sigma^*_{ij}) d\varepsilon^p_{ij} > 0$$

The first integral expresses the change of the elastic strain energy over the stress cycle considered and therefore this integral evidently becomes equal to zero. Likewise, as the elastic strains are determined entirely by Hooke's law $\varepsilon^e_{ij} = C_{ijkl}\sigma_{kl}$, the elastic strains before and after the stress cycle are equal; this implies that the second term also becomes equal to zero (provided that C_{ijkl} is constant). Denoting the plastic strain increments developing along path BC by $d\varepsilon^p_{ij}$ and letting $d\sigma_{ij}$ denote the corresponding stress increments whereas σ_{ij} denotes the stress state at point B, application of the trapezoidal rule to the last term implies that

$$W_{ex.ag.} = \frac{1}{2} \left[(\sigma_{ij} - \sigma^*_{ij}) + (\sigma_{ij} + d\sigma_{ij} - \sigma^*_{ij}) \right] d\varepsilon^p_{ij} > 0$$

i.e.

$$W_{ex.ag.} = (\sigma_{ij} - \sigma^*_{ij}) d\varepsilon^p_{ij} + \frac{1}{2} d\sigma_{ij} d\varepsilon^p_{ij} > 0 \tag{9.37}$$

If we choose $\sigma^*_{ij} = \sigma_{ij}$ then (9.37) implies

$$\boxed{d\sigma_{ij} d\varepsilon^p_{ij} > 0 \quad \textit{for associated hardening plasticity}} \tag{9.38}$$

cf. also Fig. 9.26b) As already indicated, we shall see later that (9.38) holds only for associated plasticity of a hardening material.

Let us return to (9.37) where we now know that the term $\frac{1}{2}d\sigma_{ij}d\varepsilon_{ij}^{p}$ is positive. However, if we choose σ_{ij}^{*} to be sufficiently different from σ_{ij} then the term $\frac{1}{2}d\sigma_{ij}d\varepsilon_{ij}^{p}$ becomes of second order providing

$$\boxed{(\sigma_{ij} - \sigma_{ij}^{*})d\varepsilon_{ij}^{p} \geq 0} \tag{9.39}$$

cf. also Fig. 9.26a) We shall now see that (9.39) has remarkable implications. Consider the nine-dimensional coordinate system defined by all the stress components and let the plastic strain increments be depicted in this coordinate system, then (9.39) states that the scalar product of the vectors $\sigma_{ij} - \sigma_{ij}^{*}$ and $d\varepsilon_{ij}^{p}$ is never obtuse. As σ_{ij} is located on the yield surface and as σ_{ij}^{*} is an arbitrary stress state on or within the yield surface, this implies that all points σ_{ij}^{*}, on or within the yield surface, must be located in the space opposite to the vector $d\varepsilon_{ij}^{p}$ and limited by the plane normal to $d\varepsilon_{ij}^{p}$, cf. Fig. 9.30. This implies that the yield surface must be *convex* in the stress space. Moreover, if the yield surface is assumed to be smooth, then the convexity of the yield surface and (9.39) imply that $d\varepsilon_{ij}^{p}$ must be normal to the yield surface, i.e.

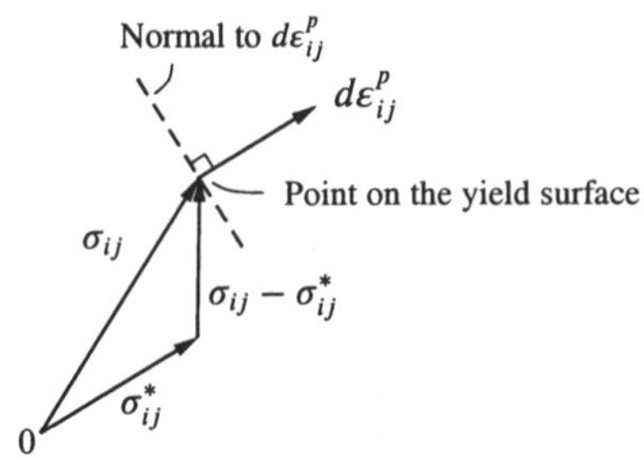

Figure 9.30: Convexity of yield surface in the stress space.

$$\boxed{d\varepsilon_{ij}^{p} = d\lambda\frac{\partial f}{\partial \sigma_{ij}}; \quad d\lambda \geq 0} \tag{9.40}$$

which is exactly the associated flow rule established previously in (9.35). Therefore, Drucker's postulate for hardening plasticity implies the following two very important points: *convexity* of the yield surface as well as the *normality principle* stated by (9.40). In addition, we shall later see that fulfillment of Drucker's postulate also ensures the *uniqueness* of the elasto-plastic boundary value problem, cf. Chapter 24. We may also note that if the yield surface has a sharp corner - as, for instance, is the case for Tresca's and Coulomb's yield criteria - then (9.39) implies that $d\varepsilon_{ij}^{p}$ is located somewhere within the 'cone' defined by the two normals at the corner, see Fig. 9.31. In Section 22.5, this property will be utilized to establish the *flow rule of Koiter*. We might finally mention that the conclusion of the yield surface being convex is in close agreement with experimental evidence, cf. Chapter 8.

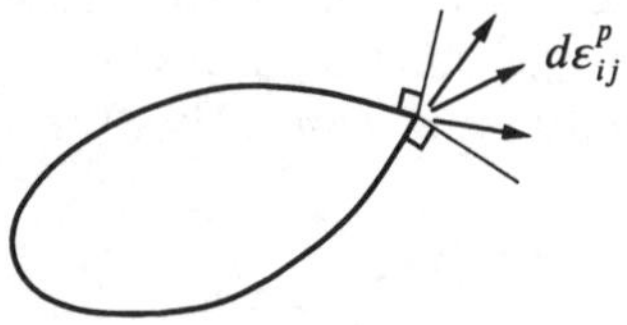

Figure 9.31: Possible directions of $d\varepsilon_{ij}^p$ when the yield surface has a corner.

The fundamental expression (9.39) was derived by Drucker's postulate for hardening plasticity and it implied convexity of the yield surface and the normality principle, i.e. the associated flow rule. It is of interest, however, that (9.39) holds even for ideal and softening plasticity, i.e. it holds even when Drucker's postulate does not hold.

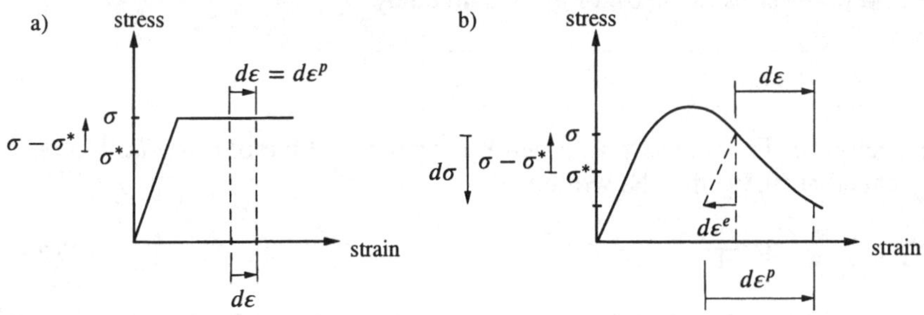

Figure 9.32: Expression (9.39) holds even for ideal and softening plasticity; The strain increment $d\varepsilon$ is imposed on the stress state given by σ.

To see this, consider Fig. 9.32 and recall that the stress σ^* is located inside or on the yield surface. Moreover, we do not consider a stress cycle, but we simply evaluate (9.39) when a strain increment $d\varepsilon > 0$ is imposed on the state with the stress σ. For ideal plasticity, cf. Fig. 9.32a), we have $\sigma - \sigma^* \geq 0$ and $d\varepsilon^p$ is positive, i.e. it follows that (9.39) holds. Considering the softening plasticity case shown in Fig. 9.32b), we have

$$d\varepsilon = d\varepsilon^e + d\varepsilon^p = \frac{d\sigma}{E} + d\varepsilon^p$$

Since $d\varepsilon^e$ is a negative quantity, $d\varepsilon^p$ is certainly a positive quantity and as $\sigma - \sigma^* \geq 0$, (9.39) holds again. Consequently, even though (9.39) was derived for hardening materials using Drucker's postulate, it holds even for ideal and softening plasticity and we are then led to the normality rule and the convexity of the yield surface also for these materials.

It is evident, however, that the manner in which one may argue for the establishment of (9.39) is open to discussion. Already von Mises (1928) observed that if the associated flow rule (9.40) is accepted, then (9.39) follows. On the other hand, (9.39) was directly postulated by Taylor (1947) and Hill (1948b) and considering crystal plasticity it was derived by Bishop and Hill (1951).

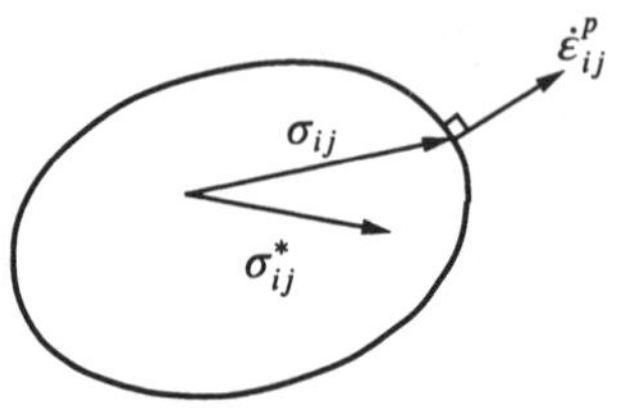

Figure 9.33: Quantities entering the postulate of maximum plastic dissipation.

Expression (9.39) has an interesting interpretation. The *rate of plastic work* related to the stress state σ_{ij} or σ_{ij}^* is defined by

$$\boxed{\dot{W}^p = \sigma_{ij}\dot{\varepsilon}_{ij}^p \; ; \qquad \dot{W}^{p*} = \sigma_{ij}^*\dot{\varepsilon}_{ij}^p} \tag{9.41}$$

respectively. The rate of plastic work is also called the *plastic dissipation.* It appears that (9.39) may be written as

$$\boxed{\dot{W}^p \geq \dot{W}^{p*}} \tag{9.42}$$

We observe that the stress state σ_{ij} is the stress state on the yield surface which in reality is related to the incremental plastic strains $\dot{\varepsilon}_{ij}^p$, cf. Fig. 9.33. Accepting the value for $\dot{\varepsilon}_{ij}^p$, (9.42) shows that the plastic dissipation relating to the real stress (σ_{ij}) is larger than or equal to the plastic dissipation relating to any other stress state (σ_{ij}^*) within or on the yield surface. Therefore, when (9.39) is expressed in terms of (9.42), it is called the *postulate of maximum plastic dissipation*; later, in Chapter 22, we will return to this important postulate and present a more stringent discussion.

We have seen that there are much stronger reasons for accepting the associated flow rule (9.35) than we originally suggested. In fact, for many workers within the field, Drucker's postulate and the postulate of maximum plastic dissipation were accepted over a span of years as being more or less laws of nature. We have shown that there are good arguments for the acceptance of the associated flow rule, but this flow rule is not a necessity. To generalize the flow rule (9.35) we may therefore write

$$\boxed{\dot{\varepsilon}_{ij}^p = \dot{\lambda}\frac{\partial g}{\partial \sigma_{ij}} \; ; \qquad \dot{\lambda} \geq 0} \tag{9.43}$$

where g is a *potential function* that may or may not be equal to the yield function f. In general, we expect g to depend on the same quantities as f, i.e. in accordance with (9.8) we have

$$\boxed{g = g(\sigma_{ij}, K_\alpha)}$$

If $g = f$, we have an *associated flow rule* and if $g \neq f$ we have a *nonassociated flow rule*. The nonassociated flow was suggested by Melan (1938) and later by Prager (1949). Clearly, an associated flow rule simplifies the theory and it should be used if the predictions are in agreement with the experimental findings. This is the case for metals and steel. For frictional materials like concrete, rocks and soils, use of a nonassociated flow rule is most often required in order to obtain realistic predictions

9.4 Consistency relation and evolution laws

Based on experimental evidence, suppose that we have determined the potential function g, which for associated plasticity is taken as the yield function. Then use of the flow rule (9.43) only determines the direction of the incremental plastic strains $\dot{\varepsilon}_{ij}^p$. However, the magnitude of $\dot{\varepsilon}_{ij}^p$ is still unknown since the plastic multiplier $\dot{\lambda}$ is as yet unknown. The next task is therefore to determine this quantity.

It is a fundamental property of plasticity theory that during plastic development, the current stress state is always located on the current yield surface. The current yield surface changes in general during plastic loading, as we have previously discussed, but, by definition, the current stress state is always located on the current yield surface during this evolution.

Having chosen the hardening rule, i.e. the hardening parameters, the current yield function is given in its general form by (9.8), i.e.

$$f(\sigma_{ij}, K_\alpha) = 0 \tag{9.44}$$

where K_α are our, as yet, unspecified hardening parameters. Since $f = 0$ during plastic loading, we can express the so-called *consistency relation* by

$$\boxed{\dot{f} = 0}$$

which with (9.44) and the chain rule lead to

$$\frac{\partial f}{\partial \sigma_{ij}}\dot{\sigma}_{ij} + \frac{\partial f}{\partial K_\alpha}\dot{K}_\alpha = 0 \tag{9.45}$$

where the summation convention is also used for the Greek letter α. The consistency relation was introduced by Prager (1949) and (9.45) tells us that during plastic loading where the stress state varies, also the hardening parameters K_α

vary in such a manner that the stress state always remains on the yield surface. According to (9.12) the hardening parameters K_α depend on the internal variables κ_α, i.e.

$$K_\alpha = K_\alpha(\kappa_\beta) \tag{9.46}$$

The choice of appropriate internal variables κ_α is given by us and also the relation between K_α and κ_α must be prescribed by us according to our experimental and other knowledge. Use of (9.46) in (9.45) provides

$$\frac{\partial f}{\partial \sigma_{ij}}\dot{\sigma}_{ij} + \frac{\partial f}{\partial K_\alpha}\frac{\partial K_\alpha}{\partial \kappa_\beta}\dot{\kappa}_\beta = 0 \tag{9.47}$$

In (9.47) the increments of the internal variables enter the consistency relation, and the next topic is therefore to establish some laws that tell us how κ_α evolutes with the plastic deformation. These laws are called the *evolution laws* and they must be specified by us based on our experimental knowledge. However, it turns out to be possible to establish the general format of these evolution equations.

A consequence of (9.46) is that $\dot{\kappa}_\alpha$ depends on the hardening parameters K_α. Moreover, since the incremental plastic strains $\dot{\varepsilon}_{ij}^p$ are the primary reason for the change of the internal variables, i.e. the change of the material, we may write

$$\dot{\kappa}_\alpha = a_\alpha(\dot{\varepsilon}_{ij}^p, K_\beta)$$

where a_α denote some functions. With (9.43) and since g depends on σ_{ij} and K_α we obtain

$$\dot{\kappa}_\alpha = a_\alpha(\dot{\lambda}, \sigma_{ij}, K_\beta)$$

Since our plasticity theory must be independent of time, we can accomplish this by writing $\dot{\kappa}_\alpha$ as an expression that is a homogeneous function of time. This is obtained by writing

$$\boxed{\dot{\kappa}_\alpha = \dot{\lambda}\, k_\alpha(\sigma_{ij}, K_\beta)} \tag{9.48}$$

where k_α denote some functions, the *evolution functions*, that must be chosen by us based on experimental or other evidence. Equation (9.48) constitutes our *evolution laws*. Following the flow rule (9.43), no plastic strains develop when $\dot{\lambda} = 0$. In accordance with expectations, (9.48) shows that the internal variables also remain unchanged when $\dot{\lambda} = 0$, cf. also (9.14).

Two sets of evolution equations have now been established: the flow rule $\dot{\varepsilon}_{ij}^p = \dot{\lambda}\partial g/\partial \sigma_{ij}$ and $\dot{\kappa}_\alpha = \dot{\lambda} k_\alpha$. It seems tempting to expect that the function $g(\sigma_{ij}, K_\alpha)$ not only serves as a potential function for $\dot{\varepsilon}_{ij}^p$, but also for $\dot{\kappa}_\alpha$. In Chapter 22 where the plasticity theory is formulated from the basic principles of thermodynamics, we will, in fact, show that these expectations become fulfilled as it turns out that

$$\dot{\varepsilon}^p_{ij} = \dot{\lambda}\frac{\partial g}{\partial \sigma_{ij}} \qquad \text{and} \qquad \dot{\kappa}_\alpha = -\dot{\lambda}\frac{\partial g}{\partial K_\alpha}$$

which means that $k_\alpha = -\partial g/\partial K_\alpha$; for associated plasticity these relations become

$$\dot{\varepsilon}^p_{ij} = \dot{\lambda}\frac{\partial f}{\partial \sigma_{ij}} \qquad \text{and} \qquad \dot{\kappa}_\alpha = -\dot{\lambda}\frac{\partial f}{\partial K_\alpha}$$

However, at the present stage we shall use the more general format given by (9.48).

Insertion of (9.48) into (9.47) leads to

$$\boxed{\frac{\partial f}{\partial \sigma_{ij}}\dot{\sigma}_{ij} - H\,\dot{\lambda} = 0} \tag{9.49}$$

where the quantity H is defined by

$$\boxed{H = -\frac{\partial f}{\partial K_\alpha}\frac{\partial K_\alpha}{\partial \kappa_\beta}k_\beta} \tag{9.50}$$

This quantity is termed the *generalized plastic modulus* and in the next chapter we will evaluate it in more detail.

If $H \neq 0$ then (9.49) provides

$$\boxed{\dot{\lambda} = \frac{1}{H}\frac{\partial f}{\partial \sigma_{kl}}\,\dot{\sigma}_{kl} \geq 0} \tag{9.51}$$

i.e. the flow rule (9.43) takes the form

$$\boxed{\dot{\varepsilon}^p_{ij} = \frac{1}{H}\frac{\partial g}{\partial \sigma_{ij}}\frac{\partial f}{\partial \sigma_{kl}}\,\dot{\sigma}_{kl}} \tag{9.52}$$

Based on our experimental and other knowledge we choose the yield function f, the potential function g, relation (9.46) and the evolution functions k_α entering the evolution law (9.48). This implies that the plastic modulus H is also known. Consequently, all quantities on the right-hand side of (9.52) are known, i.e. the incremental plastic strain $\dot{\varepsilon}^p_{ij}$ is known once the incremental stresses $\dot{\sigma}_{ij}$ are given; for evident reasons, formulation (9.52) is called a *stress driven format*. This means that we have, at last, obtained a constitutive relation for the incremental plastic strain. It is of interest that we may even reverse the procedure outlined above by specifying the generalized plastic modulus H as well as all evolution functions except for one quantity. Expression (9.50) then serves as the vehicle to determine the remaining unknown quantity in the evolution law.

In the next chapter, we will derive an expression for $\dot{\varepsilon}^p_{ij}$ that also holds when $H = 0$ and which also expresses $\dot{\varepsilon}^p_{ij}$ in terms of the incremental total strains $\dot{\varepsilon}_{ij}$. This more general formulation is called a *strain driven format.*

9.5 Preliminary loading and unloading criteria

With reference to Fig. 9.21, it is obvious that we must establish criteria which make it possible to decide whether we have plastic loading or elastic unloading. In the next chapter, we will derive such criteria which hold in general, but here we will establish some preliminary rules that hold for hardening plasticity.

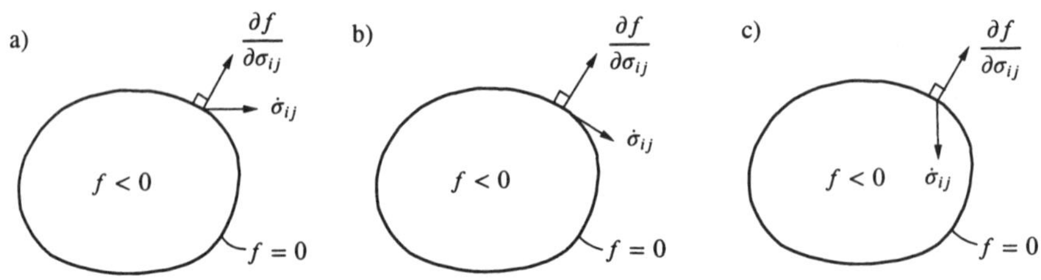

Figure 9.34: a) plastic loading; b) elastic unloading; c) neutral loading.

If the stress state is located inside the yield surface, i.e. $f < 0$, then we can only have elastic behavior. Therefore, evolution of plastic strains requires that $f = 0$. With reference to Fig. 9.34, we can conclude

$$\begin{aligned} f = 0 \quad \text{and} \quad \frac{\partial f}{\partial \sigma_{ij}}\dot{\sigma}_{ij} > 0 \quad &\Rightarrow \quad \text{plastic loading} \\ f = 0 \quad \text{and} \quad \frac{\partial f}{\partial \sigma_{ij}}\dot{\sigma}_{ij} < 0 \quad &\Rightarrow \quad \text{elastic unloading} \\ f = 0 \quad \text{and} \quad \frac{\partial f}{\partial \sigma_{ij}}\dot{\sigma}_{ij} = 0 \quad &\Rightarrow \quad \text{neutral loading} \end{aligned} \tag{9.53}$$

For neutral loading, it follows from (9.51) and (9.53) that $\dot{\lambda} = 0$ and thereby $\dot{\varepsilon}_{ij}^p = 0$, i.e. neutral loading results in a purely elastic response even though it may formally be treated as the occurrence of plasticity.

Consider the following two situations: the first where $\dot{\sigma}_{ij}$ is directed into the plastic regime, but infinitely close to neutral loading and the second where $\dot{\sigma}_{ij}$ is directed into the elastic regime, but infinitely close to neutral loading. Since $\dot{\varepsilon}_{ij}^p = 0$ holds for neutral loading, the two cases therefore result in the same response. As a result, the plasticity theory fulfills the so-called *continuity requirement* introduced by Handelman *et al.* (1947) and previously discussed in relation to nonlinear elastic models, cf. Section 5.4.

9.6 Isotropic hardening of a von Mises material

We will illustrate some of the findings above by considering isotropic hardening of a von Mises material. Associated plasticity is adopted and (9.21) provides

$$f = \sqrt{3J_2} - \sigma_y(\kappa)\ ; \quad f = 0 \quad \text{where} \quad \sigma_y(\kappa) = \sigma_{yo} + K(\kappa) \tag{9.54}$$

i.e.

$$\frac{\partial f}{\partial \sigma_{ij}} = \frac{3}{2}\frac{s_{ij}}{\sigma_y} \tag{9.55}$$

We write the general evolution law for the internal variable κ in accordance with (9.48), i.e.

$$\dot{\kappa} = \dot{\lambda}\, k(\sigma_{ij}, K) \tag{9.56}$$

The plastic modulus H is defined by (9.50). Since $dK/d\kappa = d\sigma_y/d\kappa$, we obtain with (9.54) and (9.56)

$$\boxed{H = \frac{d\sigma_y}{d\kappa}k} \tag{9.57}$$

The flow rule (9.43) gives with $g = f$ and (9.55)

$$\boxed{\dot{\varepsilon}^p_{ij} = \dot{\lambda}\,\frac{3}{2}\frac{s_{ij}}{\sigma_y}} \tag{9.58}$$

from which it is concluded that $\dot{\varepsilon}^p_{ii} = 0$, i.e. we have the plastic incompressibility that is very characteristic for metals and steel; as $\dot{\varepsilon}^p_{ii} = 0$, it follows that the plastic strains are purely deviatoric. To evaluate the plastic multiplier $\dot{\lambda}$, we multiply each side of (9.58) by itself to obtain

$$\dot{\varepsilon}^p_{ij}\,\dot{\varepsilon}^p_{ij} = \dot{\lambda}^2\,\frac{3}{2}\,\frac{3s_{ij}s_{ij}}{2\sigma_y^2} \tag{9.59}$$

Define the *effective plastic strain rate* $\dot{\varepsilon}^p_{eff}$ by

$$\boxed{\dot{\varepsilon}^p_{eff} = (\frac{2}{3}\dot{\varepsilon}^p_{ij}\dot{\varepsilon}^p_{ij})^{1/2}} \tag{9.60}$$

and it is evident that $\dot{\varepsilon}^p_{eff}$ is an invariant and that it is always non-negative. With (9.60) and (9.54), (9.59) takes the form

$$\boxed{\dot{\lambda} = \dot{\varepsilon}^p_{eff}} \tag{9.61}$$

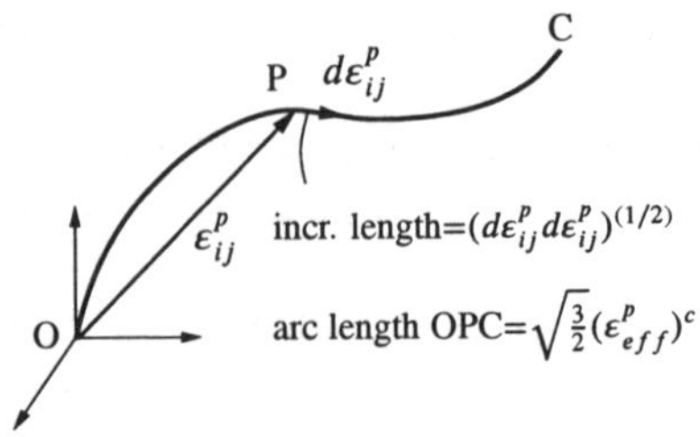

Figure 9.35: History of plastic strains ε^p_{ij} in plastic strain space.

Moreover, define the *effective stress* σ_{eff} by

$$\boxed{\sigma_{eff} = \sqrt{3J_2}} \qquad \text{i.e.} \qquad \dot{\sigma}_{eff} = \frac{3s_{ij}\dot{s}_{ij}}{2\sigma_{eff}} \tag{9.62}$$

where σ_{eff} is clearly an invariant. It follows from (9.54) that the yield criterion may then be expressed as

$$f = \sigma_{eff} - \sigma_y(\kappa) \; ; \qquad f = 0 \tag{9.63}$$

It turns out that the quantities σ_{eff} and $\dot{\varepsilon}^p_{eff}$ have particularly simple interpretations for uniaxial tension. For uniaxial tension we have $\sigma_{11} = \sigma$, $\sigma_{22} = \sigma_{33} = \sigma_{12} = \sigma_{13} = \sigma_{23} = 0$ which leads to $\sigma_{eff} = \sigma$. Moreover, we have $\dot{\varepsilon}^p_{11} = \dot{\varepsilon}^p$ and the flow rule (9.58) implies that $\dot{\varepsilon}^p_{22} = \dot{\varepsilon}^p_{33} = -\dot{\varepsilon}^p/2$ as well as $\dot{\varepsilon}^p_{12} = \dot{\varepsilon}^p_{13} = \dot{\varepsilon}^p_{23} = 0$ and (9.60) therefore leads to $\dot{\varepsilon}^p_{eff} = \dot{\varepsilon}^p$. In conclusion

$$\boxed{\sigma_{eff} = \sigma \; ; \quad \dot{\varepsilon}^p_{eff} = \dot{\varepsilon}^p \quad \textit{for uniaxial tension}} \tag{9.64}$$

The consistency relation (9.49) states that

$$\frac{\partial f}{\partial \sigma_{ij}}\dot{\sigma}_{ij} - H\dot{\lambda} = 0 \tag{9.65}$$

From (9.63) follows that

$$\frac{\partial f}{\partial \sigma_{ij}}\dot{\sigma}_{ij} = \dot{\sigma}_{eff} \tag{9.66}$$

Insertion of (9.66) in (9.65) and noting that $\sigma_{eff} = \sigma_y$ during plastic loading, cf. (9.63), we find with (9.61) that

$$\boxed{H = \frac{d\sigma_y(\kappa)}{d\varepsilon^p_{eff}}} \tag{9.67}$$

Up until now, the internal variable κ has not been specified and it is of interest that (9.67) holds irrespective of our choice of κ. However, in view of (9.67) it seems tempting to choose the internal variable κ as the effective plastic strain ε^p_{eff}, i.e. $\sigma_y = \sigma_y(\varepsilon^p_{eff})$, which also leads to $K(\kappa) = K(\varepsilon^p_{eff})$. To substantiate this choice, we obtain from definition (9.60) that

$$\boxed{\varepsilon^p_{eff} = \int_0^t \dot{\varepsilon}^p_{eff} dt \quad \text{i.e.} \quad \varepsilon^p_{eff} = \int_0^C d\varepsilon^p_{eff}\,; \quad C{=}\textit{current state}} \tag{9.68}$$

where

$$d\varepsilon^p_{eff} = (\frac{2}{3} d\varepsilon^p_{ij} d\varepsilon^p_{ij})^{1/2}$$

The plastic strain space is shown in Fig. 9.35 and at some state during the load history, we have the plastic strains ε^p_{ij} indicated by point P. The current plastic strain state corresponds to point C and it appears that path OPC describes the evolution of the plastic strains up until the current state. At an arbitrary point P on this path, the incremental arc length is given by $(d\varepsilon^p_{ij} d\varepsilon^p_{ij})^{1/2} = \sqrt{\frac{3}{2}} d\varepsilon^p_{eff}$, i.e. the incremental arc length is proportional to $d\varepsilon^p_{eff}$. Moreover, it appears that the arc length OPC $= \sqrt{\frac{3}{2}}(\varepsilon^p_{eff})^c$ where $(\varepsilon^p_{eff})^c$ is the effective plastic strain at point C. We emphasize that in accordance with (9.68), $(\varepsilon^p_{eff})^c$ is obtained by an integration along the entire path OPC; this means that arc length OPC differs, in general, from the distance of the straight line between point O and point C. In general, we therefore have $(\varepsilon^p_{eff})^c \neq \sqrt{\frac{2}{3}(\varepsilon^p_{ij})^c(\varepsilon^p_{ij})^c}$.

On the basis of this description, it seems tempting to assume that the current value of ε^p_{eff} is an expression of the plastic history. We may therefore assume that the internal variable κ equals ε^p_{eff}, i.e. the evolution law becomes

$$\boxed{\dot{\kappa} = \dot{\varepsilon}^p_{eff} \quad \text{i.e.} \quad \dot{\kappa} = \dot{\lambda} \qquad \textit{strain hardening}} \tag{9.69}$$

It is evident that κ is a non-decreasing quantity. A comparison with (9.56) shows that the evolution function $k = 1$ and (9.57) yields

$$\boxed{H = \frac{d\sigma_y}{d\varepsilon^p_{eff}}\,; \qquad \sigma_y = \sigma_y(\varepsilon^p_{eff})} \tag{9.70}$$

in accordance with (9.67). The choice of internal variable given by (9.69) is often referred to as *strain hardening* and it was suggested by Odqvist (1933).

We will now see how (9.70) can be used to calibrate the plasticity model to experimental data. Suppose that we perform a uniaxial tension test into the plastic region. From these experimental data, we can plot the results in a stress

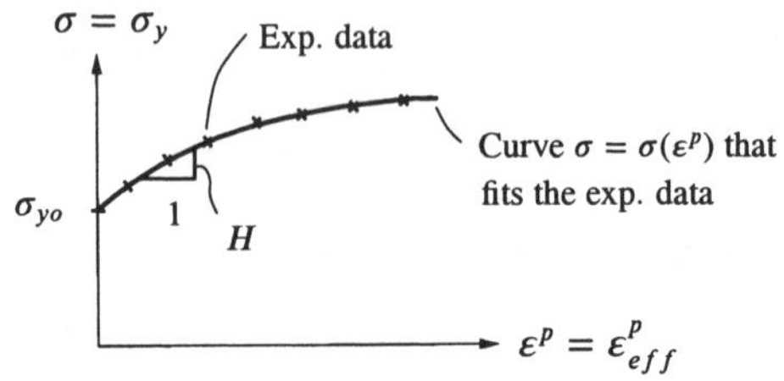

Figure 9.36: Uniaxial tension; experimental determination of the relation $\sigma_y = \sigma_y(\varepsilon^p_{eff})$.

(σ) – plastic strain (ε^p) diagram as shown in Fig. 9.36. Moreover, in accordance with Fig. 9.36 we may identify a curve given by some mathematical expression $\sigma = \sigma(\varepsilon^p)$ that fits these experimental data. For uniaxial tension, (9.63) and (9.64) show that $\sigma = \sigma_{eff} = \sigma_y$ and $\varepsilon^p = \varepsilon^p_{eff}$, i.e. the relation $\sigma = \sigma(\varepsilon^p)$ may equally well be written as $\sigma_y = \sigma_y(\varepsilon^p_{eff})$. Once this relation has been identified we may determine the plastic modulus H in accordance with (9.70) and illustrated in Fig. 9.36.

The essential point in this derivation is that we have determined the relation $\sigma_y(\varepsilon^p_{eff})$ and thereby $H = d\sigma_y/d\varepsilon^p_{eff}$ from a uniaxial experiment. However, since the expression $\sigma_y(\varepsilon^p_{eff})$ only involves invariants, it holds for any load history and we can now use this expression in the constitutive relation for any loading. In turn, this also implies that any load path can be used to determine the relation $\sigma_y(\varepsilon^p_{eff})$, but uniaxial loading is evidently very simple to accomplish and evaluate.

If the plastic modulus $H \neq 0$ (in Section 10.2 we will return to the general situation where H may even be zero), we obtain from the consistency relation (9.65) and (9.55) that

$$\boxed{\dot{\lambda} = \frac{1}{H}\frac{3s_{kl}\dot{\sigma}_{kl}}{2\sigma_y}} \tag{9.71}$$

At any state during the plastic loading, the current yield stress σ_y and the current plastic modulus H are known quantities. For any stress rate $\dot{\sigma}_{kl}$, the plastic multiplier $\dot{\lambda}$ is then determined by (9.71) and the flow rule (9.58) then provides us with the corresponding plastic strain rate $\dot{\varepsilon}^p_{ij}$. As $\dot{\varepsilon}_{ij} = \dot{\varepsilon}^e_{ij} + \dot{\varepsilon}^p_{ij}$ and as the elastic strain rates $\dot{\varepsilon}^e_{ij}$ are determined from Hooke's law, i.e.

$$\dot{\varepsilon}^e_{ij} = C_{ijkl}\dot{\sigma}_{kl}$$

where C_{ijkl} is the elastic flexibility tensor, we conclude that the entire response of the material can be identified for any stress increments $\dot{\sigma}_{kl}$.

Here we have identified the function $\sigma_y = \sigma_y(\varepsilon^p_{eff})$ and thereby the plastic modulus H, cf. (9.70), by means of a uniaxial tension test. However, as $\sigma_y =$

$\sigma_y(\varepsilon^p_{eff})$ is postulated to be a universal relation irrespective of the loading, other load paths may equally well be used even though the uniaxial tension test is especially convenient for this purpose. To illustrate this aspect, we imagine that we have performed a pure torsion test into the plastic region. Pure torsion is characterized by $\sigma_{12} = \tau$; $\sigma_{11} = \sigma_{22} = \sigma_{33} = \sigma_{13} = \sigma_{23} = 0$ and the effective stress σ_{eff} defined by (9.62) then becomes $\sigma_{eff} = \sqrt{3}\,\tau$. Moreover, for pure torsion we have $\dot{\varepsilon}^p_{12} \neq 0$ and the flow rule (9.58) provides $\dot{\varepsilon}^p_{11} = \dot{\varepsilon}^p_{22} = \dot{\varepsilon}^p_{33} = \dot{\varepsilon}^p_{13} = \dot{\varepsilon}^p_{23} = 0$ and the effective plastic strain rate (9.60) then becomes $\dot{\varepsilon}^p_{eff} = \dot{\gamma}^p/\sqrt{3}$ where $\dot{\gamma}^p = 2\dot{\varepsilon}^p_{12}$. Therefore

$$\sigma_{eff} = \sigma_y = \sqrt{3}\tau\,; \qquad \dot{\varepsilon}^p_{eff} = \frac{\dot{\gamma}^p}{\sqrt{3}} \qquad \text{for pure torsion} \tag{9.72}$$

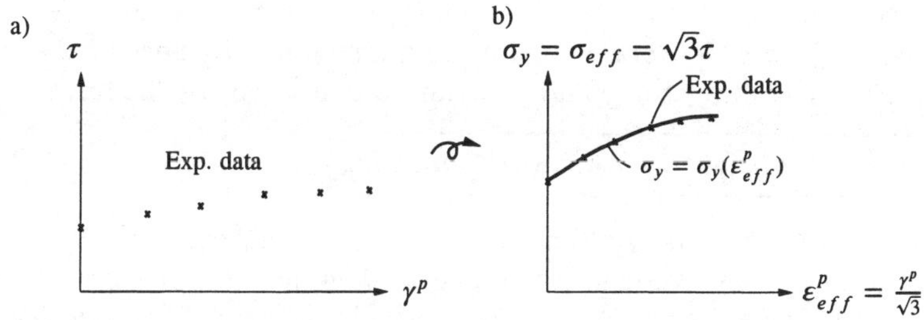

Figure 9.37: Pure torsion; experimental determination of the relation $\sigma_y = \sigma_y(\varepsilon^p_{eff})$.

From the torsion test, the experimental data illustrated in Fig. 9.37a) are obtained. Using (9.72), these data are converted into the experimental data shown in Fig. 9.37b). As shown, these data are again fitted by the curve $\sigma_y = \sigma_y(\varepsilon^p_{eff})$ which is now a known expression and we then use (9.70) to determine the plastic modulus H. It appears that any response measured in the laboratory may be used to identify the function $\sigma_y = \sigma_y(\varepsilon^p_{eff})$, but we also observe that a uniaxial tension test is particularly convenient for this purpose.

The remaining issue that needs to be addressed is the loading/unloading criteria which inform us whether the incremental response is elastic or elasto-plastic. If the current stress state is such that when inserted into the yield function f, we obtain $f < 0$ then we know that the incremental response is elastic. Therefore, for the incremental response to be elasto-plastic, it is necessary that $f = 0$ and the sign of the quantity $\dot{\sigma}_{ij}\partial f/\partial\sigma_{ij}$ then settles the question, cf. (9.53). In the present case, we obtain from (9.66) that

$$\text{if } f = 0 \quad \text{and} \quad \dot{\sigma}_{eff} = \begin{cases} > 0 & \text{plastic loading} \\ = 0 & \text{neutral loading} \\ < 0 & \text{elastic unloading} \end{cases}$$

where it is recalled that neutral loading produces no plastic strains.

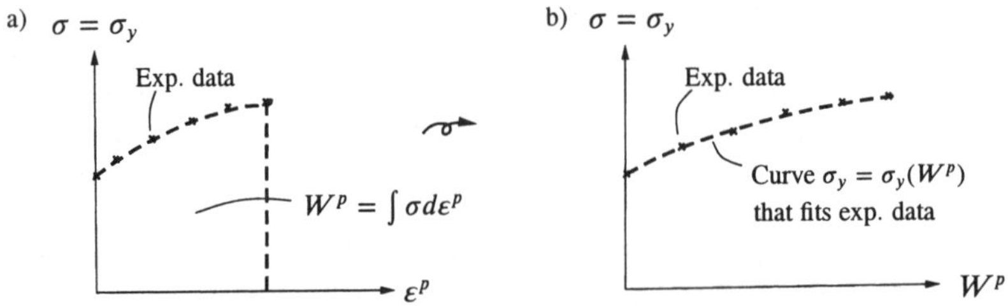

Figure 9.38: Uniaxial tension; experimental determination of the relation $\sigma_y = \sigma_y(W^p)$.

Previously, we adopted the strain hardening assumption that the internal variable κ is equal to the effective plastic strain ε^p_{eff}, cf. (9.69). Another common assumption is the *plastic work hardening* assumption usually attributed to Hill (1950) p.26, where the evolution law is assumed to be given by the format

$$\boxed{\dot{\kappa} = \dot{W}^p = \sigma_{ij}\dot{\varepsilon}^p_{ij} \qquad \textit{plastic work hardening}} \tag{9.73}$$

where $\dot{W}^p$ denotes the *rate of plastic work* or *plastic dissipation*, cf. (9.41) and where $\dot{W}^p$ is clearly an invariant. According to (9.54) and (9.73) it is postulated that $\sigma_y = \sigma_y(W^p)$, i.e. we use W^p to express the plastic history. In order to write (9.73) in the general format (9.56), we introduce the flow rule (9.58) into (9.73) to obtain

$$\dot{\kappa} = \dot{W}^p = \dot{\lambda}k \qquad \text{where} \qquad k = \frac{3s_{ij}\sigma_{ij}}{2\sigma_y} = \frac{3s_{ij}s_{ij}}{2\sigma_y} = \sigma_y \tag{9.74}$$

and where advantage was taken of (9.54). It is evident that κ is an invariant and that it is a non-decreasing quantity. Concerning plastic work hardening where (9.74) holds, the plastic modulus H given by (9.57) becomes

$$\boxed{H = \frac{d\sigma_y}{dW^p}\sigma_y \; ; \qquad \sigma_y = \sigma_y(W^p)} \tag{9.75}$$

Indeed, the same result is obtained from (9.67), which with $\kappa = W^p$ gives

$$H = \frac{d\sigma_y(W^p)}{d\varepsilon^p_{eff}} = \frac{d\sigma_y(W^p)}{dW^p}\frac{dW^p}{d\varepsilon^p_{eff}} = \frac{d\sigma_y(W^p)}{dW^p}\frac{dW^p}{d\lambda} = \frac{d\sigma_y(W^p)}{dW^p}\sigma_y$$

where it also was used that $d\varepsilon^p_{eff} = d\lambda$ and that $dW^p/d\lambda = k = \sigma_y$, cf. (9.74).

To calibrate the isotropic von Mises model using the evolution law given by the work hardening assumption (9.73), we may consider uniaxial tension. In this case (9.73) degenerates to

$$\dot{\kappa} = \dot{W}^p = \sigma\dot{\varepsilon}^p \qquad \text{for uniaxial tension}$$

From the uniaxial test, we may plot the results in the $\sigma\varepsilon^p$ - diagram shown in Fig. 9.38a) and obtain values for the invariant W^p corresponding to each of the data points. Since $\sigma = \sigma_y$ we may then plot σ_y as a function of the experimentally determined W^p-values as illustrated in Fig. 9.38b); a curve given by the expression $\sigma_y = \sigma_y(W^p)$ is then fitted to these data points and the plastic modulus H is subsequently determined by means of (9.75).

It is of considerable interest to compare the assumptions of strain hardening and work hardening. Since $\dot{\lambda} = \dot{\varepsilon}^p_{eff}$, the plastic dissipation (9.74) may be written as

$$\dot{W}^p = \sigma_y \, \dot{\varepsilon}^p_{eff} \tag{9.76}$$

Assume first that we adopt the strain hardening assumption, i.e. $\sigma_y = \sigma_y(\varepsilon^p_{eff})$. It then follows from (9.76) that

$$W^p = \int_0^{\varepsilon^p_{eff}} \sigma_y(\varepsilon^p_{eff}) d\varepsilon^p_{eff} \quad \text{i.e.} \quad W^p = W^p(\varepsilon^p_{eff}) \quad \text{or} \quad \varepsilon^p_{eff} = \varepsilon^p_{eff}(W^p)$$

This means that the strain hardening assumption $\sigma_y = \sigma_y(\varepsilon^p_{eff})$ may equally well be written in the form $\sigma_y = \sigma_y(W^p)$ which is exactly the work hardening assumption. Therefore, in this case, the two assumptions are identical. Assume next that we adopt the work hardening assumption. From $\sigma_y = \sigma_y(W^p)$ and (9.76) follow that

$$\varepsilon^p_{eff} = \int_0^{W^p} \frac{dW^p}{\sigma_y(W^p)} \quad \text{i.e.} \quad \varepsilon^p_{eff} = \varepsilon^p_{eff}(W^p) \quad \text{or} \quad W^p = W^p(\varepsilon^p_{eff})$$

and we infer that the assumption $\sigma_y = \sigma_y(W^p)$ implies the relation $\sigma_y = \sigma_y(\varepsilon^p_{eff})$ i.e. the strain hardening assumption.

We conclude that, in reality, it is immaterial whether we assume strain hardening or work hardening. However, it is emphasized that whereas this coincidence holds for isotropic hardening of a von Mises material, it is not a general conclusion that holds for arbitrary plasticity models. From the discussion above, we observe that the coincidence of strain hardening and work hardening hinges on the property that the plastic dissipation $\dot{W}^p$ defined in general by (9.73), for the isotropic von Mises model takes the particular form given by (9.76). A further discussion is given by Bland (1957).

To deform a material plastically, external work is required. The development of plastic deformations is accompanied by irreversible phenomena that require dissipation of some of the external work within the material. These irreversible phenomena may manifest themselves in terms of a reorganization of the microstructure of the material and in terms of *heat generation*. It is of considerable interest that Taylor and Quinney (1934) found from experiments with metals and steel that as much as 90-95% of the incremental plastic work $dW^p = \sigma_{ij} d\varepsilon^p_{ij}$ is transformed into heat. This supports the everyday experience that plastic deformations may result in a considerable temperature increase and a strict derivation of such heat generation will be given in Chapter 21.

Finally, we mention that isotropic hardening of a von Mises material is treated more systematically in Section 12.2.

9.7 Proportional loading of isotropic hardening von Mises material

It turns out to be of interest to investigate the response of an isotropic hardening von Mises material during increasing *proportional loading*. Proportional loading is occasionally called *radial loading*.

The stress history for proportional loading is defined as

$$\sigma_{ij}(t) = \beta(t)\sigma^*_{ij} \; ; \qquad \sigma^*_{ij} = \text{constant}$$

where σ^*_{ij} is an arbitrary constant stress state and $\beta(t)$ is a scalar function that increases with time t; moreover, $\beta(t = 0) = 0$. It follows that

$$\sigma_{kk} = \beta\sigma^*_{kk} \; ; \qquad s_{ij} = \beta s^*_{ij} \; ; \qquad \sigma_{eff} = \beta\sigma^*_{eff}$$

During plastic loading, we have $\sigma_y = \sigma_{eff}$, i.e. the flow rule (9.58) becomes

$$\dot{\varepsilon}^p_{ij} = \dot{\lambda}\frac{3s^*_{ij}}{2\sigma^*_{eff}}$$

where $\dot{\lambda} = \dot{\varepsilon}^p_{eff}$. The expression above may be integrated to obtain

$$\varepsilon^p_{ij} = \lambda\frac{3s^*_{ij}}{2\sigma^*_{eff}} \tag{9.77}$$

Multiplication of each side of (9.77) by itself then gives

$$\lambda = \varepsilon^p_{eff} = (\frac{2}{3}\varepsilon^p_{ij}\varepsilon^p_{ij})^{1/2} \tag{9.78}$$

Let us assume isotropic elasticity. In that case, Hooke's law is given by (4.85) and (4.86), i.e.

$$\varepsilon^e_{kk} = \frac{\sigma_{kk}}{3K} \; ; \qquad e^e_{ij} = \frac{s_{ij}}{2G^e}$$

where the notation G^e is used to emphasize that this shear modulus refers to the linear elastic behavior. From (9.77) and (9.78), we obtain

$$\varepsilon^p_{kk} = 0 \; ; \qquad e^p_{ij} = \varepsilon^p_{ij} = \varepsilon^p_{eff}\frac{3s^*_{ij}}{2\sigma^*_{eff}} = \varepsilon^p_{eff}\frac{3s_{ij}}{2\sigma_{eff}}$$

It then follows that

$$\varepsilon_{kk} = \varepsilon^{e}_{kk} + \varepsilon^{p}_{kk} = \frac{\sigma_{kk}}{3K} \tag{9.79}$$

and

$$e_{ij} = e^{e}_{ij} + e^{p}_{ij} = (\frac{1}{2G^{e}} + \frac{3\varepsilon^{p}_{eff}}{2\sigma_{eff}})s_{ij} \tag{9.80}$$

Let us assume that we have strain hardening, i.e. $\sigma_y = \sigma_y(\varepsilon^{p}_{eff})$ and thereby $\sigma_y = \sigma_{eff} = \sigma_{eff}(\varepsilon^{p}_{eff})$. This may be written as $\varepsilon^{p}_{eff} = \varepsilon^{p}_{eff}(\sigma_{eff})$ and as $\sigma_{eff} = \sqrt{3J_2}$, it appears that (9.79) and (9.80) can be written as

$$\boxed{\sigma_{kk} = 3K\varepsilon_{kk}\ ; \qquad s_{ij} = 2Ge_{ij}} \tag{9.81}$$

where

$$\boxed{K = constant\ ; \qquad G = G(J_2) = \frac{G^{e}}{1 + G^{e}\dfrac{3\varepsilon^{p}_{eff}(\sigma_{eff})}{\sigma_{eff}}}} \tag{9.82}$$

These constitutive relations are referred to as the *deformational plasticity theory* of Hencky (1924) and we have seen that they may be derived from the incremental formulation of an isotropic hardening von Mises material during increasing proportional loading. Moreover, a comparison of (9.81) and (9.82) with (4.102) and (4.105) shows that we have recovered a nonlinear isotropic Hooke formulation. As discussed previously in Sections 4.10 and 5.4, the major drawback of such nonlinear elasticity formulations is their unrealistic predictions during unloading.

9.8 Conditions for plastic incompressibility

We have previously indicated in (4.104) that the nonlinearity of metals and steel is entirely related to the deviatoric response whereas the volumetric response is linear. Likewise, we indicated in (4.106) that for materials like concrete, rocks and soil, the nonlinear response is related both to the deviatoric and volumetric response. With these observations, it is of interest to investigate the conditions for which a plasticity model will imply plastic incompressibility, i.e. a linear elastic volumetric response.

The general flow rule (9.43) states that

$$\dot{\varepsilon}^{p}_{ij} = \dot{\lambda}\frac{\partial g}{\partial \sigma_{ij}} \tag{9.83}$$

where the potential function is given by $g = g(\sigma_{ij}, K_\alpha)$. It follows that plastic incompressibility $\dot{\varepsilon}^p_{kk} = 0$ is obtained for

$$\boxed{\frac{\partial g}{\partial \sigma_{ii}} = 0 \quad \Rightarrow \quad \textit{plastic incompressibility}}$$

In general, we observe that

$$\frac{\partial g}{\partial \sigma_{ii}} = \frac{\partial g}{\partial \sigma_{11}} + \frac{\partial g}{\partial \sigma_{22}} + \frac{\partial g}{\partial \sigma_{33}} \neq \frac{\partial g}{\partial I_1}$$

However, for isotropic materials the potential function can be written as

$$g = g(I_1, J_2, J_3, K_\alpha)$$

This implies

$$\frac{\partial g}{\partial \sigma_{ij}} = \frac{\partial g}{\partial I_1}\frac{\partial I_1}{\partial \sigma_{ij}} + \frac{\partial g}{\partial J_2}\frac{\partial J_2}{\partial \sigma_{ij}} + \frac{\partial g}{\partial J_3}\frac{\partial J_3}{\partial \sigma_{ij}}$$

i.e.

$$\frac{\partial g}{\partial \sigma_{ij}} = \frac{\partial g}{\partial I_1}\delta_{ij} + \frac{\partial g}{\partial J_2}s_{ij} + \frac{\partial g}{\partial J_3}(s_{ik}s_{kj} - \frac{2}{3}J_2\delta_{ij}) \tag{9.84}$$

It then follows that

$$\boxed{\frac{\partial g}{\partial \sigma_{ii}} = 3\frac{\partial g}{\partial I_1}} \tag{9.85}$$

For metals and steel, it is concluded that we can assume that $g = g(J_2, J_3, K_\alpha)$ whereas concrete, rocks and soil require the inclusion of the first stress invariant I_1, i.e. $g = g(I_1, J_2, J_3, K_\alpha)$.

10 GENERAL PLASTICITY THEORY

In the previous chapter, the ingredients of the plasticity theory were introduced. The objective was to familiarize the reader with various important topics, but the exposition was tuned more towards illustrative viewpoints rather than a systematic treatment. With this background, we will now present a systematic exposition of the general plasticity theory. Reference may also be made to the textbooks of Chen and Han (1988) and Chen (1994) as well as to Hill (1950), Khan and Huang (1995), Lubliner (1990), Mroz (1966) and Stouffer and Dame (1996).

10.1 Fundamental equations

Let us first list the fundamental equations that comprise the general plasticity theory. Later on, we shall manipulate these equations in order to obtain a final framework that is suitable also from a computational point of view.

The total strains consist of the sum of the elastic and plastic strains, i.e.

$$\boxed{\dot{\varepsilon}_{ij} = \dot{\varepsilon}^e_{ij} + \dot{\varepsilon}^p_{ij}} \tag{10.1}$$

where a dot, as usual, denotes the time rate. The elastic strains are determined from Hooke's law, i.e.

$$\sigma_{ij} = D_{ijkl}(\varepsilon_{kl} - \varepsilon^p_{kl}) \tag{10.2}$$

where D_{ijkl} is the elastic stiffness tensor. We shall allow general elasticity, i.e. D_{ijkl} may even refer to anisotropic elasticity. Due to the symmetry of σ_{ij} and ε_{ij}, D_{ijkl} possesses the usual symmetry properties $D_{ijkl} = D_{jikl}$ and $D_{ijkl} = D_{ijlk}$, cf. (4.21). Moreover, we assume the symmetry property given by (4.23), i.e. we have in total that

$$\boxed{D_{ijkl} = D_{jikl} \;;\quad D_{ijkl} = D_{ijlk} \;;\quad D_{ijkl} = D_{klij}} \tag{10.3}$$

which leads to symmetry of the elastic stiffness matrix $\boldsymbol{D}$, cf. (4.38). In accordance with (4.24a) and (4.25), D_{ijkl} is positive definite, i.e.

$$a_{ij}D_{ijkl}a_{kl} > 0 \qquad \text{for any } a_{ij} \neq 0 \tag{10.4}$$

holds for any symmetric second-order tensor a_{ij}. It follows that D_{ijkl} is non-singular, i.e. the homogeneous equation system

$$D_{ijkl}a_{kl} = 0 \qquad \Rightarrow \qquad a_{kl} = 0 \tag{10.5}$$

possesses only the trivial solution $a_{kl} = 0$.

Assuming the tensor D_{ijkl} to be constant with respect to the loading, we obtain from (10.2) that

$$\boxed{\dot{\sigma}_{ij} = D_{ijkl}(\dot{\varepsilon}_{kl} - \dot{\varepsilon}^p_{kl})} \tag{10.6}$$

We assume the existence of a yield function $f(\sigma_{ij}, K_\alpha)$ so that development of plastic strains requires that

$$\boxed{f(\sigma_{ij}, K_\alpha) = 0 \qquad \textit{for development of plasticity}} \tag{10.7}$$

In this expression K_α denotes the hardening parameters, which may be scalars or higher-order tensors. Since $\alpha = 1, 2 \ldots$ we may have one, two or more hardening parameters, cf. the discussion relating to (9.6) - (9.8). To confirm with the previous terminology, (10.7) denotes the current yield surface which for $K_\alpha = 0$ reduces to the initial yield surface $F(\sigma_{ij})$. However, since we have discussed these concepts in detail in the previous chapter, we shall for convenience merely refer to (10.7) as the *yield function*. We also recall that the manner in which we go from the initial to the current yield criterion is controlled by the hardening rule, which may be chosen in the form of an isotropic, kinematic or mixed hardening rule.

For fixed hardening parameters, (10.7) describes a surface in the stress space. The sign of the function $f(\sigma_{ij}, K_\alpha)$ is chosen such that

$$\boxed{f(\sigma_{ij}, K_\alpha) < 0 \qquad \Rightarrow \qquad \textit{elastic behavior}} \tag{10.8}$$

Therefore, in order that changes may occur in the hardening parameters K_α and the plastic strains ε^p_{ij}, it is necessary that (10.7) is fulfilled.

The state of the material is described by the internal variables, which may be scalars or higher-order tensors. In analogy with the notation above, we denote the internal variables by κ_α. Since the internal variables describe the condition, i.e. the state, of the material, they are often termed state variables in the literature. In principle, the only quantities that can be directly measured, i.e. observed, are the total strains ε_{ij} and the temperature and the internal variables κ_α are therefore non-observable variables; this is the reason why κ_α are occasionally called 'hidden' variables in the literature . The internal variables κ_α

memorize the plastic loading history of the material. As an example of an internal variable, we may take the effective plastic strain. As the internal variables characterize the elasto-plastic material, we have

$$K_\alpha = K_\alpha(\kappa_\beta) \tag{10.9}$$

and the number of hardening parameters equals the number of internal variables; otherwise, the relation between K_α and κ_α will not be unique.

Like the yield function $f(\sigma_{ij}, K_\alpha)$, we assume the existence of the potential function g defined by

$$\boxed{g = g(\sigma_{ij}, K_\alpha)}$$

i.e. the potential function depends on the same parameters as the yield function. In accordance with (9.43), the flow rule is written in the following general form

$$\boxed{\dot{\varepsilon}_{ij}^p = \dot{\lambda}\frac{\partial g}{\partial \sigma_{ij}}\; ; \qquad \dot{\lambda} \geq 0} \tag{10.10}$$

If $g = f$, we have associated plasticity and if $g \neq f$, nonassociated plasticity holds. The flow rule states that the direction of $\dot{\varepsilon}_{ij}^p$ is given by the gradient $\partial g/\partial \sigma_{ij}$ whereas the plastic multiplier $\dot{\lambda}$ determines the magnitude of $\dot{\varepsilon}_{ij}^p$. If we have $\dot{\lambda} = 0$, then no plastic strains develop; otherwise, $\dot{\lambda} > 0$ ensures that $\dot{\varepsilon}_{ij}^p$ and $\partial g/\partial \sigma_{ij}$ possess the same direction.

The consistency relation states that during development of plastic strains, the yield criterion (10.7) is fulfilled, i e.

$$\boxed{\dot{f} = \frac{\partial f}{\partial \sigma_{ij}}\dot{\sigma}_{ij} + \frac{\partial f}{\partial K_\alpha}\dot{K}_\alpha = 0} \tag{10.11}$$

where the summation convention also holds for Greek letters. The consistency relation involves the time rate of the hardening parameters and from (10.9) we obtain the following evolution of K_α

$$\boxed{\dot{K}_\alpha = \frac{\partial K_\alpha}{\partial \kappa_\beta}\dot{\kappa}_\beta} \tag{10.12}$$

In turn, this expression involves the time rate of internal variables, i.e. how κ_α evolves with time. In accordance with (9.48), we assume the following evolution laws

$$\boxed{\dot{\kappa}_\alpha = \dot{\lambda} k_\alpha(\sigma_{ij}, K_\beta)} \tag{10.13}$$

where the evolution functions k_α, in general, are allowed to depend on the same variables as the yield function and the potential function. However, nothing is changed in the following exposition, if the evolution functions k_α are also allowed to depend on other variables than σ_{ij} and K_α. In accordance with our objective to formulate constitutive relations for time-independent materials, we observe that (10.13) is an expression that is homogeneous in time, i.e. the denominator dt present in both $\dot{\kappa}_\alpha = d\kappa_\alpha/dt$ and $\dot{\lambda} = d\lambda/dt$ may be canceled. This illustrates that the occurrence of time in (10.13), as well as in the other expressions, is purely artificial, enabling a convenient notation. Therefore, the evolution laws are written as (10.13) instead of the following more clumsy notation: $d\kappa_\alpha = d\lambda\, k_\alpha(\sigma_{ij}, K_\alpha)$. We also observe that (10.13) comprises a very general form of the evolution laws.

Equations (10.1) - (10.13) form the backbone of the general plasticity theory. No more assumptions or concepts are needed and the rest of this chapter describes the implications of these equations. It is emphasized that the relations above hold not only for isotropic materials, but also for anisotropic materials.

We finally notice that in order to obtain a specific plasticity model we must choose the yield function f, the potential function g, the hardening parameters K_α, the internal variables κ_α, the manner in which K_α and κ_α relate to each other (10.9) and the evolution functions k_α present in (10.13). We notice that the choice of hardening parameters K_α implies the choice of the hardening rule (isotropic, kinematic or mixed). For advanced plasticity models, all these choices are certainly not trivial and they must be based on our experimental and general knowledge of the material behavior. Often, a trial-and-error process is involved in making these choices whereby the predictions of the plasticity model are tuned to fit the pertinent experimental data.

It is of significant interest that the formulation of the general plasticity theory as presented above can also be derived entirely by means of the principles of *thermodynamics*. A detailed exposition of this thermodynamic approach will require the introduction of a number of concepts and in order to set the scene, we will postpone this fruitful, but complex treatment to Chapter 21. However, the fact that the general plasticity theory can be derived from thermodynamics means that it rests on the basic laws of nature and not just on our interpretation of experimental evidence and some reasonable assumptions. The key point in thermodynamics is the fulfillment of the *second law of thermodynamics*, i.e. the so-called *dissipation inequality*, which turns out to take the form $\sigma_{ij}\dot{\varepsilon}_{ij}^p - K_\alpha\dot{\kappa}_\alpha \geq 0$. Essentially this inequality excludes the existence of a *perpetual motion of the second kind*. If the yield function $f(\sigma_{ij}, K_\alpha)$ is a convex function, then it can be shown that the dissipation inequality is fulfilled for the following evolution laws: $\dot{\varepsilon}_{ij}^p = \dot{\lambda}\partial f/\partial\sigma_{ij}$ and $\dot{\kappa}_\alpha = -\dot{\lambda}\partial f/\partial K_\alpha$. A comparison with (10.10) and (10.13) shows that we have recovered associated plasticity and that the evolution function in (10.13) is in the form $k_\alpha = -\partial f/\partial K_\alpha$. Even nonassociated plasticity can be derived from thermodynamics and for the potential function $g = g(\sigma_{ij}, K_\alpha)$ being convex and fulfilling some rather weak requirements, we obtain

Potential function approach gives $\dot{\varepsilon}^p_{ij} = \dot{\lambda}\dfrac{\partial g}{\partial \sigma_{ij}}$ $\dot{\kappa}_\alpha = -\dot{\lambda}\dfrac{\partial g}{\partial K_\alpha}$

(10.14)

which again is contained in the formulation given by (10.10) and (10.13).

Two important observations follow from this discussion. The first observation arises from the second law given by $\sigma_{ij}\dot{\varepsilon}^p_{ij} - K_\alpha\dot{\kappa}_\alpha \geq 0$. Here, the term $\sigma_{ij}\dot{\varepsilon}^p_{ij}$ has the dimension of energy rate per unit volume and consequently, the term $K_\alpha\dot{\kappa}_\alpha$ has the same dimension. One uses the phrase that σ_{ij} is *conjugated* to $\dot{\varepsilon}^p_{ij}$ since their product is an energy rate (per unit volume). In the same sense, K_α is conjugated to $\dot{\kappa}_\alpha$ (or, in view of the minus sign in the second law, it may be more reasonable to say that $-K_\alpha$ is conjugated to $\dot{\kappa}_\alpha$). Therefore, just as it is natural to work with the stresses σ_{ij} and the plastic strains ε^p_{ij}, it is natural to work with the hardening parameters K_α and the internal variables κ_α.

The second observation is that just as $g(\sigma_{ij}, K_\alpha)$ serves as a potential for $\dot{\varepsilon}^p_{ij}$ via $\dot{\varepsilon}^p_{ij} = \dot{\lambda}\partial g/\partial\sigma_{ij}$ so does g serve as a potential for $\dot{\kappa}_\alpha$ via $\dot{\kappa}_\alpha = -\dot{\lambda}\partial g/\partial K_\alpha$. This underlines again the duality between σ_{ij} and ε^p_{ij} and between K_α and κ_α.

Here we have merely indicated some of the results that follow from thermodynamics. Since the use of thermodynamics requires the introduction of a number of abstract concepts, at the present stage we will continue to investigate plasticity from a purely mechanical point of view.

Another reason for this choice of presentation is that not all valid plasticity formulations fit into the format given by (10.14). This is not to say that they do not fulfill the second law of thermodynamics. To appreciate the difference, compare (10.13), i.e. $\dot{\kappa}_\alpha = \dot{\lambda}k_\alpha$ and the evolution law for $\dot{\kappa}_\alpha$ given by (10.14), i.e. $\dot{\kappa}_\alpha = -\dot{\lambda}\partial g/\partial K_\alpha$; it appears that the potential approach requires the evolution function k_α to be derived from the potential function $g(\sigma_{ij}, K_\alpha)$ i.e. $k_\alpha = -\partial g/\partial K_\alpha$. This is certainly a restriction and we conclude that our present approach (10.13) is somewhat more general that the potential function approach given by (10.14). Again this does not imply that models based on (10.13) do not fulfill the second law of thermodynamics; if they fulfill the inequality $\sigma_{ij}\dot{\varepsilon}^p_{ij} - K_\alpha\dot{\kappa}_\alpha \geq 0$ all formal requirements are fulfilled. The principal difference between (10.13) and (10.14) is that if the potential function g is convex and fulfills some rather weak requirements, then we know *a priori* that the formulation (10.14) fulfills the second law $\sigma_{ij}\dot{\varepsilon}^p_{ij} - K_\alpha\dot{\kappa}_\alpha \geq 0$ whereas with the formulation (10.13) we have, in principle, to check *a posteriori* that this inequality is fulfilled.

In Chapters 21 and 22 we will return to the thermodynamic approach and explore its ramifications in detail, not only for plasticity, but also for other nonlinear material behaviors.

10.2 Generalized plastic modulus - Relation between stress rates and total strain rates

The flow rule (10.10) determines the direction of the plastic strain rates, but the magnitude of $\dot{\varepsilon}^p_{ij}$ is unknown since the plastic multiplier $\dot{\lambda}$ is still unknown. To determine $\dot{\lambda}$, advantage is taken of the consistency relation and the evolution laws for $\dot{\kappa}_\alpha$.

Insert the evolution laws (10.13) into (10.12) to obtain

$$\boxed{\dot{K}_\alpha = \dot{\lambda}\frac{\partial K_\alpha}{\partial \kappa_\beta}k_\beta} \tag{10.15}$$

Insertion of (10.15) in the consistency relation (10.11) then provides

$$\boxed{\frac{\partial f}{\partial \sigma_{ij}}\dot{\sigma}_{ij} - H\dot{\lambda} = 0} \tag{10.16}$$

where the generalized plastic modulus H is defined by

$$\boxed{H = -\frac{\partial f}{\partial K_\alpha}\frac{\partial K_\alpha}{\partial \kappa_\beta}k_\beta} \tag{10.17}$$

We will see in a moment that once the yield function f, the potential function g and the generalized plastic modulus H are known then the plasticity formulation is complete. Referring to (10.17) and (10.15) it is allowable to directly specify the combined term $\partial K_\alpha/\partial \kappa_\beta k_\beta$ instead of specifying each of the separate factors $\partial K_\alpha/\partial \kappa_\beta$ and k_β. This implies that in some plasticity models the term $\partial K_\alpha/\partial \kappa_\beta k_\beta$ is directly postulated whereas in other models each of the factors $K_\alpha/\partial \kappa_\beta$ and k_β are postulated separately.

If $H \neq 0$, then (10.16) and the flow rule (10.10) give

$$\dot{\varepsilon}^p_{ij} = \frac{1}{H}\left(\frac{\partial f}{\partial \sigma_{kl}}\dot{\sigma}_{kl}\right)\frac{\partial g}{\partial \sigma_{ij}} \tag{10.18}$$

Moreover, (10.1) and (10.6) provide

$$\dot{\varepsilon}^e_{ij} = C_{ijkl}\,\dot{\sigma}_{kl} \tag{10.19}$$

where C_{ijkl} is the elastic flexibility tensor. Combination of (10.1), (10.18) and (10.19) gives

$$\dot{\varepsilon}_{ij} = C^{ep}_{ijkl}\dot{\sigma}_{kl} \tag{10.20}$$

where the *elasto-plastic flexibility tensor* C^{ep}_{ijkl} is given by

$$C^{ep}_{ijkl} = C_{ijkl} + \frac{1}{H}\frac{\partial g}{\partial \sigma_{ij}}\frac{\partial f}{\partial \sigma_{kl}} \tag{10.21}$$

Therefore, if $H \neq 0$ and if the stress rate $\dot{\sigma}_{kl}$ is given, then (10.20) determines the response completely. This formulation comprises the *stress driven* format.

However, we want to be able to determine the response in the general case where we may have $H = 0$. This turns out to be possible if - instead of a prescribed stress rate $\dot{\sigma}_{ij}$ - the total strain rate $\dot{\varepsilon}_{ij}$ is given. Moreover, this general format also has the advantage of fitting directly into our numerical formulation given in terms of the nonlinear finite element method, as we shall see later.

To obtain this general format, we insert the flow rule (10.10) into Hooke's law (10.6), i.e.

$$\dot{\sigma}_{ij} = D_{ijkl}\dot{\varepsilon}_{kl} - \dot{\lambda} D_{ijst}\frac{\partial g}{\partial \sigma_{st}} \tag{10.22}$$

Multiply this expression by $\partial f / \partial \sigma_{ij}$ and use (10.16) to obtain

$$\boxed{\dot{\lambda} = \frac{1}{A}\frac{\partial f}{\partial \sigma_{ij}} D_{ijkl}\dot{\varepsilon}_{kl} \; ; \qquad \dot{\lambda} \geq 0} \tag{10.23}$$

Here the parameter A is defined by

$$\boxed{A = H + \frac{\partial f}{\partial \sigma_{ij}} D_{ijkl}\frac{\partial g}{\partial \sigma_{kl}} \; ; \qquad A > 0} \tag{10.24}$$

To be able to derive (10.23), we must require that $A \neq 0$. However, we will later show that A is also a positive parameter.

If (10.23) is inserted into the flow rule (10.10), we find that

$$\dot{\varepsilon}^p_{ij} = \frac{1}{A}\left(\frac{\partial f}{\partial \sigma_{kl}} D_{klmn}\dot{\varepsilon}_{mn}\right)\frac{\partial g}{\partial \sigma_{ij}} \tag{10.25}$$

which proves that for given total strain rates, the plastic strain rates are known. Of more importance, however, is that insertion of (10.23) into (10.22) yields

$$\boxed{\dot{\sigma}_{ij} = D^{ep}_{ijkl}\dot{\varepsilon}_{kl}} \tag{10.26}$$

where the *elasto-plastic stiffness tensor* is given by

$$\boxed{D^{ep}_{ijkl} = D_{ijkl} - \frac{1}{A} D_{ijst}\frac{\partial g}{\partial \sigma_{st}}\frac{\partial f}{\partial \sigma_{mn}} D_{mnkl}} \tag{10.27}$$

This result comprises our objective, namely that for a given total strain rate $\dot{\varepsilon}_{ij}$, the stress rate $\dot{\sigma}_{ij}$ is determined from (10.26). This expression holds in general and we have in (10.24) emphasized that A is a positive parameter. This will be proved shortly. For evident reasons, (10.26) comprises the *strain driven* format. We also observe that D^{ep}_{ijkl} determines the current *tangential stiffness* of the material.

The establishment of the important relation (10.26) was first given by Hill (1950) for a von Mises material and later by Hill (1958) for general associated plasticity and by Mroz (1966) for general nonassociated plasticity. Often however, the establishment of the strain driven format (10.26) is attributed to Yamada *et al.* (1968) and Zienkiewicz *et al.* (1969).

Evidently, since D^{ep}_{ijkl} depends on the stress state σ_{ij}, the hardening parameters K_α and the plastic modulus H, (10.26) determines a nonlinear material response. However, we observe that D^{ep}_{ijkl} does not depend on either the stress rate $\dot{\sigma}_{ij}$ or the strain rate $\dot{\varepsilon}_{ij}$, i.e. relation (10.26) is incrementally linear.

It appears that the general format (10.26) looks very much as the incremental form of Hooke's law and it is therefore not surprising that (10.26) is of fundamental importance in nonlinear finite element calculations. Due to (10.3a,b) it follows from (10.27), that we always have

$$\boxed{D^{ep}_{ijkl} = D^{ep}_{jikl} \ ; \quad D^{ep}_{ijkl} = D^{ep}_{ijlk}}$$

It also follows from (10.27) and (10.3c) that

$$\boxed{\begin{aligned} &D^{ep}_{ijkl} = D^{ep}_{klij} \quad \textit{for associated plasticity} \\ &D^{ep}_{ijkl} \neq D^{ep}_{klij} \quad \textit{for nonassociated plasticity} \end{aligned}} \tag{10.28}$$

Similar to (4.37), we may write (10.26) in the following matrix format

$$\boxed{\dot{\boldsymbol{\sigma}} = \boldsymbol{D}^{ep}\, \dot{\boldsymbol{\varepsilon}}} \tag{10.29}$$

which is of importance in nonlinear finite element calculations.

It appears from (10.28) that associated plasticity $f = g$ implies symmetry, i.e. $\boldsymbol{D}^{ep} = \boldsymbol{D}^{ep^T}$ whereas nonassociated plasticity $f \neq g$ implies $\boldsymbol{D}^{ep} \neq \boldsymbol{D}^{ep^T}$. We conclude that nonassociated plasticity is more complicated, not only from a conceptual viewpoint, but also from a computational viewpoint. The advantage that is offered by nonassociated plasticity is the greater possibility to fit a specific plasticity model so that it complies well with the experimental evidence. In general, associated plasticity gives accurate predictions for metals and steel whereas nonassociated plasticity is often required when concrete, soil and rocks are considered.

10.3 Evaluation of plastic modulus *H*

When deriving (10.23), we implicitly assumed that the parameter $A \neq 0$. Let us now prove that A is a positive quantity as already stated by (10.24). The theory that we develop is supposed to be general, so it must also hold for an associated flow rule ($f = g$) even when the plastic modulus H is zero; according to (10.17), the plastic modulus H is zero for ideal plasticity, where the yield function does not involve any hardening parameters K_α, i.e. $\partial f/\partial K_\alpha = 0$. In that case, parameter A as given by (10.24) reduces to

$$A = \frac{\partial f}{\partial \sigma_{ij}} D_{ijkl} \frac{\partial f}{\partial \sigma_{kl}}$$

which is clearly a positive quantity, cf. (10.4). We are then led to the general requirement that

$$\boxed{A > 0}$$

always holds. Having proved this inequality, we will obtain an evaluation of the generalized plastic modulus H that, up to now, has just been defined by expression (10.17).

When using the general format (10.26), we may be interested in a situation where non-zero total strain rates $\dot{\varepsilon}_{ij} \neq 0$ imply $\dot{\sigma}_{ij} = 0$, i.e. we want to investigate possible non-trivial solutions $\dot{\varepsilon}_{kl}$ of the following homogeneous equation system

$$\dot{\sigma}_{ij} = D^{ep}_{ijkl}\, \dot{\varepsilon}_{kl} = 0 \tag{10.30}$$

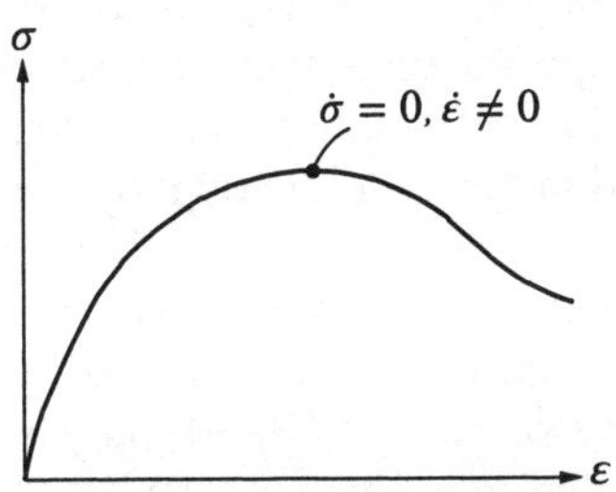

Figure 10.1: Limit point, i.e. peak stress where we have $\dot{\varepsilon} \neq 0$, but $\dot{\sigma} = 0$.

For uniaxial loading, this situation is illustrated in Fig. 10.1 and it is called a *limit point*. Since $\dot{\sigma}_{ij} = 0$, it follows from Hooke's law that $\dot{\varepsilon}^e_{ij} = 0$ i.e. $\dot{\varepsilon}_{ij} = \dot{\varepsilon}^p_{ij}$. With the flow rule (10.10), a solution to (10.30) must therefore be of the form

$$\dot{\varepsilon}_{ij} = \dot{\varepsilon}^p_{ij} = \dot{\lambda} \frac{\partial g}{\partial \sigma_{ij}} \tag{10.31}$$

To obtain a non-trivial solution we must require $\dot{\lambda} \neq 0$, i.e. insertion of (10.31) into (10.30) yields

$$D^{ep}_{ijkl} \frac{\partial g}{\partial \sigma_{kl}} = 0$$

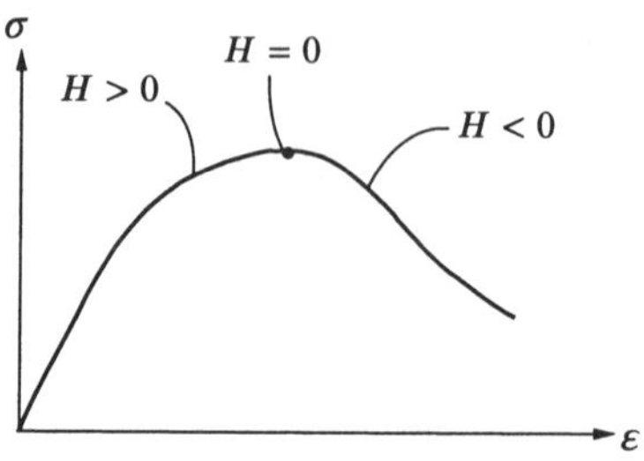

Figure 10.2: Interpretation of the plastic modulus H.

which with (10.27) and (10.24) takes the form

$$\frac{H}{A} D_{ijst} \frac{\partial g}{\partial \sigma_{st}} = 0 \tag{10.32}$$

Since D_{ijkl} is non-singular, cf. (10.5), (10.32) implies that

$$H = 0$$

We have then proved that (10.30) possesses the non-trivial solution (10.31) only when $H = 0$. Referring to the interpretation of (10.30) illustrated in Fig. 10.1, we are considering a situation where the stresses take their peak values and the tangent to the stress-strain curve is horizontal. This situation also holds for ideal plasticity and in that case the requirement $H = 0$, is indeed, not surprising, since the yield function for ideal plasticity, per definition, does not depend on any hardening parameters, cf. (9.15); since no hardening parameters K_α appear in the yield function, (10.17) shows that $H = 0$. More generally, however, the derivation above shows that (10.30) only has a non-trivial solution when $H = 0$ and that this solution is given by (10.31).

Considering (10.27) and recalling that $A > 0$, it appears that $D^{ep}_{ijkl} \to D_{ijkl}$ for $A \to \infty$, which requires $H \to \infty$. Therefore, the response of our elasto-plastic material approaches the linear elastic response when $H \to \infty$. We conclude that $H > 0$ corresponds to hardening plasticity and, consequently, $H < 0$ corresponds to softening plasticity.

The findings above are illustrated in Fig. 10.2, which may be summarized as

$$\boxed{\begin{array}{ll} H > 0 & \Rightarrow \textit{hardening plasticity} \\ H = 0 & \Rightarrow \textit{ideal plasticity} \\ H < 0 & \Rightarrow \textit{softening plasticity} \end{array}} \tag{10.33}$$

For isotropic hardening of a von Mises material, we observed in (9.67) that H denotes the slope $d\sigma/d\varepsilon^p$ for uniaxial loading. This finding is in accordance with (10.33), but we observe that $H = d\sigma/d\varepsilon^p$ is not a general interpretation. The general interpretation is given by (10.33) and Fig. 10.2.

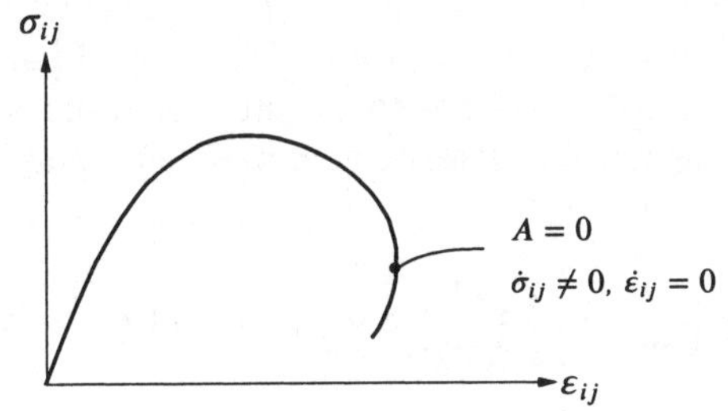

Figure 10.3: Illustration of the consequence of $A \to 0$. Note that all strain rate components $\dot{\varepsilon}_{ij} = 0$.

From (10.27) appears that formulation ceases to be valid if $A \to 0$ and we will next show that requirement $A > 0$ places restrictions on the amount of softening that can be modeled.

Against this background and as a complement to the problem posed by (10.30), it may be of interest to investigate whether it is possible to have the following situation

$$\dot{\sigma}_{ij} \neq 0 \ ; \quad \dot{\varepsilon}_{ij} = 0$$

This situation is illustrated in Fig. 10.3. According to (10.20) and (10.21) we have

$$\dot{\varepsilon}_{ij} = \dot{\varepsilon}^e_{ij} + \dot{\varepsilon}^p_{ij} = (C_{ijkl} + \frac{1}{H}\frac{\partial g}{\partial \sigma_{ij}}\frac{\partial f}{\partial \sigma_{kl}})\dot{\sigma}_{kl} \tag{10.34}$$

Since $\dot{\varepsilon}_{ij} = 0$ and $\dot{\varepsilon}^p_{ij} = \dot{\lambda}\partial g/\partial \sigma_{ij}$ hold, (10.6) gives

$$\dot{\sigma}_{kl} = -\dot{\lambda} D_{klmn}\frac{\partial g}{\partial \sigma_{mn}} \tag{10.35}$$

Use of (10.35) in (10.34) and noting that $\dot{\varepsilon}_{ij} = 0$ and $\dot{\lambda} \neq 0$ imply

$$0 = \frac{\partial g}{\partial \sigma_{ij}} + \frac{1}{H}\frac{\partial g}{\partial \sigma_{ij}}(\frac{\partial f}{\partial \sigma_{kl}} D_{klmn}\frac{\partial g}{\partial \sigma_{mn}})$$

With the definition (10.24), we obtain

$$0 = \frac{A}{H}\frac{\partial g}{\partial \sigma_{ij}}$$

and it is required that $A = 0$. It follows that the limit $A \to 0$ corresponds to the situation illustrated in Fig. 10.3. For this situation to be possible, the stress rates must fulfill (10.35). It is also of importance that all strain rate components must be zero, i.e. $\dot{\varepsilon}_{ij} = 0$, in order that the solution (10.35) be possible. For an isotropic hardening von Mises material, (10.35) implies that the incremental stresses should be purely deviatoric (for instance pure shear) and uniaxial tension will therefore not give rise to the situation discussed above.

Let us finally specialize to associated hardening plasticity. In that case we obtain with (10.18)

$$\dot{\sigma}_{ij}\dot{\varepsilon}_{ij}^{p} = \frac{1}{H}\left(\frac{\partial f}{\partial \sigma_{ij}}\dot{\sigma}_{ij}\right)\left(\frac{\partial f}{\partial \sigma_{kl}}\dot{\sigma}_{kl}\right) > 0 \quad \text{assoc. hard. plasticity}$$

in accordance with the implication of Drucker's postulate, cf. (9.38).

10.4 General loading and unloading criteria

In the previous chapter, we have derived criteria that make it possible to determine whether we have plastic loading or elastic unloading, cf. (9.53). As mentioned there, these criteria are only applicable to hardening plasticity and we will therefore now derive loading and unloading criteria that hold in general.

If the stress state is located inside the yield surface, i.e. $f < 0$, then, in accordance with (10.8) we have incrementally an elastic behavior. Therefore, evolution of plastic strains requires that $f = 0$. Let us define the *elastic stress rate* $\dot{\sigma}_{ij}^{e}$ by

$$\boxed{\dot{\sigma}_{ij}^{e} = D_{ijkl}\dot{\varepsilon}_{kl}} \tag{10.36}$$

where the term 'elastic' refers to the fact that this is the stress rate that would result for a given total strain rate provided the material responded elastically. With this definition, (10.23) takes the form

$$\dot{\lambda} = \frac{1}{A}\frac{\partial f}{\partial \sigma_{ij}}\dot{\sigma}_{ij}^{e} \geq 0 \tag{10.37}$$

where $A > 0$. From (10.37) it follows that plastic strains develop if $\partial f/\partial \sigma_{i}\dot{\rho}_{ij}^{e} > 0$. If $\partial f/\partial \sigma_{ij}\, \dot{\sigma}_{ij}^{e} < 0$, (10.37) implies $\dot{\lambda} < 0$ which cannot occur, i.e. this case must correspond to elastic unloading. Finally, in the limiting case where $\partial f/\partial \sigma_{ij}\, \dot{\sigma}_{ij}^{e} = 0$ we have neutral loading, which may formally be treated as the development of plastic strains even though the incremental response is purely elastic. We may summarize these findings into

$$\boxed{\begin{array}{lll} f = 0 \;\; \textit{and} \;\; \dfrac{\partial f}{\partial \sigma_{ij}}\dot{\sigma}_{ij}^{e} > 0 & \Rightarrow & \textit{plastic loading} \\ f = 0 \;\; \textit{and} \;\; \dfrac{\partial f}{\partial \sigma_{ij}}\dot{\sigma}_{ij}^{e} = 0 & \Rightarrow & \textit{neutral loading} \\ f = 0 \;\; \textit{and} \;\; \dfrac{\partial f}{\partial \sigma_{ij}}\dot{\sigma}_{ij}^{e} < 0 & \Rightarrow & \textit{elastic unloading} \end{array}} \tag{10.38}$$

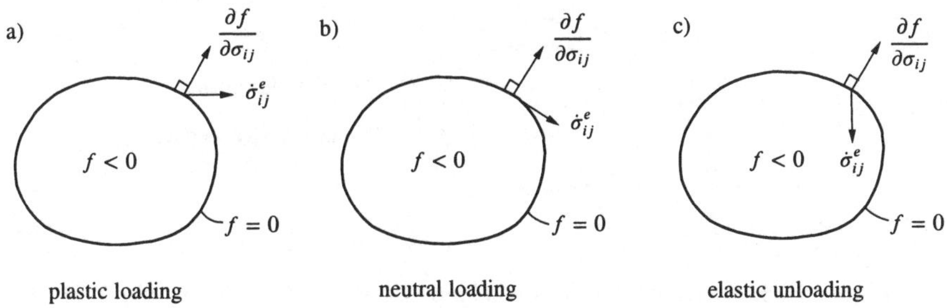

Figure 10.4: General loading/unloading criteria.

These general loading and unloading criteria are illustrated in Fig. 10.4.

We emphasize that the loading/unloading criteria (10.38) hold in general irrespective of whether we have hardening, ideal or softening plasticity. In these criteria, the elastic stress rate $\dot{\sigma}^e_{ij}$ given by (10.36) plays an important role. It is of significant interest that the term $\dot{\sigma}^e_{ij}$ also turns out to be of major importance in the numerical treatment of the constitutive equation, as will be discussed later on in Chapter 18. Moreover, it is emphasized that these loading/unloading criteria are not postulated, but they rather follow from the assumption that the plastic multiplier is non-negative, i.e. $\dot{\lambda} \geq 0$. Due to (10.36), the loading/unloading criteria (10.38) are strain driven.

Previously, we established the loading/unloading criteria given by (9.53) which formally have the same structure as the criteria (10.38). However, we claimed that (9.53) is not of general validity and we will now prove this point. From the consistency relation (10.16) we obtain with (10.37)

$$\frac{\partial f}{\partial \sigma_{ij}}\dot{\sigma}_{ij} = \frac{H}{A}\frac{\partial f}{\partial \sigma_{ij}}\dot{\sigma}^e_{ij} \tag{10.39}$$

With the definitions (10.33), evaluation of (10.39) shows that the general loading/unloading criteria (10.38) only coincide with the loading/unloading criteria (9.53) when $H > 0$, i.e. when we have hardening plasticity.

To further substantiate this conclusion, we may consider softening plasticity obtained via an isotropic 'hardening' concept where the yield surface shrinks. Figure 10.5 shows this situation and it is obvious that during the plastic deformation we have $\partial f / \partial \sigma_{ij}\, \dot{\sigma}_{ij} < 0$ in contradiction with the criteria (9.53).

If we have hardening plasticity, we may equally well use the loading criteria (9.53). However, even in this case, there are significant advantages relating to the general format (10.38) when nonlinear finite element calculations are considered. In a nonlinear finite element scheme, we know the nodal displacement increments from which the total strain increments $d\varepsilon_{ij}$ can be determined di-

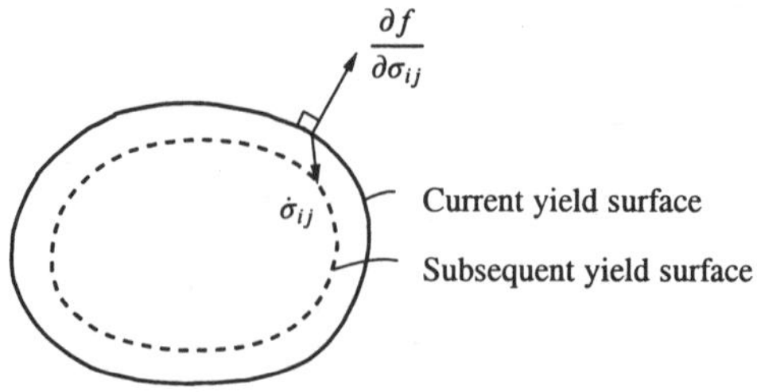

Figure 10.5: Softening plasticity modelled by an isotropic 'hardening' rule.

rectly, i.e. also the elastic stress increments $d\sigma_{ij}^e$ are known directly from (10.36) and the general criteria (10.38) can therefore be applied directly. If, however, the criteria (9.53) are used, we need to determine the stress increments $d\sigma_{ij}$ from (10.26) which require a (numerical) integration since the elasto-plastic stiffness tensor D_{ijkl}^{ep} varies with the loading. Therefore also for hardening plasticity, the general criteria (10.38) are preferable to the criteria (9.53) and we will return to this subject in Chapter 18.

The general criteria (10.38) were, in fact, introduced already by Hill (1958), but it took some time before the generality and advantages of this format were recognized in the literature.

The inability of the stress driven loading/unloading criteria (9.53) to include ideal and softening materials led Naghdi and Trapp (1975), see also Yoder and Iwan (1981), to propose a so-called *strain space plasticity*. This means that, for instance, the yield criterion is expressed as $f(\varepsilon_{ij}, K_\alpha)$ instead of $f(\sigma_{ij}, K_\alpha)$. However, due to (10.36) the loading/unloading criteria (10.38) are strain driven and since they apply to hardening, ideal and softening plasticity, there is no need to resort to the full strain space formulation mentioned above.

10.5 Plane strain

In practice, the two-dimensional problem of plane strain is of great importance and it is easy to derive the general plasticity equations also in that case. Focus will be directed to the strain driven format.

Let the plane of interest be described by the x_1x_2-coordinates, i.e. the x_3-direction denotes the out-of-plane direction. Per definition, plane strain is then characterized by

$$\dot{\varepsilon}_{13} = \dot{\varepsilon}_{23} = \dot{\varepsilon}_{33} = 0$$

Let Greek indices take the values 1, 2 whereas Latin indices, as usual, take the values 1, 2 and 3. From (10.26) and (10.27), the relation between the in-plane stress rate $\dot{\sigma}_{\alpha\beta}$ and the in-plane total strain rates $\dot{\varepsilon}_{\alpha\beta}$ is then given by

$$\boxed{\dot{\sigma}_{\alpha\beta} = D^{ep}_{\alpha\beta\gamma\delta}\dot{\varepsilon}_{\gamma\delta}} \tag{10.40}$$

where

$$\boxed{D^{ep}_{\alpha\beta\gamma\delta} = D_{\alpha\beta\gamma\delta} - \frac{1}{A}D_{\alpha\beta st}\frac{\partial g}{\partial\sigma_{st}}\frac{\partial f}{\partial\sigma_{mn}}D_{mn\gamma\delta}} \tag{10.41}$$

One may note the simultaneous occurrence of Greek and Latin indices in this latter expression. For convenience, we again list expression (10.24) for the parameter A, i.e.

$$\boxed{A = H + \frac{\partial f}{\partial\sigma_{ij}}D_{ijkl}\frac{\partial g}{\partial\sigma_{kl}}; \qquad A > 0} \tag{10.42}$$

For isotropic elasticity, we obtain from (4.110)

$$\boxed{D_{\alpha\beta\gamma\delta} = 2G[\tfrac{1}{2}(\delta_{\alpha\gamma}\delta_{\beta\delta} + \delta_{\alpha\delta}\delta_{\beta\gamma}) + \tfrac{\nu}{1-2\nu}\delta_{\alpha\beta}\delta_{\gamma\delta}]} \tag{10.43}$$

We also emphasize that the out-of-plane stress rates $\dot{\sigma}_{13}$, $\dot{\sigma}_{23}$ and $\dot{\sigma}_{33}$, in general, need to be determined from (10.26) since these stresses, in general, are non-zero and since they, in general, enter the expression for the yield function. Expression (10.26) leads to

$$\dot{\sigma}_{i3} = D^{ep}_{i3\gamma\delta}\dot{\varepsilon}_{\gamma\delta}$$

In general, also the out-of-plane plastic components $\dot{\varepsilon}^p_{13}$, $\dot{\varepsilon}^p_{23}$ and $\dot{\varepsilon}^p_{33}$ are of interest since they may enter the internal variables κ_α. From (10.25) and the flow rule follow that

$$\dot{\varepsilon}^p_{i3} = \dot{\lambda}\frac{\partial g}{\partial\sigma_{i3}} \quad \text{where} \quad \dot{\lambda} = \frac{1}{A}\frac{\partial f}{\partial\sigma_{kl}}D_{kl\gamma\delta}\dot{\varepsilon}_{\gamma\delta}$$

10.6 Plane stress

The plane of interest is given by the x_1x_2-coordinates, i.e. the x_3-direction denotes the out-of-plane direction. Per definition, plane stress is then defined by

$$\sigma_{13} = \sigma_{23} = \sigma_{33} = 0 \tag{10.44}$$

The aim is to establish the strain driven format. It will turn out that for plane stress conditions, this derivation is much more involved than the corresponding derivations for the plane strain case. The problem is that we do not know beforehand the out-of-plane strain components, implying that we cannot make direct use of the formulation (10.26). On the other hand, since the out-of-plane stress components are known, we start instead with

$$\dot{\varepsilon}_{ij} = \dot{\varepsilon}^e_{ij} + \dot{\varepsilon}^p_{ij}$$

Hooke's law then states

$$\dot{\varepsilon}_{ij} - \dot{\varepsilon}^p_{ij} = C_{ijkl}\dot{\sigma}_{kl}$$

The plane stress condition reduces this expression to

$$\dot{\varepsilon}_{ij} - \dot{\varepsilon}^p_{ij} = C_{ij\gamma\delta}\dot{\sigma}_{\gamma\delta} \tag{10.45}$$

i.e., the in-plane components are given by

$$\dot{\varepsilon}_{\alpha\beta} - \dot{\varepsilon}^p_{\alpha\beta} = C_{\alpha\beta\gamma\delta}\dot{\sigma}_{\gamma\delta}$$

or

$$\dot{\sigma}_{\alpha\beta} = D^*_{\alpha\beta\gamma\delta}(\dot{\varepsilon}_{\gamma\delta} - \dot{\varepsilon}^p_{\gamma\delta}) \tag{10.46}$$

Here $D^*_{\alpha\beta\gamma\delta}$ denotes the inverse to $C_{\alpha\beta\gamma\delta}$, i.e.

$$D^*_{\alpha\beta\theta\psi}C_{\theta\psi\gamma\delta} = \frac{1}{2}(\delta_{\alpha\gamma}\delta_{\beta\delta} + \delta_{\alpha\delta}\delta_{\beta\gamma})$$

For isotropic elasticity, $D^*_{\alpha\beta\gamma\delta}$ is given by (4.114), i.e.

$$\boxed{D^*_{\alpha\beta\gamma\delta} = 2G[\tfrac{1}{2}(\delta_{\alpha\gamma}\delta_{\beta\delta} + \delta_{\alpha\delta}\delta_{\beta\gamma}) + \tfrac{\nu}{1-\nu}\delta_{\alpha\beta}\delta_{\gamma\delta}]} \tag{10.47}$$

The difference between $D_{\alpha\beta\gamma\delta}$ given by (10.43) and $D^*_{\alpha\beta\gamma\delta}$ given by (10.47) should be observed.

From the general flow rule (10.10), we obtain

$$\dot{\varepsilon}^p_{\alpha\beta} = \dot{\lambda}\frac{\partial g}{\partial \sigma_{\alpha\beta}} \tag{10.48}$$

The plane stress conditions (10.44) reduce the general consistency relation (10.16) to

$$\frac{\partial f}{\partial \sigma_{\alpha\beta}}\dot{\sigma}_{\alpha\beta} - \dot{\lambda}H = 0 \tag{10.49}$$

where the plastic modulus H is defined as before, i.e. (10.17).

Insertion of the flow rule (10.48) into Hooke's law (10.46) gives

$$\dot{\sigma}_{\alpha\beta} = D^*_{\alpha\beta\gamma\delta}\dot{\varepsilon}_{\gamma\delta} - \dot{\lambda}D^*_{\alpha\beta\pi\theta}\frac{\partial g}{\partial \sigma_{\pi\theta}} \tag{10.50}$$

Multiplication by $\partial f/\partial\sigma_{\alpha\beta}$ and use of (10.49) then yield

$$\dot{\lambda} = \frac{1}{A^*}\frac{\partial f}{\partial\sigma_{\alpha\beta}}D^*_{\alpha\beta\gamma\delta}\dot{\varepsilon}_{\gamma\delta} \tag{10.51}$$

where A^* is defined by

$$\boxed{A^* = H + \frac{\partial f}{\partial\sigma_{\alpha\beta}}D^*_{\alpha\beta\gamma\delta}\frac{\partial g}{\partial\sigma_{\gamma\delta}} \quad ; \quad A^* > 0} \tag{10.52}$$

Finally, insertion of (10.51) into (10.50) results in

$$\boxed{\dot{\sigma}_{\alpha\beta} = D^{*ep}_{\alpha\beta\gamma\delta}\dot{\varepsilon}_{\gamma\delta}} \tag{10.53}$$

where

$$\boxed{D^{*ep}_{\alpha\beta\gamma\delta} = D^*_{\alpha\beta\gamma\delta} - \frac{1}{A^*}D^*_{\alpha\beta\pi\theta}\frac{\partial g}{\partial\sigma_{\pi\theta}}\frac{\partial f}{\partial\sigma_{\phi\psi}}D^*_{\phi\psi\gamma\delta}} \tag{10.54}$$

It is of interest to compare (10.52)-(10.54) with the corresponding formulation (10.40)-(10.42) applicable to plane strain. First of all, only Greek indices enter the plane stress formulation whereas a mixture of Greek and Latin indices enters the plane strain formulation. Second, whereas only $D^*_{\alpha\beta\gamma\delta}$ enters the plane stress case, both $D_{\alpha\beta\gamma\delta}$, $D_{\alpha\beta st}$ and D_{ijkl} enter the plane strain case. Third, the difference between A^* and A should be recognized.

For plane stress, it may be of importance to determine the out-of-plane plastic components $\dot{\varepsilon}^p_{i3}$, since they may enter the internal variables κ_α. From the flow rule (10.10) we have

$$\dot{\varepsilon}^p_{i3} = \dot{\lambda}\frac{\partial g}{\partial\sigma_{i3}} \tag{10.55}$$

where $\dot{\lambda}$ is given by (10.51). In general, we observe that $\partial g/\partial\sigma_{i3} \neq 0$ even though $\sigma_{i3} = 0$.

Moreover, the out-of-plane components $\dot{\varepsilon}_{i3}$ is determined from (10.45), i.e.

$$\dot{\varepsilon}_{i3} = \dot{\varepsilon}^p_{i3} + C_{i3\gamma\delta}\dot{\sigma}_{\gamma\delta}$$

where $\dot{\varepsilon}^p_{i3}$ is given by (10.55) and $\dot{\sigma}_{\gamma\delta}$ by (10.53).

The present derivation of the plane stress relations differs from most expositions. Here, we interpret $\partial f/\partial\sigma_{ij}$ and $\partial g/\partial\sigma_{ij}$ in the same manner as in the general three-dimensional case. This means that we use the general three-dimensional formulation of f and g, perform the differentiations $\partial f/\partial\sigma_{ij}$ and $\partial g/\partial\sigma_{ij}$ and then, finally, introduce the plane stress conditions into these expressions. In other expositions, the plane stress conditions are directly introduced into the expressions for f and g and then the differentiations are performed. In general, the two approaches differ and the advantage of the present approach is that all quantities have the same meaning, irrespective of whether the three-dimensional case, the plane strain case or the plane stress case is considered. In turn, this means that, for instance, the plastic multiplier $\dot{\lambda}$ and the plastic modulus H have the same interpretation in all these cases.

This discussion also illustrates an interesting observation, namely that the most general case, the three-dimensional case, preserves all symmetry properties of the problem whereas these properties may be less apparent when specialized conditions are considered. We are then led to the somewhat surprising conclusion that it is easier to derive general three-dimensional elasto-plasticity than plane stress elasto-plasticity.

11 PLASTIC COLLAPSE THEOREMS

In this chapter, we will present some classical *plastic collapse theorems* that have been used extensively during the years to evaluate the forces in metal forming processes and to determine the maximum load capacity of concrete structures and soil foundations. Moreover, these theorems often form the basis of various national building codes.

The body is assumed to consist of a material that may either be *stiff-ideal plastic* or *elastic-ideal plastic*, cf. Fig. 9.2. In that case the external load on the body cannot be increased indefinitely. This means that the structure possesses a certain maximum load capacity, i.e. a certain *limit load* exists. *Limit design* is concerned with the determination of this maximum load capacity - the *collapse load* - and the plastic collapse theorems take the forms of a *lower bound theorem* and an *upper bound theorem*.

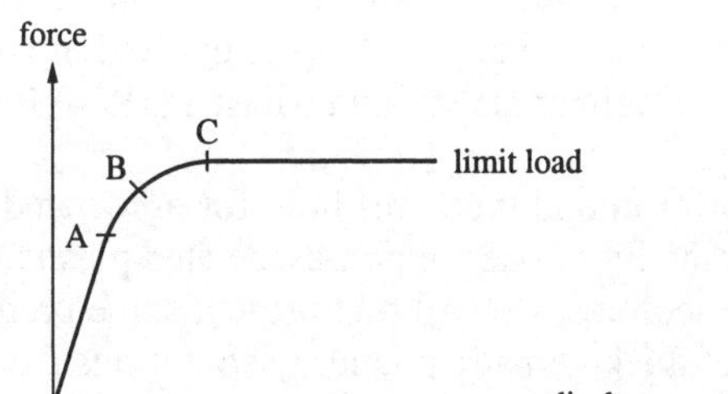

Figure 11.1: Illustration of limit load for a structure consisting of an elastic-ideal plastic material.

For an arbitrary structure consisting of an elastic-ideal plastic material, the concept of the limit load is shown in Fig. 11.1. At load stage A, plasticity is initiated and at load stage C, the stress distribution caused by plasticity is fully developed; load stage B is an intermediate load. For a beam loaded by a point force, these stages are illustrated in Fig. 11.2. From Fig. 11.1, it is apparent that, at the limit load, the external forces are constant and that the deformations can

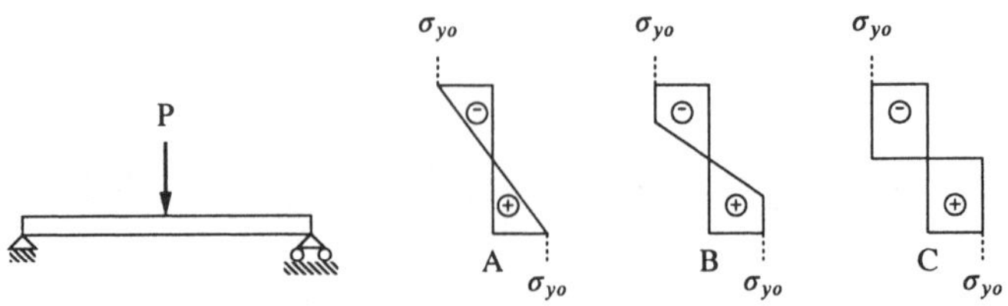

Figure 11.2: Elastic-ideal plastic beam and the stress distribution at the midsection at load stages A, B and C, cf. Fig. 11.1.

then take infinitely large values. Since small strains are assumed, we are only concerned with the initial stage of collapse.

Strict formal proofs of the lower and upper bound theorems were given by Gvozdev (1938) and they were later formulated independently by Drucker *et al.* (1952). However, the essential content of these theorems was known by intuition long before the establishment of the formal proofs. Indeed, the manner in which Coulomb (1776) investigated the collapse load of soil structures is essentially an application of the upper bound theorem. The formulation by Johansen (1943) of the so-called *yield line theory* applicable to concrete and steel plates also makes use of the upper bound theorem (an English translation is given by Johansen (1962)). On the other hand, determination of the collapse load of soil structures by means of a technique equivalent to the lower bound theorem was provided by Rankine in the mid-nineteenth century. The lower and upper bound theorems therefore belong to the classical topics in plasticity theory and we may refer to Koiter (1960), Lubliner (1990) and Nielsen (1984) for further historical remarks.

The upper and lower bound theorems hold for associated plasticity, but we will later touch upon the limit load for non-associated plasticity. Before we turn our interest to these theorems, we will first present some prerequisites.

Let us assume that the kinematic boundary conditions are given by

$$\boldsymbol{u} = \boldsymbol{u}(\boldsymbol{x}) \quad along \quad S_u \tag{11.1}$$

cf. (3.34); since, in general, $\boldsymbol{u} = \boldsymbol{u}(\boldsymbol{x}, t)$ this implies that the displacements along S_u do not vary due to the loading, i.e. $\dot{\boldsymbol{u}} = \boldsymbol{0}$ along S_u, and this situation covers most cases of practical interest. Moreover, considering static conditions and choosing the arbitrary virtual displacement v_i as $v_i = \dot{u}_i$, the principle of virtual work (3.33) takes the form

$$\int_V \dot{\varepsilon}_{ij}\sigma_{ij}dV = \int_S \dot{u}_i t_i dS + \int_V \dot{u}_i b_i dV \tag{11.2}$$

Due to (11.1), $\dot{u}_i$=0 holds along S_u and we therefore obtain

$$\int_S \dot{u}_i t_i dS = \int_{S_u} \dot{u}_i t_i dS + \int_{S_t} \dot{u}_i t_i dS = \int_{S_t} \dot{u}_i t_i dS$$

i.e. the principle of virtual work (11.2) takes the form

$$\int_V \dot{\varepsilon}_{ij}\sigma_{ij}dV = \int_{S_t} \dot{u}_i t_i dS + \int_V \dot{u}_i b_i dV \tag{11.3}$$

By this approach, the contribution to the boundary term from the reaction forces acting along S_u has been eliminated. This means that the right-hand side of (11.3) in addition to the displacement rate $\dot{u}_i$ only involves the known external load.

The formulation (11.3) was obtained from the equilibrium equations $\sigma_{ij,j} + b_i = 0$. By differentiating these equations with respect to time we obtain the incremental equilibrium equations $\dot{\sigma}_{ij,j} + \dot{b}_i = 0$. Evidently, we can derive a weak form of these latter equations by completely analogous manipulations to those that resulted in (11.3). Therefore, we obtain

$$\int_V \dot{\varepsilon}_{ij}\dot{\sigma}_{ij}dV = \int_{S_t} \dot{u}_i \dot{t}_i dS + \int_V \dot{u}_i \dot{b}_i dV \tag{11.4}$$

At the limit load, the external forces are constant, i.e. at the limit load (11.4) leads to

$$\int_V \dot{\varepsilon}_{ij}\dot{\sigma}_{ij}dV = 0 \tag{11.5}$$

For associated plasticity, we have

$$\dot{\varepsilon}_{ij} = \dot{\varepsilon}^e_{ij} + \dot{\varepsilon}^p_{ij} = C_{ijkl}\dot{\sigma}_{kl} + \dot{\lambda}\frac{\partial f}{\partial \sigma_{ij}}$$

i.e.

$$\dot{\varepsilon}_{ij}\dot{\sigma}_{ij} = \dot{\sigma}_{ij}C_{ijkl}\dot{\sigma}_{kl} + \dot{\lambda}\frac{\partial f}{\partial \sigma_{ij}}\dot{\sigma}_{ij} \tag{11.6}$$

For ideal plasticity where no hardening parameters exist, the consistency relation (10.11) gives $\dot{f} = \frac{\partial f}{\partial \sigma_{ij}}\dot{\sigma}_{ij} = 0$. Then (11.6) takes the form

$$\dot{\varepsilon}_{ij}\dot{\sigma}_{ji} = \dot{\sigma}_{ij}C_{ijkl}\dot{\sigma}_{kl} \tag{11.7}$$

and we observe that this expression holds for both a purely elastic behavior as well as for elastic-ideal plastic behavior when associated plasticity is assumed. Since the elastic flexibility tensor C_{ijkl} is positive definite, cf. (4.30), i.e. $\dot{\sigma}_{ij}C_{ijkl}\dot{\sigma}_{kl} > 0$ when $\dot{\sigma}_{ij} \neq 0$ it follows from (11.7) that (11.5) only can hold true if $\dot{\sigma}_{ij} = 0$ holds at the limit load. Since $\dot{\varepsilon}^e_{ij} = 0$ when $\dot{\sigma}_{ij} = 0$, we find

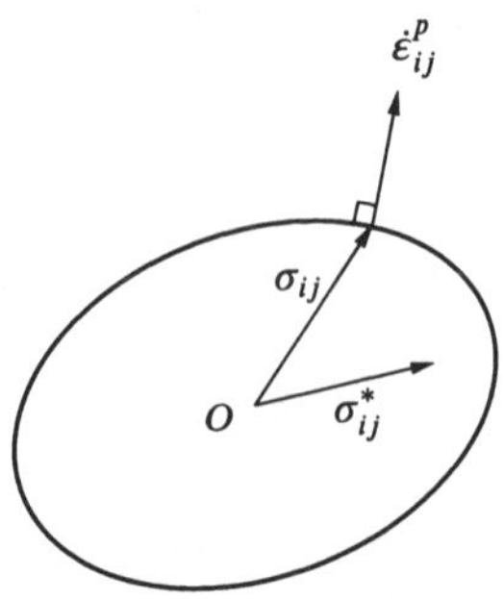

Figure 11.3: Quantities entering the postulate of maximum plastic dissipation.

that $\dot{\varepsilon}_{ij} = \dot{\varepsilon}^p_{ij}$ applies at the limit load when the material is elastic-ideal plastic. Evidently, $\dot{\varepsilon}_{ij} = \dot{\varepsilon}^p_{ij}$ also holds for stiff-ideal plasticity, i.e. we have

For stiff-ideal plasticity and elastic-ideal plasticity

$$\dot{\varepsilon}_{ij} = \dot{\varepsilon}^p_{ij} \tag{11.8}$$

holds at the limit point

Let us recall the postulate of maximum plastic dissipation given by (9.42) which reads

$$(\sigma_{ij} - \sigma^*_{ij})\dot{\varepsilon}^p_{ij} \geq 0 \tag{11.9}$$

Here σ_{ij} is the stress state on the yield surface related to the plastic strain rate $\dot{\varepsilon}^p_{ij}$ and σ^*_{ij} is any stress state within or on the yield surface; these features are illustrated in Fig. 11.3.

Since we are considering ideal plasticity, where no hardening parameters exist, the restriction on the stress state σ^*_{ij} can be expressed as

$$f(\sigma^*_{ij}) \leq 0$$

It is recalled that the postulate of maximum plastic dissipation leads to the associated flow rule. Moreover, since we are considering ideal plasticity where the convex yield surface certainly encloses the stress origin, the scalar product $\sigma_{ij}\dot{\varepsilon}^p_{ij}$ must be positive, cf. Fig. 11.3, i.e.

$$\sigma_{ij}\dot{\varepsilon}^p_{ij} > 0 \tag{11.10}$$

11.1 Lower bound theorem

To establish the lower bound theorem, we first introduce the concept of an *allowable* and *statically admissible stress field* σ_{ij}^a. It is defined as

Allowable and statically admissible stress field σ_{ij}^a :

$$\begin{aligned} &\sigma_{ij,j}^a + b_i^a = 0 \\ &f(\sigma_{ij}^a) \leq 0 \\ &t_i^a = \mu^- t_i \quad \text{is given along } S_t \\ &b_i^a = \mu^- b_i \quad \text{in } V \end{aligned} \tag{11.11}$$

The stress field σ_{ij}^a is created by the external forces t_i^a and b_i^a and is called statically admissible because it fulfills the differential equations of equilibrium and is called allowable because it does not violate the yield criterion. Moreover, this stress field σ_{ij}^a is created by external forces that are proportional to the real external forces and this proportionality is expressed by the factor μ^-. We will assume that by some means we are able to determine such a statically admissible and allowable stress field σ_{ij}^a.

Evidently, just as it is possible to make a weak formulation of the equilibrium equations for the real stress field σ_{ij}, it is possible to make a weak form of the equilibrium equations for the stress field σ_{ij}^a. Similar to (11.3), we then obtain

$$\int_V \dot{\varepsilon}_{ij}\sigma_{ij}^a dV = \int_{S_t} \dot{u}_i t_i^a dS + \int_V \dot{u}_i b_i^a dV$$

Due to (11.11), we then obtain

$$\int_V \dot{\varepsilon}_{ij}\sigma_{ij}^a dV = \mu^-[\int_{S_t} \dot{u}_i t_i dS + \int_V \dot{u}_i b_i dV] \tag{11.12}$$

Multiplication of (11.3) by μ^- and then subtraction of (11.12) give

$$\int_V (\mu^- \dot{\varepsilon}_{ij}\sigma_{ij} - \dot{\varepsilon}_{ij}\sigma_{ij}^a)dV = 0$$

As $\mu^- = (\mu^- - 1) + 1$, we obtain

$$\int_V [(\mu^- - 1)\dot{\varepsilon}_{ij}\sigma_{ij} + \dot{\varepsilon}_{ij}(\sigma_{ij} - \sigma_{ij}^a)]dV = 0 \tag{11.13}$$

Since σ_{ij}^a does not violate the yield criterion, we can as σ_{ij}^* take the stress σ_{ij}^a in Fig. 11.3 and the postulate of maximum plastic dissipation (11.9) in conjunction with (11.8) then shows that $\dot{\varepsilon}_{ij}(\sigma_{ij} - \sigma_{ij}^a) \geq 0$. Moreover, (11.10) and (11.8) show that $\dot{\varepsilon}_{ij}\sigma_{ij} > 0$. The only possibility for fulfillment of (11.13) is then $\mu^- \leq 1$. We are then led to the following important theorem

Lower bound theorem: *The external loading corresponding to an allowable and statically admissible stress field is less than or equal to the collapse load, i.e.* $\mu^- \leq 1$	(11.14)

Provided that we are able to identify an allowable and statically admissible stress field, the theorem provides a lower bound to the collapse load. The key question is then how such a stress field is identified. The simplest possibility is to perform an elastic analysis and then choose the factor μ^- such that the yield criterion is not violated. Clearly this would provide a very conservative μ^--factor and another possibility is to make some simplified analysis where the plastic effects are accounted for. As an example, consider a reinforced rectangular concrete plate. This plate is imagined to carry the load as two sets of beams at right angles to each other. It is then possible to satisfy the equilibrium differential equations for these two sets of beams when plastic effects are accounted for. If these stresses are within or on the yield surface and if they also satisfy the static boundary conditions, an allowable and statically admissible stress field has been established. This is the concept of the *strip method* proposed by Hillerborg (1974). There exists a number of approaches that often depend on the specific structure in question, and the reader may consult the specialized literature on this subject, see Chen and Han (1988), Johnson and Mellor (1983), Lubliner (1990) and Nielsen (1984). We will later give a simple example.

11.2 Upper bound theorem

Let us next introduce the concept of a *compatible displacement field* u_i^c; occasionally, such a displacement field is called an *admissible displacement field*. A compatible displacement field u_i^c is defined as one where $\dot{u}_i^c = 0$ holds along S_u and where the corresponding strain rate field is given by $\dot{\varepsilon}_{ij}^c = \frac{1}{2}(\dot{u}_{i,j}^c + \dot{u}_{j,i}^c)$. In accordance with (11.8), we will take $\dot{\varepsilon}_{ij}^c$ to be purely plastic. Moreover, the (plastic) strain rate $\dot{\varepsilon}_{ij}^c$ will be taken to be in accordance with the normality rule for associated plasticity. By this we mean that there exists some stress state σ_{ij}^c such that $\dot{\varepsilon}_{ij}^c = \dot{\lambda}(\frac{\partial f}{\partial \sigma_{ij}})_c$ where $(\frac{\partial f}{\partial \sigma_{ij}})_c$ indicates that $\frac{\partial f}{\partial \sigma_{ij}}$ is evaluated at some stress state σ_{ij}^c where $f(\sigma_{ij}^c) = 0$. We do not necessarily know this stress field σ_{ij}^c, but we do know that $\dot{\varepsilon}_{ij}^c = (\dot{\varepsilon}_{ij}^p)^c$ is in accordance with the normality rule. For von Mises ideal plasticity, for instance, where $\frac{\partial f}{\partial \sigma_{ij}} = \frac{3s_{ij}}{2\sigma_{yo}}$, cf. (9.55), we can take $\dot{\varepsilon}_{ij}^c$ as any deviatoric strain state. With these remarks, we consider the following

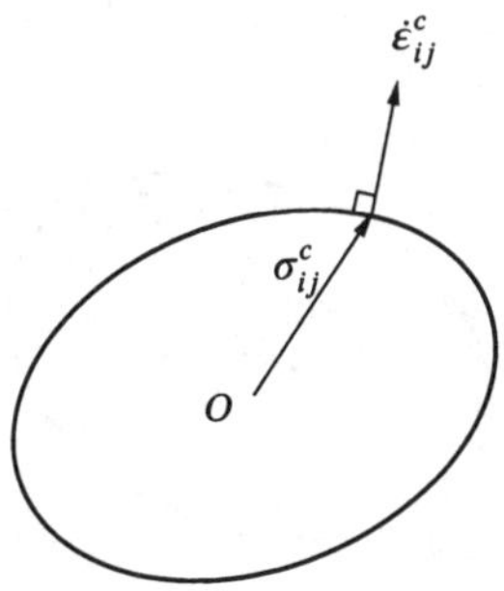

Figure 11.4: Correspondence between σ_{ij}^c and $\dot{\varepsilon}_{ij}^c$ according to the normality rule.

field

$$\boxed{\begin{array}{l} \textit{Compatible displacement field:} \\ \quad \dot{u}_i^c = 0 \text{ along } S_u \\ \quad \dot{\varepsilon}_{ij}^c = \frac{1}{2}(\dot{u}_{i,j}^c + \dot{u}_{j,i}^c) \\ \textit{Moreover, } \dot{\varepsilon}_{ij}^c \textit{ fulfills the normality rule for some} \\ \textit{stress field } \sigma_{ij}^c\textit{, i.e.} \\ \quad \dot{\varepsilon}_{ij}^c = \dot{\lambda}(\frac{\partial f}{\partial \sigma_{ij}})_c \\ \textit{where} \\ \quad f(\sigma_{ij}^c) = 0 \end{array}} \tag{11.15}$$

As the stress state σ_{ij}^c is located on the yield surface and as $\dot{\varepsilon}_{ij}^c$ is in accordance with the normality rule, we have the situation shown in Fig. 11.4. Since we are considering ideal plasticity, the convex yield surface certainly encloses the stress origin and we have

$$\sigma_{ij}^c \dot{\varepsilon}_{ij}^c > 0 \tag{11.16}$$

Moreover, as the real stress field σ_{ij} is located inside or on the convex yield surface, Fig. 11.5 illustrates the situation. It follows that the scalar product $\dot{\varepsilon}_{ij}^c(\sigma_{ij}^c - \sigma_{ij})$ is positive or zero, i.e.

$$\dot{\varepsilon}_{ij}^c(\sigma_{ij}^c - \sigma_{ij}) \geq 0 \tag{11.17}$$

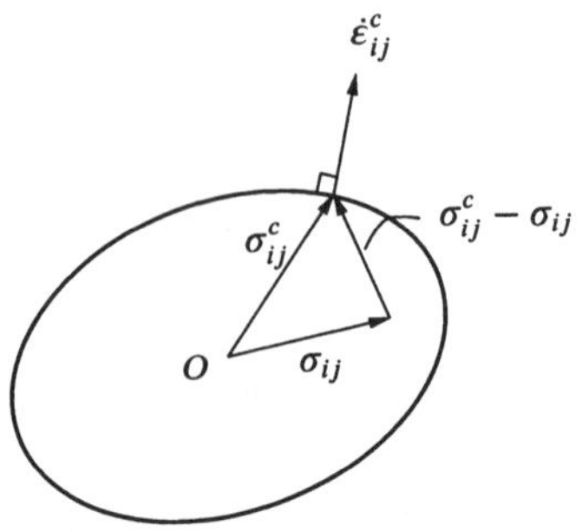

Figure 11.5: The real stress state σ_{ij} is located inside or on the yield surface.

As will be shown in a moment, we then obtain

> *Upper bound theorem:*
> *Let $\dot{u}^c_i$ and $\dot{\varepsilon}^c_{ij}$ be defined by (11.15) and consider the loads*
> $t^c_i = \mu^+ t_i$ along S_t
> $b^c_i = \mu^+ b_i$ in V
> *If μ^+ is calculated from*
> $$\mu^+ \underbrace{[\int_{S_t} \dot{u}^c_i t_i dS + \int_V \dot{u}^c_i b_i dV]}_{known} = \underbrace{\int_V \dot{\varepsilon}^c_{ij}\sigma^c_{ij} dV}_{known}$$
> then
> $\mu^+ \geq 1$

(11.18)

If μ^+ is calculated in this fashion, we have evidently determined an external load that is larger than or equal to the real collapse load that is, we have obtained an upper bound to the real collapse load. We mentioned previously that the stress field σ^c_{ij} does not necessary need to be known; however, as emphasized in (11.18) the quantity $\dot{\varepsilon}^c_{ij}\sigma^c_{ij}$ must be known up to, in principle, the unknown factor $\dot{\lambda}$ which then also enters the expression for $\dot{u}^c_i$. For von Mises ideal plasticity for instance, we have

$$\dot{\varepsilon}^c_{ij}\sigma^c_{ij} = \dot{\lambda}(\frac{\partial f}{\partial \sigma_{ij}})_c \sigma^c_{ij} = \dot{\lambda}\frac{3s^c_{ij}}{2\sigma_{yo}}\sigma^c_{ij} = \dot{\lambda}\sigma_{yo} \tag{11.19}$$

To prove the upper bound theorem, we have

$$\begin{aligned}\int_V \dot{\varepsilon}^c_{ij}\sigma_{ij} dV = \int_V \dot{u}^c_{i,j}\sigma_{ij} dV &= \int_V [(\dot{u}^c_i\sigma_{ij})_{,j} - \dot{u}^c_i\sigma_{ij,j}] dV \\ &= \int_V (\dot{u}^c_i\sigma_{ij})_{,j} dV + \int_V \dot{u}^c_i b_i dV\end{aligned}$$

where the equilibrium equations $\sigma_{ij,j} + b_i = 0$ for the real stress field were used. With the divergence theorem and as $\sigma_{ij}n_j = t_i$, it then follows that

$$\int_V \dot{\varepsilon}_{ij}^c \sigma_{ij} dV = \int_S \dot{u}_i^c t_i dS + \int_V \dot{u}_i^c b_i dV$$

and since $\dot{u}_i^c = 0$ along S_u, we obtain

$$\int_{S_t} \dot{u}_i^c t_i dS + \int_V \dot{u}_i^c b_i dV = \int_V \dot{\varepsilon}_{ij}^c \sigma_{ij} dV$$

Multiplication by μ^+ and subtracting the result from (11.18) gives

$$\int_V (\dot{\varepsilon}_{ij}^c \sigma_{ij}^c - \mu^+ \dot{\varepsilon}_{ij}^c \sigma_{ij}) dV = 0$$

Since $1 = \mu^+ + (1 - \mu^+)$, we obtain

$$\int_V [\mu^+ \dot{\varepsilon}_{ij}^c (\sigma_{ij}^c - \sigma_{ij}) + (1 - \mu^+) \dot{\varepsilon}_{ij}^c \sigma_{ij}^c] dV = 0$$

The factor μ^+ is positive and according to (11.17) and (11.16) we have $\dot{\varepsilon}_{ij}^c(\sigma_{ij}^c - \sigma_{ij}) \geq 0$ and $\dot{\varepsilon}_{ij}^c \sigma_{ij}^c > 0$. We immediately conclude that $\mu^+ \geq 1$ in accordance with the result already stipulated in (11.18).

To use the upper bound theorem, use is made of the fact that $\dot{\varepsilon}_{ij}^c$ is a plastic strain. Then the body is often assumed to deform as rigid regions separated by narrow bands - so-called *slip bands* or *shear bands* - where the yield criterion is assumed to be fulfilled. Within these slip bands the displacement field is taken in accordance with the normality rule and such that $\dot{u}_i^c = 0$ along S_u. For plates, the similar approach is called *yield line* theory, cf. Johansen (1943, 1962). The reader may consult the specialized literature for further details, Chen and Han (1988), Hill (1950), Johnson and Mellor (1983), Lubliner (1990) and Nielsen (1984).

It is evident that the lower and upper bound theorems can be used to identify the bounds for the real limit load and if the lower bound solution is equal to the upper bound solution, then the true limit load has been determined.

11.3 Simple example

In order to illustrate the use of the lower and upper bound theorems in a simple fashion, the semi-infinite body loaded by the uniform pressure p shown in Fig. 11.6 is considered; von Mises ideal plasticity as well as plane strain are assumed.

It turns out to be convenient to work with the yield shear stress τ_{yo}, which according to (8.30) is related to the yield stress in tension σ_{yo} through

$$\tau_{yo} = \frac{\sigma_{yo}}{\sqrt{3}} \tag{11.20}$$

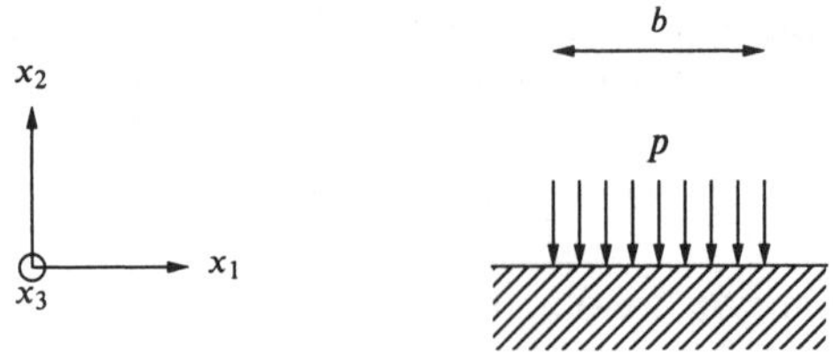

Figure 11.6: Semi-infinite body loaded by a pressure p.

The von Mises criterion (8.27) can then be written as

$$\sqrt{\frac{1}{6}[(\sigma_1-\sigma_2)^2+(\sigma_1-\sigma_3)^2+(\sigma_2-\sigma_3)^2]}-\tau_{yo}=0 \qquad (11.21)$$

To use the lower bound theorem, an allowable and statically admissible stress field needs to be established. The stress field in Fig. 11.7 is characterized by

$$\begin{aligned}\text{Regions: I}\quad & \sigma_1=-2\tau_{yo};\ \ \sigma_2=0;\ \ \sigma_3=-\tau_{yo}\\ \text{II}\quad & \sigma_1=-2\tau_{yo};\ \ \sigma_2=-4\tau_{yo};\ \ \sigma_3=-3\tau_{yo}\end{aligned} \qquad (11.22)$$

It appears that the out-of-plane stress σ_3 is taken as the mean value of σ_1 and

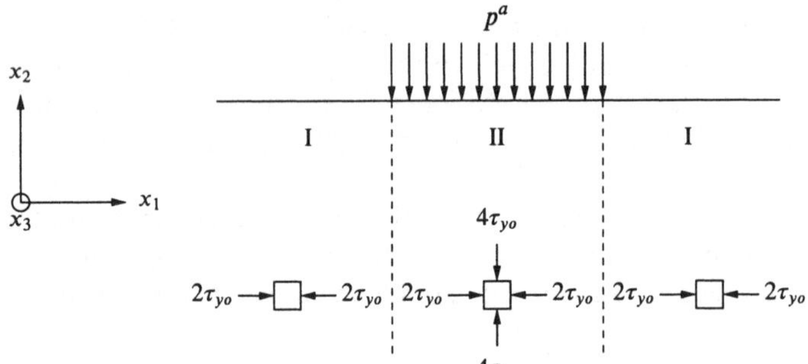

Figure 11.7: Allowable and statically admissible stress distribution.

σ_2. By inspection, these stress fields are seen to fulfill the yield criterion (11.21) so they are allowable stress fields. Since the stress fields are constant they also satisfy the equilibrium differential equations. The stress field in regions I fulfill the static boundary condition along the free surface ($t_i = 0$) and if we assume that the pressure p^a, see Fig. 11.7, creates the stress fields and that

$$p^a = 4\tau_{yo}$$

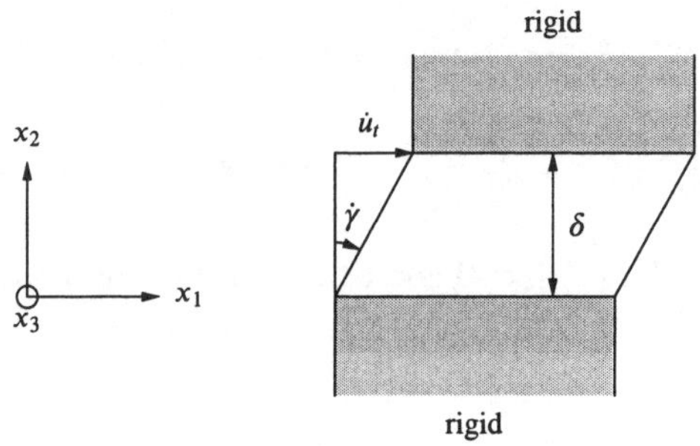

Figure 11.8: Shear band between two rigid blocks; $\dot{u}_t$ is the relative tangential velocity between these blocks.

then the static boundary condition along the pressurized surface is also fulfilled. This implies that the stress fields (11.22) in addition to being allowable are also statically admissible. The lower bound theorem (11.14) then provides

$$4\tau_{yo} \leq p_{\text{limit load}} \tag{11.23}$$

We may note that there is a discontinuity in the stress field along the border between region I and II. However, along this border the traction vector $t_i = \sigma_{ij}n_j$ for region I is opposite, but otherwise equal to the traction vector for region II. This implies that the equilibrium equations also are fulfilled along this border and the discontinuity in the stress field therefore creates no problems.

To be able to use the upper bound theorem, consider the shear band between two rigid blocks as shown in Fig. 11.8. We will take the displacements within the shear band as a compatible displacement field. From Fig. 11.8 it follows that

$$\dot{u}_1^c = \dot{\gamma}x_2 \ ; \qquad u_2^c = u_3^c = 0 \tag{11.24}$$

According to Fig. 11.8, $\dot{u}_t$ is the relative tangential velocity between the two rigid blocks, i.e. $\dot{\gamma}\delta = \dot{u}_t$ where δ is the thickness of the shear band. From (11.24), the corresponding strain rates then become

$$\dot{\varepsilon}_{12}^c = \frac{1}{2}\dot{\gamma} = \frac{\dot{u}_t}{2\delta} \ ; \qquad \dot{\varepsilon}_{ij} = 0 \quad \text{otherwise} \tag{11.25}$$

According to (11.15), the compatible displacement field is required to fulfill $\dot{u}_i^c = 0$ along S_u; however, since for the problem in question there is no S_u-surface, this requirement is fulfilled trivially.

The last requirement to the compatible displacement field is that $\dot{\varepsilon}_{ij}^c$ fulfills the normality rule, i.e. $\dot{\varepsilon}_{ij}^c = \dot{\lambda}(\frac{\partial f}{\partial \sigma_{ij}})_c$ where $f(\sigma_{ij}^c) = 0$, cf. (11.15). For von Mises ideal plasticity, we therefore require

$$\dot{\varepsilon}_{ij}^c = \dot{\lambda}\frac{3s_{ij}^c}{2\sigma_{yo}}$$

A comparison with (11.25) shows that we must have

$$\dot{\varepsilon}^c_{12} = \frac{\dot{u}_t}{2\delta} = \dot{\lambda}\frac{3s^c_{12}}{2\sigma_{yo}}; \qquad s^c_{ij} = 0 \quad \text{otherwise} \tag{11.26}$$

Finally, σ^c_{ij} must fulfill $f(\sigma^c_{ij}) = 0$ and as the stress state defined above corresponds $\sigma_1 = s^c_{12}$, $\sigma_2 = -s^c_{12}$ and $\sigma_3 = 0$, (11.21) provides

$$s^c_{12} = \tau_{yo} \tag{11.27}$$

as expected. According to (11.15), we have then demonstrated that u^c_i is a compatible displacement field. Indeed, we observe that the shearing displacement mode in Fig. 11.8 may be caused by yielding due to a shear stress alone and this motivates the terminology of the deformation mode of Fig. 11.8 being a shear band. Insertion of (11.20) and (11.27) into (11.26) then gives

$$\dot{\lambda} = \frac{1}{\sqrt{3}}\frac{\dot{u}_t}{\delta}$$

Use of this expression in (11.19) and taking advantage of (11.20) result in

$$\dot{\varepsilon}^c_{ij}\sigma^c_{ij} = \frac{\dot{u}_t}{\delta}\tau_{yo}$$

Let w be the width in the x_3-direction of the shear band. Then the plastic dissipation per unit length of the shear band becomes

$$\int_A \dot{\varepsilon}^c_{ij}\sigma^c_{ij}dA = \int_0^w \left(\int_0^\delta \frac{\dot{u}_t}{\delta}\tau_{yo}dx_2\right)dx_3 = \int_0^w \dot{u}_t\tau_{yo}dx_3 = w\dot{u}_t\tau_{yo}$$

We then obtain

$$\int_V \dot{\varepsilon}^c_{ij}\sigma^c_{ij}dV = \int_{\mathcal{L}} w\dot{u}_t\tau_{yo}d\mathcal{L} = w\dot{u}_t\tau_{yo}\mathcal{L} \tag{11.28}$$

where $\mathcal{L}$ is the length of the shear band.

Adopting the failure mechanism shown in Fig. 11.9, which corresponds to a compatible displacement field, we are now in a position to apply the upper bound theorem (11.18). Since there are no body forces and as $\mu^+ t_i = t^c_i$ we have

$$\int_{S_t} \dot{u}^c_i t^c_i dS = \int_V \dot{\varepsilon}^c_{ij}\sigma^c_{ij}dV$$

The left-hand side is the rate of external work. In view of Fig. 11.9 and (11.28) and recalling that w is the width in the x_3-direction, we obtain

$$p^c bw\frac{1}{2}\dot{u}_t = w\dot{u}_t\tau_{yo}\pi b$$

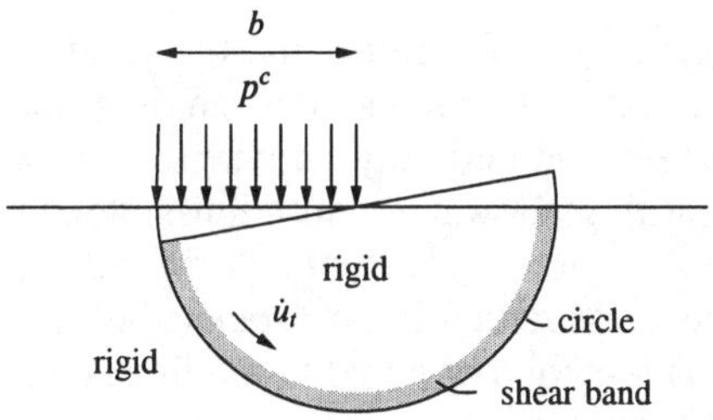

Figure 11.9: Compatible displacement field.

i.e.

$$p^c = 2\pi\tau_{yo}$$

and as this is an upper bound solution, we have

$$p_{\text{limit load}} \leq 2\pi\tau_{yo}$$

Combination with (11.23) gives

$$4\tau_{yo} \leq p_{\text{limit load}} \leq 2\pi\tau_{yo} \tag{11.29}$$

It is of interest that the mean of the lower and upper bound becomes $(2+\pi)\tau_{yo}$ and this can, in fact, be shown to be the exact solution cf. for instance Nielsen (1984). It is also shown there that the same result (11.29) is obtained if Tresca's yield criterion is used instead of the von Mises criterion. Indeed, the exact solution of the more general problem where the material follows Coulomb's yield criterion was already established by Prandtl (1920).

11.4 Nonassociated plasticity

To be able to apply the lower and upper bound theorem, the material must, in addition to being stiff-ideal plastic or elastic-ideal plastic, fulfill the normality rule, i.e. these theorems hold for associated plasticity. However, it turns out to be possible to obtain certain theorems for the limit load of nonassociated ideal plastic bodies; Lubliner (1990) gives a detailed discussion as well as relevant references. Here, we shall simply mention one important theorem which is due to Radenkovic (1961). It reads

Consider a body made of an ideal plastic material with a certain given yield criterion. *The limit load for a nonassociated flow rule* $\leq$ *the limit load for an associated flow rule*

(11.30)

The proof is straightforward. Observing that the yield criterion is assumed to be given, consider the body with a nonassociated flow rule at its limit load. The corresponding stress field can certainly be taken as allowable and admissible, cf. (11.11). Adopting this allowable and admissible stress field in the lower bound theorem (11.14), which holds for the body with an associated flow rule, it then follows that the stress field for nonassociated plasticity corresponds to an external loading that is less than or equal to the limit load for the body with an associated flow rule. This proves the correctness of statement (11.30).

12 COMMON PLASTICITY MODELS

Having in Chapter 10 derived the general format of plasticity, it is of interest to consider some specific models. We will discuss various typical plasticity models, evaluate their range of applicability and establish the tangential stiffness matrix $\boldsymbol{D}^{ep}$ that is of importance in nonlinear finite element calculations.

The exposition will mainly be based on the von Mises yield criterion, since it is the standard criterion used for metals and steel. Isotropic, kinematic and mixed hardening of a von Mises material will be discussed and their benefits and shortcomings will be illustrated, and, as an example of a plasticity model for anisotropic materials, a formulation based on Hill's criterion will be presented. Finally, isotropic hardening of a nonassociated Drucker-Prager material will be treated and taken as a simplistic prototype for frictional materials like concrete, soil and rocks.

In the present chapter we will also illustrate the difficulties in modeling of nonlinear kinematic hardening and therefore only linear kinematic hardening will be treated. The next chapter is then entirely devoted to various fairly recent approaches to capturing nonlinear kinematic hardening.

Evidently, a number of other and more advanced plasticity models exist, but from a conceptual point of view most of these models are generalizations of the models treated here and in the next chapter. With the background presented here, the reader should have sufficient information to be able to consult and evaluate the more specialized literature, whenever necessary. We may also mention that several aspects of the specific plasticity models discussed in this chapter may be used when formulating viscoplastic and creep models, cf. Chapter 15.

The present chapter deals only with smooth yield surfaces; the treatment of non-smooth yield criteria that includes, for instance, the Tresca and the Coulomb criteria will be postponed until Chapter 22.

A number of algebraic manipulations are often encountered when a specific plasticity model is considered and in order not to lose the perspective, it is appropriate to recall that all models fit into the general framework discussed in Chapter 10. In particular, the fundamental equations treated in Section 10.1 are here of importance.

Before some typical plasticity models are presented, we will discuss some general experimental characteristics for the plastic response of different groups of materials.

12.1 Experimental characteristics

As previously mentioned, the general experimental evidence for metals and steel shows that the volumetric response is linear elastic and that all nonlinearity is related to the deviatoric response. It turns out that it suffices to adopt associated plasticity theory for these materials. Since the volumetric response is linear, we have plastic incompressibility and referring to (9.85), this implies that the yield criterion does not depend on the first stress invariant $I_1 = \sigma_{kk}$.

For concrete, soil and rocks, the volumetric and deviatoric responses are coupled and both the volumetric and deviatoric responses are nonlinear. To obtain accurate predictions, it is often necessary to adopt nonassociated plasticity theory. Since the response is greatly influenced by the hydrostatic stress, the invariant $I_1 = \sigma_{kk}$ must be included in the formulation and this leads to plastic *compressibility* as well as plastic *dilatancy* depending on the particular load situation, cf. Fig. 4.10b).

Some further experimental characteristics will now be discussed. As cyclic response of metals and steel is often of importance in applications, we will first provide a qualitative discussion of some aspects related to this behavior. For more detail, the reader is referred to Chaboche (1989); Chaboche (1989), Drucker and Palgen (1981), Khan and Huang (1995), Krempl (1971), Lemaitre and Chaboche (1990) and Morrow (1965).

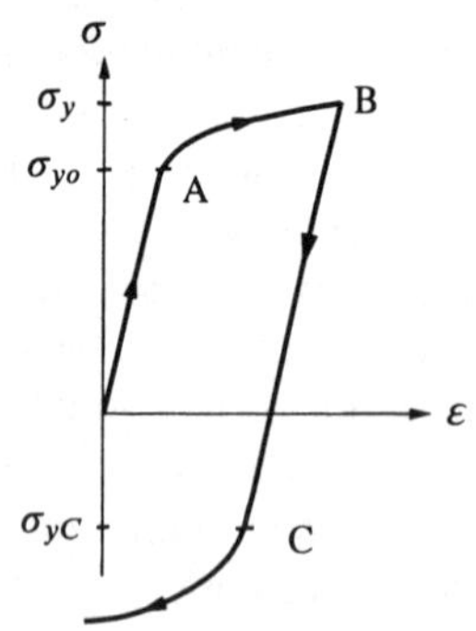

Figure 12.1: Bauschinger effect.

In Chapter 9, we discussed the Bauschinger effect, cf. Fig. 12.1. On reversing the loading from point B, plasticity is again invoked already at point C. The

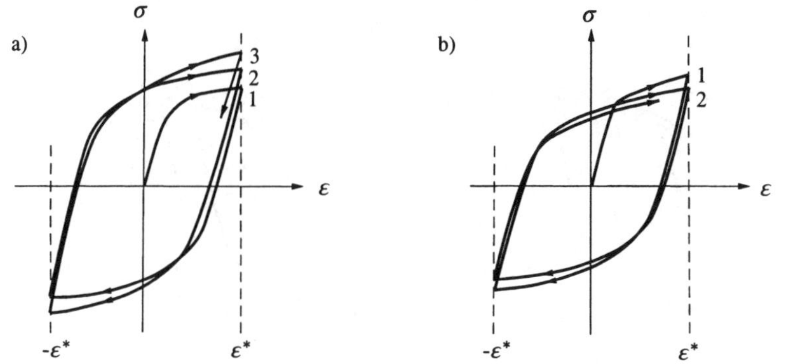

Figure 12.2: Strain cycling between ε^* and $-\varepsilon^*$; a) cyclic hardening; b) cyclic softening.

Bauschinger effect is the observation that the yield stress σ_{yc} at point C is much less than the yield stress σ_y at point B, i.e. $\sigma_{yc} < \sigma_y$.

Often, cycling between different load levels is encountered. If the material is cycled between fixed total strain values, we have *strain cycling*. In the case where the total strain cycles symmetrically between the fixed values ε^* and $-\varepsilon^*$, the situation shown in Fig. 12.2 arises. In Fig. 12.2a), the stress amplitude first increases in each cycle and this phenomenon is called *cyclic hardening*. Eventually, after several cycles, the stress-strain curve repeats itself in each cycle and we have then reached the so-called *stabilized cyclic stress-strain curve*. In Fig. 12.2b), the stress amplitude first decreases in each cycle, i.e. we have *cyclic softening* and, eventually, the stabilized cyclic stress-strain curve is obtained. Whether we have cyclic hardening or softening depends on the particular material in question. Often, a high strength steel is cyclic softening whereas a mild steel is cyclic hardening.

If the material is cycled between fixed stress values, we have *stress cycling*. In the case where the stress cycles symmetrically between the fixed values σ^* and $-\sigma^*$, the situation shown in Fig. 12.3 may be obtained. Cyclic hardening now manifests itself in the form given in Fig. 12.3a) whereas cyclic softening is shown in Fig. 12.3b).

For cycling between fixed strain values that are not symmetrically located about zero, we have, in general, the situation shown in Fig. 12.4. The mean stress is defined as the mean of the maximum and minimum stress in each cycle. It appears that the mean stress decreases with the number of cycles and one speaks of *mean stress relaxation*.

Finally, for cycling between fixed stress values that are not symmetrically located about zero, the situation shown in Fig. 12.5 arises. After each cycle, the total strain increases and one speaks of a *ratcheting effect*. Moreover, the ratcheting occurs in the 'direction' of the mean stress. Referring to Fig. 12.5, the

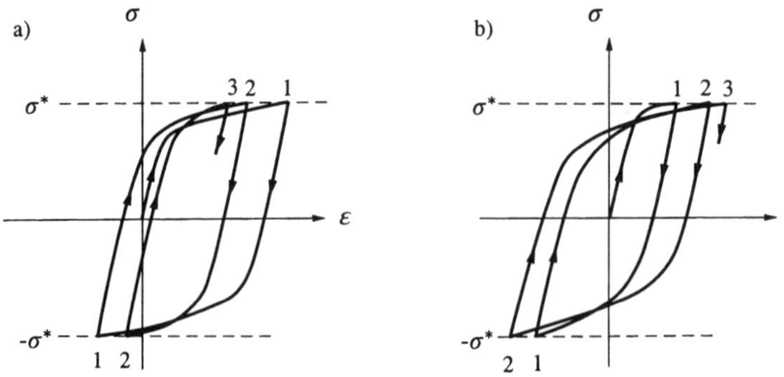

Figure 12.3: Stress cycling between σ^* and $-\sigma^*$; a) cyclic hardening; b) cyclic softening.

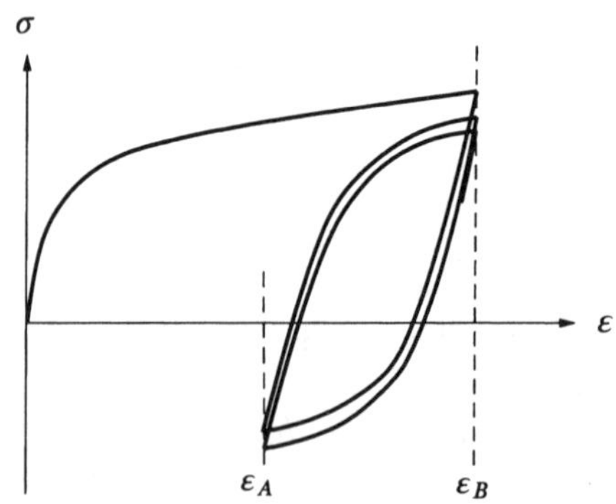

Figure 12.4: Strain cycling between the unsymmetric strain values ε_A and ε_B; mean stress relaxation.

mean stress is positive, i.e. the ratcheting results in an increasing strain. Instead of speaking of a ratcheting effect, the terminology of cyclic creep is often used in the literature, but this terminology seems not to be entirely appropriate since no creep effects – i.e. time-dependent material properities – are present.

For concrete, soil or rocks the influence of the hydrostatic stress as well as the coupling between volumetric and deviatoric response complicate the response and thereby the requirements on an accurate constitutive model. Plasticity models are therefore mostly used for load situations where significant cycling does not occur and we may therefore restrict ourselves to the illustration given in Fig. 12.6. Concrete is here loaded triaxially along the compressive meridian, cf. (8.22); first a purely hydrostatic loading is applied and then $\sigma_1 = \sigma_2$ is kept constant while σ_3 is decreased (tension is considered as a positive quantity). The effect of the hydrostatic stress is clearly observed.

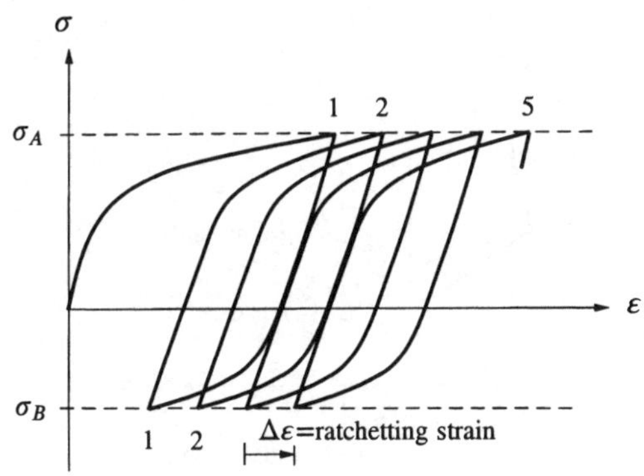

Figure 12.5: Stress cycling between the unsymmetric stress values σ_A and σ_B; ratcheting in the direction of the mean stress.

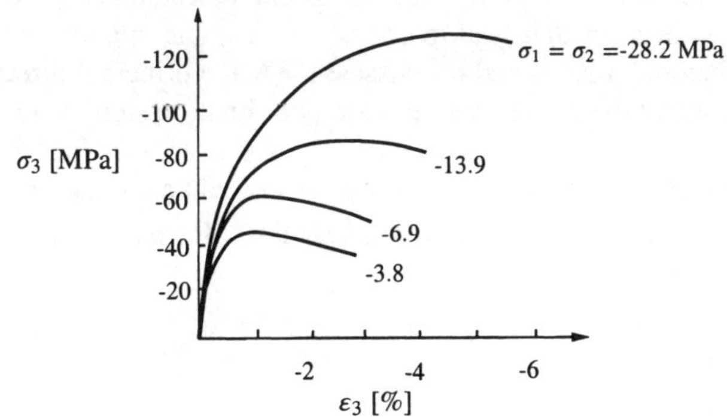

Figure 12.6: Triaxial compression of concrete along the compressive meridian, Richart *et al.* (1928); effect of hydrostatic stress.

It was mentioned above that frictional materials like concrete, soil and rocks may exhibit plastic compressibility as well as plastic dilatancy, i.e. a plastic volume change occurs. This issue complicates the modeling efforts significantly and we will now discuss various aspects of this issue.

For isotropic materials, where the potential function g may be written in the form $g = g(I_1, J_2, J_3, K_\alpha)$, (9.83) and (9.85) lead to

$$\dot{\varepsilon}_{ii}^p = \dot{\lambda} 3 \frac{\partial g}{\partial I_1}$$

i.e., frictional materials require that the first stress invariant I_1 is included in the formulation. The relation between the plastic volume change and the form of the potential surface may perhaps be more easily seen by the following arguments.

Considering isotropic materials, a general expression for $\partial g / \partial \sigma_{ij}$ is given by (9.84). If the coordinate system is chosen collinear with the principal directions

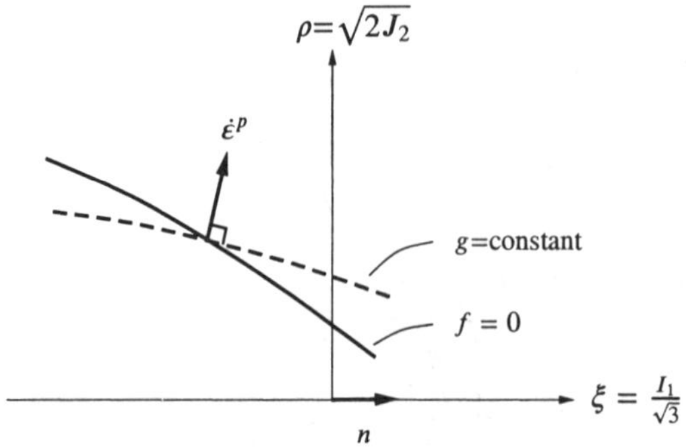

Figure 12.7: Meridian plane and illustration of $\dot{\boldsymbol{\varepsilon}}^p$ being normal to the potential surface. In the present case, a plastic volume increase will occur.

of the stress tensor then σ_{ij} becomes a diagonal matrix; consequently, both s_{ij} as well as $s_{ik}s_{kj}$ become diagonal matrices. In turn, this implies that also $\partial g/\partial\sigma_{ij}$ becomes diagonal, i.e. $\dot{\varepsilon}_{ij}^p$ and σ_{ij} possess the same principal directions.

We now adopt the Haigh-Westergaard coordinate system where the principal stresses σ_1, σ_2 and σ_3 are measured along the axes, cf. Fig. 8.4. Since the principal directions for σ_{ij} and $\dot{\varepsilon}_{ij}^p$ coincide, we can also measure the principal plastic strain rates $\dot{\varepsilon}_1^p$, $\dot{\varepsilon}_2^p$ and $\dot{\varepsilon}_3^p$ along these axes. Introduce for the moment the notation

$$\dot{\boldsymbol{\varepsilon}}^p = \begin{bmatrix} \dot{\varepsilon}_1^p \\ \dot{\varepsilon}_2^p \\ \dot{\varepsilon}_3^p \end{bmatrix} \tag{12.1}$$

The potential surface g = constant describes a surface in the Haigh-Westergaard coordinate system. According to the flow rule, the plastic strain rate $\dot{\varepsilon}_{ij}^p$ is normal to the potential surface. That is, in the Haigh-Westergaard coordinate system, the vector $\dot{\boldsymbol{\varepsilon}}^p$ defined by (12.1) is normal to the potential surface. The vector $\dot{\boldsymbol{\varepsilon}}^p$ is therefore also normal to the meridians of the potential surface as illustrated in Fig. 12.7.

Define the unit vector $\boldsymbol{n}$ in the Haigh-Westergaard coordinate system according to

$$\boldsymbol{n} = \frac{1}{\sqrt{3}}\begin{bmatrix} 1 \\ 1 \\ 1 \end{bmatrix} \tag{12.2}$$

In accordance with Fig. 8.4, this unit vector is located along the hydrostatic axis, i.e. it takes the direction shown in Fig. 12.7. We have

$$\dot{\varepsilon}_{ii}^p = \dot{\varepsilon}_{11}^p + \dot{\varepsilon}_{22}^p + \dot{\varepsilon}_{33}^p = \dot{\varepsilon}_1^p + \dot{\varepsilon}_2^p + \dot{\varepsilon}_3^p$$

which with (12.1) and (12.2) may be written as

$$\dot{\varepsilon}^p_{ii} = \sqrt{3}\boldsymbol{n}^T\dot{\boldsymbol{\varepsilon}}^p \tag{12.3}$$

As an illustration of the use of this expression, we have plastic dilatancy if the meridians of the potential surface open in the direction of the negative hydrostatic axis; this situation is illustrated in Fig. 12.7.

For metals and steel, we have plastic incompressibility $\dot{\varepsilon}^p_{ii} = 0$, i.e. the plastic strain rate $\dot{\varepsilon}^p_{ij}$ is purely deviatoric. For frictional materials, however, the plastic strain rate $\dot{\varepsilon}^p_{ij}$ also involves a volumetric component. The flow rule states that $\dot{\varepsilon}^p_{ij} = \dot{\lambda}\partial g/\partial\sigma_{ij}$ and since $\dot{\varepsilon}^p_{ij}$ for frictional materials is more complex than $\dot{\varepsilon}^p_{ij}$ for metals and steel, the requirements to the potential function g are also more complex. It is therefore to be expected that whereas for metals and steel it suffices to adopt an associated flow rule, frictional materials require, in general, a nonassociated flow rule. This expectation is supported by modeling experience.

12.2 Isotropic von Mises hardening

This model is based on associated plasticity and it is of relevance for the response of metals and steel for which significant reversed loadings do not occur. The model was treated to some extent in Section 9.6 and we will here provide the general format of this model that is applicable in nonlinear finite element schemes as well as a discussion of its prediction capabilities. For evident reasons, von Mises plasticity theory is often called *J_2-plasticity*.

Recognizing (9.20), the yield criterion is given by

$$\boxed{f(\sigma_{ij}, K) = (\frac{3}{2}s_{kl}s_{kl})^{1/2} - \sigma_y \; ; \quad f = 0 \; ; \quad \sigma_y = \sigma_{yo} + K(\kappa)} \tag{12.4}$$

i.e.

$$\frac{\partial f}{\partial \sigma_{ij}} = \frac{3}{2}\frac{s_{ij}}{\sigma_y} \; ; \qquad \frac{\partial f}{\partial K} = -1 \; ; \qquad \frac{\partial K}{\partial \kappa} = \frac{d\sigma_y}{d\kappa} \tag{12.5}$$

The flow rule states that $\dot{\varepsilon}^p_{ij} = \dot{\lambda}\partial f/\partial\sigma_{ij}$, i.e.

$$\dot{\varepsilon}^p_{ij} = \dot{\lambda}\frac{3s_{ij}}{2\sigma_y} \tag{12.6}$$

It follows that we have plastic incompressibility, i.e $\dot{\varepsilon}^p_{kk} = 0$. The evolution law (10.13) for the internal variable κ is

$$\dot{\kappa} = \dot{\lambda}k(\sigma_{ij}, K) \tag{12.7}$$

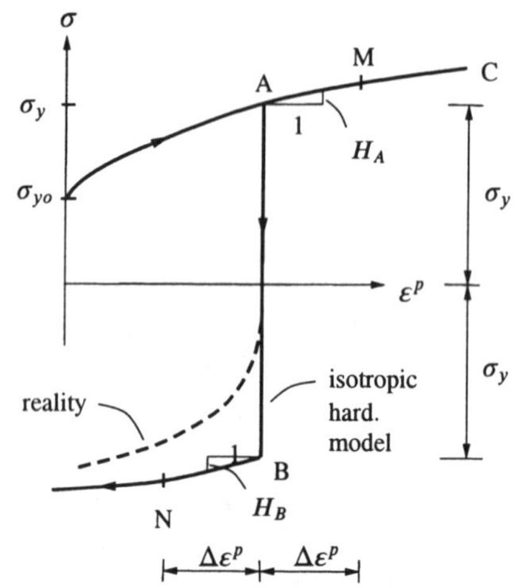

Figure 12.8: Uniaxial loading. Effects of reversed loading; $H_B = H_A$ and $H_N = H_M$.

The consistency relation $\dot{f} = 0$ yields

$$\frac{\partial f}{\partial \sigma_{ij}}\dot{\sigma}_{ij} - H\dot{\lambda} = 0$$

where the plastic modulus H according to (10.17) is defined by

$$H = -\frac{\partial f}{\partial K}\frac{\partial K}{\partial \kappa}k = \frac{d\sigma_y}{d\kappa}k \tag{12.8}$$

Introduce the definitions

$$\boxed{\sigma_{eff} = (\frac{3}{2}s_{kl}s_{kl})^{1/2} \; ; \qquad \dot{\varepsilon}^p_{eff} = (\frac{2}{3}\dot{\varepsilon}^p_{ij}\dot{\varepsilon}^p_{ij})^{1/2}} \tag{12.9}$$

i.e. the yield criterion takes the form

$$\sigma_{eff} - \sigma_y(\kappa) = 0$$

and we obtain

$$\dot{\varepsilon}^p_{eff} = \dot{\lambda} \tag{12.10}$$

We choose the strain hardening assumption

$$\dot{\kappa} = \dot{\varepsilon}^p_{eff} \quad \text{i.e.} \quad \dot{\kappa} = \dot{\lambda} \tag{12.11}$$

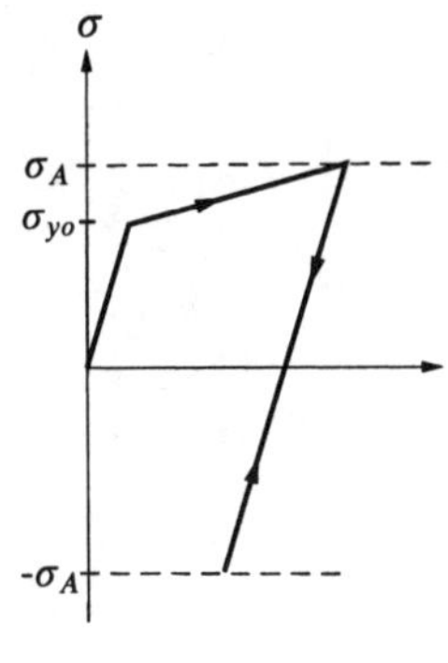

Figure 12.9: Effects of stress cycling between σ_A and $-\sigma_A$; elastic shake-down.

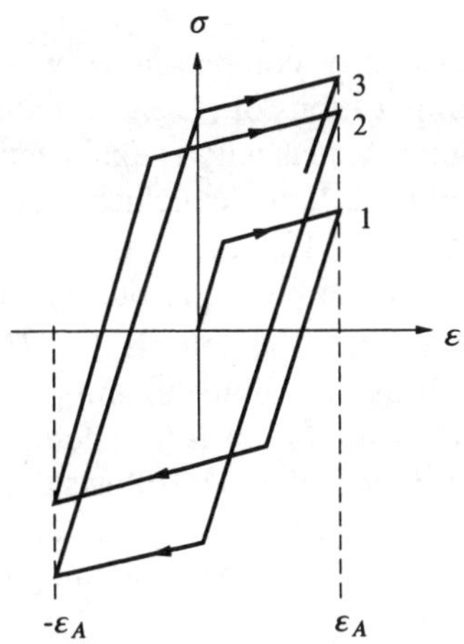

Figure 12.10: Effects of strain cycling between ε_A and $-\varepsilon_A$.

This means that $\sigma_y = \sigma_y(\kappa)$ becomes $\sigma_y = \sigma_y(\varepsilon^p_{eff})$. A comparison of (12.11b) with (12.7) shows that $k = 1$ and (12.8) then leads to the following expression for the plastic modulus

$$\boxed{H = \frac{d\sigma_y(\varepsilon^p_{eff})}{d\varepsilon^p_{eff}}} \tag{12.12}$$

Moreover, $K = K(\kappa)$ and (12.11) implies

$$\dot{K} = \frac{dK}{d\kappa}\dot{\lambda} \tag{12.13}$$

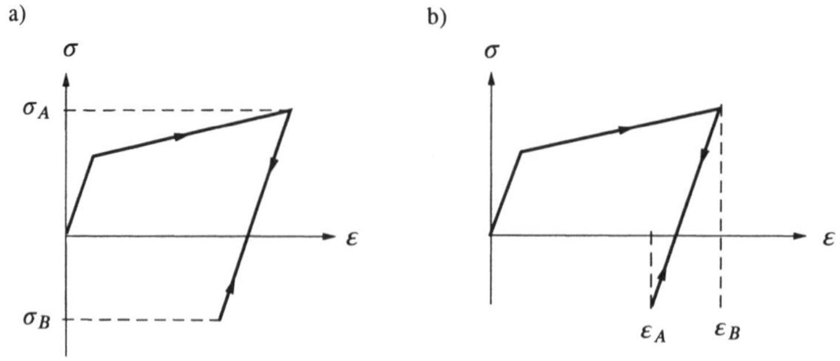

Figure 12.11: a) unsymmetric stress cycling, no ratcheting; b) unsymmetric strain cycling, no mean stress relaxation.

For uniaxial loading (tension or compression), we have $|\sigma| = \sigma_{eff} = \sigma_y$ and $|d\varepsilon^p| = d\varepsilon^p_{eff}$ and the implications of (12.12) are illustrated in Fig. 12.8. If we reverse the loading at point A, plasticity is again activated at point B where $\sigma_B = -\sigma_A$. Referring to Fig. 12.1, we conclude that the Bauschinger effect is not modeled. According to (12.12), the slope at point A is H_A and since no change occurs in the internal variable ε^p_{eff} during the elastic response from point A to B, we have the same slope at point B, i.e. $H_B = H_A$. As $\dot{\varepsilon}^p_{eff} \geq 0$, the effective plastic strain always accumulates the plastic strain contributions and for uniaxial loading, we have $\dot{\varepsilon}^p_{eff} = |\dot{\varepsilon}^p|$. Therefore, the effective plastic strains at points M and N are identical and it follows that the slopes at points M and N are the same. The real response during reversed loading is also indicated in Fig. 12.8, and it appears that the isotropic hardening model provides a rather inferior prediction.

For the purpose of evaluating cyclic loading, we adopt for simplicity linear hardening, i.e. H becomes a constant. If the loading cycles between the stress states $\pm\sigma_A$, we obtain the result illustrated in Fig. 12.9; as the response is purely elastic during cycling, one speaks of an immediate *elastic shake-down*. If the loading cycles between the strain states $\pm\varepsilon_A$, isotropic hardening predicts the cyclic hardening behavior shown in Fig. 12.10. After an infinitely large number of strain cycles, the isotropic hardening model will predict a cyclic behavior that is purely elastic.

For cycling between unsymmetric stress states, the isotropic hardening model predicts the response shown in Fig. 12.11a); no ratcheting is predicted and elastic shake-down occurs immediately. Figure 12.11b) shows cycling between unsymmetric strain states and it appears that no mean stress relaxation is predicted.

The discussion relating to Figs. 12.8-12.11 clearly emphasizes that isotropic hardening is applicable when continued plastic loading is involved whereas the effect of significant reversed loadings cannot be modeled in a proper fashion.

In order to determine the tangential elasto-plastic stiffness matrix $\boldsymbol{D}^{ep}$ of relevance in nonlinear finite element calculations, we shall assume isotropic elasticity, i.e. (4.89) provides

$$D_{ijkl} = 2G\left[\frac{1}{2}(\delta_{ik}\delta_{jl} + \delta_{il}\delta_{jk}) + \frac{\nu}{1-2\nu}\delta_{ij}\delta_{kl}\right] \tag{12.14}$$

We then obtain with (12.5)

$$\frac{\partial f}{\partial \sigma_{mn}} D_{mnkl} = 3G\,\frac{s_{kl}}{\sigma_y}\,; \qquad D_{ijst}\frac{\partial f}{\partial \sigma_{st}} = 3G\,\frac{s_{ij}}{\sigma_y}$$

and (10.24) then yields

$$A = H + 3G$$

With these expressions, (10.26) and (10.27) give

$$\dot{\sigma}_{ij} = D^{ep}_{ijkl}\dot{\varepsilon}_{kl} \tag{12.15}$$

where

$$D^{ep}_{ijkl} = D_{ijkl} - \frac{9G^2}{A}\frac{s_{ij}s_{kl}}{\sigma_y^2} \tag{12.16}$$

If we express (12.15) in a matrix form similar to (4.37), we arrive at

$$\dot{\boldsymbol{\sigma}} = \boldsymbol{D}^{ep}\dot{\boldsymbol{\varepsilon}} \tag{12.17}$$

where

$$\boldsymbol{D}^{ep} = \boldsymbol{D} - \boldsymbol{D}^{p} \tag{12.18}$$

and

$$\boldsymbol{D} = \frac{2G}{1-2\nu}\begin{bmatrix} 1-\nu & \nu & \nu & 0 & 0 & 0 \\ \nu & 1-\nu & \nu & 0 & 0 & 0 \\ \nu & \nu & 1-\nu & 0 & 0 & 0 \\ 0 & 0 & 0 & \frac{1}{2}(1-2\nu) & 0 & 0 \\ 0 & 0 & 0 & 0 & \frac{1}{2}(1-2\nu) & 0 \\ 0 & 0 & 0 & 0 & 0 & \frac{1}{2}(1-2\nu) \end{bmatrix} \tag{12.19}$$

$$\boldsymbol{D}^{p} = \frac{9G^2}{A\sigma_y^2}\begin{bmatrix} s_{11}^2 & s_{11}s_{22} & s_{11}s_{33} & s_{11}s_{12} & s_{11}s_{13} & s_{11}s_{23} \\ s_{22}s_{11} & s_{22}^2 & s_{22}s_{33} & s_{22}s_{12} & s_{22}s_{13} & s_{22}s_{23} \\ s_{33}s_{11} & s_{33}s_{22} & s_{33}^2 & s_{33}s_{12} & s_{33}s_{13} & s_{33}s_{23} \\ s_{12}s_{11} & s_{12}s_{22} & s_{12}s_{33} & s_{12}^2 & s_{12}s_{13} & s_{12}s_{23} \\ s_{13}s_{11} & s_{13}s_{22} & s_{13}s_{33} & s_{13}s_{12} & s_{13}^2 & s_{13}s_{23} \\ s_{23}s_{11} & s_{23}s_{22} & s_{23}s_{33} & s_{23}s_{12} & s_{23}s_{13} & s_{23}^2 \end{bmatrix} \tag{12.20}$$

The $\boldsymbol{D}^{ep}$-matrix is clearly symmetric. It is of interest that even though isotropic elasticity is adopted, we have a fully populated $\boldsymbol{D}^{ep}$-matrix, i.e. we have coupling effects between normal stresses and shear strains as well as between shear stresses and normal strains. Moreover, by setting $H = 0$ and $\sigma_y = \sigma_{yo}$, which implies that $A = 3G$, the formulation above also holds for ideal plasticity.

12.2.1 Plane strain

For plane strain, i.e. $\dot{\varepsilon}_{13} = \dot{\varepsilon}_{23} = \dot{\varepsilon}_{33} = 0$, (12.17)-(12.20) directly provide

$$\dot{\boldsymbol{\sigma}} = \boldsymbol{D}^{ep}\dot{\boldsymbol{\varepsilon}} \tag{12.21}$$

where

$$\boldsymbol{D}^{ep} = \boldsymbol{D} - \boldsymbol{D}^{p} \tag{12.22}$$

and

$$\boldsymbol{D} = \frac{2G}{1-2\nu}\begin{bmatrix} 1-\nu & \nu & 0 \\ \nu & 1-\nu & 0 \\ 0 & 0 & \frac{1}{2}(1-2\nu) \end{bmatrix} \tag{12.23}$$

$$\boldsymbol{D}^{p} = \frac{9G^2}{A\sigma_y^2}\begin{bmatrix} s_{11}^2 & s_{11}s_{22} & s_{11}s_{12} \\ s_{22}s_{11} & s_{22}^2 & s_{22}s_{12} \\ s_{12}s_{11} & s_{12}s_{22} & s_{12}^2 \end{bmatrix} \tag{12.24}$$

Moreover, considering the out-of-plane stresses, we obtain from (12.17)-(12.20)

$$\begin{bmatrix} \dot{\sigma}_{33} \\ \dot{\sigma}_{13} \\ \dot{\sigma}_{23} \end{bmatrix} = \frac{2G}{1-2\nu}\begin{bmatrix} \nu & \nu & 0 \\ 0 & 0 & 0 \\ 0 & 0 & 0 \end{bmatrix}\begin{bmatrix} \dot{\varepsilon}_{11} \\ \dot{\varepsilon}_{22} \\ 2\dot{\varepsilon}_{12} \end{bmatrix} - \frac{9G^2}{A\sigma_y^2}\begin{bmatrix} s_{33}s_{11} & s_{33}s_{22} & s_{33}s_{12} \\ s_{13}s_{11} & s_{13}s_{22} & s_{13}s_{12} \\ s_{23}s_{11} & s_{23}s_{22} & s_{23}s_{12} \end{bmatrix}\begin{bmatrix} \dot{\varepsilon}_{11} \\ \dot{\varepsilon}_{22} \\ 2\dot{\varepsilon}_{12} \end{bmatrix}$$

It appears that $\sigma_{13} = s_{13} = 0$ and $\sigma_{23} = s_{23} = 0$ hold in the elastic region before any plasticity occurs. The expression above therefore implies that $\dot{\sigma}_{13} = \dot{\sigma}_{23} = 0$ holds at the event that plasticity is initiated, i.e. $\sigma_{13} = \sigma_{23} = 0$ holds in general. It is concluded that

$$\sigma_{13} = \sigma_{23} = 0 \tag{12.25}$$

and

$$\dot{\sigma}_{33} = \frac{2G\nu}{1-2\nu}(\dot{\varepsilon}_{11} + \dot{\varepsilon}_{22}) - \frac{9G^2}{A\sigma_y^2}s_{33}(s_{11}\dot{\varepsilon}_{11} + s_{22}\dot{\varepsilon}_{22} + 2s_{12}\dot{\varepsilon}_{12})$$

To determine the out-of-plane plastic strain components, we obtain from (12.6)

$$\dot{\varepsilon}_{i3}^{p} = \dot{\lambda}\frac{3s_{i3}}{2\sigma_y}$$

From (12.25), we conclude

$$\varepsilon_{13}^{p} = \varepsilon_{23}^{p} = 0\,; \qquad \dot{\varepsilon}_{33}^{p} = \dot{\lambda}\frac{3s_{33}}{2\sigma_y}$$

where $\dot{\lambda} = \dot{\varepsilon}_{eff}^{p}$.

12.2.2 Plane stress

For plane stress, i.e. $\sigma_{13} = \sigma_{23} = \sigma_{33} = 0$, the formulation becomes somewhat more involved as already discussed in Section 10.6. From (10.47) and (12.5) follow that

$$D^*_{\alpha\beta\gamma\delta}\frac{\partial f}{\partial \sigma_{\gamma\delta}} = \frac{3G}{\sigma_y}(s_{\alpha\beta} + \frac{\nu}{1-\nu}\delta_{\alpha\beta}s_{\gamma\gamma}) \tag{12.26}$$

i.e., the quantity A^* given by (10.52) becomes

$$A^* = H + \frac{9G}{2\sigma_y^2}(s_{\alpha\beta}s_{\alpha\beta} + \frac{\nu}{1-\nu}s_{\gamma\gamma}s_{\beta\beta}) \tag{12.27}$$

Note, that A^* varies with the loading also for linear hardening, i.e. H=constant. This is in contrast to the general three-dimensional situation and the plane strain case, where H=constant implies that A=constant. From (12.26), we obtain

$$D^*_{\alpha\beta\pi\theta}\frac{\partial f}{\partial \sigma_{\pi\theta}}\frac{\partial f}{\partial \sigma_{\phi\psi}}D^*_{\phi\psi\gamma\delta} = \frac{9G^2}{\sigma_y^2}(s_{\alpha\beta} + \frac{\nu}{1-\nu}\delta_{\alpha\beta}s_{\theta\theta})(s_{\gamma\delta} + \frac{\nu}{1-\nu}\delta_{\gamma\delta}s_{\pi\pi})$$

With (10.54) and (4.112) it then follows that

$$\dot{\boldsymbol{\sigma}} = \boldsymbol{D}^{*ep}\dot{\boldsymbol{\varepsilon}} \tag{12.28}$$

where

$$\boldsymbol{D}^{*ep} = \boldsymbol{D}^* - \boldsymbol{D}^{*p} \tag{12.29}$$

and

$$\boldsymbol{D}^* = \frac{E}{1-\nu^2}\begin{bmatrix} 1 & \nu & 0 \\ \nu & 1 & 0 \\ 0 & 0 & \frac{1}{2}(1-\nu) \end{bmatrix} \tag{12.30}$$

$$\boldsymbol{D}^{*p} = \frac{9G^2}{A^*\sigma_y^2(1-\nu)^2}\begin{bmatrix} (s_{11}^*)^2 & s_{11}^*s_{22}^* & (1-\nu)s_{11}^*s_{12} \\ s_{22}^*s_{11}^* & (s_{22}^*)^2 & (1-\nu)s_{22}^*s_{12} \\ (1-\nu)s_{12}s_{11}^* & (1-\nu)s_{12}s_{22}^* & (1-\nu)^2s_{12}^2 \end{bmatrix} \tag{12.31}$$

where

$$s_{11}^* = s_{11} + \nu s_{22}\,; \quad s_{22}^* = s_{22} + \nu s_{11}$$

Finally, for the out-of-plane plastic strain components, (12.6) together with $\sigma_{13} = \sigma_{23} = 0$ yield

$$\varepsilon_{13}^p = \varepsilon_{23}^p = 0\,; \quad \dot{\varepsilon}_{33}^p = \dot{\lambda}\frac{3s_{33}}{2\sigma_y}$$

The plane stress condition implies $s_{33} = -(\sigma_{11} + \sigma_{22})/3$ and, in general, we therefore have $\dot{\varepsilon}_{33}^p \neq 0$. Moreover, it is recalled that $\dot{\lambda} = \dot{\varepsilon}_{eff}^p$

As previously discussed, it is worthwhile to observe that the elasto-plastic relations for plane stress are more involved than those pertinent for the general three-dimensional situation as well as for the plane strain case.

12.3 Kinematic von Mises hardening

Kinematic hardening of a von Mises material is of relevance for metals and steel where reversed loadings are of significance. First, the general properties of kinematic hardening shall be scrutinized. Referring to (9.24), kinematic hardening is, in general, described by

$$f(\sigma_{ij}, K_\alpha) = F(\sigma_{ij} - \alpha_{ij}) = 0 \tag{12.32}$$

where we have one hardening parameter given in terms of the *back-stress tensor* α_{ij}, which describes the center position of the current yield surface.

Considering the initial yield criterion $F(\sigma_{ij}) = 0$, we observed in Chapter 8 that this criterion for an isotropic material may be described, for instance, in terms of

$$F(I_1, J_2, \cos 3\theta) = 0 \tag{12.33}$$

where the stress invariants are defined by

$$I_1 = \sigma_{ii}\,; \quad J_2 = \frac{1}{2} s_{ij} s_{ij}\,; \quad \cos 3\theta = \frac{3\sqrt{3} J_3}{2 J_2^{3/2}}$$

and $J_3 = s_{ij} s_{jk} s_{ki}/3$. For metals and steel, the hydrostatic stress I_1 has no influence on yielding, i.e. (12.33) reduces to $F(J_2, \cos 3\theta) = 0$. The invariant J_2 provides information of the magnitude of the deviatoric stresses whereas the invariant $\cos 3\theta$ gives information of the direction of the deviatoric stresses. The von Mises criterion was obtained by dropping the influence of the term $\cos 3\theta$, i.e $F(J_2) = 0$ or $\sqrt{3J_2} - \sigma_{yo} = 0$, where σ_{yo} is the initial yield stress in tension.

Let us now treat (12.32) in the same manner. Define the *reduced stress tensor* $\bar{\sigma}_{ij}$ by

$$\bar{\sigma}_{ij} = \sigma_{ij} - \alpha_{ij} \tag{12.34}$$

i.e.

$$\bar{\sigma}_{ii} = \sigma_{ii} - \alpha_{ii}$$

The deviatoric part of the reduced stress tensor is defined by $\bar{s}_{ij} = \bar{\sigma}_{ij} - \delta_{ij}\bar{\sigma}_{kk}/3$ and it becomes

$$\boxed{\bar{s}_{ij} = s_{ij} - \alpha^d_{ij}} \tag{12.35}$$

where α^d_{ij} denotes the deviatoric part of the back-stress tensor α_{ij}, i.e.

$$\alpha^d_{ij} = \alpha_{ij} - \frac{1}{3}\delta_{ij}\alpha_{kk}$$

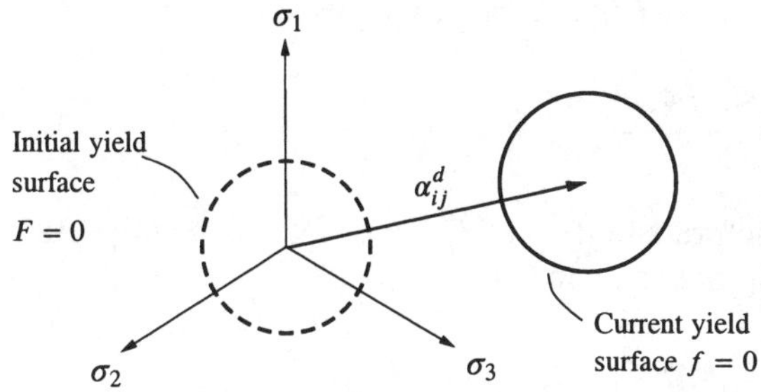

Figure 12.12: Deviatoric plane. Kinematic hardening of a von Mises material.

With (12.34), (12.32) takes the form $F(\bar{\sigma}_{ij}) = 0$, which, in analogy with (12.33), can be expressed as

$$F(\bar{I}_1, \bar{J}_2, \cos 3\bar{\theta}) = 0 \tag{12.36}$$

where

$$\bar{I}_1 = \bar{\sigma}_{ii}\,; \qquad \bar{J}_2 = \frac{1}{2}\bar{s}_{ij}\bar{s}_{ij}\,; \qquad \cos 3\bar{\theta} = \frac{3\sqrt{3}\bar{J}_3}{2\bar{J}_2^{3/2}}$$

and $\bar{J}_3 = \bar{s}_{ij}\bar{s}_{jk}\bar{s}_{ki}/3$. Arguing as previously, we ignore the influence of the invariants $\bar{I}_1$ and $\cos 3\bar{\theta}$, i.e. (12.36) reduces to $F(\bar{J}_2) = 0$ or $\sqrt{3\bar{J}_2} - \sigma_{yo} = 0$. With (12.35), we then find for kinematic hardening of a von Mises material that the current yield criterion is given by

$$\boxed{f(\sigma_{ij}, K_\alpha) = \left[\frac{3}{2}(s_{kl} - \alpha_{kl}^d)(s_{kl} - \alpha_{kl}^d)\right]^{1/2} - \sigma_{yo} = 0} \qquad K_\alpha = \{\alpha_{ij}^d\} \tag{12.37}$$

where σ_{yo} is the initial yield stress in tension and it is indicated that the hardening parameters K_α take the form of the back-stress. For $\alpha_{kl}^d = 0$, this expression evidently reduces to the initial von Mises criterion.

The von Mises surface is a cylinder in the stress space and the deviatoric part α_{ij}^d of the back-stress describes the center of this cylinder in the deviatoric plane, as illustrated in Fig. 12.12. Note that in accordance with experimental evidence, cf. Figs. 9.18 and 9.19, the current yield surface does not need to include the origin of the stress space.

After these introductory remarks, the fundamental equations for kinematic hardening of a von Mises material will be presented. From (12.37) it follows that

$$\frac{\partial f}{\partial \sigma_{ij}} = \frac{3(s_{ij} - \alpha_{ij}^d)}{2\sigma_{yo}}\,; \qquad \frac{\partial f}{\partial \alpha_{ij}^d} = -\frac{3(s_{ij} - \alpha_{ij}^d)}{2\sigma_{yo}} \tag{12.38}$$

The flow rule $\dot{\varepsilon}_{ij}^p = \dot{\lambda}\partial f/\partial\sigma_{ij}$ then provides

$$\dot{\varepsilon}_{ij}^p = \dot{\lambda}\,\frac{3(s_{ij} - \alpha_{ij}^d)}{2\sigma_{yo}} \tag{12.39}$$

and plastic incompressibility, i.e. $\dot{\varepsilon}_{kk}^p = 0$, follows readily. As usual, the effective plastic strain rate is defined by

$$\dot{\varepsilon}_{eff}^p = (\frac{2}{3}\dot{\varepsilon}_{ij}^p\dot{\varepsilon}_{ij}^p)^{1/2}$$

and the flow rule (12.39) as well as the yield criterion (12.37) then imply

$$\dot{\varepsilon}_{eff}^p = \dot{\lambda} \tag{12.40}$$

Let us now evaluate the plastic modulus H. In the present case, we have one hardening parameter in terms of the deviatoric part α_{ij}^d of the back-stress and referring to (10.15) and (10.17), we have

$$\dot{\alpha}_{ij}^d = \dot{\lambda} q_{ij} \quad \text{where} \quad q_{ij} = \frac{\partial \alpha_{ij}^d}{\partial \kappa_{kl}} k_{kl} \tag{12.41}$$

and

$$H = -\frac{\partial f}{\partial \alpha_{ij}^d} q_{ij} \tag{12.42}$$

and where q_{ij} may be considered as a combined evolution function. The consistency relation $\dot{f} = 0$ is written in the standard form as

$$\frac{\partial f}{\partial \sigma_{ij}}\dot{\sigma}_{ij} - H\dot{\lambda} = 0 \tag{12.43}$$

To obtain an interpretation of the plastic modulus H, we multiply the consistency relation (12.43) by $\dot{\lambda}$ and use the flow rule $\dot{\varepsilon}_{ij}^p = \dot{\lambda}\partial f/\partial\sigma_{ij}$ as well as (12.40) to obtain

$$\dot{\varepsilon}_{ij}^p\dot{\sigma}_{ij} - (\dot{\varepsilon}_{eff}^p)^2 H = 0 \tag{12.44}$$

Considering uniaxial stress conditions, the term $\dot{\varepsilon}_{ij}^p\dot{\sigma}_{ij}$ reduces to $\dot{\varepsilon}^p\dot{\sigma}$ and $\dot{\varepsilon}_{eff}^p$ becomes equal to $|\dot{\varepsilon}^p|$. In this case, (12.44) reduces to

$$\boxed{\frac{d\sigma}{d\varepsilon^p} = H} \tag{12.45}$$

This interpretation of H is illustrated in Fig. 12.13.

Let us now choose the specific evolution law to be used, i.e. we have to specify the evolution function q_{ij} present in (12.41). As discussed in relation to (10.17) it is allowable to directly specify the combined evolution function q_{ij}

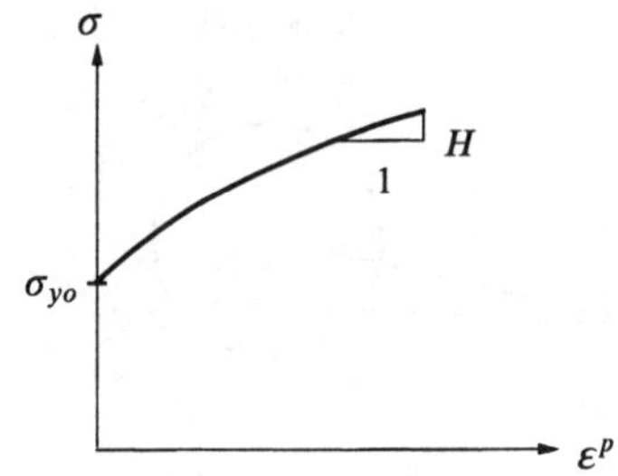

Figure 12.13: Interpretation of plastic modulus H.

instead of specifying the individual factors $\partial\alpha_{ij}^d/\partial\kappa_{kl}$ and k_{kl}. This approach is often advantageous and it is the case when kinematic hardening occurs. The classical kinematic evolution law was proposed by Melan (1938) and later by Prager (1955) and it reads

$$\boxed{\dot{\alpha}_{ij} = c\,\dot{\varepsilon}_{ij}^p \qquad \textit{Melan-Prager's evolution law}} \tag{12.46}$$

where c is a positive material parameter that may depend on the load history. We will later in this chapter discuss other kinematic evolution laws. In the present case of a von Mises material, (12.39) implies $\dot{\varepsilon}_{ii}^p = 0$, i.e. (12.46) shows that $\dot{\alpha}_{ii} = 0$, i.e. the back-stress tensor is a purely deviatoric tensor

$$\alpha_{ij}^d = \alpha_{ij} \tag{12.47}$$

Use of (12.39) into (12.46) then yields

$$\dot{\alpha}_{ij}^d = \dot{\lambda}\frac{3c}{2\sigma_{yo}}(s_{ij} - \alpha_{ij}^d) \tag{12.48}$$

As a consequence of the normality rule and Melan-Prager's evolution law (12.46), $\dot{\alpha}_{ij}^d$ is normal to the yield surface and (12.48) shows that $\dot{\alpha}_{ij}^d$ has the direction given by $s_{ij} - \alpha_{ij}^d$, cf. Fig. 12.14.

A comparison of (12.48) with (12.41) shows that the combined evolution function q_{ij} is given by

$$q_{ij} = \frac{3c}{2\sigma_{yo}}(s_{ij} - \alpha_{ij}^d)$$

Insertion of this expression as well as (12.38b) into (12.42) leads to

$$H = \frac{3}{2}c\,\frac{3(s_{ij} - \alpha_{ij}^d)(s_{ij} - \alpha_{ij}^d)}{2\sigma_{yo}^2}$$

which with the yield criterion (12.37) reduces to

$$H = \frac{3}{2}c \tag{12.49}$$

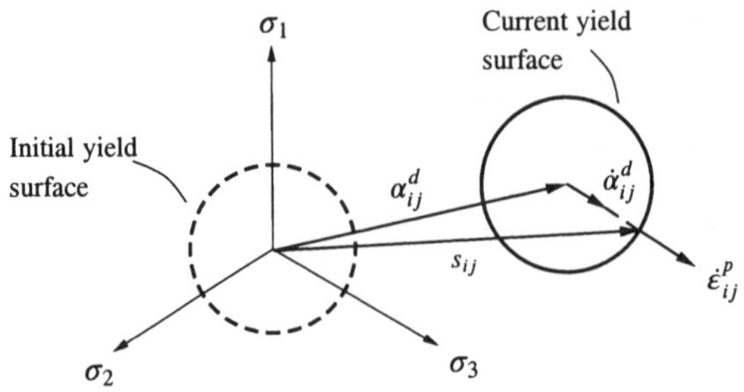

Figure 12.14: Illustration of Melan-Prager's evolution law in the deviatoric plane.

i.e. if c is constant, we have linear hardening. We emphasize that whereas (12.49) holds in general, the interpretation (12.45) only holds for uniaxial stress conditions.

Having established the fundamental elasto-plastic equations for kinematic hardening of a von Mises material, it is timely to discuss the prediction capabilities of this model.

The first issue of interest is the material parameter c appearing in the evolution law for the back-stress, cf. (12.46). The simplest choice is evidently to take c as a constant, but it is fully allowable to let c depend on some internal variable.

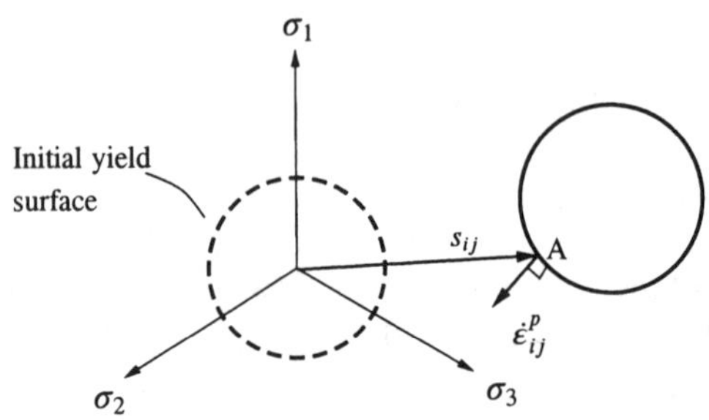

Figure 12.15: Deviatoric plane. The incremental plastic work $\dot{W}^p$ is negative when loading at point A, i.e. $\dot{W}^p = s_{ij}\dot{\varepsilon}^p_{ij} < 0$.

We have previously, in the discussion of isotropic hardening of a von Mises material, used the plastic work W^p as an internal variable, cf. (9.73). Let us therefore investigate the consequences of assuming that

$$c = c(W^p) \quad \text{where} \quad \dot{W}^p = \sigma_{ij}\dot{\varepsilon}^p_{ij}$$

In the present case, $\dot{\varepsilon}^p_{ij}$ is purely deviatoric and the incremental plastic work may therefore also be written as

$$\dot{W}^p = s_{ij}\dot{\varepsilon}^p_{ij}$$

It is of interest that whereas $\dot{W}^p$ is non-negative, i.e. $\dot{W}^p \geq 0$, for isotropic hardening of a von Mises material, cf. (9.76), this is not the case for kinematic hardening. Suppose that the current yield surface of the kinematic hardening von Mises material is located as shown in Fig. 12.15. It is evident that the scalar product $s_{ij}\dot{\varepsilon}^p_{ij}$ is negative, i.e. $\dot{W}^p < 0$, when plastic loading occurs at point A. For uniaxial stressing, the corresponding situation is shown in Fig. 12.16. According to (12.45) and (12.49), the slopes at point P and Q are identical. However, for uniaxial stressing we have $\dot{W}^p = \sigma_{ij}\dot{\varepsilon}^p_{ij} = \sigma\dot{\varepsilon}^p$; this means that $\dot{W}^p < 0$ when loading from point Q towards point R. Consequently, and as illustrated in Fig. 12.16, the slope will then increase when loading from point Q towards R. Clearly, this prediction is in contrast with experimental evidence and we can therefore reject the plastic work W^p as a suitable internal variable, when kinematic hardening is involved.

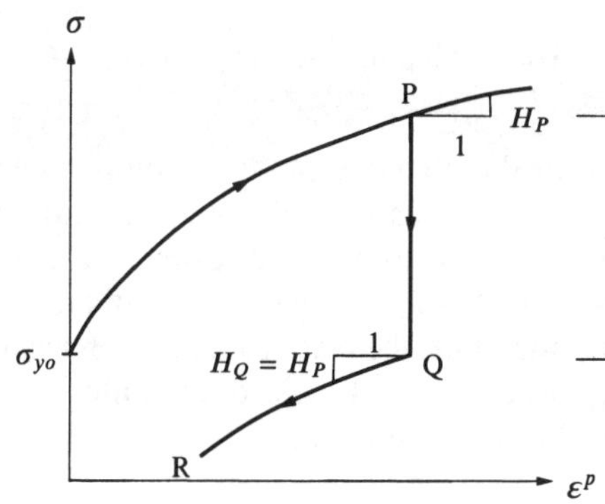

Figure 12.16: Consequence of letting $c = c(W^p)$, where $H = \frac{3}{2}c(W^p)$.

Let us next assume that the parameter c depends on the effective plastic strain, i.e.

$$c = c(\varepsilon^p_{eff}) \quad \text{where} \quad \dot{\varepsilon}^p_{eff} = (\frac{3}{2}\dot{\varepsilon}^p_{ij}\dot{\varepsilon}^p_{ij})^{1/2} = \dot{\lambda} \geq 0$$

Here, $\dot{\varepsilon}^p_{eff}$ is certainly non-negative, but even so the consequences of the assumption above are not promising.

To illustrate this, consider the uniaxial stressing shown in Fig. 12.17. The slopes at points P and Q are identical. Moreover, as $\dot{\varepsilon}^p_{eff} = |\dot{\varepsilon}^p|$ we find that the effective plastic strains ε^p_{eff} at points R and S are identical. Consequently and referring to Fig. 12.17, the slopes at points R and S are also identical and the response during reversed loading therefore becomes in inferior agreement with the real behavior of metals and steel.

The discussion above illustrates two important points. First, whereas it makes no difference for isotropic hardening of a von Mises material whether

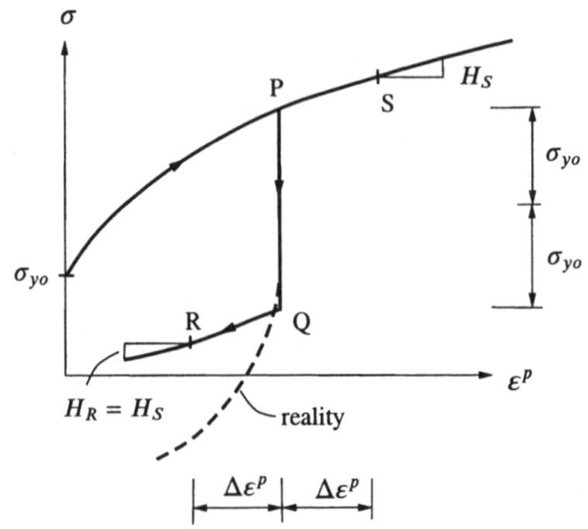

Figure 12.17: Consequence of letting $c = c(\varepsilon^p_{eff})$, where $H = \frac{3}{2}c(\varepsilon^p_{eff})$.

we adopt the plastic work or the effective plastic strain as internal variable, it makes a great difference when kinematic hardening is considered. Second, for kinematic hardening it is not trivial to obtain a realistic response for reversed loading, if the plastic modulus H is to vary with the loading.

In view of these latter difficulties, the classical approach is simply to assume that the parameter c is a constant, i.e. the plastic modulus $H = 3c/2$ is also constant and we then have linear hardening, i.e.

$$\boxed{H = \frac{3}{2}c = constant}$$

In the next chapter, we will return to more advanced kinematic models that allow for nonlinear hardening and provide a detailed discussion. For linear hardening, the response is illustrated in Fig. 12.18 and it appears that the Bauschinger effect is fairly well approximated even though the agreement with the real response is not overwhelming.

For stress cycling between the stresses σ_A and $-\sigma_A$, the linear kinematic model predicts the behavior shown in Fig. 12.19a). No cyclic hardening or softening effects can be predicted and the stabilized cyclic stress-strain curve is obtained already after one cycle. For strain cycling between ε_A and $-\varepsilon_A$, the prediction shown in Fig. 12.19b) is obtained. Since the stabilized cyclic stress-strain curve is obtained already after one cycle and since it is symmetric about the origin, there is no difference between stress cycling and strain cycling, cf. Figs. 12.19a) and 12.19b).

For cycling between unsymmetrical stress states, the result in Fig. 12.20a) is obtained; no ratcheting is predicted and the stabilized cyclic stress-strain curve is obtained already after one cycle. Similarly, Fig. 12.20b) shows unsymmetri-

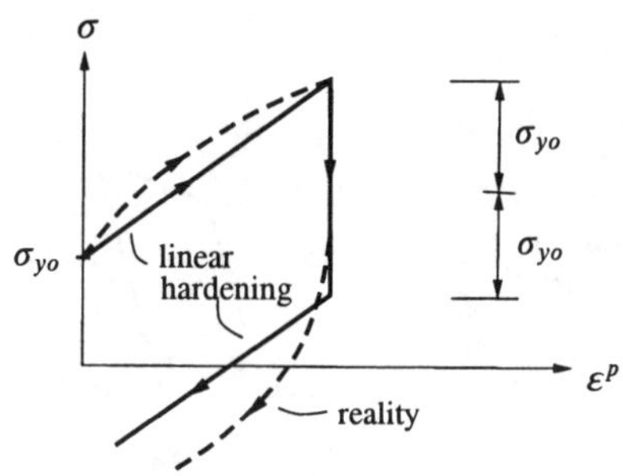

Figure 12.18: Prediction of kinematic hardening von Mises model with linear hardening, i.e. $H = \frac{3}{2}c$=constant.

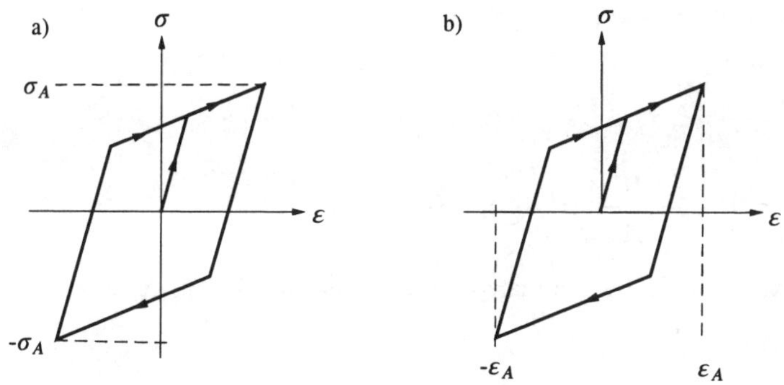

Figure 12.19: a) symmetrical stress cycling; b) symmetrical strain cycling.

cal strain cycling; no mean stress relaxation is predicted and since the stabilized cyclic stress-strain curve is obtained already after one cycle and since it is symmetric, there is no difference between stress cycling and strain cycling.

Even though linear kinematic hardening is not able to predict all important phenomena relating to cyclic behavior, it is evident that it is much more attractive than the isotropic hardening model when such loadings are of relevance.

Having illustrated that linear kinematic hardening of a von Mises material provides a first approximation when significant reversed loadings are of interest, we will now establish the tangential elasto-plastic stiffness matrix $\boldsymbol{D}^{ep}$ that is of importance in nonlinear finite element calculations.

Again, we assume elastic isotropy as given by (12.14), i.e. we obtain with (12.38a) and associated plasticity

$$\frac{\partial f}{\partial \sigma_{mn}} D_{mnkl} = \frac{3G}{\sigma_{yo}}(s_{kl} - \alpha_{kl}^{d}) \; ; \qquad D_{ijst} \frac{\partial f}{\partial \sigma_{st}} = \frac{3G}{\sigma_{yo}}(s_{ij} - \alpha_{ij}^{d})$$

From (10.24) and (12.37), it then follows that

$$A = H + 3G$$

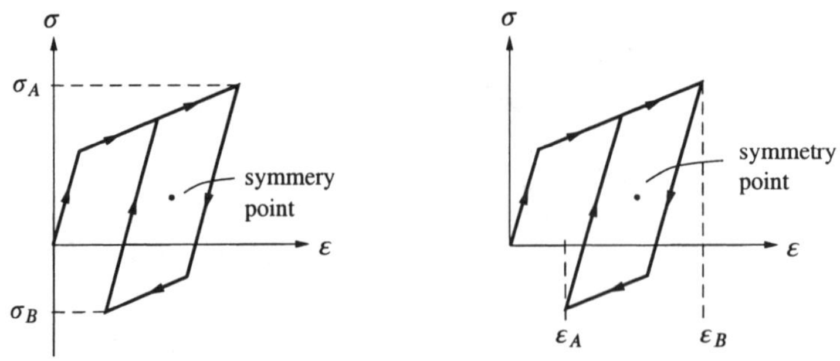

Figure 12.20: a) unsymmetrical stress cycling; b) unsymmetrical strain cycling.

Expressions (10.26) and (10.27) then lead to

$$\dot{\sigma}_{ij} = D^{ep}_{ijkl}\dot{\varepsilon}_{kl} \tag{12.50}$$

where

$$D^{ep}_{ijkl} = D_{ijkl} - \frac{9G^2}{A\sigma^2_{yo}}(s_{ij} - \alpha^d_{ij})(s_{kl} - \alpha^d_{kl})$$

If we express (12.50) in a matrix form similar to (4.37), we obtain

$$\dot{\boldsymbol{\sigma}} = \boldsymbol{D}^{ep}\dot{\boldsymbol{\varepsilon}} \tag{12.51}$$

where

$$\boldsymbol{D}^{ep} = \boldsymbol{D} - \boldsymbol{D}^{p}$$

and

$$\boldsymbol{D} = \frac{2G}{1-2\nu}\begin{bmatrix} 1-\nu & \nu & \nu & 0 & 0 & 0 \\ \nu & 1-\nu & \nu & 0 & 0 & 0 \\ \nu & \nu & 1-\nu & 0 & 0 & 0 \\ 0 & 0 & 0 & \frac{1}{2}(1-2\nu) & 0 & 0 \\ 0 & 0 & 0 & 0 & \frac{1}{2}(1-2\nu) & 0 \\ 0 & 0 & 0 & 0 & 0 & \frac{1}{2}(1-2\nu) \end{bmatrix} \tag{12.52}$$

$$\boldsymbol{D}^{p} = \frac{9G^2}{A\sigma^2_{yo}}\begin{bmatrix} \bar{s}^2_{11} & \bar{s}_{11}\bar{s}_{22} & \bar{s}_{11}\bar{s}_{33} & \bar{s}_{11}\bar{s}_{12} & \bar{s}_{11}\bar{s}_{13} & \bar{s}_{11}\bar{s}_{23} \\ \bar{s}_{22}\bar{s}_{11} & \bar{s}^2_{22} & \bar{s}_{22}\bar{s}_{33} & \bar{s}_{22}\bar{s}_{12} & \bar{s}_{22}\bar{s}_{13} & \bar{s}_{22}\bar{s}_{23} \\ \bar{s}_{33}\bar{s}_{11} & \bar{s}_{33}\bar{s}_{22} & \bar{s}^2_{33} & \bar{s}_{33}\bar{s}_{12} & \bar{s}_{33}\bar{s}_{13} & \bar{s}_{33}\bar{s}_{23} \\ \bar{s}_{12}\bar{s}_{11} & \bar{s}_{12}\bar{s}_{22} & \bar{s}_{12}\bar{s}_{33} & \bar{s}^2_{12} & \bar{s}_{12}\bar{s}_{13} & \bar{s}_{12}\bar{s}_{23} \\ \bar{s}_{13}\bar{s}_{11} & \bar{s}_{13}\bar{s}_{22} & \bar{s}_{13}\bar{s}_{33} & \bar{s}_{13}\bar{s}_{12} & \bar{s}^2_{13} & \bar{s}_{13}\bar{s}_{23} \\ \bar{s}_{23}\bar{s}_{11} & \bar{s}_{23}\bar{s}_{22} & \bar{s}_{23}\bar{s}_{33} & \bar{s}_{23}\bar{s}_{12} & \bar{s}_{23}\bar{s}_{13} & \bar{s}^2_{23} \end{bmatrix} \tag{12.53}$$

It is recalled that $\bar{s}_{ij} = s_{ij} - \alpha^d_{ij}$.

It may be of interest to observe that isotropic and kinematic von Mises hardening formally possess the same structure, cf. (12.19), (12.20) with (12.52) and (12.53). In fact, if we in (12.20) replace σ_y by σ_{y0} and s_{ij} by $\bar{s}_{ij}$ then (12.53) is obtained. This implies that the $\boldsymbol{D}^{ep}$-matrix for plane strain can be obtained directly from (12.21)-(12.24) whereas the $\boldsymbol{D}^{ep}$-matrix for plane stress can be established directly from (12.27) and (12.28)-(12.31).

12.4 Mixed von Mises hardening

We have discussed isotropic as well as kinematic hardening of a von Mises material and it seems tempting to investigate whether we can combine these models and thereby achieve additional freedom in our effort to model metals and steel in an accurate fashion.

In fact, we have previously, in (9.26) and (9.27), touched upon the concept of mixed hardening, which combines isotropic and kinematic hardening and which was introduced by Hodge (1957). Different mixed hardening formulations exist, but due to its usefulness we shall here adopt the formulation proposed by Goel and Malvern (1970), Krieg and Key (1976) and Axelsson and Samuelsson (1979). While this formulation holds for arbitrary yield functions, we will here restrict ourselves to mixed hardening of a von Mises material. The presentation will be given in fashion that is consistent with the previous exposition.

Within a von Mises concept and referring to (12.4) and (12.37), the yield criterion for mixed hardening reads

$$\boxed{f(\sigma_{ij}, K_\alpha) = [\frac{3}{2}(s_{kl} - \alpha^d_{kl})(s_{kl} - \alpha^d_{kl})]^{1/2} - \sigma_{yo} - K(\kappa) = 0} \tag{12.54}$$

where the deviatoric part of the back-stress α^d_{kl} and the parameter K are formally included in the hardening parameters K_α, i.e.

$$K_\alpha = \{\alpha^d_{ij}, K\} \tag{12.55}$$

We also note that K depends on the internal variable κ. It appears that if $K = 0$ then (12.54) reduces to purely kinematic hardening and if $\alpha^d_{ij} = 0$ then purely isotropic hardening is obtained. Let us define the quantity $\bar{\sigma}_y(\kappa)$ by

$$\boxed{\bar{\sigma}_y = \sigma_{yo} + K(\kappa)} \tag{12.56}$$

From (12.54) then follows that

$$\frac{\partial f}{\partial \sigma_{ij}} = \frac{3(s_{ij} - \alpha^d_{ij})}{2\bar{\sigma}_y}; \quad \frac{\partial f}{\partial \alpha^d_{ij}} = -\frac{3(s_{ij} - \alpha^d_{ij})}{2\bar{\sigma}_y}; \quad \frac{\partial f}{\partial K} = -1 \tag{12.57}$$

The associated flow rule then gives

$$\dot{\varepsilon}^p_{ij} = \dot{\lambda}\frac{3(s_{ij} - \alpha^d_{ij})}{2\bar{\sigma}_y} \tag{12.58}$$

which implies $\dot{\varepsilon}^p_{ii} = 0$, i.e. plastic incompressibility. This means that $\dot{\varepsilon}^p_{ij}$ is a purely deviatoric tensor. Define as usual the effective plastic strain rate by

$$\dot{\varepsilon}^p_{eff} = (\frac{2}{3}\dot{\varepsilon}^p_{ij}\dot{\varepsilon}^p_{ij})^{1/2}$$

With the flow rule (12.58) as well as the yield criterion (12.54), it then follows that

$$\dot{\varepsilon}^p_{eff} = \dot{\lambda} \tag{12.59}$$

In our mixed hardening formulation, we have two hardening parameters in terms of the parameter K and the deviatoric part α^d_{ij} of the back-stress, cf. (12.55). Referring to the general expression (10.15) for the evolution laws, we write

$$\begin{aligned} \dot{K} &= \dot{\lambda} q^{(1)} \quad \text{where} \quad q^{(1)} = \frac{\partial K}{\partial \kappa} k \\ \dot{\alpha}^d_{ij} &= \dot{\lambda} q^{(2)}_{ij} \quad \text{where} \quad q^{(2)}_{ij} = \frac{\partial \alpha^d_{ij}}{\partial \kappa_{kl}} k_{kl} \end{aligned} \tag{12.60}$$

where $q^{(1)}$ and $q^{(2)}_{ij}$ are the two combined evolution functions. The consistency relation is written in the standard format

$$\frac{\partial f}{\partial \sigma_{ij}} \dot{\sigma}_{ij} - H\dot{\lambda} = 0 \tag{12.61}$$

where the plastic modulus H, according to (10.17) and (12.60), is defined by

$$H = -\frac{\partial f}{\partial K} q^{(1)} - \frac{\partial f}{\partial \alpha^d_{ij}} q^{(2)}_{ij} \tag{12.62}$$

To obtain an interpretation of the plastic modulus H, we multiply (12.61) by $\dot{\lambda}$ and use the flow rule $\dot{\varepsilon}^p_{ij} = \dot{\lambda}\partial f/\partial \sigma_{ij}$ as well as (12.59) to obtain

$$\dot{\varepsilon}^p_{ij}\dot{\sigma}_{ij} - (\dot{\varepsilon}^p_{eff})^2 H = 0 \tag{12.63}$$

Considering uniaxial stressing, the term $\dot{\varepsilon}^p_{ij}\dot{\sigma}_{ij}$ reduces to $\dot{\varepsilon}^p\dot{\sigma}$ and $\dot{\varepsilon}^p_{eff}$ becomes equal to $|\dot{\varepsilon}^p|$. In this case, (12.63) reduces to

$$\boxed{\frac{d\sigma}{d\varepsilon^p} = H} \tag{12.64}$$

This interpretation is shown in Fig. 12.13.

The next step is to choose the specific form of the evolution functions $q^{(1)}$ and $q^{(2)}_{ij}$ present in (12.60). Inspired by the strain hardening assumption (12.13) used for isotropic hardening and the Melan-Prager evolution law (12.46) for kinematic hardening, we choose

$$\boxed{\begin{aligned}\dot{K} &= \dot{\lambda} m \frac{dK}{d\kappa}\,; \\ \dot{\alpha}_{ij} &= (1-m)c\dot{\varepsilon}^{p}_{ij}\end{aligned}} \tag{12.65}$$

where the constant parameter m is called the *mixed hardening parameter* and it is chosen by us; moreover, we have $0 \leq m \leq 1$. It appears that if $m = 0$, then a purely kinematic hardening model emerges, whereas $m = 1$ implies a purely isotropic hardening model. For m between these two values, we have mixed hardening and this suggests the terminology of m being the mixed hardening parameter. From (12.65a) and (12.59) as well as $\dot{K} = \frac{dK}{d\kappa}\dot{\kappa}$, we conclude that

$$\kappa = m\varepsilon^{p}_{eff} \tag{12.66}$$

Since $\dot{\varepsilon}^{p}_{ij}$ is purely deviatoric, (12.65b) implies

$$\alpha^{d}_{ij} = \alpha_{ij} \tag{12.67}$$

Therefore, use of the flow rule (12.58) in (12.65b) gives

$$\dot{\alpha}^{d}_{ij} = \dot{\lambda}(1-m)\frac{3c}{2\bar{\sigma}_y}(s_{ij} - \alpha^{d}_{ij}) \tag{12.68}$$

and a comparison of (12.65a) and (12.68) with (12.60a) and (12.60b), respectively, results in

$$q^{(1)} = m\frac{dK}{d\kappa}\,; \qquad q^{(2)}_{ij} = (1-m)\frac{3c}{2\bar{\sigma}_y}(s_{ij} - \alpha^{d}_{ij}) \tag{12.69}$$

The plastic modulus H is defined by (12.62). Use of (12.57b) and (12.57c) as well as (12.69) then provide

$$H = \frac{dK(\kappa)}{d\kappa}m + \frac{3}{2}c(1-m)\frac{3(s_{ij} - \alpha^{d}_{ij})(s_{ij} - \alpha^{d}_{ij})}{2\bar{\sigma}^2_y}$$

which with the yield criterion reduces to

$$H = \frac{dK(\kappa)}{d\kappa}m + \frac{3}{2}c(1-m) \tag{12.70}$$

It is emphasized that whereas this expression holds in general, the interpretation (12.64) only holds for uniaxial stress conditions.

In order to determine the material properties $dK(\kappa)/d\kappa$ and c, we rewrite (12.70) according to

$$H = [\frac{dK(\kappa)}{d\kappa} - \frac{3}{2}c]m + \frac{3}{2}c \tag{12.71}$$

It is advantageous to construct the mixed hardening model such that irrespective of our choice for the parameter m, we obtain the same response for monotonic proportional loading. For all other load histories, however, the value of m influences the response and this is particularly true when reversed loadings are involved. Let us therefore investigate the consequences of monotonic proportional loading.

For monotonic proportional loading, we have

$$s_{ij} = p(t)s_{ij}^* \tag{12.72}$$

where $p(t)$ is an increasing function of time t and s_{ij}^* is an arbitrary fixed deviatoric stress tensor. Before any plasticity is involved, we have $\alpha_{ij}^d = 0$. Referring to the evolution law (12.68), this means that the first increment $\dot{\alpha}_{ij}^d$ (when $\alpha_{ij}^d = 0$) will be proportional to s_{ij} and thereby to s_{ij}^*. This allows us to conclude that for monotonic proportional loading, α_{ij}^d will be proportional to s_{ij}^*. With this result and (12.72), it is concluded that

$$s_{ij} - \alpha_{ij}^d = q(t)s_{ij}^* \tag{12.73}$$

where $q(t)$ is a function of time. From (12.61) and the flow rule, we have

$$\dot{\varepsilon}_{ij}^p = \frac{1}{H}\frac{\partial f}{\partial \sigma_{kl}}\dot{\sigma}_{kl}\frac{\partial f}{\partial \sigma_{ij}}$$

which with (12.57a) becomes

$$\dot{\varepsilon}_{ij}^p = \frac{1}{H}\frac{3(s_{kl} - \alpha_{kl}^d)}{2\bar{\sigma}_y}\dot{\sigma}_{kl}\frac{3(s_{ij} - \alpha_{ij}^d)}{2\bar{\sigma}_y}$$

Noting that $(s_{kl} - \alpha_{kl}^d)\dot{\sigma}_{kl} = (s_{kl} - \alpha_{kl}^d)\dot{s}_{kl}$, use of (12.73) yields

$$\dot{\varepsilon}_{ij}^p = \frac{1}{H}\frac{3qs_{kl}^*\dot{p}s_{kl}^*}{2\bar{\sigma}_y}\frac{3qs_{ij}^*}{2\bar{\sigma}_y} \tag{12.74}$$

Insertion of (12.73) into the yield criterion (12.54) leads to

$$|q|(\frac{3}{2}s_{kl}^*s_{kl}^*)^{1/2} = \bar{\sigma}_y$$

Elimination of the term $s_{kl}^*s_{kl}^*$ in (12.74) then implies the result

$$\dot{\varepsilon}_{ij}^p = \frac{3}{2H}\dot{p}s_{ij}^*$$

and it is concluded that different plasticity models provide the same result for monotonic proportional loading, if the plastic modulus H is the same.

In view of (12.71), the requirement that the mixed hardening model should predict the same response for monotonic proportional loading irrespective of the value of the parameter m then leads to

$$\boxed{\frac{dK(\kappa)}{d\kappa} = \frac{3}{2}c\,; \qquad H = \frac{3}{2}c} \tag{12.75}$$

Linear hardening

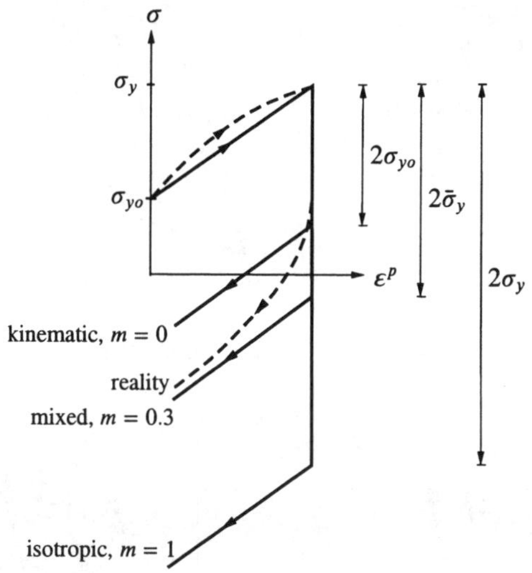

Figure 12.21: Prediction of mixed hardening von Mises model with linear hardening.

In order to proceed any further, the discussion in the previous section concerning purely kinematic hardening is recalled; it proved difficult to deal with a plastic modulus H that varies with the loading. In the following, we will therefore restrict ourselves to linear hardening, i.e.

$$\boxed{H = \frac{3}{2}c = constant} \tag{12.76}$$

Under these conditions, integration of (12.75a) yields $K = \frac{3}{2}c\kappa$, which with (12.66) and (12.56) leads to

$$\boxed{\bar{\sigma}_y = \sigma_{yo} + \frac{3}{2}cm\varepsilon^p_{eff}}$$

The predicted uniaxial response is illustrated in Fig. 12.21 from which also the interpretation of $\bar{\sigma}_y$ appears. It seems that by a proper choice of the mixed hardening parameter m, we can simulate a response during reversed loading that is in closer agreement with reality than the purely kinematic or purely isotropic model.

Let us in the following assume that the mixed hardening parameter is chosen as $m = 0.3$. For symmetric stress cycling between σ_A and $-\sigma_A$, the prediction is then shown in Fig. 12.22a). Cyclic hardening occurs and, eventually, after an infinite number of cycles, the response becomes purely elastic. For symmetric

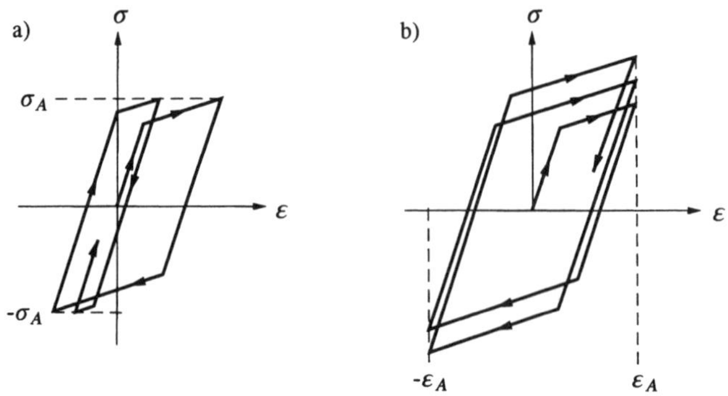

Figure 12.22: $m = 0.3$. a) symmetric stress cycling; b) symmetric strain cycling.

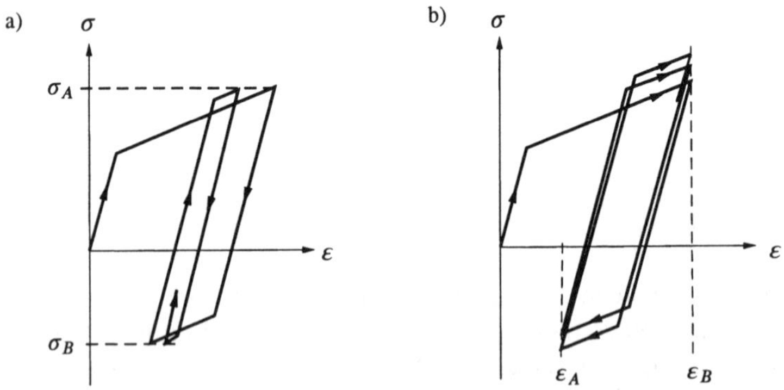

Figure 12.23: $m = 0.3$. a) unsymmetric stress cycling; b) unsymmetric strain cycling.

strain cycling between ε_A and $-\varepsilon_A$, the cyclic hardening response is shown in Fig. 12.22b) and, again, the stabilized cyclic behavior will be purely elastic. For unsymmetric stress cycling, the mixed hardening model predicts the response illustrated in Fig. 12.23a) and Fig. 12.23b) shows the situation of unsymmetric strain cycling.

In conclusion, when significant cyclic loadings are of relevance, mixed hardening offers advantages compared to purely isotropic or purely kinematic hardening, but even so, the mixed hardening model provides only a rather crude approximation to the real cyclic behavior.

Having discussed the prediction capabilities of mixed hardening, we turn to the establishment of the tangential elasto-plastic stiffness matrix $\boldsymbol{D}^{ep}$. Associated plasticity is adopted and elastic isotropy is again assumed, i.e. (12.14) together with (12.57a) yield

$$\frac{\partial f}{\partial \sigma_{mn}} D_{mnkl} = \frac{3G}{\bar{\sigma}_y}(s_{kl} - \alpha^d_{kl}) \, ; \qquad D_{ijst} \frac{\partial f}{\partial \sigma_{st}} = \frac{3G}{\bar{\sigma}_y}(s_{ij} - \alpha^d_{ij})$$

From (10.24) and (12.54), it follows that

$$A = H + 3G$$

Expressions (10.26) and (10.27) then lead to

$$\dot{\sigma}_{ij} = D^{ep}_{ijkl}\dot{\varepsilon}_{kl} \tag{12.77}$$

where

$$D^{ep}_{ijkl} = D_{ijkl} - \frac{9G^2}{A\bar{\sigma}_y^2}(s_{ij} - \alpha^d_{ij})(s_{kl} - \alpha^d_{kl})$$

If we express (12.77) in a matrix form similar to (4.37), we obtain

$$\dot{\boldsymbol{\sigma}} = \boldsymbol{D}^{ep}\dot{\boldsymbol{\varepsilon}}$$

where

$$\boldsymbol{D}^{ep} = \boldsymbol{D} - \boldsymbol{D}^{p}$$

and

$$\boldsymbol{D} = \frac{2G}{1-2\nu}\begin{bmatrix} 1-\nu & \nu & \nu & 0 & 0 & 0 \\ \nu & 1-\nu & \nu & 0 & 0 & 0 \\ \nu & \nu & 1-\nu & 0 & 0 & 0 \\ 0 & 0 & 0 & \frac{1}{2}(1-2\nu) & 0 & 0 \\ 0 & 0 & 0 & 0 & \frac{1}{2}(1-2\nu) & 0 \\ 0 & 0 & 0 & 0 & 0 & \frac{1}{2}(1-2\nu) \end{bmatrix} \tag{12.78}$$

$$\boldsymbol{D}^{p} = \frac{9G^2}{A\bar{\sigma}_y^2}\begin{bmatrix} \bar{s}_{11}^2 & \bar{s}_{11}\bar{s}_{22} & \bar{s}_{11}\bar{s}_{33} & \bar{s}_{11}\bar{s}_{12} & \bar{s}_{11}\bar{s}_{13} & \bar{s}_{11}\bar{s}_{23} \\ \bar{s}_{22}\bar{s}_{11} & \bar{s}_{22}^2 & \bar{s}_{22}\bar{s}_{33} & \bar{s}_{22}\bar{s}_{12} & \bar{s}_{22}\bar{s}_{13} & \bar{s}_{22}\bar{s}_{23} \\ \bar{s}_{33}\bar{s}_{11} & \bar{s}_{33}\bar{s}_{22} & \bar{s}_{33}^2 & \bar{s}_{33}\bar{s}_{12} & \bar{s}_{33}\bar{s}_{13} & \bar{s}_{33}\bar{s}_{23} \\ \bar{s}_{12}\bar{s}_{11} & \bar{s}_{12}\bar{s}_{22} & \bar{s}_{12}\bar{s}_{33} & \bar{s}_{12}^2 & \bar{s}_{12}\bar{s}_{13} & \bar{s}_{12}\bar{s}_{23} \\ \bar{s}_{13}\bar{s}_{11} & \bar{s}_{13}\bar{s}_{22} & \bar{s}_{13}\bar{s}_{33} & \bar{s}_{13}\bar{s}_{12} & \bar{s}_{13}^2 & \bar{s}_{13}\bar{s}_{23} \\ \bar{s}_{23}\bar{s}_{11} & \bar{s}_{23}\bar{s}_{22} & \bar{s}_{23}\bar{s}_{33} & \bar{s}_{23}\bar{s}_{12} & \bar{s}_{23}\bar{s}_{13} & \bar{s}_{23}^2 \end{bmatrix} \tag{12.79}$$

It is recalled that $\bar{s}_{ij} = s_{ij} - \alpha^d_{ij}$.

It is of interest that isotropic hardening and mixed hardening formally possess the same structure, cf. (12.19), (12.20) with (12.78) and (12.79). In fact, if we in (12.20) replace σ_y by $\bar{\sigma}_y$ and s_{ij} by $\bar{s}_{ij}$ then (12.79) is recovered. This implies that the $\boldsymbol{D}^{ep}$-matrix for plane strain can be obtained directly from (12.21)-(12.24) whereas the $\boldsymbol{D}^{ep}$-matrix for plane stress may be established directly from (12.27) and (12.28)-(12.31).

12.5 Melan-Prager's evolution law versus Ziegler's evolution law

The evolution law for the back-stress α_{ij} was previously given in the form suggested by Melan (1938) and Prager (1955), i.e.

$$\boxed{\dot{\alpha}_{ij} = c\,\dot{\varepsilon}^{p}_{ij} \qquad \textit{Melan-Prager's evolution law}} \tag{12.80}$$

cf. (12.46). With the flow rule, we obtain in general

$$\dot{\alpha}_{ij} = \dot{\lambda}\, c \frac{\partial g}{\partial \sigma_{ij}} \qquad \text{Melan-Prager} \tag{12.81}$$

which fits into the general format for evolution laws given by (10.15). For associated plasticity, this means that the direction of $\dot{\alpha}_{ij}$ is given by the normal to the yield surface at the current stress point as illustrated in Fig. 12.24a).

Another evolution law was proposed by Ziegler (1959) and it reads

$$\boxed{\dot{\alpha}_{ij} = \dot{\mu}(\sigma_{ij} - \alpha_{ij}) \qquad \textit{Ziegler's evolution law}} \tag{12.82}$$

where $\dot{\mu}$ is a non-negative quantity. This evolution law is illustrated in Fig. 12.24b). Since μ may be regarded as an internal variable, we can write the evolution law for μ in accordance with the general format given by (10.15), i.e

$$\dot{\mu} = \dot{\lambda} k \tag{12.83}$$

where k is a certain positive evolution function. Use of (12.83) in (12.82) implies

$$\dot{\alpha}_{ij} = \dot{\lambda} k(\sigma_{ij} - \alpha_{ij}) \qquad \text{Ziegler} \tag{12.84}$$

which fits into the general framework for evolution laws of hardening parameters given by (10.15).

In general, the two proposals (12.81) and (12.84) are different, but there are cases where they coincide. Assume that we have mixed hardening of a von Mises material. Since associated plasticity is assumed, Melan-Prager's evolution law (12.81) and (12.57a) give

$$\dot{\alpha}_{ij} = \dot{\lambda}\frac{3c}{2\bar{\sigma}_y}(s_{ij} - \alpha_{ij})\ ; \qquad \alpha^d_{ij} = \alpha_{ij} \tag{12.85}$$

As indicated, this evolution law implies that α_{ij} is purely deviatoric. In the von Mises yield criterion only the deviatoric part α^d_{ij} of the back-stress enters. Therefore, if Ziegler's evolution law (12.84) is adopted, we obtain

$$\dot{\alpha}^d_{ij} = \dot{\lambda} k(s_{ij} - \alpha^d_{ij}) \tag{12.86}$$

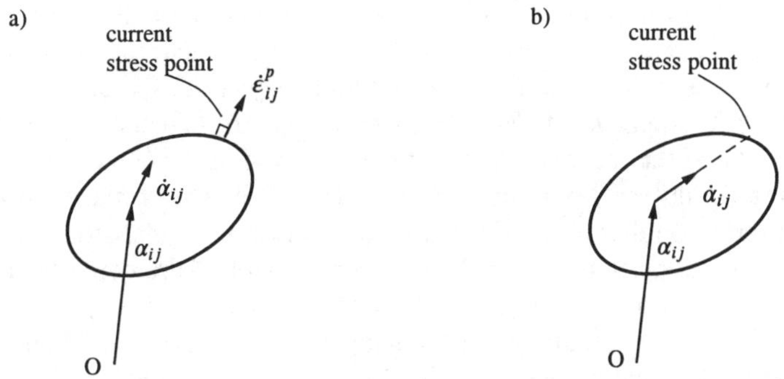

Figure 12.24: Illustration of evolution laws in the general stress space where O denotes the origin; a) Melan-Prager's rule for associated plasticity; b) Ziegler's rule.

It is noted that the hydrostatic part $\dot{\alpha}_{ii} = \dot{\lambda} k(\sigma_{ii} - \alpha_{ii})$ of the back-stress does not at all enter the mixed von Mises formulation. Moreover, a comparison of (12.85) and (12.86) reveals that the two evolution laws are identical and that we have

$$k = \frac{3c}{2\bar{\sigma}_y}$$

It is concluded that

For a mixed von Mises formulation, Melan-Prager's and Ziegler's evolution laws provide the same response	(12.87)

Despite this identity, the interpretation of the two evolution laws may be different. This may occur if the stress space considered is not the full nine-dimensional stress space defined by all the nine components of σ_{ij}, but rather some subspace. We shall first discuss these aspects in general terms and then provide a simple illustration.

For Melan-Prager's evolution law, two issues are here of importance. For associated plasticity, Melan-Prager's evolution law states that the direction of $\dot{\alpha}_{ij}$ is given by the normal to the yield surface at the current stress point, cf. Fig. 12.24a). This holds in the full nine-dimensional stress space, but not necessarily in some subspace. Moreover, for a mixed von Mises formulation the back-stress α_{ij}^d defines the center of the yield surface in the deviatoric plane. In some subspace, however, the quantity $\alpha_{ij}^d = \alpha_{ij}$ as determined by Melan-Prager's evolution law does not necessarily define the center.

For Ziegler's evolution law, however, the interpretation that α_{ij} defines the center of the yield surface is independent of whether we work in the full nine-dimensional stress space or some subspace. To see this, we observe that $\alpha_{ij} = 0$

holds initially. Assume now that in some subspace some component of σ_{ij} is always zero, say $\sigma_{13} = 0$. It follows from (12.82) that also the component α_{13} will always be zero. This implies that α_{ij} will always describe the center of the yield surface irrespective of what stress space we are considering. It follows from (12.82) that the interpretation that the direction of the center movement is given by the quantity $\sigma_{ij} - \alpha_{ij}$ holds irrespective of the stress space considered.

The observation that it is only in the full nine-dimensional stress space that Melan-Prager's evolution law predicts the movement of the center to occur in the direction given by the normal to the yield surface was first pointed out by Hodge (1957). A general analysis of this topic was given by Shield and Ziegler (1958) and further insight was provided by Clavout and Ziegler (1959) as well as by Ziegler (1959).

For a von Mises material, the two evolution laws predict the same response. For other materials, however, their predictions will differ, but at present there does not seem to exist conclusive experimental evidence that allows one to decide which of the two evolution laws is most accurate.

12.6 Orthotropic Hill plasticity

Anisotropic plasticity is of importance in many applications and orthotropy is of relevance for rolled profiles of steel and metals as well as paper, wood and ice. Formulations of orthotropic Hill plasticity have been presented, for instance, by Hill (1950), Hu (1956), Chen and Han (1988), de Borst and Feenstra (1990) and Schellekens and de Borst (1992). For ideal plasticity, the effect of anisotropy is more or less trivially considered since no hardening parameters or internal variables enter the formulation; however, for hardening or softening anisotropic plasticity the situation becomes more complex. Here, we shall present an approach for isotropic hardening of a Hill material. Hill's yield criterion was treated in Section 8.13 and the approach presented here provides a natural extension of the similar concept of an isotropic hardening von Mises material. It is recalled that Hill's criterion is of relevance for orthotropic materials.

Let us rewrite Hill's initial yield criterion given by (8.92) as

$$(\sigma_{yo}^2 \boldsymbol{s}^T \boldsymbol{P} \boldsymbol{s})^{1/2} - \sigma_{yo} = 0$$

where the constant material parameter σ_{yo} having the dimension of stress is defined by

$$\boxed{\sigma_{yo} = [\frac{3}{2(F+G+H)}]^{1/2}}$$

According to (8.96), the term $F + G + H$ is always positive. From (8.94), the symmetric matrix $\boldsymbol{P}$ is defined by

$$\boldsymbol{P} = \begin{bmatrix} \tilde{\boldsymbol{P}} & \boldsymbol{0} \\ \boldsymbol{0} & \tilde{\boldsymbol{Q}} \end{bmatrix} \tag{12.88}$$

where

$$\tilde{\boldsymbol{P}} = \begin{bmatrix} F+G & -F & -G \\ -F & F+H & -H \\ -G & -H & G+H \end{bmatrix} \; ; \quad \tilde{\boldsymbol{Q}} = \begin{bmatrix} 2L & 0 & 0 \\ 0 & 2M & 0 \\ 0 & 0 & 2N \end{bmatrix} \tag{12.89}$$

From (8.98) and considering an isotropic material, it is recalled that if

$$3F = 3G = 3H = L = M = N = \frac{3}{2\sigma_{yo}^2} \quad \Rightarrow \quad \textit{von Mises} \tag{12.90}$$

then Hill's criterion degenerates to the criterion of von Mises where σ_{yo} now, as previously, denotes the initial yield stress.

To obtain isotropic hardening of a Hill material, we write

$$\boxed{f(\sigma_{ij}, K) = (\sigma_{yo}^2 \boldsymbol{s}^T \boldsymbol{P} \boldsymbol{s})^{1/2} - \sigma_y = 0 \; ; \quad \sigma_y(\kappa) = \sigma_{yo} + K(\kappa)} \tag{12.91}$$

where K is the hardening parameter and κ is an internal variable, not yet specified.

As already indicated, we have here abandoned the tensor notation since many anisotropic yield criteria are expressed in matrix notation. As shown in Section 8.13, structural tensors can be used to describe anisotropic yield criteria and then a tensorial formulation is obtained. However, for illustration purposes we will here adopt a matrix format. The elastic stiffness is then also expressed in matrix notation and Hooke's law reads

$$\dot{\boldsymbol{\sigma}} = \boldsymbol{D}(\dot{\boldsymbol{\varepsilon}} - \dot{\boldsymbol{\varepsilon}}^p) \tag{12.92}$$

where $\boldsymbol{D}$ for the orthotropic material in question is defined by (4.55).

Adopting associated plasticity, the flow rule states that $\dot{\varepsilon}_{ij}^p = \dot{\lambda}\partial f/\partial\sigma_{ij}$ and in matrix notation this is expressed as

$$\boxed{\dot{\boldsymbol{\varepsilon}}_p = \dot{\lambda}\frac{\partial f}{\partial \boldsymbol{\sigma}}} \tag{12.93}$$

The quantity $\dot{\boldsymbol{\varepsilon}}^p$ is defined in the same manner as $\boldsymbol{\varepsilon}$, i.e. the engineering shear strains enter this column matrix, cf. (4.35). To define the meaning of $\partial f/\partial\boldsymbol{\sigma}$ some caution must be shown. To illustrate this aspect, let us express the von Mises criterion as

$$f = [\frac{3}{2}(s_{11}^2 + s_{22}^2 + s_{33}^2 + s_{12}^2 + s_{21}^2 + s_{13}^2 + s_{31}^2 + s_{23}^2 + s_{32}^2)]^{1/2} - \sigma_y = 0$$

$$\Rightarrow \quad \text{for instance} \quad \frac{\partial f}{\partial \sigma_{12}} = \frac{3 s_{12}}{2\sigma_y} \quad \text{i.e.} \quad \dot{\varepsilon}_{12}^p = \dot{\lambda}\frac{3 s_{12}}{2\sigma_y}$$

in accordance with our previous results. However, if advantage is taken of the symmetry of the stress tensor, we may write

$$f = [\frac{3}{2}(s_{11}^2 + s_{22}^2 + s_{33}^2 + 2s_{12}^2 + 2s_{13}^2 + 2s_{23}^2)]^{1/2} - \sigma_y = 0$$

$$\Rightarrow \quad \text{for instance} \quad \frac{\partial f}{\partial \sigma_{12}} = \frac{3s_{12}}{\sigma_y} \quad \text{i.e.} \quad 2\dot{\varepsilon}_{12}^p = \dot{\lambda}\frac{\partial f}{\partial \sigma_{12}}$$

It is concluded, that if formulation (12.93) is adopted then we have the following interpretation:

$$\frac{\partial f}{\partial \boldsymbol{\sigma}} = \begin{bmatrix} \frac{\partial f}{\partial \sigma_{11}} \\ \frac{\partial f}{\partial \sigma_{22}} \\ \frac{\partial f}{\partial \sigma_{33}} \\ \frac{\partial f}{\partial \sigma_{12}} \\ \frac{\partial f}{\partial \sigma_{13}} \\ \frac{\partial f}{\partial \sigma_{23}} \end{bmatrix} \quad \textit{if advantage is taken of the symmetry of the stress tensor when formulating the yield criterion} \qquad (12.94)$$

In the yield criterion (12.91) advantage was taken of the symmetry of the stress tensor, i.e. the interpretation (12.94) holds.

To determine $\partial f/\partial \boldsymbol{\sigma}$ from (12.91), we first note that the deviatoric stresses $\boldsymbol{s}$ may be written as

$$\boldsymbol{s} = (\begin{bmatrix} \boldsymbol{I} & \boldsymbol{0} \\ \boldsymbol{0} & \boldsymbol{I} \end{bmatrix} - \frac{1}{3}\begin{bmatrix} \boldsymbol{E} & \boldsymbol{0} \\ \boldsymbol{0} & \boldsymbol{0} \end{bmatrix})\boldsymbol{\sigma}$$

where $\boldsymbol{I}$, as usual, is the unit matrix whereas $\boldsymbol{E}$ is defined by

$$\boldsymbol{E} = \begin{bmatrix} 1 & 1 & 1 \\ 1 & 1 & 1 \\ 1 & 1 & 1 \end{bmatrix}$$

With the symmetric matrix $\boldsymbol{P}$ given by (12.88), we then obtain

$$\frac{\partial(\boldsymbol{s}^T\boldsymbol{P}\boldsymbol{s})}{\partial \boldsymbol{\sigma}} = 2(\frac{\partial \boldsymbol{s}}{\partial \boldsymbol{\sigma}})^T\boldsymbol{P}\boldsymbol{s} = 2(\begin{bmatrix} \boldsymbol{I} & \boldsymbol{0} \\ \boldsymbol{0} & \boldsymbol{I} \end{bmatrix} - \frac{1}{3}\begin{bmatrix} \boldsymbol{E} & \boldsymbol{0} \\ \boldsymbol{0} & \boldsymbol{0} \end{bmatrix})\begin{bmatrix} \tilde{\boldsymbol{P}} & \boldsymbol{0} \\ \boldsymbol{0} & \tilde{\boldsymbol{Q}} \end{bmatrix}\boldsymbol{s}$$

by using $\boldsymbol{I}$ and $\boldsymbol{E}$ as symmetric. Noting that $\boldsymbol{E}\tilde{\boldsymbol{P}} = \boldsymbol{0}$, we obtain

$$\frac{\partial(\boldsymbol{s}^T\boldsymbol{P}\boldsymbol{s})}{\partial \boldsymbol{\sigma}} = 2\boldsymbol{P}\boldsymbol{s}$$

The flow rule (12.93) in combination with (12.91) then leads to the following expression

$$\boxed{\dot{\boldsymbol{\varepsilon}}^p = \dot{\lambda}\frac{\sigma_{yo}^2}{\sigma_y}\boldsymbol{P}\boldsymbol{s}} \qquad (12.95)$$

It is easily checked that plastic incompressibility, i.e. $\dot{\varepsilon}^p_{11} + \dot{\varepsilon}^p_{22} + \dot{\varepsilon}^p_{33} = 0$, holds - as expected.

Let us also observe that if we define the effective stress σ_{eff} by

$$\boxed{\sigma_{eff} = (\sigma_{yo}^2 \boldsymbol{s}^T \boldsymbol{P} \boldsymbol{s})^{1/2}} \tag{12.96}$$

then the yield criterion (12.91) may be expressed as

$$\sigma_{eff} - \sigma_y = 0 \tag{12.97}$$

For an isotropic material, where (12.90) holds, we find that (12.96) reduces to $\sigma_{eff} = (\frac{3}{2} s_{ij} s_{ij})^{1/2}$, i.e. the familiar effective von Mises stress.

It turns out to be advantageous to invert the flow rule (12.95), i.e. express $\boldsymbol{s}$ in terms of $\dot{\boldsymbol{\varepsilon}}^p$. Referring to (12.88), this inversion is hampered by the fact that while the inverse of $\tilde{\boldsymbol{Q}}$ always exists, we have $\det \tilde{\boldsymbol{P}} = 0$, i.e. also $\det \boldsymbol{P} = 0$ holds. However, we note that if a matrix $\tilde{\boldsymbol{M}}$ exists such that

$$\tilde{\boldsymbol{M}} \tilde{\boldsymbol{P}} = \frac{1}{3} \begin{bmatrix} 2 & -1 & -1 \\ -1 & 2 & -1 \\ -1 & -1 & 2 \end{bmatrix} \tag{12.98}$$

then

$$\tilde{\boldsymbol{M}} \tilde{\boldsymbol{P}} \begin{bmatrix} s_{11} \\ s_{22} \\ s_{33} \end{bmatrix} = \begin{bmatrix} s_{11} \\ s_{22} \\ s_{33} \end{bmatrix} \tag{12.99}$$

This particular property hinges on the fact that $s_{11} + s_{22} + s_{33} = 0$. With $\tilde{\boldsymbol{P}}$ given by (12.89), inspection shows that the following symmetric matrix

$$\tilde{\boldsymbol{M}} = \frac{1}{3(FG + FH + GH)} \begin{bmatrix} F+H & F-G-H & -H \\ F-G-H & F+G & -G \\ -H & -G & 2F \end{bmatrix}$$

fulfills (12.98). Define the symmetric matrix $\boldsymbol{M}$ by

$$\boldsymbol{M} = \begin{bmatrix} \tilde{\boldsymbol{M}} & \boldsymbol{0} \\ \boldsymbol{0} & \tilde{\boldsymbol{Q}}^{-1} \end{bmatrix} \quad \text{where} \quad \tilde{\boldsymbol{Q}}^{-1} = \begin{bmatrix} \frac{1}{2L} & 0 & 0 \\ 0 & \frac{1}{2M} & 0 \\ 0 & 0 & \frac{1}{2N} \end{bmatrix}$$

Using (12.88) and (12.99), it then follows that

$$\boldsymbol{M} \boldsymbol{P} \boldsymbol{s} = \boldsymbol{s} \tag{12.100}$$

and (12.95) then provides

$$\boldsymbol{s} = \frac{\sigma_y}{\dot{\lambda} \sigma_{yo}^2} \boldsymbol{M} \dot{\boldsymbol{\varepsilon}}^p \tag{12.101}$$

With the results (12.100) and (12.101), it turns out to be possible to express the plastic multiplier $\dot{\lambda}$ in terms of the plastic stain rate components. For this purpose, multiply (12.95) by $\boldsymbol{s}^T$ and use (12.91) to obtain

$$\boldsymbol{s}^T \dot{\boldsymbol{\varepsilon}}^p = \dot{\lambda}\sigma_y \tag{12.102}$$

Insertion of (12.101) gives

$$\frac{\sigma_y}{\dot{\lambda}\sigma_{yo}^2}(\dot{\boldsymbol{\varepsilon}}^p)^T \boldsymbol{M} \dot{\boldsymbol{\varepsilon}}^p = \dot{\lambda}\sigma_y$$

where it was used that $\boldsymbol{M}$ is symmetric. This leads to the result sought for

$$\boxed{\dot{\lambda} = [\frac{(\dot{\boldsymbol{\varepsilon}}^p)^T \boldsymbol{M} \dot{\boldsymbol{\varepsilon}}^p}{\sigma_{yo}^2}]^{1/2}} \tag{12.103}$$

Note that apart from some constant material parameters, the plastic multiplier $\dot{\lambda}$ has now been expressed entirely in terms of the plastic strain rate components.

With these results, let us evaluate (12.103) for an isotropic material. In this case (12.90) holds, which implies

$$\boldsymbol{M} = \frac{4}{9}\sigma_{yo}^2 \begin{bmatrix} 1 & -\frac{1}{2} & -\frac{1}{2} & 0 & 0 & 0 \\ -\frac{1}{2} & 1 & -\frac{1}{2} & 0 & 0 & 0 \\ -\frac{1}{2} & -\frac{1}{2} & 1 & 0 & 0 & 0 \\ 0 & 0 & 0 & \frac{3}{4} & 0 & 0 \\ 0 & 0 & 0 & 0 & \frac{3}{4} & 0 \\ 0 & 0 & 0 & 0 & 0 & \frac{3}{4} \end{bmatrix}$$

Insertion into (12.103) and using $\dot{\varepsilon}_{11}^p + \dot{\varepsilon}_{22}^p + \dot{\varepsilon}_{33}^p = 0$ lead to

$$\dot{\lambda} = \{\frac{2}{3}[(\dot{\varepsilon}_{11}^p)^2 + (\dot{\varepsilon}_{22}^p)^2 + (\dot{\varepsilon}_{33}^p)^2 + \frac{1}{2}(2\dot{\varepsilon}_{12}^p)^2 + \frac{1}{2}(2\dot{\varepsilon}_{13}^p)^2 + \frac{1}{2}(2\dot{\varepsilon}_{23}^p)^2]\}^{1/2}$$

i.e. $\dot{\lambda} = (2\dot{\varepsilon}_{ij}^p\dot{\varepsilon}_{ij}^p/3)^{1/2}$. It appears that $\dot{\lambda}$ for an isotropic material reduces to the previously defined effective plastic strain rate $\dot{\varepsilon}_{eff}^p$, cf. (12.9).

For orthotropic materials, it therefore seems natural also to define the effective plastic strain rate by

$$\boxed{\dot{\varepsilon}_{eff}^p = \dot{\lambda}} \tag{12.104}$$

where $\dot{\lambda}$ is given by (12.103). We may note that this definition leads to the following expression for the rate of plastic work

$$\dot{W}^p = \boldsymbol{\sigma}^T \dot{\boldsymbol{\varepsilon}}^p = \boldsymbol{s}^T \dot{\boldsymbol{\varepsilon}}^p = \dot{\lambda}\sigma_y$$

where (12.102) was used. Taking advantage of (12.104) and the yield criterion (12.97), we find the following simple expression

$$\boxed{\dot{W}^p = \sigma_y \dot{\varepsilon}^p_{eff} = \sigma_{eff} \dot{\varepsilon}^p_{eff}} \tag{12.105}$$

Expression (12.105) is in complete analogy with (9.76) applicable for a von Mises material.

In matrix notation, the consistency relation (10.16) reads

$$(\frac{\partial f}{\partial \boldsymbol{\sigma}})^T \dot{\boldsymbol{\sigma}} - H\dot{\lambda} = 0 \tag{12.106}$$

where the plastic modulus H according to (10.17) is defined by

$$H = -\frac{\partial f}{\partial K}\frac{\partial K}{\partial \kappa} k \tag{12.107}$$

and

$$\dot{\kappa} = \dot{\lambda} k \tag{12.108}$$

cf. (10.13). Just as for von Mises plasticity, we shall assume *strain hardening*, i.e.

$$\dot{\kappa} = \dot{\varepsilon}^p_{eff} \tag{12.109}$$

Since $\dot{\varepsilon}_{eff} = \dot{\lambda}$ we conclude from (12.108) that $k = 1$ and in view of (12.91), we have $\partial f / \partial K = -1$, i.e. (12.107) reduces to

$$H = \frac{\partial K}{\partial \kappa} \tag{12.110}$$

From (12.91) and (12.109) we have

$$\frac{\partial K}{\partial \kappa} = \frac{d\sigma_y(\varepsilon^p_{eff})}{d\varepsilon^p_{eff}}$$

i.e. (12.110) gives

$$\boxed{H = \frac{d\sigma_y(\varepsilon^p_{eff})}{d\varepsilon^p_{eff}}} \tag{12.111}$$

in complete analogy with von Mises plasticity, cf. (12.12).

Let us next illustrate how it is possible to calibrate the model by means of a uniaxial tension experiment. For uniaxial tension, the yield criterion (12.91) reduces to

$$\sigma_y = \sigma_{yo}\sigma_{11}\sqrt{F+G} \tag{12.112}$$

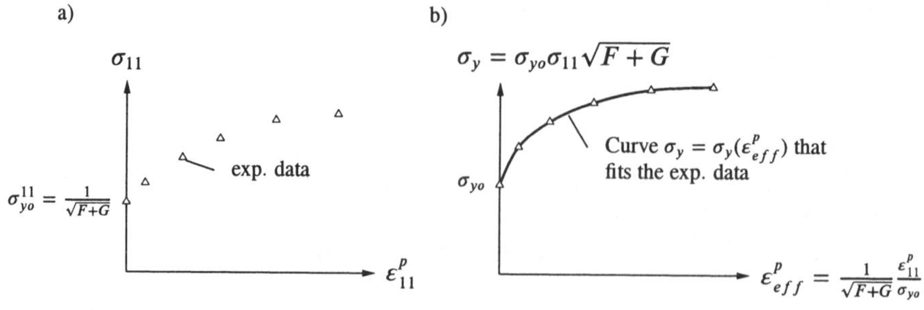

Figure 12.25: Uniaxial tension; a) original experimental data in a $\sigma_{11}, \varepsilon_{11}^{p}$-diagram, b) data transformed to a $\sigma_y, \varepsilon_{eff}^{p}$-diagram.

where initial yielding is obtained for $\sigma_y = \sigma_{yo}$, cf. (12.91), i.e. $\sigma_{11} = 1/\sqrt{F+G}$. From the flow rule (12.95), we find $2\dot{\varepsilon}_{12}^{p} = 2\dot{\varepsilon}_{13}^{p} = 2\dot{\varepsilon}_{23}^{p} = 0$ as well as

$$\begin{bmatrix} \dot{\varepsilon}_{11}^{p} \\ \dot{\varepsilon}_{22}^{p} \\ \dot{\varepsilon}_{33}^{p} \end{bmatrix} = \dot{\lambda}\frac{\sigma_{yo}^2}{\sigma_y}\sigma_{11}\begin{bmatrix} F+G \\ -F \\ -G \end{bmatrix}$$

Use of (12.112) then gives

$$\dot{\varepsilon}_{11}^{p} = \dot{\lambda}\frac{\sigma_{yo}^2}{\sigma_y}\sigma_{11}(F+G) = \dot{\lambda}\sigma_{yo}\sqrt{F+G}$$

and as $\dot{\lambda} = \dot{\varepsilon}_{eff}^{p}$, we obtain

$$\dot{\varepsilon}_{eff}^{p} = \frac{1}{\sqrt{F+G}}\frac{\dot{\varepsilon}_{11}^{p}}{\sigma_{yo}} \tag{12.113}$$

From (12.105), (12.112) and (12.113) we may note that $\dot{W}^p = \sigma_y\dot{\varepsilon}_{eff}^{p} = \sigma_{11}\dot{\varepsilon}_{11}^{p}$. However, returning to the calibration process we plot the experimental results in a stress (σ_{11})– plastic strain (ε_{11}^{p}) diagram, as shown in Fig. 12.25a). These data are converted into the data points shown in Fig. 12.25b) and approximated by the curve $\sigma_y = \sigma_y(\varepsilon_{eff}^{p})$, which is now a known expression and (12.111) is then used to determine the plastic modulus H.

Let us finally determine the elasto-plastic stiffness matrix. From Hooke's law (12.92), the flow rule (12.93) and the consistency relation (12.106) we obtain in complete analogy with the procedure described in Chapter 10 that

$$\dot{\boldsymbol{\sigma}} = \boldsymbol{D}^{ep}\dot{\boldsymbol{\varepsilon}}$$

where the elasto-plastic stiffness matrix is given by

$$\boldsymbol{D}^{ep} = \boldsymbol{D} - \frac{1}{A}\boldsymbol{D}\frac{\partial f}{\partial \boldsymbol{\sigma}}(\frac{\partial f}{\partial \boldsymbol{\sigma}})^T\boldsymbol{D}$$

and the positive parameter A is defined by

$$A = H + \left(\frac{\partial f}{\partial \boldsymbol{\sigma}}\right)^T \boldsymbol{D} \frac{\partial f}{\partial \boldsymbol{\sigma}}$$

In the present case of orthotropy, the elastic stiffness matrix $\boldsymbol{D}$ is given by (4.55). Moreover, from (12.93) and (12.95) we have

$$\frac{\partial f}{\partial \boldsymbol{\sigma}} = \frac{\sigma_{yo}^2}{\sigma_y} \boldsymbol{P s}$$

This leads to

$$A = H + \left(\frac{\sigma_{yo}^2}{\sigma_y}\right)^2 \boldsymbol{s}^T \boldsymbol{P D P s}$$

and

$$\boldsymbol{D}^{ep} = \boldsymbol{D} - \boldsymbol{D}^p$$

where

$$\boldsymbol{D}^p = \frac{1}{A}\left(\frac{\sigma_{yo}^2}{\sigma_y}\right)^2 \boldsymbol{D P s s}^T \boldsymbol{P D}$$

As a concluding remark to the above derivations, let us remember that the coordinate system was selected such that it coincides with the orthotropic material directions. If this is not the case *a priori*, the transformation rules for matrices, cf. Section 4.5, must first be used to achieve this situation.

12.7 Drucker-Prager plasticity. Frictional materials

So far, we have discussed common plasticity models for steel and metals, which all deal with various von Mises formulations; these materials are characterized by plastic incompressibility. To obtain an idea of how to model frictional materials like concrete, soil and rocks, we will consider *Drucker-Prager plasticity*. This formulation gives rise to plastic volume changes that are most characteristic for frictional materials; however, it will turn out that accurate predictions of these important plastic volume changes require much more than just a simple Drucker-Prager formulation. In this section we will provide an indication of the fundamental problems.

Referring to (8.33), isotropic hardening of a Drucker-Prager material is given by

$$f = 0; \quad \text{where} \quad f = \sigma_{eff} + \alpha I_1 - \beta - K \tag{12.114}$$

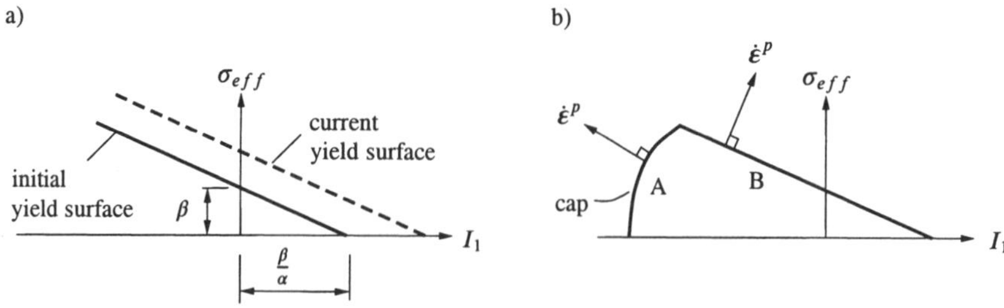

Figure 12.26: a) Isotropic hardening of Drucker-Prager material, b) Drucker-Prager yield surface supplemented by a cap.

where α and β are positive constants and the hardening parameter depends on an internal variable, i.e. $K = K(\kappa)$, cf. Fig. 12.26a). Adopting associated plasticity, we obtain

$$\dot{\varepsilon}_{ij}^p = \dot{\lambda}\frac{\partial f}{\partial \sigma_{ij}} = \dot{\lambda}(\frac{3s_{ij}}{2\sigma_{eff}} + \alpha\delta_{ij})\,; \qquad \dot{\kappa} = -\dot{\lambda}\frac{\partial f}{\partial K} = \dot{\lambda} \tag{12.115}$$

It follows that

$$\dot{\varepsilon}_{ii}^p = 3\dot{\lambda}\alpha \tag{12.116}$$

and a plastic volume increase is present. This volumetric plastic strain is only a very crude approximation of what happens in reality. Consider, for instance, purely hydrostatic compression of concrete where experimental data are shown in Fig. 4.9. It is evident that the nonlinear relation between hydrostatic stress and volumetric strain shown there can never be simulated by Drucker-Prager plasticity. According to Fig. 12.26a), purely hydrostatic compression of a Drucker-Prager material results in a linear elastic response.

To remedy this unfortunate situation, DiMaggio and Sandler (1971) added a yield surface, a so-called *cap*, to the Drucker-Prager yield surface. As shown in Fig. 12.26b), this cap closes the stress space so that purely hydrostatic compression will introduce plasticity; often, this cap is taken as part of an ellipse. According to the discussion relating to Fig. 12.7 and (12.3), it is concluded that at point A in Fig. 12.26b), we have $\dot{\varepsilon}_{ii}^p < 0$ whereas at point B, $\dot{\varepsilon}_{ii}^p > 0$ holds.

If the cap is taken to be part of an ellipse then

$$f_{cap} = 0\,; \quad \text{where} \quad f_{cap} = (I_1 - c)^2 + \sigma_{eff}^2 - q \tag{12.117}$$

where $(c, 0)$ denotes the center of the ellipse in the I_1, σ_{eff}-coordinate system. The parameters c and q in (12.117 both depend on hardening parameters.

Despite the progress achieved by introducing a cap, in practice it turns out that accurate modeling of the important plastic volume changes occurring in frictional materials cannot be captured by the model described. With associated plasticity, the plastic volume change is controlled by the form of the yield surface in the meridian plane, cf. Fig. 12.26b). This is an all too restricted format and nonassociated plasticity is therefore adopted. We have

For frictional materials, the general experience is: *Associated plasticity is used for the deviatoric response.* *Nonassociated plasticity is used for the volumetric response*	(12.118)

In Section 22.3 we will return to Drucker-Prager plasticity and find that in order to fulfill the second law of thermodynamics, both associated and nonassociated Drucker-Prager plasticity are subject to certain nontrivial restrictions.

In addition to these complications, we have previously observed in (8.25) that it is not sufficient to consider the stress invariants I_1 and J_2 as was done by the model in Fig. 12.26b); for frictional materials, the third stress invariant in terms of the quantity $\cos 3\theta$ is of major influence. Thus, frictional materials require a nonassociated formulation and the yield criterion must consider the effect of all three stress invariants.

A model for soil that contains these ingredients was proposed by Lade (1977). In Fig. 12.26b) what corresponds to the cap and the Drucker-Prager yield surface was combined by Lade (1977) into one yield surface. This idea was further developed and refined by Lade and Kim (1995) and a somewhat similar concept was suggested by Krenk (2000) and Ahadi and Krenk (2000). In fact, a number of soil models exists in the literature and the same applies to concrete where we take the opportunity to refer to Schreyer and Babcock (1985), Faruque and Chang (1990), Lubarda *et al.* (1994) together with the textbooks of Chen and Han (1988) and Chen (1994) for additional information.

13 NONLINEAR KINEMATIC HARDENING LAWS

In the previous chapter the problems with establishing a plasticity model that can predict an accurate response when significant reversed loadings occur were highlighted. A generalization of Melan-Prager's kinematic hardening law to account for nonlinear hardening behavior was discussed, and it was shown that the most natural assumptions lead to unrealistic behavior of the model, cf. Section 12.3. It was concluded that it is difficult to account for nonlinear hardening when Melan-Prager's kinematic hardening rule is used. In this chapter, other possibilities to model nonlinear kinematic hardening will be discussed and special emphasis will be given to the situation where the von Mises yield criterion is adopted.

The candidates considered in this exposition are the model of Mróz (1967), the so-called bounding surface models of Dafalias and Popov (1975, 1976) and Krieg (1975), and the nonlinear kinematic formulation of Armstrong and Frederick (1966). A review of these models is also provided by Khan and Huang (1995)

13.1 Mróz model

An intriguing model suitable when reversed loadings occur was proposed by Mróz (1967); somewhat similar models were proposed by Besseling (1953) and Iwan (1967). The *Mróz model* forms the basis of many more advanced models and we will therefore provide a detailed discussion of its ingredients.

For reasons previously discussed, kinematic hardening was restricted to linear hardening. The essential feature of the Mróz model is that the response of the material is approximated by a multilinear response and for each linear part of the response, a linear kinematic hardening model is adopted. For uniaxial loading, this idea is illustrated in Fig. 13.1; it appears that in each linear region, we have a constant plastic modulus, i.e.

$$H_n = \text{constant} \; ; \quad n = 0, 1, 2, \cdots$$

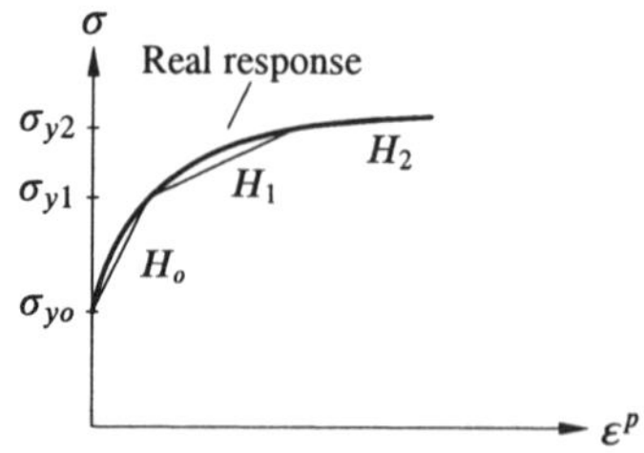

Figure 13.1: Multilinear approximation of uniaxial response with constant plastic moduli given by H_0, H_1,H_2,$\cdots$.

Figure 13.1 shows that between the constant stress values σ_{y0} and σ_{y1}, the plastic modulus is given by H_0 and between the constant stress values σ_{y1} and σ_{y2}, the plastic modulus is given by H_1, etc. We have

$$\sigma_{yn} = \text{constant} \; ; \quad n = 0, 1, 2, \cdots \tag{13.1}$$

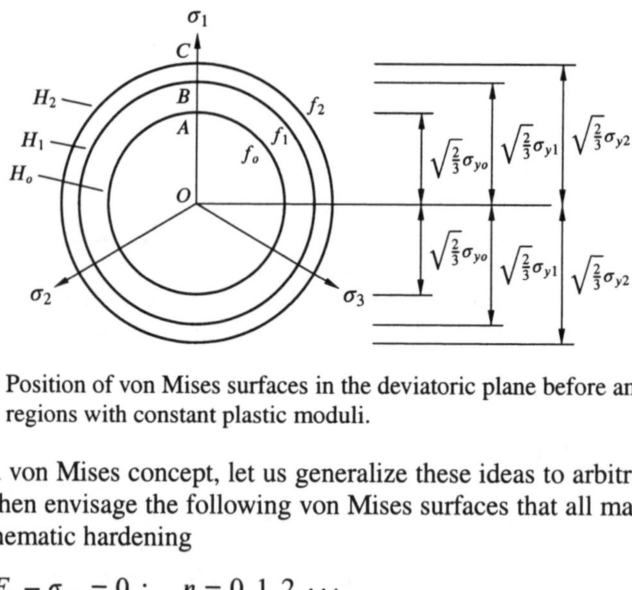

Figure 13.2: Position of von Mises surfaces in the deviatoric plane before any loading; regions with constant plastic moduli.

Within a von Mises concept, let us generalize these ideas to arbitrary stress states. We then envisage the following von Mises surfaces that all may be subjected to kinematic hardening

$$f_n = F_n - \sigma_{yn} = 0 \; ; \quad n = 0, 1, 2, \cdots \tag{13.2}$$

where the function F_n is defined by

$$F_n = [\frac{3}{2}(s_{ij} - \alpha_{ij}^{d(n)})(s_{ij} - \alpha_{ij}^{d(n)})]^{1/2} \; ; \quad n = 0, 1, 2, \cdots \tag{13.3}$$

and $\alpha_{ij}^{d(n)}$ denotes the deviatoric part of the back-stress belonging to the surface f_n with size σ_{yn}. Before any loading, all back-stresses $\alpha_{ij}^{d(n)}$ are zero and in the deviatoric plane, the surfaces f_n then take the form shown in Fig. 13.2 where

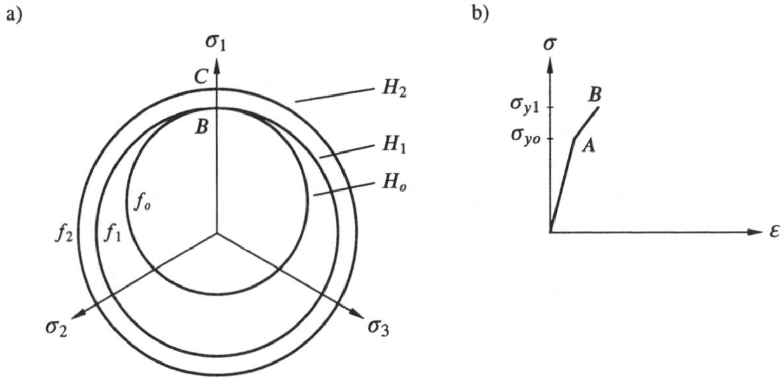

Figure 13.3: Increasing uniaxial loading; a) deviatoric plane, b) uniaxial response.

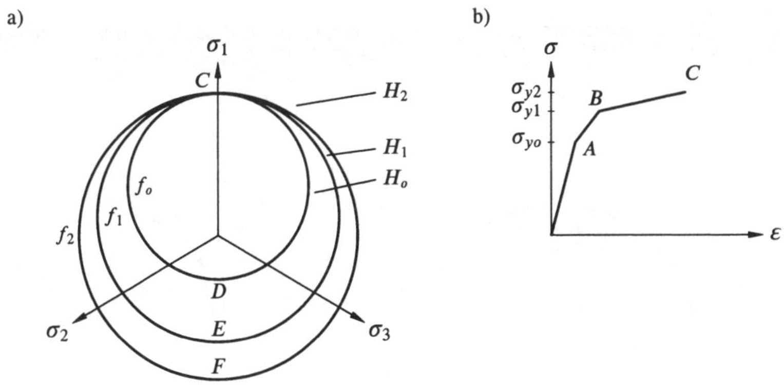

Figure 13.4: Increasing uniaxial loading; a) deviatoric plane, b) uniaxial response.

the radii are given by $\sqrt{2J_2} = \sqrt{\frac{2}{3}}\sigma_{yn}$, cf. (8.13). In the regions between the circles, different constant plastic moduli apply.

Referring to Fig. 13.2, let us now evaluate the model for uniaxial loading. When the stress point moves from the origin O along the vertical axis, it reaches the initial yield stress σ_{y0} at point A and the circle f_0 moves as a rigid body along this axis until it contacts and touches the circle f_1 at point B; all other circles remain fixed during this process and the situation when the stress has reached point B is shown in Fig. 13.3a). The corresponding stress-strain curve is shown in Fig. 13.3b) where it is recalled that between A and B, the plastic modulus is constant and given by H_0.

When the stress point moves from B towards C, the circles f_0 and f_1 translate together until point C is reached. At that state, f_0 and f_1 touch the circle f_2, which up until now remained at rest, cf. Fig. 13.4a). Between point B and

C, the plastic modulus is constant and given by H_1, i.e. we have the response shown in Fig. 13.4b). This type of process will continue for increased loading.

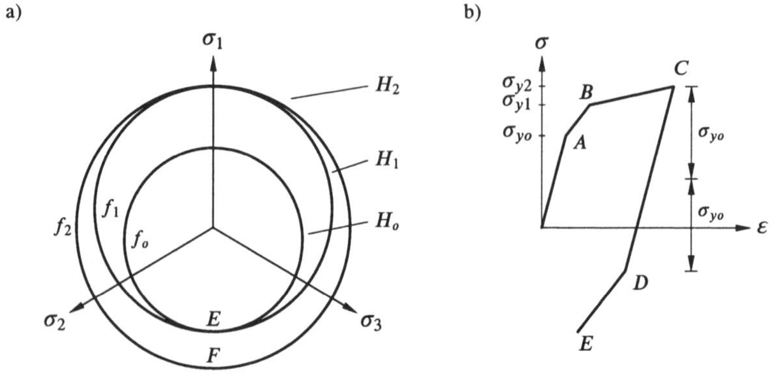

Figure 13.5: Reversed uniaxial loading; a) deviatoric plane, b) uniaxial response.

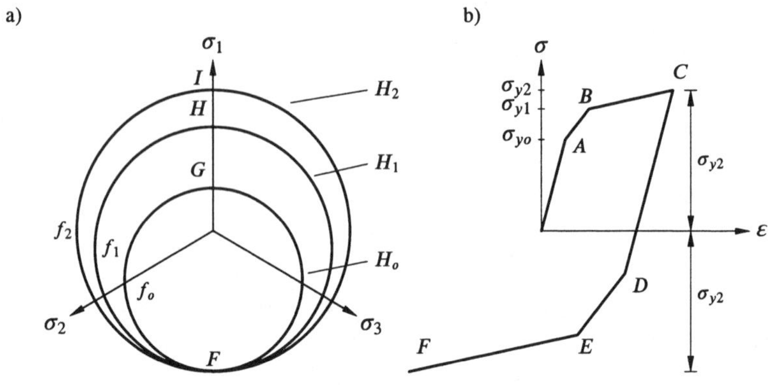

Figure 13.6: Reversed uniaxial loading; a) deviatoric plane, b) uniaxial response.

Consider now the process of reversed loading. When moving from point C towards point D, cf. Fig. 13.4a), an elastic response occurs. When the stress point reaches point D, the circle f_0 translates downwards until it reaches the circle f_1 at point E. This situation is shown in Fig. 13.5a) and since the plastic modulus between circle f_0 and f_1 is given by H_0, we have the same slope between D and E as between A and B, cf. Fig. 13.5b). Note that the stress difference between point D and E equals twice the stress difference between point A and B. When the stress point moves from point E towards F, cf. Fig. 13.5a), the two circles f_0 and f_1 move together and the relevant plastic modulus is now that applicable between circle f_1 and f_2, i.e. H_1. The situation where point F has been reached is shown in Fig. 13.6a) and the corresponding stress-strain

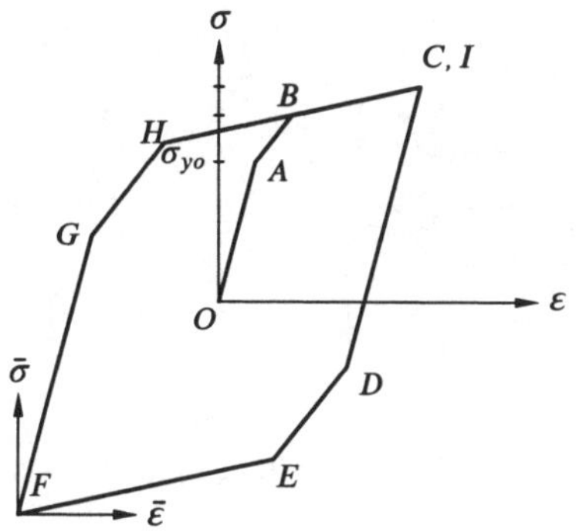

Figure 13.7: Reversed uniaxial response.

curve is given in Fig. 13.6b). Again the slope between E and F is the same as between B and C; moreover, the stress difference between E and F is twice the stress difference between B and C. At point F the stress is $\sigma = -\sigma_{y2}$ and it is recalled that at point C the stress is $\sigma = \sigma_{y2}$. Since the stress difference between $C-D$, $D-E$ and $E-F$ is twice the difference between $O-A$, $A-B$ and $B-C$, respectively, the strains at point F and at point C are related by $\varepsilon_F = -\varepsilon_C$.

When we reverse the loading from point F, cf. Figs. 13.6 and 13.7, the process is repeated, i.e. FG, GH and HI corresponds to CD, DE and EF, respectively; therefore, point I will coincide with point C. Let us next locate a $\bar{\sigma}\bar{\varepsilon}$–coordinate system at point F, cf. Fig. 13.7. It is then apparent that if the stress-strain curve $OABC$ is described by $\sigma = f(\varepsilon)$, then the curve $FGHI$ is described by

$$\boxed{\frac{1}{2}\bar{\sigma} = f(\frac{1}{2}\bar{\varepsilon}) \; ; \quad \textit{Masing's rule}} \tag{13.4}$$

This expression corresponds to the so-called *Masing's rule* and as observed by Masing (1927) it provides a fairly close approximation to the real behavior of metals and steel. If we now reverse the loading from point C, we will again follow path $CDEF$ and it is apparent that Fig. 13.7 corresponds both to symmetric stress cycling and symmetric strain cycling; it appears that no cyclic hardening or softening effects are predicted, cf. Figs. 12.2 and 12.3, and the stabilized cyclic stress-strain curve is obtained already after one cycle. However, the predicted response during reversed loading is much closer to the real behavior of metals and steel than that predicted by the previous models.

Consider the bar consisting of the three layers, cf. Fig. 13.8; each layer is assumed to exhibit a perfectly plastic response. As all layers undergo the same deformation, it is evident that the response in uniaxial loading will be identical to that shown in Fig. 13.7 and this forms the basis for the *sublayer models* proposed by Besseling (1953) as well as by Iwan (1967). It was also used by Masing (1927) to derive relation (13.4).

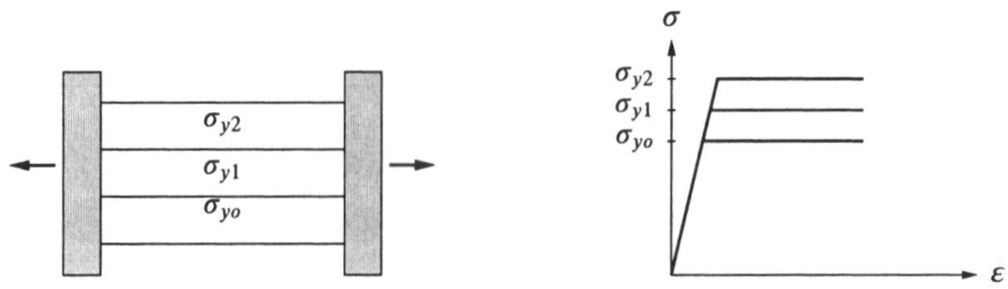

Figure 13.8: Sublayer model also resulting in the response shown in Fig. 13.7.

We will now see how the concepts above can be generalized to arbitrary non-proportional loading. We will also allow the yield functions to be written in the quite general form

$$\boxed{\begin{aligned} &f_n = F_n - (\sigma_{yn})^p = 0 \ ; \quad n = 0, 1, 2, \cdots \\ &where \quad F_n = F(\sigma_{ij} - \alpha_{ij}^{(n)}) \end{aligned}} \tag{13.5}$$

As usual, the parameters σ_{yn} have the dimension of stress. To ensure that F_n and $(\sigma_{yn})^p$ possess the same dimension, the function F is required to be a *homogeneous function of degree p* in the variables $\sigma_{ij} - \alpha_{ij}^{(n)}$. This means that for any number 'a' we have

$$\boxed{F(a(\sigma_{ij} - \alpha_{ij}^{(n)})) = a^p F(\sigma_{ij} - \alpha_{ij}^{(n)}) \ ; \quad n = 0, 1, 2, \cdots} \tag{13.6}$$

cf. for instance Sokolnikoff and Redheffer (1958) p.234. As an example, in the von Mises case where F is given by (13.3), F is a homogeneous function of degree one; in that case $p = 1$ and (13.5) reduces to (13.2).

In the general stress space, the active yield surface f_n is moving and the next yield surface f_{n+1} is fixed in the stress space, cf. Fig. 13.9. Here point O is the origin of the stress space, O_n is the center of the active yield surface f_n and O_{n+1} is the center of the next yield surface f_{n+1}.

The problem we are facing is to devise a rule for how the surface f_n is moving, i.e. we have to identify the evolution law for $\dot{\alpha}_{ij}^{(n)}$. For this purpose, the following two criteria are used:

- The active surface f_n is not allowed to intersect the surface f_{n+1} at any point. At contact, the smaller surface f_n must therefore touch the larger surface f_{n+1} tangentially.

- After tangential contact has been established, the two surfaces move together and approach the next surface f_{n+2}. The two surfaces f_n and f_{n+1} are detached only after elastic unloading and subsequent plastic reloading in a new direction.

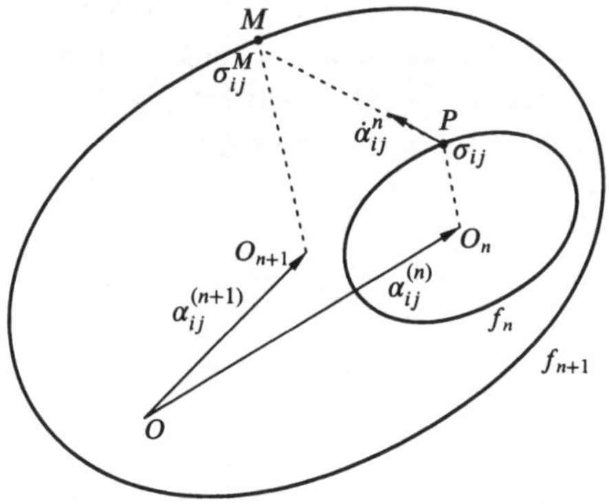

Figure 13.9: Stress space illustrating the general situation; f_n = active yield surface, f_{n+1} = next yield surface.

The notion of *nesting yield surfaces* is often used in connection with the Mróz model and this terminology hinges on the fulfillment of the criteria just discussed. Before the evolution law for $\dot{\alpha}_{ij}^{(n)}$ can be identified, some preliminary results shall be established.

Suppose that the current stress state σ_{ij} is located at point P, cf. Fig. 13.9. Define next a stress state σ_{ij}^M located on the surface f_{n+1} at point M such that the vector $O_{n+1}M$ is parallel with the vector O_nP. This means that

$$\boxed{\sigma_{ij}^M - \alpha_{ij}^{(n+1)} = a(\sigma_{ij} - \alpha_{ij}^{(n)}) \; ; \quad a > 1} \tag{13.7}$$

where a is some proportionality factor larger than one. For reasons that will become apparent in a moment, σ_{ij}^M is called the *image stress* or *mapping stress*. To determine the factor a present in (13.7), we take that the mapping point M is located on the yield surface f_{n+1} whereas the the current stress point P is located on the active surface f_n. From (13.5), we then obtain

$$\frac{F(\sigma_{ij}^M - \alpha_{ij}^{(n+1)})}{F(\sigma_{ij} - \alpha_{ij}^{(n)})} = (\frac{\sigma_{y(n+1)}}{\sigma_{yn}})^p \tag{13.8}$$

Moreover, (13.7) and (13.6) give

$$F(\sigma_{ij}^M - \alpha_{ij}^{(n+1)}) = a^p F(\sigma_{ij} - \alpha_{ij}^{(n)}) \tag{13.9}$$

Insertion into (13.8) provides

$$\boxed{a = \frac{\sigma_{y(n+1)}}{\sigma_{yn}}} \tag{13.10}$$

With this result, (13.7) determines the mapping stress σ_{ij}^M uniquely.

Intuitively, and referring to Fig. 13.9, it appears that the normal to the surface f_{n+1} at point M and the normal to the surface f_n at point P are parallel. To prove this formally, we observe that (13.7) and (13.10) imply that to each stress σ_{ij} on the surface f_n there exists a stress σ_{ij}^M on the surface f_{n+1}. From (13.5), (13.9) and (13.7), we then have

$$\frac{\partial f_{n+1}}{\partial \sigma_{ij}^M} = a^p \frac{\partial F(\sigma_{st} - \alpha_{st}^{(n)})}{\partial \sigma_{kl}} \frac{\partial \sigma_{kl}}{\partial \sigma_{ij}^M} = a^{p-1} \frac{\partial F(\sigma_{st} - \alpha_{st}^{(n)})}{\partial \sigma_{ij}}$$

Moreover, (13.5) gives

$$\frac{\partial f_n}{\partial \sigma_{ij}} = \frac{\partial F(\sigma_{st} - \alpha_{st}^{(n)})}{\partial \sigma_{ij}}$$

It appears that

$$\frac{\partial f_{n+1}}{\partial \sigma_{ij}^M} = a^{p-1} \frac{\partial f_n}{\partial \sigma_{ij}} \tag{13.11}$$

i.e. the normals at point M and at point P are parallel.

Referring to Fig. 13.9, we then conclude that if $\dot{\alpha}_{ij}^{(n)}$ is always directed from the current stress point P towards the corresponding mapping point M, then the two surfaces will contact each other at the mapping point M. On contact, the current stress point P and the mapping point M coincide. Since the normals at point P and at point M are parallel, this contact will occur tangentially without any intersection of the two surfaces. We are then led to the *kinematic evolution law of Mróz* that reads

$$\boxed{\dot{\alpha}_{ij}^{(n)} = \dot{\lambda}\, q(\sigma_{ij}^M - \sigma_{ij}) \;; \quad \textit{evolution law of Mróz}} \tag{13.12}$$

where q is a positive quantity. To be precise, Mróz (1967) did not introduce the plastic multiplier $\dot{\lambda}$ at this point. However, in order to be consistent with the theoretical framework adopted here, we will deviate somewhat from the exposition by Mróz and simply note that the resulting equations are identical. It is also observed that (13.12) fits into the framework (10.15).

With expression (13.7), we find that

$$\boxed{\sigma_{ij}^M - \sigma_{ij} = (a-1)(\sigma_{ij} - \alpha_{ij}^{(n)}) + \alpha_{ij}^{(n+1)} - \alpha_{ij}^{(n)}}$$

where all terms on the right-hand side are known quantities. Therefore, in the evolution law (13.12) it remains to identify the quantity q and we will return to this shortly. Before that we observe that if $\alpha_{ij}^{(n)} = \alpha_{ij}^{(n+1)}$ then $\sigma_{ij}^M - \sigma_{ij}$ becomes proportional to $\sigma_{ij} - \alpha_{ij}^{(n)}$, i.e. the evolution law (13.12) becomes similar to

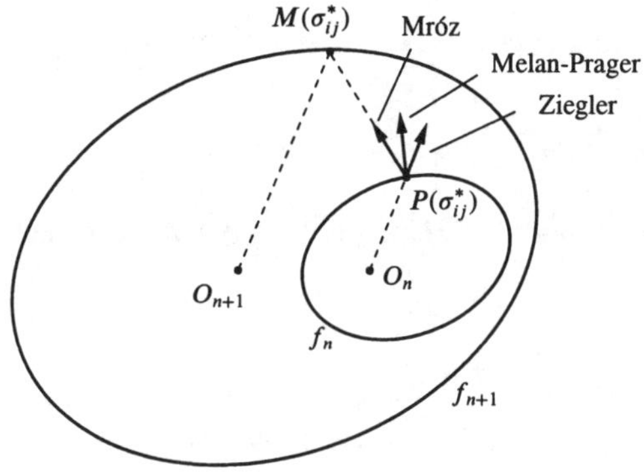

Figure 13.10: Comparison of different kinematic evolution laws; associated plasticity.

Ziegler's evolution law (12.82). In the general situation, however, the evolution laws of Melan-Prager, Ziegler and Mróz are compared in Fig. 13.10.

For clarity, let us recall that:

The largest yield surface that moves in the stress space is called the active yield surface and it is denoted by f_n

The non-associated flow rule reads

$$\boxed{\dot{\varepsilon}_{ij}^{p} = \dot{\lambda}\frac{\partial g}{\partial \sigma_{ij}}} \tag{13.13}$$

where g, as usual, denotes the potential function, which is equal to f_n for associated plasticity. Moreover, the consistency relation takes the form

$$\dot{f}_n = \frac{\partial f_n}{\partial \sigma_{ij}}\dot{\sigma}_{ij} + \frac{\partial f_n}{\partial \alpha_{ij}^{(n)}}\dot{\alpha}_{ij}^{(n)} = 0 \tag{13.14}$$

In view of (13.5), it follows that

$$\frac{\partial f_n}{\partial \sigma_{ij}} = \frac{\partial f_n}{\partial(\sigma_{kl} - \alpha_{kl}^{(n)})}\frac{\partial(\sigma_{kl} - \alpha_{kl}^{(n)})}{\partial \sigma_{ij}} = \frac{\partial f_n}{\partial(\sigma_{ij} - \alpha_{ij}^{(n)})}$$

$$\frac{\partial f_n}{\partial \alpha_{ij}^{(n)}} = \frac{\partial f_n}{\partial(\sigma_{kl} - \alpha_{kl}^{(n)})}\frac{\partial(\sigma_{kl} - \alpha_{kl}^{(n)})}{\partial \alpha_{ij}^{(n)}} = -\frac{\partial f_n}{\partial(\sigma_{ij} - \alpha_{ij}^{(n)})}$$

i.e.

$$\frac{\partial f_n}{\partial \alpha_{ij}^{(n)}} = -\frac{\partial f_n}{\partial \sigma_{ij}} \tag{13.15}$$

Insertion into (13.14) and use of (13.12) then give

$$\boxed{\frac{\partial f_n}{\partial \sigma_{ij}}\dot{\sigma}_{ij} - H_n\dot{\lambda} = 0} \tag{13.16}$$

where the plastic modulus H_n applicable between surface f_n and f_{n+1} is given by

$$H_n = q\frac{\partial f_n}{\partial \sigma_{ij}}(\sigma_{ij}^M - \sigma_{ij})$$

It appears that the quantity q present in the evolution law (13.12) is given by

$$\boxed{q = \frac{H_n}{\frac{\partial f_n}{\partial \sigma_{ij}}(\sigma_{ij}^M - \sigma_{ij})}} \tag{13.17}$$

As discussed previously, the fundamental concept of the Mróz model is the multilinear response; therefore, we have

$$\boxed{H_n = constant}$$

In the Mróz model, the plastic moduli H_n ($n = 0, 1, 2, \cdots$) are chosen. For the active yield surface in question, (13.17) then determines the corresponding quantity q and the evolution law (13.12) is then identified completely. All fundamental equations in the Mróz model have then been established.

In relation to Fig. 13.7, it was observed that the stabilized cyclic stress-strain curve was obtained already after one cycle; this means that neither softening nor hardening effects are present. As discussed by Mróz (1967), the model described above may be generalized to consider such effects. Instead of (13.1), we now allow that the yield stresses σ_{yn} may vary, i.e. we have $\sigma_{yn} = \sigma_{yn}(\kappa)$, where κ is an internal variable. For instance, we may choose $\kappa = \lambda$, i.e.

$$\sigma_{yn} = \sigma_{yn}(\lambda) \ ; \quad n = 0, 1, 2, \cdots \tag{13.18}$$

The evolution law (13.12) is unchanged, but (13.5) and (13.18) now imply the following consistency relation

$$\dot{f}_n = \frac{\partial f_n}{\partial \sigma_{ij}}\dot{\sigma}_{ij} + \frac{\partial f_n}{\partial \alpha_{ij}^{(n)}}\dot{\alpha}_{ij}^{(n)} + \frac{\partial f_n}{\partial \lambda}\dot{\lambda} = 0 \tag{13.19}$$

where

$$\frac{\partial f_n}{\partial \lambda} = -p(\sigma_{yn})^{p-1}\frac{d\sigma_{yn}(\lambda)}{d\lambda}$$

With this expression and (13.12), (13.19) becomes

$$\frac{\partial f_n}{\partial \sigma_{ij}}\dot{\sigma}_{ij} - H_n\dot{\lambda} = 0$$

where the plastic modulus H_n is defined by

$$H_n = q\frac{\partial f_n}{\partial \sigma_{ij}}(\sigma_{ij}^M - \sigma_{ij}) + p(\sigma_{yn})^{p-1}\frac{d\sigma_{yn}(\lambda)}{d\lambda}$$

The quantity q is then defined by

$$q = \frac{H_n - p(\sigma_{yn})^{p-1}\frac{d\sigma_{yn}(\lambda)}{d\lambda}}{\frac{\partial f_n}{\partial \sigma_{ij}}(\sigma_{ij}^M - \sigma_{ij})} \tag{13.20}$$

In addition to the expression $\sigma_{yn} = \sigma_{yn}(\lambda)$, we choose as before a constant value for the plastic modulus H_n and (13.20) then determines the quantity q to be used in the evolution law (13.12).

13.1.1 von Mises yield function

Let us finally illustrate the pertinent equations for associated plasticity when a von Mises concept is adopted. For simplicity, the yield stresses σ_{yn} are taken as constants. The main thing that we want to discuss is the interpretation of the plastic modulus H_n. With (13.5) where $p = 1$ and F_n is given by (13.3), the flow rule (13.13) becomes

$$\dot{\varepsilon}_{ij}^p = \dot{\lambda}\frac{3(\sigma_{ij} - \alpha_{ij}^{d(n)})}{2\sigma_{yn}} \tag{13.21}$$

As usual, define the effective plastic strain rate by

$$\dot{\varepsilon}_{eff}^p = (\frac{2}{3}\dot{\varepsilon}_{ij}^p\dot{\varepsilon}_{ij}^p)^{1/2} \tag{13.22}$$

For uniaxial stressing, we have $\dot{\varepsilon}_{22}^p = \dot{\varepsilon}_{33}^p = -\dot{\varepsilon}_{11}^p/2$, i.e. $\dot{\varepsilon}_{eff}^p = |\dot{\varepsilon}_{11}^p|$, as usual. Moreover, insertion of (13.21) into (13.22) shows that

$$\dot{\varepsilon}_{eff}^p = \dot{\lambda}$$

Multiply the consistency relation (13.16) by $\dot{\lambda}$ and use the flow rule (13.13) for associated plasticity to obtain

$$\dot{\varepsilon}_{ij}^p\dot{\sigma}_{ij} = (\dot{\varepsilon}_{eff}^p)^2 H_n$$

For uniaxial stress conditions, the term $\dot{\varepsilon}_{ij}^p\dot{\sigma}_{ij}$ reduces to $\dot{\varepsilon}_{11}^p\dot{\sigma}_{11}$, i.e.

$$\frac{d\sigma_{11}}{d\varepsilon_{11}^p} = H_n$$

just like in the previous von Mises models.

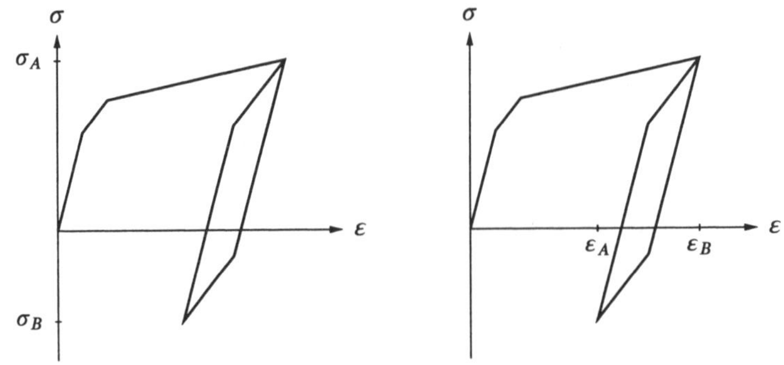

Figure 13.11: a) unsymmetrical stress cycling; b) unsymmetrical strain cycling.

It also readily follows that the elasto-plastic stiffness matrix $\boldsymbol{D}^{ep}$ is completely similar to that applicable to linear kinematic hardening, cf. (12.51).

For symmetric stress and strain cycles, the response has already been illustrated in Fig. 13.7. For cycling between unsymmetric stress states, the result in Fig. 13.11a) is obtained; no ratcheting is predicted and the stabilized cyclic stress-strain curve is already obtained after one cycle. Similarly, Fig. 13.11b) shows unsymmetrical strain cycling; no mean stress relaxation is predicted and the stabilized cyclic stress-strain curve is obtained after one cycle, and since it is symmetric, there is no difference between stress cycling and strain cycling. Despite these deficiencies, it is evident that the Mróz model is much better suited for cyclic loadings than the previously considered models. This is especially true if the multilinear approach contains many small segments such that a smooth curved response is approached. However, in a computer program this increases the storage requirements significantly, since each yield surface requires the storage of the six components of $\alpha_{ij}^{(n)}$ (as well as of σ_{yn} if this yield stress is allowed to vary).

13.2 Bounding surface models

In the Mróz model, the plastic modulus is constant in the region between two yield surfaces; this in turn leads to a multilinear response. We will now discuss an interesting formulation which makes use of only two yield surfaces and where the plastic modulus varies continuously between these two surfaces. This formulation forms the basis for many recent models and we will therefore provide a rather detailed discussion.

The concept we will discuss is the *bounding surface model* proposed simultaneously by Dafalias and Popov (1975, 1976) and Krieg (1975). Since the model of Dafalias and Popov was formulated in a more general manner, we will

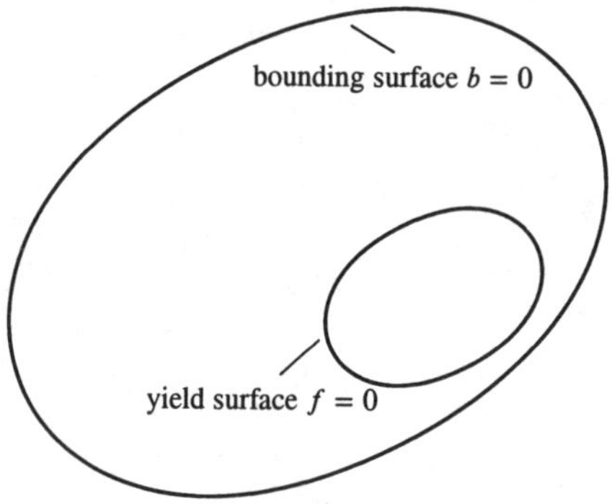

Figure 13.12: Yield surface and bounding surface in stress space.

here focus on this model.

The model contains two surfaces: a *bounding surface* and a *yield surface*, cf. Fig. 13.12. For evident reasons, the model is often called a *two-surface plasticity model.* During plastic loading, both the bounding surface and the yield surface may move and change size and they may come into contact with each other; however, the bounding surface always encloses the yield surface. The region inside the yield surface corresponds to an elastic response and plastic loading requires the current stress point to be located on the yield surface. The key issue is that it is the proximity of the two surfaces that determines the plastic modulus.

Let the yield surface f and the bounding surface b be defined by

$$\boxed{\begin{aligned} f &= F(\sigma_{ij} - \alpha_{ij}) - (\sigma_y)^p = 0 \quad \textit{yield surface} \\ b &= F(\sigma_{ij} - \beta_{ij}) - (\sigma_b)^p = 0 \quad \textit{bounding surface} \end{aligned}} \qquad (13.23)$$

Here, the back-stress α_{ij} denotes the center of the yield surface whereas the back-stress β_{ij} denotes the center of the bounding surface. The parameters σ_y and σ_b, which have the dimensions of stress, depend in general on some internal variables κ_α, i.e.

$$\boxed{\sigma_y = \sigma_y(\kappa_\alpha) \ ; \quad \sigma_b = \sigma_b(\kappa_\alpha)} \qquad (13.24)$$

In order that the function F appearing in (13.23) possess the correct dimension, it is required that F be a homogeneous function of degree p in the variables $\sigma_{ij} - \alpha_{ij}$ (or $\sigma_{ij} - \beta_{ij}$). Referring to (13.6), this means that

$$\boxed{F(a(\sigma_{ij} - \alpha_{ij})) = a^p F(\sigma_{ij} - \alpha_{ij})} \qquad (13.25)$$

where a is a quantity.

To determine the rules for how the back-stresses α_{ij} and β_{ij} change, we make use of the concepts introduced when the Mróz model was discussed. In the

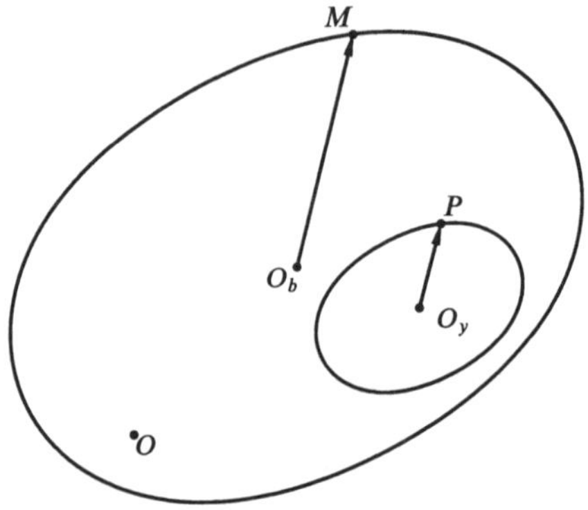

Figure 13.13: Current stress point P and mapping stress point M.

general stress space shown in Fig. 13.13, O denotes the origin of the stress space, O_y is the center of the yield surface and O_b is the center of the bounding surface. Assume that the current stress state σ_{ij} is located at point P on the yield surface. Define the mapping stress σ_{ij}^M located on the bounding surface at point M, where the point M is determined by the requirement that vector O_bM is parallel with vector O_yP. This means that

$$\boxed{\sigma_{ij}^M - \beta_{ij} = a(\sigma_{ij} - \alpha_{ij}) \ ; \quad a > 1} \tag{13.26}$$

where a is a proportionality factor. To determine this factor, we have from (13.23)

$$\frac{F(\sigma_{ij}^M - \beta_{ij})}{F(\sigma_{ij} - \alpha_{ij})} = (\frac{\sigma_b}{\sigma_y})^p \tag{13.27}$$

Moreover, (13.25) and (13.26) give

$$F(\sigma_{ij}^M - \beta_{ij}) = a^p F(\sigma_{ij} - \alpha_{ij})$$

Insertion into (13.27) provides

$$\boxed{a = \frac{\sigma_b}{\sigma_y}} \tag{13.28}$$

With this result, (13.26) determines the mapping stress σ_{ij}^M uniquely. From (13.26), we have

$$\boxed{\sigma_{ij}^M - \sigma_{ij} = (a-1)(\sigma_{ij} - \alpha_{ij}) + \beta_{ij} - \alpha_{ij}} \tag{13.29}$$

Similar to the discussion of the Mróz model, we conclude that the normals at point P and point M are parallel, cf. (13.11). Following again the discussion of the Mróz model, it is required that the bounding surface and the yield surface never intersect and that contact occurs tangentially. This is achieved if the motion of the yield surface relative to the motion of the bounding surface occurs in the direction of vector $\boldsymbol{PM}$, cf. Fig. 13.13. Therefore

$$\boxed{\dot{\alpha}_{ij} - \dot{\beta}_{ij} = \dot{\lambda}\, q^*(\sigma_{ij}^M - \sigma_{ij})} \tag{13.30}$$

where q^* is a positive quantity to be determined later. To be precise, Dafalias and Popov (1975, 1976) did not introduce the plastic multiplier $\dot{\lambda}$ at this point. However, in order to be consistent with the theoretical framework adopted here, we will deviate somewhat from the exposition of Dafalias and Popov and simply note that the resulting equations are identical.

The non-associated flow rule reads

$$\boxed{\dot{\varepsilon}_{ij}^p = \dot{\lambda}\frac{\partial g}{\partial \sigma_{ij}}} \tag{13.31}$$

where g, as usual, denotes the potential function, which is equal to f for associated plasticity. With (13.23) and (13.24), the consistency relation becomes

$$\dot{f} = \frac{\partial f}{\partial \sigma_{ij}}\dot{\sigma}_{ij} + \frac{\partial f}{\partial \alpha_{ij}}\dot{\alpha}_{ij} + \frac{\partial f}{\partial \kappa_\alpha}\dot{\kappa}_\alpha = 0$$

Similar to (13.15), we have $\partial f/\partial \alpha_{ij} = -\partial f/\partial \sigma_{ij} = -\partial F/\partial \sigma_{ij}$ and we then obtain

$$\frac{\partial F}{\partial \sigma_{ij}}\dot{\sigma}_{ij} - \frac{\partial F}{\partial \sigma_{ij}}\dot{\alpha}_{ij} - p(\sigma_y)^{p-1}\frac{\partial \sigma_y}{\partial \kappa_\alpha}\dot{\kappa}_\alpha = 0 \tag{13.32}$$

In order to proceed, we have, as usual, to specify the evolution laws for $\dot{\alpha}_{ij}$ and $\dot{\kappa}_\alpha$ and insert these expressions into the consistency relation (13.32). Most often, we have assumed that the evolution functions depend on the stresses σ_{ij} and the hardening parameters K_α, cf. (10.13). Referring to (13.23), we have two sets of hardening parameters for the yield criterion: one given by the back-stress α_{ij} and the other relating to the quantity σ_y. As usual, we may write $\sigma_y = \sigma_y(\kappa_\alpha)$ as

$$\boxed{\sigma_y = \sigma_{yo} + K(\kappa_\alpha)}$$

where σ_{yo} is some initial yield stress and K is the hardening parameter related to σ_y. In accordance with (10.13), we therefore take

$$\boxed{\dot{\kappa}_\alpha = \dot{\lambda} k_\alpha \quad where \quad k_\alpha = k_\alpha(\sigma_{ij} - \alpha_{ij}, K(\kappa_\beta))} \tag{13.33}$$

For the evolution function relating to $\dot{\alpha}_{ij}$ and in accordance with the similar discussion following (10.13), we shall allow the following more general formulation

$$\boxed{\dot{\alpha}_{ij} = \dot{\lambda} q_{ij}^{(y)} \quad where \quad q_{ij}^{(y)} = q_{ij}^{(y)}(\sigma_{ij} - \alpha_{ij}, K(\kappa_\beta), \delta, \delta_{in})} \tag{13.34}$$

where superscript (y) indicates that the back-stress α_{ij} is related to the yield surface. The variables δ and δ_{in} are so-called *discrete memory parameters*. The discrete memory parameter δ is defined as the distance between the mapping stress point σ_{ij}^M and the current stress point σ_{ij}, i.e.

$$\boxed{\delta = [(\sigma_{ij}^M - \sigma_{ij})(\sigma_{ij}^M - \sigma_{ij})]^{1/2}} \tag{13.35}$$

According to (13.29), the discrete memory parameter δ is known at every state during plastic loading. The discrete memory parameter δ_{in} is defined as δ at initiation of a plastic loading process:

$$\boxed{\begin{aligned}\delta_{in} \;&=\; \textit{initial value of } \delta\textit{, i.e. the value of } \delta \textit{ every}\\ &\;\; \textit{time a plastic loading process is initiated}\end{aligned}}$$

For continued plastic loading, δ_{in} is kept constant, i.e. δ_{in} is only updated every time an elastic response is followed by plastic loading.

Whereas the definition of δ is unique, the definition of δ_{in} is more vague and, in fact, several possibilities exist. We will return later to a more specific definition of δ_{in} and it will turn out that a consistent definition of δ_{in} that works for general three-dimensional loading is not trivial; in fact, we will see that whereas the bounding surface model of Dafalias and Popov (1975, 1976) exhibits a number of appealing properties, the main drawback relates to a reasonable definition of δ_{in}.

Before we enter into a more detailed discussion of the definition of δ_{in}, we observe that δ is only defined during plastic loading whereas δ_{in} is constant during plastic loading and it is only updated when an elastic response is followed by plastic response. Therefore, both δ and δ_{in} may vary discontinuously and this is the reason for the term: *discrete memory parameters*; they memorize the load history and they do that in a discrete manner. We emphasize that δ and δ_{in} are not internal variables since internal variables only change during plastic loading, cf. (10.13). Considering, for instance, δ_{in} this quantity takes one constant value at initiation of plasticity; after elastic unloading and subsequent plastic reloading, δ_{in} takes another constant value. This means that δ_{in} is changed because of elastic unloading and, consequently, δ_{in} cannot be an internal variable.

After this discussion of the evolution laws for $\dot{\kappa}_\alpha$ and $\dot{\alpha}_{ij}$, we insert (13.33) and (13.34) into the consistency relation (13.32) to obtain

$$\boxed{\frac{\partial F}{\partial \sigma_{ij}}\dot{\sigma}_{ij} - H\dot{\lambda} = 0} \tag{13.36}$$

where the plastic modulus H is defined by

$$\boxed{H = H^{kin} + H^{iso}} \tag{13.37}$$

and where H^{kin} and H^{iso} are defined by

$$\boxed{H^{kin} = \frac{\partial F}{\partial \sigma_{ij}} q_{ij}^{(y)} \;; \qquad H^{iso} = p(\sigma_y)^{p-1} \frac{\partial \sigma_y}{\partial \kappa_\alpha} k_\alpha} \tag{13.38}$$

It appears that the plastic modulus has been split into two parts: one related to the kinematic hardening and another related to the isotropic hardening.

The fundamental issue of the bounding surface theory of Dafalias and Popov (1975, 1976) will now be introduced. It is assumed that the plastic modulus H depends on the current distance δ as well as on the initial distance δ_{in}:

$$\boxed{H = H(\delta, \delta_{in})} \tag{13.39}$$

Moreover, the plastic modulus H is specified by us, i.e. we choose the expression indicated by (13.39); we will later return to a suitable explicit expression. In (13.37), we also choose the quantity H^{iso}, i.e. how σ_y varies with the internal variables κ_α. Moreover, we also choose the direction of $\dot{\alpha}_{ij}$ that means the direction of $q_{ij}^{(y)}$. Having chosen H, H^{iso} and the direction of $q_{ij}^{(y)}$, (13.37) and (13.38) finally determine the magnitude of $q_{ij}^{(y)}$. We may summarize this discussion by

$H = H(\delta, \delta_{in})$ and H^{iso} are chosen by us. Also the direction of $q_{ij}^{(y)}$ is specified by us. Expressions (13.37) and (13.38) then determine the magnitude of $q_{ij}^{(y)}$. In turn, the evolution law for $\dot{\alpha}_{ij}$ has then been determined completely.	(13.40)

It is of interest that this format allows any kinematic hardening rule to be chosen for $\dot{\alpha}_{ij}$. For instance, we may choose the direction of $\dot{\alpha}_{ij}$ according to Melan-Prager's, Ziegler's or Mróz's evolution laws, to which we will later return.

When the yield surface and the bounding surface are in contact with each other, $\delta = 0$ holds. The corresponding value of H turns out to be of particular interest and we use the following notation:

$$\boxed{H_o = H(\delta = 0, \delta_{in})} \tag{13.41}$$

This means that the plastic modulus applicable when the yield surface and bounding surface are in contact is the plastic modulus H_o.

According to the discussion relating to (13.40), any kinematic hardening rule can be chosen for the direction of $\dot{\alpha}_{ij}$. In general, however, the bounding surface also moves in the stress space and (13.30) relates the back-stress rates $\dot{\alpha}_{ij}$ to $\dot{\beta}_{ij}$ via the unspecified quantity q^*. Let us therefore determine this quantity.

Similar to (13.34), the evolution law for $\dot{\beta}_{ij}$ can be written in the following general format

$$\dot{\beta}_{ij} = \dot{\lambda} q_{ij}^{(b)} \tag{13.42}$$

where superscript (b) indicates that the back-stress β_{ij} is related to the bounding surface. Multiplication of (13.30) by $\partial F/\partial\sigma_{ij}$ and use of the evolution laws (13.34) and (13.42) then lead to

$$\frac{\partial F}{\partial\sigma_{ij}}q_{ij}^{(y)} - \frac{\partial F}{\partial\sigma_{ij}}q_{ij}^{(b)} = q^*\frac{\partial F}{\partial\sigma_{ij}}(\sigma_{ij}^M - \sigma_{ij}) \tag{13.43}$$

Let us define the quantity $\tilde{H}$ by

$$\tilde{H} = \frac{\partial F}{\partial\sigma_{ij}}q_{ij}^{(b)} \tag{13.44}$$

With (13.38), (13.43) then gives

$$q^* = \frac{H^{kin} - \tilde{H}}{\frac{\partial F}{\partial\sigma_{ij}}(\sigma_{ij}^M - \sigma_{ij})} \tag{13.45}$$

It appears that q^* is known once the quantity $\tilde{H}$ is known. Let us now make the assumption that $\tilde{H}$ only depends on the initial distance δ_{in} as well as on the internal variables κ_α, i.e.

$$\tilde{H} = \tilde{H}(\delta_{in}, \kappa_\alpha) \tag{13.46}$$

To determine $\tilde{H}$, we apply that the mapping stress σ_{ij}^M is located on the bounding surface, i.e. (13.23) gives $b = F(\sigma_{ij}^M - \beta_{ij}) - (\sigma_b)^p = 0$. Since σ_{ij}^M is always located on the bounding surface, the consistency relation $\dot{b} = 0$ for the bounding surface then reads

$$\frac{\partial F(\sigma_{kl}^M - \beta_{kl})}{\partial\sigma_{ij}^M}\dot{\sigma}_{ij}^M + \frac{\partial F(\sigma_{kl}^M - \beta_{kl})}{\partial\beta_{ij}}\dot{\beta}_{ij} - p(\sigma_b)^{p-1}\frac{\partial\sigma_b}{\partial\kappa_\alpha}\dot{\kappa}_\alpha = 0$$

As usual, cf. (13.15), we have $\partial F(\sigma_{kl}^M - \beta_{kl})/\partial\beta_{ij} = -\partial F(\sigma_{kl}^M - \beta_{kl})/\partial\sigma_{ij}^M$, which with (13.33) leads to

$$\frac{\partial F(\sigma_{kl}^M - \beta_{kl})}{\partial\sigma_{ij}^M}\dot{\sigma}_{ij}^M - \frac{\partial F(\sigma_{kl}^M - \beta_{kl})}{\partial\sigma_{ij}^M}\dot{\beta}_{ij} - p(\sigma_b)^{p-1}\frac{\partial\sigma_b}{\partial\kappa_\alpha}k_\alpha\dot{\lambda} = 0 \tag{13.47}$$

Moreover, similar to (13.11) we have

$$\frac{\partial F(\sigma_{kl}^M - \beta_{kl})}{\partial\sigma_{ij}^M} = a^{p-1}\frac{\partial F(\sigma_{kl} - \beta_{kl})}{\partial\sigma_{ij}} = a^{p-1}\frac{\partial F}{\partial\sigma_{ij}}$$

i.e. (13.47) takes the form

$$a^{p-1}\frac{\partial F}{\partial\sigma_{ij}}\dot{\sigma}_{ij}^M - a^{p-1}\tilde{H}\dot{\lambda} - p(\sigma_b)^{p-1}\frac{\partial\sigma_b}{\partial\kappa_\alpha}k_\alpha\dot{\lambda} = 0 \tag{13.48}$$

where advantage also was taken of (13.42) and (13.44). The expression above holds in general and it therefore also holds for the special case of contact between the yield surface and the bounding surface. On contact, we have $\dot{\sigma}_{ij}^M = \dot{\sigma}_{ij}$, i.e. $\partial F/\partial\sigma_{ij}\dot{\sigma}_{ij}^M = \partial F/\partial\sigma_{ij}\dot{\sigma}_{ij} = H\dot{\lambda}$ where (13.36) was used. On contact, the current distance $\delta = 0$ and the plastic modulus H then takes the value H_o, cf. (13.41). With these remarks and noting that $\tilde{H}$ is independent of the current distance δ, cf. (13.46), evaluation of (13.48) on contact leads to

$$\tilde{H} = H_o - p(\sigma_y)^{p-1}\frac{\partial\sigma_b}{\partial\kappa_\alpha}k_\alpha \tag{13.49}$$

where (13.28) was used. All terms on the right-hand side are known quantities that determine $\tilde{H}$ uniquely. Insertion of (13.49) into (13.45) and use of (13.37) then provide

$$\boxed{q^* = \frac{H - H_o - H^{iso} + p(\sigma_y)^{p-1}\frac{\partial\sigma_b}{\partial\kappa_\alpha}k_\alpha}{\frac{\partial F}{\partial\sigma_{ij}}(\sigma_{ij}^M - \sigma_{ij})}} \tag{13.50}$$

Once the quantity q^* has been determined, the evolution of $\dot{\beta}_{ij}$ is given by (13.30), i.e.

$$\boxed{\dot{\beta}_{ij} = \dot{\alpha}_{ij} - \dot{\lambda}\,q^*(\sigma_{ij}^M - \sigma_{ij})} \tag{13.51}$$

All quantities of interest in the Dafalias-Popov model have then been identified.

If neither the yield surface nor the bounding surface experience isotropic hardening, then the numerator of the expression for q^* reduces to $H - H_o$, which becomes zero on contact; during contact, we therefore have $\dot{\beta}_{ij} = \dot{\alpha}_{ij}$, as expected. However, if isotropic hardening is involved, we have $\dot{\beta}_{ij} \neq \dot{\alpha}_{ij}$ even during contact.

Let us return to the expression for the plastic modulus H given in general by (13.39). The explicit form of this expression is specified by us and, evidently, many choices are possible. A particularly simple and advantageous form was suggested by Dafalias and Popov (1976) and it reads

$$\boxed{H(\delta, \delta_{in}) = H_o + \frac{\delta}{\delta_{in} - \delta}h} \tag{13.52}$$

where both H_o and the positive parameter h are taken as constants. On contact, this expression provides $H = H_o$ as required; moreover, at every event where plastic loading is initiated, we have $\delta_{in} = \delta$, i.e. $H \to \infty$. This implies a smooth transition when going from an elastic to a plastic response and, as discussed later, this property is often very close to experimental evidence.

13.2.1 von Mises yield function

Let us below illustrate the Dafalias-Popov model for associated plasticity within a von Mises concept. Only kinematic hardening will be considered and (13.23) then reduces to

$$\begin{aligned} f &= [\tfrac{3}{2}(s_{ij} - \alpha_{ij}^d)(s_{ij} - \alpha_{ij}^d)]^{1/2} - \sigma_{yo} = 0 \\ b &= [\tfrac{3}{2}(s_{ij} - \beta_{ij}^d)(s_{ij} - \beta_{ij}^d)]^{1/2} - \sigma_{bo} = 0 \end{aligned} \tag{13.53}$$

where σ_{yo} and σ_{bo} are taken as constant quantities whereas α_{ij}^d and β_{ij}^d, as usual, denote the deviatoric parts of the back-stresses. The flow rule (13.31) then becomes

$$\dot{\varepsilon}_{ij}^p = \dot{\lambda}\frac{3(s_{ij} - \alpha_{ij}^d)}{2\sigma_{yo}} \tag{13.54}$$

As usual, we have

$$\dot{\lambda} = \dot{\varepsilon}_{eff}^p \quad \text{where} \quad \dot{\varepsilon}_{eff}^p = (\frac{2}{3}\dot{\varepsilon}_{ij}^p\dot{\varepsilon}_{ij}^p)^{1/2} \tag{13.55}$$

Since only kinematic hardening is considered, (13.37) gives

$$H^{kin} = H \tag{13.56}$$

Moreover, (13.50) gives

$$q^* = \frac{2\sigma_{yo}(H - H_o)}{3(s_{ij} - \alpha_{ij}^d)(s_{ij}^M - s_{ij})} \tag{13.57}$$

In general, δ is the distance between the mapping stress point σ_{ij}^M and the current stress point σ_{ij}, cf. (13.35). However, since the plastic response of a von Mises material is assumed to depend on deviatoric quantities, only, δ is redefined according to

$$\delta = [(s_{ij}^M - s_{ij})(s_{ij}^M - s_{ij})]^{1/2} \tag{13.58}$$

Melan-Prager's evolution law

Assume that Melan-Prager's evolution law (12.80) holds for the back-stress α_{ij}, i.e.

$$\dot{\alpha}_{ij} = c\,\dot{\varepsilon}_{ij}^p$$

With (13.54) and (13.34), we find that

$$\dot{\alpha}_{ij} = \dot{\alpha}_{ij}^d = \dot{\lambda}c\frac{3(s_{ij} - \alpha_{ij}^d)}{2\sigma_{yo}} \quad \text{i.e.} \quad q_{ij}^{(y)} = c\frac{3(s_{ij} - \alpha_{ij}^d)}{2\sigma_{yo}} \tag{13.59}$$

From (13.38) and (13.56) it is then concluded that

$$c = \frac{2}{3}H \tag{13.60}$$

The evolution law for $\dot{\beta}_{ij}$ is given by (13.51). According to (13.53) it is only the deviatoric part $\dot{\beta}^d_{ij}$ that is of interest. Therefore, use of (13.59), (13.60) and (13.57) in (13.51) give

$$\dot{\beta}^d_{ij} = \dot{\lambda}\frac{H}{\sigma_{yo}}(s_{ij} - \alpha^d_{ij}) - \dot{\lambda}\frac{2\sigma_{yo}(H - H_o)}{3(s_{kl} - \alpha^d_{kl})(s^M_{kl} - s_{kl})}(s^M_{ij} - s_{ij}) \tag{13.61}$$

Ziegler's evolution law

Assume next that Ziegler's evolution law (12.82) holds for the back-stress α_{ij}. Adopting the formulation given by (12.84), we then have

$$\dot{\alpha}_{ij} = \dot{\lambda}k(\sigma_{ij} - \alpha_{ij}) \qquad \text{i.e.} \qquad q^{(y)}_{ij} = k(\sigma_{ij} - \alpha_{ij})$$

From (13.38) and (13.56) follow that

$$k = \frac{H}{\sigma_{yo}}$$

i.e.

$$\alpha^d_{ij} = \dot{\lambda}\frac{H}{\sigma_{yo}}(s_{ij} - \alpha^d_{ij})$$

Comparing with (13.59) and observing (13.60), it appears that we are back to the Melan-Prager formulation; this is certainly to be expected, cf. (12.87).

Mróz' evolution law

Finally, assume that Mróz' evolution law (13.12) holds for the back-stress α_{ij}, i.e.

$$\dot{\alpha}_{ij} = \dot{\lambda}\, q(\sigma^M_{ij} - \sigma_{ij}) \qquad \text{i.e.} \qquad q^{(y)}_{ij} = q(\sigma^M_{ij} - \sigma_{ij}) \tag{13.62}$$

With (13.38) and (13.56), we then obtain

$$q = \frac{2\sigma_{yo}H}{3(s_{ij} - \alpha^d_{ij})(s^M_{ij} - s_{ij})}$$

With (13.62) as well as (13.57), the evolution law (13.51) for $\dot{\beta}_{ij}$ then gives

$$\dot{\beta}_{ij} = \dot{\lambda}\frac{2\sigma_{yo}H_o}{3(s_{kl} - \alpha^d_{kl})(s^M_{kl} - s_{kl})}(\sigma^M_{ij} - \sigma_{ij}) \tag{13.63}$$

It may be observed that both (13.61) and (13.63) imply

$$\frac{\partial F}{\partial \sigma_{ij}}\dot{\beta}_{ij} = \frac{3(s_{ij} - \alpha^d_{ij})}{2\sigma_{yo}}\dot{\beta}^d_{ij} = \dot{\lambda}H_o \tag{13.64}$$

in accordance with (13.44) and (13.49).

13.2.2 Uniaxial loading

Let us finally specialize to purely uniaxial loading. In that case, we have

$$[\sigma_{ij}] = \begin{bmatrix} \sigma & 0 & 0 \\ 0 & 0 & 0 \\ 0 & 0 & 0 \end{bmatrix} \tag{13.65}$$

For uniaxial stressing, the most direct interpretation of the foregoing von Mises formulations turns out to be provided by Mróz' evolution law. Initially, before any plasticity is involved, we have $\alpha_{ij} = \beta_{ij} = 0$ and (13.26) then shows that, initially, σ_{ij}^M is proportional to σ_{ij}. From (13.62) and (13.63) then follow that both $\dot{\alpha}_{ij}$ and $\dot{\beta}_{ij}$ are initially proportional to σ_{ij}. This allows us to conclude that σ_{ij}^M, α_{ij} and β_{ij} are always proportional to σ_{ij}, i.e.

$$[\sigma_{ij}^M] = \begin{bmatrix} \sigma^M & 0 & 0 \\ 0 & 0 & 0 \\ 0 & 0 & 0 \end{bmatrix} \; ; \; [\alpha_{ij}] = \begin{bmatrix} \alpha & 0 & 0 \\ 0 & 0 & 0 \\ 0 & 0 & 0 \end{bmatrix} \; ; \; [\beta_{ij}] = \begin{bmatrix} \beta & 0 & 0 \\ 0 & 0 & 0 \\ 0 & 0 & 0 \end{bmatrix} \tag{13.66}$$

Insertion of (13.65) and (13.66b) into the yield criterion (13.53) gives

$$\sigma - \alpha = z\sigma_{yo} \tag{13.67}$$

where

$$\boxed{z = \begin{cases} 1 & \textit{for increasing stress} \\ -1 & \textit{for decreasing stress} \end{cases}}$$

Likewise, insertion of (13.66a) and (13.66c) into the bounding criterion (13.53) gives

$$\sigma^M - \beta = z\sigma_{bo} \tag{13.68}$$

The distance δ is defined by (13.58), which with (13.65) and (13.66a) becomes

$$\boxed{\delta = z\sqrt{\frac{2}{3}}(\sigma^M - \sigma)}$$

Let us finally determine the evolution expressions. For uniaxial conditions (13.55) shows that

$$\dot{\lambda} = z\dot{\varepsilon}^p \tag{13.69}$$

where $\dot{\varepsilon}^p = \dot{\varepsilon}_{11}^p$. Multiplication of the consistency relation (13.36) by $\dot{\lambda}$ and use of the associated flow rule result in

$$\dot{\varepsilon}_{ij}^p \dot{\sigma}_{ij} - (\dot{\lambda})^2 H = 0$$

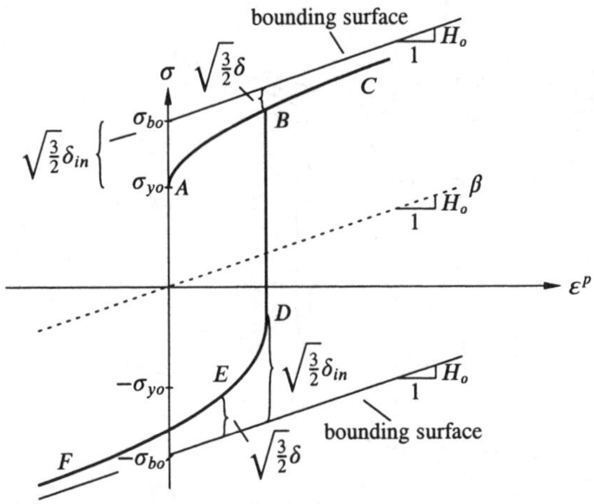

Figure 13.14: Bounding surface model.

For uniaxial conditions, this expression reduces to

$$\boxed{\dot{\sigma} = H\dot{\varepsilon}^p}$$

where the plastic modulus is given by (13.52). Expression (13.67) then shows that

$$\boxed{\dot{\alpha} = H\dot{\varepsilon}^p}$$

To determine how σ^M and β evolute with the loading, we first observe (13.65) and (13.66b) and find that

$$\frac{3(s_{ij} - \alpha_{ij}^d)}{2\sigma_{yo}}\dot{\beta}_{ij}^d = \frac{\sigma - \alpha}{\sigma_{yo}}\dot{\beta}$$

Insertion of this expression into (13.64) and taking advantage of (13.67) and (13.69) give

$$\boxed{\dot{\beta} = H_o\dot{\varepsilon}^p}$$

which with (13.68) leads to

$$\boxed{\dot{\sigma}^M = H_o\dot{\varepsilon}^p}$$

For uniaxial loading the response is shown in Fig. 13.14. For steadily increasing loading, the uniaxial curve will be described by ABC. If loading is reversed at point B, an elastic unloading will take place from B to D. At point D plastic reloading will start and the discrete memory parameter δ_{in} will be updated to the new distance. The plastic reloading will then follow DEF and this

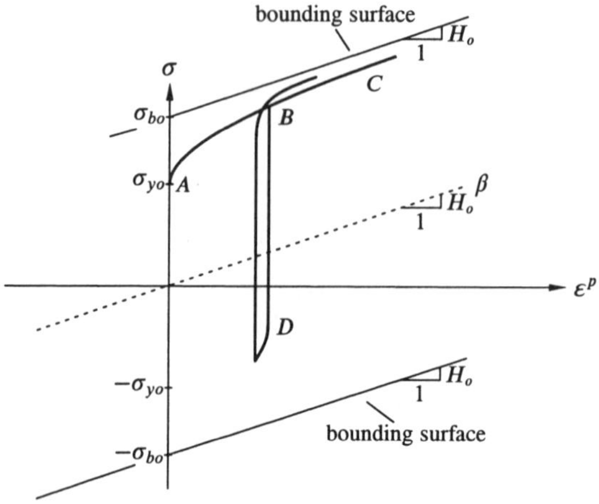

Figure 13.15: Illustration of overshooting effect.

response is in close agreement with the behavior of metals and steel; in particular the smooth transition, i.e. $H \rightarrow \infty$, when going from an elastic to a plastic response is attractive.

However, there exist some difficulties to be solved in this model. Assume, for instance, that we follow the path ABD and just touch the yield surface at point D such that the discrete memory parameter δ_{in} will be updated. Let the loading then be reversed once more. As a result δ_{in} will be updated at point B where we will have $\delta = \delta_{in}$, i.e. $H \rightarrow \infty$ implying that the stress-strain curve will be significantly above the curve BC, cf. 13.15. This so-called *overshooting* effect is clearly not acceptable and it is probably the largest deficiency in the bounding surface model.

13.3 Armstrong and Frederick model

As the last example let us consider a model introduced by Armstrong and Frederick (1966). This model has significant benefits as will be shown and it solves many of the problems relating to nonlinear kinematic hardening. The kinematic hardening model described here has been enhanced in various ways by many researchers to predict more and more complex material behavior.

For purely kinematic hardening, Armstrong and Frederick (1966) proposed the following evolution law for the back-stress tensor

$$\boxed{\dot{\alpha}_{ij} = h\left(\frac{2}{3}\dot{\varepsilon}^p_{ij} - \frac{\alpha_{ij}}{\alpha_\infty}\dot{\varepsilon}^p_{eff}\right); \quad \textit{Armstrong-Frederick evolution law}} \quad (13.70)$$

where h and α_∞ are constants, $\dot{\varepsilon}^p_{eff}$ is the effective plastic strain rate and α_{ij} is the back-stress tensor defining the center of the yield surface. We note that by letting $\alpha_\infty \to \infty$, Melan-Prager's classical kinematic hardening law (12.46) is recovered. As will be shown later, the last term in (13.70) serves as a *recall term*.

It is of interest to consider purely isotropic hardening in a format similar to (13.70) and we take

$$\dot{K} = h(1 - \frac{K}{K_\infty})\dot{\varepsilon}^p_{eff} \tag{13.71}$$

where K measures the increase of the yield surface and h and K_∞ are constants. If $K_\infty \to \infty$, we obtain linear hardening with the plastic modulus h. Integration of (13.71) with $K(\varepsilon^p_{eff} = 0) = 0$ gives

$$K = K_\infty(1 - e^{-\frac{h\varepsilon^p_{eff}}{K_\infty}}) \tag{13.72}$$

i.e. exponential hardening with the asymptotic value K_∞, cf. (9.3); it also appears that strain hardening is used to model the isotropic hardening of the yield surface. For isotropic hardening von Mises plasticity given by (12.4), the maximum value of the yield stress will then become $\sigma_{yo} + K_\infty$, where σ_{yo} is the initial yield stress. From (13.72) we can draw the conclusion that the term h/α_∞ in (13.70) serves as a damping factor that determines how fast α_{ij} should approach its asymptotic value.

13.3.1 von Mises yield function

For a von Mises material, we will illustrate mixed hardening in terms of a combination of the Armstrong-Frederick kinematic hardening model with isotropic hardening and the great potential of the Armstrong-Frederick model will then become evident. The yield function can be written as

$$f = \bar{\sigma}_{eff} - \sigma_{yo} - K \tag{13.73}$$

where

$$\bar{\sigma}_{eff} = \left(\frac{3}{2}\bar{s}_{ij}\bar{s}_{ij}\right)^{1/2} \quad ; \quad \bar{s}_{ij} = s_{ij} - \alpha^d_{ij}$$

where $\bar{s}_{ij}$ is the reduced deviatoric stress tensor.

The flow rule provides

$$\dot{\varepsilon}^p_{ij} = \dot{\lambda}\frac{\partial f}{\partial \sigma_{ij}} = \dot{\lambda}\frac{3}{2}\frac{\bar{s}_{ij}}{\bar{\sigma}_{eff}} \tag{13.74}$$

and we have $\dot{\varepsilon}^p_{ii} = 0$, i.e. plastic incompressibility, as expected. The effective plastic strain rate is as usual defined as

$$\dot{\varepsilon}^p_{eff} = \left(\frac{2}{3}\dot{\varepsilon}^p_{ij}\dot{\varepsilon}^p_{ij}\right)^{1/2}$$

and making use of the flow rule (13.74), it then follows that

$$\boxed{\dot{\varepsilon}^p_{eff} = \dot{\lambda}} \tag{13.75}$$

This means that we can interpret the plastic multiplier $\dot{\lambda}$ in (13.74) as the rate of the effective plastic strain.

To obtain mixed hardening with the isotropic hardening described by (13.71) and the kinematic hardening described by (13.70) we adopt

$$\boxed{\begin{aligned} \dot{K} &= m\,h\left(1 - \frac{K}{K_\infty}\right)\dot{\varepsilon}^p_{eff} \\ \dot{\alpha}_{ij} &= (1-m)h\left(\frac{2}{3}\dot{\varepsilon}^p_{ij} - \frac{\alpha_{ij}}{\alpha_\infty}\dot{\varepsilon}^p_{eff}\right) \end{aligned}} \tag{13.76}$$

In fact, (13.76a) and (13.76b) are similar to the expressions adopted when Melan-Prager's linear kinematic hardening rule was used to derive a mixed isotropic/kinematic von Mises model cf. (12.65). It appears that for $m = 1$ a purely isotropic hardening model is obtained and for $m = 0$ a purely Armstrong-Frederick kinematic hardening model is obtained, and for values in between, a mixed formulation emerges

Since $\dot{\varepsilon}^p_{ij}$ is deviatoric and as $\alpha_{ij} = 0$ holds initially before plasticity is introduced, it follows that the back-stress α_{ij} will always be purely deviatoric, i.e.

$$\alpha_{ij} = \alpha^d_{ij}$$

It is also observed that the evolution laws (13.76) can be written in the general format given by (10.15).

To determine the response of the above model, we start by calculating an expression for the plastic modulus. Making use of the consistency relation, $\dot{f} = 0$, written in the standard form (10.16), we have

$$\frac{\partial f}{\partial \sigma_{ij}}\dot{\sigma}_{ij} - H\dot{\lambda} = 0 \tag{13.77}$$

where H is the plastic modulus. From (13.73), the consistency relation $\dot{f} = 0$ becomes

$$\dot{f} = \frac{\partial f}{\partial \sigma_{ij}}\dot{\sigma}_{ij} + \frac{\partial f}{\partial \alpha_{ij}}\dot{\alpha}_{ij} + \frac{\partial f}{\partial K}\dot{K} = 0$$

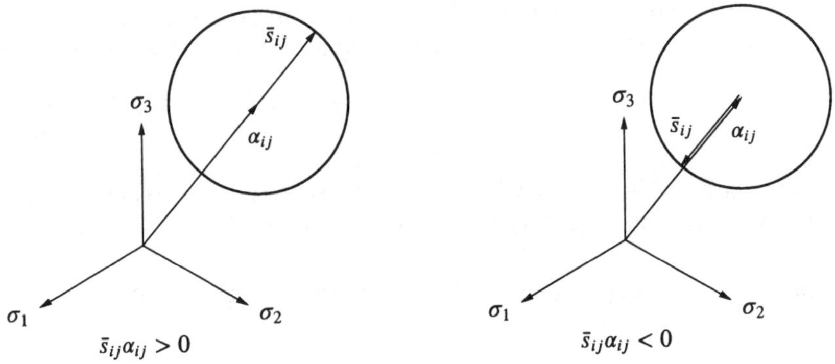

Figure 13.16: Illustration that the term $\bar{s}_{ij}\alpha_{ij}$ changes when the loading is reversed

Insertion of (13.76) as well as (13.74) and (13.75) and comparison with (13.77) then provide

$$H = -\frac{\partial f}{\partial \alpha_{ij}}(1-m)h(\frac{\bar{s}_{ij}}{\bar{\sigma}_{eff}} - \frac{\alpha_{ij}}{a_\infty}) - \frac{\partial f}{\partial K}mh(1-\frac{K}{K_\infty})$$

Determination of $\partial f/\partial \alpha_{ij}$ and $\partial f/\partial K$ and use of the yield condition then leads to

$$H = h\left[(1-m)\left(1-\frac{3}{2}\frac{\bar{s}_{ij}\alpha_{ij}}{\bar{\sigma}_{eff}\alpha_\infty}\right) + m\left(1-\frac{K}{K_\infty}\right)\right] \tag{13.78}$$

The interesting part of the plastic modulus is the term $\bar{s}_{ij}\alpha^d_{ij}$, which obviously can change sign, depending upon the loading direction, i.e. the plastic modulus will change (increase) when the loading is reversed. This is the key property of the Armstrong-Frederick model and is illustrated in Fig. 13.16. It also appears that this property is due to the *recall* term $\alpha_{ij}/\alpha_\infty\ \dot{\varepsilon}^p_{ij}$ present in (13.70); this term 'recalls' –i.e. recovers– and thereby increases the plastic modulus when the loading is reversed.

To obtain an interpretation of the plastic modulus H, we multiply (13.77) by the plastic multiplier $\dot{\lambda}$ and use the flow rule as well as(13.75) to obtain

$$\dot{\varepsilon}^p_{ij}\dot{\sigma}_{ij} = H(\dot{\varepsilon}^p_{eff})^2$$

For uniaxial conditions, the term $\dot{\varepsilon}^p_{ij}\dot{\sigma}_{ij}$ reduces to $\dot{\varepsilon}^p_{11}\dot{\sigma}_{11}$ and we then obtain

$$\frac{d\sigma_{11}}{d\varepsilon^p_{11}} = H$$

as in other von Mises models.

To analyze the model in more detail, a monotonic proportional loading situation is considered and we will determine the relation between σ_{eff} and ε^p_{eff}. We have

$$s_{ij} = p(t)s^*_{ij} \tag{13.79}$$

where s^*_{ij} is an arbitrarily fixed deviatoric stress tensor, whereas $p(t)$ serves as the loading function. Before any plasticity is involved $\alpha_{ij} = 0$ holds, i.e. (13.76) and (13.74) imply that $\dot{\alpha}_{ij}$ initially is proportional to s_{ij} and thereby to s^*_{ij}. It follows that α_{ij} will be proportional to s^*_{ij} for proportional loading, i.e.

$$\alpha_{ij} = q(t)s^*_{ij} \tag{13.80}$$

Obviously, $q(t)$ and $p(t)$ are not independent loading functions. We can take $p(t)$ and $q(t)$ as positive functions and it is evident that $p(t) \geq q(t)$. With (13.79) and (13.80) it follows that

$$\bar{s}_{ij} = (p-q)s^*_{ij}\ ; \quad \bar{\sigma}_{eff} = (p-q)\sigma^*_{ij}\ ; \quad \sigma_{eff} = p\sigma^*_{eff} \tag{13.81}$$

Integration of (13.76a) with the initial condition that $K(\varepsilon^p_{eff} = 0) = 0$ gives

$$K = K_\infty(1 - e^{-\frac{mh}{K_\infty}\varepsilon^p_{eff}}) \tag{13.82}$$

For proportional loading, the flow rule (13.74) results in

$$\dot{\varepsilon}^p_{kl} = \dot{\varepsilon}^p_{eff}\frac{3}{2\sigma^*_{eff}}s^*_{kl}$$

Insertion into the evolution equation (13.76b) gives

$$\dot{q}s^*_{ij} = (1-m)h(\frac{1}{\sigma^*_{eff}} - \frac{q}{\alpha_\infty})\dot{\varepsilon}^p_{eff}s^*_{ij}$$

from which it is concluded that

$$\dot{q}\sigma^*_{eff} = (1-m)h(1 - \frac{q\sigma^*_{eff}}{\alpha_\infty})\dot{\varepsilon}^p_{eff}$$

With the initial condition $\varepsilon^p_{eff} = 0$ implies $q = 0$, integration of this differential equation in q gives

$$q\sigma^*_{eff} = \alpha_\infty(1 - e^{-\frac{(1-m)h}{\alpha_\infty}\varepsilon^p_{eff}}) \tag{13.83}$$

For proportional loading, the yield condition (13.73) is written as

$$(p-q)\sigma^*_{eff} - \sigma_{yo} - K = 0$$

which with (13.81c) becomes

$$\sigma_{eff} = p\sigma^*_{eff} = q\sigma^*_{eff} + \sigma_{yo} + K$$

Insertion of (13.82) and (13.83) results in

$$\sigma_{eff} = \sigma_{yo} + \alpha_\infty(1 - e^{-\frac{(1-m)h}{\alpha_\infty}\varepsilon^p_{eff}}) + K_\infty(1 - e^{-\frac{mh}{K_\infty}\varepsilon^p_{eff}}) \tag{13.84}$$

It appears that when $\varepsilon^p_{eff} \to \infty$ then $\sigma_{eff} \to \sigma_{yo} + \alpha_\infty + K_\infty$.

We want a mixed formulation such that for increasing proportional loading, the response is the same irrespective of the m-value. This is achieved by setting

$$\alpha_\infty + K_\infty = \text{constant} \tag{13.85}$$

Moreover, it is required that

$$\frac{1-m}{\alpha_\infty} = \frac{m}{K_\infty} \quad \Rightarrow \quad \frac{m}{K_\infty} = \frac{1}{\alpha_\infty + K_\infty} \tag{13.86}$$

and (13.84) then reduces to the result sought

$$\sigma_{eff} = \sigma_{yo} + (\alpha_\infty + K_\infty)(1 - e^{-\frac{h}{\alpha_\infty + K_\infty}\varepsilon^p_{eff}})$$

It then follows that

$$\frac{d\sigma_{eff}}{d\varepsilon^p_{eff}} = he^{-\frac{h}{\alpha_\infty + K_\infty}\varepsilon^p_{eff}}$$

and the initial slope for $\varepsilon^p_{eff} = 0$ is given by h.

The prediction of the model in uniaxial loading is illustrated in Figs. 13.17-13.20. In these examples we have used the constraints given by (13.85) and (13.86) such that the initial loading curves are the same irrespective of the m-value. From Fig. 13.17 it appears that the model is capable of predicting a realistic nonlinear response both during loading and reversal loading. The figure also shows the interpretation of the asymptotic value $\sigma_{yo} + \alpha_\infty + K_\infty$, as well as the interpretation of H and h. It appears that by a proper choice of material parameter m, we can simulate a response that is in much closer agreement with reality than the mixed hardening von Mises model with linear hardening, cf. Fig 12.21.

The response during symmetrical loading conditions is shown in Fig. 13.18, for $m = 0$; it shows that the response of the model stabilizes after one cycle. Introducing also isotropic hardening into the model gives the response shown in Fig 13.19; due to the increase of the elastic region the response stabilizes after some cycles. Note that in the cyclic stabilized state we have that $K = K_\infty$, i.e. no additional isotropic hardening occurs. Therefore considering the symmetric

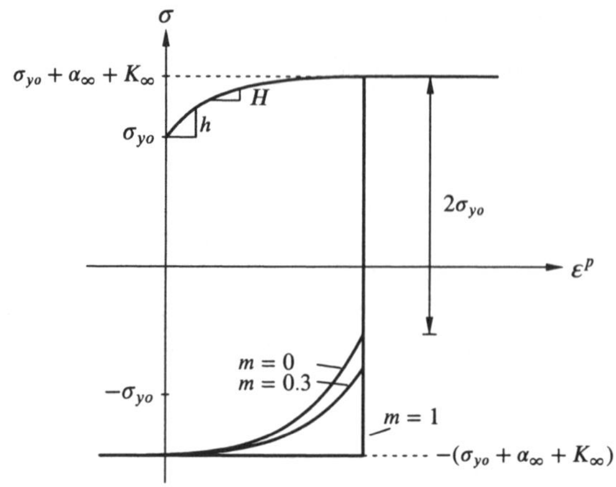

Figure 13.17: Prediction of mixed hardening von Mises model with Armstrong-Frederick's hardening rule

stress cycling case, increase of the stress amplitude after a stabilized state has been achieved will give a response where we only have kinematic hardening, and for subsequent cyclic loading the stabilized cycle is then obtained after one cycle, just like in Fig. 13.18.

To show some deficiencies of the model, we choose $m = 0$, i.e. the purely kinematic hardening case, and the responses for unsymmetric cycling are then shown in Fig. 13.20. For unsymmetric strain cycling, the model shows the desired property that the mean stress relaxes to zero, but as indicated in Fig. 13.20a it takes place within a very few cycles. Referring to Fig. 13.20b, this property of the model is more clearly evident if unsymmetric stress cycling is considered, where the model will show a very large ratcheting behavior with a constant $\Delta\varepsilon^p$-value in every cycle. For a stable material, one should expect a decreasing ratcheting behavior with $\Delta\varepsilon^p \to 0$ as the unsymmetric stress cycling is continued.

It might be in order to say something about the development and improvement of the Armstrong-Frederick model. A generalization introduced by Chaboche *et al.* (1979) and Chaboche (1989); Chaboche (1989) is based on superposition of several nonlinear kinematic hardening rules according to

$$\alpha_{ij} = \sum_{n=1}^{m} \alpha_{ij}^n \tag{13.87}$$

$$\dot{\alpha}_{ij}^n = h^n \left(\frac{2}{3}\dot{\varepsilon}_{ij}^p - \frac{\alpha_{ij}^n}{\alpha_\infty^n}\dot{\varepsilon}_{eff}^p \right) \qquad \text{(no summation over } n\text{)}$$

The reason for this generalization was to improve the large overestimate of ratcheting effects and to allow for a greater flexibility of the model. The isotropic

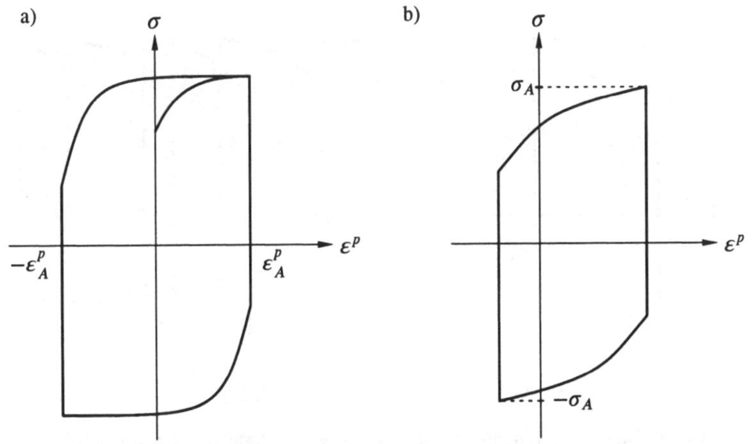

Figure 13.18: Kinematic hardening $m = 0$. a) symmetric strain cycling; b) symmetric stress cycling.

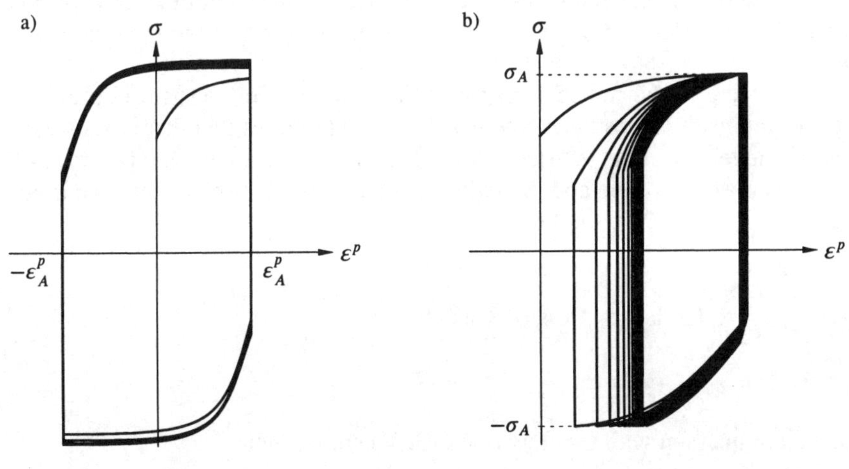

Figure 13.19: Mixed hardening $m = 0.3$. a) symmetric strain cycling; b) symmetric stress cycling.

hardening rule (13.71) could of course be generalized in the same way. Note that (13.87) can be viewed as a series expansion in exponential functions, i.e. in the limit it would be possible to include all types of curves within the model. Discussions of this generalization are also provided by Jiang and Kurath (1996).

Another improvement of the model also introduced by Chaboche *et al.* (1979) is to introduce a dependence between the saturated value of K, i.e. K_∞, and some measure of the maximum plastic strain amplitude, i.e. $K_\infty(\varepsilon_{ij}^p \varepsilon_{ij}^p)$. It fol-

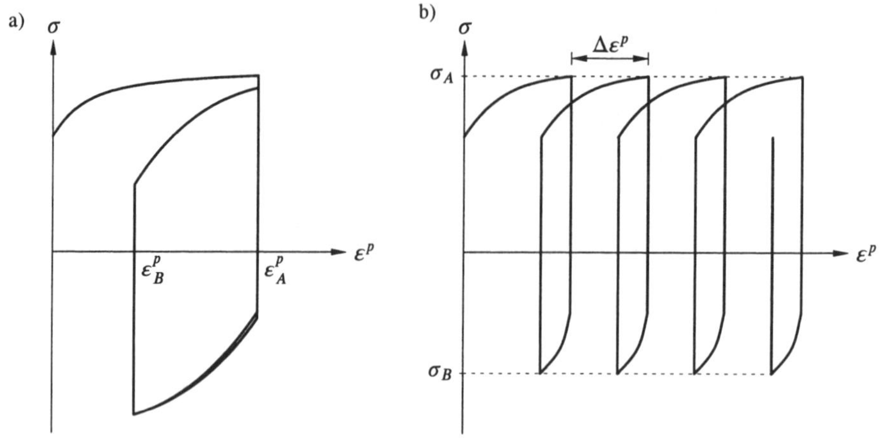

Figure 13.20: Kinematic hardening $m = 0$. a) unsymmetric strain cycling; b) unsymmetric stress cycling.

lows from the discussion above, that once we have reached the stabilized state for symmetric strain cycling, then in order to introduce additional isotropic hardening into the model K_∞ must vary with a quantity relating to the strain amplitude.

Returning to the mixed Armstrong-Frederick model considered previously, it is of interest to investigate whether this model implies that there exists a maximum range that limits the stress state. If this is the case, the yield surface must be at its maximum size and, in addition, also the back-stresses must be limited. According to (13.76), we then have

$$K = K_\infty \qquad \text{as well as} \qquad \dot{\alpha}_{ij} = 0$$

As $\dot{\varepsilon}^p_{eff} = \dot{\lambda}$, the last expression leads to

$$\dot{\varepsilon}^p_{ij} = \dot{\lambda}\frac{3}{2}\frac{\alpha_{ij}}{\alpha_\infty}$$

and a comparison with the flow rule (13.74) shows that

$$\bar{s}_{ij} = \frac{\bar{\sigma}_{eff}}{\alpha_\infty}\alpha_{ij}$$

In this expression, use of the yield criterion $f = 0$ with f given by (13.73) and observing that $K = K_\infty$ results in

$$\bar{s}_{ij} = \frac{\sigma_{yo} + K_\infty}{\alpha_\infty}\alpha_{ij} \qquad \text{i.e.} \qquad s_{ij} = (1 + \frac{\sigma_{yo} + K_\infty}{\alpha_\infty})\alpha_{ij} \tag{13.88}$$

Insertion of (13.88a) into the yield criterion $f = 0$ gives

$$\alpha_\infty = \alpha_{eff} \quad \text{where} \quad \alpha_{eff} = \left(\frac{3}{2}\alpha_{ij}\alpha_{ij}\right)^{1/2} \tag{13.89}$$

On the other hand, use of (13.88b) gives

$$\sigma_{eff} = \left(\frac{3}{2}s_{ij}s_{ij}\right)^{1/2} = \left(1 + \frac{\sigma_{yo} + K_\infty}{\alpha_\infty}\right)\alpha_{eff}$$

which with (13.89) implies

$$\boxed{\sigma_{eff,max} = \sigma_{yo} + \alpha_\infty + K_\infty} \tag{13.90}$$

where subscript 'max' is used to emphasize that this value of σ_{eff} is the largest possible value of the effective stress. This result agrees – as expected – with (13.84) when $\varepsilon^p_{eff} \to \infty$. The result (13.90) therefore represents a bounding surface as shown in Fig. 13.21.

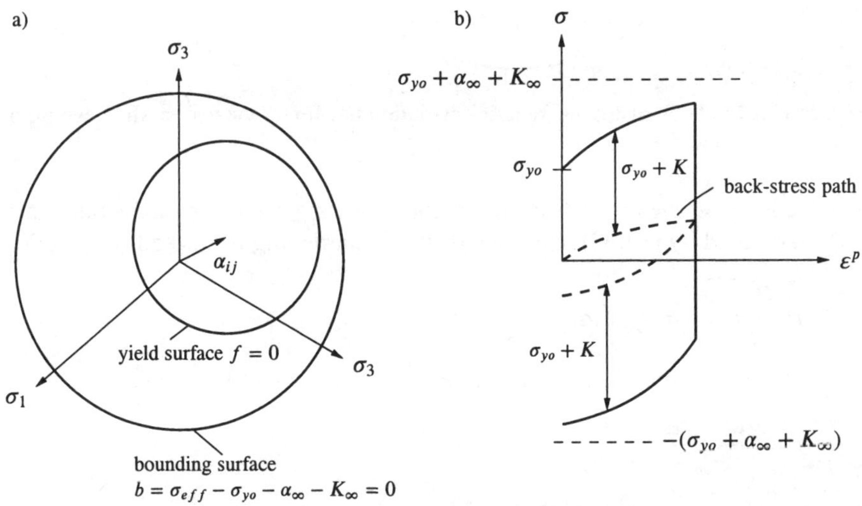

Figure 13.21: Mixed Armstrong-Frederick model with its bounding surface; a) deviatoric plane, b) uniaxial response.

The result (13.90) makes for an interesting interpretation of the Armstrong-Frederick model first pointed out by Marquis (1979) and also discussed by Chaboche (1989) and Jiang and Kurath (1996) namely that this model may be viewed also as a bounding surface model, i.e. a two-surface model, with Mróz' evolution law for the back-stresses α_{ij}.

To achieve this interpretation, isotropic hardening is ignored and the situation is then shown in Fig. 13.22. In the spirit of Mróz, the mapping stress s^M_{ij} is determined by O_bM being parallel to O_yP, i.e.

$$s^M_{ij} = a(s_{ij} - \alpha_{ij}) \tag{13.91}$$

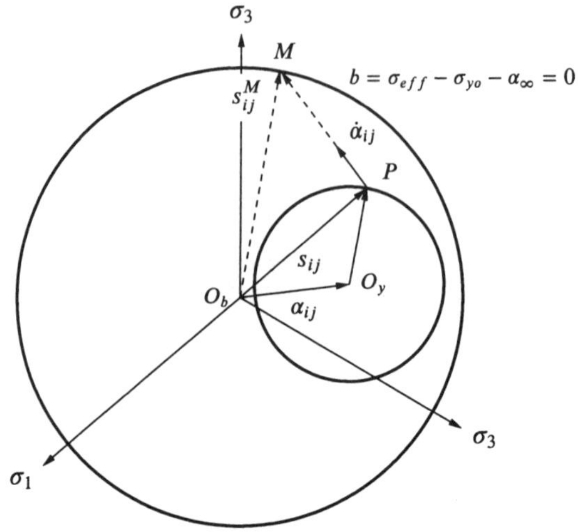

Figure 13.22: The Armstrong-Frederick evolution law for $\dot{\alpha}_{ij}$ viewed as Mróz evolution law.

where a is a positive proportionality factor. Since s_{ij}^M is located on the boundary surface defined by (13.90) and since isotropic hardening is ignored, we obtain

$$a\sqrt{\frac{3}{2}\bar{s}_{ij}\bar{s}_{ij}} = \sigma_{yo} + \alpha_\infty$$

Use of the yield condition $f = 0$ where f is defined by (13.73) gives

$$a = \frac{\sigma_{yo} + \alpha_\infty}{\sigma_{yo}}$$

Insertion of this expression into (13.91) results in

$$\alpha_{ij} = s_{ij} - \frac{\sigma_{yo}}{\sigma_{yo} + \alpha_\infty} s_{ij}^M \tag{13.92}$$

Use of the flow rule (13.74) in (13.76) with $m = 0$ gives

$$\dot{\alpha}_{ij} = h\left[\frac{s_{ij} - \alpha_{ij}}{\sigma_{yo}} - \frac{\alpha_{ij}}{\alpha_\infty}\right]\dot{\lambda}$$

Finally, insertion of (13.92) on the right-hand side provides

$$\dot{\alpha}_{ij} = \dot{\lambda}\frac{h}{\alpha_\infty}(s_{ij}^M - s_{ij})$$

This result is illustrated in Fig. 13.22 and it appears that we have recovered the evolution law of Mróz, cf. (13.12) and Fig. 13.10. The principal difference between the two formulations is that in Mróz' model the plastic modulus is chosen to be constant (within each of the nested yield surfaces) and then the evolution law for the back-stress follows from the assumption that the active yield surface approaches the next yield surface so that no intersection occurs. In the Armstrong Frederick model, the evolution law for the back-stresses is directly postulated and the plastic modulus then turns out to vary in a realistic fashion; *a posteriori*, the Armstrong-Frederick model may then be interpreted as a two-surface model with a Mróz-type evolution law for the back-stresses.

14 INTRODUCTION TO TIME-DEPENDENT MATERIAL BEHAVIOR

It is an everyday experience that even if a material is loaded by a constant load, its deformation may increase with time; a book-shelf loaded by too heavy books may increase its deflection as years goes by. As a consequence, one speaks of *time-dependent material behavior* also termed *creep behavior*. The terminology in the literature is not unique and, traditionally, when the time-dependent strains are related linearly to the stresses, one speaks of *viscoelasticity* whereas the notations of *creep* and *viscoplasticity* are often used when the time-dependent strains depend nonlinearly on the stresses.

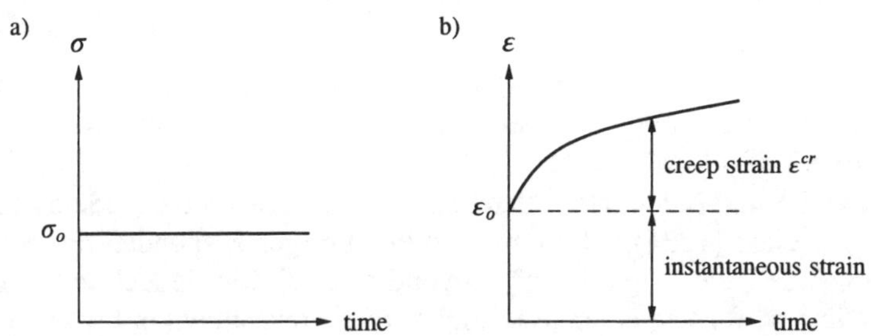

Figure 14.1: Creep test; a) stress history, b) strain history.

In this chapter, we will deal with viscoelasticity, whereas creep and viscoplasticity will be addressed in the next chapter. However, we will start with a general discussion of various experimental findings relating to time-dependent behavior.

There are three standard tests used to identify the time-dependent response of a material: the *creep test*, the *relaxation test* and the *constant strain-rate test*. In the creep test, the stress σ_0 is applied instantaneously and then kept constant, cf. Fig. 14.1a), and as a result the strain history may vary as shown in

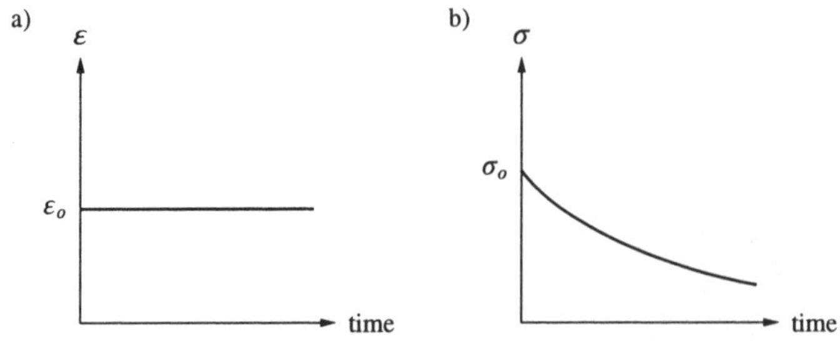

Figure 14.2: Relaxation test; a) strain history, b) stress history.

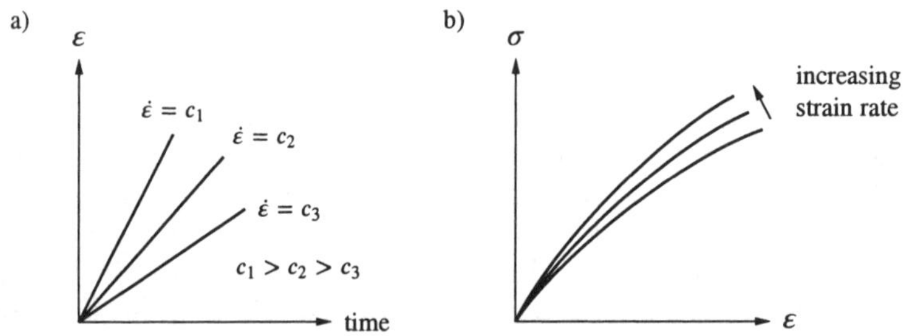

Figure 14.3: Constant strain-rate test; a) strain history for three tests, b) corresponding stress-strain responses.

Fig. 14.1b). In Fig. 14.1b), the strain ε_0 is the instantaneous strain that may be elastic (then $\varepsilon_0 = \sigma_0/E$) or elasto-plastic (then $\varepsilon_0 = \sigma_0/E + \varepsilon^p$) and with time the *creep strain* ε^{cr} develops.

Historically, the first quantitative statements about creep were made by the French engineer Vicat (1834), who observed that bridges suspended by hardened iron cables deflected significantly beyond their elastic deflections. For such cables, Vicat performed creep tests similar to the one shown in Fig. 14.1.

In the *relaxation test*, the total strain is applied instantaneously and then kept constant at the value ε_0, cf. Fig. 14.2a), and as a result the stress history may vary as shown in Fig. 14.2b). In Fig. 14.2b), the stress σ_0 is the instantaneous stress that may be a result of elastic (then $\sigma_0 = E\varepsilon_0$) or elasto-plastic response (then $\sigma_0 = E(\varepsilon_0 - \varepsilon^p)$) when enforcing the instantaneous strain ε_0. As time goes by, Fig. 14.2b) shows that the stress gradually decreases - it relaxes.

In the *constant strain-rate test*, the total strain rate $\dot{\varepsilon} = d\varepsilon/dt$ =constant is enforced on the material and the stress response is then measured so that the stress-strain relation can be established. In Fig. 14.3, the results of three such tests are shown and it appears that the larger the total strain-rate, the stiffer the material behaves. This is a characteristic property for materials that exhibit

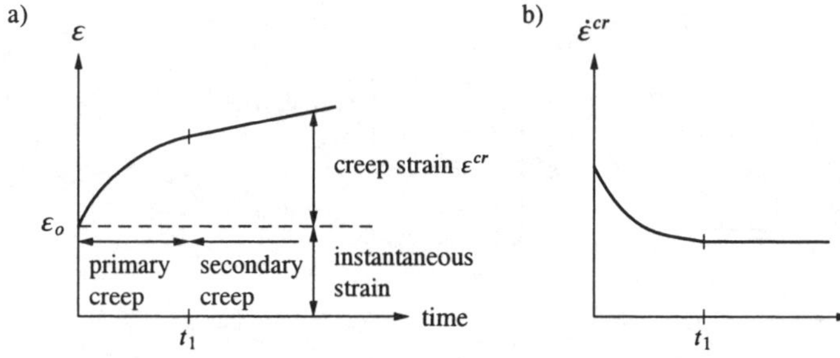

Figure 14.4: Creep test for a 'small' stress; a) strain history, b) creep strain rate $\dot{\varepsilon}^{cr}$.

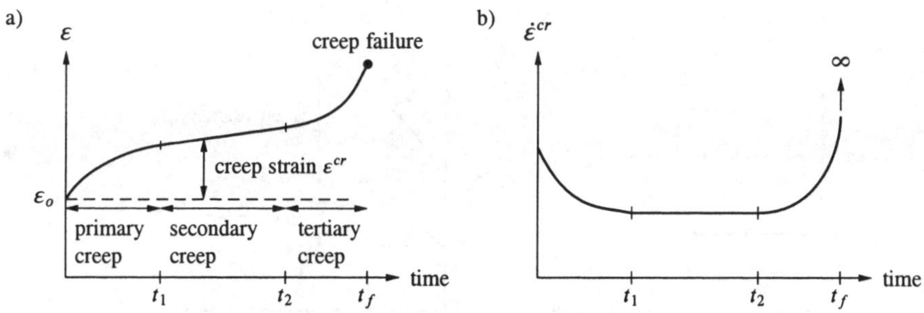

Figure 14.5: Creep test for a 'large' stress; a) strain history, b) creep strain rate $\dot{\varepsilon}^{cr}$.

creep deformation and it is concluded that for such materials, the response is *rate-dependent*. This is in striking contrast to elasto-plasticity where the response is independent of the rate applied.

If the stress applied in a creep test is not too large, the response shown in Fig. 14.4a) is obtained. Up until the time t_1, we have *primary creep* - also called *transient creep* - and after time t_1 we have *secondary creep* - also called *stationary creep*. In Fig. 14.4b), the creep strain rate $\dot{\varepsilon}^{cr} = d\varepsilon^{cr}/dt$ is shown and it appears that during primary or transient creep, $\dot{\varepsilon}^{cr}$ is decreasing whereas during secondary or stationary creep, $\dot{\varepsilon}^{cr}$ is constant. For some materials, the primary creep region is small and may be ignored; this is often the case for metals and steels exposed to high temperatures and constant load.

If the stress applied in a creep test is sufficiently large, the response shown in Fig. 14.5a) is obtained. Now we also obtain *tertiary creep* after time t_2 and in this region the creep strain rate $\dot{\varepsilon}^{cr}$ increases, cf. Fig. 14.5b), and at time t_f the material fails - *creep failure* has occurred. Modeling of the phenomenon of creep failure is very complex and we will here only be concerned with modeling of primary and secondary creep.

Let us next discuss linear and nonlinear creep response. In Fig. 14.6, the results of two creep tests are shown; one with the constant stress σ_0 and another

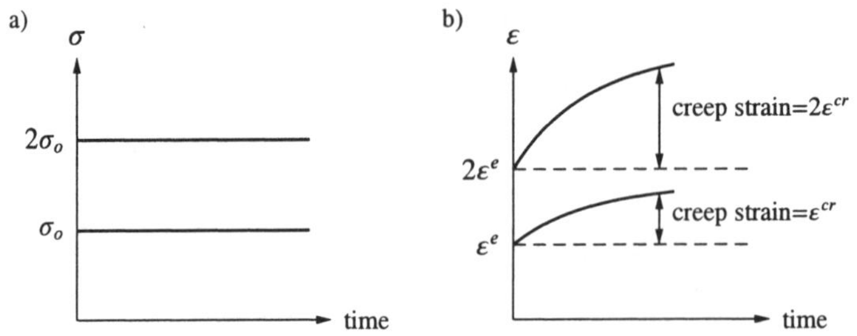

Figure 14.6: Creep test where linearity holds; if the stress is doubled the creep strain is also doubled.

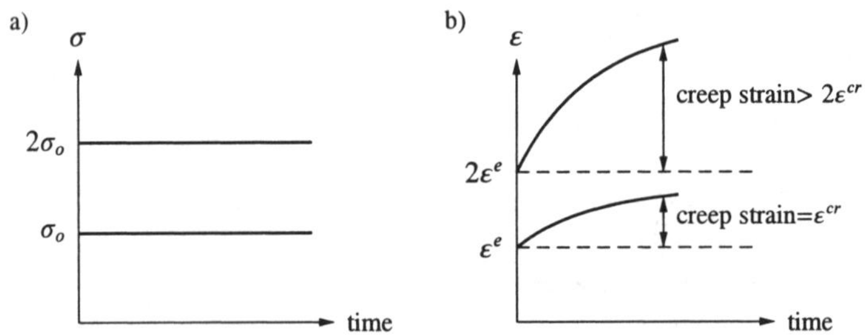

Figure 14.7: Creep test when nonlinearity holds; if the stress is doubled the creep strain is more than doubled.

test with twice the stress, i.e. $2\sigma_0$. For simplicity, the instantaneous response is assumed to be linear elastic, i.e. if it is ε^e in the first test it is $2\varepsilon^e$ in the second test. For linearity to hold, then if ε^{cr} is the creep strain in the first test at some time, the creep strain in the second test should be $2\varepsilon^{cr}$ at the same time. This linearity is characteristic for *viscoelasticity*, which will be discussed in the next section. Viscoelastic response is typical for polymers and concrete loaded not too close to their ultimate strength.

To illustrate nonlinear creep response, Fig. 14.7 is considered. Again two creep tests are performed, one with the stress σ_0 and the other with the stress $2\sigma_0$. For simplicity, the instantaneous response is again assumed to be linear elastic, i.e. the instantaneous strain is ε^e in the first test and $2\varepsilon^e$ in the second test. Now however, the creep strain in the second test is not twice the creep strain in the first test; in practice it is larger. This nonlinear creep response is typical for creep of metals and steel and we shall discuss various means to model such creep response in the next chapter.

To illustrate the phenomenon of *recovery*, the creep test in Fig. 14.8 is considered. As shown in Fig. 14.8a), the stress σ_0 is removed at time t_1 and after that

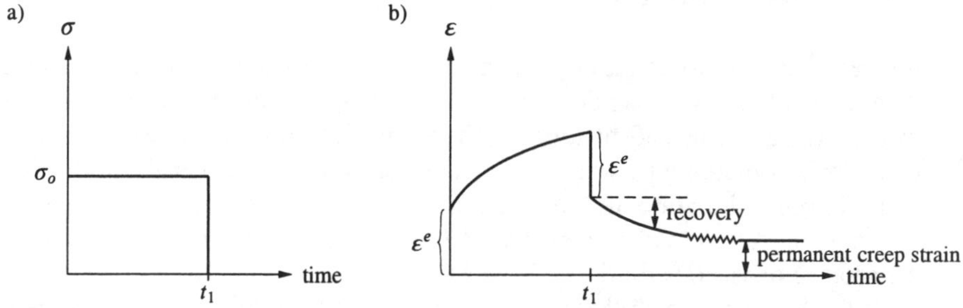

Figure 14.8: Phenomenon of recovery.

the material is completely unloaded. The corresponding strain development is shown in Fig. 14.8b) where - for simplicity - the instantaneous strain is assumed to be elastic, i.e. equal to ε^e. Figure 14.8b) shows that when the specimen is unloaded at time t_1, it responds elastically and thereby the strain decreases with the amount ε^e at time t_1; however, even though the material is unstressed after time t_1, the strain continues to decrease, i.e. the strain - or some part of it - is recovered. This phenomenon is characteristic for viscoelastic materials like polymers and concrete. The part of the total strain that is not recovered even after infinitely long time is called *permanent creep strain.*

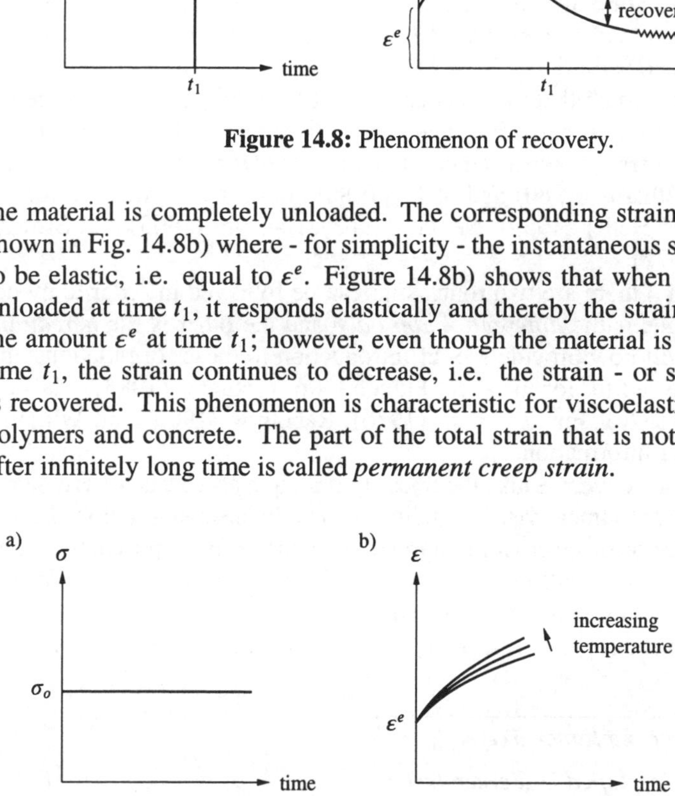

Figure 14.9: Influence of temperature.

In practice, it turns out that development of creep strains is very sensitive to temperature; the higher the temperature, the larger the creep strain. This is illustrated in Fig. 14.9 where - for simplicity - the instantaneous response is assumed to be linear elastic and where the E−modulus is assumed to be temperature-independent. We shall later return to a more detailed discussion of the influence of temperature on various materials.

14.1 Viscoelasticity

It is recalled that viscoelasticity means that if the stress in a creep test is doubled then the total strain is also doubled, cf. Fig. 14.6; this linearity is also assumed to hold for general load histories. The term viscoelasticity is used since the behavior is something between that of a viscous fluid and an elastic solid. Especially in older literature and in material science literature, viscoelastic models are also referred to as *rheological models* and strictly speaking *rheology* means the science of viscous fluids.

If the loading is such that the response - apart from being time-dependent - is elastic, the linearity property that is characteristic for viscoelasticity is closely fulfilled for polymers, cf. for instance Finnie and Heller (1959), Bartenev and Zuev (1968), Williams (1980) and Mills (1986), and for concrete, cf. for instance Finnie and Heller (1959), Neville (1963), Hannant (1969) and Browne and Blundell (1972).

It turns out that there are two routes that can be followed in order to model viscoelasticity; one is the *differential approach* and the other is the *hereditary approach*. We will now provide a brief introduction to these formulations and the reader is referred to, for instance, Flügge (1967), Hunter (1983), Malvern (1969), Pipkin (1972), Findley *et al.* (1976), Rabotnov (1980) and Williams (1980) for further information.

As emphasized above, we take the linearity principle as a basis for viscoelasticity and this leads to linear viscoelasticity that will be discussed below. Linear viscoelasticity has been successfully applied to concrete, most polymers, wood and paper. However, it will turn out that this formulation even makes for the possibility of modeling nonlinear viscoelasticity.

Let us first introduce the *creep compliance* $J(t)$ according to the following definition

The creep compliance $J(t)$
= *strain developed in a creep test*
when loaded by a unit stress

(14.1)

Since the linearity principle holds, the strain development in a creep test with the constant stress σ_0 is then $\varepsilon(t) = J(t)\sigma_0$. Comparing with Fig. 14.1, the creep compliance function $J(t)$ therefore gives information on how the strain develops with time in a creep test.

In a similar manner, the *relaxation modulus* $G(t)$ is defined by

The relaxation modulus $G(t)$
= *stress developed in a relaxation test*
when loaded by a unit strain

(14.2)

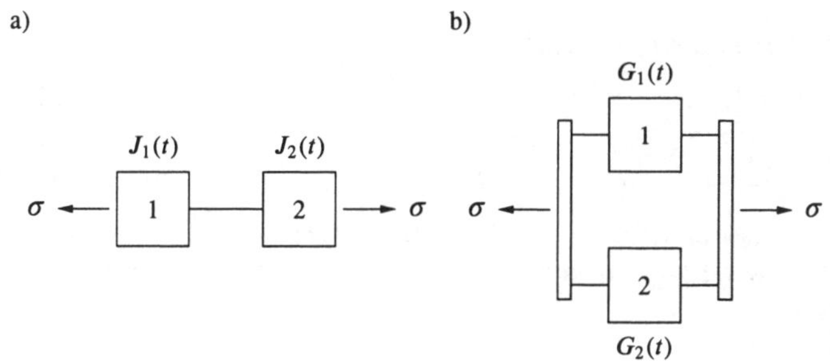

Figure 14.10: Two viscoelastic models in a) series and b) parallel.

where the relaxation modulus $G(t)$ should not be confused with the shear modulus G related to linear elasticity. Since the linearity principle holds, the stress development in a relaxation test with the constant strain ε_0 is then given by $\sigma(t) = G(t)\varepsilon_0$. Comparing with Fig. 14.2, the relaxation modulus function $G(t)$ therefore gives information on how the stress develops with time in a relaxation test. Certainly, one may expect that, in some fashion, the creep compliance $J(t)$ and the relaxation modulus $G(t)$ are related and we will establish this relation in the section dealing with the hereditary approach.

Suppose that two viscoelastic models with creep compliances $J_1(t)$ and $J_2(t)$ are placed in series, cf. Fig. 14.10a). With evident notation, we have

$$\sigma = \sigma_1 = \sigma_2\,; \qquad \varepsilon = \varepsilon_1 + \varepsilon_2$$

In a creep test with the stress σ_0, it follows that $\varepsilon = J_1(t)\sigma_1 + J_2(t)\sigma_2 = (J_1(t) + J_2(t))\sigma_0$. It is concluded that

For two models placed in series

$$J(t) = J_1(t) + J_2(t) \qquad (14.3)$$

holds

Indeed, in a relaxation test it is also possible to establish how $G(t)$ is related to $G_1(t)$ and $G_2(t)$, but in the following we will not make use of this slightly more complex relation.

Consider next that two viscoelastic models with relaxation moduli $G_1(t)$ and $G_2(t)$ are placed in parallel, cf. Fig. 14.10b). It follows that

$$\sigma = \sigma_1 + \sigma_2\,; \qquad \varepsilon = \varepsilon_1 = \varepsilon_2$$

In a relaxation test with the strain ε_0, we then obtain $\sigma = G_1(t)\varepsilon_1 + G_2(t)\varepsilon_2 =$

$(G_1(t) + G_2(t))\varepsilon_0$. Consequently

For two models placed in parallel
$G(t) = G_1(t) + G_2(t)$
holds

(14.4)

In a creep test, it is also possible to establish how $J(t)$ is related to $J_1(t)$ and $J_2(t)$, but in the following we will not make use of this slightly more complex relation.

14.2 Differential equation approach

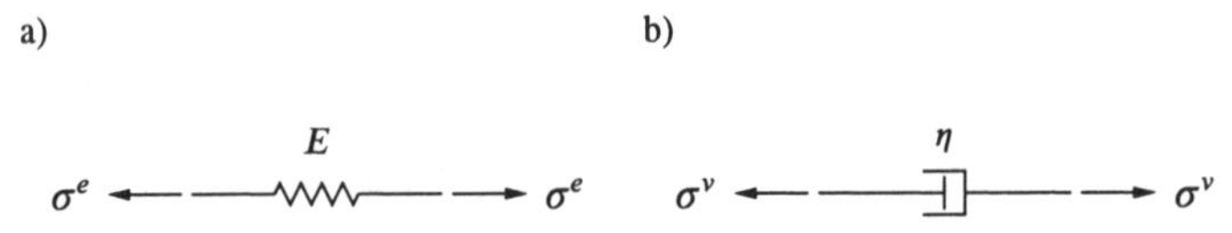

Figure 14.11: a) Linear spring; b) dashpot.

In the differential approach, viscoelastic models are constructed by various combinations of linear springs and dashpots. We have already touched upon this approach in Section 6.4 and the constitutive equations that control the spring and the dashpot shown in Fig. 14.11 are

$$\sigma^e = E\varepsilon^e\,; \qquad \sigma^v = \eta\dot{\varepsilon}^v \tag{14.5}$$

where superscript *e* and *v* refer to elastic and viscous behavior, respectively; the *viscosity coefficient* η has the dimension [Pa·s].

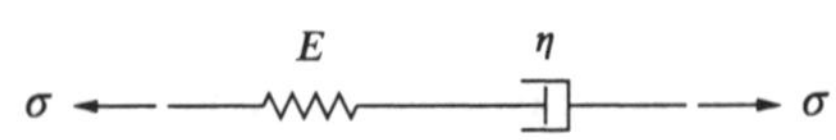

Figure 14.12: Maxwell model.

The *Maxwell model* - established by Maxwell (1868) - is shown in Fig. 14.12 and it consists of a spring and dashpot in series. It appears that

$$\dot{\varepsilon} = \dot{\varepsilon}^e + \dot{\varepsilon}^v\,; \qquad \sigma = \sigma^e = \sigma^v \tag{14.6}$$

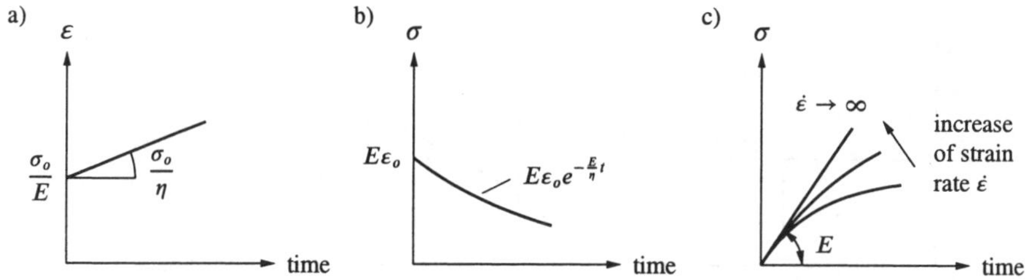

Figure 14.13: Response of Maxwell model; a) creep test, b) relaxation test and c) constant strain-rate test.

Insertion of (14.5) in (14.6a) and use of (14.6b) give the following constitutive equation for the Maxwell model

$$\dot{\varepsilon} = \frac{\dot{\sigma}}{E} + \frac{\sigma}{\eta} \qquad \textit{Maxwell model} \tag{14.7}$$

To determine the response during a creep test, (14.7) is multiplied by dt and integrated. This gives $\varepsilon = \frac{\sigma}{E} + \frac{1}{\eta}\int_0^t \sigma(\tau)d\tau = (\frac{1}{E} + \frac{t}{\eta})\sigma_0$ and a comparison with (14.1) shows that the creep compliance is $J(t) = \frac{1}{E} + \frac{t}{\eta}$. The response to a creep test is shown in Fig. 14.13a) and it appears that the Maxwell model exhibits secondary creep. For a relaxation test, we have $\varepsilon = \varepsilon_0 =$ constant and (14.7) then reduces to $\dot{\sigma} + \frac{E}{\eta}\sigma = 0$ with the solution $\sigma = Ce^{-\frac{E}{\eta}t}$ where the arbitrary constant C is determined from the condition that $t = 0$ gives $\sigma = E\varepsilon_0$, i.e. $\sigma = E\varepsilon_0 e^{-\frac{E}{\eta}t}$ as illustrated in Fig. 14.13b). Moreover, a comparison with (14.2) shows that the relaxation modulus is $G(t) = Ee^{-\frac{E}{\eta}t}$. Therefore

$$J(t) = \frac{1}{E} + \frac{t}{\eta}; \quad G(t) = Ee^{-\frac{E}{\eta}t} \qquad \textit{Maxwell model} \tag{14.8}$$

For a constant strain-rate test, we have $\varepsilon = \dot{\varepsilon}t$ where the constant $\dot{\varepsilon}$ is the strain-rate. Insertion into (14.7) and integration gives $\sigma = \dot{\varepsilon}\eta + Ce^{-\frac{E}{\eta}t}$ where the arbitrary constant C is determined from the condition that $t = 0$ gives $\sigma = 0$, i.e. $\sigma = \dot{\varepsilon}\eta(1 - e^{-\frac{E}{\eta}t})$. Since $t = \varepsilon/\dot{\varepsilon}$, the stress-strain relation becomes $\sigma = \dot{\varepsilon}\eta(1 - e^{-\frac{E}{\eta}\frac{\varepsilon}{\dot{\varepsilon}}})$, which implies $\frac{d\sigma}{d\varepsilon} = Ee^{-\frac{E}{\eta}\frac{\varepsilon}{\dot{\varepsilon}}}$ and this means that the initial slope is always given as $\frac{d\sigma}{d\varepsilon} = E$ as shown in Fig. 14.13c). It appears that the stress-strain response depends on the strain-rate and that, for an infinitely large strain-rate, the response approaches linear elasticity. Considering Fig. 14.12, this is certainly not surprising since the dashpot responds as a rigid member for a sudden application of the load.

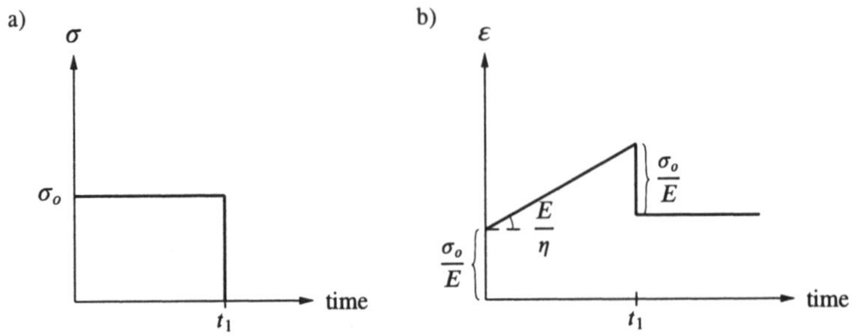

Figure 14.14: Response of Maxwell model; no recovery effect.

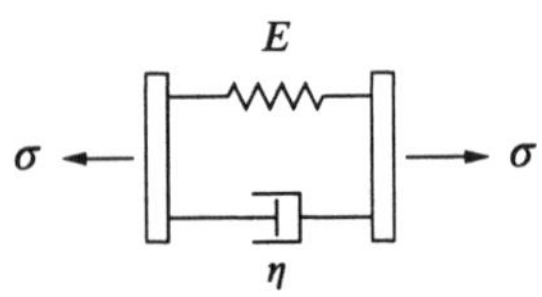

Figure 14.15: Kelvin model.

If the stress in a creep test is removed at time t_1, the Maxwell model reacts elastically during the unloading process and since $\sigma = \dot{\sigma} = 0$ when $t > t_1$, (14.7) then gives that $\dot{\varepsilon} = 0$ when $t > t_1$. We then obtain the response shown in Fig. 14.14; the Maxwell model shows no recovery, cf. Fig. 14.8.

The *Kelvin model* - established by Kelvin (1875) and also by Voigt (1892) and therefore also called the *Voigt model* - is shown in Fig. 14.15 and it consists of a spring and a dashpot in parallel. It follows that

$$\varepsilon = \varepsilon^e = \varepsilon^v; \qquad \sigma = \sigma^e + \sigma^v \tag{14.9}$$

Insertion of (14.5) into (14.9b) and use of (14.9a) give the following constitutive equation

$$\boxed{\sigma = E\varepsilon + \eta\dot{\varepsilon} \qquad \textit{Kelvin model}} \tag{14.10}$$

To determine the response in a creep test with the constant stress σ_0, (14.10) is integrated to give $\varepsilon = \frac{\sigma_0}{E} + Ce^{-\frac{E}{\eta}t}$ where C is an arbitrary constant. Since the dashpot reacts as a rigid member when a load is suddenly applied, we have the condition $\varepsilon = 0$ when $t = 0$. This condition determines C and we then obtain $\varepsilon = \frac{\sigma_0}{E}(1 - e^{-\frac{E}{\eta}t})$ as shown in Fig. 14.16a). From (14.1) it is concluded that

$$\boxed{J(t) = \frac{1}{E}(1 - e^{-\frac{E}{\eta}t}) \qquad \textit{Kelvin model}} \tag{14.11}$$

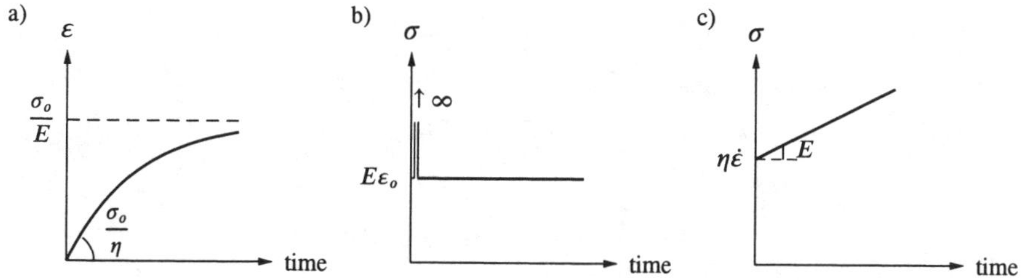

Figure 14.16: Response of Kelvin model; a) creep test, b) relaxation test and c) constant strain-rate test.

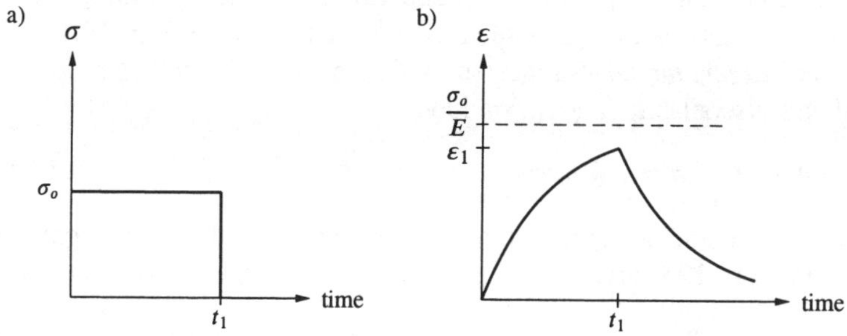

Figure 14.17: Response of Kelvin model; full recovery.

It appears from Fig. 14.16a) that the Kelvin model exhibits primary creep only. Moreover, due to the dashpot, the Kelvin model reacts as a rigid material for a sudden application of the load. Due to this peculiarity, application of the Kelvin model should be performed judiciously. This special property is also responsible for the very special response when a relaxation test is performed. According to (14.10) and in order to maintain a constant strain ε_0, the stress must be $\sigma = E\varepsilon_0$. On the other hand, in order to instantaneously deform the Kelvin model to this strain value, an infinitely large stress is required. Therefore, in a relaxation test the stress increases instantaneously to infinity and immediately after, the stress will take the value $\sigma = E\varepsilon_0$; this result is illustrated in Fig. 14.16b). For a constant strain-rate test where $\varepsilon = \dot{\varepsilon}t$ with $\dot{\varepsilon}$ being constant, (14.10) immediately gives the result shown in Fig. 14.16c). It appears that for an infinitely small strain-rate, i.e. $\dot{\varepsilon} \to 0$, the Kelvin model reacts as an elastic material.

If the stress in a creep test is removed at time t_1, the Kelvin model reacts as a rigid material during the unloading process and since $\sigma = 0$ when $t > t_1$, (14.10) implies $0 = E\varepsilon + \eta\dot{\varepsilon}$ with the solution $\varepsilon = Ce^{-\frac{E}{\eta}t}$ where the arbitrary constant C is determined by the condition $\varepsilon = \varepsilon_1$ when $t = t_1$, i.e. $\varepsilon = \varepsilon_1 e^{-\frac{E}{\eta}(t-t_1)}$ when $t \geq t_1$. This response is shown in Fig. 14.17 and with $t \to \infty$ we obtain $\varepsilon \to 0$,

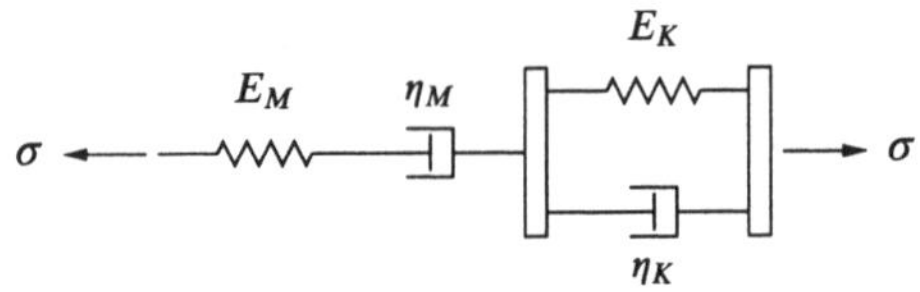

Figure 14.18: Burgers model.

i.e. the Kelvin model exhibits full recovery. Certainly, this is also evident from Fig. 14.15 since the tensile stress in the spring, when the external stress $\sigma = 0$ for $t > t_1$, will compress the dashpot until the situation $\varepsilon = 0$ has been reached.

Having discussed the simple Maxwell and Kelvin models in detail, it is evident that these models can be combined in series as shown in Fig. 14.18. We then obtain the *Burgers model* - suggested by Burgers (1935) - which represents a pretty realistic viscoelastic model. We have

$$\varepsilon = \varepsilon_M + \varepsilon_K \,; \qquad \sigma = \sigma_M = \sigma_K \tag{14.12}$$

where subscripts M and K refer to the Maxwell part and the Kelvin part, respectively. Since (14.12a) gives $\varepsilon_K = \varepsilon - \varepsilon_M$ insertion into (14.10) gives with $\sigma_K = \sigma$

$$\sigma = E_K(\varepsilon - \varepsilon_M) + \eta_K(\dot{\varepsilon} - \dot{\varepsilon}_M)$$

Differentiation with respect to time and insertion of (14.7) and using $\sigma_M = \sigma$ then give the following constitutive equation

Burgers model

$$\frac{\eta_K}{E_M}\ddot{\sigma} + \left(1 + \frac{\eta_K}{\eta_M} + \frac{E_K}{E_M}\right)\dot{\sigma} + \frac{E_K}{\eta_M}\sigma = \eta_K\ddot{\varepsilon} + E_K\dot{\varepsilon}$$

To investigate the response in a creep test, we may force the condition $\dot{\sigma} = 0$ on the constitutive equation and this leads to a linear second order differential equation in ε, which may easily be solved. This solution will involve two arbitrary constants to be determined from the initial conditions. While this procedure is certainly possible, the identification of the initial conditions turns out to be somewhat cumbersome. Indeed, this complication occurs for all viscoelastic differential equations of a higher order than one. An easier approach is to make use of (14.3), which with (14.8) and (14.11) directly gives the result

$$J(t) = \frac{1}{E_M} + \frac{t}{\eta_M} + \frac{1}{E_K}\left(1 - e^{-\frac{E_K}{\eta_K}t}\right) \qquad \textit{Burgers model} \tag{14.13}$$

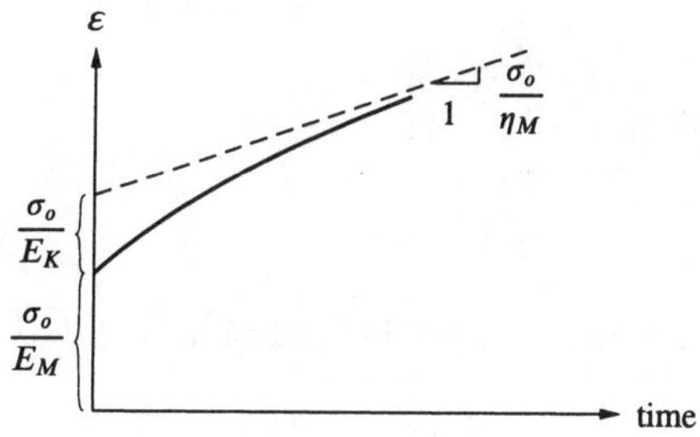

Figure 14.19: Response of Burgers model in creep test.

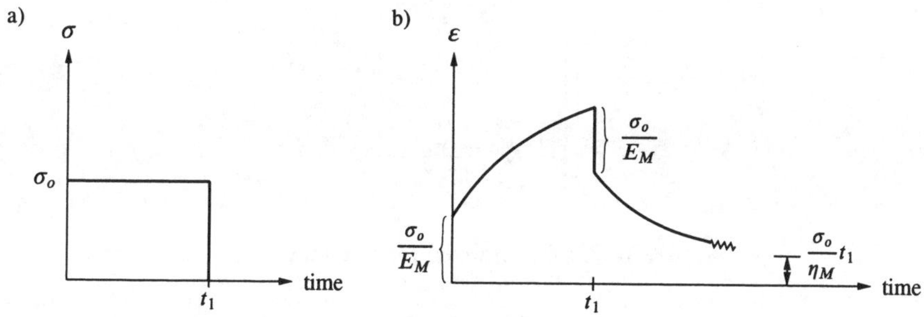

Figure 14.20: Response of Burgers model; partial recovery.

This result is shown in Fig. 14.19 and it appears that primary and - in the limit - also secondary creep occurs and that the strain rate approaches σ_0/η_M when $t \to \infty$.

If the stress in a creep test is removed at time t_1, the Maxwell part responds elastically during unloading with no further deformation when $t > t_1$, cf. Fig. 14.14, whereas the Kelvin part reacts as a rigid material during unloading and then full recovery is achieved in the Kelvin part when $t \to \infty$. Eventually, the only remaining strain is the viscous strain $\frac{\sigma_0}{\eta_M}t_1$ developed in the Maxwell part up until time t_1. Therefore, the Burgers model will show the response shown in Fig. 14.20 and only partial recovery occurs. In conclusion, Figs. 14.19 and 14.20 indicate that - except for tertiary creep - the Burgers model exhibits all the principal characteristics of a real material.

Another often used model is the *3-parameter model* shown in Fig. 14.21; this model is also called the *standard linear solid*. The constitutive equation can be obtained from (14.13) by letting $\eta_M \to \infty$. It is evident that this model will respond elastically to instantaneous loading and that it will only exhibit primary creep.

The *generalized Maxwell model* appears from Fig. 14.22. Its response in a

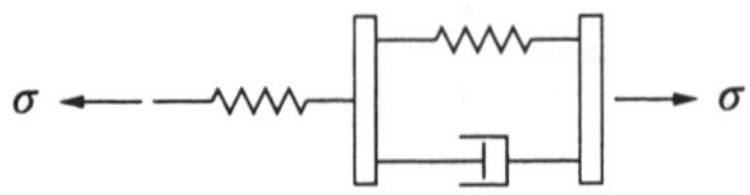

Figure 14.21: 3-parameter model (standard linear solid).

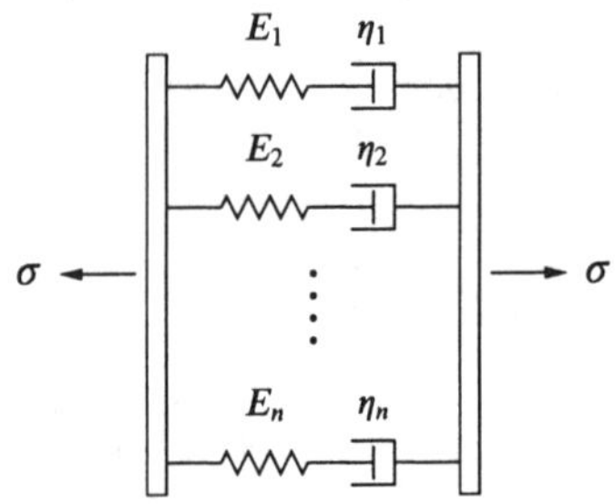

Figure 14.22: Generalized Maxwell model.

relaxation test can be inferred from (14.4) and (14.8) to provide

Generalized Maxwell model

$$G(t) = \sum_{i=1}^{n} E_i e^{-\frac{E_i}{\eta_i}t} \tag{14.14}$$

It is evident that a close approximation can be obtained to experimental data. In a creep test it can be shown that the response will also involve primary creep. In general, the response is controlled by the following equations

$$\varepsilon = \varepsilon_1 = \varepsilon_2 = \cdots = \varepsilon_n\,; \qquad \sigma = \sigma_1 + \sigma_2 + \cdots + \sigma_n$$

and from (14.7) follows that

$$\begin{aligned}
\dot{\varepsilon}_1 &= \frac{\dot{\sigma}_1}{E_1} + \frac{\sigma_1}{\eta_1} \\
\dot{\varepsilon}_2 &= \frac{\dot{\sigma}_2}{E_2} + \frac{\sigma_2}{\eta_2} \\
&\vdots \\
\dot{\varepsilon}_n &= \frac{\dot{\sigma}_n}{E_n} + \frac{\sigma_n}{\eta_n}
\end{aligned}$$

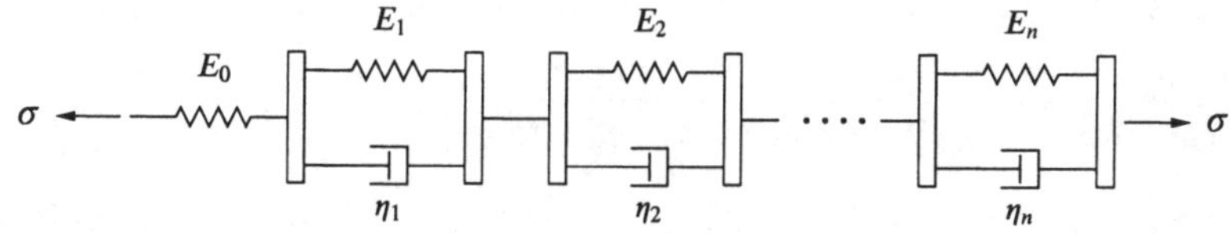

Figure 14.23: Generalized Kelvin model.

It is easily checked that there are $2n + 1$ of these constitutive equations which involve the $2 + 2n$ unknowns given by $\varepsilon, \sigma, \varepsilon_1, \varepsilon_2, \cdots, \varepsilon_n, \sigma_1, \sigma_2, \cdots, \sigma_n$. By a tedious elimination process, the result becomes one constitutive relation in terms of one higher-order differential equation in ε and σ.

The *generalized Kelvin model* is given in Fig. 14.23 where the elastic spring with the stiffness E_0 has been added in order that the response to an instantaneous loading be elastic. Its response in a creep test can be inferred from (14.3) and (14.11) to provide

Generalized Kelvin model

$$J(t) = \frac{1}{E_0} + \sum_{i=1}^{n} \frac{1}{E_i}(1 - e^{-\frac{E_i}{\eta_i}t}) \tag{14.15}$$

It appears that primary creep, only, can be modeled even though a large possibility exists for close approximations to experimental data; possible secondary creep can be modeled by adding a dashpot in series with the spring E_0. The series in (14.15) containing exponentials is called a *Dirichlet series* or - occasionally - a *Prony series*. In general, the response of the model is given by the following constitutive equations

$$\varepsilon = \varepsilon_0 + \varepsilon_1 + \varepsilon_2 + \cdots + \varepsilon_n\,; \qquad \sigma = \sigma_0 = \sigma_1 = \sigma_2 = \cdots = \sigma_n$$

and for the elastic spring and from (14.10) follow that

$$\begin{aligned}
\sigma_0 &= E_0\varepsilon_0 \\
\sigma_1 &= E_1\varepsilon_1 + \eta_1\dot{\varepsilon}_1 \\
\sigma_2 &= E_2\varepsilon_2 + \eta_2\dot{\varepsilon}_2 \\
&\vdots \\
\sigma_n &= E_n\varepsilon_n + \eta_n\dot{\varepsilon}_n
\end{aligned}$$

Again a tedious elimination process makes it possible to obtain one constitutive relation in terms of a higher-order differential equation in time of ε and σ.

Certainly, a very comprehensive and accurate model may be constructed by combining a generalized Maxwell model and a generalized Kelvin model. Irrespective of how we combine springs and dashpots, it comes as no surprise that it is always possible to write the constitutive equation in the following general form

General format of the constitutive equation

$$\sum_{i=0}^{n} a_i \frac{d^i \sigma}{dt^i} = \sum_{j=1}^{m} b_j \frac{d^j \varepsilon}{dt^j} \tag{14.16}$$

where, per definition, we have $\frac{d^0\sigma}{dt^0} = \sigma$ and $\frac{d^0\varepsilon}{dt^0} = \varepsilon$. In (14.16), a_i and b_j are some constant material parameters and, in general, n and m differ. An advantage of the differential equation approach discussed above is that each model is easy to physically understand and interpret and models can be constructed in an intuitive fashion. The drawback, however, is that the more advanced models soon become cumbersome to deal with and as already touched on in relation to the Burgers model the implication of a higher-order differential equation in time is that the initial conditions become difficult to deal with.

We have seen that an exponential term in the form $e^{-\frac{t}{t_r}}$, where t_r is a constant with the dimension of time, often emerges, cf. (14.8) and (14.11). It appears that this exponential term is unity for $t = 0$ whereas it has decreased to the value $\frac{1}{e}$ when $t = t_r$. If t_r emerges in the creep compliance function, cf. (14.11) where $t_r = \frac{\eta}{E}$, t_r is called a *retardation time* and if it emerges in the relaxation modulus, cf. (14.8) where t_r again takes the value $\frac{\eta}{E}$, it is called a *relaxation time*. If the creep compliance $J(t)$ or relaxation modulus $G(t)$ contain more exponential terms, cf. (14.15) and (14.14), it is possible to speak of a retardation or relaxation *spectrum*. In that case, a good approximation to experimental data over a large time span can be achieved if the retardation - or relaxation - times are spread uniformly; typically, a factor of 10 is chosen between these times, cf. for instance Bažant (1982) and Mills (1986).

Returning to the constitutive relation (14.16), it appears that this differential equation is a linear and the *superposition principle* then holds - as expected. Therefore, if the stress history $\sigma = \sigma_1(t)$ implies the response $\varepsilon = \varepsilon_1(t)$, then the stress history $\sigma = k\sigma_1(t)$, where k is a constant, implies the response $\varepsilon = k\varepsilon_1(t)$. Moreover, if the stress history $\sigma = \sigma_1(t)$ implies $\varepsilon = \varepsilon_1(t)$ and the stress history $\sigma = \sigma_2(t)$ implies $\varepsilon = \varepsilon_2(t)$ then the combined stress history $\sigma = \sigma_1(t) + \sigma_2(t)$ implies $\varepsilon = \varepsilon_1(t) + \varepsilon_2(t)$.

In (14.16), the material parameters a_i and b_j are constants, but we can relax this requirement and still maintain the superposition principle; as long as a_i and b_j do not depend on the stress or strain, the superposition principle will still hold. As an example, creep depends strongly on temperature and we may then let $a_i = a_i(T)$ and $b_j = b_j(T)$ where T is the temperature.

It turns out that the effect of temperature can often be evaluated in a simple and elegant fashion and for this purpose consider the relaxation modulus defined by (14.2). Since the material parameters now depend on the temperature T we have $G = G(t, T)$. Consider now relaxation tests performed at different temperatures where T_0 is a reference temperature. Since increase of temperature enhances creep deformation and plotting experimental data against logarithmic time, the results shown in Fig. 14.24a) are obtained.

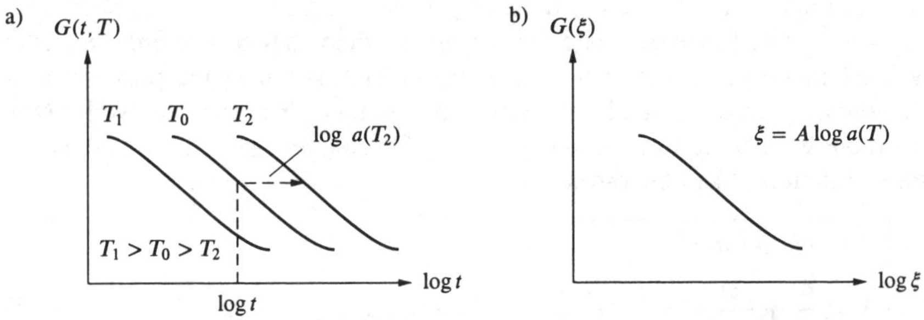

Figure 14.24: Relaxation data plotted against logarithmic times; a) original data, b) master relaxation curve.

For polymers, cf. Findley *et al.* (1976), and also for concrete, see Mukaddam and Bresler (1972), it is often observed experimentally that if the curves in Fig. 14.24a) are moved horizontally they can be brought to coincide; therefore, if the relaxation curve for the reference temperature T_0 is moved the horizontal distance $a(T_2)$, where $a(T)$ is a function of the temperature, then this curve coincides with the relaxation curve for the temperature T_2, cf. Fig. 14.24a). With this property we have

$$G_{T_0}(\log t) = G_{T_2}(\log t + \log a(T_2))$$

This implies that if the *reduced time* ξ is defined by $\xi = t\,a(T)$, where $a(T_0) = 1$, then all relaxation moduli can be written as $G(t, T) = G(\xi)$ and all curves in Fig. 14.24a) can then be expressed by one curve as shown in Fig. 14.24b); this curve is called the *master relaxation curve*. The procedure above is called the *time-temperature shift principle* and evidently the same concepts hold for the creep compliance, i.e.

Time-temperature shift principle

$$G(t, T) = G(\xi)\,; \qquad J(t, T) = J(\xi)$$

The reduced time ξ *is defined by*

$$\xi = t\,a(T)$$

where $a(T)$ *is the shift-factor and* $a(T_0) = 1$

In reality, this principle is not a principle *per se*, but it is an elegant procedure that can be adapted to many materials; it was introduced by Alfrey (1957) and Schwarzl and Staverman (1952) and expressions for the shift-factor $a(T)$ can be found, for instance, in Findley *et al.* (1976). If the temperature varies, it seems reasonable to use $\xi = \int_0^t a(T(\tau))d\tau$ where $T(\tau)$ expresses the temperature - time history.

Returning to (14.16) and in order to achieve a better correlation with experimental data, the material parameters are sometimes allowed to vary with the loading time. Since the material parameters then change - or *harden* - with time, such models are called *time-hardening models*; still the superposition principle holds. As an example, let the viscosity parameter η in the Maxwell model depend on the loading time, as suggested by Dischinger (1937). Then the constitutive relation (14.7) becomes

Dischinger model

$$\dot{\varepsilon} = \frac{\dot{\sigma}}{E} + \frac{\sigma}{\eta(t)} \tag{14.17}$$

exhibits time-hardening

The *Dischinger model* is very often used for modeling creep of concrete and by choosing a proper function for $\eta(t)$ close agreement with experimental data can be obtained; this model forms the basis in many national design codes. However, time-hardening models should be used with care and to illustrate that Fig. 14.25 is considered. In Fig. 14.25a), the stress is infinitely small up to $t = t_1$ where it is increased instantaneously to σ_0; since the loading time starts at time $t = 0$ the strain rate at $t = t_1$ according to (14.17) is $\dot{\varepsilon} = \sigma_0/\eta(t_1)$. In Fig. 14.25b), the loading up to time t_1 is now exactly zero and since the loading time therefore starts at time t_1 the strain rate following (14.17) becomes $\dot{\varepsilon} = \sigma_0/\eta(0)$. In reality, the two responses shown in Fig. 14.25a) and b) must be identical and this illustrates the problems with using loading time as a *creep hardening parameter*; we will return to this problem later.

Apart from creep hardening, another issue is that of *aging*, i.e. the material parameters change with time irrespective of the material being loaded or not. This aging effect is prominent when, for instance, glue is setting or concrete is hardening after it has been cast. Since such aging effects are strongly dependent on the temperature, it is a poor measure for aging just to use time as a parameter; some kind of *maturity concept* is more realistic. For hardening concrete, for instance, one may define an *equivalent maturity time* t_e such that t_e is the same for two concrete specimens (made of the same composition) if the aging is the same despite the two specimens having been exposed to different temperature-time histories. This maturity concept was introduced by Plowman (1956) and a review is given by Byfors (1980). The equivalent maturity time t_e is defined by

Equivalent maturity time t_e for measuring aging effects

$$t_e = \int_0^t \frac{f(\theta(\tau))}{f(\theta_0)} d\tau \tag{14.18}$$

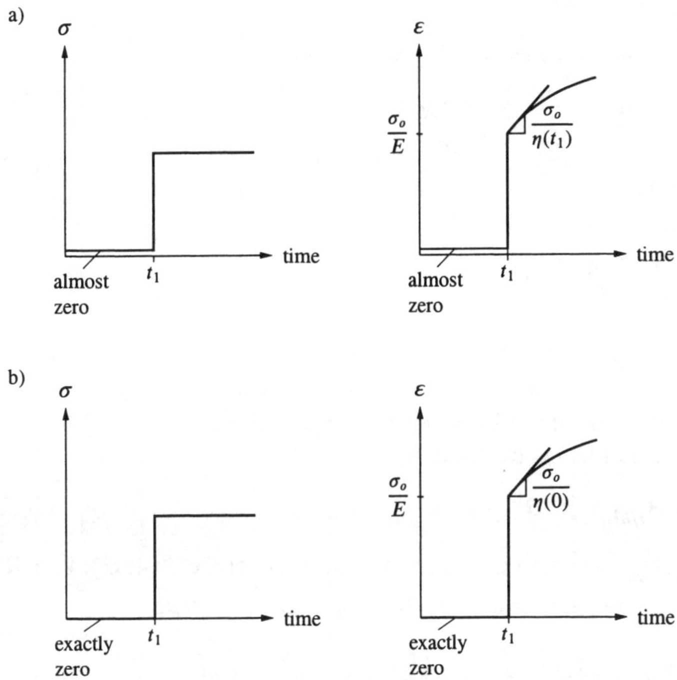

Figure 14.25: Creep tests of Dischinger's time-hardening model; a) between $t = 0$ and $t = t_1$ the stress is infinitely small, b) between $t = 0$ and $t = t_1$ the stress is exactly zero.

where f is a function of the absolute temperature θ $[K]$ and θ_0 is a reference temperature. In (14.18) $\theta = \theta(\tau)$ expresses the temperature-time history and the integration above is performed from time zero where the concrete was cast up until the current time t. As function $f(\theta)$, the *Arrhenius function* for thermal activation is often adopted, i.e. $f(\theta) = ke^{-\frac{Q}{R\theta}}$ where k is a constant, Q is the *activation energy for creep* $[J/mol]$ for the material in question and R is the *universal gas constant* = 8.314 $[J/mol\ K]$.

In view of these remarks, it then seems reasonable to model creep of aging materials by letting the material parameters in (14.16) depend on the equivalent maturity time t_e, i.e. $a_i = a_i(t_e)$ and $b_j = b_j(t_e)$.

Nonlinear viscoelasticity based on (14.16) can be achieved by letting the parameters a_i and b_i depend on the stress and/or strain and we will return to this important possibility in Section 15.3.

With this detailed discussion of various linear viscoelastic models exposed to uniaxial stress conditions, it is timely to see how these models can be generalized to three-dimensional stress conditions. Indeed this is straightforward and from (14.16) it follows directly that

General format for three-dimensional loading

$$\sum_{\alpha=0}^{n} A^{\alpha}_{ijkl} \frac{d^{\alpha}\sigma_{kl}}{dt^{\alpha}} = \sum_{\beta=0}^{m} B^{\beta}_{ijkl} \frac{d^{\beta}\varepsilon_{kl}}{dt^{\beta}}$$

which in a matrix format reads (14.19)

$$\sum_{\alpha=0}^{n} \boldsymbol{A}^{\alpha} \frac{d^{\alpha}\boldsymbol{\sigma}}{dt^{\alpha}} = \sum_{\beta=0}^{m} \boldsymbol{B}^{\beta} \frac{d^{\beta}\boldsymbol{\varepsilon}}{dt^{\beta}}$$

where A^{α}_{ijkl} and B^{β}_{ijkl} are fourth-order tensors; the corresponding matrices $\boldsymbol{A}^{\alpha}$ and $\boldsymbol{B}^{\beta}$ are obtained in a fashion similar to the discussion in Section 4.4.

As an example, consider a Maxwell model for isotropic material behavior. A comparison with (14.7) shows that

$$A^{1}_{ijkl}\dot{\sigma}_{kl} + A^{0}_{ijkl}\sigma_{kl} = B^{1}_{ijkl}\dot{\varepsilon}_{kl} \tag{14.20}$$

With A^{1}_{ijkl} and A^{0}_{ijkl} being isotropic fourth-order tensors, cf. (4.93), and B^{1}_{ijkl} being the *unit fourth-order tensor*, i.e. $B^{1}_{ijkl}\dot{\varepsilon}_{kl} = \dot{\varepsilon}_{ij}$, we obtain

$$\begin{aligned} A^{1}_{ijkl} &= C_{ijkl} = \frac{1}{2G}\Big[\frac{1}{2}(\delta_{ik}\delta_{jl} + \delta_{il}\delta_{jk}) - \frac{\nu}{1+\nu}\delta_{ij}\delta_{kl}\Big] \\ A^{0}_{ijkl} &= \frac{1+\xi}{\eta}\Big[\frac{1}{2}(\delta_{ik}\delta_{jl} + \delta_{il}\delta_{jk}) - \frac{\xi}{1+\xi}\delta_{ij}\delta_{kl}\Big] \\ B^{1}_{ijkl} &= \frac{1}{2}(\delta_{ik}\delta_{jl} + \delta_{il}\delta_{jk}) \end{aligned} \tag{14.21}$$

where G and ν as usual denote the shear modulus and Poisson's ratio, respectively, whereas η is a viscosity material parameter and ξ is a dimensionless material parameter; moreover, A^{1}_{ijkl} is recognized as the isotropic elastic flexibility tensor C_{ijkl}. Insertion of (14.21) into (14.20) provides

Three-dimensional isotropic Maxwell model

$$\frac{1}{2G}(\dot{\sigma}_{ij} - \frac{\nu}{1+\nu}\delta_{ij}\dot{\sigma}_{kk}) + \frac{1+\xi}{\eta}(\sigma_{ij} - \frac{\xi}{1+\xi}\delta_{ij}\sigma_{kk}) = \dot{\varepsilon}_{ij} \tag{14.22}$$

Apart from a redefinition of parameters, this expression corresponds exactly to (6.49), which, however, was derived by somewhat different means.

In accordance with (14.6), (14.22) may be written as $\dot{\varepsilon}^{e}_{ij} + \dot{\varepsilon}^{v}_{ij} = \dot{\varepsilon}_{ij}$ and it is concluded that the viscous strain rate is given by $\dot{\varepsilon}^{v}_{ij} = \frac{1+\xi}{\eta}(\sigma_{ij} - \frac{\xi}{1+\xi}\delta_{ij}\sigma_{kk})$ which for the choice $\xi = 1/2$ gives $\dot{\varepsilon}^{v}_{ij} = \frac{3}{2\eta}s_{ij}$, i.e. the viscous strains - or the creep strains - are incompressible if the material parameter ξ is chosen as $\xi = 1/2$ and the volume changes are then purely elastic.

Another choice of the material parameter ξ becomes apparent if (14.22) is specialized to uniaxial stress conditions. Evaluating the axial strain rate $\dot{\varepsilon}_{11}$ and the transverse strain rate $\dot{\varepsilon}_{22} = \dot{\varepsilon}_{33}$, we obtain

$$\begin{aligned} \frac{1}{E}\dot{\sigma}_{11} + \frac{1}{\eta}\sigma_{11} &= \dot{\varepsilon}_{11} \\ -\frac{\nu}{E}\dot{\sigma}_{11} - \frac{\xi}{\eta}\sigma_{11} &= \dot{\varepsilon}_{22} \end{aligned} \tag{14.23}$$

The first of these equations is in correspondence with (14.7). However, if ξ is chosen as $\xi = \nu$ then (14.23) shows that the relation $\dot{\varepsilon}_{22} = -\nu\dot{\varepsilon}_{11}$ holds not only for elastic conditions, but also during development of creep strains. This convenient situation is often supported experimentally for viscoelastic materials, for instance for concrete, cf. Hannant (1969) and Browne and Blundell (1972).

As another example, consider a Kelvin model for isotropic material behavior. A comparison of (14.19) with (14.10) shows that

$$A^0_{ijkl}\sigma_{kl} = B^0_{ijkl}\varepsilon_{kl} + B^1_{ijkl}\dot{\varepsilon}_{kl} \tag{14.24}$$

With A^0_{ijkl} being the unit fourth-order tensor whereas B^0_{ijkl} and B^1_{ijkl} are isotropic fourth-order tensors, cf. (4.89), we obtain

$$\begin{aligned} A^0_{ijkl} &= \frac{1}{2}(\delta_{ik}\delta_{jl} + \delta_{il}\delta_{jk}) \\ B^0_{ijkl} &= D_{ijkl} = 2G[\frac{1}{2}(\delta_{ik}\delta_{jl} + \delta_{il}\delta_{jk}) + \frac{\nu}{1-2\nu}\delta_{ij}\delta_{kl}] \\ B^1_{ijkl} &= \frac{\eta}{1+\nu}[\frac{1}{2}(\delta_{ik}\delta_{jl} + \delta_{il}\delta_{jk}) + \frac{\xi}{1-2\xi}\delta_{ij}\delta_{kl}] \end{aligned} \tag{14.25}$$

As before, G and ν denote the shear modulus and Poisson's ratio, respectively, whereas η is a viscosity material parameter and ξ is a dimensionless material parameter; moreover, B^0_{ijkl} is recognized as the isotropic elastic stiffness tensor D_{ijkl}. Intuitively, it seems more natural to choose the common factor in B^1_{ijkl} as $\frac{\eta}{1+\xi}$, but the choice $\frac{\eta}{1+\nu}$ turns out to be more convenient as the later calculations will show. Insertion of (14.25) into (14.24) gives

Three-dimensional isotropic Kelvin model

$$\sigma_{ij} = 2G(\varepsilon_{ij} + \frac{\nu}{1-2\nu}\delta_{ij}\varepsilon_{kk}) + \frac{\eta}{1+\nu}(\dot{\varepsilon}_{ij} + \frac{\xi}{1-2\xi}\delta_{ij}\dot{\varepsilon}_{kk}) \tag{14.26}$$

Apart from a redefinition of parameters, this expression corresponds exactly to (6.56), which, however, was derived by somewhat different means. The constitutive relation above implies $\sigma_{ii} = 3K\varepsilon_{ii} + \frac{\eta}{1+\nu}\frac{1+\xi}{1-2\xi}\dot{\varepsilon}_{ii}$ and the choice $\xi = -1$ therefore results in $\sigma_{ii} = 3K\varepsilon_{ii}$, i.e. the volumetric response is purely elastic.

Another choice of the material parameter ξ becomes apparent if (14.26) is specialized to uniaxial stress conditions where ε_{11} is the axial strain and $\varepsilon_{22} = \varepsilon_{33}$ is the transverse strain. If we again enforce the condition that $\varepsilon_{22} = -\nu\varepsilon_{11}$ should hold even during creep development, we obtain $\varepsilon_{kk} = (1-2\nu)\varepsilon_{11}$ and (14.26) evaluated for $ij = 11$ and $ij = 22$ then provides

$$\begin{aligned} \sigma_{11} &= E\varepsilon_{11} + \frac{\eta}{1+\nu}(1+\xi\frac{1-2\nu}{1-2\xi})\dot{\varepsilon}_{11} \\ 0 &= \frac{\eta}{1+\nu}\frac{\xi-\nu}{1-2\xi}\dot{\varepsilon}_{11} \end{aligned}$$

The second of these equations is fulfilled for $\xi = \nu$ and the first equation then reduces to

$$\sigma_{11} = E\varepsilon_{11} + \eta\dot{\varepsilon}_{11}$$

which corresponds to (14.10). We have then shown that the choice $\xi = \nu$ implies that $\varepsilon_{22} = -\nu\varepsilon_{11}$ holds during uniaxial stress conditions.

Above we have considered isotropic material behavior, but if the material is, say, orthotropic, it is often more convenient to work with the matrix version of (14.19). Then the matrices $\boldsymbol{A}^{\alpha}$ and $\boldsymbol{B}^{\beta}$ are either unit matrices or orthotropic matrices and the orthotropic matrices will each contain nine independent material parameters in complete similarity with the discussion in Section 4.6, cf. in particular the orthotropic matrix format given by (4.55). Orthotropic viscoelasticity is often used to model creep of wood, see Mårtensson (1992) and Ormarsson (1999), and paper, see Lif *et al.* (1999).

14.3 Hereditary approach

We will now introduce the *hereditary approach* to linear viscoelasticity, which allows greater freedom when constructing models than the differential approach that relies on the concepts of certain combinations of springs and dashpots. The essential issue in the hereditary approach is that of superposition.

Consider the uniaxial stress history in Fig. 14.26a) where the stress is increased instantaneously the constant amount $\Delta\sigma$ at time τ. If $\Delta\sigma$ were applied at time $t = 0$, the corresponding strain at time t would be given by $\Delta\varepsilon(t) = J(t)\Delta\sigma$, cf. (14.1), but now $\Delta\sigma$ is applied at time τ so the strain $\Delta\varepsilon(t)$ at time t caused by $\Delta\sigma$ applied at time τ becomes

$$\Delta\varepsilon(t) = J(t-\tau)\Delta\sigma$$

For an infinitesimal stress change $d\sigma$ applied at time τ, we then obtain $d\varepsilon(t) = J(t-\tau)d\sigma$ and integration all infinitesimal stress changes over the entire load history up until the current time t then provides

$$\varepsilon(t) = \int_{-\infty}^{t} J(t-\tau)d\sigma(\tau) \tag{14.27}$$

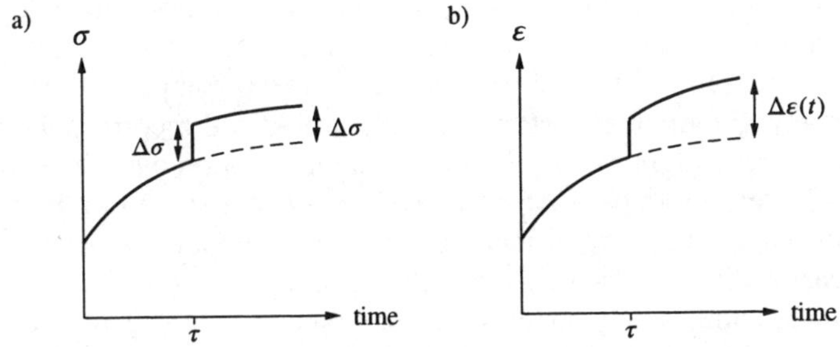

Figure 14.26: a) Stress history where the stress is increased the constant amount $\Delta\sigma$ at time τ, b) corresponding strain response.

where the notation $d\sigma(\tau)$ expresses that the infinitesimal stress changes are given as a function of the stress history $\sigma = \sigma(\tau)$. In (14.27), the lower integration limit is by tradition taken as $-\infty$, but since the stress is equal to zero up until time zero, where the loading begins, the contribution form the integral in (14.27) from $-\infty$ to zero is nil. Moreover, suppose that an instantaneous loading occurs at time zero according to

$$\sigma(t) = \begin{cases} 0 & \text{when } \; t < 0 \\ \sigma_0 + \sigma_1(t) \, ; \quad \sigma_1(0) = 0 & \text{when } \; t \geq 0 \end{cases}$$

then care should be taken in the integration. With evident notation we obtain

$$\begin{aligned} \varepsilon(t) &= \int_{-\infty}^{0^-} J(t-\tau) d\sigma(\tau) + \int_{0^-}^{0^+} J(t-\tau) d\sigma(\tau) + \int_{0^+}^{t} J(t-\tau) d\sigma(\tau) \\ &= 0 + J(t)\sigma_0 + \int_{0^+}^{t} J(t-\tau) d\sigma(\tau) \end{aligned}$$

i.e.

$$\varepsilon(t) = J(t)\sigma_0 + \int_{0^+}^{t} J(t-\tau) d\sigma(\tau)$$

With this interpretation, it appears that jumps are allowed in the stress history and the integral in (14.27) is therefore a so-called *Stieltjes integral*.

However, if the stress history $\sigma = \sigma(\tau)$ is smooth, we have $d\sigma(\tau) = \frac{d\sigma(\tau)}{d\tau} d\tau$ and (14.27) then takes the form

Hereditary approach

$$\varepsilon(t) = \int_{-\infty}^{t} J(t-\tau) \frac{d\sigma(\tau)}{d\tau} d\tau \tag{14.28}$$

where the integral is the usual *Rieman integral.* However, in the following we will adopt the format (14.28) in the sense that if the stress history exhibits jumps, then the interpretation (14.27) should be used. If the creep compliance $J(t)$ is known then for any given stress history, (14.28) provides the corresponding strain. Since the current strain $\varepsilon(t)$ is obtained as an integration over the entire loading history, the terminology of *hereditary approach* is assigned to the format (14.28). Moreover, the integral is an example of a so-called *convolution integral* and occasionally it is also called a *Duhamel integral.*

The result (14.28) hinges only on the superposition principle which is called *Boltzmann's superposition principle* and is due to Boltzmann (1874), but the specific format given by (14.28) is due to Volterra (1913) and a material obeying (14.28) is therefore also called a *Boltzmann-Volterra material.* In (14.28), knowledge of $J(t-\tau)$ and of the stress history provides the corresponding strain. However, if in (14.28) the strain $\varepsilon(t)$ and the creep compliance $J(t-\tau)$ are taken as given and the stress history $\sigma = \sigma(\tau)$ is taken as unknown, (14.28) represents an *integral equation* which, not surprisingly, turns out to be a *Volterra integral equation.* Viewed in this manner $J(t-\tau)$ comprises the so-called *kernel* and for more details see, for instance, Hildebrand (1965) and Volterra (1959).

Since (14.28) only rests on the superposition principle, the models derived in the previous section can be recast into this format. As an example, take the Maxwell model with the creep compliance given by (14.8). From (14.28) it then follows that

$$
\begin{aligned}
\varepsilon(t) &= \int_{-\infty}^{t} [\frac{1}{E} + \frac{t-\tau}{\eta}] \frac{d\sigma(\tau)}{d\tau} d\tau \\
&= (\frac{1}{E} + \frac{t}{\eta}) \int_{0-}^{t} d\sigma(\tau) - \frac{1}{\eta} \int_{0-}^{t} \tau \, d\sigma(\tau) \\
&= (\frac{1}{E} + \frac{t}{\eta})\sigma - \frac{1}{\eta}\{[\tau\,\sigma(\tau)]_{0-}^{t} - \int_{0-}^{t} \sigma(\tau)\, d\tau\} \\
&= \frac{\sigma}{E} + \frac{1}{\eta} \int_{0-}^{t} \sigma(\tau)\, d\tau
\end{aligned}
$$

Differention with respect to time then provides

$$
\dot{\varepsilon} = \frac{\dot{\sigma}}{E} + \frac{\sigma}{\eta}
$$

in accordance with (14.7).

Therefore, per definition, the format (14.28) includes the models derived in the previous section, but (14.28) is more general since we can now specify any creep compliance $J(t)$ and for a given stress history, (14.28) determines the current strain. Therefore, the creep compliance $J(t)$ ca be identified entirely by means of experimental evidence and no resort has to be taken to an interpretation in terms of springs and dashpots.

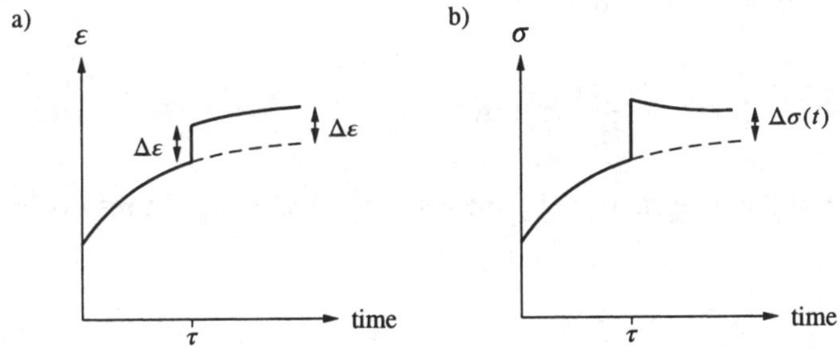

Figure 14.27: a) Strain history where the strain is increased the constant amount $\Delta\varepsilon$ at time τ, b) corresponding stress response.

Let us now assume that the strain history is known and let us derive a result analogous to (14.28). If the constant strain change $\Delta\varepsilon$ is applied at time τ, see Fig. 14.27a), the corresponding stress change $\Delta\sigma(t)$ at the current time t is given by $\Delta\sigma(t) = G(t-\tau)\Delta\varepsilon$, cf. (14.2). We are then led to

$$\sigma(t) = \int_{-\infty}^{t} G(t-\tau)\, d\varepsilon(\tau) \tag{14.29}$$

where the notation $d\varepsilon(\tau)$ expresses that the infinitesimal strain changes are given as function of the strain history $\varepsilon = \varepsilon(\tau)$. If a jump exists in the strain history, the interpretation of (14.29) is similar to (14.27), i.e. (14.29) is a Stieltjes integral. If the strain history $\varepsilon = \varepsilon(\tau)$ is smooth, (14.29) takes the format

Hereditary approach

$$\sigma(\tau) = \int_{-\infty}^{t} G(t-\tau)\frac{d\varepsilon(\tau)}{d\tau}\, d\tau \tag{14.30}$$

Again the lower integration limit is per tradition taken as $-\infty$, but the integral contributes with nil up to time zero where the strain is applied, cf. the similar discussion following (14.27). It appears that once the relaxation modulus $G(t)$ and the strain history are known, (14.30) provides the corresponding stress.

Since the format (14.30) only relies on the superposition principle, this format contains all the models discussed in the previous section. As an example, take the Maxwell model with the relaxation modulus $G(t)$ given by (14.8). From (14.30) it then follows that

$$\begin{aligned}\sigma(t) &= \int_{-\infty}^{t} E e^{-\frac{E}{\eta}(t-\tau)} \frac{d\varepsilon(\tau)}{d\tau} d\tau \\ &= E e^{-\frac{E}{\eta}t} \int_{0^-}^{t} e^{\frac{E}{\eta}\tau} \frac{d\varepsilon(\tau)}{d\tau} d\tau\end{aligned}$$

Differentiation with respect to time gives

$$\dot{\sigma} = -\frac{E^2}{\eta} e^{-\frac{E}{\eta}t} \int_{0^-}^{t} e^{\frac{E}{\eta}\tau} \frac{d\varepsilon(\tau)}{d\tau} d\tau + E\dot{\varepsilon}$$

and elimination of the integral term by means of the first equation results in

$$\dot{\sigma} = -\frac{E}{\eta}\sigma + E\dot{\varepsilon}$$

in accordance with (14.7).

However, the format (14.30) is more general than the format discussed in the previous section since it is now possible from experimental evidence to directly propose any relaxation modulus $G(t)$ and for a given strain history, (14.30) then provides the corresponding stress.

We have previously indicated that one might expect that some relation exists between the creep compliance $J(t)$ and the relaxation modulus $G(t)$; let us now identify this relation.

Consider a creep test where the constant stress σ_0 is applied instantaneously at time $t = 0$. Then (14.28) - or (14.27) - gives

$$\varepsilon(t) = J(t)\sigma_0$$

as expected. Supposing that this strain history is known, then insertion into (14.30) should provide $\sigma(t) = \sigma_0$, i.e.

$$\boxed{\int_{-\infty}^{t} G(t-\tau)\frac{dJ(\tau)}{d\tau} d\tau = 1} \tag{14.31}$$

In the integration, it is noted that both G and J are zero when $t < 0$ and the integration for $-\infty$ up to $t = 0^-$ therefore gives no contribution and most often there will be a discontinuity in $J(t)$ at time $t = 0$.

Consider next a relaxation test where the constant strain ε_0 is applied instantaneously at time $t = 0$. Then (14.30) - or (14.29) - gives

$$\sigma(t) = G(t)\varepsilon_0$$

as expected. Supposing that this stress history is known, then insertion into (14.28) should provide $\varepsilon(t) = \varepsilon_0$, i.e.

$$\boxed{\int_{-\infty}^{t} J(t-\tau)\frac{dG(\tau)}{d\tau} d\tau = 1}$$

In this integration similar arguments hold to those discussed in relation to (14.31).

The freedom with which we can choose $J(t)$ - or $G(t)$ - is a major advantage of the hereditary approach. Moreover, it is easy to generalize to three-dimensional loading and we obtain directly for (14.28) and (14.30)

Three-dimensional loading

$$\varepsilon_{ij}(t) = \int_{-\infty}^{t} J_{ijkl}(t-\tau)\frac{d\sigma_{kl}(\tau)}{d\tau}d\tau$$

or

$$\sigma_{ij}(t) = \int_{-\infty}^{t} G_{ijkl}(t-\tau)\frac{d\varepsilon_{kl}(\tau)}{d\tau}d\tau$$

which in evident matrix notation becomes

$$\begin{aligned}\boldsymbol{\varepsilon}(t) &= \int_{-\infty}^{t} \boldsymbol{J}(t-\tau)\frac{d\boldsymbol{\sigma}(\tau)}{d\tau}d\tau \\ \boldsymbol{\sigma}(t) &= \int_{-\infty}^{t} \boldsymbol{G}(t-\tau)\frac{d\boldsymbol{\varepsilon}(\tau)}{d\tau}d\tau\end{aligned} \qquad (14.32)$$

We will now illustrate a particular property relating to the hereditary approach. Suppose that $\boldsymbol{J}(t-\tau)$ and the stress history $\boldsymbol{\sigma}(\tau)$ are known; at time $t = t_1$, (14.32a) then gives

$$\boldsymbol{\varepsilon}(t_1) = \int_{-\infty}^{t_1} \boldsymbol{J}(t_1-\tau)\frac{d\boldsymbol{\sigma}(\tau)}{d\tau}d\tau$$

Consider now the strains at time $t = t_1 + \Delta t$ where Δt is a small time increment; we obtain

$$\boldsymbol{\varepsilon}(t_1+\Delta t) = \int_{-\infty}^{t_1+\Delta t} \boldsymbol{J}(t_1+\Delta t-\tau)\frac{d\boldsymbol{\sigma}(\tau)}{d\tau}d\tau$$

i.e.

$$\begin{aligned}\boldsymbol{\varepsilon}(t_1+\Delta t) = &\int_{-\infty}^{t_1} \boldsymbol{J}(t_1+\Delta t-\tau)\frac{d\boldsymbol{\sigma}(\tau)}{d\tau}d\tau \\ &+ \int_{t_1}^{t_1+\Delta t} \boldsymbol{J}(t_1+\Delta t-\tau)\frac{d\boldsymbol{\sigma}(\tau)}{d\tau}d\tau\end{aligned}$$

The important thing is that the first term on the right-hand side is not equal to $\boldsymbol{\varepsilon}(t_1)$ since the argument in the function $\boldsymbol{J}$ differs in the two cases. Therefore the first term on the right-hand side needs to be integrated from the beginning of the load history. Evidently, this complicates the application since at each time one must perform an integration over the entire load history to obtain the corresponding strain; indeed this is a consequence of the hereditary approach where the current response is a result of the entire load history. However, this complicates the (numerical) determination of the response and it may be argued that the response of a material is more dependent on its recent history than on its past; this expectation is referred to as the concept of *fading memory* which is often adopted in constitutive mechanics, cf. the discussion given by Eringen (1975b); let us now see how the above problems can be circumvented.

The generalized Kelvin model appears from Fig. 14.23 and it was shown that close predictions to experimental data can be obtained with this model. The corresponding creep compliance is given by (14.15), i.e

$$J(t) = \frac{1}{E_0} + \sum_{i=1}^{n} \frac{1}{E_i}(1 - e^{-\frac{E_i}{\eta_i}t})$$

This expression is immediately generalized to three-dimensional loading according to

$$\boldsymbol{J}(t) = \boldsymbol{C}_0 + \sum_{i=1}^{n} \boldsymbol{C}_i(1 - e^{-\frac{E_i}{\eta_i}t})$$

where $\boldsymbol{C}_0$ and $\boldsymbol{C}_i$ are constant matrices. Insertion into (14.32) provides

$$\boldsymbol{\varepsilon}(t) = \int_{-\infty}^{t} [\boldsymbol{C}_0 + \sum_{i=1}^{n} \boldsymbol{C}_i(1 - e^{-\frac{E_i}{\eta_i}(t-\tau)})] \frac{d\boldsymbol{\sigma}(\tau)}{d\tau} d\tau$$

which can be written as

Generalized Kelvin

$$\boldsymbol{\varepsilon}(t) = (\boldsymbol{C}_0 + \sum_{i=1}^{n} \boldsymbol{C}_i)\boldsymbol{\sigma}(t) - \sum_{i=1}^{n} e^{-\frac{E_i}{\eta_i}t} \tilde{\boldsymbol{\varepsilon}}_i(t) \tag{14.33}$$

where

$$\tilde{\boldsymbol{\varepsilon}}_i(t) = \boldsymbol{C}_i \int_{-\infty}^{t} e^{\frac{E_i}{\eta_i}\tau} \frac{d\boldsymbol{\sigma}(\tau)}{d\tau} d\tau$$

At time $t = t_1$, we obtain

$$\boldsymbol{\varepsilon}(t_1) = (\boldsymbol{C}_0 + \sum_{i=1}^{n} \boldsymbol{C}_i)\boldsymbol{\sigma}(t_1) - \sum_{i=1}^{n} e^{-\frac{E_i}{\eta_i}t_1} \tilde{\boldsymbol{\varepsilon}}_i(t_1) \tag{14.34}$$

and at time $t = t_1 + \Delta t$ where Δt is a small time increment, we have with $\boldsymbol{\sigma}(t_1 + \Delta t) = \boldsymbol{\sigma}(t_1) + \Delta\boldsymbol{\sigma}$

$$\begin{aligned} \boldsymbol{\varepsilon}(t_1 + \Delta t) =& (\boldsymbol{C}_0 + \sum_{i=1}^{n} \boldsymbol{C}_i)(\boldsymbol{\sigma}(t_1) + \Delta\boldsymbol{\sigma}) \\ & - \sum_{i=1}^{n} e^{-\frac{E_i}{\eta_i}(t_1 + \Delta t)} \tilde{\boldsymbol{\varepsilon}}_i(t_1 + \Delta t) \end{aligned} \tag{14.35}$$

Let the quantity $\Delta\tilde{\varepsilon}_i$ be determined as

$$\Delta\tilde{\varepsilon}_i = \boldsymbol{C}_i \int_{t_1}^{t_1+\Delta t} e^{\frac{E_i}{\eta_i}\tau} \frac{d\sigma(\tau)}{d\tau} d\tau \tag{14.36}$$

then

$$\tilde{\varepsilon}_i(t_1 + \Delta t) = \tilde{\varepsilon}_i(t_1) + \Delta\tilde{\varepsilon}_i \tag{14.37}$$

Insertion of (14.37) in (14.35) and subtraction of (14.34) provide

$$\begin{aligned}\varepsilon(t_1 + \Delta t) &= \varepsilon(t_1) + (\boldsymbol{C}_0 + \sum_{i=1}^{n} \boldsymbol{C}_i)\Delta\sigma \\ &- \sum_{i=1}^{n}(e^{-\frac{E_i}{\eta_i}(t_1+\Delta t)} - e^{-\frac{E_i}{\eta_i}t_1})\tilde{\varepsilon}_i(t_1) - \sum_{i=1}^{n} e^{-\frac{E_i}{\eta_i}(t_1+\Delta t)}\Delta\tilde{\varepsilon}_i\end{aligned}$$

It appears that once $\varepsilon(t_1)$ and $\tilde{\varepsilon}_i(t_1)$ are known, all terms on the right-hand side are trivially identified except for the very last term that involves the quantity $\Delta\tilde{\varepsilon}_i$ which involves an integration. However, this integration is given by (14.36) and the important issue is that the integration limits are t_1 and $t_1 + \Delta t$, i.e the integration should only be performed over the current time step and not over the entire loading history.

The format (14.33) therefore allows a close prediction to experimental data and it implies a computational scheme that is very simple; for that reason this format is often adopted in the literature. In essence, it was suggested by Zienkiewicz and Watson (1966) and more information and generalizations to include aging material parameters are given by Bažant (1979, 1982, 1996) and Dahlblom (1987). Applications to orthotropic materials are discussed for wood by Mårtensson (1992) and Ormarsson (1999) and for paper by Lif (2003).

Let us as an example establish the matrices $\boldsymbol{C}_0$ and $\boldsymbol{C}_i$ - or rather their tensorial counterparts - in (14.33) for isotropic materials. In analogy with (14.21) we take

$$\begin{aligned}C^0_{ijkl} &= C_{ijkl} = \frac{1}{2G}[\frac{1}{2}(\delta_{ik}\delta_{jl} + \delta_{il}\delta_{jk}) - \frac{\nu}{1+\nu}\delta_{ij}\delta_{kl}] \\ C^i_{ijkl} &= \frac{1}{2G_i}[\frac{1}{2}(\delta_{ik}\delta_{jl} + \delta_{il}\delta_{jk}) - \frac{\xi_i}{1+\xi_i}\delta_{ij}\delta_{kl}]\end{aligned}$$

For uniaxial stress conditions, where ε_{11} is the axial strain and $\varepsilon_{22} = \varepsilon_{33}$ is the transverse strain, (14.33) gives

$$\varepsilon_{11} = (\frac{1}{2G}\frac{1}{1+\nu} + \sum_{i=1}^{n}\frac{1}{2G_i}\frac{1}{1+\xi_i})\sigma_{11} - \sum_{i=1}^{n} e^{-\frac{E_i}{\eta_i}t}\int_{-\infty}^{t} e^{\frac{E_i}{\eta_i}\tau}\frac{1}{2G_i}\frac{1}{1+\xi_i}\frac{d\sigma_{11}}{d\tau}d\tau$$

$$\varepsilon_{22} = -(\frac{1}{2G}\frac{\nu}{1+\nu} + \sum_{i=1}^{n}\frac{1}{2G_i}\frac{\xi_i}{1+\xi_i})\sigma_{11} + \sum_{i=1}^{n} e^{-\frac{E_i}{\eta_i}t}\int_{-\infty}^{t} e^{\frac{E_i}{\eta_i}\tau}\frac{1}{2G_i}\frac{\xi_i}{1+\xi_i}\frac{d\sigma_{11}}{d\tau}d\tau$$

It appears that if we choose $\xi_i = \nu$ then $\varepsilon_{22} = -\nu\varepsilon_{11}$ holds not only for elastic conditions, but also during development of creep strains; as previously discussed, this convenient situation is often supported experimentally, for instance for concrete, and it implies in (14.33) the great simplification that $\boldsymbol{C}_i = \frac{G}{G_i}\boldsymbol{C}_0$.

We finally observe that the format (14.33), in general, provides a very simple expression for the strain rate, namely

$$\dot{\boldsymbol{\varepsilon}}(t) = \boldsymbol{C}_0\dot{\boldsymbol{\sigma}}(t) + \sum_{i=1}^{n} \frac{E_i}{\eta_i} e^{-\frac{E_i}{\eta_i}t} \tilde{\boldsymbol{\varepsilon}}_i(t)$$

Let us finally observe that it is tradition in the literature on linear viscoelasticity to make use of the *Laplace-transform*, which is a convenient means to transform linear ordinary differential equations like (14.16) into algebraic equations and which also transforms convolution integrals like (14.28) to simple expressions in the Laplace-transforms. For convenience and in order to emphasize the physical aspects, we have here chosen not to make use of this approach.

The hereditary approach described above is derived from Boltzmann's superposition principle, i.e. it relies on linearity. However, it is possible to adopt a hereditary format and even develop a theory for nonlinear viscoelasticity. A number of possibilities exists and one approach is to use a multiple integral representation instead of the single integral appearing in (14.28). However, the formulations soon become very involved and applications for the solution of engineering problems seem to be scarce. Comprehensive reviews of various nonlinear hereditary theories are also included in the expositions of Findley *et al.* (1976) and Rabotnov (1980).

15 CREEP AND VISCOPLASTICITY

In the following chapter, we will discuss the *classical creep* theories for steel and metals, as well as the general formulations for creep and *viscoplasticity*. In these formulations - and in contrast with viscoelasticity - it is characteristic that the time-dependent strains are nonproportionally related to the stresses.

The classical creep theories for steel and metals are of relevance when the loading is below the initial yield stress. In that case, it turns out that in addition to the elastic strain, a sufficiently high temperature - and stress - will also give rise to a creep strain, i.e.

$$\boxed{\varepsilon = \varepsilon^e + \varepsilon^{cr}} \tag{15.1}$$

The question then arises what a 'sufficiently' high temperature means, and, according to Ashby and Jones (1980), it is found for metals and steel that

$$\boxed{\frac{\theta}{\theta_M} > 0.3 - 0.4 \quad \Rightarrow \quad creep}$$

where θ is the absolute temperature $[K]$ and θ_M is the melting temperature; for ceramics it is found that a temperature of $\theta/\theta_M > 0.4 - 0.5$ is required for development of creep strains. The ratio θ/θ_M is called the *homologous temperature*. For lead $\theta_M \approx 600\,K$ and at room temperature we therefore have $\theta/\theta_M \approx 0.5$ and lead pipes in old buildings are therefore often found to exhibit significant creep deformations. For steel, $\theta_M \approx 1800\,K$ and creep is therefore of importance when the temperature is above $270 - 450\,^oC$.

General reviews of classical creep theories for steel and metals, are given by Finnie and Heller (1959), Lubahn and Felgar (1961), Odqvist and Hult (1962), Nadai (1963), Odqvist (1966), Hult (1966), Rabotnov (1969) and Stouffer and Dame (1996).

15.1 Results based on the standard creep test

For a standard creep test performed at the constant temperature θ and where the constant tensile stress σ is applied instantaneously, we have $\varepsilon^{cr} = \varepsilon^{cr}(\sigma,$ load history, θ) and as the load history is uniquely related to the loading time t, we can write

$$\varepsilon^{cr} = \varepsilon^{cr}(\sigma, t, \theta)$$

Frequently, however, it is assumed that the influence of stress, loading time and temperature can be factorized, i.e.

$$\varepsilon^{cr} = f(\sigma)g(t)h(\theta) \tag{15.2}$$

It is characteristic for creep of steel and metals that ε^{cr} for stress states of engineering importance depends nonlinearly on stress and most often a *power law* is adopted, i.e.

$$f(\sigma) \sim (\frac{\sigma}{\sigma^*})^n \tag{15.3}$$

as proposed by Norton (1929) and Bailey (1929). Here the *creep exponent* $n \geq 1$ is usually in the range of 3-8, cf. Ashby and Jones (1980), and σ^* is an arbitrary, positive reference stress that we choose with the sole purpose of making the expression dimensionless; in the literature, this power law is often called *Norton's law*. A number of other expressions is available, for instance $f(\sigma) \sim e^{B\sigma} - 1$, where the parameter B is positive, as proposed by Soderberg (1936), and $f(\sigma) \sim sinh(C\sigma)$, where the parameter C is positive, as proposed by Prandtl (1928) and Nadai (1938). For larger stresses, these two expressions coincide in the limit; moreover, for small stresses both expressions provide a linear stress dependence and we will return to this aspect later. In practice, both the power law, the exponential law and the hyperbolic sine law can - within a certain stress range - closely approximate experimental data, but the power law is the expression commonly used. For a more detailed discussion and comparison of these expressions, the reader may consult Nadai (1963), Odqvist (1966) and Rabotnov (1969).

The influence of loading time is often modeled through the simple expression

$$g(t) \sim (\frac{t}{t^*})^m \tag{15.4}$$

where $0 < m \leq 1$; it appears that $0 < m < 1$ corresponds to primary creep and $m = 1$ to secondary creep. Moreover, t^* is an arbitrary reference time we choose with the sole purpose that the expression becomes dimensionless. Expression (15.4) has the drawback that either primary creep ($0 < m < 1$) or secondary creep $m = 1$ is considered. A format which allows primary and secondary creep to be considered in one and the same expression was proposed by McVetty (1934) and it reads $\varepsilon^{cr} = f(\sigma, \theta)(1 - e^{-at}) + g(\sigma, \theta)t$, where a is a positive parameter. It appears that when $t \to \infty$ then $\dot{\varepsilon}^{cr} \to g(\sigma, \theta)$ where $g(\sigma, \theta)$ is the secondary creep rate.

Finally, the influence of temperature is most often modeled via the *law of Arrhenius*, i.e.

$$h(\theta) \sim e^{-\frac{Q}{R\theta}} \tag{15.5}$$

where R is the *universal gas constant* = 8.314 $[\frac{\text{J}}{\text{mol}\,K}]$ and Q is the *activation energy* for creep [J/mol]. The reason for this choice is that, as we will discuss later, creep development is closely related to diffusion processes and the temperature influence on diffusion processes is controlled by the law of Arrhenius, cf. for instance Ashby and Jones (1980).

Certainly, there exist other relations for $f(\sigma)$, $g(t)$ and $h(\theta)$ than those discussed above - reviews are given by Altenbach (1999), Boresi and Sidebottom (1972) and Stouffer and Dame (1996) - but the expressions we have chosen are the ones most often adopted. Combining (15.2) with (15.3)-(15.5) we obtain

$$\varepsilon^{cr} = A(\frac{\sigma}{\sigma^*})^n(\frac{t}{t^*})^m e^{-\frac{Q}{R\theta}} \tag{15.6}$$

where A is a positive dimensionless material parameter. It is recalled that σ^* and t^* are an arbitrary positive reference stress and reference time respectively, that we choose with the sole purpose of obtaining a convenient dimension of the parameter A. Once σ^* and t^* have been chosen, (15.6) involves the four material parameters A, n, m and Q. Expression (15.6) holds for a creep test where σ is a tensile stress and in order that a compressive creep test is also allowed for, (15.6) is rewritten as

$$\boxed{\begin{array}{l}\varepsilon^{cr} = A(\frac{|\sigma|}{\sigma^*})^{n-1}\frac{\sigma}{\sigma^*}(\frac{t}{t^*})^m e^{-\frac{Q}{R\theta}} \\ \textit{where } \sigma^* \textit{ and } t^* \textit{ are arbitrary positive reference quantities}\end{array}} \tag{15.7}$$

In practice, secondary creep is often of great practical interest. If the stress and temperature are kept constant over a long period, secondary creep will occur and this is the dominant creep mechanism in, for instance, a steam generator at a power station which in practice operates under stationary conditions over very long periods. Much research has therefore focused on secondary creep and how temperature and stress, for a given material, influence the creep behavior. Research within material science has identified some characteristic creep mechanisms in steel and metals that will be discussed shortly; for a more detailed discussion the reader may consult Gittus (1975), Ashby and Jones (1980), Evans and Wilshire (1993) and Stouffer and Dame (1996).

For secondary creep, it turns out that creep for small stresses may occur due to so-called *diffusional creep* where atoms and vacancies are transported by means of diffusion; in *Nabarro-Herring creep* this transport occurs through the grains whereas in *Coble creep* it occurs along the grain boundaries. It turns out that diffusional creep depends almost linearly on stress, i.e. the creep exponent in (15.6) is $n \approx 1$.

Still considering secondary creep, but now evaluating creep development for large stresses, another type of creep process becomes dominant - this is the so-called *dislocation creep* mechanism. From the field of material science, it is recalled that plastic strains in steel and metals develop due to movements of *dislocations*; a dislocation being an irregularity in the arrangement of atoms in the crystal lattice. When a dislocation meets an obstacle (dissolved solute atoms, vacancies, etc.), an increased stress is necessary to move the dislocation further and this is what happens in hardening plasticity. However, when exposed to the same stress over a long time - as it occurs in the standard creep test - another possibility for the motion of the dislocation emerges; namely that due to diffusional processes, the dislocation becomes able to 'climb' over the obstacle. Therefore, dislocation creep is also referred to as *dislocation climb* whereas plasticity (in steel and metals) is due to *dislocation glide*. It is characteristic that dislocation creep results in a strongly nonlinear dependence of the stress, and theoretical considerations within material science suggest that the creep exponent in Norton's law is $n \approx 4$.

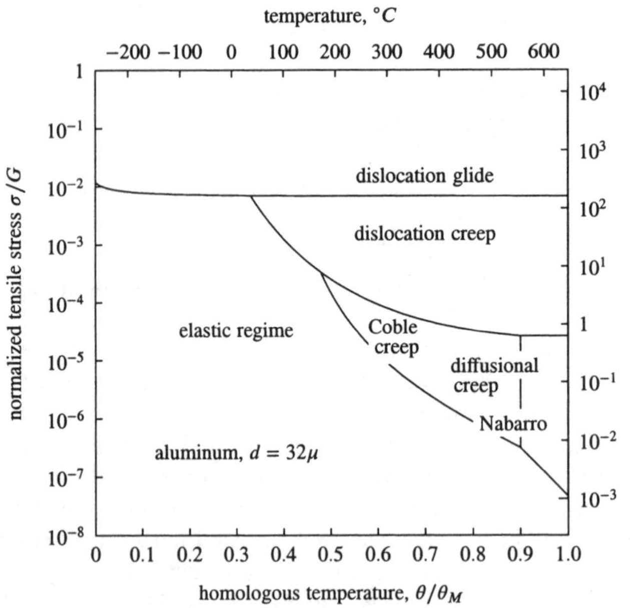

Figure 15.1: Deformation mechanism map for aluminum (grain diameter d = 32 μm), Gittus (1975); secondary creep is considered.

Information of how the stress and temperature influence the dominant deformation mechanisms for a particular material can be conveniently shown in a *deformation mechanism map*; as an example, Fig. 15.1 shows the result for aluminum. Above the curve 'dislocation glide', plasticity occurs and the dominant regions for elastic behavior, diffusional creep where $n \approx 1$ and dislocation creep where $n > 1$ appear from this figure; it is recalled that these creep regions refer to the occurrence of secondary creep data. Moreover, in Fig. 15.1 the boundary separating the elastic response from the creep regions is defined by $\dot{\varepsilon}^c = 10^{-8}[1/s]$. Deformation mechanism maps are available for a number of materials, see Gittus (1975) and Frost and Ashby (1982), and qualitatively they all have the same appearance as Fig. 15.1. Moreover, it turns out that the grain diameter d has a strong influence on the deformation mechanism map.

15.2 Uniaxial stress changes - Classical hardening rules

For a constant temperature, (15.6) can be written as

$$\varepsilon^{cr} = A(\frac{\sigma}{\sigma^*})^n(\frac{t}{t^*})^m \tag{15.8}$$

where the new parameter A has replaced the old term $Ae^{-\frac{Q}{R\theta}}$; moreover, for convenience only non-negative stresses are considered and the format (15.7) is then not needed. Expression (15.8) relates to the standard creep test where the stress is constant. Differentiation with respect to time gives

$$\dot{\varepsilon}^{cr} = \frac{mA}{t^*}(\frac{\sigma}{\sigma^*})^n(\frac{t}{t^*})^{m-1} \tag{15.9}$$

In plasticity theory it was argued that no unique relation exists between the stresses and strains, and the constitutive relations must therefore be of an incremental nature. In exactly the same fashion, we cannot expect that the creep strain for an arbitrary stress history only depends on the current stress value and loading time; if that were the case, the creep strain would then be independent of the load history. Therefore, when the stress changes during the load history, we must try to establish an expression for $\dot{\varepsilon}^{cr}$ and not for ε^{cr}.

Formally, (15.9) is identical to (15.8), but if (15.9) is now postulated to hold even when the stress is allowed to change during the load history, an expression for the creep strain rate has been achieved that holds for general uniaxial stress conditions. Since the creep strain rate changes - that is, hardens - as a function of time (unless $m = 1$ holds), (15.9) is a *time-hardening model*; this expression is easily generalized to the following format

$$\boxed{\dot{\varepsilon}^{cr} = f(\sigma, t) \quad \textit{time-hardening}} \tag{15.10}$$

where $f(\sigma, t)$ is some function. In relation to (14.17) and Fig. 14.25 we have already discussed time-hardening and argued that loading time in itself is a poor measure of hardening.

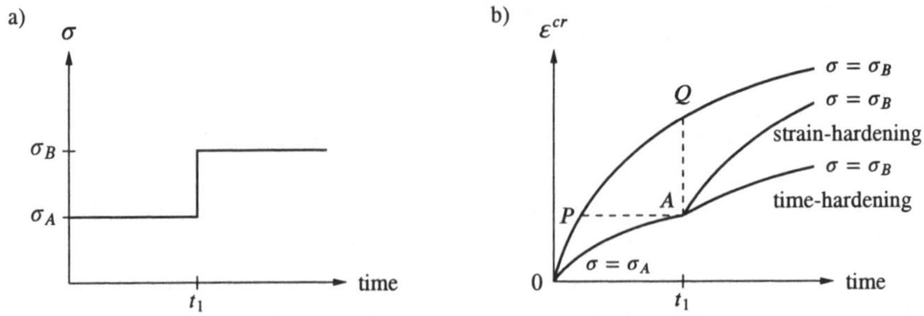

Figure 15.2: a) Stress history, b) creep strain predicted by time-hardening and strain-hardening.

However, another hardening possibility suggests itself. Eliminate the time t in (15.9) by means of (15.8) to obtain

$$\dot{\varepsilon}^{cr} = \frac{mA^{\frac{1}{m}}}{t^*}\left(\frac{\sigma}{\sigma^*}\right)^{\frac{n}{m}}(\varepsilon^{cr})^{\frac{m-1}{m}} \tag{15.11}$$

Again, this expression is identical to (15.8) for a constant stress, but now (15.11) is postulated to hold even when the stress changes during the loading history. Since the creep strain rate changes - i.e. hardens - as a function of the creep strain ε^{cr}, (15.11) is a *strain-hardening model*; it is easily generalized to

$$\boxed{\dot{\varepsilon}^{cr} = f(\sigma, \varepsilon^{c}) \quad \textit{strain-hardening}} \tag{15.12}$$

where $f(\sigma, \varepsilon^{cr})$ is a function. According to this model the creep strain ε^{cr} characterizes the state of the material and - in the spirit of Section 10.1 - it may therefore be viewed as an internal variable.

The strain-hardening model is certainly expected to be much more realistic than the time-hardening model, and the difference becomes significant when large stress changes occur. However, historically the time-hardening model has been used extensively due to its simplicity when deriving analytical solutions for structural elements.

To further substantiate the superiority of the strain-hardening model, consider the stress history in Fig. 15.2a). For a standard creep test with $\sigma = \sigma_B$, the creep strain curve OPQ in Fig. 15.2b) is obtained. For the actual stress history both models predict the same creep strain curve up to time $t = t_1$. At time $t = t_1$, the stress is suddenly increased to from $\sigma = \sigma_A$ to $\sigma = \sigma_B$ and according to the time-hardening model (15.10) the creep strain rate, i.e. the slope, at point A equals the creep strain rate at point Q. However, according to the strain-hardening model (15.12) the creep strain rate at point A equals that at point P. The expectation that the strain-hardening model is in closer agreement with experimental data than the time-hardening model is confirmed by numerous experiments for steel and metals, cf. Boresi and Sidebottom (1972) and Finnie and Heller (1959), and also for rock salt, see Ottosen and Krenk (1982).

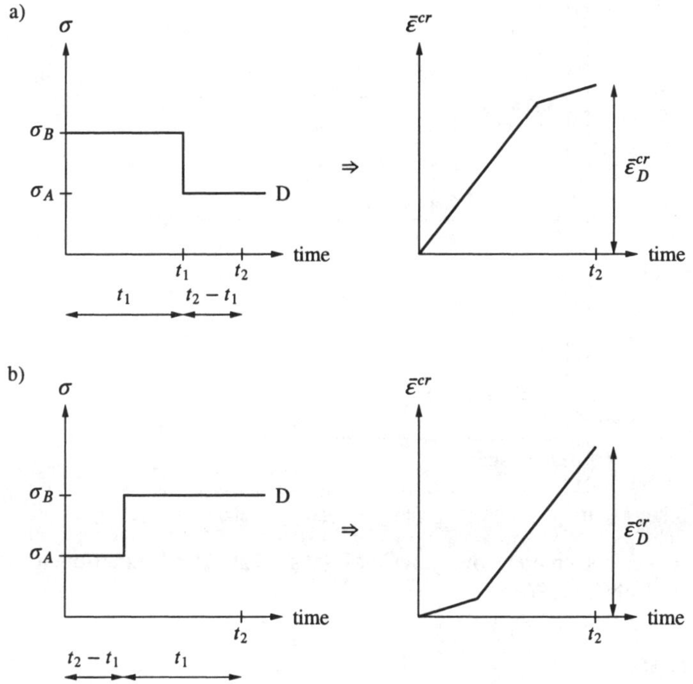

Figure 15.3: The factorized stain-hardening model exhibits no sequence effects; the load histories in a) and b) result in the same creep strain at point D (for simplicity, only secondary creep is involved).

In general, one may expect that the sequence with which the load is applied would influence the final creep strain; a small stress followed by a large stress is not expected to result in the same final creep strain as a large stress followed by a small stress, and this is confirmed by experiments, see Rabotnov (1969). The factorized strain-hardening model exhibits no such *sequence effects*. In the factorized form of the strain-hardening model, (15.12) takes the form

$$\boxed{\dot{\varepsilon}^{cr} = f(\sigma)g(\varepsilon^{cr}) \quad \textit{factorized strain-hardening model}} \tag{15.13}$$

where $f(\sigma)$ and $g(\varepsilon^{cr})$ are functions, cf. the similar structure in (15.11). Define now the variable $\bar{\varepsilon}^{cr}$ by

$$\dot{\bar{\varepsilon}}^{cr} = \frac{\dot{\varepsilon}^{cr}}{g(\varepsilon^{cr})}$$

and it is evident that the quantity $\bar{\varepsilon}^{cr}$ is related uniquely to the creep strain ε^{cr}. Then the factorized strain-hardening model (15.13) becomes

$$\dot{\bar{\varepsilon}}^{cr} = f(\sigma)$$

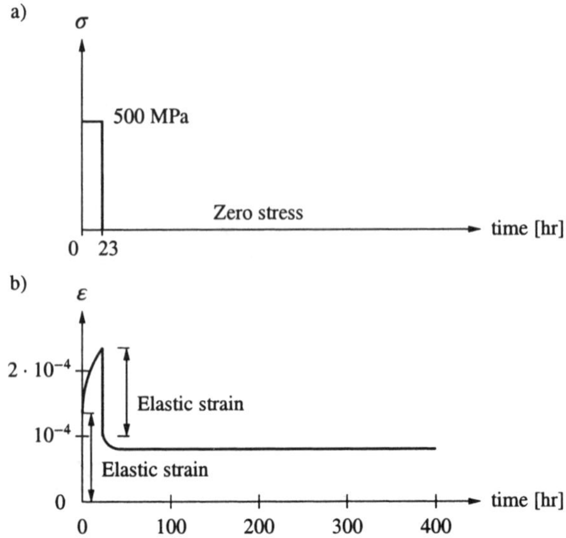

Figure 15.4: Creep and creep recovery of Cr-Mo-V steel at 425 $°C$, experimental results of Lubahn (1961).

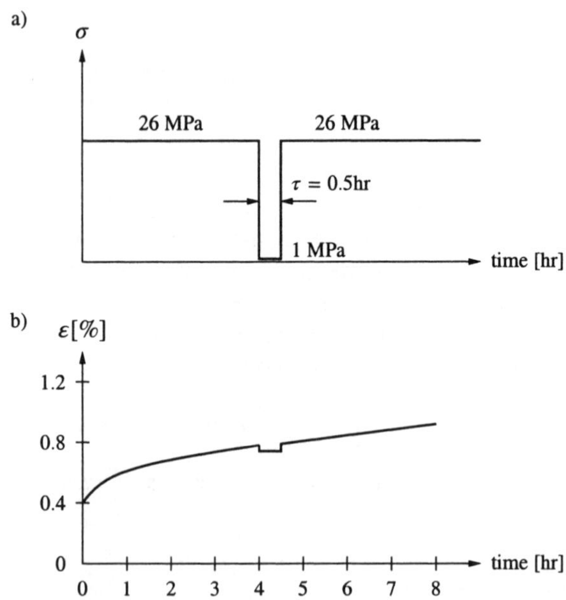

Figure 15.5: Interrupted creep test where the time of interruption is short (=0.5 hr); Al 1100 at 150 $°C$, experimental results of Wang and Onat (1968).

It appears that $\dot{\bar{\varepsilon}}^{cr}$ only depends on the current stress and not on the current value of $\bar{\varepsilon}^{cr}$ (and thereby not on the current value of ε^{cr}). It then follows that there can be no sequence effects. As an example, the factorized strain-hardening model will predict the same creep strain at point D in Fig. 15.3a) and b) where - for simplicity - only secondary creep is involved.

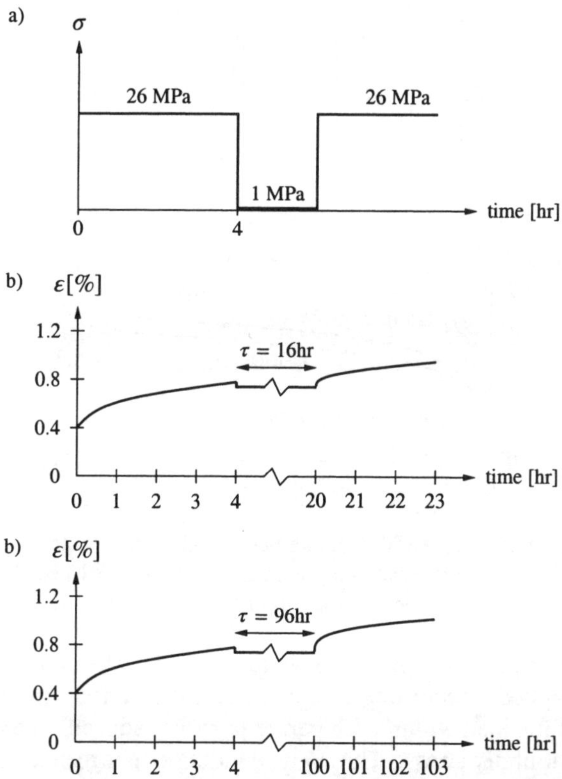

Figure 15.6: Interrupted creep test when the time of interruption is long; a) stress history, b) 16 hr interruption and c) 96 hr interruption; Al 1100 at 150 $^\circ C$, experimental results of Wang and Onat (1968).

In linear viscoelasticity, creep recovery effects are often significant, see for instance Fig. 14.20. For steel and metals, however, this recovery effect is usually modest as shown in Fig. 15.4.

Recognizing the modest recovery, it would be tempting to expect that no significant changes occur during periods of total unloading. For the short unloading period (=0.5 hr) in Fig. 15.5 the material continues to creep after reloading as it did before. However, if the unloading period is longer the material exhibits significant primary creep upon reloading, cf. Fig. 15.6. It is therefore concluded that at intermediate or high temperature, the internal state of a metal or steel changes even though it has been completely unloaded. Evidently, this change of the internal state cannot be predicted by the strain-hardening model.

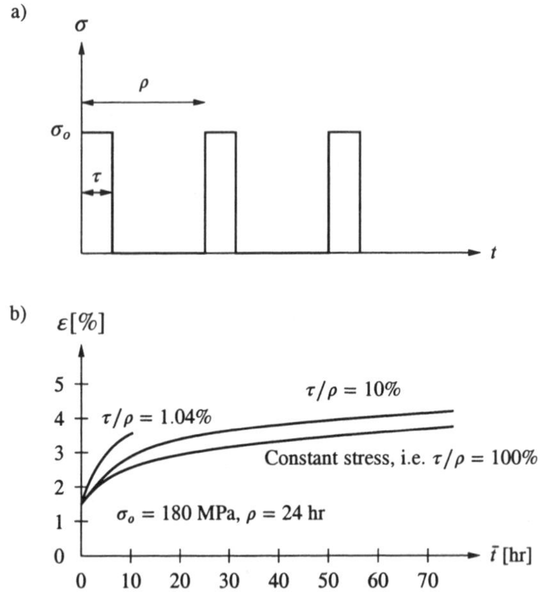

Figure 15.7: Creep of steel at 450 $^\circ C$ under periodically varying stress; a) stress history, b) experimental results reported by Onat and Fardshisheh (1972) where $\bar{t}$ is the reduced time, i.e. the total time spent under stress.

The effects discussed above become very evident when a periodically varying stress is applied. According to Fig. 15.7a), the period of the load is ρ and the duration of the load within each period is τ; the reduced time $\bar{t}$ measures the total time spent under stress. The corresponding experimental total strain ε is shown in Fig. 15.7b) and - intuitively somewhat surprisingly - it appears that the periodic loading gives rise to much larger strains than the constant load; again, this significant effect cannot be modeled by the strain-hardening model.

15.3 Multiaxial stress states

With this detailed discussion of the response for uniaxial stress conditions, it is timely to establish the multiaxial constitutive formulation. Let us as our point of departure take the strain-hardening model (15.11) written so that it applies not only to tensile stresses, but also to compressive stresses; we then have

$$\dot{\varepsilon}^{cr} = \frac{mA^{\frac{1}{m}}}{t^*}(\frac{|\sigma|}{\sigma^*})^{\frac{n}{m}-1}\frac{\sigma}{\sigma^*}(|\varepsilon^{cr}|)^{\frac{m-1}{m}} \tag{15.14}$$

Moreover, (15.1) gives

$$\dot{\varepsilon} = \frac{\dot{\sigma}}{E} + \dot{\varepsilon}^{cr} \tag{15.15}$$

A comparison of (15.14) and (15.15) with (14.7) shows that the creep model takes the form of a nonlinear Maxwell model, i.e.

$$\dot{\varepsilon} = \frac{\dot{\sigma}}{E} + \frac{\sigma}{\eta(|\sigma|, |\varepsilon^{cr}|)} \tag{15.16}$$

where

$$\frac{1}{\eta(|\sigma|, |\varepsilon^{cr}|)} = \frac{mA^{\frac{1}{m}}}{t^*\sigma^*}(\frac{|\sigma|}{\sigma^*})^{\frac{n}{m}-1}(|\varepsilon^{cr}|)^{\frac{m-1}{m}}$$

The next piece of information is that all experimental evidence shows that

$$\boxed{\begin{array}{l}\textit{For metals and steel the creep strains}\\ \textit{are incompressible, i.e. } \dot{\varepsilon}_{ii}^{cr} = 0\end{array}} \tag{15.17}$$

and we note the complete similarity with the observations for plasticity.

For an isotropic material, the three-dimensional Maxwell model is given by (14.22) where it was concluded that the parameter ξ should be chosen as $\xi = \frac{1}{2}$ if (15.17) were to hold. We are then led to

$$\dot{\varepsilon}_{ij} = C_{ijkl}\dot{\sigma}_{kl} + \dot{\varepsilon}_{ij}^{cr} \quad \text{where} \quad \dot{\varepsilon}_{ij}^{cr} = \frac{3}{2\eta}s_{ij} \tag{15.18}$$

where C_{ijkl} is the elastic flexibility tensor. It is recalled that the viscosity parameter η depends on the stresses and the creep strains, i.e. $\eta = \eta(\sigma_{ij}, \varepsilon_{ij})$. In order that (15.18) reduce to (15.16) for uniaxial stress conditions, we define, in complete analogy with plasticity theory, the effective stress σ_{eff} and the *effective creep strain* ε_{eff}^{cr} by

$$\boxed{\sigma_{eff} = (\frac{3}{2}s_{ij}s_{ij})^{1/2}; \quad \varepsilon_{eff}^{cr} = \int_0^t \dot{\varepsilon}_{eff}^{cr} dt; \quad \dot{\varepsilon}_{eff}^{cr} = (\frac{2}{3}\dot{\varepsilon}_{ij}^{cr}\dot{\varepsilon}_{ij}^{cr})^{1/2}} \tag{15.19}$$

cf. (9.60) and (9.62). The viscosity parameter η is now taken as

$$\eta = \eta(\sigma_{eff}, \varepsilon_{eff}^{cr}) \tag{15.20}$$

Since σ_{eff} and ε_{eff}^{cr} for uniaxial stress conditions reduce to $|\sigma|$ and $|\varepsilon^{cr}|$ respectively, the formulation (15.18) and (15.20) reduces exactly to (15.16) for uniaxial conditions and η becomes

$$\frac{1}{\eta(\sigma_{eff}, \varepsilon_{eff}^{cr})} = \frac{mA^{\frac{1}{m}}}{t^*\sigma^*}(\frac{\sigma_{eff}}{\sigma^*})^{\frac{n}{m}-1}(\varepsilon_{eff}^{cr})^{\frac{m-1}{m}} \tag{15.21}$$

Finally, insertion of this expression into (15.18b) and recalling that the elastic stiffness tensor D_{ijkl} is the inverse to the elastic flexibility tensor C_{ijkl}, we obtain the result

Odqvist formulation of power law creep

$$\dot{\sigma}_{ij} = D_{ijkl}(\dot{\varepsilon}_{kl} - \dot{\varepsilon}^{cr}_{kl})$$

where

$$\dot{\varepsilon}^{cr}_{ij} = \frac{mA^{\frac{1}{m}}}{t^*}\left(\frac{\sigma_{eff}}{\sigma^*}\right)^{\frac{n}{m}}(\varepsilon^{cr}_{eff})^{\frac{m-1}{m}}\frac{3s_{ij}}{2\sigma_{eff}} \tag{15.22}$$

This generalization to three-dimensional stress states was achieved by Odqvist (1934) although only secondary creep ($m = 1$) was considered. For constant multiaxial stresses, (15.22) is in close agreement with experimental data as discussed by Odqvist and Hult (1962) and Odqvist (1966).

The format (15.22) exhibits strain-hardening and it is tempting to generalize the result in a similar fashion as when the uniaxial formulation (15.11) was generalized to (15.12). We then obtain

$$\dot{\varepsilon}^{cr}_{ij} = \Lambda\frac{3s_{ij}}{2\sigma_{eff}}; \qquad \Lambda = \Lambda(\sigma_{eff}, \varepsilon^{cr}_{eff}) \tag{15.23}$$

where the function Λ with the unit [1/s] is chosen is accordance with experimental data. Further generalizations almost suggest themselves. Considering the von Mises function $f = (\frac{3}{2}s_{ij}s_{ij})^{1/2}$, the above result can be written as $\dot{\varepsilon}^{cr}_{ij} = \Lambda\partial f/\partial\sigma_{ij}$, i.e. f serves as a potential function. The formulation now looks very much like plasticity theory and in that spirit - cf. the discussion in Section 10.1 - the next generalization is to adopt a general potential function $g = g(\sigma_{ij}, K_\alpha)$ where K_α are hardening parameters and then achieve the following general format

General creep formulation

$$\dot{\sigma}_{ij} = D_{ijkl}(\dot{\varepsilon}_{kl} - \dot{\varepsilon}^{cr}_{kl})$$

where

$$\dot{\varepsilon}^{cr}_{ij} = \Lambda\frac{\partial g}{\partial\sigma_{ij}} \tag{15.24}$$

and

$$g(\sigma_{ij}, K_\alpha); \qquad \Lambda = \Lambda(\sigma_{ij}, \kappa_\alpha) \geq 0$$

The function Λ is chosen

Here κ_α are internal variables and K_α are the corresponding hardening parameters, cf. the discussion in Section 10.1. If only one internal variable κ in terms of the effective creep strain ε^{cr}_{eff} is chosen and Λ is taken as $\Lambda = \Lambda(\sigma_{eff}, \varepsilon^{cr}_{eff})$

and $g = (\frac{3}{2}s_{ij}s_{ij})^{1/2}$, then (15.24) reduces to (15.23). In general, evolution laws need to be established for $\dot{K}_\alpha$ and/or $\dot{\kappa}_\alpha$.

The general format (15.24) looks very much like the general plasticity format; however, there are important differences. First of all, no yield criterion is involved in (15.24) and secondly, the function Λ is directly prescribed by us based on experimental evidence. In plasticity theory, we have $\dot{\varepsilon}^p_{ij} = \dot{\lambda}\partial g/\partial\sigma_{ij}$, but here the plastic multiplier $\dot{\lambda}$ is determined as a consequence of the consistency relation. In creep theory, however, there is no yield criterion so the function Λ is directly chosen by us to fit the experimental data in the best possible way.

Moreover, since no yield criterion exists, the question of loading and unloading disappears. Another difference is that whereas development of plastic strains requires a change of stresses or strains, then also for constant stresses, (15.24) shows that a development of creep strains occurs.

Apart from those differences, the similarity between creep theory and plasticity theory is close and in the following we will take full advantage of this fact. As an example, *orthotropic creep theory* is achieved by taking D_{ijkl} to be orthotropic and choosing the potential function g, for instance, as the orthotropic Hill function discussed in Sections 8.13 and 12.6.

15.3.1 Bodner and Partom model

We have shown that the generalized strain-hardening model (15.22) exhibits a number of limitations when larger stress changes occur. Let us therefore turn to the general format (15.24) and investigate other specific possibilities.

As an example, we will discuss the model proposed by Bodner and Partom. A number of other authors has also contributed to the development of this model, but its origins were proposed by Bodner (1968), Bodner and Partom (1972, 1975), Bodner *et al.* (1979) and a comprehensive review is given by Bodner (1987) and Stouffer and Dame (1996).

The term *state variable approach* is occasionally used for the model of Bodner and Partom and some authors even use the notation of a *unified* model. A number of such models exists, for instance Hart (1970), Miller (1976, 1987a,b), Robinson (1978), Walker (1981) and Krempl (1996) and Krempl *et al.* (1986), all of which have been proposed within the last two or three decades; detailed reviews are given by Stouffer and Dame (1996) as well as by Miller (1987a,b) and Krausz and Krausz (1996). These models have been developed to meet the increasing demands within the gas-turbine and nuclear industry to deal with complex thermo-mechanical loadings including creep and cyclic loadings. As a typical example of such models, we will here focus on the Bodner-Partom model.

Within the general framework (15.24), the potential function g is chosen as

$$g = \frac{1}{2} s_{ij} s_{ij} \quad \text{i.e.} \quad \dot{\varepsilon}_{ij}^{cr} = \Lambda s_{ij} \tag{15.25}$$

In analogy with the definition $J_2 = \frac{1}{2} s_{ij} s_{ij} = \frac{1}{3}\sigma_{eff}^2$, the invariant D_2 is defined as

$$D_2 = \frac{1}{2} \dot{\varepsilon}_{ij}^{cr} \dot{\varepsilon}_{ij}^{cr}$$

and it appears that $D_2 = \frac{3}{4}(\dot{\varepsilon}_{eff}^{cr})^2$, cf. (15.19). Since $\sqrt{D_2} \sim \dot{\varepsilon}_{eff}^{cr}$, an expression for $\sqrt{D_2}$ is directly postulated based on general experimental evidence, i.e.

Evolution equation

$$\sqrt{D_2} = D_0 e^{-\frac{1}{2}\left(\frac{Z^2}{\sigma_{eff}^2}\right)^n} \tag{15.26}$$

where D_0 is a positive parameter with the dimension [1/s], Z is a positive variable with the dimension of stress and n is a dimensionless positive parameter. Since $\dot{\varepsilon}_{eff}^{cr}$ can be considered to be the time rate of an internal variable, i.e. $\dot{\kappa}$, so can $\sqrt{D_2}$ and it is then natural to consider Z as a *hardening parameter*. Formally, (15.26) can therefore be viewed as an evolution law in the format $\dot{\kappa} = f(\sigma_{eff}, K)$; in the literature, expression (15.26) is often called the *kinetic equation*. However, what in this situation is termed an internal variable and a hardening parameter is open for discussion and most authors within these state variable models prefer to call Z an internal variable. Since both internal variables and hardening parameters characterize the state of the material, they are *state variables* and the concept is therefore often termed the *state variable approach*, cf. Stouffer and Dame (1996).

Multiply each side of (15.25b) by itself to obtain

$$\Lambda = \frac{\sqrt{3D_2}}{\sigma_{eff}}$$

With (15.26), (15.25b) then takes the form

Bodner-Partom

$$\dot{\varepsilon}_{ij}^{cr} = \Lambda s_{ij}$$

where

$$\Lambda = \Lambda(\sigma_{eff}, Z) = \frac{\sqrt{3}D_0}{\sigma_{eff}} e^{-\frac{1}{2}\left(\frac{Z^2}{\sigma_{eff}^2}\right)^n} \geq 0 \tag{15.27}$$

Despite the views often put forward in the literature, it is evident that this model is, essentially, a creep model since even constant stresses will result in the development of strains. To evaluate the choice of the evolution law (15.26) - the kinetic equation - and to identify the role of the hardening parameter Z, the expression above is evaluated for uniaxial stress conditions to obtain

$$\dot{\varepsilon}^{cr} = \frac{2D_0}{\sqrt{3}} e^{-\frac{1}{2}(\frac{Z^2}{\sigma^2})^n} \tag{15.28}$$

Therefore $2D_0/\sqrt{3}$ is the maximum value of $\dot{\varepsilon}^{cr}$. It also appears that $\dot{\varepsilon}^{cr}$ is a decreasing function of Z^2/σ^2; when $Z^2/\sigma^2 \to \infty$ - which it does for small stresses - then $\dot{\varepsilon}^{cr} \to 0$, as expected, and when $Z^2/\sigma^2 \ll 1$ - i.e. large stresses - then $\dot{\varepsilon}^{cr} \to 2D_0/\sqrt{3}$. Therefore, Z is called in the literature the *equivalent yield stress*, since it is the threshold beyond which the (creep) deformation becomes pronounced. Moreover, the creep strain rate $\dot{\varepsilon}^{cr}$ depends nonlinearly on the stress state and this nonlinearity is controlled by the exponent n. It will be shown later that the smaller the exponent n, the more strain-rate sensitive will be the response of the material.

Consider now a standard creep test and assume that the positive variable Z increases from its initial value and approaches a constant value; then the ratio Z^2/σ^2 will increase and eventually approach a constant value. In turn, the inelastic strain rate $\dot{\varepsilon}^{cr}$ will first be large and then approach a constant, smaller value, but this is exactly what happens in a creep test: First primary creep occurs and eventually secondary creep occurs. The choice (15.26) and thereby (15.28) therefore fulfills all our expectations.

It turns out that a number of phenomena can be modeled, if Z is split into two contributions according to

$$Z = Z^I + Z^D \tag{15.29}$$

where

Z^I = *isotropic hardening parameter*
Z^D = *directional hardening parameter*

It is not surprising that the *isotropic hardening parameter* Z^I will depend on some scalar quantities whereas the so-called *directional hardening parameter* Z^D must depend on some second-order tensor much along the lines of kinematic hardening in plasticity.

The next topic is to choose an evolution law for $\dot{Z}$ such that it fulfills the requirements established above: in a creep test, the positive quantity Z should increase from its initial value and eventually reach a constant value. Let us for the time being ignore the influence of the directional hardening parameter Z^D and establish the evolution law for $\dot{Z}^I$. What can create a change in Z^I is development of creep strains, so one possibility would be to relate $\dot{Z}^I$ to $\dot{\varepsilon}^{cr}_{eff}$ (or $\sqrt{D_2}$). However, another non-negative quantity is the rate of inelastic

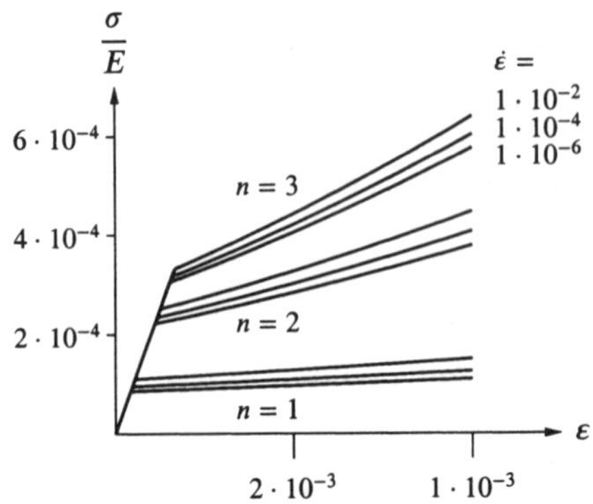

Figure 15.8: Constant strain-rate tests predicted by (15.30) for different values of the exponent n; material parameters: $m_1 E = 3.4 \cdot 10^5$, $D_0 = 10^4\,[1/s]$, $Z_0/E = 5.8 \cdot 10^{-4}$ and $Z_1/E = 3.3 \cdot 10^{-3}$ which are representative for aluminum at 200 °C where also $n = 1.43$ holds, cf. Bodner (1987).

work $\dot{W}^{cr} = \sigma_{ij}\dot{\varepsilon}^{cr}_{ij}$, which according to (15.27) becomes $\dot{W}^{cr} = \Lambda s_{ij}s_{ij} \geq 0$. A simple evolution law would then be $\dot{Z}^I = (m_1 Z_1)\dot{W}^{cr}$ where $m_1 Z_1$ is a positive material parameter (the reason for this somewhat curious notation for one material parameter will become apparent in a moment); however, since $\dot{W}^{cr} > 0$ for all non-zero stresses, the result is that Z^I increases without bounds and then only primary creep in a creep test can be modeled. To eventually obtain secondary creep in a creep test, it is required that Z^I eventually reach a constant value. This can be achieved by writing the evolution law as

$$\dot{Z}^I = m_1 Z_1 \dot{W}^{cr} - \underbrace{m_1 Z^I \dot{W}^{cr}}_{\text{dynamic recovery term}} = m_1(Z_1 - Z^I)\dot{W}^{cr} \tag{15.30}$$

where

Z_1 = *saturation value of* Z^I; Z_0 = *initial value of* Z^I
$0 < Z_0 \leq Z^I \leq Z_1$

and m_1 [1/Pa], Z_0 [Pa] and Z_1 [Pa] are positive material parameters. It appears that $\dot{Z}^I > 0$ holds in the beginning until Z^I eventually reaches the value Z_1 and then $\dot{Z}^I = 0$ whereby secondary creep is encountered in a creep test. The parameter Z_1 is therefore called the *saturation value* of Z^I and the second term on the right-hand side of (15.30) is called the *dynamic recovery term*.

For uniaxial tension, the predictions of (15.27) and (15.30) for constant strain-rate tests are shown in Fig. 15.8 where the material data are representative for aluminum at 200 °C and also $n = 1.43$ is representative. As already mentioned, the smaller the n exponent, the larger the strain-rate sensitivity of the

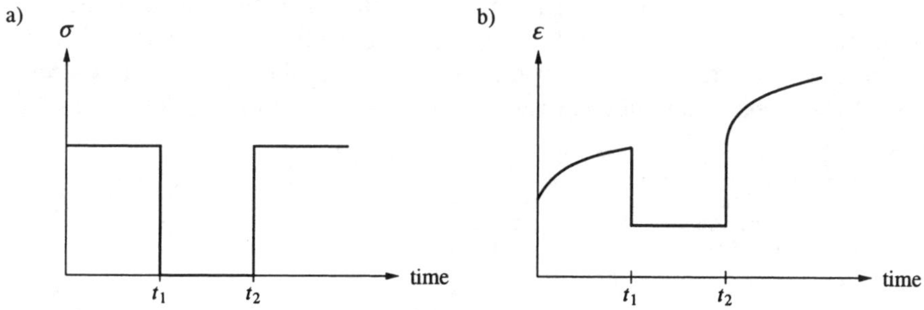

Figure 15.9: Interrupted creep test. Secondary creep is assumed to occur at time $t = t_1$, but reloading at time $t = t_2$ introduces primary creep.

material. Moreover, it is also seen that the material behaves as if there were a yield stress below which insignificant creep strains develop; this apparent yield stress increases with the exponent n.

To be able to model a realistic response for stress changes like those shown in Figs. 15.5-15.7, an additional term is introduced into the evolution law (15.30), which now takes the form

Evolution law for isotropic hardening parameter

$$\dot{Z}^I = m_1(Z_1 - Z^I)\dot{W}^{cr} - \underbrace{A_1 Z_1 \left(\frac{Z^I - Z_2}{Z_1}\right)^{r_1}}_{\textit{thermal recovery term}} \tag{15.31}$$

where $0 \leq Z_0 \leq Z^I \leq Z_1$ *and* $Z_2 = Z_0$

where the additional positive material parameters have the dimensions A_1 [1/s], Z_2 [Pa] and r_1 is dimensionless. In Bodner (1987), for instance, Z_2 is recommended to be chosen close to Z_0, but here we require $Z_2 = Z_0$. Since the exponent r_1 can be any positive number, the thermal recovery term only makes sense if $Z_0 \geq Z_2$, but if $Z_0 > Z_2$ holds then (15.31) will predict $\dot{Z}^i < 0$ even before any load is applied and to avoid this awkward situation, it is here required that $Z_2 = Z_0$.

To illustrate the influence of the *thermal recovery term* in (15.31), consider the interrupted creep test in Fig. 15.9. For simplicity, it is assumed that secondary creep has been achieved at time $t = t_1$. During the unloaded period, $\dot{W}^{cr} = 0$ holds and (15.31) then predicts $\dot{Z}^I < 0$, that is, Z^I is decreased from its value in the secondary creep regime towards a smaller Z^I-value, i.e towards the primary creep regime. Therefore, when reloaded at time $t = t_2$ primary creep is introduced and this is exactly the behavior strived for, cf. Fig. 15.6.

The thermal recovery term is occasionally termed the *static* recovery term since it is active even during unloaded periods in contrast to the dynamic recovery term that is only active when the material is loaded.

To improve the response during cyclic loading, the directional hardening parameter Z^D is introduced, cf. (15.29). To establish the evolution law for $\dot{Z}^D$, a symmetric stress tensor β_{ij} is introduced, much along the lines of a back-stress tensor for kinematic hardening in plasticity theory. The following definitions are introduced

$$u_{ij} = \frac{\sigma_{ij}}{\sqrt{\sigma_{kl}\sigma_{kl}}}; \qquad v_{ij} = \frac{\beta_{ij}}{\sqrt{\beta_{kl}\beta_{kl}}}$$

and

$$Z^D = \beta_{ij}u_{ij}$$

where $\beta_{ij}(t=0) = 0$

i.e. u_{ij} is the normalized stress tensor and v_{ij} is the normalized β_{ij}-tensor.

In complete analogy with (15.31), the following evolution law for $\dot{\beta}_{ij}$ is adopted

$$\dot{\beta}_{ij} = m_2(Z_3 u_{ij} - \beta_{ij})\dot{W}^{cr} - A_2 Z_1 \left(\frac{\sqrt{\beta_{kl}\beta_{kl}}}{Z_1}\right)^{r_2} v_{ij} \qquad (15.32)$$

where m_2 [1/Pa], Z_3 [Pa], A_2 [1/s] and r_2 (dimensionless) are additional positive material parameters. The evolution law for $\dot{Z}^D$ is then determined by $\dot{Z}^D = \dot{\beta}_{ij}u_{ij} + \beta_{ij}\dot{u}_{ij}$. Similar to (15.31), the term $m_2\beta_{ij}\dot{W}^{cr}$ is called a dynamic recovery term and the last term on the right-hand side is called a thermal recovery term.

The directional hardening parameter Z^D improves the response during cyclic loading, for a the detailed discussion of (15.31) the reader is referred to, for instance Bodner (1987), for a further evaluation and comparison with experimental data. It is of interest to note that (15.32) is similar in structure to (15.31) and a fundamental similarity exists also between (15.32) and the Armstrong-Frederick evolution law cf. (13.70).

We have seen that the Bodner-Partom model contains a number of interesting features. However, the number of material parameters is large and the identification of these parameters from experimental data is not trivial; a discussion is given by Bodner (1987) and Rowley and Thornton (1996).

15.4 Viscoplasticity

We will now turn our interest to *viscoplasticity*, which has received considerable attention during the last decades, and also refer to the comprehensive reviews presented by Perzyna (1966, 1971), Lemaitre and Chaboche (1990), Lubliner (1990), Krausz and Krausz (1996) as well as Simo and Hughes (1998). In the beginning of the 20th century, where the basic concepts for linear viscoelasticity and viscous fluids were known, it was observed that some viscous materials

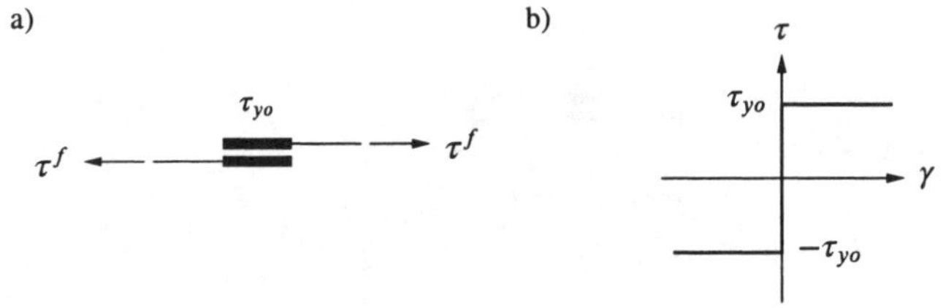

Figure 15.10: a) Saint-Venant friction element, b) response.

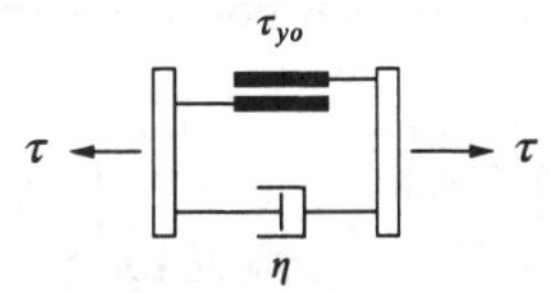

Figure 15.11: Bingham model.

could not be categorized into these two groups of materials. For instance, a large drop of oil-paint will run down the wall whereas a small drop will remain its position on the wall; it follows that a certain amount of stress is required in order that the material exhibits viscous effects.

To model this behavior, we first introduce the *friction element* shown in Fig. 15.10 where superscript 'f' refers to 'frictional'; traditionally, this friction element is attributed to Saint-Venant. If a shear stress τ^f is applied, the friction element behaves as a rigid body when $|\tau^f| < \tau_{yo}$ where τ_{yo} is the *yield shear stress*; development of a shear strain γ is only possible if $|\tau^f| = \tau_{yo}$. The constitutive model for the friction element is therefore

$$\begin{aligned}
&\dot{\gamma}^f = 0 \quad \text{if} \quad |\tau^f| < \tau_{yo} \\
&\dot{\gamma}^f \geq 0 \quad \text{if} \quad \tau^f = \tau_{yo} \\
&\dot{\gamma}^f \leq 0 \quad \text{if} \quad \tau^f = -\tau_{yo}
\end{aligned}$$

and deformations are only possible if the shear stress equals the yield shear stress.

To model the oil-paint behavior mentioned above, Bingham (1922) introduced the rigid-viscous model shown in Fig. 15.11. This *Bingham model* consists of a friction element and a viscous element in parallel. In accordance with (14.5) the constitutive law for the viscous element is $\dot{\gamma}^v = \tau^v/\eta$ where η is a viscosity parameter. To establish the constitutive law for the Bingham model, we first observe that $\tau = \tau^f + \tau^v$ and $\dot{\gamma} = \dot{\gamma}^f = \dot{\gamma}^v$. If $|\tau^f| < \tau_{yo}$ then $\dot{\gamma}^f = 0$ i.e. $\dot{\gamma} = 0$ and then $\tau^v = 0$ and thereby $|\tau| < \tau_{yo}$. If $\tau^f = \tau_{yo}$ then $\dot{\gamma}^f = \dot{\gamma} = \tau^v/\eta \geq 0$ and as $\tau^v = \tau - \tau^f = \tau - \tau_{yo} \geq 0$ we obtain $\tau \geq \tau_{yo}$ and

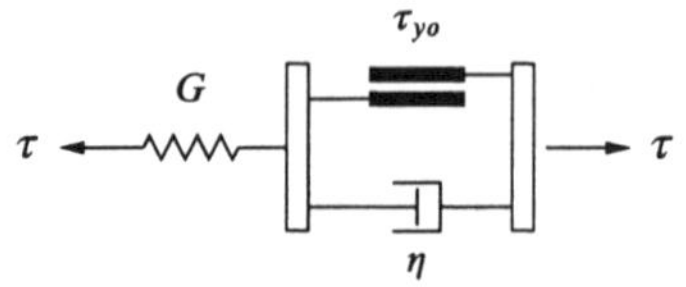

Figure 15.12: Hencky model consisting of a Bingham element and an elastic spring in series.

$\dot{\gamma} = (\tau - \tau_{yo})/\eta = (1 - \frac{\tau_{yo}}{\tau})\frac{\tau}{\eta}$. Similar considerations hold if $\tau^f = -\tau_{yo}$ and in conclusion, we obtain

$$\boxed{\begin{array}{lll} \dot{\gamma} = 0 & \text{if } |\tau| \leq \tau_{yo} & \\ \dot{\gamma} = \dfrac{1 - \frac{\tau_{yo}}{|\tau|}}{\eta}\tau & \text{if } |\tau| \geq \tau_{yo} & \textit{Bingham} \end{array}}$$

Therefore if $|\tau| \leq \tau_{yo}$ the Bingham element is rigid and if $|\tau| > \tau_{yo}$ the element exhibits secondary creep.

The *Hencky model* shown in Fig. 15.12 comprises a Bingham model and an elastic spring in series, Hencky (1925); the elastic spring is characterized by the shear modulus G. To establish the constitutive law, it is observed that $\tau = \tau^e = \tau^B$ and $\dot{\gamma} = \dot{\gamma}^e + \dot{\gamma}^B$ and it then follows that

$$\begin{aligned} \dot{\gamma} &= \frac{\dot{\tau}}{G} & \text{if } |\tau| \leq \tau_{yo} \\ \dot{\gamma} &= \frac{\dot{\tau}}{G} + \frac{1 - \frac{\tau_{yo}}{|\tau|}}{\eta}\tau & \text{if } |\tau| \geq \tau_{yo} \end{aligned} \tag{15.33}$$

The response of the Hencky model in a creep test is shown in Fig. 15.13. If the shear stress is below the yield shear stress τ_{yo}, the material is elastic and no creep strains develop whereas secondary creep occurs if $\tau > \tau_{yo}$.

To generalize the Hencky format (15.33) to isotropic three-dimensional behavior, it is first observed that the viscous strains might be thought of as being fluid-like, i.e. incompressible and it then follows that $\dot{\varepsilon}_{kk} = \dot{\sigma}_{kk}/3K$, where K is the bulk modulus. It is also observed that $J_2 = \frac{1}{2}s_{ij}s_{ij}$ for pure shear reduces to $\sqrt{J_2} = |\tau|$ and that the engineering shear strains are twice the tensorial shear strains; then an immediate generalization of (15.33) becomes

$$\begin{aligned} \dot{\varepsilon}_{ii} &= \frac{\dot{\sigma}_{ii}}{3K} & & \\ \dot{e}_{ij} &= \frac{\dot{s}_{ij}}{2G} & \text{if} \quad & \sqrt{J_2} \leq \tau_{yo} \\ \dot{e}_{ij} &= \frac{\dot{s}_{ij}}{2G} + \frac{1 - \frac{\tau_{yo}}{\sqrt{J_2}}}{2\eta}s_{ij} & \text{if} \quad & \sqrt{J_2} \geq \tau_{yo} \end{aligned}$$

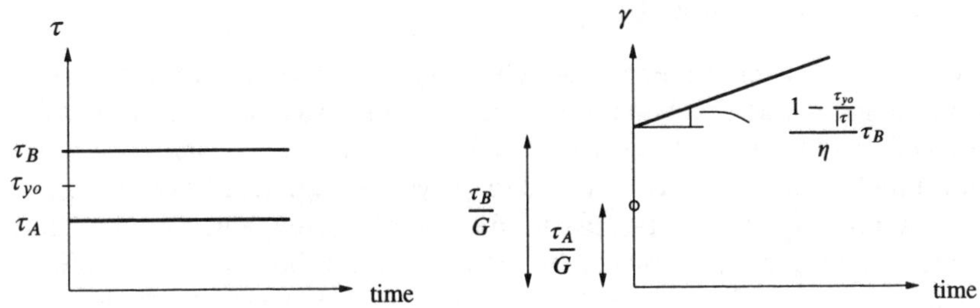

Figure 15.13: Creep test of Hencky model; one test below the yield shear stress τ_{yo} and another above τ_{yo}.

This generalization was established by Hohenemser and Prager (1932) and if the elastic response is disregarded, the formulation was given already by Hencky (1925). Observing that $\sqrt{3J_2} = \sigma_{eff}$ and that according to the yield criterion of von Mises, the initial yield stress σ_{yo} in tension is given by $\sigma_{yo} = \sqrt{3}\tau_{yo}$, the format above can also be written as

Hohenemser-Prager viscoplasticity

$$\dot{\varepsilon}_{ij} = \dot{\varepsilon}^e_{ij} + \dot{\varepsilon}^{vp}_{ij}$$

where

$$\begin{aligned} \varepsilon^e_{ij} &= C_{ijkl}\sigma_{kl} \\ \dot{\varepsilon}^{vp}_{ij} &= 0 && \textit{if} \quad \sigma_{eff} \leq \sigma_{yo} \\ \dot{\varepsilon}^{vp}_{ij} &= \frac{1-\frac{\sigma_{yo}}{\sigma_{eff}}}{2\eta} s_{ij} && \textit{if} \quad \sigma_{eff} \geq \sigma_{yo} \end{aligned} \tag{15.34}$$

It follows that when the stress state is inside or on the initial yield surface $F(\sigma_{ij}) = \sigma_{eff} - \sigma_{yo} = 0$, i.e. $F(\sigma_{ij}) \leq 0$ an elastic response occurs and if $F(\sigma_{ij}) > 0$ holds viscoplastic strains develop; it is the *excess stress* or the *overstress* $\sigma_{eff} - \sigma_{yo}$ that drives the viscoplastic development. For proportional loading, we also observe that (15.34) implies that the viscoplastic strain rates increase linearly with the stresses.

Following Hohenemser and Prager (1932) an interesting limit process will now be considered. If the viscosity parameter $\eta \to 0$ then, in order that $\dot{\varepsilon}^{vp}_{ij}$ be a finite quantity, we must have $1-\frac{\sigma_{yo}}{\sigma_{eff}} \to 0$. In that case the factor $(1-\frac{\sigma_{yo}}{\sigma_{eff}})/\eta \to 0/0 = \dot{\lambda}$ where $\dot{\lambda}$ is an undetermined quantity; therefore when $\eta \to 0$, (15.34) shows that $\dot{\varepsilon}^{vp}_{ij} \to \dot{\lambda}s_{ij}$ i.e. the classical Prandl-Reuss equations, cf. (9.31), where $\dot{\lambda}$ is the plastic multiplier and $\dot{\varepsilon}^{vp}_{ij}$ then becomes the plastic strain rates $\dot{\varepsilon}^p_{ij}$.

15.4.1 Perzyna viscoplasticity

Despite these early achievements, viscoplasticity played a minor role for a number of years and it was first when impact phenomena became an area of intensive research that the interest in viscoplasticity was revived. Constant strain-rate tests for a nickel alloy and concrete are displayed in Fig. 15.14 and it appears that the response fits into a viscoplastic framework; below a certain threshold value, the response is linear elastic and above this threshold strain-rate effects occur (note that it is allowable to choose the σ_{yo}-value as any threshold value and not just as the initial yield stress).

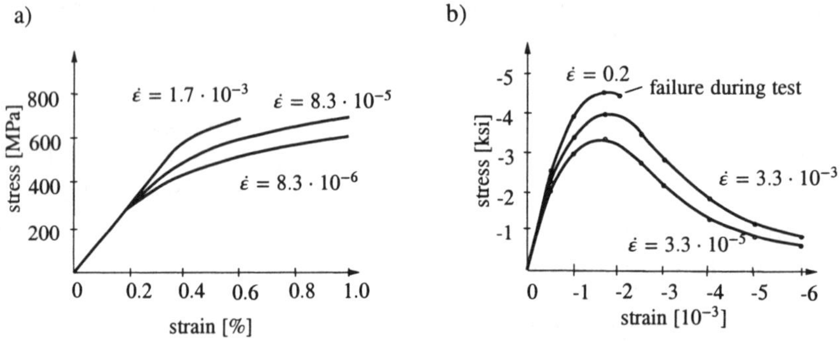

Figure 15.14: Constant strain-rate tests for uniaxial loading; a) nickel-based superalloy, B1900+Hf at 871°C , Chan *et al.* (1988), b) concrete in compression, Dilger *et al.* (1984).

The hardening effects shown in Fig. 15.14 cannot be modeled by the Hohenemser-Prager formulation (15.34) and in this model, the viscoplastic strain rate even depends linearly on strains. For uniaxial tension, Malvern (1951) suggested a model where hardening effects are considered. However, ignoring for the time being hardening effects, it is evident that (15.34) can be reformulated such that the viscoplastic strain rate depends nonlinearly on the stresses. We obtain

$$\dot{\varepsilon}_{ij}^{vp} = \begin{cases} 0 & \text{if} \quad \sigma_{eff} \leq \sigma_{yo} \\ \dfrac{\Phi(F(\sigma_{ij}))}{\eta}\dfrac{3s_{ij}}{2\sigma_{eff}} & \text{if} \quad \sigma_{eff} \geq \sigma_{yo} \end{cases}$$

where the function $\Phi(F(\sigma_{ij}) \geq 0$ is non-negative and $F(\sigma_{ij}) = \sigma_{eff} - \sigma_{yo}$. Moreover, since $\partial F/\partial\sigma_{ij} = 3s_{ij}/(2\sigma_{eff})$ the expression above can be written as

$$\dot{\varepsilon}_{ij}^{vp} = \begin{cases} 0 & \text{if} \quad \sigma_{eff} \leq \sigma_{yo} \\ \dfrac{\Phi(F(\sigma_{ij}))}{\eta}\dfrac{\partial F}{\partial\sigma_{ij}} & \text{if} \quad \sigma_{eff} \geq \sigma_{yo} \end{cases} \tag{15.35}$$

The function $\Phi(F)$ is assumed to be an increasing function of F and it is required that $\Phi(F = 0) = 0$. The formulation (15.35) was presented by Perzyna

(1963). Originally, $F(\sigma_{ij})$ was the yield function according to ideal von Mises plasticity, but we may also consider $F(\sigma_{ij})$ to be the yield function for ideal plasticity in general. With this format, it is easy to generalize so that hardening effects are included. Instead of the expression $F(\sigma_{ij})$ for the initial yield function, we use the expression $f(\sigma_{ij}, K_\alpha)$ for the current yield function, cf. the discussion related to (9.8). One may also allow for a nonassociated formulation and introduce the potential function $g = g(\sigma_{ij}, K_\alpha)$. We are then led to

Perzyna nonassociated viscoplasticity

$$\sigma_{ij} = D_{ijkl}(\varepsilon_{kl} - \varepsilon_{kl}^{vp})$$

$$\dot{\varepsilon}_{ij}^{vp} = \begin{cases} 0 & \text{if} \quad f(\sigma_{ij}, K_\alpha) \leq 0 \\ \dfrac{\Phi(f)}{\eta}\dfrac{\partial g}{\partial \sigma_{ij}} & \text{if} \quad f(\sigma_{ij}, K_\alpha) > 0 \end{cases} \tag{15.36}$$

where $f = f(\sigma_{ij}, K_\alpha) = 0$ *is the static yield surface*

and $\Phi(f) \geq 0$ *as well as* $\Phi(f = 0) = 0$

Moreover, $g = g(\sigma_{ij}, K_\alpha)$ *is the potential function*

Associated viscoplasticity is obtained by choosing the potential function as the yield function, i.e. $g = f(\sigma_{ij}, K_\alpha)$. Then (15.36) takes the format suggested by Perzyna (1966, 1971). In the present context, the surface described by $f(\sigma_{ij}, K_\alpha) = 0$ is called the *static yield surface* and we will return to that aspect in a moment.

Evidently, close similarities exist between Perzyna viscoplasticity and the time-independent plasticity theory discussed in Section 10.1. However, in the Perzyna viscoplasticity the factor $\Phi(f)/\eta$ is chosen by us whereas the corresponding factor in plasticity theory is the plastic multiplier $\dot{\lambda}$ which is determined by the consistency relation. In Perzyna viscoplasticity, the stress state is required to be located outside the static yield surface $f(\sigma_{ij}, K_\alpha)$ in order that viscoplastic strains develop whereas in time-independent plasticity the stresses can never be located outside the yield surface.

With this discussion of the similarities between viscoplasticity and plasticity in mind, we return to the general Perzyna formulation (15.36) and observe that hardening parameters K_α are involved. In analogy with plasticity theory we associate to the hardening parameters K_α some internal variables κ_α and, in general, we have the evolution equations

Evolution equations

$$K_\alpha = K_\alpha(\kappa_\beta)$$

$$\dot{\kappa}_\alpha = -\frac{\Phi(f)}{\eta}\frac{\partial g}{\partial K_\alpha} \tag{15.37}$$

A comparison of the evolution laws for $\dot{\varepsilon}_{ij}^{vp}$ and $\dot{\kappa}_\alpha$ given by (15.36) and (15.37) with the corresponding evolution laws for $\dot{\varepsilon}^p_{ij}$ and $\dot{\kappa}_\alpha$ given by (10.14) underlines the similarities and we refer to (10.14) for a further discussion.

In analogy with the discussion following (15.34), we observe from (15.36) that if the viscosity parameter $\eta \to 0$ then - in order that $\dot{\varepsilon}^{vp}_{ij}$ be a finite quantity - we must have $\Phi(f) \to 0$, i.e. $f \to 0$. As a result $\Phi(f)/\eta \to 0/0 = \dot{\lambda}$ where $\dot{\lambda}$ is an undetermined quantity and we then obtain $\dot{\varepsilon}^{vp}_{ij} \to \dot{\lambda}\partial g/\partial\sigma_{ij}$, i.e. rate independent plasticity theory with $\dot{\lambda}$ being the plastic multiplier

> *When the viscosity parameter $\eta \to 0$, Perzyna viscoplasticity reduces to rate independent plasticity theory*

This property is by Simo and Honein (1990) called the *viscoplastic regularization*

As a further illustration of Perzyna viscoplasticity, assume that the static yield function is taken as isotropic linear hardening of a von Mises material, i.e.

$$f(\sigma_{ij}, K_\alpha) = \sigma_{eff} - \sigma_{yo} - H\varepsilon^{vp}_{eff}$$

where H is a constant. In analogy with the definition of the effective plastic strain, we define the effective viscoplastic strain according to

$$\boxed{\dot{\varepsilon}^{vp}_{eff} = (\frac{2}{3}\dot{\varepsilon}^{vp}_{ij}\dot{\varepsilon}^{vp}_{ij})^{1/2}; \qquad \varepsilon^{vp}_{eff} = \int_0^t \dot{\varepsilon}^{vp}_{eff} dt} \tag{15.38}$$

For uniaxial tension, we then obtain

$$\dot{\varepsilon}^{vp} = \begin{cases} 0 & \text{if } f = \sigma - \sigma_{yo} - H\varepsilon^{vp} \leq 0 \\ \dfrac{\Phi(f)}{\eta} & \text{if } f = \sigma - \sigma_{yo} - H\varepsilon^{vp} \geq 0 \end{cases}$$

where $\Phi(f) \geq 0$.

A creep test with the constant stress $\sigma_A > \sigma_{yo}$ is shown in Fig. 15.15a). At point A, the overstress is given by $f_A = \sigma_{eff} - \sigma_{yo} - H\varepsilon^{vp} = \sigma_A - \sigma_{yo} > 0$ and we then have $\dot{\varepsilon}^{vp} > 0$. At point B, the overstress has been reduced to $f_B = \sigma_A - \sigma_{yo} - H\varepsilon^{vp}_B > 0$ and $f_B < f_A$ and thereby $\dot{\varepsilon}^{vp}_B < \dot{\varepsilon}^{vp}_A$. Finally, at point C where $f_C = \sigma_A - \sigma_{yo} - H\varepsilon^{vp}_C = 0$, the overstress has been reduced to zero and we have then reached the static yield surface and no more viscoplastic strains develop. The corresponding development over time of the viscoplastic strain is displayed in Fig. 15.15b) and it appears that the material only exhibits primary creep.

It is evident that the conclusions above can be generalized and we obtain

> *When $t \to \infty$, the stress state is located on the static yield surface*

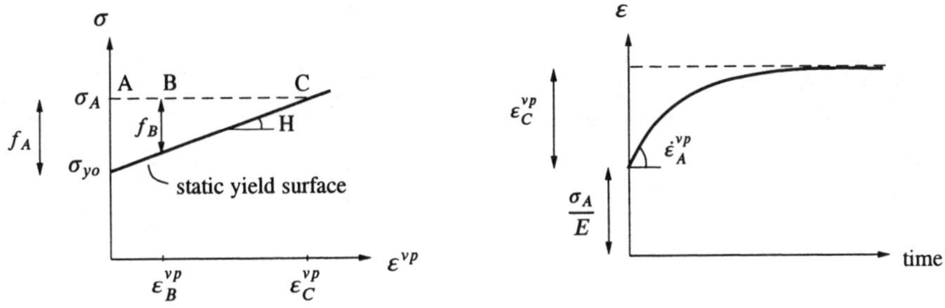

Figure 15.15: Creep test of linear hardening von Mises material; a) illustration in $\sigma-\varepsilon^{vp}$ diagram, b) development of strain over time.

This does not necessarily mean that the solution to a viscoplastic problem for $t \to \infty$ will become identical to the similar problem for plasticity. In the general case, hardening parameters K_α are involved and their development certainly depends on the loading history, and this implies that the solution to the viscoplastic problem and the corresponding plastic problem might differ. However, if the viscoplastic body is loaded by forces that increase infinitely slowly then the viscoplastic solution and the plastic solution coincide, i.e.

> *If the external loading is applied infinitely slowly, the viscoplastic solution and the plastic solution coincide*

Another extreme situation occurs if the load is applied very rapidly. The viscoplastic strain rate will then be large, but the integration of the viscoplastic strain rate over the time duration will be small when the time duration is short. Consequently, the viscoplastic material will in the limit behave as a linear elastic body when the loading rate is large.

In (15.36) the static yield surface $f(\sigma_{ij}, K_\alpha) = 0$ needs not be the same yield surface as that encountered in plasticity theory. Indeed, it is allowable in (15.36) to choose the static yield function such that $f(\sigma_{ij}, K_\alpha) \geq 0$ always; in a von Mises context one might for instance choose the static yield function as $f = \sigma_{eff}$ which implies $f \geq 0$ always. If this choice is adopted and if the factor $\Phi(f)/\eta$ in (15.36) is renamed and called Λ then the Perzyna format reduces to the general creep format (15.24); therefore

> *If the static yield function $f(\sigma_{ij}, K_\alpha)$ in Perzyna viscoplasticity is chosen as $f(\sigma_{ij}, K_\alpha) \geq 0$ then Perzyna viscoplasticity reduces to general creep theory*

It follows that Perzyna viscoplasticity contains two interesting bounds: if the time goes towards infinity, the stresses will be located on the static yield surface and if the static yield function is chosen properly, general creep theory emerges.

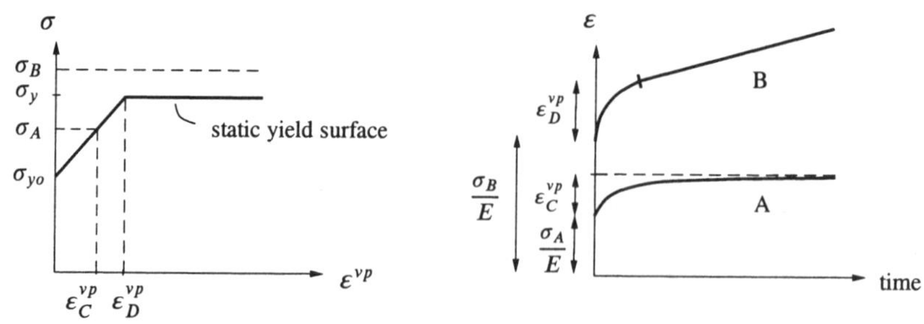

Figure 15.16: Creep tests showing both primary creep and secondary creep; a) primary creep, b) primary creep followed by stationary creep.

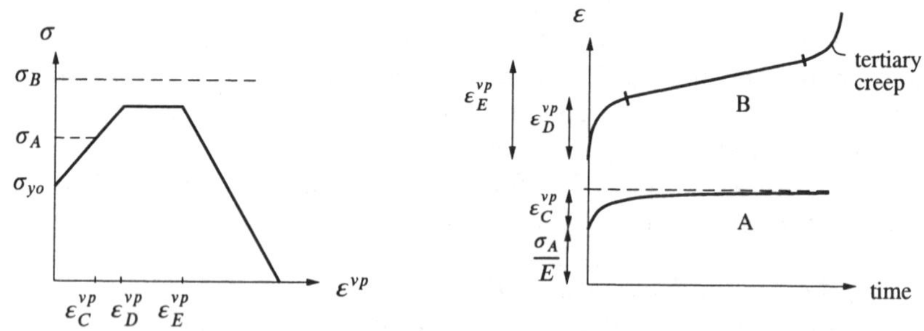

Figure 15.17: Creep tests showing primary, secondary and tertiary creep; a) primary creep, b) primary creep followed by secondary creep and tertiary creep.

Let us also illustrate that by proper choice of the static yield surface, a number of different creep characteristics can be modeled. Suppose the static yield function is chosen as shown in Fig. 15.16a) where ideal plasticity is reached with the yield stress σ_y when the viscoplastic strain has increased to the value ε_D^{vp}. For the constant stress σ_A, which is below σ_y, only primary creep is activated just like in Fig. 15.15. However, for the constant stress σ_B, which is larger than the yield stress σ_y, we first have development of primary creep and then - after some time - when the distance between the current stress and the static yield curve is constant, secondary creep develops.

In Fig. 15.17 the static yield surface is chosen such that it involves a softening branch when the viscoplastic strain is larger that ε_E^{vp}. As before, the constant stress σ_A will only give rise to primary creep, but the constant stress σ_B will first activate primary creep, then secondary creep and eventually tertiary creep, which might even involve creep failure.

Let us generalize these ideas and take the rate of the static yield function to obtain

$$\dot{f} = \frac{\partial f}{\partial \sigma_{ij}}\dot{\sigma}_{ij} + \frac{\partial f}{\partial K_\beta}\dot{K}_\beta$$

which with (15.37) becomes

$$\dot{f} = \frac{\partial f}{\partial \sigma_{ij}}\dot{\sigma}_{ij} - \frac{\partial f}{\partial K_\beta}\frac{\partial K_\beta}{\partial \kappa_\alpha}\frac{\partial g}{\partial K_\alpha}\frac{\Phi(f)}{\eta} \tag{15.39}$$

In analogy with (10.17) and (10.13) we define the generalized plastic modulus H as

$$\boxed{H = \frac{\partial f}{\partial K_\beta}\frac{\partial K_\beta}{\partial \kappa_\alpha}\frac{\partial g}{\partial K_\alpha}}$$

that is, (15.39) becomes

$$\dot{f} = \frac{\partial f}{\partial \sigma_{ij}}\dot{\sigma}_{ij} - H\frac{\Phi(f)}{\eta}$$

In a creep test, the stresses are constant and we then have $\dot{f} = -H\Phi(f)/\eta$. If $H > 0$ then $\dot{f} < 0$, i.e. primary creep, if $H = 0$ then $\dot{f} = 0$ and secondary creep develops and, finally, if $H < 0$ then $\dot{f} > 0$ and tertiary creep is present. Consequently

$H > 0 \Rightarrow$ *hardening viscoplasticity (primary creep)*
$H = 0 \Rightarrow$ *ideal viscoplasticity (secondary creep)*
$H < 0 \Rightarrow$ *softening viscoplasticity (tertiary creep)*

and these conclusions correspond to those applicable in plasticity theory, cf. (10.33); further discussions are given by Lubliner (1990) as well as by Ristinmaa and Ottosen (2000).

We will now reformulate the Perzyna equations (15.36) in a manner which leads to the interesting concept of a *dynamic yield surface* introduced by Perzyna (1963, 1966). Use of (15.36) in expression (15.38) for the effective viscoplastic strain rate provides

$$\eta\dot{\varepsilon}^{vp}_{eff}\sqrt{\frac{2}{3}\frac{\partial g}{\partial \sigma_{ij}}\frac{\partial g}{\partial \sigma_{ij}}} = \Phi(f) \tag{15.40}$$

Associated with the non-negative monotonic increasing function Φ there exists an inverse non-negative function φ with the properties

$$\boxed{\varphi(\Phi(f)) = f \quad \textit{i.e.} \quad \varphi \geq 0 \;\; \textit{and} \;\; \varphi(0) = 0} \tag{15.41}$$

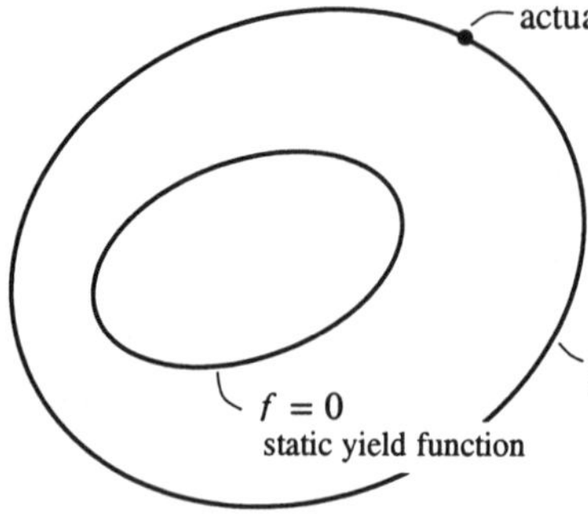

Figure 15.18: Illustration of static and dynamic yield surfaces.

The property $\phi(0) = 0$ follows directly from the previous mentioned requirement $\Phi(0) = 0$. We then obtain from (15.40)

$$\varphi(\eta\dot{\varepsilon}^{vp}_{eff}\sqrt{\frac{2}{3}\frac{\partial g}{\partial \sigma_{ij}}\frac{\partial g}{\partial \sigma_{ij}}}) = f \tag{15.42}$$

Define the dynamic yield surface f^d by $f^d = f - \varphi$. It then follows from (15.42) that $f^d = 0$ holds during viscoplastic development. If the response is elastic then $f < 0$ as well as $\dot{\varepsilon}^{vp}_{eff} = 0$ implying that $f^d = f - \varphi = f < 0$. We also observe that $f^d > 0$ can never occur. In conclusion

Dynamic yield function

$$f^d = f - \varphi$$

$$f^d \begin{cases} < 0 & \textit{elastic response} \\ = 0 & \textit{viscoplastic response} \end{cases} \tag{15.43}$$

f = *static yield function*

$$\varphi = \varphi(\eta\dot{\varepsilon}^{vp}_{eff}\sqrt{\frac{2}{3}\frac{\partial g}{\partial \sigma_{ij}}\frac{\partial g}{\partial \sigma_{ij}}})$$

This result is illustrated in Fig. 15.18 and a further discussion is provided by Ristinmaa and Ottosen (2000).

To illustrate this concept, assume that the static yield function is taken as an isotropic hardening von Mises formulation, i.e.

$$f = \sigma_{eff} - \sigma_{yo} - K(\kappa) \tag{15.44}$$

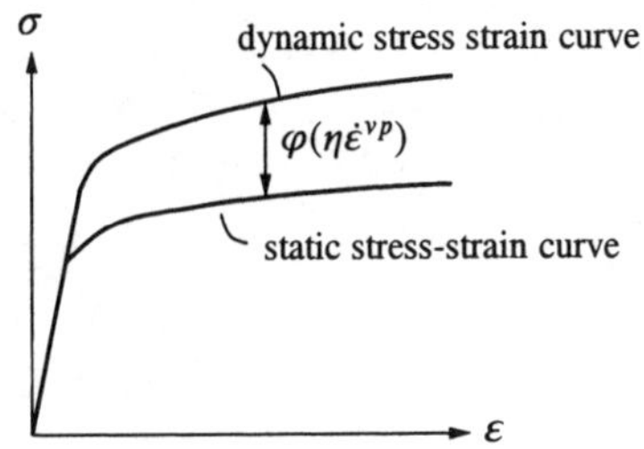

Figure 15.19: Static and dynamic stress-strain curves according to (15.47).

Then (15.36) gives

$$\dot{\varepsilon}_{ij}^{vp} = \frac{\Phi(f)}{\eta}\frac{3s_{ij}}{2\sigma_{eff}} \tag{15.45}$$

and the effective viscoplastic strain rate defined by (15.38) becomes

$$\dot{\varepsilon}_{eff}^{vp} = \frac{\Phi(f)}{\eta} \tag{15.46}$$

Use of the inverse function φ defined by (15.41) leads to

$$\varphi(\eta\dot{\varepsilon}_{eff}^{vp}) = f$$

The dynamic yield function is then given by $f^d = f - \varphi(\eta\dot{\varepsilon}_{eff}^{vp}) = 0$ which with (15.44) gives

$$\sigma_{eff} = \sigma_{yo} + K(\varepsilon_{eff}^{vp}) + \varphi(\eta\dot{\varepsilon}_{eff}^{vp}) \tag{15.47}$$

where the internal variable κ was chosen as $\kappa = \varepsilon_{eff}^{vp}$. The above expression has the uniaxial interpretation shown in Fig. 15.19.

Indeed, (15.47) makes for some interesting interpretations of various experimental findings. For steel and metals, it is observed experimentally that the initial yield stress increases with the loading rate as shown in Fig. 15.20 where the experimental data are taken from Manjoine (1944).

To fit such data, the model of Cowper and Symonds (1962) is often adopted; this model reads

$$\sigma_{yo}^d = \sigma_{yo}[1 + (\frac{\dot{\varepsilon}_{eff}^{vp}}{D})^{1/p}] \tag{15.48}$$

where σ_{yo}^d is the dynamic initial yield stress and σ_{yo}, as usual, is the static initial yield stress. Moreover, D is an arbitrary, positive reference strain rate with the sole purpose of making the term $\dot{\varepsilon}_{eff}^{vp}/D$ dimensionless; evidently when

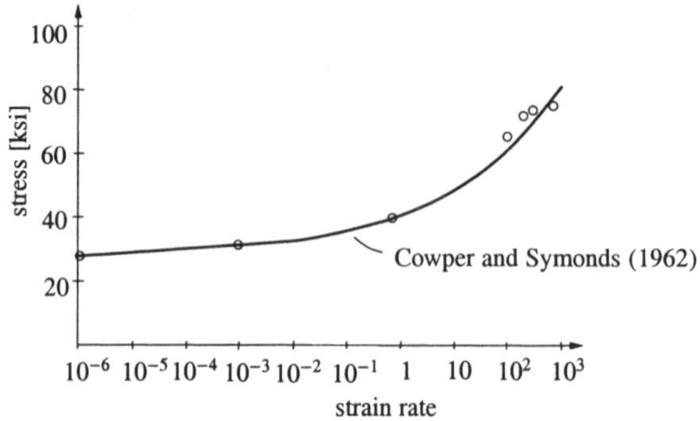

Figure 15.20: Variation of initial yield stress for mild steel with loading rate, Manjoine (1944). Curve fitting of Cowper-Symonds model (15.48) with $D = 30$ [1/s] and $p = 5$.

$\dot{\varepsilon}^{vp}_{eff} = D$ then $\sigma^d_{yo} = 2\sigma_{yo}$. The positive dimensionless parameter p in (15.48) is often in the range of 5-10. It is emphasized that the strain rate used in (15.48) is the effective viscoplastic strain rate and not the total strain rate as used by some authors.

Evaluation of (15.47) at initiation of viscoplasticity and by using $K(0) = 0$ give $\sigma_{eff} = \sigma_{yo} + \varphi(\eta\dot{\varepsilon}^{vp}_{eff})$. A comparison with (15.48) then provides

$$\varphi(\eta\dot{\varepsilon}^{vp}_{eff}) = \sigma_{yo}(\frac{\dot{\varepsilon}^{vp}_{eff}}{D})^{1/p}$$

which can be written as

$$\varphi(x) = \sigma_{yo}(\frac{x}{\eta D})^{1/p} \tag{15.49}$$

To identify the function Φ present in (15.36), we observe from (15.41) that $\varphi(\Phi(f)) = f$. Therefore if we in (15.49) adopt $x = \Phi(f)$ we obtain

$$\sigma_{yo}(\frac{\Phi(f)}{\eta D})^{1/p} = f$$

which leads to

$$\boxed{\Phi(f) = \eta D(\frac{f}{\sigma_{yo}})^p} \tag{15.50}$$

Therefore, if the function $\Phi(f)$ is chosen in this fashion the Cowper-Symonds model (15.48) is retrieved.

Led by expression (15.50), we choose

$$\boxed{\Phi(f) = \eta D(\frac{f}{\sigma_{yo} + K(\varepsilon^{vp}_{eff})})^p} \tag{15.51}$$

Insertion into (15.46) gives

$$f = [\sigma_{yo} + K(\varepsilon^{vp}_{eff})](\frac{\dot{\varepsilon}^{vp}_{eff}}{D})^{1/p}$$

and use of (15.44) results in

$$\sigma_{eff} = \sigma_{st}(\varepsilon^{vp}_{eff})[1 + (\frac{\dot{\varepsilon}^{vp}_{eff}}{D})^{1/p}] \tag{15.52}$$

where the static stress-strain curve is given by $\sigma_{st}(\varepsilon^{vp}_{eff}) = \sigma_{yo} + K(\varepsilon^{vp}_{eff})$. Also expression (15.52) is often used to fit experimental data.

Here, we have discussed the implications of the choices (15.50) and (15.51) and we refer to Perzyna (1966) for further interpretations and possibilities.

The rather detailed discussion above was based on the simple model provided by isotropic hardening von Mises formulation (15.44) and it is evident that a number of other possibilities exists; in fact, to each of the plasticity models discussed in Chapters 12 and 13 there exists a corresponding viscoplastic model. We will not pursue this line any further, but simply mention that the Armstrong-Frederick model discussed in detail in Section 13.3 forms the basis of a number of very successful models used to simulate the complex creep behavior in the gas-turbine and nuclear industry. As an example, we refer to the formulations proposed by Chaboche (1989, 1993a) as well as Chaboche and Nouailhas (1989) and also dealt with in the textbook of Lemaitre and Chaboche (1990).

15.4.2 Duvaut-Lions viscoplasticity

Having discussed Perzyna-viscoplasticity in detail, we now turn to the other major approach which is provided by Duvaut and Lions (1972).

The driving force in viscoplasticity is the overstress, which is a measure of how far we are outside the static yield surface $f = 0$. In Perzyna viscoplasticity, this measure is provided by determination of the scalar $f(\sigma_{ij}, K_\alpha)$, i.e. evaluating the static yield function at the current state.

Referring to the uniaxial case shown in Fig. 15.21a), we could equally well measure the overstress as the stress difference between the current stress and the stress at the static stress-strain curve. In the general situation shown in Fig. 15.21b), the overstress is taken as the stress difference $\sigma_{ij} - \bar{\sigma}_{ij}$ where $\bar{\sigma}_{ij}$ is a stress value at the static yield surface $f = 0$. This leads to the *Duvaut-Lions formulation* given by

$$\dot{\varepsilon}^{vp}_{ij} = \frac{1}{\eta}(\sigma_{ij} - \bar{\sigma}_{ij}) \tag{15.53}$$

where η is a constant viscosity parameter. The original Duvaut-Lions model applies to ideal viscoplasticity where no hardening parameters exist and (15.53)

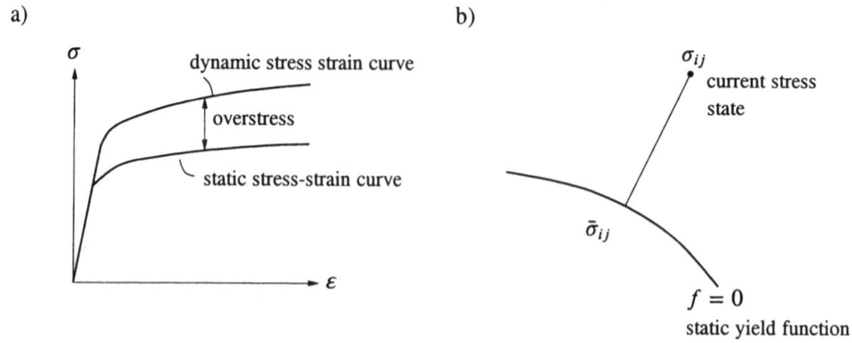

Figure 15.21: Illustration of Duvaut-Lions viscoplasticity based on overstress; a) uniaxial situation, b) overstress $\sigma_{ij} - \bar{\sigma}_{ij}$ in the general situation.

is the only evolution law. It remains to define the stress state $\bar{\sigma}_{ij}$ located on the static yield surface which for ideal viscoplasticity takes the format $f(\bar{\sigma}_{ij}, \bar{K}_\alpha = 0) = F(\bar{\sigma}_{ij}) = 0$. In the Duvaut-Lions model, the $\bar{\sigma}_{ij}$-value is taken as the *closest-point-projection* of the current stress state on the static yield surface $F(\bar{\sigma}_{ij}) = 0$. Let us discuss the concept of the closest-point-projection in more detail.

Considering a point in a space, this point can be projected on a surface such that the distance between the point and its projection on the surface is as small as possible; this constitutes the closest-point-projection. Now, any space possesses a *metric* which controls how distances are measured. In Euclidean space, the metric tensor is given by Kronecker's delta δ_{ij} such that the quadratic form $s^2 = x_i \delta_{ij} x_j$ measures the squared distance between the start and end point of the vector x_i. In the original Duvaut-Lions formulation, the $\bar{\sigma}_{ij}$-value is determined as the Euclidean closest-point-projection on the static yield surface $F(\bar{\sigma}_{ij}) = 0$.

When the Duvaut-Lions formulation is generalized to include hardening viscoplasticity, the static yield surface now also depends on the hardening parameters and it seems reasonable to measure the distance between the current state and the static yield surface not only expressed in terms of stresses, but also in terms of hardening parameters; following Simo *et al.* (1988) we will now see how this generalization can be achieved.

The static yield function is given by $f(\sigma_{ij}, K_\alpha)$. Similar to (15.37), the hardening parameters K_α are related to the internal variables κ_α through $K_\alpha = K_\alpha(\kappa_\beta)$ which leads to

$$\boxed{\dot{K}_\alpha = d_{\alpha\beta} \dot{\kappa}_\beta \quad where \quad d_{\alpha\beta} = \frac{\partial K_\alpha}{\partial \kappa_\beta}} \tag{15.54}$$

Thermodynamic considerations discussed later in Chapter 22 show that the matrix $d_{\alpha\beta}$ is symmetric. Its inverse matrix $c_{\alpha\beta}$ fulfills per definition the expression $c_{\alpha\gamma}d_{\gamma\beta} = \delta_{\alpha\beta}$ and (15.54) then leads to

$$\boxed{\dot{\kappa}_\alpha = c_{\alpha\beta}\dot{K}_\beta}$$

As usual, C_{ijkl} denotes the elastic flexibility tensor, cf. (4.26). It is now assumed that the metric tensor in the σ_{ij}, K_α - space is defined by

$$\begin{bmatrix} C_{ijkl} & 0 \\ 0 & c_{\alpha\beta} \end{bmatrix} \tag{15.55}$$

In the σ_{ij}, K_α - space, the distance s between the state σ_{ij}, K_α and the state $\bar{\sigma}_{ij}, \bar{K}_\alpha$ is, per definition, given by

$$s^2 = \frac{1}{2}(\sigma_{ij} - \bar{\sigma}_{ij})C_{ijkl}(\sigma_{kl} - \bar{\sigma}_{kl}) + \frac{1}{2}(K_\alpha - \bar{K}_\alpha)c_{\alpha\beta}(K_\beta - \bar{K}_\beta) \tag{15.56}$$

where the factor $1/2$ has been introduced for convenience; a more detailed discussion is given by Ristinmaa and Ottosen (1998).

In the discussion leading to (10.14) it was mentioned that, for plasticity, the second law of thermodynamics is expressed as $\sigma_{ij}\dot{\varepsilon}^p_{ij} - K_\alpha\dot{\kappa}_\alpha \geq 0$. It then follows that the quantities $\sigma_{ij}C_{ijkl}\sigma_{kl}$ and $K_\alpha c_{\alpha\beta}K_\beta$ are of the same dimension [Nm/m^3] and this supports the choice of the metric given by (15.55).

Let σ_{ij}, K_α be the current state and let $\bar{\sigma}_{ij}, \bar{K}_\alpha$ be some state of the static yield surface given by $f(\bar{\sigma}_{ij}, \bar{K}_\alpha) = 0$. Assuming both C_{ijkl} and $c_{\alpha\beta}$ to be constant quantities, we now determine the state $\bar{\sigma}_{ij}, \bar{K}_\alpha$ such that it becomes the closest-point-projection on the static yield surface. Therefore, for given σ_{ij}, K_α we want to determine the state $\bar{\sigma}_{ij}, \bar{K}_\alpha$ which minimizes the distance s^2 defined by (15.56) under the constraint at $f(\bar{\sigma}_{ij}, \bar{K}_\alpha) = 0$. The Lagrangian multiplier method is summarized in Appendix and (A.15) then provides

$$\begin{aligned}
&-C_{ijkl}(\sigma_{kl} - \bar{\sigma}_{kl}) + \mu\frac{\partial f}{\partial \bar{\sigma}_{ij}} = 0 \\
&-c_{\alpha\beta}(K_\beta - \bar{K}_\beta) + \mu\frac{\partial f}{\partial \bar{K}_\alpha} = 0 \\
&f(\bar{\sigma}_{ij}, \bar{K}_\alpha) = 0
\end{aligned} \tag{15.57}$$

where μ is a non-negative Lagrangian multiplier. This equation system determines the unknowns $\bar{\sigma}_{ij}, \bar{K}_\alpha$ and μ.

With these preliminaries, we are now in a position to express generalized

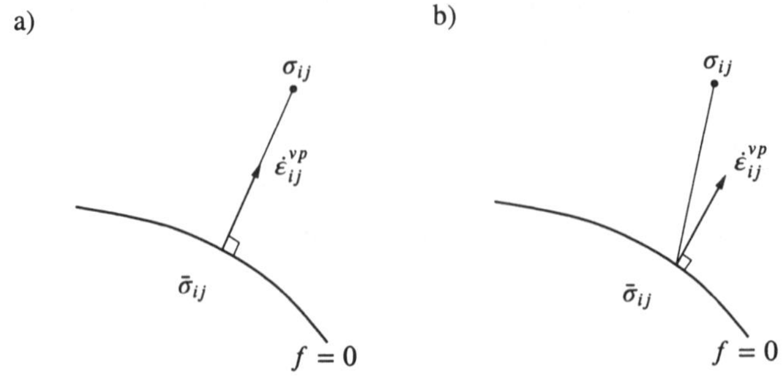

Figure 15.22: Ideal viscoplasticity. a) Duvaut-Lions formulation in stress space; b) Generalized Duvaut-Lions formulation in stress space.

Duvaut-Lions viscoplasticity according to

Generalized Duvaut-Lions viscoplasticity

$$\sigma_{ij} = D_{ijkl}(\varepsilon_{kl} - \varepsilon_{kl}^{vp})$$

if $f(\sigma_{ij}, K_\alpha) \leq 0$ *then*

$$\dot{\varepsilon}_{ij}^{vp} = 0 \quad \dot{\kappa}_\alpha = 0$$

if $f(\sigma_{ij}, K_\alpha) \geq 0$ *then* (15.58)

$$\dot{\varepsilon}_{ij}^{vp} = \Lambda C_{ijkl}(\sigma_{kl} - \bar{\sigma}_{kl})$$

$$\dot{\kappa}_\alpha = -\Lambda c_{\alpha\beta}(K_\alpha - \bar{K}_\alpha)$$

where $f = f(\sigma_{ij}, K_\alpha)$ *is the static yield function and* $\bar{\sigma}_{ij}, \bar{K}_\alpha$ *is the closest-point-projection on the static yield surface* $f(\bar{\sigma}_{ij}, \bar{K}_\alpha) = 0$

Moreover, Λ is any positive quantity that we choose and it seems natural to let Λ depend on the distance s defined in (15.56). We shall see later that the formulation (15.58) is very attractive from a computational point of view.

To further explore the differences between the original Duvaut-Lions model and the generalized Duvaut-Lions formulation, the consequences of (15.53) are illustrated in Fig. 15.22a) showing the closest-point-projection in Euclidean space as well as the direction of $\dot{\varepsilon}_{ij}^{vp}$. In Fig. 15.22b), the generalized Duvaut-Lions formulation is shown where the closest-point-projection $\bar{\sigma}_{ij}$ is determined by the metric given in (15.55). As a comparison of (15.58) and (15.57) shows, it is of interest that $\dot{\varepsilon}_{ij}^{vp}$ is still orthogonal to the static yield surface $f = 0$.

Despite the differences in formulation, it turns out that there are cases where Perzyna and generalized Duvaut-Lions viscoplasticity coincide and a detailed investigation is provided by Runesson *et al.* (1999). As an illustration, take

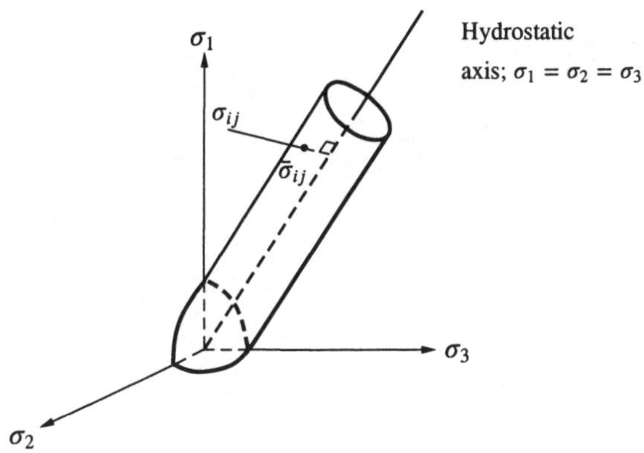

Figure 15.23: Closest point projection $\bar{\sigma}_{ij}$ of the current stress point σ_{ij} on the static yield surface in terms of a von Mises cylinder, the space is Euclidean.

the static yield function as an isotropic hardening von Mises formulation, i.e. $f = \sigma_{eff} - \sigma_{yo} - K$. Then (15.45) shows that Perzyna viscoplasticity is given by

$$\dot{\varepsilon}_{ij}^{vp} = \frac{\Phi(f)}{\eta} \frac{3s_{ij}}{2\sigma_{eff}} \tag{15.59}$$

Let the current stress state σ_{ij} be projected on the von Mises cylinder to obtain the Euclidean closest-point-projection $\bar{\sigma}_{ij}$ on the static yield surface as shown in Fig. 15.23. Since both σ_{ij} and $\bar{\sigma}_{ij}$ are located in the same deviatoric plane, we have $\sigma_{ij} - \bar{\sigma}_{ij} = s_{ij} - \bar{s}_{ij}$. Assuming isotropic elasticity, the flexibility tensor C_{ijkl} is given by (4.93) and we then obtain $C_{ijkl}(\sigma_{kl} - \bar{\sigma}_{kl}) = C_{ijkl}(s_{kl} - \bar{s}_{kl}) = (s_{ij} - \bar{s}_{ij})/2G$. Finally, we have $\bar{s}_{ij} = ks_{ij}$ where k is a proportionality factor which leads to $(s_{ij} - \bar{s}_{ij})/2G = (1-k)s_{ij}/2G$, i.e. $C_{ijkl}(\sigma_{kl} - \bar{\sigma}_{kl}) = (1-k)s_{ij}/2G$ and a comparison of (15.59) and (15.58) shows that by proper choice of, say, the Λ-quantity, the Perzyna and Duvaut-Lions formulations coincide.

16 NONLINEAR FINITE ELEMENT METHOD

As engineers, we are interested in the response of structures and thereby in the solution of boundary value problems. Having established the equations that control elasto-plastic, viscoplastic and creep behavior, it is evident that these equations are so complex that exact analytical solutions of boundary value problems cannot, in general, be established. Instead we must look for approximative solution methods and, today, the most powerful numerical means turns out to be the *finite element method*.

For linear problems, use of the finite element (FE) method is straightforward. Elasto-plastic, viscoplastic and creep problems, however, are nonlinear and this gives rise to a number of questions that must be resolved before a reliable solution can be established. These new questions can be summarized into: formulation of the nonlinear finite element method, solution of the nonlinear global equations, and integration of the constitutive equations. Here, we will first present the formulation of the nonlinear finite element method whereas the next two chapters will deal with the solution of the nonlinear global equations for static problems and numerical integration of the constitutive equations.

Let us therefore first formulate the nonlinear finite element method for general nonlinear solid mechanics. The formulation of the nonlinear FE method is very similar to that of the linear FE method, i.e. it is based on the weak formulation of the equations of motion, i.e. on the principle of virtual work. Detailed discussions of the finite element method are given, for instance, by Bathe (1996), Belytschko *et al.* (2000), Hughes (1987), Ottosen and Petersson (1992) and Zienkiewicz and Taylor (1989).

16.1 Equations of motion

In the first place, we will express the equations of motion in a finite element format. The weak form of the equations of motion is given by (3.33), i.e.

$$\int_V \rho v_i \ddot{u}_i dV + \int_V \varepsilon^v_{ij} \sigma_{ij} dV = \int_S v_i \, t_i dS + \int_V v_i \, b_i dV \tag{16.1}$$

where it is recalled that $v_i = v_i(x_i)$ is an arbitrary weight vector, t_i is the traction vector, b_i the body force, i.e. force per unit volume and $\ddot{u}_i$ is the acceleration vector. Moreover, V denotes the region, i.e. the volume, of the body whereas S is the boundary surface of the body. Finally, according to (3.32) the quantity ε^v_{ij} is defined by

$$\varepsilon^v_{ij} = \frac{1}{2}(v_{i,j} + v_{j,i}) \tag{16.2}$$

It appears that ε^v_{ij} is defined in a similar manner as the strain tensor ε_{ij}, but to emphasize that ε^v_{ij} just is a quantity defined by (16.2), we have used the superscript v.

In finite element formulations, matrix notation turns out to be particularly convenient. With the notations, cf. (4.35)

$$\boldsymbol{\varepsilon}^v = \begin{bmatrix} \varepsilon^v_{11} \\ \varepsilon^v_{22} \\ \varepsilon^v_{33} \\ 2\varepsilon^v_{12} \\ 2\varepsilon^v_{13} \\ 2\varepsilon^v_{23} \end{bmatrix}; \quad \boldsymbol{\sigma} = \begin{bmatrix} \sigma_{11} \\ \sigma_{22} \\ \sigma_{33} \\ \sigma_{12} \\ \sigma_{13} \\ \sigma_{23} \end{bmatrix}$$

$$\ddot{\boldsymbol{u}} = \begin{bmatrix} \ddot{u}_1 \\ \ddot{u}_2 \\ \ddot{u}_3 \end{bmatrix}; \quad \boldsymbol{v} = \begin{bmatrix} v_1 \\ v_2 \\ v_3 \end{bmatrix}; \quad \boldsymbol{t} = \begin{bmatrix} t_1 \\ t_2 \\ t_3 \end{bmatrix}; \quad \boldsymbol{b} = \begin{bmatrix} b_1 \\ b_2 \\ b_3 \end{bmatrix}$$

we may write the weak form (16.1) of the equations of motion as

$$\int_V \rho \boldsymbol{v}^T \ddot{\boldsymbol{u}} dV + \int_V (\boldsymbol{\varepsilon}^v)^T \boldsymbol{\sigma} dV = \int_S \boldsymbol{v}^T \boldsymbol{t} dS + \int_V \boldsymbol{v}^T \boldsymbol{b} dV \tag{16.3}$$

The *boundary conditions* of the body can be expressed as

$\boldsymbol{u} =$	*is given along* S_u
$\boldsymbol{t} =$	*is given along* S_t

that is, the displacement vector $\boldsymbol{u}$ is prescribed along the boundary surface S_u and the traction vector $\boldsymbol{t}$ is prescribed along the boundary surface S_t. The sum S_u and S_t comprises the entire boundary S as illustrated in Fig. 16.1.

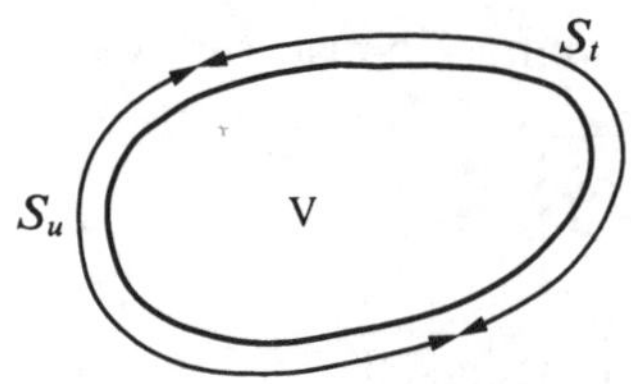

Figure 16.1: Boundary conditions

The finite element method is based on the concept that the displacement vector $\boldsymbol{u}$ throughout the body can be expressed in an approximate manner as

$$\boldsymbol{u} = \boldsymbol{N}\boldsymbol{a} \tag{16.4}$$

where $\boldsymbol{N}$ denotes the *global shape functions* and $\boldsymbol{a}$ is a column matrix that includes all the *nodal displacements* of the body. The displacement vector $\boldsymbol{u}$ depends on both position and time whereas the global shape functions only depend on position, i.e. we have

$$\boldsymbol{u} = \boldsymbol{u}(x_i, t)\ ; \qquad \boldsymbol{N} = \boldsymbol{N}(x_i)\ ; \qquad \boldsymbol{a} = \boldsymbol{a}(t) \tag{16.5}$$

It follows that

$$\ddot{\boldsymbol{u}} = \boldsymbol{N}\ddot{\boldsymbol{a}}$$

With the displacements given by (16.4) we can determine the corresponding strains, which, in a matrix form, may be expressed as

$$\boldsymbol{\varepsilon} = \boldsymbol{B}\boldsymbol{a}\ ; \qquad \boldsymbol{B} = \boldsymbol{B}(x_i) \tag{16.6}$$

where the matrix $\boldsymbol{B}$ is derived from the matrix $\boldsymbol{N}$.

The fundamental issue of the standard finite element method is that the arbitrary weight vector $\boldsymbol{v}$ is chosen in accordance with *Galerkin's method*, i.e. it is approximated in the same manner as the displacement $\boldsymbol{u}$. In analogy with (16.4) we therefore write

$$\boldsymbol{v} = \boldsymbol{N}\boldsymbol{c} \tag{16.7}$$

Since $\boldsymbol{v}$ is arbitrary and the global shape functions $\boldsymbol{N}$ are specified by us, the column matrix $\boldsymbol{c}$ is arbitrary. Like (16.5), we observe that $\boldsymbol{c}$ does not depend on position. From (16.7), we can determine the quantity $\boldsymbol{\varepsilon}^v$ similar to (16.6) i.e.

$$\boldsymbol{\varepsilon}^v = \boldsymbol{B}\boldsymbol{c} \tag{16.8}$$

Use of (16.7) and (16.8) in the weak form (16.3) of the equations of motion and noting that $\boldsymbol{c}$ is independent of position gives

$$\boldsymbol{c}^T\left[\left(\int_V \rho\boldsymbol{N}^T\boldsymbol{N}dV\right)\ddot{\boldsymbol{a}} + \int_V \boldsymbol{B}^T\boldsymbol{\sigma}dV - \int_S \boldsymbol{N}^T\boldsymbol{t}dS - \int_V \boldsymbol{N}^T\boldsymbol{b}dV\right] = 0$$

As this equation holds for arbitrary $\boldsymbol{c}$-matrices, we conclude that

$$\boxed{\boldsymbol{M}\ddot{a} + \int_V \boldsymbol{B}^T \boldsymbol{\sigma} dV = \boldsymbol{f}} \tag{16.9}$$

where the *mass matrix* $\boldsymbol{M}$ is defined by

$$\boldsymbol{M} = \int_V \rho \boldsymbol{N}^T \boldsymbol{N} dV$$

i.e. $\boldsymbol{M}$ is symmetric. Moreover, $\boldsymbol{f}$ defines the *external forces* according to

$$\boldsymbol{f} = \int_S \boldsymbol{N}^T \boldsymbol{t} dS + \int_V \boldsymbol{N}^T \boldsymbol{b} dV$$

We emphasize that expression (16.9) was derived entirely from the equations of motion without any information on the particular constitutive relation and (16.9) therefore holds for any constitutive relation.

16.2 Static conditions

Let us now specialize to static conditions, i.e. the nodal accelerations $\ddot{\boldsymbol{a}}$ are zero; the equations of motion (16.9) then reduce to the *equilibrium equations*

$$\boxed{\boldsymbol{\psi} = \boldsymbol{0}} \tag{16.10}$$

where

$$\boxed{\boldsymbol{\psi} = \int_V \boldsymbol{B}^T \boldsymbol{\sigma} dV - \boldsymbol{f}} \tag{16.11}$$

It is recalled that $\boldsymbol{f}$ is an expression for the external loading of the body given by the traction vector $\boldsymbol{t}$ and the body force $\boldsymbol{b}$. Likewise, the term $\int_V \boldsymbol{B}^T \boldsymbol{\sigma} dV$ expresses the *internal forces* that the stresses $\boldsymbol{\sigma}$ give rise to and (16.10) and (16.11) therefore states that in order that the body be in equilibrium, the external forces must be equal to the internal forces.

Equations (16.10) and (16.11) express the equilibrium equations for the body and since no consideration was made of the constitutive relation, they hold for any body in equilibrium.

Evidently, the constitutive relation must be invoked in order to solve a specific boundary value problem. For linear elasticity, this step is straightforward. In that case, we have with (16.6)

$$\boldsymbol{\sigma} = \boldsymbol{D}\boldsymbol{\varepsilon} = \boldsymbol{D}\boldsymbol{B}\boldsymbol{a} \tag{16.12}$$

where, in general, the elastic stiffness $\boldsymbol{D}$ depends on position, i.e. $\boldsymbol{D} = \boldsymbol{D}(x_i)$, but not on the loading. Introduction of (16.12) in (16.11) and use of (16.10) provide

$$\boldsymbol{K}\boldsymbol{a} = \boldsymbol{f} \quad \text{where} \quad \boldsymbol{K} = \int_V \boldsymbol{B}^T \boldsymbol{D}\boldsymbol{B} dV \tag{16.13}$$

where the elastic stiffness matrix $\boldsymbol{K}$ is constant. It appears that (16.13) is a linear equation system that - after consideration of the boundary conditions - can be solved directly of provide the current value of the nodal displacements $\boldsymbol{a}$.

For general nonlinear problems, the situation is quite different. Here we cannot express the current stresses $\boldsymbol{\sigma}$ directly in terms of the current strains $\boldsymbol{\varepsilon}$; instead, we simply know the incremental relation between the stress rate and the strain rate. For elasto-plastic problems, for instance, this relation is given by (10.29), i.e.

$$\dot{\boldsymbol{\sigma}} = \boldsymbol{D}^{ep}\dot{\boldsymbol{\varepsilon}} \tag{16.14}$$

and the current stresses $\boldsymbol{\sigma}$ must be obtained by integration of (16.14) along the actual load history.

Since the constitutive relation (16.14) is nonlinear, it is no surprise that also the global equilibrium equations (16.10) and (16.11) become nonlinear. This is the first problem we encounter when solving, for instance, elasto-plastic boundary value problems. The next problem is that at each material point, we have to integrate the constitutive equations given by (16.14).

In order to further illustrate these problems, we observe that (16.14) is given in an incremental form. It is therefore tempting to differentiate (16.10) with respect to time to obtain

$$\int_V \boldsymbol{B}^T \dot{\boldsymbol{\sigma}} dV = \dot{\boldsymbol{f}} \tag{16.15}$$

where

$$\dot{\boldsymbol{f}} = \int_S \boldsymbol{N}^T \dot{\boldsymbol{t}} dS + \int_V \boldsymbol{N}^T \dot{\boldsymbol{b}} dV$$

and where it was used that the global shape functions $\boldsymbol{N}$ and the matrix $\boldsymbol{B}$ are independent of time, cf. (16.5) and (16.6). We observe that whereas (16.10) expresses the total equilibrium of the body, (16.15) expresses the incremental equilibrium condition of the body. For elasto-plasticity, it follows from (16.14) and (16.6) that

$$\dot{\boldsymbol{\sigma}} = \boldsymbol{D}_t \boldsymbol{B}\dot{\boldsymbol{a}} \quad \text{where} \quad \boldsymbol{D}_t = \begin{cases} \boldsymbol{D} & \text{if the point behaves elastically} \\ \boldsymbol{D}^{ep} & \text{if the point behaves plastically} \end{cases} \tag{16.16}$$

Since the incremental nodal displacements $\dot{\boldsymbol{a}}$ are independent of position, use of (16.16) in (16.15) yields

$$\boldsymbol{K}_t\,\dot{\boldsymbol{a}} = \dot{\boldsymbol{f}} \tag{16.17}$$

where the *tangential stiffness matrix* $\boldsymbol{K}_t$ of the entire body is given by

$$\boldsymbol{K}_t = \int_V \boldsymbol{B}^T\,\boldsymbol{D}_t\,\boldsymbol{B}\,dV$$

Expression (16.17) comprises the global equation system that we need to solve in order to determine the incremental response of the body.

It is of importance that the tangential stiffness matrix $\boldsymbol{K}_t$ is not a constant matrix, i.e. (16.17) comprises a system of nonlinear equations. Expression (16.17) suggests that the external load $\boldsymbol{f}$ is increased in small steps and for each of these steps, the corresponding change of the nodal displacements $\boldsymbol{a}$ is determined by means of (16.17). Therefore, the solution of our boundary value problem has been reformulated into a *stepwise*, i.e. an *incremental solution procedure* that traces the response of the body with increased loading.

The first fundamental problem in the nonlinear finite element method is therefore to solve the global nonlinear equations (16.17). Moreover, this solution must have a form that ensures that the total equilibrium conditions (16.10) for the body are also satisfied. However, to use (16.10) – and considering for the moment elasto-plastic problems – we need to know the total stresses $\boldsymbol{\sigma}$ which, in turn, requires an integration of the constitutive equations (16.16) at each material point along the load path which is certainly not trivial and this comprises the second fundamental problem in the nonlinear FE method. The next chapter discusses different solution strategies for solution of the equilibrium equations whereas Chapter 18 is concerned with the integration of the constitutive equations for elasto-plasticity, viscoplasticity and creep.

In reality, a reliable solution scheme of the global equilibrium equations is based directly on (16.10) and not on (16.17). This will be discussed in detail in the next chapter and we emphasize that the format (16.17) was mainly introduced in order to provide a suitable background for a discussion of the principal problems.

17 SOLUTION OF NONLINEAR EQUILIBRIUM EQUATIONS

We have seen that the finite element method for general nonlinear problems leads to the solution of nonlinear equations. The present chapter is therefore devoted to a discussion of various methods to solve nonlinear equations. Many of these methods have their origin not only in solid and structural mechanics, but also in nonlinear optimization theory and as general references, the reader may consult Bathe (1996), Belytschko *et al.* (2000), Crisfield (1991), Fletcher (1980), Luenberger (1984), Papadrakakis (1993) and Zienkiewicz and Taylor (1991) for relevant information.

The nonlinear equations of interest here are the equilibrium equations given by (16.10) and (16.11), i.e.

$$\boxed{\boldsymbol{\psi} = \mathbf{0}} \tag{17.1}$$

where

$$\boxed{\boldsymbol{\psi} = \boldsymbol{f}_{int} - \boldsymbol{f}} \tag{17.2}$$

and $\boldsymbol{f}$ denotes the *external forces*, i.e. the load on the body, defined by

$$\boldsymbol{f} = \int_S \boldsymbol{N}^T \boldsymbol{t} dS + \int_V \boldsymbol{N}^T \boldsymbol{b} dV$$

whereas the *internal forces* $\boldsymbol{f}_{int}$ are defined by

$$\boldsymbol{f}_{int} = \int_V \boldsymbol{B}^T \boldsymbol{\sigma} dV \tag{17.3}$$

Expression (17.1) holds for any body in equilibrium and our problem is to satisfy this equation.

In order to solve the boundary value problem in question, we have to consider the response of the actual material. Here, we will for illustration purposes assume elasto-plasticity i.e.

$$\boxed{\dot{\boldsymbol{\sigma}} = \boldsymbol{D}_t \dot{\boldsymbol{\varepsilon}}} \tag{17.4}$$

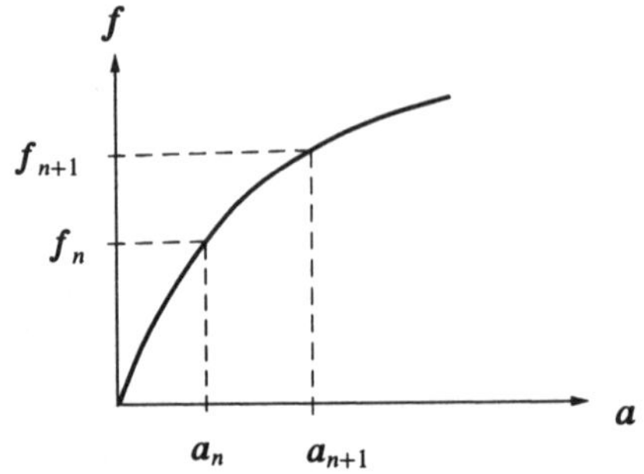

Figure 17.1: State n is known; we want to determinate state $n+1$.

where $\boldsymbol{D}_t$ is the tangential stiffness equal to the elastic stiffness if elastic loading occurs and equal to the elasto-plastic tangential stiffness if plastic loading is present. Apart from this constitutive relation that is specific for elasto-plastic problems, all aspects and results that are related to the solution of the nonlinear equilibrium equations are general and they therefore also hold for viscoplasticity and creep, for instance. The equations of equilibrium are also referred to as the *global equations*, as they hold for the entire body, whereas the constitutive relations (17.4) are referred to as the *local equations*, as they hold for each material point.

The external loading $\boldsymbol{f}$ is assumed to be known and both (17.1) and (17.4) are nonlinear equation systems. However, since the elasto-plastic response of the material depends on the entire load history, we cannot simply solve (17.1) and (17.4) by imposing the entire external load directly. Instead, we have to adopt a *stepwise*, i.e. an *incremental solution procedure*, where the external loading is increased in small steps.

With reference to Fig. 17.1, we assume that we have reached state n where everything is known. That means the nodal displacements $\boldsymbol{a}_n$, the external forces $\boldsymbol{f}_n$, the strains $\boldsymbol{\varepsilon}_n$ and the stresses $\boldsymbol{\sigma}_n$ are all known quantities. To start the process, the state n may be taken to be the state where the body is completely unloaded. The external load is now increased to $\boldsymbol{f}_{n+1}$ and we want to determine $\boldsymbol{a}_{n+1}$, $\boldsymbol{\varepsilon}_{n+1}$ and $\boldsymbol{\sigma}_{n+1}$ at state $n+1$.

Before a general approach to the solution of (17.1) is presented, it is of interest to discuss the simplest possible approach, the *Euler forward scheme*.

17.1 Euler forward scheme

In order to illustrate some important issues, we will start with the simple Euler forward scheme. Historically, this method was the first to be used for the solution of nonlinear finite element problems within solid mechanics. However, it will turn out that this method possesses a fundamental drawback, but the illustration of this drawback may serve as a prelude to a more formal and correct manner of approaching our problem.

Since the constitutive relation (17.4) is given in an incremental fashion, it is tempting to differentiate (17.1) with respect to time to obtain

$$\int_V \boldsymbol{B}^T \dot{\boldsymbol{\sigma}} dV = \dot{\boldsymbol{f}} \tag{17.5}$$

where

$$\dot{\boldsymbol{f}} = \int_S \boldsymbol{N}^T \dot{\boldsymbol{t}} dS + \int_V \boldsymbol{N}^T \dot{\boldsymbol{b}} dV \tag{17.6}$$

According to (17.4) and our finite element approximation, we have

$$\dot{\boldsymbol{\sigma}} = \boldsymbol{D}_t \dot{\boldsymbol{\varepsilon}} = \boldsymbol{D}_t \boldsymbol{B} \dot{\boldsymbol{a}}$$

Insertion of this expression into (17.5) provides

$$\boldsymbol{K}_t \dot{\boldsymbol{a}} = \dot{\boldsymbol{f}} \tag{17.7}$$

where the *tangential stiffness matrix* $\boldsymbol{K}_t$ is defined by

$$\boxed{\boldsymbol{K}_t = \int_V \boldsymbol{B}^T \boldsymbol{D}_t \boldsymbol{B} dV} \tag{17.8}$$

It is evident that $\boldsymbol{K}_t$ represents the current, i.e. the tangential stiffness of the body.

Referring to Fig. 17.1, everything is known at state n. The external load is increased to the known quantity $\boldsymbol{f}_{n+1}$ and we want to determine $\boldsymbol{a}_{n+1}$, $\boldsymbol{\varepsilon}_{n+1}$ and $\boldsymbol{\sigma}_{n+1}$ at state $n+1$. Therefore, multiplying (17.7) by dt and integrating from state n to state $n+1$, we obtain

$$\int_{\boldsymbol{a}_n}^{\boldsymbol{a}_{n+1}} \boldsymbol{K}_t d\boldsymbol{a} = \boldsymbol{f}_{n+1} - \boldsymbol{f}_n \tag{17.9}$$

The fundamental problem that we are faced with is that $\boldsymbol{K}_t$ is not a constant matrix; in fact, $\boldsymbol{K}_t$ depends on the displacements, cf. (17.8), since $\boldsymbol{D}_t$ does so. In addition, it is not even known how $\boldsymbol{K}_t$ varies from state n to state $n+1$. However, at state n everything is known, i.e. also the constitutive matrix $\boldsymbol{D}_t$ is known and the simplest thing we can do is to approximate $\boldsymbol{K}_t$ in (17.9) by $\boldsymbol{K}_t$ evaluated at state n. With the obvious notation

$$(\boldsymbol{K}_t)_n = \int_V \boldsymbol{B}^T (\boldsymbol{D}_t)_n \boldsymbol{B} dV$$

we then obtain approximately from (17.9) that

$$\boxed{(\boldsymbol{K}_t)_n(\boldsymbol{a}_{n+1} - \boldsymbol{a}_n) = \boldsymbol{f}_{n+1} - \boldsymbol{f}_n \qquad \textit{Euler forward}} \tag{17.10}$$

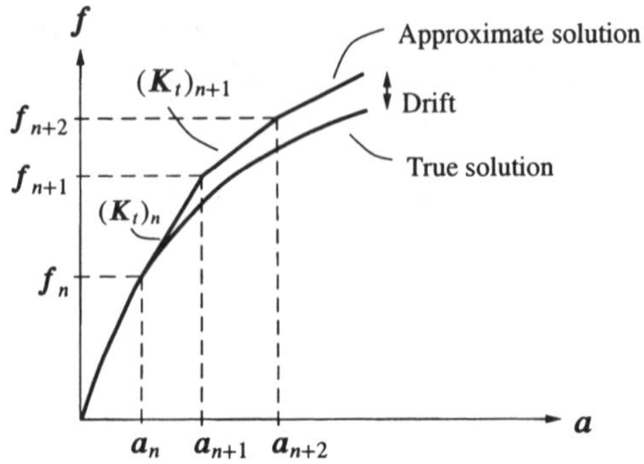

Figure 17.2: Euler forward scheme.

Except for $\boldsymbol{a}_{n+1}$, everything is known and we can therefore solve this linear equation system to provide $\boldsymbol{a}_{n+1}$. The approach obtained is termed the *Euler forward scheme*. The word 'forward' refers to the fact that we use our information about state n to determine state $n+1$ by a direct extrapolation.

When $\boldsymbol{a}_{n+1}$ has been determined from (17.10), the corresponding strain is given by $\boldsymbol{\varepsilon}_{n+1} = \boldsymbol{B}\boldsymbol{a}_{n+1}$. Moreover, the stress state $\boldsymbol{\sigma}_{n+1}$ is obtained by integration of the constitutive relation (17.4) from state n to state $n+1$. In a symbolic manner, we write

$$\boldsymbol{\sigma}_{n+1} = \boldsymbol{\sigma}_n + \int_{\boldsymbol{\varepsilon}_n}^{\boldsymbol{\varepsilon}_{n+1}} \boldsymbol{D}_t d\boldsymbol{\varepsilon} \tag{17.11}$$

For points that respond elastically this integration is trivial. However, for points that respond in a elasto-plastic manner, the integration is far from being trivial. We will deal with this problem in detail in the next chapter; at the present stage, we simply accept that this integration can be performed.

Since (17.10) is certainly an approximate solution to our original problem, the repeated use of (17.10) to trace the response of the body is bound to introduce some errors. An illustration of this effect is shown in Fig. 17.2.

In this figure, we assume that the exact solution is known at state n. Increasing the external load from $\boldsymbol{f}_n$ to $\boldsymbol{f}_{n+1}$, (17.10) then determines the nodal displacements $\boldsymbol{a}_{n+1}$. Since $(\boldsymbol{K}_t)_n$ is the tangential stiffness matrix at state n, we obtain the over-shooting effect shown in Fig. 17.2. If this process is now repeated with the state $n+1$ considered to be the known state, we obtain the response to the next step also shown in Fig. 17.2. It is apparent that use of the Euler forward scheme gives rise to a certain drift from the true solution.

This immediately raises the question whether it is possible to correct this drift and thereby improve our solution. In order to do so, we must find a quantitative expression for the drift and this problem is complicated by the fact that, in general, we do not know the true solution. The key point in this discussion

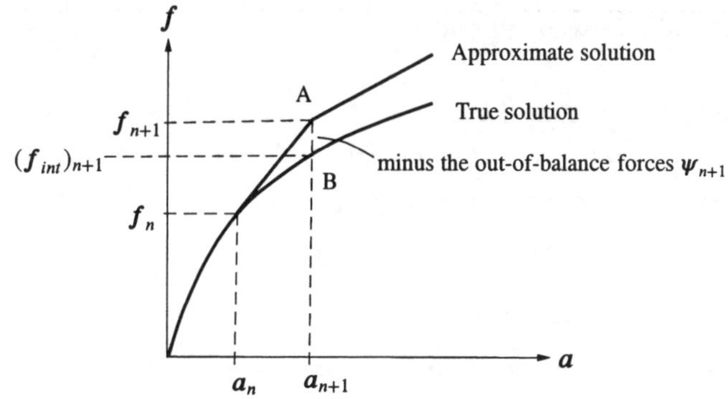

Figure 17.3: Discussion of drift and equilibrium; $(\boldsymbol{f}_{int})_{n+1}$ denotes the internal forces corresponding to the stresses $\boldsymbol{\sigma}_{n+1}$. These stresses are calculated from the nodal displacements $\boldsymbol{a}_{n+1}$.

turns out to be that of equilibrium.

Referring to (17.2), it is recalled that $\boldsymbol{f}_{int}$ represents the internal forces that the stresses give rise to. Since $\boldsymbol{f}$ denotes the known external forces, (17.1) states that equilibrium of the body requires that the external forces be equal to the internal forces. With the stresses $\boldsymbol{\sigma}_{n+1}$ determined by (17.11), we can calculate the internal forces from (17.3) according to

$$(\boldsymbol{f}_{int})_{n+1} = \int_V \boldsymbol{B}^T \boldsymbol{\sigma}_{n+1} dV$$

The internal forces $(\boldsymbol{f}_{int})_{n+1}$ correspond to the stresses $\boldsymbol{\sigma}_{n+1}$ and these stresses are calculated from the nodal displacements $\boldsymbol{a}_{n+1}$. Since the integration (17.11) may for the time being be assumed to be exact, the internal forces $(\boldsymbol{f}_{int})_{n+1}$ take the value illustrated in Fig. 17.3. It is evident that $(\boldsymbol{f}_{int})_{n+1} - \boldsymbol{f}_{n+1}$, i.e. the drift indicated in Fig. 17.2, is expressed by the fact that the Euler forward scheme does not fulfill equilibrium of the body. The drift $(\boldsymbol{f}_{int})_{n+1} - \boldsymbol{f}_{n+1}$ is also called the *out-of-balance* or *residual forces* and the equilibrium condition (17.1) states that the out-of-balance forces, i.e. the residual forces, must be zero.

We have seen that the Euler forward scheme results in a certain drift from the true solution and this drift manifests itself in the form that the residual forces are different from zero, i.e. the equilibrium conditions for the body are not fulfilled.

Against this background, it seems natural to evaluate the nonlinear equilibrium equations (17.1) in more detail and investigate how we can ensure that our solution fulfills these conditions. Before we turn to this subject, we summarize the Euler forward algorithm as shown in Box 17.1. In this box, the means to enforce the boundary conditions has not been indicated; this aspect is dealt with in Section 17.6.

Box 17.1 Euler forward algorithm

- *Initiation of quantities*

  $a_0 = \mathbf{0}\,; \quad \varepsilon_0 = \mathbf{0} \quad \sigma_0 = \mathbf{0}\,; \quad f_0 = \mathbf{0}$

- *For load step* $n = 0, 1, 2, \ldots .N_{max}$

  - *Determine new load level* $\boldsymbol{f}_{n+1}$
  - *Calculate* $\boldsymbol{K}_t = \int_V \boldsymbol{B}^T (\boldsymbol{D}_t)_n \boldsymbol{B} dV$
  - *Calculate* $\boldsymbol{a}_{n+1}$ *from* $\boldsymbol{K}_t(\boldsymbol{a}_{n+1} - \boldsymbol{a}_n) = \boldsymbol{f}_{n+1} - \boldsymbol{f}_n$
  - *Calculate* $\boldsymbol{\varepsilon}_{n+1} := \boldsymbol{B}\boldsymbol{a}_{n+1}$
  - *Determine* $\boldsymbol{\sigma}_{n+1}$ *by integration of the constitutive equations* (see next chapter)
  - *Accept quantities*

    $\boldsymbol{a}_{n+1}\,; \quad \boldsymbol{\varepsilon}_{n+1}\,; \quad \boldsymbol{\sigma}_{n+1}$

- *End load step loop*

17.2 General iteration format

We have argued for the requirement that our solution fulfills the equilibrium equations (17.1). Since these equations are nonlinear, this section is devoted to a discussion of some general aspects relating to the solution of nonlinear equations and in the next sections, we will then adopt these viewpoints to the nonlinear FE equations.

Suppose that we have some unknowns collected in the column matrix $\boldsymbol{a}$. These unknowns are determined by the following nonlinear equation system

$$\boldsymbol{a} = \boldsymbol{F}(\boldsymbol{a}) \tag{17.12}$$

To solve this equation system, a common *iteration scheme* is

$$\boldsymbol{a}^i = \boldsymbol{F}(\boldsymbol{a}^{i-1})\,; \qquad i = 1, 2, \ldots \tag{17.13}$$

where i denotes the number of iterations and where $\boldsymbol{a}^0$ is the so-called *start vector* that must be specified by us ($\boldsymbol{a}^{i-1} = \boldsymbol{a}^0$ for $i = 1$).

Let the true solution of (17.12) be denoted by $\boldsymbol{z}$ and the question then arises whether the iteration scheme (17.13) converges towards the true solution $\boldsymbol{z}$. A strict mathematical proof that defines the conditions for which the iteration scheme (17.13) converges towards the true solution, is given, for instance, by Dahlquist and Björk (1974). However, it turns out to be difficult to make practical use of these mathematical conditions and we will therefore simply state the

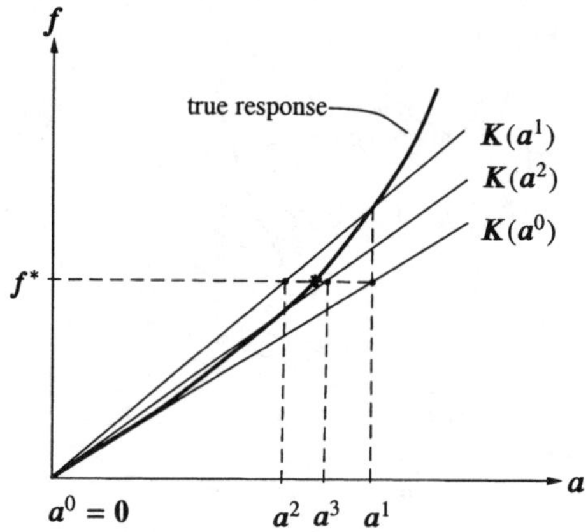

Figure 17.4: Convergent iteration scheme. True solution for external force f^* is indicated by $(*)$.

convergence theorem in the following verbal form

> *If the start vector $\boldsymbol{a}^0$ is sufficiently close to the true solution of (17.12), then the iteration scheme (17.13) will converge towards this true solution*

What in this formulation is meant by 'sufficiently close' cannot be given an explicit formulation that is applicable in practice. However, it emphasizes an important point that also arises in nonlinear finite element calculations, namely that for a given elasto-plastic problem, say, we cannot, in general, be sure that a given solution scheme will provide the solution. Instead, we may have to modify the load steps, the number of iterations and even be forced to adopt another solution scheme in order to obtain a solution. In general therefore, nonlinear finite elements calculations are far from being trivial, but a solid knowledge of the underlying theory and experience in the solution of similar problems greatly enhance the possibility of achieving a solution.

To illustrate the procedure, we will consider the following equation system

$$\boldsymbol{K}(\boldsymbol{a})\boldsymbol{a} = \boldsymbol{f}$$

where $\boldsymbol{K}$ is a nonlinear global stiffness matrix, $\boldsymbol{f}$ the external load and $\boldsymbol{a}$ the nodal displacements; the stiffness matrix $\boldsymbol{K}(\boldsymbol{a})$ can be viewed as the secant stiffness. This expression is typical for elastic problems involving geometrical nonlinearities and in this case one may often have the response shown in Fig. 17.4. Here, the structure stiffens with increasing loading and an example is the response of a laterally loaded plate with fixed supports, which stiffens due to the membrane forces created by large deformations. According to (17.12)

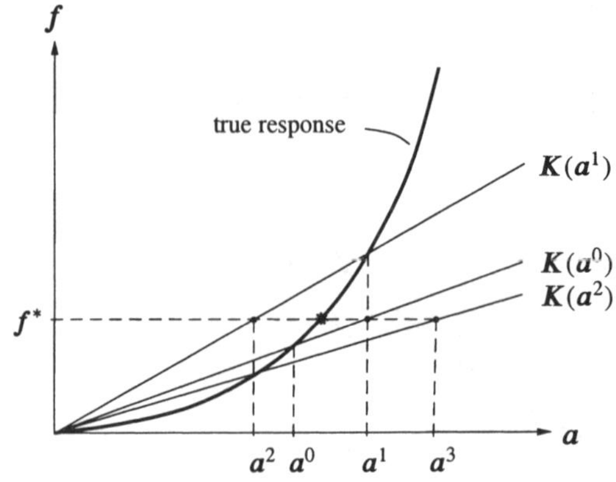

Figure 17.5: Divergent iteration scheme. True solution for external force $\boldsymbol{f}^*$ is indicated by $(*)$.

and (17.13), we adopt the following iteration scheme after the external load has been increased from zero to $\boldsymbol{f} = \boldsymbol{f}^*$

$$\boldsymbol{a}^i = [\boldsymbol{K}(\boldsymbol{a}^{i-1})]^{-1}\boldsymbol{f}^* \; ; \qquad \mathrm{i} = 1, 2, \ldots$$

Choosing the start vector as $\boldsymbol{a}^0 = \boldsymbol{0}$, then $\boldsymbol{K}(\boldsymbol{a}^0) = \boldsymbol{K}(\boldsymbol{0})$ becomes equal to the initial elastic stiffness, and we then obtain the convergent iteration scheme shown in Fig. 17.4.

To illustrate that even small changes in the response may create a divergent iteration scheme, consider the response shown in Fig. 17.5. This response is quite close to that shown in Fig. 17.4, except that the stiffening effect is larger. The starting vector $\boldsymbol{a}^0$ is now taken to be different from zero; in fact the start vector is rather close to the exact solution. Nevertheless, Fig. 17.5 shows a divergent iteration scheme and it is somewhat surprising that the qualitatively similar responses shown in Figs. 17.4 and 17.5 can give rise to completely different iteration courses.

17.3 Standard iteration format for equilibrium iterations - Iteration matrix

Having discussed the general iteration format which started with the nonlinear equation system (17.12) and resulted in the iteration scheme (17.13), we will now see how the nonlinear equilibrium equations arising in nonlinear finite element calculations can be treated in the same manner.

We have seen that the backbone in nonlinear FE analysis is the fulfillment of the nonlinear equilibrium equations (17.1). Since the stresses depend on the nodal displacements $\boldsymbol{a}$, and since we want to fulfill equilibrium at a given fixed external loading, we have from (17.1)-(17.3)

$$\boldsymbol{\psi}(\boldsymbol{a}) = \boldsymbol{0} \tag{17.14}$$

This relation may also be expressed in the form of the homogeneous equation system

$$\boldsymbol{0} = -(\boldsymbol{A}(\boldsymbol{a}))^{-1}\boldsymbol{\psi}(\boldsymbol{a}) \tag{17.15}$$

where $\boldsymbol{A}^{-1}$ is a square matrix. In the most general case, $\boldsymbol{A}^{-1}$ may also depend on the unknown $\boldsymbol{a}$-values. In order that (17.15) should provide the trivial solution $\boldsymbol{\psi} = \boldsymbol{0}$ only, we must require

$$\boxed{\det \boldsymbol{A}^{-1} \neq 0} \tag{17.16}$$

This requirement is certainly fulfilled if $\boldsymbol{A}^{-1}$ is positive definite, but even if $\boldsymbol{A}^{-1}$ is not positive definite, its use in (17.15) is acceptable as long as (17.16) holds.

Expression (17.15) may even be formulated as

$$\boldsymbol{a} = \boldsymbol{a} - (\boldsymbol{A}(\boldsymbol{a}))^{-1}\boldsymbol{\psi}(\boldsymbol{a})$$

or

$$\boldsymbol{a} = \boldsymbol{F}(\boldsymbol{a}) \quad \text{where} \quad \boldsymbol{F}(\boldsymbol{a}) = \boldsymbol{a} - (\boldsymbol{A}(\boldsymbol{a}))^{-1}\boldsymbol{\psi}(\boldsymbol{a}) \tag{17.17}$$

By these manipulations, we have retrieved formulation (17.12).

In accordance with the general iteration scheme (17.13), we then obtain from (17.17) that

$$\boldsymbol{a}^i = \boldsymbol{a}^{i-1} - (\boldsymbol{A}(\boldsymbol{a}^{i-1}))^{-1}\boldsymbol{\psi}(\boldsymbol{a}^{i-1})$$

which may be written as

Standard iteration format
$\boldsymbol{A}(\boldsymbol{a}^{i-1})(\boldsymbol{a}^i - \boldsymbol{a}^{i-1}) = -\boldsymbol{\psi}(\boldsymbol{a}^{i-1})\;; \qquad \mathrm{i} = 1, 2, \cdots$
where $\boldsymbol{A}$ *is the iteration matrix*

(17.18)

All quantities denoted by $i - 1$ are known quantities and (17.18) therefore determines the new nodal displacement estimate $\boldsymbol{a}_i$. This iteration scheme is the vehicle by which we solve (17.14) and it will be referred to as *the standard iteration format*; the matrix $\boldsymbol{A}$ is called the *iteration matrix*.

The iteration matrix $\boldsymbol{A}$ must be chosen by us and we will later see how the format (17.18) enables us to derive a very large group of iteration methods applicable to nonlinear FE calculations in a very simple and convenient manner. Before we probe further into this topic, it is worthwhile to scrutinize the restrictions that we must place on the iteration matrix $\boldsymbol{A}$ in more detail. We have already indicated restriction (17.16). Suppose that equilibrium is satisfied, i.e. $\boldsymbol{\psi}(\boldsymbol{a}^{i-1}) = \mathbf{0}$. Then the correct solution has been achieved and the iteration scheme (17.18) should then imply no further changes of the nodal displacements, i.e. $\boldsymbol{a}^i = \boldsymbol{a}^{i-1}$. However, the only possibility that $\boldsymbol{A}(\boldsymbol{a}^i - \boldsymbol{a}^{i-1}) = \mathbf{0}$ provides this solution is

$$\boxed{\det \boldsymbol{A} \neq 0} \tag{17.19}$$

Apart from the restrictions given by (17.16) and (17.19), the iteration matrix may be chosen arbitrarily, but we will experience that especially restriction (17.19) has important consequences for nonlinear finite element calculations.

Referring to our fundamental problem illustrated in Fig. 17.1, we start from state n where equilibrium is fulfilled and where the nodal displacements $\boldsymbol{a}_n$, the strains $\boldsymbol{\varepsilon}_n$, the stresses $\boldsymbol{\sigma}_n$ and the external loading $\boldsymbol{f}_n$ are all known. The external loading is then increased to $\boldsymbol{f}_{n+1}$ and we then want to determine the corresponding displacements $\boldsymbol{a}_{n+1}$, the strains $\boldsymbol{\varepsilon}_{n+1}$ and thereby also the stresses $\boldsymbol{\sigma}_{n+1}$ (obtained by an integration of the constitutive equations). The purpose of the iteration scheme (17.18) is to fulfill equilibrium at state $n + 1$. Since the external loading is given by $\boldsymbol{f}_{n+1}$, the out-of-balance forces $\boldsymbol{\psi}(\boldsymbol{a}^{i-1})$ defined by (17.2)-(17.3) become

$$\boxed{\boldsymbol{\psi}(\boldsymbol{a}^{i-1}) = \int_V \boldsymbol{B}^T \boldsymbol{\sigma}^{i-1} dV - \boldsymbol{f}_{n+1}} \tag{17.20}$$

Whereas the external loading $\boldsymbol{f}_{n+1}$ is specified by us, the stress $\boldsymbol{\sigma}^{i-1}$ must be obtained by integration of the constitutive relations (17.4). The last state where the stresses were known and where the equilibrium equations were satisfied was state n. For each point in the body, $\boldsymbol{\sigma}^{i-1}$ is therefore obtained by the following symbolic integration

$$\begin{aligned} \boldsymbol{\sigma}^{i-1} &= \boldsymbol{\sigma}_n + \boldsymbol{D}(\boldsymbol{\varepsilon}^{i-1} - \boldsymbol{\varepsilon}_n) \quad \text{if the point behaves elastically} \\ \boldsymbol{\sigma}^{i-1} &= \boldsymbol{\sigma}_n + \int_{\boldsymbol{\varepsilon}_n}^{\boldsymbol{\varepsilon}^{i-1}} \boldsymbol{D}^{ep} d\boldsymbol{\varepsilon} \quad \text{if the point behaves plastically} \end{aligned} \tag{17.21}$$

The specific manner by which the integration is performed when plastic behavior occurs is dealt with in the next chapter; at the present stage, we simply accept that this integration can be performed. In practice, the determination of the stresses $\boldsymbol{\sigma}^{i-1}$ by means of (17.21) is not performed for all points in the body, but only for the *Gauss points*.

To start the general iteration scheme (17.18) for $i = 1$, we need to specify a starting value of $\boldsymbol{a}$, i.e. we have to specify $\boldsymbol{a}^0$. The most recent known value of $\boldsymbol{a}$ is $\boldsymbol{a}_n$ and it is therefore appropriate to choose $\boldsymbol{a}^0 = \boldsymbol{a}_n$ and thereby $\boldsymbol{\varepsilon}^0 = \boldsymbol{\varepsilon}_n$ which, according to (17.21) implies $\boldsymbol{\sigma}^0 = \boldsymbol{\sigma}_n$. Likewise, as the iteration matrix

A^0 we take the one given by A_n. The starting conditions for the iteration scheme therefore become

$$a^0 = a_n \; ; \qquad \sigma^0 = \sigma_n \; ; \qquad A^0 = A_n \tag{17.22}$$

From (17.20), we then obtain

$$\psi^0 = \int_V B^T \sigma_n dV - f_{n+1} \tag{17.23}$$

By means of (17.18), we have within each load step created a series of iterations which gradually improve the solution. The quantity that controls the iterations is the out-of-balance forces $\psi(a^{i-1})$, which measure the difference between the internal forces $\int_V B^T \sigma^{i-1} dV$ and the external forces f_{n+1}. When the out-of-balance forces approach zero, the correction to the nodal displacements also approaches zero. In practice therefore, the iteration scheme (17.18) is stopped when the out-of-balance forces become smaller than a certain amount specified by the user of the FE program. We shall return to this subject in Section 17.7. When $\psi(a^{i-1})$ is small enough, we accept the solution a^{i-1}, i.e

When convergence is accepted

$$a_{n+1} = a^{i-1} \; ; \quad \varepsilon_{n+1} = \varepsilon^{i-1} \; ; \quad \sigma_{n+1} = \varepsilon^{i-1} \tag{17.24}$$

Since the objective of the iterations indicated in (17.18) is to fulfill the equilibrium equations, these iterations are called *equilibrium iterations*. Moreover, since the external load is increased incrementally and as iterations are performed within each load step, this approach is often called an *incremental-iterative approach.*

Let us return to the integration of the constitutive equations given by (17.21). The integration limits turn out to be very important and we observe that they are given by ε_n, which is the last state where the equilibrium conditions were satisfied, and ε^{i-1} which is the current value of the strains. That is, with the formulation (17.21), we deliberately ignore the intermediate states between ε_n and ε^{i-1} since these intermediate states may, for instance, involve false elastic unloading excursions. If we performed the integration of the constitutive equations by first integrating from ε_n to ε^1, then from ε^1 to ε^2 and so on, we would force the stress state σ^{i-1} to depend on some possibly false intermediate states, i.e. false load history and this pitfall is avoided by formulation (17.21). Apart from that in the next chapter, we shall see how integration of (17.21) is performed in practice.

It may come as some surprise that the only point where the constitutive equations are involved is in the integration scheme (17.21). Moreover, the iteration strategy (17.18) has the very important property that possible non-zero out-of-balance forces in one load step are automatically transferred to the next load step. That is, there is no possibility for an accumulation of non-zero out-of-balance forces.

It is of significant interest that all iteration strategies proposed in the literature and used to solve nonlinear FE equations are, in principle, embraced by the scheme (17.18) and (17.20)-(17.24). The key point is the selection of the iteration matrix $\boldsymbol{A}$ where

The choice of a specific iteration matrix $\boldsymbol{A}$ *represents the establishment of a specific solution strategy*	(17.25)

We recall that the only restrictions on the iteration matrix $\boldsymbol{A}$ are given by (17.16) and (17.19) and if (17.18) is to be an expression that is dimensionally correct, $\boldsymbol{A}$ must be a stiffness matrix; except for that, each choice of $\boldsymbol{A}$ represents a valid solution strategy. Certainly, this is not to say that all choices of $\boldsymbol{A}$ are equally promising, but the statement (17.25) is the vehicle that produces an arsenal of different methods where specific advantages and shortcomings are related to each choice of the iteration matrix $\boldsymbol{A}$. With this general framework, we will now discuss some classical and some more recent iteration strategies.

17.4 Newton-Raphson scheme

It was mentioned that the choice of different iteration matrices $\boldsymbol{A}$ in the standard iteration scheme (17.18) creates different solution strategies. The choice of $\boldsymbol{A}$ is subject to the constraints (17.16) and (17.19) and that it should represent some stiffness matrix, but apart from that we can make any choice. Evidently, not all choices create an efficient solution scheme and to obtain some kind of feeling for what is an appropriate choice, we will now identify the iteration matrix $\boldsymbol{A}$ that corresponds to the well-known *Newton-Raphson scheme.* Often this method is simply called *Newton's method*, but it was derived simultaneously by Raphson; Bićanić and Johnson (1979) give the relevant historical background.

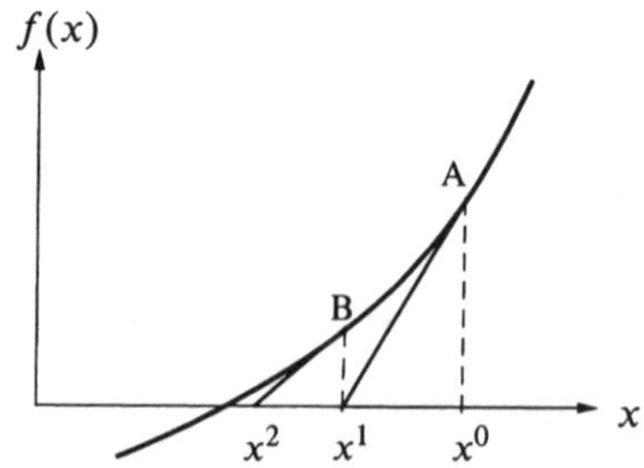

Figure 17.6: Newton-Raphson strategy for a one-dimensional problem.

For a one-dimensional problem, the essence of the Newton-Raphson strategy is illustrated in Fig. 17.6. The problem is to identify the solution to the nonlinear equation $f(x) = 0$. A starting value x^0 is guessed by us. At the corresponding point A on the curve $f(x)$, the tangent is determined and this tangent is extrapolated to obtain the next estimate x^1 for the solution. This process is then repeated so that the tangent at point B provides the next estimate x^2, etc.

It appears that the fundamental idea of the Newton-Raphson approach is to linearize the nonlinear function about a given point. This means that the nonlinear function is approximated by a Taylor expansion about the point in question and in this Taylor expansion, terms higher than the linear ones are ignored.

In our case, the nonlinear multi-dimensional function is given by the equilibrium equations, i.e.

$$\boldsymbol{\psi}(\boldsymbol{a}) = \boldsymbol{0}$$

where

$$\boldsymbol{\psi}(\boldsymbol{a}) = \int_V \boldsymbol{B}^T \boldsymbol{\sigma} dV - \boldsymbol{f} \tag{17.26}$$

and the external forces $\boldsymbol{f}$ are known and fixed whereas the stresses $\boldsymbol{\sigma}$ depend on the nodal displacements $\boldsymbol{a}$. Assume now that the approximation $\boldsymbol{a}^{i-1}$ to the true solution $\boldsymbol{a}$ has been established. Ignoring higher-order terms, a Taylor expansion of $\boldsymbol{\psi}$ about $\boldsymbol{a}^{i-1}$ yields

$$\boldsymbol{\psi}(\boldsymbol{a}^i) = \boldsymbol{\psi}(\boldsymbol{a}^{i-1}) + (\frac{\partial \boldsymbol{\psi}}{\partial \boldsymbol{a}})^{i-1}(\boldsymbol{a}^i - \boldsymbol{a}^{i-1}) \tag{17.27}$$

This expression provides the linearized approximation to the true expression for $\boldsymbol{\psi}(\boldsymbol{a}^i)$; it therefore represents the tangent to the curve at point $\boldsymbol{a}^{i-1}$. Similar to the one-dimensional case, we require $\boldsymbol{\psi}(\boldsymbol{a}^i) = \boldsymbol{0}$ and it then follows from (17.27) that

$$\boldsymbol{0} = \boldsymbol{\psi}(\boldsymbol{a}^{i-1}) + (\frac{\partial \boldsymbol{\psi}}{\partial \boldsymbol{a}})^{i-1}(\boldsymbol{a}^i - \boldsymbol{a}^{i-1}) \tag{17.28}$$

To proceed further, we have to identify the derivative $\partial \boldsymbol{\psi}/\partial \boldsymbol{a}$. Since the external loading is fixed, (17.26) implies

$$\frac{\partial \boldsymbol{\psi}}{\partial \boldsymbol{a}} = \int_V \boldsymbol{B}^T \frac{d\boldsymbol{\sigma}}{d\boldsymbol{a}} dV \tag{17.29}$$

From the constitutive relation $\dot{\boldsymbol{\sigma}} = \boldsymbol{D}_t \dot{\boldsymbol{\varepsilon}}$, cf. (17.4), and noting that the variable now is $\boldsymbol{a}$, we obtain

$$d\boldsymbol{\sigma} = \boldsymbol{D}_t d\boldsymbol{\varepsilon} = \boldsymbol{D}_t \boldsymbol{B} d\boldsymbol{a}$$

i.e.

$$\frac{d\boldsymbol{\sigma}}{d\boldsymbol{a}} = \boldsymbol{D}_t \boldsymbol{B} \tag{17.30}$$

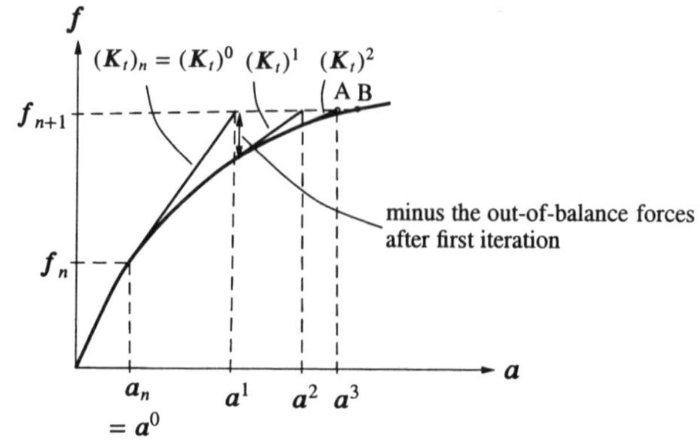

Figure 17.7: Newton-Raphson scheme showing the equilibrium iterations. Point B is the true solution and point A is the solution obtained after three iterations.

Insertion of (17.30) into (17.29) yields

$$\frac{\partial \boldsymbol{\psi}}{\partial \boldsymbol{a}} = \boldsymbol{K}_t \quad \text{where} \quad \boldsymbol{K}_t = \int_V \boldsymbol{B}^T \boldsymbol{D}_t \boldsymbol{B} dV \tag{17.31}$$

where $\boldsymbol{K}_t$ is the tangent stiffness matrix of the body, cf. (17.8).

With this result, (17.28) takes the form

$$(\boldsymbol{K}_t)^{i-1}(\boldsymbol{a}^i - \boldsymbol{a}^{i-1}) = -\boldsymbol{\psi}(\boldsymbol{a}^{i-1}) \tag{17.32}$$

and a comparison with the standard iteration scheme (17.18) shows that

$$\boxed{\boldsymbol{A} = \boldsymbol{K}_t \quad \Rightarrow \quad \textit{Newton-Raphson scheme}} \tag{17.33}$$

Therefore, this choice of the iteration matrix $\boldsymbol{A}$ results in the well-known Newton-Raphson approach.

For the first iteration $i = 1$, we find with (17.23) and the starting conditions (17.22) that

$$(\boldsymbol{K}_t)_n(\boldsymbol{a}^1 - \boldsymbol{a}_n) = \boldsymbol{f}_{n+1} - \int_V \boldsymbol{B}^T \boldsymbol{\sigma}_n dV \qquad \text{first iteration}$$

If equilibrium is fulfilled at state n then $\boldsymbol{f}_n = \int_V \boldsymbol{B}^T \boldsymbol{\sigma}_n dV$ and we immediately observe that the first iteration is, in fact, identical to the Euler forward scheme, cf. (17.10). From (17.32) follows that in each equilibrium iteration, the Newton-Raphson approach makes use of the current tangential stiffness matrix and this feature is illustrated in Fig. 17.7. The Newton-Raphson approach is one of the most often used solution schemes and the reason is the fast convergence as illustrated in Fig. 17.7. Since the Newton-Raphson method starts with a Euler forward prediction and then continues with successive corrections

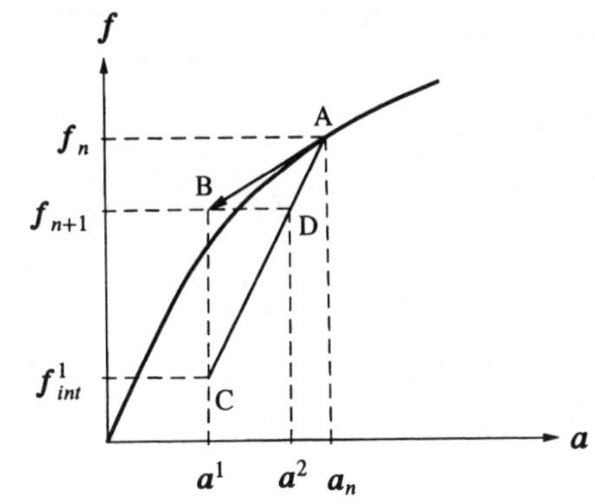

Figure 17.8: Newton-Raphson scheme during unloading of elasto-plastic body. Correct solution is obtained after two iterations.

to fulfill equilibrium, the terminology of a *predictor-corrector scheme* is often used.

In Fig. 17.7, we assumed that plastic loading occurs and it may be of interest to investigate the Newton-Raphson approach when elastic unloading occurs and Fig. 17.8 illustrates this issue. It appears that the correct response will be predicted after two iterations. First, unloading occurs along AB using the tangential stiffness at point A. Observing that, in reality, elastic unloading occurs along AC, the internal forces corresponding to the displacements $\boldsymbol{a}^1$ are given by point C. That is, at that stage the out-of-balance forces $\boldsymbol{\psi}(\boldsymbol{a}^1)$ become with (17.20) $\boldsymbol{\psi}(\boldsymbol{a}^1) = \boldsymbol{f}^1_{int} - \boldsymbol{f}_{n+1} = -|\boldsymbol{BC}|$. When these residual forces are applied in (17.32) in the next iteration where $i = 2$ and observing that the current tangential stiffness now amounts to the elastic stiffness at point C, we arrive at the correct point D after the second iteration.

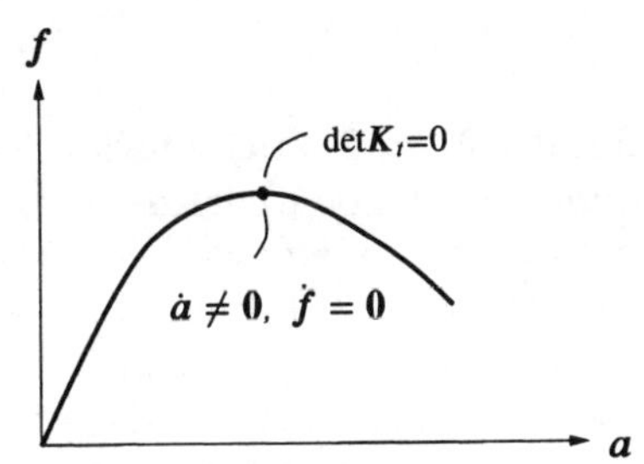

Figure 17.9: The Newton-Raphson scheme will have difficulties close to a peak load.

Since the tangential matrix $\boldsymbol{K}_t$ varies with the loading, it is of interest to investigate whether at some state in the Newton-Raphson scheme we may violate

Box 17.2 Newton-Raphson algorithm

- *Initiation of quantities*

  $\boldsymbol{a}_0 = \mathbf{0}\,; \quad \boldsymbol{\varepsilon}_0 = \mathbf{0}\,; \quad \boldsymbol{\sigma}_0 = \mathbf{0}\,; \quad \boldsymbol{f}_0 = \mathbf{0}\,; \quad \boldsymbol{f}_{int} = \mathbf{0}$
- *For load step* $n = 0, 1, 2, \ldots. N_{max}$
  - *Determine new load level* $\boldsymbol{f}_{n+1}$
  - *Initiation of iteration quantities*

    $\boldsymbol{a}^0 := \boldsymbol{a}_n$
  - *Iterate* $i = 1, 2, \ldots$ *until* $|\boldsymbol{\psi}|_{norm} = |\boldsymbol{f}_{int} - \boldsymbol{f}_{n+1}|_{norm} <$*epsilon*
    - *Calculate* $\boldsymbol{K}_t = \int_V \boldsymbol{B}^T \boldsymbol{D}_t^{i-1} \boldsymbol{B} dV$
    - *Calculate* $\boldsymbol{a}^i$ *from* $\boldsymbol{K}_t(\boldsymbol{a}^i - \boldsymbol{a}^{i-1}) = \boldsymbol{f}_{n+1} - \boldsymbol{f}_{int}$
    - *Calculate* $\boldsymbol{\varepsilon}^i := \boldsymbol{B}\boldsymbol{a}^i$
    - *Determine* $\boldsymbol{\sigma}^i$ *by integration of the constitutive equations* (see next chapter)
    - *Calculate internal forces* $\boldsymbol{f}_{int} = \int_V \boldsymbol{B}^T \boldsymbol{\sigma}^i dV$
  - *End iteration loop*
  - *Accept quantities*

    $\boldsymbol{a}_{n+1} := \boldsymbol{a}^i\,; \quad \boldsymbol{\varepsilon}_{n+1} := \boldsymbol{\varepsilon}^i\,; \quad \boldsymbol{\sigma}_{n+1} := \boldsymbol{\sigma}^i\,; \quad \boldsymbol{f}_{int}$
- *End load step loop*

---

the important restriction given by (17.19), which with (17.33) becomes

$$\det \boldsymbol{K}_t \neq 0$$

This restriction is violated if $\det \boldsymbol{K}_t = 0$ which implies that the homogeneous equation system $\boldsymbol{K}_t\dot{\boldsymbol{a}} = \mathbf{0}$ possesses a non-trivial $\dot{\boldsymbol{a}}$-solution. As illustrated in Fig. 17.9, this situation corresponds to the peak load where we have $\dot{\boldsymbol{a}} \neq \mathbf{0}$, $\dot{\boldsymbol{f}} = \mathbf{0}$ which satisfy the equation $\boldsymbol{K}_t\dot{\boldsymbol{a}} = \dot{\boldsymbol{f}}$ from (16.17) when $\det \boldsymbol{K}_t = 0$. Use of the Newton-Raphson scheme will therefore imply increasing difficulties when a peak load is approached. If the response of the body, as illustrated in Fig. 17.9, exhibits a softening branch, great difficulties are encountered when we try to trace the response over the peak load and into the softening region. This latter observation is relevant, not only for the Newton-Raphson scheme, but also for other methods and we shall return to this subject later on.

We finally observe that in every Newton-Raphson iteration, a new $\boldsymbol{K}_t$-matrix needs to be established and that, in principle, the inverse matrix $(\boldsymbol{K}_t)^{-1}$ also needs to be identified, cf. (17.32), i.e. each such iteration is costly. With this remark, we may summarize the properties of the Newton-Raphson scheme as follows:

$$
\boxed{\begin{array}{l}
* \textit{ Newton-Raphson works well both in loading} \\
\quad \textit{and unloading} \\
* \textit{ Newton-Raphson provides a fast convergence} \\
* \textit{ problems may occur close to peak points} \\
* \textit{ every Newton-Raphson iteration is costly}
\end{array}} \qquad (17.34)
$$

The Newton-Raphson algorithm is summarized in Box. 17.2; here the means to impose the boundary conditions have not been indicated, see Section 17.6.

17.5 Initial stiffness and modified Newton-Raphson schemes

It was observed that the choice $\boldsymbol{A} = \boldsymbol{K}_t$ leads to the well-known and efficient Newton-Raphson scheme. However, we have previously stressed the freedom that we have when choosing the iteration matrix $\boldsymbol{A}$ in the standard iteration scheme (17.18). Against this background, it is not surprising that a number of choices exists for the iteration matrix which are modifications of the Newton-Raphson approach. Since such modifications exist, the true Newton-Raphson method is occasionally referred to as the *full* Newton-Raphson scheme. Scrutinizing the standard iteration scheme (17.18), it follows that any $\boldsymbol{A}$-matrix must have the same dimension as the stiffness matrix of the body.

Recognizing that every Newton-Raphson iteration is costly, it is tempting to choose a constant $\boldsymbol{A}$-matrix in all iterations. The most evident choice would then be $\boldsymbol{A} = \boldsymbol{K}$ where $\boldsymbol{K}$ is the linear elastic stiffness matrix. Since $\boldsymbol{K}$ is the initial stiffness of the body, this choice leads to the *initial stiffness method*, i.e.

$$
\boxed{\boldsymbol{A} = \boldsymbol{K} \quad \Rightarrow \quad \textit{Initial stiffness scheme}}
$$

Since the elastic stiffness matrix $\boldsymbol{K}$ is positive definite, we have $det\,\boldsymbol{K} \neq 0$ and none of the restrictions given by (17.16) and (17.19) are violated.

The performance of this approach is illustrated in Fig. 17.10 and a comparison with Fig. 17.7 clearly illustrates that the initial stiffness method simplifies the calculations considerably, but at the expense of a much slower convergence. This is especially true when the structure approaches its peak load, where the force-displacement curve is flat. For elastic unloading of the body, it is evident that the initial stiffness method will provide the correct response in just one

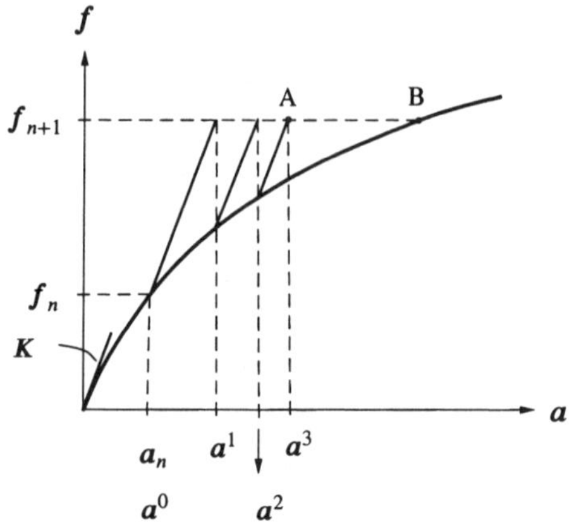

Figure 17.10: Initial stiffness method. Point B is the true solution and point A is the solution obtained after three iterations.

iteration. We are then led to the following conclusions

> * *Initial stiffness approach works well both in loading and unloading*
> * *every iteration is cheap*
> * *Initial stiffness approach converges slowly*

The initial stiffness method may be recast into other equivalent forms that have been proposed in the literature. With Hooke's law $\boldsymbol{\sigma} = \boldsymbol{D}(\boldsymbol{\varepsilon} - \boldsymbol{\varepsilon}^p) = \boldsymbol{DBa} - \boldsymbol{D\varepsilon}^p$, the internal forces become

$$\boldsymbol{f}_{int}^{i-1} = \int_V \boldsymbol{B}^T \boldsymbol{\sigma}^{i-1} dV = \boldsymbol{K}\boldsymbol{a}^{i-1} - \int_V \boldsymbol{B}^T \boldsymbol{D}(\boldsymbol{\varepsilon}^p)^{i-1} dV$$

where $\boldsymbol{K} = \int_V \boldsymbol{B}^T \boldsymbol{DB} dV$ is the elastic stiffness matrix. Insertion of the expression above into the general iteration scheme (17.18) and (17.20) with $\boldsymbol{A} = \boldsymbol{K}$ gives

$$\boldsymbol{K}\boldsymbol{a}^i = \boldsymbol{f}_{n+1} + \int_V \boldsymbol{B}^T \boldsymbol{D}(\boldsymbol{\varepsilon}^p)^{i-1} dV \tag{17.35}$$

This formulation was suggested by Argyris (1965b) under the name *initial strain method*. The reason for this terminology is that linear elasticity with initial strains is given by $\boldsymbol{\sigma} = \boldsymbol{D}(\boldsymbol{\varepsilon} - \boldsymbol{\varepsilon}^o)$ where $\boldsymbol{\varepsilon}^o$ are the initial strains, cf. (4.62). It appears that the plastic strains $\boldsymbol{\varepsilon}^p$ may be interpreted as initial strains. On the other hand, if we define the quantity $\boldsymbol{\sigma}^o$ by $\boldsymbol{\sigma}^o = -\boldsymbol{D\varepsilon}^p$ then Hooke's law can be written as $\boldsymbol{\sigma} = \boldsymbol{D\varepsilon} + \boldsymbol{\sigma}^o$, i.e. for zero strains, we have $\boldsymbol{\sigma} = \boldsymbol{\sigma}^o$ and $\boldsymbol{\sigma}^o$ may therefore be viewed as initial stresses. Replacing $\boldsymbol{D\varepsilon}^p$ by $-\boldsymbol{\sigma}^o$, the

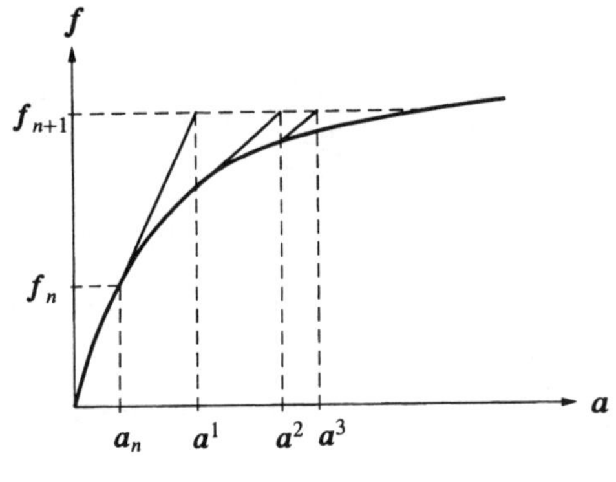

Figure 17.11: Modified Newton-Raphson. Updating of the tangential stiffness matrix after the first iteration in each load step.

scheme (17.35) was proposed by Zienkiewicz *et al.* (1969) under the name: *initial stress method.* However, as we have seen, both the initial strain and the initial stress method are identical to the initial stiffness method; it is only the physical interpretation that differs.

A compromise between full Newton-Raphson, where updating of the tangential stiffness matrix $\boldsymbol{K}_t$ occurs in every iteration, and the initial stiffness method, where no updating occurs at all, is the modification where updating is performed only once in each load step. This approach is called *modified Newton-Raphson.* Intuitively, it seems reasonable to perform this updating at the beginning of the load step. However, we experienced in Fig. 17.8 that full Newton-Raphson predicts the correct response during elastic unloading after two iterations. Therefore, the approach suggested above would cause problems during unloading. Instead, the updating of the tangential stiffness matrix is chosen to occur after the first equilibrium iteration in each load step. This approach is illustrated in Fig. 17.11.

Finally, we may mention the so-called *self-correcting procedure* of Stricklin *et al.* (1971) and further elaborated on by Stricklin and Haisler (1977). Essentially, it consists of using the full Newton-Raphson approach, but make just one equilibrium iteration in every load step. That is, the general iteration scheme (17.18) is used with $i = 1$ and with $\boldsymbol{A} = \boldsymbol{K}_t$. We have previously emphasized that the scheme (17.18) has the advantage that possible non-zero out-of-balance forces in one load step are automatically transferred to the next load step and the self-correcting procedure takes full advantage of this property.

To illustrate this issue, assume that the numerical procedure has determined point A in Fig. 17.12 to be the response at state n. For the nodal displacements $\boldsymbol{a}_n$, the corresponding internal forces are given by $(\boldsymbol{f}_{int})_n = \int_V \boldsymbol{B}^T \sigma_n dV$, i.e. point P. The out-of-balance forces are therefore given by $(\boldsymbol{f}_{int})_n - \boldsymbol{f}_n = -|\text{AP}|$. When the external loading is increased to $\boldsymbol{f}_{n+1}$, we obtain from (17.18) with

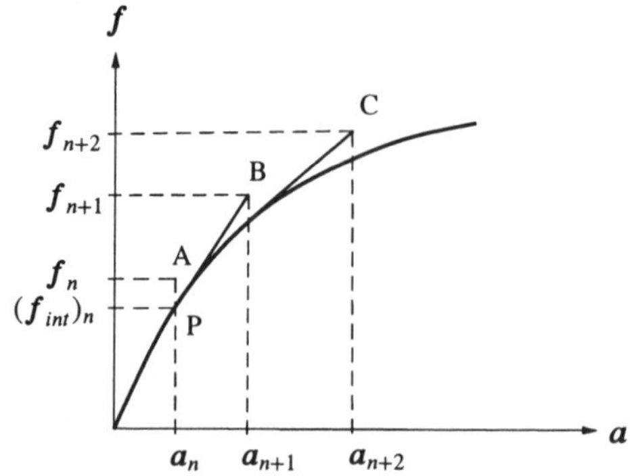

Figure 17.12: Self-correcting procedure of Stricklin *et al.* (1971); non-zero out-of-balance forces are automatically transferred to the next load step.

$\boldsymbol{A} = \boldsymbol{K}_t$ and $i = 1$

$$(\boldsymbol{K}_t)^0(\boldsymbol{a}^1 - \boldsymbol{a}^0) = -\boldsymbol{\psi}^0 \tag{17.36}$$

Since, we take just one iteration, the quantity $\boldsymbol{a}^1$ are taken as $\boldsymbol{a}_{n+1}$. With (17.22) and (17.23), (17.36) then becomes

$$(\boldsymbol{K}_t)_n(\boldsymbol{a}_{n+1} - \boldsymbol{a}_n) = -((\boldsymbol{f}_{int})_n - \boldsymbol{f}_{n+1}) = \boldsymbol{f}_{n+1} - \boldsymbol{f}_n - ((\boldsymbol{f}_{int})_n - \boldsymbol{f}_n)$$

It appears that the out-of-balance forces $(\boldsymbol{f}_{int})_n - \boldsymbol{f}_n$ at state n are automatically transferred to the next step. This feature is illustrated in Fig. 17.12 where the predicted response then is given by points A, B and C.

If elastic unloading occurs in a load step, the response in this load step will follow the tangential elasto-plastic stiffness instead of the elastic stiffness. It is only if elastic unloading continues in the next load step that the procedure will recognize the elastic unloading effects and use the elastic stiffness. Therefore the procedure will give an inferior prediction during elastic unloading and this can only be accepted as a reasonable approximation if the load steps are very small. The method is frequently used in, for instance, large strain plasticity problems where the load steps of other reasons are kept small and where the method often is referred to as the *incremental linear method.*

17.6 Consideration of boundary conditions

The essence in any equilibrium iteration is to produce a linear equation system that can be solved to provide the new nodal displacement estimate $\boldsymbol{a}^i$. In reality, this solution cannot be established before the boundary conditions of the problem have been introduced into the equation system and we have not, in fact, been precise on that point previously.

As an example, we may refer to the Newton-Raphson scheme given by (17.32) where the tangential stiffness matrix $\boldsymbol{K}_t$ defined by (17.31) enters the equation system. When establishing $\boldsymbol{K}_t$ in this manner, no considerations were taken of the kinematic boundary conditions; therefore, (17.32) also allows rigid-body motions. If $\dot{\boldsymbol{a}} \neq \boldsymbol{0}$ corresponds to a rigid-body motion, this would create no strains in the body, i.e. $\dot{\boldsymbol{\varepsilon}} = \boldsymbol{B}\dot{\boldsymbol{a}} = \boldsymbol{0}$. It follows that the homogeneous equation system $\boldsymbol{K}_t\dot{\boldsymbol{a}} = \boldsymbol{0}$ possesses a non-trivial solution $\dot{\boldsymbol{a}} \neq \boldsymbol{0}$ and $\boldsymbol{K}_t$ is therefore singular, i.e. *det* $\boldsymbol{K}_t = 0$. This means, in fact, that we cannot solve the equation system (17.32).

To obtain a solution of our boundary value problem, we therefore have to introduce the pertinent boundary conditions that, among other things, ensure that rigid-body motions are prevented.

Let us write the standard iteration scheme (17.18) and (17.20) in a simplified form as

$$\boldsymbol{A}\delta\boldsymbol{a} = \boldsymbol{f} - \boldsymbol{f}_{int} \tag{17.37}$$

where

$$\boldsymbol{A} = \boldsymbol{A}^{i-1} \; ; \qquad \delta\boldsymbol{a} = \boldsymbol{a}^{i} - \boldsymbol{a}^{i-1} \tag{17.38}$$

Moreover

$$\boldsymbol{f} = \int_S \boldsymbol{N}^T \boldsymbol{t}_{n+1} dS + \int_V \boldsymbol{N}^T \boldsymbol{b}_{n+1} dV \; ; \quad \boldsymbol{f}_{int} = \int_V \boldsymbol{B}^T \boldsymbol{\sigma}^{i-1} dV \tag{17.39}$$

where we notice that the stresses $\boldsymbol{\sigma}^{i-1}$ are known, i.e. the internal forces $\boldsymbol{f}_{int}$ are also known.

Since $\boldsymbol{A}$ may always be interpreted as a stiffness matrix of the body, it follows in accordance with the discussion above that

$$\det \boldsymbol{A} = 0$$

i.e., equation system (17.37) can first be solved after the boundary conditions have been introduced.

The boundary conditions are prescribed as

$\boldsymbol{u} =$ *is given along* S_u
$\boldsymbol{t} =$ *is given along* S_t

That is, the displacement vector $\boldsymbol{u}$ is prescribed along the boundary surface S_u and the traction vector $\boldsymbol{t}$ is prescribed along the boundary surface S_t. This is illustrated in Fig. 17.13a).

We now split the finite element nodes into two groups: one group along the boundary S_u where the displacements are prescribed and another group which contains the remaining nodes. Correspondingly, the nodal displacements $\boldsymbol{a}$ can

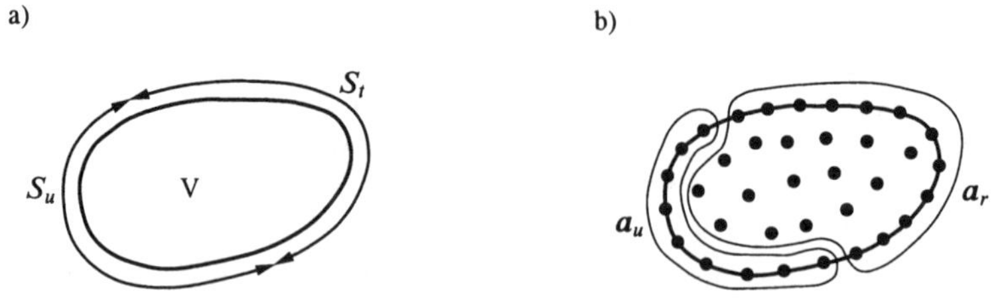

Figure 17.13: a) Along S_u the displacements are given and along S_t the traction vector is known; b) split of nodal displacements into prescribed values $\boldsymbol{a}_u$ along S_u and the remaining, i.e. unknown, values $\boldsymbol{a}_r$ in the remaining part of the body.

then be partitioned into $\boldsymbol{a}_u$ and $\boldsymbol{a}_r$, respectively, i.e. $\boldsymbol{a}_u$ are known whereas the remaining nodal displacements $\boldsymbol{a}_r$ are to be determined; this partitioning is illustrated in Fig. 17.13b). In accordance with this procedure, we may, in principle, partition (17.37) and with evident notation, we obtain

$$\begin{bmatrix} \boldsymbol{A}_{uu} & \boldsymbol{A}_{ur} \\ \boldsymbol{A}_{ru} & \boldsymbol{A}_{rr} \end{bmatrix} \begin{bmatrix} \delta \boldsymbol{a}_u \\ \delta \boldsymbol{a}_r \end{bmatrix} = \begin{bmatrix} \boldsymbol{f}_u - (\boldsymbol{f}_{int})_u \\ \boldsymbol{f}_r - (\boldsymbol{f}_{int})_r \end{bmatrix} \tag{17.40}$$

Since all the internal forces $\boldsymbol{f}_{int}$ are known, both $(\boldsymbol{f}_{int})_u$ and $(\boldsymbol{f}_{int})_r$ are known forces. However, of the external forces $\boldsymbol{f}$, only the part $\boldsymbol{f}_r$ is, in fact, known whereas the part $\boldsymbol{f}_u$ is unknown, since it corresponds to the reactions where the displacements are prescribed.

From the second row of (17.40), we find

$$\boldsymbol{A}_{rr}\delta \boldsymbol{a}_r = \boldsymbol{f}_r - (\boldsymbol{f}_{int})_r - \boldsymbol{A}_{ru}\delta \boldsymbol{a}_u$$

where the right-hand side is known. Written explicitly, using (17.38) and (17.39), we have

$$\boxed{\boldsymbol{A}_{rr}^{i-1}(\boldsymbol{a}_r^i - \boldsymbol{a}_r^{i-1}) = (\boldsymbol{f}_{n+1})_r - (\boldsymbol{f}_{int}^{i-1})_r - \boldsymbol{p}_r^{i-1}} \tag{17.41}$$

where $\boldsymbol{p}_r^{i-1}$ represents the quantity

$$\boldsymbol{p}_r^{i-1} = \boldsymbol{A}_{ru}^{i-1}(\boldsymbol{a}_u^i - \boldsymbol{a}_u^{i-1})$$

The known displacement changes along the boundary S_u are imposed in the very first integration, i.e.

$$\boldsymbol{p}_r^0 = (\boldsymbol{A}_{ru})_n[(\boldsymbol{a}_u)_{n+1} - (\boldsymbol{a}_u)_n]$$

In the next iterations, no change occurs of the displacements $\boldsymbol{a}_u$ along S_u, i.e.

$$\boldsymbol{p}_r^{i-1} = \boldsymbol{0} \quad \text{when} \quad i = 2, 3, \ldots \tag{17.42}$$

In reality, it is on the equation system (17.41) we perform the equilibrium equations and all the previous as well as the following sections should be interpreted in this manner.

The iteration scheme will always take at least one iteration and it is only if the out-of-balance forces after this first iteration are too large that the iterations will continue. Therefore, we obtain with (17.41) and (17.42) that the part of the out-of-balance forces on which we measure equilibrium is $\boldsymbol{\psi}_r^{i-1} = (\boldsymbol{f}_{int}^{i-1})_r - (\boldsymbol{f}_{n+1})_r$. Consequently

$$\boxed{\textit{Fulfillment of equilibrium is checked for the out-of-balance forces } \boldsymbol{\psi}_r^{i-1} = (\boldsymbol{f}_{int}^{i-1})_r - (\boldsymbol{f}_{n+1})_r} \tag{17.43}$$

The iterations are stopped when equilibrium has been fulfilled. In that case $\delta\boldsymbol{a}_r = \boldsymbol{a}_r^i - \boldsymbol{a}_r^{i-1} = \boldsymbol{0}$ and as also $\delta\boldsymbol{a}_u = \boldsymbol{0}$ (after the first iteration), (17.40) shows that the reaction forces along S_u, i.e. $(\boldsymbol{f}_{n+1})_u$ are given by

$$(\boldsymbol{f}_{n+1})_u = (\boldsymbol{f}_{int}^{i-1})_u \tag{17.44}$$

Since the stresses $\boldsymbol{\sigma}^{i-1}$ are known, the internal force $\boldsymbol{f}_{int}^{i-1}$ and thereby also the part $(\boldsymbol{f}_{int}^{i-1})_u$ are known. After equilibrium has been obtained, the unknown reactions $(\boldsymbol{f}_{n+1})_u$ along S_u are therefore determined by (17.44).

On the other hand, if we always determine the reaction forces by (17.44) then equilibrium along the boundary S_u is always fulfilled, i.e. $\boldsymbol{\psi}_u^{i-1} = (\boldsymbol{f}_{int}^{i-1})_u - (\boldsymbol{f}_{n+1})_u = \boldsymbol{0}$, and as $(\boldsymbol{\psi}^{i-1})^T = [(\boldsymbol{\psi}_r^{i-1})^T, (\boldsymbol{\psi}_u^{i-1})^T]$, we may instead of (17.43) check the equilibrium conditions on the total out-of-balance force $\boldsymbol{\psi}^{i-1}$.

17.7 Convergence criteria

Let us return to the general iteration scheme (17.18)-(17.24) which starts with the iteration counter $i = 1$ and then continues for increasing i-values. Hopefully, the iteration scheme converges in the sense that the out–of–balance forces $\boldsymbol{\psi}$ approach zero, which implies that the new solution $\boldsymbol{a}^i$ differs only insignificantly from the previous solution $\boldsymbol{a}^{i-1}$. In practice however, the out–of–balance forces $\boldsymbol{\psi}$ will never be exactly zero, so we have to specify some threshold for $\boldsymbol{\psi}$ that terminates the iterations. Such a threshold value is called a *convergence criterion.*

Convergence criteria may be formulated in a number of different manners and for a more detailed discussion, we refer to Bathe and Cimento (1980), Bathe (1996), Bergan and Clough (1972) and Crisfield (1991). Instead of a threshold value for $\boldsymbol{\psi}$ we may measure the improvement of our solution in each iteration in terms of $\boldsymbol{a}^i - \boldsymbol{a}^{i-1}$ and apply a convergence criterion for this quantity.

To enforce a threshold value on the out–of–balance forces $\boldsymbol{\psi}^{i-1}$, we have to measure $\boldsymbol{\psi}^{i-1}$ against something and for this purpose it seems logical to choose the external forces $\boldsymbol{f}_{n+1}$. The quantities $\boldsymbol{\psi}^{i-1}$ and $\boldsymbol{f}_{n+1}$ contain many components and in order to compare these quantities, we may use their lengths measured by the scalar product. Therefore, an often used force convergence criterion is given by

$$[(\boldsymbol{\psi}^{i-1})^T \boldsymbol{\psi}^{i-1}]^{1/2} \leq \varepsilon_{F_1} (\boldsymbol{f}_{n+1}^T \boldsymbol{f}_{n+1})^{1/2} \tag{17.45}$$

where ε_{F_1} expresses the threshold. Typical values for this threshold are $\varepsilon_{F_1} = 10^{-3} - 10^{-2}$, cf. Zienkiewicz and Taylor (1991) and Crisfield (1991), and the smaller the value ε_{F_1}, the less is the acceptable error in the out–of–balance forces. If ε_{F_1} is chosen too small, many (costly) iterations are performed without improving the solution significantly and if ε_{F_1} is chosen too large, the solution becomes inaccurate.

It is a general experience in the literature that it is difficult to recommend a specific tolerance that is always of relevance; one tolerance may work well in some case and may be inferior in other situations. For a given problem, one often starts with a rather crude tolerance and then performs calculations with narrower tolerances in an effort to judge whether a converged solution has been obtained.

In (17.45), the out-of-balance forces are compared with the total external force. When the load steps are small, the out-of-balance forces are often compared with the load step itself, i.e.

$$[(\boldsymbol{\psi}^{i-1})^T \boldsymbol{\psi}^{i-1}]^{1/2} \leq \varepsilon_{F_2} [\Delta \boldsymbol{f}^T \Delta \boldsymbol{f}]^{1/2}$$

where $\Delta \boldsymbol{f} = \boldsymbol{f}_{n+1} - \boldsymbol{f}_n$; Bathe and Cimento (1980) suggested $\varepsilon_{F_2} \approx 0.1$.

For plastic structures with a small plastic modulus H, force criteria may be misleading; the out-of-balance force may be small, but the structure is far from its correct displacement pattern. Therefore, in addition to the force convergence criteria, use is often made of some displacement convergence criteria, for instance

$$[(\delta \boldsymbol{u}^i)^T \delta \boldsymbol{u}^i]^{1/2} \leq \varepsilon_D [\boldsymbol{a}_n^T \boldsymbol{a}_n]^{1/2}$$

where $\delta \boldsymbol{u}^i = \boldsymbol{a}^i - \boldsymbol{a}^{i-1}$ and $\varepsilon_D \approx 10^{-3}$, cf. Bathe and Cimento (1980).

The convergence criteria above possess the drawback that quantities having one dimension may be added to quantities having another dimension; the components of $\boldsymbol{a}$ may involve both displacements and rotations and, likewise, the components of $\boldsymbol{\psi}$ may involve forces as well as moments. A convergence criterion in terms of an energy criterion avoids this problem since force components are here multiplied by displacements whereas moments are multiplied by rotations, i.e. it involves so-called *conjugated quantities*. Therefore an energy convergence criterion in the form

$$[(\boldsymbol{\psi}^{i-1})^T \boldsymbol{a}_n]^{1/2} \leq \varepsilon_E [\boldsymbol{f}_n^T \boldsymbol{a}_n]^{1/2}$$

is often adopted; the tolerance $\varepsilon_E = 10^{-3} - 10^{-2}$ is often used, cf. Zienkiewicz and Taylor (1991).

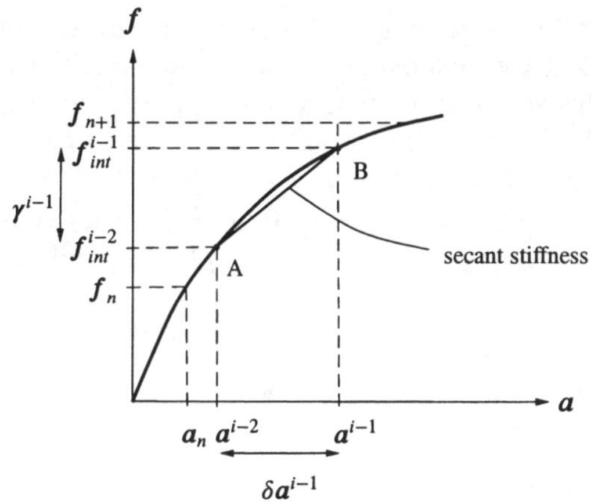

Figure 17.14: Current state given by point B and previous state given by point A; secant stiffness between these points.

17.8 Quasi-Newton methods

The Newton-Raphson scheme possesses a number of advantages among which its fast convergence is the major reason for its popularity, cf. (17.34). However, it was mentioned that every Newton-Raphson iteration requires the formulation of a new tangent stiffness matrix $\boldsymbol{K}_t$ and that, in principle, the inverse matrix $\boldsymbol{K}_t^{-1}$ also needs to be established; this means that every Newton-Raphson iteration is costly. This drawback can be obviated by the initial stiffness method where the elastic stiffness matrix is used in every iteration or in the modified Newton-Raphson method where the stiffness matrix is occasionally updated. However, these approaches have a slower convergence rate than the full Newton-Raphson scheme. It is of considerable interest that there exists an entirely different approach by which we can maintain a fast convergence without performing a costly matrix inversion in every iteration.

In the previous iteration schemes, we only took advantage of the information of the current state defined by $\boldsymbol{a}^{i-1}$ and $\boldsymbol{\psi}^{i-1}$ and with this information, we tried to establish a new estimate $\boldsymbol{a}^i$ and thereby $\boldsymbol{\psi}^i$. Essentially, this means that all previous information of the response is ignored. To illustrate this viewpoint, consider Fig. 17.14 and suppose that we increase the external loading from $\boldsymbol{f}_n$ to $\boldsymbol{f}_{n+1}$. Equilibrium iterations are then performed with, say, a Newton-Raphson procedure. At the present stage, we have then obtained the nodal displacements $\boldsymbol{a}^{i-1}$ at point B whereas the previous nodal displacements are given by $\boldsymbol{a}^{i-2}$ at point A.

In a Newton-Raphson procedure, we then determine the tangential stiffness matrix $\boldsymbol{K}_t$ at point B to obtain the new nodal displacement estimate $\boldsymbol{a}^i$. Our objective is now to establish a procedure that as closely as possible retains the fast convergence of the Newton-Raphson approach. Instead of the tangential stiffness at point B, it is then tempting to use the secant stiffness between point A and point B. This secant stiffness is a close approximation to the tangential stiffness at point B. To define this secant stiffness matrix, we observe that the internal forces at point B are given by $\boldsymbol{f}_{int}^{i-1} = \int_V \boldsymbol{B}^T \boldsymbol{\sigma}^{i-1} dV$ whereas the internal forces at point A are given by $\boldsymbol{f}_{int}^{i-2} = \int_V \boldsymbol{B}^T \boldsymbol{\sigma}^{i-2} dV$. These internal forces are illustrated in Fig. 17.14. Define the quantities $\delta \boldsymbol{a}^{i-1}$ and $\boldsymbol{\gamma}^{i-1}$ by

$$\delta \boldsymbol{a}^{i-1} = \boldsymbol{a}^{i-1} - \boldsymbol{a}^{i-2} \, ; \qquad \boldsymbol{\gamma}^{i-1} = \boldsymbol{f}_{int}^{i-1} - \boldsymbol{f}_{int}^{i-2} \tag{17.46}$$

These quantities are illustrated in Fig. 17.14. The secant stiffness $\boldsymbol{K}_s^{i-1}$ between point A and B is then defined by

$$\boldsymbol{K}_s^{i-1} \delta \boldsymbol{a}^{i-1} = \boldsymbol{\gamma}^{i-1}$$

which implies

$$\boxed{\delta \boldsymbol{a}^{i-1} = \boldsymbol{H}^{i-1} \boldsymbol{\gamma}^{i-1} \quad \textit{Quasi-Newton relation}} \tag{17.47}$$

where $\boldsymbol{H}^{i-1}$ is the inverse of $\boldsymbol{K}_s^{i-1}$. This relation is called the *quasi-Newton relation* since it defines a matrix $\boldsymbol{H}^{i-1}$ that is almost equal to the inverse of the tangential stiffness at point B that arises in the Newton-Raphson scheme. It appears that if in the general iteration scheme (17.18) we make the choice

$$(\boldsymbol{A}^{i-1})^{-1} = \boldsymbol{H}^{i-1} \tag{17.48}$$

then we will obtain a convergence that is almost as fast as the Newton-Raphson scheme.

Before we proceed further, assume that the total number of degrees of freedom in the finite element discretization is N, then the matrix $\boldsymbol{H}^{i-1}$ has the dimension $N \times N$. However, the quasi-Newton equation (17.47) - i.e. the secant relation - only comprises N equations and this implies that $\boldsymbol{H}^{i-1}$ is not defined uniquely by (17.47) unless in the trivial case where $N = 1$. To establish $\boldsymbol{H}^{i-1}$ so that it satisfies (17.47) therefore leaves us with considerable freedom.

That apart, our goal was to establish a method by which costly matrix inversions are avoided. Therefore, suppose that in the previous iteration we have obtained the matrix $\boldsymbol{H}^{i-2}$ we then want to be able to establish $\boldsymbol{H}^{i-1}$ by means of the simple updating scheme

$$\boxed{\boldsymbol{H}^{i-1} = \boldsymbol{H}^{i-2} + \boldsymbol{H}_{corr}^{i-1}} \tag{17.49}$$

where the *correction matrix* $\boldsymbol{H}_{corr}^{i-1}$ should be easy to determine. If this can be achieved, we have fulfilled our objective of devising a method with almost as good convergence properties as the Newton-Raphson method without having to perform costly inversions of some stiffness matrix.

Methods that satisfy (17.47)-(17.49) are called *quasi-Newton methods* - occasionally also called *variable matrix methods* - and they were originally derived within nonlinear optimization theory. The literature on quasi-Newton methods is very extensive and reviews are given by Dennis and Moré (1977), Fletcher (1980) and Luenberger (1984).

17.8.1 Rank one correction

To provide a simple illustration of a quasi-Newton method, suppose that we take the updating scheme (17.49) in the form

$$\boldsymbol{H}^{i-1} = \boldsymbol{H}^{i-2} + \boldsymbol{v}^{i-1}(\boldsymbol{v}^{i-1})^T \tag{17.50}$$

where $\boldsymbol{v}^{i-1}$ is a column matrix that is to be determined so that the quasi-Newton equation (17.47) is fulfilled. Insertion of (17.50) in (17.47) gives

$$\delta\boldsymbol{a}^{i-1} = \boldsymbol{H}^{i-2}\boldsymbol{\gamma}^{i-1} + \boldsymbol{v}^{i-1}((\boldsymbol{v}^{i-1})^T\boldsymbol{\gamma}^{i-1})$$

which provides

$$\boldsymbol{v}^{i-1} = \frac{1}{(\boldsymbol{v}^{i-1})^T\boldsymbol{\gamma}^{i-1}}(\delta\boldsymbol{a}^{i-1} - \boldsymbol{H}^{i-2}\boldsymbol{\gamma}^{i-1}) \tag{17.51}$$

and thereby

$$((\boldsymbol{v}^{i-1})^T\boldsymbol{\gamma}^{i-1})^2 = (\delta\boldsymbol{a}^{i-1} - \boldsymbol{H}^{i-2}\boldsymbol{\gamma}^{i-1})^T\boldsymbol{\gamma}^{i-1}$$

Insertion of (17.51) in (17.50) and use of the expression above give the result

$$\boldsymbol{H}^{i-1} = \boldsymbol{H}^{i-2} + \frac{(\delta\boldsymbol{a}^{i-1} - \boldsymbol{H}^{i-2}\boldsymbol{\gamma}^{i-1})(\delta\boldsymbol{a}^{i-1} - \boldsymbol{H}^{i-2}\boldsymbol{\gamma}^{i-1})^T}{(\delta\boldsymbol{a}^{i-1} - \boldsymbol{H}^{i-2}\boldsymbol{\gamma}^{i-1})^T\boldsymbol{\gamma}^{i-1}} \tag{17.52}$$

Since $\delta\boldsymbol{a}^{i-1}$ and $\boldsymbol{\gamma}^{i-1}$ are known, it appears that if $\boldsymbol{H}^{i-2}$ is also known then the new value of $\boldsymbol{H}^{i-1}$, i.e. the new value of the inverse of the secant stiffness matrix, is obtained by performing the simple operations in (17.52) and no costly inversion technique is necessary.

To further illustrate this approach, let us recall some results from matrix algebra, cf. for instance Strang (1980). For any square matrix $\boldsymbol{P}$, the eigenvalue problem is given by $\boldsymbol{P}\boldsymbol{c} = \lambda\boldsymbol{c}$, where $\boldsymbol{c}$ is the eigenvector and λ is the corresponding eigenvalue. The eigenvalue problem can also be written as the homogeneous equation system $(\boldsymbol{P}-\lambda\boldsymbol{I})\boldsymbol{c} = \boldsymbol{0}$ and a non-trivial $\boldsymbol{c}$-solution requires that $\det(\boldsymbol{P} - \lambda\boldsymbol{I}) = 0$; this condition is the *characteristic equation*. If the matrix $\boldsymbol{P}$ possesses an eigenvalue that is zero, then we must have $\det\boldsymbol{P} = 0$. Let the matrix $\boldsymbol{P}$ be of dimension $N \times N$, then the characteristic equation $\det(\boldsymbol{P} - \lambda\boldsymbol{I}) = 0$ becomes a polynomial of order N and we therefore have N eigenvalues. The number of non-zero eigenvalues of the matrix $\boldsymbol{P}$ is called the *rank of the matrix* and it is denoted by r. If $\det(\boldsymbol{P} - \lambda\boldsymbol{I}) = 0$ possesses no zero eigenvalues, the rank is $r = N$. If it possesses one zero eigenvalue, the rank is $r = N - 1$ and if it possesses two zero eigenvalues, the rank is $r = N - 2$ and it is then said that the matrix $\boldsymbol{P}$ has the eigenvalue zero with a *multiplicity* of two. Therefore, if the homogeneous equation system $\boldsymbol{Pc} = \boldsymbol{0}$ has q non-trivial $\boldsymbol{c}$ - solutions that are linearly independent then the rank of $\boldsymbol{P}$ is $r = N - q$ and the matrix $\boldsymbol{P}$ possesses the zero eigenvalue with a multiplicity of q.

In the present case, consider the correction matrix $\boldsymbol{H}^{i-1}_{corr}$ which according to (17.49) and (17.50) is given by $\boldsymbol{H}^{i-1}_{corr} = \boldsymbol{v}^{i-1}(\boldsymbol{v}^{i-1})^T$ and the dimension of $\boldsymbol{H}^{i-1}_{corr}$ is $N \times N$. Investigate the possibility of a non-trivial solution of the homogeneous equation system $\boldsymbol{H}^{i-1}_{corr}\boldsymbol{c} = \boldsymbol{0}$, i.e. $\boldsymbol{v}^{i-1}((\boldsymbol{v}^{i-1})^T\boldsymbol{c}) = \boldsymbol{0}$. It appears that there are $q = N - 1$ linear independent $\boldsymbol{c}$-vectors that fulfill $(\boldsymbol{v}^{i-1})^T\boldsymbol{c} = 0$, i.e. $\boldsymbol{H}^{i-1}_{corr}$ has a zero eigenvalue with a multiplicity of $q = N - 1$. The rank of $\boldsymbol{H}^{i-1}_{corr}$ is therefore $r = N - q = N - (N - 1) = 1$. Therefore, the quasi-Newton method defined by (17.50) is called a *rank one correction.*

The rank one correction quasi-Newton method defined by (17.50) was suggested by Broyden (1967) and later independently by others. It appears that if $\boldsymbol{H}^{i-2}$ is symmetric then this method has the neat property that the new update $\boldsymbol{H}^{i-1}$ is also symmetric. However, if the denominator in (17.52) is also small this may lead to numerical difficulties. Therefore, we shall not dwell on this approach any further and here merely take it as a simple illustration of a quasi-Newton method.

17.8.2 Rank two correction. BFGS-method

To establish a quasi-Newton method that is of relevance for our purpose, consider the following form

$$\boldsymbol{H}^{i-1} = (\boldsymbol{I} + \boldsymbol{v}\boldsymbol{w}^T)^T\boldsymbol{H}^{i-2}(\boldsymbol{I} + \boldsymbol{v}\boldsymbol{w}^T) \tag{17.53}$$

where the matrix $\boldsymbol{v}\boldsymbol{w}^T$ must be chosen so that (17.53) fulfills the quasi-Newton equation (17.47). Carrying out the multiplications and assuming $\boldsymbol{H}^{i-2}$ to be symmetric we obtain

$$\boldsymbol{H}^{i-1} = \boldsymbol{H}^{i-2} + (\boldsymbol{v}^T\boldsymbol{H}^{i-2}\boldsymbol{v})\boldsymbol{w}\boldsymbol{w}^T + \boldsymbol{w}(\boldsymbol{H}^{i-2}\boldsymbol{v})^T + \boldsymbol{H}^{i-2}\boldsymbol{v}\boldsymbol{w}^T \tag{17.54}$$

which is evidently of the format given by (17.49). In the quasi-Newton approach defined by (17.50), the correction matrix $\boldsymbol{H}^{i-1}_{corr}$ was found to be of rank one. Let us now determine the rank of the correction matrix in the approach defined by (17.54). Therefore, we shall investigate non-trivial $\boldsymbol{c}$-solutions of the homogeneous equation system $\boldsymbol{H}^{i-1}_{corr}\boldsymbol{c} = \boldsymbol{0}$. With (17.54) and (17.49) and defining the vector $\boldsymbol{b}$ by $\boldsymbol{b} = \boldsymbol{H}^{i-2}\boldsymbol{v}$, the following equation system is therefore considered

$$\boldsymbol{H}^{i-1}_{corr}\boldsymbol{c} = (\boldsymbol{v}^T\boldsymbol{b})\boldsymbol{w}(\boldsymbol{w}^T\boldsymbol{c}) + \boldsymbol{w}(\boldsymbol{b}^T\boldsymbol{c}) + \boldsymbol{b}(\boldsymbol{w}^T\boldsymbol{c}) = \boldsymbol{0}$$

This equation system is fulfilled if both $\boldsymbol{w}^T\boldsymbol{c} = 0$ and $\boldsymbol{b}^T\boldsymbol{c} = 0$. With $\boldsymbol{H}^{i-1}_{corr}$ having the dimension $N \times N$, these conditions can be fulfilled for $N - 2$ linear independent $\boldsymbol{c}$-vectors. Referring to our previous discussion, $\boldsymbol{H}_{corr}^{i-1}$ has therefore the eigenvalue zero with a multiplicity of $q = N - 2$ and the rank r of $\boldsymbol{H}_{corr}^{i-1}$ is then $r = N - q = N - (N - 2) = 2$. Consequently, the quasi-Newton approach defined by (17.54) is a *rank two correction*.

The next topic is to identify the vectors $\boldsymbol{v}$ and $\boldsymbol{w}$ in (17.54). Since the vector $\delta\boldsymbol{a}^{i-1}$ is one of the vectors we know, it seems natural to choose $\boldsymbol{w}$ to be proportional to $\delta\boldsymbol{a}^{i-1}$. According to (17.53), $\boldsymbol{v}$ and $\boldsymbol{w}$ only occur in the combination $\boldsymbol{v}\boldsymbol{w}^T$ or $\boldsymbol{w}\boldsymbol{v}^T$ and the length of one of these vectors can therefore be chosen arbitrarily; this will only affect the length of the other vector. Let us therefore choose $\boldsymbol{w}$ according to

$$\boldsymbol{w} = \frac{1}{b}\delta\boldsymbol{a}^{i-1} \qquad \text{where} \quad b = (\boldsymbol{\gamma}^{i-1})^T\delta\boldsymbol{a}^{i-1} \tag{17.55}$$

Since both $\delta\boldsymbol{a}^{i-1}$ and $\boldsymbol{\gamma}^{i-1}$ are known, also $\boldsymbol{w}$ is known.

With (17.55), (17.54) becomes

$$\begin{aligned}\boldsymbol{H}^{i-1} &= \boldsymbol{H}^{i-2} + (\boldsymbol{v}^T\boldsymbol{H}^{i-2}\boldsymbol{v})\frac{1}{b^2}\delta\boldsymbol{a}^{i-1}(\delta\boldsymbol{a}^{i-1})^T + \frac{1}{b}\delta\boldsymbol{a}^{i-1}(\boldsymbol{H}^{i-2}\boldsymbol{v})^T \\ &+\frac{1}{b}\boldsymbol{H}^{i-2}\boldsymbol{v}(\delta\boldsymbol{a}^{i-1})^T\end{aligned} \tag{17.56}$$

Insertion into the quasi-Newton equation (17.47) results in

$$\begin{aligned}\delta\boldsymbol{a}^{i-1} &= \boldsymbol{H}^{i-2}\boldsymbol{\gamma}^{i-1} + (\boldsymbol{v}^T\boldsymbol{H}^{i-2}\boldsymbol{v})\frac{1}{b}\delta\boldsymbol{a}^{i-1} \\ &+\frac{1}{b}\delta\boldsymbol{a}^{i-1}(\boldsymbol{v}^T\boldsymbol{H}^{i-2}\boldsymbol{\gamma}^{i-1}) + \boldsymbol{H}^{i-2}\boldsymbol{v}\end{aligned}$$

where it was assumed that $\boldsymbol{H}^{i-2}$ is symmetric. Determination of $\boldsymbol{v}$ from the expression gives

$$\boldsymbol{v} = m(\boldsymbol{H}^{i-2})^{-1}\delta\boldsymbol{a}^{i-1} - \boldsymbol{\gamma}^{i-1} \tag{17.57}$$

where

$$m = 1 - \frac{1}{b}\boldsymbol{v}^T\boldsymbol{H}^{i-2}(\boldsymbol{v} + \boldsymbol{\gamma}^{i-1}) \tag{17.58}$$

Expression (17.57) leads to

$$\boldsymbol{v}^T\boldsymbol{H}^{i-2}(\boldsymbol{v} + \boldsymbol{\gamma}^{i-1}) = m^2(\delta\boldsymbol{a}^{i-1})^T(\boldsymbol{H}^{i-2})^{-1}\delta\boldsymbol{a}^{i-1} - m\,b \tag{17.59}$$

where advantage was taken of (17.55). Insertion of (17.59) in (17.58) results in

$$m = \pm\left[\frac{b}{(\delta\boldsymbol{a}^{i-1})^T(\boldsymbol{H}^{i-2})^{-1}\delta\boldsymbol{a}^{i-1}}\right]^{1/2} \tag{17.60}$$

The vector $\boldsymbol{v}$ given by (17.57) and (17.60) has now been expressed in terms of known quantities. Moreover, this expression for $\boldsymbol{v}$ ensures that $\boldsymbol{H}^{i-1}$ given by (17.56) fulfills the quasi-Newton equation (17.47). It is easy to be convinced that irrespective of the sign we choose for m in (17.60), (17.56) provides the same updated matrix $\boldsymbol{H}^{i-1}$. In (17.60), we shall for convenience choose the positive sign.

In view of this comment, the results (17.53), (17.55), (17.57) and (17.60) can be summarized as

BFGS-method

$$\boldsymbol{H}^{i-1} = (\boldsymbol{I} + \boldsymbol{v}\boldsymbol{w}^T)^T \boldsymbol{H}^{i-2}(\boldsymbol{I} + \boldsymbol{v}\boldsymbol{w}^T)$$

where

$$\boldsymbol{v} = \left[\frac{(\boldsymbol{\gamma}^{i-1})^T \delta\boldsymbol{a}^{i-1}}{(\delta\boldsymbol{a}^{i-1})^T(\boldsymbol{H}^{i-2})^{-1}\delta\boldsymbol{a}^{i-1}}\right]^{1/2} (\boldsymbol{H}^{i-2})^{-1}\delta\boldsymbol{a}^{i-1} - \boldsymbol{\gamma}^{i-1} \qquad (17.61)$$

and

$$\boldsymbol{w} = \frac{\delta\boldsymbol{a}^{i-1}}{(\boldsymbol{\gamma}^{i-1})^T \delta\boldsymbol{a}^{i-1}}$$

This comprises the so-called *BFGS-method*, suggested by Broyden (1970), Fletcher (1970), Goldfarb (1970) and Shannon (1970). The BFGS-method can be expressed in various identical forms and the product form adopted in (17.61) was proposed by Matthies and Strang (1979). Indeed, Matthies and Strang (1979) were the first to apply the BFGS-method in a finite element context.

In the BFGS-scheme (17.61), we apparently need to establish the inverse matrix $(\boldsymbol{H}^{i-2})^{-1}$. This inverse matrix always occurs in the form $(\boldsymbol{H}^{i-2})^{-1}\delta\boldsymbol{a}^{i-1}$ and we will now show that this term is already known. From (17.46a) and the general iteration scheme (17.18), we have

$$\boldsymbol{A}^{i-2}\delta\boldsymbol{a}^{i-1} = \boldsymbol{A}^{i-2}(\boldsymbol{a}^{i-1} - \boldsymbol{a}^{i-2}) = -\boldsymbol{\psi}^{i-2}$$

and since (17.48) shows that $\boldsymbol{A}^{i-2} = (\boldsymbol{H}^{i-2})^{-1}$, we obtain

$$(\boldsymbol{H}^{i-2})^{-1}\delta\boldsymbol{a}^{i-1} = -\boldsymbol{\psi}^{i-2}$$

The out-of-balance force $\boldsymbol{\psi}^{i-2}$ is already known and this facilitates the use of the BFGS-scheme (17.61) considerably.

The product form for the updated matrix $\boldsymbol{H}^{i-1}$ given by (17.61) is of advantage for several reasons. Defining the matrix $\boldsymbol{M}$ by

$$\boldsymbol{M} = \boldsymbol{I} + \boldsymbol{v}\boldsymbol{w}^T \qquad (17.62)$$

we obtain

$$\boldsymbol{H}^{i-1} = \boldsymbol{M}^T \boldsymbol{H}^{i-2}\boldsymbol{M} \qquad (17.63)$$

The stiffness matrix in the finite element formulation is often symmetric and so is therefore its inverse. In the previous derivations, we assumed $\boldsymbol{H}^{i-2}$ to be symmetric and (17.63) then shows the interesting result that the updated matrix $\boldsymbol{H}^{i-1}$ is also symmetric. Let us next assume that $\boldsymbol{H}^{i-2}$ is positive definite. To investigate whether the updated matrix $\boldsymbol{H}^{i-1}$ is also positive definite, we choose an arbitrary vector $\boldsymbol{c}$ and calculate the quadratic form of $\boldsymbol{H}^{i-1}$. Denoting the value of the quadratic form by a, we obtain with (17.63)

$$a = \boldsymbol{c}^T \boldsymbol{H}^{i-1} \boldsymbol{c} = \boldsymbol{y}^T \boldsymbol{H}^{i-2} \boldsymbol{y} \tag{17.64}$$

where $\boldsymbol{y}$ is defined by

$$\boldsymbol{M}\boldsymbol{c} = \boldsymbol{y}$$

It appears that if $\boldsymbol{c} \neq \boldsymbol{0}$ implies $\boldsymbol{y} \neq \boldsymbol{0}$ then, since $\boldsymbol{H}^{i-2}$ was assumed to be positive definite, $a > 0$ and $\boldsymbol{H}^{i-1}$ is therefore also positive definite.

The only situation where $\boldsymbol{c} \neq \boldsymbol{0}$ implies $\boldsymbol{y} = \boldsymbol{0}$ is if $\boldsymbol{M}\boldsymbol{c} = \boldsymbol{0}$ possesses a non-trivial solution. In turn this requires that the matrix $\boldsymbol{M}$ has an eigenvalue that is zero. Let us therefore determine the eigenvalues λ of the matrix $\boldsymbol{M}$, i.e. $\boldsymbol{M}\boldsymbol{c} = \lambda\boldsymbol{c}$. With (17.62), we then obtain

$$(1 - \lambda)\boldsymbol{c} + \boldsymbol{v}(\boldsymbol{w}^T \boldsymbol{c}) = \boldsymbol{0} \tag{17.65}$$

Let us assume that $\boldsymbol{w}^T \boldsymbol{c} = 0$; with the dimension of $\boldsymbol{M}$ being $N \times N$, this condition is fulfilled by $N - 1$ linearly independent $\boldsymbol{c}$-vectors and for each of these $\boldsymbol{c}$-vectors, (17.65) is fulfilled for $\lambda = 1$. We conclude

$$\lambda = 1 \qquad \text{with multiplicity } N - 1$$

Assume next that $\boldsymbol{w}^T \boldsymbol{c} \neq 0$. Expression (17.65) then shows that the eigenvector $\boldsymbol{c}$ must be proportional to $\boldsymbol{v}$, i.e. $\boldsymbol{c} = \alpha\boldsymbol{v}$ where α is a constant. Use of $\boldsymbol{c} = \alpha\boldsymbol{v}$ in (17.65) gives $[1 - \lambda + (\boldsymbol{w}^T \boldsymbol{v})]\boldsymbol{v} = \boldsymbol{0}$, i.e. we obtain the following remaining eigenvalue

$$\lambda = 1 + (\boldsymbol{w}^T \boldsymbol{v})$$

which with $\boldsymbol{v}$ and $\boldsymbol{w}$ given by (17.61) becomes

$$\lambda = \left[\frac{(\delta \boldsymbol{a}^{i-1})^T (\boldsymbol{H}^{i-2})^{-1} \delta \boldsymbol{a}^{i-1}}{(\boldsymbol{\gamma}^{i-1})^T \delta \boldsymbol{a}^{i-1}} \right]^{1/2} \tag{17.66}$$

Since $\boldsymbol{H}^{i-2}$ and thereby $(\boldsymbol{H}^{i-2})^{-1}$ was assumed to be positive definite, the eigenvalue given by (17.66) can never be zero. In conclusion, we have proved that the matrix $\boldsymbol{M}$ has no zero eigenvalues and the discussion relating to (17.64) then shows that the updated matrix $\boldsymbol{H}^{i-1}$ is positive definite, if $\boldsymbol{H}^{i-2}$ is so. In conclusion

> *If $\boldsymbol{H}^{i-2}$ is symmetric and positive definite then the BFGS-method implies that the updated matrix $\boldsymbol{H}^{i-1}$ is symmetric and positive definite*

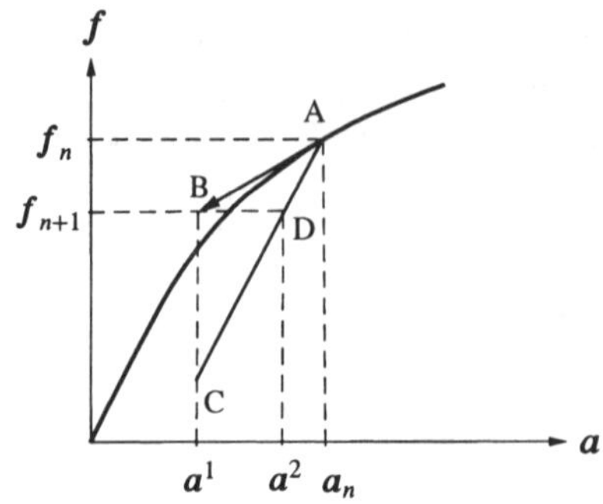

Figure 17.15: BFGS-method during unloading of elasto-plastic body. Correct solution is obtained after two iterations.

This interesting property implies, for instance, that if the very first $\boldsymbol{H}^{i-2}$-matrix is chosen as the inverse of the elastic stiffness matrix, which certainly is symmetric and positive definite, then all the following updates $\boldsymbol{H}^{i-1}$ generated by (17.61) are symmetric and positive definite.

We will now show that the BFGS-method for unloading of a elasto-plastic body provides the correct elastic unloading after two iterations; this important property is a result of the secant relation expressed by the quasi-Newton equation (17.47). In the first iteration and referring to Fig. 17.15, unloading takes place along some secant direction AB giving the first displacement estimate $\boldsymbol{a}^1$. The internal forces corresponding to $\boldsymbol{a}^1$ are given by point C. In the next iteration, the secant stiffness is determined by the last two sets of displacements and the last two sets of internal forces. Since we can assume that equilibrium is fulfilled at point A, this new secant stiffness is given by CA equal to the elastic stiffness. With this stiffness together with the residual force vector given by CB we reach, in the second iteration, exactly the correct response given by point D.

Efficient implementation of the BFGS-method evidently hinges on keeping the number of computer operations as small as possible and this requires a number of considerations. As we are here only concerned with the concepts behind the BFGS-method, the reader is referred to Bathe and Cimento (1980), Crisfield (1991), Fletcher (1980), Luenberger (1984) and Matthies and Strang (1979) for information on implementation techniques. Here we merely observe that the BFGS-method provides an attractive approach, which gives a convergence rate that is almost as fast as the Newton-Raphson without performing costly matrix inversions. The BFGS-method therefore provides an interesting compromise between the Newton-Raphson scheme and the modified Newton-Raphson scheme. The advantages within a finite element context are well documented, see Bathe (1996), Bathe and Cimento (1980), Crisfield (1991) and Matthies and Strang (1979). The use of a nonsymmetric quasi-Newton approach in finite element calculations of fluid mechanics has been investigated by Engelman *et al.* (1981).

17.9 Line search

Irrespective of the choice of iteration matrix $\boldsymbol{A}$, it turns out to be possible to reduce the number of equilibrium iterations significantly by introducing the concept of *line search*. Like many other issues relating to solution of nonlinear equations, this concept has its origin in optimization theory, see for instance Luenberger (1984).

Let us define the quantity $\boldsymbol{s}^{i-1}$ by

$$\boxed{\boldsymbol{s}^{i-1} = -(\boldsymbol{A}^{i-1})^{-1}\boldsymbol{\psi}^{i-1} \qquad \textit{search direction}} \tag{17.67}$$

Then the general iteration scheme (17.18) can be written as

$$\boldsymbol{a}^{i} = \boldsymbol{a}^{i-1} + \boldsymbol{s}^{i-1} \tag{17.68}$$

It appears that the new solution $\boldsymbol{a}^i$ is obtained by correcting the old solution $\boldsymbol{a}^{i-1}$ by the quantity $\boldsymbol{s}^{i-1}$. This quantity is called the *search direction* since it is the direction in which we search for the new solution. With this interpretation, it is tempting to accept the search direction $\boldsymbol{s}^{i-1}$, but leave open how far we shall go in this direction. Therefore, instead of (17.68), we now adopt the scheme

$$\boxed{\boldsymbol{a}^{i} = \boldsymbol{a}^{i-1} + \beta^{i-1}\boldsymbol{s}^{i-1}} \tag{17.69}$$

where the parameter β^{i-1}, the *acceleration factor*, is to be determined. With this viewpoint, we accept the search direction $\boldsymbol{s}^{i-1}$ and adjust the acceleration factor β^{i-1} in some way so that (17.69) gives us the best possible new estimate. The scheme is illustrated in Fig. 17.16

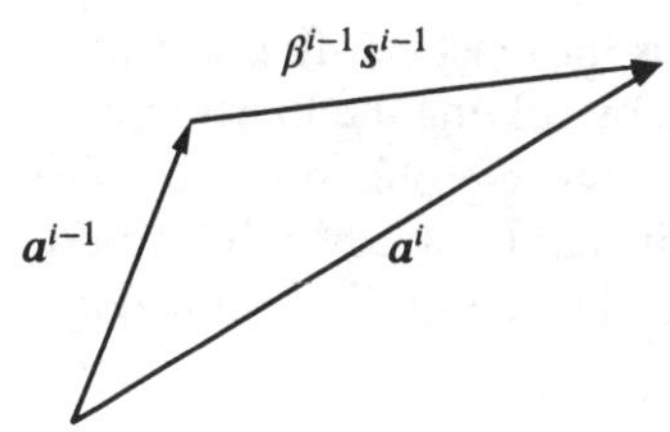

Figure 17.16: Scheme when line search is adopted

A comparison of (17.69) and (17.18) shows that instead of working with the iteration matrix $\boldsymbol{A}^{i-1}$ we now work with the iteration matrix $\frac{1}{\beta^{i-1}}\boldsymbol{A}^{i-1}$. The new scheme is therefore fully acceptable as long as $\beta^{i-1} \neq 0$ and it appears that line search can be applied in combination with any of the methods discussed previously.

However, the question remains of how to determine β^{i-1} - occasionally called the *step length* - so that the scheme (17.69) is optimal in some sense. The concept of *line search* means that we go from an old solution $\boldsymbol{a}^{i-1}$ corresponding to $\beta^{i-1} = 0$ in the search direction $\boldsymbol{s}^{i-1}$ until we obtain the best possible new solution $\boldsymbol{a}^i$. Since the iteration scheme without line search corresponds to $\beta^{i-1} = 1$, it is evident that we must have $\beta^{i-1} > 0$. The best estimate for $\boldsymbol{a}^i$ is the one which makes the out-of-balance forces $\boldsymbol{\psi}(\boldsymbol{a}^i) = \boldsymbol{0}$. In general, this condition cannot be fulfilled since it requires that the search direction be the correct one. This is just to say that we cannot expect to obtain that all components of $\boldsymbol{\psi}(\boldsymbol{a}^i)$ become zero just by adjusting one parameter β^{i-1}. Since β^{i-1} is one parameter, we must establish one condition from which β^{i-1} can be determined. Such a condition can be obtained by requiring that the new out-of-balance force $\boldsymbol{\psi}(\boldsymbol{a}^i)$ is orthogonal to a vector, i.e. the component of $\boldsymbol{\psi}(\boldsymbol{a}^i)$ in the direction of this vector is required to be zero. The question then arises which vector we shall choose when enforcing this orthogonality requirement. Referring to Fig. 17.16, three vectors are possible candidates: $\boldsymbol{a}^i$, $\boldsymbol{a}^{i-1}$ and $\boldsymbol{s}^{i-1}$. It turns out that the search direction $\boldsymbol{s}^{i-1}$ is the most natural choice, i.e. we adopt the orthogonality condition $(\boldsymbol{s}^{i-1})^T\boldsymbol{\psi}(\boldsymbol{a}^i) = 0$, which with (17.69) reads

$$\boxed{(\boldsymbol{s}^{i-1})^T\boldsymbol{\psi}(\boldsymbol{a}^{i-1} + \beta^{i-1}\boldsymbol{s}^{i-1}) = 0 \qquad \textit{line search}} \tag{17.70}$$

Since $\boldsymbol{s}^{i-1}$ and $\boldsymbol{a}^{i-1}$ are known quantities, this orthogonality condition provides one (nonlinear) relation by which β^{i-1} can be determined.

To motivate the use of the search direction $\boldsymbol{s}^{i-1}$ in the orthogonality condition (17.70), it is instructive to consider linear elasticity. In that case $\boldsymbol{\sigma} = \boldsymbol{D}\boldsymbol{\varepsilon} = \boldsymbol{D}\boldsymbol{B}\boldsymbol{a}$ and the out-of-balance forces $\boldsymbol{\psi}$ defined by (17.2) then becomes

$$\boldsymbol{\psi} = \boldsymbol{K}\boldsymbol{a} - \boldsymbol{f} \tag{17.71}$$

where $\boldsymbol{K} = \int_V \boldsymbol{B}^T\boldsymbol{D}\boldsymbol{B}dV$ is the elastic stiffness matrix. The equilibrium condition $\boldsymbol{\psi} = \boldsymbol{0}$ is then fulfilled by solving the linear equation system $\boldsymbol{K}\boldsymbol{a} - \boldsymbol{f} = \boldsymbol{0}$.

Further insight into the linear problem can be obtained by introducing the *potential energy* Π of the linear elastic body. If the body is discretized by means of finite elements, the potential energy Π is defined by

$$\Pi = \frac{1}{2}\boldsymbol{a}^T\boldsymbol{K}\boldsymbol{a} - \boldsymbol{a}^T\boldsymbol{f} \tag{17.72}$$

where the external force $\boldsymbol{f}$ is viewed as constant. It appears that $\Pi = \Pi(\boldsymbol{a})$ and therefore

$$\frac{\partial\Pi}{\partial\boldsymbol{a}} = \boldsymbol{K}\boldsymbol{a} - \boldsymbol{f}$$

A comparison with (17.71) shows that equilibrium is expressed as $\partial\Pi/\partial\boldsymbol{a} = \boldsymbol{0}$, i.e. at equilibrium the potential energy Π takes an extremum value.

Evidently, it is possible to solve the linear equation system $\boldsymbol{Ka} - \boldsymbol{f} = \boldsymbol{0}$ directly, but it is also possible to solve this linear equation system in an iterative manner using the scheme (17.69). With $\boldsymbol{a}^i$ given by (17.69), the potential energy defined by (17.72) becomes.

$$\Pi(\boldsymbol{a}^i) = \frac{1}{2}(\boldsymbol{a}^i)^T \boldsymbol{K} \boldsymbol{a}^i - (\boldsymbol{a}^i)^T \boldsymbol{f} \tag{17.73}$$

Since $\boldsymbol{a}^{i-1}$ and $\boldsymbol{s}^{i-1}$ are given quantities, $\boldsymbol{a}^i$ given by (17.69) depends only on β^{i-1} and we therefore have $\Pi = \Pi(\beta^{i-1})$. From (17.73) it then follows that

$$\frac{d\Pi}{d\beta^{i-1}} = (\boldsymbol{s}^{i-1})^T (\boldsymbol{K}\boldsymbol{a}^i - \boldsymbol{f}) \tag{17.74}$$

and

$$\frac{d^2\Pi}{d(\beta^{i-1})^2} = (\boldsymbol{s}^{i-1})^T \boldsymbol{K} \boldsymbol{s}^{i-1} > 0 \tag{17.75}$$

where it was used that $\boldsymbol{K}$ is positive definite. In accordance with (17.71), we have for linear elasticity that

$$\boldsymbol{\psi}^i = \boldsymbol{K}\boldsymbol{a}^i - \boldsymbol{f} \tag{17.76}$$

We observed above that the potential energy takes an extremum value at equilibrium. As we are performing equilibrium iterations, the condition of equilibrium has not been achieved, but with the quantities $\boldsymbol{a}^{i-1}$ and $\boldsymbol{s}^{i-1}$ given, the best possible new solution is obtained by requiring Π to be extremum, i.e. $d\Pi/d\beta^{i-1} = 0$. Therefore, (17.74) gives with (17.76)

$$(\boldsymbol{s}^{i-1})^T \boldsymbol{\psi}^i = 0 \tag{17.77}$$

According to (17.75), $d^2\Pi/d(\beta^{i-1})^2 > 0$ and the extremum property of the potential energy Π is therefore that of a minimum. Similar to (17.76), we have

$$\boldsymbol{\psi}^{i-1} = \boldsymbol{K}\boldsymbol{a}^{i-1} - \boldsymbol{f}$$

With (17.76) and (17.69), (17.77) therefore provides the following solution

$$\beta^{i-1} = -\frac{(\boldsymbol{s}^{i-1})^T \boldsymbol{\psi}^{i-1}}{(\boldsymbol{s}^{i-1})^T \boldsymbol{K} \boldsymbol{s}^{i-1}} \tag{17.78}$$

which is the best possible choice of β^{i-1}. If $\boldsymbol{\psi}^{i-1} = \boldsymbol{0}$, then the estimate $\boldsymbol{a}^{i-1}$ fulfills the equilibrium equations and it is therefore the solution sought for. Accordingly, (17.78) provides $\beta^{i-1} = 0$ when $\boldsymbol{\psi}^{i-1} = \boldsymbol{0}$. However, if $\boldsymbol{\psi}^{i-1} \neq \boldsymbol{0}$ then (17.78) gives a solution for β^{i-1} which minimizes the potential energy Π as much as possible.

It appears that for linear problems, use of the search direction $\boldsymbol{s}^{i-1}$ in the orthogonality condition is the best possible choice, cf. (17.77). It therefore seems

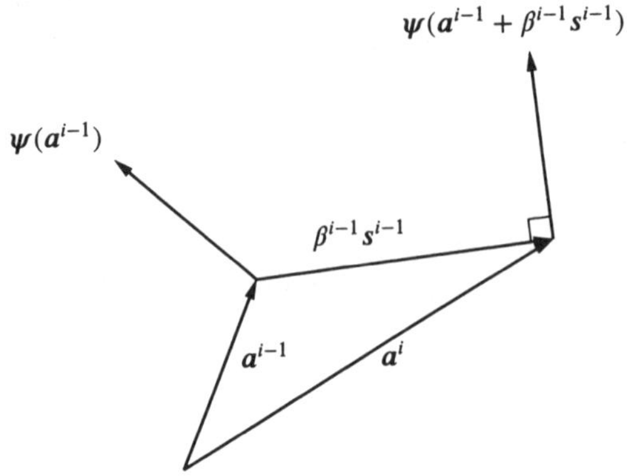

Figure 17.17: Line search for orthogonality between s^{i-1} and $\psi(a^{i-1}+\beta^{i-1}s^{i-1})$.

reasonable to adopt the same orthogonality condition also for other problems and this was already anticipated by the choice (17.70).

To further evaluate (17.70), we consider the scalar product $(s^{i-1})^T\psi(a^{i-1})$ which with (17.67) can be written as

$$(s^{i-1})^T\psi(a^{i-1}) = (\psi(a^{i-1}))^T s^{i-1} = -(\psi(a^{i-1}))^T (A^{i-1})^{-1}\psi(a^{i-1}) \quad (17.79)$$

Since the iteration matrix A^{i-1} is usually a positive definite matrix, (17.79) shows that $(s^{i-1})^T\psi(a^{i-1}) < 0$ usually holds. We then obtain the illustration shown in Fig. 17.17, where it is observed that $\psi(a^{i-1})$ corresponds to $\beta^{i-1}=0$. When β^{i-1} is changed, the out-of-balance forces change and β^{i-1} is changed such that s^{i-1} and $\psi(a^{i-1}+\beta^{i-1}s^{i-1})$ eventually become orthogonal. Therefore, when β^{i-1} is determined from (17.70) this implies that the out-of-balance force has no components in the search direction s^{i-1}.

As already mentioned, (17.70) provides for, say, plasticity problems a nonlinear relation for determination of β^{i-1}. Therefore, in principle, solution of (17.70) requires an iterative procedure and in each of these iterations, a new out-of-balance force vector needs to be determined. This in turn requires determination of the stress field, which means integration of the constitutive equations for all (Gauss) points in the body. In the next chapter, we shall treat integration of the nonlinear elasto-plastic constitutive equations in detail, but already at this stage it is evident that this integration is far from being trivial. Therefore, using a number of iterations to solve β^{i-1} from (17.70) in an accurate manner requires a computer effort that is not insignificant. Moreover, since the search direction s^{i-1} is only an approximation to the correct one, this also questions the meaningfulness of achieving an accurate β^{i-1}-solution. Finally, whereas the orthogonality expressed of (17.70) provides the best possible β^{i-1}-solution for linear problems, it can only be expected to provide a fair β^{i-1}-solution for non-linear problems.

All these arguments suggest that there is no sense in trying to solve (17.70) in an accurate manner and, instead, an approximate value for β^{i-1} is sufficient. Indeed, this is supported by numerical experience both within structural mechanics, see Crisfield (1991), and optimization theory, see Luenberger (1984). This numerical experience also shows that also use of an approximate β^{i-1}-value, in general, reduces the number of equilibrium iterations significantly and reduces the total computer effort considerably. Due to its effectiveness, line search is available in most general purpose finite element programs.

With DOF being the number of degrees of freedom for the structure and B the bandwidth of the equation system, the computational cost for solving the equilibrium equation system by means of Gauss elimination is proportional to DOF·B^2, cf. for instance Ottosen and Petersson (1992). The stresses $\boldsymbol{\sigma}^{i-1}$ are evaluated at each Gauss point in the structure and the number of Gauss points is roughly proportional to DOF=degrees of freedom. It is concluded that the advantage with the use of line search increases with the number of degrees of freedom.

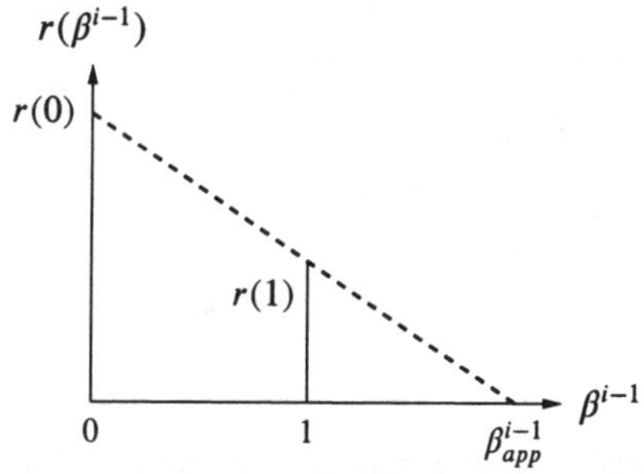

Figure 17.18: Approximate determination of β^{i-1}.

To arrive at an approximate β^{i-1}-value, we first observe that $\beta^{i-1} = 1$ corresponds to the iteration scheme without line search, cf. (17.69) and (17.18). We therefore certainly expect that $\beta^{i-1} > 0$. Define the quantity $r(\beta^{i-1})$ by

$$r(\beta^{i-1}) = -(\boldsymbol{s}^{i-1})^T \boldsymbol{\psi}(\boldsymbol{a}^{i-1} + \beta^{i-1}\boldsymbol{s}^{i-1})$$

where the minus sign is motivated by the discussion following (17.79). The quantity $\boldsymbol{\psi}(\boldsymbol{a}^{i-1})$ has already been calculated, cf. (17.67), and it is therefore easy to calculate $r(\beta^{i-1} = 0)$; calculate also r for $\beta^{i-1} = 1$. We then obtain the situation sketched in Fig. 17.18. According to (17.70) we want $r(\beta^{i-1}) = 0$ and making the linear extrapolation shown in Fig. 17.18, we obtain the following approximate value for β^{i-1}

$$\beta^{i-1}_{app} = \frac{r(0)}{r(0) - r(1)} \tag{17.80}$$

If the difference $r(0) - r(1)$ is small, β^{i-1}_{app} becomes large and, in principle, we may even encounter that β^{i-1}_{app} as predicted by (17.80) becomes negative. Since the entire motivation for the orthogonality condition (17.70) holds only strictly for linear problems, very small or very large β^{i-1}-values should be considered with caution. In practice therefore, acceptable values for β^{i-1} are restricted to a certain interval, say $0.3 \leq \beta^{i-1} \leq 3$. If (17.80) provides $\beta^{i-1}_{app} < 0.3$ then $\beta^{i-1} = 0.3$ is used and if (17.80) predicts $\beta^{i-1} > 3$ then $\beta^{i-1} = 3$ is used.

17.10 Limit points

In the previous discussion of incremental-iterative schemes, we have assumed that, for each load step, the load is increased by a certain amount and then held constant during the equilibrium iterations. This approach works well when the displacements increase with increasing load level.

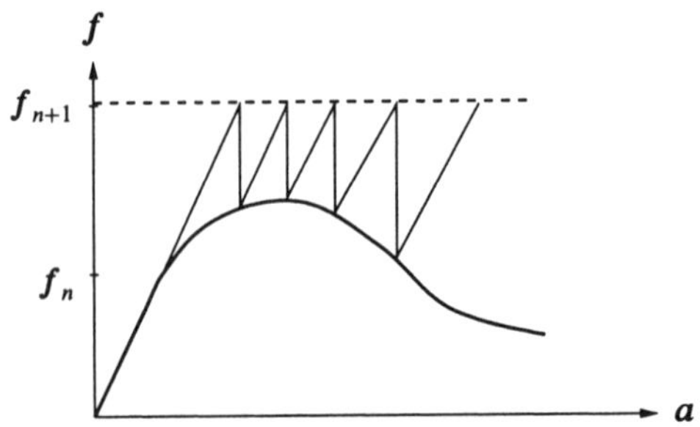

Figure 17.19: Structure displaying a softening response after the peak load. With the load $\boldsymbol{f}_{n+1}$ held fixed, equilibrium iterations will continue for ever (initial stiffness method used in the figure).

Consider now the situation where the structure displays a softening response, i.e. the structure has a maximum load capacity with a descending branch after the peak load, cf. Fig. 17.19. Evidently, the approach where the load is increased from $\boldsymbol{f}_n$ to $\boldsymbol{f}_{n+1}$ and then held constant will not be able to trace the softening branch; as soon as the load $\boldsymbol{f}_{n+1}$ exceeds the maximum load capacity of the structure, equilibrium iterations will simply continue for ever. This is illustrated in Fig. 17.19 using the initial stiffness method. Even if the response after the peak load does not display a descending branch, but exhibits a horizontal plateau (e.g. ideal plasticity), equilibrium iterations will continue for ever, if $\boldsymbol{f}_{n+1}$ exceeds the peak load.

From an engineering point of view, the peak load, i.e. the *limit load*, is often of fundamental interest. Occasionally, it is argued that once the equilibrium iterations do not converge then the limit load has been reached. However, this is a dangerous approach since equilibrium iterations may diverge for a number of other reasons. Moreover, it is often interesting to trace the descending branch after the peak load since it contains important information of the ductility or brittleness of the structural response, cf. Fig 17.20. In Fig. 17.20c) the phenomenon of *snap-back* is illustrated.

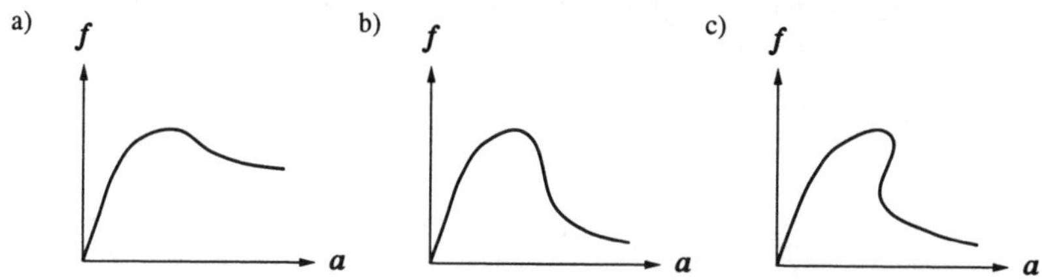

Figure 17.20: a) Ductile response; b) brittle response; c) snap-back.

The situation of *snap-through* often encountered in buckling problems is shown in Fig. 17.21. With the iterative approach discussed so far, one would only be able to trace the response OABC whereas the part ADB of the response cannot be traced.

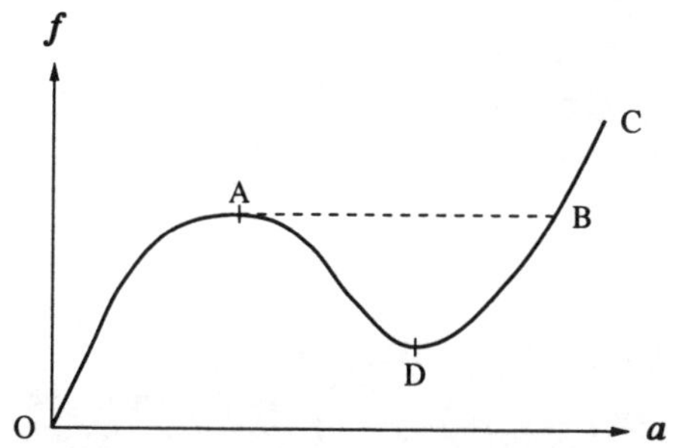

Figure 17.21: Snap-through of structure

It is evidently of importance to derive iterative schemes that are able to trace the various post-peak responses discussed above. Strategies that possess this property are called *path-following methods* and borrowing from the terminology of buckling analysis they are also often called *continuation methods*.

Whereas we previously accepted the load level and then modified the displacements until equilibrium is fulfilled, it is evident that we must now devise iterative schemes where both the displacements and the load level are adjusted, i.e.

Determination of the post-peak response requires simultaneous adjustment of the displacements and of the load level

Before entering this discussion, it is of interest to be able to characterize the different kinds of response discussed in relation to Figs. 17.20 and 17.21. Assume for simplicity that the external loading is proportional, i.e.

$$\boldsymbol{f} = \lambda \boldsymbol{f}_{ref} \tag{17.81}$$

where $\boldsymbol{f}_{ref}$ is a constant *reference load* and the proportionality factor $\lambda = \lambda(t)$, where t is the time, is called the *load parameter*. According to (17.2), the out-of-balance forces $\boldsymbol{\psi}$ are then given by

$$\boldsymbol{\psi} = \boldsymbol{f}_{int} - \lambda \boldsymbol{f}_{ref}$$

where the internal forces $\boldsymbol{f}_{int}$ depend on the stresses and thereby on the nodal displacements $\boldsymbol{a}$. The equilibrium condition is therefore given by

$$\boldsymbol{\psi} = \boldsymbol{0}\,; \quad \boldsymbol{\psi} = \boldsymbol{\psi}(\boldsymbol{a}, \lambda)$$

During equilibrium we have $\dot{\boldsymbol{\psi}} = \boldsymbol{0}$, which leads to

$$\frac{\partial \boldsymbol{\psi}}{\partial \boldsymbol{a}}\dot{\boldsymbol{a}} + \frac{\partial \boldsymbol{\psi}}{\partial \lambda}\dot{\lambda} = \boldsymbol{0}$$

i.e.

$$\boxed{\boldsymbol{K}_t\dot{\boldsymbol{a}} - \dot{\lambda}\boldsymbol{f}_{ref} = \boldsymbol{0}} \tag{17.82}$$

where use was made of (17.31).

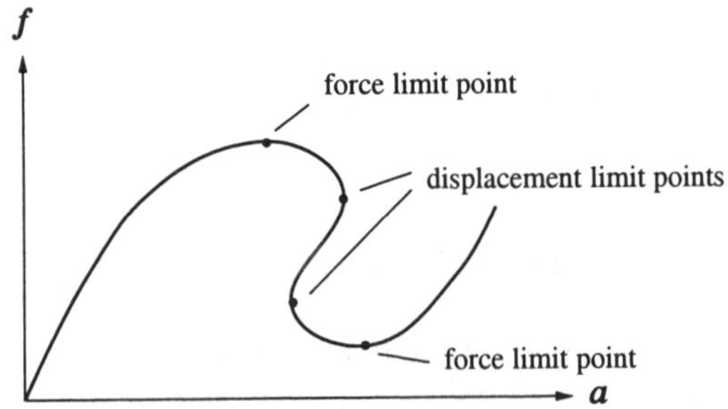

Figure 17.22: Illustration of force and displacement limit points

Let us make the following definitions:

$$\boxed{\begin{array}{l} \textit{Force limit point: } \dot{\lambda} = 0 \textit{ and } \dot{\boldsymbol{a}} \neq \boldsymbol{0} \\[1ex] \textit{Displacement limit point: } \dot{\lambda} \neq 0 \textit{ and } \dot{\boldsymbol{a}} = \boldsymbol{0} \end{array}} \tag{17.83}$$

These *limit points* are illustrated in Fig. 17.22 and we may note that in the literature such limit points are often called *turning points*.

To determine a force limit point, (17.82) and (17.83a) lead to $\boldsymbol{K}_t\dot{\boldsymbol{a}} = \boldsymbol{0}$ and a non-trivial solution requires $\det \boldsymbol{K}_t = 0$, i.e.

$$\boxed{\det \boldsymbol{K}_t = 0 \quad \Rightarrow \quad \textit{force limit point}}$$

Let us next assume that $\det \boldsymbol{K}_t \neq 0$; then (17.82) implies

$$\dot{\boldsymbol{a}} = \dot{\lambda}\boldsymbol{K}_t^{-1}\boldsymbol{f}_{ref}$$

A comparison with (17.83b) shows that the only possibility for a displacement limit point is that $\boldsymbol{f}_{ref}$ fulfills $\boldsymbol{K}_t^{-1}\boldsymbol{f}_{ref} = \boldsymbol{0}$, which requires $\det \boldsymbol{K}_t^{-1} = 0$, i.e.

$$\boxed{\det \boldsymbol{K}_t^{-1} = 0 \;\text{ and }\; \boldsymbol{K}_t^{-1}\boldsymbol{f}_{ref} = \boldsymbol{0} \quad \Rightarrow \quad \textit{displacement limit point}}$$

17.11 Bergan's minimum residual force method

In the previous iteration schemes, the load was increased step-wise and after each load increase, the load was held constant and, via equilibrium iterations, the nodal displacements were then adjusted in an iterative fashion until equilibrium was satisfied.

An elegant and entirely different approach, which allows force limit points to be passed as well as tracing of the post-peak response, was proposed by Bergan (1979, 1980). For a certain load level, the standard iteration scheme (17.18) is adopted, i.e.

$$\boldsymbol{A}^{i-1}(\boldsymbol{a}^i - \boldsymbol{a}^{i-1}) = -\boldsymbol{\psi}^{i-1} \tag{17.84}$$

where

$$\boldsymbol{\psi}^{i-1} = \boldsymbol{f}_{int}^{i-1} - \lambda^{i-1}\boldsymbol{f}_{ref}$$

and proportional loading according to (17.81) was assumed. It follows that

$$\boldsymbol{\psi}^0 = (\boldsymbol{f}_{int})_n - \lambda^0\boldsymbol{f}_{ref} \tag{17.85}$$

where it was used that $\boldsymbol{f}_{int}^0 = \int_V \boldsymbol{B}^T\boldsymbol{\sigma}^0 dV = \int_V \boldsymbol{B}^T\boldsymbol{\sigma}_n dV = (\boldsymbol{f}_{int})_n$, cf. (17.22). Moreover, in (17.85) we set

$$\lambda^0 = \lambda_n + \Delta\lambda^* \tag{17.86}$$

where $\boldsymbol{f}_n = \lambda_n\boldsymbol{f}_{ref}$ and the quantity $\Delta\lambda^*$ is specified by us.

From (17.84), the displacements $\boldsymbol{a}^i$ can now be determined; the corresponding stresses $\boldsymbol{\sigma}^i$ can now also be obtained by integration of the constitutive equations. With the stress field $\boldsymbol{\sigma}^i$ known throughout the body, the corresponding internal forces can be determined in the usual manner by means of

$$\boldsymbol{f}_{int}^i = \int_V \boldsymbol{B}^T\boldsymbol{\sigma}^i dV \tag{17.87}$$

The new residual forces then become

$$\boldsymbol{\psi}^i = \boldsymbol{f}^i_{int} - \lambda^i \boldsymbol{f}_{ref} \tag{17.88}$$

where the new load parameter λ^i is viewed as unknown. The key point in *Bergan's minimum residual force method* is then to adjust the load parameter λ^i so that the residual force $\boldsymbol{\psi}^i$ becomes minimum.

Define the quantity b by

$$b = (\boldsymbol{\psi}^i)^T \boldsymbol{\psi}^i$$

With both $\boldsymbol{f}_{ref}$ and $\boldsymbol{f}^i_{int}$ being known and fixed and λ^i the only unknown, we obtain

$$\frac{db}{d\lambda^i} = -2(\boldsymbol{\psi}^i)^T \boldsymbol{f}_{ref}$$

as well as $d^2b/(d\lambda^i)^2 = 2\boldsymbol{f}^T_{ref}\boldsymbol{f}_{ref} > 0$. Therefore, the minimum of the quantity b with respect to λ^i is given by $db/d\lambda^i = 0$, i.e.

$$\boxed{(\boldsymbol{\psi}^i)^T \boldsymbol{f}_{ref} = 0} \tag{17.89}$$

Insertion of (17.88) gives the new load parameter λ^i according to

$$\boxed{\lambda^i = \frac{(\boldsymbol{f}^i_{int})^T \boldsymbol{f}_{ref}}{\boldsymbol{f}^T_{ref}\boldsymbol{f}_{ref}}} \tag{17.90}$$

In view of the orthogonality expressed by (17.89) and the definitions of residual forces given by (17.88), we obtain the graphical illustration of the method shown in Fig. 17.23. After $\boldsymbol{a}^i$ has been determined from (17.84), these nodal displacements are accepted and the corresponding fixed internal forces $\boldsymbol{f}^i_{int}$ are then given by (17.87). The new load parameter λ^i is now adjusted according to (17.90) until the new external force $\lambda^i \boldsymbol{f}_{ref}$ and the new residual force $\boldsymbol{\psi}^i$ become orthogonal. Each iteration can therefore be viewed as a *two-step procedure*.

With this new load parameter, we perform a new iteration, i.e. the iteration number i is increased by one and the procedure is repeated until convergence – according to the procedures discussed in Section 17.6 – has been obtained. In that case, we put $\boldsymbol{a}_{n+1} = \boldsymbol{a}^i$ and $\lambda_{n+1} = \lambda^i$.

Bergan's minimum residual method is simple and very effective. In (17.84), we can take any iteration matrix; for instance, $\boldsymbol{A}^{i-1}$ may be taken as the tangential stiffness matrix $\boldsymbol{K}^{i-1}_t$ or the constant initial stiffness matrix $\boldsymbol{K}$. This latter choice is often adopted since it is cheap and since it allows force limit points to be passed and post-peak response to be traced. These features are illustrated in the one-dimensional case shown in Fig. 17.24.

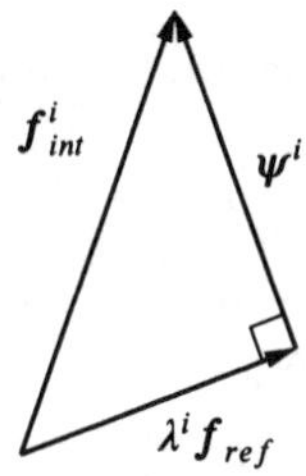

Figure 17.23: Orthogonal property obtained by Bergan's minimum residual method.

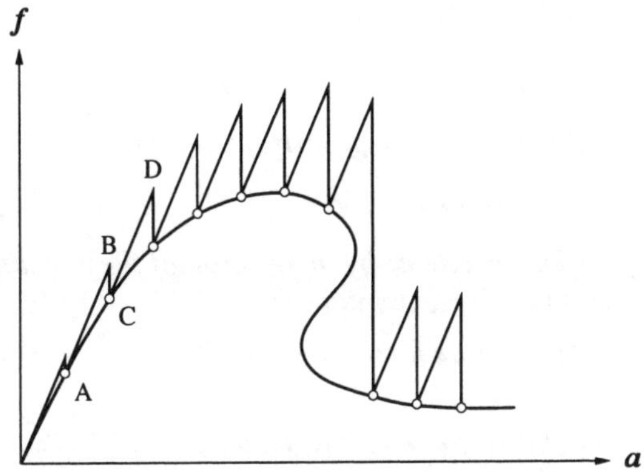

Figure 17.24: One-dimensional system. Bergan's method used with the initial stiffness matrix; after equilibrium is fulfilled, the imposed load increment is the same in all load steps. The predictions are indicated by (o).

Assume that point A has been reached. The load is then increased by some amount specified by us ($\Delta\lambda^*$ in (17.86)) to point B and the corresponding displacement is determined by (17.84) using the initial stiffness. These displacements are accepted and the corresponding internal force is determined by (17.87); this internal force corresponds to point C. The load is now adjusted so that the residual force, i.e. BC, becomes minimum. In the present one-dimensional case, this means reducing the load from B to C, i.e. the new residual force becomes in the one-dimensional case exactly equal to zero. Therefore in one iteration, exact equilibrium has been obtained. The loading is now increased by an amount specified by us ($\Delta\lambda^*$ in (17.86)) which brings us to point D and the procedure is repeated; in the case shown in Fig. 17.24 all specified load steps $\Delta\lambda^*$ are taken as equal.

While exact fulfillment of equilibrium in just one iteration is not a general property, but holds only for the one-dimensional problems, Fig. 17.24 illustrates that Bergan's method allows us to pass the force limit point and to trace the post-peak response. However, the figure also illustrates that Bergan's method cannot trace a snap-back response.

Box 17.3 Bergan's minimum residual force method

- *Initiation of quantities*
  $\boldsymbol{a}_0 = \mathbf{0}\ ; \quad \boldsymbol{\varepsilon}_0 = \mathbf{0}\ ; \quad \boldsymbol{\sigma}_0 = \mathbf{0}\ ; \quad \boldsymbol{f}_0 = \mathbf{0}\ ; \quad \boldsymbol{f}_{int} = \mathbf{0}$
- *For load step* $n = 0, 1, 2, \dots . N_{max}$
  - *Determine new load level* $\lambda \boldsymbol{f}_{ref}$
  - *Initiation of iteration quantities*
    $\boldsymbol{a}^0 := \boldsymbol{a}_n$
  - *Iterate* $i = 1, 2, \dots$ *until* $|\boldsymbol{\psi}|_{norm} = |\boldsymbol{f}_{int} - \lambda \boldsymbol{f}_{ref}|_{norm} <$ *epsilon*
    - *Calculate* $\boldsymbol{K} = \int_V \boldsymbol{B}^T \boldsymbol{D} \boldsymbol{B} dV$
    - *Calculate* $\boldsymbol{a}^i$ *from* $\boldsymbol{K}(\boldsymbol{a}^i - \boldsymbol{a}^{i-1}) = \lambda \boldsymbol{f}_{ref} - \boldsymbol{f}_{int}$
    - *Calculate* $\boldsymbol{\varepsilon}^i := \boldsymbol{B}\boldsymbol{a}^i$
    - *Determine* $\boldsymbol{\sigma}^i$ *by integration of the constitutive equations* (see next chapter)
    - *Calculate internal forces* $\boldsymbol{f}_{int}$
    - *Calculate load parameter* $\lambda = \dfrac{\boldsymbol{f}_{int}^T \boldsymbol{f}_{ref}}{\boldsymbol{f}_{ref}^T \boldsymbol{f}_{ref}}$
  - *End iteration loop*
  - *Accept quantities*
    $\boldsymbol{a}_{n+1} := \boldsymbol{a}^i\ ; \quad \boldsymbol{\varepsilon}_{n+1} := \boldsymbol{\varepsilon}^i\ ; \quad \boldsymbol{\sigma}_{n+1} := \boldsymbol{\sigma}^i\ ; \quad \boldsymbol{f}_{int}$
- *End load step loop*

With the use of the initial stiffness matrix $\boldsymbol{K}$, each iteration using Bergan's method is in practice as cheap as iterations using the standard initial stiffness method (Bergan's method only requires the additional computation of λ^i given by (17.90), which is inexpensive). However, the standard initial stiffness method is not able to pass the force limit point and to trace the post-peak response. In addition to these features Bergan's method improves the convergence speed very significantly, as can be seen by comparing Figs. 17.10 and 17.24.

Here, we have formulated Bergan's method for proportional loading expressed by $\boldsymbol{f} = \lambda \boldsymbol{f}_{ref}$, but it is straightforward to generalize the method to non-proportional loading. In that case, the loading is considered as piece-wise linear and the above approach only requires modest modifications.

The algorithm for Bergan's method in combination with the initial stiffness matrix $\boldsymbol{K}$ is summarized in Box. 17.3; here the means to enforce the kinematic boundary conditions have not been indicated.

It is also noticed that a closely related method that improves the possibilities for passing displacement limit points was proposed by Krenk (1995) and Krenk and Hededal (1995). Instead of the orthogonality relation (17.89), this method makes use of the orthogonality relation $(\boldsymbol{\psi}^i)^T(\boldsymbol{a}^i - \boldsymbol{a}^{i-1}) = 0$ to determine the new load parameter λ^i.

Finally, we mention the very successful *arc-length* method where a combination of load and displacement control is adopted. This method facilitates the passage of limit points and a detailed discussion is provided, for instance by Crisfield (1991).

18 INTEGRATION OF CONSTITUTIVE EQUATIONS

In the previous chapter, a detailed discussion was devoted to the solution of the nonlinear equilibrium equations that arise in the FE scheme. Returning to the standard iteration scheme given by (17.18), the out-of-balance forces $\boldsymbol{\psi}(\boldsymbol{a}^{i-1})$ given by (17.20) depend on the stress state $\boldsymbol{\sigma}^{i-1}$. Therefore, the determination of the stress state $\boldsymbol{\sigma}^{i-1}$ is fundamental for the iteration scheme.

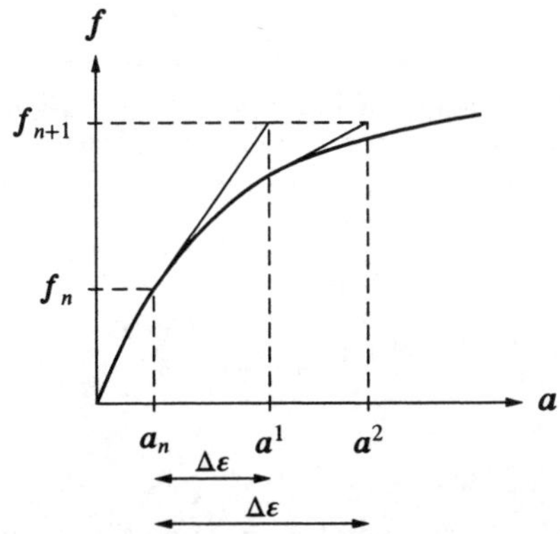

Figure 18.1: Illustration of integration limits.

Previously, we merely assumed, *a priori*, that this integration could be performed in an accurate and reliable manner. The present chapter is devoted to a discussion of how this integration is performed in practice. Recalling that $\boldsymbol{\varepsilon}_n = \boldsymbol{B}\boldsymbol{a}_n$ and $\boldsymbol{\varepsilon}^{i-1} = \boldsymbol{B}\boldsymbol{a}^{i-1}$, it is emphasized that

> *The integration of the constitutive equations is always performed from the last accepted equilibrium state* $\boldsymbol{a}_n$ *to the current state* $\boldsymbol{a}^{i-1}$

as discussed in relation to (17.24). With the notation $\Delta\boldsymbol{\varepsilon} = \boldsymbol{\varepsilon}^{i-1} - \boldsymbol{\varepsilon}_n$, these integration limits are illustrated in Fig. 18.1.

For convenience, the integration limits $\boldsymbol{\varepsilon}_n$ and $\boldsymbol{\varepsilon}^{i-1}$ will now be denoted by 1 and 2 respectively. This implies that state 1 is the known state and we want to determine the stress $\sigma_{ij}^{(2)}$. While everything is known about state 1, the only thing we know about state 2 is the total strains $\varepsilon_{ij}^{(2)}$, which are provided by the solution of the global FE equations.

In the following, we will discuss numerical integration of the constitutive equations arising in elasto-plasticity and viscoplasticity. Emphasis is given to fundamental issues and for further discussion and results, we refer to Bathe (1996), Belytschko *et al.* (2000), Crisfield (1997), Simo and Hughes (1998) and Zienkiewicz and Taylor (1991).

18.1 Elasto-plasticity

First, we will consider elasto-plastic material behavior. It was shown in Section 10.1 that the fundamental equations are given by

$$\begin{aligned} \sigma_{ij} &= D_{ijkl}(\varepsilon_{kl} - \varepsilon_{kl}^{p}) \\ K_\alpha &= K_\alpha(\kappa_\beta) \end{aligned} \tag{18.1}$$

where according to (10.14) the evolution laws for the non-associated case are given as

$$\dot{\varepsilon}_{ij}^{p} = \dot{\lambda}\frac{\partial g}{\partial \sigma_{ij}} \quad \text{and} \quad \dot{\kappa}_\alpha = -\dot{\lambda}\frac{\partial g}{\partial K_\alpha} \tag{18.2}$$

Moreover, for plastic behavior we require that

$$f(\sigma_{ij}, K_\alpha) = 0 \tag{18.3}$$

must hold. Evidently, it is the integration of the evolution equations (18.2) which is the topic of interest and the crucial point is that these evolution equations shall be solved on condition that $f = 0$ holds.

The question arises whether it is possible to integrate the constitutive equations exactly such that closed-form solutions can be obtained. Indeed, for some simple models this is possible and we refer to Krieg and Krieg (1977) for the elastic-perfectly plastic von Mises material. Later Yoder and Whirley (1984) extended the solution to linear and kinematic hardening and Ristinmaa and Tryding (1993) treated the Tresca and Coulomb materials. However, for more advanced models closed-form solutions cannot be obtained.

Two different ways to perform the integration numerically emerge. These two methods are here named the *indirect* and the *direct method* and they differ in how the yield condition during plastic loading is treated in the numerical solution procedure. In the literature, indirect methods are frequently referred to as explicit methods and direct methods as implicit methods, e.g. Zienkiewicz and Taylor (1991). It is felt that these names are somewhat misleading since both in the indirect as well as in the direct method explicit and implicit methods exist. As will be shown, the more natural names adopted here emerge from how the yield condition is treated.

In the indirect method, the consistency relation $\dot{f} = 0$ is used to reformulate the evolution equations whereas in the direct method, the condition $f = 0$ will be directly enforced.

Before we pursue this discussion, we will evaluate whether the strain increment from $\varepsilon_{ij}^{(1)}$ to $\varepsilon_{ij}^{(2)}$ results in development of plastic strains or not.

18.1.1 Loading and unloading criteria

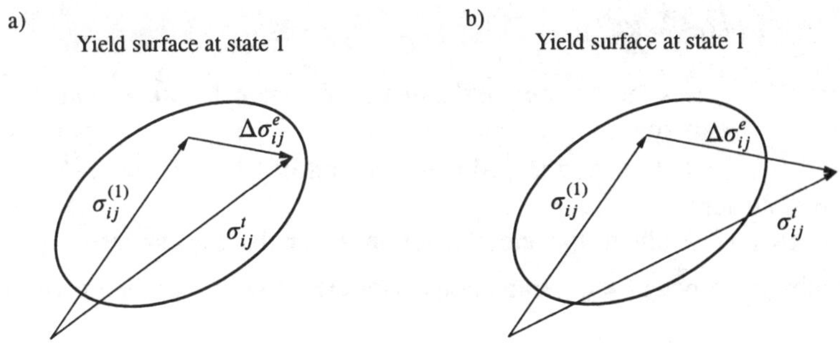

Figure 18.2: Yield surface at state 1 depicted in the stress space; illustration of the stresses $\sigma_{ij}^{(1)}$, the elastic stress increments $\Delta\sigma_{ij}^{e}$ and the trial stresses σ_{ij}^{t}; a) elastic response; b) elasto-plastic response.

From (18.1a) evaluated at state 2 and state 1, we obtain

$$\sigma_{ij}^{(2)} = D_{ijkl}(\varepsilon_{kl}^{(2)} - \varepsilon_{kl}^{p(2)}); \qquad \sigma_{ij}^{(1)} = D_{ijkl}(\varepsilon_{kl}^{(1)} - \varepsilon_{kl}^{p(1)})$$

Subtraction of these two equations gives

$$\sigma_{ij}^{(2)} = \sigma_{ij}^{(1)} + D_{ijkl}\Delta\varepsilon_{kl} - D_{ijkl}\Delta\varepsilon_{kl}^{p} \tag{18.4}$$

where

$$\Delta\varepsilon_{kl} = \varepsilon_{kl}^{(2)} - \varepsilon_{kl}^{(1)}; \qquad \Delta\varepsilon_{kl}^{p} = \varepsilon_{kl}^{p(2)} - \varepsilon_{kl}^{p(1)}$$

In analogy with (10.36), we introduce the following notation

$$\Delta\sigma_{ij}^{e} = D_{ijkl}\Delta\varepsilon_{kl} \tag{18.5}$$

i.e. $\Delta\sigma_{ij}^e$ is the stress change that would occur if the step $\Delta\varepsilon_{kl}$ is purely elastic. Moreover, define the so-called *trial stresses* σ_{ij}^t by

$$\boxed{\begin{array}{l}\textit{Trial stresses}\\ \sigma_{ij}^t = \sigma_{ij}^{(1)} + \Delta\sigma_{ij}^e\end{array}} \tag{18.6}$$

The stresses $\sigma_{ij}^{(1)}$, $\Delta\sigma_{ij}^e$ and σ_{ij}^t are illustrated in Fig. 18.2. With (18.5) and (18.6), (18.4) may be written as

$$\boxed{\sigma_{ij}^{(2)} = \sigma_{ij}^t - D_{ijkl}\Delta\varepsilon_{kl}^p} \tag{18.7}$$

It appears that if the step is purely elastic, then $\sigma_{ij}^{(2)} = \sigma_{ij}^t$.

With $\sigma_{ij}^{(1)}$ and $K_\alpha^{(1)}$ being the stresses and the hardening parameters at state 1, we introduce the notation

$$f^{(1)} = f(\sigma_{ij}^{(1)}, K_\alpha^{(1)}) \tag{18.8}$$

where $f^{(1)}$ is the value of the yield function for state 1. Since state 1 was accepted as a valid solution, we must have $f^{(1)} \leq 0$. If $f^{(1)} < 0$, then the stress state $\sigma_{ij}^{(1)}$ is located inside the yield surface and if $f^{(1)} = 0$ then $\sigma_{ij}^{(1)}$ is located on the yield surface.

Let us now evaluate the yield function f for the trial stresses σ_{ij}^t and the hardening parameters $K_\alpha^{(1)}$. In analogy with (18.8) we adopt the notation

$$f^t = f(\sigma_{ij}^t, K_\alpha^{(1)})$$

It is recalled that during elastic response where $\dot{\lambda} = 0$ holds, no change of the hardening parameters occurs. If σ_{ij}^t denotes a stress state inside or on the yield surface valid for state 1, we will have $f^t \leq 0$. In this case, the trial stresses σ_{ij}^t are, in fact, the correct stresses at state 2, i.e. $\sigma_{ij}^{(2)} = \sigma_{ij}^t$ and we can then proceed directly in the iteration scheme of the global FE equations, i.e.

$$\boxed{f^t \leq 0 \quad \Rightarrow \quad \textit{elastic response} \quad \Rightarrow \quad \sigma_{ij}^{(2)} = \sigma_{ij}^t\ ;\ \ K_\alpha^{(2)} = K_\alpha^{(1)}} \tag{18.9}$$

The situation (18.9) is illustrated in Fig. 18.2a).

Obviously, the situation where $f^t > 0$ means that the trial stress σ_{ij}^t is located outside the yields surface, cf. Fig. 18.2b). Since no stress state can be located outside the yield surface, we conclude that

$$\boxed{f^t > 0 \quad \Rightarrow \quad \textit{elasto-plastic response}} \tag{18.10}$$

Therefore, the trial stresses σ_{ij}^t that were determined assuming that the material responded in an elastic manner cannot be the stresses at state 2. In this case,

Box 18.1 Check for elasto-plastic response

· *Calculate trial stresses*

$$\sigma_{ij}^t = \sigma_{ij}^{(1)} + D_{ijkl}\Delta\varepsilon_{kl}$$

· *If* $f(\sigma^t, K_\alpha^{(1)}) \leq 0$

Elastic response

$$\sigma_{ij}^{(2)} = \sigma_{ij}^t \; ; \;\; K_\alpha^{(2)} = K_\alpha^{(1)}$$

· *Else*

Elasto-plastic response, cf. Box 18.3 or Box 18.4

$$\sigma_{ij}^{(2)} = \sigma_{ij}^t - D_{ijkl}\Delta\varepsilon_{kl}^p \quad where \quad \Delta\varepsilon_{kl}^p = \int_1^2 \frac{\partial g}{\partial \sigma_{kl}} d\lambda$$

we conclude that plastic strains will develop when the step $\Delta\varepsilon_{ij}$ is applied. The check for elasto-plastic loading is summarized in Box 18.1.

It is of interest that in this numerical approach, the elastic stress changes $\Delta\sigma_{ij}^e$ play a central role for the unloading/loading criteria given by (18.9) and (18.10) just like the elastic stress rates $\dot{\sigma}_{ij}^e$ do in the theoretical unloading/loading criteria, cf. (10.38).

When (18.10) is fulfilled, we know that plastic strains develop during the step $\Delta\varepsilon_{ij}$, but we do not know whether the entire step is elasto-plastic or if only part of the step creates plastic strains. This brings us to the concept of *contact stresses* to be discussed next.

18.1.2 Contact stresses

In the step $\Delta\varepsilon_{ij}$, the loading criterion (18.10) is assumed to hold and we therefore have to determine that part of the step which results in development of plastic strains. The situation is illustrated in Fig. 18.3).

During the step $\Delta\varepsilon_{ij}$, we will assume that the total strain varies linearly, i.e.

$$\varepsilon_{ij} = (1-\gamma)\varepsilon_{ij}^{(1)} + \gamma\varepsilon_{ij}^{(2)} \; ; \qquad 0 \leq \gamma \leq 1 \tag{18.11}$$

When $\gamma = 0$ we have $\varepsilon_{ij} = \varepsilon_{ij}^{(1)}$ whereas $\gamma = 1$ gives $\varepsilon_{ij} = \varepsilon_{ij}^{(2)}$. When determining the trial stress σ_{ij}^t, it was assumed that the material behaves in a linear elastic manner. Since the elastic stiffness tensor is constant for linear elasticity, the stress path between $\sigma_{ij}^{(1)}$ and σ_{ij}^t is a straight line in the stress space as illustrated in Fig. 18.3. This line will penetrate the yield surface valid at state 1 at

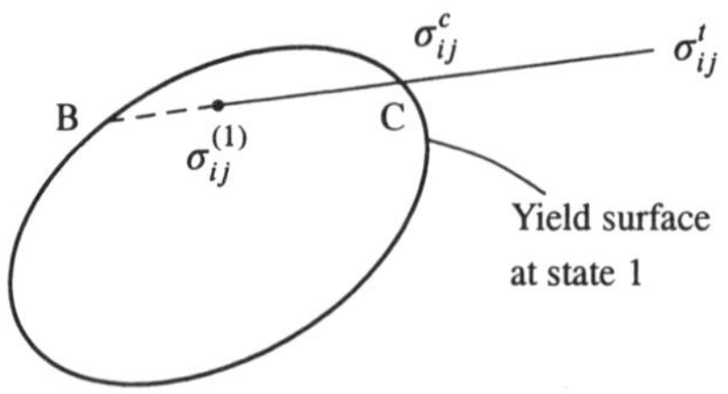

Figure 18.3: Stress space. The straight line between $\sigma_{ij}^{(1)}$ and σ_{ij}^t penetrates the yield surface at state 1 at the contact point with the contact stresses σ_{ij}^c.

the *contact point* C with the *contact stresses* σ_{ij}^c. As the stresses vary linearly between $\sigma_{ij}^{(1)}$ and σ_{ij}^t, the contact stresses are given by

$$\sigma_{ij}^c = (1-\gamma^c)\sigma_{ij}^{(1)} + \gamma^c \sigma_{ij}^t \; ; \qquad 0 \le \gamma^c \le 1 \tag{18.12}$$

where the parameter γ^c is to be determined. As the material behaves elastically from $\sigma_{ij}^{(1)}$ to σ_{ij}^c, no changes occur in the hardening parameters. The contact stresses must therefore fulfill the yield criterion with the hardening parameters $K_\alpha^{(1)}$, i.e.

$$f^c = f(\sigma_{ij}^c, K_\alpha^{(1)}) = 0$$

Insertion of (18.12) provides

$$\boxed{f((1-\gamma^c)\sigma_{ij}^{(1)} + \gamma^c \sigma_{ij}^t,\; K_\alpha^{(1)}) = 0} \tag{18.13}$$

It is recalled that $\sigma_{ij}^{(1)}$, σ_{ij}^t and $K_\alpha^{(1)}$ are known quantities and (18.13) then provides one equation with one unknown, the parameter γ^c. In general, the function f is nonlinear and any of the methods derived in the previous chapter for a set of nonlinear equations may evidently also be applied to the nonlinear equation given by (18.13); we may therefore refer to the standard iteration format given by (17.18). This scheme requires a first initial guess for γ^c, i.e. a starting value for γ^c. Recognizing the interpretation of $f^{(1)}$ given by (18.8) and f^t given by (18.10), we may interpolate linearly between these values to obtain

$$\text{starting value for} \quad \gamma^c = -\frac{f^{(1)}}{f^t - f^{(1)}}$$

where it is recalled that $f^t > 0$ and $f^{(1)} \le 0$. The contact problem to be solved is summarized in Box 18.2

Having determined the γ^c-value, the total strains ε_{ij}^c at the contact point are given by

$$\varepsilon_{ij}^c = (1-\gamma^c)\varepsilon_{ij}^{(1)} + \gamma^c \varepsilon_{ij}^{(2)} = \varepsilon_{ij}^{(1)} + \gamma^c \Delta\varepsilon_{ij}$$

Box 18.2 Calculation of contact stresses and strains.

· *Solve γ^c from*

$$f(\sigma_{ij}^c, K_\alpha^{(1)}) = 0$$

where

$$\sigma_{ij}^c = (1-\gamma^c)\sigma_{ij}^{(1)} + \gamma^c \sigma_{ij}^t$$

· *Calculate strains*

$$\varepsilon_{ij}^c = (1-\gamma^c)\varepsilon_{ij}^{(1)} + \gamma^c \varepsilon_{ij}^{(2)}$$

---

that is, the strain increment $\gamma^c \Delta\varepsilon_{ij}$ will result in elastic response whereas the remaining increment $(1-\gamma^c)\Delta\varepsilon_{ij}$ will give rise to elasto-plastic response. Therefore, (18.7) takes the form

$$\boxed{\sigma_{ij}^{(2)} = \sigma_{ij}^t - D_{ijkl}\Delta\varepsilon_{kl}^p \quad where \quad \Delta\varepsilon_{kl}^p = \int_c^2 \frac{\partial g}{\partial \sigma_{kl}}\, d\lambda} \tag{18.14}$$

Exact solution of the contact stresses

In some cases, an exact analytical solution of (18.13) is possible. This is the case for isotropic hardening of a von Mises and a Drucker-Prager material as well as for Tresca and Coulomb materials, cf. Bicanic (1989). For convenience, we shall only solve (18.13) in case of isotropic von Mises hardening. We therefore consider

$$(\frac{3}{2} s_{ij}^c s_{ij}^c)^{1/2} - \sigma_y(\kappa^{(1)}) = 0 \tag{18.15}$$

Due to (18.12) we have

$$s_{ij}^c = (1-\gamma^c)s_{ij}^{(1)} + \gamma^c s_{ij}^t = s_{ij}^{(1)} + \gamma^c s_{ij}^* \tag{18.16}$$

where

$$s_{ij}^* = s_{ij}^t - s_{ij}^{(1)}\ ; \qquad J_2^* = \frac{1}{2} s_{ij}^* s_{ij}^*\ ; \qquad J_2^{(1)} = \frac{1}{2} s_{ij}^{(1)} s_{ij}^{(1)}$$

Use of (18.16) in (18.15) then leads to

$$(\gamma^c)^2 J_2^* + \gamma^c s_{ij}^{(1)} s_{ij}^* + J_2^{(1)} - \frac{1}{3}\sigma_y^2(\kappa^{(1)}) = 0 \tag{18.17}$$

This quadratic equation provides two solutions for γ^c corresponding to the fact that line (18.12) will intersect the yield surface at two points, the contact point C and the point B, cf. Fig. 18.3. Obviously, the true solution of (18.17) will be the larger of the two solutions provided.

18.1.3 Indirect consideration of the yield condition

As already mentioned in the introduction, in the indirect method the consistency relation $\dot{f} = 0$ is used to reformulate the evolution equations given by (18.2). Enforcing $\dot{f} = 0$ and using (18.1) as well as (18.2), allow us to determine the plastic multiplier as

$$\dot{\lambda} = \frac{1}{A}\frac{\partial f}{\partial \sigma_{ij}} D_{ijkl}\dot{\varepsilon}_{kl}$$

cf. also (10.23); the parameter A is then given by (10.24). Insertion of the expression above into the evolution equations (18.2) gives

$$\begin{aligned} \dot{\varepsilon}_{ij}^{p} &= \frac{1}{A}\frac{\partial g}{\partial \sigma_{ij}}\frac{\partial f}{\partial \sigma_{mn}} D_{mnkl}\dot{\varepsilon}_{kl} \\ \dot{\kappa}_{\alpha} &= -\frac{1}{A}\frac{\partial g}{\partial K_{\alpha}}\frac{\partial f}{\partial \sigma_{mn}} D_{mnkl}\dot{\varepsilon}_{kl} \end{aligned} \tag{18.18}$$

With the assumption that the total strain varies linearly, cf. (18.11), the contact strain is calculated as

$$\varepsilon_{ij}^{c} = (1-\gamma^{c})\varepsilon_{ij}^{(1)} + \gamma^{c}\varepsilon_{ij}^{(2)} \tag{18.19}$$

where γ^c is identified from (18.13); thus we have $\gamma \in [\gamma^c, 1]$ during plastic development. In this interval, it turns out to be of advantage to introduce a new variable $z \in [0, 1]$ such that

$$\gamma = (1-\gamma^{c})z + \gamma^{c}$$

Insertion into (18.11) and taking advantage of (18.19), the strains in the elasto-plastic regime vary according to

$$\varepsilon_{ij} = (1-z)\varepsilon_{ij}^{c} + z\varepsilon_{ij}^{(2)} \qquad z \in [0, 1]$$

Moreover, from the above relation and (18.19) we find that

$$\frac{d\varepsilon_{kl}}{dz} = \varepsilon_{kl}^{(2)} - \varepsilon_{kl}^{c} = (1-\gamma^{c})\Delta\varepsilon_{kl}$$

With these introductory remarks we are now ready to define our problem. Using the result above, (18.18) can be written as

$$\begin{aligned} \frac{d\varepsilon_{ij}^{p}}{dz} &= (1-\gamma^{c})\frac{1}{A}\frac{\partial g}{\partial \sigma_{ij}}\frac{\partial f}{\partial \sigma_{mn}} D_{mnkl}\Delta\varepsilon_{kl} \\ \frac{d\kappa_{\alpha}}{dz} &= -(1-\gamma^{c})\frac{1}{A}\frac{\partial g}{\partial K_{\alpha}}\frac{\partial f}{\partial \sigma_{mn}} D_{mnkl}\Delta\varepsilon_{kl} \end{aligned} \tag{18.20}$$

Box 18.3 Stress calculation, indirect method.

- *Calculate contact stresses and strains*
  cf. Box 18.2
- *Use any ODE-solver to solve*

$$\frac{d\varepsilon_{ij}^p}{dz} = (1-\gamma^c)\frac{1}{A}\frac{\partial g}{\partial \sigma_{ij}}\frac{\partial f}{\partial \sigma_{mn}} D_{mnkl}\Delta\varepsilon_{kl}$$

$$\frac{d\kappa_\alpha}{dz} = -(1-\gamma^c)\frac{1}{A}\frac{\partial g}{\partial K_\alpha}\frac{\partial f}{\partial \sigma_{mn}} D_{mnkl}\Delta\varepsilon_{kl}$$

with the integration limits $z \in [0, 1]$ *and the initial conditions*

$$\varepsilon_{ij}^p(z=0) = \varepsilon_{ij}^{p(1)}$$

$$\kappa_\alpha(z=0) = \kappa_\alpha^{(1)}$$

To obtain the complete set of constitutive equations, two additional relations must be supplemented

$$\begin{aligned} \sigma_{ij} &= D_{ijkl}(\varepsilon_{kl} - \varepsilon_{kl}^p) \\ K_\alpha &= K_\alpha(\kappa_\beta) \end{aligned} \tag{18.21}$$

A glance at (18.20) reveals that we have obtained an initial-value problem with $6 + \alpha$ ordinary differential equations where the initial conditions are given by the values at the contact point. It is then clear that any solver for ordinary differential equations (ODEs) can be used in this integration procedure which is summarized in Box 18.3.

To facilitate a simple and natural notation that corresponds to the notation often found in textbooks dealing with ordinary differential equations, we introduce the following definitions

$$\boldsymbol{y} = \begin{bmatrix} \boldsymbol{\varepsilon}^p \\ \boldsymbol{\kappa} \end{bmatrix} \qquad \text{and} \qquad \boldsymbol{f} = (1-\gamma^c)\begin{bmatrix} \frac{1}{A}\frac{\partial g}{\partial \boldsymbol{\sigma}}(\frac{\partial f}{\partial \boldsymbol{\sigma}})^T \boldsymbol{D}\Delta\boldsymbol{\varepsilon} \\ -\frac{1}{A}\frac{\partial g}{\partial \boldsymbol{K}}(\frac{\partial f}{\partial \boldsymbol{\sigma}})^T \boldsymbol{D}\Delta\boldsymbol{\varepsilon} \end{bmatrix}$$

It then follows that (18.20) can be written as

$$\frac{d\boldsymbol{y}}{dz} = \boldsymbol{f}(\boldsymbol{y}) \quad \text{with the initial conditions} \quad \boldsymbol{y}(0) = \boldsymbol{y}^c = \begin{bmatrix} \boldsymbol{\varepsilon}^{p(1)} \\ \boldsymbol{\kappa}^{(1)} \end{bmatrix} \tag{18.22}$$

With the solution strategy described above, the result is the plastic strains $\varepsilon_{ij}^{p(2)}$ and the internal variables $\kappa_\alpha^{(2)}$ estimated for the strains $\varepsilon_{ij}^{(2)}$. From Hooke's law

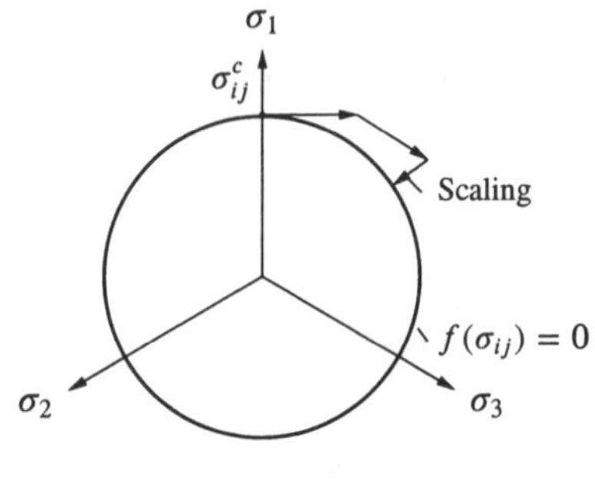

Figure 18.4: Illustration of the drift in Euler's method, ideal von Mises plasticity.

(18.21a), it is then easy to determine the corresponding stresses $\sigma_{ij}^{(2)}$ at state 2 and from (18.21b) we obtain the hardening parameters $K_\alpha^{(2)}$ at state 2.

There exists a large number of methods that can be used to solve (18.22). Since it is beyond the scope of this work to introduce and analyze the implication of all methods, only a very small fraction of the possible methods will be touched upon here.

In *single-step* methods, the entire step length $\Delta\epsilon$ is applied in (18.22) and the most simple approach to solving these differential equations is the *Euler forward* method. A more accurate method, which involves the midpoint, has also been used; in the literature it goes under the names *second-order Runge-Kutta*, *explicit midpoint method* and *leap-frog method*, cf. Dahlquist and Björk (1974) and Zienkiewicz and Taylor (1991).

In *sub-stepping* methods, the entire step $\Delta\epsilon$ is divided into sub-steps and any of the single-step methods can now be applied to each sub-step. A general discussion is provided by Gear (1971) and in a finite element context one may refer to Nayak and Zienkiewicz (1972), Bathe (1996) and Ottosen and Gunneskov (1986). The methods are rather straightforward to work with, but their intrinsic drawback is that eventually the stress state will drift away from the yield surface. Then, various scaling procedures are used by which the stresses are scaled back so that they fulfill the yield condition $f = 0$, cf. Potts and Gens (1985); this drift is illustrated in Fig. 18.4.

Adaptive sub-stepping methods are also available, cf. Dahlquist and Björk (1974). Here an error control is used to adjust the sub-step length. This type of integration technique belongs to the *embedded Runge-Kutta* methods, cf. Dormand and Prince (1980). Interesting results have been obtained for Tresca and Coulomb materials by Sloan (1987) and Sloan and Booker (1992) as well as for the delicate problem of combined plasticity and damage mechanics, cf. Wallin and Ristinmaa (2001).

In practice, however, these advanced methods are computationally expensive and most applications make use of the direct method to fulfill the yield condition; we will now address this issue in detail.

18.1.4 Direct consideration of the yield condition - Return methods

Any numerical integration scheme applied to (18.2) will imply that the plastic strains $\varepsilon_{ij}^{p(2)}$ and the hardening parameters $K_\alpha^{(2)}$ at state 2 can only be determined approximately, i.e. the stresses $\sigma_{ij}^{(2)}$ are only determined approximately. However, irrespective of these approximations we now introduce the requirement that the yield condition must be satisfied. This requirement is indeed fundamental, since the entire plasticity theory relies on the concept that the yield condition is fulfilled. By analogy, we may refer to the solution schemes for the nonlinear global FE equations where, irrespective of the other approximations introduced, we require that the equilibrium equations are fulfilled.

Let us therefore derive an integration scheme that fulfills the yield condition. First let us recall relation (18.14),

$$\sigma_{ij}^{(2)} = \sigma_{ij}^{t} - D_{ijkl}\Delta\varepsilon_{kl}^{p} \quad \text{where} \quad \Delta\varepsilon_{kl}^{p} = \int_c^2 \frac{\partial g}{\partial \sigma_{kl}}\, d\lambda \tag{18.23}$$

which is important for the following discussion.

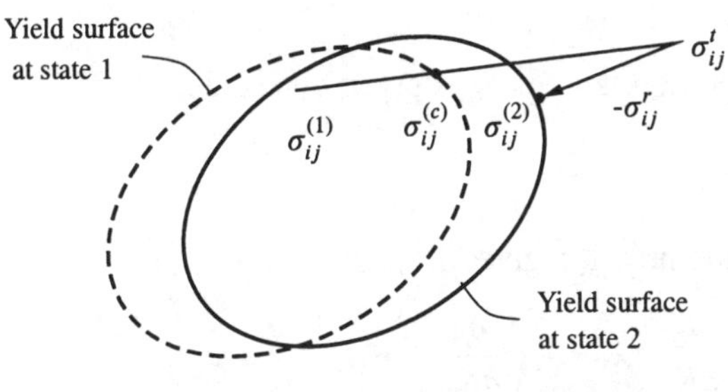

Figure 18.5: Illustration of return method.

We recall since state 1 and $\Delta\varepsilon_{kl}$ are known, $\Delta\sigma_{ij}^{e}$ and the trial stress σ_{ij}^{t} are known, cf. (18.6). Let us define the quantity σ_{ij}^{r}

$$\sigma_{ij}^{r} = D_{ijkl}\Delta\varepsilon_{kl}^{p}$$

which allows (18.23) to be written as

$$\sigma_{ij}^{(2)} = \sigma_{ij}^{t} - \sigma_{ij}^{r} \tag{18.24}$$

The format (18.24) illustrates why the integration schemes now considered are called *return methods*. First, we calculate the trial stress σ_{ij}^{t} and then we return towards the yield surface by the amount $-\sigma_{ij}^r$ and achieve the sought stress state $\sigma_{ij}^{(2)}$ as illustrated in Fig. 18.5; this stress state is required to be located at the yield surface at state 2. The problem, however, is to determine σ_{ij}^r. For this purpose, the flow rule (18.23b) is considered with its specific integration limits

$$\Delta\varepsilon_{ij}^p = \int_{\lambda^{(1)}}^{\lambda^{(1)}+\Delta\lambda} \frac{\partial g}{\partial\sigma_{ij}} d\lambda \tag{18.25}$$

where $\lambda^{(2)} = \lambda^{(1)} + \Delta\lambda$ and where it is recalled that $\lambda^c = \lambda^{(1)}$ since no plastic strains develop along the path from state 1 to the contact point. The problem is that we only know the quantity $\partial g/\partial\sigma_{ij}$ at the contact point, but we may solve (18.25) approximately by writing

$$\Delta\varepsilon_{ij}^p = \int_c^2 \frac{\partial g}{\partial\sigma_{ij}} d\lambda \approx \left(\frac{\partial g}{\partial\sigma_{ij}}\right)^* \Delta\lambda \quad \text{where} \quad g = g(\sigma_{ij}, K_\alpha) \tag{18.26}$$

where the quantity $(\partial g/\partial\sigma_{ij})^*$ denotes $\partial g/\partial\sigma_{ij}$ evaluated at some state along the integration path from the contact point to state 2.

In addition to the plastic strains, we also need to determine the hardening parameters K_α that enter the yield function and the potential function. From (18.1b), it is recalled that the hardening parameters K_α depend on the internal parameters κ_α through

$$K_\alpha = K_\alpha(\kappa_\beta) \tag{18.27}$$

Moreover, the evolution law for κ_α is given by (18.2b), i.e.

$$\dot{\kappa}_\alpha = -\dot{\lambda}\frac{\partial g}{\partial K_\alpha}$$

Integration of this expression gives

$$\Delta\kappa_\alpha = -\int_c^2 \frac{\partial g}{\partial K_\alpha} d\lambda \approx -\left(\frac{\partial g}{\partial K_\alpha}\right)^* \Delta\lambda \tag{18.28}$$

where $\Delta\kappa_\alpha = \kappa_\alpha^{(2)} - \kappa_\alpha^{(1)}$ and the quantity $(\partial g/\partial K_\alpha)^*$ denotes $\partial g/\partial K_\alpha$ evaluated at some state along the integration path from the contact point to state 2. When $\Delta\kappa_\alpha$ has been determined from (18.28), (18.27) is used to determine the value $K_\alpha^{(2)}$ and we obtain

$$K_\alpha^{(2)} = K_\alpha(\kappa_\beta^{(1)} - \Delta\lambda \left(\frac{\partial g}{\partial K_\alpha}\right)^*)$$

Finally, the key-point in the integration method outlined above is that the yield condition is fulfilled, i.e. $\sigma_{ij}^{(2)}$ and $K_\alpha^{(2)}$ must therefore fulfill the yield criterion and

$$f(\sigma_{ij}^{(2)}, K_\alpha^{(2)}) = 0$$

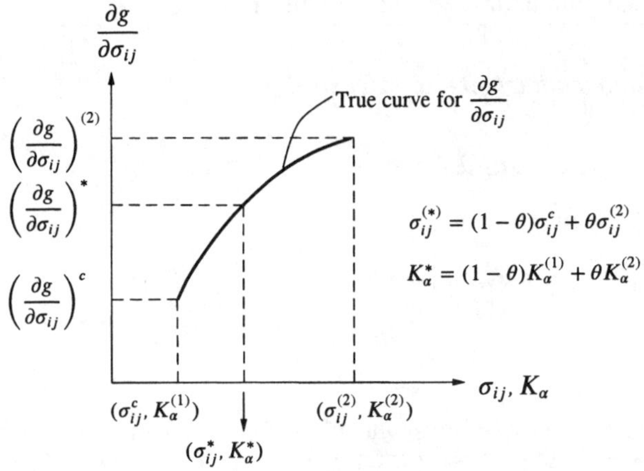

Figure 18.6: Illustration of generalized mid-point rule; note that $K_\alpha^{(1)} = K_\alpha^{(c)}$.

To determine $(\partial g/\partial\sigma_{ij})^*$ and $(\partial g/\partial K_\alpha)^*$ present in (18.26) and (18.28) respectively, two methods are available: the *generalized trapezoidal rule* and the *generalized mid-point rule*, cf. Ortiz and Popov (1985). Both methods employ information relating to the contact point and state 2. For convenience we will only focus on the generalized mid-point rule.

In the generalized mid-point rule, we interpolate σ_{ij} and K_α linearly between the contact point and state 2 and obtain

$$\boxed{\begin{aligned} \sigma_{ij}^* &= (1-\theta)\sigma_{ij}^c + \theta\sigma_{ij}^{(2)} \\ K_\alpha^* &= (1-\theta)K_\alpha^{(1)} + \theta K_\alpha^{(2)} \end{aligned} \qquad \textit{generalized mid-point rule}} \tag{18.29}$$

where $0 \le \theta \le 1$ and we note that $K_\alpha^{(c)} = K_\alpha^{(1)}$. With these values, we then determine $(\partial g/\partial\sigma_{ij})^*$ and $(\partial g/\partial K_\alpha)^*$ according to

$$\boxed{\begin{aligned} \left(\frac{\partial g}{\partial\sigma_{ij}}\right)^* &= \left.\frac{\partial g}{\partial\sigma_{ij}}\right|_{\sigma_{ij}^*, K_\alpha^*} \\ \left(\frac{\partial g}{\partial K_\alpha}\right)^* &= \left.\frac{\partial g}{\partial K_\alpha}\right|_{\sigma_{ij}^*, K_\alpha^*} \end{aligned} \qquad \textit{generalized mid-point rule}} \tag{18.30}$$

i.e. the quantities $\partial g/\partial\sigma_{ij}$ and $\partial g/\partial K_\alpha$ are evaluated for the variables given by σ_{ij}^* and K_α^*. The generalized mid-point rule is illustrated in Fig. 18.6.

We are now in a position to summarize the final equations of interest. Insertion of (18.26) into Hooke's law (18.23) gives

$$\sigma_{ij}^{(2)} = \sigma_{ij}^t - D_{ijkl}\left(\frac{\partial g}{\partial\sigma_{kl}}\right)^* \Delta\lambda \tag{18.31}$$

Box 18.4 Stress calculation generalized mid-point rule.

· *Calculate contact stresses and strains*

cf. Box 18.2

· *Solve σ_{ij}, K_α and $\Delta\lambda$ from*

$$\sigma^*_{ij} = (1-\theta)\sigma^c_{ij} + \theta\sigma^{(2)}_{ij}$$
$$K^*_\alpha = (1-\theta)K^{(1)}_\alpha + \theta K^{(2)}_\alpha$$

$$\sigma_{ij} = \sigma^t_{ij} - \Delta\lambda D_{ijkl}\left(\frac{\partial g}{\partial\sigma_{kl}}\right)^*$$

$$K_\alpha = K_\alpha(\kappa^{(1)}_\beta - \Delta\lambda\left(\frac{\partial g}{\partial K_\alpha}\right)^*)$$

subject to the constraint

$$f(\sigma_{ij}, K_\alpha) = 0$$

From (18.28) follows that

$$\kappa^{(2)}_\alpha = \kappa^{(1)}_\alpha - \left(\frac{\partial g}{\partial K_\alpha}\right)^* \Delta\lambda \tag{18.32}$$

whereas (18.27) gives

$$K^{(2)}_\alpha = K_\alpha(\kappa^{(2)}_\beta) \tag{18.33}$$

Moreover, the state $\sigma^{(2)}_{ij}$, $K^{(2)}_\alpha$ should fulfill the yield condition, i.e.

$$f(\sigma^{(2)}_{ij}, K^{(2)}_\alpha) = 0 \tag{18.34}$$

In the general case where $(\partial g/\partial\sigma_{ij})^*$ and $(\partial g/\partial K_\alpha)^*$ depend on both $\sigma^{(2)}_{ij}$ and $K^{(2)}_\alpha$ (i.e. $\kappa^{(2)}_\alpha$), equations (18.31)-(18.34) will provide a set of nonlinear equations that can be solved numerically to provide solutions for the unknowns $\sigma^{(2)}_{ij}$, $\kappa^{(2)}_\alpha$ and $\Delta\lambda$. (It is readily checked that the number of unknowns equals the number of equations=7+2α); these equations are summarized in Box 18.4.

Significant simplifications arise if we choose $\theta = 0$ and it appears that $(\partial g/\partial\sigma_{ij})^* = (\partial g/\partial\sigma_{ij})^c$ and $(\partial g/\partial K_\alpha)^* = (\partial g/\partial K_\alpha)^c$ only depend on the known quantities σ^c_{ij} and $K^{(1)}_\alpha$. This means that (18.31) and (18.33) may be inserted into (18.34) to provide one nonlinear equation with one unknown, the quantity $\Delta\lambda$. Unfortunately, the scheme $\theta = 0$ turns out to be unstable for large steps, cf. Ortiz and Popov (1985). Since the conditions at the known contact state are extrapolated to obtain the solution, one speaks of an *explicit scheme*: the method is also called a *Euler forward scheme*.

If $\theta = 1$ is chosen then we still have to work with the equations (18.31)-(18.34) simultaneously, but since $(\partial g/\partial\sigma_{ij})^* = (\partial g/\partial\sigma_{ij})^{(2)}$ and $(\partial g/\partial K_\alpha)^* = (\partial g/\partial K_\alpha)^{(2)}$ only depend on $\sigma_{ij}^{(2)}$ and $K_\alpha^{(2)}$ (i.e. $\kappa_\alpha^{(2)}$) and not on σ_{ij}^c we have established a scheme where determination of the contact stresses is unnecessary. This simplification is very convenient and since $\theta = 1$ turns out to provide a scheme that always is stable, cf. Ortiz and Popov (1985), it is very often used in practice.

Apart from the choice $\theta = 0$, we see that our scheme relies on the evaluation of the functions $(\partial g/\partial\sigma_{ij})^*$ and $(\partial g/\partial K_\alpha)^*$ at a state that is not known, *a priori*, and these schemes are therefore called *implicit schemes*. The choice $\theta = 1$ is the *fully implicit scheme* and this method is also called a *backward Euler scheme*.

We have advocated the choice $\theta = 1$, i.e. the fully implicit scheme, even though it generally results in a rather complex solution strategy. Some of the reasons for this choice have already been mentioned, but in addition, this scheme turns out to be very accurate. Let us therefore summarize the properties of this scheme

The fully implicit scheme is stable and accurate and it does not depend on the contact stresses

For the fully implicit scheme, consider now the solution procedure for the equation system defined by (18.31)-(18.34). Switching to a matrix format and using the definitions in Section 4.4 we find that the above set of equations is equivalent to

$$\begin{aligned} \boldsymbol{R}_\sigma &= \boldsymbol{\sigma} + \Delta\lambda \boldsymbol{D}\frac{\partial g}{\partial \boldsymbol{\sigma}} - \boldsymbol{\sigma}^t \\ \boldsymbol{R}_{\boldsymbol{K}} &= \boldsymbol{K} - \hat{\boldsymbol{K}}(\boldsymbol{\kappa}^{(1)} - \Delta\lambda\frac{\partial g}{\partial \boldsymbol{K}}) \\ R_f &= f \end{aligned} \tag{18.35}$$

where the superscript (2) was suppressed and $\hat{\boldsymbol{K}}$ denotes the function whereas $\boldsymbol{K}$ is the variable. Defining a residual vector as $\boldsymbol{V} = [\boldsymbol{R}_\sigma, \boldsymbol{R}_{\boldsymbol{K}}, R_f]^T$ and the vector containing the unknowns as $\boldsymbol{S} = [\boldsymbol{\sigma}, \boldsymbol{K}, \Delta\lambda]^T$, it is evident that $\boldsymbol{V}(\boldsymbol{S}) = \boldsymbol{0}$ defines the solution sought for.

Adopting the Newton-Raphson method, the iterative solution procedure is defined as

$$\boldsymbol{S}^{(i)} = \boldsymbol{S}^{(i-1)} - \left[\frac{\partial \boldsymbol{V}}{\partial \boldsymbol{S}}^{(i-1)}\right]^{-1} \boldsymbol{V}^{(i-1)}$$

which is found by considering a Taylor series expansion of $\boldsymbol{V}$, cf. (17.28). The iteration procedure is stopped when the norm of $\boldsymbol{V}$ is sufficiently small. The iteration matrix can easily be identified as

$$\frac{\partial \boldsymbol{V}^{(i-1)}}{\partial \boldsymbol{S}} = \begin{bmatrix} \boldsymbol{I}_6 + \Delta\lambda \boldsymbol{D}\dfrac{\partial^2 g}{\partial \boldsymbol{\sigma}\partial \boldsymbol{\sigma}} & \Delta\lambda \boldsymbol{D}\dfrac{\partial^2 g}{\partial \boldsymbol{\sigma}\partial \boldsymbol{K}} & \boldsymbol{D}\dfrac{\partial g}{\partial \boldsymbol{\sigma}} \\ \Delta\lambda \boldsymbol{d}\dfrac{\partial^2 g}{\partial \boldsymbol{K}\partial \boldsymbol{\sigma}} & \boldsymbol{I}_\alpha + \Delta\lambda \boldsymbol{d}\dfrac{\partial^2 g}{\partial \boldsymbol{K}\partial \boldsymbol{K}} & \boldsymbol{d}\dfrac{\partial g}{\partial \boldsymbol{K}} \\ \dfrac{\partial f}{\partial \boldsymbol{\sigma}} & \dfrac{\partial f}{\partial \boldsymbol{K}} & 0 \end{bmatrix} \tag{18.36}$$

where $\boldsymbol{I}_6$ denotes the 6×6 unit matrix and $\boldsymbol{I}_\alpha$ the $\alpha \times \alpha$ unit matrix; moreover, $\boldsymbol{d} = \partial \hat{\boldsymbol{K}}/\partial \boldsymbol{\kappa}$. In the derivation above, a new matrix $\partial g^2/\partial \boldsymbol{\sigma}\partial \boldsymbol{\sigma}$ emerges and a glance at the expression reveals that it should be defined in the same way as $\boldsymbol{D}^{-1}$, i.e.

$$\frac{\partial^2 g}{\partial \boldsymbol{\sigma}\partial \boldsymbol{\sigma}} = \begin{bmatrix} \dfrac{\partial^2 g}{\partial\sigma_{11}\partial\sigma_{11}} & \dfrac{\partial^2 g}{\partial\sigma_{11}\partial\sigma_{22}} & \dfrac{\partial^2 g}{\partial\sigma_{11}\partial\sigma_{33}} & \dfrac{\partial^2 g}{\partial\sigma_{11}\partial\sigma_{12}} & \dfrac{\partial^2 g}{\partial\sigma_{11}\partial\sigma_{13}} & \dfrac{\partial^2 g}{\partial\sigma_{11}\partial\sigma_{23}} \\ \dfrac{\partial^2 g}{\partial\sigma_{22}\partial\sigma_{11}} & \dfrac{\partial^2 g}{\partial\sigma_{22}\partial\sigma_{22}} & \dfrac{\partial^2 g}{\partial\sigma_{22}\partial\sigma_{33}} & \dfrac{\partial^2 g}{\partial\sigma_{22}\partial\sigma_{12}} & \dfrac{\partial^2 g}{\partial\sigma_{22}\partial\sigma_{13}} & \dfrac{\partial^2 g}{\partial\sigma_{22}\partial\sigma_{23}} \\ \dfrac{\partial^2 g}{\partial\sigma_{33}\partial\sigma_{11}} & \dfrac{\partial^2 g}{\partial\sigma_{33}\partial\sigma_{22}} & \dfrac{\partial^2 g}{\partial\sigma_{33}\partial\sigma_{33}} & \dfrac{\partial^2 g}{\partial\sigma_{33}\partial\sigma_{12}} & \dfrac{\partial^2 g}{\partial\sigma_{33}\partial\sigma_{13}} & \dfrac{\partial^2 g}{\partial\sigma_{33}\partial\sigma_{23}} \\ \dfrac{\partial^2 g}{\partial\sigma_{12}\partial\sigma_{11}} & \dfrac{\partial^2 g}{\partial\sigma_{12}\partial\sigma_{22}} & \dfrac{\partial^2 g}{\partial\sigma_{12}\partial\sigma_{33}} & \dfrac{\partial^2 g}{\partial\sigma_{12}\partial\sigma_{12}} & \dfrac{\partial^2 g}{\partial\sigma_{12}\partial\sigma_{13}} & \dfrac{\partial^2 g}{\partial\sigma_{12}\partial\sigma_{23}} \\ \dfrac{\partial^2 g}{\partial\sigma_{13}\partial\sigma_{11}} & \dfrac{\partial^2 g}{\partial\sigma_{13}\partial\sigma_{22}} & \dfrac{\partial^2 g}{\partial\sigma_{13}\partial\sigma_{33}} & \dfrac{\partial^2 g}{\partial\sigma_{13}\partial\sigma_{12}} & \dfrac{\partial^2 g}{\partial\sigma_{13}\partial\sigma_{13}} & \dfrac{\partial^2 g}{\partial\sigma_{13}\partial\sigma_{23}} \\ \dfrac{\partial^2 g}{\partial\sigma_{23}\partial\sigma_{11}} & \dfrac{\partial^2 g}{\partial\sigma_{23}\partial\sigma_{22}} & \dfrac{\partial^2 g}{\partial\sigma_{23}\partial\sigma_{33}} & \dfrac{\partial^2 g}{\partial\sigma_{23}\partial\sigma_{12}} & \dfrac{\partial^2 g}{\partial\sigma_{23}\partial\sigma_{13}} & \dfrac{\partial^2 g}{\partial\sigma_{23}\partial\sigma_{23}} \end{bmatrix} \tag{18.37}$$

where advantage is taken of the symmetry of the stress tensor when formulating the potential function g, cf. the discussion relating to (12.94). In that respect, a comparison of the expression above with (4.36) is illuminating. Evidently the matrix defined in (18.37) is symmetric.

The above method applies in general, but for certain models the equation set can be reduced considerably such that only one nonlinear scalar function needs to be considered in the numerical solution procedure. The steps to obtain this nonlinear equation are, usually, first to apply (18.35a) and (18.35b) and derive explicit expressions for $\boldsymbol{\sigma} = \boldsymbol{\sigma}(\Delta\lambda)$ and $\boldsymbol{K} = \boldsymbol{K}(\Delta\lambda)$. These expressions are then used in the yield criterion, (18.35c), which now provides a nonlinear scalar function in $\Delta\lambda$. This approach will be illustrated below for von Mises plasticity.

Isotropic von Mises hardening - Radial return method

Let us first recall the fundamental equations for isotropic hardening of a von Mises material. Referring to Section 12.2, the yield criterion is given by

$$f = (\frac{3}{2}s_{ij}s_{ij})^{1/2} - \sigma_y(\kappa) = 0; \qquad \sigma_y = \sigma_{yo} + K(\kappa)$$

where κ is an internal variable, σ_y is the current yield stress and σ_{yo} the initial yield stress. Moreover, the effective stress is defined by

$$\sigma_{eff} = \left(\frac{3}{2}s_{ij}s_{ij}\right)^{1/2}$$

and the yield criterion may then be written as

$$\sigma_{eff} - \sigma_y(\kappa) = 0 \tag{18.38}$$

We then obtain

$$\frac{\partial f}{\partial \sigma_{ij}} = \frac{3s_{ij}}{2\sigma_{eff}} \tag{18.39}$$

and the flow rule then provides

$$\dot{\varepsilon}^p_{ij} = \dot{\lambda}\frac{3s_{ij}}{2\sigma_{eff}}$$

The effective plastic strain rate is defined by

$$\dot{\varepsilon}^p_{eff} = (\frac{2}{3}\dot{\varepsilon}^p_{ij}\dot{\varepsilon}^p_{ij})^{1/2}$$

and it follows that $\dot{\varepsilon}^p_{eff} = \dot{\lambda}$. Strain hardening is adopted, i.e. the evolution law becomes

$$\dot{\kappa} = \dot{\lambda} \quad \text{where} \quad \dot{\lambda} = \dot{\varepsilon}^p_{eff} \tag{18.40}$$

and the plastic modulus H is given by

$$H = \frac{d\sigma_y(\varepsilon^p_{eff})}{d\varepsilon^p_{eff}}$$

where $\sigma_y(\varepsilon^p_{eff})$ is a known function determined by experimental evidence. Finally, isotropic elasticity is assumed, i.e.

$$D_{ijkl} = 2G\left[\frac{1}{2}(\delta_{ik}\delta_{jl} + \delta_{il}\delta_{jk}) + \frac{\nu}{1-2\nu}\delta_{ij}\delta_{kl}\right] \tag{18.41}$$

We will now apply the fully implicit integration scheme derived in the previous section, i.e. $\theta = 1$. From (18.26) and (18.39), we then obtain

$$\Delta\varepsilon^p_{ij} = \frac{3}{2}\frac{s^{(2)}_{ij}}{\sigma^{(2)}_{eff}}\Delta\lambda \tag{18.42}$$

where $\sigma^{(2)}_{eff}$ denotes σ_{eff} evaluated at state 2. From (18.31) and (18.41) follow that

$$\sigma^{(2)}_{ij} = \sigma^t_{ij} - 3G\frac{s^{(2)}_{ij}}{\sigma^{(2)}_{eff}}\Delta\lambda \tag{18.43}$$

i.e.

$$\boxed{\sigma^{(2)}_{kk} = \sigma^t_{kk}} \tag{18.44}$$

This result is certainly not surprising, since the hydrostatic response of a von Mises material is purely elastic. Expressions (18.43) and (18.44) lead to

$$s^{(2)}_{ij} = \frac{s^t_{ij}}{1 + 3G\dfrac{\Delta\lambda}{\sigma^{(2)}_{eff}}} \tag{18.45}$$

Multiplying each side by itself results in

$$\sigma^{(2)}_{eff} = \frac{\sigma^t_{eff}}{1 + 3G\dfrac{\Delta\lambda}{\sigma^{(2)}_{eff}}}$$

i.e.

$$\sigma^{(2)}_{eff} = \sigma^t_{eff} - 3G\Delta\lambda \tag{18.46}$$

where σ^t_{eff} denotes σ_{eff} evaluated at the trial state. Insertion into (18.45) gives the result

$$\boxed{s^{(2)}_{ij} = (1 - 3G\frac{\Delta\lambda}{\sigma^t_{eff}})s^t_{ij}} \tag{18.47}$$

It appears that once the increment $\Delta\lambda$ has been identified, the expression above determines the deviatoric stress $s^{(2)}_{ij}$. Let us therefore determine the increment $\Delta\lambda$.

We first observe that the evolution law (18.40) provides

$$\boxed{\varepsilon^{p(2)}_{eff} = \varepsilon^{p(1)}_{eff} + \Delta\lambda} \tag{18.48}$$

Box 18.5 Radial return algorithm for isotropic von Mises plasticity.

· *Given:* $\varepsilon_{ij}^{(1)}$, $\varepsilon_{ij}^{p(1)}$, $\varepsilon_{eff}^{p(1)}$, $\sigma_{ij}^{(1)}$, *and* $\Delta\varepsilon_{ij}$

· *Calculate*

$$\sigma_{ij}^{t} = \sigma_{ij}^{(1)} + D_{ijkl}\Delta\varepsilon_{kl}$$

$$\sigma_{eff}^{t} = (\frac{3}{2}s_{ij}^{t}s_{ij}^{t})^{1/2}$$

· *Determine* $\Delta\lambda$ *from* $\sigma_{eff}^{t} - 3G\Delta\lambda - \sigma_y(\varepsilon_{eff}^{p(1)} + \Delta\lambda) = 0$

· *Calculate*

$$\varepsilon_{eff}^{p(2)} = \varepsilon_{eff}^{p(1)} + \Delta\lambda$$

$$\sigma_y^{(2)} = \sigma_y(\varepsilon_{eff}^{p(2)})$$

$$\sigma_{ij}^{(2)} = s_{ij}^{(2)} + \frac{1}{3}\sigma_{kk}^{(2)}\delta_{ij} \quad \textit{where} \quad s_{ij}^{(2)} = \frac{\sigma_y^{(2)}}{\sigma_{eff}^{t}}s_{ij}^{t}\,; \quad \sigma_{kk}^{(2)} = \sigma_{kk}^{t}$$

$$\varepsilon_{ij}^{p(2)} = \varepsilon_{ij}^{p(1)} + \Delta\varepsilon_{ij}^{p} \qquad \textit{where} \quad \Delta\varepsilon_{ij}^{p} = \frac{3}{2}\frac{\Delta\lambda}{\sigma_{eff}^{t}}s_{ij}^{t}$$

The yield criterion (18.38) evaluated at state 2 takes the form

$$\sigma_{eff}^{(2)} - \sigma_y^{(2)} = 0 \qquad \text{where} \qquad \sigma_y^{(2)} = \sigma_y(\varepsilon_{eff}^{p(2)}) \tag{18.49}$$

where $\sigma_y(\varepsilon_{eff}^{p(2)})$ is a known function. Insertion of (18.46) and (18.48) into the yield criterion (18.49) then gives the result

$$\boxed{\sigma_{eff}^{t} - 3G\Delta\lambda - \sigma_y(\varepsilon_{eff}^{p(1)} + \Delta\lambda) = 0} \tag{18.50}$$

In this equation everything but $\Delta\lambda$ is known and a (numerical) solution of (18.50) then provides the quantity $\Delta\lambda$; the strategies discussed in the previous chapter, and in particular the Newton-Raphson approach, can be adopted here. It appears that (18.50) states that the stress state $\sigma_{ij}^{(2)}$ together with the internal variable $\varepsilon_{eff}^{p(2)}$ satisfy the yield criterion and this is exactly the property we originally required by our solution scheme.

Solving (18.46) for $\Delta\lambda$, insertion into (18.47) gives

$$s_{ij}^{(2)} = \frac{\sigma_y^{(2)}}{\sigma_{eff}^{t}}s_{ij}^{t} \tag{18.51}$$

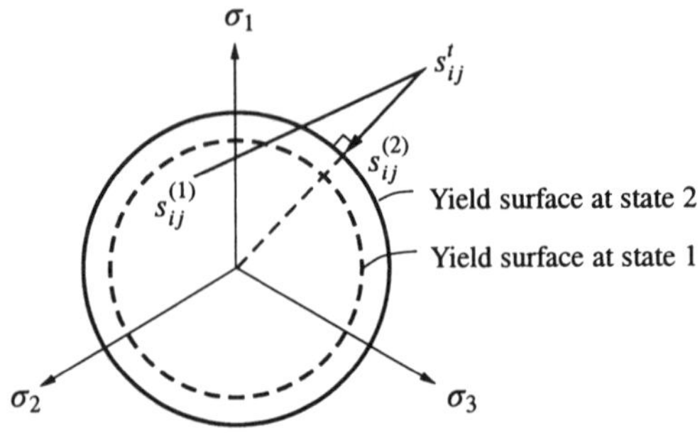

Figure 18.7: Illustration of radial return method for isotropic von Mises hardening.

Finally, inserting this expression into (18.42) we conclude that

$$\boxed{\Delta\varepsilon_{ij}^p = \frac{3}{2}\frac{\Delta\lambda}{\sigma_{eff}^t}s_{ij}^t} \tag{18.52}$$

that is, all quantities of interest can be determined in a very straightforward manner. Moreover, we recall that the yield condition is fulfilled and that we do not have to bother about determination of the contact stresses. Indeed, irrespective of the contact stresses, the results above hold.

The numerical scheme we have discussed was established by Wilkins (1964) and further investigated by Krieg and Krieg (1977) and provides a very accurate solution. The method is called the *radial return method.* This name follows directly from (18.51) showing that in the deviatoric plane, the stress $s_{ij}^{(2)}$ is located along the radial line given by the trial stress s_{ij}^t. This solution procedure is illustrated in Fig. 18.7, which supports the terminology of radial return method: first we estimate the trial stress s_{ij}^t and then we return radially to the stress state $s_{ij}^{(2)}$. The radial return algorithm is summarized in Box 18.5

Linear hardening

If $\sigma_y(\varepsilon_{eff}^p)$ is a simple function, (18.50) even allows an analytical solution for $\Delta\lambda$. This is the case for linear hardening, where

$$\sigma_y = \sigma_{yo} + H\varepsilon_{eff}^p$$

and the plastic modulus H is constant. It follows that $\sigma_y(\varepsilon_{eff}^{p(1)} + \Delta\lambda) = \sigma_{yo} + H\varepsilon_{eff}^{p(1)} + H\Delta\lambda = \sigma_y^{(1)} + H\Delta\lambda$ and insertion into (18.50) gives the solution

$$\Delta\lambda = \frac{\sigma_{eff}^t - \sigma_y^{(1)}}{3G + H} \tag{18.53}$$

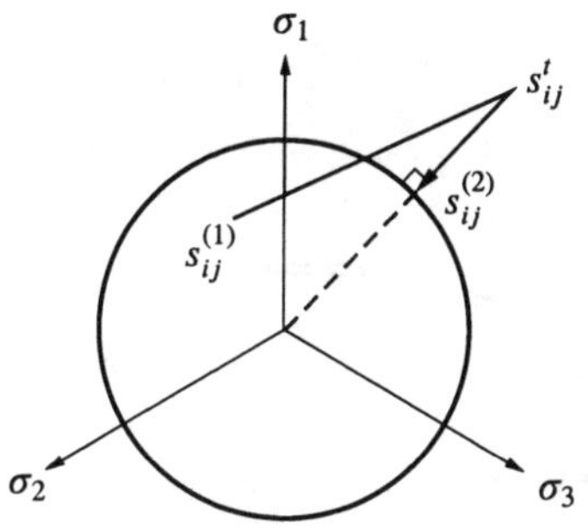

Figure 18.8: Illustration of radial return method for ideal von Mises plasticity.

Ideal von Mises plasticity

For ideal von Mises plasticity, the results simplify even more. Since σ_y is now a constant i.e. $\sigma_y = \sigma_{yo}$ and $H = 0$, (18.53) provides directly the explicit solution

$$\Delta\lambda = \frac{1}{3G}(\sigma_{eff}^t - \sigma_{yo})$$

and (18.51) becomes

$$s_{ij}^{(2)} = \frac{\sigma_{yo}}{\sigma_{eff}^t} s_{ij}^t$$

The plastic strains are still given by (18.52). Figure 18.8 shows the procedure for ideal plasticity.

18.1.5 Algorithmic tangent stiffness

It was mentioned previously in Section 17.4 that Newton-Raphson equilibrium iterations converge quickly. We will now scrutinize this property in more detail and find that in order to fully utilize this potential, the numerical procedure used for integration of the constitutive equations must be accounted for when establishing the tangential stiffness.

The essential property of the Newton-Raphson method is its *quadratic convergence*. A general discussion of this issue is given, for instance by Dahlquist and Björk (1974), and here we will merely mention the essential ingredients. Consider the nonlinear equation $f(x) = 0$ with the true solution $x = \alpha$. The error relating to the estimate x_i is defined by

$$\varepsilon_i = x_i - \alpha$$

It turns out that for a Newton-Raphson procedure, we have

Newton-Raphson procedure
For $i \to \infty \quad |\varepsilon_{i+1}| = C|\varepsilon_i|^2$ *i.e. quadratic convergence*

(18.54)

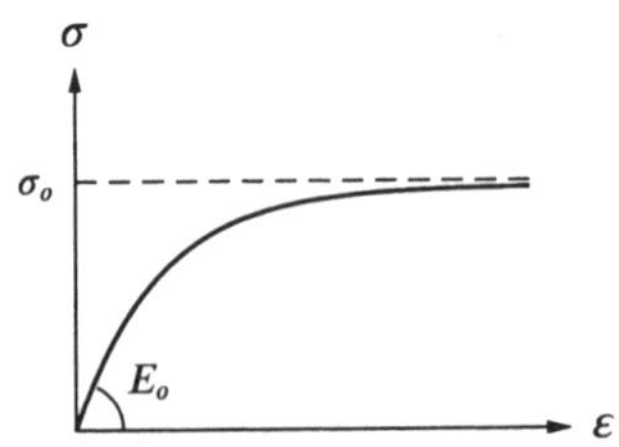

Figure 18.9: Uniaxial stress-strain curve.

where C is a positive constant (the so-called asymptotic error constant). It is evident that if the exponent is larger than two then a faster convergence is achieved whereas an exponent of less than two implies a slower convergence.

In order to maintain the result (18.54), the tangent used in the Newton-Raphson procedure must be the true tangent; if not, the quadratic convergence is lost. However, when a numerical algorithm is used for integration of the constitutive relations, it is the resulting relations that are used. Therefore, the constitutive relations as we see them are the ones that the numerical integration algorithm provides and in order that the Newton-Raphson procedure should maintain its quadratic convergence, the tangent stiffness must be derived from this algorithm. This tangent stiffness is called the *algorithmic tangent stiffness* or the *consistent tangent stiffness* and this important concept was introduced by Nagtegaal (1982) and Runesson and Booker (1982) and its ramifications were further investigated by Runesson and Samuelsson (1985) and Simo and Taylor (1985).

Let us first illustrate the problem by considering the simple uniaxial stress-strain relation given by

$$\dot{\sigma} = E_t \dot{\varepsilon} \quad \text{where} \quad E_t = E_0(1 - \frac{\sigma}{\sigma_0}) \tag{18.55}$$

Evidently, this constitutive relation can be integrated analytically to provide $\sigma = \sigma_0(1 - e^{-\frac{E_0}{\sigma_0}\varepsilon})$ and it is illustrated in Fig.18.9. Let us now integrate (18.55) numerically by means of the fully implicit method. We integrate from the last accepted equilibrium state defined by $\sigma^{(1)}$ and $\varepsilon^{(1)}$ up until the current state $\sigma^{(2)}$ and $\varepsilon^{(2)}$. We then obtain

$$\sigma^{(2)} - \sigma^{(1)} = E_t^{(2)}(\varepsilon^{(2)} - \varepsilon^{(1)})$$

i.e.

$$\sigma^{(2)} = \sigma^{(1)} + E_0(1 - \frac{\sigma^{(2)}}{\sigma_0})(\varepsilon^{(2)} - \varepsilon^{(1)}); \qquad \varepsilon^{(2)} - \varepsilon^{(1)} = \Delta\varepsilon \tag{18.56}$$

With $\sigma^{(1)}$ and $\varepsilon^{(1)}$ being constant quantities, we have $\sigma^{(2)} = g(\varepsilon^{(2)})$ and we will now establish the corresponding tangent stiffness.

This tangent stiffness can be identified if we consider an increment in $d\varepsilon^{(2)}$ and then observe the corresponding increment in $d\sigma^{(2)}$. Now this notation is somewhat awkward and, instead, we use the notation $\dot{\varepsilon}^{(2)}$ and $\dot{\sigma}^{(2)}$; occasionally we may use the terminology of differentiation which then should be understood in the manner described above. Thus from (18.56) we obtain

$$\dot{\sigma}^{(2)} = E_{ats}\dot{\varepsilon}^{(2)} \quad \text{where} \quad E_{ats} = \frac{E_0(1 - \frac{\sigma^{(2)}}{\sigma_0})}{1 + \frac{E_0}{\sigma_0}\Delta\varepsilon}$$

where E_{ats} denotes the algorithmic tangent stiffness. This tangent stiffness is the one that our numerical model experiences and it is the one that should be used in the Newton-Raphson procedure in order to maintain quadratic convergence. The so-called *continuum tangent stiffness* E_t is defined by (18.55) and we obtain

$$\frac{E_{ats}}{E_t} = \frac{1}{1 + \frac{E_0}{\sigma_0}\Delta\varepsilon}$$

Consequently, when $\Delta\varepsilon \to 0$ the algorithmic tangent stiffness approaches the continuum tangent stiffness, but otherwise $E_{ats}/E_t < 1$. For metals, we typically have $E_0/\sigma_0 \approx 500$ and with the strain increment $\Delta\varepsilon$ in the range $10^{-4} - 10^{-3}$ we obtain $E_{ats}/E_t \approx 0.67 - 0.95$. In conclusion, for strain increments encountered in practice, the algorithmic tangent stiffness is significantly smaller than the continuum tangent stiffness and the algorithmic tangent stiffness should be used in the Newton-Raphson procedure to maintain its quadratic convergence.

Derivation of D_{ats}

With the above motivating example let us now tackle the general equations of elasto-plasticity. From the discussion above and from Box 17.2, it is evident that the algorithmic tangent stiffness is defined from

$$\dot{\boldsymbol{\sigma}}^{(2)} = \boldsymbol{D}_{ats}\dot{\boldsymbol{\varepsilon}}^{(2)} \quad \text{where} \quad \boldsymbol{D}_{ats} = \boldsymbol{D}_t^{i-1}$$

The numerical integration scheme most often used in practice is the fully implicit scheme, i.e. $\theta = 1$ in (18.29) and (18.30), which will be adopted here.

First, let us summarize the equations used in the integration of the constitutive relations; the stress-strain relation where the flow rule is used, the evolution law for the internal variable (for simplicity only one internal variable is assumed to exist) and the yield function. In matrix notation we then find from (18.7),

(18.2) and (18.3)

$$\begin{aligned}
&\boldsymbol{\sigma}^{(2)} = \boldsymbol{\sigma}^t - \Delta\lambda \boldsymbol{D}\left(\frac{\partial g}{\partial \boldsymbol{\sigma}}\right)^{(2)} \\
&\Delta\kappa = -\Delta\lambda\left(\frac{\partial g}{\partial K}\right)^{(2)} \qquad K^{(2)} = K(\kappa^{(2)}) \\
&f(\boldsymbol{\sigma}^{(2)}, K^{(2)}) = 0
\end{aligned} \tag{18.57}$$

where $\Delta\lambda = \lambda^{(2)} - \lambda^{(1)}$ and $\Delta\kappa = \kappa^{(2)} - \kappa^{(1)}$. Moreover, the gradient $\partial g/\partial \boldsymbol{\sigma}$ is defined according to (12.94). In the following, for simplicity we will make the restriction that the evolution function $\partial g/\partial K_\alpha$ only depends on the hardening parameter K.

To achieve a neat notation, the superscript (2) will be omitted and recalling that all quantities at state (1) are fixed, we obtain

$$\dot{\boldsymbol{\sigma}}^{(2)} = \dot{\boldsymbol{\sigma}}; \quad (\dot{\Delta\boldsymbol{\varepsilon}}) = \dot{\boldsymbol{\varepsilon}}^{(2)} = \dot{\boldsymbol{\varepsilon}}; \quad (\dot{\Delta\lambda}) = \dot{\lambda}^{(2)} = \dot{\lambda}; \quad (\dot{\Delta\kappa}) = \dot{\kappa}^{(2)} = \dot{\kappa}$$

Since $\boldsymbol{\sigma}^t = \boldsymbol{\sigma}^{(1)} + \boldsymbol{D}\Delta\boldsymbol{\varepsilon}$, a differentiation of Hooke's law (18.57a) therefore results in

$$\boldsymbol{D}^{-1}\dot{\boldsymbol{\sigma}} = \dot{\boldsymbol{\varepsilon}} - \dot{\lambda}\frac{\partial g}{\partial \boldsymbol{\sigma}} - \Delta\lambda\frac{\partial^2 g}{\partial \boldsymbol{\sigma}\partial \boldsymbol{\sigma}}\dot{\boldsymbol{\sigma}} \tag{18.58}$$

Due to the existence of the last term, it can already be realized here that the derivations will not lead to the elasto-plastic continuum stiffness, cf. also (10.22).

From (18.58) we then obtain

$$\dot{\boldsymbol{\sigma}} = \boldsymbol{D}^a\dot{\boldsymbol{\varepsilon}} - \dot{\lambda}\boldsymbol{D}^a\frac{\partial g}{\partial \boldsymbol{\sigma}} \tag{18.59}$$

where

$$\boldsymbol{D}^a = (\boldsymbol{D}^{-1} + \Delta\lambda\frac{\partial^2 g}{\partial \boldsymbol{\sigma}\partial \boldsymbol{\sigma}})^{-1} \tag{18.60}$$

The format obtained in (18.59) is of the same form as (10.22), however, the elastic stiffness is now modified and depends on the step length $\Delta\lambda$. It is also noted that if $\Delta\lambda = 0$ then (18.59) and (10.22) become identical.

The final task is then to obtain an expression for the plastic multiplier, which is achieved by differentiation of the yield function at state (2). From (18.57c) follows

$$(\frac{\partial f}{\partial \boldsymbol{\sigma}})^T\dot{\boldsymbol{\sigma}} + \frac{\partial f}{\partial K}\dot{K} = 0 \tag{18.61}$$

The last term requires a differentiation of K. From (18.57b) we obtain

$$\dot{K} = \frac{dK}{d\kappa}\dot{\kappa} = -\frac{dK}{d\kappa}(\dot{\lambda}\frac{\partial g}{\partial K} + \Delta\lambda\frac{\partial^2 g}{\partial K\partial K}\dot{K})$$

and solving for $\dot{K}$ results in

$$\dot{K} = d^a \dot{\lambda} \quad \text{where} \quad d^a = -(1 + \Delta\lambda \frac{dK}{d\kappa} \frac{\partial^2 g}{\partial K \partial K})^{-1} \frac{dK}{d\kappa} \frac{\partial g}{\partial K}$$

Use of this expression as well as (18.59) in (18.61) provides

$$\dot{\lambda} = \frac{1}{A^a} (\frac{\partial f}{\partial \boldsymbol{\sigma}})^T \boldsymbol{D}^a \dot{\boldsymbol{\varepsilon}} \tag{18.62}$$

where

$$A^a = (\frac{\partial f}{\partial \boldsymbol{\sigma}})^T \boldsymbol{D}^a \frac{\partial g}{\partial \boldsymbol{\sigma}} - \frac{\partial f}{\partial K} d^a \tag{18.63}$$

Finally, insertion of (18.62) into (18.59) provides the result sought for

Algorithmic tangent stiffness

$$\dot{\boldsymbol{\sigma}} = \boldsymbol{D}_{ats} \dot{\boldsymbol{\varepsilon}} \quad \text{where} \quad \boldsymbol{D}_{ats} = \boldsymbol{D}^a - \frac{1}{A^a} \boldsymbol{D}^a \frac{\partial g}{\partial \boldsymbol{\sigma}} (\frac{\partial f}{\partial \boldsymbol{\sigma}})^T \boldsymbol{D}^a \tag{18.64}$$

Where $\boldsymbol{D}_{ats}$ defined above is the algorithmic tangent stiffness. Evidently it has the same format as $\boldsymbol{D}^{ep}$, but with the difference that it depends on $\Delta\lambda$, i.e. the size of the plastic load increment. In essence, the step $\Delta\lambda$ implies the following changes: $\boldsymbol{D} \rightsquigarrow \boldsymbol{D}^a$ and $A \rightsquigarrow A^a$. For a von Mises material, we will now identify these changes.

Isotropic linear hardening von Mises material

Let us now consider a von Mises material with isotropic linear hardening and derive an explicit form of the algorithmic stiffness. Certainly, we could use the general result (18.64), but it turns out to be simpler to adopt the following direct procedure.

The deviatoric stresses are given by (18.47), i.e.

$$s_{ij}^{(2)} = (1 - 3G \frac{\Delta\lambda}{\sigma_{eff}^t}) s_{ij}^t \tag{18.65}$$

where

$$s_{ij}^t = s_{ij}^{(1)} + 2G(e_{ij}^{(2)} - e_{ij}^{(1)}) \tag{18.66}$$

Moreover, the hydrostatic part of the stresses is not influenced by plasticity, i.e.

$$\sigma_{kk}^{(2)} = 3K \varepsilon_{kk}^{(2)} \tag{18.67}$$

Since linear hardening is assumed, we obtain from (18.53)

$$\Delta\lambda = \frac{\sigma_{eff}^t - \sigma_y^{(1)}}{H + 3G}$$

Use of this expression in (18.65) gives

$$s_{ij}^{(2)} = \frac{H + 3G\frac{\sigma_y^{(1)}}{\sigma_{eff}^t}}{H + 3G} s_{ij}^t \tag{18.68}$$

From (18.66) we obtain

$$\dot{s}_{ij}^t = 2G(\dot{\varepsilon}_{ij}^{(2)} - \frac{1}{3}\delta_{ij}\dot{\varepsilon}_{kk}^{(2)}) \tag{18.69}$$

Moreover

$$\sigma_{eff}^t = (\frac{3}{2}s_{kl}^t s_{kl}^t)^{1/2} \quad \text{i.e.} \quad \dot{\sigma}_{eff}^t = \frac{3G s_{kl}^t \dot{\varepsilon}_{kl}^{(2)}}{\sigma_{eff}^t} \tag{18.70}$$

Differentiation of (18.68) with respect to time and use of (18.69) and (18.70b) give

$$\dot{s}_{ij}^{(2)} = \frac{H + 3G\beta}{H + 3G} 2G(\dot{\varepsilon}_{ij}^{(2)} - \frac{1}{3}\delta_{ij}\dot{\varepsilon}_{kk}^{(2)}) - \frac{9G^2\beta}{H + 3G}\frac{s_{ij}^t s_{kl}^t}{(\sigma_{eff}^t)^2}\dot{\varepsilon}_{kl}^{(2)} \tag{18.71}$$

where the factor β is defined by

$$\beta = \frac{\sigma_y^{(1)}}{\sigma_{eff}^t}$$

It follows that $0 < \beta \leq 1$ and for infinitely small steps $\beta \to 1$. From (18.71) and (18.67) we finally achieve the result aimed at

Linear isotropic hardening von Mises material

$$\dot{\sigma}_{ij}^{(2)} = D_{ijkl}^{ats}\dot{\varepsilon}_{kl}^{(2)}$$

where

$$D_{ijkl}^{ats} = \frac{H + 3G\beta}{A^a} 2G[\frac{1}{2}(\delta_{ik}\delta_{jl} + \delta_{il}\delta_{jk}) - \frac{1}{3}\delta_{ij}\delta_{kl}] + K\delta_{ij}\delta_{kl} - \frac{9G^2\beta}{A^a}\frac{s_{ij}^t s_{kl}^t}{(\sigma_{eff}^t)^2}$$

where $A^a = A = H + 3G$

A comparison with (12.16) and (4.89) shows that the algorithmic stiffness approaches the continuum stiffness when $\beta \to 1$, as expected. Moreover, it appears that the factor β decreases the shear modulus from G to βG and it also decreases the plastic part of the stiffness tensor. It is also observed that in this specific case $A^a = A = H + 3G$.

18.2 Viscoplasticity

Having considered the integration of the constitutive relations for elasto-plasticity, the integration of the viscoplastic constitutive relations follows almost directly. As discussed in Chapter 15, creep modeling is a special case of viscoplasticity in which the elastic region disappears. Therefore, although not pursued here, the numerical treatment of creep is similar to that adopted for viscoplasticity.

Here we will focus on a method that will fit into the algorithms considered in Chapter 17. Other methods for solution of rate-dependent problems exist; these methods usually lead to both an additional force vector, due to the linearization of the constitutive relations, and a modified tangent stiffness, i.e. rate tangent stiffness; we refer to Zienkiewicz and Taylor (1991) and references therein and also to Peirce *et al.* (1984) for further information.

In Section 15.4 two different viscoplastic models were considered, the Perzyna model, (15.36), and the Duvaut-Lions model, (15.58). Except for the elasto-viscoplastic loading condition, the numerical treatment of these two models differs and will therefore be treated separately.

We have the constitutive relations

$$\begin{aligned} \sigma_{ij} &= D_{ijkl}(\varepsilon_{kl} - \varepsilon_{kl}^{vp}) \\ K_\alpha &= K_\alpha(\kappa_\beta) \end{aligned} \tag{18.72}$$

Moreover, both models take advantage of a static yield function $f(\sigma_{ij}, K_\alpha)$. If $f \leq 0$ elastic response occurs whereas viscoplastic response is obtained if $f > 0$. Assuming that the loading is elastic allows for the definition of a trial stress and in correspondence with (18.6) we have

$$\sigma_{ij}^t = \sigma_{ij}^{(1)} + \Delta\sigma_{ij}^e \quad \text{where} \quad \Delta\sigma_{ij}^e = D_{ijkl}\Delta\varepsilon_{kl} \tag{18.73}$$

where $\Delta\varepsilon_{kl} = \varepsilon_{kl}^{(2)} - \varepsilon_{kl}^{(1)}$ and where the superscript (1) indicates the value of the quantity at the last known state in equilibrium and (2) the new sought state. Note that the total strains are known at state (2). By analogy with (18.7) we also have

$$\sigma_{ij}^{(2)} = \sigma_{ij}^t - D_{ijkl}\Delta\varepsilon_{kl}^{vp} \tag{18.74}$$

With (18.73) the loading condition used in the numerical treatment can be written as

$$f^t = f(\sigma_{ij}^t, K_\alpha^{(1)}) = \begin{cases} \leq 0 & \text{elastic response} \\ > 0 & \text{viscoplastic response} \end{cases}$$

If $f < 0$ the sought stresses are given by (18.73), i.e. $\sigma_{ij}^{(2)} = \sigma_{ij}^t$ and the new hardening parameters are given as $K_\alpha^{(2)} = K_\alpha^{(1)}$. It appears that the loading condition used in the numerical treatment becomes the same as for elasto-plasticity.

18.2.1 Perzyna viscoplasticity

First let us summarize the constitutive relations for Perzyna viscoplasticity. In addition to (18.72) we have from (15.36) and (15.37) the evolution laws

$$\begin{aligned} \dot{\varepsilon}_{ij}^{vp} &= \frac{\Phi(f)}{\eta}\frac{\partial g}{\partial \sigma_{ij}} \\ \dot{\kappa}_{\alpha} &= -\frac{\Phi(f)}{\eta}\frac{\partial g}{\partial K_{\alpha}} \end{aligned} \tag{18.75}$$

As indicated above it will be assumed that viscoplastic loading occurs.

In the numerical treatment, the fully implicit method will be used, i.e. the backward Euler rule. The corresponding numerical format of (18.75) is then given by

$$\begin{aligned} \Delta\varepsilon_{ij}^{vp} &= \int_{t^{(1)}}^{t^{(2)}} \frac{\Phi(f)}{\eta}\frac{\partial g}{\partial \sigma_{ij}}dt = \Delta\lambda\left(\frac{\partial g}{\partial \sigma_{ij}}\right)^{(2)} \\ \Delta\kappa_{\alpha} &= -\int_{t^{(1)}}^{t^{(2)}} \frac{\Phi(f)}{\eta}\frac{\partial g}{\partial K_{\alpha}}dt = -\Delta\lambda\left(\frac{\partial g}{\partial K_{\alpha}}\right)^{(2)} \end{aligned}$$

where $t^{(1)}$ and $t^{(2)}$ denote the time at the last known state in equilibrium and the current time respectively. Moreover, the definition

$$\Delta\lambda = \frac{\Phi(f^{(2)})}{\eta}\Delta t \tag{18.76}$$

was introduced. An interesting observation can now be made. According to (15.41) there exists an inverse function φ to Φ and (18.76) can then be written as

$$f^d = f^{(2)} - \varphi(\eta\frac{\Delta\lambda}{\Delta t}) \; ; \qquad f^d = 0$$

which is the numerical counterpart of the dynamic yield criterion, cf. (15.43).

The set of equations used in the numerical solution procedure can therefore be summarized as

$$\begin{aligned} \sigma_{ij}^{(2)} &= \sigma_{ij}^{t} - D_{ijkl}\Delta\varepsilon_{kl}^{vp} \\ K_{\alpha}^{(2)} &= K_{\alpha}(\kappa_{\beta}^{(2)}) \end{aligned} \tag{18.77}$$

with the evolution laws

$$\begin{aligned} \Delta\varepsilon_{ij}^{vp} &= \Delta\lambda\left(\frac{\partial g}{\partial \sigma_{ij}}\right)^{(2)} \\ \Delta\kappa_{\alpha} &= -\Delta\lambda\left(\frac{\partial g}{\partial K_{\alpha}}\right)^{(2)} \end{aligned} \tag{18.78}$$

and the dynamic yield criterion

$$f^{(2)} - \varphi(\eta \frac{\Delta\lambda}{\Delta t}) = 0 \tag{18.79}$$

For certain models it is possible to reduce the above set of equations dramatically, but for anisotropy, for instance, such a reduction is not possible. Let us therefore consider the general solution procedure for the equation system above. Switching to a matrix format using the definitions in Section 4.4 we find that the above set of equations is equivalent to

$$\begin{aligned} \boldsymbol{R}_\sigma &= \boldsymbol{\sigma} + \Delta\lambda \boldsymbol{D}\frac{\partial g}{\partial \boldsymbol{\sigma}} - \boldsymbol{\sigma}^t \\ \boldsymbol{R}_K &= \boldsymbol{K} - \hat{\boldsymbol{K}}(\boldsymbol{\kappa}^{(1)} - \Delta\lambda \frac{\partial g}{\partial \boldsymbol{K}}) \\ R_f &= f - \varphi(\eta \frac{\Delta\lambda}{\Delta t}) \end{aligned} \tag{18.80}$$

where the superscript (2) was suppressed and $\hat{\boldsymbol{K}}$ denotes the function whereas $\boldsymbol{K}$ is the variable. Defining a residual vector as $\boldsymbol{V} = [\boldsymbol{R}_\sigma, \boldsymbol{R}_K, R_f]^T$ and as this vector depends on the unknowns $\boldsymbol{S} = [\boldsymbol{\sigma}, \boldsymbol{K}, \Delta\lambda]^T$, it is evident that $\boldsymbol{V}(\boldsymbol{S}) = \boldsymbol{0}$ defines the solution sought for.

Adopting the Newton-Raphson method, the iterative solution procedure is defined as

$$\boldsymbol{S}^{(i)} = \boldsymbol{S}^{(i-1)} - \left[\frac{\partial \boldsymbol{V}^{(i-1)}}{\partial \boldsymbol{S}}\right]^{-1} \boldsymbol{V}^{(i-1)} \tag{18.81}$$

which is found by considering a Taylor series expansion of $\boldsymbol{V}$, cf. (17.28). The iteration procedure is stopped when the norm of $\boldsymbol{V}$ is sufficiently small. The iteration matrix can easily be identified as

$$\frac{\partial \boldsymbol{V}^{(i-1)}}{\partial \boldsymbol{S}} = \begin{bmatrix} \boldsymbol{I}_6 + \Delta\lambda \boldsymbol{D}\frac{\partial^2 g}{\partial \boldsymbol{\sigma}\partial \boldsymbol{\sigma}} & \Delta\lambda \boldsymbol{D}\frac{\partial^2 g}{\partial \boldsymbol{\sigma}\partial \boldsymbol{K}} & \boldsymbol{D}\frac{\partial g}{\partial \boldsymbol{\sigma}} \\ \Delta\lambda \boldsymbol{d}\frac{\partial^2 g}{\partial \boldsymbol{K}\partial \boldsymbol{\sigma}} & \boldsymbol{I}_\alpha + \Delta\lambda \boldsymbol{d}\frac{\partial^2 g}{\partial \boldsymbol{K}\partial \boldsymbol{K}} & \boldsymbol{d}\frac{\partial g}{\partial \boldsymbol{K}} \\ \frac{\partial f}{\partial \boldsymbol{\sigma}} & \frac{\partial f}{\partial \boldsymbol{K}} & -\varphi' \frac{\eta}{\Delta t} \end{bmatrix} \tag{18.82}$$

where $\boldsymbol{I}_6$ denotes the 6 × 6 unit matrix, $\boldsymbol{I}_\alpha$ the $\alpha \times \alpha$ unit matrix, Moreover, $\boldsymbol{d} = \partial \hat{\boldsymbol{K}} / \partial \boldsymbol{\kappa}$ and $\varphi' = d\varphi(x)/dx$. It is interesting to compare (18.82) with its elasto-plastic counterpart (18.36) which reveals that the only difference is the term $\varphi'\eta/\Delta t$ and for $\eta = 0$, the two systems coincide, as expected.

As mentioned previously, it turns out that for certain models the equation system can be reduced to a single nonlinear equation with one unknown. In this approach, expression (18.80a) and (18.80b) can be reduced to obtain $\boldsymbol{\sigma} = \boldsymbol{\sigma}(\Delta\lambda)$ and $\boldsymbol{K} = \boldsymbol{K}(\Delta\lambda)$. These two relations are then inserted into the dynamic yield criterion (18.80c) which then, usually, becomes a nonlinear scalar equation in $\Delta\lambda$. In the following, we will consider a situation where this reduction process is possible.

Isotropic von Mises hardening

The solution procedure (18.81) holds in general, but as mentioned above there are cases where the problem can be reduced to the solution of one nonlinear equation in the unknown $\Delta\lambda$.

Let us for example consider an isotropic hardening von Mises material where the static yield function is defined as

$$f = \left(\frac{3}{2}s_{ij}s_{ij}\right)^{1/2} - \sigma_y(\varepsilon_{eff}^{vp}) \; ; \qquad \sigma_y = \sigma_{yo} + K(\varepsilon_{eff}^{vp}) \tag{18.83}$$

where the internal variable κ has been chosen as the effective viscoplastic strain ε_{eff}^{vp}, i.e. we have

$$\dot{\kappa} = \dot{\varepsilon}_{eff}^{vp} = \left(\frac{2}{3}\dot{\varepsilon}_{ij}^{vp}\dot{\varepsilon}_{ij}^{vp}\right)^{1/2} = \frac{\Phi(f)}{\eta}$$

This result was achieved by using the flow rule (18.75a) which here becomes

$$\dot{\varepsilon}_{ij}^{vp} = \frac{\Phi(f)}{\eta}\frac{3s_{ij}}{2\sigma_{eff}} \; ; \qquad \sigma_{eff} = \left(\frac{3}{2}s_{ij}s_{ij}\right)^{1/2}$$

The task is now to make use of the set of equations defined by (18.77)-(18.79) and we will make an effort to reduce the set of equations as much as possible. For the elasto-plastic problem we eventually ended up with only one nonlinear equation, that is the yield criterion with $\Delta\lambda$ as the unknown. As will be shown for the viscoplastic model, the same reduction process leads to a single nonlinear equation, namely the dynamic yield criterion which has to be solved for the unknown quantity $\Delta\lambda$. By and large, the derivation follows the elasto-plastic situation. By using

$$D_{ijkl} = 2G[\frac{1}{2}(\delta_{ik}\delta_{jl} + \delta_{il}\delta_{jk}) + \frac{\nu}{1-2\nu}\delta_{ij}\delta_{kl}]$$

insertion of (18.78a) into (18.77a) provides

$$\sigma_{ij}^{(2)} = \sigma_{ij}^{t} - 3G\frac{s_{ij}^{(2)}}{\sigma_{eff}^{(2)}}\Delta\lambda \tag{18.84}$$

for isotropic elasticity. From (18.84) it follows that

$$s_{ij}^{(2)} = \frac{s_{ij}^{t}}{1 + \dfrac{3G\Delta\lambda}{\sigma_{eff}^{(2)}}}; \qquad \sigma_{kk}^{(2)} = \sigma_{kk}^{t} \tag{18.85}$$

Each side of (18.85a) is now multiplied by itself and this results in

$$\sigma_{eff}^{(2)} = \sigma_{eff}^{t} - 3G\Delta\lambda \tag{18.86}$$

i.e. the same set of equations is obtained as found for elasto-plasticity, cf. (18.44)-(18.46). Thus to calculate the stresses we have only to determine the value of $\Delta\lambda$ which will be determined by the dynamic yield function.

With (18.86), the static yield function (18.83) can be expressed as

$$f = \sigma_{eff}^{t} - 3G\Delta\lambda - \sigma_y(\varepsilon_{eff}^{vp(1)} + \Delta\lambda)$$

and insertion into (18.79) provides

$$\sigma_{eff}^{t} - 3G\Delta\lambda - \sigma_y(\varepsilon_{eff}^{vp(1)} + \Delta\lambda) - \varphi(\eta\frac{\Delta\lambda}{\Delta t}) = 0 \tag{18.87}$$

which is a scalar equation in the unknown $\Delta\lambda$. Typically, the Newton-Raphson method is used in the numerical solution of this nonlinear equation. It is interesting to note that if $\eta = 0$ is considered, i.e. rate-independent plasticity, (18.87) reduces to (18.50).

As a first example assume that the function φ is linear, i.e. $\varphi(x) = x$ which corresponds, essentially, to Hohenemser-Prager viscoplasticity (15.34) and with hardening also being considered. From (18.87) we then find

$$\sigma_{eff}^{t} - 3G\Delta\lambda - \sigma_y(\varepsilon_{eff}^{vp(1)} + \Delta\lambda) - \eta\frac{\Delta\lambda}{\Delta t} = 0$$

For linear hardening where $\sigma_y = \sigma_{yo} + H\varepsilon_{eff}^{vp}$, a closed-form solution can be obtained and it is given by

$$\Delta\lambda = \frac{\sigma_{eff}^{t} - \sigma_y^{(1)}}{3G + H + \frac{\eta}{\Delta t}}$$

Thus if $\eta = 0$ the above relation reduces to (18.53).

Let us finally consider the Cowper-Symonds model where use of (15.49) in (18.87) gives

$$\sigma_{eff}^{t} - 3G\Delta\lambda - \sigma_y(\varepsilon_{eff}^{vp(1)} + \Delta\lambda) - \left(\frac{1}{D}\frac{\Delta\lambda}{\Delta t}\right)^{1/p} \sigma_{yo} = 0$$

Note that $1/D$ has here assumed the role of η.

Having determined $\Delta\lambda$ the stresses are given by (18.85) and the plastic strains are found from (18.78a) as

$$\varepsilon_{ij}^{vp(2)} = \varepsilon_{ij}^{vp(1)} + \Delta\lambda \frac{3s_{ij}^t}{2\sigma_{eff}^t}$$

Algorithmic tangent stiffness for Perzyna viscoplasticity

In order to keep the algebra reasonably simple, we assume that only one hardening parameter K exists and that the quantity $\partial g/\partial K$ does not depend on the stresses $\boldsymbol{\sigma}$.

It is then interesting to note that the only difference between (18.80) and (18.35) relates to the static and dynamic yield functions. Thus for deriving the algorithmic tangent stiffness for Perzyna viscoplasticity we can take advantage of the derivations for the algorithmic tangent stiffness for elasto-plasticity.

Differentiation of the dynamic yield function (18.80c) results in

$$(\frac{\partial f}{\partial \boldsymbol{\sigma}})^T \dot{\boldsymbol{\sigma}} + \frac{\partial f}{\partial K}\dot{K} - \varphi' \frac{\eta}{\Delta t} \Delta\dot{\lambda} = 0$$

and apart from the last term, this expression corresponds to the consistency relation (18.61) in plasticity. Therefore, similar to (18.62) we obtain

$$\Delta\dot{\lambda} = \frac{1}{A^{vp}} (\frac{\partial f}{\partial \boldsymbol{\sigma}})^T \boldsymbol{D}^a \dot{\boldsymbol{\varepsilon}}$$

where

$$A^{vp} = A^a + \varphi' \frac{\eta}{\Delta t}$$

Compared with elasto-plasticity, this is the only change and we then obtain the algorithmic tangent stiffness $\boldsymbol{D}_{ats}$ for viscoplasticity directly from (18.64) according to

Algorithmic tangent stiffness
$\dot{\boldsymbol{\sigma}} = \boldsymbol{D}_{ats} \dot{\boldsymbol{\varepsilon}}$ where $\boldsymbol{D}_{ats} = \boldsymbol{D}^a - \frac{1}{A^{vp}} \boldsymbol{D}^a \frac{\partial g}{\partial \boldsymbol{\sigma}} (\frac{\partial f}{\partial \boldsymbol{\sigma}})^T \boldsymbol{D}^a$

It emphasized that the only change compared with the elasto-plastic format (18.64) is that A^a is replaced by A^{vp}.

18.2.2 Duvaut-Lions viscoplasticity

As the starting point we will summarize the constitutive relations for Duvaut-Lions viscoplasticity. From (15.58) we have the evolution laws

$$\begin{aligned} \dot{\varepsilon}_{ij}^{vp} &= \Lambda C_{ijkl}(\sigma_{kl} - \bar{\sigma}_{kl}) \\ \dot{\kappa}_\alpha &= -\Lambda c_{\alpha\beta}(K_\beta - \bar{K}_\beta) \end{aligned} \tag{18.88}$$

where $(\bar{\sigma}_{ij}, \bar{K}_\alpha)$ is the closest-point-projection of (σ_{ij}, K_α) on the static yield surface $f(\bar{\sigma}_{ij}, \bar{K}_\alpha) = 0$. Moreover, $c_{\alpha\beta}$ is the inverse matrix to $d_{\alpha\beta} = \partial K_\alpha / \partial \kappa_\beta$ and $c_{\alpha\beta}$, as well as C_{ijkl}, are considered as constant quantities. The solution to the closest-point-projection is given by (15.57) and summarized here

$$\begin{aligned} -C_{ijkl}(\sigma_{kl} - \bar{\sigma}_{kl}) + \mu \frac{\partial f}{\partial \bar{\sigma}_{ij}} &= 0 \\ -c_{\alpha\beta}(K_\beta - \bar{K}_\beta) + \mu \frac{\partial f}{\partial \bar{K}_\alpha} &= 0 \\ f(\bar{\sigma}_{ij}, \bar{K}_\alpha) &= 0 \end{aligned} \tag{18.89}$$

where μ is a Lagrangian multiplier. Before proceeding it turns out that instead of (18.72) it is of advantage to use (15.54), i.e.

$$\dot{K}_\alpha = d_{\alpha\beta} \dot{\kappa}_\beta \tag{18.90}$$

as a more structured set of equations then will be obtained; here $d_{\alpha\beta}$ is the inverse of $c_{\alpha\beta}$.

Applying the backward Euler rule to (18.88) results in

$$\begin{aligned} \Delta \varepsilon_{ij}^{vp} &= \Lambda^{(2)} \Delta t C_{ijkl}(\sigma_{kl}^{(2)} - \bar{\sigma}_{kl}) \\ \Delta \kappa_\alpha &= -\Lambda^{(2)} \Delta t c_{\alpha\beta}(K_\beta^{(2)} - \bar{K}_\beta) \end{aligned}$$

Taking advantage of the above relations in (18.74) and (18.90), where the backward Euler rule is applied, provides after some manipulations

$$\begin{aligned} \sigma_{ij}^{(2)} &= \frac{1}{1 + \Lambda^{(2)} \Delta t}(\sigma_{ij}^t + \Lambda^{(2)} \Delta t \bar{\sigma}_{ij}) \\ K_\alpha^{(2)} &= \frac{1}{1 + \Lambda^{(2)} \Delta t}(K_\alpha^{(1)} + \Lambda^{(2)} \Delta t \bar{K}_\alpha) \end{aligned} \tag{18.91}$$

Insertion into (18.89) results in

$$\begin{aligned} \bar{\sigma}_{ij} &= \sigma_{ij}^t - \Delta\lambda D_{ijkl} \frac{\partial f}{\partial \bar{\sigma}_{kl}} \\ \bar{K}_\alpha &= K_\alpha^{(1)} - \Delta\lambda d_{\alpha\beta} \frac{\partial f}{\partial \bar{K}_\beta} \\ f(\bar{\sigma}_{ij}, \bar{K}_\alpha) &= 0 \end{aligned} \tag{18.92}$$

where

$$\Delta\lambda = (1 + \Lambda^{(2)}\Delta t)\mu$$

Following Simo *et al.* (1988) and considering rate-independent plasticity, an interesting observation can now be made. Let $\bar{\sigma}_{ij}$ and $\bar{K}_\alpha$ be considered as the final stress state and final hardening parameters respectively, it then appears that (18.92) expresses the fully implicit scheme. The solution of the above system provides $(\bar{\sigma}_{ij}, \bar{K}_\alpha, \Delta\lambda)$ and, finally, from (18.91) the current state is obtained. For a more detailed discussion of the numerical implication as well as for the derivation of the algorithmic tangential stiffness matrix we refer to Simo *et al.* (1988), Simo and Hughes (1998) and Runesson *et al.* (1999).

Isotropic von Mises hardening

Considering associated plasticity and linear hardening, the static yield function is given by

$$f = \left(\frac{3}{2}s_{ij}s_{ij}\right)^{1/2} - \sigma_y(\varepsilon^{vp}_{eff})\ ; \qquad \sigma_y = \sigma_{yo} + H\varepsilon^{vp}_{eff} \tag{18.93}$$

where the internal variable is taken as $\kappa = \varepsilon^{vp}_{eff}$ the effective viscoplastic strain.

As discussed above, when $(\bar{\sigma}_{ij}, \bar{K}_\alpha)$ is replaced with $(\sigma^{(2)}_{ij}, K^{(2)}_\alpha)$, (18.92) is the same set of equations that is found in rate-independent plasticity for the fully implicit scheme. From (18.47) the solution can therefore be written down directly

$$\bar{s}_{ij} = (1 - 3G\frac{\Delta\lambda}{\sigma^t_{eff}})s^t_{ij}$$

$$\bar{\sigma}_{kk} = \sigma^t_{kk}$$

$$\bar{K} = K^{(1)} + H\Delta\lambda$$

Insertion into the static yield criterion (18.93) results in

$$f = \sigma^t_{eff} - 3G\Delta\lambda - (\sigma_{yo} + K^{(1)} + H\Delta\lambda) = 0$$

with the solution

$$\Delta\lambda = \frac{\sigma^t_{eff} - (\sigma_{yo} + K^{(1)})}{3G + H}$$

The final state is then achieved directly from (18.91). If we, for instance, choose $\Lambda = G/\eta$ as well as $H = 0$, that is, ideal plasticity, it can be shown that the model corresponds to Hohenemser-Prager viscoplasticity, cf. (15.34).

19 SOLUTION OF DYNAMIC FINITE ELEMENT EQUATIONS

We will now turn our interest to dynamic problems involving an elasto-plastic or viscoplastic response. Within the framework already established, it turns out that it is easy to include dynamics caused by the inertia effects. The only new issue that we have to introduce is that of different *time integration schemes*. These schemes are the same as those encountered in linear dynamic problems and, therefore, we will not discuss these time integration schemes in detail, but simply present some information that motivates these schemes to a sufficient degree. The interested reader may consult Bathe (1996), Hughes (1983, 1987) as well as Argyris and Mlejnek (1991), Belytschko *et al.* (2000) and Zienkiewicz and Taylor (1991) for a detailed discussion of time integration schemes.

Moreover, for convenience, damping effects will be ignored. The extension to consider damping is straightforward and as we are here interested in the fundamental numerical procedures, damping is ignored.

19.1 Introduction

According to (16.9), the FE discretization of the equations of motion is given by

$$\boldsymbol{M}\ddot{\boldsymbol{a}} + \boldsymbol{\psi}(\boldsymbol{a}) = \boldsymbol{0} \tag{19.1}$$

where the mass matrix $\boldsymbol{M}$ is defined by

$$\boldsymbol{M} = \int_V \rho \boldsymbol{N}^T \boldsymbol{N} dV$$

and the out-of-balance forces $\boldsymbol{\psi}$ are defined by

$$\boldsymbol{\psi}(\boldsymbol{a}) = \int_V \boldsymbol{B}^T \boldsymbol{\sigma} dV - \boldsymbol{f} \tag{19.2}$$

For a given time, the external forces are denoted by $\boldsymbol{f}$ and since the stresses $\boldsymbol{\sigma}$ for a given time depend on the nodal displacements $\boldsymbol{a}$, we have $\boldsymbol{\psi} = \boldsymbol{\psi}(\boldsymbol{a})$.

Relation (19.1) comprises a set of nonlinear differential equations. First we will transform them into a set of nonlinear algebraic equations; thereafter, these nonlinear algebraic equations will be solved in an iterative manner similar to that discussed in Chapter 17.

To transform the nonlinear differential equations (19.1) into a set of nonlinear algebraic equations, the time integration scheme proposed by Newmark (1959) is adopted. We assume that all quantities have been determined at time t_n and we want to proceed to time t_{n+1} and determine all relevant quantities. With evident notation, the *Newmark time integration scheme* then comprises the following approximations

Newmark time integration scheme

$$\boldsymbol{a}_{n+1} = \boldsymbol{a}_n + \Delta t\dot{\boldsymbol{a}}_n + \tfrac{\Delta t^2}{2}[(1-2\beta)\ddot{\boldsymbol{a}}_n + 2\beta\ddot{\boldsymbol{a}}_{n+1}] \qquad (19.3)$$

$$\dot{\boldsymbol{a}}_{n+1} = \dot{\boldsymbol{a}}_n + \Delta t[(1-\gamma)\ddot{\boldsymbol{a}}_n + \gamma\ddot{\boldsymbol{a}}_{n+1}]$$

where β and γ are certain parameters and Δt is the time step, i.e. $\Delta t = t_{n+1} - t_n$. Depending on particular choice of β and γ, we recover several special strategies like the central difference method, the trapezoidal rule etc. For a detailed discussion of the entire subclass of methods, we may refer to Bathe (1996) and Hughes (1983, 1987).

To substantiate the Newmark scheme (19.3), we adopt the *trapezoidal rule*, which provides the following approximations for $\boldsymbol{a}_{n+1}$ and $\dot{\boldsymbol{a}}_{n+1}$

$$\boldsymbol{a}_{n+1} = \boldsymbol{a}_n + \frac{\Delta t}{2}(\dot{\boldsymbol{a}}_n + \dot{\boldsymbol{a}}_{n+1})\ ; \quad \dot{\boldsymbol{a}}_{n+1} = \dot{\boldsymbol{a}}_n + \frac{\Delta t}{2}(\ddot{\boldsymbol{a}}_n + \ddot{\boldsymbol{a}}_{n+1}) \qquad (19.4)$$

Insertion of (19.4b) into (19.4a) yields

$$\boldsymbol{a}_{n+1} = \boldsymbol{a}_n + \Delta t\dot{\boldsymbol{a}}_n + \frac{\Delta t^2}{2}(\frac{1}{2}\ddot{\boldsymbol{a}}_n + \frac{1}{2}\ddot{\boldsymbol{a}}_{n+1}) \qquad (19.5)$$

It appears that (19.5) and (19.4b) correspond to (19.3a) and (19.3b) respectively, for $\beta = 1/4$ and $\gamma = 1/2$. Therefore, the Newmark scheme (19.3) can be viewed as a generalization of the trapezoidal rule.

19.2 Explicit scheme

Let us assume that

$$\boxed{\beta = 0\ ; \quad \gamma = \frac{1}{2}} \qquad (19.6)$$

In that case (19.3) reduces to

$$\begin{aligned} \boldsymbol{a}_{n+1} &= \boldsymbol{a}_n + \Delta t \dot{\boldsymbol{a}}_n + \frac{\Delta t^2}{2}\ddot{\boldsymbol{a}}_n \\ \dot{\boldsymbol{a}}_{n+1} &= \dot{\boldsymbol{a}}_n + \frac{\Delta t}{2}(\ddot{\boldsymbol{a}}_n + \ddot{\boldsymbol{a}}_{n+1}) \end{aligned} \tag{19.7}$$

Let us write the equations of motion at the current time t_n, i.e. (19.1) becomes

$$\boldsymbol{M}\ddot{\boldsymbol{a}}_n + \boldsymbol{\psi}(\boldsymbol{a}_n) = \boldsymbol{0} \tag{19.8}$$

where

$$\boldsymbol{\psi}(\boldsymbol{a}_n) = \int_V \boldsymbol{B}^T \boldsymbol{\sigma}_n dV - \boldsymbol{f}_n$$

Let us now rewrite the time integration scheme such that the acceleration $\ddot{\boldsymbol{a}}$ is expressed in terms of displacements $\boldsymbol{a}$. Determination of $\ddot{\boldsymbol{a}}_n$ from (19.7a) yields

$$\ddot{\boldsymbol{a}}_n = \frac{2}{\Delta t^2}(\boldsymbol{a}_{n+1} - \boldsymbol{a}_n) - \frac{2}{\Delta t}\dot{\boldsymbol{a}}_n \tag{19.9}$$

i.e.

$$\ddot{\boldsymbol{a}}_{n+1} = \frac{2}{\Delta t^2}(\boldsymbol{a}_{n+2} - \boldsymbol{a}_{n+1}) - \frac{2}{\Delta t^2}\dot{\boldsymbol{a}}_{n+1} \tag{19.10}$$

Insertion of (19.9) and (19.10) into (19.7b) results in

$$\dot{\boldsymbol{a}}_{n+1} = \frac{1}{2\Delta t}(\boldsymbol{a}_{n+2} - \boldsymbol{a}_n)$$

and thereby

$$\dot{\boldsymbol{a}}_n = \frac{1}{2\Delta t}(\boldsymbol{a}_{n+1} - \boldsymbol{a}_{n-1}) \tag{19.11}$$

Then (19.9) and (19.11) imply

$$\boxed{\ddot{\boldsymbol{a}}_n = \frac{1}{\Delta t^2}(\boldsymbol{a}_{n+1} - 2\boldsymbol{a}_n + \boldsymbol{a}_{n-1})} \tag{19.12}$$

With the parameter choice (19.6), we obtain (19.12) which is, in fact, the *central difference approximation* to $\ddot{\boldsymbol{a}}_n$.

Insertion of (19.12) into (19.8) then provides the following scheme

$$\boxed{\boldsymbol{M}\boldsymbol{a}_{n+1} = \boldsymbol{M}(2\boldsymbol{a}_n - \boldsymbol{a}_{n-1}) + \Delta t^2(\boldsymbol{f}_n - \int_V \boldsymbol{B}^T \boldsymbol{\sigma}_n dV)} \tag{19.13}$$

The the nodal values $\boldsymbol{a}_n$ and $\boldsymbol{a}_{n-1}$ are known and, by the methods discussed in Chapter 18, we can determine the stress state $\boldsymbol{\sigma}_n$. It follows that all terms are known at the right-hand side of (19.13) and this expression then provides the solution $\boldsymbol{a}_{n+1}$ at time t_{n+1}.

It is of considerable interest that even though we are facing a dynamic (elasto-plastic or viscoplastic) problem that certainly is a nonlinear problem, (19.13) provides the solution $\boldsymbol{a}_{n+1}$ directly without any iterations. This very fortunate situation is enhanced further if the mass matrix $\boldsymbol{M}$ is *lumped* i.e. diagonal. In that case the inversion of $\boldsymbol{M}$ is trivial and (19.13) gives the solution directly without in reality solving any equation system.

Unfortunately, we have to pay a price for these appealing solution properties and our price is that of *conditional stability*. Instability means that any small error introduced by truncation of numbers in the computer will jeopardize the solution in the sense that errors will accumulate with time and very quickly make the solution meaningless. On the other hand, conditional stability means that in order to maintain stability, the time step Δt must be within a certain limit. In our case, it turns out that we must have

$$\boxed{\Delta t \leq \frac{T_s}{\pi} \quad \Rightarrow \quad \textit{stability}} \tag{19.14}$$

where T_s is the shortest period of the finite element assemblage that comprises the discretized body, cf. for instance Bathe (1996) and Hughes (1983, 1987).

As a result, the scheme (19.13) is very efficient for situations with a short load duration; the effects of impacts and explosions as well as wave propagation are problems that are eminently well suited for the solution scheme (19.13). For longer load durations, like for instance earthquakes, the scheme (19.13) is not advantageous since the time step limitation (19.14) implies an excessive number of time steps to be considered. We will return to this aspect in the next section.

An interesting application of (19.13) is even static problems. However, in order to fulfill the restriction (19.14) and maintain a reasonable number of time steps, it is often necessary to modify the mass matrix in an appropriate manner, i.e. to increase the mass density ρ artificially and thereby increasing the smallest period T_s.

In accordance with the notation in Chapter 18, the solution scheme (19.13) is an *explicit scheme* since from the known state at t_n we extrapolate to obtain the response at t_{n+1}. This means that we enforce the equations of motion at time t_n, cf. (19.8). In addition, (19.13) provides an explicit solution where no iterations are involved, but, in principle, that has nothing to do with the method being explicit. In general, one may have an explicit scheme that only provides a solution that is implicitly given. We have previously encountered the opposite situation in Chapter 18 where the implicit return scheme (i.e. $\theta = 1$) for the isotropic von Mises material results in an explicit solution in terms of the radial return method, cf. Section 18.1.4.

19.3 Implicit schemes

We will now assume the parameter $\beta \neq 0$. From (19.3a) we then obtain

$$\ddot{\boldsymbol{a}}_{n+1} = c_1(\boldsymbol{a}_{n+1} - \boldsymbol{a}_n) - c_2\dot{\boldsymbol{a}}_n - c_3\ddot{\boldsymbol{a}}_n \tag{19.15}$$

where

$$c_1 = \frac{1}{\beta\Delta t^2}; \quad c_2 = \frac{1}{\beta\Delta t}; \quad c_3 = \frac{1-2\beta}{2\beta} \tag{19.16}$$

All quantities are assumed to be known at the current time t_n. We will now derive *implicit schemes* and we therefore write the equations of motion (19.1) at time t_{n+1}, i.e.

$$\boldsymbol{M}\ddot{\boldsymbol{a}}_{n+1} + \boldsymbol{\psi}(\boldsymbol{a}_{n+1}) = \boldsymbol{0} \tag{19.17}$$

Insertion of (19.15) yields

$$\boldsymbol{v}(\boldsymbol{a}_{n+1}) = \boldsymbol{0} \tag{19.18}$$

where the column matrix $\boldsymbol{v}$ is given by

$$\boldsymbol{v}(\boldsymbol{a}_{n+1}) = \boldsymbol{M}[c_1(\boldsymbol{a}_{n+1} - \boldsymbol{a}_n) - c_2\dot{\boldsymbol{a}}_n - c_3\ddot{\boldsymbol{a}}_n] + \boldsymbol{\psi}(\boldsymbol{a}_{n+1}) \tag{19.19}$$

Since the quantities $\boldsymbol{a}_n$, $\dot{\boldsymbol{a}}_n$ and $\ddot{\boldsymbol{a}}_n$ are known and fixed quantities and since we are considering a given fixed time t_{n+1}, it follows that $\boldsymbol{v} = \boldsymbol{v}(\boldsymbol{a}_{n+1})$.

By making use of the Newmark time integration scheme, we have transformed the nonlinear differential equations (19.1) into the nonlinear algebraic equations (20). This implies that we can take advantage of the methods described in detail in Chapter 17. In analogy with (17.14) and (17.15), we can therefore write (19.18) as

$$-(\boldsymbol{A}(\boldsymbol{a}_{n+1}))^{-1}\boldsymbol{v}(\boldsymbol{a}_{n+1}) = \boldsymbol{0} \tag{19.20}$$

where the iteration matrix $\boldsymbol{A}^{-1}$ is nonsingular, i.e. $\det \boldsymbol{A}^{-1} \neq 0$ and the components of $\boldsymbol{A}^{-1}$ are finite quantities; otherwise the square matrix $(\boldsymbol{A}(\boldsymbol{a}_{n+1}))^{-1}$ is arbitrary. We may also write (19.20) in the form

$$\boldsymbol{a}_{n+1} = \boldsymbol{F}(\boldsymbol{a}_{n+1}) \tag{19.21}$$

where

$$\boldsymbol{F}(\boldsymbol{a}_{n+1}) = \boldsymbol{a}_{n+1} - (\boldsymbol{A}(\boldsymbol{a}_{n+1}))^{-1}\boldsymbol{v}(\boldsymbol{a}_{n+1})$$

For convenience, let us drop the subscript $n+1$, that is, the equation above becomes

$$\boldsymbol{F}(\boldsymbol{a}) = \boldsymbol{a} - (\boldsymbol{A}(\boldsymbol{a}))^{-1}\boldsymbol{v}(\boldsymbol{a}) \tag{19.22}$$

As usual, cf. (17.13) the general iteration scheme for (19.21) is $\boldsymbol{a}^i = \boldsymbol{F}(\boldsymbol{a}^{i-1})$ for $i = 1, 2, \cdots$. Together with (19.21), (19.22) and (19.19), it then follows that our iteration scheme for dynamic nonlinear problems becomes

$$\boxed{\begin{aligned}&\boldsymbol{a}^i = \boldsymbol{a}^{i-1}\\ &-(\boldsymbol{A}(\boldsymbol{a}^{i-1})^{-1})\{\boldsymbol{M}[c_1(\boldsymbol{a}^{i-1} - \boldsymbol{a}_n) - c_2\dot{\boldsymbol{a}}_n - c_3\ddot{\boldsymbol{a}}_n] + \boldsymbol{\psi}(\boldsymbol{a}^{i-1})\}\\ &i = 1, 2\cdots\end{aligned}} \tag{19.23}$$

where the out-of-balance forces $\boldsymbol{\psi}$ according to (19.2) are given by

$$\boldsymbol{\psi}(\boldsymbol{a}^{i-1}) = \int_V \boldsymbol{B}^T \boldsymbol{\sigma}^{i-1} dV - \boldsymbol{f}$$

The starting values for $i = 1$ are as usual taken as the last accepted values, i.e.

$$\boldsymbol{a}^0_{n+1} = \boldsymbol{a}_n \; ; \qquad \boldsymbol{\sigma}^0_{n+1} = \boldsymbol{\sigma}_n$$

Moreover, the stresses $\boldsymbol{\sigma}^{i-1}$ are obtained by a numerical integration of the constitutive equations; this was discussed in the previous chapter.

It appears that all terms on the right-hand side of (19.23) are known quantities and (19.23) can then be solved to provide the new values for the nodal displacements, i.e. $\boldsymbol{a}^i$.

Convergence is obtained for $\boldsymbol{a}^{i-1} \to \boldsymbol{a}^i$ and in that case (19.23) reduces to $\boldsymbol{v}(\boldsymbol{a}^i) \to 0$ where $\boldsymbol{v}$ is given by (19.19). In reality, $\boldsymbol{v}(\boldsymbol{a}^i) = \boldsymbol{0}$ is an expression for fulfillment of the equations of motion given by (19.17). Therefore, a criterion for convergence may not only involve $\boldsymbol{a}^{i-1} \to \boldsymbol{a}^i$, but it may also express the fact that $\boldsymbol{v}^i$ is close to zero. We therefore obtain the following possible convergence criteria:

$$(\boldsymbol{a}^i - \boldsymbol{a}^{i-1})^T(\boldsymbol{a}^i - \boldsymbol{a}^{i-1}) < \alpha_1 \boldsymbol{a}_n^T \boldsymbol{a}_n \tag{19.24}$$

and/or

$$\boldsymbol{v}^T(\boldsymbol{a}^i)\boldsymbol{v}(\boldsymbol{a}^i) < \alpha_2 NORM \tag{19.25}$$

and/or

$$\begin{aligned}&(\boldsymbol{a}^i - \boldsymbol{a}^{i-1})^T(\boldsymbol{M}\ddot{\boldsymbol{a}}^i + \boldsymbol{f}_{n+1} - \textstyle\int_V \boldsymbol{B}^T\boldsymbol{\sigma}^i dV)\\ &< \alpha_3(\boldsymbol{a}^1 - \boldsymbol{a}_n)^T(\boldsymbol{M}\ddot{\boldsymbol{a}}_n + \boldsymbol{f}_{n+1} - \textstyle\int_V \boldsymbol{B}^T\boldsymbol{\sigma}_n dV)\end{aligned} \tag{19.26}$$

where α_1, α_2 and α_3 are certain *tolerances* specified by us and *NORM* may be $||\boldsymbol{f}||$ or $||\boldsymbol{M}\ddot{\boldsymbol{a}}||$.

The convergence criteria (19.24) and (19.25) possess the drawback that quantities having one dimension may be added to quantities having another dimension; for instance, the components of $\boldsymbol{a}$ may involve both displacements and rotations and, likewise, the components of $\boldsymbol{v}$ may involve forces as well as moments. In that sense, (19.26) is more appealing since it is an energy criterion where each force component is multiplied by the corresponding displacement component and each moment component is multiplied by the corresponding rotation component; accordingly, (19.26) is a consistent criterion.

Typical values for the tolerances α_1,α_2 and α_3 have been discussed by Bathe (1996) as well as Belytschko *et al.* (2000). It appears that the tolerances are of the following order: $\alpha_1 \simeq 10^{-3}$, $\alpha_2 \simeq 10^{-3}$ and $\alpha_3 \simeq 10^{-7}$.

As previously discussed, the scheme (19.23) is implicit and it appears that it is stable, irrespective of the time step, if the parameter values in the Newmark approximation fulfill the following bounds

$$\boxed{\gamma \geq \frac{1}{2}\ ; \quad \beta \geq \frac{1}{4}(\gamma + \frac{1}{2})^2 \quad \Rightarrow \quad \textit{unconditional stability}} \tag{19.27}$$

cf. Bathe (1996) and Hughes (1983, 1987). The parameter values in the most often applied implicit Newmark scheme are taken as $\beta = 1/4$ and $\gamma = 1/2$, which according to (19.27) implies unconditional stability. Moreover following (19.4) and (19.5), this scheme is equivalent to the trapezoidal rule.

Different expressions for the iteration matrix $\boldsymbol{A}$ present in the scheme (19.23) may be chosen, but in order to obtain some kind of feeling for an appropriate choice, it is instructive to derive the Newton-Raphson approach. In that case, a Taylor series expansion is made of the quantity $\boldsymbol{v}$ entering the nonlinear equation (19.18). Making the Taylor series expansion about the state $\boldsymbol{a}^{i-1}$, we obtain

$$\boldsymbol{v}(\boldsymbol{a}^i) = \boldsymbol{v}(\boldsymbol{a}^{i-1}) + (\frac{\partial \boldsymbol{v}}{\partial \boldsymbol{a}})^{i-1}(\boldsymbol{a}^i - \boldsymbol{a}^{i-1}) \cdots$$

Ignoring higher order terms and assuming that $\boldsymbol{a}^i$ is close to the correct value, i.e. $\boldsymbol{v}(\boldsymbol{a}^i) = 0$, imply

$$\boldsymbol{a}^i = \boldsymbol{a}^{i-1} - [(\frac{\partial \boldsymbol{v}}{\partial \boldsymbol{a}})^{i-1}]^{-1}\boldsymbol{v}(\boldsymbol{a}^{i-1}) \tag{19.28}$$

From (19.19) we have

$$\boldsymbol{v}(\boldsymbol{a}) = \boldsymbol{M}[c_1(\boldsymbol{a} - \boldsymbol{a}_n) - c_2\dot{\boldsymbol{a}}_n - c_3\ddot{\boldsymbol{a}}_n] + \boldsymbol{\psi}(\boldsymbol{a}) \tag{19.29}$$

i.e.

$$\frac{\partial \boldsymbol{v}}{\partial \boldsymbol{a}} = c_1\boldsymbol{M} + \int_V \boldsymbol{B}^T\frac{d\boldsymbol{\sigma}}{d\boldsymbol{a}}dV$$

Since $\dot{\boldsymbol{\sigma}} = \boldsymbol{D}_t\dot{\boldsymbol{\varepsilon}} = \boldsymbol{D}_t\boldsymbol{B}\dot{\boldsymbol{a}}$, we find that

$$\frac{\partial \boldsymbol{v}}{\partial \boldsymbol{a}} = c_1\boldsymbol{M} + \boldsymbol{K}_t \tag{19.30}$$

where

$$K_t = \int_V B^T D_t B dV$$

Insertion of (19.30) into (19.28) and use of (19.29) then yield

$$\begin{aligned} a^i =& a^{i-1} \\ & - [c_1 M + K_t^{i-1}]^{-1} \{ M[c_1(a^{i-1} - a_n) - c_2 \dot{a}_n - c_3 \ddot{a}_n] \\ & + \psi(a^{i-1}) \} \end{aligned}$$

A comparison with (19.23) shows that the following choice of the iteration matrix A

$$\boxed{A^{i-1} = c_1 M + K_t^{i-1} \quad \Rightarrow \quad \textit{Newton-Raphson}}$$

provides the Newton-Raphson approach. However, in accordance with Chapter 17 we may adopt other choices for the iteration matrix A. Since the mass matrix M is constant, we may adopt the modified Newton-Raphson method where the stiffness matrix entering A is only updated once in every time step, the initial stiffness method where the stiffness matrix entering A is taken as the elastic stiffness K or a quasi-Newton method like the BFGS-approach.

If the parameters β and γ in the Newmark scheme fulfill (19.27), the time integration is always stable. However, considering a given mode in the response, the *accuracy* of the time integration scheme decreases as the ratio $\Delta t/T$ increases; here T is the period of the actual mode considered. This means that higher order modes (short period T) are integrated with a lower accuracy than higher order modes (long period T).

As discussed in the previous section, the implicit method (19.23) is especially suited to problems with long load durations, i.e. structural dynamics problems like modeling the effect of earthquakes. In that case, the contribution of higher order modes to the response is often of minor importance. On the other hand, due to the time restriction (19.14) the explicit scheme is geared to events of short duration as, for instance, wave propagation and impact, where higher order modes are often of importance. Moreover, the explicit scheme is much simpler to program than the implicit scheme, and evaluation of new concepts is therefore conveniently performed in an explicit environment.

20 BASIC PRINCIPLES OF THERMODYNAMICS

The equations of motion, which involve forces or stresses, are based entirely upon Newton's second law. The concept of a traction vector acting on a surface is based on Newton's third law that relates action and reaction. The concept of strains relies entirely upon certain kinematic considerations. Thus, both stresses and strains are based upon certain laws of nature that we accept as *axioms*, i.e. postulates that everybody accepts as being true.

When relating stresses and strains via the constitutive equations, we are leaving the firm background that the laws of nature provide. In fact, constitutive theories are our imagination of how materials behave when exposed to certain load conditions. We have discussed a number of such constitutive theories and even though we have relied on certain experimental evidence and have tried to be as consistent as possible, it is evident that the constitutive modeling considered so far is based to a certain extent on a framework that is not as firmly rooted in fundamental principles as we would like it to be.

In order to obtain a more fundamental basis for constitutive modeling, it seems intriguing to investigate whether *thermodynamics* may provide this basis. Thermodynamics deals with relations between heat and mechanical work and it is based on two fundamental laws of nature, the *first* and *second law of thermodynamics*; these laws are accepted as axioms just like, for instance, Newton's second law.

It turns out to be possible to derive constitutive theories from thermodynamics and in this chapter we will first discuss some fundamental results of thermodynamics. In the next chapters, we will then discuss in detail various ramifications for constitutive modeling. It appears that the constitutive theories discussed previously can also be derived from thermodynamical principles. This not only supports the relevance of our previous discussion, but it also makes for a completely new approach to derive other and more advanced constitutive theories. Moreover, it turns out that, with this thermodynamic basis, it is possible to combine various material behaviors in a straightforward manner. As an example, we may establish a constitutive theory that combines plasticity and damage.

Without the thermodynamic basis such a combined model is rather difficult to establish, but with the thermodynamic formulation the establishment turns out to be pretty straightforward.

One may ask why we have chosen not to start with the thermodynamic formulation right from the beginning of this exposition. The answer is threefold: first of all, the thermodynamic formulation is rather abstract whereas the previous discussion of various constitutive theories has relied heavily on simple experimental evidence and direct physical considerations; secondly, when the thermodynamic formulation is used to establish once more the previously discussed models (as well as providing a number of new theories) it will - hopefully - enlighten the reader; thirdly - and this may be the most important point - with the background obtained up to now, the reader will have a much better background to understand and fully appreciate why the thermodynamic formulation takes the specific forms that we will discuss.

In this chapter, we will introduce various concepts like temperature, heat, entropy, etc. that enter thermodynamics. For evident reasons, we will confine our attention to mechanical and heat processes and exclude electric, magnetic, chemical and ion diffusion processes. For introductions to traditional thermodynamics that do not focus on the implications for constitutive modeling of solid materials, we may refer, for instance, to Sears (1959), Schmidt (1963) and Çengel and Boles (1994). In the book of Kestin (1979), some attention is given to solid materials. Even though we will take advantage of of these viewpoints, for evident reasons we will focus here on a thermodynamic formulation that is of direct relevance for establishing constitutive theories for solid materials. Within this scope, there exist many different approaches to thermodynamics and the reader may, for instance, consult Truesdell and Toupin (1960), de Groot and Mazur (1962), Truesdell (1969), Malvern (1969), Oden (1972), Eringen (1975a), Lavenda (1978), Lemaitre and Chaboche (1990), Maugin (1992) and Haupt (2000). Most of these expositions are quite formal and abstract. Here we have chosen to adopt the simplest possible approach that, hopefully, appeals to the physical insight of the reader so that new concepts like heat, temperature, entropy, etc. not only become quantities that the reader may manipulate with in various equations, but also become concepts of physical relevance.

In the next chapters, we will then take full advantage of thermodynamics and illustrate its extreme importance for constitutive modeling.

20.1 Temperature - Absolute temperature

One of the fundamental characteristics of a body is how hot or cold it is. The familiar concept of *temperature* quantifies this degree of hotness or coldness and the temperature T is simply the real number indicated on a thermometer. If the body gets hotter, its temperature rises; if it gets colder, its temperature falls. However, the *scale* with which we measure the temperature is arbitrary -

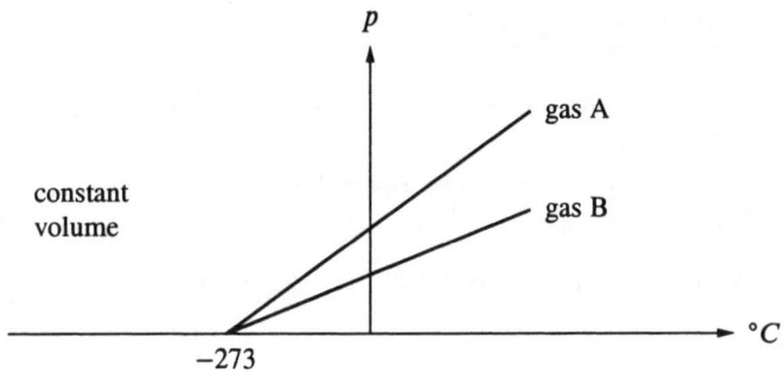

Figure 20.1: Relation between gas pressure and gas temperature for constant volume.

familiar examples are degrees Fahrenheit and degrees Celsius. In spite of that, experience shows that all temperature scales are related uniquely to each other; this implies that irrespective of the temperature scale used, we measure the same phenomenon.

> *Temperature may be viewed as an expression of how fast the molecules vibrate*

It is an experimental fact that high temperature results in powerful vibrations of the molecules of the body whereas low temperature results in small vibrations of the molecules. It is therefore not surprising that experimental evidence shows that, regardless of the scale used for a thermometer, there is a temperature below which no body can be cooled. At this temperature, no molecular vibrations occur and we therefore have:

> *Irrespective of the temperature scale used, there exists a temperature below which no body can be cooled*

To substantiate this viewpoint, consider a gas that occupies a certain fixed volume. Experience then shows that there is a relation between the pressure p and the temperature T. If this relation is measured in the laboratory, it turns out to be linear, cf. Fig. 20.1. This figure also illustrates that irrespective of what gas we consider, there is a certain temperature at which all gases exhibit a pressure that is nil; this temperature turns out to be -273^oC. As a negative gas pressure cannot exit, we are again led to the conclusion that there exists a temperature below which no body can be cooled.

Therefore, irrespective of the temperature scale used, the temperature is bounded below. If we assign the temperature value zero to the greatest lower bound, then the temperature is said to be *absolute* and, instead of the notation T for temperature, we use the notation θ to indicate that the temperature measure is absolute, i.e.

$$\boxed{\theta > 0 \quad \textit{absolute temperature}} \tag{20.1}$$

We conclude that the absolute temperature approaches the value zero at the lowest possible temperature.

As already discussed, the lower bound is about -273^oC. If we choose to measure temperature changes by means of degrees Celsius (^oC) then we can construct an absolute temperature measure by means of so-called *Kelvin degrees* $[K]$ according to

$$K = 273 + ^oC \ ; \quad K > 0 \tag{20.2}$$

Let us recall that degrees Fahrenheit (^oF) are related to degrees Celsius by $(^oF - 32)5/9 = ^oC$. Therefore, using degrees Fahrenheit as a measure of temperature changes and noting that -273^oC corresponds to $9(-273)/5 + 32 = -459^oF$, we may construct another absolute temperature measure ^oR, the *Rankine scale*, by setting $^oR = 459 + {}^oF$, $^oR > 0$, but (20.2) is the one usually adopted.

We shall later, in Section 20.10, show that strict thermodynamical considerations also infer the existence of an absolute temperature.

20.2 Heat and heat flow

The incremental mechanical work done by a point force $\mathbf{f}$, when the point moves the distance $d\mathbf{u}$ is given by the scalar product $d\mathbf{u}^T\mathbf{f}$. Mechanical work is therefore energy that is produced by the displacement of forces.

Experience shows that there is another form of energy, namely that related to *heat*. If two bodies at different temperatures are brought into contact with each other, heat will flow from the hotter body into the colder body. We therefore have:

> *Heat flow is a transportation of energy and*
> *heat flow requires a temperature difference*

That heat, in fact, is energy that may even be transformed to mechanical work is illustrated in Fig. 20.2. Here a gas is contained in a cylinder with a weightless piston and a mass M is placed on the piston. In Fig. 20.2a), a Bunsen burner is used to heat the system and due to the increase of gas temperature, the gas expands so that after some time the piston has moved the distance u, cf. Fig. 20.2b). The pressure p is unchanged, since the only mechanical loading is due to the mass M. Due to the heating, the system has performed the mechanical work uMg, where g is the gravity acceleration.

That mechanical work may also be transformed to heat is evident for everybody who has inflated a bicycle tyre by using an ordinary hand pump.

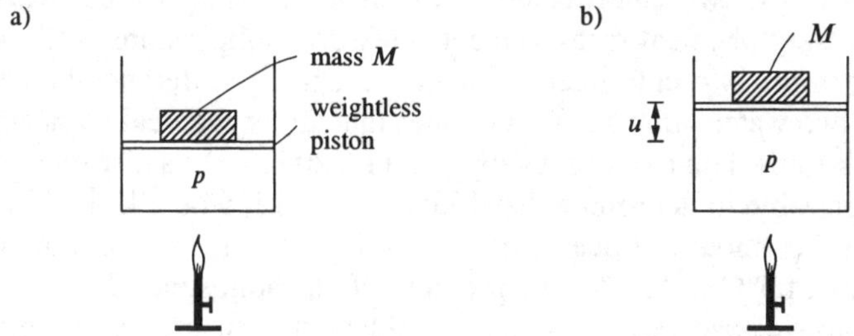

Figure 20.2: Heating of a cylinder containing a gas with the constant pressure p; a) start of heating; b) end of heating.

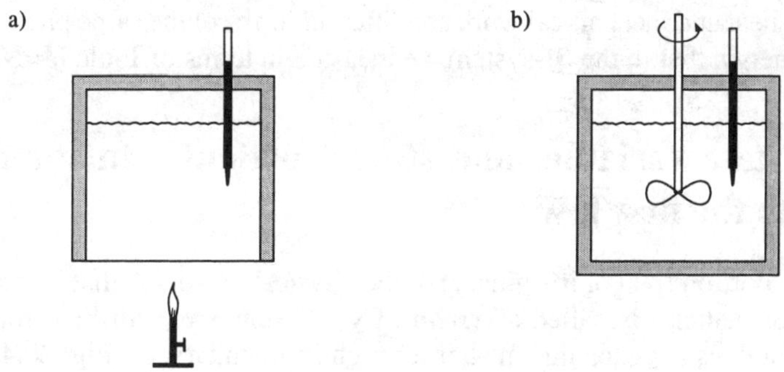

Figure 20.3: Joule's experiment; a) temperature increase by heating; b) same temperature increase by performing mechanical work.

Therefore, heat and mechanical work are of the same nature and both phenomena are referred to as *energy*. Even though heat represents energy, heat can only be measured in terms of a transportation of a certain amount of energy and heat therefore manifests itself in terms of heat flow. Often, in some process between two states, the word heat is used for the total flow of heat that occurs between state 1 and state 2. Heat flow is not to be confused with *heat flux*, which refers to the transportation of heat per unit time.

Heat or heat flow represents energy, but the question remains how we measure heat and how this measurement is related to our usual (SI) concepts of energy. For this purpose, we refer to the celebrated experiment performed by Joule in 1843, see Fig. 20.3. In Fig. 20.3a), a certain amount of water is heated by a Bunsen burner and the temperature then increases. In Fig. 20.3b), mechanical work is supplied to the same amount of water by means of a propeller; due to turbulence and friction in the water, the water temperature then increases. The pressure in the two calorimeters is kept at the same constant level and the initial temperature of the two water samples is the same. When the temperature increase in the two calorimeters is the same, the mechanical work input is measured. Any physical measurement (pressure, temperature, volume, etc.) of the water in the two calorimeters cannot reveal any difference between the states of the two water samples. This implies that all mechanical work supply in Fig. 20.3b) is equivalent to all heat supply in Fig. 20.3a). As a historical remark, it was then possible to determine that 1 kcal = 4.19 kJ, where kcal=kilocalorie and 1 kcal is the amount of heat necessary to increase the temperature of 1 kg of water from 14.5^oC to 15.5^oC at a pressure of one atmosphere.

With the above observation, the amount of heat flow in any process may now, in principle, be measured; the heat flow in the process in question is applied to the calorimeter in Fig. 20.3a) and by the test set-up in Fig. 20.3b), we are able to measure the amount of heat flow.

We conclude from Joule's experiment that we are able to measure heat; moreover heat and mechanical work are different forms of the same phenomenon, namely energy, that in the SI-system we measure in terms of Joule [J=Nm].

20.3 State variables and state functions - Introduction to the first law

A specific portion (real or imaginary) of the physical universe - that is, a specific quantity of matter - is called a *system*. By a system, we will here refer to a system that does not exchange matter through its boundary, cf. Fig. 20.4.

In order to describe the system, we need the concept of *state variables*. For simplicity, we will here assume that homogeneous conditions hold for the system. The following definition is used:

State variables are quantities that characterize the state of the system	(20.3)

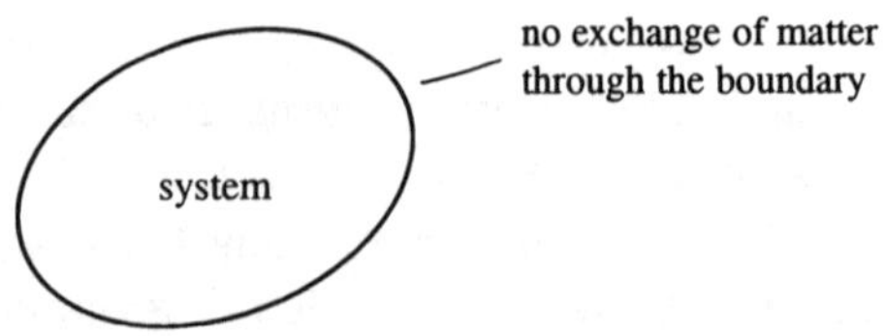

Figure 20.4: Illustration of system.

If in rigid body mechanics we consider a particle as a system, then the current position vector x_i is a state variable. If we consider a gas in a cylinder, state variables are given by the pressure p, the volume V and the temperature θ. In general, however, the identification of the state variables of a given system is not trivial. Moreover, the choice of state variables for one and the same system depends upon what we are interested in. Consider as an example, a steel bar loaded in uniaxial tension; the steel bar is painted blue. If we are a painter, then the color blue would be a state variable whereas as engineers we may select, for instance, the strain, the stress and the temperature as state variables.

Having established the concept of state variables, we will now introduce the concept of a *state function* according to the following definition:

A state function is a function that only depends on the state of the system and not on the manner in which this state is achieved	(20.4)

As an illustration of a state function, consider a person walking on a mountain from point 1 to point 2. The altitude difference $H_2 - H_1$ is certainly a state function, since it is independent of the specific path the person chooses. However, the walking distance between point 1 and point 2 is not a state function, since it depends on the specific path chosen. Since the state variables characterize the state, we conclude that

State function = function of state variables	(20.5)

In order to arrive at *the first law of thermodynamics*, we first introduce the concept of a *perpetual motion of the first kind*:

A perpetual motion of the first kind is a machine operating in cycles which, in any number of complete cycles, will produce more mechanical work than is absorbed in the form of heat

That the machine operates in cycles means that after one cycle, the machine is in the same condition, i.e. the same state, as before the cycle. One manner in which the first law of thermodynamics is expressed is the following:

A perpetual motion of the first kind is impossible	(20.6)

We will later return to the first law of thermodynamics and evaluate it in greater detail, but already now we will emphasize that the first law of thermodynamics is a postulate. However, all experience supports the first law of thermodynamics and it is therefore accepted as an *axiom* just like Newton's second law. One special occasion where (20.6) becomes evident is where the perpetual motion of the first kind just produces mechanical work without absorbing any heat. There also exists a perpetual motion of the *second* kind and it will be discussed later.

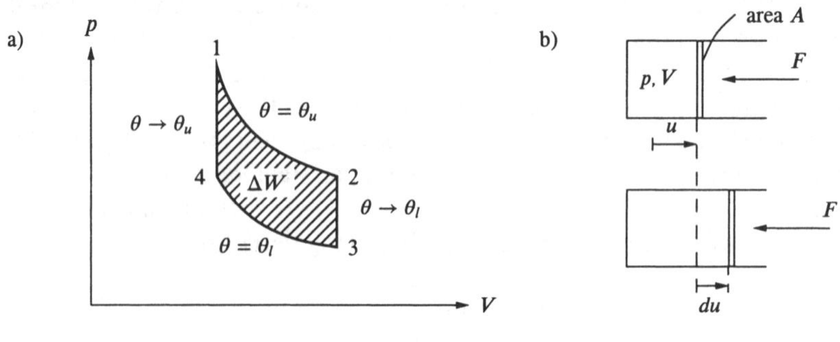

Figure 20.5: a) Complete cycle of gas in a cylinder; b) external force F acting on the piston.

Against this background, we will now prove the important result that neither the mechanical work input nor the heat input to a system are state functions. Referring to Fig. 20.5, let us consider a system in the form of a gas contained in a cylinder with a frictionless piston. Let p, V and θ denote the gas pressure, volume and absolute temperature, respectively. These quantities are taken as state variables that completely define the state of the system. We now perform a cycle via the states 1-2-3-4-1, cf. Fig. 20.5a), so that when returning to state 1, the state is the same as before the cycle.

To gain insight, we may assume that the gas follows the *ideal gas law* given by

$$pV = mR\theta \tag{20.7}$$

where m is the mass of the gas and R denotes the gas constant for the specific gas in question. Along curve 1-2 the gas expands *isothermally*, i.e. at constant temperature. This temperature is denoted by θ_u, where subscript u refers to 'upper'; therefore curve 1-2 is described by pV =constant. This isothermal expansion can be established by bringing the cylinder in contact with a large reservoir having the temperature θ_u. Along curve 2-3, the change is *isochoric*, that is, the volume V is kept constant. The fall in pressure p is obtained by decreasing the temperature from θ_u to θ_l, cf. (20.7); subscript l refers to 'lower'. Along curve 3-4, the gas is compressed isothermally at the temperature θ_l. Finally, the cycle is completed by curve 4-1 where the gas is compressed isochorically by increasing the temperature from θ_l to θ_u.

Referring to Fig. 20.5b), the incremental mechanical work supplied to the system is $-F\,du$ (minus sign because the force F and the displacement u are measured positively in opposite directions). The incremental mechanical work supplied to the surroundings (i.e. the incremental mechanical work produced by the system) is then given by $F du = pA du = pdV$, i.e. the mechanical work produced by the system during the cycle is given by the area ΔW shown in Fig. 20.5a). Therefore

$$\text{Mechanical work produced during the cycle} = \Delta W > 0 \tag{20.8}$$

Since the system after the cycle has returned to its original state, we conclude that

$$\boxed{\textit{Mechanical work is not a state function}} \tag{20.9}$$

cf. the definition of a state function given by (20.4).

Assume now that heat is a state function. Then, when the cycle has been completed, the total heat input must be zero. However, referring to (20.8) we have then created a machine that during one cycle only produces mechanical work without receiving any heat. This is a perpetual motion of the first kind and according to (20.6) such a machine is impossible. It is therefore concluded that

$$\boxed{\textit{Heat is not a state function}} \tag{20.10}$$

20.4 First law of thermodynamics

With these introductory remarks, we are now able to present the first law of thermodynamics in a general form suitable for our purposes.

We consider a region of a body, i.e. a system, and we will first formulate the rate with which mechanical work is performed on the body. With t_i being the traction vector acting on the boundary surface S, b_i the body force per unit volume acting over the region V and $\dot{u}_i$ the displacement rate, we have

$$\begin{aligned} \frac{\delta W}{dt} &= \text{rate of mechanical work input} \\ &= \text{mechanical power input} \\ &= \int_S \dot{u}_i t_i dS + \int_V \dot{u}_i b_i dV \end{aligned} \tag{20.11}$$

We have here chosen to denote this mechanical power input by $\delta W/dt$ instead of dW/dt. If the latter notation was used, this would suggest that the quantity W (the mechanical work) exists in the form of a function that depends on some variables, i.e. that W is a state function. But we have just shown that W is not a state function, cf. (20.9), and this issue is emphasized by the notation $\delta W/dt$. Put another way, dW is not a perfect differential and this is emphasized by writing the incremental mechanical work as δW. If dW were a perfect differential, then there would exist a function W that depends on some variables and W would then be a state function in contrast with our previous findings.

Let us next determine the rate of heat input to the body. Let q_i be the *heat flux vector*; this vector has the direction of the heat flow and its length expresses the heat per unit time which passes through a unit surface area perpendicular to the direction of heat flow. Let n_i be the unit vector normal to the boundary and directed outwards. Moreover, let r denote the amount of *heat supply* per unit time and unit volume; this quantity represents heat sources within the body (inductive heating, for instance). We then have

$$\begin{aligned} \frac{\delta Q}{dt} &= \text{rate of heat input} \\ &= \int_V r dV - \int_S q_i n_i dS \end{aligned} \tag{20.12}$$

Again we use the notation $\delta Q/dt$ to indicate that heat is not a state function, cf. (20.10), i.e. no function Q exists.

The *kinetic energy* K of the body is defined by

$$K = \frac{1}{2}\int_V \rho \dot{u}_i \dot{u}_i dV \tag{20.13}$$

and it is certainly a state function depending on the velocity $\dot{u}_i$ and the mass density ρ. Since we assume small strains, the mass density ρ can be considered as a constant, i.e.

$$\dot{K} = \int_V \rho \dot{u}_i \ddot{u}_i dV \tag{20.14}$$

We are now in a position to formulate the first law of thermodynamics. It states that

$$\boxed{\dot{K} + \dot{U} = \frac{\delta W}{dt} + \frac{\delta Q}{dt} \qquad \textit{global form of first law}} \tag{20.15}$$

where U is defined to be the so-called *internal energy* of the body. Moreover, it is postulated that

The internal energy is a state function	(20.16)

Expression (20.15) is referred to as the *global form* of the first law since it applies to the entire body in question.

The first law of thermodynamics is a postulate which, however, is supported by all experience; it is therefore accepted as an *axiom*. The first law may be viewed as the *principle of conservation of energy*, since it states that the sum of the rate of mechanical work input and the rate of heat input equals the sum of the rate of kinetic energy and rate of internal energy. It is of significant importance that whereas neither mechanical work input nor heat input are state functions, the sum of their rates creates $\dot{K} + \dot{U}$. Here the kinetic energy K is certainly a state function, cf. (20.13), whereas the internal energy U is postulated to be a state function. Therefore, the first law of thermodynamics is not only given by (20.15), but it also defines U to be a state function.

Let us write (20.15) as

$$dK + dU = \delta W + \delta Q \tag{20.17}$$

We then observe, as expected, that the first law excludes the existence of a perpetual motion of the first kind, cf. definition (20.6); integrating (20.17) over one cycle and observing that K and U are state functions, the left-hand side becomes zero, which yields $0 = \Delta W + \Delta Q$ and this expression excludes the existence of a perpetual motion of the first kind. The quantitative formulation of the first law was originally provided by Clausius in 1850.

Expression (20.15) formulates the first law in a global form that applies to the entire body. It turns out to be important to obtain a *local form* of the first law which holds at every point in the body. For this purpose, we first define the *specific internal energy u* per unit mass by

$$\boxed{U = \int_V \rho u dV} \tag{20.18}$$

The specific internal energy u is a state function just like the total internal energy U of the body is a state function.

Since the traction vector is given by $t_i = \sigma_{ij}n_j$, use of Gauss's divergence theorem in (20.11) implies that

$$\frac{\delta W}{dt} = \int_V (\dot{u}_i\sigma_{ij})_{,j}dV + \int_V \dot{u}_i b_i dV \tag{20.19}$$

Likewise, use of Gauss's divergence theorem in (20.12) gives

$$\frac{\delta Q}{dt} = \int_V r dV - \int_V q_{i,i}dV \tag{20.20}$$

Then insertion of (20.14) and (20.18)-(20.20) into (20.15) provides

$$\int_V [\dot{u}_i(\rho\ddot{u}_i - \sigma_{ij,j} - b_i) + \rho\dot{u} - \dot{u}_{i,j}\sigma_{ij} - r + q_{i,i}]dV = 0$$

Taking advantage of the equations of motion, cf. (3.29), we then obtain

$$\int_V (\rho\dot{u} - \dot{\varepsilon}_{ij}\sigma_{ij} - r + q_{i,i})dV = 0$$

Since this expression holds for arbitrary regions V, we conclude that

$$\boxed{\rho\dot{u} = \dot{\varepsilon}_{ij}\sigma_{ij} + r - q_{i,i} \qquad \textit{local form of first law}} \tag{20.21}$$

This is the *local form* of the first law of thermodynamics that holds at every point in the body.

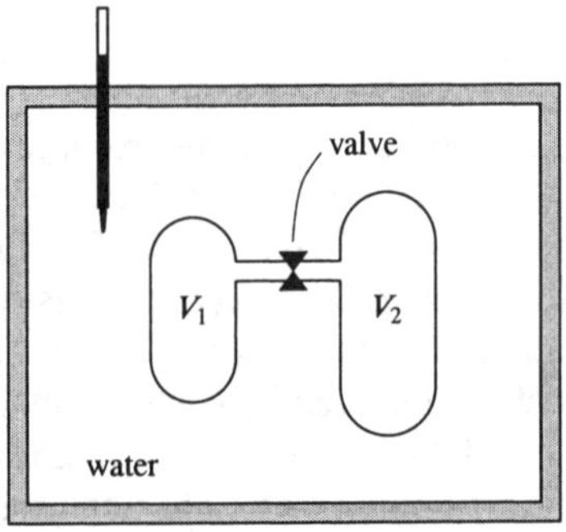

Figure 20.6: Joule's internal energy experiment.

20.5 Ideal gases

In order to introduce the second law of thermodynamics, we shall later use the concept of a *Carnot cycle*. For that purpose, we need information of certain processes performed with *ideal gases*. Ideal gases are described by the *ideal gas law*

$$\boxed{pV = mR\theta\,; \quad \textit{ideal gas law}} \tag{20.22}$$

where p=pressure, V=volume, m=mass of gas, R=gas constant for the gas in question and θ is the absolute temperature. The ideal gas law was formulated preliminarily by Boyle in 1662 and by Mariotte in 1676 and given in its final form by Gay-Lussac in 1802. The ideal gas law applies with close accuracy for gases at not too high a pressure.

The state variables for the ideal gas are p, V and θ and since they are related via (20.22), one may choose, for instance, V and θ. This means that the total internal energy U of a certain amount of gas can be written in general as

$$U = U(V, \theta) \tag{20.23}$$

Note that the total internal energy U refers to a certain amount of matter.

Referring to (20.23) and in order to investigate the internal energy of ideal gases in more detail, Joule performed in 1845 the experiment shown in Fig. 20.6. Two vessels with volumes V_1 and V_2 are placed in a calorimeter filled with water. Initially, the vessel with volume V_1 is occupied by an ideal gas whereas the other vessel is evacuated. The valve connecting the two vessels is then opened and eventually the pressure will be the same in the two vessels. The experiment shows that no temperature change occurs in the calorimeter. Considering the gas as a system, we therefore have $\delta Q = 0$. Moreover, since the vessel walls can be regarded as rigid, the gas has performed no mechanical work on the surroundings, i.e. $\delta W = 0$. According to the first law (20.17) and waiting until all kinematic energy has disappeared, the total internal energy U is then unchanged. Before opening of the valve, the gas volume is V_1 and after opening of the valve, the gas volume is $V_1 + V_2$. Referring to (20.23) and since the total internal energy is unchanged, we conclude that

$$\boxed{U = U(\theta)\,; \quad \textit{ideal gas}} \tag{20.24}$$

Note that the total internal energy U refers to all the particles that constitute the amount of gas considered. This gas amount may be compressed or expanded, but if the temperature is unchanged the total internal energy is also unchanged.

Let us next write the first law (20.17) in a form suitable for gases. Ignoring body forces and since the traction vector is given by $t_i = -pn_i$, (20.11) becomes

$$\frac{\delta W}{dt} = -\int_S \dot{u}_i p n_i dS = -p\int_S \dot{u}_i n_i dS = -p\dot{V}$$

For static conditions, (20.17) then becomes

$$\boxed{dU = -pdV + \delta Q\,; \quad \textit{for gas}} \tag{20.25}$$

Let us next introduce the *specific heat capacity c* of a material. There are different measures of specific heat capacity, but here we will consider the specific heat capacity c_V at constant volume. It is defined as

The specific heat capacity c_V is the amount of heat that must be supplied without change of volume in order to increase the temperature of one unit mass one degree	(20.26)

Note that according to this definition of c_V, the volume is kept constant and this is indicated by the subscript V. Since V is constant, (20.25) implies that $dU = \delta Q$ and according to definition (20.26) we then have

$$dU = \delta Q = mc_V d\theta \qquad \text{when } V = \text{constant} \tag{20.27}$$

Following (20.18) and (20.24), we have $U = mu(\theta)$ where u is the internal energy per unit mass. Therefore, we obtain

$$dU = m\frac{du}{d\theta}d\theta$$

and a comparison with (20.27) shows that

$$dU = mc_V d\theta \quad \text{where} \quad c_V = \frac{du}{d\theta}\,; \quad \text{ideal gas} \tag{20.28}$$

Since (20.24) holds in general for ideal gases, (20.28) also holds in general for ideal gases. However, the material parameter c_V is measured at constant volume as indicated by (20.27). For ideal gases, c_V can be considered as constant with close accuracy.

For later purposes, we will consider *isothermal processes*, i.e. θ = constant, for ideal gases and *adiabatic processes*, i.e. $\delta Q = 0$, for ideal gases.

For isothermal processes where θ = constant, (20.28) yields $dU = 0$ and the first law (20.25) then provides

$$\delta Q = pdV$$

Elimination of the pressure p by means of the gas law (20.22) gives

$$\boxed{\delta Q = mR\theta\frac{dV}{V} \quad \textit{isothermal, ideal gas}} \qquad (20.29)$$

For adiabatic processes where $\delta Q = 0$, the first law (20.25) gives $dU = -pdV$ which with (20.28) results in

$$mc_V d\theta = -pdV$$

Elimination of the pressure p by means of the gas law (20.22) gives

$$\boxed{\frac{c_V}{R}\frac{d\theta}{\theta} = -\frac{dV}{V} \quad \textit{adiabatic, ideal gas}} \qquad (20.30)$$

It should be recalled that the only purpose of this section is the establishment of relations (20.29) and (20.30). Advantage will be taken of these results later.

20.6 Reversible and irreversible processes

Let us next introduce the concept of *reversible* and *irreversible processes*. By definition, we have

A process is reversible if both the system and all its surroundings can be brought back to their initial conditions	(20.31)

This implies that a reversible process is defined as a process that can be reversed without leaving any trace on the system or its surroundings. Moreover

A process that is not reversible is irreversible

It should be pointed out that a system may be restored to its initial state following a process, irrespective of whether the process is reversible or irreversible. But for reversible processes, this restoration is made without leaving any trace on the surroundings whereas for irreversible processes, the surroundings will have changed their conditions.

Friction is a familiar form of irreversibility. When two bodies in contact are forced to move relative to each other, a friction force develops which opposes the motion and some mechanical work is needed to overcome this friction force. During the process, this mechanical work is eventually converted to heat and is transferred to the bodies in contact, as evidenced by a temperature increase at the interface. When the direction of the motion is reversed, the bodies can be restored to their initial position, but the interface will not cool during this reversed motion. On the contrary, additional mechanical work is needed to overcome the friction force that also opposes the reversed motion and this additional mechanical work will also be converted to heat and imply a temperature increase of the interface. As a result, the initial position of the bodies has been restored and even the heat developed during the process may be transferred to the surroundings so that the system is in its initial condition. However, the surroundings are not in their initial conditions, since they have received a certain amount of heat; the process is evidently irreversible.

Another form of irreversibility is heat transfer through a finite temperature difference. Consider as an example isothermal expansion or compression of the ideal gas where the heat exchange δQ is given by (20.29). From Fourier's law $\delta Q/\Delta t \sim \Delta\theta$ and a constant temperature difference $\Delta\theta$ between gas and surroundings, we may write in a symbolic form

$$\Delta Q = \text{constant } \Delta\theta\Delta t \tag{20.32}$$

where Δt is the time duration of the heat transfer. When the gas expands, (20.29) shows that $\delta Q > 0$, i.e. the gas temperature is lower than the temperature of the reservoir. When the gas is compressed, (20.29) provides $\delta Q < 0$, i.e. the gas temperature is higher than the reservoir. Since the reservoir temperature is assumed to be constant, the same gas temperature during expansion and compression can only be achieved by letting $\Delta\theta \to 0$, but since ΔQ is a finite quantity, (20.32) shows that we must require $\Delta t \to \infty$. This implies that if the heat transfer during expansion and compression is to be the same, the process must proceed infinitely slowly. A reversible heat transfer can therefore only be achieved if the temperature differences are infinitely small. Conversely, any finite temperature difference will result in an irreversible process.

It appears that reversible processes cannot be achieved in practice. However, reversible processes are important since they may be viewed as theoretical, idealized processes.

An evident example of an irreversible process is Joule's experiment illustrated in Fig. 20.6. Initially, the gas occupies the volume V_1; after opening of the valve, the gas occupies the volume $V_1 + V_2$ without leaving any trace on the surroundings. However, the opposite process cannot be imagined; it is not possible to bring the gas back to occupy its initial volume V_1 without changing the condition of the surroundings.

We may conclude that reversible processes are only possible if all changes

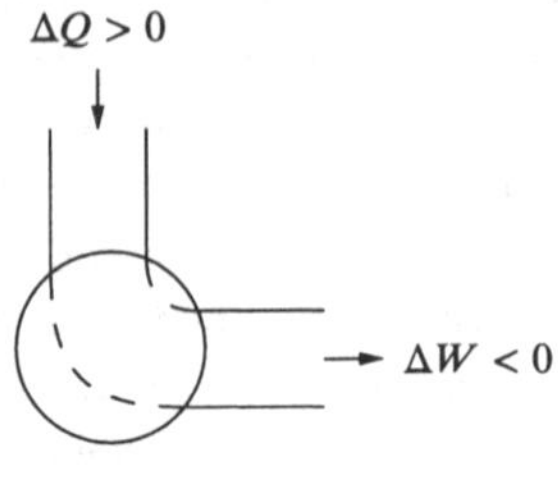

Figure 20.7: Principal sketch of perpetual motion of the second kind.

(of temperature, pressure, etc.) are infinitely small. This implies that a reversible process requires a uniform temperature and no unbalanced forces.

20.7 Introduction to the second law

The second law of thermodynamics can be formulated in various ways. Here we will adopt the most simple and direct physical approach. Just like the first law of thermodynamics was based on the impossibility of a perpetual motion of the first kind, we will base the *second law of thermodynamics* on the impossibility of a *perpetual motion of the second kind*. Let us make the following definition:

A perpetual motion of the second kind is a machine operating in cycles which, in any number of complete cycles, will receive heat from a single heat reservoir and produce an equivalent amount of mechanical work	(20.33)

Such a machine is illustrated in Fig. 20.7. The perpetual motion of the second kind works in cycles; therefore, after one cycle all conditions in the machine are unchanged and from the first law (20.17), we then have $0 = \Delta W + \Delta Q$ where ΔQ is the amount of heat received during one cycle and ΔW is the mechanical work produced during one cycle. Note that per definition, cf. (20.17), ΔW is positive if mechanical work is supplied to the system; in turn, $\Delta W < 0$ then means that the system produces work. We observe that a perpetual motion of the second kind does not violate the first law of thermodynamics. However, it is postulated that

A perpetual motion of the second kind is impossible	(20.34)

This formulation is the famous *Kelvin statement of the second law* from 1851 (William Thomson became Lord Kelvin when he was raised to the peerage); there exist other formulations of the second law - for instance the one proposed

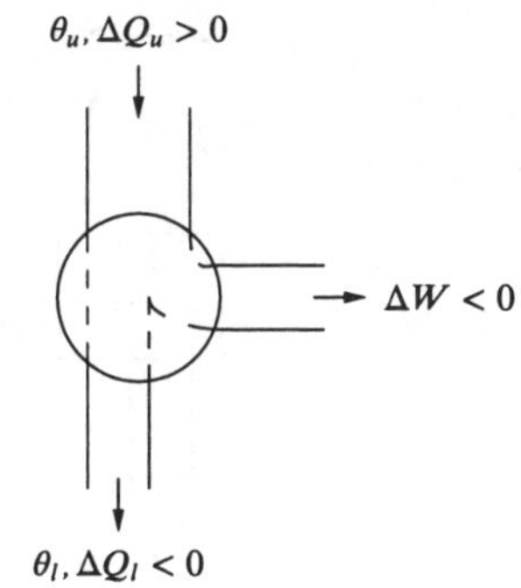

Figure 20.8: Principal sketch of heat engine.

by Clausius in 1850 - but we will later prove that the Kelvin statement and the Clausius statement are identical.

Whereas it is possible to convert mechanical energy completely into heat, cf. for instance the experiment of Joule shown in Fig. 20.3, it is an everyday experience that an amount of heat cannot be converted entirely into mechanical work and this is another, looser formulation of the Kelvin statement.

The impossibility of a perpetual motion of the second kind does not rely as much upon the fact that such a machine has never been constructed, but rather that it has a number of consequences that are in agreement with all experimental evidence and the second law of thermodynamics is therefore accepted as an *axiom*.

According to (20.33) and (20.34), it is impossible to construct a machine that operates in cycles and only receives heat. A machine must therefore also be able to reject heat. A principle sketch of an engine, a *heat engine*, therefore takes the form shown in Fig. 20.8. Here, in one cycle the amount of heat ΔQ_u (> 0) is supplied to the engine, which produces the mechanical work ΔW (where $\Delta W < 0$) while it rejects the amount of heat ΔQ_l (where $\Delta Q_l < 0$). Note that we measure heat and mechanical work as positive quantities when they are supplied to a system. The temperature θ_u at the reservoir providing the amount of heat ΔQ_u and the temperature θ_l at the reservoir receiving the amount of heat ΔQ_l are conveniently taken as constant temperatures since the heat engine is assumed to operate in cycles; that means, after one cycle, all conditions are assumed to be unchanged.

From the first law (20.17), we have

$$0 = \Delta W + \Delta Q_u + \Delta Q_l \tag{20.35}$$

The *efficiency* η of the heat engine is defined as

$$\boxed{\textit{efficiency } \eta = \frac{-\Delta W}{\Delta Q_u}} \tag{20.36}$$

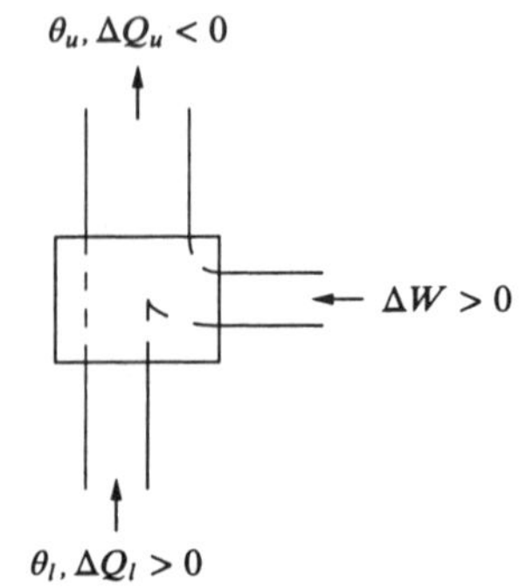

Figure 20.9: Principal sketch of refrigerator.

where it is noted that $-\Delta W > 0$. Combination of (20.35) and (20.36) yields

$$\boxed{\eta = 1 + \frac{\Delta Q_l}{\Delta Q_u}} \tag{20.37}$$

where $\Delta Q_l < 0$ and $\Delta Q_u > 0$.

From (20.36) appears that the efficiency η is a non-negative quantity. If $\eta = 1$, (20.37) shows that $\Delta Q_l = 0$ and the heat engine then turns into a perpetual motion of the second kind that only receives heat from one reservoir (ΔQ_u) and produces an equivalent amount of mechanical work (ΔW). But according to the Kelvin statement (20.34) such a machine is impossible and we conclude that

$$\boxed{0 \leq \eta < 1} \tag{20.38}$$

Evidently, when constructing a heat engine the aim is to achieve an efficiency that is as large as possible, but this efficiency is always smaller that unity. We shall later return to this subject.

A machine that operates in cycles and that represents the opposite of a heat engine is a *refrigerator*, cf. Fig. 20.9. Here, the mechanical work ΔW is supplied to the refrigerator, which receives the amount of heat ΔQ_l from the reservoir having the low temperature θ_l and rejects the amount of heat ΔQ_u to the reservoir having the high temperature θ_u.

Let us next turn to another formulation of the second law. It is an everyday experience that heat does not, by itself, flow from a cold region towards a warmer region. This is the content of the *Clausius statement of the second law* from 1850, which, in a more precise formulation, reads

> *It is impossible to construct a machine operating in cycles which, in any number of complete cycles, only results in a transfer of heat from a reservoir at one temperature to a reservoir at a higher temperature* (20.39)

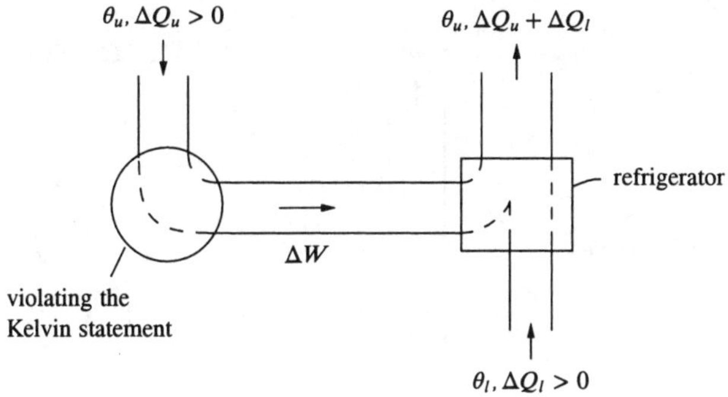

Figure 20.10: Refrigerator is powered by a perpetual motion of the second kind.

We will now prove that the Kelvin statement (20.34) and the Clausius statement (20.39) are equivalent. The proof consists of showing that if it were possible to violate one statement, the other statement would also be violated.

Suppose first that we have a heat engine that violates the Kelvin statement, i.e. it receives the amount of heat ΔQ_u (> 0) and produces an equivalent amount of mechanical work ΔW (< 0), where $\Delta W + \Delta Q_u = 0$. As shown in Fig. 20.10, this mechanical work is used to drive a refrigerator, which also receives the amount of heat ΔQ_l (> 0) at the reservoir having the temperature θ_l. Note that the mechanical work ΔW is considered as a positive quantity when it is supplied to the system (the machine) and as a negative quantity when it is produced by the system. The refrigerator rejects the amount of heat $\Delta Q_u + \Delta Q_l$ to the high-temperature reservoir. The combined result of this machinery is to take the amount of heat ΔQ_l from the low-temperature reservoir and deliver it to the high-temperature reservoir. We have then proved that violation of the Kelvin statement leads to violation of the Clausius statement.

Suppose next that we have a machine that violates the Clausius statement, i.e. it receives the amount of heat ΔQ_l from the low-temperature reservoir and transfers all this heat to the high-temperature reservoir. Referring to Fig. 20.11, this amount of heat ΔQ_l at the high-temperature reservoir as well as the additional amount of heat ΔQ_{add} is supplied to the heat engine that produces the mechanical work ΔW and rejects the amount of heat ΔQ_l to the low-temperature reservoir. The combined result of this machinery is to take the amount of heat ΔQ_{add} from the high-temperature reservoir and produce an equivalent amount of mechanical work ΔW. But this is a perpetual motion of the second kind, which violates the Kelvin statement.

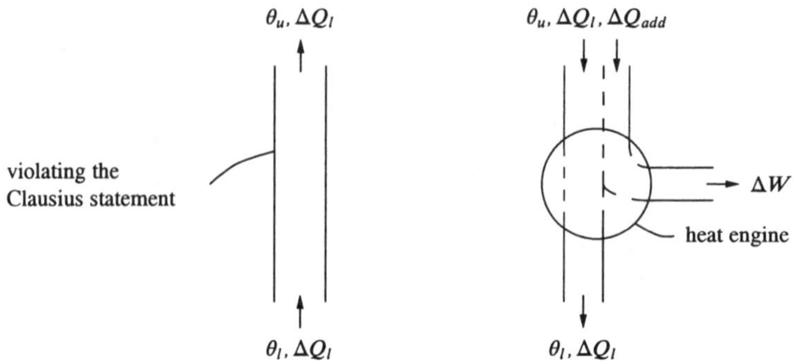

Figure 20.11: Combined effect of machine that violates the Clausius statement with a heat engine.

We have therefore proved that

The Kelvin statement and the Clausius statement of the second law are equivalent statements

20.8 Efficiency of various heat engines

Let us next prove the following statement

The efficiency of an irreversible heat engine is always less than the efficiency of a reversible heat engine operating between the same reservoirs

With evident notation, this statement may be written as

$$\boxed{\eta_{irr} < \eta_{rev}} \tag{20.40}$$

To prove (20.40), consider an irreversible heat engine and a reversible heat engine working between the same reservoirs, as shown in Fig. 20.12. The first law yields

$$0 = \Delta W^{irr} + \Delta Q_u^{irr} + \Delta Q_l^{irr} \quad ; \quad 0 = \Delta W^{rev} + \Delta Q_u^{rev} + \Delta Q_l^{rev} \tag{20.41}$$

The two engines are operated such that they deliver the same mechanical work, i.e.

$$\Delta W^{irr} = \Delta W^{rev} = \Delta W \tag{20.42}$$

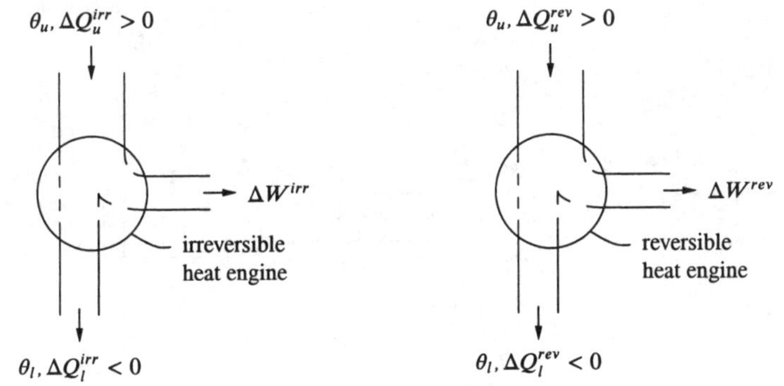

Figure 20.12: Irreversible and reversible heat engine working between the same reservoirs.

From (20.41) then follows that

$$\Delta Q_u^{irr} - \Delta Q_u^{rev} + \Delta Q_l^{irr} - \Delta Q_l^{rev} = 0 \tag{20.43}$$

Since the heat engine to the right in Fig. 20.12 is reversible, it can be reversed and thereby operate as a refrigerator. The irreversible heat engine is now coupled to this refrigerator and due to (20.42), all the mechanical work produced by the irreversible heat engine is used to drive the refrigerator, cf. Fig. 20.13.

Assume that the high-temperature reservoir receives heat from the combined machine, i.e.

$$\Delta Q_u^{irr} - \Delta Q_u^{rev} < 0 \quad \text{(assumption)} \tag{20.44}$$

Due to (20.43), this implies that $\Delta Q_l^{irr} - \Delta Q_l^{rev} > 0$, i.e. the low-temperature reservoir supplies heat to the combined machine. We are then led to a situation where the only result of the combined machine is a transfer of heat from a low-temperature reservoir to a high-temperature reservoir. But according to the Clausius statement (20.39) such a machine is impossible.

Assume next that

$$\Delta Q_u^{irr} - \Delta Q_u^{rev} = 0 \quad \text{(assumption)} \tag{20.45}$$

i.e. the total heat transfer to the high-temperature reservoir is nil. Due to (20.43), the total heat transfer from the low-temperature reservoir is then also nil. Within the combined machine, a number of processes takes place. However, after one cycle all conditions of the combined machine (the system) as well as of the reservoirs (the surroundings) have been restored. According to (20.31), this means that the processes are reversible, but this is in contradiction with the assumption that the combined machine involves an irreversible heat engine.

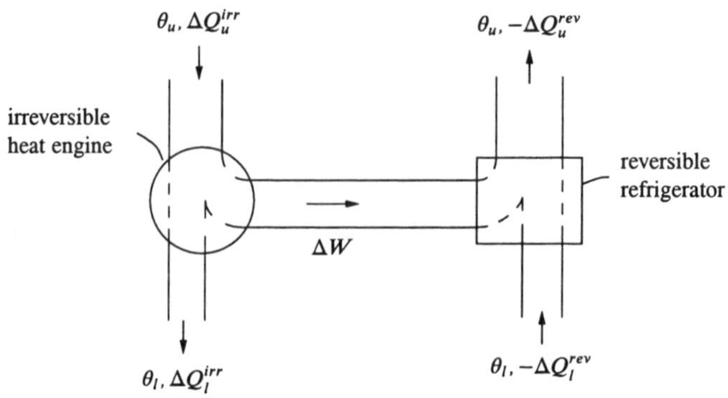

Figure 20.13: Irreversible heat engine driving a (reversible) refrigerator.

Since also (20.45) must be rejected, we are left with

$$\Delta Q_u^{irr} - \Delta Q_u^{rev} > 0 \quad \Rightarrow \quad \frac{\Delta Q_u^{rev}}{\Delta Q_u^{irr}} < 1 \tag{20.46}$$

where it was used that $\Delta Q_u^{irr} > 0$. Use of (20.37) and (20.43) yield

$$\eta_{irr} = \frac{\Delta Q_u^{irr} + \Delta Q_l^{irr}}{\Delta Q_u^{irr}} = \frac{\Delta Q_u^{rev} + \Delta Q_l^{rev}}{\Delta Q_u^{irr}} = \frac{\Delta Q_u^{rev}}{\Delta Q_u^{irr}} \eta_{rev} < \eta_{rev}$$

where advantage was taken of (20.46). This completes the proof of (20.40).

We will also prove that

All reversible heat engines that operate between the same reservoirs have the same efficiency	(20.47)

This statement is quite surprising, since it says that irrespective of how the cycles are devised and irrespective of the working medium used in the heat engines, the efficiency of reversible heat engines is the same, when the two reservoirs are the same.

To prove (20.47), Fig. 20.12 is considered again, but now the irreversible heat engine to the left is replaced by any reversible heat engine. The two heat engines are operated so that they deliver the same mechanical work $\Delta W_{left}^{rev} = \Delta W_{right}^{rev}$, cf. (20.42). The reversible heat engine to the right is again reversed so that it operates as a refrigerator that is driven by the heat engine to the left, cf. Fig. 20.13. Again we assume condition (20.44), which now reads $\Delta Q_u^{left} - \Delta Q_u^{right} < 0$, and again we conclude that this assumption violates the Clausius statement. Since we can choose arbitrarily which of the reversible heat engines that is to work as a refrigerator, we are left with the possibility that ΔQ_u^{left} −

$\Delta Q_u^{right} = 0$. In combination with (20.42) ($\Delta W_{left}^{rev} = \Delta W_{right}^{rev}$), it follows that $\eta_{rev}^{left} = \eta_{rev}^{right}$ and statement (20.47) have then been proved.

For any two given reservoirs, restriction (20.38), i.e. $0 \leq \eta < 1$, holds and we have found that the largest efficiency is provided by a reversible heat engine. Any attempt to construct an efficient heat engine is therefore confined to reduce the irreversibilities as much as possible.

20.9 The Carnot process

It was shown that all reversible heat engines possess the same efficiency. In order to determine this efficiency, we may select any reversible process and the so-called *Carnot process* turns out to be especially convenient for this purpose. The reason is that the Carnot process is devised such that the requirement of reversibility can be fulfilled, at least in principle.

The Carnot process was suggested by Carnot in 1824 and it consists of the following cycle:

> *Carnot process:*
> 1-2 *reversible isothermal expansion*
> 2-3 *reversible adiabatic expansion*
> 3-4 *reversible isothermal compression*
> 4-1 *reversible adiabatic compression*

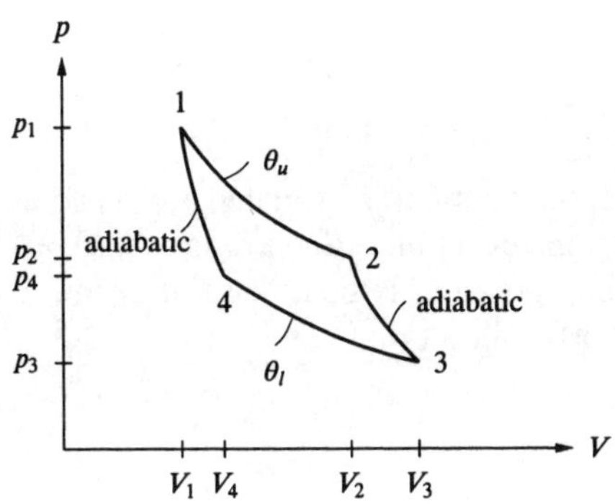

Figure 20.14: Carnot process with ideal gas; reversibility is assumed.

To illustrate this process, we may select as the working medium an ideal gas in a cylinder with a frictionless piston, cf. Fig. 20.14.

Initially in process 1-2, the gas temperature is θ_u and the cylinder are brought into contact with the high-temperature reservoir having the temperature θ_u; an isothermal expansion is then enforced. In order that the process be reversible, the difference between the temperature θ_u and the gas temperature should be infinitely small; this requires an infinitely slow expansion. Moreover, a reversible process requires that the piston moves in the cylinder in a frictionless manner. During this isothermal expansion, the gas receives heat from the high-temperature reservoir according to (20.29) and we obtain

$$\Delta Q_u = mR\theta_u ln\frac{V_2}{V_1} \tag{20.48}$$

In process 2-3, the cylinder is insulated and an expansion occurs. In order that the process be reversible, no friction is allowed and the expansion should proceed infinitely slowly. According to (20.30), the temperature decreases and the expansion is continued until the temperature θ_l has been reached. From (20.30), we obtain

$$-ln\frac{V_3}{V_2} = \int_{\theta_u}^{\theta_l} \frac{c_V}{R}\frac{d\theta}{\theta}$$

or

$$lnV_3 = lnV_2 + \int_{\theta_l}^{\theta_u} \frac{c_V}{R}\frac{d\theta}{\theta} \tag{20.49}$$

where we have changed the integration direction.

In process 3-4, the cylinder is brought into contact with the low-temperature reservoir having the temperature θ_l. A reversible isothermal compression occurs and (20.29) yields that heat is transferred to the low-temperature reservoir, where

$$\Delta Q_l = mR\theta_l ln\frac{V_4}{V_3} \tag{20.50}$$

Finally in process 4-1, the cylinder is insulated and an adiabatic compression occurs; the process is again reversible. According to (20.30), the temperature then increases and the compression is continued until the temperature θ_u has been reached. From (20.30) follows that

$$-ln\frac{V_1}{V_4} = \int_{\theta_l}^{\theta_u} \frac{c_V}{R}\frac{d\theta}{\theta}$$

or

$$lnV_1 = lnV_4 - \int_{\theta_l}^{\theta_u} \frac{c_V}{R}\frac{d\theta}{\theta} \tag{20.51}$$

From (20.37), (20.48) and (20.50), we obtain

$$\eta_{rev} = 1 + \frac{\theta_l}{\theta_u}\frac{lnV_4 - lnV_3}{lnV_2 - lnV_1}$$

and insertion of (20.49) and (20.51) yields the result

$$\boxed{\eta_{rev} = 1 - \frac{\theta_l}{\theta_u}} \tag{20.52}$$

Therefore, the efficiency of a reversible heat engine is entirely determined by the temperatures of the reservoirs. We have obtained this conclusion by considering a reversible process in terms of a Carnot process using an ideal gas as working medium, but since all reversible heat engines have the same efficiency, (20.52) holds for all reversible heat engines, irrespective of the specific reversible cycle and irrespective of the working medium.

20.10 Thermodynamic temperature scale

The concept of temperature T and absolute temperature θ was discussed in Section 20.1. However, up to now the measurement of temperature hinges on the fact that some properties of a medium change with its coldness or hotness; such a measurement may, for instance, be based on expansion or contraction of mercury in a capillary, as used in a traditional thermometer. As another example, if a vessel of constant volume is filled with ideal gas, then the (absolute) temperature is proportional to the gas pressure, see Fig. 20.1. This means that by measuring the pressure, we can measure the temperature. Such a thermometer is called a *constant-volume gas thermometer* and the corresponding temperature is called the *ideal gas temperature scale*, which, as we have seen, is identical to the absolute temperature, i.e.

Constant-volume gas thermometer ⇒ *ideal gas temperature scale = absolute temperature* θ

However, irrespective of the thermometer we adopt so far, the temperature measurement, in itself, relies on the properties of the substances used in the thermometer.

Theoretically, it would be of importance if the temperature measurement could be performed in such a manner that it is independent of the properties of some substances - at least in principle. Such a temperature scale is called a *thermodynamic temperature scale* and we have

A temperature scale that is independent of the properties of the substances that are used to measure temperature is called a thermodynamic temperature scale

It turns out to be possible to achieve a thermodynamic temperature scale and for that purpose, we first combine (20.37) and (20.52) to obtain

$$\boxed{\frac{-\Delta Q_l}{\Delta Q_u} = \frac{\theta_l}{\theta_u} \qquad \textit{reversible machine}} \tag{20.53}$$

To derive this relation, we used that for the same two reservoirs, all reversible heat engines have the same efficiency irrespective of their construction, working process and working medium. By specifically choosing a reversible process in terms of a Carnot process using an ideal gas as working medium, (20.53) was then derived. Therefore, (20.53) holds for all reversible machines.

The important point is that the ratio $-\Delta Q_l/\Delta Q_u$ is independent of the nature of the working substances, since the efficiency η_{rev} is so. Since the energy of amounts of heat can be measured directly by means of calorimetry, cf. the Joule test set-up shown in Fig. 20.3, without having defined a temperature scale, the ratio $-\Delta Q_l/\Delta Q_u$ can be measured. If we then consider (20.53) as a matter of definition, then we can measure the temperature ratio θ_l/θ_u only by means of calorimetry. Running a reversible heat engine and since the ratio $-\Delta Q_l/\Delta Q_u$ is independent of the involved substances also the temperature ratio θ_l/θ_u is independent of the involved substances. The thus-defined temperature ratio θ_l/θ_u therefore establishes a thermodynamic temperature scale. In principle, these considerations were put forward by Kelvin in 1854.

Since (20.53) has been shown to be satisfied if temperatures are measured by an ideal gas temperature scale = absolute temperature, we conclude that

Ideal gas temperature scale = absolute temperature = thermodynamic temperature

To be precise, (20.53) only defines the ratio θ_l/θ_u. To identify the thermodynamic temperature scale uniquely, one more piece of information is needed. In 1954, the *triple point of water* (the state at which ice, water and vapor exist in equilibrium) was assigned the value 273 K (273.16 K to be precise) and we are then back to (20.2).

From (20.52), we have

$$\theta_l = \theta_u(1 - \eta_{rev})$$

and since $0 \leq \eta_{rev} < 1$, cf. (20.38), we are led to the conclusion that the lowest attainable temperature θ_l is always greater than zero. This observation supports our previous conclusion given by (20.1).

20.11 Entropy - Clausius's inequality

Relation (20.53) holds when a reversible cycle is performed and it may be written as

$$\frac{\Delta Q_u}{\theta_u} + \frac{\Delta Q_l}{\theta_l} = 0 \qquad \text{reversible cycle}$$

To indicate that a cycle is performed, this expression may be written as

$$\oint \frac{\delta Q}{\theta} = 0 \qquad \textit{reversible cycle} \tag{20.54}$$

This relation is of significant interest. It was shown previously that heat is not a state function, cf. (20.10), therefore during a cycle we have, in general, $\oint \delta Q \neq 0$. Indeed, this was the motivation for denoting an infinitely small amount of heat by δQ and not dQ, since the latter notation would suggest that there exists a function Q that depends on some variables and then Q would be a state function, in contrast to (20.10). Referring to the discussion following (20.11) and (20.12), an infinitely small amount of heat is not a perfect differential and this is emphasized by using the notation δQ.

Any function, say $z = z(y_\alpha)$, where z depends on some variables y_α, will have the property that during any cycle, we have $\oint dz = 0$. If the variables y_α are state variables, then $z = z(y_\alpha)$ would be a state function. However, whereas heat is not a state function, (20.54) implies that there exists a state function, the *entropy S of the system*, defined by

$$dS = \frac{\delta Q}{\theta} \qquad \text{reversible process} \tag{20.55}$$

To check that S is a state function, we calculate the change of S during a reversible cycle and due to (20.54), it follows that $\oint dS = 0$, in accordance with our expectations.

To discuss the concepts of a state function and a perfect differential in more detail, consider the following expression

$$\text{infinitely small quantity} = c_\alpha dy_\alpha \tag{20.56}$$

where the functions c_α depend on the variables y_α. Following, for instance, Kreyszig (1962) we have

A necessary and sufficient condition for (20.56) being a perfect differential is that

$$\frac{\partial c_\alpha}{\partial y_\beta} = \frac{\partial c_\beta}{\partial y_\alpha} \tag{20.57}$$

We then conclude that (20.56) may be written as

$$\begin{aligned} da &= c_\alpha dy_\alpha && \text{if perfect differential} \\ \delta a &= c_\alpha dy_\alpha && \text{if not a perfect differential} \end{aligned}$$

Any function $z = z(y_\alpha)$ fulfills requirement (20.57) since

$$dz = \frac{\partial z}{\partial y_\alpha} dy_\alpha \qquad \text{i.e.} \qquad \frac{\partial}{\partial y_\beta}\left(\frac{\partial z}{\partial y_\alpha}\right) = \frac{\partial}{\partial y_\alpha}\left(\frac{\partial z}{\partial y_\beta}\right)$$

On the other hand, consider for instance the expression

$$\delta a = -y_2 dy_1 + y_1 dy_2 \tag{20.58}$$

Since $\partial(-y_2)/\partial y_2 = -1 \neq \partial(y_1)/\partial y_1 = 1$, (20.57) is not fulfilled and the quantity δa is not a perfect differential, as already indicated by the notation δa. However, if (20.58) is multiplied by the factor $1/y_2^2$, we obtain

$$\frac{\delta a}{y_2^2} = -\frac{1}{y_2}dy_1 + \frac{y_1}{y_2^2}dy_2 \tag{20.59}$$

It is now observed that

$$\frac{\partial}{\partial y_2}(-\frac{1}{y_2}) = \frac{1}{y_2^2} = \frac{\partial}{\partial y_1}(\frac{y_1}{y_2^2}) = \frac{1}{y_2^2}$$

i.e. (20.57) is fulfilled and, indeed, (20.59) may be integrated to provide

$$db = \frac{\delta a}{y_2^2} \quad \text{where} \quad b = -\frac{y_1}{y_2}$$

In the considered case, we could convert expression (20.58) into a perfect differential by multiplication with a factor, in the present case the factor $1/y_2^2$. Such factor is called an *integrating factor*.

With this discussion, it appears that the factor $1/\theta$ is the integrating factor which for a reversible process transforms the quantity δQ into the perfect differential dS, cf. (20.55).

After these preliminary remarks, we introduce the *entropy S of the body* in the following manner, as suggested by Clausius in 1865:

$$\text{\textit{Clausius's inequality}} \quad dS \begin{cases} = \dfrac{\delta Q}{\theta} & \text{\textit{reversible process}} \\ > \dfrac{\delta Q}{\theta} & \text{\textit{irreversible process}} \end{cases} \tag{20.60}$$

Moreover, irrespective of whether the process is reversible or irreversible, we assume that

The entropy is a state function (20.61)

Equations (20.60) and (20.61) express the second law of thermodynamics in a mathematical form. Just like the first law, the second law is a postulate, but since all experimental evidence supports this postulate, it is accepted as an axiom.

To illustrate the use of (20.60) and (20.61) and to show that they contain the various forms of the second law that we discussed previously, consider first a reversible heat engine having performed one cycle in a Carnot process, cf. Fig. 20.14. After one cycle, the system (the machine) has returned to its initial state. Since the entropy S of the system is a state function, it therefore takes the same value before and after the cycle. Therefore, (20.60) yields

$$0 = \oint dS = \oint \frac{\delta Q}{\theta} = \frac{\Delta Q_u}{\theta_u} + \frac{\Delta Q_l}{\theta_l} \quad \text{i.e.} \quad \frac{\Delta Q_l}{\Delta Q_u} = -\frac{\theta_l}{\theta_u}$$

Insertion into (20.37) gives

$$\eta_{rev} = 1 - \frac{\theta_l}{\theta_u}$$

in accordance with (20.52).

Consider next an irreversible heat engine having performed one cycle in a process similar to that of Carnot, but without the assumption of reversibility. Again, the entropy S is a state function and it therefore takes the same value before and after the cycle. Consequently, (20.60) gives

$$0 = \oint dS > \oint \frac{\delta Q}{\theta} = \frac{\Delta Q_u}{\theta_u} + \frac{\Delta Q_l}{\theta_l} \quad \text{i.e.} \quad \frac{\Delta Q_l}{\Delta Q_u} < -\frac{\theta_l}{\theta_u}$$

Insertion into (20.37) then yields

$$\eta_{irr} = 1 + \frac{\Delta Q_l}{\Delta Q_u} < 1 - \frac{\theta_l}{\theta_u} = \eta_{rev}$$

in accordance with (20.40).

It is of interest that whereas the first law postulates the existence of a state function, the internal energy, the second law postulates the existence of another state function, the entropy. The word 'entropy' originates from Greek and means 'change'.

The first law does not discriminate whether a process goes in one or another direction; it simply performs the book-keeping of the various energy changes. On the other hand, the second law is discriminative in the sense that it informs us whether a process is possible or not; a process is only possible if it satisfies Clausius's inequality (20.60).

20.12 Maximum entropy at thermodynamic equilibrium

Since the second law selects the processes that are possible, it indicates a trend in nature. It may therefore be used to determine a state of *thermodynamic equilibrium*. By definition, we have

> *Consider a system where no change in the boundary conditions occurs, i.e. no mechanical work input and no heat input.*
> *The system is then in thermodynamic equilibrium if no changes of the state variables can occur* (20.62)

Thus thermodynamic equilibrium implicitly refers to static conditions. Since $\delta W = \delta Q = 0$, the first law (20.15) implies that $dU = 0$, i.e. the internal energy is constant. A necessary condition for thermodynamic equilibrium is therefore that the internal energy is constant. With $\delta Q = 0$, the second law (20.60) gives $dS \geq 0$; therefore, if $\delta W = \delta Q = 0$ then the only processes that are allowable are those which imply that $dS \geq 0$. If it is the case that the entropy takes a maximum value, then any change of the state variables would imply that S decreases, which is impossible, and we therefore have thermodynamic equilibrium. Consequently:

> *If $\delta W = \delta Q = 0$ and if the entropy has a maximum value, thermodynamic equilibrium exists* (20.63)

This statement is due to Gibbs and it was formulated in 1876.

20.13 The second law and the Clausius-Duhem inequality

It is timely to establish a form of the second law that is of relevance for constitutive mechanics. First of all, we define the *specific entropy* s per unit mass, i.e.

$$\boxed{S = \int_V \rho s dV} \tag{20.64}$$

The specific entropy s is a state function just like the total entropy S of the body is a state function.

Dividing Clausius's inequality (20.60) by dt, we obtain

$$\dot{S} \geq \frac{1}{\theta}\frac{\delta Q}{dt} \tag{20.65}$$

This expression holds for a homogenous system (body) where the temperature is constant. For a body within which the temperature is allowed to vary we obtain with (20.12)

$$\boxed{\dot{S} \geq \int_V \frac{r}{\theta} dV - \int_S \frac{q_i}{\theta} n_i dS \quad \text{global form of second law} = \text{Clausius-Duhem inequality}} \tag{20.66}$$

This is the *global form* of the second law and following the terminology suggested by Truesdell and Toupin (1960), it forms the celebrated *Clausius-Duhem inequality* in its global form. The volume term in (20.66) is often called the *entropy source term* whereas the boundary term in (20.66) often is called the *entropy flux term*.

An important point here is that the temperatures in the previous discussion were the temperatures of the reservoirs, i.e. the temperatures at the boundary of the system. This observation motivates the boundary term $\int_S q_i n_i/\theta \, dS$ in

(20.66). However, in the previous discussion no reference was made to the temperature within the system (within the heat engine) and one may therefore question the volume term $\int_V r/\theta\, dV$ in (20.66). If the temperature is constant within the body (the system), (20.66) certainly reduces to (20.65). In general, however, the temperature is not uniform and with the just mentioned remarks of the volume term in (20.66) in mind, it is not evident that the total entropy S obeys inequality (20.66). A further discussion of this delicate problem goes beyond the scope of the present exposition and the interested reader may, for instance, consult Gurtin and Williams (1966), Truesdell (1969) and Hutter (1977). Here, we adopt (20.66) as an axiom and it is the global form of the second law.

According to the divergence theorem of Gauss, we have

$$\int_S \frac{q_i}{\theta} n_i dS = \int_V (\frac{q_i}{\theta})_{,i} dV \tag{20.67}$$

Insertion of (20.64) and (20.67) into (20.66) and observing that due to the assumption of small strains, the mass density ρ can be considered as constant, we obtain

$$\int_V [\rho \dot{s} - \frac{r}{\theta} + (\frac{q_i}{\theta})_{,i}] dV \geq 0$$

As this expression holds for arbitrary regions V, it is concluded that

$\rho \dot{s} - \frac{r}{\theta} + \frac{q_{i,i}}{\theta} - \frac{q_i \theta_{,i}}{\theta^2} \geq 0$	*local form of second law* *= Clausius-Duhem inequality*

(20.68)

This is the *local form* of the second law, i.e. the local form of the Clausius-Duhem inequality that holds at every point in the body.

It is recalled that the equality sign holds for reversible processes whereas the inequality sign holds for irreversible processes. The essence of constitutive mechanics has now been boiled down to fulfillment of the Clausius-Duhem inequality. If our suggestions for constitutive models fulfill this inequality, the models satisfy the second law and thereby fulfill all formal physical requirements. Whether the suggested models provide a close approximation to the actual behavior of the materials in question is another issue where thermodynamics cannot give any help. Therefore

Any constitutive model that fulfills the Clausius-Duhem inequality, fulfills all formal requirements

One may wonder whether fulfillment of the first law also places restrictions on constitutive modeling. However, as we shall see in the next chapter the first law turns into the *heat equation*, which places no restrictions on constitutive modeling, but which instead determines the temperature distribution within the body.

20.14 Various approaches to thermodynamics

The essence of this chapter is the first law in its global and local forms, cf. (20.15) and (20.21), as well as the second law, i.e. the Clausius-Duhem inequality, in its global and local forms, cf. (20.66) and (20.68). Moreover, the specific internal energy u and the specific entropy s are postulated to be state functions, i.e. functions that only depend on the current values of those variables, the state variables, that characterize the state, cf. (20.3) and (20.4).

To achieve to this point, the essential principles of the classic thermodynamics were presented and they served as a physical motivation for these main results. This is in contrast with the mainstream of textbooks within constitutive modeling where the first and second laws are simply presented as principles that must be satisfied. Here, we have taken a somewhat less abstract viewpoint by trying to motivate these laws by various physical arguments and the hope is that a quantity like the entropy will appeal to the physical insight of the reader and not just become an abstract quantity that the reader is able to manipulate in various expressions.

However, it would be unfair not to mention that our approach may be criticized and that other approaches to thermodynamics exist. While the scientific community accepts the first law as truly being an axiom, arguments may be raised against the second law and this manifests itself in different formulations of the second law.

For reversible processes, we could argue strictly that the factor $1/\theta$ is the integrating factor that transforms the quantity δQ into the perfect differential $dS = \delta Q/\theta$, cf. (20.55) and (20.60); this implies that the entropy S certainly is a state function. However, as mentioned in Section 20.6, reversible processes are only possible if all changes (of temperature, pressure, etc.) are infinitely small. This implies that a reversible process requires a uniform temperature and no unbalanced forces. For reversible processes to occur, the body must therefore be subjected to conditions that are in equilibrium and this is the reason why this part of thermodynamics often is termed *thermostatics*. In that case, we can prove with confidence that the entropy is a state function.

However, for irreversible processes the body is not in equilibrium (in the manner defined above) and it may then be questioned whether the entropy is a state function, cf. for instance Meixner (1970) and Hutter (1977). This is the reason for two schools of thought: the so-called *irreversible thermodynamics* and the so-called *rational thermodynamics*.

The school of irreversible thermodynamics focusses on the fact that under equilibrium (reversible) conditions, the entropy is a state function. Therefore, in irreversible thermodynamics the irreversible processes are thought of as being close to equilibrium and only slight, i.e. linear, deviations are allowed.

Within the school of rational thermodynamics with which names like Truesdell, Coleman, Toupin, Eringen and Noll are associated, it is simply accepted that the entropy is a state function also for irreversible processes far from equilibrium and the second law is then taken in the form of the Clausius-Duhem inequality. This line of thought then investigates the consequences for constitutive theories when they are to fulfill this inequality. The result is a field theory of great beauty and with very important consequences.

Here, we will adopt in the following chapters the viewpoint of the school of rational thermodynamics, that means accept the first and second laws as stated by (20.16), (20.21) and (20.61), (20.68), respectively. However, instead of simply stating these laws, as it is common in rational thermodynamics, we have tried to provide a physical motivation for these laws. In the present chapter, we have also deviated from the school of rational thermodynamics in another aspect by first introducing the concept of temperature as something we can measure and accept and then obtaining the entropy as a derived, i.e. related, concept. This means that we have used the temperature as a so-called *primitive quantity*. Within rational thermodynamics, the entropy is introduced first as a primitive quantity and the temperature then becomes a derived quantity. However, irrespective of which approach is adopted, the resulting relations become the same, but the present approach has been adopted since it seems to be more appealing to the physical intuition of the reader.

21 THERMODYNAMIC FRAMEWORK FOR CONSTITUTIVE MODELING

In this chapter, we will establish the framework for deriving constitutive models that is provided by thermodynamics. In the next chapters, we shall then take full advantage of this framework and demonstrate that we may retrieve a number of theories discussed previously and also establish a number of new constitutive models.

Let us first recall the first and second laws. From (20.21), the first law in its local form is given by

$$\boxed{\rho\dot{u} = \dot{\varepsilon}_{ij}\sigma_{ij} + r - q_{i,i} \qquad \textit{first law}} \tag{21.1}$$

where u is the specific internal energy, q_i is the heat flux vector and r is the heat supply per unit time and unit volume. Moreover, the internal energy u is a state function, i.e. a function that only depends on the current values of the state variables. The state variables are variables that characterize the state of the material and, as previously discussed in Section 20.3, it is not trivial to identify these variables; we shall later return to this important point.

The second law expressed in terms of the local form of the Clausius-Duhem inequality is given by (20.68), i.e.

$$\boxed{\rho\dot{s} - \frac{r}{\theta} + \frac{q_{i,i}}{\theta} - \frac{q_i\theta_{,i}}{\theta^2} \geq 0 \qquad \begin{array}{l}\textit{second law}\\ = \textit{Clausius-Duhem inequality}\end{array}} \tag{21.2}$$

where s is the specific entropy and θ is the absolute temperature. The entropy s is a state function.

The term $q_{i,i} - r$ appearing in (21.2) can be eliminated by means of the first law (21.1) and the Clausius-Duhem inequality then turns into what is called the *dissipation inequality*

$$\boxed{\begin{array}{l} \textit{Dissipation inequality in terms of internal energy } u \\ \gamma \geq 0 \\ \textit{where} \\ \gamma \equiv \theta\rho\dot{s} - \rho\dot{u} + \dot{\varepsilon}_{ij}\sigma_{ij} - \dfrac{q_i\theta_{,i}}{\theta} \end{array}} \tag{21.3}$$

where it is emphasized that the internal energy u enters the dissipation inequality. It is recalled that the inequality sign holds for irreversible processes whereas the equality sign holds for reversible processes. The quantity γ defined by (21.3) has the dimension of energy rate per unit volume, i.e. [J/s m^3], and it is zero for reversible processes and positive for irreversible processes; this suggests the terminology of γ being the *dissipation*. Moreover, as emphasized in relation to (20.68), any constitutive model that fulfills (21.3) fulfills all formal requirements enforced by thermodynamics.

In addition to the state functions u and s, it turns out to be convenient to introduce another state function proposed by Helmholtz in 1882, namely *Helmholtz' free energy* ψ per unit mass defined by

$$\boxed{\begin{array}{l} \textit{Helmholtz' free energy} \\ \psi = u - s\theta \\ \psi \textit{ is a state function} \end{array}} \tag{21.4}$$

It is evident that the absolute temperature θ can be considered as a state variable, since it certainly is a quantity that - together with other quantities - characterizes the state of the body, cf. definition (20.3). Moreover, since u and s are state functions, it follows from the definition of ψ that it is a state function, cf. (20.5).

From (21.3) and (21.4), we then obtain the following alternative formulation of the dissipation inequality

$$\boxed{\begin{array}{l} \textit{Dissipation inequality in terms of free energy } \psi \\ \gamma \geq 0 \\ \textit{where} \\ \gamma \equiv -\rho(\dot{\psi} + s\dot{\theta}) + \dot{\varepsilon}_{ij}\sigma_{ij} - \dfrac{q_i\theta_{,i}}{\theta} \end{array}} \tag{21.5}$$

21.1 Thermo-elastic materials

In Section 20.3, we introduced the concept of state variables and it was noticed that the choice of state variables is not trivial; indeed it depends upon what we are interested in.

In order to approach this problem, it seems reasonable to consider the simplest possible situation, which certainly is that of a reversible process. In Section 20.6, we found that a reversible process requires infinitely small temperature gradients. Assume therefore that the temperature does not vary over the body and that a reversible process occurs. Intuitively, we would expect that the material then is to respond in an elastic manner and, indeed, our expectations are fulfilled as will appear shortly. Since $\gamma = 0$ and $\theta_{,i} = 0$, (21.5) reduces to

$$\dot{\psi} = -s\dot{\theta} + \frac{1}{\rho}\sigma_{ij}\dot{\varepsilon}_{ij} \tag{21.6}$$

Since the free energy ψ is a state function, the expression above suggests that

$$\boxed{\psi = \psi(\theta, \varepsilon_{ij}) \qquad \textit{thermo-elastic material}} \tag{21.7}$$

Indeed, we obtain

$$\dot{\psi} = \frac{\partial\psi}{\partial\theta}\dot{\theta} + \frac{\partial\psi}{\partial\varepsilon_{ij}}\dot{\varepsilon}_{ij} \tag{21.8}$$

and a comparison with (21.6) supports expression (21.7). Moreover, subtracting (21.8) from (21.6) gives

$$0 = (-s - \frac{\partial\psi}{\partial\theta})\dot{\theta} + (\frac{1}{\rho}\sigma_{ij} - \frac{\partial\psi}{\partial\varepsilon_{ij}})\dot{\varepsilon}_{ij} \tag{21.9}$$

Since we can control the spatially uniform temperature and the strains independently, it is possible to assign arbitrary values to $\dot{\theta}$ and $\dot{\varepsilon}_{ij}$. Expression (21.9) then implies

$$\boxed{s = -\frac{\partial\psi}{\partial\theta}; \qquad \sigma_{ij} = \rho\frac{\partial\psi}{\partial\varepsilon_{ij}}} \tag{21.10}$$

Since the entropy s is obtained from ψ by differentiation with respect to the temperature θ, use is made of the phrase that ψ serves as a *potential function* for the entropy. Likewise, the stresses are obtained from ψ by differentiation with respect to the strains and the free energy ψ then serves as a potential function for the stresses.

We found that the quantities θ and ε_{ij} appeared in a natural manner as state variables for the free energy ψ and they are therefore the *natural* or *canonical state variables* for ψ when considering a thermo-elastic material. We might also mention that as both the temperature θ and the strains ε_{ij} can be measured, i.e. observed, θ and ε_{ij} are examples of *observable state variables*; we shall later encounter state variables that are not observable.

To demonstrate that (21.7) and thereby (21.10) define the response of a thermo-elastic material - as already anticipated - let us consider the special case where $\dot{\theta} = 0$, that is, the temperature does not change with time (we have previously assumed that the temperature does not change with position, i.e. $\theta_{,i} = 0$). Since θ is constant, (21.7) reduces to $\psi = \psi(\theta = \text{constant}, \varepsilon_{ij})$. Moreover, writing $\rho\psi(\theta = \text{constant}, \varepsilon_{ij}) = W(\varepsilon_{ij})$ and observing that the mass density ρ for small strains can be considered as constant, it follows from (21.10) that $\sigma_{ij} = \partial W/\partial \varepsilon_{ij}$. A comparison with (4.5) and (4.8) reveals that we then have identified the strain energy W per unit volume, i.e. we have identified a situation where hyper-elasticity occurs. In conclusion

For constant temperature

$$\rho\psi(\theta = \textit{constant}, \varepsilon_{ij}) = W(\varepsilon_{ij}) = \textit{strain energy}$$

$$\sigma_{ij} = \frac{\partial W}{\partial \varepsilon_{ij}} \qquad \textit{hyper-elasticity} \tag{21.11}$$

The requirement of constant temperature is identical with isothermal conditions. Therefore, during isothermal conditions $\rho\psi$ becomes equal to the strain energy W. If the surroundings are kept at a constant temperature, isothermal conditions can be achieved, if the material is loaded so slowly that any temperature difference between the material and the surroundings is allowed to disappear.

The strain energy for elastic materials can be recovered and perform mechanical work. It follows that for isothermal conditions, Helmholtz' free energy ψ is the amount of energy that is 'free' to perform mechanical work.

Let us next consider the case where the temperature is allowed to vary, i.e. $\psi = \psi(\theta, \varepsilon)$. It is expected that this formulation would turn into thermo-elasticity and, indeed, this is the situation. Let us expand the free energy $\psi(\theta, \varepsilon_{ij})$ in a Taylor series about the reference state where $\theta = \theta_o$ and $\varepsilon_{ij} = 0$, i.e.

$$\begin{aligned}\psi(\theta, \varepsilon_{ij}) = \psi_o &+ (\frac{\partial \psi}{\partial \varepsilon_{ij}})_o \varepsilon_{ij} + (\frac{\partial \psi}{\partial \theta})_o (\theta - \theta_o) + \frac{1}{2}\varepsilon_{ij}(\frac{\partial^2 \psi}{\partial \varepsilon_{ij}\partial \varepsilon_{kl}})_o \varepsilon_{kl} \\ &+ \frac{1}{2}(\frac{\partial^2 \psi}{\partial \theta^2})_o (\theta - \theta_o)^2 + (\frac{\partial^2 \psi}{\partial \varepsilon_{ij}\partial \theta})_o (\theta - \theta_o)\varepsilon_{ij}\end{aligned} \tag{21.12}$$

where the notation $(\;)_o$ indicates that the quantity in question is evaluated at the reference state where $\theta = \theta_o$ and $\varepsilon_{ij} = 0$. From (21.10) and (21.12), it then follows that

$$\sigma_{ij} = \rho\frac{\partial \psi}{\partial \varepsilon_{ij}} = \rho(\frac{\partial \psi}{\partial \varepsilon_{ij}})_o + \rho(\frac{\partial^2 \psi}{\partial \varepsilon_{ij}\partial \varepsilon_{kl}})_o \varepsilon_{kl} + \rho(\frac{\partial^2 \psi}{\partial \varepsilon_{ij}\partial \theta})_o (\theta - \theta_o)$$

Let us assume that no stresses exist at the reference state. It follows that $(\partial\psi/\partial\varepsilon_{ij})_o = 0$ and the expression above reduces to

$$\sigma_{ij} = D_{ijkl}\varepsilon_{kl} - \beta_{ij}(\theta - \theta_o) \tag{21.13}$$

where D_{ijkl} and β_{ij} are defined by

$$D_{ijkl} = \rho(\frac{\partial^2 \psi}{\partial \varepsilon_{ij} \partial \varepsilon_{kl}})_o; \quad \beta_{ij} = -\rho(\frac{\partial^2 \psi}{\partial \varepsilon_{ij} \partial \theta})_o \tag{21.14}$$

Since the elastic stiffness tensor D_{ijkl} is positive definite, cf. (4.24) and (4.25), it is always possible to determine the quantity α_{ij} from the equation

$$D_{ijkl}\alpha_{kl} = \beta_{ij} \tag{21.15}$$

and with C_{ijkl} being the elastic flexibility tensor, we obtain

$$\alpha_{ij} = C_{ijkl}\beta_{kl}$$

From (21.13) and (21.15) follow

$$\sigma_{ij} = D_{ijkl}(\varepsilon_{kl} - \varepsilon^o_{kl}) \quad \text{where} \quad \varepsilon^o_{ij} = \alpha_{ij}(\theta - \theta_o) \tag{21.16}$$

A comparison with (4.62) and (4.64) reveals that we have recovered a description of thermo-elasticity, where ε^o_{ij} becomes the *thermal strains* and α_{ij} the *tensor of thermal expansion coefficients*. If the material is isotropic, then the tensor α_{ij} reduces to $\alpha\delta_{ij}$, where α is the *thermal expansion coefficient*.

According to (21.14), both D_{ijkl} and α_{ij} appearing in (21.16) are constant quantities, but it is easy to obtain a formulation where they may depend upon the temperature. For this purpose and recalling that the term $(\partial\psi/\partial\varepsilon_{ij})_o$ is zero, we may generalize (21.12) and with evident notation postulate

$$\boxed{\begin{aligned} \rho\psi(\theta, \varepsilon_{ij}) &= C + A(\theta)(\theta - \theta_o) + \tfrac{1}{2}B(\theta)(\theta - \theta_o)^2 \\ &\quad + \tfrac{1}{2}\varepsilon_{ij}D_{ijkl}(\theta)\varepsilon_{kl} - \beta_{ij}(\theta)(\theta - \theta_o)\varepsilon_{ij} \end{aligned}} \tag{21.17}$$

From (21.10), it then follows that

$$\boxed{\begin{array}{l} \textit{Thermo-elasticity with temperature-dependent properties} \\ \sigma_{ij} = D_{ijkl}(\theta)(\varepsilon_{kl} - \varepsilon^o_{kl}) \quad \textit{where} \quad \varepsilon^o_{ij} = \alpha_{ij}(\theta)(\theta - \theta_o) \\ \textit{or} \\ \sigma_{ij} = D_{ijkl}(\theta)\varepsilon^e_{kl} \quad \textit{where} \quad \varepsilon^e_{ij} = \varepsilon_{ij} - \varepsilon^o_{ij} \end{array}} \tag{21.18}$$

Moreover

$$\boxed{D_{ijkl} = \rho\frac{\partial^2 \psi}{\partial \varepsilon_{ij} \partial \varepsilon_{kl}} \quad \textit{and} \quad \alpha_{ij} = C_{ijkl}\beta_{kl}} \tag{21.19}$$

We have found that thermo-elasticity follows from Helmholtz' free energy taken in the form of $\psi = \psi(\theta, \varepsilon_{ij})$. However, it turns out to be possible to obtain another format for elasticity by specifying the internal energy u. For that purpose, we return to definition (21.4), which after differentiation, becomes

$$\dot{u} = \dot{\psi} + \dot{s}\theta + s\dot{\theta}$$

With (21.4) and (21.6), it follows that

$$\dot{u} = \theta \dot{s} + \frac{1}{\rho}\sigma_{ij}\dot{\varepsilon}_{ij} \tag{21.20}$$

Since the internal energy u is a state function, this expression leads to

$$\boxed{u = u(s, \varepsilon_{ij}) \qquad \textit{elasticity}} \tag{21.21}$$

Differentiation and comparison with (21.20) yield

$$\boxed{\theta = \frac{\partial u}{\partial s}; \qquad \sigma_{ij} = \rho \frac{\partial u}{\partial \varepsilon_{ij}}} \tag{21.22}$$

In accordance with the terminology discussed previously and since the temperature is obtained from u by differentiation with respect to the entropy, u serves as a potential function for the temperature θ; likewise, u serves as a potential function for the stresses.

When considering the internal energy u, we found that the quantities s and ε_{ij} appeared in natural manner as the state variables and they are therefore the natural or canonical state variables for u when considering an elastic material. While the state variable ε_{ij} can be measured, i.e. ε_{ij} is an observable state variable, we cannot determine the entropy s directly just by making some kind of measurement and s is therefore an example of a state variable that is not observable.

Another interesting point is that u is the *Legendre transformation* of ψ and, likewise, ψ is the Legendre transformation of u. To illustrate the last property, we accept (21.21) and (21.22). Instead of having s and ε_{ij} as the state variables, we can - without knowing the explicit form of $u(s, \varepsilon_{ij})$ - switch to another description where θ and ε_{ij} serve as state variables. For this purpose, the free energy ψ is defined by (21.4). Differentiation gives

$$\dot{\psi} = \dot{u} - \dot{s}\theta - s\dot{\theta}$$

and use of (21.21) and (21.22) results in

$$\dot{\psi} = -s\dot{\theta} + \frac{1}{\rho}\sigma_{ij}\dot{\varepsilon}_{ij}$$

which, as before, implies $\psi = \psi(\theta, \varepsilon_{ij})$ and thereby also relations (21.10). We may also refer to the similar discussion relating to (4.11).

Two additional state functions are often introduced in thermodynamics. If, instead of the description $u = u(s, \varepsilon_{ij})$, we want a formulation with the state variables s and σ_{ij}, the *enthalpy* h is defined by the Legendre transformation $h = u - \frac{1}{\rho}\sigma_{ij}\varepsilon_{ij}$ and it follows that $h = h(s, \sigma_{ij})$. Likewise, if, instead of $\psi = \psi(\theta, \varepsilon_{ij})$, we want a description with the state variables θ and σ_{ij}, *Gibb's free energy* ϕ is defined by the Legendre transformation $\phi = \psi - \frac{1}{\rho}\sigma_{ij}\varepsilon_{ij}$ and it follows that $\phi = \phi(\theta, \sigma_{ij})$. Indeed it is easy to show that Gibb's free energy ϕ for θ = constant is equal to $-C$ where C is the complementary energy defined by (4.11). However, for constitutive modeling it suffices in most cases to deal with the state functions already introduced, namely: the entropy s, the internal energy u and Helmholtz' free energy ψ.

It is evident that (21.7) and (21.21) are different descriptions of the same phenomena and just as we have shown that (21.7) corresponds to elasticity, (21.21) must also provide a description of elasticity. To illustrate this point, we recall that a reversible process is considered. If, in addition, it is assumed that the heat supply r is zero and that no temperature gradients exist and thereby that the heat flux vector q_i is zero, then (21.2) implies that $\rho\dot{s} = 0$, i.e. the entropy is constant; this is called an *isentropic process*. From (21.21) and (21.22), it then appears that

For constant entropy

$$\rho u(s = \text{constant}, \varepsilon_{ij}) = W(\varepsilon_{ij}) = \text{strain energy}$$

$$\sigma_{ij} = \frac{\partial W}{\partial \varepsilon_{ij}} \quad \text{hyper-elasticity} \tag{21.23}$$

It follows that we have identified another situation where hyper-elasticity occurs. The conditions $r = 0$ and $\theta_{,i} = 0$ are equivalent to an adiabatic process (no heat input); this situation can be achieved if the material is loaded very quickly since heat exchange with the surroundings then has no time to take place. Under these conditions, (21.23) shows that a hyper-elastic response can occur. On the other hand, if isothermal conditions are considered, (21.11) shows that a hyper-elastic response can occur. Isothermal conditions can be achieved in practice if the material is loaded very slowly.

To obtain the result that (21.7) or (21.21) describes an elastic behavior, it was assumed that no temperature gradients exist and that a reversible process occurs. Suppose now that temperature gradients may exist and therefore that an irreversible process is allowed. Also for these conditions, we maintain the definitions of elasticity given by (21.7) and (21.21). If (21.7) is adopted and advantage is taken of (21.10), then the dissipation inequality (21.5) reduces to $\gamma = -q_i\theta_{,i}/\theta \geq 0$. Likewise, if (21.21) is adopted and advantage is taken of (21.22), the dissipation inequality (21.3) also reduces to $\gamma = -q_i\theta_{,i}/\theta \geq 0$. It is concluded that

For elastic materials, i.e. $\psi = \psi(\theta, \varepsilon_{ij})$ *or* $u = u(s, \varepsilon_{ij})$, *the dissipation inequality reduces to*

$$\gamma = -\frac{q_i\theta_{,i}}{\theta} \geq 0 \tag{21.24}$$

This inequality is often termed *Fourier's inequality*. Since the temperature $\theta > 0$, we obtain that the only restriction on elasticity provided by the second law is $q_i\theta_{,i} \leq 0$. This restriction simply states that the scalar product of the heat flux vector q_i and the temperature gradient $\theta_{,i}$ is non-positive, i.e. heat flows from warm regions towards cold regions. The statement (21.24) also shows that when elastic materials are considered, the only source to irreversibility is the existence of temperature gradients.

We have found that both $\psi = \psi(\theta, \varepsilon_{ij})$ and $u = u(s, \varepsilon_{ij})$ provide descriptions of elasticity; however, whereas the temperature θ is an observable state variable that we can control and, in practice, assign arbitrary values, this is not the case for the state variable s, which is not observable. This means that we cannot, in practice, create experimental conditions by which the entropy s is assigned arbitrary values. Therefore, even though $\psi = \psi(\theta, \varepsilon_{ij})$ and $u = u(s, \varepsilon_{ij})$, in principle, provide equivalent descriptions of elasticity, it is evident that the format $\psi = \psi(\theta, \varepsilon_{ij})$ is the most convenient. It is no surprise that also for inelastic behavior, a description in terms of the free energy ψ is the most advantageous format. We conclude that

> *A description in terms of Helmholtz' free energy ψ is more convenient than a description in terms of the internal energy u*

21.2 Inelastic materials - Internal variables

After the detailed discussion of thermo-elasticity and the state variables that characterize thermo-elasticity, it is timely to consider the description of inelastic materials. Let us first note that, in general, we can write

$$\boxed{\varepsilon_{ij} = \varepsilon_{ij}^{e} + \varepsilon_{ij}^{ie} + \varepsilon_{ij}^{o}}$$

where ε_{ij}^{e} are the elastic strains, ε_{ij}^{ie} the *inelastic strains* and ε_{ij}^{o} are the thermal strains; the inelastic strains may be due to plasticity, creep, viscoplasticity, etc. Moreover, the elastic strains ε_{ij}^{e} are defined by Hooke's law i.e. $\varepsilon_{ij}^{e} = C_{ijkl}\sigma_{kl}$ where C_{ijkl} is the elastic flexibility tensor.

When choosing the state variables for inelastic materials, it is evident that the strains must enter in some form. To approach this problem, it is observed that there are cases where an arbitrary inelastic material responds in an elastic manner; for plastic materials this occurs during unloading and for time-dependent materials an elastic response follows if the loading rate is sufficiently fast.

As an example, consider the hardening plasticity response shown in Fig. 21.1. An elastic behavior occurs between point O and point A and an elastic behavior is also present during unloading between point B and point C. If the elasticity between O and A is to be described, the free energy ψ is taken as $\psi = \psi(\theta, \varepsilon_{ij})$, cf. (21.7). However, if the elasticity between B and C is to be described, it follows from Fig. 21.1 that we should now take ψ as $\psi = \psi(\theta, \varepsilon_{ij} - \varepsilon_{ij}^{ie})$ where $\varepsilon_{ij}^{ie} = \varepsilon_{ij}^{p}$.

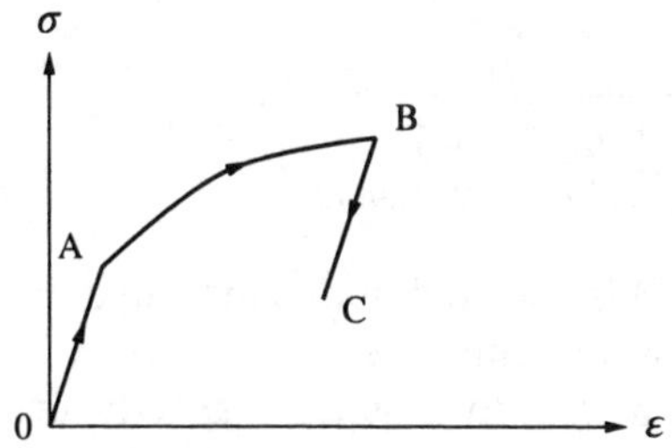

Figure 21.1: Modeling of elastic behavior for hardening plasticity.

This example motivates that while the total strain tensor ε_{ij} is a proper state variable for elasticity, the corresponding state variable for inelastic behavior is $\varepsilon_{ij} - \varepsilon_{ij}^{ie}$, i.e. $\psi = \psi(\theta, \varepsilon_{ij} - \varepsilon_{ij}^{ie})$. It is not surprising that this format is not sufficiently general to model all types of inelasticity and we therefore introduce an additional set of state variables, κ_α, where $\alpha = 1, 2, \cdots$, that characterize the inelastic material. We are then led to the following format

$$\boxed{\psi = \psi(\theta, \varepsilon_{ij} - \varepsilon_{ij}^{ie}, \kappa_\alpha) \qquad \textit{inelastic material}} \tag{21.25}$$

The additional set of state variables given by κ_α is also called *internal variables* or *hidden variables*, i.e.

$$\boxed{\kappa_\alpha = \textit{state variables=internal variables= hidden variables}}$$

We do not know the number of internal variables beforehand, and, as indicated, we may have one, two or more internal variables. Moreover, at this point we do not know the type of internal variables, which may be scalars or higher-order tensors. However, we collect all these internal variables into the notation κ_α and use the following definition

$$\boxed{\kappa_\alpha = \textit{internal variables} \;\; (\alpha = 1, 2, \dots)}$$

This discussion of internal variables is completely similar to that encountered in plasticity theory, cf. the discussion relating to (9.10) and (9.11).

Of the state variables introduced, the only quantities we can measure directly, without making use of our imagination of how materials behave, are the observable state variables θ and ε_{ij}. Therefore, the internal variables κ_α (and, in principle, also ε_{ij}^{ie}) are not observable state variables and this is the reason for the somewhat peculiar terminology of 'hidden' variables. We will follow the trend in the literature and exclusively use the word internal variables since the terminology of 'hidden' variables may act as a psychological block to acceptance. As an example of an internal variable, we may take the effective plastic strain.

Indeed our choice of internal variables influences the type of inelasticity we can simulate. This aspect will be illustrated in detail in the following chapters.

If we adopt the definition

> *The internal variables κ_α and ε_{ij}^{ie} change only during inelastic response*

then it is evident that (21.25) is able to describe the situation where the inelastic material responds in an elastic manner, cf. for instance Fig. 21.1. That the format (21.25), in fact, enables us to simulate plasticity, creep, viscoplasticity, etc. is illustrated in the following chapters.

It is a characteristic feature of inelasticity that the material response depends on the previous load history; in that sense the inelastic material possesses a memory. The format (21.25) implies that the load history is monitored through the internal variables κ_α and ε_{ij}^{ie} and this concept is referred to as the *internal variable concept*. This approach was introduced by Onsager (1931a,b), Meixner (1953), Biot (1954) and Ziegler (1958) and later generalized by Coleman and Gurtin (1967) and further explored by Valanis (1968), Lubliner (1972) and Halphen and Nguyen (1975) as well as many others. The review articles by Germain *et al.* (1983) and Reddy and Martin (1994) provide a detailed discussion of the internal variable formalism and contain a number of relevant references.

In this exposition, we will adopt the internal variable approach and only mention that there exists another strategy to represent the memory of the material, the so-called *functional approach*. We may recall that a *functional* can be viewed as a function of a function. As an example, consider the quantity I defined by

$$I(y) = \int_a^b F(x, y(x))dx$$

where $F(x, y)$ and $y = y(x)$ are known functions. Just like the function $y = y(x)$ can be thought of as a rule which assigns a single number $y(x)$ to every number x, then $I(y)$ as defined above assigns a single number $I(y)$ to every function y; in that sense, the quantity $I(y)$ can be viewed as a function of a function and the quantity $I(y)$ is an example of a functional.

For illustration purposes only, assume that we have chosen the functions a and b_{ij}. Consider then the expression

$$\psi = \int_{-\infty}^{t} \frac{\partial \theta(t')}{\partial t'} a(t - t')dt' + \int_{-\infty}^{t} \frac{\partial \varepsilon_{ij}(t')}{\partial t'} b_{ij}(t - t')dt'$$

where t denotes the present time and t' is an integration variable. Then the free energy ψ would be a functional of θ and ε_{ij} and ψ would then depend, not only on the present values of θ and ε_{ij}, but also on the history of θ and ε_{ij}.

With this illustration in mind, let us assume that the free energy ψ is given by

$$\boxed{\psi = \mathcal{F}(\theta, \varepsilon_{ij}) \quad \textit{functional approach}} \tag{21.26}$$

where $\mathcal{F}(\theta, \varepsilon_{ij})$ denotes a functional of θ and ε_{ij} i.e. it mimics the entire history of θ and ε_{ij}. The format (21.26) then implies that the free energy ψ not only depends on the current values of θ and ε_{ij}, but also on their histories.

The functional approach illustrated in principle by (21.26) was proposed by Coleman and Noll (1960, 1961) and further explored by Coleman (1964) and it has been used successfully for modeling of viscoelasticity. In general however, when considering, for instance, elasto-plasticity and damage mechanics, the internal variable concept seems to be much more fruitful and following the general trend in the literature, we will here exclusively deal with the internal variable concept.

With these remarks, let us return to the general expression for the free energy ψ given by (21.25), i.e.

$$\psi = \psi(\theta, \varepsilon_{ij} - \varepsilon_{ij}^{ie}, \kappa_\alpha) \tag{21.27}$$

Since θ, $\varepsilon_{ij} - \varepsilon_{ij}^{ie}$ and κ_α are assumed to be the state variables that determine the state, it is only a small restriction to also assume that the stress state σ_{ij} and the entropy s depend on the same state variables, i.e.

$$\sigma_{ij} = \sigma_{ij}(\theta, \varepsilon_{ij} - \varepsilon_{ij}^{ie}, \kappa_\alpha)\,; \qquad s = s(\theta, \varepsilon_{ij} - \varepsilon_{ij}^{ie}, \kappa_\alpha) \tag{21.28}$$

These expressions also represent another viewpoint, namely the *axiom of equipresence*. It states that at the onset, all relations for the material are considered as relations that contain the same list of variables until the contrary is deduced, see for instance Eringen (1975a) and Truesdell and Toupin (1960). However, even though the word 'axiom' is commonly used in the literature, it seems to us to be somewhat too strong a statement and to quote Eringen (1975a) "it is rather a precautionary measure that is valid in all proper scientific methods".

Insertion of (21.27) into the dissipation inequality (21.5) gives

$$\gamma \equiv (\sigma_{ij} - \rho\frac{\partial\psi}{\partial\varepsilon_{ij}})\dot{\varepsilon}_{ij} - \rho(s + \frac{\partial\psi}{\partial\theta})\dot{\theta} + \rho\frac{\partial\psi}{\partial\varepsilon_{ij}}\dot{\varepsilon}_{ij}^{ie} - \rho\frac{\partial\psi}{\partial\kappa_\alpha}\dot{\kappa}_\alpha - \frac{q_i\theta_{,i}}{\theta} \geq 0 \tag{21.29}$$

According to (21.27) and (21.28) all terms in (21.29) except the term $q_i\theta_{,i}/\theta$ are independent of the temperature gradient $\theta_{,i}$. Since we can choose the temperature gradient $\theta_{,i}$ arbitrarily, it is concluded that (21.29) implies that

$$\boxed{\begin{aligned} &\gamma \equiv \gamma_{mech} + \gamma_{ther} \geq 0 \\ &\textit{where} \\ &\gamma_{mech} \equiv (\sigma_{ij} - \rho\frac{\partial\psi}{\partial\varepsilon_{ij}})\dot{\varepsilon}_{ij} - \rho(s + \frac{\partial\psi}{\partial\theta})\dot{\theta} + \rho\frac{\partial\psi}{\partial\varepsilon_{ij}}\dot{\varepsilon}_{ij}^{ie} - \rho\frac{\partial\psi}{\partial\kappa_\alpha}\dot{\kappa}_\alpha \geq 0 \\ &\gamma_{ther} \equiv -\frac{q_i\theta_{,i}}{\theta} \geq 0 \end{aligned}} \tag{21.30}$$

That is, the non-negative dissipation γ can be split into two parts: the non-negative quantity γ_{mech} and the non-negative quantity γ_{ther}. It is evident that γ_{mech} refers to the irreversibility, i.e. the dissipation, of the material as such and it is therefore called the *mechanical dissipation*; in the literature it is often called the *intrinsic* or *internal dissipation*. Likewise, γ_{ther} refers to the irreversibility, i.e. the dissipation, relating to the existence of temperature gradients, i.e. heat flow, and it is therefore called the *thermal dissipation*.

In fact, we have previously discussed in relation to (21.24) the thermal dissipation inequality where it was called Fourier's inequality and which simply states that heat must flow from warm regions towards cold regions. We will assume that the constitutive equation for the heat flux vector q_i is such that Fourier's inequality is always fulfilled; indeed we will later evaluate the consequences for *Fourier's law* given by $q_i = -k_{ij}\theta_{,j}$ when this law is to fulfill Fourier's inequality, cf. the discussion relating to (21.59). We therefore have

The thermal dissipation inequality = Fourier's inequality

$$\gamma_{ther} \equiv -\frac{q_i\theta_{,i}}{\theta} \geq 0 \tag{21.31}$$

In order to fulfill (21.30), we are then left with the requirement $\gamma_{mech} \geq 0$. According to (21.30) and the discussion above, we emphasize that reversibility as such requires that both $\gamma_{mech} = 0$ and $\gamma_{ther} = 0$; however, the material responds reversibly if $\gamma_{mech} = 0$. This is readily shown for thermo-elasticity where $\varepsilon_{ij}^{ie} = \kappa_\alpha = 0$ in combination with (21.10) implies $\gamma_{mech} = 0$.

Let us return to the general expression for γ_{mech} given by (21.30). For thermo-elasticity where $\varepsilon_{ij}^{ie} = \kappa_\alpha = 0$, (21.30) reduces to

$$\gamma_{mech} \equiv (\sigma_{ij} - \rho\frac{\partial\psi}{\partial\varepsilon_{ij}})\dot{\varepsilon}_{ij} - \rho(s + \frac{\partial\psi}{\partial\theta})\dot{\theta} = 0$$

It is recalled that ε_{ij} and θ are observable state variables that we can control independently of each other. Therefore, since neither σ_{ij}, s nor ψ depend on $\dot{\varepsilon}_{ij}$ and $\dot{\theta}$ and as $\dot{\varepsilon}_{ij}$ and $\dot{\theta}$ can be chosen arbitrarily, we immediately conclude from the expression above that we must have $\sigma_{ij} = \rho\,\partial\psi/\partial\varepsilon_{ij}$ and $s = -\partial\psi/\partial\theta$, in accordance with (21.10).

In the general case where the variables κ_α and ε_{ij}^{ie} are also present, the implication of $\gamma_{mech} \geq 0$ as given by (21.30) is much more complex. The reason is that even though $\dot{\varepsilon}_{ij}$ and $\dot{\theta}$ still can be chosen arbitrarily and independently of each other, the rates $\dot{\varepsilon}_{ij}^{ie}$ and $\dot{\kappa}_\alpha$ may, in general, depend on $\dot{\varepsilon}_{ij}$ and $\dot{\theta}$ and this complicates the picture considerably.

However, for elasticity we found that $\sigma_{ij} = \rho\,\partial\psi/\partial\varepsilon_{ij}$. It is therefore tempting to assume that (21.30) implies that

$$\sigma_{ij} = \rho\frac{\partial\psi}{\partial\varepsilon_{ij}} + b_{ij} \tag{21.32}$$

Since the unknown quantity b_{ij}, at this point, is completely arbitrary there is, in fact, no loss of generality when assuming the format (21.32). In analogy with (21.28), we assume

$$b_{ij} = b_{ij}(\theta, \varepsilon_{ij} - \varepsilon_{ij}^{ie}, \kappa_\alpha) \tag{21.33}$$

For a purely elastic behavior, where $\dot{\varepsilon}_{ij}^{ie} = \dot{\kappa}_\alpha = 0$, we found that (21.30) implies $\sigma_{ij} = \rho\, \partial\psi/\partial\varepsilon_{ij}$. However, even when we have an inelastic material, its response may, in the limit, manifest itself as elasticity; for elasto-plasticity the response approaches elasticity when neutral loading is approached and for creep and viscoplasticity, the response approaches elasticity when the loading rate is sufficiently rapid. In these cases, we have $\dot{\varepsilon}_{ij}^{ie} \to 0$ and $\dot{\kappa}_\alpha \to 0$ and we must then have $\sigma_{ij} = \rho\, \partial\psi/\partial\varepsilon_{ij}$. From (21.32), it is concluded that $b_{ij} \to 0$ must hold when $\dot{\varepsilon}_{ij}^{ie} \to 0$ and $\dot{\kappa}_\alpha \to 0$. However, in view of (21.33) the quantity b_{ij} is independent of $\dot{\varepsilon}_{ij}^{ie}$ and $\dot{\kappa}_\alpha$ and the only manner in which these requirements can be fulfilled is to require that $b_{ij} = 0$, i.e.

$$\sigma_{ij} = \rho \frac{\partial\psi}{\partial\varepsilon_{ij}} \tag{21.34}$$

Likewise, for thermo-elasticity (21.30) implies that $s = -\partial\psi/\partial\theta$, cf. (21.10). In analogy with (21.28) there is no loss of generality to assume that for inelasticity we have

$$s = -\frac{\partial\psi}{\partial\theta} + c$$

where the unknown function c has the form

$$c = c(\theta, \varepsilon_{ij} - \varepsilon_{ij}^{ie}, \kappa_\alpha)$$

Arguing in exactly the same manner as above, we are led to the conclusion that $c = 0$, i.e.

$$s = -\frac{\partial\psi}{\partial\theta} \tag{21.35}$$

The result that (21.34) and (21.35) hold not only for thermo-elasticity but also for general inelasticity is called *Coleman's relations*. They were derived by Coleman and Gurtin (1967) using a method which in the literature is referred to as *Coleman's method*. This method is completely different from the one adopted here. Indeed, it has been argued by Lubliner (1972, 1982) that Coleman's method, in fact, does not provide the unique solution given by (21.34) and (21.35) and that other solutions are possible. Following Lubliner, it is only when the set of possible solutions is evaluated in relation to the elastic response occurring in the limit also for inelastic materials that the solutions (21.34) and (21.35) emerge. The arguments presented here bear similarities to those of

Lubliner, but they are essentially different and the reader may consult the cited works of Lubliner as well as the original method suggested by Coleman and Noll to appreciate the differences.

According to (21.32) and (21.33) the argumentation above was based on the assumption that the stresses only depend on the current state. However, there are cases where the stresses also depend on the rate of some quantities and the Kelvin model, cf. (14.9) and (14.5), provides such an example. To generalize the results above we therefore write

$$\sigma_{ij} = \rho \frac{\partial \psi}{\partial \varepsilon_{ij}} + \sigma_{ij}^{v} \tag{21.36}$$

where σ_{ij}^{v} is the *viscous stress* that depends on the rate of some quantities, i.e. $\sigma_{ij}^{v} = \sigma_{ij}^{v}(\dot{\varepsilon}_{ij}, \dot{\varepsilon}_{ij}^{ie}, \dot{\kappa}_{\alpha})$ with the property $\sigma_{ij}^{v}(0,0,0) = 0$. As already touched upon, a viscous stress will appear in some viscoelastic formulations whereas it will be zero in all other cases. With the results (21.36) and (21.35) inserted into the mechanical dissipation inequality in (21.30), we are led to the following important conclusions

> *For inelasticity we have*
>
> $$\psi = \psi(\theta, \varepsilon_{ij} - \varepsilon_{ij}^{ie}, \kappa_{\alpha})$$
>
> *where*
>
> $$\sigma_{ij} = \rho \frac{\partial \psi}{\partial \varepsilon_{ij}} + \sigma_{ij}^{v}; \qquad s = -\frac{\partial \psi}{\partial \theta}$$
>
> *and* $\sigma_{ij}^{v} = \sigma_{ij}^{v}(\dot{\varepsilon}_{ij}, \dot{\varepsilon}_{ij}^{ie}, \dot{\kappa}_{\alpha})$ *denotes the viscous stress.*
> *The mechanical dissipation inequality becomes* (21.37)
>
> $$\gamma_{mech} \equiv \sigma_{ij}\dot{\varepsilon}_{ij}^{ie} + \sigma_{ij}^{v}(\dot{\varepsilon}_{ij} - \dot{\varepsilon}_{ij}^{ie}) - K_{\alpha}\dot{\kappa}_{\alpha} \geq 0$$
>
> *where the conjugated thermodynamic forces* K_{α} *are defined by*
>
> $$K_{\alpha} \equiv \rho \frac{\partial \psi}{\partial \kappa_{\alpha}}$$

Note the definition of the *conjugated thermodynamic forces*. From the expression for γ_{mech} which has the unit [Nm/(m^{3}s)], i.e. energy density rate, we see that the stress σ_{ij} is the 'force' that is energy conjugated to the 'flux' $\dot{\varepsilon}_{ij}^{ie}$; in the same fashion the 'thermodynamic force' K_{α} is energy conjugated to the 'flux' $\dot{\kappa}_{\alpha}$ (or rather to $-\dot{\kappa}_{\alpha}$). It is recalled that elasticity occurs when $\gamma_{mech} = 0$ and this is evidently fulfilled when $\sigma_{ij}^{v} = \dot{\varepsilon}_{ij}^{ie} = \dot{\kappa}_{\alpha} = 0$. However, reversibility as such requires that $\gamma_{mech} = 0$ as well as $\gamma_{ther} = 0$, cf. (21.30).

With the main result (21.37), and for simplicity excluding the viscous stress σ_{ij}^{v} the only remaining topic is to choose expressions for $\dot{\varepsilon}_{ij}^{ie}$ and $\dot{\kappa}_{\alpha}$ that fulfill the mechanical dissipation inequality. Expressions for $\dot{\varepsilon}_{ij}^{ie}$ and $\dot{\kappa}_\alpha$ are called *evolution laws* and we will now enter a discussion of proper choices of these evolution laws.

21.3 Choice of evolution laws - Fulfillment of the mechanical dissipation inequality

Any constitutive relation must fulfill the mechanical dissipation inequality and excluding for the moment those viscoelastic formulations which contain a viscous stress, the mechanical dissipation inequality appears from (21.37) as

$$\gamma_{mech} \equiv \sigma_{ij}\dot{\varepsilon}_{ij}^{ie} - K_\alpha \dot{\kappa}_\alpha \geq 0 \qquad (21.38)$$

The question is now how we can choose expressions for the fluxes $\dot{\varepsilon}_{ij}^{ie}$ and $\dot{\kappa}_\alpha$ so that the mechanical dissipation inequality is fulfilled. These expressions for $\dot{\varepsilon}_{ij}^{ie}$ and $\dot{\kappa}_\alpha$ are called *evolution laws* and in this section we will provide a general discussion of various ways to establish such evolution laws.

Let us first write the mechanical dissipation inequality in a slightly more convenient form. For this purpose, define the sets A_Θ and $\dot{a}_\Theta$ by

$$\boxed{A_\Theta = \{\sigma_{ij},\, K_\alpha\}\,; \qquad \dot{a}_\Theta = \{\dot{\varepsilon}_{ij}^{ie},\, -\dot{\kappa}_\alpha\}} \qquad (21.39)$$

i.e. the set A_Θ comprises all σ_{ij} and K_α components, i.e. all the 'forces', whereas the set $\dot{a}_\Theta$ comprises all $\dot{\varepsilon}_{ij}^{ie}$ and $-\dot{\kappa}_\alpha$ components, i.e. all the 'fluxes'. The minus sign in front of $\dot{\kappa}_\alpha$ turns out to be convenient since the definitions (21.39) allow the mechanical dissipation inequality (21.38) to be written as

$$\boxed{\gamma_{mech} \equiv A_\Theta \dot{a}_\Theta \geq 0} \qquad (21.40)$$

Below, we will consider the dissipation inequality in this general form and ignore, for the moment, that in our case the components of A_Θ and $\dot{a}_\Theta$ are given by (21.39). As an illustration of this generality, we may write the thermal dissipation inequality (21.31) as $\gamma_{ther} \equiv A\dot{a}$ where $A = \{\theta_{,i}/\theta\}$ and $\dot{a} = \{-q_i\}$. Since we could equally well have chosen $A = \{\theta_{,i}\}$ and $\dot{a} = \{-q_i/\theta\}$ this aspect illustrates that the definition of forces and fluxes is not unique.

If all forces are zero, it seems reasonable to assume that also all fluxes are zero, i.e.

$$\dot{a}_\Theta = 0 \quad \text{if} \quad A_\Theta = 0 \qquad (21.41)$$

Below, we will discuss various approaches for establishment of the evolution laws for $\dot{a}_\Theta$ so that the dissipation inequality (21.40) is fulfilled; for related viewpoints see for instance Halphen and Nguyen (1975), Germain *et al.* (1983), Eve *et al.* (1990) and Nguyen (2000).

21.3.1 Direct approach

It is certainly possible to directly postulate some evolution laws for $\dot{a}_\Theta$ and then *a posteriori*, check that the dissipation inequality (21.40) is fulfilled. The drawback in this approach is that this *a posteriori* check must be performed for each material model and no general information can be derived for a group of material models

Direct approach:
Postulate some evolution laws for $\dot{a}_\Theta$. Check a posteriori that the dissipation inequality is fulfilled

21.3.2 Onsager approach

The most simple specific expression that can be adopted for the relation between the fluxes and forces is linear, i.e.

$$\dot{a}_\Theta = L_{\Theta\Upsilon} A_\Upsilon$$

where the coefficient matrix $L_{\Theta\Upsilon}$ is taken as constant; this expression certainly fulfills (21.41) and insertion into (21.40) provides

$$\gamma_{mech} \equiv A_\Theta L_{\Theta\Upsilon} A_\Upsilon > 0 \quad \text{when} \quad A_\Theta \neq 0$$

where $\geq$ has been replaced by $>$ since we are considering irreversible processes. It appears that the coefficient $L_{\Theta\Upsilon}$ must be positive definite. Let us furthermore assume that $L_{\Theta\Upsilon}$ is symmetric. We are then led to

Onsager's relations:

$$\dot{a}_\Theta = L_{\Theta\Upsilon} A_\Upsilon$$

where

$$L_{\Theta\Upsilon} = L_{\Upsilon\Theta} \quad \textit{and} \quad L_{\Theta\Upsilon} \quad \textit{is positive definite} \tag{21.42}$$

These relations where proposed by Onsager (1931a,b) and the symmetry property $L_{\Theta\Upsilon} = L_{\Upsilon\Theta}$ is referred to as the *Onsager reciprocal relations*. Onsager arrived at the reciprocal relations by statistical mechanics considerations and they have been widely used in the literature and confirmed experimentally for a number phenomena within heat, diffusion and electricity, see Fung (1965), and in physical chemistry. However, they have also been criticized, cf. for instance Eringen (1975a) and Truesdell (1969), since the definition of forces and fluxes, as already touched upon, is not unique. In our situation, we can safely state that the Onsager relations are not fundamental laws of nature since they cannot be used to describe general plasticity and viscoplasticity; however, the Onsager relations describe with close accuracy a number of other phenomena and in that sense they can be viewed as relevant material descriptions for some fields of applications.

21.3.3 Potential approach

In order to generalize Onsager's linear theory and obtain a nonlinear theory, let us assume that there exists a function ϕ such that

$$\boxed{\dot{a}_\Theta = \dot{\lambda}\frac{\partial \phi}{\partial A_\Theta}; \qquad \dot{\lambda} \geq 0 \qquad \textit{Potential formulation}} \tag{21.43}$$

It is easily shown that this format contains Onsager's linear theory as a special case. In that case, choose the function ϕ as

$$\phi = \frac{1}{2} A_\Theta L_{\Theta\Upsilon} A_\Upsilon$$

where nothing is said about the symmetry of $L_{\Theta\Upsilon}$. Splitting $L_{\Theta\Upsilon}$ into its symmetric and antisymmetric parts according to

$$L_{\Theta\Upsilon} = L^s_{\Theta\Upsilon} + L^a_{\Theta\Upsilon}$$

it then follows that

$$\phi = \frac{1}{2} A_\Theta L_{\Theta\Upsilon} A_\Upsilon = \frac{1}{2} A_\Theta L^s_{\Theta\Upsilon} A_\Upsilon \tag{21.44}$$

Choosing $\dot{\lambda} = 1$ and inserting (21.44) into (21.43) then provide (21.42).

Returning to the general situation, insertion of (21.43) into (21.40) gives

$$\boxed{\gamma_{mech} \equiv \dot{\lambda} A_\Theta \frac{\partial \phi}{\partial A_\Theta} \geq 0} \tag{21.45}$$

The function ϕ is called the *dissipation potential* and it evidently depends on the forces A_Θ, but we may also allow it to depend on some other variables Z_q, i.e.

$$\boxed{Z_q = \text{some variables } (q = 1, 2, \ldots)} \tag{21.46}$$

We then obtain

$$\phi = \phi(A_\Theta, Z_q)$$

Within the potential approach there are various ways to ensure that (21.45) is fulfilled.

Homogeneous potential function

One route is to assume that ϕ is a homogeneous function of degree n in A_Θ, i.e.

ϕ is a homogeneous function of A_Θ of degree n if
$\phi(kA_\Theta, Z_q) = k^n\phi(A_\Theta, Z_q)$

As an example, $\phi = (A_1^3+A_2^3)Z$ is homogeneous of degree 3 since $\phi(kA_\theta, Z_q) = ((kA_1)^3 + (kA_2)^3)Z = k^3(A_1^3 + A_2^3)Z = k^3\phi(A_\Theta, Z_q)$; as another example, ϕ given by (21.44) is homogeneous of degree 2. By Euler's theorem, see for instance Sokolnikoff and Redheffer (1958), it then follows that

$$A_\Theta\frac{\partial\phi}{\partial A_\Theta} = n\phi \tag{21.47}$$

From (21.43) and (21.47) we then obtain

The evolution laws
$\dot{a}_\Theta = \dot{\lambda}\frac{\partial\phi}{\partial A_\Theta}; \quad \dot{\lambda} \geq 0$
where $\phi = \phi(A_\Theta, Z_q)$ and ϕ is a homogeneous function of A_Θ of degree n implies
$\gamma_{mech} \equiv \dot{\lambda}A_\Theta\frac{\partial\phi}{\partial A_\Theta} = \dot{\lambda}n\phi \geq 0$
and is fulfilled if $\phi \geq 0$

As a relevant example, ideal von Mises plasticity is given by

$$f = F - \sigma_{yo} \quad \text{where} \quad F = \sigma_{eff} = (\frac{3}{2}s_{ij}s_{ij})^{1/2}$$

where f is the yield function, cf. (12.4). For ideal plasticity there are no internal variables, i.e. $\kappa_\alpha = 0$ and thereby no conjugated thermodynamic forces K_α, cf. the definition of K_α given by (21.37). It then follows from (21.39) that $A_\Theta = \{\sigma_{ij}\}$ and $\dot{a}_\Theta = \{\dot{\varepsilon}_{ij}^p\}$. Choose the dissipation function ϕ as

$$\phi(A_\Theta, Z_q) = \phi(\sigma_{ij}) = F = \sigma_{eff}$$

Since F is a homogeneous function of σ_{ij} of degree 1 we have $\sigma_{ij}\partial F/\partial\sigma_{ij} = F = \sigma_{eff}$, i.e. the dissipation inequality becomes $\gamma_{mech} \equiv \dot{\lambda}A_\Theta\partial\phi/\partial A_\Theta = \dot{\lambda}\sigma_{ij}\partial F/\partial\sigma_{ij} = \dot{\lambda}\sigma_{eff} \geq 0$ which certainly is fulfilled. Moreover, the evolution law $\dot{a}_\Theta = \dot{\lambda}\partial\phi/\partial A_\Theta$ becomes

$$\dot{\varepsilon}_{ij}^p = \dot{\lambda}\frac{\partial F}{\partial\sigma_{ij}} = \dot{\lambda}\frac{\partial f}{\partial\sigma_{ij}}$$

which is exactly the associated flow rule.

Convex potential function

We will now present a very powerful approach for establishment of the evolution laws so that the dissipation inequality is fulfilled.

Let us assume that ϕ is a convex function in A_Θ. From Appendix (A.5), it then follows that

$$\phi(A_\Theta^{(2)}, Z_q) - \phi(A_\Theta^{(1)}, Z_q) \geq (\frac{\partial \phi}{\partial A_\Theta})^{(1)}(A_\Theta^{(2)} - A_\Theta^{(1)}) \tag{21.48}$$

where $A_\Theta^{(1)}$ and $A_\Theta^{(2)}$ are two arbitrary sets of A_Θ and $(\ )^{(1)}$ means that the quantity within the parenthesis should be evaluated at the point $A_\Theta^{(1)}$. Choosing $A_\Theta^{(2)} = 0$ we obtain

$$(\frac{\partial \phi}{\partial A_\theta})^{(1)} A_\Theta^{(1)} \geq \phi(A_\Theta^{(1)}, Z_q) - \phi(0, Z_q)$$

Since $A_\Theta^{(1)}$ denotes an arbitrary set of A_Θ, we write this expression as

$$\frac{\partial \phi}{\partial A_\theta} A_\Theta \geq \phi(A_\Theta, Z_q) - \phi(0, Z_q) \tag{21.49}$$

In (21.48), choose next $A_\Theta^{(1)} = 0$ to obtain

$$\phi(A_\Theta^{(2)}, Z_q) - \phi(0, Z_q) \geq (\frac{\partial \phi}{\partial A_\theta})_{A_\Theta=0} A_\theta^{(2)} \tag{21.50}$$

Due to (21.41) and (21.43), we have

$$(\frac{\partial \phi}{\partial A_\Theta})_{A_\Theta=0} = 0$$

and since $A_\Theta^{(2)}$ in (21.50) denotes an arbitrary set of A_Θ, we can write (21.50) as

$$\phi(A_\Theta, Z_q) - \phi(0, Z_q) \geq 0$$

A comparison of this expression with (21.49) shows that

$$A_\Theta \frac{\partial \phi}{\partial A_\Theta} \geq 0$$

i.e. the dissipation inequality is fulfilled. We are then led to the following very powerful conclusion

> *If the dissipation function $\phi = \phi(A_\Theta, Z_q)$ is a convex function of A_Θ and if*
>
> $$\phi(A_\Theta, Z_q) - \phi(0, Z_q) \geq 0$$
>
> *then the evolution laws* (21.51)
>
> $$\dot{a}_\Theta = \dot{\lambda} \frac{\partial \phi}{\partial A_\Theta}; \quad \dot{\lambda} \geq 0$$
>
> *fulfill the dissipation inequality*

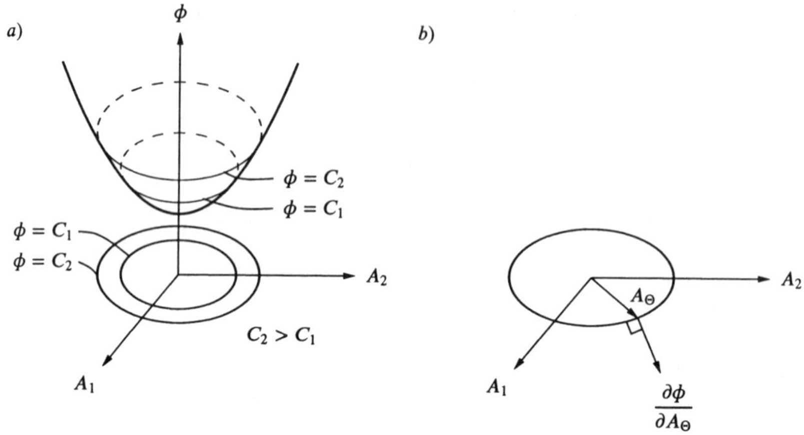

Figure 21.2: Illustration that the evolution laws $\dot{a}_\Theta = \dot{\lambda}\partial\phi/\partial A_\Theta$ fulfill the dissipation inequality when ϕ is a convex function.

We observe that the requirement $\phi(A_\Theta, Z_q) - \phi(0, Z_q) \geq 0$ follows from the requirement $(\partial\phi/\partial A_\Theta)_{A_\Theta=0} = 0$ (as well as convexity of ϕ). Moreover, when evaluating the inequality $\phi(A_\Theta, Z_q) - \phi(0, Z_q) \geq 0$, the quantity $\phi(A_\Theta, Z_q)$ should be evaluated at the current state where the evolution law $\dot{a}_\Theta = \dot{\lambda}\partial\phi/\partial A_\Theta$ is employed. The fundamental result (21.51) is due to Edelen (1972) and the proof given here follows closely that of Eringen (1975a).

If A_Θ merely consists of two components, a simple graphical illustration of (21.51) is possible. Since ϕ is a convex function and as $(\partial\phi/\partial A_\Theta)_{A_\Theta=0} = 0$, it follows that ϕ has a minimum at $A_\Theta = 0$, cf. Fig. 21.2a). Moreover, the vector $\partial\phi/\partial A_\Theta$ is normal to the contour curves of ϕ in the A_Θ-space and $\partial\phi/\partial A_\Theta$ is directed outwards (towards increasing values of ϕ). That the scalar product $A_\Theta\partial\phi/\partial A_\Theta$ and thereby the dissipation inequality is non-negative are then evident from Fig. 21.2b)

Let us for the moment choose $\dot{\lambda} = 1$; then the evolution laws become

$$\dot{a}_\Theta = \frac{\partial \phi}{\partial A_\Theta} \tag{21.52}$$

Define the *pseudo-dissipation function* ϕ^* by

$$\phi^* = \phi - \dot{a}_\Theta A_\Theta$$

It follows that

$$\dot{\phi}^* = \frac{\partial \phi}{\partial A_\Theta}\dot{A}_\Theta + \frac{\partial \phi}{\partial Z_q}\dot{Z}_q - \ddot{a}_\Theta A_\Theta - \dot{a}_\Theta \dot{A}_\Theta$$

which with the result (21.52) reduces to

$$\dot{\phi}^* = -A_\Theta \ddot{a}_\Theta + \frac{\partial \phi}{\partial Z_q}\dot{Z}_q$$

It follows that

$$\phi^* = \phi^*(\dot{a}_\Theta, Z_q) \quad \text{and} \quad A_\Theta = -\frac{\partial \phi^*}{\partial \dot{a}_\Theta}; \quad \frac{\partial \phi^*}{\partial Z_q} = \frac{\partial \phi}{\partial Z_q} \tag{21.53}$$

Whereas (21.52) determines the fluxes by means of the forces, (21.53b) expresses the forces in terms of the fluxes. In recognition of the previous discussion following (21.22), it is evident that ϕ^* is the Legendre transformation of ϕ. We also note that what we have here termed the dissipation function is called by some authors the pseudo-dissipation function and *visa versa*, cf. Lemaitre and Chaboche (1990) and Maugin (1992); however, since the function ϕ turns out to be the one that is most convenient to work with for constitutive modeling, our choice of terminology seems justified.

Returning to the essential result (21.51) and focusing on plasticity/viscoplasticity, let us choose the dissipation function $\phi(A_\Theta, Z_q)$ where $A_\Theta = \{\sigma_{ij}, K_\alpha\}$ as the potential function $g(\sigma_{ij}; K_\alpha)$ previously introduced for nonassociated plasticity and viscoplasticity. Since $A_\Theta = \{\sigma_{ij}, K_\alpha\}$ and $\dot{a}_\Theta = \{\dot{\varepsilon}^p_{ij}, -\dot{\kappa}_\alpha\}$, the evolution laws given by (21.51) become

Potential function approach

$$\dot{\varepsilon}^p_{ij} = \dot{\lambda}\frac{\partial g}{\partial \sigma_{ij}}; \qquad \dot{\kappa}_\alpha = -\dot{\lambda}\frac{\partial g}{\partial K_\alpha} \tag{21.54}$$

which corresponds exactly to the evolution laws postulated for plasticity in Chapter 10. Now however, we know exactly from (21.51) what requirements must be posed on the potential function g in order that these evolution laws be physically meaningful. Replacing $\dot{\lambda}$ by the Λ or $<\Phi(f)>/\eta$ we obtain the evolution equations for creep and Perzyna viscoplasticity, respectively, cf. (15.24) and (15.36).

Recalling that associated plasticity is obtained if the potential function g is chosen as the yield function f, we will now establish associated plasticity by a completely different approach.

The approach to use the convex function ϕ as a dissipation function and derive the evolution equations by (21.54) was pioneered by Halphen and Nguyen (1975). According to their notation, a material with these evolution equations is termed a *generalized standard material.*

Postulate of maximum dissipation

For plasticity, the dissipation inequality (21.37) reads

$$\gamma_{mech} \equiv \sigma_{ij}\dot{\varepsilon}^p_{ij} - K_\alpha\dot{\kappa}_\alpha \geq 0$$

where the yield function f fulfills $f \leq 0$. In accordance with (10.7) and (10.8), it seems natural to choose the thermodynamic forces K_α as the hardening parameters, i.e.

$$f(\sigma_{ij}, K_\alpha) \leq 0$$

The key point is to ensure that $\gamma_{mech} \geq 0$. It then seems natural to investigate the case where γ_{mech} takes the maximal possible value, i.e. we adopt *the postulate of maximum dissipation*; in that case we evidently ensure that $\gamma_{mech} \geq 0$. Maximization of γ_{mech} is equivalent to minimization of $-\gamma_{mech}$. We are then led to the following problem

Postulate of maximum dissipation: *For given fluxes* $\dot{\varepsilon}^p_{ij}$ *and* $\dot{\kappa}_\alpha$, *find those stresses* σ_{ij} *and conjugated forces* K_α *that minimize* $-\gamma_{mech}$ *under the constraint that* $f(\sigma_{ij}, K_\alpha) \leq 0$	(21.55)

In accordance with the Appendix A, this problem is a minimization problem with a constraint in terms of an inequality. Following (A.16), we define the Lagrange function $\mathcal{L}(\sigma_{ij}, K_\alpha, \dot{\lambda})$ by

$$\mathcal{L}(\sigma_{ij}, K_\alpha, \dot{\lambda}) = -\gamma_{mech} + \dot{\lambda} f = -\sigma_{ij}\dot{\varepsilon}^p_{ij} + K_\alpha \dot{\kappa}_\alpha + \dot{\lambda} f(\sigma_{ij}, K_\alpha)$$

where $\dot{\lambda}$ is a *Lagrange multiplier* and f is assumed to be a convex function of σ_{ij} and K_α. In accordance with (A.17), it follows that $\partial\mathcal{L}/\partial\sigma_{ij} = -\dot{\varepsilon}^p_{ij} + \dot{\lambda}\partial f/\partial\sigma_{ij} = 0$ and $\partial\mathcal{L}/\partial K_\alpha = \dot{\kappa}_\alpha + \dot{\lambda}\partial f/\partial K_\alpha = 0$. From (A.17) and (A.15) we then obtain the following *Kuhn-Tucker relations*

For plasticity theory where $f(\sigma_{ij}, K_\alpha) \leq 0$ *and* f *is assumed to be a convex function of* σ_{ij} *and* K_α, *the postulate of maximum dissipation leads to the associated evolution equations*

$$\dot{\varepsilon}^p_{ij} = \dot{\lambda}\frac{\partial f}{\partial \sigma_{ij}}; \qquad \dot{\kappa}_\alpha = -\dot{\lambda}\frac{\partial f}{\partial K_\alpha}$$

as well as (21.56)

$$\dot{\lambda} \geq 0$$

$$\dot{\lambda} = 0 \qquad \textit{for elastic behavior}$$

and

$$\dot{\lambda} f = 0$$

It appears that the postulate of maximum dissipation for plasticity leads to associated plasticity, cf. Chapter 10, and that we certainly ensure that the mechanical dissipation inequality is fulfilled. Moreover, the quantity $\dot{\lambda}$, which was originally introduced as a Lagrange multiplier, turns out to be the plastic multiplier.

The postulate of maximum dissipation is not a principle in the sense of being a law of nature even though some literature tends to suggest that; instead, it may be viewed simply as a means to fulfill the dissipation inequality. However, it may be possible to appeal to some kind of physics behind this postulate. We found in (20.62) and (20.63) that at thermodynamic equilibrium where no mechanical work input and no heat input are supplied, then the entropy is maximum. On the other hand, it seems reasonable to assume that the process to achieve that state is characterized by maximum entropy production and thereby maximum dissipation. However, this is an assumption - a postulate - and not a strict result; if it were a law of nature, nonassociated plasticity would not exist.

Another route of argumentation relates to (9.41) where the postulate of maximum dissipation was also encountered and where it was shown to lead to the associated flow rule $\dot{\varepsilon}^p_{ij} = \dot{\lambda}\partial f/\partial\sigma_{ij}$. We shall provide a further discussion of this aspect in the following chapter.

It is of significant interest that the postulate of maximum dissipation without any further assumptions by the Kuhn-Tucker relations leads to the conclusion that $\dot{\lambda} \geq 0$ when $f = 0$ and $\dot{\lambda} = 0$ when $f < 0$. Previously, in Chapter 10 we had to assume $\dot{\lambda}$ as being non-negative.

In this section, we have discussed various ways of establishing evolution laws that fulfill the mechanical dissipation inequality. The main results are (21.51), which for plasticity and viscoplasticity lead to the nonassociated evolution laws given by (21.54), and the postulate of maximum dissipation (21.55), which for plasticity leads to the associated evolution laws (21.56). We will later return to these results and further explore their implications. However, in the present chapter we will return to a discussion of the ramifications of the first law of thermodynamics, which deserves some attention.

21.4 Heat equation

We have shown that fulfillment of the second law of thermodynamics implies the main result given by (21.31) and (21.37). In particular, we observe the restrictions for constitutive modeling given by $\gamma_{ther} \geq 0$ and $\gamma_{mech} \geq 0$. On the other hand, the first law of thermodynamics places no restrictions on constitutive modeling; it simply turns into the *heat equation*, which determines the temperature distribution within the body.

Before attention is given to a derivation of the heat equation, we return to the thermal dissipation inequality (21.31), i.e. *Fourier's inequality*, which may also be written as

$$-q_i\theta_{,i} \geq 0 \qquad (21.57)$$

As discussed in Section 20.6, any temperature gradient gives rise to irreversibility and (21.57) can therefore be expressed as

$$-q_i\theta_{,i} > 0 \quad \text{when} \quad \theta_{,i} \neq 0 \tag{21.58}$$

This inequality simply states that heat flows from hot regions towards cold regions.

The reason for heat flow is the existence of a temperature gradient and the simplest constitutive equation that relates the heat flux vector q_i and the temperature gradient $\theta_{,i}$ is a linear relation. This leads to *Fourier's law*, which dates back to 1822, and which reads

$$\boxed{q_i = -k_{ij}\theta_{,j} \qquad \textit{Fourier's law}} \tag{21.59}$$

where k_{ij} denotes the *tensor of thermal conductivities*. Referring to (21.42), we may consider Fourier's law as an application of the Onsager relations. We will now discuss the pertinent properties of k_{ij} so that inequality (21.58) is fulfilled.

In general, we may assume that k_{ij} depends on the same variables as the free energy ψ, i.e.

$$k_{ij} = k_{ij}(\theta, \varepsilon_{kl} - \varepsilon_{kl}^{ie}, \kappa_\alpha)$$

In practice, however, the tensor k_{ij} is often assumed to be constant or to depend on the temperature θ alone. Insertion of (21.59) into (21.58) yields

$$\theta_{,i}k_{ij}\theta_{,j} > 0 \quad \text{when} \quad \theta_{,i} \neq 0 \tag{21.60}$$

In general, k_{ij} may be split into a symmetric part k_{ij}^s and an antisymmetric part k_{ij}^a according to

$$k_{ij} = k_{ij}^s + k_{ij}^a$$

where

$$k_{ij}^s = \frac{1}{2}(k_{ij} + k_{ji})\,; \qquad k_{ij}^a = \frac{1}{2}(k_{ij} - k_{ji})$$

Then (21.60) becomes

$$\theta_{,i}k_{ij}^s\theta_{,j} > 0 \quad \text{when} \quad \theta_{,i} \neq 0$$

and it follows that the symmetric part k_{ij}^s is positive definite.

This conclusion follows from the second law. However, from the first law and following Truesdell (1969), it turns out to be possible to evaluate the antisymmetric part k_{ij}^a. Referring to the first law (21.1), the heat flux vector q_i only appears in the form of the divergence given by $q_{i,i}$. Using (21.59), we obtain

$$q_{i,i} = -\frac{\partial k_{ij}}{\partial x_i}\theta_{,j} - k_{ij}\theta_{,ij} = -\frac{\partial k_{ij}}{\partial x_i}\theta_{,j} - k_{ij}^s\theta_{,ij} \tag{21.61}$$

where advantage was taken of $k^a_{ij}\theta_{,ij} = -k^a_{ji}\theta_{,ij} = -k^a_{ij}\theta_{,ji} = -k^a_{ij}\theta_{,ij}$, i.e. $k^a_{ij}\theta_{,ij} = 0$. For a homogeneous material, the thermal conductivity tensor k_{ij} does not change with position, i.e. $\partial k_{ij}/\partial x_i = 0$. In that case (21.61) reduces to

$$q_{i,i} = -k^s_{ij}\theta_{,ij}$$

i.e. for homogeneous materials, it is only the symmetric part k^s_{ij} of k_{ij} that influences the divergence $q_{i,i}$. It seems hard to accept that the principal properties of k_{ij} (symmetric or not) are to depend on whether the material is taken in a homogeneous form or in an inhomogeneous form. It therefore seems fully acceptable to assume that k_{ij} is always symmetric and this is in accordance with general experimental evidence. We are then led to the following conclusions

$$\boxed{k_{ij} \textit{ is symmetric and positive definite}} \tag{21.62}$$

We observe that both (21.59) and (21.62) are in agreement with the Onsager relations (21.42). For isotropic materials, k_{ij} reduces to $k_{ij} = k\delta_{ij}$, with k being the *thermal conductivity* and (21.62) implies that $k > 0$, as expected. With Fourier's law and the conclusions above, the inequality $\gamma_{ther} \geq 0$ is always fulfilled, as already anticipated in (21.31). Therefore, the requirements imposed by the second law have been reduced to fulfillment of the mechanical dissipation inequality $\gamma_{mech} \geq 0$ alone.

The heat equation will turn out to involve a new material parameter - the *specific heat capacity*. Before attention is turned to solids, and to fully appreciate the difference between various specific heat capacities, it is instructive to first discuss this subject for ideal gases. In fact, for a gas we have already touched upon this subject when we introduced the specific heat capacity c_V at constant volume. According to (20.26), we have

Gases: *The specific heat capacity c_V is the amount of heat that must be supplied without change of volume in order to increase the temperature of one unit mass one degree*	(21.63)

i.e.

$$\delta Q = mc_V d\theta \qquad \text{when } V = \text{constant} \tag{21.64}$$

where m is the mass of the gas in question. For gases, the global form of the first law is given by (20.25), i.e.

$$dU = -pdV + \delta Q \tag{21.65}$$

According to (20.24), we have for ideal gases $U(\theta) = mu(\theta)$, i.e.

$$dU = m\frac{du}{d\theta}d\theta \qquad \text{always holds for ideal gases} \tag{21.66}$$

When V is constant, (21.64) and (21.65) give $dU = mc_V d\theta$ and a comparison with (21.66), which always holds, reveals that

$$dU = mc_V d\theta \qquad \text{always holds for ideal gases} \tag{21.67}$$

where $c_V = du/d\theta$, cf. (20.28).

For a gas, let us next define the specific heat capacity c_p by

> *Gases:*
> *The specific heat capacity c_p is the amount of heat that must be supplied without change of pressure in order to increase the temperature of one unit mass one degree* (21.68)

i.e.

$$\delta Q = mc_p d\theta \qquad \text{when } p = \text{constant}$$

Insertion of this expression and of (21.67) into (21.65) gives

$$mc_V d\theta = -pdV + mc_p d\theta \tag{21.69}$$

For constant pressure, the ideal gas law $pV = mR\theta$ implies

$$pdV = mRd\theta$$

Insertion into (21.69) results in

$$\boxed{c_p - c_V = R \qquad \textit{ideal gas}}$$

The specific heat capacity c_p is therefore larger than c_V and the physical reason is that when c_p is measured, heat is supplied, but in order to maintain a constant pressure the gas must expand and thereby perform mechanical work on the surroundings. Therefore when c_p is measured, the heat input is used not only to increase the gas temperature, but also to perform mechanical work on the surroundings. On the other hand, when c_V is measured, the volume is constant and all the heat input goes into an increase of the gas temperature. The difference between c_p and c_V is quite significant for ideal gases and typically, we have $c_p/c_V \approx 1.3 - 1.7$, cf. Schmidt (1963).

Against this background, let us now turn to the specific heat capacities of solid materials. In analogy with (21.63) and (21.68), two definitions are available

> *Solids:*
> *The specific heat capacity c_ε is the amount of heat that must be supplied without change of strains in order to increase the temperature of one unit mass one degree*

and

Solids: *The specific heat capacity c_σ is the amount of heat that must be supplied without change of stresses in order to increase the temperature of one unit mass one degree*

Evidently, when c_ε or c_σ are measured, the body is assumed to be in a state of homogeneous conditions. By the definitions above, we therefore have

$$\frac{\delta Q}{dt} = mc_\varepsilon \dot{\theta} \qquad \text{when } \varepsilon_{ij} = \text{constant} \tag{21.70}$$

and

$$\frac{\delta Q}{dt} = mc_\sigma \dot{\theta} \qquad \text{when } \sigma_{ij} = \text{constant} \tag{21.71}$$

where m is the mass of the body. Under homogeneous conditions, the first law (21.1) may with (20.12) be written as

$$\rho \dot{u} = \dot{\varepsilon}_{ij}\sigma_{ij} + \frac{1}{V}\frac{\delta Q}{dt}$$

where V is the total volume of the body. Since $m = \rho V$, we obtain

$$\rho \dot{u} = \dot{\varepsilon}_{ij}\sigma_{ij} + \frac{\rho}{m}\frac{\delta Q}{dt}$$

According to (21.4), we have $u = \psi + s\theta$, which leads to

$$\rho(\dot{\psi} + \dot{s}\theta + s\dot{\theta}) = \dot{\varepsilon}_{ij}\sigma_{ij} + \frac{\rho}{m}\frac{\delta Q}{dt} \tag{21.72}$$

For a general inelastic material, the free energy ψ is given by (21.37), i.e.

$$\psi = \psi(\theta, \varepsilon_{ij} - \varepsilon_{ij}^{ie}, \kappa_\alpha)$$

Differentiation and use of the results given in (21.37) then imply

$$\dot{\psi} = -s\dot{\theta} + \frac{\partial \psi}{\partial \varepsilon_{ij}}\dot{\varepsilon}_{ij} - \frac{\partial \psi}{\partial \varepsilon_{ij}}\dot{\varepsilon}_{ij}^{ie} + \frac{\partial \psi}{\partial \kappa_\alpha}\dot{\kappa}_\alpha \tag{21.73}$$

where

$$s = -\frac{\partial \psi}{\partial \theta}$$

It follows that

$$\dot{s} = -\frac{\partial^2 \psi}{\partial \theta^2}\dot{\theta} - \frac{\partial^2 \psi}{\partial \varepsilon_{ij}\partial \theta}\dot{\varepsilon}_{ij} + \frac{\partial^2 \psi}{\partial \varepsilon_{ij}\partial \theta}\dot{\varepsilon}_{ij}^{ie} - \frac{\partial^2 \psi}{\partial \kappa_\alpha \partial \theta}\dot{\kappa}_\alpha \tag{21.74}$$

Insertion of this expression and of (21.73) into (21.72) results in

$$-\rho\theta\frac{\partial^2\psi}{\partial\theta^2}\dot{\theta} - \rho\theta\frac{\partial^2\psi}{\partial\varepsilon_{ij}\partial\theta}\dot{\varepsilon}_{ij} + \rho(\theta\frac{\partial^2\psi}{\partial\varepsilon_{ij}\partial\theta} - \frac{\partial\psi}{\partial\varepsilon_{ij}})\dot{\varepsilon}_{ij}^{ie}$$
$$+ (\rho\frac{\partial\psi}{\partial\varepsilon_{ij}} - \sigma_{ij})\dot{\varepsilon}_{ij} + \rho(\frac{\partial\psi}{\partial\kappa_\alpha} - \theta\frac{\partial^2\psi}{\partial\kappa_\alpha\partial\theta})\dot{\kappa}_\alpha = \frac{\rho}{m}\frac{\delta Q}{dt} \tag{21.75}$$

When the strains are constant, i.e. $\dot{\varepsilon}_{ij} = 0$, combination of (21.70) and (21.75) gives

$$-\theta\frac{\partial^2\psi}{\partial\theta^2}\dot{\theta} + (\theta\frac{\partial^2\psi}{\partial\varepsilon_{ij}\partial\theta} - \frac{\partial\psi}{\partial\varepsilon_{ij}})\dot{\varepsilon}_{ij}^{ie}$$
$$+ (\frac{\partial\psi}{\partial\varepsilon_{ij}} - \frac{\sigma_{ij}}{\rho})\dot{\varepsilon}_{ij} + (\frac{\partial\psi}{\partial\kappa_\alpha} - \theta\frac{\partial^2\psi}{\partial\kappa_\alpha\partial\theta})\dot{\kappa}_\alpha = c_\varepsilon\dot{\theta} \tag{21.76}$$

Likewise, if the stresses are constant, combination of (21.71) and (21.75) gives

$$-\theta\frac{\partial^2\psi}{\partial\theta^2}\dot{\theta} - \theta\frac{\partial^2\psi}{\partial\varepsilon_{ij}\partial\theta}\dot{\varepsilon}_{ij} + (\theta\frac{\partial^2\psi}{\partial\varepsilon_{ij}\partial\theta} - \frac{\partial\psi}{\partial\varepsilon_{ij}})\dot{\varepsilon}_{ij}^{ie}$$
$$+ (\frac{\partial\psi}{\partial\varepsilon_{ij}} - \frac{\sigma_{ij}}{\rho})\dot{\varepsilon}_{ij} + (\frac{\partial\psi}{\partial\kappa_\alpha} - \theta\frac{\partial^2\psi}{\partial\kappa_\alpha\partial\theta})\dot{\kappa}_\alpha = c_\sigma\dot{\theta} \tag{21.77}$$

where the condition that the stresses are constant remains to be enforced.

It appears that, in principle, both measurements of c_ε and c_σ are dependent on the inelastic processes occurring in the material, as manifested by the presence of the factors $\dot{\varepsilon}_{ij}^{ie}$ and $\dot{\kappa}_\alpha$ in (21.76) and (21.77). In practice, however, c_ε and c_σ are measured when the material responds in a thermo-elastic manner. Therefore

In practice, c_ε and c_σ are measured during thermo-elastic conditions

In that case and since the viscous stress now is zero, i.e $\rho\partial\psi/\partial\varepsilon_{ij} = \sigma_{ij}$, (21.76) reduces to

$$c_\varepsilon = -\theta\frac{\partial^2\psi}{\partial\theta^2} \tag{21.78}$$

which in the following we will take as the definition of c_ε. Likewise, (21.77) becomes

$$-\theta\frac{\partial^2\psi}{\partial\theta^2}\dot{\theta} - \theta\frac{\partial^2\psi}{\partial\varepsilon_{ij}\partial\theta}\dot{\varepsilon}_{ij} = c_\sigma\dot{\theta} \tag{21.79}$$

Thermo-elasticity is defined by $\psi = \psi(\theta, \varepsilon_{ij})$ as well as $\sigma_{ij} = \rho\partial\psi/\partial\varepsilon_{ij}$. For constant stresses, it then follows that

$$\frac{\partial^2\psi}{\partial\theta\partial\varepsilon_{ij}}\dot{\theta} + \frac{\partial^2\psi}{\partial\varepsilon_{ij}\partial\varepsilon_{kl}}\dot{\varepsilon}_{kl} = 0$$

	E [Pa]	ν	α $[\frac{1}{K}]$	ρ $[\frac{\text{kg}}{\text{m}^3}]$
Steel	$2 \cdot 10^{11}$	0.3	$1.1 \cdot 10^{-5}$	$7.8 \cdot 10^3$
Aluminum alloys	$0.7 \cdot 10^{11}$	0.3	$2.3 \cdot 10^{-5}$	$2.7 \cdot 10^3$
Concrete	$0.3 \cdot 10^{11}$	0.15	$1.0 \cdot 10^{-5}$	$2.4 \cdot 10^3$

Table 21.1: Approximative material parameters at room temperature.

Use of (21.19) then gives

$$\dot{\varepsilon}_{ij} = -\rho C_{ijkl} \frac{\partial^2 \psi}{\partial \theta \partial \varepsilon_{kl}} \dot{\theta}$$

With this expression and (21.78) inserted into (21.79), we obtain the result

$$\boxed{c_\sigma - c_\varepsilon = \rho\theta \frac{\partial^2 \psi}{\partial \theta \partial \varepsilon_{ij}} C_{ijkl} \frac{\partial^2 \psi}{\partial \theta \partial \varepsilon_{kl}}} \tag{21.80}$$

Since the flexibility tensor C_{ijkl} is positive definite, we conclude, as expected, that $c_\sigma \geq c_\varepsilon$.

To evaluate (21.80) in more detail, we assume that D_{ijkl} and β_{ij} appearing in (21.17) are temperature independent; we then obtain

$$\frac{\partial^2 \psi}{\partial \theta \partial \varepsilon_{ij}} = -\frac{1}{\rho} \beta_{ij} \quad \text{temperature-independent properties}$$

Insertion of this expression and of (21.19) into (21.80) then results in

$$\boxed{c_\sigma - c_\varepsilon = \frac{\theta}{\rho} \alpha_{ij} D_{ijkl} \alpha_{kl} \qquad \textit{temperature-independent properties}} \tag{21.81}$$

If we, furthermore, assume isotropy, where D_{ijkl} is given by (4.89), and $\alpha_{ij} = \alpha \delta_{ij}$, where α is the thermal expansion coefficient, (21.81) reduces to

$$\boxed{c_\sigma - c_\varepsilon = \frac{9K\alpha^2}{\rho} \theta \qquad \begin{array}{l} \textit{isotropy} \\ \textit{temperature-independent properties} \end{array}} \tag{21.82}$$

where K is the bulk modulus.

Some approximative material parameters are given in Tables 21.1 and 21.2. From the values of Table 21.2 it appears that, at room temperature, the difference between c_σ and c_ε, in practice, is negligible. Moreover, when measuring the

	$9K\alpha^2/\rho$ $[\frac{\text{Nm}}{\text{kg}K^2}]$	c_σ $[\frac{\text{Nm}}{\text{kg}\,K}]$	$3K\alpha$ $[\frac{\text{N}}{\text{m}^2\,K}]$	ρc_σ $[\frac{\text{N}}{\text{m}^2\,K}]$	$\frac{c_\sigma - c_\varepsilon}{c_\sigma}$ [%]
Steel	$2.3 \cdot 10^{-2}$	480	$5.5 \cdot 10^6$	$3.7 \cdot 10^6$	1.4
Aluminum alloys	$10 \cdot 10^{-2}$	960	$5.8 \cdot 10^6$	$2.6 \cdot 10^6$	3.1
Concrete	$0.54 \cdot 10^{-2}$	900	$1.9 \cdot 10^6$	$2.2 \cdot 10^6$	0.2

Table 21.2: Approximative material parameters at room temperature.

specific heat capacity of a solid, the specimen is not exposed to mechanical loading, i.e. $\sigma_{ij} = 0$. In conclusion:

In practice, c_σ is the specific heat capacity that is measured. However, the $c_\sigma-$ and $c_\varepsilon-$values are very close to each other (21.83)

Very often the specific heat capacity only varies slowly with temperature; if, therefore, c_ε as given by (21.78) is assumed to be constant, the free energy can be obtained by integration and the result becomes

$$\begin{aligned}\psi(\theta, \varepsilon_{ij} - \varepsilon_{ij}^{ie}, \kappa_\alpha) &= K_1 + c_\varepsilon(\theta - \theta \ln \frac{\theta}{\theta^*}) \\ &\quad + \theta p(\varepsilon_{ij} - \varepsilon_{ij}^{ie}, \kappa_\alpha) + q(\varepsilon_{ij} - \varepsilon_{ij}^{ie}, \kappa_\alpha)\end{aligned}$$

where K_1 is an arbitrary constant and θ^* an arbitrary constant temperature whereas p and q denote arbitrary functions. Considering for simplicity thermo-elasticity and by proper choices of the functions p and q, we obtain

$$\begin{aligned}\psi(\theta, \varepsilon_{ij}) =& \rho K_1 + \rho c_\varepsilon(\theta - \theta \ln \frac{\theta}{\theta^*}) \\ &+ \frac{1}{2}\varepsilon_{ij} D_{ijkl} \varepsilon_{kl} - \beta_{ij}(\theta - \theta_o)\varepsilon_{ij}\end{aligned} \tag{21.84}$$

which holds for thermo-elasticity with constant material parameters; it is evident that formulation (21.17) can be adapted to coincide with the expression above.

With this discussion of the specific heat capacity, we now finally turn to a derivation of the heat equation from the first law. From the first law (21.1) and (21.4), we have

$$\rho(\dot{\psi} + \dot{s}\theta + s\dot{\theta}) = \dot{\varepsilon}_{ij}\sigma_{ij} + r - q_{i,i} \tag{21.85}$$

With $\dot{\psi}$ and $\dot{s}$ given by (21.73) and (21.74) respectively, insertion of these expressions into (21.85) results in

$$-\rho\theta\frac{\partial^2\psi}{\partial\varepsilon_{ij}\partial\theta}\dot{\varepsilon}_{ij}+\rho(\theta\frac{\partial^2\psi}{\partial\varepsilon_{ij}\partial\theta}-\frac{\partial\psi}{\partial\varepsilon_{ij}})\dot{\varepsilon}^{ie}_{ij}+(\rho\frac{\partial\psi}{\partial\varepsilon_{ij}}-\sigma_{ij})\dot{\varepsilon}_{ij}$$
$$+\rho(\frac{\partial\psi}{\partial\kappa_\alpha}-\theta\frac{\partial^2\psi}{\partial\theta\partial\kappa_\alpha})\dot{\kappa}_\alpha-\rho\theta\frac{\partial^2\psi}{\partial\theta^2}\dot{\theta}=r-q_{i,i}$$

With c_ε defined by (21.78) and the heat flux vector q_i given by Fourier's law (21.59), we finally obtain

Coupled heat equation

$$\rho c_\varepsilon\dot{\theta}=(k_{ij}\theta_{,j})_{,i}+r$$
$$+\rho\theta\frac{\partial^2\psi}{\partial\varepsilon_{ij}\partial\theta}(\dot{\varepsilon}_{ij}-\dot{\varepsilon}^{ie}_{ij})+\rho\frac{\partial\psi}{\partial\varepsilon_{ij}}\dot{\varepsilon}^{ie}_{ij} \quad (21.86)$$
$$+(\sigma_{ij}-\rho\frac{\partial\psi}{\partial\varepsilon_{ij}})\dot{\varepsilon}_{ij}-\rho(\frac{\partial\psi}{\partial\kappa_\alpha}-\theta\frac{\partial^2\psi}{\partial\theta\partial\kappa_\alpha})\dot{\kappa}_\alpha$$

It appears that determination of the temperature distribution within the body is coupled not only to the strain rate $\dot{\varepsilon}_{ij}$, but also to the inelastic rates $\dot{\varepsilon}^{ie}_{ij}$ and $\dot{\kappa}_\alpha$. Therefore (21.86) comprises a coupled problem and it can only be solved if the equations for $\dot{\varepsilon}_{ij}$, $\dot{\varepsilon}^{ie}_{ij}$ and $\dot{\kappa}_\alpha$ are solved simultaneously.

Let us consider a different format of the coupled heat equation which is based on the entropy. Taking advantage of (21.73) in (21.85) leads to

Coupled heat equation

$$\rho\theta\dot{s}=\gamma_{mech}+r-q_{i,i}$$

where (21.87)

$$\gamma_{mech}=\rho\frac{\partial\psi}{\partial\varepsilon_{ij}}\dot{\varepsilon}^{ie}_{ij}+(\sigma_{ij}-\rho\frac{\partial\psi}{\partial\varepsilon_{ij}})\dot{\varepsilon}_{ij}-\rho\frac{\partial\psi}{\partial\kappa_\alpha}\dot{\kappa}_\alpha$$

This neat alternative formulation of the heat equation is occasionally preferred in numerical computations; in the following, however, we choose to work with the temperature instead of its conjugated quantity, the entropy.

It may be of interest to evaluate (21.86) for thermo-elastic behavior. In that case $\dot{\varepsilon}^{ie}_{ij}=\dot{\kappa}_\alpha=0$, $\sigma_{ij}=\rho\partial\psi/\partial\varepsilon_{ij}$ and we obtain

Coupled heat equation for thermo-elasticity

$$\rho c_\varepsilon\dot{\theta}=(k_{ij}\theta_{,j})_{,i}+r+\rho\theta\frac{\partial^2\psi}{\partial\varepsilon_{ij}\partial\theta}\dot{\varepsilon}_{ij} \quad (21.88)$$

and the solution of this equation is still coupled to the development of $\dot{\varepsilon}_{ij}$. If, furthermore, both D_{ijkl} and β_{ij} in (21.17) and (21.19) are assumed to be independent of the temperature, we have

$$\rho\frac{\partial^2\psi}{\partial\varepsilon_{ij}\partial\theta} = -\beta_{ij} = -D_{ijkl}\alpha_{kl}$$

and (21.88) turns into

$$\rho c_\varepsilon\dot{\theta} = (k_{ij}\theta_{,j})_{,i} + r - \theta\dot{\varepsilon}_{ij}D_{ijkl}\alpha_{kl}$$

If, in addition, isotropy is assumed this expression reduces to

Coupled heat equation for isotropic thermo-elasticity and temperature-independent values of E, ν and α (21.89)

$$\rho c_\varepsilon\dot{\theta} = (k\theta_{,i})_{,i} + r - 3K\alpha\theta\dot{\varepsilon}_{ii}$$

where, for isotropy, we have $k_{ij} = k\delta_{ij}$ with k being the thermal conductivity. It appears that it is only the volumetric part of the strain rate that contributes to the coupling between strains and temperature. This coupled heat equation for thermo-elasticity was derived by Duhamel already in 1837.

Expression (21.89) may be used to determine the thermo-elastic response for specimens subjected to a homogeneous stress state and loaded either during isothermal or adiabatic conditions. Since an isotropic material is considered, (21.18) reduces to

$$s_{ij} = 2Ge_{ij}; \qquad \sigma_{kk} = 3K[\varepsilon_{kk} - 3\alpha(\theta - \theta_o)] \tag{21.90}$$

Let the surroundings of the specimen be kept at a constant temperature. The isothermal condition is then obtained by applying the forces on the specimen so slowly that any temperature difference between the specimen and the surroundings is allowed to disappear. The mechanical loading is obtained by prescribing the stresses; since $\theta = \theta_o$ (21.90) gives

$$\varepsilon_{kk} = \frac{\sigma_{kk}}{3K} \tag{21.91}$$

Consider next the adiabatic condition where no heat input occurs which is obtained by the conditions $r = 0$ and $\theta_{,i} = 0$. This situation can be achieved if the specimen is loaded very quickly since heat exchange with the surroundings then has no time to take place. In that case (21.89) provides

$$\dot{\theta} = -\frac{3K\alpha}{\rho c_\varepsilon}\theta\dot{\varepsilon}_{kk} \tag{21.92}$$

Recalling that (21.89) was derived on condition that the material parameters are temperature independent, integration of the expression above gives

$$\ln\frac{\theta}{\theta_o} = -\frac{3K\alpha}{\rho c_\varepsilon}\varepsilon_{kk} \tag{21.93}$$

We shall later see that the temperature change is very small and we may then write

$$\theta = \theta_o + \Delta\theta \quad \text{where} \quad |\Delta\theta| \ll 1$$

Under these conditions, a Taylor expansion gives

$$\ln\frac{\theta}{\theta_o} = \ln(1 + \frac{\Delta\theta}{\theta_o}) \approx \frac{\Delta\theta}{\theta_o} = \frac{\theta - \theta_o}{\theta_o} \approx \frac{\theta - \theta_o}{\theta}$$

and then (21.93) becomes

$$\theta - \theta_o = -\frac{3K\alpha}{\rho c_\varepsilon}\theta\varepsilon_{kk} \tag{21.94}$$

Insertion into (21.90) provides

$$\varepsilon_{kk} = \frac{\sigma_{kk}}{3K(1 + \frac{9K\alpha^2\theta}{\rho c_\varepsilon})} \tag{21.95}$$

which can be written as

$$\varepsilon_{kk} = \frac{\sigma_{kk}}{3K_{ad}} \quad \text{where} \quad K_{ad} = K(1 + \frac{9K\alpha^2\theta}{\rho c_\varepsilon}) \tag{21.96}$$

where K_{ad} is the apparent bulk modulus during adiabatic loading. During isothermal loading we have (21.91), i.e. $K_{iso} = K$ and (21.96b) may then be written as

$$\boxed{\frac{K_{ad}}{K_{iso}} = 1 + \frac{9K\alpha^2\theta}{\rho c_\varepsilon}}$$

where $K_{iso} = K$. It appears that the material behaves in a stiffer manner during adiabatic loading than during isothermal loading. However, this effect is most often extremely small and for steel we find from Table 21.2 - and recalling that $c_\sigma \approx c_\varepsilon$ - that at room temperature, we have $K_{ad}/K_{iso} \approx 1.014$.

Let us return to (21.94) and demonstrate that the temperature change for thermo-elastic materials during adiabatic conditions is indeed very small. Combination of (21.94) and (21.95) and taking advantage of (21.82) give

$$\theta - \theta_o = -\frac{\alpha\theta\sigma_{kk}}{\rho c_\sigma}$$

Since $\theta - \theta_o$ is very small, we have $\theta \approx \theta_o$, i.e.

Kelvin's expression for thermo-elasticity

$$\boxed{\theta - \theta_o = -\frac{\alpha\theta_o\sigma_{kk}}{\rho c_\sigma}} \tag{21.97}$$

This expression was established by Lord Kelvin already in 1855 and it is called the *Kelvin effect*. For hydrostatic tension, it implies a fall in temperature whereas hydrostatic compression gives rise to a temperature increase and it may be considered as the counterpart to the same phenomenon for gases. However, this effect is very modest. As an example, consider uniaxial loading where $\sigma_{kk} = \sigma$; referring to Tables 21.1 and 21.2, we find for steel at room temperature that a sudden application of a tensile stress of 200 MPa results in a temperature drop of about $0.17\,^{o}C$.

We have demonstrated that straining of a thermo-elastic material in itself only influences the temperature to a very modest degree. In general however, when thermo-elastic behavior is considered the material is not only exposed to mechanical loading, but also to significant temperature loadings that, in turn, influence the stress and strain state, and the problem of coupling effects then becomes more complex. In order to evaluate the coupling effect apparent from (21.89) in such cases, we rewrite this expression according to

$$\rho c_\varepsilon \dot{\theta}(1 + M) = (k\theta_{,i})_{,i} + r \tag{21.98}$$

where the coupling term M is given by

$$M = \frac{9K\alpha^2\theta}{\rho c_\varepsilon}\frac{\dot{\varepsilon}_{ii}}{3\alpha\dot{\theta}}$$

The term $\dot{\varepsilon}_{ii}/(3\alpha\dot{\theta})$ is a natural quantity, since it is the ratio of the total volumetric strain rate to the volumetric thermal strain rate. We can ignore the coupling effect if $M \ll 1$, i.e.

$$\frac{\dot{\varepsilon}_{ii}}{3\alpha\theta} \ll \frac{\rho c_\varepsilon}{9K\alpha^2\theta}$$

According to Table 21.2, at room temperature the term $\rho c_\varepsilon/(9K\alpha^2\theta)$ is 71, 33 and 569 for steel, aluminum and concrete, respectively. We can safely replace these restrictions by a stronger condition and - traditionally - one chooses $\dot{\varepsilon}_{ii}/(3\alpha\theta) \ll 20$; in that case the coupling term M in (21.98) can be ignored. In conclusion

If $\frac{\dot{\varepsilon}_{ii}}{3\alpha\dot{\theta}} \ll 20$ *then*
the uncoupled heat equation
$$\rho c_\varepsilon = (k\theta_{,i})_{,i} + r \tag{21.99}$$
holds for thermo-elasticity with close accuracy

The condition $\dot{\varepsilon}_{ii}/(3\alpha\dot{\theta}) \ll 20$ does not seem to be a severe restriction, since we intuitively expect the total volumetric strain rate $\dot{\varepsilon}_{ii}$ to be of the same order of magnitude as the volumetric thermal strain rate $3\alpha\dot{\theta}$. Indeed - and as already discussed - this restriction turns out to be fulfilled in most cases of practical interest and the reader is referred to Boley and Weiner (1960) for a further evaluation of this issue.

After this discussion of thermo-elasticity, we return to the general coupled heat equation given by (21.86). If the loading is performed very quickly, no heat exchange can occur with the surroundings of the body. Moreover, this rapid loading does not allow any heat flow to take place. Disregarding heat flow can be obtained by ignoring the heat flux vector $q_i = -k_{ij}\theta_{,j}$, i.e. by putting $k_{ij}\theta_{,j} = 0$. Likewise, since no heat exchange with the surroundings is possible, we put $r = 0$. These conditions are called *adiabatic heating* and (21.86) then reduces to

Adiabatic heating

$$\rho c_\varepsilon \dot{\theta} = \sigma_{ij}\dot{\varepsilon}^{ie}_{ij} + (\sigma_{ij} - \rho\frac{\partial\psi}{\partial\varepsilon_{ij}})\dot{\varepsilon}_{ij} + \rho\theta\frac{\partial^2\psi}{\partial\varepsilon_{ij}\partial\theta}(\dot{\varepsilon}_{ij} - \dot{\varepsilon}^{ie}_{ij}) - \rho(\frac{\partial\psi}{\partial\kappa_\alpha} - \theta\frac{\partial^2\psi}{\partial\theta\partial\kappa_\alpha})\dot{\kappa}_\alpha \quad (21.100)$$

For a given stress state σ_{ij} and a given expression for the free energy ψ, then once the quantities $\dot{\varepsilon}_{ij}$, $\dot{\varepsilon}^{ie}_{ij}$ and $\dot{\kappa}_\alpha$ are known, the right-hand side of this expression is known and (21.100) then determines directly the temperature rate $\dot{\theta}$. That is, the temperature change can be determined without solving any field problem that involves the boundary conditions of the body; instead the problem is solved directly at each point in the body. For isotropic thermo-elasticity with temperature-independent material properties (21.100) reduces to (21.92). It should also be emphasized that (21.100) for an inhomogeneously loaded body allows the development of an inhomogeneous temperature distribution over the body. This inhomogeneous temperature field is allowed to exist since the loading was assumed to be so rapid that heat flow within the body has no time to take place.

It is noted, however, that since $\dot{\varepsilon}_{ij}$, $\dot{\varepsilon}^{ie}_{ij}$ and $\dot{\kappa}_\alpha$, in general, depends on $\dot{\theta}$, expression (21.100) still defines a coupled problem in the numerical sense (but it is not a coupled boundary value problem). Numerical solution strategies for this problem as well as the general case of the coupled heat equation (21.86) will be discussed later on in Chapter 23.

Expression (21.100) determines the temperature change that results from mechanical loading alone. For thermo-elasticity, we have seen that this effect is very modest, cf. the discussion of (21.97). For inelastic materials, however, the effect may be very significant and we shall return to this aspect in Chapter 23.

With evident notation, (21.100) may be written as

$$\rho c_\varepsilon \dot{\theta} = \sigma_{ij}\dot{\varepsilon}^{ie}_{ij} + Y$$

From the experiments of Taylor and Quinney (1934) for steel and metals it is found that

$$\rho c_\varepsilon \dot{\theta} = \eta\sigma_{ij}\dot{\varepsilon}^{ie}_{ij} \quad \textit{Adiabatic heating} \quad (21.101)$$

where $\eta \approx 0.90 - 0.95$. It appears that $Y \approx -(0.05 - 0.10)\sigma_{ij}\dot{\varepsilon}_{ij}^{ie}$ and since Y has the dimension of energy rate [Nm/(m^{3}s)], it follows that the material absorbs a certain energy that does not manifest itself in terms of a temperature increase. Instead, this absorbed energy is used to transform the microstructure of the material by phase changes, pile-up of dislocations etc. The absorbed energy therefore transforms into *latent heat* that can be recovered when the material is heated, i.e. annealed, cf. the discussion of Taylor and Quinney (1934) and Bever *et al.* (1973).

In many applications in the literature, the factor η in (21.101) is treated as a material constant. However, if (21.100) is written in the form (21.101) it is evident that η must depend on the evolution laws for $\dot{\kappa}_\alpha$ and $\dot{\varepsilon}_{ij}^{ie}$ and on the stresses σ_{ij} and the thermodynamic forces $K_\alpha(= \rho\partial\psi/\partial\kappa_\alpha)$ and - despite the traditional attitude in the literature - it is not surprising that η for a given material will not become a constant. In recent years, there has been an increasing interest for theoretical determination of the η-factor and detailed considerations are presented by Chaboche (1993b), Kamlah and Haupt (1998) and Rosakis *et al.* (2000).

21.5 Properties of state functions at thermodynamical equilibrium

We will now consider a body at a state where no mechanical work input or heat input occur. Following (20.62), the body is then in thermodynamical equilibrium, if no changes of the state variables can occur. In (20.63) and following Gibbs, it was then concluded that when the internal energy is constant, thermodynamic equilibrium exists when the entropy attains a maximum value. We will now express these properties in precise mathematical terms by which we can characterize the properties of the state functions at thermodynamical equilibrium; in particular we want to identify which properties the free energy ψ must possess in order to achieve thermodynamic equilibrium.

According to (21.4), we have

$$u = \psi + s\theta$$

Differentiation with respect to time and use of (21.37) results in

$$\dot{u} = \theta\dot{s} + \frac{\partial\psi}{\partial\varepsilon_{ij}}(\dot{\varepsilon}_{ij} - \dot{\varepsilon}_{ij}^{ie}) + \frac{\partial\psi}{\partial\kappa_\alpha}\dot{\kappa}_\alpha \tag{21.102}$$

The internal energy is a state function and the expression above suggests that

$$u = u(s, \varepsilon_{ij} - \varepsilon_{ij}^{ie}, \kappa_\alpha)$$

Differentiation of the expression and comparison with (21.102) provides

$$\frac{\partial u}{\partial s} = \theta\,; \qquad \frac{\partial u}{\partial\varepsilon_{ij}} = \frac{\partial\psi}{\partial\varepsilon_{ij}}; \qquad \frac{\partial u}{\partial\kappa_\alpha} = \frac{\partial\psi}{\partial\kappa_\alpha}$$

Since $\theta > 0$, (21.102) may be solved for $\dot{s}$ to obtain

$$\dot{s} = \frac{1}{\theta}\dot{u} - \frac{1}{\theta}\frac{\partial\psi}{\partial\varepsilon_{ij}}(\dot{\varepsilon}_{ij} - \dot{\varepsilon}_{ij}^{ie}) - \frac{1}{\theta}\frac{\partial\psi}{\partial\kappa_\alpha}\dot{\kappa}_\alpha \tag{21.103}$$

The entropy s is a state function and we must therefore have

$$s = s(u, \varepsilon_{ij} - \varepsilon_{ij}^{ie}, \kappa_\alpha)$$

i.e.

$$\dot{s} = \frac{\partial s}{\partial u}\dot{u} + \frac{\partial s}{\partial\varepsilon_{ij}}(\dot{\varepsilon}_{ij} - \dot{\varepsilon}_{ij}^{ie}) + \frac{\partial s}{\partial\kappa_\alpha}\dot{\kappa}_\alpha \tag{21.104}$$

A comparison of this expression with (21.103) shows that

$$\frac{\partial s}{\partial u} = \frac{1}{\theta}; \qquad \frac{\partial s}{\partial\varepsilon_{ij}} = -\frac{1}{\theta}\frac{\partial\psi}{\partial\varepsilon_{ij}}; \qquad \frac{\partial s}{\partial\kappa_\alpha} = -\frac{1}{\theta}\frac{\partial\psi}{\partial\kappa_\alpha} \tag{21.105}$$

With these preliminaries we now consider a situation where the internal energy u is constant and where the entropy s attains a maximum value at thermodynamic equilibrium. Therefore $\dot{u} = \dot{s} = 0$ and (21.104) then provides

$$\frac{\partial s}{\partial\varepsilon_{ij}}(\dot{\varepsilon}_{ij} - \dot{\varepsilon}_{ij}^{ie}) + \frac{\partial s}{\partial\kappa_\alpha}\dot{\kappa}_\alpha = 0 \tag{21.106}$$

Moreover, following Appendix (A.12) the condition that $s = s(u, \varepsilon_{ij} - \varepsilon_{ij}^{ie}, \kappa_\alpha)$ for constant u attains a maximum value can be expressed as

$$\begin{aligned}(\dot{\varepsilon}_{ij} - \dot{\varepsilon}_{ij}^{ie})&\frac{\partial^2 s}{\partial\varepsilon_{ij}\partial\varepsilon_{kl}}(\dot{\varepsilon}_{kl} - \dot{\varepsilon}_{kl}^{ie}) \\ &+ 2(\dot{\varepsilon}_{ij} - \dot{\varepsilon}_{ij}^{ie})\frac{\partial^2 s}{\partial\varepsilon_{ij}\partial\kappa_\alpha}\dot{\kappa}_\alpha + \dot{\kappa}_\alpha\frac{\partial^2 s}{\partial\kappa_\alpha\partial\kappa_\beta}\dot{\kappa}_\beta < 0\end{aligned} \tag{21.107}$$

This is simply to say that the entropy s at thermodynamic equilibrium is a concave function with respect to the variables $\varepsilon_{ij} - \varepsilon_{ij}^{ie}$ and κ_α.

Having identified in mathematical terms the properties of the entropy s, we will now identify the corresponding properties for Helmholtz' free energy ψ. Insertion of (21.105b) and (21.105c) into (21.106) and (21.107) gives the results sought for

At thermodynamic equilibrium

$$\frac{\partial\psi}{\partial\varepsilon_{ij}}(\dot{\varepsilon}_{ij} - \dot{\varepsilon}_{ij}^{ie}) + \frac{\partial\psi}{\partial\kappa_\alpha}\dot{\kappa}_\alpha = 0 \tag{21.108}$$

and

$$\boxed{\begin{aligned}&\textit{At thermodynamic equilibrium}\\&(\dot{\varepsilon}_{ij}-\dot{\varepsilon}^{ie}_{ij})\frac{\partial^2\psi}{\partial\varepsilon_{ij}\partial\varepsilon_{kl}}(\dot{\varepsilon}_{kl}-\dot{\varepsilon}^{ie}_{kl})\\&\quad+2(\dot{\varepsilon}_{ij}-\dot{\varepsilon}^{ie}_{ij})\frac{\partial^2\psi}{\partial\varepsilon_{ij}\partial\kappa_\beta}\dot{\kappa}_\beta+\dot{\kappa}_\alpha\frac{\partial^2\psi}{\partial\kappa_\alpha\partial\kappa_\beta}\dot{\kappa}_\beta>0\end{aligned}} \tag{21.109}$$

which according to Appendix (A.8) means that Helmholtz' free energy ψ is a convex function with respect to the variables $\varepsilon_{ij}-\varepsilon^{ie}_{ij}$ and κ_α.

From (21.37) we have $s=-\partial\psi/\partial\theta$ and differentiation gives

$$\dot{s}=-[\frac{\partial^2\psi}{\partial\theta^2}\dot{\theta}+\frac{\partial^2\psi}{\partial\theta\partial\varepsilon_{mn}}(\dot{\varepsilon}_{mn}-\dot{\varepsilon}^{ie}_{mn})+\frac{\partial^2\psi}{\partial\theta\partial\kappa^\gamma}\dot{\kappa}^\gamma]$$

and as $\dot{s}=0$ holds at thermodynamic equilibrium we obtain

$$\dot{\theta}=-\frac{\theta}{c_\varepsilon}[\frac{\partial^2\psi}{\partial\theta\partial\varepsilon_{mn}}(\dot{\varepsilon}_{mn}-\dot{\varepsilon}^{ie}_{mn})+\frac{\partial^2\psi}{\partial\theta\partial\kappa^\gamma}\dot{\kappa}^\gamma]$$

However, by using (21.108) in this expression we obtain

$$\boxed{\dot{\theta}=0\quad \textit{at thermodynamic equilibrium}}$$

We will now prove that the free energy ψ not only is a convex function in $\varepsilon_{ij}-\varepsilon^{ie}_{ij}$ and κ_α, but also that it attains a minimum at thermodynamic equilibrium. Since $\psi=\psi(\theta,\varepsilon-\varepsilon^{ie}_{ij},\kappa_\alpha)$ we obtain with the expression above as well as (21.108)

$$\boxed{\dot{\psi}=0\qquad \textit{at thermodynamic equilibrium}}$$

This result combined with (21.109) shows that ψ attains a minimum value at thermodynamic equilibrium. According to (21.78) the specific heat capacity c_ε is defined by $c_\varepsilon=-\theta\partial^2\psi/\partial\theta^2$ and as $\theta>0$ and we certainly expect the heat capacity to be positive, we conclude that ψ is a concave function of the absolute temperature θ.

Let us finally specialize to thermo-elasticity, where the internal variables ε^{ie}_{ij} and κ_α do not exist and where (21.109) reduces to

$$\dot{\varepsilon}_{ij}\frac{\partial^2\psi}{\partial\varepsilon_{ij}\partial\varepsilon_{kl}}\dot{\varepsilon}_{kl}>0 \tag{21.110}$$

We do not expect any peculiarities connected with thermo-elasticity; therefore, it seems reasonable to expect that thermo-elastic materials are stable in a thermodynamic sense and we therefore expect the expression above to be fulfilled.

Since $\partial^2\psi/\partial\varepsilon_{ij}\partial\varepsilon_{kl} = D_{ijkl}/\rho$, cf. (21.19), we conclude that (21.110) is fulfilled for

$$\boxed{D_{ijkl} \quad \textit{is positive definite}}$$

That the elastic stiffness tensor D_{ijk} is positive definite is precisely what we anticipated already in Chapter 4, cf. (4.24) and (4.25), and for isotropic elasticity this is fulfilled when $E > 0$ and $-1 < \nu < \frac{1}{2}$, cf. (4.96).

22 PLASTICITY, VISCOPLASTICITY AND VISCOELASTICITY

Having established the general framework provided by thermodynamics, we will now illustrate its use on plasticity, viscoplasticity and viscoelasticity. The reader may consult the textbooks of Lemaitre and Chaboche (1990), Maugin (1992), Simo and Hughes (1998) and Nguyen (2000) for related viewpoints. We will first present the general plasticity equations and then give examples of the establishment of some specific and typical plasticity models.

22.1 Fundamental equations of plasticity

For plasticity theory where the viscous stress σ_{ij}^{v} disappears, the results (21.37) read

$$\boxed{\begin{aligned} &\psi = \psi(\theta, \varepsilon_{ij} - \varepsilon_{ij}^{p}, \kappa_\alpha) \\ &\textit{and} \\ &\gamma_{mech} \equiv \sigma_{ij}\dot{\varepsilon}_{ij}^{p} - K_\alpha \dot{\kappa}_\alpha \geq 0 \\ &\textit{where} \\ &\sigma_{ij} = \rho\frac{\partial \psi}{\partial \varepsilon_{ij}}\,; \qquad K_\alpha \equiv \rho\frac{\partial \psi}{\partial \kappa_\alpha} \end{aligned}} \tag{22.1}$$

We certainly know that at least the stress σ_{ij} must enter the yield function and the question is what other variables are needed to monitor hardening/softening behavior. According to the mechanical dissipation inequality, the 'forces' σ_{ij} and K_α are conjugated to the 'fluxes' $\dot{\varepsilon}_{ij}^{p}$ and $-\dot{\kappa}_\alpha$ respectively. It therefore seems natural to take these forces as the variables in the yield function f and we then have

$$\boxed{f = f(\sigma_{ij}, K_\alpha) \leq 0}$$

A comparison with Chapter 10 shows that the conjugated thermodynamic forces K_α for plasticity theory take the form of being the *hardening parameters*.

Once Helmholtz' free energy ψ has been chosen, both the stresses σ_{ij} and the forces K_α are given in accordance with (22.1). Disregarding for the moment thermo-plasticity, which we will deal with in the next chapter, and considering the expression for $\rho\psi$ given by (21.17) that is valid for thermo-elasticity, a natural choice for plasticity is

$$\boxed{\rho\psi(\theta, \varepsilon_{ij}-\varepsilon_{ij}^p, \kappa_\alpha) = \rho h(\theta) + \frac{1}{2}(\varepsilon_{ij}-\varepsilon_{ij}^p)D_{ijkl}(\varepsilon_{kl}-\varepsilon_{kl}^p) + \rho\psi^p(\kappa_\alpha)} \tag{22.2}$$

Here, $h(\theta)$ is an arbitrary function of the temperature and $\psi^p(\kappa_\alpha)$ is an arbitrary function of the internal variables; moreover, D_{ijkl} is the elastic stiffness tensor which is considered to be a constant. It appears that the free energy function ψ is split into three separate parts and that no coupling effects are present between the three sets of variables: θ, $\varepsilon_{ij} - \varepsilon_{ij}^p$ and κ_α, cf. Lubliner (1972). From (22.2) and (22.1) follow that

$$\boxed{\sigma_{ij} = D_{ijkl}(\varepsilon_{kl} - \varepsilon_{kl}^p)\,; \qquad K_\alpha = \rho\frac{\partial\psi^p(\kappa_\beta)}{\partial\kappa_\alpha}} \tag{22.3}$$

where Hooke's law is recovered by (22.3a).

Having determined the 'forces' σ_{ij} and K_α, the next topic is to establish evolution laws for the 'fluxes' $\dot{\varepsilon}_{ij}^p$ and $\dot{\kappa}_\alpha$. The only formal requirement is that these evolution laws must be such that the dissipation inequality (22.1) is fulfilled. A natural route is to require that γ_{mech} is as large as possible; we then evidently ensure that $\gamma_{mech} \geq 0$. This leads to the postulate of maximum dissipation discussed in Section 21.3. As the mathematical literature on extremum properties is by tradition concerned with minimization problems, we change our requirement of γ_{mech} being maximal to $-\gamma_{mech}$ being minimal. Moreover, we will treat the fluxes $\dot{\varepsilon}_{ij}^p$ and $\dot{\kappa}_\alpha$ as given - although arbitrary - and consider the forces σ_{ij} and K_α as the variables. In agreement with (21.55) we then have

> *Postulate of maximum dissipation:*
>
> *For given fluxes $\dot{\varepsilon}_{ij}^p$ and $\dot{\kappa}_\alpha$, find those stresses σ_{ij} and conjugated forces K_α that minimize $-\gamma_{mech}$ under the constraint that $f(\sigma_{ij}, K_\alpha) \leq 0$. The yield function f is assumed to be a convex function in σ_{ij} and K_α* (22.4)

In accordance with the Appendix, we are faced with a minimization problem with a constraint in terms of an inequality. Following (A.16), we define the Lagrange function $\mathcal{L}(\sigma_{ij}, K_\alpha, \dot{\lambda})$ by

$$\mathcal{L}(\sigma_{ij}, K_\alpha, \dot{\lambda}) = -\gamma_{mech} + \dot{\lambda}f = -\sigma_{ij}\dot{\varepsilon}_{ij}^p + K_\alpha\dot{\kappa}_\alpha + \dot{\lambda}f(\sigma_{ij}, K_\alpha)$$

where $\dot{\lambda}$ is a Lagrange multiplier. In conformity with (A.17) and (A.15), it follows that $\partial\mathcal{L}/\partial\sigma_{ij} = -\dot{\varepsilon}_{ij}^p + \dot{\lambda}\partial f/\partial\sigma_{ij} = 0$ and $\partial\mathcal{L}/\partial K_\alpha = \dot{\kappa}_\alpha + \dot{\lambda}\partial f/\partial K_\alpha = 0$.

From these expressions and (A.15), the following *Kuhn-Tucker relations* are then obtained

Postulate of maximum dissipation implies:
The associated evolution equations

$$\dot{\varepsilon}_{ij}^{p} = \dot{\lambda}\frac{\partial f}{\partial \sigma_{ij}}; \qquad \dot{\kappa}_{\alpha} = -\dot{\lambda}\frac{\partial f}{\partial K_{\alpha}}$$

as well as the loading/unloading criteria (22.5)

$$\dot{\lambda} \geq 0$$

$$\dot{\lambda} = 0 \quad \textit{for elastic behavior}$$

and

$$\dot{\lambda} f = 0$$

It appears that the postulate of maximum dissipation leads to the associated evolution laws and that the quantity $\dot{\lambda}$, which was originally introduced as a Lagrange multiplier, turns out to be the *plastic multiplier*. Moreover, the Kuhn-Tucker relations imply that $\dot{\lambda} \geq 0$ when $f = 0$ and $\dot{\lambda} = 0$ when $f < 0$. Previously, in Chapter 10 we had to assume these properties for $\dot{\lambda}$, but now they are mathematical consequences of the postulate of maximum dissipation. We also recall from Chapter 10 that these properties for $\dot{\lambda}$ lead to the general loading/unloading criteria given by (10.38).

To obtain the evolution equations for $\dot{\varepsilon}_{ij}^{p}$ and $\dot{\kappa}_{\alpha}$, another fruitful route is to use the results given by (21.51) where the dissipation function $\phi = \phi(A_{\Theta}, Z^{p})$ was introduced and where $A_{\Theta} = \{\sigma_{ij}, K_{\alpha}\}$, i.e. the set A_{Θ} comprises all σ_{ij} and K_{α} components and Z^{p} are simply some variables, cf. (21.39) and (21.46). Then the evolution equations are given by $\dot{a}_{\Theta} = \dot{\lambda}\partial\phi/\partial A_{\Theta}$ where $\dot{a}_{\Theta} = \{\dot{\varepsilon}_{ij}^{p}, -\dot{\kappa}_{\alpha}\}$, i.e. the set $\dot{a}_{\Theta}$ comprises all $\dot{\varepsilon}_{ij}^{p}$ and $-\dot{\kappa}_{\alpha}$ components, cf. (21.39). Since $\phi = \phi(A_{\Theta}, Z^{p})$ and $A_{\Theta} = \{\sigma_{ij}, K_{\alpha}\}$ we may choose the dissipation function ϕ as the *potential function* $g(\sigma_{ij}, K_{\alpha})$. The evolution equations $\dot{a}_{\Theta} = \dot{\lambda}\partial\phi/\partial A_{\Theta}$ then become $\dot{\varepsilon}_{ij}^{p} = \dot{\lambda}\partial g/\partial\sigma_{ij}$ and $-\dot{\kappa}_{\alpha} = \dot{\lambda}\partial g/\partial K_{\alpha}$. With (21.51), we then obtain

If the potential function $g = g(\sigma_{ij}, K_{\alpha})$ is a convex function in σ_{ij} and K_{α} and if

$$g(\sigma_{ij}, K_{\alpha}) - g(0,0) \geq 0$$

then the nonassociated evolution equations (22.6)

$$\dot{\varepsilon}_{ij}^{p} = \dot{\lambda}\frac{\partial g}{\partial \sigma_{ij}}; \qquad \dot{\kappa}_{\alpha} = -\dot{\lambda}\frac{\partial g}{\partial K_{\alpha}}; \qquad \dot{\lambda} \geq 0$$

fulfill the dissipation inequality

Referring to the discussion of (21.51), we observe that the requirement $g(\sigma_{ij}, K_{\alpha}) - g(0,0) \geq 0$ is implied by the requirements $(\partial g/\partial\sigma_{ij})_{\sigma_{ij}=0, K_{\alpha}=0} = 0$ and

$(\partial g/\partial K_\alpha)_{\sigma_{ij}=0,K_\alpha=0} = 0$ (as well as convexity of g, that is, g is minimum at $g(0,0)$). It appears that we have retrieved nonassociated plasticity and that associated plasticity emerges if we choose $g = f$. However, in contrast to Chapter 10 we now know exactly the requirements that must be posed on the potential function g in order that the nonassociated evolution laws are physically meaningful (i.e. the dissipation inequality is fulfilled). This route of establishing the evolution equations only leave us with the information that $\dot{\lambda} \geq 0$; however, it is natural to impose the same further requirements on $\dot{\lambda}$ as in associated plasticity, i.e. $\dot{\lambda} \geq 0$ when $f = 0$ and $\dot{\lambda} = 0$ when $f < 0$.

It is recalled that (22.5) and (22.6) are convenient mathematical tools by which we can identify the precise requirements that ensure *a priori* that the dissipation inequality is fulfilled. However, in principle, it is possible to relax these requirements and simply write $\dot{\varepsilon}^p_{ij} = \dot{\lambda} h_{ij}$ and $\dot{\kappa}_\alpha = -\dot{\lambda} b_\alpha$ ($\dot{\lambda} \geq 0$) where the functions h_{ij} and b_α then should fulfill $\sigma_{ij}h_{ij} + K_\alpha b_\alpha \geq 0$. The drawback of this more general approach is that it is not possible to identify, *a priori*, the properties that h_{ij} and b_α should possess in order to fulfill this inequality. Therefore, for each specific model, i.e. each specific choice of h_{ij} and b_α, one must, *a posteriori*, check that the inequality is fulfilled. With these remarks, in the following we will restrict ourselves to the formulations given by (22.5) and (22.6).

Both for associated and nonassociated plasticity we have in accordance with the previous discussion $\dot{\lambda} f = 0$, cf. (22.5). Differentiation gives $\ddot{\lambda} f + \dot{\lambda}\dot{f} = 0$. For development of plasticity to occur, we must have $f = 0$ which implies $\dot{\lambda}\dot{f} = 0$ and since development of plastic strains (and of the internal variables) requires $\dot{\lambda} > 0$ we are left with $\dot{f} = 0$. According to Chapter 10, we have then recovered the *consistency relation* in an elegant fashion. With $f = f(\sigma_{ij}, K_\alpha)$, the consistency relation becomes

$$\dot{f} = \frac{\partial f}{\partial \sigma_{ij}}\dot{\sigma}_{ij} + \frac{\partial f}{\partial K_\alpha}\dot{K}_\alpha = 0 \tag{22.7}$$

From (22.1) and (22.6), we have

$$\dot{K}_\alpha = \rho\frac{\partial^2 \psi^p(\kappa^\gamma)}{\partial \kappa_\alpha \partial \kappa_\beta}\dot{\kappa}_\beta = -\dot{\lambda}\rho\frac{\partial^2 \psi^p(\kappa^\gamma)}{\partial \kappa_\alpha \partial \kappa_\beta}\frac{\partial g}{\partial K_\beta}$$

Insertion of this expression in (22.7) leads to

$$\boxed{\dot{f} = \frac{\partial f}{\partial \sigma_{ij}}\dot{\sigma}_{ij} - H\dot{\lambda} = 0} \tag{22.8}$$

where the *generalized plastic modulus* H is defined by

$$\boxed{H \equiv \frac{\partial f}{\partial K_\alpha}\rho\frac{\partial^2 \psi^p(\kappa^\gamma)}{\partial \kappa_\alpha \partial \kappa_\beta}\frac{\partial g}{\partial K_\beta}} \tag{22.9}$$

in complete analogy with Section 10.2.

Assuming the elastic stiffness D_{ijkl} to be constant, (22.3a) gives

$$\dot{\sigma}_{ij} = D_{ijkl}(\dot{\varepsilon}_{kl} - \dot{\varepsilon}^p_{kl}) = D_{ijkl}\dot{\varepsilon}_{kl} - \dot{\lambda} D_{ijkl}\frac{\partial g}{\partial \sigma_{kl}}$$

and insertion into the consistency relation (22.8) provides

$$\boxed{\dot{\lambda} = \frac{1}{A}\frac{\partial f}{\partial \sigma_{ij}} D_{ijkl}\dot{\varepsilon}_{kl} \quad where \quad A = H + \frac{\partial f}{\partial \sigma_{ij}} D_{ijkl}\frac{\partial g}{\partial \sigma_{kl}} \geq 0} \qquad (22.10)$$

in accordance with (10.23) and where, as usual, it is required that $A > 0$.

It appears that all fundamental equations of general plasticity previously discussed in Chapter 10 have now been recovered in a fashion that fulfills all formal thermodynamical requirements. All the subsequent manipulations of these fundamental equations are therefore completely identical to those discussed in Chapter 10; thus, they will not be repeated here.

The only remaining topic is the choice of the function $\psi^p(\kappa_\alpha)$, present in (22.3), and thereby also the choice of the internal variables κ_α. Some typical examples will be presented in Section 22.3, but before that specific discussion it is worthwhile to return to a discussion of the postulate of maximum dissipation.

22.2 Further discussion of the postulate of maximum dissipation

The postulate of maximum dissipation was used to derive associated plasticity and it was emphasized in Section 21.3 that this postulate is not a law of nature. Indeed it may simply be viewed as a convenient mathematical means to ensure that the mechanical dissipation inequality is fulfilled. On the other hand, it may be possible to appeal to some kind of physics behind this postulate. Let us consider a body where no mechanical work input and no heat input are supplied (i.e. $\delta W = \delta Q = 0$). Following (20.63) it was shown that if $\delta W = \delta Q = 0$ and if the entropy has a maximum value then thermodynamical equilibrium exists. Suppose that we consider a process where the body is carried through such stages of maximum entropy and thermodynamic equilibrium. Then it seems reasonable to assume that a process of maximum entropy production occurs and, consequently, this process is one of maximum dissipation. However, this kind of process is an assumption and not a strict result; otherwise nonassociated plasticity would not exist.

Before we enter a further discussion of the postulate of maximum dissipation, a few results about the concepts of a *convex function* and a *convex set of points* need to be recalled.

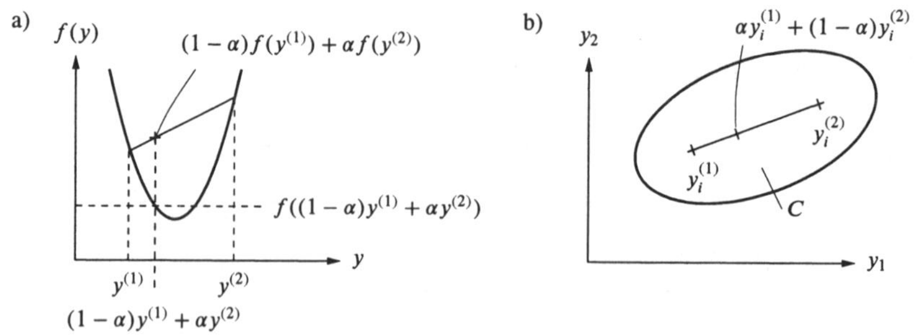

Figure 22.1: a) One-dimensional convex function, b) two-dimensional convex set of points.

A function $f(y_i)$ of the N variables y_i, i.e. $i = 1, 2, \cdots, N$ is *convex* if and only if

Convex function:

$$f((1-\alpha)y_i^{(1)} + \alpha y_i^{(2)}) \leq (1-\alpha)f(y_i^{(1)}) + \alpha f(y_i^{(2)})$$

where $0 \leq \alpha \leq 1$ and where $y_i^{(1)}$ and $y_i^{(2)}$ are two arbitrary points, cf. Appendix (A.1). Consider next a *set of points* C where both $y_i^{(1)}$ and $y_i^{(2)}$ belong to C, i.e. $y_i^{(1)}$ and $y_i^{(2)} \in C$. A set of points C is *convex* if and only if

Convex set of points C :

$$\textit{If } y_i^{(1)} \textit{ and } y_i^{(2)} \in C \textit{ then } \alpha y_i^{(1)} + (1-\alpha)y_i^{(2)} \in C \tag{22.11}$$

where $0 \leq \alpha \leq 1$. A one-dimensional example of a convex function is shown in Fig. 22.1a) and a two-dimensional example of a convex set of points is illustrated in Fig. 22.1b).

Consider an arbitrary function $f(y_i)$ and define the set of points C by those points y_i that fulfill $f(y_i) \leq K = constant$, i.e. the set of points is bounded by a contour curve of the function f. It then turns out that

Let $f(y_i)$ be a convex function.
The set of points C defined by

$$f(y_i) \leq K = constant \tag{22.12}$$

is then a convex set of points

To prove this, we first use that f is a convex function, i.e.

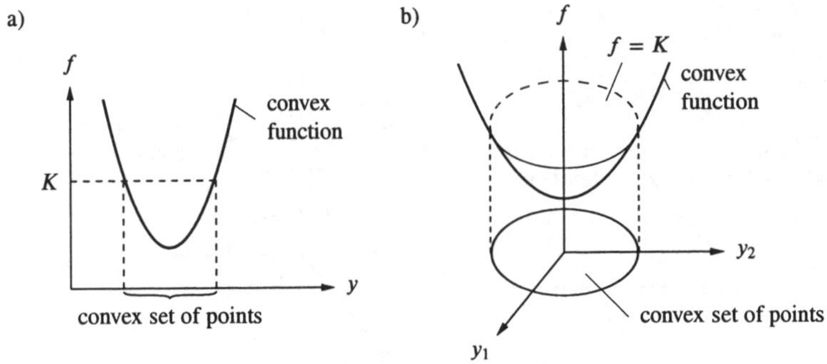

Figure 22.2: Convexity of the function f implies convexity of the set of points C defined by $f(y_i) \leq K = constant$; a) one-dimensional case, b) two-dimensional case.

$$f(\alpha y_i^{(1)} + (1-\alpha)y_i^{(2)}) \leq \alpha f(y_i^{(1)}) + (1-\alpha)f(y_i^{(2)})$$

From the definition of the set C, we have $f(y_i^{(1)}) \leq K$ and $f(y_i^{(2)}) \leq K$ and insertion in the expression above gives

$$f(\alpha y_i^{(1)} + (1-\alpha)y_i^{(2)}) \leq \alpha f(y_i^{(1)}) + (1-\alpha)f(y_i^{(2)}) \leq \alpha K + (1-\alpha)K = K$$

Consequently, the point $\alpha y_i^{(1)} + (1-\alpha)y_i^{(2)}$ also belongs to the set C. Following (22.11) we then conclude that the set C is convex.

The result (22.12) is illustrated in Figs. 22.2a) and 23.2b). While convexity of a function f implies convexity of the set of points C defined by $f(y_i) \leq K = constant$, the opposite is not true; simple examples are given in Fig. 22.3.

With these preliminaries, let us next prove the following

If the yield function $f(\sigma_{ij}, K_\alpha)$ is a convex function in σ_{ij} and K_α, and if

$$\dot{\varepsilon}_{ij}^p = \dot{\lambda}\frac{\partial f}{\partial \sigma_{ij}}; \qquad \dot{\kappa}_\alpha = -\dot{\lambda}\frac{\partial f}{\partial K_\alpha}; \qquad \dot{\lambda} \geq 0$$

then

$$(\sigma_{ij} - \sigma_{ij}^*)\dot{\varepsilon}_{ij}^p - (K_\alpha - K_\alpha^*)\dot{\kappa}_\alpha \geq 0 \qquad (22.13)$$

where

$$f(\sigma_{ij}, K_\alpha) = 0 \quad and \quad f(\sigma_{ij}^*, K_\alpha^*) \leq 0$$

where the evolution laws are evaluated at the state (σ_{ij}, K_α), i.e. $\dot{\varepsilon}_{ij}^p$ and $\dot{\kappa}_\alpha$ belong to the state (σ_{ij}, K_α). It appears from (22.13) that the point (σ_{ij}, K_α) is

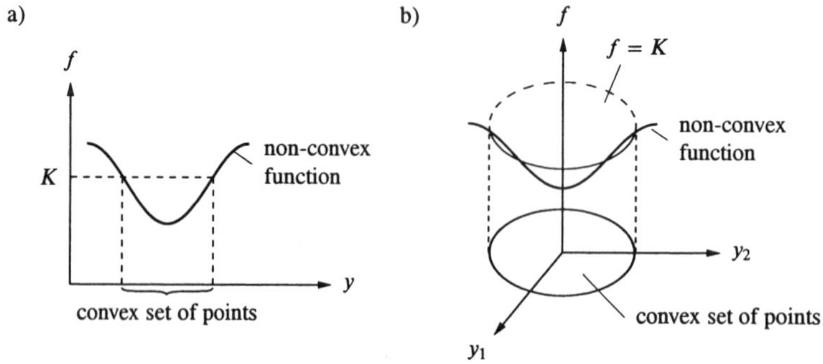

Figure 22.3: Convexity of the set of points C defined by $f(y_i) \leq K = constant$ does not imply convexity of the function f.

located on the yield surface whereas the point $(\sigma^*_{ij}, K^*_\alpha)$ is located on or inside the yield surface. Rewrite (22.13) as $\sigma_{ij}\dot{\varepsilon}^p_{ij} - K_\alpha\dot{\kappa}_\alpha \geq \sigma^*_{ij}\dot{\varepsilon}^p_{ij} - K^*_\alpha\dot{\kappa}_\alpha$ which, with (22.1) and evident notation, can be written as $\gamma_{mech} \geq \gamma^*_{mech}$. Then (22.13) can be interpreted as: the dissipation related to the real state (σ_{ij}, K_α) is larger than or equal to the dissipation related to any other state $(\sigma^*_{ij}, K^*_\alpha)$ within or on the yield surface. Indeed, this is just another way of expressing the postulate of maximum dissipation.

To prove (22.13), we accept that f is a convex function in σ_{ij} and K_α. From Appendix (A.5) we then have

$$f(\sigma^*_{ij}, K^*_\alpha) - f(\sigma_{ij}, K_\alpha) \geq \frac{\partial f}{\partial \sigma_{ij}}(\sigma^*_{ij} - \sigma_{ij}) + \frac{\partial f}{\partial K_\alpha}(K^*_\alpha - K_\alpha)$$

where $\partial f/\partial \sigma_{ij}$ and $\partial f/\partial K_\alpha$ are evaluated at the point (σ_{ij}, K_α). Multiplication by $\dot{\lambda}\ (\geq 0)$, use of the evaluation equations as well as $f(\sigma_{ij}, K_\alpha) = 0$ give

$$\dot{\lambda} f(\sigma^*_{ij}, K^*_\alpha) \geq \dot{\varepsilon}^p_{ij}(\sigma^*_{ij} - \sigma_{ij}) - \dot{\kappa}_\alpha(K^*_\alpha - K_\alpha)$$

Therefore

$$(\sigma_{ij} - \sigma^*_{ij})\dot{\varepsilon}^p_{ij} - (K_\alpha - K^*_\alpha)\dot{\kappa}_\alpha \geq -\dot{\lambda} f(\sigma^*_{ij}, K^*_\alpha) \geq 0$$

since $\dot{\lambda} \geq 0$ and $f(\sigma^*_{ij}, K^*_\alpha) \leq 0$; this is exactly the result stated in (22.13).

Let us next prove

$$\boxed{\begin{array}{l} \textit{Let } \dot{\varepsilon}_{ij}^{p} \textit{ and } \dot{\kappa}_{\alpha} \textit{ be given and let} \\ \quad (\sigma_{ij}-\sigma_{ij}^{*})\dot{\varepsilon}_{ij}^{p}-(K_{\alpha}-K_{\alpha}^{*})\dot{\kappa}_{\alpha}\geq 0 \\ \textit{and } \ f(\sigma_{ij},K_{\alpha})=0\,; \quad f(\sigma_{ij}^{*},K_{\alpha}^{*})\leq 0 \quad \textit{then} \\ \quad \dot{\varepsilon}_{ij}^{p}=\dot{\lambda}\dfrac{\partial f}{\partial \sigma_{ij}}\,; \quad \dot{\kappa}_{\alpha}=-\dot{\lambda}\dfrac{\partial f}{\partial K_{\alpha}}\,; \quad \dot{\lambda}\geq 0 \\ \textit{and the set defined by} \\ \quad f(\sigma_{ij},K_{\alpha})\leq 0 \textit{ is a convex set of points} \end{array}} \tag{22.14}$$

where $\partial f/\partial\sigma_{ij}$ and $\partial f/\partial K_\alpha$ are evaluated at the point (σ_{ij}, K_α).

Let us first prove the associated evolution laws. Without losing generality, we can write

$$\dot{\varepsilon}_{ij}^{p}=\dot{\lambda}\frac{\partial f}{\partial \sigma_{ij}}+p_{ij}(\dot{\lambda},\sigma_{kl},K_{\beta})\,; \qquad \dot{\kappa}_{\alpha}=-\dot{\lambda}\frac{\partial f}{\partial K_{\alpha}}+q_{\alpha}(\dot{\lambda},\sigma_{kl},K_{\beta}) \tag{22.15}$$

where p_{ij} and q_α are some general functions and since $\dot{\varepsilon}_{ij}^{p}$ and $\dot{\kappa}_\alpha$ are given, p_{ij} and q_α do not depend on $(\sigma_{ij}^*, K_\alpha^*)$. Then the expression for maximum dissipation given by (22.14) becomes

$$\begin{aligned} &\dot{\lambda}[\frac{\partial f}{\partial \sigma_{ij}}(\sigma_{ij}-\sigma_{ij}^{*})+\frac{\partial f}{\partial K_{\alpha}}(K_{\alpha}-K_{\alpha}^{*})] \\ &\qquad +(\sigma_{ij}-\sigma_{ij}^{*})p_{ij}-(K_{\alpha}-K_{\alpha}^{*})q_{\alpha}\geq 0 \end{aligned} \tag{22.16}$$

From $f = f(\sigma_{ij}, K_\alpha)$ we obtain

$$df=\frac{\partial f}{\partial \sigma_{ij}}d\sigma_{ij}+\frac{\partial f}{\partial K_{\alpha}}dK_{\alpha} \tag{22.17}$$

The arbitrary state $(\sigma_{ij}^*, K_\alpha^*)$ is located within or on the yield surface. Thus, it is always possible to choose $(\sigma_{ij}^*, K_\alpha^*)$ such that

$$\sigma_{ij}-\sigma_{ij}^{*}=-d\sigma_{ij}\,; \qquad K_{\alpha}-K_{\alpha}^{*}=-dK_{\alpha} \tag{22.18}$$

Suppose that $(\sigma_{ij}^*, K_\alpha^*)$ is chosen on the yield surface; an illustration of σ_{ij}, σ_{ij}^* and $d\sigma_{ij}$ is shown below in Fig. 22.4a). Since we are moving tangentially to the yield surface we have $df = 0$ and (22.17) and (22.18) then provide

$$\frac{\partial f}{\partial \sigma_{ij}}(\sigma_{ij}-\sigma_{ij}^{*})+\frac{\partial f}{\partial K_{\alpha}}(K_{\alpha}-K_{\alpha}^{*})=0$$

Insertion of this expression and (22.18) into (22.16) gives

$$-d\sigma_{ij}p_{ij}+dK_{\alpha}q_{\alpha}\geq 0 \tag{22.19}$$

Figure 22.4: The same σ_{ij}-value applies, but σ^*_{ij} is chosen such that $(d\sigma_{ij})^b = -(d\sigma_{ij})^a$.

Since f is assumed to be a smooth surface, it is always possible to choose σ^*_{ij} and K^*_α such that $d\sigma_{ij}$ and dK_α given by (22.18) take the opposite values to those they had previously; an illustration of σ_{ij} and σ^*_{ij} and the new $d\sigma_{ij}$ is illustrated in Fig. 22.4b). In that case (22.19) takes the form

$$-(-d\sigma_{ij}p_{ij} + dK_\alpha q_\alpha) \geq 0 \tag{22.20}$$

and a comparison of (22.19) and (22.20) shows that $p_{ij} = 0$ and $q_\alpha = 0$; i.e. (22.15) reduces to the associated evolution laws $\dot{\varepsilon}^p_{ij} = \dot{\lambda}\partial f/\partial\sigma_{ij}$ and $\dot{\kappa}_\alpha = -\dot{\lambda}\partial f/\partial K_\alpha$. Moreover, (22.16) reduces to

$$\dot{\lambda}[\frac{\partial f}{\partial\sigma_{ij}}(\sigma_{ij} - \sigma^*_{ij}) + \frac{\partial f}{\partial K_\alpha}(K_\alpha - K^*_\alpha)] \geq 0 \tag{22.21}$$

We next have to prove that $\dot{\lambda} \geq 0$. For this purpose we choose the point $(\sigma^*_{ij}, K^*_\alpha)$ to be inside the yield surface. With (22.18) and since $(\sigma^*_{ij}, K^*_\alpha)$ is located inside the yield surface we have $df < 0$, i.e. (22.17) and (22.18) imply

$$-\frac{\partial f}{\partial\sigma_{ij}}(\sigma_{ij} - \sigma^*_{ij}) - \frac{\partial f}{\partial K_\alpha}(K_\alpha - K^*_\alpha) < 0$$

From this inequality and (22.21), we conclude that $\dot{\lambda} \geq 0$.

Finally, let us prove that the postulate of maximum dissipation given by (22.14) implies that the set of points C defined by $f(\sigma_{ij}, K_\alpha) \leq 0$ is a convex set of points. For the particular points (σ_{ij}, K_α) and $(\sigma^*_{ij}, K^*_\alpha)$ present in the postulate of maximum dissipation given by (22.14) we have $f(\sigma_{ij}, K_\alpha) = 0$ and $f(\sigma^*_{ij}, K^*_\alpha) \leq 0$; both these points therefore belong to the set of points C defined by $f(\sigma_{ij}, K_\alpha) \leq 0$. Assume that this set of points is not convex. Following (22.11) we can then have a situation where

$$f(\alpha\sigma^*_{ij} + (1-\alpha)\sigma_{ij}, \alpha K^*_\beta + (1-\alpha)K_\beta) > 0$$

where $0 \leq \alpha \leq 1$. From this expression and since $f(\sigma_{ij}, K_\beta) = 0$ we obtain for $\alpha \neq 0$

$$\frac{f(\sigma_{ij} + \alpha(\sigma_{ij}^* - \sigma_{ij}),\ K_\beta + \alpha(K_\alpha^* - K_\beta)) - f(\sigma_{ij}, K_\beta)}{\alpha} > 0$$

Letting $\alpha \to 0$, the left-hand side of this expression is the so-called *directional derivative* of f when moving in the direction of $\sigma_{ij}^* - \sigma_{ij}$, $K_\beta^* - K_\beta$, see the discussion of (A.3). From (A.3), we then obtain

$$\frac{\partial f}{\partial \sigma_{ij}}(\sigma_{ij}^* - \sigma_{ij}) + \frac{\partial f}{\partial K_\beta}(K_\beta^* - K_\beta) > 0$$

where $\partial f/\partial \sigma_{ij}$ and $\partial f/\partial K_\beta$ are evaluated at the state (σ_{ij}, K_β). Multiplication by $-\dot{\lambda}$ (≤ 0) and use of the evolution laws give

$$\dot{\varepsilon}_{ij}^p(\sigma_{ij} - \sigma_{ij}^*) - \dot{\kappa}_\beta(K_\beta - K_\beta^*) < 0$$

However, this expression is in contradiction with the postulate of maximum dissipation given by (22.14); thus we have proved that the set of points defined by $f(\sigma_{ij}, K_\alpha) \leq 0$ must be convex. This concludes the proof of the results given in (22.14).

The assumptions and results in (22.14) are very close to the previous formulation of the postulate of maximum dissipation given by (22.4) and the resulting consequences given by the Kuhn-Tucker relations (22.5); however, there are subtle differences. In both cases, we obtain the associated evolution laws. In (22.4), however, the yield function is *a priori* assumed to be a convex function and as a result of the Kuhn-Tucker relations, we obtain $\dot{\lambda} \geq 0$ if $f = 0$ and $\dot{\lambda} = 0$ if $f < 0$. In (22.14), on the other hand, we *a priori* presuppose $f = 0$ to hold for plasticity to develop, but we can now instead prove that the set of points defined by $f(\sigma_{ij}, K_\alpha) \leq 0$ is convex.

While (22.4) and the resulting Kuhn-Tucker relations (22.5) are now used extensively in modern more advanced texts, the classical approach is that provided by (22.14). Indeed, we have seen some reference to (22.14) in relation to (9.39) - which emerged as a consequence of Drucker's postulate - where this expression was viewed as an example of the postulate of maximum dissipation. The expression in question reads

$$(\sigma_{ij} - \sigma_{ij}^*)\dot{\varepsilon}_{ij}^p \geq 0 \tag{22.22}$$

A comparison with (22.14) shows that (22.22) lacks the influence of the hardening parameters K_α and of the fluxes $\dot{\kappa}_\alpha$. However, it was proved in Chapter 9 that (22.22) leads to the associated flow rule $\dot{\varepsilon}_{ij}^p = \dot{\lambda}\partial f/\partial \sigma_{ij}$ whereas no information can be derived from (22.22) regarding the evolution laws for $\dot{\kappa}_\alpha$. However, one may state that if the postulate of maximum dissipation is accepted then the expression given by (22.14) makes physical sense whereas (22.22) cannot be justified (unless of course for ideal plasticity where no hardening parameters K_α and thereby no internal variables κ_α exist).

For further viewpoints and discussions, the reader is referred to the works of Moreau (1970) and Eve *et al.* (1990).

22.3 Examples of typical plasticity models

With the choice of Helmholtz' free energy ψ given by (22.2), the stresses σ_{ij} and hardening parameters K_α are obtained by (22.3). Moreover, the evolution laws are either given by (22.5) for associated plasticity or by (22.6) for nonassociated plasticity. The only remaining topic is the choice of the function $\rho\psi^p(\kappa_\alpha)$ present in (22.2) and thereby also the choice of the internal variables κ_α.

In this section, various choices of $\rho\psi^p(\kappa_\alpha)$ will be provided. Only some typical and illustrative cases will be presented, but they should enable the reader to easily make suitable generalizations; a number of specific models is also presented by Lemaitre and Chaboche (1990) and by Maugin (1992).

For ideal plasticity, no hardening parameters K_α and consequently no internal parameters exist, i.e.

$$\boxed{\psi^p(\kappa_\alpha) = 0 \quad \Rightarrow \quad \textit{ideal plasticity}}$$

Isotropic hardening of von Mises material

For associated plasticity, consider next isotropic hardening of a von Mises material. According to (12.4) the yield function is then given by

$$\begin{aligned} &f = \sigma_{eff} - \sigma_{yo} - K \\ &\text{where} \quad \sigma_{eff} = (\tfrac{3}{2}s_{kl}s_{kl})^{1/2} \quad \text{and} \quad \sigma_y = \sigma_{yo} + K \end{aligned} \tag{22.23}$$

where s_{ij} is the deviatoric stress tensor, σ_{yo} is the initial yield stress and σ_y is the current yield stress. It follows that

$$\dot{\varepsilon}^p_{ij} = \dot{\lambda}\frac{\partial f}{\partial \sigma_{ij}} = \dot{\lambda}\frac{3s_{ij}}{2\sigma_{eff}} \quad \text{i.e.} \quad \dot{\lambda} = \dot{\varepsilon}^p_{eff} = (\frac{2}{3}\dot{\varepsilon}^p_{ij}\dot{\varepsilon}^p_{ij})^{1/2} \tag{22.24}$$

Since only one hardening parameter K exists, we have only one internal variable κ, cf. the evolution law $\dot{\kappa}_\alpha = -\dot{\lambda}\partial f/\partial K_\alpha$, which becomes

$$\dot{\kappa} = -\dot{\lambda}\frac{\partial f}{\partial K} = \dot{\lambda} = \dot{\varepsilon}^p_{eff} \tag{22.25}$$

It appears that the effective plastic strain ε^p_{eff} emerges as a natural internal variable for isotropic von Mises plasticity. This neat result is a consequence of the thermodynamic formulation whereas in Section 9.6 we had to posit various intuitive, but reasonable arguments for this choice. From (22.23) and (22.9) the plastic modulus H becomes

$$H = \rho \frac{d^2\psi^p(\kappa)}{d\kappa^2}$$

Since $K \equiv \rho d\psi^p / d\kappa$ we obtain with $\kappa = \varepsilon^p_{eff}$

$$H = \frac{dK}{d\varepsilon^p_{eff}} = \frac{d(\sigma_{yo} + K)}{d\varepsilon^p_{eff}} = \frac{d\sigma_y}{d\varepsilon^p_{eff}}$$

This result is in agreement with (12.12).

If linear hardening is assumed then

$$\sigma_y = \sigma_{yo} + H\varepsilon^p_{eff}$$

where H is constant. Since $\sigma_y = \sigma_{yo} + K$, it follows that

$$K = \rho \frac{d\psi^p}{d\varepsilon^p_{eff}} = H\varepsilon^p_{eff}$$

i.e.

$$\rho\psi^p(\varepsilon^p_{eff}) = \frac{1}{2}H(\varepsilon^p_{eff})^2; \quad \varepsilon^p_{eff} = \kappa \tag{22.26}$$

If a Taylor expansion of $\psi^p(\varepsilon^p_{eff})$ is made about the point $\varepsilon^p_{eff} = 0$, we obtain $\rho\psi^p = A + B\varepsilon^p_{eff} + \frac{1}{2}H(\varepsilon^p_{eff})^2$ where higher order terms are ignored. Then $K = \rho d\psi^p / d\varepsilon^p_{eff} = B + H\varepsilon^p_{eff}$ and as $\varepsilon^p_{eff} = \kappa = 0$ must imply $K = 0$ - otherwise the yield condition for $\varepsilon^p_{eff} = 0$ does not coincide with the initial yield condition - $B = 0$ holds; moreover, the constant term A is of no influence. It then appears that if nothing is known *a priori*, the most simple and straightforward approach is to make a Taylor expansion of ψ^p and this leads to linear hardening.

If power law hardening is assumed, we have

$$\sigma_y = \sigma_{yo} + k\sigma_{yo}(\varepsilon^p_{eff})^n; \quad 0 < n \leq 1$$

cf. (9.4). Since $\sigma_y = \sigma_{yo} + K$, it follows that

$$K = \rho \frac{d\psi^p(\varepsilon^p_{eff})}{d\varepsilon^p_{eff}} = k\sigma_{yo}(\varepsilon^p_{eff})^n$$

i.e.

$$\rho\psi^p(\varepsilon^p_{eff}) = \frac{k\sigma_{yo}}{n+1}(\varepsilon^p_{eff})^{n+1}; \quad \varepsilon^p_{eff} = \kappa$$

Consider finally exponential hardening where

$$\sigma_y = \sigma_{yo} + K_\infty(1 - e^{-\frac{\varepsilon^p_{eff}}{\varepsilon_o}}) \tag{22.27}$$

cf. (9.3). Consequently

$$K = \rho\frac{d\psi^p(\varepsilon^p_{eff})}{d\varepsilon^p_{eff}} = K_\infty(1 - e^{-\frac{\varepsilon^p_{eff}}{\varepsilon_o}})$$

i.e.

$$\rho\psi^p(\varepsilon^p_{eff}) = K_\infty(\varepsilon^p_{eff} + \varepsilon_o e^{-\frac{\varepsilon^p_{eff}}{\varepsilon_o}}); \qquad \varepsilon^p_{eff} = \kappa$$

The associated evolution laws follow from the postulate of maximum dissipation, provided that the yield function $f = f(\sigma_{ij}, K_\alpha)$ is a convex function in σ_{ij} and K_α, cf. (22.4). For isotropic von Mises hardening, let us therefore prove formally that f is a convex function. According to the Appendix (A.8) the function $f = \sigma_{eff} - \sigma_{yo} - K$ is convex in σ_{ij} and K if and only if its *Hessian* is positive semi-definite. This implies that an arbitrary quadratic form I of the Hessian must be non-negative. With $(\bar{\sigma}_{ij}, \bar{K}_\alpha)$ being an arbitrary point we therefore require

$$\begin{aligned} I &= \begin{bmatrix} \bar{\sigma}_{ij} & \bar{K} \end{bmatrix} \begin{bmatrix} \frac{\partial^2 f}{\partial\sigma_{ij}\partial\sigma_{kl}} & \frac{\partial^2 f}{\partial\sigma_{ij}\partial K} \\ \frac{\partial^2 f}{\partial K\partial\sigma_{ij}} & \frac{\partial^2 f}{\partial K^2} \end{bmatrix} \begin{bmatrix} \bar{\sigma}_{kl} \\ \bar{K} \end{bmatrix} \\ &= \begin{bmatrix} \bar{\sigma}_{ij} & \bar{K} \end{bmatrix} \begin{bmatrix} A_{ijkl} & 0 \\ 0 & 0 \end{bmatrix} \begin{bmatrix} \bar{\sigma}_{kl} \\ \bar{K} \end{bmatrix} \geq 0 \end{aligned}$$

where

$$A_{ijkl} = \frac{3}{2\sigma^2_{eff}}\left\{\sigma_{eff}\left[\frac{1}{2}(\delta_{ik}\delta_{jl} + \delta_{il}\delta_{jk}) - \frac{1}{3}\delta_{ij}\delta_{kl}\right] - s_{ij}\frac{3s_{kl}}{2\sigma_{eff}}\right\}$$

It follows that $I = \bar{\sigma}_{ij}A_{ijkl}\bar{\sigma}_{kl}$, i.e.

$$I = \frac{3}{2\sigma^2_{eff}}\left[\sigma_{eff}\bar{s}_{kl}\bar{s}_{kl} - \bar{s}_{ij}s_{ij}\frac{3s_{kl}\bar{s}_{kl}}{2\sigma_{eff}}\right] \geq 0 \tag{22.28}$$

It is always possible to write $\bar{s}_{ij}$ as $\bar{s}_{ij} = as_{ij} + bs^\perp_{ij}$ where the quantity $s^\perp_{ij}$ fulfills $s_{ij}s^\perp_{ij} = 0$ and a and b are constants. Then $\bar{s}_{ij}\bar{s}_{ij} = a^2 s_{ij}s_{ij} + b^2 s^\perp_{ij}s^\perp_{ij}$ and $\bar{s}_{ij}s_{ij} = as_{ij}s_{ij}$; with these expressions and $\sigma^2_{eff} = \frac{3}{2}s_{ij}s_{ij}$, (22.28) reduces to

$$I = \frac{3}{2\sigma_{eff}}b^2 s^\perp_{ij}s^\perp_{ij} \geq 0$$

which certainly is fulfilled. It follows that f is a convex function in σ_{ij} and K, i.e. the prerequisites for the Kuhn-Tucker relations (and thereby the associated evolution laws) are fulfilled; accordingly the postulate of maximum dissipation (22.4) ensures that $\gamma_{mech} \geq 0$.

However, in the present case of isotropic von Mises plasticity it is much easier to simply check *a posteriori* that $\gamma_{mech} \geq 0$ holds. With (22.24) and (22.25), we obtain

$$\gamma_{mech} \equiv \sigma_{ij}\dot{\varepsilon}_{ij}^{p} - K_\alpha \dot{\kappa}_\alpha = \dot{\lambda}(\sigma_{ij}\frac{3s_{ij}}{2\sigma_{eff}} - K) = \dot{\lambda}(\sigma_{eff} - K) \geq 0$$

which with the yield condition $f = 0$ becomes

$$\gamma_{mech} = \dot{\lambda}\sigma_{yo} \geq 0$$

that certainly is fulfilled. It is of interest that this expression for γ_{mech} holds irrespective of the particular hardening that is chosen.

A summary of these results reads

Isotropic von Mises hardening:

$$f = \sigma_{eff} - \sigma_{yo} - K$$

$$\dot{\varepsilon}_{ij}^{p} = \dot{\lambda}\frac{\partial f}{\partial \sigma_{ij}} = \dot{\lambda}\frac{3s_{ij}}{2\sigma_{eff}}$$

$$\dot{\kappa} = -\dot{\lambda}\frac{\partial f}{\partial K} = \dot{\lambda} = \dot{\varepsilon}_{eff}^{p}$$

$$\gamma_{mech} = \dot{\lambda}\sigma_{yo} \geq 0; \quad H = \rho\frac{d^2\psi^p}{d\varepsilon_{eff}^{p2}} \tag{22.29}$$

Linear hardening $\rho\psi^p = \frac{1}{2}(\varepsilon_{eff}^{p})^2 H$

Power law hardening $\rho\psi^p = \frac{k\sigma_{yo}}{n+1}(\varepsilon_{eff}^{p})^{n+1}$

Exponential hardening $\rho\psi^p = K_\infty(\varepsilon_{eff}^{p} + \varepsilon_o e^{-\frac{\varepsilon_{eff}^{p}}{\varepsilon_o}})$

Isotropic hardening of von Mises material – Nonassociated formulation

In the presentation above, it was more or less implied that there exists only one manner in which isotropic hardening of a von Mises material with a given stress-strain response can be achieved. This is not the case. Indeed there exists a number of formulations that lead to an identical stress-strain response; however, the mechanical dissipation connected with these formulations differs and this can be utilized when heat generation is modeled. The key point is that a nonassociated formulation is adopted.

Take the plastic part of Helmholtz' free energy according to

$$\rho\psi^p = \frac{1}{2}c\kappa^2 \tag{22.30}$$

where c is a constant. It follows that

$$K = \rho\frac{\partial\psi}{\partial\kappa} = c\kappa \tag{22.31}$$

As before the yield function is given by

$$f = \sigma_{eff} - \sigma_{yo} - K\,; \qquad \sigma_y = \sigma_{yo} + K$$

Nonassociated plasticity is adopted according to

$$g = f(\sigma_{ij}, K) + g^*(K) \tag{22.32}$$

The evolution equations then become

$$\begin{aligned} \dot{\varepsilon}^p_{ij} &= \dot{\lambda}\frac{\partial g}{\partial\sigma_{ij}} = \dot{\lambda}\frac{\partial f}{\partial\sigma_{ij}} = \dot{\lambda}\frac{3s_{ij}}{2\sigma_{eff}} \\ \dot{\kappa} &= -\dot{\lambda}\frac{\partial g}{\partial K} = -\dot{\lambda}(-1 + \frac{\partial g^*}{\partial K}) \end{aligned} \tag{22.33}$$

It appears that the potential function has been chosen in such a way that the flow rule corresponds exactly to associated plasticity; it is only the evolution law for the internal variable that is influenced by the formulation being nonassociated.

As a specific example, suppose that we want to establish a model that exhibits exponential hardening. According to (22.27), we then require

$$K = K_\infty(1 - e^{-\frac{\varepsilon^p_{eff}}{\varepsilon_0}})$$

Let us assume that

$$g^* = \frac{K^2}{2K_\infty} \tag{22.34}$$

Taking advantage of (22.33b) leads to

$$\dot{\kappa} = \dot{\lambda}(1 - \frac{K}{K_\infty}) \tag{22.35}$$

From (22.33a) follows as usual that $\dot{\varepsilon}^p_{eff} = \dot{\lambda}$ and insertion of the expression above into (22.31) provides

$$\dot{K} = c\,\dot{\varepsilon}^p_{eff}(1 - \frac{K}{K_\infty})$$

Using $K(0) = 0$ an integration results in

$$K = K_\infty(1 - e^{-\frac{c\varepsilon^p_{eff}}{K_\infty}})$$

Thus by choosing $c = K_\infty/\varepsilon_o$ the nonassociated formulation given by (22.30), (22.32) and (22.34) results in the same stress-strain response as the associated formulation given by (22.29); even so, the mechanical dissipation relating to these two models differs.

By way of demonstration, let us determine the mechanical dissipation for the nonassociated formulation. From (22.33) and (22.35) we obtain

$$\gamma_{mech,nonas} = \sigma_{ij}\dot{\varepsilon}^p_{ij} - K\dot{\kappa} = \dot{\lambda}(\sigma_{eff} - K + \frac{K^2}{K_\infty})$$

Since $\sigma_{eff} = \sigma_{yo} + K$, it follows that

$$\gamma_{mech,nonas} = \dot{\lambda}(\sigma_{yo} + \frac{K^2}{K_\infty})$$

A comparison with (22.29) shows that $\gamma_{mech,nonas} \geq \gamma_{mech,ass}$. Since the mechanical dissipation enters the heat equation, we have now obtained a situation by which we can adjust the model so that for a given stress-strain response it exhibits different heat generation properties.

Linear kinematic hardening of von Mises material

Consider next linear kinematic hardening von Mises plasticity and, again, our purpose is to identify the quantity $\rho\psi^p$ present in (22.2). According to (12.37), we have

$$f(\sigma_{ij}, K_{ij}) = [\frac{3}{2}(s_{kl} - K^d_{kl})(s_{kl} - K^d_{kl})]^{1/2} - \sigma_{yo}$$

where K^d_{ij} is the deviatoric part of K_{ij} and $K^d_{ij} = \alpha^d_{ij}$ is the deviatoric *back-stress tensor*. Associated plasticity is assumed, i.e.

$$\dot{\varepsilon}^p_{ij} = \dot{\lambda}\frac{\partial f}{\partial \sigma_{ij}} = \dot{\lambda}\frac{3(s_{ij} - K^d_{ij})}{2\sigma_{yo}} \quad \text{i.e.} \quad \dot{\lambda} = \dot{\varepsilon}^p_{eff}$$

and the evolution law for the internal variable becomes

$$\dot{\kappa}_{ij} = -\dot{\lambda}\frac{\partial f}{\partial K_{ij}} = -\dot{\lambda}\frac{\partial f}{\partial K^d_{mn}}\frac{\partial K^d_{mn}}{\partial K_{ij}} = \dot{\lambda}\frac{3(s_{ij} - K^d_{ij})}{2\sigma_{yo}} \qquad (22.36)$$

It appears that κ_{ij} is a purely deviatoric quantity and that

$$\kappa_{ij} = \varepsilon^p_{ij}$$

From (22.1) the hardening parameters K_{ij} are obtained as

$$K_{ij} = \rho \frac{\partial \psi^p(\kappa_{kl})}{\partial \kappa_{ij}}$$

i.e.

$$\dot{K}_{ij} = \rho \frac{\partial^2 \psi^p(\kappa_{mn})}{\partial \kappa_{ij} \partial \kappa_{kl}} \dot{\kappa}_{kl} = \rho \frac{\partial^2 \psi^p(\kappa_{mn})}{\partial \kappa_{ij} \partial \kappa_{kl}} \dot{\varepsilon}^p_{kl} \tag{22.37}$$

To obtain linear *Melan-Prager kinematic hardening*, cf. (12.46) and (12.47), we require

$$\dot{K}_{ij} = \dot{K}^d_{ij} = \dot{\alpha}^d_{ij} = c\dot{\varepsilon}^p_{ij}$$

where c is a constant. A comparison with (22.37) shows that

$$\rho \frac{\partial^2 \psi^p(\kappa_{mn})}{\partial \kappa_{ij} \partial \kappa_{kl}} = c\frac{1}{2}(\delta_{ik}\delta_{jl} + \delta_{il}\delta_{jk})$$

As $\kappa_{ij} = \varepsilon^p_{ij} = 0$ must imply $K_{ij} = \rho \partial \psi^p / \partial \kappa_{ij} = 0$ - otherwise the initial yield condition makes no sense - we obtain

$$\rho \psi^p = \frac{1}{2} c \kappa_{ij} \kappa_{ij} \tag{22.38}$$

Moreover, from (22.9) and as associated plasticity is adopted, we obtain with (22.36) and (22.38)

$$\begin{aligned} H &= \frac{\partial f}{\partial K_{ij}} \rho \frac{\partial^2 \psi^p}{\partial \kappa_{ij} \partial \kappa_{kl}} \frac{\partial f}{\partial K_{kl}} \\ &= \frac{3(s_{ij} - K^d_{ij})}{2\sigma_{yo}} c\frac{1}{2}(\delta_{ik}\delta_{jl} + \delta_{il}\delta_{jk}) \frac{3(s_{kl} - K^d_{kl})}{2\sigma_{yo}} \end{aligned}$$

i.e.

$$H = \frac{3}{2} c$$

in accordance with (12.49). Just like linear isotropic hardening can be obtained by making a Taylor expansion of $\psi^p(\varepsilon^p_{eff})$, cf. the discussion following (22.26), it is of interest that simply by making a Taylor expansion of $\psi^p(\kappa_{ij})$ we are led to Melan-Prager's kinematic hardening rule, cf. (22.38).

Linear mixed von Mises hardening

Consider next linear mixed von Mises hardening. From (12.54) we then have

$$f(\sigma_{ij}, K_{ij}, K) = [\frac{3}{2}(s_{kl} - K^d_{kl})(s_{kl} - K^d_{kl})]^{1/2} - \sigma_{yo} - K$$

where $K^d_{ij} = \alpha^d_{ij}$. It follows that

$$\dot{\varepsilon}^p_{ij} = \dot{\lambda}\frac{\partial f}{\partial \sigma_{ij}} = \dot{\lambda}\frac{3(s_{ij} - K^d_{ij})}{2\bar{\sigma}_{yo}} \quad \text{i.e.} \quad \dot{\lambda} = \dot{\varepsilon}^p_{eff}$$

where

$$\bar{\sigma}_{yo} = \sigma_{yo} + K$$

Moreover, with the internal variables κ_{ij} conjugated to K_{ij} and the internal variable κ conjugated to K, the associated evolution laws give

$$\dot{\kappa}_{ij} = -\dot{\lambda}\frac{\partial f}{\partial K_{ij}} = \dot{\lambda}\frac{(s_{ij} - K^d_{ij})}{2\bar{\sigma}_{yo}} = \dot{\varepsilon}^p_{ij}; \quad \dot{\kappa} = -\dot{\lambda}\frac{\partial f}{\partial K} = \dot{\lambda} = \dot{\varepsilon}^p_{eff} \tag{22.39}$$

From (22.1) and as $\rho\psi^p(\kappa_{ij}, \kappa)$

$$K_{ij} = \rho\frac{\partial \psi^p(\kappa_{mn}, \kappa)}{\partial \kappa_{ij}}; \quad K = \rho\frac{\partial \psi^p(\kappa_{ij}, \kappa)}{\partial \kappa}$$

These expressions imply that

$$\begin{aligned} \dot{K}_{ij} &= \rho\frac{\partial^2 \psi^p}{\partial \kappa_{ij}\partial \kappa_{kl}}\dot{\kappa}_{kl} + \rho\frac{\partial^2 \psi^p}{\partial \kappa_{ij}\partial \kappa}\dot{\kappa} \\ \dot{K} &= \rho\frac{\partial^2 \psi^p}{\partial \kappa\partial \kappa_{ij}}\dot{\kappa}_{ij} + \rho\frac{\partial^2 \psi^p}{\partial \kappa^2}\dot{\kappa} \end{aligned} \tag{22.40}$$

Since mixed linear hardening is considered, we require with (12.65), (12.67) and (12.75) that

$$\begin{aligned} \dot{K}_{ij} &= \dot{K}^d_{ij} = \dot{\alpha}^d_{ij} = (1 - m)c\dot{\varepsilon}^p_{ij} = (1 - m)c\dot{\kappa}_{ij} \\ \dot{K} &= m\frac{3}{2}c\dot{\lambda} = m\frac{3}{2}c\dot{\kappa} \end{aligned} \tag{22.41}$$

where the *mixed hardening parameter* m is in the range $0 \leq m \leq 1$; for $m = 0$ we obtain a purely linear kinematic hardening whereas $m = 1$ implies a purely linear isotropic hardening. A comparison of (22.40) and (22.41) gives

$$\rho\psi^p = \frac{1}{2}(1 - m)c\kappa_{ij}\kappa_{ij} + \frac{3}{4}mc\kappa^2$$

Indeed this expression is not surprising since it can also be obtained by writing $\rho\psi^p(\kappa_{ij}, \kappa)$ as $\rho\psi^p(\kappa_{ij}, \kappa) = (1 - m)p(\kappa_{ij}) + mq(\kappa)$ and then making a Taylor expansion of the functions $p(\kappa_{ij})$ and $q(\kappa)$.

For the plastic modulus H, we obtain with (22.9)

$$H = \frac{\partial f}{\partial K_{ij}}\rho\frac{\partial^2\psi^p}{\partial\kappa_{ij}\partial\kappa_{kl}}\frac{\partial f}{\partial K_{kl}} + \frac{\partial f}{\partial K}\rho\frac{\partial^2\psi^p}{\partial\kappa^2}\frac{\partial f}{\partial K} = \frac{3}{2}c$$

in accordance with (12.76).

Let us finally calculate the mechanical dissipation. From (22.1) and (22.39) and the yield condition $f = 0$ follow that

$$\begin{aligned} \gamma_{mech} &= \sigma_{ij}\dot{\varepsilon}_{ij}^p - K_{ij}\dot{\kappa}_{ij} - K\dot{\kappa} \\ &= \dot{\lambda}[\sigma_{ij}\frac{3(s_{ij} - K_{ij}^d)}{2\bar{\sigma}_{yo}} - K_{ij}\frac{3(s_{ij} - K_{ij}^d)}{2\bar{\sigma}_{yo}} - K] \\ &= \dot{\lambda}(\bar{\sigma}_{yo} - K) = \dot{\lambda}\sigma_{yo} \geq 0 \end{aligned} \tag{22.42}$$

and this inequality is certainly fulfilled. It is of interest that γ_{mech} is independent of the mixed hardening parameter m and that (22.42) coincides with expression (22.29) valid for arbitrary isotropic von Mises hardening.

In summary, we have

Linear mixed von Mises hardening:

$$f = [\frac{3}{2}(s_{ij} - K_{ij}^d)(s_{ij} - K_{ij}^d)]^{1/2} - \bar{\sigma}_{yo}\,; \qquad \bar{\sigma}_{yo} = \sigma_{yo} + K$$

$$\rho\psi^p = \frac{1}{2}(1 - m)c\kappa_{ij}\kappa_{ij} + \frac{3}{4}mc\kappa^2\,; \qquad 0 \leq m \leq 1$$

This implies

$$\begin{aligned} \dot{\varepsilon}_{ij}^p &= \dot{\lambda}\frac{\partial f}{\partial\sigma_{ij}} = \dot{\lambda}\frac{3(s_{ij} - K_{ij}^d)}{2\bar{\sigma}_{yo}}\,; \qquad \dot{\lambda} = \dot{\varepsilon}_{eff}^p \\ \dot{\kappa}_{ij} &= -\dot{\lambda}\frac{\partial f}{\partial K_{ij}} = \dot{\varepsilon}_{ij}^p \\ \dot{\kappa} &= -\dot{\lambda}\frac{\partial f}{\partial K} = \dot{\lambda} \end{aligned} \tag{22.43}$$

Moreover

$$\gamma_{mech} = \dot{\lambda}\sigma_{yo} \geq 0$$

$$\dot{K}_{ij} = \dot{K}_{ij}^d = (1 - m)c\dot{\varepsilon}_{ij}^p\,; \qquad \dot{K} = \dot{\lambda}\frac{3}{2}mc$$

$$H = \frac{3}{2}c$$

Armstrong-Frederick kinematic hardening

Let us next turn to Armstrong-Frederick kinematic hardening, which, within a kinematic hardening concept, allows nonlinear hardening to be modeled in an elegant fashion. The yield function is again given by

$$f = [\frac{3}{2}(s_{kl} - K_{kl}^d)(s_{kl} - K_{kl}^d)]^{1/2} - \sigma_{yo} \tag{22.44}$$

Following Armstrong and Frederick (1966) and in accordance with (13.70), it is required that

$$\dot{K}_{ij} = \dot{K}_{ij}^d = \dot{\alpha}_{ij}^d = h(\frac{2}{3}\dot{\varepsilon}_{ij}^p - \frac{K_{ij}^d}{\alpha_\infty}\dot{\varepsilon}_{eff}^p) \tag{22.45}$$

where h and α_∞ are constant parameters and $\alpha_\infty \geq 0$; it appears that the Armstrong-Frederick formulation degenerates to Melan-Prager hardening for $\alpha_\infty \to \infty$ and $h = 3c/2$. According to (22.43), we have seen that an associated format leads to Melan-Prager hardening and to obtain (22.45) a nonassociated format must therefore be adopted. However, as the plastic flow rule is still to be given in the form $\dot{\varepsilon}_{ij}^p = \dot{\lambda}\partial f/\partial\sigma_{ij}$, we are led to the conclusion that the evolution law for $\dot{\varepsilon}_{ij}^p$ is to follow an associated format whereas the evolution law for $\dot{\kappa}_{ij}$ must follow a nonassociated format. The result is a potential function of the form

$$g(\sigma_{ij}, K_{ij}) = f(\sigma_{ij}, K_{ij}) + g^*(K_{ij}) \tag{22.46}$$

With the general nonassociated formulation given by (22.6), use of (22.44) and (22.46) give

$$\dot{\varepsilon}_{ij}^p = \dot{\lambda}\frac{\partial g}{\partial\sigma_{ij}} = \dot{\lambda}\frac{\partial f}{\partial\sigma_{ij}} = \dot{\lambda}\frac{3(s_{ij} - K_{ij}^d)}{2\sigma_{yo}} \quad \text{i.e.} \quad \dot{\lambda} = \dot{\varepsilon}_{eff}^p \tag{22.47}$$

Moreover

$$\begin{aligned} \dot{\kappa}_{ij} &= -\dot{\lambda}\frac{\partial g}{\partial K_{ij}} = -\dot{\lambda}\frac{\partial f}{\partial K_{ij}} - \dot{\lambda}\frac{\partial g^*}{\partial K_{ij}} \\ &= \dot{\varepsilon}_{ij}^p - \dot{\lambda}\frac{\partial g^*}{\partial K_{ij}} \end{aligned} \tag{22.48}$$

and (22.1) gives

$$K_{ij} = \rho\frac{\partial\psi^p(\kappa_{mn})}{\partial\kappa_{ij}}$$

i.e.

$$\dot{K}_{ij} = \rho\frac{\partial^2\psi^p}{\partial\kappa_{ij}\partial\kappa_{kl}}\dot{\kappa}_{kl}$$

Insertion of (22.48) and use of (22.47b) provide

$$\dot{K}_{ij} = \rho \frac{\partial^2 \psi^p}{\partial \kappa_{ij} \partial \kappa_{kl}} \dot{\varepsilon}^p_{kl} - \rho \frac{\partial^2 \psi^p}{\partial \kappa_{ij} \partial \kappa_{kl}} \frac{\partial g^*}{\partial K_{kl}} \dot{\varepsilon}^p_{eff} \tag{22.49}$$

It appears that if we choose

$$\rho \psi^p = \frac{1}{3} h \kappa_{ij} \kappa_{ij}$$

and

$$g^* = \frac{3}{2\alpha_\infty} K^d_{ij} K^d_{ij} \tag{22.50}$$

then (22.49) reduces to (22.45). It may be of interest to observe that the most simple expression for $g^*(K_{ij})$ is obtained by a straightforward Taylor expansion which, in fact, is the one given by (22.50).

The plastic modulus H defined by (22.9) becomes

$$H = \frac{\partial f}{\partial K_{ij}} \rho \frac{\partial^2 \psi^p}{\partial \kappa_{ij} \partial \kappa_{kl}} \frac{\partial g}{\partial K_{kl}}$$

$$= -\frac{3(s_{ij} - K^d_{ij})}{2\sigma_{yo}} \frac{h}{3} \frac{1}{2} (\delta_{ik}\delta_{jl} + \delta_{il}\delta_{jk}) [-\frac{3(s_{kl} - K^d_{kl})}{2\sigma_{yo}} + \frac{3}{\alpha_\infty} K^d_{kl}]$$

i.e.

$$H = h[1 - \frac{3(s_{ij} - K^d_{ij}) K^d_{ij}}{2\alpha_\infty \sigma_{yo}}]$$

in accordance with (13.78).

Let us finally determine the mechanical dissipation. From (22.47), (22.48) and (22.50), we obtain

$$\gamma_{mech} = \sigma_{ij} \dot{\varepsilon}^p_{ij} - K^d_{ij} \dot{\kappa}_{ij}$$

$$= \dot{\lambda}[\sigma_{ij} \frac{3(s_{ij} - K^d_{ij})}{2\sigma_{yo}} - K^d_{ij} \frac{3(s_{ij} - K^d_{ij})}{2\sigma_{yo}} + \frac{3}{\alpha_\infty} K^d_{ij} K^d_{ij}] \geq 0$$

which with the yield condition $f = 0$ gives

$$\gamma_{mech} = \dot{\lambda}(\sigma_{yo} + \frac{3}{\alpha_\infty} K^d_{ij} K^d_{ij}) \geq 0$$

which certainly is fulfilled.

In conclusion

Armstrong-Frederick kinematic hardening:

$$f = [\frac{3}{2}(s_{ij}-K_{ij}^d)(s_{ij}-K_{ij}^d)]^{1/2}-\sigma_{yo}$$

$$g = f + \frac{3}{2\alpha_\infty}K_{ij}^d K_{ij}^d$$

$$\rho\psi^p = \frac{h}{3}\kappa_{ij}\kappa_{ij}$$

This implies

$$\dot{\varepsilon}_{ij}^p = \dot{\lambda}\frac{\partial g}{\partial \sigma_{ij}} = \dot{\lambda}\frac{\partial f}{\partial \sigma_{ij}} = \dot{\lambda}\frac{3(s_{ij} - K_{ij}^d)}{2\sigma_{yo}}; \qquad \dot{\lambda} = \dot{\varepsilon}_{eff}^p$$

$$\dot{\kappa}_{ij} = -\dot{\lambda}\frac{\partial g}{\partial K_{ij}} = \dot{\varepsilon}_{ij}^p - \dot{\lambda}\frac{3}{\alpha_\infty}K_{ij}^d; \qquad \kappa_{ij} = \kappa_{ij}^d$$

Moreover

$$\gamma_{mech} = \dot{\lambda}(\sigma_{yo} + \frac{3}{\alpha_\infty}K_{ij}^d K_{ij}^d) \geq 0$$

$$K_{ij} = \rho\frac{\partial \psi^p}{\partial \kappa_{ij}} = \frac{2h}{3}\kappa_{ij}$$

$$i.e.\ \dot{K}_{ij} = h(\frac{2}{3}\dot{\varepsilon}_{ij}^p - \frac{1}{\alpha_\infty}K_{ij}^d\dot{\lambda}) \quad and \quad K_{ij} = K_{ij}^d$$

$$H = h[1 - \frac{3(s_{ij} - K_{ij}^d)K_{ij}^d}{2\alpha_\infty\sigma_{yo}}]$$

We emphasize that this formulation of Armstrong-Frederick hardening is an interesting example of a format that is intrinsically nonassociated. However, the potential function is such that the plastic flow rule turns out to be associated; thus, only the evolution law for the internal variables is nonassociated.

Isotropic Drucker-Prager hardening

As a simplistic prototype for soil and concrete mechanics consider finally isotropic Drucker-Prager hardening. From (12.114), we have

$$f = \sigma_{eff} + \alpha I_1 - \beta - K$$

where - for practical purposes - both α and β are non-negative parameters. Nonassociated plasticity is adopted and the potential function is taken as

$$g = \sigma_{eff} + \alpha^* I_1 - \beta^* - K$$

where α^* is a non-negative parameter and β^* is a parameter. It appears that associated plasticity is obtained if $\alpha^* = \alpha$ and $\beta^* = \beta$. The plastic flow rule becomes

$$\dot{\varepsilon}_{ij}^p = \dot{\lambda}\frac{\partial g}{\partial \sigma_{ij}} = \dot{\lambda}(\frac{3s_{ij}}{2\sigma_{eff}} + \alpha^*\delta_{ij}) \tag{22.51}$$

i.e. $\dot{\varepsilon}_{ii}^p = 3\dot{\lambda}\alpha^*$ and

$$\dot{\varepsilon}_{eff}^p = (\frac{2}{3}\dot{\varepsilon}_{ij}^p\dot{\varepsilon}_{ij}^p)^{1/2} = \dot{\lambda}\sqrt{1 + 2(\alpha^*)^2}$$

The evolution law for the internal variable κ becomes

$$\dot{\kappa} = -\dot{\lambda}\frac{\partial g}{\partial K} = \dot{\lambda} \tag{22.52}$$

and it is observed that the internal variable is proportional to the effective plastic strain.

Using (22.51) and (22.52), the mechanical dissipation takes the form

$$\begin{aligned} \gamma_{mech} &= \sigma_{ij}\dot{\varepsilon}_{ij}^p - K\dot{\kappa} = \dot{\lambda}[\sigma_{ij}(\frac{3s_{ij}}{2\sigma_{eff}} + \alpha^*\delta_{ij}) - K] \\ &= \dot{\lambda}(\sigma_{eff} + \alpha^* I_1 - K) \geq 0 \end{aligned}$$

Use of the yield condition $f = 0$ implies

$$\gamma_{mech} = \dot{\lambda}[\beta - I_1(\alpha - \alpha^*)] \geq 0 \tag{22.53}$$

Since the model primarily intends to work for compressive stresses, (22.53) must be fulfilled for arbitrary negative values of I_1 and we therefore require

$$\alpha \geq \alpha^* \tag{22.54}$$

This restriction is shown in Fig. 22.5

Leaving aside the specific Drucker-Prager material in question and considering frictional materials in general, it is found experimentally, see for instance Ko and Scott (1968) and Lade and Duncan (1973), that the frictional angle (α) is larger than the dilatancy angle (α^*). The theoretical result (22.54) is in neat agreement with this experimental evidence and it underlines the significance of nonassociated plasticity for frictional materials.

Let us next confine ourselves to associated plasticity. In that case $\alpha^* = \alpha$ and (22.53) reduces to

$$\gamma_{mech} = \dot{\lambda}\beta \geq 0$$

For $\beta > 0$ this expression is certainly fulfilled. However, if a cohesionless material like sand is considered, $\beta = 0$ holds and we always obtain $\gamma_{mech} = 0$. Since

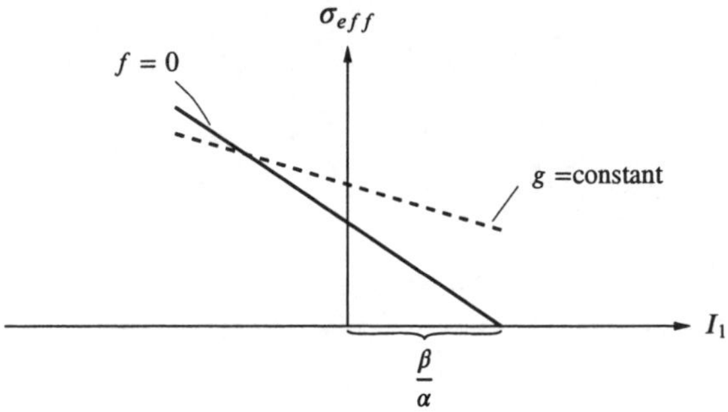

Figure 22.5: Illustration of the restriction $\alpha \geq \alpha^*$.

$\gamma_{mech} > 0$ implies an irreversible material behavior whereas $\gamma_{mech} = 0$ implies a reversible material behavior, the model in question will imply the development of plastic strains and even so it will be reversible! This evident contradiction cannot be accepted and it is concluded that associated isotropic hardening of a cohesionless Drucker-Prager material violates the physical implication of the second law of thermodynamics. If we furthermore restrict ourselves to ideal plasticity, the rejection of such a model has long been known in the literature without, however, referring to strict thermodynamical arguments. Instead, it was observed, see for instance Vermeer and de Borst (1984) that the rate of plastic work for associated ideal plasticity of a cohesionless Drucker-Prager material becomes $\dot{W}^p = \sigma_{ij}\dot{\varepsilon}_{ij}^p = \dot{\lambda}\sigma_{ij}[3s_{ij}/(2\sigma_{eff}) + \alpha\delta_{ij}] = \dot{\lambda}(\sigma_{eff} + \alpha I_1)$ and as the yield condition reads $f = \sigma_{eff} + \alpha I_1 = 0$, we obtain $\dot{W}^p = 0$. A graphical illustration of this result is shown in Fig. 22.6 where the observation that $\dot{\varepsilon}_{ij}^p$ is orthogonal to the yield surface immediately implies that the scalar product $\sigma_{ij}\dot{\varepsilon}_{ij}^p = 0$. Here we can extend this rejection also to include the rejection of isotropic hardening/softening of a cohesionless associated Drucker-Prager material; we additionally note that these rejections are intimately connected with the existence of straight meridians in the Drucker-Prager criterion, cf. Fig. 22.6.

For a cohesionless nonassociated formulation, (22.53) reduces to $\gamma_{mech} = -\dot{\lambda}I_1(\alpha - \alpha^*) \geq 0$, which is only fulfilled for $I_1 < 0$, i.e. we must exclude hardening and only ideal and softening plasticity are allowed.

In general, from (22.53) it is concluded that $\gamma_{mech} \geq 0$ requires

$$I_1 \leq \frac{\beta}{\alpha - \alpha^*}$$

From (80) and $f = 0$ the maximum value of I_1 is given by $I_{1,max} = (\beta + K)/\alpha$ and if the model is to be thermodynamically acceptable this $I_{1,max}$-value must

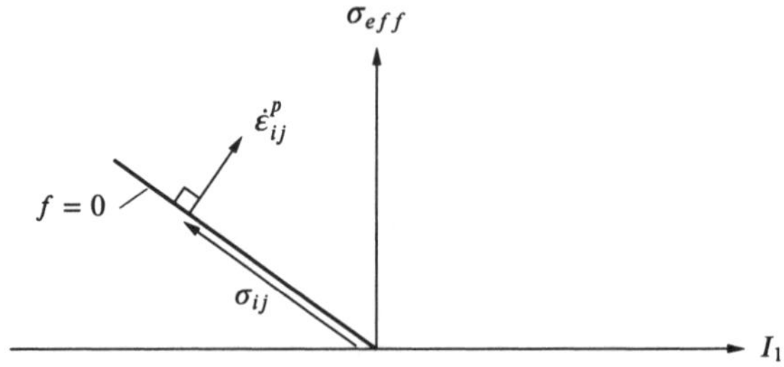

Figure 22.6: Illustration that $\dot{W}^p = \sigma_{ij}\dot{\varepsilon}^p_{ij} = 0$ for associated ideal plasticity of a cohesionless Drucker-Prager material.

be less than or equal to the limit given above. This leads to the following restriction

$$K(\alpha - \alpha^*) \leq \beta\alpha^* \qquad (22.55)$$

Considering in the following a Drucker-Prager material with cohesion (i.e. $\beta > 0$) and as $\alpha^* \geq 0$ then (22.55) is certainly fulfilled for softening plasticity (i.e. $K < 0$) and for ideal plasticity (i.e. $K = 0$); however, hardening nonassociated plasticity places a restriction on the amount of hardening.

With these observations, (22.9) gives the following expression for the generalized plastic modulus

$$H = \frac{\partial f}{\partial K}\rho\frac{d^2\psi^p(\kappa)}{d\kappa^2}\frac{\partial g}{\partial K} = \rho\frac{d^2\psi^p}{d\kappa^2}$$

Let us in particular assume linear hardening where H is constant. Then the expression above gives

$$\rho\psi^p = \frac{1}{2}H\kappa^2$$

Moreover

$$K = \rho\frac{d\psi^p}{d\kappa} = H\kappa$$

Since the hardening parameter K for linear hardening can take arbitrarily large positive values, (22.55) will, for nonassociated plasticity, be violated at some state; this formulation must thus be rejected.

In conclusion

Drucker-Prager isotropic hardening:

$$f = \sigma_{eff} + \alpha I_1 - \beta - K$$

$$g = \sigma_{eff} + \alpha^* I_1 - \beta^* - K$$

where $\alpha \geq 0$, $\beta \geq 0$ *and* $\alpha^* \geq 0$

$$\dot{\varepsilon}^p_{ij} = \dot{\lambda}\frac{\partial g}{\partial \sigma_{ij}} = \dot{\lambda}(\frac{3s_{ij}}{2\sigma_{eff}} + \alpha^* \delta_{ij})\,; \qquad \dot{\varepsilon}^p_{eff} = \dot{\lambda}\sqrt{1 + 2(\alpha^*)^2}$$

$$\dot{\kappa} = -\dot{\lambda}\frac{\partial g}{\partial K} = \dot{\lambda}$$

$$\gamma_{mech} = \dot{\lambda}[\beta - I_1(\alpha - \alpha^*)] \geq 0$$

It is required that $\alpha \geq \alpha^*$

Material without cohesion: associated plasticity is not allowed; nonassociated plasticity valid only for ideal and softening plasticity

Material with cohesion: Associated plasticity is always allowed; nonassociated plasticity is valid for ideal and softening plasticity, but restrictions apply for hardening plasticity and linear hardening is not allowed

$$H = \rho\frac{d^2\psi(\kappa)}{d\kappa^2}\,; \qquad K = \rho\frac{d\psi^p(\kappa)}{d\kappa}$$

Linear hardening: $\rho\psi^p = \frac{1}{2}H\kappa^2$

22.4 Use of plastic work as internal variable

We have seen that when an internal variable takes the form of a scalar, this internal variable often turns out to be the effective plastic strain or a quantity proportional to the effective plastic strain; examples are isotropic and mixed von Mises hardening and isotropic Drucker-Prager hardening. Therefore, the natural manifestation of a scalar internal variable seems to be that of effective plastic strain.

In Section 9.6, isotropic von Mises hardening was introduced and two choices for the internal variable were discussed: *strain hardening* where κ was chosen as the effective plastic strain ε^p_{eff} and *work hardening* where κ was chosen as the plastic work W^p. It was shown that the two assumptions lead to the same model, since there exists a one-to-one relation between ε^p_{eff} and W^p. However, it was emphasized that whereas this conclusion holds for isotropic von Mises plasticity, it is not a general conclusion that holds for arbitrary plasticity models.

With the format of the specific plasticity models discussed above, we observed that the natural interpretation of the scalar internal variable κ is that of ε^p_{eff} (or proportional to ε^p_{eff}). However, in many specific plasticity models used in the literature - and especially for those plasticity models that are derived in the traditional way without making use of thermodynamics - the plastic work W^p is often used as an internal variable. Thus, let us establish a framework which provides W^p as the natural interpretation of the scalar internal variable κ. To do so, we will take advantage of the results presented by Ristinmaa (1999).

In soil and concrete constitutive modeling, the yield function and potential function are often written as

$$f(\sigma_{ij}, K) = p(K)F(\sigma_{ij}) - c_1\,; \qquad g(\sigma_{ij}, K) = q(K)G(\sigma_{ij}) - c_2 \tag{22.56}$$

i.e. only one scalar hardening parameter K and therefore only one scalar internal variable κ is involved; moreover, c_1 and c_2 are constants. Formulation (22.56) applies, for instance, to the soil and concrete model of Lade and Kim (1995). The characteristic feature of (22.56) is that the influence of σ_{ij} and K appears in a factorized form. Adopting nonassociated plasticity, it follows from (22.6) that

$$\dot{\varepsilon}^p_{ij} = \dot{\lambda}\frac{\partial g}{\partial \sigma_{ij}} = \dot{\lambda} q(K)\frac{\partial G(\sigma_{kl})}{\partial \sigma_{ij}}\,; \qquad \dot{\kappa} = -\dot{\lambda}\frac{\partial g}{\partial K} = -\dot{\lambda}\frac{dq(K)}{dK}G(\sigma_{ij}) \tag{22.57}$$

Assume that the function $G(\sigma_{ij})$ is homogeneous in σ_{ij} of degree n. From Euler's theorem (21.47), we then have

$$\sigma_{ij}\frac{\partial G}{\partial \sigma_{ij}} = nG \tag{22.58}$$

The rate of plastic work then becomes

$$\dot{W}^p = \sigma_{ij}\dot{\varepsilon}^p_{ij} = \dot{\lambda} q \sigma_{ij}\frac{\partial G}{\partial \sigma_{ij}} = \dot{\lambda} q n G \tag{22.59}$$

Since we want

$$\dot{\kappa} = \dot{W}^p$$

insertion of (22.57b) and (22.59) gives

$$-\frac{dq(K)}{dK} = nq$$

with the solution

$$q = e^{-nK} \tag{22.60}$$

With (22.56), (22.58) and (22.60), we conclude that the internal variable becomes $\kappa = W^p$. Indeed, this formulation was adopted by Lade and Kim (1995) using traditional plasticity theory whereas we have here followed Ristinmaa (1999) for its thermodynamic formulation. As another implication of thermodynamics, the dissipation inequality becomes

$$\gamma_{mech} = \sigma_{ij}\dot{\varepsilon}_{ij}^{p} - K\dot{\kappa} = \dot{\lambda}(qnG + K\frac{dq}{dK}G) \geq 0$$

by using (22.59) and (22.57b). With (22.60), we arrive at

$$\gamma_{mech} = \dot{\lambda}Gne^{-nK}(1 - K) \geq 0$$

and it is required that the hardening parameter K fulfills $K < 1$. It turns out that in the formulation of Lade and Kim (1995), this restriction is fulfilled.

22.5 Corner plasticity

Many yield and potential surfaces contain *corners* and examples are given by the criteria of Tresca and Coulomb. Considering for simplicity associated plasticity we have used the format $\dot{\varepsilon}_{ij}^{p} = \dot{\lambda}\partial f/\partial\sigma_{ij}$, but this certainly presumes that the yield surface is smooth. Considering ideal plasticity and adopting the postulate of maximum plastic dissipation, we have $(\sigma_{ij} - \sigma_{ij}^{*})\dot{\varepsilon}_{ij}^{p} \geq 0$ where σ_{ij} is the current stress state located on the yield surface, σ_{ij}^{*} is an arbitrary point on or inside the yield surface and the plastic strain rate $\dot{\varepsilon}_{ij}^{p}$ is related to the current stress state σ_{ij}. Regarding Fig. 22.7 this implies that $\dot{\varepsilon}_{ij}^{p}$ is located somewhere within the 'cone' defined by the two normals at the corner; as a result, $\dot{\varepsilon}_{ij}^{p}$ can be written as a linear combination of these two normals.

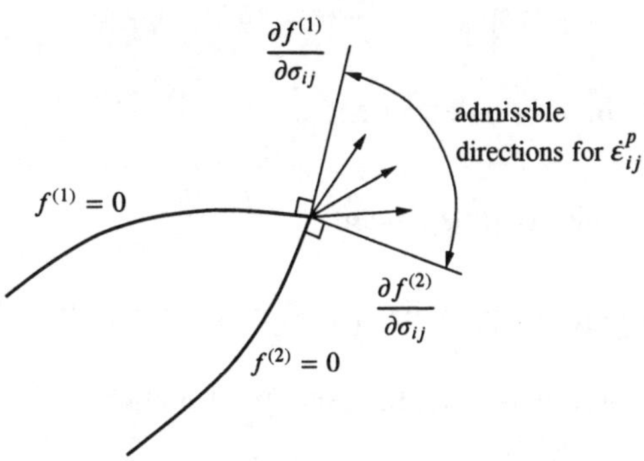

Figure 22.7: Stress space with two yield functions. Admissible directions for $\dot{\varepsilon}_{ij}^{p}$ at a corner.

Generalizing these ideas and assuming that F_{max} is the number of yield surfaces that meet at a corner, we are led to

Koiter's flow rule

$$\dot{\varepsilon}^p_{ij} = \sum_{I=1}^{F_{max}} \dot{\lambda}^I \frac{\partial f^I}{\partial \sigma_{ij}}$$

as suggested by Koiter (1953, 1960); here f^I, $I = 1, 2, \ldots, F_{max}$ denote the yield functions and $\dot{\lambda}^I$ the corresponding plastic multipliers.

Following Ottosen and Ristinmaa (1996) we will now generalize Koiter's flow rule both for associated and nonassociated plasticity. Each of the yield functions depends on the stresses and the hardening parameters

$$f^I = f^I(\sigma_{ij}, K_\alpha)\,; \qquad I = 1, 2, \ldots, F_{max}$$

As usual, the mechanical dissipation is given by $\gamma_{mech} = \sigma_{ij}\dot{\varepsilon}^p_{ij} - K_\alpha \dot{\kappa}_\alpha$. Adopting the postulate of maximum dissipation we are faced with the following problem: for given $\dot{\varepsilon}^p_{ij}$ and $\dot{\kappa}_\alpha$, minimize the quantity $-\gamma_{mech}$ subject to the constraints $f^I \leq 0$. From the Kuhn-Tucker relations given by Appendix (A.15), we obtain

Associated plasticity

$$\dot{\varepsilon}^p_{ij} = \sum_{I=1}^{F_{max}} \dot{\lambda}^I \frac{\partial f^I}{\partial \sigma_{ij}}\,; \qquad \dot{\kappa}_\alpha = -\sum_{I=1}^{F_{max}} \dot{\lambda}^I \frac{\partial f^I}{\partial K_\alpha}$$

where

$$\dot{\lambda}^I \geq 0 \quad \textit{and} \quad \dot{\lambda}^I f^I = 0$$

Following Appendix (A.13), a prerequisite for this result is that the point is a *regular point* meaning that

$$\begin{aligned} &\frac{\partial f^I}{\partial \sigma_{ij}} \quad \text{are linearly independent} \\ &\frac{\partial f^I}{\partial K_\alpha} \quad \text{are linearly independent} \end{aligned} \tag{22.61}$$

Moreover, according to Appendix (A.18) it is required that f^I are convex functions.

For nonassociated plasticity, the potential functions are given by

$$g^\Phi = g^\Phi(\sigma_{ij}, K_\alpha)\,; \qquad \Phi = 1, 2, \ldots, G_{max}$$

where G_{max} is the number of potential functions meeting at the corner; in general, we will allow the number of yield surfaces F_{max} to differ from the number of potential surfaces G_{max} and later we will return to this issue. It is required that the potential functions are convex and that $g^{\Phi}(\sigma_{ij}, K_\alpha) - g^{\Phi}(0,0) \geq 0$. Following (21.51) it is then evident that the following evolution laws fulfill the dissipation inequality

> *Nonassociated plasticity*
>
> $$\dot{\varepsilon}_{ij}^{p} = \sum_{\Phi=1}^{G_{max}} \dot{\lambda}^{\Phi} \frac{\partial g^{\Phi}}{\partial \sigma_{ij}}; \qquad \dot{\kappa}_\alpha = -\sum_{\Phi=1}^{G_{max}} \dot{\lambda}^{\phi} \frac{\partial g^{\phi}}{\partial K_\alpha} \tag{22.62}$$
>
> *where*
>
> $\dot{\lambda}^{\phi} \geq 0$

In accordance with (22.2) we choose Helmholtz' free energy according to

$$\rho\psi(\theta, \varepsilon_{ij} - \varepsilon_{ij}^{p}, \kappa_\alpha) = \rho h(\theta) + \frac{1}{2}(\varepsilon_{ij} - \varepsilon_{ij}^{p}) D_{ijkl}(\varepsilon_{kl} - \varepsilon_{kl}^{p}) + \rho\psi^{p}(\kappa_\alpha)$$

This leads to

$$\sigma_{ij} = \rho\frac{\partial \psi}{\partial \varepsilon_{ij}} = D_{ijkl}(\varepsilon_{kl} - \varepsilon_{kl}^{p})$$

$$K_\alpha = \rho\frac{\partial \psi}{\partial \kappa_\alpha} = \rho\frac{\partial \psi^{p}}{\partial \kappa_\alpha}$$

and it follows that

$$\dot{\sigma}_{ij} = D_{ijkl}(\dot{\varepsilon}_{kl} - \dot{\varepsilon}_{kl}^{p}); \qquad \dot{K}_\alpha = d_{\alpha\beta}\dot{\kappa}_\beta \tag{22.63}$$

where

$$\boxed{D_{ijkl} = \rho\frac{\partial^2 \psi}{\partial \varepsilon_{ij}\partial \varepsilon_{kl}}; \qquad d_{\alpha\beta} = \rho\frac{\partial^2 \psi^{p}}{\partial \kappa_\alpha \partial \kappa_\beta}}$$

With these fundamental equations, we will now derive the incremental relation $\dot{\sigma}_{ij} = D_{ijkl}^{ep}\dot{\varepsilon}_{kl}$. Differentiation of the yield function f^I gives

$$\dot{f}^I = \frac{\partial f^I}{\partial \sigma_{ij}}\dot{\sigma}_{ij} + \frac{\partial f^I}{\partial K_\alpha}\dot{K}_\alpha$$

Insertion of (22.63b) gives

$$\dot{f}^I = \frac{\partial f^I}{\partial \sigma_{ij}}\dot{\sigma}_{ij} + \frac{\partial f^I}{\partial K_\alpha} d_{\alpha\beta}\dot{\kappa}_\beta$$

and the evaluation law (22.62b) then provides

$$\boxed{\begin{aligned}&\dot{f}^I = \frac{\partial f^I}{\partial \sigma_{ij}}\dot{\sigma}_{ij} - \sum_{\Phi=1}^{G_{max}} H^{I\Phi}\dot{\lambda}^{\Phi}\\ &\textit{where}\\ &H^{I\Phi} = \frac{\partial f^I}{\partial K_\alpha} d_{\alpha\beta} \frac{\partial g^{\Phi}}{\partial \kappa_\beta}\\ &\textit{is the matrix of plastic moduli}\end{aligned}} \tag{22.64}$$

Insertion of the flow rule (22.62a) into Hooke's law (22.63a) gives

$$\dot{\sigma}_{ij} = D_{ijkl}\dot{\varepsilon}_{kl} - D_{ijst} \sum_{\Phi=1}^{G_{max}} \frac{\partial g^{\Phi}}{\partial \sigma_{st}} \dot{\lambda}^{\Phi} \tag{22.65}$$

and with this expression (22.64) becomes

$$\dot{f}^I = \dot{a}^I - \sum_{\Phi=1}^{G_{max}} A^{I\Phi}\dot{\lambda}^{\Phi} \tag{22.66}$$

where

$$\dot{a}^I = \frac{\partial f^I}{\partial \sigma_{ij}} D_{ijkl}\dot{\varepsilon}_{kl}$$

and

$$\boxed{A^{I\Phi} = H^{I\Phi} + \frac{\partial f^I}{\partial \sigma_{ij}} D_{ijkl} \frac{\partial g^{\Phi}}{\partial \sigma_{kl}}} \tag{22.67}$$

When plastic loading occurs, the consistency relation states that $\dot{f}^I = 0$ which with (22.66) gives

$$\sum_{\Phi=1}^{G_{max}} A^{I\Phi}\dot{\lambda}^{\Phi} = \dot{a}^I \quad \text{or} \quad \boldsymbol{A}\dot{\boldsymbol{\lambda}} = \dot{\boldsymbol{a}} \tag{22.68}$$

To be able to derive a strain driven format, this equation must provide a unique $\dot{\lambda}^{\Phi}$-solution. The augmented matrix $\boldsymbol{T}$ is defined by $\boldsymbol{T} = [\boldsymbol{A}, \dot{\boldsymbol{a}}]$ and it is of dimension $F_{max} \times (G_{max} + 1)$. From Ayres (1962), for instance, a unique $\dot{\lambda}^{\Phi}$-solution requires that $\text{rank}\,\boldsymbol{T} = \text{rank}\,\boldsymbol{A} = G_{max}$; thus, it follows that $\dot{a}^I$ can be expressed by linear combinations of the columns in $A^{I\Phi}$. We therefore require

$$\text{rank}\,\boldsymbol{A} = G_{max} \tag{22.69}$$

This requirement can never be achieved if $F_{max} < G_{max}$ and we conclude

$$\boxed{F_{max} \geq G_{max}} \tag{22.70}$$

That is, the number of yield surfaces must be larger than or equal to the number of potential surfaces. If these requirements above are fulfilled then (22.68) has a unique $\dot{\lambda}^{\Phi}$-solution.

Due to (22.69) the $F_{max} \times G_{max}$ matrix $\boldsymbol{A}$ has a *left inverse* $\underline{\boldsymbol{A}}$ such that

$$\underline{\boldsymbol{A}}\boldsymbol{A} = \boldsymbol{I} \quad \text{or} \quad \sum_{I=1}^{F_{max}} \underline{A}^{\Theta I} A^{I\Phi} = \delta^{\Theta\Phi}$$

cf. Ayres (1962). As a result, (22.68) provides the solution

$$\dot{\boldsymbol{\lambda}} = \underline{\boldsymbol{A}}\dot{\boldsymbol{a}} \quad \text{or} \quad \dot{\lambda}^{\Phi} = \sum_{I=1}^{F_{max}} \underline{A}^{\Phi I} \dot{a}^{I}$$

It is of interest that even though the left inverse $\underline{\boldsymbol{A}}$, is not, in general, uniquely determined, the $\dot{\lambda}^{\Phi}$-solution given above is a unique solution, cf. Eves (1980). Insertion of the result above into Hooke's law (22.65) gives the result sought for

$$\boxed{\begin{aligned} &\dot{\sigma}_{ij} = D^{ep}_{ijkl}\dot{\varepsilon}_{kl} \\ &\textit{where} \\ &D^{ep}_{ijkl} = D_{ijkl} - D_{ijst} \sum_{\Phi=1}^{G_{max}} \frac{\partial g^{\Phi}}{\partial \sigma_{st}} \sum_{I=1}^{F_{max}} \underline{A}^{\Phi I} \frac{\partial f^{I}}{\partial \sigma_{mn}} D_{mnkl} \end{aligned}}$$

which for $F_{max} = G_{max} = 1$ evidently reduces to expression (10.27) valid for smooth yield and potential functions. It also appears that $\boldsymbol{D}^{ep}$ for associated plasticity becomes a symmetric matrix.

In addition to requirement (22.69), let us now investigate the matrix $A^{I\Phi}$ in more detail. Observing (22.70), it is then evident that it is always possible to choose the numbering of the yield functions such that the $A^{I\Phi}$-matrix can be written as

$$A^{\Phi I} = \begin{bmatrix} P^{\Theta\Phi} \\ Q^{J\Phi} \end{bmatrix}; \qquad \text{rank}\, P^{\Theta\Phi} = G_{max} \tag{22.71}$$

where $P^{\Theta\Phi}$ is of dimension $G_{max} \times G_{max}$ whereas $Q^{J\Phi}$ is of dimension $(F_{max} - G_{max}) \times G_{max}$. With (22.69) we immediately observe that det $\boldsymbol{P} \neq 0$ and we will now show that the eigenvalues of $\boldsymbol{P}$ are positive.

Consider as a special case ideal plasticity where no hardening parameters exist. In that case the matrix of plastic moduli $\boldsymbol{H}$ is zero and if we furthermore specialize to associated plasticity, $\boldsymbol{A}$ given by (22.67) reduces to the following symmetric matrix

$$A^{IJ} = \frac{\partial f^I}{\partial \sigma_{ij}} D_{ijkl} \frac{\partial f^J}{\partial \sigma_{kl}}$$

Since D_{ijkl} is positive definite and as $\partial f^I / \partial \sigma_{ij}$ are linearly independent, cf. (22.61), it follows that also A^{IJ} is positive definite, i.e. it possesses only positive eigenvalues. As the theory we want to establish is to be general, we therefore assume - in recognition of (22.71) - that

$$\boxed{\textit{The eigenvalues of } P^{\Theta\Phi} \textit{ must be positive}} \qquad (22.72)$$

In the special case of associated plasticity, Sewell (1973) and Simo *et al.* (1988) argued that A^{IJ} should be positive definite and under these conditions, (22.72) reduces to that requirement. We may also note when $F_{max} = G_{max} = 1$, the requirement above reduces to the result $A > 0$ valid for conventional plasticity, cf. (10.24).

Having determined the properties of the matrix $A^{I\Phi}$, it also turns out to be important to evaluate the properties of the plastic moduli matrix. For this purpose, let us return to (22.64) and investigate the possibilities for a stress driven format. The consistency relation gives

$$\sum_{\Phi=1}^{G_{max}} H^{I\Phi} \dot{\lambda}^{\Phi} = \frac{\partial f^I}{\partial \sigma_{ij}} \dot{\sigma}_{ij} \qquad (22.73)$$

A stress driven format requires that this equation provides a unique $\dot{\lambda}^{\Phi}$-solution and it is concluded that

Stress driven format requires rank $H^{I\Phi} = G_{max}$

Equivalent to (22.71) it is always possible to write

$$H^{I\Phi} = \begin{bmatrix} R^{\Theta\Phi} \\ S^{J\Phi} \end{bmatrix}; \qquad \begin{array}{l} \text{stress driven format requires} \\ \text{rank } R^{\Theta\Phi} = G_{max} \end{array} \qquad (22.74)$$

where $R^{\Theta\Phi}$ is of dimension $G_{max} \times G_{max}$ and $S^{J\Phi}$ is of dimension $(F_{max} - G_{max}) \times G_{max}$.

A *limit point* is defined according to

$$\dot{\sigma}_{ij} = 0 \quad \text{and} \quad \dot{\varepsilon}_{ij} \neq 0 \;\Rightarrow\; \text{limit point}$$

Since $\dot{\sigma}_{ij} = 0$, (22.73) gives $\sum_{\Phi=1}^{G_{max}} H^{I\Phi} \dot{\lambda}^{\Phi} = 0$ and it is concluded that

$$\text{limit point} \;\Leftrightarrow\; \text{rank } H^{\Theta\Phi} < G_{max} \qquad (22.75)$$

Consider now the special case of ideal plasticity. In that case no hardening parameters exist and we have

$$H^{\Theta\Phi} = 0 \quad \Leftrightarrow \quad \text{ideal plasticity}$$

That is, all components of $H^{I\Phi}$ are zero. We now observe a difference between corner plasticity and conventional plasticity. For conventional plasticity, ideal plasticity ($H = 0$) and the existence of a limit are identical statements, cf. Section 10.3. For corner plasticity, a glance at the two last equations reveals a fundamental difference between ideal plasticity and the existence of a limit point.

Considering now associated hardening plasticity, then during plastic loading we expect all the components of the quantity $\partial f^I/\partial\sigma_{ij}\dot{\sigma}_{ij}$ to be positive. Since the plastic multipliers $\dot{\lambda}^I$ are also positive during plastic loading, we obtain $\sum_{I=1}^{F_{max}} \dot{\lambda}^I \partial f^I/\partial\sigma_{ij}\dot{\sigma}_{ij} > 0$. From (22.73) it then follows that

$$\sum_{I=1}^{F_{max}} \sum_{J=1}^{F_{max}} \dot{\lambda}^I H^{IJ} \dot{\lambda}^J > 0 \quad \text{for associated hardening plasticity}$$

and the matrix of plastic moduli is therefore positive definite. Generalizing this result and observing (22.74) we arrive at the following definitions

Hardening plasticity $\Leftrightarrow$ $R^{\Theta\Phi}$ *has only positive eigenvalues*

With reference to (22.75), the existence of a limit point is reformulated as

Limit point $\Leftrightarrow$ $R^{\Theta\Phi}$ *has at least one zero eigenvalue*

We are then left with the following conclusion

Softening plasticity $\Leftrightarrow$	$R^{\Theta\Phi}$ *has at least one negative eigenvalue and no eigenvalues are zero*

It is recalled that ideal plasticity exists if all the components of $\boldsymbol{H}$ are zero. With these definitions, it follows that hardening plasticity allows a stress driven format, i.e. (22.73) can be solved for $\dot{\lambda}^{\Phi}$.

Suppose that two yield surfaces meet at a corner and that hardening of one yield surface is independent of what happens with the other yield surface, then *independent hardening* occurs and Fig. 22.8 shows an example. On the other hand, if hardening of one yield surface depends on what happens with the other yield surface, then *dependent hardening* occurs as illustrated in Fig. 22.9.

For associated plasticity where two yield surfaces meet at a corner, two classic formats of the plastic moduli matrix are given by

$$\boldsymbol{H} = \begin{bmatrix} k & k \\ k & k \end{bmatrix} \quad \text{Taylor hardening} \tag{22.76}$$

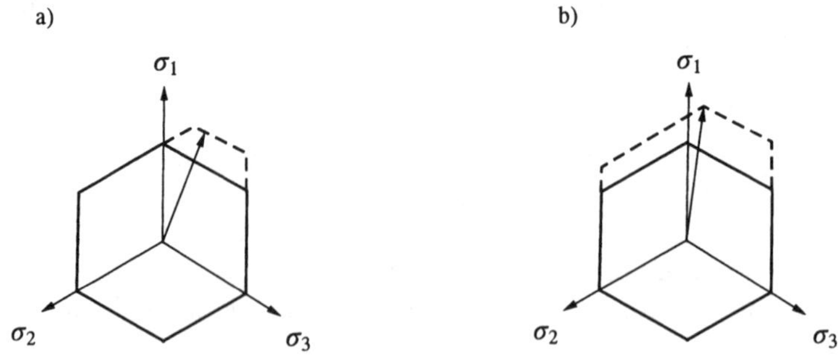

Figure 22.8: Tresca criterion with independent hardening; a) loading along a smooth yield surface, b) corner loading.

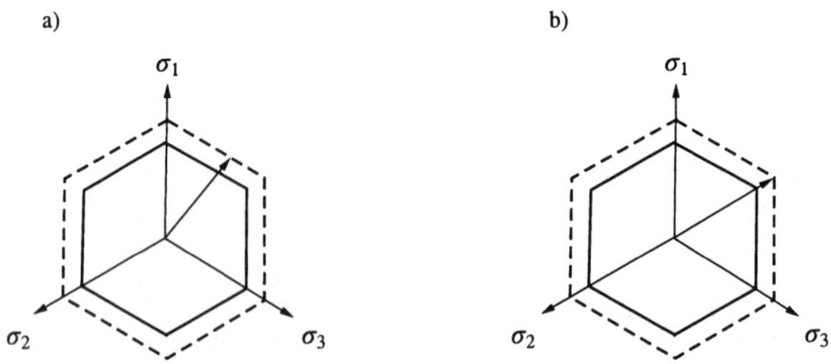

Figure 22.9: Tresca criterion with dependent hardening; a) loading along a smooth yield surface, b) corner loading.

and

$$\boldsymbol{H} = \begin{bmatrix} H_{11} & 0 \\ 0 & H_{22} \end{bmatrix} \quad \text{Koiter hardening}$$

where (22.76) is due to Taylor (1938). Koiter hardening implies independent hardening whereas Taylor hardening may be involved for dependent hardening. It is of interest that Taylor hardening does not allow a stress driven format, cf. (22.74); for further discussions, we refer to Mandel (1965), Hill (1966) and Ottosen and Ristinmaa (1996).

With the discussion above, we have established a firm basis for general nonassociated corner plasticity. It turns out, however, that in the most general case the loading/unloading criteria become rather complex and a discussion is provided by Ottosen and Ristinmaa (1996). The issue of numerical integration is treated by Simo *et al.* (1988), Pramono and Willam (1989) and Simo and Hughes (1998) for the associated case.

22.6 Viscoplasticity

As discussed in Chapter 15 there exist two major concepts when formulating viscoplasticity: the Perzyna and the Duvaut-Lions format. We will now illustrate how these formulations can be given a thermodynamic interpretation.

Perzyna viscoplasticity

The general nonassociated format is easily obtained. Let the potential function $g(\sigma_{ij}, K_\alpha)$ possess the property $g(\sigma_{ij}, K_\alpha) - g(0,0) \geq 0$; then, according to (21.51) and (21.54) the dissipation inequality is fulfilled for the following evolution laws

$$\dot{\varepsilon}_{ij}^{vp} = \Lambda \frac{\partial g}{\partial \sigma_{ij}}; \qquad \dot{\kappa}_\alpha = -\Lambda \frac{\partial g}{\partial K_\alpha} \tag{22.77}$$

where Λ is a non-negative function. Choosing this quantity as $\Lambda = \Phi(f)/\eta$ we have recovered the format (15.36).

For Perzyna viscoplasticity, the postulate of maximum dissipation was originally restricted to rely on subtle regularization and penalty techniques that only hold in the limit when viscoplasticity degenerates to inviscid plasticity, cf. Simo and Honein (1990). However, by using the concept of the dynamic yield surface, Ristinmaa and Ottosen (2000) demonstrated that this allows the postulate of maximum dissipation to be adopted in a straightforward manner as will be shown next.

According to (15.43) the dynamic yield function is given by

$$f^d(\sigma_{ij}, K_\alpha, \dot{\varepsilon}_{eff}^{vp}) = f(\sigma_{ij}, K_\alpha) - \varphi(\eta \dot{\varepsilon}_{eff}^{vp}) \tag{22.78}$$

where $f^d < 0$ implies elastic behavior, $f^d = 0$ results in development of viscoplasticity and $f^d > 0$ cannot occur; moreover, $f(\sigma_{ij}, K_\alpha)$ denotes the static yield function and the effective viscoplastic strain rate is defined by $\dot{\varepsilon}_{eff}^{vp} = (2\dot{\varepsilon}_{ij}^{vp}\dot{\varepsilon}_{ij}^{vp}/3)^{1/2}$.

The mechanical dissipation inequality is given by $\gamma_{mech} = \sigma_{ij}\dot{\varepsilon}_{ij}^{vp} - K_\alpha\dot{\kappa}_\alpha \geq 0$ and the postulate of maximum dissipation can accordingly be formulated as: For given fluxes $\dot{\varepsilon}_{ij}^{vp}$ and $\dot{\kappa}_\alpha$ find those stresses σ_{ij} and conjugated forces K_α that minimize $-\gamma_{mech}$ under the constraint that $f^d(\sigma_{ij}, K_\alpha, \dot{\varepsilon}_{eff}^{vp}) \leq 0$. Assuming the dynamic yield function to be convex, Appendix (A.15) and (A.18) provide the Kuhn-Tucker relations

$$\dot{\varepsilon}_{ij}^{vp} = \Lambda \frac{\partial f^d}{\partial \sigma_{ij}}; \qquad \dot{\kappa}_\alpha = -\Lambda \frac{\partial f^d}{\partial K_\alpha}$$

where $\Lambda \geq 0$, $f^d \leq 0$ and $\Lambda f^d = 0$. With the dynamic yield function defined by (22.78), we obtain

$$\dot{\varepsilon}_{ij}^{vp} = \Lambda \frac{\partial f}{\partial \sigma_{ij}}; \qquad \dot{\kappa}_\alpha = -\Lambda \frac{\partial f}{\partial K_\alpha} \tag{22.79}$$

From (22.79a) and the definition of the effective viscoplastic strain rate, we find

$$\dot{\varepsilon}_{eff}^{vp} = \Lambda (\frac{2}{3} \frac{\partial f}{\partial \sigma_{ij}} \frac{\partial f}{\partial \sigma_{ij}})^{1/2} \tag{22.80}$$

When viscoplasticity develops $f^d = 0$ holds and (22.78) then provides

$$f(\sigma_{ij}, K_\alpha) = \varphi(\eta \dot{\varepsilon}_{eff}^{vp}) \quad \Rightarrow \quad \Phi(f) = \eta \dot{\varepsilon}_{eff}^{vp} \tag{22.81}$$

where Φ is the inverse function of φ. A comparison of (22.80) and (22.81b) shows that the non-negative function Λ is given by

$$\Lambda = \frac{\Phi(f)}{\eta(\frac{2}{3} \frac{\partial f}{\partial \sigma_{ij}} \frac{\partial f}{\partial \sigma_{ij}})^{1/2}} \tag{22.82}$$

In most cases, for instance, when the static yield function is chosen in terms of the von Mises, Drucker-Prager or Coulomb criterion, the quantitity $\partial f / \partial \sigma_{ij}$ $\partial f / \partial \sigma_{ij}$ becomes a constant and (22.79) and (22.82) then coincide exacly with associated Perzyna viscoplasticity.

Duvaut-Lions viscoplasticity

To establish a thermodynamic framework for Duvaut-Lions viscoplasticity is somewhat more complex; in fact, it turns out that Duvaut-Lions viscoplasticity apparently can only be formulated as nonassociated viscoplasticity. We will follow the proposal by Ristinmaa and Ottosen (1998), which rests on the concept of an *additive split* of the conjugated forces.

The mechanical dissipation inequality is given by

$$\gamma_{mech} = \sigma_{ij} \dot{\varepsilon}_{ij}^{vp} - K_\alpha \dot{\kappa}_\alpha \geq 0 \tag{22.83}$$

With $f = f(\sigma_{ij}, K_\alpha)$ being the static yield function, it was shown in Section 15.4.2 that the closest-point-projection $\bar{\sigma}_{ij}, \bar{K}_\alpha$ on the static yield surface was given by (15.57), i.e.

$$\begin{aligned} &-C_{ijkl}(\sigma_{kl} - \bar{\sigma}_{kl}) + \mu \frac{\partial f}{\partial \bar{\sigma}_{ij}} = 0 \\ &-c_{\alpha\beta}(K_\beta - \bar{K}_\beta) + \mu \frac{\partial f}{\partial \bar{K}_\alpha} = 0 \\ &f(\bar{\sigma}_{ij}, \bar{K}_\alpha) = 0 \end{aligned} \tag{22.84}$$

Define the quantities σ^*_{ij}, K^*_α according to

$$\sigma_{ij} = \sigma^*_{ij} + \bar{\sigma}_{ij}\,; \qquad K_\alpha = K^*_\alpha + \bar{K}_\alpha \tag{22.85}$$

It appears that we have made an additive split of the conjugated forces. Insertion into the mechanical dissipation inequality (22.83) gives

$$\gamma_{mech} = \sigma^*_{ij}\dot{\varepsilon}^{vp}_{ij} - K^*_\alpha\dot{\kappa}_\alpha + \bar{\sigma}_{ij}\dot{\varepsilon}^{vp}_{ij} - \bar{K}_\alpha\dot{\kappa}_\alpha \geq 0$$

This inequality is certainly fulfilled provided

$$\gamma_{mech,1} = \sigma^*_{ij}\dot{\varepsilon}^{vp}_{ij} - K^*_\alpha\dot{\kappa}_\alpha \geq 0\,; \qquad \gamma_{mech,2} = \bar{\sigma}_{ij}\dot{\varepsilon}^{vp}_{ij} - \bar{K}_\alpha\dot{\kappa}_\alpha \geq 0 \tag{22.86}$$

Each of these inequalities will now be fulfilled separately using the potential function approach. The static yield fuction is convex and it must fulfill $f(\bar{\sigma}_{ij}, \bar{K}_\alpha) - f(0,0) \geq 0$. According to (21.51) and (21.54) we then comply with the inequality $\gamma_{mech,2} \geq 0$ for

$$\dot{\varepsilon}^{vp}_{ij} = \Lambda_2\frac{\partial f}{\partial \bar{\sigma}_{ij}}\,; \qquad \dot{\kappa}_\alpha = -\Lambda_2\frac{\partial f}{\partial \bar{K}_\alpha} \tag{22.87}$$

where Λ_2 is a non-negative quantity. Considering the other inequality, we adopt as potential function the same function as that which defines the distance between the current state and the closest-point-projection. According to (15.56), we then choose the potential function $G(\sigma^*_{ij}, K^*_\alpha)$ as

$$G(\sigma^*_{ij}, K^*_\alpha) = \frac{1}{2}\sigma^*_{ij}C_{ijkl}\sigma^*_{kl} + \frac{1}{2}K^*_\alpha c_{\alpha\beta}K^*_\beta \tag{22.88}$$

The elastic flexibility tensor is certainly positive definite and $c_{\alpha\beta}$ is also positive definite for hardening viscoplasticity, as shown in Ristinmaa and Ottosen (1998). There it is proved that the potential function introduced fulfills our requirements even when softening viscoplasticity is involved. Therefore, we obtain

$$\dot{\varepsilon}^{vp}_{ij} = \Lambda_1\frac{\partial G}{\partial \sigma^*_{ij}} = \Lambda_1 C_{ijkl}\sigma^*_{kl}\,; \qquad \dot{\kappa}_\alpha = -\Lambda_1\frac{\partial G}{\partial K^*_\alpha} = -\Lambda_1 c_{\alpha\beta}K^*_\beta \tag{22.89}$$

The two sets of evolution equations given by (22.87) and (22.89) must be identical. Indeed, if Λ_2 is chosen as $\Lambda_2 = \mu\Lambda_1$ then, in view of (22.84), it appears that this equality is achieved. Thus, from (22.89) and making the definition $\Lambda = \Lambda_1$ we obtain

$$\dot{\varepsilon}^{vp}_{ij} = \Lambda C_{ijkl}(\sigma_{kl} - \bar{\sigma}_{kl})\,; \qquad \dot{\kappa}_\alpha = -\Lambda c_{\alpha\beta}(K_\beta - \bar{K}_\beta) \tag{22.90}$$

which is exactly the Duvaut-Lions formulation, cf. (15.58).

22.7 Viscoelasticity

Having illustrated how plasticity and viscoplasticity are placed within a thermodynamic framework, it is of interest to consider viscoelasticity; information is also provided by the textbooks of Lemaitre and Chaboche (1990), Maugin (1992) and Simo and Hughes (1998). An introduction to viscoelasticity was given in Chapter 14 and we will here investigate Maxwell and Kelvin viscoelasticity in their generalized forms. It turns out to be useful to return to the mechanical dissipation inequality (21.5), which reads

$$\gamma_{mech} = -\rho(\dot{\psi} + s\dot{\theta}) + \sigma_{ij}\dot{\varepsilon}_{ij} \geq 0 \tag{22.91}$$

In the framework discussed so far, it was emphasized in relation to (21.37) that viscoelasticity may involve the occurrence of a viscous stress. However, before we pursue this observation, it will first be shown that Maxwell viscoelasticity and generalized Maxwell viscoelasticity fit into the thermodynamic framework already discussed.

Maxwell models

A very simple form of viscoelasticity is given by Maxwell viscoelasticity for which we assume

$$\psi = \psi(\theta, \varepsilon_{ij} - \varepsilon_{ij}^{v}) \tag{22.92}$$

where ε_{ij}^{v} denotes the *viscous strains*. From (21.37) it then follows that

$$\sigma_{ij} = \rho\frac{\partial\psi}{\partial\varepsilon_{ij}} \tag{22.93}$$

and the dissipation inequality becomes

$$\boxed{\gamma_{mech} = \sigma_{ij}\dot{\varepsilon}_{ij}^{v} \geq 0} \tag{22.94}$$

In order to fulfill this inequality and as linear viscoelasticity is aimed at, we take the potential function Φ according to

$$\Phi = \frac{1}{2}\sigma_{ij}B_{ijkl}\sigma_{kl}$$

where the constant tensor B_{ijkl} is assumed to be positive definite. From this expression, the evolution equation becomes

$$\dot{\varepsilon}_{ij}^{v} = \frac{\partial\Phi}{\partial\sigma_{ij}} = B_{ijkl}\sigma_{kl} \tag{22.95}$$

Insertion into (22.94) immediately reveals that the dissipation inequality is fulfilled; we may note that the evolution law (22.95) can also be viewed as being an Onsager relation, cf. (21.42).

To be specific, take Helmholtz' free energy according to

$$\rho\psi(\theta, \varepsilon_{ij} - \varepsilon_{ij}^{v}) = A(\theta) + \frac{1}{2}(\varepsilon_{ij} - \varepsilon_{ij}^{v})D_{ijkl}(\varepsilon_{kl} - \varepsilon_{kl}^{v})$$

The stresses are given by (22.93) and a time differentiation and use of (22.95) result in

$$C_{ijkl}\dot{\sigma}_{kl} + B_{klmn}\sigma_{mn} = \dot{\varepsilon}_{ij}$$

where C_{ijkl} is the flexibility tensor; this result coincides with the findings given by (14.20) and (14.22). In conclusion

Maxwell viscoelasticity
$\psi = \psi(\theta, \varepsilon_{ij} - \varepsilon_{ij}^{v})$
$\sigma_{ij} = \rho\dfrac{\partial\psi}{\partial\varepsilon_{ij}}; \quad \dot{\varepsilon}_{ij}^{v} = B_{ijkl}\sigma_{kl}$
where B_{ijkl} *is positive definite*

(22.96)

Since Maxwell viscoelasticity is a very simplistic model for viscoelastic behavior, it is of interest to consider generalized Maxwell viscoelasticity, cf. (14.14) and the discussion following that expression. With the result (22.96), this generalization can be achieved in a straightforward manner. Take the free energy according to

$$\psi = \sum_{k=1}^{n} \psi^{(k)} \quad \text{where} \quad \psi^{(k)} = \psi^{(k)}(\theta, \varepsilon_{ij} - \varepsilon_{ij}^{v(k)}) \tag{22.97}$$

Define $\sigma_{ij}^{(k)}$ by

$$\sigma_{ij}^{(k)} = -\rho\frac{\partial\psi^{(k)}}{\partial\varepsilon_{ij}^{v(k)}} \tag{22.98}$$

The stress tensor is still given by (22.93) and it follows that

$$\sigma_{ij} = \rho\frac{\partial\psi}{\partial\varepsilon_{ij}} = \sum_{k=1}^{n} \sigma_{ij}^{(k)} \tag{22.99}$$

Insertion of (22.97) into the dissipation inequality (22.91) and observing that $\partial\psi/\partial\theta = -s$ still holds, result in

$$\rho\sum_{k=1}^{n} \frac{\partial\psi^{(k)}}{\partial\varepsilon_{ij}^{v(k)}}(\dot{\varepsilon}_{ij} - \dot{\varepsilon}_{ij}^{v(k)}) + \sigma_{ij}\dot{\varepsilon}_{ij} \geq 0$$

Due to (22.98) and (22.99), this expression takes the form

$$\sum_{k=1}^{n} \sigma_{ij}^{(k)} \dot{\varepsilon}_{ij}^{v(k)} \geq 0 \tag{22.100}$$

Take the potential function Φ according to

$$\Phi = \sum_{k=1}^{n} \Phi^{(k)} \quad \text{where} \quad \Phi^{(k)} = \frac{1}{2}\sigma_{ij}^{(k)} B_{ijkl}^{(k)} \sigma_{kl}^{(k)}$$

where the constant tensors $B_{ijkl}^{(k)}$ are assumed to be positive definite. From (21.51), the evolution equations then become

$$\dot{\varepsilon}_{ij}^{v(k)} = \frac{\partial \Phi^{(k)}}{\partial \sigma_{ij}^{(k)}} = B_{ijkl}^{(k)} \sigma_{kl}^{(k)} \tag{22.101}$$

which evidently fulfill the dissipation inequality (22.100).

To be specific, take Helmholtz' free energy according to

$$\rho\psi = A(\theta) + \sum_{k=1}^{n} \frac{1}{2}(\varepsilon_{ij} - \varepsilon_{ij}^{v(k)}) D_{ijkl}^{(k)} (\varepsilon_{kl} - \varepsilon_{kl}^{v(k)})$$

From (22.99) follows that

$$\sigma_{ij} = \sum_{k=1}^{n} D_{ijkl}^{(k)} (\varepsilon_{kl} - \varepsilon_{kl}^{v(k)}) \tag{22.102}$$

and (22.98) provides

$$\sigma_{ij}^{(k)} = D_{ijkl}^{(k)} (\varepsilon_{kl} - \varepsilon_{kl}^{v(k)}) \tag{22.103}$$

A comparison of (22.101)-(22.103) with the expressions following (14.14) shows that we have achieved the three-dimensional generalized Maxwell model. It is also noted that the three-dimensional formulation of the standard model illustrated in Fig. 14.21 is obtained by keeping only two terms in the summation (i.e. $n = 2$).

Kelvin models

The considerations above demonstrate that Maxwell and generalized Maxwell viscoelasticity fit into the usual thermodynamic framework without any *viscous stresses*. However, when Kelvin viscoelasticity is addressed, then the viscous stresses mentioned in (21.37) enter the formulation. Indeed, formulation (14.9) for the one-dimensional Kelvin model already suggests that the total stresses consist of the elastic stresses and some viscous stresses. We have also seen in (14.5) that these viscous stresses depend on some strain rates.

Let us begin with the simplest possible Kelvin formulation and take Helmholtz' free energy in the form

$$\psi = \psi(\theta, \varepsilon_{ij}) \tag{22.104}$$

According to (21.37) the stresses are now given by

$$\sigma_{ij} = \rho \frac{\partial \psi}{\partial \varepsilon_{ij}} + \sigma_{ij}^{v}(\dot{\varepsilon}_{kl}) \tag{22.105}$$

where the viscous stresses depend on the strain rate. Again we have $s = -\partial\psi/\partial\theta$ and insertion of (22.104) and (22.105) into the dissipation inequality (22.91) result in

$$\boxed{\gamma_{mech} = \sigma_{ij}^{v}\dot{\varepsilon}_{ij} \geq 0} \tag{22.106}$$

Take the dissipation function (pseudo-dissipation function) as

$$\Phi = \frac{1}{2}\dot{\varepsilon}_{ij} A_{ijkl} \dot{\varepsilon}_{kl} \tag{22.107}$$

where the constant tensor A_{ijkl} is assumed to be positive definite. With the evolution law

$$\sigma_{ij}^{v} = \frac{\partial \Phi}{\partial \dot{\varepsilon}_{ij}} = A_{ijkl}\dot{\varepsilon}_{kl} \tag{22.108}$$

it appears immediately that the dissipation inequality is fulfilled.

To be specific, choose the free energy as

$$\rho\psi(\theta, \varepsilon_{ij}) = A(\theta) + \frac{1}{2}\varepsilon_{ij} D_{ijkl} \varepsilon_{kl} \tag{22.109}$$

From (22.105) and (22.108), we then obtain

$$\sigma_{ij} = D_{ijkl}\varepsilon_{kl} + A_{ijkl}\dot{\varepsilon}_{kl} \tag{22.110}$$

This result coincides with the findings given by (14.24)-(14.26). In conclusion

$$\boxed{\begin{array}{l} \textit{Kelvin viscoelasticity} \\ \psi = \psi(\theta, \varepsilon_{ij}) \\ \sigma_{ij} = \rho \dfrac{\partial \psi}{\partial \varepsilon_{ij}} + \sigma_{ij}^{v}; \qquad \sigma_{ij}^{v} = A_{ijkl}\dot{\varepsilon}_{kl} \\ \textit{where } A_{ijkl} \textit{ is positive definite} \end{array}} \tag{22.111}$$

In general, this Kelvin model is all too simple to be able to simulate viscoelastic materials in an accurate fashion. However, with the result (22.111) it is easy to derive the formulation of generalized Kelvin viscoelasticity introduced in relation to (14.15). Let the free energy be given by

$$\psi = \sum_{k=0}^{n} \psi^{(k)} \quad \text{where} \quad \psi^{(k)} = \psi^{(k)}(\theta, \varepsilon_{ij}^{(k)}) \tag{22.112}$$

Insertion into the dissipation inequality (22.91) and noting that $s = -\partial\psi/\partial\theta$, we obtain

$$-\rho \sum_{k=0}^{n} \frac{\partial \psi^{(k)}}{\partial \varepsilon_{ij}^{(k)}} \dot{\varepsilon}_{ij}^{(k)} + \sigma_{ij}\dot{\varepsilon}_{ij} \geq 0 \tag{22.113}$$

Let the quantities $\varepsilon_{ij}^{(k)}$ fulfill

$$\varepsilon_{ij} = \sum_{k=0}^{n} \varepsilon_{ij}^{(k)} \tag{22.114}$$

Inserting this expression in (22.113), and take

$$\sigma_{ij} = \rho \frac{\partial \psi^{(k)}}{\partial \varepsilon_{ij}^{(k)}} + \sigma_{ij}^{v(k)} \quad \text{where} \quad \sigma_{ij}^{v(0)} = 0 \tag{22.115}$$

Then the dissipation inequality (22.113) becomes

$$\sum_{k=1}^{n} \sigma_{ij}^{v(k)} \dot{\varepsilon}_{ij}^{(k)} \geq 0 \tag{22.116}$$

Take the dissipation function as

$$\Phi = \sum_{k=1}^{n} \Phi^{(k)} \quad \text{where} \quad \Phi^{(k)} = \frac{1}{2} \dot{\varepsilon}_{ij}^{(k)} A_{ijkl}^{(k)} \dot{\varepsilon}_{kl}^{(k)} \tag{22.117}$$

where all the constant tensors $A_{ijkl}^{(k)}$ are assumed to be positive definite. With the following evolution laws

$$\sigma_{ij}^{v(k)} = \frac{\partial \Phi^{(k)}}{\partial \dot{\varepsilon}_{ij}^{(k)}} = A_{ijkl}^{(k)} \dot{\varepsilon}_{kl}^{(k)} \quad k = 1, 2, \ldots, n \tag{22.118}$$

the dissipation inequality becomes fulfilled.

To be specific, Helmholtz' free energy is now taken as

$$\rho\psi(\theta, \varepsilon_{ij}^{(k)}) = A(\theta) + \sum_{k=0}^{n} \frac{1}{2} \varepsilon_{ij}^{(k)} D_{ijkl}^{(k)} \varepsilon_{kl}^{(k)} \tag{22.119}$$

and (22.115) then provides

$$\sigma_{ij} = D_{ijkl}^{(k)} \varepsilon_{kl}^{(k)} + \sigma_{ij}^{v(k)} \quad \text{where} \quad \sigma_{ij}^{v(0)} = 0 \tag{22.120}$$

Otherwise, the viscous stresses are given by (22.118). A comparison of (22.114), (22.118) and (22.120) with the expressions following (14.15) shows that we have now achieved the three-dimensional generalized Kelvin model.

23 THERMO-PLASTICITY

If the material is loaded so that not only plastic strains develop, but also the temperature is changed, then *thermo-plasticity* is encountered. This phenomenon was first studied in detail by Prager (1958) for rigid hardening plasticity and later extended to elasto-plasticity by Boley and Weiner (1960) and Naghdi (1960). Attempts to formulate thermo-plasticity within a thermodynamic framework were initiated by Dillon (1963), in the review paper by Perzyna and Sawczuk (1973) and by Raniecki and Sawczuk (1975); later efforts are reported, for instance, by Lehmann (1984), Gnoneim (1990), Allen (1991), Simo and Miehe (1992), Wu and Glockner (1996) and by Celentano *et al.* (1996). The latter papers use an approach somewhat similar to the one we will adopt.

The most evident implication of a change of temperature is that the total strains now consist of the elastic, plastic and thermal strains, i.e.

$$\varepsilon_{ij} = \varepsilon_{ij}^{e} + \varepsilon_{ij}^{p} + \varepsilon_{ij}^{o}$$

where the elastic strains, per definition, are defined by Hooke's law, i.e. $\varepsilon_{ij}^{e} = C_{ijkl}\sigma_{kl}$. However, change of temperature also influences the material parameters and it is worthwhile to evaluate this effect in some detail.

We also emphasize that plastic strains – or more generally, dissipative mechanisms – will result in a heat generation and thereby thermal strains; this phenomenon will also be considered here.

23.1 Change of material parameters with temperature

It is of interest to obtain knowledge about the variation of different material parameters when the temperature changes. The intention of this review is not to provide a detailed catalogue of the temperature dependence of different material parameters for various materials, but simply to provide typical illustrations of the effect of temperature. The reader is referred, for instance, to review contributions by Browne and Blundell (1972), Eibl *et al.* (1974), Schneider *et al.* (1981), Schneider (1988) and Vecchi (1998).

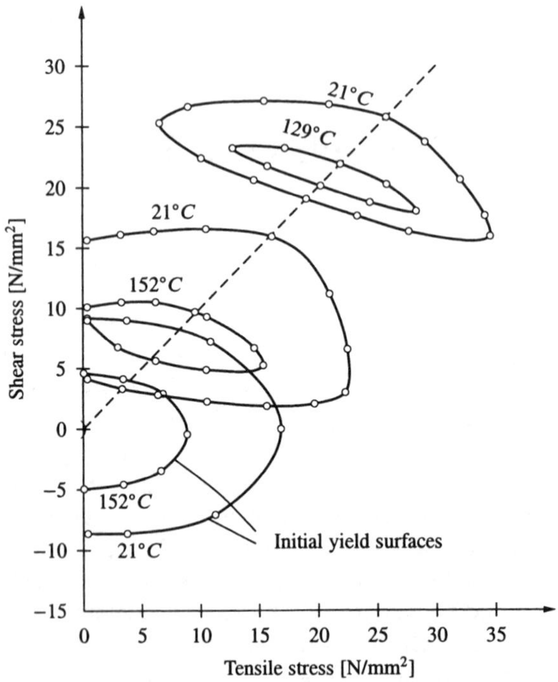

Figure 23.1: Effect of temperature on the yield surfaces; pure aluminum tested in combined torsion and tension by Phillips and Tang (1972).

The most characteristic feature of the temperature influence is that the yield stress changes with temperature; in practice it decreases with increasing temperature. This effect is evident from, for instance, the experimental results on pure aluminum of Phillips and Tang (1972) shown in Fig. 23.1 where the contraction of the yield surface with increasing temperature is apparent; further investigations are reported by Phillips (1974). For uniaxial stress conditions, the softening effect due to the temperature change shown in Fig. 23.1 is illustrated in Fig. 23.2 for aluminum 2024-T4 and carbon steel.

The elastic parameters are also influenced by the temperature; Fig. 23.3a) shows the decrease of the E-modulus with increasing temperature for carbon steel and pure aluminum and Fig. 23.3b) for quartzitic concrete, where the data are normalized by E_0, i.e. the E-modulus at 21^oC.

In Fig. 23.3a) the experimental data are interpolated linearly by

$$E(\theta) = E_0 + (\theta - \theta_0)E_1 \tag{23.1}$$

where $\theta_0 = 294K = 21^oC$ and E_0 is Young's modulus at this reference temper-

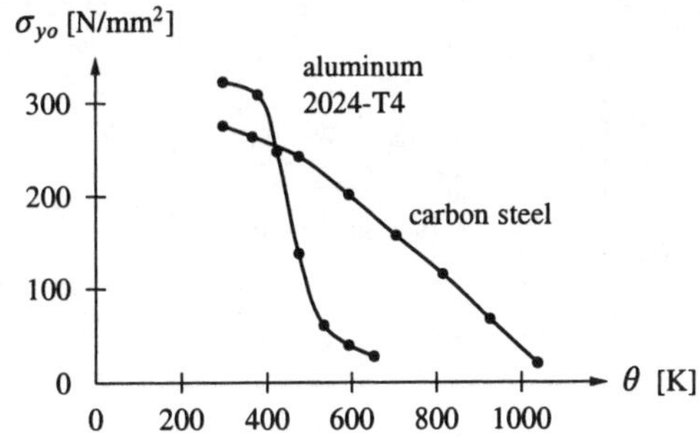

Figure 23.2: Effect of temperature on initial yield stress; experimental data of Eshbach and Souders (1975).

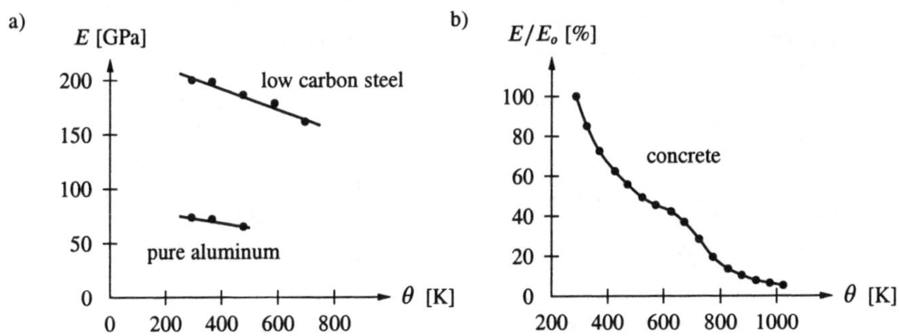

Figure 23.3: Effect of temperature on E-modulus; a) linear approximations and experimental data of Eshbach and Souders (1975), b) for quartzitic concrete, Schneider *et al.* (1981).

ature; the coefficients E_0 and E_1 are given in Table 23.1.

	E_0[GPa]	E_1[GPa/K]
carbon steel	202	-0.096
pure aluminum	73	-0.043

Table 23.1: Coefficients for E-modulus in (23.1).

The fact that mechanical parameters like σ_{yo} and E in general decrease with increasing temperature is often referred to in the literature as the phenomenon of *thermal softening*.

The influence of temperature on Poisson's ratio ν is shown in Fig. 23.4 and

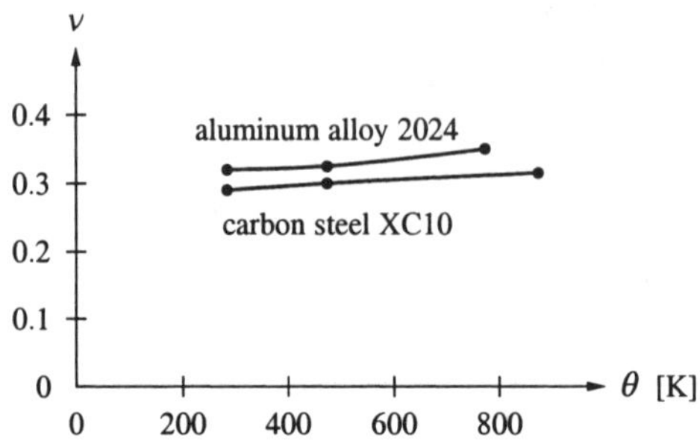

Figure 23.4: Effect of temperature on Poisson's ratio ν; experimental data reported by Metals Handbook (1978)

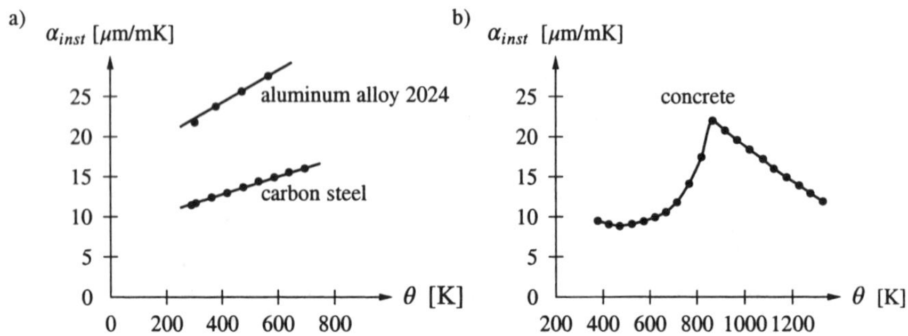

Figure 23.5: Effect of temperature on the instantaneous thermal expansion coefficient α_{inst}; a) linear approximations and experimental data for aluminum alloy and carbon steel by Metals Handbook (1978), b) for quartzitic concrete; Schneider *et al.* (1981).

it appears that the effect is very modest. Noting that the exact magnitude of ν is only of secondary importance for the response, it is concluded that in most applications ν can be taken as constant.

The temperature effect on the instantaneous thermal expansion coefficient α_{inst} (defined as $\dot{\varepsilon}^0 = \alpha_{inst}\dot{\theta}$) is shown in Fig. 23.5a) for carbon steel and aluminum alloy 2024. It appears that the experimental data are in close agreement with the linear approximation given by

$$\alpha_{inst}(\theta) = \alpha_0 + (\theta - \theta_0)2\alpha_1 \tag{23.2}$$

where $\theta_0 = 294K = 21^oC$ and the factor 2 ahead of α_1 is introduced for convenience, see (23.3). The coefficients α_0 and α_1 are given in Table 23.2.

	α_0 [10^{-6}/K]	α_1 [10^{-6}/K^2]
carbon steel	11.6	0.006
2024 al.alloy	22.1	0.010

Table 23.2: Coefficients for thermal expansion coefficient in (23.2) and (23.3).

The usual thermal expansion coefficient α is defined by $\varepsilon^0 = (\theta - \theta_0)\alpha$, i.e. $\dot{\varepsilon}^0 = \frac{d}{d\theta}[(\theta - \theta_0)\alpha]\dot{\theta}$ and it appears that $\alpha_{inst} = \frac{d}{d\theta}[(\theta - \theta_0)\alpha]$; from (23.2) it then follows that

$$\alpha(\theta) = \alpha_0 + (\theta - \theta_0)\alpha_1 \tag{23.3}$$

For quartzitic concrete, the variation of the thermal expansion coefficient is illustrated in Fig. 23.5b).

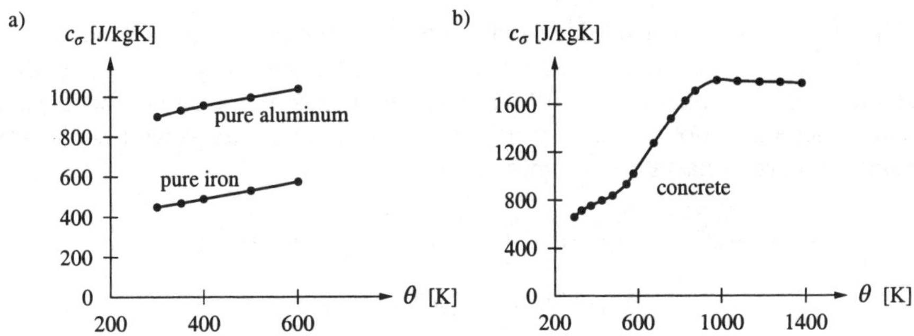

Figure 23.6: Effect of temperature on specific heat capacity; a) aluminum and iron, CRC (1995), b) quartzitic concrete, Schneider *et al.* (1981).

For the specific heat capacity c_σ, the temperature dependence for pure iron and pure aluminum is shown in Fig. 23.6a) (recall from (21.83) that $c_\varepsilon \approx c_\sigma$ holds with close accuracy) and for quartzitic concrete in Fig. 23.6b).

The variation of the thermal conductivity k with temperature is illustrated in Fig. 23.7a) for carbon steel and pure aluminum and in Fig. 23.7b) for granitic concrete.

Finally, for the mass density ρ, Perry's Chemical Engineering Handbook (1984) indicates that for Al, Cu and Fe the mass density only decreases by about 4% when the temperature is raised from room temperature to melting point. The mass density can therefore be considered in most applications as constant.

It appears that of all material parameters, it is the initial yield stress that exhibits the greatest temperature dependence and that the effect of temperature on Poisson's ratio can be ignored for most cases of practical interest.

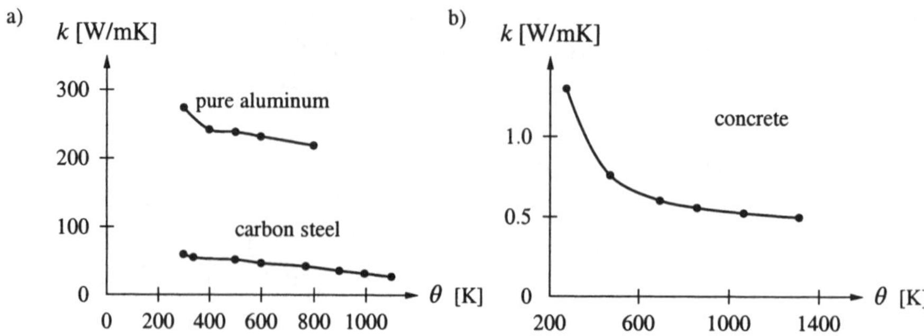

Figure 23.7: Effect of temperature on thermal conductivity k; a) carbon steel, Metals Handbook (1978) and pure aluminum, Perry's Chemical Engineering Handbook (1984), b) granitic concrete, Betonghandbok (1982).

23.2 Equations of thermo-plasticity

With the discussion above of the influence of temperature on the various material parameters, it is timely to establish a form of Helmholtz' free energy ψ that results in thermo-plasticity. If all material parameters are taken as constants, a natural generalization of (22.2) valid for isothermal plasticity as well as (21.84) valid for thermo-elasticity becomes

$$\begin{aligned}\rho\psi(\theta, \varepsilon_{ij}-\varepsilon_{ij}^{p}, \kappa_\alpha) =& \rho c_\varepsilon(\theta - \theta \ln \frac{\theta}{\theta^*}) + \frac{1}{2}(\varepsilon_{ij} - \varepsilon_{ij}^{p}) D_{ijkl}(\varepsilon_{kl} - \varepsilon_{kl}^{p}) \\ &- \beta_{ij}(\varepsilon_{ij} - \varepsilon_{ij}^{p})(\theta - \theta_0) + \rho\psi^{p}(\kappa_\alpha)\end{aligned} \tag{23.4}$$

where θ^* is an arbitrary constant temperature and θ_0 is the reference temperature at which no thermal strains exist. In the expression above, the tensor β_{ij} is related to the thermal expansion tensor α_{ij} by (21.19)

$$\beta_{ij} = D_{ijkl}\alpha_{kl}$$

It is easily verified that $\theta\partial^2\psi/\partial\theta^2 = -c_\varepsilon$ in accordance with definition (21.78); moreover, from (21.37) and using the expression above we obtain

$$\sigma_{ij} = D_{ijkl}(\varepsilon_{kl} - \varepsilon_{kl}^{p} - \varepsilon_{kl}^{o}) \quad \text{where} \quad \varepsilon_{ij}^{o} = \alpha_{ij}(\theta - \theta_0)$$

that is Hooke's law. Finally, the conjugated thermodynamic forces become

$$K_\alpha = \rho\frac{\partial\psi}{\partial\kappa_\alpha} = \rho\frac{\partial\psi^{p}(\kappa_\beta)}{\partial\kappa_\alpha}$$

These results were based on the assumption that D_{ijkl}, β_{ij} (and thereby α_{ij}) and c_ε are constants; moreover, the forces K_α depend only on the internal variables κ_α. In view of the previous discussion which showed that these quantities in general depend on the temperature, it is natural to seek a proper generalization of (23.4).

Let us instead of (23.4) take Helmholtz' free energy in the format

$$\begin{aligned}\rho\psi(\theta, \varepsilon_{ij} - \varepsilon^p_{ij}, \kappa_\alpha) =& \rho h(\theta) + \frac{1}{2}(\varepsilon_{ij} - \varepsilon^p_{ij})D_{ijkl}(\theta)(\varepsilon_{kl} - \varepsilon^p_{kl}) \\ &- \beta_{ij}(\theta)(\varepsilon_{ij} - \varepsilon^p_{ij})(\theta - \theta_0) + \rho\psi^p(\theta, \kappa_\alpha)\end{aligned} \tag{23.5}$$

where $h(\theta)$, $D_{ijkl}(\theta)$, $\beta_{ij}(\theta)$ and $\psi^p(\theta, \kappa_\alpha)$ depend on the temperature. We then obtain Hooke's law

$$\sigma_{ij} = \rho\frac{\partial\psi}{\partial\varepsilon_{ij}} = D_{ijkl}(\theta)(\varepsilon_{kl} - \varepsilon^p_{kl} - \varepsilon^o_{kl}) \quad \text{where} \quad \varepsilon^o_{ij} = \alpha_{ij}(\theta)(\theta - \theta_0) \tag{23.6}$$

and

$$K_\alpha = \rho\frac{\partial\psi^p(\theta, \kappa_\beta)}{\partial\kappa_\alpha} \tag{23.7}$$

Moreover, assuming the mass density to be independent of the temperature, we find

$$\boxed{c_\varepsilon = -\theta\,\frac{\partial^2\psi}{\partial\theta^2} = -\theta\,\frac{d^2h(\theta)}{d\theta^2} + \tilde{c}_\varepsilon} \tag{23.8}$$

where

$$\begin{aligned}\tilde{c}_\varepsilon =& -\frac{\theta}{2\rho}(\varepsilon_{ij} - \varepsilon^p_{ij})\frac{d^2D_{ijkl}}{d\theta^2}(\varepsilon_{kl} - \varepsilon^p_{kl}) \\ &+ \frac{\theta}{\rho}(\varepsilon_{ij} - \varepsilon^p_{ij})\frac{d^2}{d\theta^2}[\beta_{ij}(\theta - \theta_0)] - \theta\frac{\partial^2\psi^p}{\partial\theta^2}\end{aligned} \tag{23.9}$$

We will later show that the contribution of $\tilde{c}_\varepsilon$ is slight, i.e. $\tilde{c}_\varepsilon/c_\varepsilon \approx 0$.

Let us next consider the dissipation inequality and the evolution laws. Irrespective of whether we have isothermal plasticity or thermo-plasticity, the mechanical dissipation inequality is given by (21.37)

$$\gamma_{mech} \equiv \sigma_{ij}\dot{\varepsilon}^p_{ij} - K_\alpha\dot{\kappa}_\alpha \geq 0 \tag{23.10}$$

If we accept the postulate of maximum dissipation as a means to fulfill this inequality then σ_{ij} and K_α are the natural variables in the yield function, i.e. $f = f(\sigma_{ij}, K_\alpha)$. However, we know that the temperature influences both the conjugated forces, see (23.7), and the initial yield stress σ_{yo}, cf. Figs. 23.1 and 23.2. This is accomplished by writing

$$\boxed{f = f(\sigma_{ij}, K_\alpha, \theta) \leq 0} \tag{23.11}$$

For an isotropic hardening von Mises model - for instance - this expression reads

$$f = \sigma_{eff} - \sigma_{yo}(\theta) - K \leq 0$$

where we may have $K = K(\theta, \kappa)$ but since $f = f(\sigma_{ij}, K, \theta)$ it follows that $\partial f/\partial\theta = -d\sigma_{yo}/d\theta$.

To fulfill the dissipation inequality (23.10) we may adopt the nonassociated format and introduce a potential function $g = g(\sigma_{ij}, K_\alpha, \theta)$ and with (22.6) we obtain

$$\dot{\varepsilon}_{ij}^p = \dot{\lambda}\frac{\partial g}{\partial \sigma_{ij}}; \quad \dot{\kappa}_\alpha = -\dot{\lambda}\frac{\partial g}{\partial K_\alpha} \qquad (23.12)$$

with the restrictions $\dot{\lambda} \geq 0$ $f \leq 0$ and $\dot{\lambda} f = 0$; moreover, associated thermo-plasticity is retrieved if $g = f$.

As for isothermal plasticity, differentiation of $\dot{\lambda} f = 0$ leads to $\ddot{\lambda} f + \dot{\lambda}\dot{f} = 0$. Development of plasticity requires $f = 0$ and $\dot{\lambda} > 0$ and we are then led to the consistency relation $\dot{f} = 0$ for development of plasticity.

With (23.11), we have

$$\dot{f} = \frac{\partial f}{\partial \sigma_{ij}}\dot{\sigma}_{ij} + \frac{\partial f}{\partial K_\alpha}\dot{K}_\alpha + \frac{\partial f}{\partial \theta}\dot{\theta} \qquad (23.13)$$

From $K_\alpha = \rho\partial\psi^p(\theta, \kappa_\beta)/\partial\kappa_\alpha$ it follows that

$$\dot{K}_\alpha = \rho\frac{\partial^2\psi^p}{\partial\kappa_\alpha\partial\kappa_\beta}\dot{\kappa}_\beta + \rho\frac{\partial^2\psi^p}{\partial\kappa_\alpha\partial\theta}\dot{\theta}$$

which with (23.12b) becomes

$$\dot{K}_\alpha = -\rho\frac{\partial^2\psi^p}{\partial\kappa_\alpha\partial\kappa_\beta}\frac{\partial g}{\partial K_\beta}\dot{\lambda} + \frac{\partial K_\alpha}{\partial\theta}\dot{\theta}$$

Insertion of this expression into (23.13) provides

$$\dot{f} = \frac{\partial f}{\partial \sigma_{ij}}\dot{\sigma}_{ij} - H\dot{\lambda} + S\dot{\theta} \qquad (23.14)$$

where the generalized plastic modulus H is defined in the usual way as

$$H = \frac{\partial f}{\partial K_\alpha}\rho\frac{\partial^2\psi^p}{\partial\kappa_\alpha\partial\kappa_\beta}\frac{\partial g}{\partial K_\beta}$$

cf. (22.9); as previously $H > 0$, $H = 0$, $H < 0$ corresponds to hardening, ideal and softening thermo-plasticity respectively. Moreover, in (23.14) the quantity S is defined as

$$S = \frac{\partial f}{\partial\theta} + \frac{\partial f}{\partial K_\alpha}\frac{\partial K_\alpha}{\partial\theta}$$

and it is observed that $S = df/d\theta$. It appears that S depends on how much the yield function parameters and the forces K_α change with temperature. Therefore, the *sensitivity parameter* S expresses how the yield function changes with temperature. To evaluate what the sign of S means, suppose that the state $(\sigma_{ij}, K_\alpha, \theta)$ initially corresponds to a point located inside the yield surface, i.e. $f < 0$. Consider then a situation where $\dot{\sigma}_{ij} = 0$ and as also $\dot{\lambda} = 0$ hold, (23.14) gives $\dot{f} = S\dot{\theta}$. If $S > 0$ then $\dot{f} > 0$ for $\dot{\theta} > 0$ and for a sufficiently elevated temperature we will be in a situation where $f = 0$ is achieved. It is concluded that $S > 0$ corresponds to the situation where the yield surface contracts with increasing temperature. Evidently $S = 0$ implies that the yield function is independent of the temperature and, finally, $S < 0$ must then be the situation where the yield surface expands with increasing temperature. With reference to Fig. 23.1, the normal situation is that the yield surface contracts with increasing temperature, i.e. $S > 0$ holds in most cases. In conclusion

For change of temperature

$$S \begin{cases} >0 \;\Rightarrow\; \textit{yield surface contracts} \\ =0 \;\Rightarrow\; \textit{yield surface remains constant} \\ <0 \;\Rightarrow\; \textit{yield surface expands} \end{cases} \tag{23.15}$$

where

$$S = \frac{\partial f}{\partial \theta} + \frac{\partial f}{\partial K_\alpha}\frac{\partial K_\alpha}{\partial \theta}$$

Let us now derive the incremental response for a thermo-plastic material as well as the loading/unloading criteria. Differentiation of Hooke's law (23.6) provides

$$\dot{\sigma}_{ij} = D_{ijkl}(\dot{\varepsilon}_{kl} - \dot{\varepsilon}^p_{kl}) - P_{ij}\dot{\theta} \tag{23.16}$$

where

$$P_{ij} = \beta_{ij} + \frac{d\beta_{ij}}{d\theta}(\theta - \theta_0) - \frac{dD_{ijkl}}{d\theta}(\varepsilon_{kl} - \varepsilon^p_{kl})$$

Since both β_{ij} and D_{ijkl} are symmetric in i and j, it follows that P_{ij} is a symmetric tensor. Use of (23.12a) in (23.16) implies

$$\dot{\sigma}_{ij} = D_{ijkl}\dot{\varepsilon}_{kl} - D_{ijkl}\frac{\partial g}{\partial \sigma_{kl}}\dot{\lambda} - P_{ij}\dot{\theta} \tag{23.17}$$

Insertion of this expression into (23.14) and noting that $\dot{f} = 0$ holds then provide

$$\frac{\partial f}{\partial \sigma_{ij}}D_{ijkl}\dot{\varepsilon}_{kl} - A\dot{\lambda} + (S - \frac{\partial f}{\partial \sigma_{ij}}P_{ij})\dot{\theta} = 0 \tag{23.18}$$

where the quantity A is defined in the usual way as

$$A = H + \frac{\partial f}{\partial \sigma_{ij}} D_{ijkl} \frac{\partial g}{\partial \sigma_{kl}} > 0$$

cf. (22.10) and where, again, it is required that $A > 0$. From (23.18) then follows

$$\dot{\lambda} = \frac{1}{A}[\frac{\partial f}{\partial \sigma_{ij}} D_{ijkl}\dot{\varepsilon}_{kl} + (S - \frac{\partial f}{\partial \sigma_{ij}} P_{ij})\dot{\theta}] \tag{23.19}$$

To obtain the incremental response, the expression above is inserted into (23.17) to obtain

Incremental constitutive relation

$$\dot{\sigma}_{ij} = D^{ep}_{ijkl}\dot{\varepsilon}_{kl} - \beta^{ep}_{ij}\dot{\theta}$$

where

$$\begin{aligned}
D^{ep}_{ijkl} &= D_{ijkl} - \frac{1}{A} D_{ijst} \frac{\partial g}{\partial \sigma_{st}} \frac{\partial f}{\partial \sigma_{mn}} D_{mnkl} \\
\beta^{ep}_{ij} &= P_{ij} + \frac{1}{A} D_{ijkl} \frac{\partial g}{\partial \sigma_{kl}} (S - \frac{\partial f}{\partial \sigma_{mn}} P_{mn}) \\
P_{ij} &= \beta_{ij} + \frac{d\beta_{ij}}{d\theta}(\theta - \theta_0) - \frac{d\, D_{ijkl}}{d\theta}(\varepsilon_{kl} - \varepsilon^{p}_{kl}) \\
A &= H + \frac{\partial f}{\partial \sigma_{ij}} D_{ijkl} \frac{\partial g}{\partial \sigma_{kl}} > 0 \\
H &= \frac{\partial f}{\partial K_\alpha} \rho \frac{\partial^2 \psi^p}{\partial \kappa_\alpha \partial \kappa_\beta} \frac{\partial g}{\partial K_\beta}
\end{aligned} \tag{23.20}$$

The similarity in structure between (23.20) and (10.26) is evident. Moreover, it appears that D^{ep}_{ijkl} is the usual elasto-plastic stiffness tensor and that β^{ep}_{ij} is a symmetric tensor.

It is of interest to consider (23.17) for purely thermo-elastic behavior where $\dot{\lambda} = 0$; we then obtain $\dot{\sigma}_{ij} = D_{ijkl}\dot{\varepsilon}_{kl} - P_{ij}\dot{\theta}$. In analogy with (10.36), we then define the *thermo-elastic stress rate* $\dot{\sigma}^{te}_{ij}$ as

$$\boxed{\dot{\sigma}^{te}_{ij} \equiv D_{ijkl}\dot{\varepsilon}_{kl} - P_{ij}\dot{\theta}} \tag{23.21}$$

where the term 'thermo-elastic' refers to the fact that this is the stress rate that would result for a given total strain rate and a given temperature rate provided that the material responds thermo-elastically. With (23.21), (23.19) takes the form

$$\dot{\lambda} = \frac{1}{A}(\frac{\partial f}{\partial \sigma_{ij}} \dot{\sigma}^{te}_{ij} + S\dot{\theta}) \tag{23.22}$$

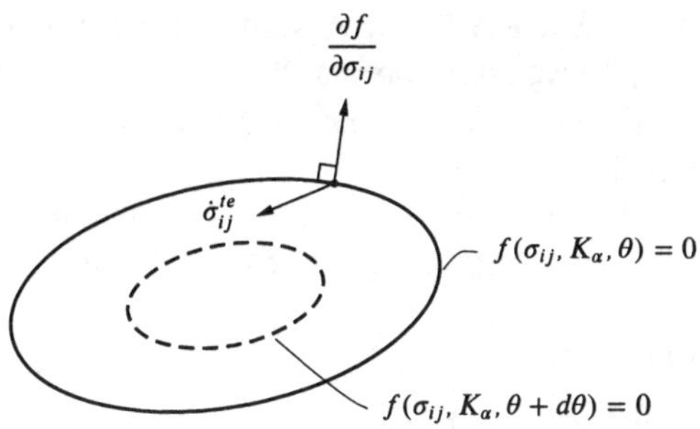

Figure 23.8: Illustration with $S > 0$ (i.e. the yield surface contracts with increasing temperature) where even $\partial f/\partial\sigma_{ij}\ \dot{\sigma}_{ij}^{te} < 0$ may result in thermo-plastic loading for increasing temperature.

To obtain the general loading/unloading criteria, it is first observed that $f < 0$ implying that a thermo-elastic response occurs. Thermo-plasticity requires that $f = 0$ and $\dot{\lambda} \geq 0$. If $\partial f/\partial\sigma_{ij}\ \dot{\sigma}_{ij}^{te} + S\dot{\theta} > 0$, then (23.22) implies $\dot{\lambda} > 0$ and we then have thermo-plastic loading. If $\partial f/\partial\sigma_{ij}\ \dot{\sigma}_{ij}^{te} + S\dot{\theta} < 0$, (23.22) implying $\dot{\lambda} < 0$, which is impossible, and this case must therefore correspond to thermo-elastic unloading. Finally, in the limiting case where $\partial f/\partial\sigma_{ij}\ \dot{\sigma}_{ij}^{te} + S\dot{\theta} = 0$, we have neutral thermo-plasticity loading, which is formally treated as thermo-plasticity even though the incremental response is purely thermo-elastic. These findings are summarized as

$$\begin{array}{lll} f = 0 \ \ and \ \ \dfrac{\partial f}{\partial \sigma_{ij}}\dot{\sigma}_{ij}^{te} + S\dot{\theta} > 0 & \Rightarrow & thermo\text{-}plastic\ loading \\ f = 0 \ \ and \ \ \dfrac{\partial f}{\partial \sigma_{ij}}\dot{\sigma}_{ij}^{te} + S\dot{\theta} = 0 & \Rightarrow & neutral\ loading \\ f = 0 \ \ and \ \ \dfrac{\partial f}{\partial \sigma_{ij}}\dot{\sigma}_{ij}^{te} + S\dot{\theta} < 0 & \Rightarrow & thermo\text{-}elastic\ unloading \end{array}$$

The similarity with the general loading/unloading criteria for isothermal plasticity given by (10.38) is striking. If $S = 0$, i.e. the yield function is unaffected by temperature changes, then $\dot{\lambda} = \frac{1}{A}\frac{\partial f}{\partial\sigma_{ij}}\dot{\sigma}_{ij}^{te}$ whereas for isothermal plasticity we have $\dot{\lambda} = \frac{1}{A}\frac{\partial f}{\partial\sigma_{ij}}\dot{\sigma}_{ij}^{e}$ where the elastic stress rate $\dot{\sigma}_{ij}^{e}$ is defined by $\dot{\sigma}_{ij}^{e} \equiv D_{ijkl}\dot{\varepsilon}_{kl}$. If $S > 0$, which is normally the case, the yield surface contracts with increasing temperature and even if $\partial f/\partial\sigma_{ij}\ \dot{\sigma}_{ij}^{te} < 0$ we may still have a situation where $\partial f/\partial\sigma_{ij}\ \dot{\sigma}_{ij}^{te} + S\dot{\theta} > 0$ for increasing temperature and thermo-plastic loading then occurs; this situation is illustrated in Fig. 23.8.

As a further illustration, consider a situation where thermo-plastic loading

$\dot{f} = 0$ holds. Suppose that the stresses are held constant, i.e. $\dot{\sigma}_{ij} = 0$; then (23.14) gives $-H\dot{\lambda} + S\dot{\theta} = 0$. Assuming hardening behavior, $H > 0$, we then obtain $\dot{\lambda} = \frac{S}{H}\dot{\theta}$. Again it appears that $S > 0$ (contraction of the yield surface with increasing temperature) implies $\dot{\lambda} > 0$ and thereby development of plastic strains if $\dot{\theta} > 0$.

Let us now turn our interest to the energy equation. From (21.86) and $K_\alpha = \rho\partial\psi/\partial\kappa_\alpha$ and with $\rho\psi$ given by (23.5) we obtain

Heat equation

$$\rho c_\varepsilon \dot{\theta} = r - q_{i,i} + Q$$

where

$$Q = \gamma_{mech} - \theta P_{ij}(\dot{\varepsilon}_{ij} - \dot{\varepsilon}^p_{ij}) + \theta\frac{\partial K_\alpha}{\partial\theta}\dot{\kappa}_\alpha \tag{23.23}$$

$$\gamma_{mech} = \sigma_{ij}\dot{\varepsilon}^p_{ij} - K_\alpha\dot{\kappa}_\alpha$$

where γ_{mech} is the mechanical dissipation, P_{ij} is defined in (23.20) and the heat flux vector q_i is related to the temperature gradient $\theta_{,i}$ by the constitutive equation in terms of Fourier's law, i.e.

$$q_i = -k_{ij}\theta_{,j} \tag{23.24}$$

Insertion of the evolution equations for $\dot{\varepsilon}^p_{ij}$ and $\dot{\kappa}_\alpha$ given by (23.12) and (23.19) into (23.23) leads, after a little algebra, to

Heat equation for thermo-plasticity

$$Q^m_{kl}\dot{\varepsilon}_{kl} + [\rho c_\varepsilon + L(\frac{\partial f}{\partial\sigma_{mn}}P_{mn} - S)]\dot{\theta} = r - q_{i,i}$$

where

$$Q^m_{kl} = \theta P_{kl} - L\frac{\partial f}{\partial\sigma_{mn}}D_{mnkl} \tag{23.25}$$

$$L = \frac{1}{A}[(\theta P_{ij} + \sigma_{ij})\frac{\partial g}{\partial\sigma_{ij}} + (K_\alpha - \theta\frac{\partial K_\alpha}{\partial\theta})\frac{\partial g}{\partial K_\alpha}]$$

It appears that the symmetric tensor Q^m_{kl} expresses the mechanical coupling term that produces heat due to strain rates.

23.2.1 Specific heat

It appears that due to the contribution $\tilde{c}_\varepsilon$ in (23.8), c_ε now not only depends on the temperature, but $c_\varepsilon = c_\varepsilon(\theta, \varepsilon_{ij} - \varepsilon^p_{ij}, \kappa_\alpha)$. Intuitively, one may expect that the contribution $\tilde{c}_\varepsilon$ is slight and, accordingly, it is natural to seek the most general circumstances for which $\tilde{c}_\varepsilon = 0$. From (23.9) follows that $\tilde{c}_\varepsilon = 0$, if and only if

$$\begin{aligned} D_{ijkl}(\theta) &= D^0_{ijkl} + (\theta - \theta_0)D^1_{ijkl} \\ \beta_{ij} &= D_{ijkl}\alpha_{kl} = \text{constant} \\ \psi^p(\theta, \kappa_\alpha) &= \psi^p_o(\kappa_\alpha) + (\theta - \theta_0)\tilde{\psi}^p(\kappa_\alpha) \end{aligned} \tag{23.26}$$

where D^0_{ijkl} and D^1_{ijkl} are constant quantities. In that case, it is allowed that c_ε is an arbitrary function of the temperature alone. For an isotropic material in which Poisson's ratio is taken as constant whereas the E-modulus may vary with the temperature, we can, in view of (4.89), write D_{ijkl} as $b(\theta)D^0_{ijkl}$ where D^0_{ijkl} is constant and where $b(\theta) = E(\theta)/E_0$. According to (23.26a), we then have

$$D_{ijkl}(\theta) = b(\theta)D^0_{ijkl} = [1 + (\theta - \theta_0)b_1]D^0_{ijkl} \tag{23.27}$$

where b_1 is a constant, which is in agreement with (23.1). With the result above, (23.26b) gives

$$\alpha_{ij} = \frac{1}{b(\theta)}C^0_{ijkl}\beta_{kl} = \frac{\alpha^0_{ij}}{b(\theta)} \tag{23.28}$$

Since isotropic materials are considered (where $\alpha_{ij} = \alpha\delta_{ij}$), the expression above implies $E\alpha/(E_0\alpha_0) = 1$. This relation exhibits the right tendency: as E decreases, α increases, cf. Figs. 23.3a) and 23.5a). For carbon iron, (23.1) and (23.3) as well as Tables 23.1 and 23.2 imply that $E\alpha/(E_0\alpha_0)$ varies from 1 to 0.98 in the range $\theta = 294 - 700K$ and for aluminum $E\alpha/(E_0\alpha_0)$ varies from 1 to 0.93 in the range $\theta = 294 - 600K$. However, in general, the E-modulus must be zero when the material is in its liquid phase and $E\alpha/(E_0\alpha_0) = 1$ would then imply that the thermal expansion coefficient should be infinitely large. This absurdity emphasizes that (23.27) and (23.28) are far from being general.

In general, it is therefore concluded that the contribution $\tilde{c}_\varepsilon$ present in (23.8) is not zero and it is of interest to evaluate its magnitude; isotropy is assumed. For simplicity take ψ^p as a linear function of the temperature so that its contribution to $\tilde{c}_\varepsilon$ disappears. Moreover, based on Figs. 23.3a) and 23.4 it is a fair approximation to assume the variation of E to be linear and ν to be constant. Since we then have $d^2D_{ijkl}/d\theta^2 = 0$, (23.9) reduces to

$$\tilde{c}_\varepsilon \approx \frac{\theta}{\rho}(\varepsilon_{ij} - \varepsilon^p_{ij})\frac{d^2}{d\theta^2}[D_{ijkl}\alpha_{kl}(\theta - \theta_0)] = \frac{\theta}{\rho}\frac{\varepsilon_{ii} - \varepsilon^p_{ii}}{1 - 2\nu}\frac{d^2}{d\theta^2}[E\alpha(\theta - \theta_0)]$$

For uniaxial stress conditions, $\varepsilon_{ii} - \varepsilon^p_{ii} = \varepsilon_{ii} = (1-2\nu)\varepsilon^e_{11}$ holds. Finally, with not only E, but also α taken as a linear function of the temperature, cf. Fig. 23.5a), the expression above reduces to

$$\tilde{c}_\varepsilon \approx 2\frac{\theta}{\rho}[\frac{dE}{d\theta}\frac{d\alpha}{d\theta}(\theta - \theta_0) + \frac{dE}{d\theta}\alpha + E\frac{d\alpha}{d\theta}]\varepsilon^e_{11}$$

Consider aluminum, for instance, and take the linear variations of E and α given by (23.1) and (23.3) and Tables 23.1 and 23.2; moreover $\rho \approx 2700$ kg/m^3 and $\varepsilon_{11}^e \approx 10^{-3}$. Then for the temperature in the range $\theta = 294 - 600K$, we obtain at most $\tilde{c}_\varepsilon \approx -0.3J/kgK$ and as c_ε for aluminum is about 900 J/kg$K$, cf. Fig. 23.6a), this gives $\tilde{c}_\varepsilon / c_\varepsilon \approx -3 \cdot 10^{-4}$ and $\tilde{c}_\varepsilon$ indeed turns out to be very small.

23.3 Isotropic hardening von Mises plasticity

As an illustration of thermo-plasticity, we will consider isotropic hardening von Mises plasticity and derive the incremental constitutive relation as well as the corresponding heat equation. For simplicity, only the initial yield stress σ_{yo} is assumed to be temperature dependent and Helmholtz' free energy is given by (23.4). The yield function is then given by

$$f = f(\sigma_{ij}, K, \theta) = (\frac{3}{2} s_{ij} s_{ij})^{1/2} - \sigma_y(\theta, K)$$

where $\sigma_y(\theta, K) = \sigma_{yo}(\theta) + K(\kappa)$. Isotropy is assumed, i.e.

$$D_{ijkl} = 2G[\frac{1}{2}(\delta_{ik}\delta_{jl} + \delta_{il}\delta_{jk}) + \frac{\nu}{1-2\nu}\delta_{ij}\delta_{kl}]\,; \qquad \alpha_{ij} = \alpha\delta_{ij}$$

The evolution laws for associated plasticity provide

$$\dot{\varepsilon}_{ij}^p = \dot{\lambda}\frac{\partial f}{\partial \sigma_{ij}} = \dot{\lambda}\frac{3s_{ij}}{2\sigma_y}\,; \qquad \dot{\kappa} = -\dot{\lambda}\frac{\partial f}{\partial K} = \dot{\lambda}$$

and it follows that $\dot{\varepsilon}_{eff}^p = (\frac{2}{3}\dot{\varepsilon}_{ij}^p\dot{\varepsilon}_{ij}^p)^{1/2} = \dot{\lambda}$. It also follows from (23.20) that

$$A = H + \frac{\partial f}{\partial \sigma_{ij}} D_{ijkl} \frac{\partial f}{\partial \sigma_{kl}} = H + 3G$$

where $H = \rho\partial^2\psi^p(\kappa)/\partial\kappa^2$ is temperature independent. Moreover

$$D_{ijkl}^{ep} = D_{ijkl} - \frac{9G^2}{A}\frac{s_{ij}s_{kl}}{\sigma_y^2} \tag{23.29}$$

holds as usual, cf. (12.16).

The consistency relation $\dot{f} = 0$ is given by (23.14) and we obtain

$$\dot{\sigma}_{eff} - H\dot{\lambda} + S\dot{\theta} = 0$$

At yield, $\sigma_{eff} = \sigma_y$ holds and for isothermal conditions, the usual condition

$$\frac{\partial \sigma_y}{\partial \varepsilon_{eff}^p} = H$$

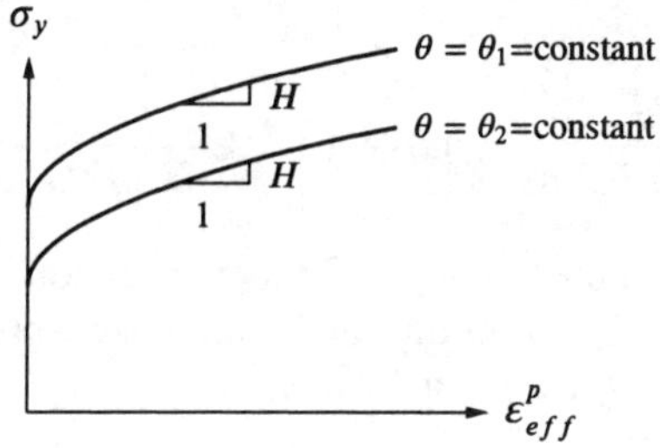

Figure 23.9: Relation between σ_y and ε^p_{eff}; only the initial yield stress σ_{yo} is influenced by the temperature.

holds, cf. (12.12). Since H is temperature independent, the response shown in Fig. 23.9 is obtained. As the initial yield stress σ_{yo} only depends on temperature, the two curves in Fig. 23.9 are identical apart from a shift along the σ_y-axis.

To determine β^{ep}_{ij} present in the incremental relation $\dot{\sigma}_{ij} = D^{ep}_{ijkl}\dot{\varepsilon}_{kl} - \beta^{ep}_{ij}\dot{\theta}$, we first calculate β_{ij} according to

$$\beta_{ij} = D_{ijkl}\alpha_{kl} = \frac{E}{1-2\nu}\alpha\delta_{ij} = 3K\alpha\delta_{ij}$$

Since only the initial yield stress σ_{yo} is assumed to depend on the temperature, (23.20) provides

$$P_{ij} = \beta_{ij} + \frac{d\beta_{ij}}{d\theta}(\theta - \theta_0) - \frac{dD_{ijkl}}{d\theta}(\varepsilon_{kl} - \varepsilon^p_{kl}) = \beta_{ij} = 3K\alpha\delta_{ij}$$

It then follows that

$$\frac{\partial f}{\partial \sigma_{mn}}P_{mn} = 0$$

Moreover

$$S = \frac{\partial f}{\partial \theta} + \frac{\partial f}{\partial K_\alpha}\frac{\partial K_\alpha}{\partial \theta} = \frac{\partial f}{\partial \theta} = -\frac{d\sigma_{y0}}{d\theta}$$

We then obtain from (23.20)

$$\beta^{ep}_{ij} = P_{ij} + \frac{1}{A}D_{ijkl}\frac{\partial f}{\partial \sigma_{kl}}(S - \frac{\partial f}{\partial \sigma_{mn}}P_{mn}) = 3K\alpha\delta_{ij} - \frac{3G}{A}\frac{s_{ij}}{\sigma_y}\frac{d\sigma_{yo}}{d\theta}$$

With this expression, as well as (23.29), the incremental constitutive relation becomes

$$\boxed{\dot{\sigma}_{ij} = (D_{ijkl} - \frac{9G^2}{A}\frac{s_{ij}s_{kl}}{\sigma_y^2})\dot{\varepsilon}_{kl} - (3K\alpha\delta_{ij} - \frac{3G}{A}\frac{s_{ij}}{\sigma_y}\frac{d\sigma_{yo}}{d\theta})\dot{\theta}}$$

in accordance with Ray and Utko (1989).

With these results, it is trivial to identify the heat equation and from (23.25) we obtain

$$(\theta 3K\alpha\delta_{ij} - \frac{3G}{H+3G}\frac{\sigma_{yo}}{\sigma_y}s_{ij})\dot{\varepsilon}_{ij} + (\rho c_\varepsilon + \frac{\sigma_{yo}}{H+3G}\frac{d\sigma_{yo}}{d\theta})\dot{\theta} = r - q_{i,i}$$

For adiabatic loading where $r - q_{i,i} = 0$ this expression may be compared with the Kelvin expression (21.92) for thermo-elasticity. It appears that plastic effects create a significant heat generation.

23.4 Field equations - Finite element formulation

Having evaluated in detail the mechanical constitutive equation and the heat equation, we will now discuss the field equations of thermo-plasticity as well as their finite element formulation.

Considering first the mechanical part of the problem, the equations of motion - i.e. the balance equations - read

$$\boxed{\sigma_{ij,j} + b_i = \rho\ddot{u}_i} \tag{23.30}$$

where b_i is the body force per unit volume and u_i is the displacement vector. The kinematic relation is

$$\boxed{\varepsilon_{ij} = \frac{1}{2}(u_{i,j} + u_{j,i})} \tag{23.31}$$

The second balance equation that needs to be considered is given by the energy equation. From (23.23) we have

$$\boxed{\rho c_\varepsilon\dot{\theta} = r - q_{i,i} + Q} \tag{23.32}$$

where the heat flux vector q_i is related to the temperature gradient $\theta_{,i}$ by the constitutive equation in terms of Fourier's law, i.e.

$$\boxed{q_i = -k_{ij}\theta_{,j}} \tag{23.33}$$

The mechanical field equations consist of (23.30) and (23.31) and the thermal field equations are given by (23.32) and (23.33). It appears that, in general, these field equations are coupled: as the stresses depend on the temperature, we cannot solve the mechanical field equations without knowledge of the temperature and - on the other hand - the thermal field equations cannot be solved without knowledge of the strains and stresses. Instead of (23.32) we could use (23.25) which directly links strain rates and temperature changes. However, (23.25) applies only to thermo-plasticity whereas (23.32) also holds for viscoplasticity and creep if $\dot{\varepsilon}_{ij}^{p}$ is replaced by $\dot{\varepsilon}_{ij}^{vp}$ and $\dot{\varepsilon}_{ij}^{c}$ respectively. Consequently, by adopting (23.32) we obtain a very general approach which not only applies to plasticity, but also to viscoplasticity and creep.

To derive the finite element formulation of the field equations, we begin with the equations of motion given by (23.30); first the corresponding weak form is obtained and then the FE-approximations are introduced. As this task has already been performed in Section 3.8, we will only repeat the final relations. From (16.9) we have

$$\boxed{\begin{aligned} &\boldsymbol{M}\ddot{\boldsymbol{a}} + \boldsymbol{\psi} = \boldsymbol{0} \\ &\textit{where} \\ &\boldsymbol{M} = \int_V \rho \boldsymbol{N}^T \boldsymbol{N}\, dV \\ &\boldsymbol{\psi} = \int_V \boldsymbol{B}^T \boldsymbol{\sigma}\, dV - \boldsymbol{f} \\ &\boldsymbol{f} = \int_S \boldsymbol{N}^T \boldsymbol{t}\, dS + \int_V \boldsymbol{N}^T \boldsymbol{b}\, dV \end{aligned}} \tag{23.34}$$

where $\boldsymbol{M}$ is the *mass matrix*, $\boldsymbol{\psi}$ represents the *out-of-balance forces* and $\boldsymbol{f}$ denotes the *external forces*.

To obtain the weak form of the heat equation (23.32), we proceed in a similar manner as in Section 3.8. Therefore, (23.32) is multiplied by the arbitrary weight function v and integrated over the region V, i.e.

$$\int_V v\rho c_\varepsilon \dot{\theta}\, dV = \int_V vr\, dV - \int_V vq_{i,i}\, dV + \int_V vQ\, dV \tag{23.35}$$

From the divergence theorem of Gauss, it follows that

$$\begin{aligned} \int_V vq_{i,i}\, dV &= \int_V (vq_i)_{,i}\, dV - \int_V v_{,i}q_i\, dV \\ &= \int_S vq_n\, dS - \int_V v_{,i}q_i\, dV \end{aligned}$$

where the flux q_n is defined by $q_n = q_i n_i$; here n_i is the unit vector normal to the boundary S of the body and directed outwards. Use of the expression above in (23.35) gives

$$\int_V v\rho c_\varepsilon \dot{\theta}\, dV = \int_V vr\, dV - \int_S vq_n\, dS + \int_V v_{,i}q_i\, dV + \int_V vQ dV \tag{23.36}$$

To rewrite this expression in matrix form, define the following matrices

$$\boldsymbol{q} = \begin{bmatrix} q_1 \\ q_2 \\ q_3 \end{bmatrix}; \quad \nabla = \begin{bmatrix} \frac{\partial}{\partial x_1} \\ \frac{\partial}{\partial x_2} \\ \frac{\partial}{\partial x_3} \end{bmatrix} \tag{23.37}$$

where ∇ is the *gradient operator*, i.e. $v_{,i} = \nabla v$. Similarly to the finite element formulation of the mechanical part, the temperature within the body is approximated by

$$\theta(x_i, t) = \boldsymbol{N}_\theta(x_i)\boldsymbol{a}_\theta(t)$$

where $\boldsymbol{N}_\theta$ denotes the global shape function matrix and $\boldsymbol{a}_\theta$ is a column matrix that contains all the nodal values of the temperature. It follows that

$$\dot{\theta} = \boldsymbol{N}_\theta \dot{\boldsymbol{a}}_\theta \tag{23.38}$$

and that the temperature gradient $\theta_{,i} = \nabla\theta$ is given by

$$\nabla\theta = \boldsymbol{B}_\theta \boldsymbol{a}_\theta \quad \text{where} \quad \boldsymbol{B}_\theta = \nabla \boldsymbol{N}_\theta$$

In evident notation, Fourier's law (23.33) then reads

$$\boldsymbol{q} = -\boldsymbol{k}\nabla\theta = -\boldsymbol{k}\boldsymbol{B}_\theta \boldsymbol{a}_\theta \tag{23.39}$$

Adopting again Galerkin's method, the arbitrary weight function v is chosen as

$$v = \boldsymbol{N}_\theta \boldsymbol{c}_\theta = \boldsymbol{c}_\theta^T \boldsymbol{N}_\theta^T; \qquad \nabla v = \boldsymbol{B}_\theta \boldsymbol{c}_\theta \tag{23.40}$$

where $\boldsymbol{c}_\theta$ is a constant, but arbitrary column matrix. Insertion of (23.40) into (23.36), where $v_{,i}q_i = (\nabla v)^T \boldsymbol{q}$, and $\boldsymbol{c}_\theta$ is independent of the coordinates lead to

$$\boldsymbol{c}_\theta^T \Big[\int_V \boldsymbol{N}_\theta^T \rho c_\varepsilon \dot{\theta}\, dV - \int_V \boldsymbol{N}_\theta^T r\, dV + \int_S \boldsymbol{N}_\theta^T q_n\, dS - \int_V \boldsymbol{B}_\theta^T \boldsymbol{q}\, dV - \int_V \boldsymbol{N}_\theta^T Q\, dV \Big] = 0$$

and since $\boldsymbol{c}_\theta$ is arbitrary, it is concluded that

$$\int_V \boldsymbol{N}_\theta^T \rho c_\varepsilon \dot{\theta}\, dV - \int_V \boldsymbol{N}_\theta^T r\, dV + \int_S \boldsymbol{N}_\theta^T q_n\, dS - \int_V \boldsymbol{B}_\theta^T \boldsymbol{q}\, dV - \int_V \boldsymbol{N}_\theta^T Q\, dV = \boldsymbol{0}$$

Finally, insertion of (23.38) and (23.39) leads to the following finite element formulation

$$\boxed{\begin{aligned} &\boldsymbol{C}\dot{\boldsymbol{a}}_\theta + \boldsymbol{K}_\theta \boldsymbol{a}_\theta - \boldsymbol{Q} - \boldsymbol{f}_\theta = \boldsymbol{0} \\ &\textit{where} \\ &\boldsymbol{C} = \int_V \boldsymbol{N}_\theta^T \rho c_\varepsilon \boldsymbol{N}_\theta\, dV \\ &\boldsymbol{K}_\theta = \int_V \boldsymbol{B}_\theta^T \boldsymbol{k} \boldsymbol{B}_\theta\, dV \\ &\boldsymbol{Q} = \int_V \boldsymbol{N}_\theta^T Q\, dV \\ &\boldsymbol{f}_\theta = \int_V \boldsymbol{N}_\theta^T r\, dV - \int_S \boldsymbol{N}_\theta^T q_n\, dS \end{aligned}} \tag{23.41}$$

It appears that $\boldsymbol{C}$ is the *capacity matrix*, $\boldsymbol{K}_\theta$ is the *thermal stiffness matrix* (or *conductivity matrix*) and $\boldsymbol{f}_\theta$ is the *thermal external force*; moreover, $\boldsymbol{Q}$ expresses the amount by which the strain rate produces heat.

We will now discuss numerical techniques to solve the mechanical balance equation (23.34) and the thermal balance equation (23.41). For simplicity we will ignore inertia effects, and the equations of motion (23.34) then reduce to the static equilibrium equations given by

$$\boxed{\boldsymbol{\psi} = \boldsymbol{0}}$$

The cases of uncoupled thermo-plasticity, coupled thermo-plasticity and adiabatic heating will now be dealt with.

23.5 Solution of uncoupled thermo-plasticity

For thermo-elasticity, we have shown in relation to (21.99) that very often the mechanical response influences the heat equation only insignificantly. For thermo-plasticity, if the external heat supply expressed by $\boldsymbol{f}_\theta$ in (23.41) is significantly larger than the heat generation due to the mechanical response, we can ignore this latter part and a situation of uncoupled thermo-plasticity has arisen.

23.5.1 Thermal field equations

Since the heat generation due to the mechanical response is ignored, we have

$$\boldsymbol{Q} = \boldsymbol{0}$$

and (23.41) then reduces to the uncoupled heat equation

$$\boxed{\boldsymbol{C}\dot{\boldsymbol{a}}_\theta + \boldsymbol{K}_\theta \boldsymbol{a}_\theta - \boldsymbol{f}_\theta = \boldsymbol{0}} \qquad (23.42)$$

Since the specific heat capacity c_ε as well as conductivity matrix $\boldsymbol{k}$ may depend on the temperature, $\boldsymbol{C}$ and $\boldsymbol{K}_\theta$ may also depend on the temperature

$$\boxed{\boldsymbol{C} = \boldsymbol{C}(\theta); \quad \boldsymbol{K}_\theta = \boldsymbol{K}_\theta(\theta)}$$

As a result, (23.42) comprises in general a set of nonlinear ordinary differential equations. Like the discussion in Chapter 19, we will first transform this set of nonlinear ordinary differential equations into a set of nonlinear algebraic equations.

This transformation is achieved by introducing a *time integration scheme* and we will here adopt the *generalized Euler scheme* according to which

$$\boldsymbol{a}_{\theta,n+1} = \boldsymbol{a}_{\theta,n} + \Delta t[(1-\gamma)\dot{\boldsymbol{a}}_{\theta,n} + \gamma\dot{\boldsymbol{a}}_{\theta,n+1})] \qquad (23.43)$$

where $\Delta t = t_{n+1} - t_n$ is the time step. In this expression, $\boldsymbol{a}_{\theta,n}$ and $\dot{\boldsymbol{a}}_{\theta,n}$ are known quantities at state n and we seek $\boldsymbol{a}_{\theta,n+1}$ and $\dot{\boldsymbol{a}}_{\theta,n+1}$ at the new state $n+1$; moreover, γ is a parameter we choose in the range $0 \leq \gamma \leq 1$.

Explicit scheme

To obtain the *explicit scheme*, the *Euler forward scheme*, we choose $\gamma = 0$ and (23.43) then gives

$$\boxed{\dot{\boldsymbol{a}}_{\theta,n} = \frac{1}{\Delta t}(\boldsymbol{a}_{\theta,n+1} - \boldsymbol{a}_{\theta,n})} \tag{23.44}$$

The balance equation (23.42) is now evaluated at state n, i.e.

$$\boldsymbol{C}_n\dot{\boldsymbol{a}}_{\theta,n} + \boldsymbol{K}_{\theta,n}\boldsymbol{a}_{\theta,n} - \boldsymbol{f}_{\theta,n} = \boldsymbol{0}$$

where $\boldsymbol{C}_n$, $\boldsymbol{K}_{\theta,n}$ and $\boldsymbol{f}_{\theta,n}$ are the respective quantities evaluated at the known state n. Insertion of (23.44) gives

$$\boxed{\boldsymbol{C}_n\boldsymbol{a}_{\theta,n+1} = (\boldsymbol{C}_n - \Delta t\boldsymbol{K}_{\theta,n})\boldsymbol{a}_{\theta,n} + \Delta t\boldsymbol{f}_{\theta,n}} \tag{23.45}$$

Since all terms on the right-hand side of this expression are known, $\boldsymbol{a}_{\theta,n+1}$ can be determined directly. This direct solution also holds for nonlinear heat problems where $\boldsymbol{C}$ and $\boldsymbol{K}_\theta$ depend on the temperature. This very fortunate situation is further enhanced if the capacity matrix $\boldsymbol{C}$ is taken as *lumped*, i.e. diagonal. In that case the inversion of $\boldsymbol{C}_n$ is trivial, and (23.45) then gives the solution directly without, in reality, solving any equation system.

This situation is quite similar to the explicit method discussed for dynamic mechanical problems, cf. (19.13). However, we have to pay a corresponding price for this simplicity namely that of *conditional stability*. Instability means that any minor error introduced by truncation of numbers in the computer will jeopardize the solution in the sense that errors will accumulate with time and very quickly render the solution meaningless. On the other hand, conditional stability means that in order to maintain stability, the time step Δt must be within a certain limit. In the present case, it turns out that we must have

$$\boxed{\Delta t \leq \frac{T_s}{\pi} \quad \Rightarrow \quad \textit{stability}}$$

where T_s is the shortest period for the *eigenmodes* of the finite element assemblage that comprises the discretized body, cf. for instance Bathe (1996) or Hughes (1977).

Implicit scheme

To obtain an *implicit scheme*, we choose $\gamma > 0$ and (23.43) then gives

$$\dot{\boldsymbol{a}}_{\theta,n+1} = \frac{1}{\gamma\Delta t}(\boldsymbol{a}_{\theta,n+1} - \boldsymbol{a}_{\theta,n}) - \frac{1-\gamma}{\gamma}\dot{\boldsymbol{a}}_{\theta,n}$$

However, whereas $\gamma \geq 1/2$ implies *unconditional stability* for linear problems, only $\gamma = 1$, i.e. the *backward Euler scheme*, results in unconditional stability for nonlinear problems, cf. Hughes (1977). Thus, $\gamma = 1$ will be assumed in the following and the relation above then reduces to

$$\boxed{\dot{\boldsymbol{a}}_{\theta,n+1} = \frac{1}{\Delta t}(\boldsymbol{a}_{\theta,n+1} - \boldsymbol{a}_{\theta,n})} \tag{23.46}$$

We might mention that a time integration scheme (the 'α-method') closely related to (23.43) exists, which is unconditionally stable for $\gamma \geq 1/2$ even for nonlinear problems, see Hughes (1977) and Bathe (1996). From (23.46) and since $\theta = \boldsymbol{N}_\theta \boldsymbol{a}_\theta$ we also have

$$\dot{\theta}_{n+1} = \frac{1}{\Delta t}(\theta_{n+1} - \theta_n) \tag{23.47}$$

The balance equation (23.42) is now evaluated at the state $n+1$, i.e.

$$\boldsymbol{C}_{n+1}\dot{\boldsymbol{a}}_{\theta,n+1} + \boldsymbol{K}_{\theta,n+1}\boldsymbol{a}_{\theta,n+1} - \boldsymbol{f}_{\theta,n+1} = \boldsymbol{0}$$

Dropping the subscript $n+1$ and insertion of (23.46) provides the sought for residual format for the heat equation

$$\boxed{\begin{aligned} &\boldsymbol{\psi}_\theta(\boldsymbol{a}_\theta) = \boldsymbol{0} \\ &\textit{where} \\ &\boldsymbol{\psi}_\theta(\boldsymbol{a}_\theta) = \frac{1}{\Delta t}\boldsymbol{C}(\boldsymbol{a}_\theta - \boldsymbol{a}_{\theta,n}) + \boldsymbol{K}_\theta \boldsymbol{a}_\theta - \boldsymbol{f}_\theta \end{aligned}} \tag{23.48}$$

Whereas $\boldsymbol{a}_{\theta,n}$ and the external force $\boldsymbol{f}_\theta = \boldsymbol{f}_{\theta,n+1}$ are known quantities, $\boldsymbol{a}_\theta$, $\boldsymbol{C}$ and $\boldsymbol{K}_\theta$ are unknown. Clearly, (23.48) is a nonlinear equation system in $\boldsymbol{a}_\theta$ and an iterative solution method must be applied; a natural procedure is to adopt the *Newton-Raphson scheme*, see Chapter 17. Therefore, assume that the approximation $\boldsymbol{a}_\theta^{i-1}$ ($i = 1, 2, \ldots$) to the true solution $\boldsymbol{a}_\theta$ has been established. Ignoring higher-order terms, a Taylor series expansion of $\boldsymbol{\psi}_\theta$ about $\boldsymbol{a}_\theta^{i-1}$ gives

$$\boldsymbol{\psi}_\theta(\boldsymbol{a}_\theta^i) = \boldsymbol{\psi}_\theta(\boldsymbol{a}_\theta^{i-1}) + \left(\frac{\partial \boldsymbol{\psi}_\theta}{\partial \boldsymbol{a}_\theta}\right)^{i-1}(\boldsymbol{a}_\theta^i - \boldsymbol{a}_\theta^{i-1})$$

According to the Newton-Raphson method, we require $\boldsymbol{\psi}(\boldsymbol{a}_\theta^i) = \boldsymbol{0}$, i.e.

$$\boldsymbol{0} = \boldsymbol{\psi}_\theta(\boldsymbol{a}_\theta^{i-1}) + \left(\frac{\partial \boldsymbol{\psi}_\theta}{\partial \boldsymbol{a}_\theta}\right)^{i-1}(\boldsymbol{a}_\theta^i - \boldsymbol{a}_\theta^{i-1}) \tag{23.49}$$

In order to evaluate the partial derivative above, it is first noted that (23.41) implies

$$\frac{1}{\Delta t}\boldsymbol{C}(\boldsymbol{a}_\theta - \boldsymbol{a}_{\theta,n}) = \int_V \boldsymbol{N}_\theta^T \dot{\theta} \rho c_\varepsilon \, dV \tag{23.50}$$

where $\dot{\theta} = \dot{\theta}_{n+1}$ is defined by (23.47). By using $c_\varepsilon = c_\varepsilon(\theta)$, two useful results can now be established

$$\frac{\partial \dot{\theta}}{\partial \boldsymbol{a}_\theta} = \frac{1}{\Delta t}\boldsymbol{N}_\theta\,; \qquad \frac{\partial c_\varepsilon}{\partial \boldsymbol{a}_\theta} = \frac{dc_\varepsilon}{d\theta}\frac{\partial \theta}{\partial \boldsymbol{a}_\theta} = \frac{dc_\varepsilon}{d\theta}\boldsymbol{N}_\theta$$

From the results above it then follows that

$$\frac{1}{\Delta t}\frac{\partial[\boldsymbol{C}(\boldsymbol{a}_\theta - \boldsymbol{a}_{\theta,n})]}{\partial \boldsymbol{a}_\theta} = \tilde{\boldsymbol{C}} + \frac{1}{\Delta t}\boldsymbol{C} \tag{23.51}$$

where

$$\boxed{\tilde{\boldsymbol{C}} = \int_V \boldsymbol{N}_\theta^T \rho \dot{\theta} \frac{dc_\varepsilon}{d\theta}\boldsymbol{N}_\theta\, dV} \tag{23.52}$$

Likewise, from (23.41) we have

$$\boldsymbol{K}_\theta \boldsymbol{a}_\theta = \int_V \boldsymbol{B}_\theta^T \boldsymbol{k} \nabla\theta\, dV \quad \text{where} \quad \nabla\theta = \boldsymbol{B}_\theta \boldsymbol{a}_\theta$$

Since $\boldsymbol{k} = \boldsymbol{k}(\theta)$, observation of (23.37) gives

$$\frac{\partial(\boldsymbol{k}\nabla\theta)}{\partial \boldsymbol{a}_\theta} = \frac{d\boldsymbol{k}}{d\theta}\nabla\theta\frac{\partial\theta}{\partial \boldsymbol{a}_\theta} + \boldsymbol{k}\frac{\partial\nabla\theta}{\partial \boldsymbol{a}_\theta} = \frac{d\boldsymbol{k}}{d\theta}\nabla\theta\boldsymbol{N}_\theta + \boldsymbol{k}\boldsymbol{B}_\theta$$

The above expressions then imply

$$\frac{\partial(\boldsymbol{K}_\theta \boldsymbol{a}_\theta)}{\partial \boldsymbol{a}_\theta} = \tilde{\boldsymbol{K}}_\theta + \boldsymbol{K}_\theta \tag{23.53}$$

where

$$\boxed{\tilde{\boldsymbol{K}}_\theta = \int_V \boldsymbol{B}_\theta^T \frac{d\boldsymbol{k}}{d\theta}\nabla\theta\boldsymbol{N}_\theta\, dV} \tag{23.54}$$

Finally, use of (23.51) and (23.53) in (23.48b) provides

$$\frac{\partial \boldsymbol{\psi}_\theta}{\partial \boldsymbol{a}} = \tilde{\boldsymbol{C}} + \frac{1}{\Delta t}\boldsymbol{C} + \tilde{\boldsymbol{K}}_\theta + \boldsymbol{K}_\theta$$

and insertion into (23.49) yields the sought for Newton-Raphson scheme

$$\boxed{\begin{aligned} &(\tilde{\boldsymbol{C}}^{i-1} + \frac{1}{\Delta t}\boldsymbol{C}^{i-1} + \tilde{\boldsymbol{K}}_\theta^{i-1} + \boldsymbol{K}_\theta^{i-1})(\boldsymbol{a}_\theta^{i} - \boldsymbol{a}_\theta^{i-1}) = -\boldsymbol{\psi}_\theta^{i-1} \\ &\textit{where} \\ &\boldsymbol{\psi}_\theta^{i-1} = \frac{1}{\Delta t}\boldsymbol{C}^{i-1}(\boldsymbol{a}_\theta^{i-1} - \boldsymbol{a}_{\theta,n}) + \boldsymbol{K}_\theta^{i-1}\boldsymbol{a}_\theta^{i-1} - \boldsymbol{f}_{\theta,n+1} \end{aligned}} \tag{23.55}$$

where $\boldsymbol{a}_{\theta,n}$ and $\boldsymbol{f}_{\theta,n+1}$ are constant given quantities. As in Chapter 17 we choose as starting vectors ($i = 1$) the values at the last accepted state n, that is, $\boldsymbol{a}_\theta^0 = \boldsymbol{a}_{\theta,n}$

and the iterations are stopped when the convergence criterion/criteria similar to those discussed in Section 17.7 are satisfied and then $\boldsymbol{a}_{n+1} = \boldsymbol{a}^i$.

It is evident that the computational effort to establish $\tilde{\boldsymbol{K}}_\theta$ is significant. Moreover, as $\tilde{\boldsymbol{C}}$ and $\tilde{\boldsymbol{K}}_\theta$ are corrections to $\boldsymbol{C}$ and $\boldsymbol{K}_\theta$ respectively, which vary slowly with temperature, cf. Figs. 23.6 and 23.7, it seems reasonable to ignore them in (23.55) and thereby obtain a modified Newton-Raphson scheme without significant loss of convergence speed, cf. Celentano *et al.* (1994). The slow temperature variation of $\boldsymbol{C}$ and $\boldsymbol{K}_\theta$ also motivates accepting constant values and only occasionally making an update; this is, in fact, frequently adopted, cf. Bathe (1996). If the contributions $\tilde{\boldsymbol{C}}$ and $\tilde{\boldsymbol{K}}$ (which is unsymmetric) are ignored, the coefficient matrix in (23.55) becomes $\frac{1}{\Delta t}\boldsymbol{C}^{i-1} + \boldsymbol{K}_\theta^{i-1}$, which is symmetric and positive definite, cf. the definitions in (23.41). These are precisely the prerequisites for application of the quasi-Newton method in terms of the BFGS-method, cf. Section 17.8, and this route was pursued by Argyris *et al.* (1981).

23.5.2 Mechanical field equations

Having discussed the solution of the uncoupled heat equation in some detail, we next turn to the mechanical part of the problem. From (23.6) $\boldsymbol{\sigma}$ depends both on the strains and the temperature, that is, the stresses depend both on the nodal displacements $\boldsymbol{a}$ and the nodal temperatures $\boldsymbol{a}_\theta$. Assume that $\boldsymbol{a}_n$ and $\boldsymbol{a}_{\theta,n}$ at state n are known. The external force is then changed to $\boldsymbol{f}_{n+1}$. Again dropping for convenience the subscript $n+1$ on quantities that depend on $\boldsymbol{a}_{n+1}$ and $\boldsymbol{a}_{\theta,n+1}$, and as only static problems are considered, we have from (23.34)

$$\boldsymbol{\psi}(\boldsymbol{a}, \boldsymbol{a}_\theta) = \boldsymbol{0} \tag{23.56}$$

where the out-of-balance forces $\boldsymbol{\psi}$ are given by

$$\boldsymbol{\psi}(\boldsymbol{a}, \boldsymbol{a}_\theta) = \int_V \boldsymbol{B}^T \boldsymbol{\sigma}\, dV - \boldsymbol{f} \tag{23.57}$$

where $\boldsymbol{f}$ denotes the known external forces at state $n+1$.

As the heat problem is solved independently to provide the temperature changes, the temperature field is now known when the mechanical problem is considered. Since the mechanical problem has already been treated in Chapter 17 we will here simply reiterate the equations for a Newton-Raphson scheme. From (17.26) and (17.32) the iteration scheme is given by

$$(\boldsymbol{K}_t)^{i-1}(\boldsymbol{a}^i - \boldsymbol{a}^{i-1}) = -\boldsymbol{\psi}(\boldsymbol{a}^{i-1}, \boldsymbol{a}_{\theta,n+1})$$

where

$$\boldsymbol{\psi}(\boldsymbol{a}^{i-1}, \boldsymbol{a}_{\theta,n+1}) = \int_V \boldsymbol{B}^T \boldsymbol{\sigma}^{i-1}\, dV - \boldsymbol{f}_{n+1}$$

and where

$$\boldsymbol{K}_t = \int_V \boldsymbol{B}^T \boldsymbol{D}_t \boldsymbol{B} dV$$

The starting vector for $\boldsymbol{a}^{i-1}$ for $i = 1$ is taken as $\boldsymbol{a}^0 = \boldsymbol{a}_n$, that is the value at the last accepted state n.

It appears that the only way in which the temperature field influences the mechanical field equations is through the stresses which, in turn, determine the internal forces $\int_V \boldsymbol{B}^T \boldsymbol{\sigma} dV$; this is the topic of the next section.

23.5.3 Numerical treatment of the constitutive relations

Evidently, we need to integrate the thermo-plastic constitutive relations to obtain the current stresses. Fortunately, it turns out that many of the results in Chapter 18 can be used. Here we will use methods that directly fulfill the yield criterion.

The first task is to consider whether plastic strains will develop during the load increment. Adopting the approach utilized in Section 18.1.1, a trial stress is defined assuming that no plastic strains will develop during the increment. Denote the current state and the last known state in equilibrium as state 2 and state 1 respectively. Evaluating (23.6) at state 2 and state 1 then results in

$$\sigma_{ij}^{(2)} = D_{ijkl}(\varepsilon_{kl}^{(2)} - \varepsilon_{kl}^{p(2)}) - \beta_{ij}(\theta^{(2)} - \theta_o)$$

and

$$\sigma_{ij}^{(1)} = D_{ijkl}(\varepsilon_{kl}^{(1)} - \varepsilon_{kl}^{p(1)}) - \beta_{ij}(\theta^{(1)} - \theta_o)$$

where for simplicity it was assumed that D_{ijkl} and β_{ij} are constant. Subtracting the above equations

$$\boxed{\sigma_{ij}^{(2)} = \sigma_{ij}^t - D_{ijkl}\Delta\varepsilon_{kl}^p} \tag{23.58}$$

where the trial stress is defined as

$$\boxed{\sigma_{ij}^t = \sigma_{ij}^{(1)} + \Delta\sigma_{ij}^{te} \qquad where \qquad \Delta\sigma_{ij}^{te} = D_{ijkl}\Delta\varepsilon_{kl} - \beta_{ij}\Delta\theta} \tag{23.59}$$

and

$$\Delta\varepsilon_{kl} = \varepsilon_{kl}^{(2)} - \varepsilon_{kl}^{(1)} \qquad \Delta\varepsilon_{kl}^p = \varepsilon_{kl}^{p(2)} - \varepsilon_{kl}^{p(1)} \qquad \Delta\theta = \theta^{(2)} - \theta^{(1)}$$

From (23.58) it can then be concluded that if no plastic strains develop, we have $\sigma_{ij}^{(2)} = \sigma_{ij}^t$.

The yield function is now evaluated for the trial stresses according to

$$f^t = f(\sigma_{ij}^t, K_\alpha^{(1)}, \theta^{(2)})$$

It then follows that

$$\boxed{\begin{array}{ll} f^t \le 0 \;\Rightarrow\; \textit{thermo-elastic response}\ \sigma_{ij}^{(2)} = \sigma_{ij}^t \\ f^t > 0 \;\Rightarrow\; \textit{thermo-plastic response} \end{array}}$$

For a thermo-plastic response, the constitutive relations must be integrated numerically. In the uncoupled approach, the temperature is known when the integration of the constitutive relations is considered. This allows for the use of the methods described in Section 18.1.4. Let us summarize the set of equations considered in the integration procedure

$$\begin{aligned}
\boldsymbol{\sigma} &= \boldsymbol{\sigma}^t - \Delta\lambda \boldsymbol{D}\frac{\partial g}{\partial \boldsymbol{\sigma}} \\
K &= K(\kappa^{(1)} - \Delta\lambda \frac{\partial g}{\partial K}, \theta) \\
f &= f(\boldsymbol{\sigma}, K, \theta) = 0
\end{aligned} \tag{23.60}$$

where, for simplicity, it is assumed that only one hardening parameter exists $K = K(\kappa, \theta)$. Moreover, in (23.60) the superscript 2 was omitted.

A comparison of (23.60) with (18.35) reveals a complete similarity. In fact, by replacing $\boldsymbol{\sigma}^t$, which is a known quantity, in (18.35) with (23.59) and considering θ as a given quantity, the system in (18.35) can be directly used for the calculation of stresses. As an example, the solution for a thermo-plastic von Mises material directly follows the derivations in Section 18.1.4 and we will therefore not repeat the integration procedures here. Accordingly, the algorithmic tangent stiffness is unchanged and the results given in Section 18.1.5 hold.

23.6 Solution of coupled thermo-plasticity

We will now discuss solution strategies in the general case of thermo-plasticity where heat flow is accounted for and the heat generation due to plastic dissipation is significant. This problem has been the center of increasing interest during recent years due to its importance in metal-forming processes and development of shear bands; the latter topic is, for instance, treated by Lemonds and Needleman (1986a,b). In Section 23.8 an alternative solution method will be discussed: the staggered solution scheme.

Define $\tilde{\boldsymbol{a}}$ and $\tilde{\boldsymbol{\psi}}$ by

$$\tilde{\boldsymbol{a}} = \begin{bmatrix} \boldsymbol{a} \\ \boldsymbol{a}_\theta \end{bmatrix}; \quad \tilde{\boldsymbol{\psi}} = \begin{bmatrix} \boldsymbol{\psi} \\ \boldsymbol{\psi}_\theta \end{bmatrix}$$

that is, the nodal quantities $\tilde{\boldsymbol{a}}$ and the out-of-balance forces $\tilde{\boldsymbol{\psi}}$ include both the mechanical and thermal parts. In the spirit of the general iteration format for equilibrium iterations provided by the '$\boldsymbol{A}$-matrix' approach, cf. (17.18), we then have

$$\boldsymbol{A}^{i-1}(\tilde{\boldsymbol{a}}^{i} - \tilde{\boldsymbol{a}}^{i-1}) = -\tilde{\boldsymbol{\psi}}(\tilde{\boldsymbol{a}}^{i-1}) \tag{23.61}$$

where $i = 1, 2, \ldots$ and $\tilde{\boldsymbol{a}}^0 = \tilde{\boldsymbol{a}}_n$. The only restrictions placed on the iteration matrix $\boldsymbol{A}$ are given by (17.16) and (17.19), i.e.

$$\det \boldsymbol{A} \neq 0 \quad \text{and} \quad \det \boldsymbol{A}^{-1} \neq 0$$

To find information on relevant choices of the $\boldsymbol{A}$-matrix, we will adopt the Newton-Raphson scheme.

23.6.1 Thermal field equations

Adopt the backward Euler scheme

$$\dot{\boldsymbol{a}}_{n+1} = \frac{1}{\Delta t}(\boldsymbol{a}_{n+1} - \boldsymbol{a}_n)\,; \qquad \dot{\boldsymbol{a}}_{\theta,n+1} = \frac{1}{\Delta t}(\boldsymbol{a}_{\theta,n+1} - \boldsymbol{a}_{\theta,n})$$

Evaluation of the coupled heat equation given by (23.41) at state $n + 1$, and dropping the subscript $n + 1$ on quantities that depend on $\boldsymbol{a}_{n+1}$ and $\boldsymbol{a}_{\theta,n+1}$, we obtain

$$\boxed{\begin{aligned}&\boldsymbol{\psi}_\theta(\boldsymbol{a}, \boldsymbol{a}_\theta) = \boldsymbol{0}\\ &\textit{where}\\ &\boldsymbol{\psi}_\theta(\boldsymbol{a}, \boldsymbol{a}_\theta) = \frac{1}{\Delta t}\boldsymbol{C}(\boldsymbol{a}_\theta - \boldsymbol{a}_{\theta,n}) + \boldsymbol{K}_\theta \boldsymbol{a}_\theta - \boldsymbol{Q} - \boldsymbol{f}_\theta\end{aligned}} \tag{23.62}$$

The quantities $\boldsymbol{a}_n$, $\boldsymbol{a}_{\theta,n}$ and $\boldsymbol{f}_{\theta,n+1}$ are known quantities, i.e. the thermal out-of-balance force $\boldsymbol{\psi}_\theta$ is a function of $\boldsymbol{a}$ and $\boldsymbol{a}_\theta$ as already indicated above.

For the continued treatment, the quantity $\boldsymbol{Q}$ needs some attention. From (23.41) we have

$$\boldsymbol{Q} = \int_V \boldsymbol{N}_\theta^T Q \, dV \tag{23.63}$$

where Q is evaluated at state $n + 1$. Moreover, from (23.23) we can identify

$$Q = \boldsymbol{\sigma}^T \dot{\boldsymbol{\varepsilon}}^p - \boldsymbol{K}^T \dot{\boldsymbol{\kappa}} - \theta \boldsymbol{P}^T (\dot{\boldsymbol{\varepsilon}} - \dot{\boldsymbol{\varepsilon}}^p) + \theta(\frac{\partial \boldsymbol{K}}{\partial \theta})^T \dot{\boldsymbol{\kappa}} \tag{23.64}$$

where the different rates are evaluated at state $n+1$ and approximated using the backward Euler scheme, i.e.

$$\dot{\boldsymbol{\varepsilon}} = \frac{1}{\Delta t}(\boldsymbol{\varepsilon}_{n+1} - \boldsymbol{\varepsilon}_n)\,; \quad \dot{\boldsymbol{\varepsilon}}^p = \frac{1}{\Delta t}(\boldsymbol{\varepsilon}^p_{n+1} - \boldsymbol{\varepsilon}^p_n)\,; \quad \dot{\boldsymbol{\kappa}} = \frac{1}{\Delta t}(\boldsymbol{\kappa}_{n+1} - \boldsymbol{\kappa}_n) \tag{23.65}$$

Then we can write (23.64) as

$$Q = \frac{\Delta Q}{\Delta t} \tag{23.66}$$

where

$$\Delta Q = \boldsymbol{\sigma}^T \Delta \boldsymbol{\varepsilon}^p - \boldsymbol{K}^T \Delta \boldsymbol{\kappa} - \theta \boldsymbol{P}^T (\Delta \boldsymbol{\varepsilon} - \Delta \boldsymbol{\varepsilon}^p) + \theta (\frac{\partial \boldsymbol{K}}{\partial \theta})^T \Delta \boldsymbol{\kappa} \tag{23.67}$$

With these results we are ready to consider the solution method for the non-linear equation system (23.62); the Newton-Raphson scheme will be adopted. Therefore, assume that the approximation $\boldsymbol{a}^{i-1}$ and $\boldsymbol{a}_\theta^{i-1}$ $(i = 1, 2, \ldots)$ to the true solution $\boldsymbol{a}_{n+1}$ and $\boldsymbol{a}_{\theta,n+1}$ has been established. Ignoring higher order terms, a Taylor series expansion of $\boldsymbol{\psi}_\theta$ about $\boldsymbol{a}^{i-1}$, $\boldsymbol{a}_\theta^{i-1}$ gives

$$\begin{aligned}\boldsymbol{\psi}_\theta(\boldsymbol{a}^i, \boldsymbol{a}_\theta^i) =& \boldsymbol{\psi}_\theta(\boldsymbol{a}^{i-1}, \boldsymbol{a}_\theta^{i-1}) \\ &+ (\frac{\partial \boldsymbol{\psi}_\theta}{\partial \boldsymbol{a}})^{i-1}(\boldsymbol{a}^i - \boldsymbol{a}^{i-1}) + (\frac{\partial \boldsymbol{\psi}_\theta}{\partial \boldsymbol{a}_\theta})^{i-1}(\boldsymbol{a}_\theta^i - \boldsymbol{a}_\theta^{i-1})\end{aligned}$$

According to the Newton-Raphson method, we require $\boldsymbol{\psi}_\theta(\boldsymbol{a}^i, \boldsymbol{a}_\theta^i) = \boldsymbol{0}$, i.e.

$$\boldsymbol{0} = \boldsymbol{\psi}_\theta(\boldsymbol{a}^{i-1}, \boldsymbol{a}_\theta^{i-1}) + (\frac{\partial \boldsymbol{\psi}_\theta}{\partial \boldsymbol{a}})^{i-1}(\boldsymbol{a}^i - \boldsymbol{a}^{i-1}) + (\frac{\partial \boldsymbol{\psi}_\theta}{\partial \boldsymbol{a}_\theta})^{i-1}(\boldsymbol{a}_\theta^i - \boldsymbol{a}_\theta^{i-1}) \tag{23.68}$$

Similar to the analysis given in Section 23.5.1, it is easily shown that

$$\frac{\partial \boldsymbol{\psi}_\theta}{\partial \boldsymbol{a}_\theta} = \tilde{\boldsymbol{C}} + \frac{1}{\Delta t}\boldsymbol{C} + \tilde{\boldsymbol{K}}_\theta + \boldsymbol{K}_\theta - \frac{1}{\Delta t}\boldsymbol{H} \tag{23.69}$$

where $\tilde{\boldsymbol{C}}$ and $\tilde{\boldsymbol{K}}$ are defined by (23.52) and (23.54) respectively and

$$\boxed{\boldsymbol{H} = \int_V \boldsymbol{N}_\theta^T \frac{\partial \Delta Q}{\partial \boldsymbol{a}_\theta}\, dV} \tag{23.70}$$

In the same fashion, it is easily shown that

$$\frac{\partial \boldsymbol{\psi}_\theta}{\partial \boldsymbol{a}} = \tilde{\boldsymbol{G}} - \frac{1}{\Delta t}\boldsymbol{G} \tag{23.71}$$

where

$$\boxed{\tilde{\boldsymbol{G}} = \int_V \boldsymbol{N}_\theta^T \dot{\theta} \rho \frac{\partial c_\varepsilon}{\partial \boldsymbol{a}} dV \; ; \qquad \boldsymbol{G} = \int_V \boldsymbol{N}_\theta^T \frac{\partial \Delta Q}{\partial \boldsymbol{a}}\, dV} \tag{23.72}$$

These partial derivatives will be addressed later on when the stress calculations are considered. We also note that according to (23.8), $c_\varepsilon = -\theta d^2 h(\theta)/d\theta^2 + \tilde{c}_\varepsilon$ where $\tilde{c}_\varepsilon$ depends on the deformation; however, it was shown in Section 23.2.1

that $\tilde{c}_\varepsilon / c_\varepsilon \approx 0$ so we expect the matrix $\tilde{\boldsymbol{G}}$ to be very small compared with the other matrices entering the problem.

Finally, insertion of (23.71) and (23.69) into the Newton-Raphson scheme (23.68) results in

$$\boldsymbol{T}_a^{i-1}(\boldsymbol{a}^i - \boldsymbol{a}^{i-1}) + \boldsymbol{T}_\theta^{i-1}(\boldsymbol{a}_\theta^i - \boldsymbol{a}_\theta^{i-1}) = -\boldsymbol{\psi}_\theta^{i-1} \tag{23.73}$$

where

$$\boxed{\begin{aligned} \boldsymbol{T}_a &= \tilde{\boldsymbol{G}} - \frac{1}{\Delta t}\boldsymbol{G} \\ \boldsymbol{T}_\theta &= \tilde{\boldsymbol{C}} + \frac{1}{\Delta t}\boldsymbol{C} + \tilde{\boldsymbol{K}}_\theta + \boldsymbol{K}_\theta - \frac{1}{\Delta t}\boldsymbol{H} \end{aligned}} \tag{23.74}$$

23.6.2 Mechanical field equations

From (23.34) we have $\boldsymbol{\psi}(\boldsymbol{a}, \boldsymbol{a}_\theta) = \boldsymbol{0}$ and again we seek a Newton-Raphson scheme for this nonlinear equation system. Assume that the approximation $\boldsymbol{a}^{i-1}$ and $\boldsymbol{a}_\theta^{i-1}$ to the state $n+1$ are known and make a Taylor series expansion about the state $\boldsymbol{a}^{i-1}$, $\boldsymbol{a}_\theta^{i-1}$, i.e.

$$\begin{aligned} \boldsymbol{\psi}(\boldsymbol{a}^i, \boldsymbol{a}_\theta^i) &= \boldsymbol{\psi}(\boldsymbol{a}^{i-1}, \boldsymbol{a}_\theta^{i-1}) \\ &\quad + \left(\frac{\partial \boldsymbol{\psi}}{\partial \boldsymbol{a}}\right)^{i-1}(\boldsymbol{a}^i - \boldsymbol{a}^{i-1}) + \left(\frac{\partial \boldsymbol{\psi}}{\partial \boldsymbol{a}_\theta}\right)^{i-1}(\boldsymbol{a}_\theta^i - \boldsymbol{a}_\theta^{i-1}) \end{aligned}$$

and setting $\boldsymbol{\psi}(\boldsymbol{a}^i, \boldsymbol{a}_\theta^i) = \boldsymbol{0}$ gives

$$\begin{aligned} \boldsymbol{0} &= \boldsymbol{\psi}(\boldsymbol{a}^{i-1}, \boldsymbol{a}_\theta^{i-1}) \\ &\quad + \left(\frac{\partial \boldsymbol{\psi}}{\partial \boldsymbol{a}}\right)^{i-1}(\boldsymbol{a}^i - \boldsymbol{a}^{i-1}) + \left(\frac{\partial \boldsymbol{\psi}}{\partial \boldsymbol{a}_\theta}\right)^{i-1}(\boldsymbol{a}_\theta^i - \boldsymbol{a}_\theta^{i-1}) \end{aligned} \tag{23.75}$$

To obtain the Newton-Raphson scheme we must obviously evaluate the above partial derivatives. For the time being, let us observe that it is possible to evaluate these partial derivative using the algorithmic relation

$$d\boldsymbol{\sigma} = \boldsymbol{D}_{ats} d\boldsymbol{\varepsilon} - \boldsymbol{\beta}_{ats} d\theta$$

i.e.

$$d\boldsymbol{\sigma} = \boldsymbol{D}_{ats}\boldsymbol{B} d\boldsymbol{a} - \boldsymbol{\beta}_{ats}\boldsymbol{N}_\theta d\boldsymbol{a}_\theta$$

The establishment of this relation will be performed later on in this section. From (23.34) it then follows that

$$\frac{\partial \boldsymbol{\psi}}{\partial \boldsymbol{a}} = \boldsymbol{K}_t; \qquad \frac{\partial \boldsymbol{\psi}}{\partial \boldsymbol{a}_\theta} = -\boldsymbol{L}$$

where

$$\boxed{\boldsymbol{K}_t = \int_V \boldsymbol{B}^T \boldsymbol{D}_{ats} \boldsymbol{B}\, dV \quad \boldsymbol{L} = \int_V \boldsymbol{B}^T \boldsymbol{\beta}_{ats} \boldsymbol{N}_\theta\, dV}$$

Insertion into (23.75) results in

$$(\boldsymbol{K}_t)^{i-1}(\boldsymbol{a}^i - \boldsymbol{a}^{i-1}) - \boldsymbol{L}^{i-1}(\boldsymbol{a}_\theta^i - \boldsymbol{a}_\theta^{i-1}) = -\boldsymbol{\psi}(\boldsymbol{a}^{i-1}, \boldsymbol{a}_\theta^{i-1}) \tag{23.76}$$

where

$$\boldsymbol{\psi}(\boldsymbol{a}^{i-1}, \boldsymbol{a}_\theta^{i-1}) = \int_V \boldsymbol{B}^T \boldsymbol{\sigma}^{i-1}\, dV - \boldsymbol{f}_{n+1}$$

The starting vectors for $\boldsymbol{a}^{i-1}$ and $\boldsymbol{a}_\theta^{i-1}$ are taken as $\boldsymbol{a}^0 = \boldsymbol{a}_n$ and $\boldsymbol{a}_\theta^0 = \boldsymbol{a}_{\theta,n}$ i.e. the values at the last accepted state n.

From (23.73) and (23.76) it is apparent that the Newton-Raphson formulation of the coupled problem becomes

$$\begin{bmatrix} (\boldsymbol{K}_t)^{i-1} & -\boldsymbol{L}^{i-1} \\ \boldsymbol{T}_a^{i-1} & \boldsymbol{T}_\theta^{i-1} \end{bmatrix} \begin{bmatrix} \boldsymbol{a}^i - \boldsymbol{a}^{i-1} \\ \boldsymbol{a}_\theta^i - \boldsymbol{a}_\theta^{i-1} \end{bmatrix} = - \begin{bmatrix} \boldsymbol{\psi}^{i-1} \\ \boldsymbol{\psi}_\theta^{i-1} \end{bmatrix}$$

where

$$\begin{aligned} \boldsymbol{\psi}^{i-1} &= \boldsymbol{\psi}(\boldsymbol{a}^{i-1}, \boldsymbol{a}_\theta^{i-1}) = \int_V \boldsymbol{B}^T \boldsymbol{\sigma}^{i-1}\, dV - \boldsymbol{f}_{n+1} \\ \boldsymbol{\psi}_\theta^{i-1} &= \boldsymbol{\psi}_\theta(\boldsymbol{a}^{i-1}, \boldsymbol{a}_\theta^{i-1}) \\ &= \boldsymbol{C}\frac{1}{\Delta t}(\boldsymbol{a}_\theta^{i-1} - \boldsymbol{a}_{\theta,n}) + \boldsymbol{K}_\theta^{i-1}\boldsymbol{a}_\theta^{i-1} - \boldsymbol{Q}^{i-1} - \boldsymbol{f}_{\theta,n+1} \end{aligned} \tag{23.77}$$

A comparison with (23.61) shows that the iteration matrix $\boldsymbol{A}$ for the Newton-Raphson method is given as the coefficient matrix in (23.77).

Unfortunately the coefficient matrix in (23.77) turns out to be unsymmetric and the bandwidth is very broad. In addition, several of the individual matrices are complex to derive. This is especially true for the matrices $\tilde{\boldsymbol{G}}$, $\boldsymbol{G}$ and $\boldsymbol{H}$, but even the more simple matrices $\tilde{\boldsymbol{C}}$ and $\tilde{\boldsymbol{K}}_\theta$ call for a substantial computational effort; it is noted that $\tilde{\boldsymbol{C}}$ and $\tilde{\boldsymbol{K}}_\theta$ are zero when the specific heat capacity c_ε and the conductivity matrix $\boldsymbol{k}$ are temperature independent. However, the contributions from $\tilde{\boldsymbol{G}}$, $\boldsymbol{H}$, $\tilde{\boldsymbol{C}}$ and $\tilde{\boldsymbol{K}}_\theta$ can be considered as slight and if they are ignored (23.77) reduces to

$$\begin{bmatrix} (\boldsymbol{K}_t)^{i-1} & -\boldsymbol{L}^{i-1} \\ -\frac{1}{\Delta t}\boldsymbol{G}^{i-1} & \frac{1}{\Delta t}\boldsymbol{C}^{i-1} + \boldsymbol{K}_\theta^{i-1} \end{bmatrix} \begin{bmatrix} \boldsymbol{a}^i - \boldsymbol{a}^{i-1} \\ \boldsymbol{a}_\theta^i - \boldsymbol{a}_\theta^{i-1} \end{bmatrix} = - \begin{bmatrix} \boldsymbol{\psi}^{i-1} \\ \boldsymbol{\psi}_\theta^{i-1} \end{bmatrix} \tag{23.78}$$

for $i = 1, 2, \ldots$. As usual, the starting vectors are taken as $\boldsymbol{a}^0 = \boldsymbol{a}_n$, $\boldsymbol{a}_\theta^0 = \boldsymbol{a}_{\theta,n}$ where n is the last state where the mechanical and thermal balance equations are fulfilled. Although solution of the fully coupled equation system (23.77) without introducing any simplifications is computationally very demanding, it was performed e.g. by Slużalec (1988).

Algorithmic tangent stiffness

We note that in the iteration scheme both $\boldsymbol{\varepsilon}^{(2)}$ and $\theta^{(2)}$ are known when the stresses are to be calculated. Accordingly, this topic has already been treated in Section 23.5.3 and will therefore not be repeated here.

Let us instead establish the algorithmic tangent stiffness. In order not to complicate the algebra, it will be assumed that $\partial g/\partial\boldsymbol{\sigma}$ only depends on $\boldsymbol{\sigma}$. In that case (23.60) becomes similar to (18.57). We will therefore utilize the same ideas as presented in Section 18.1.5; however, we now have to account for the temperature in the derivations.

Differentiation of (23.60a) and use of (23.59) result in

$$\dot{\boldsymbol{\sigma}} = \boldsymbol{D}\dot{\boldsymbol{\varepsilon}} - \dot{\lambda}\boldsymbol{D}\frac{\partial g}{\partial\boldsymbol{\sigma}} - \Delta\lambda\boldsymbol{D}\frac{\partial^2 g}{\partial\boldsymbol{\sigma}\partial\boldsymbol{\sigma}}\dot{\boldsymbol{\sigma}} - \boldsymbol{\beta}\dot{\theta}$$

which leads to

$$\dot{\boldsymbol{\sigma}} = \boldsymbol{D}^a\dot{\boldsymbol{\varepsilon}} - \dot{\lambda}\boldsymbol{D}^a\frac{\partial g}{\partial\boldsymbol{\sigma}} - \boldsymbol{P}^a\dot{\theta} \tag{23.79}$$

where

$$\boldsymbol{D}^a = (\boldsymbol{D}^{-1} + \Delta\lambda\frac{\partial^2 g}{\partial\boldsymbol{\sigma}\partial\boldsymbol{\sigma}})^{-1}$$
$$\boldsymbol{P}^a = \boldsymbol{D}^a\boldsymbol{D}^{-1}\boldsymbol{\beta}$$

It is interesting to note that $\boldsymbol{D}^a$ is identical to $\boldsymbol{D}^a$ given in (18.60). Considering the yield condition (23.60c), a differentiation provides

$$(\frac{\partial f}{\partial\boldsymbol{\sigma}})^T\dot{\boldsymbol{\sigma}} + \frac{\partial f}{\partial K}\dot{K} + \frac{\partial f}{\partial\theta}\dot{\theta} = 0 \tag{23.80}$$

Eventually, let us consider (23.60b). A straightforward differentiation results in

$$\dot{K} = d^a\dot{\lambda} + d^a_\theta\dot{\theta} \tag{23.81}$$

where

$$d^a = -(1 + \Delta\lambda\frac{\partial K}{\partial\kappa}\frac{\partial^2 g}{\partial K^2})^{-1}\frac{\partial K}{\partial\kappa}\frac{\partial g}{\partial K}$$
$$d^a_\theta = -(1 + \Delta\lambda\frac{\partial K}{\partial\kappa}\frac{\partial^2 g}{\partial K^2})^{-1}(\frac{\partial K}{\partial\kappa}\Delta\lambda\frac{\partial^2 g}{\partial K\partial\theta} - \frac{\partial K}{\partial\theta})$$

Insertion of (23.81) and (23.79) into (23.80) provides

$$\dot{\lambda} = \frac{1}{A^a}(\frac{\partial f}{\partial\boldsymbol{\sigma}})^T\boldsymbol{D}^a\dot{\boldsymbol{\varepsilon}} + \frac{1}{A^a}[S^a - (\frac{\partial f}{\partial\boldsymbol{\sigma}})^T\boldsymbol{P}^a]\dot{\theta} \tag{23.82}$$

where

$$A^a = (\frac{\partial f}{\partial \sigma})^T \boldsymbol{D}^a \frac{\partial g}{\partial \sigma} - \frac{\partial f}{\partial K} d^a$$
$$S^a = \frac{\partial f}{\partial K} d^a_\theta + \frac{\partial f}{\partial \theta}$$

Finally, insertion of (23.82) into (23.79) provides the result sought for

Algorithmic constitutive relation

$$\dot{\boldsymbol{\sigma}} = \boldsymbol{D}_{ats}\dot{\boldsymbol{\varepsilon}} - \boldsymbol{\beta}_{ats}\dot{\theta}$$

where

$$\boldsymbol{D}_{ats} = \boldsymbol{D}^a - \frac{1}{A^a}\boldsymbol{D}^a \frac{\partial g}{\partial \sigma}(\frac{\partial f}{\partial \sigma})^T \boldsymbol{D}^a \qquad (23.83)$$

$$\boldsymbol{\beta}_{ats} = \boldsymbol{P}^a + \frac{1}{A^a}\boldsymbol{D}^a \frac{\partial g}{\partial \sigma}[S^a - (\frac{\partial f}{\partial \sigma})^T \boldsymbol{P}^a]$$

This relation is the algorithmic version of (23.20).

In (23.70) and (23.72) the matrices $\tilde{\boldsymbol{G}}$, $\boldsymbol{G}$ and $\boldsymbol{H}$ were defined, but their specific content was left open. We have already argued that $\tilde{\boldsymbol{G}}$ is expected to be small compared with other matrices, but it can be obtained by a linearization of c_ε; the derivations are lengthy but straightforward and will not be pursued here. To identify the matrices $\boldsymbol{G}$ and $\boldsymbol{H}$ a linearization of (23.67) provides

$$(\dot{\Delta Q}) = \bar{\boldsymbol{Q}}_\varepsilon^T \dot{\boldsymbol{\varepsilon}} + \bar{B}\dot{\theta} = \bar{\boldsymbol{Q}}_\varepsilon^T \boldsymbol{B}\dot{\boldsymbol{a}} + \bar{B}\boldsymbol{N}_\theta \dot{\boldsymbol{a}}_\theta$$

The specific formats for $\boldsymbol{Q}_\varepsilon$ and B are straightforward to establish. From the above it is then possible to identify

$$\frac{\partial \Delta Q}{\partial \boldsymbol{a}} = \bar{\boldsymbol{Q}}_\varepsilon^T \boldsymbol{B}\,; \qquad \frac{\partial \Delta Q}{\partial \boldsymbol{a}_\theta} = \bar{B}\boldsymbol{N}_\theta$$

and the matrices $\boldsymbol{G}$ and $\boldsymbol{H}$ defined by (23.72) and (23.70) then become

$$\boldsymbol{G} = \int_V \boldsymbol{N}_\theta^T \bar{\boldsymbol{Q}}_\varepsilon^T \boldsymbol{B} dV\,; \qquad \boldsymbol{H} = \int_V \boldsymbol{N}_\theta^T \bar{B}\boldsymbol{N}_\theta dV$$

23.7 Adiabatic heating

If the loading is performed very rapidly, no heat exchange can occur with the surroundings of the body. Moreover, this rapid loading does not allow any heat flow to take place. Disregarding heat flow can be done by ignoring the heat flux vector $q_i = -k_{ij}\theta_{,j}$; likewise, since no heat exchange with the surroundings is possible, we set $r = 0$. These conditions are called *adiabatic heating* and the heat equation (23.23b), cf. also (21.100), then reduces to

$$\boxed{\begin{aligned} &E = \rho c_\varepsilon \dot{\theta} - Q = 0 \\ &\textit{where} \\ &Q = \sigma_{ij}\dot{\varepsilon}^p_{ij} - K_\alpha \dot{\kappa}_\alpha - \theta P_{ij}(\dot{\varepsilon}_{ij} - \dot{\varepsilon}^p_{ij}) + \theta \frac{\partial K_\alpha}{\partial \theta}\dot{\kappa}_\alpha \end{aligned}} \tag{23.84}$$

It appears that it is only the mechanical coupling that produces heat and thereby change of temperature. It can also be noted that given the strain rate, plastic strain rate and the rate of the internal variables the change in temperature can be obtained without solving any boundary value problem. This implies that (23.84) should be treated together with the constitutive relations, i.e. on the material level. Adiabatic heating has been considered e.g. by Nemat-Nasser (1988), Batra and Wright (1988) and Wriggers *et al.* (1992).

The gain of not having to solve a boundary value problem must be paid for on the constitutive level, which now becomes more elaborate. Therefore, let us first consider the calculation of stresses and later return to the finite element formulation for the mechanical problem.

Before considering the algorithm for calculating the stresses, it is noted that (23.84) is in rate form. To obtain a suitable format for our purposes, (23.84) is integrated over the loading step. With the fully implicit method we obtain

$$\begin{aligned} &E = \rho c_\varepsilon \Delta\theta - \Delta Q = 0 \\ &\text{where} \\ &\Delta Q = \boldsymbol{\sigma}^T \Delta\boldsymbol{\varepsilon}^p - K\Delta\kappa - \theta \boldsymbol{P}^T(\Delta\boldsymbol{\varepsilon} - \Delta\boldsymbol{\varepsilon}^p) + \theta\frac{\partial K}{\partial \theta}\Delta\kappa \end{aligned} \tag{23.85}$$

where a matrix formulation was adopted and where it was assumed that only one internal variable exists. Let us also for simplicity assume that $\boldsymbol{D}$ and $\boldsymbol{\beta}$ are constant quantities; this is in analogy with the assumptions in Section 23.5.3. With this in mind let us now recall the set of equations that needs to be considered for the calculation of stresses

$$\begin{aligned} &\boldsymbol{R}_\sigma = \boldsymbol{\sigma} + \Delta\lambda \boldsymbol{D}\frac{\partial g}{\partial \boldsymbol{\sigma}} - \boldsymbol{\sigma}^t \\ &\boldsymbol{R}_K = K - \hat{K}(\kappa^{(1)} - \Delta\lambda\frac{\partial g}{\partial K}, \theta) \\ &R_f = f(\boldsymbol{\sigma}, K, \theta) \\ &R_E = \rho c_\varepsilon \Delta\theta - \Delta Q(\boldsymbol{\sigma}, K, \theta, \Delta\lambda) \end{aligned} \tag{23.86}$$

where it is assumed that only one hardening parameter $K = K(\kappa, \theta)$ exists. Defining the residual vector as $\boldsymbol{V} = [\boldsymbol{R}_\sigma, \boldsymbol{R}_K, R_f, R_E]^T$ and the vector containing the unknowns as $\boldsymbol{S} = [\boldsymbol{\sigma}, \boldsymbol{K}, \Delta\lambda, \theta]^T$, it is evident that $\boldsymbol{V}(\boldsymbol{S}) = \boldsymbol{0}$ defines the solution sought for.

The solution of (23.86) can be found by adopting the Newton-Raphson method and the iterative procedure becomes

$$\mathbf{0} = \boldsymbol{S}^{(i-1)} - \left[\frac{\partial \boldsymbol{V}}{\partial \boldsymbol{S}}^{(i-1)}\right]^{-1} \boldsymbol{V}^{(i-1)}$$

which is found by considering a Taylor series expansion of $\boldsymbol{V}$. The iteration procedure is stopped when the norm of $\boldsymbol{V}$ is sufficiently small. The iteration matrix is straightforward to identify and will therefore not be considered here.

As the last issue consider the algorithmic stiffness that will be used in the mechanical equilibrium iterations. The derivation follows the lines discussed in Section 23.6.2 and only one additional relation needs to be regarded, namely E given in (23.85). According to (23.83), the result from the first three relations in (23.86) is

$$\dot{\boldsymbol{\sigma}} = \boldsymbol{D}_{ats}\dot{\boldsymbol{\varepsilon}} - \boldsymbol{\beta}_{ats}\dot{\theta} \tag{23.87}$$

where it was assumed that $\partial g/\partial \boldsymbol{\sigma}$ only depends on $\boldsymbol{\sigma}$. The task is then to relate $\dot{\theta}$ to $\dot{\boldsymbol{\varepsilon}}$ using (23.86). To maintain a compact formulation, a differentiation of (23.86d) provides after a little algebra

$$\boldsymbol{Q}_{\varepsilon}^{T}\dot{\boldsymbol{\varepsilon}} + B\dot{\theta} = 0$$

where $\boldsymbol{Q}_{\varepsilon}$ and B can be identified if advantage is taken of (23.81), (23.82) and (23.87). The above relation is the algorithmic counterpart of (23.25) for $r = 0$ as well as $q_{i,i} = 0$.

Use of these expressions in (23.87) results in the algorithmic tangent stiffness for adiabatic conditions, namely

Algorithmic tangent stiffness for adiabatic conditions

$$\dot{\boldsymbol{\sigma}} = \boldsymbol{D}_{ats}^{ad}\dot{\boldsymbol{\varepsilon}}$$

where

$$\boldsymbol{D}_{ats}^{ad} = \boldsymbol{D}_{ats} + \frac{1}{B}\boldsymbol{\beta}_{ats}\boldsymbol{Q}_{\varepsilon}^{T}$$

In analogy with (23.76), it is then evident that a Newton-Raphson approach to the equilibrium equations (23.56) and (23.57) results in

Iteration format for adiabatic conditions

$$(\boldsymbol{K}_t)^{i-1}(\boldsymbol{a}^i - \boldsymbol{a}^{i-1}) = -\boldsymbol{\psi}^{i-1}$$

where

$$\boldsymbol{\psi}^{i-1} = \int_V \boldsymbol{B}^T\boldsymbol{\sigma}^{i-1}\,dV - \boldsymbol{f}_{n+1}$$

$$\boldsymbol{K}_t = \int_V \boldsymbol{B}^T\boldsymbol{D}_{ats}^{ad}\boldsymbol{B}\,dV$$

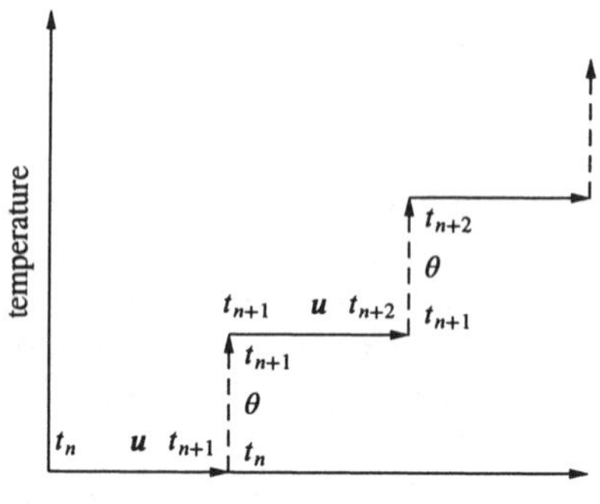

Figure 23.10: Illustration of the isothermal staggered approach.

The elimination of the temperature as described above results in a simple and convenient scheme.

23.8 Staggered solution scheme

As previously discussed, the computational effort relating to the solution of the fully coupled equation system (23.77) is substantial and one approach to avoid this complication is to make use of a so-called *staggered solution scheme*. Considering another type of coupled response, this approach was originately introduced by Park *et al.* (1977) and a review is given by Park and Felippa (1983). The procedure is as follows: the temperature field derived in the previous time step, i.e. at time t_n, is used as an input in the mechanical field equations (23.77a) and iterations are then performed - without changing the temperature field - until the mechanical balance equations are fulfilled at time t_{n+1}. Now the corresponding displacements at time t_{n+1} are used as input in the thermal field equations (23.77b) and iterations are then performed - without changing the displacement field - until the thermal balance equations are fulfilled at time t_{n+1}. Then the next time step Δt is performed. This approach is said to be based on the *isothermal split* and the process is illustrated in Fig. 23.10. The 'staggered' route is the motivation for the terminology of a 'staggered' solution scheme.

This solution scheme is the one most often adopted in coupled thermo-plasticity and it was used in the pioneering work of Argyris *et al.* (1981) as well as in a number of other investigations, cf. for instance Simo and Miehe (1992), Slużalec (1992), Tuğcu (1995), Celentano *et al.* (1996) and Håkansson *et al.* (2005). One advantage of the staggered scheme is that it permits different time steps to be used in the mechanical and in the thermal field equations, thereby considering the different time scales that control these two fields. However, the main disadvantage of the staggered scheme described above is that it is only conditionally stable, i.e. the time step cannot be taken arbitrarily large. This was shown by Armero and Simo (1992) who also concluded that the staggered scheme based on the isothermal split is not suitable for strongly coupled problems; instead they proposed alternative staggered schemes in terms of a so-called *adiabatic split* and an *isentropic split*, which are unconditionally stable. The reader is referred to Armero and Simo (1992) and Armero and Simo (1993) for further details.

Let us consider the staggered solution scheme based on the isothermal split and adopt the backward Euler scheme; the index $n+1$ is as usual omitted. In the first step, the mechanical step, the temperature is kept constant and we are left with the mechanical problem

$$\sigma_{ij,j} + b_i = 0$$

alone. Since we have a purely mechanical problem, the solution procedures described in Chapter 17 can be used directly. For calculation of the stresses, and as the temperature is unchanged, the algorithms described in Chapter 18 are applicable.

In step 2, the thermal part, we consider

$$\rho c_\varepsilon \dot{\theta} = r - q_{i,i} + Q$$

and the configuration is kept constant. It should be noted that owing to the temperature changes this will have an effect on the constitutive relations as these depend on the temperature. In conclusion, even though the configuration (total strains) is kept fixed, changes in the constitutive relations must be considered, i.e. new stresses must be calculated.

It turns out that we have already carried out most of the derivation when we considered the fully coupled problem in Section 23.6. From this derivation we can extract our sought for relations by imposing that the configuration is fixed. Based on (23.62) we therefore obtain the finite element formulation for the thermal problem

$$\boxed{\begin{aligned} &\boldsymbol{\psi}_\theta(\boldsymbol{a}_\theta) = \boldsymbol{0} \\ &\textit{where} \\ &\boldsymbol{\psi}_\theta(\boldsymbol{a}_\theta) = \frac{1}{\Delta t}\boldsymbol{C}(\boldsymbol{a}_\theta - \boldsymbol{a}_{\theta,n}) + \boldsymbol{K}_\theta \boldsymbol{a}_\theta - \boldsymbol{Q} - \boldsymbol{f}_\theta \end{aligned}} \tag{23.88}$$

Adopting the Newton-Raphson scheme and keeping the configuration fixed, the solution procedure is found from (23.73)

$$\boldsymbol{T}_\theta^{i-1}(\boldsymbol{a}_\theta^i - \boldsymbol{a}_\theta^{i-1}) = -\boldsymbol{\psi}_\theta^{i-1}$$

where

$$\boldsymbol{T}_\theta = \tilde{\boldsymbol{C}} + \frac{1}{\Delta t}\boldsymbol{C} + \tilde{\boldsymbol{K}}_\theta + \boldsymbol{K}_\theta - \frac{1}{\Delta t}\boldsymbol{H}$$

with $\tilde{\boldsymbol{C}}$, $\tilde{\boldsymbol{K}}_\theta$ and $\boldsymbol{H}$ defined by (23.52), (23.54) and (23.70) respectively.

As mentioned previously, the stresses need to be calculated in the thermal step. Once more it can be concluded that this topic has already been dealt with in Section 23.5.3. In this integration of the stresses, the total strain increment $\Delta\boldsymbol{\varepsilon}$ is kept fixed and equal to the value found from step 1, but now we change the temperature. This will lead to a new value of the trial stress given by (23.59), and the integration procedure then follows the lines described in Section 23.5.3.

Let us finally consider the adiabatic split. As the name indicates one step is an adiabatic step similar to that dealt with in Section 23.7. Therefore, in the first step we consider

$$\sigma_{ij,j} + b_i = 0\ ; \qquad \rho c_\varepsilon \frac{\tilde{\theta} - \theta_n}{\Delta t} = Q$$

where the last relation is the heat equation for adiabatic heating. The temperature field $\tilde{\theta}$ obtained influences the displacement increment. In the next step the configuration is fixed and the temperature field is achieved by the solution of the following heat equation

$$\rho c_\varepsilon \frac{\theta - \tilde{\theta}}{\Delta t} = r - q_{i,i}$$

and it appears that the $\tilde{\theta}$-field may be considered as the initial temperature field. This second step turns out to be a linear problem if c_ε is constant, otherwise an iterative scheme must be adopted. This can be found by considering (23.88) and adopting $\tilde{\boldsymbol{C}} = \boldsymbol{0}$, $\boldsymbol{Q} = \boldsymbol{0}$ as well as $\boldsymbol{H} = \boldsymbol{0}$. As the derivation of this iterative scheme is straightforward, it will not be pursued further.

24 UNIQUENESS AND DISCONTINUOUS BIFURCATIONS

When solving boundary value problems, it is evidently of great importance to know whether the problem in question has a *unique* solution or not. Unique solutions are most often encountered in solid mechanics even though Euler buckling provides a trivial example of loss of uniqueness and where, above the Euler load, not only the *fundamental solution*, but also the *bifurcated solution* are possible. In this specific case, the finding is a consequence of equilibrium being related to the deformed configuration and not to the undeformed configuration.

Here, we will be concerned with uniqueness and bifurcations entirely related to the material behavior in itself; elasto-plasticity will be assumed. With the exceptions of the textbooks of Nguyen (2000) and Lubarda (2002), these topics are mostly addressed in the journal literature; before entering into a systematic analysis, we will present a simple example.

24.1 Simple illustration - Tension bar

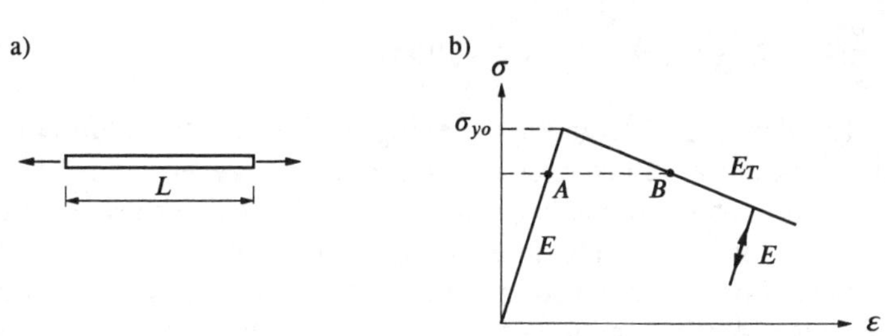

Figure 24.1: a) Homogeneous bar loaded in tension; b) bilinear stress-strain curve.

Consider the homogeneous bar of length L which is loaded uniaxially by a prescribed axial elongation into its postpeak regime, Fig. 24.1a); the material exhibits the bilinear stress-strain curve shown in Fig. 24.1b). Letting ε denote the total axial strain and ε_t the total transverse strain, then

$$\dot{\varepsilon} = \dot{\varepsilon}^e + \dot{\varepsilon}^p; \qquad \dot{\varepsilon}_t = \dot{\varepsilon}_t^e + \dot{\varepsilon}_t^p \tag{24.1}$$

The incremental Hooke's law states that

$$\dot{\varepsilon}^e = \frac{\dot{\sigma}}{E}; \qquad \dot{\varepsilon}_t^e = -\nu\dot{\varepsilon}^e = -\frac{\nu}{E}\dot{\sigma} \tag{24.2}$$

For the plastic behavior, we adopt associated plasticity with the yield function f in terms of the maximum tensile stress criterion, i.e.

$$f(\sigma_{ij}, K) = \sigma - \sigma_y(K); \qquad \sigma = \sigma_1 \tag{24.3}$$

where $\sigma_1 \geq \sigma_2 \geq \sigma_3$ are the principal stresses and σ_y is the current yield stress. The plastic strain rates are given by $\dot{\varepsilon}_{ij}^p = \dot{\lambda}\partial f/\partial\sigma_{ij}$ which in the present case become

$$\dot{\varepsilon}^p = \dot{\lambda}; \qquad \dot{\varepsilon}_t^p = 0 \tag{24.4}$$

According to Fig. 24.1b), we have

$$\dot{\varepsilon} = \frac{\dot{\sigma}}{E_T} \tag{24.5}$$

where E_T is a constant. Use of this expression as well as (24.4) and (24.2) in (24.1) gives

$$\dot{\varepsilon}^p = (\frac{1}{E_T} - \frac{1}{E})\dot{\sigma}; \qquad \dot{\varepsilon}_t = -\frac{\nu}{E}\dot{\sigma} \quad \text{softening regime} \tag{24.6}$$

For elastic unloading, we obtain from (24.2) and (24.1)

$$\dot{\varepsilon} = \frac{\dot{\sigma}}{E}; \qquad \dot{\varepsilon}_t = -\frac{\nu}{E}\dot{\sigma} \quad \text{elastic regime} \tag{24.7}$$

Evidently, when the bar is in the softening regime, a valid solution is to assume that the plastic zone occupies the entire bar. However, let us investigate the possibility that other solutions are realizable. For this purpose assume that the bar has been loaded up to its maximum load capacity σ_{yo}, see Fig. 24.1b). When the bar is further elongated, assume that two regions emerge in the bar as shown in Fig. 24.2: one region of length L_s exhibiting strain softening corresponding to point B in Fig. 24.1b) and another region of length $L - L_s$ is elastically unloading corresponding to point A in Fig. 24.1b). In each region, the strain state is assumed to be uniform and no dynamic effects are involved.

From equilibrium follows that the same axial stress rate $\dot{\sigma}$ applies in both regions. Referring to (24.6) and (24.7) we therefore have the same transverse

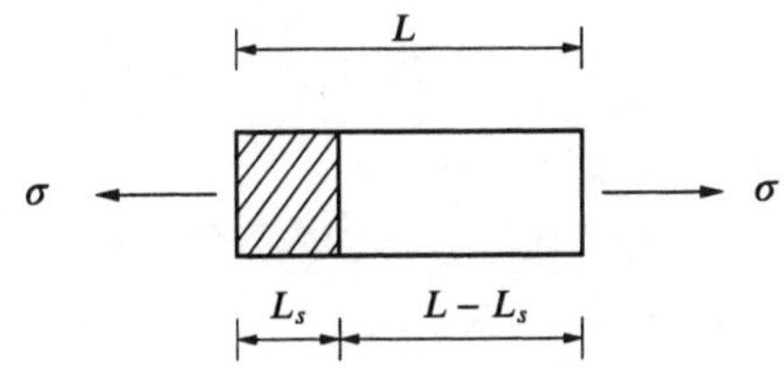

Figure 24.2: Tension bar in softening with one region in elastic unloading and the other exhibiting strain softening.

strain rates in the two regions; consequently, no shear stresses exist along the interface of the two regions. The deformation mode shown in Fig. 24.2 therefore fulfills all field equations and since we have said nothing about the length L_s we have lost uniqueness of the solution. Moreover, as the axial strain rate differs in the two regions we have obtained a situation of a (discontinuous) bifurcation.

If hardening plasticity occurs, i.e. $E_T > 0$, then it is evident that the situation shown in Fig. 24.2 cannot exist. This follows directly from equilibrium, which requires that the stress in both regions must exceed the initial yield stress σ_{yo}, cf. Fig. 24.1b). Therefore, hardening plasticity only allows one solution where the plastic region occupies the entire bar. For softening plasticity, however, uniqueness is lost and we have the situation in Fig. 24.2 where the length L_s is arbitrary. Let us evaluate this case more detailed.

For constant size L_s of the softening region, the elongation rate $\dot{\delta}$ of the bar becomes with (24.5) and (24.7a)

$$\dot{\delta} = (\frac{L_s}{E_T} + \frac{L - L_s}{E})\dot{\sigma} \tag{24.8}$$

If the prescribed elongation is assumed to always increase, we have $\dot{\delta} > 0$ and as $\dot{\sigma} \leq 0$ for $E_T \leq 0$ it follows that

$$L_{min} < L_s \leq L \quad \text{where} \quad L_{min} = \frac{L}{1 - \frac{E}{E_T}} \tag{24.9}$$

If L_s is written as $L_s = kL_{min}$ where k is a dimensionless positive number, (24.8) takes the format

$$\frac{\dot{\sigma}}{\sigma_{yo}} = \frac{1}{1-k}\frac{\dot{\delta}E}{L\sigma_{yo}}; \qquad L_s = kL_{min} \tag{24.10}$$

which is illustrated in Fig. 24.3. It appears that $k > 1$ implies stability in the sense that the bar can be exposed to an increasing elongation, whereas $k < 1$ results in a snap-back response; these aspects were discussed by Bažant (1976)

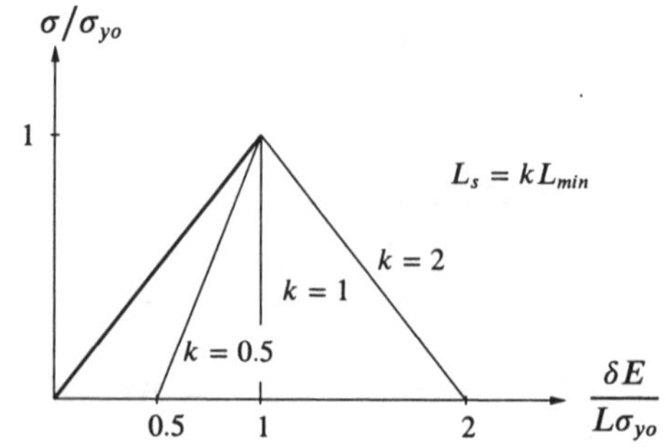

Figure 24.3: Behavior of bar for different k-values; $L_s = kL_{min}$.

and Sture and Ko (1978). Moreover, this stability criterion was combined by Ottosen (1986) with Gibbs' conditions for thermodynamic equilibrium, cf. Section 20.12. It was then shown that cracking of brittle materials like concrete will be characterized by a *cohesive zone* with a constitutive relation between stress and crack opening displacement much along the lines of the *fictitious crack model* of Hillerborg *et al.* (1976).

24.2 Equations of plasticity theory

Let us for convenience summarize the equations of plasticity theory. It is assumed that the stress state is located on the yield surface and then

$$\frac{\partial f}{\partial \sigma_{ij}} D_{ijkl} \dot{\varepsilon}_{kl} \begin{cases} < 0 \Rightarrow & \text{elastic unloading} \\ = 0 \Rightarrow & \text{neutral loading} \\ > 0 \Rightarrow & \text{plastic loading} \end{cases} \tag{24.11}$$

When plastic loading occurs, the incremental stress-strain relation is given by

$$\dot{\sigma}_{ij} = D^{ep}_{ijkl} \dot{\varepsilon}_{kl} \tag{24.12}$$

where

$$D^{ep}_{ijkl} = D_{ijkl} - \frac{1}{A} D_{ijst} \frac{\partial g}{\partial \sigma_{st}} \frac{\partial f}{\partial \sigma_{mn}} D_{mnkl} \tag{24.13}$$

and

$$A = H + \frac{\partial f}{\partial \sigma_{ij}} D_{ijkl} \frac{\partial g}{\partial \sigma_{kl}} > 0 \tag{24.14}$$

The consideration of the following generalized eigenvalue problem turns out to be important

$$D^{ep}_{ijkl} z_{kl} = \lambda D_{ijkl} z_{kl} \tag{24.15}$$

where z_{ij} is the eigenvector and λ the eigenvalue. Use of (24.13) in this expression gives

$$(1 - \lambda) D_{ijkl} z_{kl} - \frac{1}{A} D_{ijst} \frac{\partial g}{\partial \sigma_{st}} \frac{\partial f}{\partial \sigma_{mn}} D_{mnkl} z_{kl} = 0$$

It appears that a solution is given by

$$\lambda = 1 \quad \text{when} \quad \frac{\partial f}{\partial \sigma_{mn}} D_{mnkl} z_{kl} = 0 \tag{24.16}$$

This means that the eigenvalue $\lambda = 1$ holds when neutral loading occurs. On the other hand, if $\lambda \neq 1$ then the expression above shows that the eigenvector $z_{ij} \sim \partial g / \partial \sigma_{ij}$. With (24.14) this leads to

$$\frac{1}{A}(H - \lambda A) D_{ijkl} \frac{\partial g}{\partial \sigma_{kl}} = 0$$

and it is concluded that

$$\lambda = \frac{H}{A} \quad \text{when} \quad z_{ij} = \frac{\partial g}{\partial \sigma_{ij}} \tag{24.17}$$

With these results, it is useful to consider the conditions for a *limit point*. A limit point is defined by $\dot{\sigma}_{ij} = 0$ when $\dot{\varepsilon}_{ij} \neq 0$ and a limit point therefore corresponds to a peak stress. In view of the generalized eigenvalue problem (24.15) and the result (24.17) we conclude that

$$\boxed{\textit{A limit point } \dot{\sigma}_{ij} = 0;\, \dot{\varepsilon}_{ij} \neq 0 \textit{ requires } H^{lp} = 0} \tag{24.18}$$

where superscript 'lp' underlines that it is the plastic modulus at the limit point.

For later purposes, some further properties of the tangential stiffness tensor will now be investigated. Since D_{ijkl} is a positive definite tensor, there exists a positive definite tensor B_{ijkl} with the properties

$$B_{ijmn} B_{mnkl} = D_{ijkl} \tag{24.19}$$

cf. Strang (1980). It follows that B_{ijkl} possesses minor and major symmetries just like D_{ijkl}. Define the following quantities

$$p_{ij} = B_{ijmn} \frac{\partial f}{\partial \sigma_{mn}}; \quad q_{ij} = B_{ijst} \frac{\partial g}{\partial \sigma_{st}} \tag{24.20}$$

Then D^{ep}_{ijkl} given by (24.13) can be written as

$$D^{ep}_{ijkl} = D_{ijkl} - \frac{1}{A} B_{ijst} q_{st} p_{mn} B_{mnkl}$$

The symmetric part $D^{ep,s}_{ijkl}$ of D^{ep}_{ijkl} is then defined by

$$D^{ep,s}_{ijkl} = D_{ijkl} - \frac{1}{2A}(B_{ijst} q_{st} p_{mn} B_{mnkl} + B_{ijst} p_{st} q_{mn} B_{mnkl}) \tag{24.21}$$

With these introductory remarks we will consider the following generalized eigenvalue problem

$$\boxed{D^{ep,s}_{ijkl} z_{kl} = \omega D_{ijkl} z_{kl}} \tag{24.22}$$

which will later prove useful when uniqueness properties are discussed. In analogy with (24.20) we define

$$\psi_{ij} = B_{ijkl} z_{kl}$$

and the eigenvalue problem then takes the form

$$(1-\omega)\psi_{ij} - \frac{1}{2A}(q_{ij} p_{st} \psi_{st} + p_{ij} q_{st} \psi_{st}) = 0 \tag{24.23}$$

If we consider the symmetric second-order tensors p_{ij}, q_{ij} and ψ_{ij} in the six-dimensional space, we obtain the following solution to the eigenvalue problem

$$\boxed{\omega_2 = \omega_3 = \omega_4 = \omega_5 = 1 \quad \text{when} \;\; p_{st}\psi_{st} = 0 \;\; \text{and} \;\; q_{st}\psi_{st} = 0} \tag{24.24}$$

i.e. $\omega = 1$ is an eigenvalue with multiplicity of four.

In view of (24.23) the last two eigenvectors are given by

$$\psi_{ij} = \alpha p_{ij} + \beta q_{ij}$$

where α and β are arbitrary constants. Insertion into (24.23) results in

$$\begin{aligned} &[(1-\omega)\alpha - \frac{1}{2A}(\alpha q_{mn} p_{mn} + \beta q_{mn} q_{mn})] p_{ij} \\ &+ [(1-\omega)\beta - \frac{1}{2A}(\alpha p_{mn} p_{mn} + \beta q_{mn} p_{mn})] q_{ij} = 0 \end{aligned} \tag{24.25}$$

Introduce the following quantities

$$m_{ij} = \frac{p_{ij}}{|p|}; \quad n_{ij} = \frac{q_{ij}}{|q|} \quad \Rightarrow \quad m_{ij} n_{ij} = cos\theta \tag{24.26}$$

and it follows that $m_{ij}m_{ij} = 1$ and $n_{ij}n_{ij} = 1$. Then (24.25) takes the form

$$[(1-\omega)\alpha - \frac{1}{2A}(\alpha|q||p|cos\theta + \beta|q|^2)]p_{ij}$$
$$+ [(1-\omega)\beta - \frac{1}{2A}(\alpha|p|^2 + \beta|q||p|cos\theta)]q_{ij} = 0 \tag{24.27}$$

The coefficients to p_{ij} and q_{ij} must be zero leading to a homogeneous equation system in α and β; a nontrivial solution requires that the determinant to the coefficient matrix be zero and this gives

$$\omega = 1 - \frac{|p||q|}{2A}(cos\theta \pm 1)$$

Since

$$A = H + |p||q|cos\theta$$

we finally obtain

$$\left.\begin{matrix}\omega_1\\ \omega_6\end{matrix}\right\} = \frac{1}{A}[H + \frac{1}{2}|p||q|(cos\theta \pm 1)] \tag{24.28}$$

and it follows that $\omega_1 \geq 1$ and $\omega_6 \leq 1$; these results will be of importance when uniqueness is investigated.

24.3 Uniqueness of elasto-plastic materials

It is of great importance to identify the conditions for which a boundary value problem possesses a unique solution. For this purpose we will follow the concept originately proposed by Kirchhoff (1859) for linear elastic bodies and later applied by Melan (1938) to elasto-plastic problems.

Assume firstly that the loading has brought the body to a certain state and secondly that uniqueness exists; then consider the response for a further incremental loading. From (3.29) we have

$$\dot{\sigma}_{ij,j} + \dot{b}_i = 0$$

Similar to (3.33), the weak format of this equation is given by

$$\int_V v_{i,j}\dot{\sigma}_{ij}\, dV = \int_S v_i\dot{t}_i\, dS + \int_V v_i\dot{b}_i\, dV \tag{24.29}$$

where v_i is an arbitrary weight vector. For a given incremental loading, assume that there exist two different solutions to the problem. Similar to (24.29) we then obtain with evident notation

$$\int_V v_{i,j}\dot{\sigma}_{ij}^{(2)}\,dV = \int_S v_i\dot{t}_i^{(2)}\,dS + \int_V v_i\dot{b}_i^{(2)}\,dV$$

$$\int_V v_{i,j}\dot{\sigma}_{ij}^{(1)}\,dV = \int_S v_i\dot{t}_i^{(1)}\,dS + \int_V v_i\dot{b}_i^{(1)}\,dV$$

Subtraction of these expressions gives

$$\int_V v_{i,j}(\dot{\sigma}_{ij}^{(2)}-\dot{\sigma}_{ij}^{(1)})\,dV = \int_S v_i(\dot{t}_i^{(2)}-\dot{t}_i^{(1)})\,dS + \int_V v_i(\dot{b}_i^{(2)}-\dot{b}_i^{(1)})\,dV \tag{24.30}$$

The two solutions correspond to the same loading so $\dot{b}_i^{(2)} = \dot{b}_i^{(1)}$ holds. Moreover, along the outer boundary the displacements are prescribed along S_u and the traction vector is prescribed along the boundary S_t; therefore $\dot{t}_i^{(2)} = \dot{t}_i^{(1)}$ holds along S_t. With these remarks (24.30) becomes

$$\int_V v_{i,j}(\dot{\sigma}_{ij}^{(2)} - \dot{\sigma}_{ij}^{(1)})\,dV = \int_{S_u} v_i(\dot{t}_i^{(2)} - \dot{t}_i^{(1)})\,dS$$

The weight vector v_i is arbitrary and it is now chosen as $v_i = \dot{u}_i^{(2)} - \dot{u}_i^{(1)}$ which implies that $v_i = 0$ along S_u. The expression above then reduces to

$$\int_V v_{i,j}(\dot{\sigma}_{ij}^{(2)} - \dot{\sigma}_{ij}^{(1)})\,dV = 0$$

Introduce the notation

$$[\dot{\varepsilon}_{ij}] = \dot{\varepsilon}_{ij}^{(2)} - \dot{\varepsilon}_{ij}^{(1)} \qquad [\dot{\sigma}_{ij}] = \dot{\sigma}_{ij}^{(2)} - \dot{\sigma}_{ij}^{(1)} \tag{24.31}$$

which leads to

If two solutions exist, they fulfill

$$\int_V [\dot{\varepsilon}_{ij}][\dot{\sigma}_{ij}]\,dV = 0 \tag{24.32}$$

If both solutions correspond to elastic unloading or neutral loading then $[\dot{\sigma}_{ij}] = D_{ijkl}[\dot{\varepsilon}_{kl}]$ leading to

$$\int_V [\dot{\varepsilon}_{ij}]D_{ijkl}[\dot{\varepsilon}_{kl}]\,dV = 0 \tag{24.33}$$

Since we required the strain energy to be positive and thereby D_{ijkl} to be positive definite, cf.(4.40), the expression above implies $[\dot{\varepsilon}_{ij}] = 0$ meaning that solutions to linear elastic boundary value problems are unique.

Before uniqueness of elasto-plastic boundary value problems are considered, we will derive an intermediate result. Assume that both solutions correspond to plastic loading then

$$\boxed{I_{pp} = [\dot{\varepsilon}_{ij}][\dot{\sigma}_{ij}] = [\dot{\varepsilon}_{ij}]D^{ep}_{ijkl}[\dot{\varepsilon}_{kl}]} \tag{24.34}$$

where I refers to the integrand in (24.32) and the subscript 'pp' indicates that both solutions respond plastically. Use of (24.13) implies that we also have

$$I_{pp} = [\dot{\varepsilon}_{ij}]D_{ijkl}[\dot{\varepsilon}_{kl}] - \frac{1}{A}[\dot{\varepsilon}_{ij}]D_{ijst}\frac{\partial g}{\partial \sigma_{st}}\frac{\partial f}{\partial \sigma_{mn}}D_{mnkl}[\dot{\varepsilon}_{kl}] \tag{24.35}$$

Suppose that one solution (2) corresponds to plastic loading and the other solution (1) corresponds to elastic unloading/neutral loading, then

$$I_{pe} = [\dot{\varepsilon}_{ij}][\dot{\sigma}_{ij}] = [\dot{\varepsilon}_{ij}](D^{ep}_{ijkl}\dot{\varepsilon}^{(2)}_{kl} - D_{ijkl}\dot{\varepsilon}^{(1)}_{kl}) \tag{24.36}$$

Insertion of D_{ijkl} as determined by (24.13) results in

$$I_{pe} = [\dot{\varepsilon}_{ij}]D^{ep}_{ijkl}[\dot{\varepsilon}_{kl}] - \frac{1}{A}[\dot{\varepsilon}_{ij}]D_{ijst}\frac{\partial g}{\partial \sigma_{st}}\underbrace{\frac{\partial f}{\partial \sigma_{mn}}D_{mnkl}\dot{\varepsilon}^{(1)}_{kl}}_{\leq 0}$$

Assume that $[\dot{\varepsilon}_{ij}]D_{ijst}\frac{\partial g}{\partial \sigma_{st}} \geq 0$, then, since solution (1) was assumed to unload elastically, a comparison of the expression above with (24.34) shows that

$$\boxed{I_{pp} \leq I_{pe}} \tag{24.37}$$

Next assume that $[\dot{\varepsilon}_{ij}]D_{ijst}\frac{\partial g}{\partial \sigma_{st}} < 0$, then insertion of D^{ep}_{ijkl} as given by (24.13) into (24.36) gives

$$I_{pe} = [\dot{\varepsilon}_{ij}]D_{ijkl}[\dot{\varepsilon}_{kl}] - \frac{1}{A}\underbrace{[\dot{\varepsilon}_{ij}]D_{ijst}\frac{\partial g}{\partial \sigma_{st}}}_{<0}\underbrace{\frac{\partial f}{\partial \sigma_{mn}}D_{mnkl}\dot{\varepsilon}^{(2)}_{kl}}_{>0}$$

Since the elastic stiffness D_{ijkl} is positive definite, $I_{pe} > 0$ holds. Consequently, assume that we have reached a situation where $I_{pp} = 0$ which means loss of uniqueness, but at the same tine we have $I_{pe} > 0$. With this result, conclusion (24.37) is obtained once again.

Expression (24.37) shows that loss of uniqueness involving a situation of plastic loading and elastic unloading can never precede a situation of loss of uniqueness where both solutions exhibit plastic loading. Thus, the case of plastic loading/plastic loading is the most critical situation and this was used by Hill (1958, 1978) to introduce a so-called *linear comparison solid* which is a solid that only exhibits plastic loading.

We therefore conclude that as long as I_{pp} given by (24.34) is positive for $|\dot{\varepsilon}_{ij}| \neq 0$ then uniqueness exists, i.e.

$$\boxed{\begin{array}{l}\textit{Uniqueness exists if}\\ [\dot{\varepsilon}_{ij}]D^{ep}_{ijkl}[\dot{\varepsilon}_{kl}] > 0\end{array}} \tag{24.38}$$

We also observe that in this quadratic expression, it is only the symmetric part $D^{ep,s}_{ijkl}$ of the tangential stiffness tensor that is of importance. When one of the eigenvalues of $D^{ep,s}_{ijkl}$ becomes zero then uniqueness is lost; considering the eigenvalue problem (24.22) with the solutions (24.24) and (24.28) it is concluded that this occurs when $\omega_6 = 0$. We then obtain the result

$$\boxed{\begin{array}{l}\textit{Loss of uniqueness:}\\ \quad H^{lu} = \dfrac{1}{2}|p||q|(1 - cos\theta)\\ \textit{where}\\ \quad |p| = (\dfrac{\partial f}{\partial \sigma_{ij}}D_{ijkl}\dfrac{\partial f}{\partial \sigma_{kl}})^{1/2}; \quad |q| = (\dfrac{\partial g}{\partial \sigma_{ij}}D_{ijkl}\dfrac{\partial g}{\partial \sigma_{kl}})^{1/2}\\ \quad cos\theta = \dfrac{1}{|p||q|}\dfrac{\partial f}{\partial \sigma_{ij}}D_{ijkl}\dfrac{\partial g}{\partial \sigma_{kl}}\end{array}} \tag{24.39}$$

where H^{lu} denotes the value of the plastic modulus when loss of uniqueness occurs.

For associated plasticity where $g = f$ and thereby $cos\theta = 1$ hold, we obtain $H^{lu} = 0$, i.e. uniqueness is assured in the hardening regime and this is in accordance with the previous discussion in the introductory example. For nonassociated plasticity, however, $cos\theta < 1$ and loss of uniqueness then occurs in the hardening regime where $H^{lu} > 0$. The result given in (24.39) was established by Maier and Hueckel (1979) as well as by Raniecki and Bruhns (1981) and the analysis presented here shares much in common with the approach suggested by Runesson and Mroz (1989).

It is of interest to compare the conditions for uniqueness, namely D^{ep}_{ijkl} being positive definite, with those of thermodynamic equilibrium as given by (21.109). For isothermal plasticity, Helmholtz' free energy ψ is given by (22.2) and we obtain

$$\rho\frac{\partial^2 \psi}{\partial \varepsilon_{ij}\partial \varepsilon_{kl}} = D_{ijkl}; \quad \rho\frac{\partial^2 \psi}{\partial \varepsilon_{ij}\partial \kappa_\alpha} = 0$$

Insertion of these expressions into (21.109) gives

$$(\dot{\varepsilon}_{ij} - \dot{\varepsilon}^p_{ij})D_{ijkl}(\dot{\varepsilon}_{kl} - \dot{\varepsilon}^p_{kl}) + \dot{\kappa}_\alpha \rho\frac{\partial^2 \psi^p}{\partial \kappa_\alpha \partial \kappa_\beta}\dot{\kappa}_\beta > 0$$

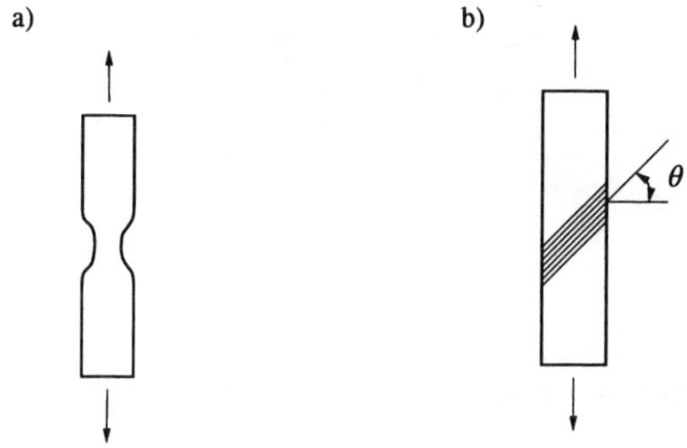

Figure 24.4: a) Continuous bifurcation; b) discontinuous bifurcation i.e. a shear band.

Since $\dot{\sigma}_{ij} = D^{ep}_{ijkl}\dot{\varepsilon}_{kl} = D_{ijkl}(\dot{\varepsilon}_{kl} - \dot{\varepsilon}^p_{kl})$ we obtain

$$\dot{\varepsilon}_{ij} D^{ep}_{ijkl}\dot{\varepsilon}_{kl} - \dot{\varepsilon}^p_{ij} D_{ijkl}(\dot{\varepsilon}_{kl} - \dot{\varepsilon}^p_{kl}) + \dot{\kappa}_\alpha \rho \frac{\partial^2 \psi^p}{\partial \kappa_\alpha \partial \kappa_\beta}\dot{\kappa}_\beta > 0 \tag{24.40}$$

For associated plasticity, the evolution laws are given by

$$\dot{\varepsilon}^p_{ij} = \dot{\lambda}\frac{\partial f}{\partial \sigma_{ij}}; \qquad \dot{\kappa}_\alpha = -\dot{\lambda}\frac{\partial f}{\partial K_\alpha}$$

Insertion into (24.40) and making use of (24.14) provide

$$\dot{\varepsilon}_{ij} D^{ep}_{ijkl}\dot{\varepsilon}_{kl} - \dot{\lambda}\frac{\partial f}{\partial \sigma_{ij}} D_{ijkl}\dot{\varepsilon}_{kl} + (\dot{\lambda})^2 A > 0$$

and since $\partial f/\partial \sigma_{ij} D_{ijkl}\dot{\varepsilon}_{kl} = A\dot{\lambda}$ we obtain $\dot{\varepsilon}_{ij} D^{ep}_{ijkl}\dot{\varepsilon}_{kl} > 0$. We have then reached the interesting observation that for associated plasticity, the conditions for uniqueness and thermodynamic equilibrium coincide.

24.4 Discontinuous bifurcations

When uniqueness is lost, more solutions are possible and apart from the fundamental solution, *bifurcations* are possible; these bifurcations can manifest themselves in terms of continuous bifurcations or *discontinuous bifurcations* as illustrated in Fig. 24.4.

The discontinuous bifurcations are also called *shear bands*, i.e. narrow bands across which a discontinuity occurs in the rate of the displacement gradient; for metals and steel, they can take the form of *Lüder's bands* as discussed, for instance, by Nadai (1950). In the pioneering works of Rudnicki and Rice (1975)

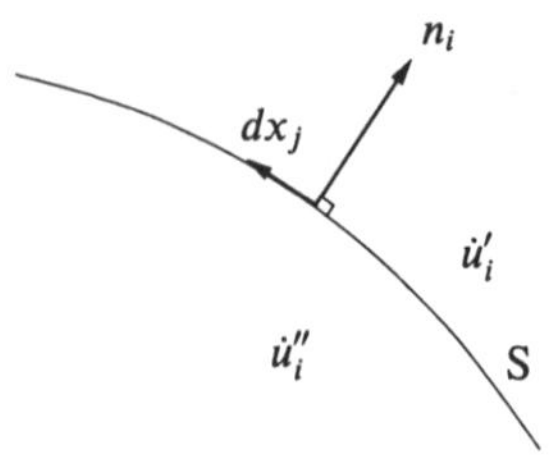

Figure 24.5: Singular surface i.e. characteristic surface.

and Rice (1976), the phenomenon was attributed directly to the properties of the constitutive model; however, the effects of large strains were also considered in these works.

The concept of a characteristic surface, across which a discontinuity of the rate of deformation gradient is permitted was already considered by Hill (1950) and Thomas (1961) and the classical argument for *localization* then implies a shear band bounded by two characteristic surfaces. We will see that the predictions of the directions of the characteristic surfaces made by plasticity theory are in close accuracy with experimental data; however, the width of the shear band is left unspecified. This is simply to say that conventional continuum theory lacks information on a length scale and this aspect has been used as an argument for invoking non-local continuum theories which contain a length scale that reflects the internal structure of the material, cf. Aifantis (1984), Belytschko *et al.* (1986), de Borst (1991), Strömberg and Ristinmaa (1996), Bassani (2001), Fleck and Hutchinson (2001) and Gurtin (2002).

Assume that the current state is characterized by continuous displacements, stresses and strains. With increased loading, we will consider the possibility that discontinuous bifurcations of the displacement rate $\dot{u}_i$ and the rate of the displacement gradient $\dot{u}_{i,j}$ can occur across a fixed *singular surface* S within the body, cf. Fig. 24.5. It is assumed that the difference between the value of $\dot{u}_i$ for the bifurcated and fundamental fields is preserved along S, i.e. $[\dot{u}_i]$ = constant along S, where the bracket denotes the difference of the two fields. It will appear that the strain rate and the stress rate then become discontinuous across S; note that here we have also allowed the velocity $\dot{u}_i$ to become discontinuous across S.

Let the orientation of this singular, or characteristic, surface S be defined by the unit normal vector n_i and denote the position of S by x_i. The assumption that $[\dot{u}_i]$ = constant along S implies that

$$d[\dot{u}_i] = [\dot{u}_{i,j}]\, dx_j = 0$$

where $[\dot{u}_{i,j}] = \partial[\dot{u}_i]/\partial x_j$ and dx_j is an arbitrary differential vector tangential to S, cf. Fig. 24.5. The general solution to the expression above is easily seen to become

$$[\dot{u}_{i,j}] = c_i n_j \quad \Rightarrow \quad [\dot{\varepsilon}_{ij}] = \frac{1}{2}(c_i n_j + n_i c_j) \tag{24.41}$$

where c_i is an arbitrary vector.

It follows from equilibrium considerations that the traction rate across the singular surface S must be unique

$$[\dot{\sigma}_{ij}] n_j = 0 \tag{24.42}$$

Assume that the material at both sides of the surface S responds plastically. As the field before bifurcation was assumed to be continuous, the tangential stiffness tensor takes the same value on either side of the surface S. Therefore, insertion of $[\dot{\sigma}_{ij}] = D^{ep}_{ijkl}[\dot{\varepsilon}_{kl}]$ into (24.42) and taking advantage of (24.41) results in

Discontinuous bifurcation condition

$$Q^{ep}_{il} c_l = 0$$

where the elasto-plastic acoustic tensor Q^{ep}_{il} *is defined by*

$$Q^{ep}_{il} = n_j D^{ep}_{ijkl} n_k \tag{24.43}$$

Other names for the *acoustic tensor* are *characteristic stiffness tensor* and *polarization tensor* and we will later comment upon this terminology; we observe that Q^{ep}_{il} is symmetric if D^{ep}_{ijkl} is likewise.

Consider now the situation where the material on one side of the surface S responds plastically ($\dot{\varepsilon}''_{ij}$) and the material on the other side unloads elastically ($\dot{\varepsilon}'_{ij}$). Then (24.43) leads to

$$n_j(D^{ep}_{ijkl}\dot{\varepsilon}''_{kl} - D_{ijkl}\dot{\varepsilon}'_{kl}) = 0 \tag{24.44}$$

Elimination of D_{ijkl} by means of (24.13) gives

$$Q^{ep}_{il} c_l = \frac{1}{A} n_j D_{ijst} \frac{\partial g}{\partial \sigma_{st}} \frac{\partial f}{\partial \sigma_{mn}} D_{mnkl} \dot{\varepsilon}'_{kl} \tag{24.45}$$

Expressions (24.43) and (24.45) are the classical bifurcation conditions discussed by Rice (1976).

We will now show that plastic/plastic bifurcation determined by (24.43) always occurs before the plastic/elastic bifurcation given by (24.45); this situation is analogous with that encountered in uniqueness, cf. (24.37), and we can therefore concentrate in the following on the plastic/plastic bifurcation. Before that, we must prove this result. For this purpose introduce the notations

$$a_i = n_j D_{ijst} \frac{\partial f}{\partial \sigma_{st}}; \qquad b_i = n_j D_{ijst} \frac{\partial g}{\partial \sigma_{st}} \tag{24.46}$$

and

$$\alpha = \frac{\partial f}{\partial \sigma_{mn}} D_{mnkl} \dot{\varepsilon}'_{kl} \leq 0\,; \qquad \beta = \frac{\partial f}{\partial \sigma_{mn}} D_{mnkl} \dot{\varepsilon}''_{kl} > 0 \tag{24.47}$$

where $\dot{\varepsilon}'_{ij}$ corresponds to elastic unloading and $\dot{\varepsilon}''_{ij}$ corresponds to plastic loading. Then (24.45) can be written as

$$Q^{ep}_{il} c_l = \frac{\alpha}{A} b_i \tag{24.48}$$

Elimination of D^{ep}_{ijkl} in (24.44) by means of (24.13) results in

$$Q_{il} c_l = \frac{\beta}{A} b_i \tag{24.49}$$

where

$$\boxed{Q_{il} = n_j D_{ijkl} n_k} \tag{24.50}$$

is the elastic acoustic tensor; as D_{ijkl} is symmetric and positive definite, so is Q_{il}. Determination of b_i from (24.49) and insertion into (24.48) lead to

$$Q^{ep}_{il} c_l = \frac{\alpha}{\beta} Q_{il} c_l \tag{24.51}$$

Consider this expression as a generalized eigenvalue problem with the eigenvalue α/β. According to (24.47), $\alpha/\beta < 0$ and as (24.43) corresponds to a zero eigenvalue, it follows immediately that plastic/plastic bifurcation occurs prior to plastic/elastic bifurcation.

The trivial solution $c_i = 0$ of the bifurcation condition (24.43) implies that the solution is continuous, cf. (24.41) whereas a nontrivial solution exists if Q^{ep}_{il} is singular. By analogy with the terminology used to classify scalar second-order partial differential equations, smooth solutions imply *ellipticity* and discontinuous solutions can only exist if ellipticity is lost. Therefore, when Q^{ep}_{il} becomes singular, ellipticity is lost, cf. Knops and Payne (1971). Clearly, only bifurcation solutions with the discontinuity defined by (24.41) are associated with loss of ellipticity and for linear elasticity where the acoustic tensor is positive definite, ellipticity always exists.

To identify when the elasto-plastic acoustic tensor becomes singular, we consider the following generalized eigenvalue problem

$$Q^{ep}_{il} z_l = \Lambda Q_{il} z_l \tag{24.52}$$

With the definition of the elasto-plastic acoustic tensor given by (24.43), we obtain

$$Q^{ep}_{il} = Q_{il} - \frac{1}{A} b_i a_l$$

and insertion into (24.52) results in

$$(1-\Lambda)Q_{il}z_l = \frac{1}{A}b_i(a_l z_l) \tag{24.53}$$

Since the elastic acoustic tensor is positive definite and symmetric, it possesses an inverse S_{ki} that is also positive definite and symmetric, i.e.

$$\boxed{S_{ki}Q_{il} = \delta_{kl}} \tag{24.54}$$

Therefore, multiplication of (24.53) by S_{ki} gives

$$(1-\Lambda)z_k = \frac{1}{A}S_{ki}b_i(a_l z_l) \tag{24.55}$$

It follows directly that

$$\Lambda_1 = \Lambda_2 = 1\,; \quad \text{when} \quad a_l z_l = 0 \tag{24.56}$$

is an eigenvalue of multiplicity of two. If $\Lambda \neq 1$ (24.55) shows that z_k must be proportional to $S_{ki}b_i$ and insertion of this eigenvector gives

$$\Lambda_3 = 1 - \frac{1}{A}a_i S_{il} b_l \tag{24.57}$$

From the results (24.56) and (24.57) appear that $\Lambda_3 = 0$ is the only possibility for O_{il}^{ep} being singular. With (24.14) and (24.46) we then obtain

$$\boxed{H(n_i) = -\frac{\partial f}{\partial \sigma_{ij}}D_{ijkl}\frac{\partial g}{\partial \sigma_{kl}} + n_j D_{ijst}\frac{\partial f}{\partial \sigma_{st}}S_{il}\frac{\partial g}{\partial \sigma_{mn}}D_{mnkl}n_k} \tag{24.58}$$

In this expression, all quantities except n_i are known and $H = H(n_i)$ then holds. The task is now to determine that shear band direction n_i which provides the maximum value of H. Evidently, this corresponds to the situation when a discontinuous bifurcation first becomes possible; this value is called H^{db} and we have

$$\boxed{\begin{array}{l}\textit{Discontinuous bifurcations occur for}\\ \quad H^{db} = \max H(n_i)\\ \textit{which corresponds to loss of ellipticity}\end{array}} \tag{24.59}$$

For nonassociated Drucker-Prager plasticity, analytical solutions to this problem were given by Rudnicki and Rice (1975) and later generalizations were established by Ottosen and Runesson (1991a), Runesson *et al.* (1991), Ottosen and Runesson (1991c), Bigoni and Hueckel (1991), Neilsen and Schreyer (1993) and Schreyer and Neilsen (1996). As shown by Rizzi *et al.* (1995, 1996) and Ekh and Runesson (2000), it is possible to rephrase elastic-damage models

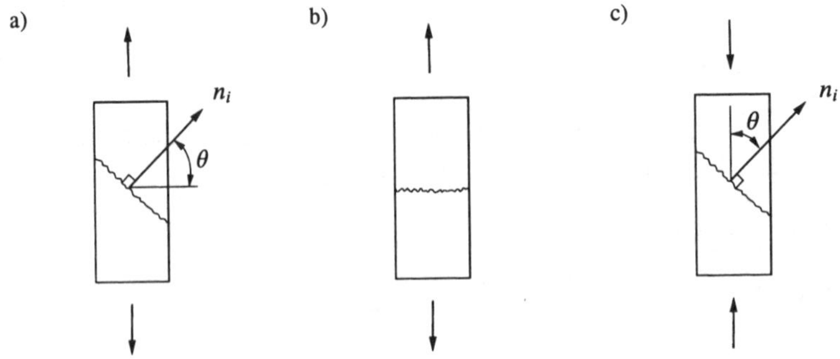

Figure 24.6: Predicted bifurcation directions for associated plasticity; a) von Mises material, b) Rankine material in tension and c) Coulomb material in compression.

in a plasticity-like format so that the results above can be used for bifurcation analysis in damage mechanics. In a finite element context, bifurcation results have been incorporated by the concepts proposed by Ortiz *et al.* (1987) and Larsson *et al.* (1993, 1998); we also refer to the regularization techniques used in localization analysis and discussed by Needleman (1988), Needleman and Tvergaard (1992) and Tomita (1994).

In order to illustrate a few of these results, the coordinate system is chosen collinear with the principal stress directions so that the x_1-axis is in the direction of the largest principal stress σ_1 where $\sigma_1 \geq \sigma_2 \geq \sigma_3$; the angle θ is measured from the unit vector n_i normal to the shear band to the σ_3-direction. For a von Mises material in plane stress, we then obtain

$$\tan^2\theta = -\frac{s_1}{s_2}; \qquad H^{db} = 0 \tag{24.60}$$

valid when $\sigma_1 \geq \sigma_2/2$ and $\sigma_1 \leq 2\sigma_2$, cf. Runesson *et al.* (1991). For uniaxial tension this gives $\theta = 54.7^0$ which is the classical result obtained by Nadai (1950) and Thomas (1961), cf. Fig. 24.6a). It is of considerable interest that experimental data reported by Nadai (1950) shows that $\theta \approx 57^0$. This demonstrates that the bifurcation analysis predicts shear band directions that are in close accuracy with experimental data. However, no information is given about the thickness of the shear band and, as discussed previously, such information requires the use of non-local theories. Returning to (24.60), it is observed that pure shear ($s_1 = -s_2$) results in $\theta = 45^0$.

For associated Coulomb plasticity, Ottosen and Runesson (1991c) derived the following results

$$\tan^2\theta = \frac{1+\sin\phi}{1-\sin\phi}; \qquad H^{db} = 0 \tag{24.61}$$

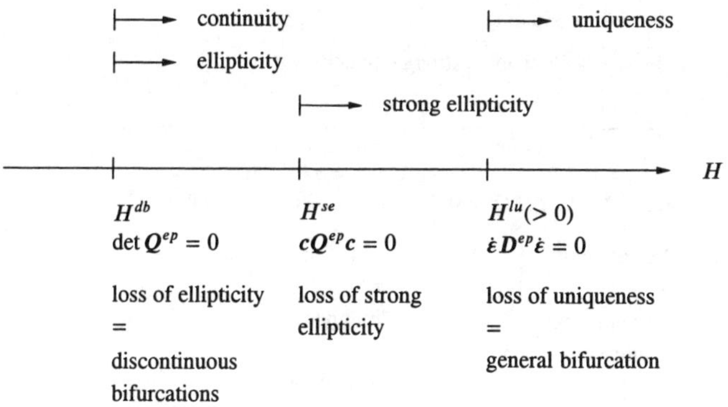

Figure 24.7: Identification of events for nonassociated plasticity.

where the angle θ for uniaxial compression is shown in Fig. 24.6c) and ϕ denotes the friction angle. It follows that $\theta = \pm(45^o + \phi/2)$ and it is of considerable interest that this is exactly the angle predicted by Mohr's failure mode criterion (8.52). The results for nonassociated Coulomb plasticity are also presented by Ottosen and Runesson (1991c). The fact that these results differ from (24.61), underlines that Mohr's failure mode criterion is a postulate that is in agreement with bifurcation analysis for associated Coulomb plasticity. If the friction angle is chosen as $\phi = 90^o$ then the Coulomb criterion reduces to the Rankine criterion and (24.61) provides $\theta = 90^o$. Recalling that θ measures the angle from the normal vector n_i to the σ_3-direction, this means that the bifurcation band is orthogonal to the maximum principal stress as illustrated in Fig. 24.6b); this is in close agreement with the emergence of cracks in brittle materials like concrete and ceramics.

As we have previously defined the concept of ellipticity, we now introduce the condition of *strong ellipticity* according to

Ellipticity is defined by
$Q_{il}^{ep} c_l \neq 0$
and strong ellipticity is defined by
$c_i Q_{il}^{ep} c_l > 0$

(24.62)

cf. Knops and Payne (1971) and Bigoni and Zaccaria (1992).

Let H^{se} be the plastic modulus associated with loss of strong ellipticity. A comparison of (24.38) and (24.62) shows that uniqueness implies strong ellipticity, whereas the converse is not true; therefore $H^{lu} \geq H^{se}$. Moreover, loss of ellipticity $Q_{il}^{ep} c_l = 0$ implies $c_i Q_{il}^{ep} c_l = 0$, i.e. loss of strong ellipticity. On the other hand, loss of strong ellipticity does not necessarily mean loss of ellipticity.

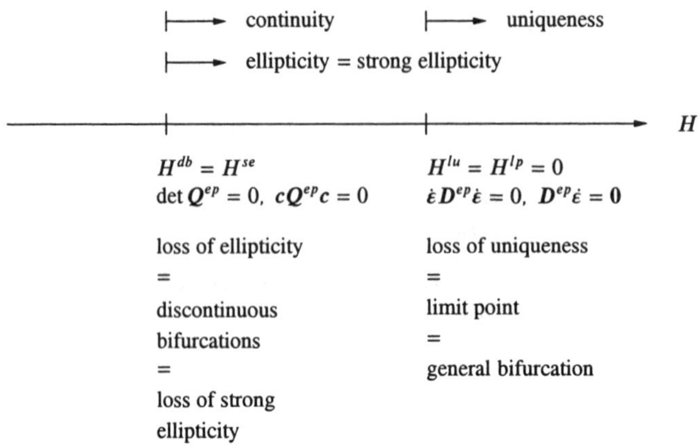

Figure 24.8: Identification of events for associated plasticity.

Thus, $H^{se} \geq H^{db}$ stating that strong ellipticity is a stronger requirement than ellipticity, as expected. With these observations as well as (24.39), we have arrived at the conclusions

$$\boxed{\begin{array}{l}\textit{Nonassocitated plasticity}\\ H^{lu} \geq H^{se} \geq H^{db}; \qquad H^{lu} > H^{lp} = 0\end{array}} \tag{24.63}$$

For associated plasticity where loss of uniqueness occurs when $H^{lu} = 0$, cf. (24.39) and where the conditions for ellipticity and strong ellipticity coincide, we obtain

$$\boxed{\begin{array}{l}\textit{Associtated plasticity}\\ 0 = H^{lp} = H^{lu} \geq H^{db} = H^{se}\end{array}} \tag{24.64}$$

These conclusions are illustrated in Figs. 24.7 and 24.8.

24.5 Acceleration waves

Propagation of *acceleration waves* in solid bodies relates directly to the important issues of static discontinuous bifurcations, plane wave propagation and stability. Development within this fundamental field was initiated by Hadamard (1903) where elastic bodies were studied. Hill (1961, 1962) and Mandel (1962, 1964) extended this work to elasto-plasticity and further progress was made by Rice (1976); a comprehensive treatment is also given by Truesdell and Noll (1965). It turns out that propagation of acceleration waves leads to an eigenvalue problem where analytical solutions for associated plasticity were given by

Hill (1962); analytical solutions for nonassociated plasticity were established by Ottosen and Runesson (1991b). Before the problem can be formulated, some preliminary results need to be deduced.

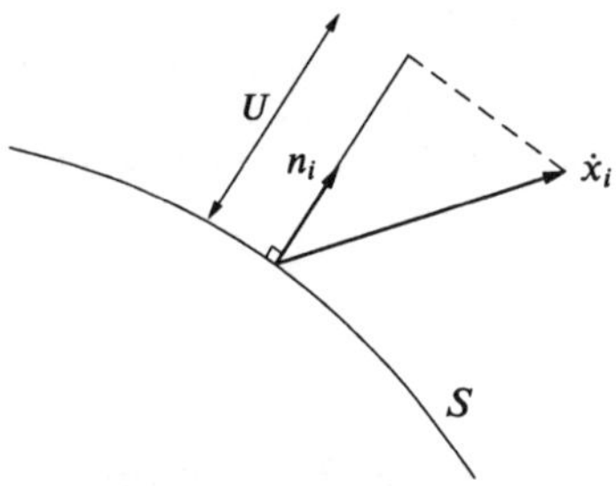

Figure 24.9: Surface S moving through the body at wave speed U.

For a given time t, the increment of the function $f_i(x_k, t)$ is given by $df_i = \partial f_i/\partial x_k\, dx_k$. Let $ds = |dx_k|$ denote the length of dx_k and $s_k = dx_k/ds$ is then the unit vector in the direction of dx_k. It follows that

$$\frac{df_i}{ds} = \frac{\partial f_i}{\partial x_k} s_k \tag{24.65}$$

Let us now study the motion of a surface S which moves through the body, cf. Fig. 24.9. At a given time, it is assumed that f_i is constant along the surface S. This means that $df_i = 0$ holds along S and in analogy with (24.41), it is concluded that

$$\frac{\partial f_i}{\partial x_k} = c_i n_k \tag{24.66}$$

where n_k is the unit vector normal to the surface S and c_i is an arbitrary vector. With this result and choosing in (24.65) s_k as n_k, we obtain

$$\frac{df_i}{dn} = c_i \tag{24.67}$$

which means that (24.66) can be written as

$$\frac{\partial f_i}{\partial x_k} = \frac{df_i}{dn} n_k \tag{24.68}$$

If, instead of the vector f_i, the second-order tensor f_{ij} is considered, we obtain in a similar manner

$$\frac{\partial f_{ij}}{dx_k} = \frac{df_{ij}}{dn} n_k \tag{24.69}$$

According to Fig. 24.9, the surface S moves through the body. A particle on this surface is at the velocity $\dot{x}_i$ and the component of the velocity in the direction of the unit normal vector n_i is by definition the so-called *wave speed* U

$$U = \dot{x}_i n_i \tag{24.70}$$

Assume now that the function $f_i(x_k, t)$ is constant along the surface S. Differentiation with respect to time gives

$$\dot{f}_i + \frac{\partial f_i}{\partial x_k} \dot{x}_k = 0 \tag{24.71}$$

where $\dot{f}_i = \partial f_i / \partial t$. Use of (24.68) in (24.71) gives

$$\dot{f}_i + U \frac{df_i}{dn} = 0 \tag{24.72}$$

Likewise, if the function $f_{ij}(x_k, t)$ is constant along the surface S, we obtain

$$\dot{f}_{ij} + U \frac{df_{ij}}{dn} = 0 \tag{24.73}$$

With these preliminary results, we are ready to investigate the situation where the surface S in Fig. 24.9 is moving through the body. The displacements, velocity, displacement gradient, strains, and stresses are assumed to vary continuously across the surface S

$$[u_i] = [\dot{u}_i] = 0\,; \qquad [u_{i,j}] = [\varepsilon_{ij}] = [\sigma_{ij}] = 0$$

The conditions for which the strain rate, stress rate and the acceleration become discontinuous across the moving surface S will now be investigated.

Since $[\dot{u}_i] = 0$ and $[\sigma_{ij}] = 0$ hold along the surface, use of (24.66) and (24.67) gives

$$\frac{\partial [\dot{u}_i]}{\partial x_k} = c_i n_k\,; \qquad c_i = \frac{d[\dot{u}_i]}{dn} \quad \Rightarrow \quad [\dot{\varepsilon}_{ij}] = \frac{1}{2}(c_i n_j + n_i c_j) \tag{24.74}$$

whereas (24.69) results in

$$\frac{\partial [\sigma_{ij}]}{\partial x_k} = \frac{d[\sigma_{ij}]}{dn} n_k \tag{24.75}$$

Similar to (24.72) and (24.73) we have

$$[\ddot{u}_i] + U \frac{d[\dot{u}_i]}{dn} = 0\,; \qquad [\dot{\sigma}_{ij}] + U \frac{d[\sigma_{ij}]}{dn} = 0 \tag{24.76}$$

Multiplication of the last equation by n_j and use of (24.75) give

$$[\dot{\sigma}_{ij}]n_j + U\frac{\partial[\sigma_{ij}]}{\partial x_j} = 0 \tag{24.77}$$

Since the fields on both sides of the surface S satisfy the equations of motion and since also the body force b_i and mass density ρ are identical, we obtain

$$\frac{\partial[\sigma_{ij}]}{\partial x_j} = \rho[\ddot{u}_i]$$

Use of this expression in (24.77) provides

$$[\dot{\sigma}_{ij}]n_j + U\rho[\ddot{u}_i] = 0 \tag{24.78}$$

With (24.74b), (24.76a) reads

$$[\ddot{u}_i] + Uc_i = 0$$

and insertion into (24.78) gives

$$[\dot{\sigma}_{ij}]n_j = \rho U^2 c_i \tag{24.79}$$

It is of interest to compare this expression with (24.42) applicable to the static situation. Since $[\dot{\sigma}_{ij}] = D^{ep}_{ijkl}[\dot{\varepsilon}_{kl}]$ insertion into the expression above and use of (24.74c) give the following eigenvalue problem

Condition for acceleration waves	
$Q^{ep}_{il}c_l = \rho U^2 c_i$	(24.80)

which certainly reduces to the condition for static discontinuous bifurcations (24.43) when the wave speed $U = 0$.

Since the eigenvalue problem (24.80) only allows certain eigenvectors c_i, these eigenvectors are said to be polarized and this is the reason why Q^{ep}_{il} is occasionally termed the polarization tensor.

The eigenvalue problem above determines the condition for acceleration waves, i.e. the condition that stress rates and strain rates vary discontinuously across a surface S that moves through the body. However, the same condition also controls propagation of *plane waves*. By definition, a plane wave in direction n_m is given by

$$u_i = c_i f(n_m x_m \pm Ut) \tag{24.81}$$

where c_i, n_m and U are constants and f denotes an arbitrary function; U is the *phase speed*. It follows that

$$\ddot{u}_i = U^2 c_i \frac{\partial^2 f}{\partial(n_m x_m \pm Ut)^2}; \qquad u_{k,lj} = c_k n_l n_j \frac{\partial^2 f}{\partial(n_m x_m \pm Ut)^2} \tag{24.82}$$

Let the material be stressed to a certain state in static equilibrium, i.e

$$\sigma_{ij,j} + \rho b_i = 0 \tag{24.83}$$

Assuming that the body forces are unchanged, we now investigate the existence of small vibrations about this equilibrium state. The additional small stresses and displacements caused by the vibrations are denoted by σ^*_{ij} and u_i respectively. Thus

$$(\sigma_{ij} + \sigma^*_{ij})_{,j} + \rho b_i = \rho \ddot{u}_i$$

Subtraction of (24.83) gives

$$\sigma^*_{ij,j} = \rho \ddot{u}_i$$

In the original equilibrium configuration, it is assumed that the material is in a homogeneous state. Consequently, the elasto-plastic stiffness tensor D^{ep}_{ijkl} is constant throughout the body and the expression above provides

$$D^{ep}_{ijkl} u_{k,lj} = \rho \ddot{u}_i$$

By using (24.82) in this expression, we are back to the acceleration wave condition given by (24.80). This demonstrates that even though acceleration waves and plane waves are physically distinct phenomena, they are controlled by the same equations. As the investigation of plane waves was based on small vibrations about an already stressed state, this is equivalent to the so-called acoustic approximation in fluid mechanics; for this reason, Q^{ep}_{il} is often referred to as the acoustic tensor.

With the interpretation of (24.80) in terms of plane waves, it is possible to draw certain conclusions about *stability*. If the eigenvalues ρU^2 are real and positive, both acceleration waves and plane waves exist. Since the amplitude of the function f in (24.81) is small, u_i will remain small; this signals a stable situation.

However, if the eigenvalue ρU^2 is real but negative, then the corresponding acceleration wave does not exist, but plane waves will still be possible. To see this, it is noted that any linear combination of solutions of the form (24.81) is a valid plane wave solution. Suppose that $U^2 = -\alpha^2$, i.e. $U = \pm i\alpha$, where α is positive. Since ρU^2 is real, the corresponding eigenvector c_i is also real and $U = \pm i\alpha$ corresponds to the same eigenvalue and therefore also to the same eigenvector. Choosing f in (24.81) as a sine-function, the following plane wave is possible

$$u_i = c_i[sin(n_m x_m + i\alpha t) + sin(n_m x_m - i\alpha t)]$$

Using Euler's formula, we obtain

$$u_i = c_i(e^{\alpha t} + e^{-\alpha t}) sin(n_m x_m) \tag{24.84}$$

Thus, this solution implies that the displacement increases with time; it indicates that any small disturbance can grow infinitely large with time and it certainly signals an unstable situation. For a fixed x_m-position, (24.84) shows that the displacement vector increases with time without any oscillations. Consequently, it is common to use the terminology of *divergence instability* in accordance with the notation adopted in aerodynamics, cf. Rice (1976) and Leipholz (1972).

For linear elasticity as well as associated plasticity, the acoustic tensor is symmetric and as a result the eigenvalues are always real. Before the static bifurcation condition (24.43) is fulfilled, the acoustic tensor only possesses positive eigenvalues. However, after this condition has been passed, a negative eigenvalue exists and the material then exhibits divergence instability.

For nonassociated plasticity, the acoustic tensor is nonsymmetric and, in principle, it may possess complex eigenvalues ρU^2; accordingly, U comprises both a real and an imaginary part. The corresponding u_i-solution can then be shown to consist of oscillations with increasing amplitude. Borrowing again the terminology from aerodynamics, this situation is called *flutter instability*, cf. Rice (1976) and Leipholz (1972). However, it was shown by Ottosen and Runesson (1991b) that for a very broad group of nonassociated plasticity, flutter instability cannot occur, cf. also the arguments presented by Loret (1992), Bigoni and Zaccaria (1994) and Bigoni (1995).

A CONVEXITY - MINIMIZATION OF FUNCTION SUBJECT TO CONSTRAINTS

In this Appendix, we will discuss certain mathematical tools that are employed in the thermodynamic treatment of evolution laws. When establishing the evolution laws, use is made of properties relating to *convex functions*. In plasticity, the evolution laws may be established by invoking the postulate of maximum dissipation. However, this postulate is to be fulfilled on condition that the stress state is located inside or on the yield surface. This constraint condition complicates the necessary mathematical tools considerably, in particular because the constraint is given in terms of an inequality. These topics are dealt with in the mathematical literature concerned with *nonlinear optimization* and for further information, we may refer, for instance, to Luenberger (1984) and Strang (1986).

A.1 Convex function

In the discussion of evolution laws that fulfill the dissipation inequality, extensive use is made of convex functions and we will therefore provide a strict mathematical definition of convexity of a function.

Considering a one-dimensional problem with the variable given by y, a *convex* function is illustrated in Fig. A.1a) whereas a *concave* function is shown in Fig. A.1b). Referring to Fig. A.2, it appears that for a one-dimensional problem, convexity may be expressed as

$$f((1-\alpha)y^{(1)} + \alpha y^{(2)}) \leq (1-\alpha)f(y^{(1)}) + \alpha f(y^{(2)})$$

where $y^{(1)}$ and $y^{(2)}$ are two arbitrary points and $0 \leq \alpha \leq 1$.

Let us generalize these results and consider a function $f(y_i)$ of N variables,

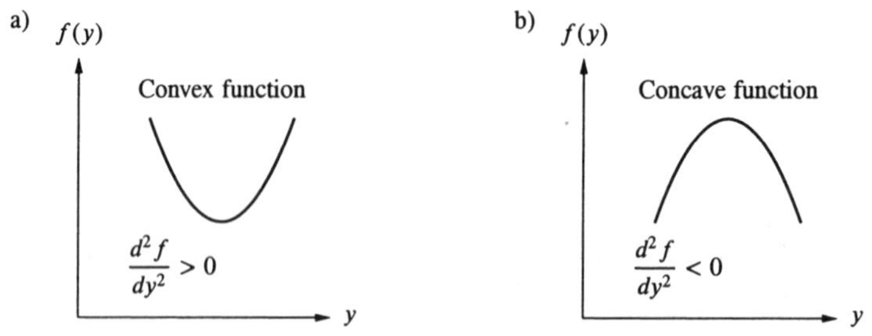

Figure A.1: One-dimensional illustration of: a) convex function and b) concave function.

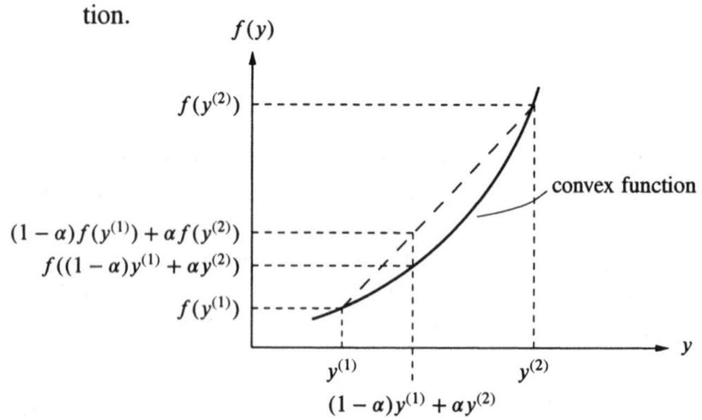

Figure A.2: Illustration of convexity.

i.e. $i = 1, 2, \cdots, N$. By definition, we then have

$$\boxed{\begin{array}{l} \textit{The function } f(y_i) \textit{ is convex, if, for any two} \\ \textit{points } y_i^{(1)} \textit{ and } y_i^{(2)}, \textit{ we have} \\ f((1-\alpha)y_i^{(1)} + \alpha y_i^{(2)}) \leq (1-\alpha)f(y_i^{(1)}) + \alpha f(y_i^{(2)}) \\ \textit{where } 0 \leq \alpha \leq 1 \end{array}} \tag{A.1}$$

The function f is said to be *strictly convex*, if the strict inequality holds whenever $y_i^{(1)} \neq y_i^{(2)}$ and $0 < \alpha < 1$.

It turns out to be advantageous to express this convexity property in a different form. From (A.1) we obtain for $0 < \alpha \leq 1$

$$\frac{f(y_i^{(1)} + \alpha \Delta y_i) - f(y_i^{(1)})}{\alpha} \leq f(y_i^{(2)}) - f(y_i^{(1)}) \tag{A.2}$$

where $\Delta y_i = y_i^{(2)} - y_i^{(1)}$. To evaluate the left-hand term we make a Taylor expansion about point $y_i^{(1)}$, i.e.

$$f(y_i^{(1)} + \alpha \Delta y_i) = f(y_i^{(1)}) + (\frac{\partial f}{\partial y_i})^{(1)} \alpha \Delta y_i + \mathcal{O}(\alpha^2)$$

where $(\ )^{(1)}$ means that the quantity within the parenthesis is to be evaluated at point $y_i^{(1)}$. From the expression above, it follows that

$$\frac{f(y_i^{(1)} + \alpha \Delta y_i) - f(y_i^{(1)})}{\alpha} = (\frac{\partial f}{\partial y_i})^{(1)} \Delta y_i + \mathcal{O}(\alpha)$$

Letting $\alpha \to 0$, we obtain

$$[\frac{f(y_i^{(1)} + \alpha \Delta y_i) - f(y_i^{(1)})}{\alpha}]_{\alpha \to 0} = (\frac{\partial f}{\partial y_i})^{(1)} \Delta y_i \tag{A.3}$$

It may be observed that the left-hand term in the mathematical literature is called the *directional derivative* of f when moving in the direction Δy_i; the terminology of *Gateaux* or *Frechet derivative* is also often used, see Vainberg (1964).

Use of (A.3) in (A.2) implies

$$(\frac{\partial f}{\partial y_i})^{(1)} \Delta y_i \le f(y_i^{(2)}) - f(y_i^{(1)}) \tag{A.4}$$

We have then shown that (A.1) implies (A.4); let us next prove that (A.4) implies (A.1). For this purpose, we accept (A.4) and choose $y_i^{(1)}$ as $(1-\alpha)y_i^{(3)} + \alpha y_i^{(4)}$ and alternatively $y_i^{(2)}$ as $y_i^{(2)} = y_i^{(3)}$ or $y_i^{(2)} = y_i^{(4)}$. This leads to

$$\begin{aligned} -(\frac{\partial f}{\partial y_i})^{(1)} \alpha (y_i^{(4)} - y_i^{(3)}) &\le f(y_i^{(3)}) - f((1-\alpha)y_i^{(3)} + \alpha y_i^{(4)}) \\ (\frac{\partial f}{\partial y_i})^{(1)} (1-\alpha)(y_i^{(4)} - y_i^{(3)}) &\le f(y_i^{(4)}) - f((1-\alpha)y_i^{(3)} + \alpha y_i^{(4)}) \end{aligned}$$

Multiplying the first expression by $(1-\alpha)$ and the second by α and adding, give the result

$$f((1-\alpha)y_i^{(3)} + \alpha y_i^{(4)}) \le (1-\alpha) f(y_i^{(3)}) + \alpha f(y_i^{(4)})$$

which is in complete agreement with (A.1). We have then proved that

The function $f(y_i)$ is convex, if and only if, for any two points $y_i^{(1)}$ and $y_i^{(2)}$, we have

$$f(y_i^{(2)}) - f(y_i^{(1)}) \ge (\frac{\partial f}{\partial y_i})^{(1)} (y_i^{(2)} - y_i^{(1)}) \tag{A.5}$$

In the one-dimensional case, this property is illustrated in Fig. A.3.

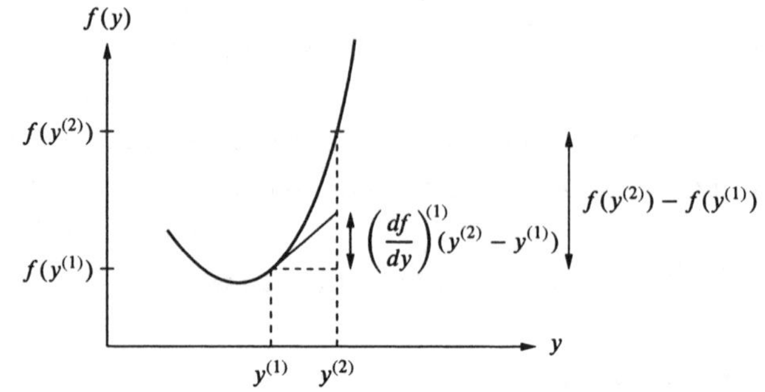

Figure A.3: Illustration of $f(y_i^{(2)}) - f(y_i^{(1)}) \geq (\frac{\partial f}{\partial y_i})^{(1)}(y_i^{(2)}) - y_i^{(1)})$.

It is also possible to express convexity in yet another form. We first make the following definition

$$\text{\textit{The Hessian } } F_{ij} \text{ \textit{of} } f(y_i) \text{ \textit{is defined by} } F_{ij} = \frac{\partial^2 f}{\partial y_i \partial y_j} \tag{A.6}$$

It appears that the Hessian is a symmetric matrix. Now a Taylor expansion about $y_i^{(1)}$ and use of (A.6) give

$$f(y_i^{(2)}) - f(y_i^{(1)}) = (\frac{\partial f}{\partial y_i})^{(1)}(y_i^{(2)} - y_i^{(1)})$$
$$+ \frac{1}{2}(y_i^{(2)} - y_i^{(1)})F_{ij}((1-\theta)y_m^{(1)} + \theta y_m^{(2)})(y_j^{(2)} - y_j^{(1)}) \tag{A.7}$$

For some θ-value in the range $0 \leq \theta \leq 1$, this result is exact. Use of (A.5) in (A.7) leads to

$$(y_i^{(2)} - y_i^{(1)})F_{ij}((1-\theta)y_i^{(1)} + \theta y_i^{(2)})(y_j^{(2)} - y_j^{(1)}) \geq 0$$

Since $y_i^{(2)} - y_i^{(1)}$ denotes arbitrary quantities, we conclude that F_{ij} must be positive semi-definite. On the other hand, if the Hessian F_{ij} is assumed to be positive semi-definite then (A.7) implies (A.5). We are then led to the following conclusion

$$\text{\textit{The function} } f(y_i) \text{ \textit{is convex, if and only if, the Hessian} } F_{ij} = \frac{\partial^2 f}{\partial y_i \partial y_j} \text{ \textit{is positive semi-definite}} \tag{A.8}$$

Referring back to the one-dimensional convex function $f(y)$ shown in Fig. A.1a), we see that convexity requires $d^2f/dy^2 \geq 0$ and (A.8) is the generalization of this result. With respect to Fig. A.1a), we may remark that the limit case $d^2f/dy^2 = 0$ corresponds to $f(y)$ being a straight line.

A.2 Unconstrained minimum

Consider the function $f(y_i)$ of N unknowns, i.e. $i = 1, 2, \cdots, N$. Assume that $f(y_i)$ at some point attains an extremum. At that point, we have per definition

$$df = \frac{\partial f}{\partial y_i} dy_i = 0 \tag{A.9}$$

In the present case, where we are considering unconstrained problems, the quantities dy_i can take any values. From (A.9) we therefore conclude that

$$\boxed{\frac{\partial f}{\partial y_i} = 0 \quad \Rightarrow \quad \textit{extremum point}} \tag{A.10}$$

This result evidently generalizes the trivial result when we have one variable, only.

Let us next devise a method by which we can decide whether the extremum is a minimum or not. Assume that a minimum point exists and that it is located at the point y_i^*. By definition, we have

$$f(y_i) \geq f(y_i^*) \quad \text{where} \quad y_i \neq y_i^* \tag{A.11}$$

A Taylor expansion of $f(y_i)$ about the point y_i^* provides

$$f(y_i) = f(y_i^*) + (\frac{\partial f}{\partial y_i})^*(y_i - y_i^*) + \frac{1}{2}(\frac{\partial^2 f}{\partial y_i \partial y_j})^*(y_i - y_i^*)(y_j - y_j^*)$$

where higher order terms were ignored. Due to (A.10), this reduces to

$$f(y_i) = f(y_i^*) + \frac{1}{2}(y_i - y_i^*)(\frac{\partial^2 f}{\partial y_i \partial y_j})^*(y_j - y_j^*)$$

For (A.11) to be fulfilled, it follows that the Hessian $F_{ij} = \partial^2 f / \partial y_i \partial y_j$ at the minimum point must be positive semi-definite. We conclude that

$$\boxed{\begin{array}{l} \textit{Necessary and sufficient conditions for } y_i^* \textit{ being} \\ \textit{a minimum point for the function } f(y_i)\textit{:} \\ \star \quad \dfrac{\partial f}{\partial y_i} = 0 \textit{ at point } y_i^* \\ \star \quad \textit{the Hessian } F_{ij} = (\dfrac{\partial^2 f}{\partial y_i \partial y_j})^* \textit{ is positive semi-definite} \end{array}} \tag{A.12}$$

If the function $f(y_i)$ is convex, cf. (A.8), it suffices to consider the condition $\partial f / \partial y_i = 0$ in order to identify the minimum point.

A.3 Inequality constrained minimum - Kuhn-Tucker relations

The problem of finding the minimum of a function where the variables are subject to some constraints, in particularly inequalities, is relevant for a number of applications. In plasticity, for instance, we want to maximize the mechanical dissipation γ_{mech} (minimize the quantity $-\gamma_{mech}$) subject to the condition that the yield criterion $f(\sigma_{ij}, K_\alpha) \leq 0$ is fulfilled. The solution to this problem is provided by the famous *Kuhn-Tucker relations* which are treated in detail, for instance, by Luenberger (1984) and Strang (1986); here, we will merely present the results.

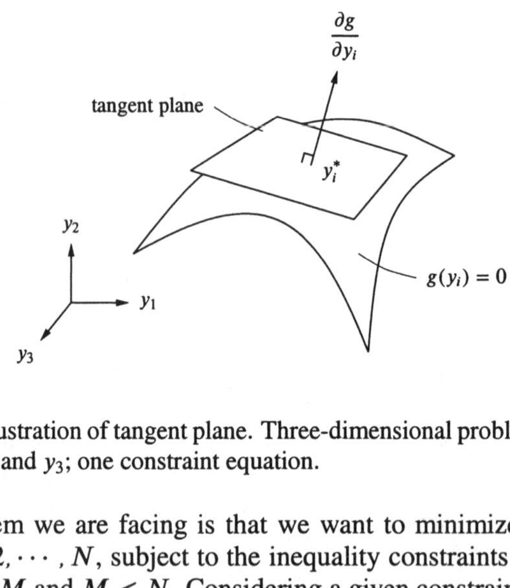

Figure A.4: Illustration of tangent plane. Three-dimensional problem with variables y_1, y_2 and y_3; one constraint equation.

The problem we are facing is that we want to minimize a function $f(y_i)$, where $i = 1, 2, \cdots, N$, subject to the inequality constraints $g^I(y_i) \leq 0$, where $I = 1, 2, \cdots, M$ and $M < N$. Considering a given constraint, it is classified as *active* if $g^I(y_i) = 0$ and *inactive* if $g^I(y_i) < 0$. A *regular point* is defined by

If at some point y_i, all the column matrices $\partial g^I/\partial y_i$ for the active constraints are linearly independent, then this point is called a regular point	(A.13)

For a given point y_i^* consider the expression

$$\left(\frac{\partial g^I}{\partial y_i}\right)^*(y_i - y_i^*) = 0 \quad \text{for } I = 1, 2, ..., M \tag{A.14}$$

For each value of I, the gradient $\partial g^I/\partial y_i$ is orthogonal to the corresponding surface $g^I = 0$. It follows that the y_i−values that fulfill (A.14) define the so-

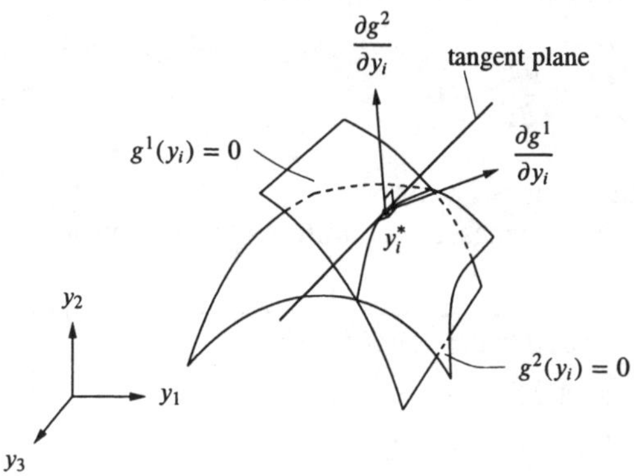

Figure A.5: Illustration of tangent plane. Three-dimensional problem with variables y_1, y_2 and y_3; two constraint equations.

called *tangent plane* at the position y_i^*. These tangent planes are illustrated in Figs. A.4 and A.5.

We then have

Problem: Determine minimum point y_i^* of the function $f(y_i)$, where $i = 1, 2, \cdots, N$, subject to the constraints $g^I(y_i) \leq 0$, where $I = 1, 2, \cdots, M$ and $M < N$.

Necessary conditions = Kuhn-Tucker relations:

The solution y_i^* must be a regular point. Moreover

$$\left(\frac{\partial f}{\partial y_i}\right)^* + \sum_{I=1}^{M} \lambda^I \left(\frac{\partial g^I}{\partial y_i}\right)^* = 0 \qquad \text{(A.15)}$$

$\lambda^I \geq 0$ *for active constraints*

$\lambda^I = 0$ *for inactive constraints*

$\lambda^{(I)} g^{(I)} = 0$ *(no summation)*

which were derived by Kuhn and Tucker (1951). The result above can be interpreted in the following elegant manner. Define the Lagrange function $\mathcal{L}(y_i, \lambda^I)$ by

$$\mathcal{L}(y_i, \lambda^I) = f(y_i) + \sum_{I=1}^{M} \lambda^I g^I(y_i) \qquad \text{(A.16)}$$

and determine the extremum point for $\mathcal{L}(y_i, \lambda^I)$ with respect to y_i as if the variables y_i are unconstrained. We then obtain

$$\frac{\partial \mathcal{L}}{\partial y_i} = \frac{\partial f}{\partial y_i} + \sum_{I=1}^{M} \lambda^I \frac{\partial g^I}{\partial y_i} = 0 \qquad (A.17)$$

which is exactly the result we want.

The results above are necessary conditions for a minimum and we also have the results

Necessary and sufficient conditions:
In addition to (A.15), we must have

$$(y_i - y_i^*) L_{ij}^* (y_j - y_j^*) \geq 0$$

where

$$L_{ij}^* = \left(\frac{\partial^2 f}{\partial y_i \partial y_j}\right)^* + \lambda^I \left(\frac{\partial^2 g^I}{\partial y_i \partial y_j}\right)^*$$

and y_i is located on the tangent plane defined by

$$\left(\frac{\partial g^I}{\partial y_i}\right)^* (y_i - y_i^*) = 0 \quad \textit{for all active constraints}$$

This means that the quantity L_{ij}^* is positive semi-definite.

These conditions can be strengthened slightly. Since $\lambda^I \geq 0$ then, if the functions $f(y_i)$ and $g^I(y_i)$ are convex functions, i.e. the Hessians $F_{ij} = \partial^2 f / \partial y_i \partial y_j$ and $G_{ij}^I = \partial^2 g^I / \partial y_i \partial y_j$ are both positive semi-definite, this implies that L_{ij}^* is always positive semi-definite. We conclude that

If the functions $f(y_i)$ and $g^I(y_i)$ are convex functions,
the condition $(y_i - y_i^*) L_{ij}^* (y_j - y_j^*) \geq 0$ (A.18)
is fulfilled

BIBLIOGRAPHY

Ahadi, A. and Krenk, S. (2000). Characteristic state plasticity for granular materials, Part II: Model calibration and results. *International Journal of Solids and Structures*, **37**, 6361–6380.

Aifantis, E. C. (1984). On the microstructural origin of certain inelastic models. *Journal of Engineering Materials and Technology*, **106**, 326–330.

Alfrey, T. J. (1957). *Mechanical Behavior of High Polymers*. Interscience Publishers.

Allen, D. H. (1991). Thermomechanical coupling in inelastic solids. *Applied Mechanics Reviews*, **44**, 361–373.

Altenbach, H. (1999). Classical and non-classical creep models. In H. Altenbach and J. J. Skrzypek, editors, *Creep and Damage in Materials and Structures*, pages 45–95. Springer-Verlag.

Argyris, J. and Mlejnek, H.-P. (1991). *Dynamics of Structures*. North-Holland.

Argyris, J. H. (1965a). Three-dimensional anisotropic and inhomogeneous media. Matrix analysis for small and large displacements. *Ingenieur-Archiv*, **34**, 33–55.

Argyris, J. H. (1965b). Elasto-plastic matrix displacement analysis of three-dimensional continua. *Journal of Aeronautical Society*, **69**, 633–635.

Argyris, J. H., Vaz, L. E., and Willam, K. J. (1981). Integrated finite-element analysis of coupled thermoviscoplastic problems. *Journal of Thermal Stresses*, **4**, 121–153.

Armstrong, P. J. and Frederick, C. O. (1966). *A mathematical representation of the multiaxial Bauschinger effect*. Berkeley Nuclear Laboratories, Report, RD/B/N731.

Armero, F. and Simo, J. C. (1992). A new unconditionally stable fractional step method for non-linear coupled thermomechanical problems. *International Journal for Numerical Methods in Engineering*, **35**, 737–766.

Armero, F. and Simo, J. C. (1993). A priori stability estimate, and unconditionally stable product formula algorithms for nonlinear coupled thermoplasticity. *Computer Methods in Applied Mechanics and Engineering*, **98**, 41–104.

ASCE Commite on Concrete and Masonry Structures (1982). *Finite Element Analysis of Reinforced Concrete Structures*. ASCE Special Publication.

Ashby, M. F. and Jones, D. R. H. (1980). *Engineering Materials 1. An Introduction to their Properties and Applications*. Pergamon Press.

Axelsson, K. (1979). *On Constitutive Modelling in Metal Plasticity. With special emphasis on aniotropic strain hardening and finite element formulation*. Publ.79:2, Dept. of Struct. Mech., Chalmers University of Technology, Sweden.

Axelsson, K. and Samuelsson, A. (1979). Finite element analysis of elastoplastic materials displaying mixed hardening. *International Journal for Numerical Methods in Engineering*, **14**, 211–225.

Ayres, F. J. (1962). *Theory and Problems of Matrices*. Schaum Publishing Company.

Bailey, R. W. (1929). *Creep of steel under single and compound stresses, and the use of high initial temperature in steam power plants*, volume 3, pages 1089–1095. Trans. World Power Conference. Tokyo.

Balmer, G. G. (1949). *Shearing strength of concrete under high triaxial stress-computation of Mohr's envelope as a curve*. Structural Research Laboratory Report No. Sp-23, United States Department of Interior, Bureau of Reclamation.

Baltov, A. and Sawczuk, A. (1965). A rule of anisotropic hardening. *Acta Mechanica*, **1**, 81–92.

Barlat, F., Lege, D. J., and Brem, J. C. (1991). A six-component yield function for anisotropic materials. *International Journal of Plasticity*, **7**, 693–712.

Bartenev, G. M. and Zuev, J. S. (1968). *Strength and Failure of Visco-elastic Materials*. Pergamon Press.

Bassani, J. L. (2001). Incompatibility and a simple gradient theory of plasticity. *Journal of the Mechanics and Physics of Solids*, **49**, 1983–1996.

Bathe, K. J. (1996). *Finite Element Procedures*. Prentice-Hall, Englewood Cliffs.

Bathe, K. J. and Cimento, A. (1980). Some practical procedures for the solution of nonlinear finite element equations. *Computer Methods in Applied Mechanics and Engineering*, **22**, 59–85.

Batra, R. C. and Wright, T. W. (1988). A comparison of solutions for adiabatic shear banding by forward-difference and crank-nicolson methods. *Communications in Applied Numerical Methods*, **4**, 741–748.

Bauschinger, J. (1886). *Über die Veränderung der Elastizitätsgrenze und der Festigkeit des Eisens und Stahls durch Strecken und Quetschen, durch Erwärmen und Abkühlen und durch oftmal wiederholte Beanspruchung*. Mitteilungen 15 aus dem mechanisch - technischen Laboratorium der K. polytechnischen Schule, München.

Bažant, Z. P. (1976). Instability, ductility and size effect in strain-softening concrete. *Journal of the Engineering Mechanics Division, ASCE*, **102**, 331–344.

Bažant, Z. P. (1979). *Inelasticity and failure of concrete*. Swedish Cement and Concrete Research Institute, Stockholm, Sweden.

Bažant, Z. P. (1982). Mathematical models for creep and shrinkage of concrete. In Z. P. Bažant and F. H. Wittmann, editors, *Creep and Shrinkage in Concrete Structures*, pages 163–256. John Wiley & Sons.

Bažant, Z. P. (1996). Mathematical modelling of creep and shrinkage of concrete at high temperature. In Z. P. Bažant and M. F. Kaplan, editors, *Concrete at High Temperatures. Material Properties and Mathematical Models*, pages 249–320. Longman Group Limited.

Belytschko, T., Bazant, Z., Hyun, Y.-W., and Chang, T.-P. (1986). Strain-softening materials and finite-element solutions. *Computers & Structures*, **23**, 163–180.

Belytschko, T., Liu, W. K., and Moran, B. (2000). *Nonlinear Finite Elements for Continua and Structures*. John Wiley & Sons.

Bergan, P. G. (1979). Solution algorithms for nonlinear structural problems. In *International Conference on Engineering Applications of the Finite Element Method*, pages 13.1–13.39. Norway, A. S. Computas.

Bergan, P. G. (1980). Solution algorithms for nonlinear structural problems. *Computers & Structures*, **12**, 497–509.

Bergan, P. G. and Clough, R. W. (1972). Convergence criteria for iterative processes. *AIAA Journal*, **10**, 1107–1108.

Bernstein, B. (1960). Hypoelasticity and elasticity. *Archive for Rational Mechanics and Analysis*, **6**, 90–104.

Besseling, I. F. (1953). *A theory of plastic flow for anisotropic hardening in plastic deformation of an initially isotropic material*. National Aerospace Laboratory, Amsterdam, Report S410.

Betonghandbok (1982). *Material*. Edited by G. Möller, N. Petersons and P. Samuelsson. Svensk Byggtjänst, Sweden.

Bever, M. B., Holt, D. L., and Titchener, A. L. (1973). *The Stored Energy of Cold Work*. Pergamon Press.

Bićanić, N. and Johnson, W. W. (1979). Who was "-Raphson"? *International Journal for Numerical Methods in Engineering*, **14**, 148–152.

Bicanic, N. P. (1989). Exact evaluation of contact stress state in computational elasto-plasticity. *Engineering Computations*, **6**, 67–73.

Bigoni, D. (1995). On flutter instability in elastoplastic constitutive models.

International Journal of Solids and Structures, **32**, 3167–3189.

Bigoni, D. and Hueckel, T. (1991). Uniqueness and localization - I. Associate and non-associative elastoplasticity - II: Coupled elastoplasticity. *International Journal of Solids and Structures*, **28**, 197–214.

Bigoni, D. and Zaccaria, D. (1992). Loss of strong ellipticity in non-associative elastoplasticity. *Journal of the Mechanics and Physics of Solids*, **40**, 1313–1331.

Bigoni, D. and Zaccaria, D. (1994). On the eigenvalues of the acoustic tensor in elastoplasticity. *European Journal of Mechanics. A/Solids*, **13**, 621–638.

Bingham, E. C. (1922). *Fluidity and Plasticity*. McGraw-Hill Book Company.

Biot, M. A. (1954). Theory of stress-strain relations in anisotropic viscoelasticity and relaxation phenomena. *Journal of Applied Physics*, **25**, 1385–1391.

Bishop, J. F. W. and Hill, R. (1951). A theory of the plastic distortion of a polycrystalline aggregate under combined stress. *Philosophical Magazine*, **42**, 414–427.

Bland, D. R. (1957). The two measures of workhardening. In *9th International Congress of Applied Mechanics*, volume 8, pages 45–50.

Bodner, S. R. (1968). Constitutive equations for dynamic material behavior. In U. S. Lindholm, editor, *Mechanical Behavior of Materials under Dynamic Loads*, pages 176–190. Springer-Verlag.

Bodner, S. R. (1987). Review of unified elastic-viscoplastic theory. In A. K. Miller, editor, *Unified Constitutive Equations for Creep and Plasticity*, pages 273–301. Elsevier Applied Science Publishers.

Bodner, S. R. and Partom, Y. (1972). A large deformation elastic-viscoplastic analysis of a thick-walled spherical shell. *Journal of Applied Mechanics*, **39**, 751–757.

Bodner, S. R. and Partom, Y. (1975). Constitutive equations for elastic-viscoplastic strain-hardening mateials. *Journal of Applied Mechanics*, **42**, 385–389.

Bodner, S. R., Partom, I., and Partom, Y. (1979). Uniaxial cyclic loading of elastic-viscoplastic materials. *Journal of Applied Mechanics*, **46**, 805–810.

Boehler, J.-P. (1975). Sur les formes invariants dans le sous-groupe orthotrope de révolution des transformations orthogonales de la relation entre deux tenseurs symétriques du second ordre. *Zeitschrift für angewandte Mathematik und Mechanik*, **55**, 609–611.

Boehler, J.-P. (1977). Irreducible representations for isotropic scalar functions. *Zeitschrift für angewandte Mathematik und Mechanik*, **57**, 323–327.

Boehler, J.-P. (1978). Lois de comportement anisotrope des milieux continus. *Journal de Méchanique*, **17**, 153–190.

Boehler, J.-P. (1979). A simple derivation of representations for non-polynomial constitutive equations for some cases of anisotropy. *Zeitschrift für angewandte Mathematik und Mechanik*, **59**, 157–167.

Boehler, J.-P. (1987a). Introduction to the invariant formulation of anisotropic constitutive equations. In J. P. Boehler, editor, *Applications of Tensor Functions in Solid mechanics*, pages 13–30. Springer-Verlag.

Boehler, J.-P. (1987b). Representations for isotropic and anisotropic non-polynomial tensor functions. In J. P. Boehler, editor, *Applications of Tensor Functions in Solid mechanics*, pages 31–53. Springer-Verlag.

Boehler, J.-P. (1987c). Anisotropic linear elasticity. In J. P. Boehler, editor, *Applications of Tensor Functions in Solid mechanics*, pages 55–65. Springer-Verlag.

Boehler, J.-P. and Sawczuk, A. (1976). Application of representation theorems to describe yielding of transversely isotropic solids. *Mechanics Research Communications*, **3**, 277–283.

Boley, A. B. and Weiner, J. H. (1960). *Theory of Thermal Stresses*. John Wiley & Sons.

Boltzmann, L. (1874). *Zur Theorie der elastischen Nachwirkung*. Sitzungsberichte Akademie Wissenschaften der Wien, Wissenschaftligen Abhandlungen, Vol.70, pages 275-306.

Boresi, A. P. and Sidebottom, O. M. (1972). Creep of metals under multiaxial states of stress. *Nuclear Engineering and Design*, **18**, 415–456.

Bridgman, P. W. (1952). *Studies in Large Plastic Flow and Fracture with Special Emphasis on the Effects of Hydrostatic Pressure*. McGraw-Hill.

Browne, R. D. and Blundell, R. (1972). The behaviour of concrete in prestressed concrete pressure vessels. *Nuclear Engineering and Design*, **20**, 429–475.

Broyden, C. (1967). Quasi-Newton methods and their application to function minimization. *Mathematical Computation*, **21**, 368–381.

Broyden, C. (1970). The convergence of a class of double rank minimization algorithms. Parts I and II. *Journal of Institute of Mathematical Application*, **6**, 76–90 and 222–236.

Burgers, J. M. (1935). *First and Second Report on Viscosity and Plasticity*. Academy of Sciences at Amsterdam, the Netherlands.

Byfors, J. (1980). *Plain Concrete at Early Stages*. Swedish Cement and Concrete Research Institute, Fo.80:3, Stockholm, Sweden.

Çengel, Y. A. and Boles, M. A. (1994). *Thermodynamics. An Engineering Approach*. McGraw-Hill.

Celentano, D., Oñate, E., and Oller, S. (1994). A temperature-based formulation for finite element analysis of generalized phase-change problems. *Inter-*

national Journal for Numerical Methods in Engineering, **37**, 3441–3465.

Celentano, D., Oller, S., and Oñate, E. (1996). A coupled thermomechanical model for the solidification of cast metals. *International Journal of Solids and Structures*, **33**, 647–673.

Chaboche, J. L., Dang-Van, K., and Cordier, G. (1979). *Modelization of the Strain memory Effect on the Cyclic Hardening of 316 Stainless Steel*. SMIRT-5, Division L. Berlin.

Chaboche, J. L. (1989). Time-independent constitutive theories for cyclic plasticity. *International Journal of Plasticity*, **2**, 247–302.

Chaboche, J. L. (1989). Constitutive equations for cyclic plasticity and cyclic viscoplasticity. *International Journal of Plasticity*, **5**, 149–188.

Chaboche, J. L. (1993a). Cyclic viscoplastic constitutive equations, part I: A thermodynamically consistent formulation. *Journal of Applied Mechanics*, **60**, 813–821.

Chaboche, J. L. (1993b). Cyclic viscoplastic constitutive equations, part II: Stored energy - Comparison between models and experiments. *Journal of Applied Mechanics*, **60**, 822–828.

Chaboche, J. L. and Nouailhas, D. (1989). A unified constitutive model for cyclic viscoplasticity and its applications to various stainless steels. *Journal of Engineering Materials and Technology*, **111**, 424–430.

Chan, K. S., Bodner, S. R., and Lindholm, U. S. (1988). Phenomenological modeling of hardening and thermal recovery in metals. *Journal of Engineering Materials and Technology*, **110**, 1–8.

Chen, W. F. (1975). *Limit Analysis and Soil Plasticity*. Elsevier.

Chen, W. F. (1982). *Plasticity in Reinforced Concrete*. McGraw-Hill Book Company.

Chen, W. F. (1994). *Constitutive Equations for Engineering Materials. Vol. 2: Plasticity and Modeling*. Elsevier Science B.V.

Chen, W. F. and Han, D. J. (1988). *Plasticity for Structural Engineers*. Springer-Verlag.

Chen, W. F. and Saleeb, A. F. (1982). *Constitutive Equations for Engineering Materials, Vol.1. Elasticity and Modeling*. John Wiley & Sons.

Chinn, J. and Zimmerman, R. M. (1965). Behavior of plain concrete under various high triaxial compression loading conditions. Technical Report No. WL TR 64-163 (AD 468460), Air Force Weapons Laboratory, New Mexico.

Clavout, C. and Ziegler, H. (1959). Über einige verfestungsregeln. *Ingenieur-Archiv*, **28**, 13–26.

Coleman, B. D. (1964). Thermodynamics of materials with memory. *Archive for Rational Mechanics and Analysis*, **17**, 1–46.

Coleman, B. D. and Gurtin, M. E. (1967). Thermodynamics with internal state variables. *The Journal of Chemical Physics*, **47**, 597–613.

Coleman, B. D. and Noll, W. (1960). An approximation theorem for functionals with applications to continuum mechanics. *Archive for Rational Mechanics and Analysis*, **6**, 355–370.

Coleman, B. D. and Noll, W. (1961). Foundations of linear viscoelasticity. *Reviews of Modern Physics*, **33**, 239–249.

Coon, M. D. and Evans, R. J. (1971). Recoverable deformation of cohesionsless soils. *Journal of the Soil Mechanics and Foundations Division, ASCE*, **97**, 375–391.

Coon, M. D. and Evans, R. J. (1972). Incremental constitutive laws and their associated failure criteria with application to plain concrete. *International Journal of Solids and Structures*, **8**, 1169–1183.

Cosserat, E. and Cosserat, F. (1909). *Théories des corps déformables*. A. Herman et fils, Paris.

Coulomb, C. A. (1776). Essai sur une application des règles de maximis & minimis à quelques problèmes de statique, relatifs a l'archiecture. *Mémoires de Mathèmatique & de Physique présentés à l'Academie Royale des Sciences*, **7**, 343–382.

Cowin, S. C. and Mehrabadi, M. M. (1995). Anisotropic symmetries of linear elasticity. *Applied Mechanics Reviews*, **48**, 247–285.

Cowper, G. R. and Symonds, P. S. (1962). Unpublished work communicated to Bodner, S.R and Symonds, P.S. Experimental and theoretical investigation of the plastic deforrmation of cantilever beams subjected to impulsive loading. *Journal of Applied Mechanics*, **29**, 719–728.

CRC (1995). *Handbook of Chemistry and Physics. 76th edition.* CRC Press.

Crisfield, M. A. (1991). *Non-linear Finite Element Analysis of Solids and Structures, Essentials*, volume 1. John Wiley & Sons.

Crisfield, M. A. (1997). *Non-linear Finite Element Analysis of Solids and Structures, Advanced Topics*, volume 2. John Wiley & Sons.

Dafalias, Y. F. and Popov, E. P. (1975). A model of nonlinearly hardening materials for complex loading. *Acta Mechanica*, **21**, 173–192.

Dafalias, Y. F. and Popov, E. P. (1976). Plastic internal variables formalism of cyclic plasticity. *Journal of Applied Mechanics*, **43**, 645–651.

Dafalias, Y. F. and Rashid, M. M. (1989). The effect of plastic spin on anisotropic material behavior. *International Journal of Plasticity*, **5**, 227–246.

Dahlblom, O. (1987). *Constitutive modelling and finite element analysis of concrete structures with regard to environmental influence*. Report TVSM-

1004, Div. of Structural Mechanics, Lund University, Sweden.

Dahlquist, G. and Björk, Å. (1974). *Numerical Methods*. Prentice Hall. Englewood Cliffs.

de Borst, R. (1991). Simulation of strain localization: A reappraisal of the Cosserat continuum. *Engineering Computations*, **8**, 317–332.

de Borst, R. and Feenstra, P. H. (1990). Studies in anisotropic plasticity with reference to the Hill criterion. *International Journal for Numerical Methods in Engineering*, **29**, 315–336.

de Groot, S. and Mazur, P. (1962). *Non-equilibrium Thermodynamics*. North-Holland.

Dennis, J. E. and Moré, J. (1977). Quasi-Newton methods, motivation and theory. *SIAM Review*, **19**, 46–89.

Desai, C. S. and Siriwardane, H. J. (1984). *Constitutive Laws for Engineering Materials, with emphasis on geologic materials*. Prentice-Hall.

Dilger, W. H., Koch, R., and Kowalczyk, R. (1984). Ductility of plain and confined concrete under different strain rates. *Journal of the American Concrete Institute*, **81**, 73–81.

Dillon, O. W. (1963). Coupled thermoplasticity. *Journal of the Mechanics and Physics of Solids*, **11**, 21–33.

DiMaggio, F. L. and Sandler, I. S. (1971). Material model for granular soils. *Journal of the Engineering Mechanics Division, ASCE*, **97**, 935–950.

Dischinger, F. (1937). Untersuchungen über die knicksicherheit, die elastische verformung und das kriechen des betons bei logenbrüchen. *Der Bauingenieur*, **18**, 487–520.

Dormand, J. R. and Prince, J. R. (1980). A family of embedded Runge-Kutta formulae. *Journal of Computational and Applied Mathematics*, **6**, 19–26.

Dougill, J. W. (1976). On stable progressively fracturing solids. *Zeitschrift für angewandte Mathematik und Physik*, **27**, 423–437.

Drucker, D. C. (1951). A more fundamental approach to plastic stress-strain relations. In *Proceedings of the 1th US National Congress of Applied Mechanics, ASME*, pages 487–491. New York.

Drucker, D. C. (1964). On the postulate of stability of material in the mechanics of continua. *Journal de Méchanique*, **3**, 235–249.

Drucker, D. C. and Palgen, L. (1981). On stress-strain relations suitable for cyclic and other loadings. *Journal of Applied Mechanics*, **48**, 479–485.

Drucker, D. C. and Prager, W. (1952). Solid mechanics and plastic analysis or limit design. *Quarterly of Applied Mathematics*, **10**, 157–165.

Drucker, D. C., Prager, W., and Greenberg, H. J. (1952). Extended limit design theorems for continuous media. *Quarterly of Applied Mathematics*, **9**, 381–

389.

Duvaut, G. and Lions, J. L. (1972). *Les inéquations en mécanique et en physique*. Dunod, Paris.

Edelen, D. G. B. (1972). A nonlinear onsager theory of irreversibility. *International Journal of Engineering Science*, **10**, 481–490.

Eibl, J., Waubke, N. V., Klingsch, W., Schneider, U., and Rieche, G. (1974). *Studie zur Erfassung specieller Betoneigenschaften im Reaktordruckbehälterbau*. Deutscher Ausschuss für Stahlbeton, Heft 237.

Eibl, J., Aschl, H., Bobrowski, J., Cedolin, L., Garas, F. K., Gerstle, K. H., Hilsdorf, H., Kotsovos, M. D., Ottosen, N. S., Wastiels, J., and Willam, K. J. (1983). *Concrete under multiaxial states of stress. Constitutive equations for practical design*. Comite Euro-International du Beton, CEB-Bulletin No. 156.

Ekh, M. and Runesson, K. (2000). Bifurcation results for plasticity coupled to damage with MCR-effect. *International Journal of Solids and Structures*, **37**, 1975–1996.

Engelman, M. S., Strang, G., and Bathe, K.-J. (1981). The application of quasi-Newton methods in fluid mechanics. *International Journal for Numerical Methods in Engineering*, **17**, 707–718.

Ericksen, J. L. (1960). Transversely isotropic fluids. *Kolloid Zeitschrift*, **173**, 117–122.

Eringen, A. C. (1975a). Thermodynamics of continua. In A. C. Eringen, editor, *Continuum Physics*, volume 2, pages 89–127. Academic Press.

Eringen, A. C. (1975b). Constitutive equations for simple materials. General theory. In A. C. Eringen, editor, *Continuum Physics*, volume 2, pages 131–172. Academic Press.

Eringen, A. C. (1999). *Microcontinuum Field Theories. I: Foundations and Solids*. Springer-Verlag.

Eshbach, O. W. and Souders, M. (1975). *Handbook of Engineering Fundamentals, 3rd edition*. John Wiley & Sons.

Evans, R. J. and Pister, K. S. (1966). Constitutive equations for a class of nonlinear elastic solids. *International Journal of Solids and Structures*, **2**, 427–445.

Evans, R. W. and Wilshire, B. (1993). *Introduction to Creep*. The Institute of Materials, London.

Eve, R. A., Reddy, B. D., and Rockafeller, R. T. (1990). An internal variable theory of elastoplasticity based on the maximum plastic work inequality. *Quarterly of Applied Mathematics*, **48**, 59–83.

Eves, H. (1980). *Elementary Matrix Theory*. Dover Publications.

Faruque, M. O. and Chang, C. J. (1990). A constitutive model for pressure sensitive materials with particular reference to plain concrete. *International Journal of Plasticity*, **6**, 29–43.

Findley, W. N., Lai, J. S., and Onaran, K. (1976). *Creep and Relaxation of Nonlinear Viscoelastic Materials*. North-Holland Publishing Company.

Finnie, I. and Heller, W. R. (1959). *Creep of Engineering Materials*. McGraw-Hill Book Company.

Fleck, N. and Hutchinson, J. W. (2001). A reformulation of strain gradient plasticity. *Journal of the Mechanics and Physics of Solids*, **49**, 2245–2271.

Fletcher, R. (1970). A new approach to variable metric algorithms. *Computer Journal*, **13**, 317–322.

Fletcher, R. (1980). *Practical Methods of Optimization*. Vol 1. John Wiley & Sons.

Flügge, W. (1967). *Viscoelasticity*. Blaisdell, Waltham.

Frost, H. J. and Ashby, M. F. (1982). *Deformation Mechanism Maps*. Pergamon Press.

Fung, Y. C. (1965). *Foundations of Solid Mechanics*. Prentice-Hall.

Gear, G. W. (1971). *Numerical Initial value Problems in Ordinary Differential equations*. Prentice Hall, Englewood Cliffs.

Germain, P., Nguyen, Q. S., and Suquet, P. (1983). Continuum thermodynamics. *Journal of Applied Mechanics*, **50**, 1010–1020.

Gerstle, K. H., Aschl, H., Bellotti, R., Bertacci, P., Kotsovos, M. D., Ko, H. Y., Linse, D., Newman, J. B., Rossi, P., Schickert, G., Taylor, M. A., Traina, L. A., Winkler, H., and Zimmerman, R. M. (1980). Behavior of concrete under multiaxial stress states. *Journal of the Engineering Mechanics Division, ASCE*, **106**, 1383–1403.

Gittus, J. (1975). *Creep, Viscoelasticity and Creep Fracture in Solids*. Applied Science Publishers.

Gnoneim, H. (1990). Analysis and application of a complex thermoviscoplasticity theory. *Journal of Applied Mechanics*, **57**, 828–835.

Goel, R. P. and Malvern, L. E. (1970). Biaxial plastic simple waves with combined kinematic and isotropic hardening. *Journal of Applied Mechanics*, **37**, 1100–1106.

Goldfarb, D. (1970). A family of variable metric methods derived by variational means. *Mathematical Computation*, **24**, 23–26.

Green, A. E. and Adkins, J. E. (1960). *Large Elastic Deformations*. Oxford, Clarendon Press.

Green, S. J. and Swanson, S. R. (1973). Static constitutive relations for concrete. Technical Report No. AFWL-TR-72-244 (AD 761820), Air Force Weapons Laboratory, New Mexico.

Gurtin, M. E. (1981). *An Introduction to Continuum Mechanics.* Academic Press.

Gurtin, M. E. (2002). A gradient theory of single-crystal viscoplasticity that accounts for geometrically necessary dislocations. *Journal of the Mechanics and Physics of Solids*, **50**, 5–32.

Gurtin, M. E. and Williams, W. O. (1966). On the Clausius-Duhem inequality. *Zeitschrift für angewandte Mathematik und Physik*, **17**, 626–633.

Gvozdev, A. A. (1938). "Opredelenie velichiny razrushayushchei nagruzki dlya staticheski neopredelimykh sistem, preterperayuschchikh plasticheskie deformatssii", moscow/leningrad, akademia nauk sssr, 19-38. English translation (1960), "The determination of the value of the collapse load for statically indeterminate systems undergoing plastic deformation". *International Journal of Mechanical Science*, **1**, 322–333.

Hadamard, J. (1903). *Leçons sur la Propagation des Ondes et les Equations de l'Hydrodynamique.* Herman et fils, Paris. Also published by Chelsea Publishing Company, New York (1949).

Haigh, B. F. (1920). The strain-energy function and the elastic limit. *Engineering, London*, **109**, 158–160.

Håkansson, P., Wallin, M., and Ristinmaa, M. (2005). Comparison of isotropic hardening and kinematic hardening in thermoplasticity. *International Journal of Plasticity*, **21**, 1435–1460.

Halphen, B. and Nguyen, Q. S. (1975). Sur les matériaux standards généralisés. *Journal de Méchanique*, **14**, 39–63.

Handelman, G. H., Lin, C. C., and Prager, W. (1947). On the mechanical behaviour of metals in the strain-hardening range. *Quarterly of Applied Mathematics*, **4**, 397–407.

Hannant, D. J. (1969). Creep and creep recovery of concrete subjected to multiaxial compressive stresses. **66**, 391–394.

Hart, E. W. (1970). A phenomenological theory for plastic deformation of polycrystalline metals. *Acta Metallurgica Materials*, **18**, 599–610.

Haupt, P. (2000). *Continuum Mechanics and Theory of Materials.* Springer-Verlag.

Hencky, H. Z. (1924). Zur theorie plastischer deformationen und der hierdurch im material hervorgerufenen nachspannungen. *Zeitschrift für angewandte Mathematik und Mechanik*, **4**, 323–334.

Hencky, H. Z. (1925). Über langsame stationäre strömungen in plastischen massen mit rücksicht auf die vorgänge beim walzen, pressen und ziehen von metallen. *Zeitschrift für angewandte Mathematik und Mechanik*, **5**, 115–124.

Hildebrand, F. B. (1965). *Methods of Applied Mathematics.* Prentice-Hall. Second edition.

Hill, R. (1948a). A theory of the yielding and plastic flow of anisotropic materials. *Proceedings of the Royal Society of London*, **A193**, 281–297.

Hill, R. (1948b). A variational principle of maximum plastic work in classical plasticity. *Quarterly Journal of Mechanics and Applied Mathematics*, **1**, 18–28.

Hill, R. (1950). *The Mathematical Theory of Plasticity*. Oxford University Press.

Hill, R. (1958). A general theory of uniqueness and stability in elastic-plastic solids. *Journal of the Mechanics and Physics of Solids*, **6**, 236–249.

Hill, R. (1961). Discontinuity relations in mechanics of solids. In I. N. Sneddon and R. Hill, editors, *Progress in Solid Mechanics*, volume 2, pages 245–276. North-Holland, Amsterdam.

Hill, R. (1962). Acceleration waves in solids. *Journal of the Mechanics and Physics of Solids*, **10**, 1–16.

Hill, R. (1966). Generalized constitutive relations for incremental deformation of metal crystals by multislip. *Journal of the Mechanics and Physics of Solids*, **14**, 95–102.

Hill, R. (1978). Aspects of invariance in solid mechanics. In C.-S. Yih, editor, *Advances in Applied Mechanics*. Academic Press.

Hill, R. (1993). A user-friendly theory of orthotropic plasticity in sheet metals. *International Journal of Mechanical Science*, **35**, 19–25.

Hillerborg, A. (1974). *Dimensionering av armerade betongplattor enligt strimlemetoden, (The strip method)*. Almqvist & and Wiksell, Stockholm.

Hillerborg, A., Modéer, M., and Petersson, P.-E. (1976). Analysis of crack formation and crack growth in concrete by means of fracture mechanics and finite elements. *Cement and Concrete Research*, **6**, 773–782.

Hodge, P. G. J. (1957). Piecewise linear hardening. In *9th International Congress on Applied Mechanics*, volume 8, pages 65–72. University of Brussels, Belgium.

Hoffman, O. (1967). The brittle strength of orthotropic materials. *Journal of Composite Materials*, **1**, 200–206.

Hohenemser, K. and Prager, W. (1932). Über die ansätze der mechanik isotroper kontinua. *Zeitschrift für angewandte Mathematik und Mechanik*, **12**, 216–226.

Horgan, C. O. (1973). On the strain-energy density in linear elasticity. *Journal of Engineering Mathematics*, **7**, 231–234.

Hosford, W. F. and Backofen, W. A. (1964). Strength and plasticity of textured metals. In W. A. Backofen, J. Burke, L. Coffin, N. Reed, and V. Weiss, editors, *Fundamentals of Deformation Processing*, pages 259–298. Syracuse

University Press, Syracuse.

Hsieh, S. S., Ting, E. C., and Chen, W. F. (1982). A plasticity-fracture model for concrete. *International Journal of Solids and Structures*, **18**, 181–197.

Hu, L. W. (1956). Studies on plastic flow of anisotropic metals. *Journal of Applied Mechanics*, **12**, 444–450.

Huber, M. T. (1904). Wlaściwa praca odkształcenia jako miara wytężenia materyalu. *Czasopismo techiczne, Lemberg*, **22**, 38–40 and 80–81.

Hughes, T. J. R. (1977). Unconditionally stable algorithms for nonlinear heat conduction. *Computer Methods in Applied Mechanics and Engineering*, **10**, 135–139.

Hughes, T. J. R. (1983). Analyis of transient algorithms with particular reference to stability behaviour. In T. Belytscho and T. J. R. Hughes, editors, *Computational Methods for Transient Analysis*, volume 1, pages 67–155. North-Holland, Amsterdam.

Hughes, T. J. R. (1987). *The Finite Element Method. Linear Static and Dynamic Finite Element Analysis*. Prentice-Hall, Englewood Cliffs.

Hult, J. (1966). *Creep in Engineering Structures*. Blaisdell Publications Company.

Hunter, S. C. (1983). *Mechanics of Continuous Media*. Ellis Horwood Limited. Second edition.

Hutter, K. (1977). Review article. The foundations of thermodynamics, its basic postulates and implications. A review of modern thermodynamics. *Acta Mechanica*, **27**, 1–54.

Ivey, H. J. (1961). Plastic stress-strain relations and yield surfaces for aluminium alloys. *Journal of Mechanical Engineering Science*, **3**, 15–31.

Iwan, W. D. (1967). On a class of models for the yielding behaviour of continuous and composite systems. *Journal of Applied Mechanics*, **34**, 612–617.

Jaunzemis, W. (1967). *Continuum Mechanics*. The MacMillan Company.

Jeager, J. C. and Cook, N. G. W. (1976). *Fundamentals of Rock Mechanics*. Chapman and Hall. Second edition.

Jiang, Y. and Kurath, P. (1996). Characteristics of the Armstrong-Frederick type plasticity models. *International Journal of Plasticity*, **12**, 387–415.

Johansen, K. W. (1943). *Brudlinieteorier, (Yield line theories)*. Gjellerup, Copenhagen.

Johansen, K. W. (1962). *Yield Line Theories*. Cement and Concrete Association, London.

Johnson, W. and Mellor, P. B. (1983). *Engineering Plasticity*. Ellis Horwood.

Kamlah, M. and Haupt, P. (1998). On the macoscopic description of stored energy and self heating during plastic deformation. *International Journal of*

Plasticity, **13**, 893–911.

Karafillis, A. P. and Boyce, M. C. (1993). A general anisotropic yield criterion using bounds and a transformation weigthing tensor. *Journal of the Mechanics and Physics of Solids*, **41**, 1859–1886.

Kelvin, S. (1875). *Elasticity*. Encyclopedia Britannica, 9th edition.

Kestin, J. (1979). *A Course in Thermodynamics*. revised printing, Vol.1 and 2, McGraw-Hill Book Comp.

Khan, A. S. and Huang, S. (1995). *Continuum Theory of Plasticity*. John Wiley & Sons.

Kirchhoff, G. (1859). Über das gleichwicht und die bewegung eines unendlich dünnen elastischen stabes. *Journal für die reine und angewandte Mathematik (Crelle's Journal)*, **56**, 285–313.

Knops, R. J. and Payne, L. E. (1971). *Uniqueness Theorems in Linear Elasticity*. Springer-Verlag.

Ko, H.-Y. and Scott, R. F. (1968). Deformation of sand at failure. *Journal of the Soil Mechanics and Foundations Division, ASCE*, **94**, 883–898.

Koiter, W. T. (1953). Stress-strain relations, uniqueness and variational theorems for elasto-plastic materials with a singular yield surface. *Quarterly of Applied Mathematics*, **11**, 350–354.

Koiter, W. T. (1960). General theorems for elastic-plastic solids. In I. N. Sneddon and R. Hill, editors, *Progress in Solid Mechanics*, pages 165–221. North-Holland Publishing Company.

Koiter, W. T. (1964). Couple-stresses in the theory of elasticity, I and II. *Proceedings of the Koninklijke Nederlandse Akademie van Wetenschappen*, **67**, 17–44.

Kotsovos, M. D. (1980). A mathematical model of the deformational behaviour of concrete under generalised stress based on the fundamental material properties. *Materials and Structures*, **13**, 289–298.

Krausz, A. S. and Krausz, K. (1996). *Unified Constitutive Laws of Plastic Deformation*. Academic Press.

Krempl, E. (1971). Cyclic plasticity: Some properties of the hysteresis curve of structural metals at room temperature. *Journal of Basic Engineering*, **93**, 317–323.

Krempl, E. (1996). A small-strain viscoplasticity theory based on overstress. In A. S. Krausz and K. Krausz, editors, *Unified Constitutive Laws of Plastic Deformation*, pages 281–318. Academic Press.

Krempl, E., McMahon, J. J., and Yao, D. (1986). Viscoplasticity based on overstress with a differential growth law for the equilibrium stress. *Mechanics of Materials*, **5**, 35–48.

Krenk, S. (1995). An orthogonal residual procedure for nonlinear finite-element equations. *International Journal for Numerical Methods in Engineering*, **38**, 823–839.

Krenk, S. (1996). Family of invariant stress surfaces. *Journal of Engineering Mechanics, ASCE*, **122**, 201–208.

Krenk, S. (2000). Characteristic state plasticity for granular materials, Part I: basic theory. *International Journal of Solids and Structures*, **37**, 6343–6360.

Krenk, S. and Hededal, O. (1995). A dual orthogonality procedure for nonlinear finite-element equations. *Computer Methods in Applied Mechanics and Engineering*, **123**, 95–107.

Kreyszig, E. (1962). *Advanced Engineering Mathematics*. John Wiley & Sons.

Krieg, R. D. (1975). A practical two surface plasticity theory. *Journal of Applied Mechanics*, **42**, 641–646.

Krieg, R. D. and Key, S. W. (1976). Implementation of a time independent plasticity theory into structural computer programs. In J. A. Stricklin and K. J. Saczalski, editors, *Constitutive equations in viscoplasticity: Computational and engineering aspects*, pages 125–136. American Society of Mechanical Engineers, New York.

Krieg, R. D. and Krieg, D. B. (1977). Accuracies of numerical solution methods for the elastic-perfectly plastic model. *Journal of Pressure Vessel Technology*, **99**, 510–515.

Kuhn, H. W. and Tucker, A. W. (1951). Nonlinear programming. In J. Neyman, editor, *Proceedings of the Second Berkeley Symposium on Mathematical Statistics and Probability*, pages 481–492. University of California Press, Berkeley and Los Angeles.

Kupfer, H. (1973). Das verhalten des Betons under mehrachsiger Kurzzeitbelastung under besonderer Berücksichtigung der zweiachsigen Beanspruchung. *Deutscher Ausschuss für Stahlbeton, Heft 229.*

Kupfer, H., Hilsdorf, H. K., and Rüsch, H. (1969). Behavior of concrete under biaxial stresses. *Journal of the American Concrete Institute*, **66**, 656–666.

Lade, P. V. (1977). Elasto-plastic stress-strain theory for cohesionless soil with curved yield surfaces. *International Journal of Solids and Structures*, **13**, 1019–1035.

Lade, P. V. and Duncan, J. M. (1973). Cubical triaxial tests on cohesionless soil. *Journal of the Soil Mechanics and Foundations Division, ASCE*, **99**, 793–812.

Lade, P. V. and Kim, K. K. (1995). Simple hardening constitutive model for soil, rock and concrete. *International Journal of Solids and Structures*, **32**, 1963–1978.

Larsson, R., Runesson, K., and Ottosen, N. S. (1993). Discontinuous displacement approximation for capturing plastic localization. *International Journal for Numerical Methods in Engineering*, **36**, 2087–2105.

Larsson, R., Steinmann, P., and Runesson, K. (1998). Finite element embedded localization band for finite strain plasticity based on regularized strong discontinuity. *Mechanics of Cohesive-Frictional Materials*, **4**, 171–194.

Lavenda, B. H. (1978). *Thermodynamics of Irreversible Processes*. Dover Publications.

Lehmann, T. (1984). General frame for definition of constitutive laws for large non-isothermic-elastic-plastic and elastic-viscoplastic deformations. In T. Lehmann, editor, *The Constitutive Law in Thermoplasticity*, pages 379–463. CISM Courses and Lectures No.281, Springer-Verlag.

Leipholz, H. H. E. (1972). Dynamic stability of elastic systems. In H. H. E. Leipholz, editor, *Stability*, pages 243–279. Study No.6. University of Waterloo, Ontario.

Lekhnitskii, S. G. (1981). *Theory of Elasticity of a Anisotropic Body*. Mir Publishers, Moscov.

Lemaitre, J. and Chaboche, J.-L. (1990). *Mechanics of Solid Materials*. Cambridge University Press.

Lemonds, J. and Needleman, A. (1986a). Finite element analyses of shear localization in rate and temperature dependent solids. *Mechanics of Materials*, **5**, 339–361.

Lemonds, J. and Needleman, A. (1986b). An analysis of shear band development incorporating heat conduction. *Mechanics of Materials*, **5**, 363–373.

Lévy, M. (1870). Mémoire sur les équations générales des mouvementes intérieurs des corps solides ductiles, au delà des limites où l'élasticité pourrait les ramener à leur premier état. *Comptes Rendus des Seanes de l'Académie des Sciences*, **70**, 1323–1325.

Lianis, G. and Ford, H. (1957). An experimental investigation of the yield criterion and the stress-strain law. *Journal of the Mechanics and Physics of Solids*, **5**, 215–222.

Lif, J. (2003). *Analysis of the Time and Humidity-Dependent Mechanical Behaviour of Paper Webs at Offset Printing Press Condition*. Doctoral Thesis No. 55, Solid Mechanics, Royal Institute of Technology, Stockholm, Sweden.

Lif, J., Östlund, S., and Fellers, C. (1999). Applicability of anistropic viscoelasticity of paper at small deformations. *Mechanics of Time-Dependent Materials*, **2**, 245–267.

Lode, W. (1926). Versuche über den einfluss der mittleren hauptspannung auf das fliessen der metalle eisen, kupfer und nickel. *Zeitschrift für Physik*, **36**, 913–939.

Loret, B. (1992). Does deviation from deviatoric associativity lead to the onset of flutter instability? *Journal of the Mechanics and Physics of Solids*, **40**,

1363–1375.

Love, A. E. H. (1944). *A Treatise on the Mathematical Theory of Elasticity*. Dover Publications. Fourth edition.

Lubahn, J. D. (1961). Deformation phenomena. In J. E. Dorn, editor, *Mechanical Behaviour of Materials at Elevated Temperatures*, pages 319–392. McGraw-Hill.

Lubahn, J. D. and Felgar, R. P. (1961). *Plasticity and Creep of Metals*. John Wiley & Sons.

Lubarda, V. A. (2002). *Elastoplasticity Theory*. CRC Press LLC.

Lubarda, V. A., Krajcinovic, D., and Mastilovic, S. (1994). Damage model for brittle elastic solids with unequal tensile and compressive strengths. *Engineering Fracture Mechanics*, **49**, 681–697.

Lubliner, J. (1972). On the thermodynamic foundations of non-linear solid mechanics. *International Journal of Non-Linear Mechanics*, **7**, 237–254.

Lubliner, J. (1982). Normality rules in large-deformation plasticity. *Mechanics of Materials*, **5**, 29–34.

Lubliner, J. (1990). *Plasticity Theory*. MacMillan Publishing Company.

Ludwik, P. (1909). *Elemente der technologischen Mechanik*. Springer-Verlag.

Luenberger, D. G. (1984). *Linear and Nonlinear Programming*. Addison-Wesley Publishing Company. Second edition.

Maier, G. and Hueckel, T. (1979). Nonassociated and coupled flow rules of elastoplasticity for rock-like materials. *International Journal of Rock Mechanics and Mining Sciences*, **16**, 77–92.

Malvern, L. E. (1951). The propagation of longitudinal waves of plastic deformation in a bar of material exhibiting a strain-rate effect. *Journal of Applied Mechanics*, **18**, 203–208.

Malvern, L. E. (1969). *Introduction to the Mechanics of a Continous Media*. Prentice-Hall.

Mandel, J. (1962). Ondes plastiques dans un milieu indéfini a trois dimensions. *Journal de Méchanique*, **1**, 3–30.

Mandel, J. (1964). Propagation des surfaces de discontinuité dans un milieu élastoplastique. In H. Kolsky and W. Prager, editors, *Stress Waves in Anelastic Solids*, pages 331–340. IUTAM Symposium, Providence, April 1963, Springer-Verlag.

Mandel, J. (1965). Generalization de la theorie de plasticite de W. L. Koiter. *International Journal of Solids and Structures*, **1**, 273–295.

Manjoine, M. J. (1944). Influence of rate of strain and temperature on yield stresses of mild steel. *Journal of Applied Mechanics*, **11**, A211–A218.

Marquis, D. (1979). *Sur un Modèle de plasticité rendant compte du comporte-*

ment cyclique. 3éme Congrès Francais de Mechanique, Nancy, France.

Masing, G. (1927). *Wissenschaftliche Veröffentlichungen aus dem Siemens-Konzern*. Band III.

Matthies, H. and Strang, G. (1979). The solution of nonlinear finite element equations. *International Journal for Numerical Methods in Engineering*, **14**, 1613–1623.

Maugin, G. A. (1980). The method of virtual power in continuum mechanics - Application to coupled fields. *Acta Mechanics*, **35**, 1–70.

Maugin, G. A. (1992). *The Thermodynamics of Plasticity and Fracture*. Cambridge University Press.

Maxwell, J. (1868). On the dynamical theory of gases. *Philosophical Magazine*, **35**, 129–145 and 185–217.

McVetty, P. G. (1934). Working stresses for high temperature service. *Mechanical Engineering*, **56**, 149–154.

Meixner, J. (1953). Die thermodynamische theorie der relaxationserscheinungen und ihr zusammenhang mit der nachwirkungstheorie. *Kolloid-Zeitschrift*, **134**, 3–19.

Meixner, J. (1970). On the foundation of thermodynamics of processes. In E. B. Stuart, B. Gal-Or, and A. J. Brainard, editors, *A Critical Review of Thermodynamics*, pages 37–47. Mono Book Company, Baltimore.

Melan, E. (1938). Zur plastizität des räumlichen kontinuums. *Ingenieur-Archiv*, **IX**, 116–126.

Mendelsohn, A. (1968). *Plasticity: Theory and Application*. The Macmillan Company.

Metals Handbook (1978). 9th edition, American Society for Metals.

Michno, M. J. and Findley, W. N. (1976). A historical perspective of yield suface investigations for metals. *International Journal of Non-Linear Mechanics*, **11**, 59–82.

Miller, A. K. (1976). An inelastic constitutive model for monotonic, cyclic and creep deformation: Part 1. equations development and analytical procedures; part 2. application to 304 stainless steel. *Journal of Engineering Materials and Technology*, **98H**, 97–113.

Miller, A. K. (1987a). The matmod equations. In A. K. Miller, editor, *Unified Constitutive Equations for Creep and Plasticity*, pages 139–219. Elsevier Applied Science Publishers.

Miller, A. K. (1987b). *Unified Constitutive Equations for Creep and Plasticity*. Elsevier Applied Science Publishers.

Mills, H. J. (1986). *Plastics: Microstructure, Properties and Applicatins*. Edward Arnold (Publishers).

Mindlin, R. D. (1964). Micro-structure in linear elasticity. *Archive for Rational Mechanics and Analysis*, **16**, 51–78.

Mindlin, R. D. and Tiersten, H. F. (1962). Effects of couple-stresses in linear elasticity. *Archive for Rational Mechanics and Analysis*, **11**, 415–448.

Mohr, O. (1882). Über die Darstellung des Spannungszustandes und des Deformationszustandes eines Körperelementes und über die Anwendung derselben in der Fastigkeitslehre. *Zivilingenieur*, **28**, 113–156.

Mohr, O. (1900). Welche umstände bedingen elastizitätsgrenze und den bruch eines materials? *Zeitschrift des Vereines Deutscher Ingenieure*, **44**, 1524–1530, 1572–1577.

Moreau, J. J. (1970). Sur les lois de frottement, de plasticité et de viscosité. *Comptes Rendus de l'Académie des Sciences. Séries A*, **271**, 608–611.

Morrow, J. (1965). Cyclic plastic strain energy and fatigue of metals. In *Internal Friction, Damping and Cyclic Plasticity*, pages 45–87. ASTM, STP, no. 378.

Mroz, Z. (1966). On forms on constitutive laws for elastic-plastic solids. *Archiwum Mechaniki Stosowanej*, **18**, 3–35.

Mróz, Z. (1967). On the description of anisotropic workhardening. *Journal of the Mechanics and Physics of Solids*, **15**, 163–175.

Mårtensson, A. (1992). *Mechanical behaviour of wood exposed to humidity variations*. Report TVBK-1006, Depart. of Structural Engineering, Lund University, Sweden.

Mukaddam, M. A. and Bresler, B. (1972). *Behavior of Concrete under Variable Temperature and Loading*. ACI Special Publ. No. 34, *Concrete for Nuclear Reactors*, Detroit.

Murakami, S. and Sawczuk, A. (1981). A unified approach to constitutive equations of inelasticity based on tensor function representations. *Nuclear Engineering and Design*, **65**, 33–47.

Nadai, A. (1938). The influence of time upon creep. the hyperbolic sine creep law. In *Stephen Timoshenko Anniversary Volume*, pages 155–170. The MacMillan Company.

Nadai, A. (1950). *Theory of Flow and Fracture of Solids*, volume 1. McGraw-Hill Book Company. Second edition.

Nadai, A. (1963). *Theory of Flow and Fracture of Solids*, volume 2. McGraw-Hill Book Company.

Naghdi, P. M. (1960). Stress-strain relations in plasticity and thermoplasticity. In E. H. Lee and P. S. Symonds, editors, *Plasticity*, pages 121–169. Proceedings of the Second Symposium in Naval Structural Mechanics (Providence 1960), Pergamon Press.

Naghdi, P. M. and Trapp, J. A. (1975). Significance of formulating plasticity theory with reference to loading surface in strain space. *International Journal of Engineering Science*, **13**, 785–797.

Nagtegaal, J. C. (1982). On the implementation of inelastic constitutive-equations with special reference to large deformation problems. *Computer Methods in Applied Mechanics and Engineering*, **33**, 469–484.

Nayak, G. C. and Zienkiewicz, O. C. (1972). Note on the "alpha"-constant stiffness method for the analysis of nonlinear problems. *International Journal for Numerical Methods in Engineering*, **4**, 579–582.

Needleman, A. (1988). Material rate dependence and mesh sensitivity in localization problems. *Computer Methods in Applied Mechanics and Engineering*, **67**, 69–85.

Needleman, A. and Tvergaard, V. (1992). Analysis of plastic flow localization in metals. *Applied Mechanics Reviews*, **45**, S3–S18.

Neilsen, M. K. and Schreyer, H. L. (1993). Bifurcations in elastic-plastic materials. *International Journal of Solids and Structures*, **30**, 521–544.

Nemat-Nasser, S. (1988). Micromechanics of failure at high strain rate: Ttheory, experiments and computations. *Computers & Structures*, **30**, 95–104.

Neville, A. M. (1963). *Properties of Concrete*. Pitman Paperbacks.

Newmark, N. M. (1959). A method of computation for structural dynamics. *Journal of Engineering Mechanics, ASCE*, **85**, 67–94.

Nguyen, Q. S. (2000). *Stability and Nonlinear Solid Mechanics*. Wiley & Sons.

Nielsen, M. P. (1984). *Limit Analysis and Concrete Plasticity*. Prentice-Hall.

Norton, F. H. (1929). *Creep of Steel at High Temperatures*. McGraw-Hill Book Company.

Oden, J. T. (1972). *Finite elements of nonlinear continua*. McGraw-Hill Book Company.

Odqvist, F. K. G. (1933). Die verfestigung von flusseienahnlichen körpern. eine beitrag zur plastizitätstheorie. *Zeitschrift für angewandte Mathematik und Mechanik*, **13**, 360–363.

Odqvist, F. K. G. (1934). Creep stresses in a rotating disc. In *Proceedings of IV International Congress of Applied Mechanics*, pages 228–229. Cambridge, England.

Odqvist, F. K. G. (1966). *Mathematical Theory of Creep and Creep Relaxation*. Oxford University Press.

Odqvist, F. K. G. and Hult, J. (1962). *Kriechfestigheit metallischer Werkstoffe*. Springer-Verlag.

Onat, E. T. and Fardshisheh, F. (1972). *Representation of Creep of Metals*. ORNL-4783, U.S. Atomic Energy Commission.

Onsager, L. (1931a). Reciprocal relations in irreversible thermodynamics, I.

Physical Review, **37**, 405–426.

Onsager, L. (1931b). Reciprocal relations in irreversible thermodynamics, II. *Physical Review*, **38**, 2265–2279.

Ormarsson, S. (1999). *Numerical Analysis of Moisture-Related Distorsions in Sawn Timber*. Publication 99.7, Depart. of Structural Mechanics, Chalmers University of Technology, Göteborg, Sweden.

Ortiz, M. and Popov, E. P. (1985). Accuracy and stability of integration algorithms for elastoplastic constitutive relations. *International Journal for Numerical Methods in Engineering*, **21**, 1561–1576.

Ortiz, M., Leroy, Y., and Needleman, A. (1987). A finite element method for localized failure analysis. *Computer Methods in Applied Mechanics and Engineering*, **61**, 189–214.

Ottosen, N. S. (1975). *Failure and Elasticity of Concrete*. Risö-M-1801, Danish Atomic Energy Commission, Research Establishment Risö, Engineering Department.

Ottosen, N. S. (1977). A failure criterion for concrete. *Journal of the Engineering Mechanics Division, ASCE*, **103**, 527–535.

Ottosen, N. S. (1979). Constitutive model for short-time loading of concrete. *Journal of the Engineering Mechanics Division, ASCE*, **105**, 127–141.

Ottosen, N. S. (1980). Discussion of constitutive model for short-time loading of concrete. *Journal of the Engineering Mechanics Division, ASCE*, **106**, 1441–1443.

Ottosen, N. S. (1986). Thermodynamic consequences of strain softening in tension. *Journal of Engineering Mechanics, ASCE*, **112**, 1152–1164.

Ottosen, N. S. (1989). Failure criteria of concrete. In *Review of Constitutive Models for Concrete*, pages 130–166. Contract No. 3301-87-12 EL ISP DK, Report to Commison of the European Communities, Joint Research Centre, ISPRA.

Ottosen, N. S. and Gunneskov, O. (1986). Numerical subincremental method for determination of elastic-plastic-creep behaviour. *International Journal for Numerical Methods in Engineering*, **21**, 2237–2256.

Ottosen, N. S. and Krenk, S. (1982). Mechanics of gas and oil cavities in rock salt. *Bygningsstatiske Meddelelser*, **53**, 1–56.

Ottosen, N. S. and Petersson, H. (1992). *Introduction to the Finite Element Method*. Prentice Hall.

Ottosen, N. S. and Ristinmaa, M. (1996). Corners in plasticity - Koiter's theory revisited. *International Journal of Solids and Structures*, **33**, 3697–3721.

Ottosen, N. S. and Runesson, K. (1991a). Properties of discontinuous bifurcation solutions in elasto-plasticity. *International Journal of Solids and Struc-*

tures, **27**, 401–421.

Ottosen, N. S. and Runesson, K. (1991b). Acceleration waves in elasto-plasticity. *International Journal of Solids and Structures*, **28**, 135–159.

Ottosen, N. S. and Runesson, K. (1991c). Discountinuous bifurcations in a nonassociated Mohr material. *Mechanics of Materials*, **12**, 255–265.

Papadrakakis, M. (1993). Solving large-scale nonlinear problems in solid and structural mechanics. In M. Papadrakakis, editor, *Solving Large-Scale Problems in Mechanics*, pages 183–223. John Wiley & Sons.

Park, K. C. and Felippa, C. A. (1983). Partitioned analysis of coupled systems, chapter 3. In T. Belytschko and T. J. R. Hughes, editors, *Computational Methods for Transient Analysis*, pages 157–219. North-Holland.

Park, K. C., Felippa, C. A., and DeRuntz, J. A. (1977). Stabilization of staggered solution procedures for fluid-structure interaction analysis. In T. Belytschko and T. L. Geers, editors, *Computational Methods for Fluid-Structure Interaction Problems*, pages 95–124. ASME Applied Mechanics Symposia Series, AMD-Vol.26.

Paul, B. (1968). Macroscopic criteria for plastic flow and brittle fracture. In B. Liebowitz, editor, *Fracture: An Advanced Treatise*, volume II, pages 315–496. Academic Press.

Pedersen, P. (1995). Simple transformations by proper contracted forms: Can we change the usual practice? *Communications in Numerical Methods and Engineering*, **11**, 821–829.

Peirce, D., Shih, C. F., and Needleman, A. (1984). A tangent modulus method for rate dependent solids. *Computers & Structures*, **18**, 875–887.

Perry's Chemical Engineering Handbook (1984). McGraw-Hill. 6th edition.

Perzyna, P. (1963). The constitutive equations for rate sensitive plastic materials. *Quarterly of Applied Mathematics*, **20**, 321–332.

Perzyna, P. (1966). Fundamental problems in viscoplasticity. In *Advances in Applied Mechanics*, volume 9, pages 243–377. Academic Press.

Perzyna, P. (1971). Thermodynamic theory of viscoplasticity. In *Advances in Applied Mechanics*, volume 11, pages 313–354. Academic Press.

Perzyna, P. and Sawczuk, A. (1973). Problems of thermoplasticity. *Nuclear Engineering and Design*, **24**, 1–55.

Phillips, A. (1974). The foundations of thermoplasticity - experiments and theory. In J. L. Zeman and F. Ziegler, editors, *Topics in Applied Continuum Mechanics*, pages 1–21. Springer-Verlag.

Phillips, A. and Tang, J. L. (1972). The effect of loading path on the yield surface at elevated temperatures. *International Journal of Solids and Structures*, **8**, 463–474.

Pipkin, A. C. (1972). *Lectures on viscoelastic theory*. Springer-Verlag.

Plowman, J. M. (1956). Maturity and the strength of concrete. *Magazine of Concrete Research*, **8**, 13–22.

Podgorski, J. (1985). General failure criterion for isotropic media. *Journal of Engineering Mechanics, ASCE*, **111**, 188–201.

Popovicz, S. (1970). A review of stress-strain relationships for concrete. *Journal of the American Concrete Institute*, **67**, 243–248.

Potts, D. M. and Gens, A. (1985). A critical assessment of methods of correcting for drift from the yield surface in elastoplastic finite element analysis. *International Journal for Numerical and Analytical Methods in Geomechanics*, **9**, 149–159.

Prager, W. (1945). Strain hardening under combined stresses. *Journal of Applied Physics*, **16**, 837–840.

Prager, W. (1949). Recent developments in mathematical theory of plasticity. *Journal of Applied Physics*, **20**, 235–241.

Prager, W. (1955). The theory of plasticity: A survey of recent achievements. *Institution of Mechanical Engineers*, **169**, 41–57.

Prager, W. (1958). Non-isothermal plastic deformation. In *Proceedings, Konnickl Nederl. Akademie Van Wetenschappen Te Amsterdam, Series B, Vol.61*.

Prager, W. (1961). *Introduction to Mechanics of Continua*. Dover Publications.

Pramono, E. and Willam, K. (1989). Implicit intergration of composite yield surfaces with corners. *Engineering Computations*, **6**, 186–197.

Prandtl, L. (1920). Über die härte plastischer körper. In *Nachrichten der Gesellschaft der Wissenschaften Göttingen, Mathematisch-Physkalische Klasse, Göttingen*, pages 74–85.

Prandtl, L. (1924). Spannungsverteilung in plastischen körpern. In *Proceedings of the first International Congress on Applied Mechanics, Delft*, pages 43–54.

Prandtl, L. (1928). Ein gedankenmodell zur kinetischen theorie der festen körper. *Zeitschrift für angewandte Mathematik und Mechanik*, **8**, 85–106.

Rabotnov, Y. N. (1969). *Creep Problems in Structural Members*. North-Holland Publishing Company.

Rabotnov, Y. N. (1980). *Elements of Hereditary Solid Mechanics*. Mir Publishers, Moscow.

Radenkovic, D. (1961). Théorémes limites pour un matériau de coulomb á dilatation non standardisée. *Comptes Rendus l'Académie des Sciences Paris*, **252**, 4103–4104.

Ramberg, W. and Osgood, W. R. (1943). *Description of Stress-Strain Curves by Three Parameters*. National Advisory Committee for Aeronautics, Technical Notes No.902, Washington.

Raniecki, B. and Bruhns, O. T. (1981). Bounds to bifurcation stresses in solids with non-associated flow law at finite strain. *Journal of the Mechanics and Physics of Solids*, **29**, 153–172.

Raniecki, B. and Sawczuk, A. (1975). Thermal effects in plasticity. Part I. Coupled theory. *Zeitschrift für angewandte Mathematik und Mechanik*, **55**, 333–341.

Rankine, W. J. M. (1858). *A Manual of Applied Mechanics*. Griffin, London.

Rathkjen, A. (1986). Failure criteria for reinforced materials. *Bygningsstatiske Meddelelser*, **57**, 1–36.

Ray, S. K. and Utko, S. (1989). A numerical model for the thermo-elasto-plastic behaviour of a material. *International Journal for Numerical Methods in Engineering*, **28**, 1103–1114.

Reddy, B. D. and Martin, J. B. (1994). Internal variable formulations and problems in elastoplasticity: Constitutive and algorithmic aspects. *Applied Mechanics Reviews*, **47**, 429–456.

Reiner, M. (1945). A mathematical theory of dilatancy. *American Journal of Mathematics*, **67**, 350–361.

Reuss, W. (1930). Berücksichtigung der elastischen formänderungen in der plastizitätstheorie. *Zeitschrift für angewandte Mathematik und Mechanik*, **10**, 266–274.

Rice, J. R. (1976). The localization of plastic deformation. In W. T. Koiter, editor, *Theoretical and Applied Mechanics. Proceedings of the 14the IUTAM Congress, Delft, The Netherlands, 30 Aug.-4 Sept*, pages 207–220. North-Holland Publishing Company.

Richart, F. E., Brandtzaeg, A., and Brown, R. L. (1928). *A Study of the Failure of Concrete under Combined Compressive Stresses*. Bulletin No. 185, University of Illinois, Engineering Experiment Station, Urbana, Il.

Ristinmaa, M. (1999). Thermodynamic formulation of plastic work hardening materials. *Journal of Engineering Mechanics, ASCE*, **125**, 152–155.

Ristinmaa, M. and Ottosen, N. S. (1998). Viscoplasticity based on an additive split of the conjugated forces. *European Journal of Mechanics. A/Solids*, **17**, 207–235.

Ristinmaa, M. and Ottosen, N. S. (2000). Consequences of dynamic yield function in viscoplasticity. *International Journal of Solids and Structures*, **37**, 4601–4622.

Ristinmaa, M. and Tryding, J. (1993). Exact integration of constitutive equations in elasto-plasticity. *International Journal for Numerical Methods in Engineering*, **36**, 2525–2544.

Rivlin, R. S. and Ericksen, J. L. (1955). Stress-deformation relations for isotropic materials. *Journal of Rational Mechanics and Analysis*, **4**, 323–425.

Rizzi, E., Carol, I., and Willam, K. (1995). Localization analysis of elastic degradation with application to scalar damage. *Journal of Engineering Mechanics, ASCE*, **121**, 541–554.

Rizzi, E., Maier, G., and Willam, K. (1996). On failure indicators in multi-dissipative materials. *International Journal of Solids and Structures*, **33**, 3187–3214.

Robinson, D. N. (1978). *A unified creep plasticity model for structural metals at high temperature*. ORNL-5969, U.S. Atomic Energy Commission. Oak Ridge National Laboratory.

Rosakis, P., Rosakis, A., Ravichandran, G., and Hodowany, J. (2000). A thermodynamic internal variable model for the partition of plastic work into heat and stored energy in metals. *Journal of the Mechanics and Physics of Solids*, **48**, 581–607.

Rowley, M. A. and Thornton, E. A. (1996). Constitutive modeling of the visco-plastic response of hastelloy-x and aluminium alloy 8009. *Journal of Engineering Materials and Technology*, **118**, 19–27.

Rudnicki, J. W. and Rice, J. R. (1975). Conditions for the localization of deformation in pressure-sensitive dilatant materials. *Journal of the Mechanics and Physics of Solids*, **23**, 371–394.

Runesson, K. and Booker, J. R. (1982). On mixed and displacement finite element methods in prefect elasto-plasticity. In *Proc. of the Fourth Int. Conf. in Austrailia on Finite Element Methods*, pages 85–89.

Runesson, K. and Mroz, Z. (1989). A note on non-associated plastic flow rules. *International Journal of Plasticity*, **5**, 639–658.

Runesson, K. and Samuelsson, A. (1985). Aspects on numerical techniques in small deformation plasticity. In J. Middleton and G. N. Pande, editors, *NUMETA 85. Numerical Methods in Engineering: Theory and Applications*, volume 1, pages 337–347. A. A. Balkema, Rotterdam.

Runesson, K., Ottosen, N. S., and Perić, D. (1991). Discontinuous bifurcations of elastic-plastic solutions at plane stress and plane strain. *International Journal of Plasticity*, **7**, 99–121.

Runesson, K., Ristinmaa, M., and Mähler, L. (1999). A comparison of viscoplasticity formats and algorithms. *Mechanics of Cohesive-Frictional Materials*, **4**, 75–98.

Saint-Venant, B. (1870). Mémorie sur l'etablissement des équations des mouvements intérieurs opérés dans les corps solides ductiles au delà des limites où l'élasticité pourrait les ramener à leur premier état. *Comptes Rendus l'Académie des Sciences*, **70**, 473–480.

Sargin, M. (1971). *Stress-Strain Relationships for Concrete and the Analysis of Structural Concrete Sections*. University of Waterloo, Canada, Solid Mechanics Division, SM Study No.4, pages 23-46.

Sawzcuk, A. (1984). On consistent formulations of stress-strain relations for concrete at combined load and heating. *Verba Volant, Scripta Manent Festschrift for prof. Masonnet*, pages 325–331.

Schellekens, J. C. J. and de Borst, R. (1992). Application of anisotropic softening plasticity to mixed-mode delamination in composites. In D. R. J. Owen, E. Oñate, and E. Hinton, editors, *Proceedings of Third International Conference on Computational Plasticity*, pages 1671–1683.

Schickert, G. and Winkler, H. (1977). Results of tests concerning strength and strain of concrete subjected to multiaxial compressive stresses. *Deutscher Ausschuss für Stahlbeton, Heft 277, Berlin.*

Schleicher, F. (1926). Der spannungszustand an der fliessgrenze (plasticitätsbedingung). *Zeitschrift für angewandte Mathematik und Mechanik*, **6**, 199–216.

Schmidt, E. (1963). *Einführung in die Technische Thermodynamik und in die Grundlagen der chemischen thermodynamik*. Springer-Verlag. Zehnten Auflage.

Schneider, U. (1988). Concrete at high temperatures - A general review. *Fire Safety Journal*, **13**, 55–68.

Schneider, U., Diederichs, U., and Ehm, C. (1981). Effect of temperature on steel and concrete for PCRV's. *Nuclear Engineering and Design*, **67**, 245–258.

Schreyer, H. L. and Babcock, S. M. (1985). A third-invariant plasticity theory for low-strength concrete. *Journal of Engineering Mechanics, ASCE*, **111**, 545–558.

Schreyer, H. L. and Neilsen, M. K. (1996). Discontinuous bifurcation states for associated smooth plasticity and damage with isotropic elasticity. *International Journal of Solids and Structures*, **33**, 3239–3256.

Schwarzl, F. and Staverman, A. J. (1952). Time-temperature dependence of linear viscoelastic behavior. *Journal of Applied Physics*, **28**, 838–843.

Scott, R. F. (1963). *Principles of Soil Mechanics*. Addison-Wesley.

Sears, F. W. (1959). *In introduction to thermodynamics, the kenetic theory of gases, and statistical mechanics*. Addison-Wesley Publishing Company. Second edition.

Segel, L. A. (1987). *Mathematics Applied to Continuum Mechanics*. Dover Publications.

Sewell, M. L. (1973). A plastic flow rule at a yield vertex. *Journal of the Mechanics and Physics of Solids*, **22**, 469–490.

Shannon, D. F. (1970). Conditioning of quasi-Newton methods for function minimization. *Mathematical Computation*, **24**, 647–656.

Shield, R. T. and Ziegler, H. (1958). On Prager's hardening rule. *Zeitschrift für angewandte Mathematik und Physik*, **IXa**, 260–276.

Simo, C. and Honein, T. (1990). Variational formulation, discrete conservation-laws, and path-domain independent integrals for elastovisoplasticity. *Journal of Applied Mechanics*, **57**, 488–497.

Simo, C. and Taylor, R. L. (1985). Consistent tangent operators for rate-independent elasto-plasticity. *Computer Methods in Applied Mechanics and Engineering*, **48**, 101–118.

Simo, J. C. and Hughes, T. J. R. (1998). *Computational Inelasticity*. Springer-Verlag.

Simo, J. C. and Miehe, C. (1992). Associate coupled thermoplasticity at finite strains: Formulation, numerical analysis and implementation. *Computer Methods in Applied Mechanics and Engineering*, **98**, 41–104.

Simo, J. C., Kennedy, J. G., and Govindjee, S. (1988). Non-smooth multisurface plasticity and viscoplasticity - Loading unloading conditions and numerical algorithms. *International Journal for Numerical Methods in Engineering*, **26**, 2161–2185.

Sloan, S. W. (1987). Substepping schemes for the numerical integration of elastoplastic stress-strain relations. *International Journal for Numerical Methods in Engineering*, **24**, 893–911.

Sloan, S. W. and Booker, J. R. (1992). Integration of Tresca and Mohr-Coulomb constitutive relations in plane strain elastoplasticity. *International Journal for Numerical Methods in Engineering*, **33**, 163–196.

Slużalec, A. (1988). An analysis of thermal effects of coupled thermo-plasticity in metal forming processes. *Communications in Applied Numerical Methods*, **4**, 675–685.

Slużalec, A. (1992). Temperature rise in elastic-plastic metal. *Computer Methods in Applied Mechanics and Engineering*, **96**, 293–302.

Smith, G. F. (1971). On isotropic functions of symmetric tensors, skew-symmetric tensors and vectors. *International Journal of Engineering Science*, **9**, 899–916.

Smith, G. F. and Rivlin, R. S. (1957). The anisotropic tensors. *Quarterly of Applied Mathematics*, **15**, 308–314.

Soderberg, C. R. (1936). The interpretation of creep tests for machine design. *Transactions of the American Society of Mechanical Engineers*, **58**, 733–743.

Sokolnikoff, I. S. (1946). *Mathematical Theory of Elasticity*. McGraw-Hill Book Company.

Sokolnikoff, I. S. (1951). *Tensor Analysis. Theory and Applications*. John Wiley

& Sons.

Sokolnikoff, I. S. and Redheffer, R. M. (1958). *Mathematics of Physics and Modern Engineering*. McGraw-Hill Book Company.

Spain, B. (1965). *Tensor Calculus*. Oliver and Boyd.

Spencer, A. J. M. (1971). Theory of invariants. In A. C. Eringen, editor, *Continuum Physics*, volume 1, pages 239–353. Academic Press.

Spencer, A. J. M. (1980). *Continuum Mechanics*. Longman Group Limited.

Stouffer, D. C. and Dame, L. T. (1996). *Inelastic Deformation of Metals*. John Wiley & Sons.

Strang, G. (1980). *Linear Algebra and its Applications*. Academic Press.

Strang, G. (1986). *Introduction to Applied Mathematics*. Wellesley-Cambridge Press.

Stricklin, J. A. and Haisler, W. E. (1977). Formulation and solution procedures for nonlinear structural analysis. *Computers & Structures*, **7**, 125–136.

Stricklin, J. A., Haisler, W. E., and von Riesmann, W. A. (1971). Self-correcting initial value formulations in nonlinear structural mechanics. *AIAA Journal*, **9**, 2066–2067.

Strömberg, L. and Ristinmaa, M. (1996). FE-formulation of a nonlocal plasticity theory. *Computer Methods in Applied Mechanics and Engineering*, **136**, 127–144.

Sture, S. and Ko, H.-Y. (1978). Strain-softening of brittle geological materials. *International Journal for Numerical and Analytical Methods in Geomechanics*, **2**, 237–253.

Swanson, S. R. and Brown, W. S. (1971). An observation of loading path independence of fracture in rock. *International Journal of Rock Mechanics and Mining Sciences*, **8**, 277–281.

Taylor, G. I. (1938). Plastic strain in metals. *Journal of Institute of Metals*, **62**, 307–324.

Taylor, G. I. (1947). A connexion between the criterion of yield and the strain ratio relationship in plastic solids. *Proceedings of the Royal Society of London Series A*, **191**, 441–446.

Taylor, G. I. and Quinney, H. (1931). The plastic distortion of metals. *Philosophical Transaction of the Royal Society, London*, **A230**, 323–362.

Taylor, G. I. and Quinney, H. (1934). The latent energy remaining in a metal after cold working. *Proceedings of the Royal Society of London Series A*, **143**, 307–326.

Thelandersson, S. (1987). Modeling of combined thermal and mechanical action in concrete. *Journal of Engineering Mechanics, ASCE*, **113**, 893–906.

Thomas, T. Y. (1961). *Plastic Flow and Fracture in Solids*. Academic Press.

Tomita, Y. (1994). Simulations of plastic instabilities in solid mechanics. *AMR*, **47**, 171–205.

Tresca, H. (1864). Mémoire sur l'écoulement des corps solids soumis á de fortes pressions. *Comptes Rendus de l'Académie des Sciences*, **59**, 754–758.

Truesdell, C. (1955a). Hypo-elasticity. *Journal of Rational Mechanics and Analysis*, **4**, 83–133 and 1019–1020.

Truesdell, C. (1955b). The simplest rate theory of pure elasticity. *Communications on Pure and Applied Mathematics*, **8**, 123–132.

Truesdell, C. (1969). *Rational Thermodynamics: A course of lectures on selected topics*. McGraw-Hill.

Truesdell, C. and Noll, W. (1965). The non-linear field theories of mechanic. In S. Flügge, editor, *Handbuch der Physik*, volume III/3. Springer-Verlag.

Truesdell, C. and Toupin, R. A. (1960). The classical field theories. In S. Flügge, editor, *Handbuch der Physik, Principles of Classical Mechanics and Field Theory*, volume III/1. Springer-Verlag.

Tryding, J. (1994). A modification of the Tsai-Wu failure criterion for the biaxial strength of paper. *Tappi Jounal*, **77**, 132–134.

Tsai, S. T. and Wu, E. M. (1971). A general theory of strength for anisotropic materials. *Journal of Composite Materials*, **5**, 58–80.

Tuğcu, P. (1995). Heat conduction effects on strain localization in plane-strain tension. *International Journal for Numerical Methods in Engineering*, **38**, 2083–2099.

Vainberg, M. M. (1964). *Variational Methods for the Study of Nonlinear Operators*. Holden-Day, San Francisco.

Valanis, K. C. (1968). The viscoelastic potential and its thermodynamic foundations. *Journal of Mathematics and Physics*, **47**, 262–275.

Vecchi, M. (1998). *Heat generation due to plastic deformations*. Lic. thesis, Solid Mechanics, Lund Institute of Technology, Sweden, LUTFD 2/(TFHF-1019).

Vermeer, P. A. and de Borst, R. (1984). Non-associated plasticity for soils, concrete and rock. *Heron (Delft)*, **29**, 1–64.

Vicat, L. J. (1834). Note sur l'allongement progressif du fil de fer soumis á diverses tensions. *Annales du ponts et chaussées, Mémories et documents relatifs à l'art des constructions et au service de l'ingénieur*.

Voigt, W. (1892). Über innere reibung fester köper, insbesondere der metalle. *Annalen der Physik und Chemie*, **47**, 671–693.

Voigt, W. (1928). *Lehrbuch der kristalphysik*. Teubuer, Leipzig.

Volterra, V. (1913). *Leçons sur les Fonctions de Lignes*. Gauthier-Villard, Paris.

Volterra, V. (1959). *Theory of Functionals and of Integral and Integro-*

differential Equations. Dover Publications.

von Kármán, T. (1911). Festigkeitsversuche unter allseitigem druck. *Zeitschrift des vereins deutscher Ingenieure*, **55**, 1749–1757.

von Mises, R. (1913). Mechanik der festen körper im plastisch deformablen zustand. *Nachrichten der Gesellschaft der Wissenschaften in Göttingen. Mathematisch-Physiskalische Klasse, Göttingen*, pages 582–592.

von Mises, R. (1928). Mechanik der plastischen formänderung von kristallen. *Zeitschrift für angewandte Mathematik und Mechanik*, **8**, 161–185.

Walker, K. P. (1981). *Research and Development Program for Nonlinear Structural Modeling with Advanced Time-Temperature Dependent Constitutive Relationships*. NASA CR 165533.

Wallin, M. and Ristinmaa, M. (2001). Accurate stress updating algorithm based on constant strain rate assumption. *Computer Methods in Applied Mechanics and Engineering*, **190**, 5583–5601.

Wang, C.-C. (1970). A new representation theorem for isotropic functions: An answer to Professor G.F. Smith's criticism of my papers on representations for isotropic functions, part 1, scalar-valued isotropic functions, part 2, vector-valued isotropic functions, symmetric tensor-valued isotropic functions and skew-symmetric tensor-valued isotropic functions. *Archive for Rational Mechanics and Analysis*, **36**, 166–197 and 198–223.

Wang, T. T. and Onat, E. T. (1968). Nonlinear mechanical behavior of 1100 aluminium at 300F. *Acta Mechanica*, **5**, 54–70.

Westergaard, H. M. (1920). On the resistance of ductile materials to combined stresses in two and three directions perpendicular to one another. *Journal of the Franklin Institute*, **189**, 627–640.

Wilkins, M. L. (1964). Calculation of elastic-plastic flow. In B. Alder, S. Fernbach, and M. Rotenberg, editors, *Methods of Computational Physics*, volume 3, pages 211–263. Academic Press.

Willam, K. J. and Warnke, E. P. (1974). *Constitutive Model for the Triaxial Behaviour of Concrete*. International Association of Bridge and Structural Engineers Seminar on Concrete Structures subjected to Triaxial Stresses, Paper III-1, Bergamo, Italy, May 17-19.

Williams, J. G. (1980). *Stress Analysis of Polymers*. Ellis Horwood Limited. Second edition.

Wriggers, P., Miehe, C., Kleiber, M., and Simo, J. C. (1992). On the coupled thermomechanical treatment of necking problems via finite element methods. *International Journal for Numerical Methods in Engineering*, **33**, 869–883.

Wu, Z. and Glockner, P. G. (1996). Thermo-mechanical coupling applied to plastics. *International Journal of Solids and Structures*, **33**, 4431–4448.

Yamada, Y., Yishimura, N., and Sakurai, T. (1968). Plastic stress-strain matrix and its application for the solution of elastic-plastic problems by the finite element method. *International Journal of Mechanical Science*, **10**, 343–354.

Yoder, P. J. and Iwan, W. F. (1981). On the formulation of strain-space plasticity with multiple loading surfaces. *Journal of Applied Mechanics*, **48**, 773–778.

Yoder, P. J. and Whirley, R. G. (1984). On the numerical implementation of elastoplastic models. *Journal of Applied Mechanics*, **51**, 283–288.

Yu, M.-H. (2002). Advances in strength theories for materials under complex stress states in the 20th century. *Applied Mechanics Reviews*, **55**, 169–218.

Zheng, Q.-S. (1994). Theory of representations for tensor functions - a unified invariant approach to constitutive equations. *Applied Mechanics Reviews*, **47**, 545–587.

Ziegler, H. (1958). An attempt to generalize Onsager's principle, and its significance for rheological problems. *Zeitschrift für angewandte Mathematik und Physik*, **9b**, 748–763.

Ziegler, H. (1959). A modification of Prager's hardening rule. *Quarterly of Applied Mathematics*, **XVII**, 55–65.

Zienkiewicz, O. C. and Watson, M. (1966). Some creep effects in stress analysis with particular reference to concrete pressure vessels. *Nuclear Engineering and Design*, **4**, 406–412.

Zienkiewicz, O. C. and Taylor, R. (1989). *The Finite Element Method, Vol. 1*. McGraw-Hill Book Company. Fourth edition.

Zienkiewicz, O. C. and Taylor, R. (1991). *The Finite Element Method, Vol. 2*. McGraw-Hill Book Company. Fourth edition.

Zienkiewicz, O. C., Valliappan, S., and King, I. P. (1969). Elasto-plastic solutions of engineering problems. Initial-stress finite element approach. *International Journal for Numerical Methods in Engineering*, **1**, 75–100.

INDEX

The Collaborative Law Companion

Neil Denny

Family Law

Published by Family Law
A publishing imprint of Jordan Publishing Limited
21 St Thomas Street
Bristol BS1 6JS

British Library Cataloguing-in-Publication Data

A catalogue record for this book is available from the British Library.

ISBN 978 1 84661 250 3

Typeset by Letterpart Ltd, Reigate, Surrey

Printed in Great Britain by CPI Antony Rowe, Chippenham, Wiltshire

For Jo, Charlotte and Henry

CONTENTS

INTRODUCTION

COLLABORATIVE LAW IS SIMPLE, NOT EASY

There is the most simple of processes and there is strength within that simplicity.

Even though there is often resentment and pain it is not unusual to see respect and compassion.

On one level, collaborative law is underwhelming. It can be presented as little more than a series of round-table meetings between separating couples and their advisers. At its lowest level, it struggles to rise above that. But when the process works most effectively, the outcomes can be astounding.

Collaborative law meetings can also be incredibly moving. Couples find that they are able to communicate frankly with one another. They may do so for the first time in months or years. They can rediscover an ability to co-operate and to commit to a joint endeavour, such as a shared plan for caring for the children, committing to the collaborative law process itself or agreeing how they speak to others about how the relationship has come to an end.

So then, collaborative law is simple. It is an important point to keep hold of.

If we are about to step into a room with our estranged partners, with lawyers, financial advisers, therapists or counsellors, then we need to know and remind ourselves that it is natural, normal even, to do so.

There are few things more normal than sitting together to talk about a problem.

The conventional divorce or separation process is quickly shown to be far from normal.

Communication within the conventional legal separation process becomes abstracted. It is made bizarre.

One partner speaks to a lawyer. That lawyer listens and takes notes. He or she writes a letter to the other lawyer according to what they think they have heard.

The other lawyer will read that letter and interpret it. They send a copy to their own client together with a covering letter containing their own interpretation.

The receiving partner will now perceive their partner's concerns via the letter from their own lawyer.

AN ALTERNATIVE TO CHINESE WHISPERS

The original concerns of the first partner will have been moderated or adopted by the first lawyer. Emphasis will have been changed depending upon how it will impact the overall case. A particular issue might be isolated with a view to maximising what can be achieved either in terms of settlement or positioning at a final hearing, should matters get that far.

The message is then reshaped once more by the reaction of the receiving lawyer and yet again by how he or she presents the communication to their own client. The childhood game of Chinese whispers springs to mind.

By the time the receiving spouse or partner receives the message that the first had wanted to communicate, it has passed through no less than three or four channels, each one introducing interference.

COLLABORATIVE LAW IS NATURAL

The process of talking and debating within a collaborative law meeting is much more natural.

At the end of a collaborative case I asked a couple who had been able to resolve everything in two or three meetings how they thought the collaborative law process had worked for them.

They looked rather blankly at each other. The question made no sense.

'How else would we have got through all of this?', they inquired back.

The process of communicating directly with one another was so obvious and in keeping with normal daily activities, whether at home, at work or in the wider community, that they had no alternative to measure it against.

SAFEGUARDING THE DIFFERENCES BETWEEN COLLABORATIVE LAW AND CONVENTIONAL LAW

Collaborative law is simple, and yet so powerful. There is so much that goes on within the process that progress may pass unnoticed. Indeed, it might be a measure of the successful collaborative law case if it does not draw attention to itself.

In the example provided above, the emerging, now separated couple might have felt that the process itself was transparent, even invisible.

As the parties collaborate, they resolve the matters that are relevant to them unhindered by court-led timetables or prescribed forms.

For all of that, we should ensure that we maintain an awareness of the differences between collaborative and conventional approaches.

By bearing those differences in mind we can ensure that we use the collaborative law process as effectively as possible. We remind ourselves of the potential outcomes that can be achieved. We can guard ourselves against the patterns of conventional separation and resist the seductive lure of conflict.

There is another reason why we need to recognise, safeguard, even celebrate the differences between collaborative and conventional law.

MAKING COLLABORATIVE LAW DISTINCTIVE

Laurie S Coltri writes in her book *Conflict Diagnosis and Alternative Dispute Resolution* (PrenticeHall, 2004) about how dominant social norms, such as legalistic or adversarial approaches, hamper innovative dispute resolution processes. It is too simplistic to argue that they do so out of vested interests, but that helps to provide an example of how such a dynamic might come about.

Conventional practitioners will try to seduce collaborative clients with the lure of 'Look at what you could have won', pulling back the curtain on the biggest prizes to be won through a litigious approach.

We see explicit attacks where collaborative practitioners find themselves having to answer to conventional law suits for malpractice or other legalistic arguments. These might be because the collaborative settlement does not match the biggest prizes that might have been available. Other cases might see collaborative practitioners being sued for having permitted a client to collaborate at all.

DISCREDITING COLLABORATIVE LAW

There are some incredibly strident critiques of collaborative law, emerging from conventional practice.

Consider these two excerpts from different articles;

> 'As a remedy for the burned-out lawyer and the terrified divorce client the Collaborative Law movement has struck a chord; for many it has become a religion.'

From 'Drinking the Kool-Aid', www.wbdlaw.com/CM/Articles/ Drinking%20the%20Kool%20Aid%20(Walling%20Firm).pdf

> 'Couples who are facing a divorce are not likely to hire lawyers who are more akin to mental health therapists and advocate that everyone get naked, jump in a pile, eat potluck and sign agreements mutually beneficial to everyone. If the couple is fair minded, we don't need collaborative law to reach settlement. If they are not fair minded, it is my opinion they are more likely to reach settlement if they have a court date and a lawyer who is prepared to try their case.'

http://ricefamilylaw.com/blog/2009/09/16/ collaborative-law-divorce-not-a-good-fit-for-north-carolina/

These are fairly typical of the derisory tones in which collaborative law is discussed within the conventional legal practice. The suggestion is that collaborative is the easy, soft way out 'For the terrified divorce client' or the 'burned out lawyer'.

Note how the language is exaggerated. It is both dismissive and deafening in its attempt to discredit. These are the characteristics of conventional litigation. Diminish the opposing argument and drown it out.

The mocking tones within both articles are unattractive. They seek to humiliate collaborative practitioners and those who choose the collaborative process to help their families when going through separation.

Note also the strong re-assertion of the dominant conventions of litigation: 'If they are not fair-minded … they are more likely to reach a settlement if they have a court date and a lawyer who is prepared to try their case.'

ATTEMPTS TO RECLAIM THE COLLABORATIVE LANDSCAPE

If it is not possible for conventional practice to discredit the collaborative process then there will be a much more subtle corruption under the guise of a creeping assimilation back into the mainstream.

The differences between the new process and conventional practice become minimised.

DISMISSING INNOVATION

Conventional lawyers will dismiss or attempt to marginalise innovative aspects of collaborative practice. This can be seen with the disqualification clause within the collaborative law participation agreement.

If either party chooses to stop collaborating and apply to court then the existing lawyers cannot continue acting. They are said to be disqualified. It is a simple provision, but the impact of it upon the dynamic of the collaborative team is powerful and should not be underestimated. It is more than just a deterrent to going to court. It shifts the whole basis of negotiation, power and communication, as we shall see within this book.

Some conventional lawyers, even some who have trained in collaborative law, argue that if people are minded to co-operate they should not need this disqualification clause.

Others will say: 'Let's get rid of the disqualification clause. We'll stop calling it collaborative law. Let's call it collaborative-lite, or co-operative law. In fact let's make this new process more like the way we have always done it. We'll have round table negotiations but we'll make it so that we can carry on as usual if our clients want to go to court.'

THE COLLABORATIVE LAW CORAL REEF

Think of a coral reef. It is ancient, massive and almost entirely comprised of a dominant dull brown coral. Suddenly, a new species of coral is planted on this reef by the ocean currents. This new species stands out with its distinctive bright colours.

Before long, the dominant coral starts to re-colonise. It creeps over the edges of this new species, shrinking it from the outside in. At the same time it poisons the new species from below, so that its colour fades and withers, becoming brown like the rest of the reef that it rests upon.

Within a season or two, the coral looks exactly like the rest of the reef. It has become indistinguishable, except upon the closest scrutiny.

The collaborative process risks becoming re-colonised and reduced from its innovative, vivid state to nothing more than a series of round-table meetings, which lawyers have been doing for years. We need to recognise and preserve the differences between collaborative and conventional practice for these reasons.

SLIPPING BACK INTO CONVENTIONAL PRACTICE

There is also a risk of re-convergence or slipping back between newly acquired collaborative skills and what went before.

If collaborative practitioners lose sight of the important differences between collaborative and conventional practice, then they risk losing their own vividness.

Many collaborative practitioners neglect to develop their collaborative skills. They lose curiosity about what the couple are experiencing or what their wishes are. They slip backwards into conventional legal practice. Perhaps they start to put forward positions within collaborative meetings or take on a driving role within meetings. They resume speaking on behalf of their clients instead of ensuring that their clients are contributing fully.

We need to safeguard and celebrate those differences, therefore, and remind ourselves of them, frequently.

INTRODUCING THE COLLABORATIVE LAW SPECTRUM

Collaborative law is a relatively new process. All collaborative lawyers need to be trained and qualified through Resolution, the England and Wales family lawyers association, but not all collaborative lawyers respond to this training the same.

The fact that no two collaborative lawyers act in an identical fashion is not a problem. That indicates the flexibility of the collaborative process. What needs to be of greater concern, particularly for the client, and if the collaborative legal community is going to grow, is the degree of commitment to be found between collaboratively trained lawyers.

This book establishes a spectrum upon which to place collaboratively trained lawyers.

At one end of the spectrum will be the sceptics of the collaborative law process. These lawyers have trained 'just in case' someone wants them to collaborate at some stage in the future. They have no desire to collaborate and no belief in the opportunities that collaborative practice makes available to families.

The far end of spectrum stretches up to the most motivated and effective of collaborative lawyers. These practitioners are strongly committed to seeing collaborative law thrive. They are convinced about the benefits of the collaborative outcome, and equally disillusioned with the damage caused by conventional outcomes to the client and the client's family.

They are the proudest corals on the reef and are often seen as leaders of the collaborative movement, whether nationally or locally.

In between the extremes are a mix of collaborative lawyers who fall somewhere between passivity and activity. They will have all understood and passed the collaborative training, but how much have they have changed their practice?

It is one of the goals of this book to inspire collaborative practitioners to become more active in adopting collaborative practice within their day-to-day work.

FIVE SOURCES OF CONFLICT IN SEPARATION

A key characteristic of family breakdown is conflict. That conflict may arise from one of the following drivers:

- *Factual.* Who did or said what? Who is to blame?
- *Legal.* What are my entitlements?
- *Communication.* What do you mean by that?
- *Displaced conflict.* For example, a parent who is anxious about their financial provision might seek to create a new conflict around child contact, or some other issue.
- *Limited resources.* Time with the children or dividing income and financial assets.

These drivers will be experienced as much within collaborative law cases as they will within conventional cases. What will be different will be how we respond to them.

Collaborative law is neither a conflict free nor an easy way of resolving the issues arising from the end of a relationship. The collaborative law process enables us to respond to those inevitable conflicts in different ways and this is where we start to see the reasoning behind the participation agreement and the disqualification cause.

RESPONDING TO CONFLICT WITHIN CONVENTIONAL SEPARATION

Conventional litigation draws its strength and authority from the legal sanctions available to the lawyer and his or her client. This means that within conventional cases there is always the consideration 'What would a judge say?'

If there is a dispute then the lawyer's consideration is 'Can I prove my client's version of events? What has been said by the other person to support my case? How can I discredit what the other party is saying?', and the ensuing dialogue or correspondence will reflect that to varying degrees.

Conventional legal conflict leads us to consider our position against relevant sections of law or previously decided cases. In doing so we consider steps that we can lead a judge to take to maximise our share of resources, namely time with children or financial resources.

This forced communication fuels entrenched positions complete with an attack–defence–counter-attack dynamic. The Chinese whispers dynamic explored above can easily create acrimony. Each partner can be horrified at the tone of communication, incorrectly thinking that because it has been made on behalf of their partner, it accurately reflects their partner's stance.

RESPONDING TO CONFLICT WITHIN COLLABORATIVE LAW

Within collaborative law we can respond to conflict differently.

This is enabled by two distinguishing features, namely that we are communicating face to face, and the unique impact of the participation agreement, complete with the aforementioned disqualification clause.

THE ADVANTAGES OF WORKING FACE TO FACE

When we are working face to face within collaborative meetings we are aware of how people are responding. We can deal with conflict as it arises. We can ask questions to clarify issues or whether those issues are important. The Chinese whispers of clients instructing lawyers, lawyers interpreting, twice, and then reporting to the other client becomes obsolete.

Misunderstood communication can be resolved by the simplest of questions being asked across the room within a collaborative meeting – 'What do you mean?' or 'Tell me more about that'.

If we are angry, anxious or upset within collaborative law then we can say as much rather than seek to express that hurt elsewhere, such as by stopping a child visiting a parent or grandparent. Within collaborative law meetings, it is clear when these emotions are felt and we can address them, explicitly, either right there or by referring to suitable support professionals, such as therapists, coaches or counsellors.

THE REDUCED ROLE OF FACTUAL AND LEGAL CONFLICT IN COLLABORATIVE LAW

The seemingly central issue of factual conflict – who said or did what – becomes less critical within collaborative law. This can sound like some kind of heresy to conventional lawyers.

There is no need to refer backwards to what has happened in the past to justify what we are asking a judge to order in the present. Instead the commitment to collaborate sees the couple looking forward to design an outcome that will benefit the whole family.

That is not to say that the factual issues or what has happened is irrelevant. It may well be necessary still to acknowledge where things have gone wrong, but the role of these factors is greatly reduced.

The role of legal consideration in collaborative law is radically different to conventional cases. This merits greater attention elsewhere in this book and we will revert to it later. The thrust of collaborative law is to design a solution that fits the specific family or couple as opposed to looking at solutions that lawyers have carved out, or a judge has previously imposed, in the hope that it fits this current family also.

Legal entitlement is still considered within collaborative law, but entitlements are not a prescription or a foregone conclusion. Instead, what a court might order sometimes becomes little more than a curiosity. In approaching legal entitlements in this way we see that the legal outcome is often far from clear, or that the outcome is unattractive for various reasons. The legal arena and 'What a judge might say' is no longer the deciding factor. What the family needs and desires is what matters most.

IS COLLABORATIVE LAW BETTER THAN CONVENTIONAL APPROACHES TO LAW?

The temptation here is to be conservative and say that collaborative law is no better than conventional approaches. It would be easy, and conflict-averse, to say that both are as good as each other, but for my own part I cannot say that.

Collaborative law is better.

The fact that collaborative law might not be suitable for some families does not mean that conventional law becomes any better for those families.

It just becomes more appropriate.

It can still be just as damaging and have repercussions for many years to come. Parents will still be pitched against parents, spouse against spouse and the excellent work carried out by all conscientious family lawyers to minimise those impacts will still need to be maintained.

Collaborative law will not be appropriate for everyone, but if it was it would be better.

This enthusiasm for collaborative law is not my own folly. It is an increasingly held view, even among leading judiciary. Consider the words of Justice Coleridge in the case of *S v P (Settlement by Collaborative Law Process)* [2008] 2 FLR 2040:

> 'Every conceivable encouragement should be given for parties to negotiate by this method.'

To offer families an opportunity to create their own solutions complete with careful representation and support is a powerful thing.

Both partners contribute to the solution that they design and so there is no resentment once the lawyers are discharged. The chances of sabotage or non-compliance, it is to be hoped, will be greatly reduced.

Within collaborative law, couples are free to consider a whole range of solutions, some creative, some less so. These solutions can go far beyond the limited considerations shaped by what a court can or cannot do.

HOW TO USE *THE COLLABORATIVE LAW COMPANION*

This book is written for lawyers who have already trained in collaborative law skills and for the curious. Its aim is to inspire collaborative lawyers once again to be curious about the potential that can be achieved through the collaborative process. I hope it will encourage many to push on to a greater level of effectiveness and commitment to collaborative law work.

I hope that it will stimulate debate and discussion within the collaborative law profession itself and with non-collaborative sectors.

Regional practice groups, or PODs as they are known, might want to start up book clubs to work through this material, or dip into the book from time to time, to access material that can be used to start discussions amongst local colleagues.

As the title suggests, it is also designed as a companion to collaborative practice.

It can feel like we are swimming against the tide of deeply entrenched social norms of litigation and adversarial practice.

Collaborative lawyers, as we have seen, might find themselves being mocked and derided by the mainstream profession as being idealistic, liberal or burnt out.

Ignore them.

We are criticised for thinking that we know better.

We do not know better, but we do know a process that we believe offers a far better way to help families going through separation.

Keep hold of that and do all that you can to serve those families that you help.

The book serves as a refresher and a provocation to consider new approaches to working that reflect the collaborative law shift.

I sincerely hope that it will help you in building your collaborative practice and skills.

This book is also written for the individuals and couples who choose to use collaborative law to explain how and why collaborative law can help your family. It aims to reveal some of the training that your collaborative lawyer will have gone through and why they do some of the things they do in the way that they do.

When I completed my collaborative training in 2006 I was amazed. Lawyers spend hundreds of pounds on learning this material and in making considerable shifts from our well-entrenched ideas of what it used to mean to resolve family disputes. It takes several days out of the office and then practice and further reading really to consolidate that shift. Yet we place our clients into those collaborative meetings with no training at all and expect them to be effective.

This book will go some way to helping you to be as effective as you can be within those meetings. When that happens, you take back the responsibility for how your family will look instead of leaving it to the lawyers and the judicial system.

DOES COLLABORATIVE LAW HAVE A TRANSFORMATIVE AGENDA?

A final source of criticism against collaborative law is that it has a transformative agenda. This means that it strives to bring about change in the individuals concerned as they pass through the process of separating and resolving disputes. The criticism is that this goes beyond acting in the client's best interests. A lawyer-led transformative agenda can say more about the lawyer's own belief system and motivations.

For my own part, I believe that collaborative law does have the power to transform. It demonstrates to families that, even though we have fallen out of love, and even though we might disagree on a whole range of issues, and even though I might be hurting very badly indeed, despite all of these things, we can still commit to working together towards a better outcome.

Collaborative law, and other alternative dispute resolution processes, model new ways of dealing with conflict. We might, as we pass through these processes, learn new insights that will colour our own conflict responses in future. As we do so we might, just might, also model to our children, friends and colleagues that there are alternatives to litigation.

That would be transformation indeed, but the transformation is a by-product of collaborative law, not the goal.

The goal is quite simply the best interests of you, the client, and your family.

This potential for transformation is held up, pilloried by critics of the collaborative law process as if it were a bad thing.

Transformation is nothing more than change. The families we work with are going through massive change anyway. Very often those changes are destructive.

Collaborative law offers these families a more hopeful change for them and their futures.

That is not a bad thing.

Hope is not a dirty word.

CHAPTER 1

THE SEVEN ASPECTS OF SEPARATION

1.1 BROADENING THE SEPARATION CONSIDERATIONS

Constance Ahrons' excellent book *The Good Divorce* (HarperCollins, 1994) refers to the difference between the emotional and legal aspects of divorce. She asserts that people are too ready to rush to the formal aspects of the legal process before taking care of the emotional elements.

There is a risk that it is easier to hand over the responsibility for the legal elements of separation and divorce to the lawyers. Doing so feels like a relief for many individuals going through separation. It feels as though something is being done by somebody. It feels like progress and many people believe that as soon as the legal matters are resolved the individual can then move on.

The consequences are dysfunctional legal divorce processes which reflect the unresolved emotional aspects, resulting in heightened conflict, power struggles and damage.

This distinction between the legal and emotional aspects of divorce and separation is important. It challenges us to be more mindful about a wider range of issues that need to be considered.

1.2 THE SEVEN ASPECTS OF SEPARATION

In the next three chapters we will break down these two aspects into seven areas for consideration. In doing so it is our intention to create a checklist of sorts that we might refer to from time to time as we progress through the resolution of issues.

The checklist can help us to ensure that everything is being taken into account. It might also help us when we feel as though progress is getting stuck. At those times, consideration of the seven aspects of separation might reveal an area that has been neglected, or might stimulate ideas on areas that we can concentrate upon to restart communication.

The shift from conventional legal representation to collaborative law gives lawyers and their clients the space to look at a far broader range of concerns. If we take that opportunity then we can start to provide dispute resolution processes and outcomes that will better serve our clients and their families.

Conventional legal divorce and separation tends to focus on legal entitlements and remedies. When that happens then we risk overlooking six other important aspects of divorce and separation.

The seven aspects of separation and divorce are therefore as follows:

- Legal
- Financial
- Parental
- Emotional
- Uncoupling
- Social
- Narrative

They can be depicted as shown in Figure 1.1.

This diagram shows how the seven aspects can be seen to interrelate.

The legal, financial and parenting issues are all closely grouped and overlap to some extent.

We need to do more to keep them distinct. If we do not, then we risk assuming that all financial and parenting considerations have been satisfied when we have only dealt with the legal questions.

We shall look at how these three considerations work together in this chapter.

Chapter 2 will examine how our clients' thoughts about legal issues, financial and parenting matters are likely to be shaped and influenced by the emotions being felt by the individuals concerned. That is reflected in the diagram by showing the emotional element as holding the first three, as if in a tight sheath. Again, collaborative law enables everyone involved to have a better understanding of how we might be feeling. The impact of those emotions needs to be appreciated.

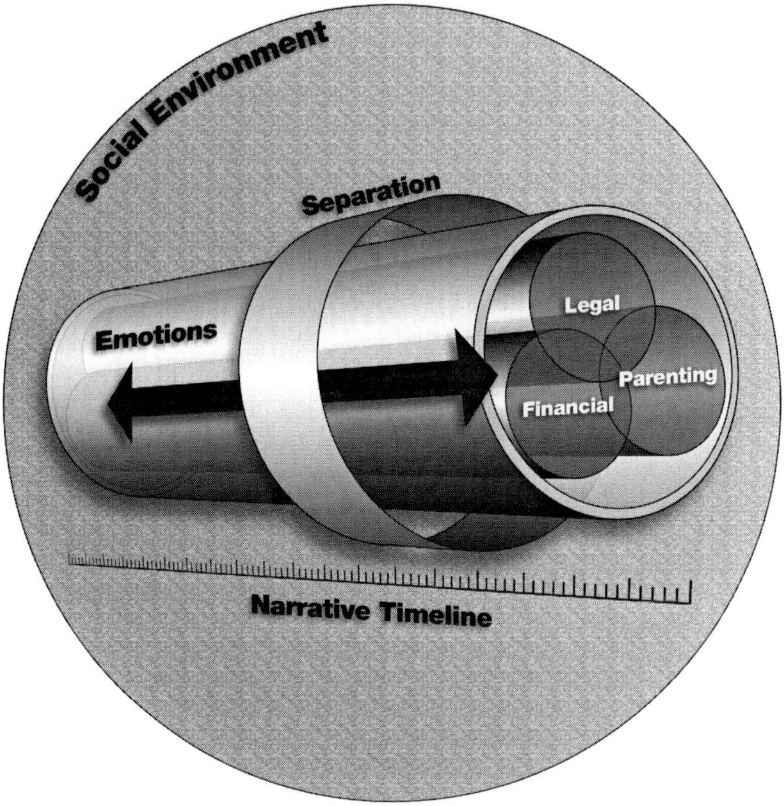

Figure 1.1: Visualising the seven aspects

The actual separation, that is to say the act of the couple moving out of one home into two, might also need to be managed. This might have already happened before lawyers are instructed. If so, is there fallout that needs to be considered? If the couple have not yet separated, then the couple have an opportunity within collaborative law to manage this 'Uncoupling' rather than leave it to chance and possible chaos.

The separation is depicted as a movable band that can be moved up or down the diagram. It might be placed at the very start of the process or further up if it has yet to take place. The separation might also be a very thin band if it is possible to deal with it quickly. Alternatively, it might need to be very broad if the plans required are complex and will take time to implement.

Collaborative law offers much more opportunity for practitioners and individuals to consider what is going on as the relationship comes to an end. Relationships end not only in our clients' lives and homes, but in the middle of their own communities, social groups and broader families. To what extent do the various social groups shape and distort the separation and how can the couple be prepared for that?

Finally, and we will explore this last point in Chapter 3, understanding the role of narrative within separation and divorce can provide couples with new opportunities to make sense of what they are going through. With the help of effective collaborative law skills, our clients can control the story that their relationship and separation will tell. Doing so gives them control over their separation and enables them to shape their family's future. The narrative model is a powerful tool in the effective collaborative practitioner's box. It is for this reason that it has a whole chapter dedicated to it.

This chapter will now look at the first three elements of separation. We will see how a broader consideration will help families to make more sense of what is going on. We will also encourage practitioners to move further away from the narrow considerations of conventional legal practice and so increase the assistance and insights that they can bring to their clients' families.

1.3 THE LEGAL SEPARATION

The legal elements of divorce or separation are the most obvious, particularly to practitioners. This is the result of the social and cultural expectations that position lawyers at the very centre of the divorce or separation process.

Conventional wisdom says that if you are getting divorced, then you had better get a lawyer.

1.4 The lawyer as gateway

The lawyer has been the gateway through which separating couples must pass to access advice, information and resolution. It has been a privileged position for the profession, but one which might not have served separating families well.

An individual who is worried about his or her relationship would make an initial appointment to meet with a divorce lawyer.

They would sit on one side of the desk and explain their situation.

Perhaps they are concerned that a divorce might be necessary at some stage and they want some information about what this will entail.

The lawyer, keen to reassure the client that they are in safe hands, very quickly advises on legal process and relevant statutes. He or she delivers a compelling and cogent breakdown of what lies ahead, including timescales, costs and a range of possible outcomes. The client, suitably informed, proceeds to instruct the lawyer to deliver the solutions he or she has outlined.

The lawyer then proceeds to go about delivering the outcomes they have outlined, over the course of several months, invoices and then closes the file.

The client emerges legally divorced with a collection of court forms and orders and is left to deal with everything else.

The legal aspects of the divorce will have been dealt with appropriately, but there has been no consideration of broader financial or parenting issues, or of the client's emotional state.

When lawyers act as the conventional gateway to separation the legal considerations will take centre stage. It is possible to lose sight of the other aspects of separation. These can be entirely marginalised within the legal process either by intent or by a simple lack of awareness.

1.5 Legal processes can block emotional considerations

Within a conventional separation the lawyer may entirely disregard emotional issues. Lawyers call this 'Remaining objective' or 'Focussed on the outcome'. That does not need not to be a criticism of the lawyer's role. It might represent his or her instructions. Remember Constance Ahrons' clients handing over responsibility to the lawyer? That act can be a means of bypassing the emotional fallout.

Many clients will not permit their lawyers to consider the emotional considerations. It is just too painful for them to do so. In some cases the legal issues are used as a sort of catharsis through which the hurt and pain can be played out by someone else, namely the hired lawyer.

Lawyers are of course trained to focus on the legal process of separation.

As a result, legal advisers will concentrate on their field and may refuse to engage with the other elements to varying degrees. This might be complete disregard in the worst cases, through to carefully selected referrals to other service providers at best.

Some lawyers will build their own mythology and reputation as being entirely dispassionate. It might speak of strength to some clients. To others it might provide reassurance that this separation is going to be dealt with in a 'Business-like' fashion. It suggests that emotions are not going to feature and that may well appear to be an easier way forward.

There is also the cost consideration acting as a limit and focus on what gets considered and discussed between lawyer and client.

The lawyer is paid for his or her legal skills. Even if the lawyer sees that there are other issues at stake as the case progresses, the lawyer might feel that they do not have the mandate from the client to explore these other arenas.

A conventional divorce lawyer could spend hours upon hours, all charged at their hourly rate, listening to and empathising with a client. That, however, may not be in the client's best interests.

Is this the best use of the client's funds?

Would that element of the costs be better spent by referring the client to a counsellor or therapist instead?

Does the lawyer have permission from the client to recommend a counsellor or therapist?

And what if the client does not take that offer up but still finds it impossible to cope with the changes going on in their life?

What is the lawyer to do then? Repeat recommendations or referrals to other supporting professionals or just carry on regardless?

1.6 The lawyer's role as legal expert

The lawyer's conventional role is to work as a guide and pilot to the client as they navigate the complex legalistic channels that lie ahead.

As guide, the lawyer uses their knowledge and communication skills to make this legalistic environment understandable. They explain what is happening and what comes next.

As pilot, the lawyer navigates a route to a destination that the client has requested or one that the lawyer has recommended through his or her legal advice. Conventionally, this journey is mapped out through lawyers' correspondence, court applications and hearings. Strange statutes and case-law act as landmarks that the lawyer alone can read and understand.

Too often, at the end of the lawyer's involvement, the client has a legal outcome but arrives dishevelled, hungry and thirsty because no one has been looking after their other needs.

1.7 The collaborative law mandate

When we work collaboratively then we have the mandate from our clients to be curious about a broader range of concerns. There is also the opportunity to bring in other collaborative professionals to manage those concerns with the client.

The active collaborative lawyer has the permission to ask and comment upon what the client is going through instead of only being mandated to deal with the legal answers. Hopefully the lawyer will have spent time developing a network of preferred counselling and therapeutic practitioners that they can refer clients on to. There is no need for the lawyer to step into therapeutic roles, but he or she should have an awareness about how those other professionals work.

The financial and parental aspects of separation will inevitably be part of the legal adviser's remit and consideration. See Figure 1.1 with the large overlap between legal, financial and parenting, but also see below for more about the areas that do not get covered.

Many conventional practitioners already consider a broader range of issues than purely the legalistic ones. That does not shift the mandate they have from their clients.

1.8 What would a judge say?

The deafening question in too many conventional cases is 'What would a judge say?'

This single consideration dominates conventional practice.

It shapes how we relate to our clients, what advice we give them and how far we are prepared to negotiate and compromise.

The result is that the whole management of a case can be shaped by the anticipation of a final contested hearing in court.

- What evidence do we need?

- What do we need to challenge?

- How can we prove or disprove certain assertions?

Everything is measured against the yardstick of judicial opinion.

If we put forward a proposal, then do we think that we could get that kind of outcome from a judge at a final hearing?

If we receive a proposal, then the question is: 'How does that compare to what a judge might order? How sure are we of that? What are the variables and does that, therefore, make this offer attractive?'

This consideration conditions the language that we use with our clients and when communicating with our client's partner. Creative thinking is stilted and we tend to think in terms of previously decided cases or what we believe we can get a judge to agree with.

Of course we deal with financial issues, and parenting issues as well, but all couched in terms of the legal backdrop and the authority of the judge, statute and case-law. The problem is that these perfectly proper considerations might fail to address fully the complete range of issues that a client is struggling with, such as:

- broader financial concerns;
- practical concerns about co-parenting after separation;
- their emotional state;
- what other friends and families are thinking; and
- what this separation tells others about the client and their partner.

1.9 THE FINANCIAL ASPECTS OF SEPARATION

In its purest form, the financial aspects of divorce or separation can be seen as a question of provision and division.

- How are both parties going to be provided for?
- How are assets, both current and future, going to be divided?

For example, in divorce cases in England and Wales finances are determined following consideration of eight provisions listed at s 25 of the Matrimonial Causes Act 1973. Those factors, shared here for the benefit of non-lawyer readers, are:

(1) the income, earning capacity, property and other financial resources which each of the parties to the marriage has or is likely to have in the foreseeable future, including in the case of earning capacity any increase in that capacity which it would in the opinion of the court be reasonable to expect a party to the marriage to take steps to acquire;

(2) the financial needs, obligations and responsibilities which each of the parties to the marriage has or is likely to have in the foreseeable future;

(3) the standard of living enjoyed by the family before the breakdown of the marriage;

(4) the age of each party to the marriage and the duration of the marriage;

(5) any physical or mental disability of either of the parties to the marriage;

(6) the contributions which each of the parties has made or is likely in the foreseeable future to make to the welfare of the family, including any contribution by looking after the home or caring for the family;

(7) the conduct of each of the parties, if that conduct is such that it would in the opinion of the court be inequitable to disregard it; and

(8) in the case of proceedings for divorce or nullity of marriage, the value to each of the parties to the marriage of any benefit which, by reason of the dissolution or annulment of the marriage, that party will lose the chance of acquiring.

All of the above are subject to the court's primary concern, namely the welfare of any child of the marriage.

Within conventional cases there would be full consideration of these factors. The available and identifiable future assets would be distributed. Having done so, a conventional approach to divorce would conclude that the financial aspects have been taken care of.

It is a mistake, however, to assume that adherence to this purely legalistic approach will result in ideal outcomes.

In one contested case, I obtained an outcome for a client, following a contested hearing, which retained 90% of the matrimonial assets for the wife. When measured against legal criteria, this represented an excellent result.

The client was, however, mortified.

She knew that the outcome of these financial considerations would leave her husband with next to nothing. She was concerned about how this would impact on their ability to work together as parents. She was also worried about how this outcome would impact upon the family, as a whole, emotionally and socially.

What would the paternal grandparents think? Would they want to see the children?

What story was this divorce, and the outcome, telling to the children and what would be the children's experiences as they grew up, seeing their father living in the conditions that he had been reduced to.

Strict adherence to a legalistic approach had not satisfied any of those concerns.

In another case we might have seen a similarly excellent legal outcome and yet leave a client with no ability to now manage his or her future financial obligations.

1.10 The zero sum approach. More for me, less for you

Conventionally, lawyers looking to negotiate or litigate financial disputes do so as a 'Zero sum exercise'.

This states that when we add up what one party has gained with what the other party has lost, then the total will be zero.

For every pound the client's partner receives, your client will get one pound less.

Conventionally, lawyers have not had any difficulty with this notion. They are instructed to act in their client's best interests. Best interests can easily mean getting more for your own client and leaving it to the other partner to protect their own interests.

It is easy to see how this drives the adversarial nature of conventional approaches.

1.11 Positions-based negotiating

The zero sum approach is also called a 'Positions-based negotiating system'.

Here, our positions are those lists of demands and payments that we insist upon for our client. Our position is that we will take the house, or a given lump sum, combined with maintenance payments of a certain level for a certain duration.

When we make proposals we build up a tower of such demands. That is our position.

Our client's partner will have their own tower representing their position. This is their own list of requirements.

What then follows is that each party, via their lawyers, sets about attacking the other party's position.

We send over correspondence discrediting their contribution or demand for a certain asset. We employ our legalistic knowledge and fire across withering letters strengthened with previous decided cases and statutes that help us to undermine the other's position.

The hope is that we land enough hits to reduce the other person's tower. If we can show that their argument has been poorly constructed, or if we take down it down a few pegs, then our own position becomes the dominant one.

While we are attacking the other's position, we have to defend our own tower from similar incoming strikes.

This is of course an entirely destructive process.

Each partner perceives that they are coming under attack from the other, or from their lawyer.

Massive resources can be spent on trying to disprove one claim or another. Very quickly it becomes more important to repel every allegation or comment that is made. We become more fixated on increasingly small details rather than concentrating on demonstrating why a certain level of financial provision is being sought.

The emergence of greater studies into effective negotiation, such as that carried out by the Harvard Negotiation Project over the last 30 years, has helped to create a new discipline in better understanding how the way that we negotiate can actually build in opportunities to create value for the benefit of both parties.

1.12 Interests-based negotiating

The biggest single shift is in the move from positions-based negotiating to interests-based negotiation. Interests-based negotiation is not so concerned with what the parties' stated positions are, but with what needs and interests each party is trying to address.

1.13 The two sisters and the orange

Two sisters are fighting over the last orange in fruit bowl. Tired of squabbling they agree to the perfect compromise and cut the orange in two.

They both then use the half orange they have received.

Later that evening they tell their mother about their squabble and their compromise. They are both pretty sore that they did not get what they wanted, namely the whole orange.

Their mother listened to them carefully and then asked each of them a simple question:

> 'What did you need the orange for?'
>
> 'I wanted to make a cake', answered the first sister. 'I needed the pith from the orange peel to flavour the batter. Now you can hardly taste the orange in it at all.'
>
> 'I was making a glass of juice', answered the other sister. 'I only had half of what I wanted and I am still thirsty.'
>
> 'You foolish girls', said their mother. 'Don't you see? If only you had each taken just a moment to understand why each of you wanted the orange, instead of trying to get as much of it as you could, then you could have both got everything you wanted.'
>
> 'The cake could have used the whole of the pith, and you could have had the juice of the full orange.'

1.14 Collaborative law is interests based

Collaborative law is much more concerned with understanding each party's interests.

Lists of financial demands are frowned upon and dialogue revolves, instead, around what is important to each party and why.

The clients are encouraged to be creative in their considerations and their lawyers need to ensure that they have an adequate grasp of interests-based negotiating to ensure that they can assist their clients and the collaborative process.

This requires collaborative practitioners to be more curious and keep asking questions. If one party is pursuing a certain asset or provision then we need to ask questions about that.

> 'Why is that important to you?'
>
> 'What would that mean to you?'
>
> 'What need would be met if you were to keep the house?'
>
> 'How might we make that work?'

> 'How will that work for your husband/wife?'

By asking these questions we avoid getting stuck into positional traps which would just see us arguing about what has been sought. Instead we maintain a curious and respectful stance. It is likely that as we do so, these questions actually continue to develop new insights and a whole range of alternatives that might satisfy an identified need or interest.

1.15 Interests can break financial impasse

In a collaborative case, John and Barbara are finding it difficult to move away from their respective positions.

John has in mind that he will pay his wife no more than £25,000 as a one-off lump sum instead of ongoing maintenance payments.

Barbara has promised to herself that she will not settle for less than £75,000.

Both have stated that their offers represent their last say on the matter. This gap between the two brought the last collaborative session to an ill-tempered end and has now carried forward into the next session a couple of weeks later.

Barbara, still adopting a positional stance, says she will split the difference and take £50,000. John objects and is about to walk out of the meeting, explaining that this is all a waste of time and money.

John feels he cannot lose face by backing down even though Barbara has reduced her request. In fact, John is angrier than ever as he is now convinced that Barbara had gone in high with a request for £75,000 only to put extra pressure on him. Going in high is a very conventional positional-based negotiating tactic. Unfortunately the result here is that John trusts Barbara even less.

One of the lawyers remembers the need to be curious in order to avoid positional traps such as these. He resolves not to defend or oppose the request for £50,000 and instead to ask questions.

> 'Barbara, tell me what this £50,000 means to you? What considerations are driving the money you are asking for?'

The problem very quickly settles.

Barbara explains that she was going to use the ongoing maintenance payments to repay her brother who had lent her a lot of money to get out of personal debt several years ago. At the time, she had not wanted to discuss it with her husband for fear that he would become angry and insist she returned to work, rather than stay at home to care for the children as they had previously jointly agreed.

He can see that the loan from the brother-in-law was properly paid and agrees that, yes, of course it should be repaid. He asks how much is still owed and Barbara thinks it is about £50,000.

Barbara accepts that she had asked for £75,000 to 'anchor' her position, expecting to get knocked down to £50,000.

John finds that he is able to agree with the reasoning and agrees to pay up to £50,000 to see that his brother-in-law is paid back, and the issue is resolved.

Remember, only a few moments earlier, John felt he was not going to be held ransom to positional demands and ultimatums. He is only able to agree once the couple shift from positions and proposed halfway compromises to curiosity about what is driving the other person's demands.

When we develop a fuller understanding of why something is being sought, then we are often able to agree with the reasoning. If Barbara had just kept asking for £50,000, they might have never agreed.

Within conventional cases, demands are often driven by percentage claims. These in turn are informed by previously decided cases. A wife and mother in a certain position might be entitled to a certain percentage share, for example. Such an outcome might have been appropriate for a previous family, but is it really appropriate for this family?

Conventional cases also get stuck on the zero game of maximising one party's share over and above the other.

Within collaborative cases we can be far more creative about how we are going to explore and settle financial claims. Having the space and the permission from clients to ask more curious questions about needs and interests can help families resolve these disputes better.

1.16 Considering both parties' financial knowledge

Are both parties familiar with financial products, such as investments, mortgages, insurance and pensions? Again, a purely legalistic approach to resolving financial matters upon divorce or separation may well see a successful outcome whereby one party receives a comprehensive wealth portfolio, but sees that person left completely stranded in understanding what to do with it now.

A lawyer can easily overlook the fact that one partner has no working knowledge of financial matters. There are many cases where the less financially aware party goes along with a legal process trusting his or her lawyers to deal with the matters appropriately. He or she might have no idea about the value or nature of the pension, unit trust or other financial product that they are negotiating upon or have received.

There are probably many court cases where the less financially astute party sits in bewildered silence as the lawyers and accountants discuss financial matters about which he or she knows nothing.

Within collaborative cases we have the opportunity to ensure that both parties have an adequate level of knowledge about the financial aspects of the separation. This is important if both partners are going to be able to make informed contributions to negotiations and the settlement of matters.

Assistance can also be brought in to help a partner who is receiving and having to manage assets or income. This does not only apply in high net worth cases. In many cases where there is a much more normal level of asset or income one or both parties might benefit from some assistance with household budgeting and management.

Separation in many cases has far broader considerations beyond the purely legalistic, zero sum considerations of what a judge would say and who might get what percentage. Within collaborative law, the couple can more fully explore the wide range of considerations and start to design a settlement for their own specific circumstances instead of going along with what a judge felt was suitable for a different family in a previous case.

1.17 PARENTING ASPECTS OF SEPARATION

Just as the financial aspects of separation cannot be fully satisfied by a legalistic approach alone, so the parenting issues need greater attention also.

A legalistic approach to resolving residence and contact issues is played out in accordance with the Children Act 1989 and relevant case-law.

Either party can apply to the court to resolve, or impose, a contact or residence arrangement. In other situations the court might be asked to impose a prohibited steps or a specific issue order, where one party is looking to the court to compel or prohibit some stated action.

1.18 A court system under pressure

The court system in England and Wales is under great pressure as a result of the number of children cases it has to deal with.

The court has inadequate resources so far as time and funding are concerned. At the time of writing, the Ministry of Justice is proposing that a far greater number of private child law cases should be diverted away from the courts to mediation services.

It is anticipated that anyone applying to the court, as of April 2011, regarding financial or parenting issues will have to show that they have attended a suitability assessment meeting with a mediator, or provide reasons why not, before their application will be issued.

There is a recognition within the court and legal profession that many children law cases at court do not deal with questions of law at all, but only questions of balancing the children's time between parents.

Very long delays can often be experienced in the preparation of CAFCASS reports or waiting for an available court appointment, depending on whereabouts in the country you, or your children, are based.

1.19 A legalistic approach to parenting issues

In typical child law disputes, an application is made by one of the parents. In the broadest of summaries, there will be an initial court hearing to determine what information, such as reports and statements, is going to be required to enable the court to make an order that might be in the child's best interests.

There is often a lengthy delay while those documents are prepared.

The parents' statements set out their respective positions. Again these are similar to the position-based arguments we saw above when looking at legalistic approaches to resolving financial matters. Each party will list their demands, safeguards and conditions, liberally sprinkled with criticisms of the other parent. These statements will persuade a court, they hope, that their position is the dominant and most compelling argument.

These statements are often flavoured by how the relationship has come to an end. The demands can be driven by a desire to punish the other person. A certain adopted position might be an attempt to correct a perceived power imbalance in the relationship or even obtain retribution.

Here we can start to see how narratives, or separation stories, start to emerge.

A more substantive hearing takes place once the statements and reports have been completed. The parents will be cross-examined by each others' lawyers or barristers. This is often a very damaging process which does nothing to enhance the chances of the couple being able to co-operate, or work together as parents in the future.

The court will also hear from professionals reporting on what might be in the child's best interests and the suitability of each parent's proposals, and even the suitability of each parent.

A judge will then impose an order for the family to abide by. In doing so, the judge inevitably favours one parent's position over the other's, or occasionally balances between the two, before sending the couple back out to manage their ongoing relationship as parents of the children concerned.

1.20 The legalistic parenting aftermath

These are now two parents who for the last 6 to 9 months have worked against each other in preparing their own statements and reading what has been said about them by the other. It is obvious that this legalistic platform is not fit for the purpose of building an effective post-separation co-parenting relationship.

In many cases, the final court hearing proves to be anything but final. It is not unusual to see patterns of defiance, breaches of court orders, sabotage or goading continue, sometimes for the whole duration of the children's childhood. Some families experience a repetitive cycle of court order, breach, application, subsequent court order and so on.

The anticipation that a legal approach will resolve parenting issues is a myth.

The resulting disempowerment, resentment and anger remains. There is little scope for parents to work together before, during or after the court process.

1.21 A more nuanced approach

The full scope of parenting concerns upon the ending of a relationship is much more nuanced than the range of concerns that a court, or conventional legal advisers, has the mandate to cover.

Parents need to be able to work together for the benefit of the children. How they are supposed to do this cannot be listed in the paragraphs of a court order, no matter how detailed it is.

For example, it has to be desirable for there to be a degree of consistency between the upbringing, rules and expectations a child experiences between one parent's home and the other. That is not to say that they have to be identical, but the respective expectations should stem from the same values.

One parent may put the children to bed at 8 o'clock, whereas the other has them in bed by 7.30. That might be because one parent has to collect from after-school club or child-minders. As long as the parents can communicate and agree on a need to have the children settled and put to bed so that they are refreshed and ready for the next day, then problems need not arise.

Even though there are differences, each parent can explain why those differences exist in a way which supports the other parent's care for the children.

Doing so enables the parents to maintain unity and consistent boundaries in front of the children.

In too many conventional cases, issues such as this become new arguments for conflicted and hurting parents to engage in. Perhaps a father, still bruised from the court hearing, determines that he will set the rules in his own household. As a result the children stay up until 10, playing videogames, watching TV or similar. The lines are drawn for a new argument.

The mother objects to the arrangements; the father sees this as a new imposition and interference. His desire to choose what happens in his own home, in his time with the children, and probably a desire to get back at his ex-partner, leads to new positions being adopted once again.

It is only a matter of time until the situation deteriorates.

If a conventional legalistic approach was used to resolve matters the first time round, then the couple will revert to a similar legalistic approach this time round. The precedent has already been set.

As a result the previous disagreement which passed through the court system sets the tone for future disagreements. The family finds itself once again going through a series of applications and court hearings. The conventional approach to resolving parenting, residence and contact disputes does not serve families well.

1.22 Using collaborative law to resolve future disputes

Future parenting disagreements will be inevitable. Within a collaborative approach, consideration can be built into any agreement in anticipation of future difficulties.

There can be discussions, either with the clients and their advisers or, even better, with a parenting coach or expert, about what they are going to do when they cannot agree on situations. Within conventional litigated cases the only precedent that is set is that if they cannot agree, then go back to the lawyers and look to the court for imposed sanctions. In collaborative cases the precedent that is set is that if you cannot agree then keep talking until you are able to design a solution.

Within the collaborative law process parents can agree to meet with a counsellor, a parenting coach or family therapist. Perhaps they can agree to convene a future mediated or collaborative meeting to discuss a specific issue. It might even be possible for one parent to say: 'You have the final say and whatever you decide upon, I agree to back you up on it.'

This last suggestion sounds outlandish, but only through the lens of a broken relationship.

Many parents, while the relationship is working, agree to support one or the other and their parenting decision without question.

Does it have to be different after the relationship has broken down, and why? Is this a question that could be explored within the collaborative process? There is no reason why it cannot be. Either collaborative lawyer could ask the question: 'How would you resolve differences in your parenting styles while things were still working within the marriage/relationship?'

The legal response to child or parenting disputes is woefully inadequate when it comes to resolving all of the parenting issues and particularly the ongoing questions of working together as parents post separation. Alternative methods need to be found to provide a safe environment to explore the full range of issues.

Some other topics that parents ought to anticipate and be prepared to have to discuss in the future are set out below:

- Introducing the children to new partners.
- New children, either ones born within subsequent relationships or children that a new partner might already have.
- Which school should the children attend?

- Behaviour, medical, religious and spiritual issues.

- Respecting the other parent.

- Handling the children's growing pains and changing needs.

Again, it is shown how the limited considerations within conventional cases fall far short of the full range of issues that need to be addressed. A shared commitment to collaborate between the separating couple creates a safe space in which the full range of parenting issues can be explored, together with professional assistance if needed. We move beyond the mere distribution of the children's time and residence and genuinely talk about how parents are going to continue to work together to meet their children's best interests.

By adhering to a purely legalistic approach we risk reducing financial and parenting issues to little more than their respective metrics of pounds and pence or days and weeks. Considerations of identity, roles and relationship, particularly within the parenting sphere, can be overlooked almost entirely. There needs to be a broader platform of consideration if couples are going to be able to continue to work together to any extent post separation.

CHAPTER 2

THE SOCIAL EMOTIONAL ASPECTS OF SEPARATION

2.1 THE EMOTIONAL ASPECT

So far we have looked at how the legal aspects of separation only partly address the financial and parenting aspects. Broader financial and parenting considerations are not recognised at all or are minimised within a conventional approach to separation or divorce.

The emotional dynamics that clients pass through within separation may be similarly neglected. This might be because the client has ruled them out of bounds. Some clients make it clear that they do not want their emotions to be taken into account. Alternatively the lawyer might have an objective and determinedly focused approach on the things that they can impact, namely the court outcome, or the line of correspondence with the other partner's lawyer.

We do not need to provide a comprehensive breakdown of the potential emotional challenges that a client might face, but we do need to be more aware of the type of issues that may occur.

2.2 THE FIVE KUBLER-ROSS STAGES OF GRIEVING

A good starting point for practitioners and clients is the widely recognised Kubler-Ross five-stage model of grieving. This model has also been applied to other situations where individuals experience large change, such as redundancy and the loss of a spouse or relationship through separation.

The five stages are:

- Denial
- Anger
- Bargaining
- Depression

- Acceptance

People might not experience all five stages and might not experience the stages in the order they are set out above. One stage may be returned to several times rather than being experienced once and then somehow 'overcome'.

People might experience wild alternating swings between one stage and another such that two stages essentially run at the same time.

It is almost inevitable that each partner will be at different stages of the grieving cycle. One partner might already have accepted that the relationship is over while the other partner is still struggling with the very earliest reactions to what has happened.

Within collaborative law we have more time and a safer space in which to explore the impact of this emotional dynamic. By sharing this model with our clients we can draw their mind to the question of how they are coping in a more directed fashion than simply asking the rather bland question, 'How are you?'

2.3 Denial

Denial, in the Kubler-Ross model, leads individuals to minimise a partner's action in leaving, or belittle the reasons they give for leaving. It can, of course, be much more fundamental and one partner might simply refuse to accept that the relationship is over, or problematic, at all.

A partner's departure might be attributed to a mid-life crisis or a moment of silliness. As a result the party who has been left in a relationship remains convinced that reconciliation, or a return to happier times, is possible.

Denial can also lead clients to place blame elsewhere. Perhaps it is the other lawyer's fault, or the sister, brother, a colleague at work, almost anyone. If they could just get them away from the influencing party then things would be alright.

Denial is not always absolute.

A client might accept their contribution to the end of the relationship but then trivialise the importance of that contributing behaviour or minimise the difficulties in bringing about any change.

Denial prevents individuals from taking responsibility for any part in what has happened or from taking an active role in resolving matters that now need to be addressed. That can make it very difficult to move forward.

Consider a wife who instructs solicitors to write to her husband explaining that the marriage has come to an end. She stresses, in the most conciliatory terms possible, that she would like to negotiate a settlement.

The husband, very much in denial, ignores the letter. After all, she will be back shortly. Her decision to instruct a lawyer is a silly little nonsense that does not merit a response.

A second and third letter are written over the coming weeks, asking for a dialogue, suggesting mediation or a collaborative law approach to resolving matters. These letters also go unanswered. The husband remains convinced that his wife will 'snap out of it' and return home once she realises that she has made a mistake and cannot get by without him.

What happens instead is that his wife and her solicitor have run out of options. The wife is now frustrated and angry that her actions and her ability to determine a course of action for herself are being trivialised by her husband, who remains steeped in denial.

The fourth letter is written. This letter explains that as the husband has demonstrated that he is not going to co-operate, his wife has no alternative but to issue an application to the court to initiate divorce proceedings. Maybe this fourth letter includes a draft divorce petition for his attention. Maybe a financial application is also required to deal with interim issues, such as maintenance payments.

Denial, in this typical scenario, has driven a case that was capable of being resolved by mediation, or collaboratively, into the court process and all that such a process might entail.

Denial, although it is the first of the Kubler-Ross grief phases, can continue long into the conduct of conventional cases. When denial persists, then clients prevaricate at every step of the process. The client fails to meet each and every deadline. They do not accept that a hearing is required or that a document really needs to be completed because, after all, they still believe that reconciliation is inevitable.

Within collaborative practice we have the opportunity to test denial. This is one of the strengths of working face to face. If there is any doubt about one partner's determination to see an issue through then we can ask that partner to reaffirm such intent and invite the partner in denial to repeat back what they have heard and understood their partner to have said.

We can also work more closely with counsellors or therapists to help clients move through this and other stages of grief.

2.4 Anger

The impact and potential for damage here should be self-evident.

We might see violent, harassing or pestering actions either towards the other person or people associated with them.

Real damage can be done to the prospects of parents working together in a post-separation co-parenting relationship. Even where there are no children, we might see the legacy of a largely successful relationship being wiped out by anger-motivated actions as that relationship comes to an end.

Communication becomes poisonous. The language used becomes loaded with hatred, insults and insinuations. The goal here is to exact revenge on the other person and those who might be seen as their allies, namely the other lawyer, supporting members of the other person's family or friends.

Anger can easily drive a conventional litigious approach. It leads many individuals to see the ensuing disputes as a battlefield.

When in anger, any settlement is impossible.

Settlement does not provide the balm that anger requires – retribution. Instead, settlement might be seen as approving one partner's act of leaving, or even rewarding it.

Consider the angry client's question, when a properly measured settlement is suggested to them, with a view to bringing litigation to an end.

> 'Why should I? Why should he [or she] get anything? They left me. They should have thought about that before they went. I haven't done anything wrong. Why should I be punished in this way?'

The angry client, convinced by anger that they have been wronged, turns away from opportunities to settle and instead commits to months of litigation, and thousands upon thousands of pounds in legal costs.

Those costs can become almost irrelevant. Anger can be so convincing in its hold that we are prepared to contemplate almost any cost in the firm belief that those costs will somehow bring about a just conclusion.

Within conventional approaches to family resolution, anger is quite normal. Despite practitioners' efforts to moderate legal correspondence, the role of anger can still be seen in the decisions that are made and the actions that are taken.

Anger can make it difficult, if not impossible, to manage a client. Pointing out to such a client that their instructions might be unwise can be fraught with problems. Lawyers may anticipate that if they challenge a client's stance too much that they will be sacked. Most lawyers will have experienced the angry client who lays down the ultimatum that: 'If you don't do what I am telling you to do then I'll get me another lawyer who will.'

Again, the problem is that we may not have the mandate from clients within conventional cases to consider non-legal issues.

Within collaborative law we do have that permission. As a result we can afford to be more challenging with our clients. We can speak out more easily about what we might be seeing within our client's actions. We can share material, such as this book, to show clients what might be going on for them as they pass through separation.

The active and effective collaborative lawyer will be watching out for times when the client is struggling with denial or anger and will be more confident in speaking out those concerns.

Again, a referral to a therapist, counsellor or coach might also help the client to move forward.

Any of these steps can act as a safeguard to ensure that anger does not distort the client's intentions and instructions. That in turn enables the collaborative effort between the couple and their advisers to keep on track, focussed on creating a solution that will fit this family and their circumstances. The alternative, an unvoiced quest for retribution and justification, is unhelpful and unattractive.

2.5 Bargaining

Bargaining is the third of the Kubler-Ross stages of grief. Remember though that these stages might not be experienced in the order they are set out.

Bargaining arrives with an air of dread realisation that change is happening. The denial has gone. Anger might have revealed itself to be counterproductive in pushing one partner even further away.

There might now be an acceptance that the partner has left and is sincere in his or her desire to leave.

This realisation now prompts efforts to get them back. Those efforts might be just as unrealistic as the initial denial was.

The bargaining client will promise anything to avoid the situation becoming permanent. They will promise change or a commitment to the most convoluted arrangements if only some degree of the relationship with their partner, the retention of some asset or the preservation of the status quo in living together in the same home as their child can be achieved.

Care needs to be taken that proposals or proposed action are realistic.

Is it realistic, for example, to agree to collaborate? Is this agreement itself taken in the belief that doing so will bring the other person back?

Is it realistic to forego claims on a pension in the hope that everything will be alright, or to sign over the shares in the limited company, to agree to further borrowing against the family home or to agree to a contact schedule which falls far short of what the client previously sought?

What is the motivation in agreeing to these terms? Is it the belief that agreement is the absence of arguing and the belief that if only we were not arguing then we could get back together?

The idea that reaching a bargain, or an early settlement, will bring an end to all of the grieving and the hurt is often made and often with unfortunate results. For examples look at case-law such as *Edgar v Edgar* and *Camm v Camm*, both cases where wives entered into bargains to settle matrimonial financial claims on terms that were not in their best interests.

Once the bargaining period of grief has passed then those bargains can look like very poor deals with hindsight. It is likely to prove very difficult to reverse them in many circumstances.

Again, the nature of the relationship between the conventionally instructed lawyer and the collaborative practitioner grants the latter the right to challenge motive.

In conventional cases, if one party is bargaining and is inclined to agree to a bad deal, then that is something that the other partner may be all too happy to go along with. In collaborative cases it will be appropriate to test the proposals and motives. Doing so ensures that the final settlement, whenever it is concluded, truly serves the family's needs. This can help the whole family to preserve a degree of mutual regard as they pass into their post-separation relationship.

2.6 Depression

Depression may lead our clients to be disinterested, detached and apathetic. In those circumstances they will be indifferent to what is happening. They might also be passive within meetings, either within the collaborative process or in discussions between the lawyer and the client.

Depression should almost be expected within our clients as they pass through separation. The question is only one of degree.

Depression causes real difficulties within conventional cases. The indifference seen in depressed clients can make it difficult to obtain instructions or co-operation. Deadlines slip in a similar way to that seen when one party is stuck in denial.

The depressed client will have no interest in the process of separating. Even simple tasks such as completing forms or providing requested documents can become impossible. Frustration quickly grows on the part of the practitioners and the other partner, who may well want to progress matters at a quicker rate.

Remember that depression is characterised not only by low periods, but also contrasting periods of exuberance and almost manic activity. Within the collaborative model we can watch out for these swings and manage them better.

This can be achieved by reality testing within collaborative meetings, and also by working closely with counsellors and therapists. The collaborative model is much more open to bringing in these other support services so that the parties going through the separation can discuss these emotional aspects with the appropriate person.

2.7 Acceptance

The final stage in the Kubler-Ross model is acceptance.

It is a great tragedy of many conventional legal cases that acceptance might never be attained. Consider the child contact case where litigation continues for many years after the initial court decision. Or a financial case where resentment is reignited every month that a maintenance payment has to be paid.

When orders and decisions are imposed by a court, or are agreed to only under the threat of the court imposing an unattractive order, then it is not surprising that resentment should remain. Also, as we have touched upon above, when a settlement has been agreed to prematurely before grief has been fully resolved, then the perspective on any deal that has been agreed can be very different when looked at from a different stage in the grief process.

One of the attractions of the legal process is the handing over of responsibility for the conduct of the case to the lawyers. The emotional process is far too painful, as we discussed at the start of this chapter. As a result, clients might fast forward straight to the formal, or the legal, process, mistaking the resolution of legal process as being the resolution of their own emotional responses.

There are two downsides to this strategy.

The first is that the stages of denial, anger, bargaining and depression are never fully processed. Instead everything becomes abstracted, and delegated, through lawyers and legal argument. In this sense, the whole act of instructing lawyers to 'Deal with it' could be seen in terms of denial, not of the end of the relationship, but denial of the grieving process itself.

A second downside comes from the loss of control and involvement in the outcome, especially if the case goes all the way to a final hearing where the judge will then be forced to impose an order. Change management theory universally talks about the need to involve the people affected by the change within the change process itself. Processing change requires individuals to wrestle with the information and to experience the five stages of grief as they do so.

The dispute resolution process that clients use at the end of a relationship should enable that to happen.

Conventional lawyer-led dispute resolution, such as litigation or lawyer-led negotiation, does not allow this to happen.

The highly effective conventional lawyer is even more effective in taking these concerns off the client's shoulders. He or she offers the promise that 'I can deal with that for you'. And such promises sound like music to the hurting client. But such a process delivers legalistic solutions to clients rather than fully immersing them in the resolution itself. This can be greatly disorientating for clients, who might speak about feeling detached from the process, almost as if they are watching the legal machine fix their situation. Consider the client who comes out of a court hearing without having had to say a word but having only observed what the lawyers and judges are discussing on their behalf.

The collaborative law process, on the other hand, deeply involves the couple in the resolution. As they collaborate with one another then they are intricately involved in the process of identifying needs. They wrestle with possible solutions, debating, arguing even, about how seemingly competing interests can be aligned.

As they do so they will also be processing their own responses to the drastic changes their lives are going through. We will see denial, anger, bargaining, depression and, powerfully, acceptance.

When people are part of creating their own solutions in the face of change, as hard and uncomfortable as that may be, then they have invested in the solution that they create. They are said to have taken ownership of the solution. In practical terms, this means that they can accept the outcome better, even though they may never have wished for it.

In conventional cases, too often, clients are simply expected to accept the outcome, whatever it might be, without having passed through the other stages of the emotional separation. The result can be lasting resentment, sabotage of the current settlement or order and future litigation and costs.

2.8 UNCOUPLING

Constance Ahrons' book *The Good Divorce* (HarperCollins, 1994) talks about separation in terms of uncoupling. Her goal in doing so is to change the language of separation. If we can move away from talking about broken relationships, failed marriages, split families and move towards a more positive language, then we can reduce the stigma and damage caused by separation.

Ahrons refers not to broken or split families, but to bi-nuclear families, a riff on the commonly accepted notion of the nuclear family. Where there was one nucleus, the family within the one home, there are now, quite simply, two.

The separation itself is referred to by Ahrons as an uncoupling. This is as an alternative to the idea of couples tearing themselves apart in a destructive fashion, fuelled by rage or litigation. Instead, the uncoupling is referred to in almost biological terms.

Consider the single cell amoeba which evolves from one into two cells. Note how it does so in a way which sustains future growth and independence.

The process is a gentle, organic progression.

The single cell wall starts to converge in the middle. Picture the soft figure of eight shape that it creates as the wall comes increasingly close and separation of the two parts is imminent. Now picture the wall pinching together softly and the two halves of the once single cell coming apart. The single cell has now become two, each constituent part starting out in its own right.

This process of uncoupling is often overlooked.

A couple might blow itself apart from one another in the most awful fashion, wounding everyone and their ability to function effectively in the future. The cell walls are not given an opportunity to create a new shape and instead are left torn and ragged.

Within a legalistic process there is no room to consider these aspects. Again, as lawyers we do not have the mandate to challenge our clients to manage the separation itself. Very often, separation has already taken place spontaneously before lawyers are appointed. As lawyers operating within the conventions of the legal-based separation, we have no role.

Within collaborative law, however, the question of the separation itself is often discussed within the collaborative law meetings. It becomes possible to manage that process to enable both partners to remain effective post separation. This can be the case whether the parties have actually physically separated or not. The only difference will be whether we need to look backwards to what has already happened and what impact that might have had, or look forward and consider what plans need to be put in place.

2.9 CONSIDERATIONS ABOUT SEPARATING

2.10 Physical separation

How and when are we, as a couple, or a family, going to move out of one home and into two?

Are we going to live together, but separately in the same house instead? How?

What timescale are we working to?

What are the cost implications?

Do we need to find money for a rental or house purchase deposit and, if so, where will that come from?

How are we going to tell the children what is happening? Are we going to do that together? How and when?

What do I need to take with me? What do I need to leave?

What items do we need two of?

2.11 Relational separation

How are we going to relate to one another?

What will we call each other? Do we abandon the old pet names? What names shall we use instead?

How will we support each other? Are we allowed to offer support to one another?

How will we support each other in our roles as parents to our children?

How are we going to talk about the things that we need to talk about?

What ground rules do we need to establish regarding communicating with one another?

How do we manage our separation in terms of relationship-based physicality – touching, stroking, kissing? What is no longer appropriate?

How can we ask for help from another?

How can we signal discomfort or disagreement?

How will we manage new relationships and informing each other, if at all, and telling the children?

Can I call you or e-mail?

Can I still speak to your parents?

2.12 Organisational separation

How can we let each other know what is going on?

Can we set up an online diary system?

How frequently do you want me to forward your mail?

What do I do if I need help with something that you usually do?

What do we need to sign and file with banks, utilities and other subscription and service providers?

Do we need to separate bank accounts or open new ones with new payments?

How will we work together to maintain consistency in our care for the children?

How will we arrange activities, holidays, birthdays, Christmas and other events?

How will we deal with maintaining discipline?

Who will attend school events? Can we both attend and, if so, together or apart?

What about clothes and haircuts?

2.13 Planning for the actual separation

The collaborative law process gives a forum in which all of these questions, and many more, can be addressed. The result is that the uncoupling can be anticipated and planned. Those arrangements can be communicated to the children, and other onlookers such as friends and extended family, in a way that the couple themselves control.

The manner in which the couple separate can establish a steady platform for the couple to continue to build their post-separation relationship upon. It can also communicate the couple's values and their own hopes for their future relationship to the children and others.

Separating in this uncoupling way is a powerful demonstration of one of the key concepts of collaborative law work. Although a couple may have fallen out of love and although there is pain and sadness they can still work together and support one another.

When parents are able to work together in this way their children see that it is still alright to maintain their own relationships with both parents. For example, if the parent who is not going to be living with the children can assist with the move into a new home, maybe decorating the children's bedrooms, or being shown around the new rooms after the moving day, then that can give the children their parent's approval and endorsement that it is okay for them to be happy there.

This is a far cry from the arrangements in many conventional cases where one parent is forbidden from returning to the family home, or entering a new home.

This demonstration of working together and the giving of permission to maintain relationships with both partners brings us onto the sixth of our seven aspects of separation.

2.14 THE SOCIAL ASPECTS OF SEPARATION

Couples going through separation do not do so in isolation.

They represent parts of larger families, with grandparents, in-laws, brothers and sisters all having an opinion on and playing a part in the process of separation.

Consider also colleagues, friends and other social groups, such as church, sports teams or hobby-based groups.

All of these people will have their own perception of the couple or family. They all have their own relationship history with one or both of the couple.

It will not be unusual for friends and family to feel embarrassed about continuing to talk to both. Any one of a couple's social connections may be inclined to support one or the other party.

We also need to be aware that each of these groups will have their own views on separation or divorce generally. Some may be more or less disapproving of relationships coming to an end.

Couples might want to consider these other social groups and how they will manage and respond to the comments that other people will make.

Families are an obvious starting point. It might be useful to think about how a couple's own parents are going to be told about the end of the relationship. Will they do this together, maybe over a dinner? Will they each tell their own parents separately?

The collaborative law process, with its high level of communication and co-operation, can help couples think about all of these things.

2.15 Grandparents, for example

Is there a need to explain to grandparents that they should not villainise the other partner?

How can the grandparents be encouraged to not blame the other partner, who will remain, of course, as the parent to their grandchildren?

If a mother and father are determined to work together to co-parent their children effectively post-separation then that could be undone if the grandparents are not able to speak of the other parent in similarly supportive terms.

It would be natural for a grandparent to offer well-intentioned support for their own son or daughter. Very often that can involve, again quite unwittingly, protecting their son or daughter from any semblance of blame. As a result, the other partner becomes the object of blame.

'She was never good enough …'

'I always knew he would …'

'You know what his problem was …'

'You weren't to know that she was always …'

'If only he had …'

If allowed to go unchecked this pattern of behaviour can increase the difficulty that a separating couple will have in working together.

The comments might also directly influence the partner who is hearing them from their own supporters. It is not impossible to imagine a complete rewriting of the marriage or relationship taking place.

It can also lead to torn loyalties. A husband, or a wife, for example, might find it hard to ask their own parents to stop putting down their spouse.

2.16 Managing friends' and families' reactions to collaborative law

Collaborative law, again, gives a forum for the couple to explore these issues and discuss how they will deal with this social separation, or divorce, just as they deal with the legal, financial, parenting, emotional and uncoupling aspects.

Consideration will also need to be given to how other people will react to the couple's decision to use the collaborative law process.

A couple's decision to turn away from litigated dispute resolution processes can be a very strong statement. It can be an outward demonstration to friends and family members about the couple's own shared values and hopes for their post-separation future.

Many friends, colleagues and family members may need some explanation about how and why the couple are able to still trust each other. Supporters will often feel that the collaborative process is letting the other party get off lightly. Those friends and family members may well have denial or anger motives of their own.

Think about, therefore, who these supporters might be. How are they likely to feel about watching their friend collaborate and continue to work with their partner? Is there work needed to get them on board as well?

The decision to collaborate can give permission to friends and families to maintain some ongoing relationship with both partners post separation.

It can say: 'We are still working together and still respect each other now we are apart. Because we can, you can.'

2.17 A DIFFERENT KIND OF DIVORCE PARTY

A market in divorce parties has recently emerged as a means to celebrate or mark the end of a relationship. Their reputation suggests single-sex, fiercely loyal events similar to a hen or stag do.

They seem to speak more of 'Good riddance' than anything else.

One client spoke of an alternative model for a collaborative divorce or separation party which was very different.

He proposed an event where the couple hosts an evening or afternoon party and invites friends and family. The message is one of gratitude for their support, but also to reconcile the two different camps that might have evolved supporting each partner.

Such an event would set a new precedent for the extended families and friends so that when events happen in the future then people could still come together. The event also recognises that the relationships that these friends had were not just with one partner or the other, but that there were also rich relationships among themselves.

Those friendships were likely to be affected by a lingering sense that they were friend and supporter to one or the other partner and had found themselves in different camps to former friends and colleagues.

The separation party here reconciles the couple with their full spectrum of friends. It also starts to rebuild trust between each partner and the other partner's friends and family. Finally, those friends and wider family members can reconcile their own relationships with one another, which may well have been damaged by their unilateral support for one partner or the other.

These first two chapters have looked at the much broader considerations that separation brings about. Separation is shown to be a much more complex issue than the purely legal.

Separation does have legal aspects to it, but there are also very important financial and parenting aspects that do not get considered within a narrow legalistic approach. The emotional and social considerations, as well as the practical need to think carefully about the act of separating itself, can also be given greater consideration. The temptation to delegate the resolution of the legal separation or divorce to lawyers for the conduct of matters to the professionals and assume that this takes care of everything needs to be resisted.

The alternative is not an easy way forward. The collaborative law process, with its specific training of lawyers and its willingness to work with other professionals, can greatly help couples going through separation to wrestle with the more emotive issues. Doing so enables the couple to preserve mutual regard for one another, and deal with their own self-care, in ways that conventional practices do not.

The final aspect of separation, the narrative aspect, is more abstract than the six we have looked at above and merits its own chapter accordingly.

CHAPTER 3

THE SEPARATION NARRATIVE

The seventh aspect of separation, the separation narrative, is the most abstract component within our model. It feels rather more philosophical than the previous six and therefore requires a little more attention. It is strongly recommended to all readers that the notions of narrative which follow are understood. Adopting a narrative approach can greatly help couples and their families as they pass through separation.

If we are able to share these insights with the couple themselves then there is the chance to make a radical impact on how they will respond to the conflict discussions and negotiations that they have yet to resolve.

This chapter will show how a narrative approach can assist families and their advisers move towards separation in new and productive ways.

3.1 NARRATIVE AND WHAT WE MEAN BY STORIES

When we talk about this thing called narrative we are talking about stories.

When we refer to the word story, or narrative, we are not seeking to belittle the importance or the difficulty of what the family is going through. Instead, we want to find a way to talk honestly and openly about those difficult and painful experiences. We need to do so in a way that will enable us all to approach these experiences safely, effectively and in creative ways. This is a central priority for collaborative practice.

Referring to narrative and stories can bring about powerful opportunities to challenge our collective and individual understanding of what is happening.

Essentially, a narrative approach enables us to change our understanding of what is happening, how the situation impacts upon us and how we relate to it. The alternative is that we simply react instinctively to what happens without really thinking through what our options might be. The risk is that our reactions might be counterproductive, make the situation even worse or continue to jeopardise trust and communication.

3.2 THE THINGS THAT HAPPEN AND THE THINGS WE UNDERSTAND

John Winslade and Gerald Monk, in their 2001 book *Narrative Mediation* (Jossey Bass) helpfully review work carried out by Jerome Bruner.

Here we are introduced to the notions of the *landscape of action* and the *landscape of meaning*. Think of these two landscapes as parallel realities operating alongside or one on top of the other.

It is beyond the remit of this practical book to ponder philosophical concepts but the following discussions on narrative and Bruner's landscape ideas are the exception. There is value in the practical benefits that can be accessed as a result of narrative-type considerations and the thinking behind these twin landscapes.

3.3 The landscape of action

The landscape of action refers to the things that happen.

These are the things that are said and done and our reactions to them. The things that happen are events within a story. These events combine to create the scenes and episodes that we experience. Together these events build up a story which we tell and retell back to ourselves and to other people.

3.4 The landscape of meaning

The landscape of meaning is rather different. As Winslade and Monk put it:

> 'Human beings are meaning making creatures. They dwell on, think about, interpret, reach conclusions about, have emotional responses to, develop attitudes toward and positions about and react against many of the plot events. All of these responses take place on the landscape of meaning.'
>
> *Narrative Mediation, p 164.*

Things happen. We then interpret or create meaning from the things that happen. Our understanding is a reaction to what we have observed.

That understanding creates this second plane, or landscape, namely the landscape of meaning.

The events, which happen in the landscape of action, shape our understanding. Our understanding builds up our own landscape of meaning.

3.5 How action and meaning determine each other

Clients retell what has happened within their separation so far. We can call this their separation story.

When they are telling their separation stories, clients tell us the narratives as they are seen on their own landscape of meaning.

As the listener, our own understanding of what has happened is entirely shaped by how the client tells their story. As far as we are concerned, the landscape of action, that is to say, the things that have happened, is created by the client.

If the teller of the story, the narrator, shifted their narrative then the landscape of action is similarly changed.

When we tell someone what has happened to us, we are in fact telling them what understanding we have about the things that happened. Our retelling is not the thing that happened itself, but only one interpretation of what happened. Our interpretation is very often a reaction to what has happened.

This is not to say that the separation stories of those going through separation are untrue or dishonest. Their understanding of what has happened is not made up. It is a sincere, important and often painful record of how they have experienced what is going on.

3.6 THE DIFFERENCE BETWEEN REACTIONS AND RESPONSES

There is an important distinction between reactions and responses.

Reactions are usually automatic reflexes. They might be shaped by our nature, previous experience or society's expectations and conventions.

They are usually unconscious or subconscious but they are always automatic.

Responses, on the other hand, tend to be deliberately and consciously chosen from a range of choices that we see as being available to us.

If I reacted angrily to the end of my relationship then I will speak to friends and colleagues, quite naturally, out of that anger and hurt. I will be telling a certain version of what has happened.

I define the things that have happened in a way that is largely influenced by me and my own reactions to what is going on.

As I tell and retell my separation story of hurt and blame I shape how other people will understand what has happened. Their perception of the events is directly coloured and defined by my retelling of what has happened. Remember, my landscape of meaning has become their only understanding of the events that took place on the landscape of action.

What if I was able to choose a response to the things that have happened, as opposed to simply reacting automatically?

What if my response to the end of the separation is to conclude that the relationship had been an intense and beautiful experience? What if I recognised that the relationship had run its course and had now come to an end?

How would that response change the separation story that I retell to other people, and indeed myself? How does the new separation story compare with the reactive story that I might have otherwise told?

The narrative is no longer a purely angry one telling of the hurt caused to me, listing the wrongs committed and attributing blame and responsibility. If I now retell my new separation story, my perception of what has happened is very different. Family and friends who now hear this alternative narrative will be hearing a radically different story. Their understanding of what has happened will be very different.

By shifting my separation story from a reaction to what has happened to a more responsive selection, I will greatly change my understanding and the understanding of the lawyers, friends and families I speak to as I pass through this separation process.

Of course, this exaggerated and over-simplified example is likely to have little bearing to the experiences of many people going through separation. It is used only to demonstrate the point that a different response leads to a very different narrative about what has happened.

This difference between reaction and response can also be seen in the immediacy of a collaborative meeting.

A wife makes a statement that touches a nerve for her husband.

Instead of reacting with 'See! I knew that was what you thought!', a more carefully considered response might be 'What is it that made you say that?'

The original event is changed from an attack into nothing more than a strange comment that arouses curiosity. The impact of the event has been changed dramatically. The events and dialogue that follow will also have been changed.

The event is no longer interpreted as the continuation of a sustained campaign by one individual against the other, but is now simply a comment made that merits clarification.

3.7 CHANGING THE NARRATIVE

There is an additional layer to this argument. As we develop different responses to the things that happen then we change our own perception of what has happened. This can again be demonstrated quite easily.

3.8 HINDSIGHT IS A WONDERFUL THING

It might help to consider this rather abstract, and not straightforward, concept by likening it to hindsight.

Hindsight is a wonderful thing.

We may think that we know what is going on as it is happening. In fact, we are often convinced that we know what is going on.

When we later discover a new fact, after the event has happened, we are sometimes confronted with a very different understanding.

That new fact might be a comment made by someone that we did not previously know about. It might be something that had happened that we had not noticed or did not previously know about.

These later discoveries create new understandings of what was going on. When we now look back at what happened we can see that what was going on was very different from what we first assumed.

Consider a murder-mystery story or film.

The events are played out in front of us. We might be led to believe that one character or another has committed the crime. We can be completely convinced about this, only to be presented with a new detail or clue which is only revealed in the dramatic closing courtroom scene.

The understanding provided by this new fact twists the meaning we have derived from the whole film and turns the narrative on its head.

3.9 THE STORY SO FAR

So, we have defined what we mean by narrative and story. We have established the idea of the separation story as being that version of events that our clients and friends tell and retell to explain what is happening.

We have looked at how the things that happen lead us to react. We have also seen how we draw a reflexive or automatic interpretation from those events. It is that interpretation that we or our clients tell and retell as our separation story.

We have looked at the difference between the plane of action and the plane of meaning and how they influence each other.

We then looked at the difference between reacting to events and being able to choose a response instead. Once we are able to move away from the reaction then we have the chance better to develop our understanding of what is going on by selecting better responses, rather than automatic reflex reactions.

These changes can be retrospective. If we can get to a different understanding of what has been happening, then we will start to perceive what has been going on differently. What we thought was happening might be shown not to have been happening at all. Perhaps we realise that one partner or the other was not trying to turn a child away from its parent, or we realise that one partner was not deliberately trying to get rid of money or assets. We used the metaphor of the murder-mystery story to show how what happened was not as we originally perceived it, but might have been quite different.

3.10 NARRATIVE THEORY IN COLLABORATIVE LAW

Narrative theory enables us, and our clients, to create hindsight moments. The object in being able to do so is to enable us to reshape a new understanding of what has happened. But why would we want to this?

An example of this might be seen in the attitude of a spouse who is determined to punish his or her partner. They perceive everything that happens as evidence of their partner's infidelity, selfishness or whichever other attributes they choose to assign to their spouse.

They are consumed with the notion of revenge. We can imagine the narrative or separation story that this husband or wife will be telling.

We can also imagine how listeners might respond with horror and sympathy in equal measure.

The result is that the client, and probably the whole family, gets caught in a predictable spiral of events–perception–reaction–event.

Something happens, namely the event. This is the landscape of action.

That is perceived and interpreted in a certain light dictated by reaction. This is the landscape of meaning, and this interpreted meaning shapes the story that is told and retold.

Maybe the client instructs lawyers to send a certain kind of letter, make an emergency application to the court, pay a visit or send embarrassing photographs to a new partner or family member.

The cycle becomes self-perpetuating and destructive. Every subsequent event is taken as justification that the initial reaction was correct. We become more and more convinced that we are being wronged by the other person, and so as our conviction grows, so do our reactions. Acrimony, litigation and legal costs all escalate.

3.11 BREAKING THE SPIRAL

The key to breaking these spirals lies in enabling a shift to take place from the automatic reactive interpretation of what is going on to a more manageable responsive interpretation.

The predictable spiral of event–perception–reaction–event can be disrupted by replacing that reaction with a selected response.

3.12 DOMINANT NARRATIVES AND DOMINANT DISCOURSES

The challenge within narrative practice is how can we loosen the influence that our reactions have over us so that we might create room for an alternative response to emerge?

Winslade and Monk, in *Narrative Mediation*, talk about 'Dominant discourses' within divorce and separation.

We have explored how the narratives or separation stories we tell are shaped by our reactive interpretations of what we see happening.

Dominant discourses dictate how society expects, in this case, a divorce or a separation to progress.

They are well-established storylines that throw us into predetermined positions in relation to one another. These are the roles and separation stories that we might subconsciously expect individuals' narratives to conform to.

These separation stories are so dominant that as soon as we see events transpiring that seem to fit in with a dominant story, then we assume that the relevant story applies to us.

There are many dominant divorce or separation narratives that may well shape how we react as we approach and pass through separation. A few of them are explored below.

3.13 *Kramer v Kramer*

The *Kramer v Kramer* story, taken from the 1979 film with Dustin Hoffman and Meryl Streep, establishes or reinforces a very prominent divorce narrative about child law, contact and residence.

When we hear of parents separating, there may well be an assumption of a certain kind of story, or drama, having to be passed through.

When separation arises, and there are children of that relationship, clients might reactively cast themselves into very well-worn, gender based roles and act them out accordingly. They will, of course, perceive the other parent through the lens of this dominant narrative and interpret everything that they say and do in that light.

3.14 It must be somebody's fault

We still have fault-based divorce for couples who want to divorce immediately and without the need for waiting for 2 or 5 years. The couple is forced into the telling of certain narratives.

There is the adultery-based divorce story. Note how the language is determined by the law that attaches itself here. Adultery is not a part of everyday language. The imposition of a legal cultural device establishes this thing called adultery as the language with which the separation story will be told.

Very little information is required as far as the court process is concerned, but crucially what is required is the identification of which partner committed this thing called adultery.

Much more elaborate adultery narratives play out within the social circles in which the family lives however.

- Who was it?
- What was the attraction?
- What happened?
- How was it discovered?

The alternative basis for an immediate application for divorce is the allegation of 'Unreasonable behaviour'. Again, note how the relevant statute defines the language to be used with its unreasonable behaviour moniker.

The applicant is forced to tell a separation story of unreasonable behaviour. Although such divorce applications can be very bland and low level, the temptation to list paragraph after paragraph of indignant allegations of behaviour can often prove to be irresistible.

3.15 Society's need to blame

Society also has its own narratives. Constance Ahrons challenges the manner in which society refers to families that have parted. The language used is that of failed relationships or broken families indicating society's dominant narrative that something must have gone wrong. Someone must be to blame.

Recent separation language talks about certain characters who are called 'the leaver and the leavee'. Those labels assign all kinds of values, judgments and assumptions against each of them.

If you are the leaver then it might be assumed that you are in control, strong and more happy about the end of the relationship than the leavee.

The leavee might be seen in terms of being the victim.

There will be implicit judgments about who was right or wrong, determined by the reasons for the departure.

3.16 'She's out to ruin me'

Consider the stereotypically male fable of the scorned wife out to ruin her husband.

This well-established divorce narrative can make it very difficult to perceive the things that are happening in anything other than a reactive and defensive light. As soon as a wife requests information about what resources are available for her family, or even as soon as she mentions going to a lawyer, the dominant narrative can convince many husbands that their wives are out to ruin them and get all they can.

3.17 The lawyer narrative

Lawyers also fall into dominant narratives. They might be seen as the defender or gladiator, being sent into the court arena by the client to act upon their instructions.

Individual lawyers will have their own narratives, within their communities, built up around the feedback of previous clients or maybe those whom the lawyer has previously acted against.

Consider the discussion between a husband and his colleague.

The husband has received an initial letter from his estranged wife's lawyer. The letter is conciliatory and polite. It recommends considering mediation or using the collaborative law process.

The husband shares this letter within his own social circle, over a coffee with his work colleague. That colleague reads the letter and then retells his own separation story from his divorce a couple of years earlier.

He tells the story of how his ex-wife had used this very same lawyer and how she did 'a real lynch job' on him.

> 'Took me to the cleaner's, she did.'

What impact does this lawyer-based narrative have upon how the husband will interpret and respond to this initial correspondence? How might it influence how he also perceives his wife and her decision to instruct this particular lawyer?

At this moment in time, the husband will have fallen into an unhelpful dominant narrative. If that is not challenged then he will interpret every letter and encounter with that lawyer in a certain reactive way.

3.18 Courts are for criminals

The dominant narrative about courts speaks of criminality, crooks, innocence and guilt.

The courts are for judges to listen to evidence, reach a judgment and hand down appropriate admonishments or punishments.

This narrative gets in the way when couples find themselves having to apply to or respond to the family courts. Judgments within child or financial matters are seen as either vindications or punishments by the party who is more or less helped by the judgment accordingly.

As a result it becomes very difficult to respond to a court order as anything other than condemnation. Likewise, the one partner is seen as having run a campaign of criminalising the other. It is hardly surprising that we see resistance and resentment ensue.

3.19 Happily ever after

Our children will be subject to dominant narratives as well. Consider the 'Happily ever after' narrative. Many married couples also subscribe to this narrative and it is one suggested reason for the high level of divorce and separation.

If the couple does not live happily ever after then, where this has been the foundation of the relationship, the marriage can come to an end in less happy times. Other relationships which were not built on such a strong adherence to this narrative might be better suited to withstand periods of hardship and disharmony.

For children though, the narrative is more elaborate. If the 'Happily ever after' narrative falters there is an alternative ready to fall into place. It is seen in the unattractive spectre of stepfathers and stepmothers handed down by years of fairy tales.

3.20 NARRATIVES WORK LIKE LENSES

These narratives and many more are well established within our culture. As soon as separation events start to unfold we tend to perceive what is happening through the filter or the lens of the narrative that is most dominant in our minds and in our particular society.

The result is that we might perceive events through the lens of Dustin Hoffman in *Kramer v Kramer*, or a conventional courtroom drama.

We cast ourselves in the roles of friends or colleagues who have previously shared their own lurid separation stories with us, about how they were 'done over', 'robbed' or 'ruined' by an ex-spouse.

These stories are so ingrained that we can, in an instant, fast forward our situations and see ourselves superimposed onto these storied outcomes.

If we or our clients have heard a story of a friend who was forced to sleep on the streets, broken and abandoned by his family, robbed by the legal process and society, then we can easily interpret what is happening in our separation story in such a way that it conforms with the dominant narrative. We buy into the narrative plot and outcomes.

Hardly surprising then that we react strongly, thinking that we are fighting against that outcome.

The awful irony is that our struggles to avoid such an outcome can often result in the outcome we are trying to avoid being brought into reality.

This happens because we react too aggressively, or assertively. We then provoke an even more strident reaction from the other partner or their lawyer.

3.21 SELF-FULFILLING NARRATIVES

Imagine a husband who is caught up within the 'She's out to ruin me' narrative.

As a result he does not disclose a bank account or two.

> 'After all, if she knew about them she would go after them as well, because she is out to ruin me.'
>
> 'And anyway, I will need those funds after she has inevitably ruined me.'
>
> 'She is going to get the lot', says the narrative, 'I'll be destitute. Therefore, if I keep these two accounts secrets, well, that is only fair and prudent in the circumstances. They're just a drop in the ocean.'

See how the narrative creates its very own logic?

The bank account is later discovered and any remaining trust on the wife's part is lost.

There are now no prospects of reaching a negotiated settlement.

The wife is now committed to a new dominant narrative that she had heard many times before but which had never resonated with her own circumstances. It is the 'My spouse is a lying cheat and cannot be trusted' story. She also signs up to the 'My lawyer is my defender and gladiator' and 'Judgment as punishment and vindication' narratives for good measure. These newly subscribed narratives will drive her actions over the coming months and ensuing litigation.

The greatest irony of all is that she may well now declare 'That does it. I *am* going to ruin him', whereas before there had been no contemplation of that at all.

She therefore pushes on to the final court hearing believing that her lawyer will protect her and reveal the truth, and that the judge will see that truth and rule in her favour. The judgment will represent, she believes, justice and what is right. That is her entitlement, the vindication of what she has said all along.

And the husband?

He is also vindicated in one sense. He has seen the fulfilment of the narrative he had subscribed to, namely that his wife was out to ruin him.

In confusing the criminal and family law court narratives he might complain that: 'I'm being treated like a criminal. I'm being punished but I've not committed any offence.' Other than not disclosing the account, of course, but his 'I'm innocent' narrative clouds that observation so as to make that entirely invisible. Remember that his actions will have been entirely justified in his eyes, through the lens of the 'She's out to ruin me' narrative that had such a hold over him.

The financial settlement decided upon by the judge will now form the conclusion of that narrative and will be told and retold by him as his own separation story for many years to come.

3.22 CONFLICT-SOAKED NARRATIVES

Winslade and Monk talk about these dominant stories as 'Soaking' our own perceptions of the things that are happening. The language that we use reflects this.

The relationships that we describe with the other partner, their lawyer and even how we perceive ourselves can illustrate our conflict-soaked perception.

Perhaps we catch ourselves talking in ways that suggest victimhood or villainy. We might attribute negative motives to the other party. Perhaps we talk in absolutist terms, suggesting that the other partner always or never does this or that.

We need to identify where narratives exist and are having a detrimental role.

A narrative approach to divorce and separation recognises these dominant storylines and the impact they have on our lives as explicit.

If we can identify which narratives we are subscribing to then we can challenge ourselves, or our clients, to question where assumptions are coming from.

We might ask why a particular word has been used or why the client is talking about their partner in a certain way.

When we realise how the conflict in the situation is driving us to tell our separation stories then we can loosen the grip that the narrative has over us. Our reactions to what is happening can be recognised for what they are – unthinking reflexes – and we can create space for alternative narratives to take hold. We will explore some of those alternative narratives below.

3.23 UNIQUE OUTCOMES

Another way of undermining the influence, or loosening the grip, that these stories can have on our lives is to hunt out what Winslade and Monk call 'Unique outcomes'. These are details within the divorce or separation story that do not conform to the assumed narrative.

An example would be parents who are locked in a contact dispute.

A father is convinced that the mother is trying to alienate his children from him.

As we talk further, however, we are told of examples where that has not happened. There might be a non-scheduled telephone call where the mother has got the children to make contact just to say hello or to tell him that they love him.

Alternatively, the parent with care of the children might have suggested an extra night's contact.

These unique outcomes threaten the cohesion of the dominant narrative. They suggest that the assumed conflict narrative is perhaps not so convincing. Perhaps our conclusions are not as reliable or as absolute as we had imagined and portrayed so far.

This is a delicate moment.

The father could recognise that there is hope and that the current conflict is not always at play. The question then develops into a discussion about:

> 'What was different on that occasion?'
>
> 'How did you feel about your wife at that time and how was that different to how you felt at other times?'

The space can be leveraged to make it even greater, and to offer the father in this example a way to respond to the conflict.

> 'Which would you prefer? To see your wife and the mother of your children as always trying to keep you apart, or to see more occasions where the children have extra contact with you?'
>
> 'How might we achieve that? Was there anything else different, anything that you had done differently, before the telephone call from the children?'

The danger is that the incident, this unique outcome, is lost. It can quickly become twisted to fit the dominant narrative. Once that has happened, then these optimistic exemptions to the conflict are now held up as further proof of the dominant storyline.

For example, the spontaneous telephone call must have been mother using the children to soften up the father ahead of some request or other: 'She probably wants more money, or to buy new trainers on the next contact weekend.'

The additional overnight contact is rejected and the offer is retold in the context of the mother treating the father as an unpaid babysitter.

It will not be long before this narrative evolves into the 'Unfit mother/father' dynamic which will need to be fully played out within the courts through statements, reports and cross-examination.

The unique outcome has been quickly explained away, and poisoned. We need to grasp them as soon as we can and hold them up, not as evidence of our assumed narratives, but as critical flaws in the narrative myths that we are subscribing to.

There is a mediator I know who previously worked as a gold miner in Africa. He was often required to mediate between rebel militias and government officials. We talked about the mining technique he was using. He and his staff would dive into murky rivers and search amongst the silt on the bottom. Every now and then they would glimpse a tiny nugget glinting amongst the gloom and dirt. If they did not grab the nugget immediately, then it would be covered once again by the mud or possibly taken away by the strong currents, never to be found again.

Unique outcomes are just like those nuggets. When we see them, we should grab them immediately. If we do not they will be lost forever, hidden once more amongst the grime. They are incredibly valuable to us in our role as collaborative lawyers.

If we can hold them up for all to see then we can show that the dominant narrative is not as convincing as it once seemed. These nuggets, these unique outcomes, are like glitches in the matrix. They signify that there might be something else going on in our clients' lives other than that destructive narrative to which they had previously subscribed.

3.24 ALTERNATIVE NARRATIVES

As the grip that the dominant conflict-soaked narratives had on our client's lives is loosened, it becomes possible to entertain alternative narratives.

> 'Neither of us was perfect and we did our best.'
>
> 'There is enough for all of us.'
>
> 'Our children will never have to choose between us.'

'Our children will grow up without regrets or sadness about their childhood or us as their parents.'

'We did what was right for us all.'

'We still value what we had together even though it is now over.'

'I am sad and hurting but I know that I will recover.'

'My partner is not a villain. I am not a victim. We have got a lot to sort out.'

'My partner didn't want things to turn out this way either.'

'We can still work together as parents.'

'We can work together to sort this out.'

'My solicitor is there to help me, not to do it for me.'

'I can learn from this awful experience.'

'No matter how other people think I should react, I am going to stay true to my values.'

'I still love him/her but I know that has to be different.'

'It will be alright.'

'This divorce is awful. We will not let it ruin everything that has gone before.'

'Our separation will amaze people who thought that we would fight one another all the way.'

'Our children will be proud of us. We will be proud of each other.'

Any one of these alternative narratives can keep things in a very different perspective. We, and our clients, will be able to respond very differently.

Once we loosen the grip that dominant conflict-soaked narratives have on our clients' experiences then we reduce the likelihood that our clients will simply react. Instead they have the time and opportunity to consider other narratives, other perceptions, that may serve them and their families better.

3.25 SUMMARY

We have looked at how the things that happen can be said to take place on the landscape of action. We and our clients interpret and create our understanding, and the understanding of other people, on the landscape of meaning.

We need to check ourselves and our clients' separation stories. Are they caught up in a narrative, or a generic conflict-soaked storyline? If so, how is that affecting them and the dialogue between them and their partner?

We need to stay alert for unique outcomes and acknowledge them as undermining the less helpful narrative. We should be particularly wary of warping these incidents to fit the destructive narrative.

3.26 OTHER NARRATIVE ELEMENTS

3.27 Plot

In any book, film or story there are various components, namely plot, characterisation, language and outcome.

A narrative approach enables us to challenge assumptions within any of those categories.

What plots are our clients anticipating?

Is that helpful?

What might be a more helpful plot?

What other plots could they imagine as we pass through this separation?

What plot would our clients choose?

What do need they need to do and we need to do to change to those preferred plots?

3.28 Characterisation

Who are the characters in this separation story?

What are their names?

What other names do we give them, perhaps when we are by ourselves or talking to our closest friends?

How are they being portrayed? What characteristics are being emphasised? Which ones are being denied?

Are we casting individual characters as villains and victims?

Is that helpful?

What happens if we recast them back into their roles as husband and wife, mother and father?

How would that impact on how they work together and communicate with one another?

How would our clients want their children to perceive their parents?

3.29 Language

What language are we using?

Why did we use a certain adjective or adverb to describe someone or what someone did?

Ask the question: 'Why did you put it that way?'

How does that reflect how you feel about him or her or what happened? Is that helping us to resolve these matters?

How would you like to be able to describe them? Try it.

3.30 Outcomes

What outcome are we expecting?

What outcome are we headed for?

What outcome would we like instead?

What outcome are we guessing the other partner wants?

What alternatives might they want if we asked them?

3.31 HOW COLLABORATIVE LAW PROCESS DISRUPTS DOMINANT NARRATIVES

Collaborative law has the scope and flexibility to build these narrative considerations into its process, just as there is scope to build in a full consideration of the other six divorce or separation considerations explored in the previous two chapters.

Litigious, court-based family dispute resolution does not have that scope. Instead, litigation is entirely soaked in certain narratives. The plot is predetermined by statutory timetables and procedures.

In contentious cases the relationships between the various protagonists, or characters, are well defined and rarely shift.

Within collaborative law, however, we can disrupt the conflict-soaked narrative and encourage our clients to create new approaches as they move towards resolution and their self-defined outcomes.

Our clients are given the opportunity to choose what story they want their relationship, and separation, to tell.

By choosing collaborative law, they will have already written the first page of that alternative narrative. As collaborative lawyers it is part of our role to help them to co-author their own radically different story.

Together.

CHAPTER 4

CHANGING THE WAY YOU WORK

A common frustration among collaborative lawyers is that they would like more collaborative clients. They have invested large sums in qualifying as a collaborative lawyer and are keen to use those new skills.

Those lawyers who experience these frustrations all understand and support the collaborative values and process. To use the common parlance of collaborative law, they have 'experienced' or 'got' the paradigm shift.

4.1 WHAT IS THE PARADIGM SHIFT?

The paradigm shift, referred to by collaborative practitioners and trainers, means being able to leave behind the learned and deeply entrenched ways of working and thinking as a conventional lawyer.

The shift suggests that there is a moment of realisation where the previously conventional practitioner looks up and says 'Aha! I get it'.

It is the shift between feeling the need to argue a point and vigorously defend counter-arguments. There is also a shift away from the perception that lawyers are the experts and paternalistic guardians of the client.

There might be the realisation that those who are going through separation are capable of working with one another, with varying degrees of support, and the realisation that we can collaborate with lawyers who we were previously working against for many years.

4.2 MIND THE GAP

There can be a considerable gap between the client's expectations in coming to meet with a lawyer and what a collaborative trained lawyer offers.

Having experienced the paradigm shift, it can then be frustrating when clients do not share a practitioner's enthusiasm for this process.

This should not be a surprise.

Clients often have had little or no exposure to collaborative law or alternative dispute resolution processes until they first meet with a lawyer. Their only expectations are those conflict-soaked narratives handed down through media or friends and family who have previously experienced separation and divorce and subsequently told and retold their own separation stories.

It may be that the client's expectations are wholly unrealistic. That does not matter. What matters is that the expectations are there and if the experience of the first meeting with the lawyer does not meet the expectation then the gap between the two can be critical.

4.3 WHY CLIENTS CHOOSE NOT TO COLLABORATE

There can be many reasons why a client does not choose to collaborate. It must be remembered at all times that this is their choice. There are steps, however, that should be taken to ensure that if the client is choosing one way or another, they make an informed decision.

4.4 SOMETIMES COLLABORATIVE LAW IS NOT APPROPRIATE

There might be problems with the nature of the relationship between the couple. In Chapter 8 we will explore the TAP model, which considers whether collaborative law is even appropriate when there are:

- Threats of violence or oppressive behaviour
- Anger or other strong emotions that might get in the way
- Positional-based negotiating or litigating stances.

More often, a client's expectations about the nature of a lawyer-client relationship make the decision to subscribe to a conventional litigation-based relationship an automatic choice.

There are steps that collaborative lawyers can take to manage this gap between conventionally based expectations and the decision to collaborate.

Practitioners should ask themselves two questions.

First, is their own conduct perpetuating and affirming conventional expectations about how clients and lawyers work together and relate to one another?

Secondly, if the practitioner is not perpetuating those conventional expectations, are they taking enough care to enable the client to make the transition from their expectations to their experience upon meeting with the collaborative lawyer?

4.5 GUARDING AGAINST COLLABORATING WHEN IT IS INAPPROPRIATE

Collaborative lawyers do need to be careful not to enter unsuitable couples into the collaborative law process.

There is no point in trying to make an unsuitable case fit the collaborative profile.

If the process breaks down then the clients will have wasted time and limited resources from which to pay legal fees. The lawyer also loses out on the future conduct of the case as a result of the disqualification clause.

Lawyers need to check that they are not trying to collaborate a case just because they themselves believe in the process. If the client does not believe in the benefits then there has to be a risk that they will drop out of the collaborative process sooner or later. If the client does not sincerely see the need, or share the will, to collaborate, then they will drop out as soon as things go awry and they can excuse themselves from the process.

4.6 DISCOURSE AND THE LAWYER-CLIENT RELATIONSHIP

The theory of discourse suggests that lawyers and clients fall into a certain pattern of relating to one another. Those relationships are immediately coloured with a range of assumptions about that relationship.

The lawyer may well be perceived as being the expert in process and law. That presumption in turn renders the client as being less of an expert to varying degrees. At the extreme, the client may readily step into a position within the relationship where they look to the lawyer to determine everything.

There might be a quite conscious desire on the part of the client to subscribe to such a relationship. It is the 'You deal with it for me' approach that might sometimes be selected by the client.

Within the conventional lawyer-client relationship, the lawyer is endowed with the role of the active driver within the process. He or she takes responsibility for progress and decisions. They are the expert.

The role of the client is passive. They do as they are asked. It is not for them to volunteer information or suggestions but to follow the lawyer's lead.

4.7 DISRUPTING THE DISCOURSE

This relationship dynamic needs to be deliberately challenged.

Our first meetings with the client can be used to start to disrupt those assumptions. We need to move away from the conventional expectation of what it means to work with a lawyer and start to rewrite how this separation story is going to develop.

If, when we first meet with our clients, we use the same approaches as we use within conventional cases, then we should not be surprised that the outcome is the same, namely that clients offer up to us the conduct of their matters. If we want our clients to choose alternative approaches, such as collaborative law, then we need to prepare them to take a more central role.

We need to shift our approach to how we relate to our clients and demonstrate that we are forsaking or rejecting the conventional position of expert and calling our clients into more active roles.

A different approach to the first appointment can make it more likely that the client will choose a collaborative approach by disrupting the discourse-based relationships that we might otherwise fall into. Exploring an alternative first appointment here also enables us to model new ways for client and lawyer to work together collaboratively.

4.8 ADOPTING A COACHING APPROACH TO CLIENT MEETINGS

The approach proposed in this chapter is designed around tools which are widely used within the coaching profession. The tools have been developed explicitly to enable the client to retain control over their own processes and can be transferred into the collaborative lawyer-client relationship.

This new approach is demonstrated below in the context of an initial meeting, but the methods can be returned to and utilised at any stage of the collaborative law case and even within the four-, or more, way collaborative meetings themselves. We will see how the methods, used within an initial meeting, create an experience that calls the client to step into a position within the lawyer-client relationship that enables and challenges them to determine for themselves what approach they would like to adopt to resolve the matters arising from their separation.

If individuals are encouraged, through our questioning, then they will design a dispute resolution process for themselves that will often look like the collaborative model or can be serviced by using collaborative processes.

4.9 FIXED FEE OR FREE FIRST MEETINGS

The client arrives at the lawyer's office. Maybe there is a fixed time period available, an initial free 30 minutes, for example, or a fixed fee, hour-long interview. Bear in mind that such arrangements, while commonplace, may well be deeply flawed.

They create a sense of time scarcity, as if to say 'I've got 30 minutes to get through all of this', for the lawyer and the client alike. This is a fallacy. The first meeting can last as long as it takes, but even then the consideration is 'I've got 30 minutes to get through all of this before he/she starts charging me'.

Each law practice will take its own view on what first appointment incentive they offer, or indeed if they offer any. I have previously advocated both. On the fixed fee front, I would also give the client a book, as well as the initial interview. We would meet for 30–60 minutes on a fixed fee arrangement. In return though, they would receive a copy of a relevant book, such as a published divorce guide. The result is that the client, in comparing a free half hour with a competitor, or a fixed fee half hour with another practice, knows they were getting something of tangible value from the fixed fee appointment.

You might take the view, for example, that it would be appropriate to share this very book with clients who choose to use the collaborative process, and give the book as a 'gift' in return for the fixed fee appointment.

The other benefit of this approach is that the clients perceive that the lawyer is opening up access to information. This again disrupts the discursive presumptions that the lawyer is the gatekeeper to a realm of privileged information that they cannot access other than through the lawyer. It is an invaluable way of building trust between lawyer and client.

4.10 THE CONVENTIONAL FIRST MEETING

The lawyer and client exchange some pleasantries and then, very quickly, get down to the business of collecting relevant information. The lawyer will have a fixed list of questions, dates and details he or she requires. That list might be set down onto a new client questionnaire sheet, or the list might be entirely memorised following years of first client meetings just like these.

And this is how the meeting goes.

> 'Names?
>
> Addresses?
>
> Ages?
>
> Occupations?
>
> Children? Dates of birth and schools please.
>
> Who do the children live with?
>
> What is happening with contact?
>
> Date of marriage?
>
> Date of separation?
>
> Has your wife or husband instructed a solicitor yet? Who?
>
> What has happened?
>
> Why are you here?
>
> What has gone wrong?
>
> How much do you each earn?
>
> Have you or he/she got a business?
>
> How much do you think that might be worth?
>
> How much is the house worth?
>
> Is it in joint names?
>
> Tenant in common or joint tenancy? [Cue explanation on the difference between the two.]
>
> How much is on the mortgage?
>
> Who pays it?
>
> Any other debts?
>
> How much?
>
> Got a pension?

How much?

Any recent or imminent inheritances?'

We discuss options, namely mediation, collaborative law and conventional representation with litigation if necessary.

We quickly discuss legal fees and then ask the bewildered client what they would like to do next. They do not know, so we offer to send a letter out to their partner, with their approval, of course, and see what happens.

What has happened is that by asking the range of prescribed questions, in accordance with our own agenda and what we need to know, we have already fallen into the discursive roles of lawyer and client.

Tell me what I need to know in response to my mainly closed questions and then listen to me, as lawyer, set out your options.

The client's presumptions about what it means to meet and work with a lawyer are reaffirmed.

They fall into the expected role of client, providing the information requested by the lawyer.

When the lawyer then presents options for the client to choose from, the client, shell-shocked by the list of questions, is not in a position to select. The dynamic of lawyer as expert and active driver has already been selected.

As a result the client asserts that they want the lawyer to deal with matters for them in a conventional fashion.

On occasion, clients will select mediation or collaborative law, but those are the exceptions.

4.11 A RADICALLY DIFFERENT FIRST CLIENT APPOINTMENT

We will be looking at three specific models which are frequently used by coaches in helping clients to identify what is important to them and also in helping those clients cope with, resolve and overcome problems they have identified. The three tools are called the Wheel, GROW and Experiential learning models.

4.12 THE WHEEL

4.13 Implementation

The Wheel is widely used within coaching and some forms of psychotherapy. There is a very full discussion on this model in the book *Co-Active Coaching*, by Laura Whitworth, Karen Kimsey-House, Henry Kimsey-House and Phillip Sandahl (Davies-Black Publishing, 2nd edn, 2007).

Its inclusion here is to demonstrate how such a model can broaden the otherwise narrow considerations of both lawyer and client into a fuller awareness of what might be important.

Practitioners will have to decide whether or not to use the Wheel tool explicitly within first client meetings.

An alternative might be for practitioners to provide clients with an exercise sheet to consider and complete prior to coming to the first meeting. This would be an excellent way of preparing the client's expectations of the range of topics that will be discussed at that meeting. The expectations of lawyer or client as expert will already be challenged as a result.

Less confident collaborative practitioners might simply want to keep the Wheel model in mind as they meet with their clients.

4.14 How the Wheel works

The Wheel tool provides the client with a large diagram of an eight-spoke wheel. Along each of these spokes are ten markers representing scores of one to ten, with ten being placed on the outside rim of the circle.

Each spoke, and the wedge clockwise of each spoke, is labelled differently. These labels can be anything that the client chooses, or that the lawyer would like them to consider.

A sample diagram might therefore have the following eight sectors:

- Finances
- Children
- Post-separation relationship
- Friends
- Personal safety

- Emotions
- Legal costs
- Immediate issues

Alternatively you might want to use the seven aspects of separation from Chapters 1–3 as guideline labels:

- Legal process
- Parenting issues
- Financial issues
- Emotional divorce
- Social aspects – family and friends' expectations and reactions
- The uncoupling – the process of separating
- The divorce story we tell

and leave the eighth for the client to label, such as:

- Housing
- Lifetime financial planning
- Career
- CD collection
- Ability to negotiate
- Managing anger
- Safety

The eight issues, however they might have been labelled, are marked on a scale of one to ten to reflect how comfortable (ten) or anxious (one) the client is about that issue. This is done by placing a mark against the spoke at one through to ten.

To help with visualising this exercise, lines can be drawn in each wedge at the marked scores. Where a client has given a low mark then the wedge will be smaller. The wedges will be larger, by which we mean they will be closer to the circumference of the wheel, to reflect those areas where the client is feeling more comfortable.

This Wheel exercise can help to draw the client's mind to areas which might not be obvious. If the client is feeling angry or upset, for example, then they may be stuck in a single-minded desire for retribution, proving right and wrong, or only maximising the financial share they receive.

By having to consider eight segments we make these new areas for consideration explicit. Doing so invites our clients to give at least some focus on an array of areas that might have, so far, been disregarded, instead of being blinded by the one most pressing or alarming concern.

4.15 How can we use the Wheel?

We can go on to ask questions about sections showing low or high scores. This might reveal areas which are still working well for our client and their partner despite everything. If we come across areas where the couple are working well together then we should ask more questions about that in keeping with the narrative approach discussed above.

We might reveal the unique outcomes that we explored and re-affirm that although the couple are separating there are still areas where they share values and are co-operating.

It is important to make these explicit.

These areas might provide clues as to how we can approach the other partner and invite them to collaborate.

As future discussions get difficult, then these areas might be safe refuges to which we can retreat to remind the couple about their co-operation.

Recognising these areas of high co-operation can also help to challenge and undermine conflict-soaked separation stories. In the narrative sense we might ask a client to explain how they are able to work together in this area despite everything else that is going on.

Likewise, if a client has marked a high score reflecting their own confidence about a particular issue then questions can be asked about the basis for that confidence. Again, this can reveal lessons or possible ways to approach areas about which the client is concerned.

Do not be afraid of asking about the low-scoring areas also. The questions we ask here can give a better understanding of the extent and nature of the client's needs.

4.16 Using the Wheel at other stages

This Wheel model can be re-used throughout the collaborative process. For example, if the couple get stuck in negotiations about a pension, they could do a quick exercise called the Pension Wheel. Each client would have a blank wheel, as before, and would have to name eight considerations relating to the pension issue, and then score the importance of each consideration. The result is that the sticking point is expanded upon – suddenly it is eight considerations, instead of one headline consideration – and everyone can see very quickly what the priorities are and also where the shared concerns meet.

Practitioners might want to use the Wheel as a tool for testing proposals and possible settlements by inviting clients, either individually or as a couple, to mark the proposal against a range of considerations. The considerations here, for example, might be:

- Clarity
- Sustainability
- Partner #1's satisfaction/acceptability
- Partner #2's satisfaction/acceptability
- Children's best interests
- Short-term provision
- Longer-term provision
- Future-proofed (are details included regarding any future change in circumstances and how will future queries or disputes be addressed?)

Again, consider any low scores. Do they suggest that we need to refine the settlement further?

4.17 THE GROW MODEL – A FRAMEWORK FOR QUESTIONS

The GROW model of questioning is discussed in many books on coaching, but in particular by Sir John Whitmore in his book *Coaching for Performance* (Nicholas Brealey Publishing, 4th revised edn, 2009).

GROW is an acronym standing for:

- Goals

- Reality
- Options
- Will

GROW presents a framework around which we can build a curiosity-based dialogue with our clients. It is an excellent basis to assist with the initial client meeting and stands in stark contrast to the standardised new client questionnaire approach.

We should explain to our clients that we want to understand, in our clients' own words, what they hope to get out of the situation. We should then explain that we will start to explore how we might go about achieving those objectives.

We should make it clear that if they become uncomfortable, or if we touch upon an issue that they are not yet able to talk about, they can tell us to stop, or to move on to a different issue.

We might reassure our clients that we can always go back to the standard new client questionnaire at any time, which will give them a safer, and perhaps much easier, way of giving information.

One approach can be to show the client that we have a new client questionnaire with us and to make a deliberate gesture of pushing it to one side. As we do so we can make it explicit to the client that we could just go through the questionnaire. We can give the client the option to complete that much more restrictive process. At the same time we might assert that it would be much more helpful if we could put the questionnaire to the side for the time being and, instead, have the client tell us about what is most important to them, not what is most important to us as the lawyer.

Again, the disruption to the assumed lawyer-client relationship is clear.

Most people do not get the opportunity to talk about themselves and what they are feeling, especially at this time of separation. Friends and family members might have got tired of hearing about what our clients are going through. Others will simply find it too painful to approach and therefore shut down such discussions, particularly if they experiencing conflicting loyalties between both partners. When we offer our clients the opportunity to speak for themselves and to be heard and understood, then the opportunity is usually welcomed.

Once we have got permission from our clients to ask these more general questions then we can start the dialogue. As experienced lawyers we should be used to the need to be mindful about how the client is feeling and how fast or slow they want to go.

It will have become apparent already that this model is not suited for a free 30-minute interview. That is an inherent problem in conventional legal practice, not in the application of these models.

4.18 The 'G' in GROW stands for Goals

This seems fairly straightforward but it has a few levels.

By goals we mean the client's goals and not ours. This is not about our desire to have more collaborative cases, to sign up a new divorce client, to get through a new client interview checklist or whatever else we might have in mind. This is about helping our clients to identify and name their own goals.

Typically, clients will start with fairly large, generalist goals. They might be stated as follows:

> 'I want to get through this as quickly as possible.'
>
> 'I want an amicable divorce.'
>
> 'I just want all of this hurt to be over.'
>
> 'I want to make sure that the children are provided for.'

We need to ask questions to understand further and define these goals.

4.19 Using questions to define goals

Try to use open questions, namely those which cannot be answered simply with a yes or a no response.

Try to rephrase such closed questions as what, how or when questions.

Try to avoid asking why questions. These can be interpreted as being too direct and confrontational.

Asking these questions further disrupts conventional lawyer-client relationships. We are demonstrating our curiosity. We are positioning the client as the expert in all of this and, as we do so, we encourage them to take on more active roles in their own separation and dispute resolution.

> 'What do you mean by "Understand"? Or by "What has happened?"?'
>
> 'You said "Amicably" … What does that mean to you?'

> 'How might we define amicable?
>
> 'What does amicability look like?'
>
> 'What is important to you about making sure we deal with this amicably?'
>
> 'Tell me more about that ...'
>
> 'Providing for the children in what way?'

These are called probing questions. They are designed to dig deeper into the areas that the client has identified, to get a clearer understanding of why that issue is important. Very often, the first expression of a goal, anxiety or issue can cover up something quite different several layers down.

We might also consider broadening, as well as deepening, the goals.

> 'Is there anything else?'
>
> 'What else is going on for you right now?'
>
> 'If I asked your best friend/mother/doctor/boss what was most important, what might they say?'
>
> 'If your children write a book about your divorce in 20 years' time, how would you like them explain what is happening now, do you think?'
>
> 'What do you think would be the largest/best/happiest chapter in that book? And the worst?'

The purpose of these questions is to challenge the client to think more widely about issues that might be important instead of becoming too single-minded. Just because one issue is very prominent does not mean it is the only issue. Asking these questions will also develop the client's own understanding of his or her initial thoughts.

Do not forget to refer back to the Wheel exercise for prompts as to areas that you might explore.

> 'I notice that you marked separating as 10 on the Wheel. Tell me about that.'
>
> 'You say that legal costs are a real worry. What are your thoughts on that?'
>
> 'If we did a Wheel exercise just to deal with "Immediate issues" what names might you give to the eight segments?'

Also, revisit the last chapter on narrative. Consider asking questions that might make the conflict-soaked separation story more explicit.

'I notice that you referred to your husband as a … How else might you have referred to him?'

'What do you call him in front of the children?'

'I can imagine how that must have felt for you at the time, but in your own words, how would you describe that?'

'How would you have preferred to have seen things turn out?'

'What other way might we approach your wife?'

'What could we call that situation?'

'If you had to make up another explanation for that, what might it be?'

All of these goal questions will provide important information and insights into what the client's concerns are.

They take the client beyond the immediate and reactive considerations arising out of fear, hurt and anger and create opportunities for clients to speak, possibly for the first time, about some very difficult issues.

4.20 Silence and the client's struggle

We should expect clients to surprise themselves at what gets uncovered. We should also expect clients to struggle to answer some of these questions. There may well be long silences as they do so.

There will be the temptation to rush in and fill the silence.

Resist that temptation.

If we fill in the silences then we rescue them from answering. This behaviour reaffirms and invites a reversion to client passivity. It repositions the lawyer as the active driver, the client as passive passenger.

There has emerged, over the last 10 years or so, a presumption that clients are bewildered and almost incapable of speaking and thinking for themselves. That drives lawyers to frame and restrict the scope of questions carefully, to make it easy for the client.

What is emerging now is an appreciation that clients are capable and have the resilience to tell their own stories.

Of course we need to be sensitive to how our clients are feeling. We can ask them or go by observation. If we see distress then we should point that out. We should ask the client if they are able to go ahead with resolving the issues arising from the separation.

The temptation, when the client becomes distressed, is to presume that it is the collaborative process that is too difficult at that moment in time and to revert to a conventional approach instead.

That disregards the fact that the emotional unreadiness will still be there within the conventional process.

Distress on the part of the client does not prove that collaborative law is unsuitable for the client. It demonstrates that the client is not ready for any process and that other support is required. That may well mean turning away the client for the time being and referring them on, instead, to a counsellor, coach or therapist.

4.21 Consider immediate and longer-term goals

We already know what our immediate goals are in this first meeting. We want to:

- Understand our client and their concerns
- Establish a rapport and an appropriate level of trust with them
- Be in a position to have some idea as to the way forward
- Work with a new client.

Surprisingly we do not need to know, at least not at this first meeting, the value of the house, the amount owed on the mortgage or, even, the date the couple were married or when they started living together. It is peculiar that we should spend so much time at the first meeting dealing with these details as we work through the restrictive new client questionnaire.

Again, put away the client questionnaire at this first meeting. If we are going to work together then we can always ask our clients to complete those details in their own time, and without us having to charge our hourly rates to go through that process. If we are not going to be working together then we are wasting our time on collating this information anyway.

Ask specific questions instead about the client's immediate concerns.

> 'So, by the time we finish our meeting here today, what would you like to have achieved?'
>
> 'If there was one thing today that you really need to resolve, that is really pressing on your mind, what would it be?'

It may well be that the client at this stage reverts to questions about entitlement and likely settlement.

We are not going to be in a position to provide specific answers in that regard, and that is okay. Remember that within this process we have had next to no information about the family, child arrangements or finances.

Instead we can acknowledge the immediate concerns and start to identify the steps that should be taken to answer those concerns.

4.22 Feed back what you have heard

Once we think we have a good grasp of our client's goals then we should feed back what we have heard.

It is okay if we have got something wrong. We are giving our clients the chance to correct us. This reassurance can lead us to resist the temptation to make copious notes as our clients are talking to us. Instead we can concentrate on what is being said and what is going on for our client.

When we feed back what we have heard we can achieve three things:

- Confirm our understanding
- Correct our understanding
- Check client understands their own position.

There is no reason why these goals cannot be recorded. They may well help in preparing the anchor statement that the couple will share at the first collaborative meeting. This anchor statement is explored fully in Chapter 8.

4.23 The 'R' in GROW stands for Reality

Having established a goal, how realistic is that? What do we need to get to that goal?

Discussing reality does not mean doubting what the client has said, but rather taking stock of where our client is at that moment.

We are trying to gauge where the client wants to get to and where they are leaving from on that journey.

How far removed is our client from being able to conduct matters amicably or from being able not to get upset at their wife or husband's behaviour? Ask them.

If a client is stating a desire to 'Get this all over and done with', then the reality enables us to focus on the need to engage with the dispute resolution process, whichever approach might be selected by our client.

> 'I can imagine why you would want all of this over and done with. What do you think it will take to get everything sorted out between the two of you? What kind of timescale have you got in mind to get this all over and done with?'
>
> 'Okay, so you want to have contact with your son without your wife making it difficult. What is happening at the moment?'
>
> 'You say that she states that you are not able to look after your 6-month-old son, is that right? How much time do you currently spend with your son? What aspects of parenting did you deal with when you were together?'
>
> 'If I asked her about areas that need to be developed, what areas do you think she would say?'
>
> 'How far have we got to go, do you think, before she will be reassured?'
>
> 'How soon would you like to be able to start seeing your son? And how long do you think it might take us if we have to go through the courts?'
>
> 'What else might we be able to do to move things forward, do you think?'

If the goals are around financial issues such as maintenance or security then these reality questions will start to shape awareness of what kind of provision might be needed, but also what the client themselves can do to help themselves, either fiscally or within the resolution process.

If it becomes apparent, for example, that they have no experience of dealing with household budgeting, or no awareness of financial products such as mortgages, pensions and insurance, then an early referral to a financial adviser or finance coach might assist.

Having established, tested and refined a range of goals, this reality phase gauges where we are in relation to those goals and gets the client to start to consider how to bridge the gap between where we are and where we need to get to.

4.24 The 'O' in GROW stands for Options

It is here that the client may well start to create their own dispute resolution process.

We will have our own views on what resolution process is appropriate, and it is likely that we will have our own preferences on which process we would like to use. It is important, however, that we stick with our client's agenda and their earlier stated goals.

Keep asking questions rather than recommending one process or another. Questions here should elicit options for what could happen next.

John Whitmore writes about this option generation becoming easily stunted by negative considerations. We are quick to dismiss ideas under a barrage of reasons why something would not work.

We could dismiss all suggestions on this basis. We are very good at thinking of 'Why nots'. We need to keep thinking of 'What ifs'.

The reason why not to do something or try something might include cost, time, the other partner's likely reaction and such like. John Whitmore suggests overcoming such dead-ends with a question along the lines of 'And if they did agree to go along with that idea? What might happen then?' Simply leapfrog the reason why not and keep the discussion creative to move beyond the possible pitfalls. He writes:

> 'By this process, which temporarily sidesteps the censorship of the rational mind, more creative thought is unleashed and perhaps the obstacle is found to be less insurmountable than it seemed.'

Coaching for Performance, p 80.

Bearing in mind that the vast majority of clients claim to want an amicable separation, then the client's suggested options are likely to be geared around dialogue and communication.

Very few clients ever suggest an application to court, particularly at these early stages of proceedings.

Instead we might get options such as these:

> 'If only I could talk to him/her …'
>
> 'If only I could get him/her to see that …'
>
> 'If I could somehow make sure that there was enough to provide for …'
>
> 'I could get some help with …'

The client is starting to design for themselves a collaborative approach to dispute resolution.

Once the client has gathered an array of ideas – and do resist the temptation simply to take the first suggestion – then we can afford to be a bit more directive.

At this stage we might discuss the various options, namely self-help, mediation, collaborative law or litigation, and observe our client's reactions to each of them.

4.25 The 'W' in GROW stands for Will

- What will the client do next?
- What will they have us do?
- What will each of us be responsible for?

Here we can start to establish expectations of one another and how we will work together. Notice how we are essentially contracting to work with one another on a certain basis. It is a lawyer-client relationship, however, where responsibility is distributed between the client and lawyer, rather than the conventional dynamic where all is allocated to the lawyer.

This is not to say that the lawyer does not act in the client's best interest. Of course they must still do that. It reflects, however, that the client, far from being incapable or lost, does have the resources and the knowledge to take an active part in their own dispute resolution.

This part of the discussion can be used to ensure that this new level of responsibility is understood and that the client is prepared to take that on. This needs them to agree that they will agree to do many things. Some of those things will be procedural. Others might be more about managing the relationship and communication.

Areas to explore might be as follows:

> 'How will you and I let each other know if we have questions or concerns?'
>
> 'How will you react when things get tough in negotiations?'
>
> 'How will you let me know when you are upset, off the pace or just plain tired?'
>
> 'How will you deal with the concerns of other family members and friends who might encourage you to adopt more conventional "hard" tactics?'

Other 'Wills' to explore, again as listed by John Whitmore include:

> 'When will you [or I] do whatever we have committed to?'
>
> 'Will this step achieve what you want?'
>
> 'Who will you tell about what is going on?'

'Who will you tell when you need help?'

'Who will you get support from and when and how will you seek it out?'

In all of this, we are also testing and strengthening the resolve, or again the will, of ourselves and our client to see the newly designed process through.

4.26 SUMMARISING THE WHEEL AND GROW

This chapter has taken the challenge that many collaboratively trained lawyers will be familiar with, namely the gap between their desire to work collaboratively on a greater proportion of their cases and clients' decisions to use conventional approaches, as opposed to collaborative law.

We have taken that challenge and used it as an introduction to considering discourse and the nature of the assumed relationship between lawyer and client and the positions that they adopt in relation to one another.

We then went on to explore how we need to change how we relate to clients with a view to disrupting convention and assumed roles. When we do that, we place our clients into more active roles within their own dispute resolution.

As our clients wrestle with and answer questions, they determine for themselves what they want to achieve and create options for how they might move things forward.

We find that many clients will not have spoken about litigation or solicitors' correspondence at this stage, but rather will express a desire for dialogue and for 'working things out'. At that stage, the client will be thinking in terms that mirror the collaborative approach and now we can suggest the collaborative process as an option.

If we persist with conventional practices at the first appointment then the conventional outcome of those first appointments will be perpetuated.

4.27 SEPARATION AS CHANGE

As clients, and indeed collaborative lawyers, come to grips with the collaborative process in practice, they will experience radical changes. For the client, the changes are taking place in their daily lives as they approach and pass through separation. They may also experience changes in their expectations of what is going to happen.

Both the client and the lawyer will experience change as they pass through the collaborative process and its pattern of four-, or more, way meetings, homework and preparation.

4.28 THE CYCLE OF EXPERIENTIAL DEVELOPMENT

There are several academic models which look at how adults learn, process and deal with life changes such as separation and divorce. The model used in the following section is taken from Geo Roberts' book *Impact!* (Xulon Press, 2010).

This model of change and development is not so much a tool to be used, as with the Wheel or the GROW framework, for questioning. Instead, it should be seen as a lens through which the challenges that we and our clients will continue to face can be understood.

The cycle of experiential development suggests that as we encounter and pass through change we pass through four quadrants of a circle and then go around again, and again.

4.29 Quadrant one – Attitude

Quadrant one is the recognition that an individual needs to, wants to, can do and will change. It requires there to be an attitude reflecting those four beliefs.

The model asserts that all four beliefs – need to, can do, want to and will do – need to be present, although not necessarily in equal measure. Without any one of them, then the will to do something will not emerge or does not persist.

For example, the absence of a realisation that there is any need to change would be prohibit any change at all.

A client might know that there needs to be change – the partner has left already or filed for a divorce perhaps – but might not believe that they can change. This will result in despair and might drive a client to instruct a lawyer in a conventional manner as they will not accept that they can do anything about their own situation. This can play into the conventional lawyer-as-guardian relationship.

The extent to which any individual wants change may well be dependent on how far they have passed through the grieving process of denial, anger, bargaining, despair and acceptance. If the changes in their life are still very new and raw, then there will not be too much want at the moment.

Suggesting that a client wants to change is likely to be clumsy. Very often neither party actually desires change. Want in this sense is more of a recognition or acceptance that there has to be change.

For example, a spouse might be deeply upset about the end of the relationship and yet still sense that they want to deal with matters in the best way possible for the benefit of the family. They might not want the separation but do want to deal with it as best as they can.

Finally, if there is no will to change, despite the recognition of need to change, the belief that they could change and even a want to change, then progress will become frustrated.

We can use these four attitudes when progress and co-operation is faltering. Where is the obstacle? Is it in the need, can, want or will attitude?

In some circumstances it might not be appropriate to take instructions for the time being, whether on a collaborative or conventional basis. A referral out to a counsellor or therapist might provide greater support, even if that means that we lose a client for the time being.

4.30 Quadrant two – Knowledge

The second quadrant is often very painful. Geo Roberts' model calls this the 'Knowledge quadrant'. It is where we know the need to change, the need to participate in the ongoing process of separation and resolution.

What happens is that our attitudes from quadrant one lead us to step out of our comfort zone and take action within this separation. Very often, when we step out of our comfort zone, we will experience failure or challenges.

It might better be called the 'Quadrant of Falling Over'.

As a collaborative practitioner you might experience this yourself when your first collaborative meeting or first meeting as part of a new team does not go very smoothly.

You might experience it the first time you try to meet with a client with no new client questionnaire and try to implement the GROW framework instead.

Clients might experience it the first time they try a collaborative meeting, or the first time one parent has to care for a child overnight without the other parent being present.

Problems crop up. We will trip and stumble, fall and get hurt.

We might feel a renewed anxiety as a result of our fumbling first attempts. The inner voice or the dominant narrative speaks up, chastising us, telling us that this was always a bad idea and why did we not listen in the first place.

Friends and family make similar 'I told you so' noises and hold up these first faltering steps as proof that the decision to collaborate, to trust the other partner or to try to take responsibility was misguided.

The real danger here is that our experiences of falling over lead us quickly to revert to the status quo behaviour.

The first collaborative meeting results in one of the couple becoming tearful and walking out. As a result, everyone concludes that the process has failed and so resolves to settle the issue at court instead.

Your first interview with a client using the GROW model discussed above becomes difficult. You run out of suitable questions to ask, or find yourself reverting to standard questions eliciting prescribed information such as dates and values. You conclude that alternative approaches to taking instructions are impossible and so reach for your new client checklist with the next new client.

One parent is caring for the young child overnight for the first time since the couple have separated. It is not easy. The young child is tearful and misses the other parent and her pink fluffy rabbit which has been left behind.

The parent resolves that this is too challenging.

The parent who usually cares for the child concludes that this is evidence that the child is too young, and he or she always said this was going to happen.

One parent or the other retracts and insists on non-staying contact as a response, with all of the predictable consequences and applications.

One of the couple, who previously has never had to deal with budgeting or domestic matters, fails in the first month and quickly becomes convinced that they cannot do this alone. Their previous resolve to be independent and learn new skills to deal with these issues alone is quickly abandoned.

Desperation returns with the conviction that 'I'll never manage'.

We should expect to be challenged and we should prepare our clients to be challenged.

It needs to be clear to our clients that we will experience tough episodes, walkouts, tears, anger, clumsy comments and slips back into conventional practice.

If we anticipate these in advance then when we encounter them we will not be distracted by them, but rather feel that we are making progress.

Perhaps the best thing we can do within the second quadrant, the quarter of falling over, is to get into it, do what it is that we have to do, see if we fall over and move on as quickly as possible.

Google had a motto which stated 'Fail Quickly'. This means that when we fall, when our efforts to do things differently do not work, we need to recognise that and move on quickly.

There can be a temptation to fall into self-pity or recrimination. We might indulge in the 'I knew this was never going to work' or even the more blame-loaded 'I told you this was not going to work'. We can look at ourselves and our efforts and draw out all kinds of self-limiting conclusions: 'I'm not good enough, I'm just not cut out for this, this is not safe, I need someone else to do this for me ...'

'Fail quickly' essentially says move on, not move backwards, and that is also what the circle of experiential learning and development tells us.

4.31 Quadrant three – Practice

The third quadrant calls to practice what we have resolved to do and have done. What worked well? What did not work well? How did the first collaborative law meeting work? Was there anything I did, or my colleague did, that you would change?

It can be helpful to talk about these questions with another person. The lawyers may well do so in conversations directly between themselves in between four-way meetings.

For the couple themselves, they can each get feedback from their own lawyer, especially if that lawyer has an element of coaching skills awareness.

An external collaborative or personal coach, therapist or counsellor can also help here.

The point of the feedback at this stage is to keep looking forward, albeit it with the perspective of new lessons learned from the quadrant of falling over.

How else might we have responded to one or another issue? What other choices did we have available to us that we had not recognised at the time, but that we might be able to bring to mind next time?

Having reviewed and listened to feedback we can now move into the final quarter, named, by Geo Roberts, as the quarter of skills and habits.

We have tried the new approach and we have reviewed it, drawing lessons from our experiences on what we might do better, or at least differently.

4.32 Quadrant four – Skills and habits

Quadrant four is about internalising those lessons so that next time we find ourselves in a similar situation, we can respond in the new way as opposed to the same way as last time. Having resolved to step into the process and having experienced the challenge of being outside of our comfort zone, and having thought about what worked well and what might have gone wrong, this quarter now sees us learning from those experiences.

We internalise the learning from our experiences. We are better prepared to manage change and so we resolve to go round again.

We reaffirm our original quarter one beliefs, namely that we have to deal with these changes and reaffirm our resolve that, no matter how hard, we will and we can respond to change and that we want to do so in ways that are important to us.

And so the next collaborative meeting comes around and it goes better than the first one. It is still not easy, but it is better.

We start to see results. Areas of agreement, co-operation and collaboration emerge. We go round and round that cycle, growing in confidence and growing in our ability to collaborate.

The attitudes of needing, wanting and believing in trying to change will all enable the client to change how they deal with their separation in a powerfully different way.

The family will be transformed from one on the brink of destroying itself through litigation to a family which has created for itself a new basis of relationship, co-operation and understanding. That and the lessons learnt through the collaborative process will put the family in good stead for the future.

4.33 SUMMARY

This chapter has shown how new approaches to the initial client meeting, and a shift in the way that lawyers speak to clients, can better help lawyers and couples going through the collaborative law process.

The Wheel challenges us to consider a broader range of considerations to include, but not limited to, the most pressing issues.

It can also be used to break down sticking points into a more varied range of considerations, which in turn might unblock later discussions.

The GROW model of questioning should be used to explore the real goals, reality, options and will of the client. Again, this model can be used at the outset when the lawyer and client first meet. In doing so we may find that many clients design their own process, which either reflects the collaborative process or can be accommodated by it.

The GROW framework can be used repeatedly and it may well be necessary to loop back and review goals, or options, from time to time.

Finally, the cycle of experiential learning and development can be used to build in expectation of and resilience to challenges that we will meet and overcome as we commit to collaborating together.

Collaborative practitioners can benefit from accessing these tools themselves to ensure that they continue to develop their own skills and collaborative law practices.

CHAPTER 5

BUILDING THE COLLABORATIVE TEAM

5.1 GETTING BOTH PARTNERS ON BOARD

In the previous chapter we looked at how a client-centred approach to taking initial instructions can help clients to identify for themselves how they would like to resolve matters arising from their separation. Many clients will self-select a collaborative-type approach if they are asked about what is important to them now and in the future and how they would like to be able to resolve issues.

Identifying the client's wishes is only one part of the process of establishing a collaborative resolution process. There still remains the not straightforward process of getting the client's partner to agree. This chapter will consider that challenge.

The chapters that follow will look at who else we might draft into the collaborative team to ensure that both partners are fully supported in this process, and how a team approach can help to progress matters more efficiently. We will also review, in the next chapter, what criteria clients use to select their own collaborative lawyers.

5.2 COLLABORATION BY DESIGN

The collaborative process is not rigid. It is a collection of values rather than a collection of rules, timetables and hoops to jump through.

The process is flexible. It can be stretched or squeezed as needs and, to some extent, budgets require, as long as the participation agreement containing a disqualification clause is present and signed up to. See Chapter 8 for more details.

Collaborative practice can be bent and twisted to accommodate almost anything and anyone. We can see examples of collaborative law cases which have pulled together an array of diverse practitioners, or very straightforward cases which have resolved all matters in a single meeting between lawyers and the separating couple.

It is partly this flexibility that gives collaborative practice its reputation as being a client-centred approach to family dispute resolution. The process is not completely without structure and we explore the collaborative cycle of preparation, meeting and follow-up in Chapters 7, 8 and 9. Within that structure, though, the clients can access any resources that are felt to be appropriate. It is one of the collaborative practitioner's roles to highlight what options and which other practitioners are available and the benefit that can be gleaned by using them.

5.3 FIXED TEAMS, FULL TEAMS AND CLIENT DESIGN

Parts of the collaborative law profession are becoming more vocal in talking of the need for 'full team' collaborative practice. This full team means having the following people involved:

- each of the couple;
- one lawyer each;
- a mental health professional (counsellor, therapist or coach) each;
- a neutral independent financial adviser; and
- a collaborative child expert.

The 'full team' therefore consists of no less than eight people, six of whom will be charging fees.

The phrase 'full team' is unhelpful as it suggests that anything less than the cast listed above is somehow incomplete. There is an implied criticism of any alternative.

Some collaborative lawyers believe that a full-team approach is the only suitable approach. In such cases, some clients may find themselves being presented with a whole package of coaches, financial advisers and therapists from the outset. That client may, or may not, want such an approach. Collaborative lawyers need to be careful to consider the client's wishes in this. They should continue to help clients identify their needs and other professionals who might be introduced to meet those needs which the lawyer cannot.

In some cases, a large team will be appropriate. In this regard the image of the 'full team' as defined above is helpful in demonstrating how a large collaborative team might be comprised rather than prescribing how it should be comprised.

5.4 RESISTANCE TO LARGE OR FULL COLLABORATIVE TEAMS

It is not helpful to start imposing large cast lists of professionals onto the collaborative process. Doing so may well engender resistance from individuals and practitioners alike for several reasons.

An obvious consideration has to be cost.

Does the family have the means to pay for all of these professionals? If not, which ones are critical? It is recognised that not all of the full team members need to be available from one collaborative meeting to the next. It may well be that one member has a very limited contribution to make. Not all members of the team therefore will have an equal cost attributed to them.

Another problem is the risk that the client gets lost amongst the crowd and that their voice is not heard above the collection of well-meaning but uninvited guests. This consideration is more easily met by careful management by the whole team. Indeed, it will be the explicit purpose of the coach or therapist, for example, to identify barriers that the client might be meeting in trying to speak up and speak out.

There are going to be other points of resistance that need to be borne in mind. The couple might not be ready to open themselves up to what they see as scrutiny from therapists, counsellors or coaches. They may, quite simply, not be comfortable about having to work with what are referred to as mental health professionals within the international collaborative law community.

5.5 MENTAL HEALTH PROFESSIONALS

This book will use the phrase 'mental health professionals' as a label encompassing the very wide range of professionals called therapists, counsellors and coaches. The unifying characteristic is that these professionals are brought in to the collaborative team to help the clients, or the lawyers, understand the social, communication and behavioural dynamics that might be encountered as the couple pass through separation. It must be up to the client to select which of those professions they want to access.

Information should be made available about the range of services available and clients should be encouraged to make informed choices about which services they want. That will mean collaborative lawyers having to communicate the role and benefits that each class of professional can bring. The chapters in this book might help.

5.6 DEVELOPING A NETWORK OF NON-LEGAL COLLABORATIVE PRACTITIONERS

Active and effective collaborative lawyers develop local networks of mental health professionals, financial advisers and accountants. Doing so helps them to understand what each other professional offers. A lawyer who has not grasped the full benefits of collaborative law may find it difficult to communicate value in the process to a client. A collaborative lawyer who has not grasped the role and benefits of the various other professionals who might assist within the collaborative law process will similarly find it difficult to communicate the same to their clients.

To this end, the better collaborative lawyers invest time in getting to know the other professionals that they might refer to in the future. They learn more about how the other professionals work within their own industry and also as an individual.

Clients should expect to be offered contact details for these other collaborative professionals. If they are not then they are probably not receiving as full a service from their lawyer as another might be able to provide.

Clients should insist that if their lawyers have not provided details they do so upon request. If the lawyer has no details then they should get some from their local colleagues or their local collaborative practice groups.

Each of the professions featured in the following chapters has an important role to play in supporting those going through separation.

It is recognised that conventional family law cases bring in other professionals as expert witnesses and individual advisers within their specific fields. This is not a practice that is exclusive to collaborative law approaches.

It is likely, however, that there is greater openness to considering such interventions within the broader considerations that we see in collaborative law. Very often the involvement of another professional within collaborative law is seen as complementing the collaborative effort. Within conventional or litigious cases, the involvement of other professionals can be resented either for representing an additional cost or because their involvement has come about as a result of disputed representations made by one partner or the other.

5.7 HAVING OTHER PROFESSIONALS AT THE FIRST CLIENT MEETING

Just as there are those who very strongly encourage a full-team approach, so there are those lawyers who insist that certain other professionals must be present at the very first client meeting. There is anecdotal evidence of some practitioners insisting on having a coach, or a child expert, sit in on the first client meeting.

The permission and the level of intervention that a client signs up for in meeting with a lawyer is very different from the permission that a client gives a therapist, counsellor or coach. They are different mandates and different kinds of working relationships. We cannot assume that clients will be comfortable in having an initial meeting, or any subsequent meeting, with mental health professionals present.

The client must be free to give proper permission to bring in other professionals.

For a lawyer to say 'I will not see you without this coach, or this therapist' jeopardises the choice that the client might otherwise be able to make.

Of course, the client can choose to instruct other lawyers and, of course, each lawyer can choose to practice as they wish, so long as it is within professional boundaries.

Problems arise though if there is a scarcity of collaborative lawyers within a region. Similarly, any given lawyer might have been strongly recommended to a client. In these circumstances the client may well feel that they have no choice and that they will have to do as the lawyer stipulates.

Do not forget the nature of narratives and discourse, discussed at length in Chapter 3.

A client looking to instruct a lawyer may perceive that the natural way of relating to that lawyer is to do what they say.

This book and the collaborative process is very much about levelling out that relationship and handing back responsibility and self-determination to the client and the couple. A lawyer's insistence on the inclusion of certain other professionals might preserve a rather conventional perception of the lawyer's role as the client's paternalistic guardian or rescuer. That might go on to reinforce certain narratives and assumed relationships between the client, adviser and process. The result can be a reduced effectiveness or level of participation on the part of the client.

If there is a risk that a client conforms to expectations of what it means to work with a lawyer, by falling into a submissive and unhelpfully deferential position within the relationship, then how might that be compounded when the client has to meet with not one but two professionals?

Clear advice should be given as to the merits of having other practitioners join the team. It is inevitable that such advice will often go so far as a clear steer towards instructing other professionals. Surely, as long as the client has got adequate information upon which to make an informed choice as to whether or not they want more professionals involved, then the lawyer has done what is required of them. The choice, ultimately, has to be the client's.

5.8 OTHER PROFESSIONALS RECOMMENDING EARLY INTERVENTION

Some financial neutrals and mental health professionals suggest that they should be in the room whenever a family lawyer first meets with a client going through separation.

Their argument states that lawyers and clients may not know when the other professionals, particularly mental health professionals, have got a role to play. The concern is that lawyers may miss an opportunity to make a timely referral.

This should not be a concern.

The effective and more active collaborative lawyers and their local collaborative colleagues in other professions should already be working with one another to increase their mutual familiarity with one another's work. That networking and cross-education will highlight when referrals are appropriate and beneficial to the client.

The counter-suggestion that a family lawyer should always sit in when a therapist or coach first meets with a client reveals this proposal as being unrealistic and unnecessary.

What is important is that there is an early identification of possible needs and an early recommendation or referral is made.

These questions of full teams and the obligatory inclusion of other collaborative professionals from the outset are areas of ongoing debate within the international collaborative law community.

Clients should discuss it with their lawyers and find a way forward which is comfortable for the client, having been put in a position where they can give fully informed consent.

5.9 GETTING THE OTHER PARTNER TO COLLABORATE

Before we can start to address the question of which other professionals to bring in, we need to overcome a much more pressing challenge.

It is not uncommon to find that whatever is suggested or recommended by one partner, or their lawyer, to the other partner is dismissed out of hand.

This happens precisely because it has been recommended by their spouse or lawyer.

Sadly, many cases go all the way to contested final court hearings for this very reason. Every time a solution is proposed, even a sensible, generous proposal, it is dismissed on the basis that there must be a catch or something in it for the proposer.

5.10 The 'No' strategy

In some cases it becomes possible to imagine a 'No' strategy. Here, any request for clarification or request to agree a proposal, a settlement or a way forward is met with a 'No' response.

Such a strategy might be driven by conflict-soaked narratives that suggest that the other partner must be manipulative or villainous. Such separation stories which cast the other partner as villains of the piece make it impossible to trust anything they say or do.

Going along with any suggestion would be seen as sleeping with the enemy. In such circumstances the suggestion of working together on an aspect of the case, in a co-operative sense, is inconceivable. Even if such co-operation was in both parties' best interests it remains unconscionable. In these conflict-soaked situations collaboration does not have the meaning we have been working with here. Instead, collaboration takes on the connotations of the collaborateurs who acquiesced and assisted the invasion of European countries in the Second World War. Any notion or suggestion of collaboration takes on connotations of treachery.

This same dynamic can be seen just as much when trying to enrol the other partner to the collaborative process. Suspicion is aroused and dominant narratives start to take a grip.

> 'What is in it for them? Is this a trick? What is he/she up to? I've got to protect myself. I don't understand what is going on. I had best get a tough lawyer who can protect me from these tricks.'

The opportunity to collaborate can be quickly lost.

So what can be done to increase the likelihood of getting both partners to agree to collaborate?

5.11 WHEN COUPLES JOINTLY AGREE TO USE THE COLLABORATIVE PROCESS

How might couples, who are still working together or speaking to one another, choose to use the collaborative law process?

It is to be hoped that as the collaborative process continues to gain in popularity, collaborative law will become as much a part of the general public's considerations as lawyers, mediators or counselling are.

Collaborative law is not there yet. It simply does not have that level of resonance or awareness among the wider public.

The vast majority of couples will have never heard about collaborative law until they happen to instruct one of the 1,500 or so collaboratively trained lawyers in England and Wales, or maybe come across the term collaborative law as they start to research matters online.

5.12 The need to raise the profile of collaborative law

Incidentally, the profile of collaborative law on internet search engines such as Google needs to be raised massively.

At the time of writing, an online search for 'How do I get divorced' shows no findings for collaborative law at all on its first three pages. Mediation only gets one mention. Instead, a couple searching such a question will be directed by Google to cheap 'Do it yourself' divorce packages and discussions about financial entitlement. That is a problem.

As soon as an individual or a couple starts to research online, the risk is that they are immediately thrown into the statutory process and legal rights. That early exposure may well shape the expectations of many couples, who then find themselves on a one-way track to a cheap divorce outcome with no support and no exploration of other alternatives.

The discourse and conventions of separation establish themselves very quickly.

If collaborative law is to raise its profile then it needs to do so to such an extent that collaborative law and divorce become synonymous. When people search 'divorce' they need to find collaborative law quickly. There is still a long way to go in this regard.

If both partners have accepted that their relationship has come to an end then there will be the opportunity to discuss and agree upon which of the available dispute resolution methodologies they would like to use to resolve outstanding matters.

5.13 COUPLES REACHING JOINT AGREEMENTS

A couple who are able to communicate with one another can consider whether they want to attend mediation, use lawyers conventionally or retain lawyers to work with them within the collaboration process.

Few couples who are still able to communicate to any effective level choose to litigate against one another. It is a damning indictment of current conventions that this is how too many cases still end up.

Couples will often express a wish to resolve matters amicably. Even though they will have very often started at a level where there is still some effective communication between them, too often, even with the best will of both partners and their instructed lawyers, matters quickly deteriorate. It would be interesting to study how the transition is made from couples who are still speaking, and committed to resolving matters amicably at the outset, and then fall into litigation. That study is beyond the scope of this book.

5.14 Self-negotiated settlements

Some couples are able to successfully negotiate between themselves without any professional assistance. This is called the 'Kitchen table' approach, as if negotiations are conducted around the said table.

Sometimes these home-based negotiations are facilitated by a well-meaning family member, or trusted family adviser, such as a financial adviser or family accountant.

This is an interesting development that family lawyers should be acutely aware of. It is an indication that couples are starting to turn to other professionals than lawyers to help them resolve matters. The shift away from lawyers being the sole gatekeepers to the dispute resolution marketplace is already underway.

These negotiations can become problematic if they are conducted without a full knowledge of the steps that are required to make such agreements binding and enforceable. There are many cases where couples believe they have separated successfully in the sense that it has been amicable and they have saved legal fees. They have resolved the distribution of the house and other assets and made arrangements relating to income and maintenance. Ignorance of the need to record agreements in a court-approved order to terminate future claims is often fatal though. Unsuspecting parties can find themselves subjected to new claims from their long-estranged spouse many years later. Examples can be seen when a former matrimonial home is sold and it transpires that it is still held in joint names. The proceeds of sale, now much more than they would have been when the parties first separated, fall to be redistributed once again. In other cases lottery winners have found themselves having to share their winnings, again many years after believing that all matters had been resolved.

Another problem can arise when both parties understand that they have reached a full agreement and only then take legal advice. The lawyer may be obliged to point out an omission or change certain terms to make them compliant with tax or other regulations. If the negotiations were at an advanced stage, it can prove to be very difficult to back-pedal and revisit parts of the negotiated agreement.

Care needs to be taken that legal advice is obtained in a timely fashion even in self-negotiated settlements to ensure that agreements reached are recorded in ways that will be binding both now and in the future.

5.15 Mediation

The couple may realise that there are tough matters that they need help to resolve. Maybe one or both of them has an inadequate understanding of the issues at hand, or they recognise that discussions get derailed into less helpful arguments. They resolve therefore to seek some assistance with the outstanding matters and so approach a mediator either directly or via lawyers.

Mediation may well be appropriate and such services are now fairly visible to families, or at least will become apparent following online searches. Not all couples will feel inclined to use mediation, however. There might be concerns on one part or the other about balance in negotiating power, vulnerability or their own ability to negotiate and represent themselves within a typical mediation meeting. In such a typical family mediation structure, there would be one mediator addressing both clients, either together in the same room or shuttling between them in separate rooms.

Some clients report that not having their lawyer with them is a hindrance, and mediators themselves express frustration that a couple can reach an agreement within a mediation process, only to see it dissolve once one client or the other has reported back to their lawyer on what has been agreed.

Mediation can be more flexible than this, and some mediators are increasingly prepared to mediate with clients attending together with their lawyer or adviser if need be. Mediation should not be dismissed out of hand therefore.

It would be quite wrong to allow the suggestion that mediation cannot deal with negotiating power imbalances remains without challenge.

Mediators are able to recognise such dynamics and imbalances and are trained to employ strategies to level out imbalances. The danger then is that the dominant party may perceive that the less comfortable partner is getting preferential treatment, or that the mediator is biased by helping him or her.

As indicated above, it is probably unlikely that a couple who are able to speak to one another about how they might resolve issues would choose each to instruct lawyers to then engage in court applications at this stage.

One or the other might, however, hire a lawyer to work on their behalf. They may make an initial appointment to meet with a lawyer to get some idea of what their entitlement might be.

Conventional first client meetings can quickly shift the direction of resolution between the couple. The lawyer might speak in conventional terms and unwittingly trigger anxieties on the part of the client. Even without such triggers, as soon as we get embroiled in discussing entitlements and likely outcomes, then it becomes very difficult for clients to move away from a positional understanding of their rights.

A lawyer might perceive or provoke a need to write immediately to the other party and advise their client as such. When the other partner learns that their spouse has instructed a lawyer, communication and relationships can take an unwanted turn.

We have already explored the discursive relationship and relative positions that can quickly establish themselves between a client and their own lawyer. The positions adopted between a client's partner and the already-instructed lawyer are even more problematic, as explored below.

5.16 Couples agreeing to instruct collaborative lawyers

If the couple have had access to literature and information about collaborative law then they may jointly feel that they would like to use such a method.

It is to be hoped that this book itself will provide couples with a reference point in their considerations on whether they think collaborative law can work for them. There are other books that can be used also. One partner might read through such a book and choose to share it with their partner, agreeing to discuss it together once they have done so.

There is also an increasing amount of literature, video and other material online, although as stated above, nowhere near enough, explaining how the collaborative process works online. A quick search on YouTube for collaborative law reveals many videos with lawyers explaining how the process works and how it might be best suited for any given family. Couples might consider this between them or one partner might share a link with the other to consider.

5.17 THE EMERGENCE OF THE DISPUTE RESOLUTION PROCESS BROKER

As the provision of legal services faces modernisation in the coming years, a new breed of family law adviser may well emerge. This person could be seen as sort of process broker.

At the moment, solicitors' conduct rules prohibit solicitors from advising both partners in a couple. That makes it difficult for a couple to meet together with a single adviser to find out more about their choices.

It is thought that some lawyers are offering this service strictly on the basis that no legal advice is given to either client at these early stages. It is not clear whether such practice is permissible under the existing codes of conduct.

If professionals were to set up as process brokers, however, then there may well be a role for them to meet with a couple and explain the range of options available to that couple. The process broker essentially positions themselves as a new gatekeeper to the full range of family separation service providers and can signpost the couple to processes and practitioners that best fit the couple's wishes, needs and values.

To be effective they would probably need to forego their own legal practice if they are to reassure clients that their referrals and recommendations are not coloured by self-interest.

There are already lawyers in the United States who fulfil this purpose. It is not thought that any such service yet exists in England and Wales, but their emergence is inevitable via the private or state sector.

Alternatively, a couple who are still communicating well might simply discuss between them which solicitors each of them will instruct. We consider what criteria the couple might want to look out for when selecting a collaborative lawyer in the following chapter.

5.18 WHEN THE COUPLE IS NOT TALKING

We should not assume that just because a couple cannot communicate directly between themselves that they cannot work together with the assistance, safeguards and boundaries, provided by the collaborative law process.

For that reason, one estranged partner might still decide, as they research their separation options, that they would like to use the collaborative process to resolve outstanding issues.

Again, it is to be hoped that an increasing availability of collaborative law material online will result in an increased profile for collaborative law services. In years to come, it will not be the role of the person who has 'left' to bring collaborative law to the knowledge of the partner they might have just parted from.

It is a tall order to expect someone who is likely to still be hurting to agree with anything that their now, or soon to be, ex-partner is suggesting. As collaborative law raises its profile it will hopefully be an automatic consideration of the partner who has been told that the other is leaving.

If both partners remain living within the same home together, then it might be possible to leave material around. One partner might want to use material from collaborative law websites or elsewhere. Consider sending copies of leaflets or brochures through, either by e-mail or within a gentle letter such as the one below.

> 'Dear …
>
> I have been thinking about how we are going to sort out matters relating to our children/the house/finances/the divorce/our future …
>
> It is important to me, and I think to our family, that we are able to work things out together without acrimony and getting caught up in blame. I have been reading about something called Collaborative Law which can enable us to talk things through in a series of meetings.
>
> You and I will each have our own lawyer but they act very differently to normal lawyers. For example, they have to commit, with us, that if they advise us to go to court to sort things out, or if we choose to go to court ourselves, that they have to stop acting for us. That means that there is no point in them threatening court action.
>
> They will also commit to helping us reach an agreement.
>
> Let me know what you think. I have enclosed some details [*either hard copies or weblinks*].

Regards …'

5.19 ENLISTING THE HELP OF THIRD PARTIES

It might be possible to enlist the help of a third party to encourage the other partner to agree to collaborate, although this is less than ideal. Care needs to be taken not to get embroiled in politicising, that is to say, trying to win supporters to push through a preferred argument or proposal. Such steps if dealt with badly can come across as being manipulative and can diminish trust. The key has to be being discrete, and not becoming a nuisance in efforts to get a message across via any third party.

One partner might consider asking another family member to help, particularly a sympathetic member of the partner's family. That person might then agree to speak to the partner, or deliver appropriate literature about the collaborative process.

If the selected family member can see merit in the collaborative process then they can be left to communicate the value that they perceive to the other partner.

It might be that a doctor, counsellor or other trusted professional might intervene. For example, financial advisers, accountants, and even school staff are finding themselves increasingly falling into informal intervener roles. They may well be able to pass on a message or try to influence the other partner.

5.20 THE CHALLENGE FOR LAWYERS WRITING TO THE OTHER PARTNER

The experience of receiving a solicitor's letter should not be underestimated.

As soon as the letter arrives, whether expected or not, the recipient's pulse will be racing. The letter, still unopened, represents a new shift in the dissolution of this relationship. That dissolution may or may not be welcomed by the recipient.

The letter may remain unopened for several days. As long as the contents remain unknown then the recipients may well fear the worst, compounding their agitation and reluctance to read it.

Even before the recipient therefore reads what the lawyer has to say, the letter has already been framed in terms of anxiety and bad news. Several narratives will have been created by the receiving partner which will cloud the interpretation of what is written, whenever it may be read.

There are relational and discursive issues at play here as well.

The solicitor's letter represents the intrusion of an external party and one that is deemed to have some authority. The recipient may be feeling disempowered as a result.

It represents a shift in the relationship between the partners themselves. Whereas once the partners would discuss matters between themselves, the other partner has now chosen to discuss the matters with somebody else.

The recipient has also been thrown into a relationship with this lawyer who has now written to him or her. How they interpret that relationship will colour how the letter is interpreted once it is read.

The dominant narratives raise their heads once again and the receiving partner will perceive the receipt of this letter in the terms of whatever narrative they are currently swayed by.

It is likely that many people, particularly if not accustomed to dealing with lawyers, will automatically vest the lawyer's role in the relationship as being the dominant or more powerful party. The lawyers, after all, have the education and training to enable them to operate in these separation or divorce issues. They will have varying degrees of experience and they may carry a certain reputation before them.

The recipient of such a letter might therefore be feeling less dominant, weaker and threatened as a result of receiving that correspondence.

Bearing this in mind, consider the following conventional letter that might be written as an opening letter to a client's partner, and then compare it with the very different letter that follows.

5.21 TWO DIFFERENT LETTERS INVITING COLLABORATION

First letter

'Dear Mr Jones

I have been instructed by your wife with regards to your marriage and recent separation. I want to reassure you that my client has made it clear that she would like to resolve all outstanding matters relating to housing, financial provision and contact with the children, Harry and Ben, amicably.

I have advised my client about a new process called collaborative law and she has indicated that she would like to use this process, together with you, to resolve the matters listed above.

I enclose a leaflet setting out how the collaborative law process works, for your information.

Given the circumstances I am obliged to recommend that you do seek independent legal advice. If you would like to use the collaborative process then it will be necessary for you to instruct a collaboratively trained lawyer to represent you. There are details of other local collaborative practitioners who you should contact within the leaflet.

I look forward to hearing from you, or your instructed solicitor, within the next 10 days.'

Second letter

'Dear Mr Jones

I have been approached by Julie following your recent separation. Julie has been keen to ensure that I understand that it is a priority for her to be able to work with you through the issues that will need to be addressed between you over the coming months.

One of the initial concerns will be just how Julie, yourself, your own adviser and I can best communicate with each other, to enable you to resolve everything. I recognise that solicitor's correspondence can often be seen as being antagonistic, and it is certainly an expensive and slow process.

You may feel, as Julie does, that it would be better to have a series of round table meetings to enable you and Julie to talk things through. This can minimise the risk of misunderstandings arising between us and can also save time.

If you agree that this sounds like an option that you would like to explore then we can go on to consider how such meetings would be facilitated. For example, how would the meetings work, how would we conduct ourselves within them and how can we ensure that all relevant issues are discussed?

There is a process called collaborative law. I have enclosed a leaflet about that for you to consider. You might feel that this will enable you and Julie to work together and avoid much of the need for correspondence and court proceedings.

I reiterate that Julie does want to find a way to work together with you. Between us, we have an immediate opportunity here to start building that co-operation as we determine a way to enable you and Julie to resolve matters.

May I recommend that you discuss this letter with your own solicitor? If you have not yet instructed a solicitor then I recommend that you instruct one who is affiliated with Resolution, the family solicitors group. You will see that the enclosed leaflet also lists several other lawyers, who are known to me and practice collaborative law locally. I look forward to working together with you, your instructed solicitor and Julie with a view to having these issues resolved.'

5.22 Compare and contrast

These two letters are very different in tone and effect although the subject matter and suggestion remains the same.

Collaborative law is put forward in both letters as a solution.

In the first letter it comes across as a prescribed method. The letter says that 'I have advised …' and such a recommendation can arouse resistance.

Furthermore, the letter then goes on to talk about the process, this thing called collaborative law, without any preamble as to what it is or what the point of it is. We considered the need to ensure that our first meeting with our own client follows a natural progression where we first disrupt assumptions about the lawyer-client relationship before preparing the ground so that collaborative law now fits and makes sense. So we need to do the same when first inviting the client's partner.

The letter also slips into referring to Julie as 'my client'. This causes problems as Julie is reduced to being a client, as opposed to the person that her husband has a pre-existing relationship with.

The letter concludes with an authoritative tone. 'I am obliged to …', 'It will be necessary for you to …' and 'You should …' before finally concluding with a request for a reply within a fixed number of days.

The recipients of these letters will dislike being told what to do. Remember, letters such as these represent an intrusion into the family dynamic. The family members themselves hold varying degrees of authority and influence within the family relationships. We need to be more respectful of that.

The content and aim of the second letter remains the same. It proposes collaborative law as a desirable process to the other partner, but the way in which the letter works is very different.

We have stripped out the depersonalising references to 'our client'.

We explicitly recognise that lawyer's letters can be seen as being antagonistic.

We then offer back the power to choose an appropriate method of dispute resolution to the recipient and Julie by way of respectful curiosity. 'You may feel, as Julie does, that it would be better to have a series of round table meetings …' We are effectively enlisting the recipient to join us in designing a dispute resolution process that will meet their goals, as well as those of their partner.

In the first letter, the lawyer has simply presented collaborative law as a solution that he or she has advised the client upon. In the second letter we have not even mentioned collaborative law yet. Instead we are asking whether the recipient agrees with his wife that a 'series of round table meetings' would be helpful.

It is hard to imagine this more generalised proposal being rejected where there remains a degree of civility and effective communication between the couple.

Having raised the issue of round table meetings we can go on to mention collaborative law by name. Now, when we mention collaborative law, it is given meaning and context by the preceding talk about round table meetings and benefits.

In the first letter, we have simply handed down a collaborative law solution without explaining what it is, other than to say 'Read the leaflet'. Here, in the second letter, we flag up the benefits of a certain approach before we name the process. The benefits are the primary consideration, not 'What is collaborative law?'

Couples are unlikely to be interested in what the process is called. They are only going to be interested in how it will help them and their family.

The phrase 'Between us, we have an immediate opportunity here to start building that co-operation' is a deliberate attempt to break down any notions that the recipient's partner's lawyer is calling the shots or is exploiting a dominant relationship position.

This paragraph puts the lawyer on an equal footing with the recipient and demonstrates a collaborative intent not only once the process has started, but even in these preliminary discussions about how we are going to even have those discussions. This letter itself is already modelling a collaborative approach.

The letter acknowledges that the lawyer cannot do it on his or her own and invites the other party to join us in collaboration, instead of telling the other person that 'This is how we are going to do it'.

The second letter therefore aims deliberately to break down presumptions of hierarchy or power relationships between the lawyer and the recipient. It does this by adopting an invitational, curious stance that inquires about the recipient's wishes, acknowledges their possible thoughts about receiving a solicitor's letter in the first place and concludes by asking the other party for their co-operation, rather than assuming it will be given. The letter is still directive and clearly guides towards the collaborative process, in accordance with the client's wishes, but it is not dictatorial.

5.23 OTHER APPROACHES TO ENGAGING THE OTHER PARTNER

5.24 Online resources

We have seen how poorly collaborative law ranks among general divorce inquiries on Google. That does not need to be a problem when one party has already instructed a lawyer, and that lawyer is trying to obtain the partner's agreement to collaborate.

The other partner will research the lawyer that has been instructed by their spouse.

Active collaborative lawyers will ensure that their website profiles clearly link through to details about the collaborative process. These details might be on an article that you have written and posted online. Maybe you want to link through to a blog or even a short video that you have recorded and put onto the internet.

Writing articles or creating other online resources about collaborative law will demonstrate to anyone who looks that you are committed to collaborative law work. You also have the opportunity to spell out the benefits of a collaborative law approach. Another advantage of preparing and distributing such material online is that you increase your chances of being found by search engines when people look for 'collaborative law' or 'collaborative practitioners', or other words and phrases that separating couples might search for online.

5.25 Printed literature

Be sure to have presentable and clear literature to send out or provide to clients for their own use or for them to pass on to their partners. It should be possible to work together with other local collaborative practitioners to develop a leaflet or flyer which contains details of all the contributing practitioners. Speak to your local collaborative practice group about this. Also speak to other neighbouring groups who have already got a local practitioner's leaflet to see if you can borrow their design as a template.

5.26 Promotional work and networking

Collaborative practitioners should continue to promote their collaborative work publicly as well. This can include press releases, either as individual firms or again on behalf of all local practitioners.

Also consider approaching local referrers of work and those other professionals who serve the public. Think about presenting to:

- Accountants
- Financial advisers
- Estate agents
- School staff
- Spiritual and pastoral care providers
- Coaches
- Therapists
- Counsellors
- GPs

By raising your profile and that of collaborative law amongst your local communities you will make it more likely that your client's partner will either have heard of collaborative law or may be considering it as an option already.

5.27 WHAT IF THE OTHER PARTNER HAS ALREADY INSTRUCTED SOLICITORS?

If the other party has instructed solicitors then much will depend on whether that lawyer is collaboratively trained or not. If they are then the client's permission to telephone should be obtained as soon as is reasonably possible to explore whether or not they agree that matters can be dealt with under the collaborative process.

If the response is 'No' then you might want to explore why not.

What concerns are there that apparently make collaborative law inappropriate?

Is there anything that can be done to overcome those reservations?

How might the couple be able to build upon trust or communication levels, in order to get them to a stage where collaborative law can work?

It might be possible for the lawyers to meet to discuss the situation and their client's respective concerns, provided of course that the client has authorised them to do so.

5.28 WHAT IF THE PARTNER'S LAWYER IS NOT COLLABORATIVELY TRAINED?

If the client's partner has already instructed a non-collaborative lawyer then it is likely that efforts to use the collaborative process will fail.

While all lawyers have a duty to advise clients on all possible methods of dispute resolution, even those that they do not offer themselves, it seems unlikely that the non-collaborative lawyer will be motivated to deliver a compelling argument as to why the collaborative process should be taken up. Doing so would mean the lawyer having to terminate their working relationship with that particular client and refer him or her on to a local competitor.

If there is a collaborative practitioner within the non-collaborative lawyer's own firm then there might be the possibility of an in-house referral.

5.29 CAN YOU COLLABORATE WITH A NON-COLLABORATIVE TRAINED LAWYER?

The rules surrounding collaborative law, set up by Resolution, the family lawyers group, stipulate that collaborative lawyers can only collaborate with other collaborative lawyers. The reason for this is that the training goes to great lengths to switch lawyers' attitudes from litigious-based arguments where they are putting forward positions and moving instead to a more interests-based negotiating model. This paradigm switch cannot be underestimated and is explored more deeply in the next chapter.

5.30 THE DIFFERENCE BETWEEN ROUND TABLE MEETINGS AND COLLABORATIVE LAW

There is no reason why a collaborative lawyer and a non-collaborative lawyer could not have a series of round table meetings within which matters are negotiated.

The distinctions between round table meetings and collaborative meetings need to be kept in mind though, not least the absence of the disqualification clause.

The disqualification clause within collaborative work means that parties are free to explore a full range of options. It enables us to move away from position-based arguments, where we might go in high on our opening bid with a view to being reduced when it comes to a later compromise, to a much deeper conversation about what is important and how those interests can be met. This is enabled in the knowledge that if one party, or either of the lawyers, elects to go to court then both lawyers have to stand down.

As a result there is less incentive simply to have a round table meeting designed to flush out information about the other party's preferences and priorities.

Another benefit of the disqualification clause is that the lawyers are stripped of their nuclear option, the 'See you in court' option. As a result there is little point in adopting hard negotiating tactics. This in turn emphasises the collaborative nature of the negotiations as opposed to simply setting up competing arguments demanding various provisions.

5.31 SUMMARY

This chapter has started to look at the first steps in pulling together the collaborative team.

We also explored who is involved and who is responsible for determining who is in and who is out of the team. We looked at some of the existing tensions as to whether it is the lawyers who are responsible for bringing the team together or whether the couple should retain control over the composition of the team.

We will explore the nature of the other potential team members in the following chapters.

We then went on to explore how we can best encourage the other partner to agree to use the collaborative process.

The biggest aid to increasing participation by the other partner will be for the collaborative law profession to raise its profile to such an extent that collaborative law simply becomes as normal a consideration upon separation as applying to the court, mediation or going to a conventional lawyer. Collaborative practitioners need to do much more yet to get their profession's profile to the required level.

In the meantime, clients might try to encourage their partners themselves where there is a higher degree of ongoing communication.

Alternatively, it might fall to the first instructed lawyer to encourage a collaborative approach from the client's partner. We explored the importance of recognising the response of the client's partner at a deeper level than that which is often seen.

Finally, we explored what can be done when trying to encourage a partner to collaborate when they have already instructed their own lawyers, and the very real difficulties that can be presented when a non-collaborative lawyer has been chosen.

CHAPTER 6

CHOOSING A LAWYER

This chapter explores what criteria clients might use to select a collaborative lawyer. We look more closely at the spectrum of commitment to collaborative work and community. This spectrum serves to demonstrate the criteria that might be used by clients and also provides suggestions on how a collaborative practitioner can develop their own profile and position upon that spectrum.

Before we can do this we need to understand the radical differences, or the paradigm shift, between conventional and collaborative practice better.

We will go on to look at what difference a lawyer's investment in their own training and personal development can make and how that can help the collaborative process work more effectively. We will also look at the question of cost and to what extent that should be used a reason for selecting a collaborative lawyer.

6.1 INTRODUCING THE COLLABORATIVE SPECTRUM

There is a wide spread of lawyer commitment to collaborative practice, ranging from the least to the most committed and highly motivated of practitioners. We can give them labels, as arbitrary as those labels are, to help distinguish between different levels.

We shall use the following labels:

- Sceptic
- Passive
- Active
- Leaders

6.2 REVISITING THE 'PARADIGM SHIFT'

The paradigm shift presented by collaborative law training is felt to varying degrees by those who access it. The four classes of practitioner in the spectrum above reflect the extent to which that shift is experienced and also acted upon by each of the practitioners.

A paradigm is a fundamental world view.

It is the lens through which we observe and make sense of our world.

When paradigms shift we are talking about much more than just changing our mind about an issue. Our whole understanding of the issue, and maybe ourselves, changes profoundly.

This paradigm shift can be further illustrated by looking at how it impacts on several aspects of conventional and collaborative law practice.

6.3 The shift from litigation to collaboration

The shift from litigation to collaboration is the most obvious shift.

Conventional legal representation, even if it is not carried out through the courts but through conventional round table negotiations or lawyer correspondence, is all carried out against the backdrop of court intervention.

Consider, for example, the frequent question that a client will ask of their legal advisers:

> 'What am I entitled to?'

This will have various variations such as 'What would I get at court?' or 'What is normal? Is this normal?' and relative questions such as 'Is this a good deal/bad deal?'

Legal advice is shaped inevitably by reference to previously decided cases. We use the core consideration of 'what would a judge say?' as the acid test in considering reasonability. The question of entitlement becomes 'what do I think I can get at court?', bearing in mind what relevant statutes and previously decided cases suggest.

This court-led paradigm dictates legal advice and representation. It colours every aspect of the conduct of that case, including non-court based correspondence and dialogue. With every letter that is written we tend to position ourselves with one eye on the judge at a final hearing that may or may not happen. We think that we might want to be able to refer back to these letters to show how reasonable we had been should we ever get to that final hearing. We imagine throwing down our letters and saying 'I told you so', perhaps in support for a claim for costs.

The paradigm shift from litigation to collaboration is massive.

At its most basic level, the shift is a move from 'what would the judge say?' to 'what needs can we identify and agree upon, and what solutions are this couple able to create for themselves and their family?'

6.4 The shift in working relationships between lawyers

There is also the challenge to lawyers to trust one another and work together to enable their respective clients to operate and move through the collaborative process. It is remarkable how great a shift that feels at the outset and how profound a change it has on lawyers' working relationships.

Many lawyers report that once they have worked collaboratively with another lawyer, future dealings with that lawyer, even in conventional cases, is markedly different.

6.5 The shift in the role of law in collaborative law

The role of law in collaborative law is sometimes over-simplified to suggest that law, whether written in statute or established through previously decided cases, has no bearing to collaborative law. That is not entirely accurate.

The questions 'what am I entitled to?' or 'what would a judge say?' are still asked by both clients and practitioners. These questions demand answers.

Simply refusing to provide an answer would create an artificial blind spot, with both lawyers shrugging their shoulders dumbly and saying 'I don't know'.

The alternative response 'it's not relevant' is just as unconvincing.

The questions demand to be answered and we need to be careful how we do so.

In a conventional case, one lawyer would present an interpretation of the relevant law, current facts and previously decided cases in a way that both shapes and supports the argument being put forward.

The other lawyer does the same, but with a different interpretation. They might have selected different facts or given the available facts a different weighting.

There always has to be a difference in how the law and facts are interpreted. If there was not, there would be no dispute in the first place and matters would be settled.

Each lawyer in a conventional case then sets about attacking the other lawyer's position. That is done by revealing new law, new cases or new facts. We see attacks on credibility in what they or their client is saying. We see allegations being made with a view to proving that one party or another is unfit or no longer entitled. The destructive nature of the process is obvious.

The danger is that by having identified some kind of authoritative accepted standard, namely what a judge might say, even if it is flawed and not necessarily appropriate for this family, we narrow the considerations of the parties. The risk is that whatever solutions the couple go on to explore between themselves, that they will always be measured against that standard. The determining factor risks being reduced to the following consideration:

> 'Is this a good deal, comparatively, or would I be better off in the court?'

Within the collaborative paradigm, these legal discussions take on a very different angle.

The simplest iteration of legal debate within collaborative law would suggest the following:

> 'My interpretation of what a judge would order is x, y and z but that outcome could not be guaranteed. Even if it was ordered then this is only one of any number possible outcomes that you and your partner might otherwise design between yourselves.
>
> You are not obliged to follow what a judge might order. That outcome is what was imposed by a court onto a particular family in a previously decided case based on the facts relevant to that family.
>
> We cannot tell if it was the best outcome for that family and we cannot guarantee it is the best possible outcome for your family.
>
> You are the experts in your own family and what you want and need. You can decide your own outcome for yourself and we are here to help you to do that.'

Legal entitlement in this situation is reduced to just one of a range of possible, plausible outcomes. The influential power of legal entitlement is said to be diminished accordingly.

A more helpful but more complex response might see each lawyer presenting an interpretation of what a court outcome might look like. There is an irony here in that the lawyers can demonstrate their disagreement, ironically, by agreeing with one another, as we shall see below.

One lawyer, for example, explains that his or her view is that the court would order maintenance payments from the husband to the wife for the rest of her life, or until she remarried or by further order of the court.

The other lawyer responds that in his or her opinion the court would actually want to see maintenance reducing down to nil within a few years.

The lawyers are seemingly in disagreement and this is, so far, entirely conventional. The lawyers can further flip the 'what am I entitled to?' paradigm further by agreeing with the other lawyer's point of view.

> 'I think that the court would order maintenance to stop within a few years. That is not to say that I think your argument is wrong and I can think of several cases where I have successfully argued what you have just set out.
>
> You see, the law does not give us black and white answers. I can think of cases where I have had both of these outcomes.
>
> Bearing that in mind, where does that leave the question of "what would the judge order?"
>
> Is it still as important do you think? Why don't we keep on exploring options and other ways that we might both be able to make sure that everyone is provided for? We can always come back to this once we have reached an agreement …'

We can see again how the rule of law is being disempowered.

Legal entitlement and the idea that there is a definitive legal outcome are shown to be unreliable. By seeing lawyers disagreeing and yet accepting that the other lawyer's argument also has merit, couples can conclude that trying to second-guess what a court outcome might be is not going to help them. The conversation is brought back to the couple's own considerations and what is important to them. At the same time, the lawyers are modelling excellent collaborative skills in being able to disagree with one another and yet keep on communicating.

6.6 The shift from delegated responsibility to mutual responsibility

As we saw when we were exploring conventional discourse and the discursive relationships and positions between lawyer and client, clients can delegate many things to the lawyers, including responsibility. A dynamic quickly arises in many cases where the client hands over their concerns to the lawyer with the simple request, 'fix it'.

When a lawyer is retained to work on behalf of a client, the client is delegating responsibility for all kinds of unexpected things, including communication between the couple (the 'speak to my lawyer' syndrome), case management and decision-making.

Following this delegation, clients can very quickly develop a sense of victimisation within family law cases. They will have become the passive subjects of the conventional law process, subjected to the advice of the lawyer and the requirements placed upon them by the relevant court process.

From this stance many clients feel that almost every letter or utterance by the other lawyer is a direct attack upon them and that everything said has to be denied. Indignation is a common feature of family law cases, fuelling the client's exhortations to their lawyer to respond in certain terms. The lawyer, because of professional courtesy, is forbidden from doing so and the client's frustration is exacerbated.

Even if the client intervenes and does something such as e-mailing their partner with some choice thoughts, or removing funds, then they will still cling to a passive position and claim that they had no choice. They were forced to it by their partner's actions or they felt they had to do something to protect their position.

This kind of comment demonstrates the absence of self-responsibility.

As conflict within the litigation spirals it becomes impossible for clients to recognise how they themselves may be contributing to those escalations. The lawyer has a very limited mandate from the client to point that out to them.

The lawyer, under the paradigm of delegated responsibility, is instead expected to manage the case, come what may.

The paradigm shift in this field of delegated responsibility is very powerful.

The couple are positioned within collaborative practice as central components of the dispute resolution process itself. They are the experts within their family, rather than external experts reporting on them. They are expected to speak out and speak up to contribute within collaborative meetings. Contrast this with the clients' experiences of sitting at the back of a courtroom at procedural hearings making next to no contribution.

The clients take back responsibility for managing themselves, their relationship and the tone of the collaborative communications. Again, contrast this with conventional cases where the lawyer is largely primed with information and then delegated the responsibility to communicate with the estranged partner's lawyer, as if in their place.

This reclaimed responsibility might not be easy for the couple to take back.

They may struggle in the early stages, for example, to contribute within a collaborative meeting. Many meetings are less than effective because one partner does not speak up about the things that they need to communicate.

The conventional expectation that 'my lawyer will speak for me' simply does not work within collaborative meetings.

Lawyers do need to watch out for this to make sure that non-contributing members of the process are encouraged and enabled to speak for themselves. In this sense, the lawyer's role is one of facilitating their client's contribution rather than making that contribution on their behalf.

Clients also need to recognise their retained responsibility to look after themselves.

If depression arises or if they are feeling victimised then they need to speak out. If discussions are running ahead too fast, or if they disagree with something that someone has said, then they must let the others in the room know. There is no point in meeting together to explore settlements if, when a proposal is put forward with which we do not agree, we say nothing.

Delegated responsibility within collaborative law shifts to a shared responsibility.

The lawyers are still responsible for their own client's best interests.

That is not necessarily limited to getting the largest financial settlement or the greatest amount of time spent with a child of the family. The clients may now bring a whole range of other criteria that they want to honour in reaching what they would see as being the best possible outcome. In addition to ensuring that their client is not at risk, the lawyers also need to look after their client's own ability to contribute within the collaborative sessions in a way that is appropriate.

Lawyers have the responsibility to indentify the need for other practitioners to play a role within the collaborative team, perhaps to help with mental health issues, repetitive patterns of communication leading to conflict, or to address a need for education on, say, financial matters.

It remains the client's responsibility to hear those recommendations and, where appropriate, take up or reject those recommendations. Again, we explore those other professionals' roles in the next chapter.

We see a much greater balance in the responsibility for completing tasks. Conventionally this is delegated to the lawyer to complete forms, make inquiries, obtain updated information and the like. Within collaborative law, it is expected that the clients, either as a couple or as individuals, will take back much of this work for themselves. It is one of the ways that costs can be saved by collaborative working.

This mutual responsibility can also result in day-to-day contact between the client and the lawyer. In conventional cases there is a much higher level of incidental contact, by telephone or mail, between lawyer and client.

This is seen particularly where the relationship between the couple has been completely abdicated to the lawyers. It then becomes necessary, in the eyes of the clients, to report every incident, e-mail and comment for the lawyers to consider and respond to if required.

That incidental contact between the client and the lawyer reduces with collaborative law cases. Instead it is often possible to diarise exactly when the client and lawyer will come together to prepare for the next meeting and, again, to diarise that meeting. Often, no contact is required between these sessions, which can also result in reduced costs.

It may well be that the reason for this is that the collaborative client is empowered by retaining a greater degree of responsibility. There is no need, accordingly, to report every incident or comment to the lawyer. The client is more able to process the issue themselves, or take up a referral to one of the other professionals, such as a counsellor, coach or therapist, who can better assist with coping mechanisms and resilience.

6.7 The paradigm shift and the lawyer's role

The lawyer's role within collaborative law is close to being unrecognisable compared with their conventional role.

The mandate or the permission that a lawyer is granted by a client in collaborative cases is broader than that granted within conventional cases. As a collaborative lawyer we have authority to point out or share observations and recommendations with our clients that otherwise we would not.

This in turn stems from the greater parity between lawyer and client, as opposed to the conventional hierarchical relationship. That sounds like a contradiction but it is not.

In conventional cases the lawyer can find him- or herself cast as the client's protector. They are responsible for ensuring the client's outcome and that the client's case is put forward and argued. The lawyer becomes a sounding board and support. The lawyer within a conventional family law case can still recommend mental health professionals who the client can work with, but if the client refuses to do so, the lawyer does not have the mandate to repeatedly encourage the client to seek help.

The conventional hierarchical relationship is built upon the lawyer's position as a lawyer. His or her authority within the relationship is absolute but it is built upon a very narrow plinth, namely that of legal expertise. It can be difficult in those circumstances for a lawyer to speak with authority outside of that scope. If a lawyer advises upon the need to access therapeutic help, for example, it becomes very easy for the client to reject the same with the comment 'that is not what I pay you for'.

Where the collaborative lawyer and client share responsibility and status these problems of authority are not present. Such recommendations are now not handed down from the narrow authoritative plinth that the client had previously placed the lawyer on. Instead, the suggestion is made from a more equal partner within this resolution process. The collaborative lawyer need not be the client's friend, but the relationship has a more social and compassionate concern than the conventional.

Within collaborative law the lawyer can perhaps best be seen as a facilitator. He or she facilitates the client's own contributions to the process. He or she facilitates the legal effectiveness of any agreed outcomes through the drafting of documents and filing of the same at court. He or she also facilitates a fuller consideration of all relevant issues by the collaborative team, and helps to anticipate future problems.

The lawyer should also take an interest in equipping the client with the skills and awareness to be as effective as they can be within the collaborative process and meetings. That might require a degree of training, coaching or other therapeutic support being made available to the client.

An example can illustrate this paradigm shift in the role of the lawyer.

In conventional cases, if an allegation is made against a client then the instinct as lawyer is to rebut that allegation strongly and to defend and protect the client. It is likely that the lawyer may feel some indignation on behalf of their client. It is seen and felt in the 'how dare they?' response when receiving the other partner's statement or latest questionnaire. Lawyers can take on the conflicts that the clients themselves have been wrestling with within their marriages and relationships.

The conventional inter-lawyer relationship quickly comes to resemble that of the couple themselves.

Collaborative lawyers are able to deal with similar situations very differently.

If an allegation is raised within the collaborative session, for example, the effective collaborative lawyer will seek to elicit more information rather than shutting down the allegation.

> 'That's interesting. What makes you say that?'
>
> 'When this thing happened, what did that say to you? What were your thoughts?'
>
> 'What that happened, what do you think his/her intentions might have been?'
>
> 'I wonder what other explanation there could there be for that. What do you think?'
>
> 'What do you think Jayne's explanation for that sequence of events might be, if I were to ask her?'

This approach can quickly exhaust the frustration behind such allegations without simply rejecting the allegation. The person making the allegation knows that their concern has been heard and, very often, can determine alternative explanations for themselves. The effective collaborative lawyer will be able to resolve these conflicts by respectful and curious questioning of the complainant, rather than seeking to discredit them. A great deal of antagonism, costly correspondence and even court applications can be avoided as a result.

6.8 A crisis in lawyer's identity

The paradigm shift in the role of the lawyer can be startling for many newly trained collaborative lawyers. There is anecdotal evidence that suggests that many lawyers can feel a real crisis of identity following training, so much so that a degree of post-training coaching or mentoring might be helpful to ease that transition.

The challenges that lawyers typically express relate to a sense that their new roles, as greater equals with their newly empowered clients, reveals the potential damage that the conventional practice can do to a family passing through separation.

Some lawyers turn their back on conventional litigation such is their disaffection with their former roles and the impact those roles had on families. They determine for themselves only to carry out collaborative work, possibly supplemented with mediation work. Other lawyers are more comfortable straddling the roles of conventional and collaborative lawyer. No judgment is made here against either camp.

Having explored this thing called the paradigm shift more deeply we can now move on to look at the spectrum of collaborative lawyers. The lawyer's positioning on the spectrum may be determined by the extent to which they have experienced the paradigm shift. This is not to say that one lawyer understands collaborative practice more than another, but rather that the more collaboratively motivated lawyer has been more moved by that paradigm shift.

6.9 BREAKING DOWN THE COLLABORATIVE SPECTRUM

To recap, the four arbitrary classes of collaborative practitioner can be labelled as follows:

- Sceptic
- Passive
- Active
- Leader

6.10 The sceptic

The collaboratively trained sceptic is the least motivated of collaborative lawyers. They are not inclined to practice collaboratively at all, which in turns begs the question, why did they invest a substantial sum in their training and time out of the office? There is an easy answer to this question.

They did not want to miss out.

As collaborative practice really gathered momentum in England and Wales from 2004 onwards there was a view that many cases would be resolved using this methodology. Such was the demand for lawyers wanting to be trained that there were lists of lawyers waiting to be notified of slots on collaborative law training courses becoming available.

The spectre of a client leaving a non-trained lawyer's office and going up the street to his or her collaboratively trained competitor took hold. The picture struck fear into the hearts of some lawyers who had no wish to change their practice whatsoever.

Those sceptics therefore obtained the minimum level of training and the appropriate certificate just in case a client one day asked for it.

The collaboratively trained sceptic does not invest time in any further skills training, nor in creating or contributing to the preparation of collaborative law material, whether for the collaborative practice community or for their own firm.

They tend not to attend local collaborative practice groups, even though Resolution, which governs the collaborative law profession, has a mandatory requirement to attend at least four such meetings each year.

As Resolution starts to enforce this requirement, then the collaboratively trained sceptic will face a tough choice. He or she might attend regional collaborative law meetings in order to keep that certificate valid. If they do so, then they will probably not contribute to the discussions at all, save to wince or argue. Some of these practitioners can be quite vocal in denouncing collaborative law altogether.

In many cases, these practitioners would rather let their collaborative training lapse completely.

The problem is that the unwitting client might inadvertently instruct the collaboratively trained sceptic. In that event there is the risk that they are served well enough by him or her in conventional terms but are denied access to the collaborative process that they had meant to employ.

There is no doubting the legal skills of the sceptic, but as far as a collaborative process and outcome goes, the results are likely to be far less than optimal.

The lack of additional skills training, curiosity or commitment makes it likely that conventional legalistic and litigious behaviours will surface, even within the collaborative meetings. Expect to see positional arguments being put forward, practitioners adopting a representational, protective stance over their client, speaking for their client, rather than facilitating their client's full contribution.

The other collaborative practitioner will have a real challenge in managing the collaboration between the lawyers. It might prove difficult, if not impossible, for example, to get the sceptic to sign up to the participation agreement, or to moderate his or her behaviour within meetings. Correspondence passing between lawyers is likely to be conventional in tone and impact.

Thankfully this class of collaborative practitioner lies at one end of the spectrum. It is a small class and will be dwindling. Greater compliance checks by Resolution on regular updating training and practice group attendance will see more sceptics choosing to let their minimal collaborative credentials lapse.

Other sceptics may well become motivated to move up the spectrum by demonstrating greater commitment to the collaborative law process and community. That might come from a conscious decision to do so, by way of refresher or advanced skills training, or it may simply be a result of the paradigm shift falling into place.

6.11 The passive collaborative lawyer

The passive collaborative practitioner offers a full range of conventional, collaborative and possibly mediation practices.

When meeting with clients they will likely continue to work as they have always worked, completing a printed or a well-worn mental checklist of dates and details. They will go so far as to point out to the client that they can choose mediation, collaborative law, or they offer to send out a first letter to the other party and 'see where that leads us'.

The passive practitioner may still be waiting for their first case, despite having trained many months ago. In some cases, this lack of collaborative work can drag the passive practitioner down the spectrum into disaffection and scepticism. The passive practitioner's natural curiosity leads them to attend their local practice group's meetings occasionally and this can serve as an antidote to scepticism.

These practitioners understand the collaborative process. They long to do more collaborative work but they might be passive in communicating the collaborative values and benefits to their clients, particularly at the crucial initial meeting. Their passivity is also to be seen in not taking deliberate steps to adjust their practice in response to the collaborative process but to carry on as they always have done.

Learning and development professionals talk about the 'internalisation' of a concept. This happens when an idea has so impacted on a person that it becomes part of their outlook. Passive practitioners, while understanding collaborative law, are not so moved by the notion as to allow it to shift all aspects of their practice. Where clients then come into contact with them they will have their own paradigms and preconceptions of what it means to work with a lawyer. When they have nothing to challenge or replace that coming back from the lawyer then the conventional presumptions will be allowed to dictate the approach taken within that case.

Practitioners at this level will provide a very good collaborative experience when asked to do so. They are likely to conduct themselves in accordance with the collaborative process. There might be lapses into conventional practice such as positional bargaining or more partisan methods of communicating and framing issues, simply out a lack of habit forming practice.

It is easier for the passive practitioner to move up the spectrum. A good way to do so will be to continue their collaborative training. This does not need to be necessarily expensive. Reading through readily available and inexpensive negotiation and communication books can certainly help. The reading list provided in Appendix 4 would be a good starting point.

If the passive practitioner further consolidates their reading with their own writing, perhaps by way of an online blog or by writing other materials to post online, then they will better internalise the collaborative way of working. That will then be apparent as they talk with new clients and referrers. What is more, they will be enhancing their own reputations as active collaborative practitioners, contributing more fully to the canon of writing and other material available online or within the practitioner's community. Posting material online will also raise the practitioner's profile so that they increase their chances of being found when clients are searching for suitable representatives.

6.12 The active collaborative lawyer

The active practitioner has a steady number of collaborative cases. He or she is passionate about the potential that collaborative law has in enabling couples to separate with the least damage. That enthusiasm is clear when they meet with new clients. They usually still offer the full range of dispute resolution processes to clients, together with the pros and cons of each, but their natural preference for collaborative law reassures clients that this process is safe and worth exploring.

The active practitioner is a frequent attendee at local practice groups. He or she works hard at developing links with other collaborative lawyers and other financial and mental health professionals. They will often be spontaneous in promoting collaborative law in almost any forum and need little prompting to write an article for the local press, host a presentation for local referrers or update an online blog or website.

They regularly read around communication and soft skills issues and have a desire constantly to increase their skills. They invest time and money in pursuing further insights and training, possibly from diverse sources and other professions.

They are likely to be more comfortable about the reform of family law and may be quite vocal in expressing concerns about the damaging nature of conventional family law.

Once a practitioner reaches a certain level of collaborative practice then there is probably no going back. After all, collaborative lawyers will frequently state 'I wish I could do more of my cases collaboratively'. No-one ever says 'I wish I could do more of my cases litigiously'.

6.13 Collaborative leaders

Collaborative leaders are everywhere. It is likely that each local collaborative practice group has one or two leaders. They will probably be the people who were so motivated as to set up the practice group in the first place.

Other leaders can be identified from their roles in pioneering collaborative law approaches within England and Wales or their region. They may well have a reputation that spreads beyond their own town or city.

They are likely to have a more strident view not only on the benefits of collaborative practice but will also campaign on the need for greater penetration of the methodology into family dispute resolution.

The term 'leader' need not be too literal. They do not need to lead a group at all, but rather they are perceived as being innovative, outspoken and passionate about their work. As a result, a larger proportion of their client base is made up of collaborative clients. Some leaders in the field have resolved only to practice collaboratively.

They continue to read and invest in their self-development and are just as passionate about sharing that newly acquired information with colleagues, such is their desire to see the general level of collaborative skills and practice develop across their community.

6.14 WHAT USE IS THE SPECTRUM?

The spectrum is not meant to be taken too seriously. It is entirely arbitrary. There are no exams or milestones to pass between one level and another. There is no badge or extra certificate.

The spectrum is designed to illustrate, though, that each collaborative lawyer will have a different level of the following criteria, and it is these criteria that clients can try to discern as they meet with potential representatives.

- *Belief.* To what extent does the practitioner believe in the benefits of a collaborative approach?

- *Contribution.* To what extent do they contribute to the collaborative community, its development and the resources, whether training or publicity material, available to it?

- *Motivation.* Measured by spontaneous self-development or investment in further collaborative skills training, and to what end?

- *Participation.* Does that practitioner meet with the local practice group to share knowledge, information and best practice? Have they developed a network of local practitioners, such as mental health workers, to whom they can refer clients?

6.15 CONSIDERATIONS WHEN SELECTING A COLLABORATIVE LAWYER

When it comes to selecting a collaborative lawyer, clients should consider the spectrum above. Do not be afraid of asking questions against the criteria, questions such as:

> 'Have you got any material I can read or take away?'
>
> 'How do you refer to other collaborative professionals? Who do you refer to? How often?'
>
> 'I'd like to know more. Are there any good books you can recommend?'
>
> 'I have heard about local practice groups. How often do you attend?'
>
> 'What further training have you had beyond the core skills training? Was it good, and if so why?'

These questions will give an indication of the degree to which the lawyer will believe, contribute and participate within the collaborative practice. Their own level of motivation will also become apparent.

There are those who say that overly motivated collaborative lawyers, likely to be the active or leader practitioners, may try too hard to make every case fit a collaborative-shaped hole. We explore screening for suitability in a later chapter.

Other general considerations that a client might consider are set out below.

> 'Was the lawyer interested in my agenda or their own?'
>
> 'What kind of questions did they ask? Open questions designed to find out more about my family and me, or only those questions to elicit dates, names and figures?'
>
> 'How did they respond when I asked them questions?'
>
> 'How did they talk about my partner? Were they against them or more neutral?'
>
> 'Did they tell me what I was entitled to or what I should ask for?'

'Did they talk about my family specifically, or talk about "In cases like these …"?'

'Did they explain all of the available options – self negotiation, mediation, collaborative, lawyer-led negotiation and litigation, or leave one or more out?'

'Do I believe this lawyer will be supportive of me in trying to resolve my case or do I feel that he or she thinks they know the answer already?'

'Did this lawyer want to help me or did they want to do it all for me?'

'Can they or are they prepared to recommend other lawyers that my partner could approach?'

'Did they ask me how I am feeling about matters or recommend support services?'

6.16 Environment

Clients will take a view on office space.

- What do the rooms that we are going to be meeting in look like?
- Are they comfortable?
- Are they smart enough? Are they *too* smart or showy?
- How will they feel about spending difficult meetings in those rooms?
- How will they feel about bringing their spouse or partner into their lawyer's offices?
- What will that say about them?

I once worked for a law firm that held client appointments in a boiler room. Admittedly, it was a large boiler room, large enough to take a table, but the boiler on the wall could not be denied. It was hideous.

There were no windows and no daylight or fresh air unless you wedged open the fire escape door that led out into a back alley.

I gave up my office, with its large window, and had that room professionally decorated, with three walls painted cream and the fourth painted a calming, sage green.

I removed my desk and put in a low coffee table instead, with drawers in it to hold relevant leaflets and details of therapists and counsellors.

In one corner was a bookshelf, filled with the kind of books that are listed in Appendix 4.

Finally, we invested in four comfortable brown leather tub chairs.

The result was a room that clients wanted to be in, and wanted to share with their partners. It cost less than £2,000 to create, but it spoke volumes about our investment in creating the right environment for our clients.

6.17 Is cost a consideration?

It sounds glib to suggest that cost should not be a consideration in selecting collaborative law as a process or in selecting one practitioner over another.

Collaborative law should not be promoted, or selected, as being a cheaper option, not least because the cost might be similar to conventional cases in many cases of simple to middle complexity. It is inconceivable, however, that the most expensive collaborative case would ever cost more than a fraction of the hundreds of thousands of pounds spent on the most expensive litigated cases.

If collaborative law is seen as being the cheaper option then clients run the risk of trying to shoehorn themselves into the process in order to save a few hundred or a few thousand pounds. If the clients are ill-suited to the process then that effort will be a waste of time and money. If the process fails several meetings later then it may become necessary to duplicate large parts of work, and cost, after switching to a more conventional approach.

In the words of another collaborative lawyer and author, Forrest Mosten: 'For the cost conscious, the collaborative process will never be cheap enough.'

Meetings get squeezed and rushed.

A point that is not agreed is not disputed within the meetings because, if it was, the meeting might last another 30 minutes. Instead, a husband or wife tries to talk about the contentious issue in between meetings, over the breakfast table or by phone and e-mail, causing a nuisance and leading the other partner to feel bullied. The collaborative effort fails accordingly.

Recommendations that counselling, therapeutic or even financial advice is brought into the collaborative process get rejected purely in order to keep the cost down.

Likewise, selecting one lawyer over another because he or she is marginally cheaper would be inappropriate. The relationship between client and collaborative lawyer needs a high level of trust and security.

Clients should, within reason, select the lawyer with whom they feel most comfortable. There might always be grounds to negotiate on fees within the collaborative meetings or directly with the lawyer.

For all of that, costs will be, of course, an inevitable consideration. Perhaps the question should not be 'how cheaply can I get all of this sorted out for?' but rather: 'can we, as a family, afford this process and can we afford the alternatives?'

6.18 SUMMARY

Clients and practitioners alike should consider the spectrum of collaborative practitioners. It is not a definitive or measured spectrum but it does serve the purpose of highlighting certain considerations.

Clients might be able to use it to get a feel for how committed their lawyer is to collaborative practice and particularly the more nuanced aspects of understanding relationships and the assumptions and dynamics within them.

Doing so may well ensure that the client is able to select a collaborative lawyer who matches their own level of expectation.

Lawyers can use the spectrum as a means of challenging themselves and their development as collaborative practitioners. This can form the start of a renewed effort to build a collaborative or broader alternative dispute resolution practice, or to improve their own profile.

CHAPTER 7

THE COLLABORATIVE TEAM MEMBERS

This chapter explores and encourages the roles of non-lawyer professionals from the financial sector and what Americans refer to as 'mental health professionals' within collaborative practice.

We examine how they can assist and how collaborative professionals might work together.

Working with other collaborative professionals can raise tensions within the legal profession. These tensions arise out of adherence to entrenched conventional practice and assumptions.

7.1 WHOSE CLIENT IS IT, ANYWAY?

Several years ago a family therapist had been introduced to a collaborative lawyers' practice group. He explained how his involvement in collaborative cases would help individuals to make progress in getting through the grieving process. He explained how he could help couples to communicate better with each other. As a result, he suggested, the couple would be able to reach agreement on more issues and do so more quickly.

One lawyer responded: 'If you are doing that, then what is the role for us?'

Family lawyers have traditionally had a privileged position as the sole gatekeepers to the family dispute resolution marketplace. That resulted in a perception, to some extent, that family law clients somehow belonged to the lawyers.

It was for the lawyers to decide on which other professionals were used to assist the parties. Other professionals, such as accountants, financial advisers and mental health professionals were obliged to court these lawyers to access those clients.

The question that was beginning to be asked, although it could never be allowed to be expressed, was: 'Whose client is this, anyway?'

As soon as the question is asked then it is revealed as an offensive approach to family law work. Such an approach is hideously lawyer-centred and yet it is the sentiment that underlies the 'What is the role for us?' question.

As soon as we challenge ourselves to think 'Whose client is this?' then it should become obvious that the answer is 'None of ours'.

Lawyers, therapists and financial advisers do not have any right or privilege over any given client. A more client-centred approach is required.

What could that mean?

Clients are not commodities that enter lawyers' offices and feature in their careers for a period of several months. Instead, the client, through circumstance and separation, is obliged to select lawyers and other professionals that can serve them and their family at these difficult times.

Clients do not enter professional's lives. We enter theirs.

In part, we have been working on this already by repositioning the client as being more of an equal within the relationship between client and professional. We have explored deliberate steps that can be taken to surrender a dominant position within that relationship while still preserving our appointed role as advisers.

In collaborative practice, the client is enabled and encouraged to step back into a position of self-determination and effectiveness, together with their partner. They are not passive. They are certainly much more than files in a filing cabinet waiting to be processed.

They are competent and experts in their own lives. They are shown to have resources and resilience that a conventional approach denies them. Instead of the client saying 'act for me' the client is now saying 'work with me'.

Increasingly the client will also be saying: 'I have already resolved this and that with my partner directly, or with another trusted professional. Now we need you to work with us …'

The dynamics are changing. There is no room for 'What is the role for us?' defensiveness. The collaborative paradigm is about collaborating. Who might we collaborate with and to what end?

7.2 COLLABORATIVE FINANCIAL PROFESSIONALS

Independent financial advisers and accountants are the most obvious financial professionals who might have a role to play within collaborative law work. They already play a role within conventional family law work, but their contribution here is slightly different.

We will also highlight the role of financial coaches and debt advisers.

7.3 Independent financial advisers

Independent financial advisers, often referred to as IFAs, have one of the most central non-lawyer roles within collaborative law. Resolution, the family law group responsible for governing the development of collaborative law practice in England and Wales, has devised an accreditation programme, complete with study, examination and a minimum 200 hours per year commitment to practising in family law.

When collaborating with another lawyer it is obligatory only to work with other collaboratively trained lawyers. That same obligation does not apply to working with IFAs, and it is possible to work with IFAs even though they are not accredited.

The role of the IFA in collaborative law, particularly accredited IFAs, is for them to work as a financial neutral.

7.4 What is a financial neutral?

This role sees the accredited IFA adopting a neutral role, where they are not acting for either party, but rather advising on how family finances can be arranged and divided in the most effective fashion.

The financial neutral is explicitly prohibited from subsequently carrying out the work required to implement what has been recommended and agreed upon by the couple. That work is referred to a different IFA, who subsequently implements the agreement. Note, however, that this prohibition may soon be removed by Resolution.

It is thought that there are only about 80 IFAs who have met Resolution's stringent accreditation requirements. That number has been kept low while Resolution is reviewing its entry examination process.

Active collaborative lawyers will have got to know who their local accredited IFAs are and will have spent time with them, either within the practice groups or by way of face-to-face discussions, to enable them to know when the financial neutral can assist.

7.5 How does a financial neutral help collaborative law?

The financial neutral carries out several roles, greatly reducing the time, cost and nuisance factor that can often be experienced within separation dispute resolution.

7.6 Collating and updating

Many financial neutrals will take on the role of obtaining up-to-date valuations and statements for the family's financial assets. The couple simply sign an authority enabling the financial neutral to request the said documents on behalf of the couple. The couple will list all of their assets, accounts and investments and the financial neutral then writes to each institution to request the up-to-date documents.

The nuisance factor that the couple would otherwise have had to tolerate as they chase up replies is avoided. Of course the financial neutral will be charging for their work on an hourly or fixed fee basis.

Financial neutrals should be clear with their clients on how they are going to charge for their work. As with the lawyers, agreements should be signed recording those details.

Some couples may not need this level of assistance. It may be that they have the time on their hands to enable them to do this work for themselves. The consideration is whether the costs or time saved in leaving it to the financial neutral to collate this information is more valuable to the couple concerned.

Lawyers have previously carried out this work and might still do in many cases. However, this exercise of collating and updating is not a legal exercise. It is likely that it can be completed at a more reasonable cost by the financial neutral or their administrative support staff. The other benefit is that all of the original documents are being delivered straight to the financial neutral rather than being shuttled between lawyers and then out to the financial neutral.

7.7 Reporting

The financial neutral then compiles a report on what assets are available. He or she will also highlight what considerations may need to be taken into account when considering the financial information that is now available.

There may be requirements to leave funds invested for a certain period of time. Certain assets might be inappropriately invested and could be better placed elsewhere.

7.8 Pensions

These reports will often report on pensions that one or both parties have.

Pensions can often be the largest assets within marriages, after the matrimonial home.

The financial neutral will be able to explain what the pensions are worth and what benefits those pensions might produce at some stage in the future. It may well be that the financial neutral him- or herself might have to call in the assistance of an actuary to help with those valuations but, again, that degree of service coordination can be left to them.

Pension reports can also indicate who should receive what percentage of the pensions. The financial neutral, together with actuarial assistance, will calculate what percentage share is required to ensure that each party would receive an equal income from the pensions that they each receive.

Clients need to be made aware that women will tend to receive less income for each pound of pension asset that they have once they retire. This is because women are deemed to live slightly longer than men on average. As a result the pension providers have to spread women's pension funds more thinly over a longer period. This means that women will often require more than male partners if their income is to be matched

7.9 Neutral financial advice

The financial neutral can now set about giving his or her advice. Exactly when that advice is given may well vary from case to case.

The couple might want to resolve children matters first, for example. In this event, provided the financial neutral has been brought in at a very early stage, he or she can be left to collate the financial information while the couple continue their parenting discussions over the next couple of collaborative meetings.

Even then, the parties might only want to see the financial report, setting out assets, investments and pensions to start with.

We need to remember that the collaborative process is client-centred. It is very easy for the financial neutral, or the lawyers, to start making recommendations at this stage but that temptation must be resisted. Similar care needs to be taken with the tone in which any report is written.

The paradigm shift for the financial neutrals is just as important as it is for the lawyers. Clients will expect that shift to have been made and to be reflected within the report writing. This is one reason why using an accredited collaborative IFA is strongly recommended.

Reports can easily slip into making recommendations, or assuming that a certain division of assets will be adopted by the couple without further question or consideration. When the report is written in a way which is seen as being directive then it can jar with the rest of the collaborative process. Clients notice this and can get alarmed that the tone has changed.

In some cases the clients, who themselves might have taken some time to get comfortable with the collaborative dynamic and their own responsibilities of self-determination and resolution, can suddenly get very protective of the process.

If a report goes so far as to hand down a solution then the couple can feel that their authority to determine outcomes for themselves within the collaborative process has been taken away from them. Even though the report might be perfectly correct in its conclusions, if the tone is too directive it can be offensive to the collaborative effort. Clients are right to complain if the lawyers or financial neutrals slip into such directive voices. This is supposed to be their process and the outcomes are supposed to be their outcomes.

Some critics might suggest that this process of withholding our own conclusions is artificial. If we know the answer, or a range of answers, why would we not tell the clients? The reason is in the clients' motivation for choosing the collaborative process.

Many collaborative clients, once they grasp the collaborative methodology, place value in designing their solutions themselves. Doing so gives them ownership of it. If we slip back into handing down predetermined solutions then the clients' courage and determination in deciding to collaborate is undermined or even dismissed entirely.

7.10 What is the role of neutral financial advice?

The role of financial advice is similar to that of legal advice. Of course there will be a range of solutions that are apparent to the financial neutral. One of those solutions might even be preferred by the adviser as being the optimum solution.

What is important is that if advice goes so far as to recommend certain divisions of assets, income and pensions, such recommendations are presented as suggestions or possible examples and not as the answer.

The clients need to be reassured that these suggestions are not definitive or prescriptive, although they can be guided to what would represent the most profitable or efficient division or investment.

It might be better to report on the data that has been obtained without making any recommendations or putting forward any permutations at this moment in time.

Instead, the next collaborative meeting is spent considering the report. The couple can then be given the time, and appropriate guidance, to consider this information and how it relates to the family's identified current and future needs.

7.11 The role of collaborative lawyers within neutral finance advice

The role of the lawyers in this process will be to consider whether suggested solutions can be implemented so that they will be binding now and in the future.

The lawyers also retain the responsibility to ensure that their clients' needs have been fully considered. That is not to say that a lawyer must get the largest share of assets for their client. Such an outcome may be unsustainable in the long term or undermine other values that the clients have expressed within their anchor statements.

Certainly the lawyer will not allow a client to stumble into a bad deal.

Meeting the client's best interests in the collaborative sense will mean reality testing possible solutions using questions such as:

> 'Will that enable you to pay for …?'
>
> 'How far would that proposal go to satisfy your need for …?
>
> 'You said that you wanted to reach a fair division of assets. How close are we to reaching that level fairness, do you think?'
>
> 'How do you think you will feel if you have these assets and you see your husband/wife/partner with these other assets?'
>
> 'How long do you think you might be able to keep going with that level of support?'
>
> 'If your children were grown up and were sitting with us in this room, what kind of arrangement do you think they would want to see the two of you reaching?'
>
> 'Let's look 10 years into the future. If you have this level of assets now, how will you be able to build upon that over the next ten years? And your husband/wife/partner?'

This method can be seen as pulling the clients into a fuller realisation of the implications and suitability of their suggested outcomes without pushing them into positions.

The financial neutral should be careful to adopt similar pull methods of advising as they go about their role of ensuring that all of the implications of their decisions are considered.

Financial neutrals can very often highlight more efficient ways of managing assets and will have specific knowledge on certain investment types. They can sometimes create value by showing alternative asset splits that might reduce, for example, tax that would be paid between the couple.

7.12 Future financial planning

7.13 Lifetime planning

Financial neutrals have the advantage of access to tools that can very clearly map out future wealth growth, or reduction. This enables clients to future-test their plans to see where they would each be at retirement age or any other trigger date they wanted to look at.

These plans can be incorporated into very clear graphs which can reveal large disparities that would otherwise be invisible to the clients and to their respective lawyers. Those disparities can, in turn, help clients to consider proposals more carefully to avoid future resentment or unfairness emerging between the couple.

7.14 Inheritance planning

Financial neutrals can also play a role where families are concerned about preserving legacies for their children, or, of course, continuing to work together to build a legacy or decent inheritance for their children.

It might be possible to explore options and new investments as a vehicle for such future financial planning. If that is appropriate then the couple's respective commitment to contributing to such plans can be incorporated within the collaborative agreement and later within the court order or deed of separation that will record the settlement.

This notion of building future family financial planning into the separation process is still relatively novel to many families. It is a further example of the paradigm shift between conventional and collaborative law.

The former law is too often about dividing the assets and maximising one party's share at the expense of the other. Collaborative law has the flexibility to incorporate a far broader array of considerations in accordance with the separating couple's wishes.

7.15 Planning for future changes

Couples who are going through separation may want to consider between themselves how they can use the pain, and cost, of separating now to ensure that the family continues to be provided for in the future. Those considerations might anticipate how commitments and plans entered into now will be accommodated as and when one or both partners have new relationships, possibly new families and the new commitments that will arrive with them. This can be very useful to avoid future disputes.

7.16 The role of accountants within collaborative law

Accountants will be more familiar with working on behalf of both parties within family separation cases as a result of the push for jointly instructed experts within the Civil Procedure Rules.

These litigation conduct rules compelling and governing the instruction of joint experts can be found at Part 35. Rule 35.2 in particular states that:

(1) It is the duty of experts to help the court on matters within their expertise.

(2) This duty overrides any obligation to the person from whom experts have received instructions or by whom they are paid.

The accountant's position within collaborative law cases is similar. He or she is not expected to put forward a viewpoint that is biased to either party. Their duty can be seen to be towards the couple and helping them, and other collaborative professionals working with them, to understand whatever accountancy issues are at hand.

7.17 Where can accountants help?

If one of the partners within the separating couple own or have a holding within a company or firm, or if there is a family business then it is likely that an accountant might be asked to work with the collaborative team to:

- explain how well a business appears to be doing;
- analyse and explain a set of accounts;
- test a company's performance;
- identify inconsistencies or irregularities that can be addressed;

- challenge company spending, profit distribution or retention; and
- comment on liquidity or viability of selling the company.

7.18 Anticipate accountancy involvement before committing to collaborate

The business-owning spouse or partner should be made aware that the involvement of an independent jointly instructed accountant is likely. This should ideally be anticipated at the very outset and probably even before the participation agreement is signed.

This factor can form part of the screening process when ascertaining whether a potential client is suitable for the collaborative process.

If the partner is not prepared to co-operate fully with such inquiries then there will be serious doubts as to whether this case is capable of being resolved by the collaborative process.

Business owners often resist external inquiries on the basis that the accounts are straightforward. For the business-owning partner the accounts will speak for themselves.

Other objections might suggest that the existing accountants can deal with any queries, to save costs. This latter suggestion might be appropriate if both parties are prepared to agree on it.

Business owners need to recognise, however, that their own level of familiarity with their own business accounts is not going to be matched by their partner or any financial neutral or either of the lawyers.

The instruction of the joint accountant is designed to ensure that everybody understands how the business is doing.

7.19 Conflict between the accountant and the business-owning party

The inquiries that an independent accountant makes can bring the accountant into conflict with the partner who runs the relevant business. This will need careful management.

Remember that collaborative law is not to be seen as a conflict-free way of resolving family separation. Instead, collaborative work does throw up possible conflicts, such as these. They should be anticipated before they arise.

The business owner in such a situation should expect that his or her accounts will be considered and that inquiries will be made about them.

They should expect to feel uncomfortable when such questions are asked. The goal in such preparation work is that when the questions arrive it is recognised that they are not attacks that need to be defended. Instead we simply need clear answers to reassure us and enable a fuller understanding.

The business-owning partner should be coached, or prepared, to recognise when they are feeling under attack and given tools to avoid these inquiries deteriorating into conventional conflict-led patterns of perceived attack, defence and counter attack. The following kind of exchange is easy to imagine:

> Accountant: 'Julie, can you help me with this. Your entertaining costs for last year were £8,000. The previous year, they were only £1,300.'
>
> Julie: 'What are you trying to say? That I'm fiddling the books? Is that it?'
>
> Accountant: 'No, not at all. I am just trying to understand what the difference was and what the costs might be, looking forward.'
>
> Julie: 'You think I have fiddled the books to save a poxy £8,000, don't you? You know what? I thought we were supposed to working together to sort all of this out, not having a go at me all the time. That does it, I've had enough. [To husband:] If this is the way you want it, if you would rather disbelieve everything I say then take me to court, but I am not going to sit here and have those allegations thrown at me. I'm out of here.' [Exit]

With prior preparation it is to be hoped that although the questions will still need to be asked and although they may still feel as uncomfortable, the response would be different.

> Accountant: 'Julie, can you help me with this. Your entertaining costs for last year were £8,000. The previous year, they were only £1,300.'
>
> Julie: 'Hmm. You know, when you say that, I feel like you are suggesting that I have done this deliberately.'
>
> Accountant: 'No, not at all. I am just trying to understand what the difference was and what the costs might be, looking forward.'
>
> Julie: 'I know, I know that's right, and I appreciate that we all need to understand what's going on here. The truth is I cannot recall right now. Do you mind if I call my own accountant and get that information, or shall I bring it to the next meeting? Is there anything else we need to clarify while I am at it?'

In the absence of collaborative law training from Resolution for accountants, it will be necessary for collaborative practitioners to ensure that their local accountants fully understand the collaborative law dynamic. This should minimise misunderstandings and ensure that they work and report in ways that fit in with the collaborative profile.

7.20 Financial coaches

When couples separate it may well be that one partner has very little knowledge about how the family finances have been run. They might have never dealt with household budgeting before.

Their knowledge of the different kinds of investments and financial products can be far lower than that of their partner.

Very often, this lack of familiarity can make it impossible for the less familiar partner to negotiate around the financial assets.

A finance coach can be brought in to help one partner, or both, to become more comfortable with understanding these issues, enabling them to contribute fully to discussions about finances, and also enabling them to be more self-reliant in the future.

It may well be that the couple will agree that the financial neutral can work with the partner who needs help to understand budgeting, financial products or investments. Alternatively, that partner might be introduced to someone they can work with outside of the collaborative meetings, so that they can build up their knowledge before a future meeting where the finances are going to be discussed.

Lawyers need to ascertain their client's own level of financial understanding and make such recommendations accordingly. Do not underestimate the extent to which a lack of confidence in these matters can inhibit either partner's effective contribution to the discussions.

7.21 Debt advisers

It is a sad reality that many relationships struggle as a result of onerous debt liabilities.

Monthly repayments on high levels of borrowing can see couples having to hold down several jobs and still have no money, or time, to care for themselves, each other or their relationship.

There is no reason why, in cases where there is unmanageable debt, a debt adviser, or debt counsellor, cannot be invited to a collaborative meeting to discuss what other approaches might be available to help control the debt burden.

7.22 MENTAL HEALTH PROFESSIONALS

The phrase 'mental health professional' is used in this book to refer collectively to the following service providers:

- General practitioners;
- Psychologists;
- Psychiatrists;
- Psychotherapists;
- Counsellors;
- Coaches;
- Holistic therapists.

All of the above can be incorporated into the collaborative team or, alternatively, one partner or both can work with the relevant professional outside of the meetings themselves.

7.23 REMOVING THE STIGMA OF MENTAL HEALTH CARE

Broaching the issue of mental health care can feel uncomfortable for the lawyer and for the client with whom the issue is being raised. One way to normalise the issue might be to ask whether a client has spoken to their GP about what is going on for them right now. Explain that it is important to watch out for the sometimes imperceptible creep of depression.

Suggesting that they speak to their GP raises the possibility that depression might take a hold to some extent or other. It might be helpful to suggest that some degree of depression is to be expected when facing change such as separation.

We can normalise the idea of mental health care in this very low-key way using familiar non-threatening terms such as depression and talk of GPs. It is then much easier to go on to offer more specific details of therapists, counsellors or coaches who might also be able to assist.

7.24 THE LAWYER'S ROLE AND RESPONSIBILITY AS GATEKEEPER

Lawyers still hold a privileged position as gatekeepers to the arena of separation service providers, albeit one that may not be as exclusive as it once was. The people who come to us looking for help will often be startled, hurting and lost on the outside of this walled marketplace. In many cases they will be completely unaware of the array of support and resolution services available to them on the inside. If they are unaware of what lies within they are unlikely to ask for those services.

They come to us as the assumed and trusted gatekeepers.

We need to recognise that they might be doing so only to pass right through our offices and into that broader marketplace that lies within. It is important that we do not abuse our position as gatekeeper. It can be too easy to use the gateway as a closed net in which clients unwittingly become caught.

Our position as gatekeepers means that we will sometimes be cast as both conflict and needs diagnostician.

It remains the lawyer's role to act in the client's best interest. Sometimes that might mean not acting at all, at least for the time being, but only referring the client on to another service provider. We have a responsibility therefore to ensure that we do broach the support offered by the various mental health professionals, as defined above, and make referrals or recommendations accordingly.

Consider the following:

A large tattooed man meets with a divorce lawyer for the first time on an initial free half-hour appointment.

He is bewildered following his wife's departure at the weekend. The lawyer takes some details in a conventional fashion, completing his new client checklist, and opens up the dialogue regarding depression, as discussed above.

The large man, sitting uncomfortably at this lawyer's desk, listens intently.

The lawyer carries on.

> '… and so, it might make sense to speak to your doctor about what is going on for you right now, so you can keep an eye open for depression.'

The man continues to listen and the lawyer goes a little bit further.

> '... and, if you think it would help, I can also let you have the names of a few local counsellors or therapists who you might be able to talk things through with.'

There is a silent moment before this large man crumples, sobbing with his face hidden in hands. After a few moments he composes himself and looks up.

> 'Thank you', he says. 'That would be very helpful. I had no idea where I could get that kind of information from. That is exactly what I was looking for.'

This real-life scenario demonstrates that it is crucial, therefore, that clients are advised about suitable mental health professionals and offered easy access to suitable contacts. It is important that the choice is theirs.

An alternative outcome to the above scenario might have seen a more closed gateway simply assuming that here is a man who needs a divorce and signing him up to a conventional lawyer-client divorce retainer. In such a scenario, only the legalistic aspect of his separation would have been considered. The other six aspects would have been neglected, either in part or in their entirety.

7.25 THE DIFFERENCES BETWEEN VARIOUS MENTAL HEALTH PROFESSIONALS

We will explore the differences between the professionals below.

Any referral to any mental health professional is designed to provide care, safeguarding and possibly education for the partner concerned. It may be that the input of the professionals will not directly impact the collaborative process or any agreement that is reached. Their role might be less direct than a financial neutral's assistance or the findings of a joint accountant might be.

The mental health professionals' involvement will, instead, enable the individual to be as effective as they can be as they contribute to the collaborative meetings.

7.26 THE NEED FOR MENTAL HEALTH CARE

It is essential that both parties take care of themselves as they pass through the emotional aspects of separation.

As big a supporter as I am of the active promotion of health care support within collaborative law, I continue to believe that it is not for the lawyer to dictate who plays a part in the collaborative team. It should, instead, be for the client to make informed choices on who should be involved and to what extent. As we have already explored, that is not a universally held belief within the collaborative profession. There are those who insist on a 'full team' approach or insist on having certain other professionals sit in with them if they are going to agree to meet with a client.

Lawyers may well face situations where they have to stop acting for a client in the absence of specific assistance or support.

It may become apparent that a client is simply not able to take part in and contribute to the collaborative process. Clients might be sullen and stubbornly refuse to speak.

When asked to comment on a proposal or suggestion a client might simply shrug or even turn away in silence. They might acquiesce with a disinterested 'whatever …' response.

7.27 THE NEED TO BE SENSITIVE TO CLIENT'S CONDITION

Similarly clients, and their lawyers, should be observant about their actions, demeanour and appearance. We should create a collaborative atmosphere where we have permission to ask one another about 'What is going on for you, right now? How are you feeling about this? I notice you haven't said anything for the last 30 minutes', or even observations such as 'You look like this is proving really difficult for you. Have I understood that right?'

We need to be able to acknowledge expressions of sadness or distress and make appropriate inquiries about that.

If the emotional state is so high then we should take a break or reconvene another day. We should also be bold in suggesting emotional or mental health support.

If we as collaborative practitioners put our clients into situations that are overly distressing to them without offering support then we will not be acting in their best interests.

There might be occasions where we have to say to a client 'I do not think it is right that we continue with this process right now' and instead refer the client to meet with a suitable mental health professional. As collaborative lawyers, that might mean that we lose a client.

We will also meet resistance from the partner who may become agitated or frustrated at delay. Remember that clients may be at different stages of the grieving process. One will be more ready than the other to move on with the progress of matters. Each client will have differing needs when it comes to emotional and mental health support accordingly. This will require careful management, particularly by the lawyer acting for the client who is further down the separation line.

If support is not offered, or if we push on to conclude an agreement even though a client is distressed, then that agreement could be challenged at a later date. We as lawyers might even face allegations that we were negligent in allowing that situation to exist. That is something that everybody will want to avoid.

7.28 Safeguarding and protection

Some clients may benefit from accessing mental health professionals to safeguard against things going wrong. Examples would include individuals who recognise that they tend to react angrily to certain triggers.

In other cases the lawyers might recognise that a couple get stuck on a particular aspect of their relationship, a certain emotive issue or children arrangements. There might be a power imbalance, or a pattern of communication, that is proving impossible for the lawyers alone to rectify.

In these circumstances it can be appropriate to explore involving an appropriate counsellor, coach or therapist who can help them to ensure that they do not spend time within collaborative meetings discussing the same hot point for hours on end.

Where one partner is too quiet or finds it difficult to speak out for his or her own interests, then, again, counselling, therapy or coaching might be considered to enable them to be more confident and resilient.

If there has been domestic violence then it might be necessary, if collaborative law is appropriate at all, for there to be regular counselling for one or both partners to safeguard against further incidents. Domestic violence is discussed further when considering screening for the suitability of the collaborative process.

7.29 Education

As the couple meet with relevant mental health professionals it is inevitable that as well as accessing care and safeguarding benefits there will also be a degree of education. The individual is likely to learn about themselves and their own patterns of response. They will learn about how certain situations lead them to act.

They may well learn techniques to enable them to keep calm or look after themselves better, even after the collaborative process has finished.

The introduction to this book explored the difficult question of whether or not collaborative law is a transformative process. By that we mean: 'Does resolving a separation by collaborative process transform the individual, as a result of going through this process?'

We can see that the involvement of suitable mental health professionals can usher in such transformations. The point remains that transformation is not the purpose of collaborative law, but may be a side-product of it, and particularly so when external mental health professionals are accessed. When that happens there is real value in the process and collaboration with other professionals. It may well be a value that cannot be quantified.

7.30 GENERAL PRACTITIONERS

The most common mental health practitioner will be the family doctor or GP. This will be the first port of call when one starts to become aware of feeling depressed or stressed, when suffering from sleeplessness, loss of appetite, loss of weight or anxiety.

It is possible to use GPs as an initial recommendation when broaching the issue of external mental health care. Doing so frames our concerns in very normal terms.

'Have you spoken to your doctor about that?' is non-threatening, whereas going straight in with 'I've got the number of a psychotherapist who can help you' feels very different.

If a client goes to meet with a GP in between collaborative sessions, then the GP might refer them on to other service providers including other mental health professionals. It would be appropriate to make a note where one partner has indicated that they are going to meet with a GP to follow up, outside of the collaborative meeting, to inquire how that meeting went. This can take place when the collaborative lawyer and client meet to review the previous meeting and prepare for the next one.

Has a referral been made to another professional? Who? Would it be helpful for that professional to give an indication on whether it is appropriate that collaborative meetings continue for the time being? Could the professional be invited to feed into the collaborative team process?

Is there immediate work that needs to be completed before the collaborative process can continue? How can we ensure that the collaborative process does not make any condition or anxiety worse, or that any treatment is not undermined or jeopardised by further meetings for the time being?

If there are serious concerns about whether it is appropriate to continue then the client can be asked to authorise a very short letter asking the GP, or other mental health professional if a referral has been made, as to the suitability of continuing with collaborative law for the time being.

7.31 PSYCHOLOGISTS, PSYCHIATRISTS AND PSYCHOTHERAPISTS

There is a great deal of confusion about the differences between psychologists, psychiatrists and psychotherapists.

Psychologists work in the field of study of the human mind and how it works. It looks at the theory of thinking and how thinking interacts with doing and being.

Psychology therefore tends to look at why we do things in a certain way, and the thought patterns behinds those actions.

Psychology is not necessarily a medical discipline.

Psychiatry, on the other hand, is a medical discipline.

Psychiatrists are doctors who are engaged in diagnosing and treating mental health issues and disorders.

Finally, psychotherapists offer therapy for various psychological issues. Psychotherapy is not necessarily a medical issue. A psychotherapist does not have to be medically qualified. There are various psychotherapeutic services available such as:

- Cognitive behavioural therapy (CBT);
- Systemic therapy;
- Family therapy;
- Integrative psychotherapy;
- Narrative therapy.

It is most likely that collaborative practitioners would refer a client to a psychotherapist rather than a psychologist, and certainly more so than to a psychiatrist, although a GP might refer him to a psychiatrist.

Active collaborative practitioners will have taken time to meet with local psychotherapists better to understand how they work and where their involvement can benefit clients within the collaborative process.

Psychotherapists who are looking to work with families going through separation would also be well advised to ensure that they network with lawyers and educate them about the benefits that family law clients can derive from psychotherapeutic work.

7.32 WHAT IS THE DIFFERENCE BETWEEN PSYCHOTHERAPY AND COUNSELLING?

One distinction is that therapists work in accordance with a theoretical background.

For example, a systemic therapist works in accordance with their understanding of systemic theory.

They would then use their expertise in that theory to help clients understand their issues from, in this example, a systemic point of view.

A counsellor tends to look at the person and the problem and works with the person to enable them to overcome the problem.

Another way of putting that is that psychotherapists are experts in a field of psychological or development theory and use that expertise to resolve psychological issues.

Counsellors do not rely as much on their own expertise, but rather build upon the client's own expertise in their own lives and their life experiences. They then enable the client to make sense of what has happened to them and subsequently build their own solutions, and the resolve to implement those solutions.

The words psychotherapist and counsellor may well be interchangeable and many psychotherapists will call themselves counsellors as it feels less imposing.

7.33 WHAT IS THE DIFFERENCE BETWEEN COUNSELLING AND COACHING?

The roles here are similar. Both counsellors and coaches enable clients to develop their own solutions to enable them to be more effective by overcoming identified obstacles. In both situations, the client is the expert in their own life and is encouraged, through structured dialogue and careful questioning and listening techniques, to help themselves.

The difference between counsellors and coaches, to put it very simply, is that counsellors tend to look back and coaches look forward.

Counsellors will want to know about what has happened. How has the client arrived at this stage in his or her life? How have they got stuck with a specific issue? How can the client better relate to what has happened in a better way? Having understood that, how else might the client respond in a way that will be more productive and helpful to them?

Coaches will want to know what is happening for the client now, and what the client would like to happen instead. Having understood those goals, as opposed to having to understand a recent history of events, how might the client go about reaching his or her goals so that they can be more productive in their life?

To summarise, therefore, a divorce counsellor would want to look at what has happened in the relationship and how the client can better understand, and moderate, their responses to what has happened.

A divorce coach would be curious about: 'Where do we go now? Given that the relationship appears to be coming to an end, what are your goals in the immediate, medium and long term?'

Again, we do not need to be too precise with our terms of reference. What is important is that collaborative practitioners recognise the benefits that each mental health professional in their region is able to offer. Network with the service providers and understand their skills better so that you have a full range of possible services to offer your clients.

7.34 CHILD AND PARENTING SPECIALISTS

With the emergence of collaborative law and its mediation predecessor, the voice and position of the child of the relationship has become much more central.

Families are increasingly searching out the input of child and parenting specialists. These professionals can help the parents to understand the child's experience of divorce and separation better.

Even the court service has now introduced a Parenting Information Course to equip couples applying to the court for contact and residence issues to refocus on their children and their duties to the children as parents. These duties bring about a different tone than when parents are more concerned about perceived rights as parents.

Within collaborative law, the child expert can help to draw out and amplify the children's concerns. Doing so presents a new dynamic to the parents' separation that may have recently been lost amongst the anxiety and noise of other separation considerations.

The child expert may well work in a coach-like capacity. Doing so encourages the parents to consider the impact of their own conduct, and the collaborative arrangements that they might make, upon the children.

Parents can learn new skills and strategies to work together as co-parents who are now living apart. This might include exploring ways of communicating effectively, managing conflict and designing arrangements that can work for both parents and the children. There will also be material to explore children's development and how the parents can work together to continue to offer a loving relationship to the children while still maintaining appropriate boundaries regarding discipline. This transition from parents living together to co-parents who are now living apart from one another is crucial and one that should not be underestimated.

The expert might also be asked to work directly with the children to help them manage and come to terms with the changes that they are experiencing.

Again, active collaborative lawyers should seek out local providers of child and parenting expertise. They should then connect with them to learn more about their specific approaches and how they will work with their clients so that they can make confident recommendations in future cases.

Not every case will need a child and parenting expert, but many more cases will benefit from this input than currently access it.

Another source of child and parenting expertise can be found within the material produced by Resolution it its *Parenting After Parting* range of material or in another book by Constance Ahrons, *We're Still Family* (Harper Paperbacks, reprint edn, 2005), in which children who have now grown into adults reflect on their own experiences of their parents' separations.

7.35 WHAT ABOUT HOLISTIC THERAPIES?

Holistic therapies offer non-conventional therapeutic services for the whole being, to include the physical, the non-physical and the spiritual.

The perceived value of holistic therapies varies dramatically from individual to individual and from specific therapy to specific therapy.

An adherence to one holistic therapy or another may well require a level of faith in the effectiveness of it, or faith in an underlying spiritual or values system that underpins any given holistic approach.

As a result, the decision as to whether to recommend holistic approaches must be left to the individual practitioner using their careful discretion with each client.

7.36 OTHER EXPERTS

Carefully consider seeking involvement of other professionals, whose roles should be self-explanatory:

- Estate agents/chartered surveyors;
- Other niche valuation experts;
- Child and parenting specialists;
- Parenting experts;
- Conflict coaches;
- Spiritual/pastoral mentors where faith issues arise.

7.37 MANAGING THE COLLABORATIVE TEAM

Resolution, the family law group of England and Wales, has guidelines relating to the instruction of mental health professionals, which it calls 'family consultants', such as those discussed above. The purpose of the guidelines is to provide some rigour to the relationship between the clients, lawyers and other professionals.

7.38 SHARING INFORMATION BETWEEN COLLABORATIVE TEAM MEMBERS

When working with other professionals there should be a written agreement specifying, for example, what information is to be shared. Therefore, if a husband is to meet with, say, an anger management expert, is that anger management expert expected to provide feedback to the whole team?

Clients need to be reminded that any professional who learns of threats to the safety of other individuals, and especially children, or indications that another crime is about to be committed, may be forced to share such information with the appropriate authorities regardless of other duties of confidentiality.

7.39 QUALIFICATION AND EXPERIENCE GUIDELINES

The Resolution guidelines set out a requirement that family consultants (as they refer to mental health professionals) must have been educated to degree level and have professional qualifications in psychology, social work, marriage and family therapy, clinical counselling or mediation.

Furthermore they should be in 'good standing' with their relevant professional regulatory body, and have 3 years' post-qualification experience related to couples and family work, be insured and be willing to be trained in certain topics relating to collaborative and family law.

In practice these can only be guidelines. It will remain a duty of the lawyers themselves to check on ability and qualifications. The best way to do that will be by getting to know local practitioners and their work.

7.40 POSSIBLE CONFLICT WITHIN THE COLLABORATIVE TEAM

Collaborative practitioners need to keep in touch with each other while working together, subject to any limits on information sharing. Doing so will avoid possible misunderstandings about who is responsible for what and the conflicts that can ensue.

The professionals also need to take responsibility for how they respond to conflicts that arise. They have a duty to their clients to model effective collaboration and conflict communication skills. This may well mean investing time in their further personal development so that their own ability to manage conflict is as good as it can be.

It will do no good to have collaborative lawyers, or cross-disciplinary professionals, say a lawyer and a therapist, bickering, sulking or storming out.

Instead, practitioners need to be prepared to recognise when things are starting to go wrong and talk about the issue with their colleague, as uncomfortable as that might be.

When we come to look at the collaborative cycle below, we see that there are plenty of opportunities for the professionals to provide feedback to one another about what worked well, what did not work so well and what they would like to change at the next meeting.

Use those opportunities rather than shirking the responsibility to do so.

Watch out for tensions discussed earlier around the 'Whose client is it anyway?' question. Ask the client or the couple for feedback on how they feel things are working out and who they would like to work with on a particular stage of the collaborative process.

If conflicts persist, despite best efforts, then consider speaking to another senior collaborative practitioner or local mediator to see if they would be prepared to intervene in the situation.

Clients should also feel that they have permission to speak out when they perceive there is conflict between the professionals. Clients should have permission to point out that they feel the professionals are getting hung up on an issue and need to re-focus on helping the couple collaborate.

7.41 EFFECTIVE COLLABORATIVE TEAMS

Open communication and trust is central to the effectiveness of these teams.

Collaborative teams need to be able to trust one another entirely. The conduct of individual professionals will be very important. They should ensure that they deliver what they promise within the timescales promised and should be prepared to be held to account to do so.

With this mutual accountability comes a responsibility on each of the professionals to speak up when there are concerns or doubts about each party's role and conduct.

These discussions do not need to take place within the four-way meetings but it may sometimes be appropriate to do so. Ask the couple if they would like to contribute to such discussions or if they are happy for these conversations take place off stage.

If the individual collaborative team members are clear on what each other's role is, and if they honour their own responsibilities and boundaries of involvement, then collaborative teams will work much more smoothly.

7.42 SUMMARY

This chapter has looked at the full range of professionals who might be invited to work within the collaborative team or operate outside of the team with either one or both of the parties.

We have considered the valuable role that each of these professionals can play and some of the benefits for the parties. Any one of the professionals, used appropriately, will enable the parties to move forward with their collaborative discussions more effectively and with a fuller understanding of the relevant facts, processes or emotional and psychological aspects of their situation.

Care needs to be taken not to overload the team with too many professionals, but also not to miss clients' needs for other assistance that the lawyers cannot provide.

CHAPTER 8

PREPARING FOR AND ATTENDING THE FIRST FOUR-WAY MEETING

In this chapter we will look at preparing for and attending the very first collaborative meeting.

We shall presume that this meeting will be held between the two partners and their respective lawyers in what tends to be called a four-way meeting.

A standard agenda for the first meeting might look as follows

(1) Meeting and introductions

(2) Anchor statements

(3) Discuss and sign participation agreement

(4) Immediate issues to be addressed

(5) Identifying longer term issues

(6) Preparing for the next meeting

There will have been a considerable amount of preparation before the first four-way meeting.

Each client and their respective lawyer will have met twice, at least.

8.1 THE FIRST CLIENT-LAWYER MEETING – A REVIEW

We have explored how a first meeting might go between a collaborative lawyer and a potential collaborative client. That format for a first meeting is likely to have prompted the client to determine that the collaborative approach is attractive to them.

If the lawyer has opted to use a questioning format for the meeting, such as the GROW framework examined in Chapter 4, then it is likely that the lawyer and client will have established a deep appreciation of the client's values and concerns.

As we suggested there, much of what was identified and explored can be used to build a platform upon which to go into the first collaborative meeting together with the client's partner and the other lawyer.

8.2 THE SECOND CLIENT-LAWYER MEETING – ORIENTATION

Once both partners have agreed to collaborate it is likely that a date will be set for the first four-way meeting.

Prior to that meeting, the lawyer and client will meet together once again. This second meeting is likely to be an orientation meeting. Its purpose is to prepare the client so that he or she has some idea about what they will encounter within the four-way meeting.

The client is likely to be feeling nervous as the first four-way meeting approaches. The client will have little knowledge about what will happen, who says what and what their own role will be. These practical questions will also collide with more emotive worries such as:

- How will I cope?
- Will I be able to keep calm?
- Will I get upset or angry and what happens if I do?
- Will I get it wrong?
- What will I say?
- How will my partner react?

One way of reducing the anxiety is to reduce the level of unknown issues. We will not be able to answer every question, not least because the collaborative process and collaborative meetings tend to be very responsive. They are flexible enough to spend more or less time on a given topic, or even to accommodate entirely new lines of discussion if they are appropriate.

It will be possible, at least, to identify what the standard agenda for a first meeting would look like so that the client will have some idea about the issues that will be discussed.

8.3 SCREENING FOR SUITABILITY

Collaborative law has a fortunate problem. It can be so attractive both to lawyers and clients that there is sometimes a rush to collaborate even when the process is not suitable for a particular family.

We should be slow to assume that the collaborative process is the best way forward where there are issues of Trust, Anger and Positional negotiating. In such circumstances we might have to take the view that we need to turn off the collaborative T.A.P (TAP).

As we have already seen, there is much to commend collaborative law to couples facing separation.

- They can consider a far broader range of issues, namely the legal, financial, parenting, emotional, uncoupling, social and narrative aspects of divorce or separation, rather than dealing only with the legal.
- The couple retains a central role in designing the best possible outcome for their family.
- Matters are kept out of court.
- Matters are less acrimonious.
- Neutral experts can help with financial, emotional or communication problems.
- Parents may well be able to work together better post separation.
- The collaborative law process might reduce legal fees.

These attractions can be so powerful that clients, or lawyers, may be blinded to the situations where collaborative law, as beneficial as it can be, is not appropriate.

Clients and practitioners need to consider the question of suitability carefully. There should be a screening process, whether formal or informal, to question why the collaborative law process is being entered into and to ascertain whether the client, and their partner, is able to work effectively within the process. This process of screening can take place at one or more of several opportunities.

Some practitioners ask clients to complete questionnaires before they even see the lawyer. This questionnaire might be sent to the client in advance of the meeting, or it might be completed while the client waits at the lawyer's office or meeting rooms.

The questionnaire has a dual purpose.

It warms up the client to think about collaborative issues before meeting with the lawyer. In that sense, it has a disruptive element, dislodging those presumptions about the lawyer-client relationship and framing the meeting that will follow in a different context.

It also works as a test. If the client's responses indicate that there are certain issues, such as Trust, Anger or Positional stances (TAP) then the lawyer can be alerted to the possible unsuitability of the collaborative law process.

Most collaborative lawyers do not yet use a questionnaire such as this. It can feel intrusive and impersonal. These lawyers use their own discernment instead when meeting with their clients at the first, and possibly the second, meetings. They will still be looking for those TAP indicators and also assessing the client's demeanour. In addition to the TAP issues there may also be concerns about the client's emotional state as explored within the grieving process.

8.4 TAP STANDS FOR TRUST, ANGER AND POSITIONS

- *Trust.* Are there trust issues, or the threat of violence or oppression within the collaborative process?

- *Anger.* Is anger, revenge, hate or a similar negative emotion motivating one party or the other?

- *Positions.* Are hard negotiating positions being held onto or are the parties more interested in interest-based negotiating?

If any of the TAP considerations are present then collaborative law might still work, but the highlighted problems, and possible solutions, will need to be carefully considered between the lawyer and client, and among the full collaborative team.

8.5 Trust and threats within collaborative law

Can we trust both partners, and other suggested collaborative practitioners, to do all that is required of them? This might mean:

- Disclosing information in a full and timely fashion

- Sincerely working together within the collaborative process

- Committing to the collaborative law process, and not giving up

We have already considered, in Chapter 7, for example, the problems that can be encountered when a business owner is expected to open up his or her business to the scrutiny of an accountancy expert within the collaborative process.

Are they going to be prepared to do that and to provide unobstructed access and full answers to questions that might be raised?

If the business-owning partner is reticent then this might be an indication that the collaborative process is not suitable. That is not to say that the business-owning partner is trying to be deceitful. It is likely that the business owner has several good reasons why the existing accounts should be accepted at face value. The problem is that if the business owner is unable to accept that the joint, collaborative accountancy expert is required to provide full transparency, then he or she is not buying into the collaborative ethic.

Within collaboration, our actions are not limited to doing as little as we can get away with, or disclosing the bare minimum. That approach may be more in keeping with conventional round table meetings comprising of positional and strategic negotiating. With the collaborative approach though, we are all required to do what is needed to enable the whole group to fully understand the complete picture.

We need to consider these questions at the outset, such as 'Will you be prepared to give a jointly instructed accountant instructions to prepare a report?' rather than risk getting half way through the collaborative process and then realising that we have a problem.

Trust might have been undermined by virtue of previous attempts to mislead. If there have already been attempts to hide assets then future conduct and the chance of future deceptions will need to be very carefully considered.

8.6 Safety within collaborative law

Trust, within the TAP model, should also remind us all to consider the issue of safety. Can we trust all involved to behave appropriately? Is there a history, or risk, of actual violence, violence being threatened or other coercive behaviour?

Collaborative lawyers need to make sure that they are not placing their clients into a system that could continue a pattern of threatening or oppressive behaviour that may have been experienced during the relationship.

To some extent, the clients themselves have a responsibility to highlight problems or concerns. The reality is, however, that any partner who is being oppressed or coerced may simply go along with what the abusive or controlling partner is suggesting.

This acquiescence might be seen either in agreeing to go along with the collaborative law process in the first place or in going along with what is being said within the collaborative meetings themselves.

For the practitioners, this will mean sensitive questioning and observations, either through the written questionnaire or through dialogue. The practitioners should keep vigilant throughout the collaborative meetings that follow.

Many experienced lawyers will have developed observational skills that will highlight concerns about violence, oppression or coercion, but it might not be adequate to rely on those alone.

Within collaborative law, the involvement of a therapist or counsellor can be very useful in ensuring that all of the family dynamics and patterns are fully appreciated within the collaborative law process. The appointment of a therapist can help to bring concerns about violence or coercive behaviour to the surface, where they can then be identified.

If there are trust or threat issues, this does not mean that the couple should be turned away from collaborative solutions. It will, however, require honest conversations setting out what those concerns are and how they can be addressed.

If one partner, or collaborative practitioner, is not willing, or is unable, to have that difficult conversation at the outset, then that may well indicate that other difficult conversations later on in the process will be similarly unsuccessful.

Managing domestic violence is a very important and specialized area. A full treatise on it is beyond the scope of this book, but an excellent essay can be found online by searching for it by its title 'The Uniform Collaborative Law Act and Intimate Partner Violence: A Roadmap for Collaborative (and Non-collaborative) Lawyers'. The report is written by Nancy Ver Steegh and is hosted at www.hofstra.edu.

It also needs to be recognised that if specific threats are made, or if it becomes clear that there is a specific threat of harm or violence, this might be a criminal matter. In those circumstances, an obligation to disclose details of such criminal threats may well overrule the otherwise very high level of confidentiality.

Of course trust does not need to be absolute. It is inevitable that a couple facing the end of their relationship will have profound trust issues between them. That is not the problem in itself.

If it is possible to talk about, recognise, openly acknowledge and address trust issues then collaborative law practice can work very well indeed.

Couples can use collaborative practice not only to resolve the current issues arising out of the separation but also to create a new basis of relationship, boundaries and expected behaviour that can lead to far greater post-separation co-operation than would be seen if the couple were simply cast into the conventional legal system.

8.7 Anger and other motivations in collaborative law

We use Anger within the TAP acronym for two reasons.

The first is that anger is such a raw emotion that it is often one of the most easy to spot.

The second is that the driving nature of anger is such that it acts as a good flag and symbol for motives generally.

The Kubler-Ross five stages of grieving also lists anger as being a stage that needs to be passed through.

Anger is not a complete barrier, therefore, to taking part in the collaborative process. Many practitioners will have had many successful collaborative law outcomes where one party or the other has experienced anger either at the outset or during the process.

There are cases, though, where anger can be problematic. If anger is not recognised, not appreciated or acknowledged, then it can sabotage the collaborative efforts.

If either party is being driven to act in anger-led ways, then that is likely to be very destructive.

Anger might lead a client to seek retribution through the terms and outcomes of the negotiated settlement. Anger can lead us to think that we have an entitlement as a result of the other person's behaviour.

Anger may also lead us to try to wound the other party, not necessarily physically, but in the things that we say and do. Sarcastic remarks directed at the other person are natural enough within four-way meetings. Nobody is perfect after all. We should have enough permission and trust to pull one another up if we are communicating in unhelpful ways. It is only if these comments continue that they undermine trust and the ability to work together.

8.8 What other motives might there be?

As we have already seen, despair and depression might lead to apathy on the part of one client or another. This can lead to apathy and acquiescence.

'Oh, give him whatever he wants …' is not a collaborative stance. It is a passive, non-participatory stance. Remember, the collaborative process requires everyone to take an active role within it.

Collaborative law can be mistakenly seen as being the easy way out for the conflict-averse client.

Collaborative law, however, needs the clients, assisted by their representatives and other collaborative practitioners, such as counsellors, coaches and the like, to be able to engage in the collaborative dialogue, even to argue and debate.

If a client is expressing a desire to use the collaborative process because it avoids the difficult stuff, then care should be taken that their expectations are realistic. We will still have difficult conversations to get through and there will still be tears and anguish. Make sure that the client appreciates that.

8.9 Collaborative law as a cheaper option

Some clients might select collaborative law out of the perception that it is the cheaper option.

Collaborative law might be cheaper than conventional, contested legal disputes, but we should be slow to sign clients up to the process where this is the main or only motivation.

If a client is sold the collaborative process on the basis that it will be cheaper, then there could be problems if complications arise that add to the cost. A client might feel that they have been misled, which would have obvious consequences on their level of satisfaction and the practitioner's own reputation.

As discussed elsewhere, if cost is a motivation then there is the risk that a client is trying to shoehorn themselves into the process for that reason, even though they know that they might not believe in the underlying collaborative principles.

8.10 What can we do about unsuitable motivations?

Where the unsuitable motivation is anger then it will often be possible to work together with a counsellor, coach or therapist to move one or both parties on from their current positions. This of course requires a degree of recognition on their own part and a desire to move forward.

Where there is despair and depression then, again, a referral to a therapist, counsellor or coach might be suitable.

Bear in mind that it might be appropriate to hold off beginning the collaborative process until these issues have been discussed and the client can contribute more effectively to any collaborative meeting.

Where money and costs considerations are the motive then it would make sense to have a very frank conversation about how costs might be incurred. Can all members of the collaborative team come together to discuss fees and expected budgets? It is helpful to overestimate the likely cost and conclude matters with budget to spare, rather than set up an overly optimistic impression of what the costs might be, only to have to revisit those costs at a later date.

Ask the question, again, at the outset 'What will you do if this matter does not resolve within £5,000, £10,000 or £15,000 worth of fees?'

8.11 Positional negotiating as a barrier to collaboration

Collaborative law is more than a regulated series of round table meetings. There is a radical difference between conventional negotiations and collaborative law discussions.

Conventional negotiation is based on competing positions and a contest of who can get most of what they have asked for. The collaborative approach is about individuals committing to work together to create not a position but a mutually beneficial outcome.

A client making such demands is likely to render the collaborative process unsuitable.

Other signals can be a client refusing to sign the participation agreement until they have been able to sort out a key preliminary issue. Sometimes this is dressed up as a trial run. The more positional partner suggests that he or she would like to use the collaborative process but needs to see it in action first. He or she goes on to suggest that the first four-way meeting might be used to explore how the process works to resolve one of the issues between the two partners.

If it works, and an agreement can be reached, then that partner promises to sign up to the participation agreement at the end of that session.

What is telling is that they then go on to propose that the key issue for them, whether it is the retention of a pension, the family home, lifetime maintenance or maybe weekly staying contact, is used as the trial issue.

What is happening here in reality is not a trial run, but the positional party saying: 'If I get my way on this issue, then I will sign up to the participation agreement.'

Such approaches should be rejected. They use the promise of agreeing to collaborate once they have reached an adequate concession on a key issue. This in itself is positional horse-trading and indicates that the collaborative philosophy is not understood.

If positional negotiating tactics emerge then there might be an opportunity to speak with the client or the other lawyer and explore how and why such behaviour does not fit. It will be the lawyer's responsibility to call whether or not they enter into the collaborative process.

If positions cannot be abandoned in order to make room for more collaborative conversations about mutual interests then we need to be very careful as to whether we sign up that client to a collaborative law approach.

It might be possible to recommend a book, such as Ury and Fisher's *Getting to Yes* (Penguin Books, 1983), to a client to see if that helps to dislodge the adherence to positions. It might also be possible to have a discussion directly or introduce the client to a coach or therapist who might be able to get the client to consider not what he or she wants, but why they want those things.

8.12 WHY IS SCREENING SO IMPORTANT – THE CLIENT'S PERSPECTIVE

The primary concern has to be the client's obvious need and right to be kept safe and not to be placed into a situation which might be damaging, perpetuate ongoing abuse or coercion, or in which he or she is unable to operate for any other reason, whether mental health reasons or lack of knowledge and information.

The disqualification clause within collaborative law will have a very onerous impact if collaborative law turns out, at a later date, to be inappropriate for this family. Both clients would have to stop the collaboration process and instruct new lawyers.

A large portion of the fees incurred in the collaborative process will now need to be replicated by the new lawyers.

Clients only have finite resources with which to resolve those matters that require attention at the end of a relationship. They may well face very real difficulties if the process breaks down and they are faced with additional legal costs as a result.

Clients need to be sure, therefore, that the collaborative process can work. Any concerns about collaborative law or the disqualification clause need to be discussed openly. The outcome of these discussions should be recorded.

Consider recording such matters in writing, such as on the client care letter that clients are asked to sign.

The client has to be in a position where they have all of the relevant information to make a fully informed decision as to whether the collaborative law approach is suitable for them. This includes having an appreciation of the potential for the process to break down. It is essential that the disqualification clause and the risk that a portion of the costs might need to be replicated are fully understood.

A client who seeks to argue that they had not been so informed may well be successful if they later try to sue the collaborative lawyer.

8.13 WHY IS SCREENING SO IMPORTANT – THE LAWYER'S PERSPECTIVE

As lawyers build up their collaborative practice, they will also be building up their own collaborative reputation.

Consider a lawyer who has a string of cases that he or she has agreed to collaborate on. These cases repeatedly break down. What is more, it is apparent to the other lawyer that the client was never prepared to be collaborative, or was wrestling with his or her own anger or entrenched in positional, hard negotiating modes.

The trust that the lawyer commands will become greatly diminished.

The next time that lawyer makes contact with a collaborative lawyer, he or she needs the other lawyer to agree to collaborate.

When the other lawyer is doing that, they trust the first lawyer to have taken a prudent view on whether their client is capable of working in the collaborative process. This is important because it is not only the original lawyer who stands to be dismissed if the process breaks down, but both lawyers.

A collaborative lawyer might earn an unwelcome reputation as someone who fails to screen for suitability. As a result, agreeing to collaborate with him or her becomes something of a shot in the dark. It might work but we have no confidence in him or her to have properly assessed his or her own client's suitability.

In time, people will refuse to take a risk with that lawyer. When the invitation to collaborate comes in, other lawyers in the area might now be saying 'No thank you'.

8.14 THE ORIENTATION MEETING IN DETAIL

Having dealt with screening, the orientation meeting between the lawyer and client now starts to prepare the client for the first four-way meeting.

Looking at the agenda that will be used for that meeting gives a helpful structure for this orientation. The client should be taken through each item on the agenda so that there are no surprises once the meeting commences.

8.15 Looking ahead to the first four-way meeting agenda

The agenda for the first collaborative four-way meeting might look as follows:

(1) Meeting and introductions

(2) Anchor statements

(3) Discuss and sign participation agreement

(4) Immediate issues to be addressed

(5) Identifying longer-term issues

(6) Preparing for the next meeting

We look at each of the above below.

8.16 Meeting and introductions

This is a very important part for the lawyers to get right. We need to come across as being professional and courteous to our respective client's partner. We need to act in ways that start to dissemble those assumptions that they will have about us.

It is helpful to have some introductory-level conversation about the weather, the journey into the office or anything. This helps to relax everybody to some extent and re-humanises both lawyers in the eyes of the partners. An important consideration is to make sure that at this introductory level, you speak directly to your client's partner. This serves to demolish the conventional expectation that one lawyer speaks to one partner, and the second lawyer speaks to the other.

Take time to anticipate this period of meeting and introductions within the orientation meeting. What should you call your client's partner? Is there anything I should really avoid? How is he or she likely to feeling and likely to react to me? Are they worried? Angry?

8.17 Anchor statements

The next item on the agenda for the four-way meeting, and therefore the next item to be discussed within this orientation meeting also, is the anchor statement.

The anchor statement is a short statement. Each partner will prepare one and share it with the rest of the collaborative team at this early stage in the first four-way meeting. The purpose of the anchor statement is to explain why each client has chosen the collaborative process and to share as much or as little as the client wants to about various things that might be important to them.

Clients can feel very self-conscious about preparing and reading out these statements. Lawyers too can find them to be very difficult. There might be a temptation to gloss over this step in the collaborative process. Try to resist that temptation.

At these early stages the anchor statement gives the meetings an intimate feeling. Effective statements open up vulnerability within the collaborative team as each partner shares their greatest hopes for this process.

Anchor statements tend to be about the future and the outcomes that each party is hoping for. We do not mean what the final deal might look like, but what the hopes might be for the couple's future relationship, their ability to co-parent after separation and such like.

Focusing on the future at the very start of the collaborative process starts to establish a habit of looking forward instead of looking backwards. It also sets an aspirational tone to the discussions as each partner is identifying what they would like to see happen.

The reading out of anchor statements also establishes that this process is determined by the clients. It calls them into those active roles that they need to take up for the collaborative process to be effective. Part of that is in the clients taking the responsibility to identify and name their own needs and wishes. Another part is the need for clients to be vocal in speaking out within the meeting.

Different clients will be able to deal with anchor statements better than others. If a client is adamant that they cannot and will not read out their statement then this might indicate unsuitability for the collaborative process. We need our clients to be active and vocal.

Such reticence displayed at the orientation meeting might signal a need to do some preparatory work before we convene within the four-way meeting. Consider and explore whether working with a coach or therapist to do some quick work on confidence and assertiveness will be helpful.

8.18 Expect upsets

An important part of the orientation meeting prior to the first four-way meeting will be getting used to the idea that we should expect upsets.

The anchor statements can be very moving.

These statements might be the first time in several years that the couple have been able to speak candidly to one another about what they had hoped for and needed from the relationship.

Sometimes parents will want to share photos of the children while they read out the statements and this can be very helpful to acknowledge that although the children are not with us in the room, that they are certainly a part of everybody's considerations.

Another emotional aspect is that as the partners each read out their statement, there is often a realisation that each of them wants similar things.

8.19 Helping clients to write their anchor statements

Consider revisiting the Wheel diagram discussed in Chapter 4 to prepare an anchor statement.

See also the seven-step exercise in Appendix 3 setting out a full process for writing an anchor statement.

What is most important? What are the greatest worries?

What hopes does the client have for the future in terms of future relationship between the couple, or between any children and the other parent?

Have clients consider the question 'What story do I want my separation to tell?'

When they look back upon their relationship and subsequent separation, how do they want to feel about the time they had together, especially during the better times? How do they want to be feeling about this separation?

If they have children, how would they like their children to look back on their childhood when they are grown adults?

Is there a need to be able to work together as parents after separation?

How would the client like friends and family to relate to both them and their partner in the future?

Perhaps they have seen their own friends and family go through separation? What was their experience of their friends' separations or divorce? Was there anything there that they felt was positive and that they would like to achieve? Was there anything they saw that they would like to avoid?

There might be strong moral, ethical or spiritual beliefs held by one or both partners. It can be helpful to identify these in anchor statements to make those considerations explicit within this process.

Try not to be too prescriptive in how the client writes their statement. It is their statement and it is a highly personal seal that will be stamped upon the whole collaborative team and process.

Encourage the client, within the orientation meeting, to make time available to think about the statement. Avoid rushing it. Allow time to get familiar with what they have written so that they will be more comfortable when it comes to sharing it at the start of the first four-way meeting.

8.20 The anchor in the anchor statement

The anchor statement acts as a reference point as the meetings and negotiations progress.

Their real value might only be seen if discussions are starting to get rocky. If a ship finds itself in a storm it will throw down its anchor to save it from being blown off course. And so it is with the anchor statement. They keep our clients, and ourselves, steady.

There might be times when it is tempting to go off track, to start arguing or going backwards and looking at who did what and who was to blame. At these times it can be helpful to read out the anchor statements again and ask one or both clients:

> 'Bearing in mind what you wanted to achieve in your anchor statement, how well are we managing to work in accordance with those wishes?'

Or

> 'If that statement represented the best that you were hoping for, how close are we to getting there? Can you think of a time when we were closer than we are now? What was the difference with how things were at that time and right now? Which did you prefer? How do you think we can get back to that?'

Or

> 'At which stage in our discussion so far do you think you were closest, and what was working well at the time?'

We can get one partner to read out the other partner's anchor statement and ask similar questions to get them to think about each other's needs and wishes. Try a similar question as one from above:

> 'How close do you think Jean feels to achieving those goals at the moment? What might help her, do you think? Is there anything that you can do that might help her?'

If the discussions are really fraught then we as lawyers can take the heat off our clients by reflecting the questions back on ourselves:

> 'How can I and my colleague here [the other partner's lawyer] help you both to get back to those best intentions you listed in your statement?'

> 'What do you need from us right now to help you to really achieve what you set out to do?'

Or

> 'Is there anyone else, do you think, that we might be able to bring in to this process to help us get back to what you wanted to achieve? Maybe a counsellor, or a communications or conflict coach?'

8.21 Anchor statements act as measuring sticks

Sooner or later within the collaborative process, we are going to start exploring solutions and possible settlements. When we get to that stage it is important that we do not just accept the first settlement that is mentioned. We need to be able to check that the settlement is appropriate. This is especially important if one or both parties are still working through the emotional issues such as bargaining or depression which can lead to a tendency to accept anything.

The anchor statements can be revisited and read alongside the proposals that are being discussed. In this sense the anchor statements provide the criteria against which the effectiveness of proposals can be measured.

We might ask similar questions as before:

> 'How well does this proposal match what you wanted to achieve at the start?'

Or

> 'John, you said you wanted to make sure that Jean had a fair deal. Have we been able to achieve that with this proposal?'

In summary, the anchor statements are so important. They can feel a bit strange, phoney even, when we first encounter them and read them out, but they are very useful in keeping everything on track, especially if we encounter stormy waters further along in the dispute resolution process.

Use them.

8.22 Anchor statements at the first meeting

We have gone into considerable depth here about anchor statements to enable clients to prepare theirs, and prepare to read theirs at the first meeting.

They will only take a few moments to read through at the meeting itself.

We should be prepared ahead of that first meeting to have some discussion about the statements once they have been read out.

Lawyers might think about questions they can ask their own client about their partner's statement:

> 'How does that sound to you?'

> 'Any thoughts while you were listening to that?'

'You seem to be quite moved by that ...?

'Is there anything there that surprised you?'

The purpose here is to make it explicit that one partner has heard and understood the other. This again establishes the start of a habit and platform basis of the couple listening to and responding to one another.

8.23 The participation agreement

As we are in our orientation meeting, we can look ahead to the next item that will be on the agenda for the first four-way meeting. This will be the signing of the participation agreement by both partners and their lawyers.

The participation agreement is a largely standardised document that all collaborative law cases are governed by.

It sets out rules for how each of us will conduct ourselves, relate to and speak to one another. It has specific clauses covering various aspects of the process that are self-evident. A copy of a standard participation agreement can be found at Appendix 1.

The orientation meeting should be used to provide an advance, standardised copy of the agreement to your client. You might go through it there and then or invite them to read through it in their own time. Make sure that you do this. You do not want to get to the first four-way meeting only to have your client say that they have never seen the participation agreement before.

8.24 Is the participation agreement necessary?

Some critics argue that if a couple is really committed to working together that they do not need the restrictions of a formal agreement to enable them to do so.

That is naive.

Almost every client, when they first approach separation, has a desire to see matters resolved amicably. Almost none would choose a court process at the outset. Unfortunately though, too many of these couples get caught in a current which leads them into making court applications before any settlement is reached.

The participation agreement, therefore, is needed even though there is a commitment to working together.

The agreement strengthens and records that commitment. It makes this shared commitment explicit and sets it down in writing. The commitment is no longer simply a vague notion or a hope. It has been converted into something physical, namely the document, that the lawyers and clients are about to agree and sign.

If it feels unnecessary at the outset, when everyone is being very polite and co-operative, then we must remember that there will be real disagreements that we will have to work through. The participation agreement is helpful in reminding us that we cannot just down tools and apply to the court without repercussions.

It holds us accountable.

The participation agreement must be signed. Collaborative law can be many things and can be incredibly flexible. The only characteristic that it must feature, however, is the signed participation agreement, including a disqualification clause.

8.25 Discussing the participation agreement

In most cases the participation agreement will be prepared with all parties' names inserted and distributed by one of the lawyers prior to the first collaborative meeting. Hopefully both partners will have had a standard sample agreement, even if not the fully completed and personalised version, to read through prior to the first meeting.

Some practitioners read through the whole participation agreement at the start of the first collaborative meeting. Others take the view that is a bit dull and long-winded.

Alternative approaches might be to ask clients whether they have read the agreement and understood it. More expansive questions might be asked to ascertain if they have any queries. You might try asking what the most important paragraph was to them and why.

The lawyers can then pull out a few clauses that they want to address explicitly before signing up to the agreement.

It is prudent to spend some considerable time on the disqualification clause to make sure that everyone understands the implications of that and is prepared to agree to the participation agreement bearing those provisions in mind. This helps to ensure that the agreement is committed to by everyone with fully informed consent.

8.26 The disqualification clause

The disqualification clause states as follows:

'If a court is involved in the family break-up otherwise than to process the divorce or to make an agreed order:

A) Husband's and Wife's instructions to Lawyer 1 and Lawyer 2 respectively will immediately lapse;
B) Husband and Wife will promptly place themselves on the court record, unless they instruct new lawyers to do so;
C) Lawyers 1 and 2 will on request:
   a) give their respective former clients the completed document needed to put that client on the court record which that client will file at court immediately;
   b) help their respective former clients and any new advisers that they appoint (whether or not they are instructed to act under a collaborative agreement) by providing 1) copies of the papers that represent the disclosure in the case; and 2) a bundle of privileged documents showing the stage reached in the negotiations/discussions;
   c) For the avoidance of doubt, any expert's report will form part of the disclosure unless there is agreement to the contrary reached at the time or subsequently to the end of the collaborative process, between the parties or their representative;

   and Lawyers 1 and 2 will be entitled to charge for so doing; but
D) Neither Lawyers 1 or 2, nor other members of their firms may be involved (as lawyers or witnesses) in any court proceedings between Husband and Wife;
E) neither Husband nor Wife nor anyone on their behalf will:
   a) use in evidence Lawyer's 1 or 2's notes except to the extent that they record relevant financial and factual disclosure;
   b) call Lawyer 1 or 2 to give evidence in any court;
   c) refer a judge to the negotiations conducted under this agreement if that judge may impose a final order on any matter that has been the subject of the negotiations; but
F) any financial and/or factual disclosure made within the process may be used in evidence.'

The disqualification clause specifies that if either party chooses to apply to court rather than continue collaborating, the lawyers currently instructed by each of the parties have to stop acting for them. No colleagues from their firms will be able to act for the couple either.

The rest of the disqualification clause specifies what is to happen to the bundles of papers and what can and cannot be used in court in future proceedings between the couple.

This disqualification clause is very important. It is this clause that gives collaborative law the safety within which a couple can be as open with one another as they like. They can discuss any possible settlement. The lawyers are also able to be far more open with one another, because neither of them, nor any of their colleagues, will be able to use things that are discussed within the collaborative meetings to their advantage, or their client's advantage, in any future court proceedings.

8.27 The difference between collaborative law meetings and conventional round table negotiations

The nature of conventional round table negotiations tends to be more guarded and secretive.

Non-collaborative, round table meetings still tend to use a positional type of negotiation, where one party might go in high, attacking the other person's proposals and expecting to be knocked down a bit on their own proposal.

Remember, in conventional round table meetings, parties do not reveal their bottom line and information tends to be held back so that parties are not revealing their hand, or what is most important to them.

Collaborative negotiations are much more open. The talk is not driven by strategically high bids or bottom lines.

Within a collaborative meeting, for example, one partner might say: 'I am not interested in keeping the house at all. It's too expensive to run and we can all move into a slightly cheaper property a couple of miles away.'

Information such as this would never be revealed within a non-collaborative round table meeting. If the negotiation did not work out, you would not want the other person to know that you were prepared to sell the house. Instead, you would want them to think that you still wanted to keep the house.

If the other person then knocked you down a bit, say, at a final contested court hearing, and the judge ordered that you had to buy a slightly smaller house a few miles away, that is what you were content with all along and you would not, therefore, have lost anything.

8.28 Signing the participation agreement and disqualification clause

Ideally, four copies of the participation agreement will be signed by both partners and lawyers near the start of the first four-way meeting. Everyone then retains a copy, signed by everybody, for their own records.

If this first meeting is being used only to see if the couple can work together then the agreement should still be discussed but does not need to be signed. It may well be that simply discussing the agreement will give a good indication of whether everyone will be able to collaborate or not.

Substantial legal, financial or other disputed issues should not be discussed at that first meeting or until the agreement has been signed. Remember the earlier warning not to use an initial four-way meeting as a trial run at trying to resolve a sample issue.

If the first meeting is only a trial one then there will be more than enough content and potential to test the couple's ability to stay in the same room and co-operate without having to stray into substantive matters. If collaborative law is going to fail then it will be quite apparent.

If the first trial meeting finishes without the agreement being signed, make sure it is signed at the very start of the next meeting at the latest.

If one party does not want to sign the agreement then think very carefully about how meetings are to continue, and on what basis the meetings are progressing. The failure to have a signed participation agreement could be the basis for a claim against a lawyer in the future if there is any confusion or misunderstanding on this point.

8.29 Immediate issues to be discussed

The next item on the agenda will be to consider what pressing issues need to be resolved. These are likely to be issues which, once agreed, will provide a platform upon which the parties can rely while the longer-term issues are identified and sorted out.

Financial matters will relate to what money can be provided on a monthly basis to enable both parties to make ends meet.

The question of meeting the ongoing costs of the collaborative process itself may well arise.

There might be organisational issues relating to changing bank accounts and direct debit payments.

Non-financial matters might see very early discussions taking place about arrangements to see the children regularly.

The common strand between all of these items is that their resolution will alleviate the immediate anxieties that would otherwise antagonise the parties. If that were allowed to happen then it is likely that the couple would be ineffective within this and future collaborative meetings.

Instead of being able to explore matters creatively and with mutual regard, the parties would simply revert time and time again to the pressing issue. They would be so concerned about resolving that matter that it would become impossible to move on.

These immediate concerns should be identified before the first collaborative four-way meeting. They might have already been identified at the very first client meeting as part of the goals that the client has identified.

Even if this has happened then repeat those concerns back to the client at the orientation meeting. Are they still pressing? Is there anything else?

Also ask the client what pressing issues their partner is likely to raise. Doing so again breaks down the partisan conventions inherent in legal proceedings. We are asking our client in a very simple fashion to start using empathy and to perceive matters from their partner's viewpoint. This can be very helpful. If the partner does then raise those same immediate concerns it becomes harder for the client to resent them. They had themselves anticipated the same issues and, in doing so, acknowledged that there was a concern.

Take care when discussing these matters with clients that issues that are very emotive, and undoubtedly important, are not confused with the issues that are really pressing.

Be bold in stating that non-urgent matters need to be explored but will need to wait. We cannot resolve everything in the first collaborative meeting and there will be time in future meetings to give those other important issues the time they require.

8.30 Sharing immediate concerns between lawyers

Do not forget to tell the client that you will be meeting with their partner's lawyer prior to the four-way meeting.

Explain that the purpose of this meeting is to share what the immediate concerns are going to be.

Some clients can feel concern at this lawyer-to-lawyer meeting. Those concerns stem from conventional expectations. Does such a meeting give the game away? Would it not be better to keep quiet until we get there?

The collaborative process is designed to avoid bandit litigation tactics such as this. We do not need surprises or either party hijacking the four-way meetings with unannounced requests. Doing so will greatly undermine trust and co-operation within the collaborative team.

The lawyers will carefully consider how the immediate topics can best be broached. Will one of the lawyers invite the client with the concern to state their concern? Will it be necessary at this very early stage for the lawyer to raise the issue, given that one or both partners might not yet feel confident enough to speak about contentious issues – it should be expected that it might take one or two meetings before one or both clients feel safe enough to speak out entirely frankly.

Once the lawyers have shared this information with one another there will be an opportunity to let both partners know. That might be done by a telephone call prior to the first collaborative meeting or when each lawyer meets with their client, as they will, 30 minutes or so immediately before the first meeting is due to start.

Bear in mind that early notice might be appropriate so that both parties can consider how they might resolve the pressing concerns and come to the meeting prepared.

As a side note, if the lawyer is to broach the issue then try to do so in a way that continues to invite and encourage the client to step into that space.

> 'Moving on, Luke, when you and I met earlier this week to talk about what matters needed resolving immediately, I explained that Jean was anxious to sort out a regular time for John to meet with the children. Jean, can you tell us a bit about how you think that will help both of you and, of course, the children?'

If the client takes up this invitation then allow the conversation to flow. It can be very rewarding for everybody in the collaborative meeting to see such a high level of involvement at such an early stage. It speaks volumes about the level of perceived safety within the room and the within the newly formed collaborative team.

If the client does not take up the invitation then avoid forcing the issue. Although we advocated silence as a good thing in an earlier chapter, this was in the safer setting of the lawyer-client meeting. Long, pregnant pauses while everyone is waiting for one partner to say something, particularly at this early stage, is likely to be acutely embarrassing and counterproductive. You may well convince the client that he or she does not want to return for any future collaborative meetings.

Instead, step back in and explain the concern. Keep checking in with your client with safe closed questions.

> 'Is that right? Have I understood that correctly?'

> 'I think that is what you were telling me, isn't it?'

Throw in some more open questions as well to see if they are taken up.

> 'How else did you think this was going to help, remind me?'

> 'What else would you like to say?'

8.31 Identifying longer-term issues

The identification of longer-term issues will make up the rest of the first collaborative four-way meeting. This may well be the most free-form part of the meeting.

Both clients should have been prepared in advance to consider what their concerns are and how they might be able to identify those within the four-way meeting.

It is important that we keep the full breadth of separation considerations in mind here. We do not need to restrict ourselves to the purely legal aspects as if considering what the final settlement might look like. Instead, encourage your client to think in terms of the financial and broader parenting concerns and not just lump sums, maintenance and contact.

How might the couple get to a position, instead, where they can be independent of one another and yet still work together?

What is happening regarding the separation? Have they yet separated? Are they yet to separate? What do they need to put in place to achieve that? How might they co-operate with new homes, or communicating the changes to their children, or their own parents?

Are there any concerns about what other people might say? Does the couple want to explore how they can present a unified and optimistic, or positive, message to friends and family, who will now be starting to represent partisan supporters?

There might be issues arising from the anchor statements that merit greater attention.

Consider a fuller discussion within the four-way meeting about other collaborative professionals. Who else might the client want to have access to? Does the couple think it would be helpful to appoint a coach or therapist, and if so, to help with which issues?

It will be obvious that not everything that is discussed at this stage can be anticipated. Much of it will come about in a quite organic fashion as the conversation between the two partners, their lawyers and any other collaborative professionals develops, but clients should still be encouraged to think about which issues they would like to explore.

8.32 Preparing for the meeting – setting homework

The penultimate aspect of the first four-way meeting will be to determine who needs to do what prior to the next session. Essentially, we are setting homework.

Homework might include the following:

- Obtaining financial documents such as bank statements and asset valuations;

- Obtaining other financial information such as cost of housing and sample housing particulars or replacement car costs, information about income or newly available tax credit allowances;

- Checking work rotas or getting school holiday dates to help facilitate conversations about contact arrangements;

- Obtaining details of suitable therapists or counsellors that either partner might access individually or together;

- Agreeing to meet with another identified collaborative professional again either individually or jointly or both.

Make sure that these steps are understood and accepted by the person with responsibility for them. It is important that these homework steps are completed once they have been committed to. Trust within the team can be damaged when one partner, or lawyer, does not complete the tasks that they have said they would.

8.33 FEEDBACK ON THE SESSION

Asking for and giving feedback on the collaborative sessions is very important but it is rarely done. If more practitioners included some preparation for this within the run-up to the first collaborative meeting then it would become much easier.

Prepare your clients, therefore, to expect to be asked questions such as those that follow. Make it clear that their honest feedback is required so that the next four-way meeting will be better. Remember the cycle of experiential development? That requires that once we have tried to implement what we said we would do, namely once we have come together within the collaborative meeting, we should elicit immediate feedback on the three Ws:

- What worked well?

- What didn't work so well?
- What would you do differently next time?

Asking these questions gives permission to the clients to let off steam about issues that might have been annoying them throughout the session but which they have felt inhibited to mention. Remember the discursive relationships that clients and lawyers can still fall into?

The lawyers may well ask each other these questions as well. Doing so presents an ideal opportunity to continue to build a rapport and trust with the other partner, especially if you can feed back on comments made by, or the general involvement of, the other partner.

When answering what did not work so well try to be aware of your own contributions. How could you have improved what you have done and brought to the collaborative effort?

It might be more appropriate to make comments on improving the clients' contributions and participation when answering the 'What would you do differently?' question.

Consider other questions such as:

> 'How do you think that went?'
>
> 'Did it meet your expectations? In what way?'
>
> 'Has this been helpful today?'
>
> 'What progress do you think we have been able to make?'
>
> 'What are you each going to do now between this meeting and the next one?'
>
> 'How can Luke and I be of more assistance between this meeting and the next one? Is there anything more – or less – that you would like us to be doing?'

I repeat the encouragement to take steps to make this kind of feedback more explicit at the end of every meeting. The questions feel weird when we first ask them.

If you are the more experienced collaborative lawyer then take the initiative in asking the questions. By doing so you will be demonstrating how the questions sound to your less experienced colleague. I know that I was very grateful to a local colleague when he led this kind of feedback session the first time that I collaborated with him. I had studied the collaborative model and knew that this was a part of the exercise but my own embarrassment at asking these very direct questions had inhibited me so far.

Once I had seen feedback in action I realised that the clients appreciate being asked for their thoughts. They were full in their responses and gave pointers on how the next session could work even better.

When that next session finished I was happy to step in and initiate the feedback questions, and I have done ever since.

8.34 NOW DO IT

This chapter has looked at the work that the lawyer and client would do as they prepare for the first collaborative four-way meeting.

This work will commence as soon as the client and lawyer first meet, particularly if the lawyer follows the suggestions in Chapter 4 regarding the different approach to a collaborative first client and lawyer meeting.

Do try to make time available to meet again with the client once the collaborative process has been agreed with the other partner. His or her second meeting should be seen as an orientation meeting.

Be sure to satisfy yourself on screening issues – be sure that the client wants to collaborate and is in a position to do so. Watch out for the TAP considerations. Make sure that you explicitly discuss the important aspects of the participation agreement and the disqualification clause in particular.

If the client is not yet ready to contribute to the collaborative meetings then consider whether there needs to be some preliminary work done with a coach or therapist. If so, put back the first collaborative meeting and get instructions from your client to authorise you to explain to the other partner what is happening.

Anticipate the first collaborative meeting. Use a standard agenda and go through the various headings discussing the issues set out above in this chapter.

Doing this will ensure that the first meeting is as comfortable as it can be for your client. It will never be easy, but at least there will be no surprises. This orientation and prior preparation will also enable the meeting to be as productive as it can be.

Having done all of this, there is now only one thing left to do, and that is attend the meeting itself. We will look at the four-way meeting itself in the next chapter.

CHAPTER 9

MANAGING PROCESS AND CONFLICT

In the previous chapter we looked at how to prepare before attending the first collaborative four-way meeting. This was done on the basis that the client and lawyer would have already met for an initial appointment and that there was then a subsequent orientation meeting.

We proposed that the basis of this meeting could be to work through the typical agenda for the first collaborative four-way meeting. This would enable the client, and the lawyer, to anticipate issues that are likely to arise.

This orientation meeting also provides a further opportunity to make sure that the collaborative process is suitable for this couple, and that the client can cope with and contribute to the collaborative law process.

This chapter will look at what happens next as we commit to the full collaborative process. We explore the cycle of meetings and the rhythm that collaborative law cases experience.

We also explore the management of the process. This includes managing the meetings – note-keeping, who says what when and the use of visual aids such as flip charts. Management will also mean managing impasse, walkouts and conflict in many cases.

9.1 THE COLLABORATIVE CYCLE – THE FOUR-WAY MEETINGS

Collaborative law takes on its own fluid rhythm.

There will be a series of collaborative four-, or more, way meetings where everyone concerned comes together. These can be seen as peak moments within the cycle. They generate a great deal of noise and anticipation. They are the markers by which the progress of the collaborative resolution can be measured.

There will then be a degree of residual work in the days that immediately follow the four-way meetings. The lawyers agree and exchange minutes.

The couple will have shorter- as well as longer-term tasks to complete between the end of one four-way meeting and the start of the next.

Do not forget, the importance of getting documents prepared by the time they are agreed to be completed by is crucial to minimise frustration and delay and avoid undermining trust.

9.2 The lawyer-to-lawyer debrief

The two lawyers may well meet together without the clients immediately after the four-way meeting, or in the days following, to reflect on what has happened so far.

What worked well, what did not and how might the process and meetings be improved?

These lawyer-to-lawyer meetings might feel uncomfortable if either lawyer wants to raise an issue of feedback for the other to consider. Something might have been said or done that was unhelpful, for example. In other circumstances, one of the lawyers, or both, might have inadvertently slipped into conventional patterns of negotiating or representing their clients in a positional way.

In these circumstances, the lawyers need to be able to listen and acknowledge the feedback being given. It may be that they do not agree with it. They will almost certainly feel pangs of defensiveness and a desire to rebut, excuse or otherwise explain away what is being said. Do not do so. This will only lead to deterioration in the collaborative relationship.

Ask questions, instead, for specific details about what was said or done. Ask the lawyer who is feeding back 'Would it have been more helpful if I …' or 'What would have been a better way of putting that question, do you think?'

We explore managing conflict in more depth below and there may be other considerations there for you to bear in mind. These feedback sessions, if handled competently, can provide rich learning for both lawyers and have a profound impact on their relationship and reputation.

9.3 The lawyer-to-client debrief

Each lawyer and their client might have a similar feedback discussion as well, despite having had the full feedback session at the end of the last four-way session. While it is to be hoped that both clients felt able to give feedback within the main meeting, the reality is that one party or another might have been anxious about doing so, particularly if the feedback was about their partner or their partner's lawyer.

9.4 Advice given privately in lawyer-client sessions

Be aware that the client might be inclined to try to develop the issues that were discussed collaboratively within the meeting at these private lawyer-client meetings.

This should be resisted.

The thrust of collaborative law is that the bulk of the work is done within the collaborative four-way meeting. Transparency and trust is maintained accordingly. It can be very difficult to maintain that trust if it appears in a future session that there has been a considerable amount of advice given between meetings.

If your client wants to discuss issues which form the substance of the collaborative sessions then explain that you will ensure that these concerns are explored in the next session.

If the client insists on discussing substantive matters privately then it will be necessary to ask whether or not this case is suitable for collaborative law. There should be no need to hide anything within this process. When we do so, we risk slipping back into conventional modes of strategic, positional or selective dialogue. This is contrary to the collaborative effort and commitment.

Strongly consider why any advice should not be given within the collaborative meeting room or at least later repeated within the collaborative meeting. If there are reasons to keep these discussions secret, is it still appropriate to collaborate?

We have seen above how the lawyer-to-lawyer and lawyer-to-client debriefing meetings represent the immediate aftermath of the four-way meetings.

There may now be a trough of relative quiet before the next meeting is upon us.

9.5 Diarising the next meeting

The date of the next four-way meeting should have already been diarised at the end of the one that precedes it.

Try to avoid finishing one meeting and not fixing a date for the next one. This can lead to great delay. One party or another might find it difficult, particularly after a challenging session, to leave the meeting and then gear up again for a next meeting. Committing to the next date before everyone leaves the room makes it more likely that the next meeting will happen.

Both parties should be getting on with their respective, or shared, tasks in this period, such as gathering information on school holiday dates or documents and valuations of assets and the like. One or both clients might have resolved to meet with one of the other collaborative professionals detailed in Chapter 7.

As the next four-way meeting approaches then the activity starts to build up once again. We find ourselves in the same process as before the first collaborative four-way meeting.

9.6 Lawyer-to-client orientation meeting

The client and lawyer will meet again. They will review the work that the client has been doing, and any tasks that the lawyer might have had to fulfil.

They will anticipate the meeting ahead and what might discussed.

The lawyer should make sure there have been no developments on issues that have already been discussed or within the client's lives generally. Examples might be new partners or an illness in the family.

Check again for difficult issues that might arise and explain to the client that these will be shared with their partner's adviser to minimise conflict and surprise.

Check on the client's health, particularly if it had been said that they were going to visit a GP or if health issues are already known.

Remember to continue looking out for suitability issues. Are there any of the TAP indicators that might suggest threats of violence or coercion? Are there any trust issues emerging? Is anger or any other emotion playing a prevalent role?

The biggest potential shift often is that a client has adopted a very determined position since the last four-way meeting.

It is inevitable that the client, either alone or together with his or her partner, will have been thinking about what was said in the previous meeting. Clients frequently come to the later orientation meetings believing that everything is fixed. They have either worked out a solution and ask the lawyer to present that at the next meeting, or even explain that they cannot see the point in having the next meeting. Alternatively they might report, sincerely, that they and their partner have reached an agreement between themselves.

9.7 The flight into health

There is a recognised phenomenon within therapeutic circles called the flight into health.

When this happens a patient responds almost immediately to therapeutic interventions. There is some debate as to whether these are genuine recoveries or a denial-based response. The response, it is argued, is an attempt to convince the self and the therapist that no further work is required and that everything is now fine. This might in turn be an effort to avoid further treatment or discomfort.

If the client sees the collaborative lawyer, or any other collaborative team member, in these practitioner terms, then an expression of 'Everything is alright now. We have sorted it out between us. Thank you!' could well fall into this category.

Suggest that the proposals can be discussed and tested for effectiveness at the next meeting.

These proposals should be held lightly. If they become the core focus of the collaborative discussion then we will greatly restrict the issues that the couple can explore.

It should be explained that the proposal is a still just a proposal. It might even be the proposal that will settle matters, but that the couple should take some effort to explore what other possibilities there are.

If the proposal becomes the central topic for discussion then the collaborative effort will have reverted to a positional basis where one partner and then the next argues whether the proposal should be improved or otherwise tweaked.

Further examination of a stated agreement often reveals flaws.

The agreement might be unsustainable in the medium to long term. It is often much more partisan than it is being held out to be. Exploring the proposals in a collaborative meeting often reveals that one party is much more pleased with it than the other. Indeed, very often one party can be seen to be almost entirely passive in acquiescing to the terms. This might be for one of the Kubler-Ross five stages of grieving reasons. More likely is that one partner has indicated agreement to get some peace from his or her well-meaning but insistent partner.

9.8 Lawyer-to-lawyer orientation meeting

As before, the lawyers will meet to explain sensitive and important issues that are going to be raised in the upcoming four-way meeting. This can be an excellent time to explain what, if any, legal advice has been given and how that can be acknowledged within the four-way meeting. Remember to have made sure that your client recognises this will happen.

9.9 The next four-way meeting

We have now come full circle as we reconvene for a second, third or more four-way meeting.

The agenda will have been agreed in advance as a result of the lawyer-to-lawyer orientation or correspondence passing between the lawyers. Alternatively, in more relaxed cases the agenda might be agreed at the start of the meeting.

The lawyer and client, as before, should arrange to meet a few moments before the meeting and the lawyer should take care to ensure that the client is comfortable. Check again for any developments and if there are any then let the other lawyer know, again, to avoid surprises.

These later meetings will feel very different to the first.

If there were not other professionals at the first four-way meeting then there may well be now. It might take a bit of time for that new presence to be accepted in the room without people being self-conscious.

The couple themselves will have a better anticipation of what they should expect within the collaborative process. As a result the lawyers can expect their clients to be more vocal and spontaneous.

For long periods of time within these subsequent meetings the couple will be able to direct their own conversation. These instances can be remarkably powerful. The lawyers' role should be to observe and listen. Notice the language that is being used and issues that one party might raise that are not responded to fully, or at all, by the other.

Be slow to interrupt as long as the couple are making progress. At these moments, the couple is creating a new way of communicating with one another that will serve them well post separation. They will be learning new forms of resolving their disputes for themselves. These are valuable moments.

If the discussion falters then consider picking up on the points that have not been fully explored. Take care at this time to use the words and phrases used by the couple. This is still their conversation. If we start to use different words to the ones they used then we risk taking the conversation away from them.

At other times it will be necessary for the lawyers to drive the conversation.

9.10 USING QUESTIONS EFFECTIVELY

As we saw in Chapter 4, we should use questions when encouraging our clients to contribute to the collaborative meetings. Doing so reaffirms their active role and their positions as experts in their own lives.

Consider the kind of questions that you use carefully.

9.11 Closed questions

Use closed questions sparingly. These are questions which will or could elicit a yes or no, or some other either/or response.

> 'Have you thought about future housing needs?'
>
> 'Is that a job you can see yourself doing?'
>
> 'Is it important that the children see both parents?'
>
> 'Is that okay with you?'

With practice these questions can be turned into open questions.

9.12 Open questions

Open questions make it very difficult to give a yes or no answer. Open questions typically start with words like who, what, how and when. The questions above would now sound like this.

> 'What are your future housing needs?'
>
> 'How would you feel about doing that job?'
>
> 'What would be the benefits of the children seeing both parents?'
>
> 'How do you feel about that?'

You might also want to consider two other types of questions, namely probing questions and broadening questions.

9.13 Probing questions

These are designed to go deeper into a response that has already been given. Always bear in mind that the answer provided to a previous question might be fairly superficial. A client might feel anxious about the full impact of the answer they have in mind. The answer they give might therefore be an attempt to answer the question that has been asked but only at a level they feel comfortable with.

Consider using probing questions in these circumstances.

> 'What happened then?'
>
> 'What was the most important thing about that for you?'
>
> 'What would help you to feel differently about that?'
>
> 'When did you think that?'

9.14 Broadening questions

These questions encourage clients to broaden their considerations. The result is different from probing questions.

Probing questions can be seen to take us down further and deeper into a client's understanding of an issue.

Broadening questions help to open up, once again, the range of answers being given.

> 'Which other houses might be appropriate? And which others?'
>
> 'What other jobs do you think you could do? Give me five others.'
>
> 'What else?'

9.15 LISTENING

It is crucial that we listen well to what is being said.

Do not be afraid to feed back regularly on what you think has been said. If you have misunderstood then the clients will be happy to correct you. You will also be demonstrating that you are trying to understand the conversation from their point of view.

If a complex point is being made by the other lawyer or one of the other collaborative professionals within the meeting then ask a question as a means of repeating the issue to enable it to be fully appreciated.

When we are listening we should be sure to respect the clients' agenda. This is their case and everything should be about them and their family. There can be tendency in normal dialogue to understand what people are saying in accordance with our own world view or experiences that we have had. This level of listening is reflected in comments such as 'That happened to me once', or even 'I understand' or 'I know'.

Try to demonstrate that you are listening to the other person and understanding it from their position. Keep on asking probing or broadening questions.

> 'And how did that make you feel?'
>
> 'And then what did you do?'
>
> 'Why was that important to you?'
>
> 'What else might be a way forward for you?'

Do not talk about yourself but talk about them.

More experienced or confident collaborative lawyers and other professionals might go beyond this straightforward listening and try even harder to understand what is going on. One way of doing this can be by observing your client and the shifts in their expressions, emotions and body language. It might be that you have an almost gut feeling that there is something different in the room.

If an adequate rapport has been established then it can be very helpful to comment on these shifts when you notice them.

> 'You looked very upset when she said that. What were you thinking?'
>
> 'What was it that made you hesitate there?'
>
> 'What were you going to say before you stopped yourself?'
>
> 'I notice that you are much more animated now. Why might that be?'
>
> 'It seems to me that things have got a bit heavy in here. I wonder what might be going on.'

9.16 KEEPING NOTES AND MINUTES

Keeping notes and minutes within collaborative meetings can be a big problem.

If one lawyer or professional is keeping notes then he or she can find it difficult to make as full a contribution to the meeting as the other lawyer.

It is likely that the lawyers will switch note-taking responsibilities from one meeting to the next.

An alternative approach can be to keep a note of the briefest comments or structure that the meeting follows. At the end of the session, as part of the wrapping-up and feedback, there can be a discussion along the lines of 'Let's recap and agree what we have discussed then …', enabling everybody to have a fuller role in what is noted. It might be appropriate to use the flipchart, if there is one, to do this where everybody can see.

There is probably little need for elaborate minutes except in the most detailed of cases.

In most situations it will be appropriate for the lawyer responsible for preparing the notes to dictate or write up a brief attendance note recording what has been discussed. That should be sent to the other lawyer as soon as possible for approval and any amendments or additions and, once agreed, sent out to the clients for their own records.

These notes are records of without prejudice conversations. They cannot therefore be used if the collaborative process comes to an end and the couple find themselves in court proceedings. Beware, however, that factual matters are likely to be open to inspection in future court proceedings.

9.17 FLIPCHARTS

Collaborative professionals should get accustomed to using flipcharts to help them to record and share information.

Using flipcharts gives a very visual dynamic to collaborative meetings that can help to clarify issues. If, for example, the couple are listing ideas on an issue then it makes sense to record them on the flipchart.

One of the professionals or clients might have to explain a complex point. Using diagrams can greatly help to explain less straightforward matters in a way that is memorable. More experienced practitioners may well have diagrams or models that they use in almost every case to explain concepts such as joint ownership of a house or the division of income, capital and pensions.

It is a good idea to remember illustrations that work. And those that do not.

Be sure to make it clear that the clients can get up and write on the flipchart if that will help them. This is another way of levelling out the relationship between lawyer and client and repositioning them as equal within that relationship.

If a meeting is getting rather bogged down and stuffy and the level of energy in the room is dropping, then the mere act of one person getting up from the table and starting to share something on the flipchart can change the dynamic. All of a sudden everybody is looking up. The act of writing on the flipchart alone has disrupted a slow part of the session.

Using flipcharts is also an excellent way of getting through impasse.

9.18 DEALING WITH IMPASSE

Any collaborative lawyer will encounter impasse sooner or later. Thankfully, not every client has to.

Impasse is the situation when discussions and negotiations have come to the point where they are stuck. It can feel like a desperate place to be. There is a sense that no one is prepared to move. It might feel like we have reached the end of the collaborative road and that the process and our efforts have failed.

Contrary to these sentiments, impasse is very often capable of being resolved.

Adopting a different approach can help to unblock the discussion and get things moving again.

An example would be to use the flipchart to recap what has been agreed.

Write a list of everything that we have been able to agree upon so far. This is very powerful as it focuses everybody's minds on how much progress has been made to date. At the moment of impasse we are all fixated, instead, on what we cannot agree. Having reminded ourselves of several points we have agreed upon it is almost as if we get back into the habit of reaching agreement. The thing we are stuck on can suddenly seem less of a problem.

Impasse can be a good time to go back to the probing questions seen above.

> 'What is it about this issue that we might be getting stuck on?'
>
> 'What is important to you about this? [Turning to partner] And to you?'

It might be helpful to backtrack from the point that we are stuck at by asking: 'How did we get stuck here? What were we talking about?'

This invites the parties to explain, usually, what each other said and did. Very often at this stage we will see where one or both parties has misinterpreted or felt hurt or angry with the other person. 'Well we were doing fine until he had to go and mention her again.'

We might call these wound points and they indicate quite precisely where the issue is. Again these will need careful management. These are likely to be conflict areas and we explore conflict management within collaborative law below.

Some practitioners will offer to take the pressure off the clients by suggesting the impasse is a problem on the part of the lawyers.

> 'I'm sorry. I have got us all stuck in this cul-de-sac. If there was one thing that [other lawyer's name] and I could do for each of you to get us moving again, what would it be?'

Try other 'If we could …' questions such as:

> 'If we were not stuck on this issue then what might we be discussing now instead?'
>
> 'If you could say one thing to get all of this moving again, what might it be?'
>
> 'If I could do one thing now, what would it be?'
>
> 'If you were not feeling and thinking this way now, how else might you (or how else would you like to) be feeling or thinking?'

Remember the purpose of these strange-sounding questions is to bypass the critical thought part of our minds and enable positive and creative considerations to take centre stage.

If the parties remain stuck then try shifting the narratives by getting one partner to explain the impasse from their partner's perspective. This exercise needs care. Do not allow either client to use this exercise as a means of criticising their partner

Invite the clients to consider what will happen if they cannot overcome this impasse. Ask questions about those alternatives to focus on the need to move forwards.

Finally, consider taking each client back to their anchor statements as discussed in Chapter 8. The statements are said to reflect the clients' biggest hopes for the end of their relationship and their futures post separation.

Have each client read their own or each other's statement. The statements will remind the clients of why they committed to the collaborative process in the first place.

Get each partner to reflect back to one another about what the other partner has said. Ask them how they feel about each other's statement, or how they feel about their own statement.

Do those statements still apply?

Do they still want to resolve matters or are they happy to leave things where they are?

Ask them how we might be able to keep exploring matters so we can be sure that we have all done everything that we can to honour each other's anchor statements.

9.19 USING EXTERNAL SOURCES TO RESOLVE IMPASSE

Consider bringing in an external source to resolve impasse.

There is a growing trend of mediators working with collaborative teams to get over a current obstacle.

Another service provider could be a collaboratively trained barrister. These barristers will take instructions from the whole collaborative team.

They might be asked to advise on what would a judge say so far as distribution of an asset is concerned, or they might be asked to provide an opinion on how a narrow point of law might be interpreted.

Requesting a barrister's opinion on possible distribution of assets or time in a child care matter is problematic.

The barrister is being asked to adopt a hypothetical judiciary role. Their intervention risks clients deferring to a higher authority to resolve their problems.

This can clash with the collaborative law notion that the couple are able to create their own solutions. As soon as an individual is granted a greater authority then the couple, and the lawyers, can very quickly snap back into conventional positions. Ordering out for an opinion on terms of settlement suggests that the couple have reached an impasse and given up on their own collaborative creativity to overcome that impasse.

It is as though they are throwing up their hands and saying: 'We are stuck. What shall we do? Let's get someone else to sort it out for us.'

This is no different from the conventional discourse where the client absolves responsibility for resolving matters to the lawyers entirely. It is almost inevitable that anything suggested by the barrister becomes the new yardstick by which any alternatives are measured. If an alternative suggestion is made then the party who was to benefit better by the barrister's opinion might find it hard to now pretend that they had never heard that opinion.

There is another role for barristers within collaborative law which is in less conflict with the collaborative ethos. There is anecdotal evidence of collaborative teams referring out specific questions relating to the interpretation of law, as opposed to requesting a determination of entitlement and division. An example might be a collaborative team needing an expert opinion on what an element of tax law, or a lesser known aspect of statute, might mean and what is required to ensure that the collaborative settlement is compliant with that.

It may well be that as barristers enjoy greater direct access to clients they will start to take on the role of collaborative lawyer and help the couple to determine their own outcomes as opposed to handing one down upon them.

9.20 REINVENTING THE WHEEL

Consider using a Wheel exercise to break an impasse.

See Chapter 4 for an example and use the flipchart to draw the eight-spoked wheel. Have the couple define what each spoke will refer to. A dialogue can now follow exploring how comfortable, how problematic or how important each issue is.

The merit in doing this exercise is that the collaborative team is challenged to consider eight aspects of the issue that they have got stuck upon. This can break away from single-minded considerations that might be blocking progress. It can also help in identifying which aspects are really serious and which aspects show a greater deal of consensus between the couple.

The revived conversation can easily highlight several new avenues to pursue and enable dialogue to resume.

Consider breaking off the current meeting if matters are becoming very strained and asking the couple each to complete the Wheel, with the same eight labels, and then discuss them at a later meeting.

9.21 CHECKLISTING CONFLICT

Impasse might be an indication that something has not been fully appreciated.

You might find it helpful to work through a list of possible conflict points.

Canadian conflict resolution expert and author, Gary Furlong, talks about breaking down the interests that underlie positions as falling into three categories, namely Results, Process and Psychological. This could form the basis of impasse-breaking questions or considerations.

Alternatively, revisit the seven aspects of separation. If we have reached an impasse, could it be that one of these aspects is being overlooked?

Also consider the dominant narratives that we listed in Chapter 3. Have we fallen into one of those? Can we create our own name for the prevalent theme or narrative that we might have created for ourselves? If we are able to give this conflict a name then we will have externalised it and made it easier to challenge.

9.22 NINE ROOTS OF CONFLICT

Also consider using the nine roots of conflict list introduced below. This suggests that conflict stems from one or more of the nine following points.

(1) Resources

(2) Law/entitlement

(3) Facts

(4) Behaviour

(5) Communication

(6) Power

(7) Values

(8) Character types

(9) Displaced conflict

9.23 Resources

Resource conflicts arise from a consideration of 'What is available?' In impasse situations we might discern that the couple are reverting to zero sum thinking. They might be looking at what the other party is getting as a direct reduction of what they are getting.

Ask if this is a concern. Recall situations when discussions were more creative or less concerned about what one party was retaining. Ask what has changed and why.

Try to move discussions away from resources and the distribution of them. Instead ask about what needs or interests the resources satisfy. Recall the section on the two sisters and the orange in Chapter 1.

Bear in mind that our clients are separating in the middle of a range of social groups and influences. Are there people who are not in the room, such as family, friends, colleagues or others, who are warning the client to adopt more conventional practices such as holding out for more or playing dirty? Are they arousing fear and casting doubt on the collaborative commitment? If so, revisit anchor statements. Ask questions that will get the client to recall their earlier goals and ambitions. Ask if they want to recommit to those goals and what the benefits would be if they were to do so. You might also ask what the consequences might be if the zero sum approach was pursued.

9.24 Law/entitlement

Is a sense of entitlement creeping into the discussions? Ask questions to determine where that sense might come from.

It might be that a client has again heard comments from another person outside of the collaborative. Maybe a mutual friend had been able to use a particular argument in conventional proceedings that resulted in a higher award, or more time being spent with the children.

Entitlement can also derive from a sense of injustice. If one party feels that they have been wronged then this will need careful consideration. How can that client be encouraged to let go of that sense of being wronged, rather than holding onto a sense of hurt?

Consider referring the client or the couple to counsellors, coaches or therapists to help the client to process this emotion so that it is no longer an obstacle.

Alternatively, the lawyers might feel confident enough to ask questions about the issue.

A narrative approach would see the lawyer asking questions about this sense of injustice. Give it a name or a label:

> 'What might we call this feeling of injustice? How does it make you feel when it arrives? How did you feel differently when it was not with you? How

> well does this sense of being wronged serve you? Or your family? Does it make agreement more or less likely? Would you like to spend more or less time feeling these things?'

If there is an issue of legal entitlement then the dialogue will be slightly different.

We saw above how the role of law can be shown to be arbitrary by the collaborative lawyers exploring contrary interpretations of the same law. If the lawyers each argue a certain stance and are still able to accept that the other lawyer's interpretation has some merit, then the presumption that law creates an absolute entitlement is greatly diminished.

Even if the legal entitlement remains, it is only one possible outcome for this family. Just because it is the statutory entitlement does not mean that it is the best outcome for this family. There might be any number of other possible permutations and arrangements. Note how this option does not remain once a barrister's opinion has been brought in. Both lawyers will have implicitly surrendered the weight within their respective interpretations by deferring to the external opinion that has been sought.

9.25 Facts

Facts-driven impasse might indicate one of two problems.

Either the collaborative team is looking backwards and debating exactly who did and said what, or there are current issues with trust between the parties and whether current assertions and disclosures are true or not.

Looking backwards does not serve the couple well.

It would be glib to say that the parties should never look backwards. On occasion it might be necessary to acknowledge one party or the other's interpretation as to what has happened and how that has led the client to feel. Care should be taken, however, that this does not lead to the sense of entitlement seen above.

If there are questions and concerns in this regard then the clients should be encouraged to speak them out so they can be explored. Those discussions, in turn, should explore whether past facts are relevant now, beyond a need to be acknowledged. Will those past facts determine future settlement and, if so, how?

If there are concerns about current facts then there is a fundamental problem with trust.

Trust is the first TAP consideration. If our clients cannot trust one another then there has to be a consideration as to whether the collaborative process is appropriate.

Query where the distrust is stemming from and how trust might be revived, if it can be at all. Mistrust may be nothing more than a misunderstanding. The client might be falling into the grips of a dominant narrative as explored in Chapter 3. That might be being encouraged from supporters and other well-intentioned parties who are not within the collaborative team.

It may well be that one party has demonstrated that he or she cannot be trusted within the collaborative process. The need to be able to trust one another is an imperative within collaborative law.

If it is the case that one party has been shown to be evasive or misleading, then there will have to be consideration as to whether the collaborative process can be continued. If it cannot then lawyers need to accept that they must advise their clients to instruct other lawyers.

9.26 Behaviour

When reaching impasse, running through the nine roots of conflict might flag up that there is something that somebody within the collaborative team has said or done that is causing immense irritation.

Perhaps one of the lawyers, in an effort to be civil, is overly enthusiastic every time their own client's partner speaks. Maybe one of the couple rolls his or her eyes repeatedly, or clenches their jaw.

Encourage the clients to address these irritations and encourage a non-defensive stance to be adopted.

We do not need to accept that feedback such as this is necessarily accurate, but we should watch out for it in future in case it is.

Behaviours are likely to be unintentional and few people would continue with such behaviours once they have been pointed out.

Behaviours might not be as explicit as the things that we do. They might be found in attitudes instead. Again it might be appropriate to refer to a coach, counsellor or therapist to work on these issues and to enable the collaborative process to progress smoothly.

Be aware that attitudes and behaviour might indicate that there are aspects to be considered within the emotional aspect as explored in Chapter 2. Ask yourself whether these issues relate to the Kubler-Ross model of grieving and its five stages of denial, anger, bargaining, depression and, less likely to be problematic, acceptance.

9.27 Communication

Are there communication issues that are driving conflict and impasse?

Look out here for patterns of communication that repeat themselves. These might be figures of speech, ways of referring to one another or non-verbal communication such as posture or facial expressions.

In other conflicts, cultural differences in communication can drive conflict. That might be less likely in collaborative law, at least between the couple themselves as they are likely to have developed a shared culture of communication. These cultural differences can still arise, however, as external professionals, including the lawyers, are brought in to help with discussions.

Try to find and identify areas where communication is not working or is leading to misinterpretation or crossed meanings.

9.28 Power

Conflict can often persist because one partner feels that they are at a disadvantage. This can indicate that there is an issue with power hierarchies within the collaborative team.

One partner might feel disadvantaged in relation to the other.

This might stem from an imbalance in knowledge or ability or an imbalance in the relationships between individuals within the collaborative.

What can be done to level out these dynamics?

If the question is one of knowledge or ability then consider asking the financial neutral, or a children or parenting expert, to spend some time developing awareness and familiarity with the relevant issue. If an appropriate expert has not been added to the collaborative team yet then consider bringing them in. Remember to discuss the costs involved in doing so and how they might be met.

If the question is one of social dynamics within the group then it will be the responsibility of everyone concerned to stay aware of that dynamic and make sure that it stays in check.

If the problem is one of confidence then consider asking a coach or therapist to get involved to increase skills in this specific area.

When dealing with power conflicts it is important that the imbalance is recognised not only by the partner in the lesser position but also by the person who is felt be in the more dominant position. If that cannot be acknowledged then it can prove impossible to challenge the power issue. That cannot be allowed to happen. The result is that the settlement will become skewed, if any agreement can be reached at all. Alternatively, the less confident partner might simply leave the collaborative effort and refer themselves to a higher level of representation, such as a conventional lawyer, to act for them in conventional ways as defender and advocate.

9.29 Values

Are different values within the collaborative team leading to impasse?

These values might differ between the couple themselves or between individuals who are assisting the couple within the collaborative team.

Consider again how those values might be different.

A common one, for example, can be a collaborative lawyer's own passion for the collaborative process to rub against the couple's need for no more than a solution.

There might be religious or moral values that are at odds.

Value systems and assumptions within the family might start to come to the surface. These might be based around gender roles or approaches to parenting, finance and other issues.

Once such issues have been recognised then it might be possible to reach an understanding in relation to them. Values can be so deeply entrenched, however, that it might be necessary to use a family or couples therapist to explore ways of bridging these differences and developing strategies for coping with them in the future.

9.30 Character types

The penultimate conflict source can be something which is no more complex than each individual in the team being just the kind of person that they are.

Businesses have used personality profiling for many years. There are several proprietary models for testing what strengths and characteristics individuals display, and also what their weaker points may be. Systems of varying complexity can then be built up around these different personalities to enable different types to relate to one another more effectively.

One model might, for example, refer to different personality types as having colours.

One colour, blue, might indicate an analytical, methodical approach to problem solving and inter-personal relationships.

Another, green, would suggest a strong need for inclusion and harmony.

A third, red, would see a rather driven and outcome-focused approach, demanding on measurable progress and results, while a fourth type, yellow, is highly creative and spontaneous.

There is no reason why personality profiling cannot play a role within collaborative teams. Each individual would then have an awareness of their own tendencies as well as those of the others. That in turn would lead to a recognition of the need to communicate differently with each other.

A more analytical individual, for example, is not going to respond very well to the spontaneous, creative type's latest idea if there are three lines of discussion which have not yet been resolved.

The results-driven partner can become easily frustrated if collaborative meetings feel as though they are drifting or are without set objectives.

The individual who seeks inclusion and harmony can be even more conflict averse than others and might find it difficult to speak out.

Many collaborative coaches will be able to advise on these types of systems. It is not thought that they are being used within collaborative law yet, but there is no reason why they should not be either at the outset to help develop the team's mutual understanding or at times of impasse.

9.31 Displaced conflict

Displaced conflict sees a partner's irritation at an issue being repressed and emerging in relation to a different topic.

A partner who is under pressure in relation to their work, for example, might suddenly become difficult within collaborative meetings. There could be any number of external pressures which might impact on conduct within meetings in this fashion. This underlines the importance of preparing fully for the four-way meetings.

Ensure you know what else is going on in your client's and their partner's lives. Are there health, employment, money or broader family issues? If so, should these be brought to the attention of the collaborative team at the meetings?

The conflict can also be displaced within the collaborative process itself.

For example, one parent might avoid debating contact arrangements for fear that the parent with whom the child lives decides to leave the area. The frustration in feeling that this issue is effectively too risky and therefore out of bounds for discussion might lead that parent to be difficult in relation to another issue.

It is for this reason that the broadening questions we explored above can be very helpful. Ask if there are other things going on which are having an impact on the issue at hand.

Care should be taken to make sure that the stated source of impasse or conflict is not overlooked or belittled in a search for an alternative displaced source.

9.32 WALKOUTS

Some of the toughest situations in collaborative law meetings are walkouts.

Walkouts should not be seen as failures of the process or the collaborative team to make things work but rather as natural responses to the pressures presented by this process.

How people respond to walkouts will vary, and may well be an example of different personality types and the dynamics they create.

It is easy to blame one another, or even ourselves, when someone chooses to walk out. Very often the person walking out will make their own attribution in this sense perfectly clear. Such recriminations should be avoided.

The lawyer representing the partner who has left should go and speak to their client.

Make it clear that it is perfectly acceptable to want to leave the room. Ascertain whether they are able to continue after a short break or if they need to bring the current collaborative four-way meeting to an end.

Even if they want to end the current session it can be helpful, if it is appropriate and safe to do so, to gather everybody together in the same room if only to agree on tasks, recap on what has been agreed and set the next date. This helps to re-establish the notion of all working together and brings the current meeting to a collaborative close.

Failure to do so will leave the couple with the walkout and a broken collaborative team representing the current status of the process between the end of this meeting and the start of the next.

Very often, these steps will not be necessary and the partner who has left can rejoin the group after a short break. Even in these situations the meeting should be brought to a close sooner rather than later.

The walkout should be discussed thoroughly when the lawyer and client debrief after the meeting. Find out what happened to trigger the walkout. Ask how things could have been handled differently. Also try to agree that if pressure is mounting that there is a form of early warning sign. The client might agree, for example, simply to ask to take a 10-minute refresher, or commit to saying out loud when they are feeling uncomfortable. The collaborative effort should enable these issues to be spoken out loud and respected.

The partner who has not walked out can find it very difficult to manage with such incidents, particularly if they keep happening.

Assistance should be sought if walkouts are repeated to understand what is going on. Again consider therapeutic or counselling assistance. Would it help for that partner to have their counsellor or therapist sit in on the meetings to manage those pressures?

9.33 MANAGING CONFLICT IN MEETINGS

Conflict is inevitable within the collaborative process. As we have repeatedly said, collaborative law is not a conflict-free way of resolving separation issues, but rather a process that enables those conflicts to be acknowledged and addressed in a safe and effective way.

At the outset, conflict should be explicitly permitted within the collaborative process. Make it clear that if there are issues they should be spoken about as early as possible.

A big problem with conflict is found in conflict avoidance. It is a natural tendency for many people to avoid the difficult conversations. We know when we are feeling anxious about something often before we know what it is. We might experience a surge of adrenalin or a racing pulse. We can also watch out for clenched fists or jaws as well as fairly obvious facial expressions and agitated body language.

It can be helpful to acknowledge these symptoms in ourselves, or when we see them in one another. The earlier that a conflict point can be expressed and explored the easier it is to resolve it.

When we avoid conflict, we might find that a considerable amount of time and effort has been invested in pursuing a certain line of inquiry by the time the conflict is finally raised. Speaking out at a late stage about something that we disagree with can invalidate what has taken place already. That in turn will lead to frustration and foreseeable responses to the conflict now being raised at all.

Conflict issues are more fluid as they start to develop. As a result they are easier to deal with and control. Positions become harder as discussions move forward. Both partners, for example, might be looking ahead to resolution on certain terms on the basis of the collaborative discussion. Raising a contrary view at a late stage, as a result of having avoided the conflict discussion, can then make it very difficult to work flexibly, or fluidly, with the well-progressed plans and dialogues that have already been established.

9.34 MODERATION

When we deal with conflict issues we need to be moderate in our conduct and speech.

- Keep calm.
- Speak slowly and quietly. Watch out for the tone of your voice and try hard to keep the pitch and volume down.
- Model calm behaviour and others in the team are more likely to follow your lead.

William Ury, a leading conflict and negotiation expert and author, had a technique whereby he would pinch the palm of his hand. This physical act would cut through his rising feelings of anger or frustration and remind him of the need to be calm and act calm.

Moderate others' actions and words as well. As we have said above, call out when you see conflict indicators in one another. 'I noticed you clenched your jaw when I mentioned that. What were you thinking?'

9.35 REFRAMING

We can moderate one another's comments as well. A good way to do this is by reframing.

When conflict is expressed in an unhelpful way, possibly in a way that attacks or blames the other partner, then acknowledge the anger or frustration but quickly go on to restate what has been said.

'You never cared about my work anyway. You never wanted me to have my own life!'

'I can hear how frustrated you are about this Jane. Will it be helpful to talk about how Brad can support you in your own work?'

The reframing has honoured the point that Jane wants to explore but turns it into a progressive, forward-looking conversation. This will be a better basis for the collaborative team to pursue.

9.36 CHALLENGE THE PROBLEM, NOT THE PERSON

It is very easy for people in conflict to blame the other person for problems. Problems are said to be characteristics of the other person. So a couple that has long struggled with money may well complain about one partner or the other 'Always spending' or 'Never making enough'. These tendencies to spend or the stated inability to earn enough are features of the other person.

If we were to reframe these problems then we might name the problem 'The money problem', or 'The budgeting problem' or 'The income issue'.

The problem is assigned its own label rather than being cast upon the other partner as an inherent characteristic of theirs.

Signs that this is happening are the words 'Always' and 'Never'. These words should be avoided as they indicate that it is impossible for the other partner to change their actions and that they are incapable of determining their own actions.

9.37 MIND YOUR LANGUAGE

My first book, *Conversational Riffs; Creating Meaning Out of Conflict* (Sunmaker, 2009), explored the pattern of attack–defence–counter-attack within conventional arguments. We will all be familiar with those arguments that simply go to and fro and get nowhere. Eventually they might settle on an uneasy compromise, or the conflict just gets shelved while everybody licks their wounds and reassembles more resources, namely time, money and energy, to renew the effort.

This is an entirely destructive pattern and does nothing to help families post separation. *Conversational Riffs* suggested that we should avoid attack–defence–counter-attack and suggested alternatives to enable us to do so.

9.38 Using invitations instead of attacks

Instead of the attack, consider an invitation. Invite the other person to discuss an issue. Consider explicitly building in time so that an immediate response is not required.

> 'Peter, at the next meeting, will you tell us about the business? Will you also let us have the accounts a week before? Thank you.'

9.39 Encourage more information instead of being defensive

Even if we feel that we are coming under attack, and we will from time to time, do not rush to defend yourself. Instead encourage more information.

> 'Go on, tell me more about that.'

> 'What was it that I said that led you to …?'

> 'Help me to understand that point.'

Bear in mind that we cannot possibly start addressing what has been said unless we fully understand it. Also, the act of being able to speak and be fully understood can be enough to resolve a problem.

9.40 Acknowledgement

Acknowledge the other person's grievance. Let them know that it is reasonable for them to hold that point of view even though we might not agree with it. So often in conflict we try to destabilise the other person's position and simply get embroiled in exchanging insults, attacking one another's premise or credibility. That is unhelpful. If we can acknowledge that the other person holds their point of view reasonably then that frees us up to explore the substance of what is being said.

9.41 Say yes. Often

Agree with the other people in the collaborative team as often as possible. It will be very easy and tempting to create a library of 'No's. If we explicitly agree then we will safeguard against the contagion of the contentious points. We make a habit of reaching consensus which confounds the expectation of disagreement and debate.

This habit of yes makes collaboration so much easier.

Yes is, in fact, the first rule of collaboration.

Imagine a comedy improvisation group. The audience throws up suggestions for an improvised sketch and the actors are expected to collaborate with one another immediately.

The rule that they will play by is simple. Whatever the previous actor has said or done the response from the next actor will be 'Yes, and …'.

The yes approves what has already been said as having value and relevance. It can then be built upon. Even if the second actor wants to take the improvisation in a different direction his or her answer is still 'Yes, and …' and they will then go on to add to what has gone before.

Notice how 'Yes, but …' denies what has happened previously. Avoid it.

Collaborative law can be very much like this.

One client might say something that we do not agree with. We will feel the temptation to defend, or to say 'No'. We might try, even, a 'Yes, but …'.

When we do so, however, we lurch into a pit of conventional conflict in which we can easily become trapped. We argue about who did what, when and why instead of building on what has already gone before, keeping our eyes ahead of us on what we can create, collaboratively, together.

9.42 ATTRIBUTION THEORY

There is a fascinating but unhelpful conflict dynamic called Attribution Theory.

This sees people in conflict attributing negative motives and characteristics against the people they are in conflict with. Their own actions, however, are endowed with big noble motives and positive characteristics.

A conflicted couple might speak about each other and themselves as follows.

> 'Oh, that's him! He is always working on some big project or other. He is never at home for me or the children and would rather be out with the boys on the few nights that he does have spare. He doesn't care about us at all, only himself.'
>
> 'Me? Well, I'm their mother, aren't I? Everything I do is for them. It is like I have to make up for his being away all the time. They are the most important thing in the world to me, and so I always think about them first. I would do anything for them. I want them to have the very best chances in life.'
>
> 'Oh, well that's her. When I met her she was the woman of my dreams. She was fun and we could just laugh and laugh. And then she started to turn

> away from me. Since the children came it is like she has been punishing me for something. She never speaks to me, never wants to come along to a work's function. She just talks about the children all the time and blocks me out.'
>
> 'Me? Well I do my best. I think I am a good provider for the family. I want them all to have the best things in life, you know, and so I work hard to give them that, not that she ever shows any gratitude. Sometimes I wonder why I bother working all those hours and networking in the evenings, but then I remember that I love her, and, of course the children.'

Consider sharing the attribution model with the whole team. You might ask for permission to point out when people attribute or use always/never statements.

The attribution theory can also lead to a dynamic known as the drama triangle.

9.43 THE DRAMA TRIANGLE

The drama triangle suggests that we experience conflict such as separation or smaller conflicts within the collaborative process as stories, not dissimilar to the notion seen in narrative theory we looked at in Chapter 3.

Within this story we, and anyone else involved, can be positioned as the victim, the villain or the rescuer. Everyone is positioned by the way that we perceive and describe them and ourselves.

The person who is telling the story will tend to see themselves as the victim of the piece. Assuming victimhood in this sense is seen as being blameless. It is a passive position where, rather than having done anything wrong, wrong things are done to us.

This role of victimhood can be very attractive.

As we grow up we might have got used to people helping us out or rescuing us. It may well be seen in the dominant narratives that we explored earlier and the idea of the lawyer as being responsible and capable of fixing the situation. It can lead, however, to a person becoming increasingly passive. They can become inactive within the collaborative process and seek to defer to their lawyer.

The lawyer should resist the temptation to play the rescuer role, not least because victims can quickly turn on rescuers. Rescuers can also transgress into villainy very easily.

Villains tend to be presented as the party who puts pressure onto people or coerces them into doing something.

In reality, all of the characters within the drama triangle model are manipulative.

The victim manipulates the rescuer into acting. The rescuer will assert, especially if they come under attack, that 'I was only trying to help' and the villain tries to manipulate the whole model by trying to reposition him- or herself as the victim of the piece, who is now under attack from both the original stated victim and the rescuer.

The roles within the drama triangle are not fixed. Instead the various characters whirl around the points of the triangle as if jostling for position. Resolution becomes very difficult among the chaos and fluctuations.

Thankfully, there is an answer and within the field of family and couples separation the answer is collaborative law.

As this book has shown, collaborative law is very simple, although not easy.

By rejecting the privileged conventional positions of lawyer as hero, defender and saviour we are already resisting the call to play rescuer.

Instead, once we step outside of those conventional roles and the relative power hierarchy found within conventional lawyer-client relationships, then we hand back active participation to our clients. As a result they are able to step out of the victim role. The role they step into is not the rescuer role of the drama triangle but something powerfully different.

It is the role of collaborative participant, committing to work together with their partner in the best interests of their family. That role strips away the villain label and enables the couple to form a new kind of relationship that may well serve them much better, post separation, than the conventional lawyer-led processes would.

9.44 REACHING SETTLEMENT AND ENDING THE COLLABORATIVE CASE

As the cycle of collaborative meetings progresses we will soon reach an agreement.

Be sure to test it.

How does it measure against the couple's anchor statements?

Consider using the Wheel model to measure it against the seven aspects of separation.

Future-test the agreement. Ask those difficult questions about what will happen when new partners or new children arrive? How will future disputes or queries be managed?

The lawyers will be responsible for drafting the document that will record that agreement. Depending on the couple's circumstances this might be a court order or a deed of settlement.

Consider dealing with the drafting within the collaborative room itself. At the very least, have a draft prepared and sent out in advance of the meeting. Have a laptop available so that the very wording can be fine-tuned to reflect the couple's wishes. It can then be printed off, signed and filed as appropriate.

Some collaborative guides encourage having a closing meeting. This might be the meeting where the agreement is signed. Wine and lunch is suggested in some places, although in England and Wales this is not common practice. The risk of mistaking what will still be a very sad occasion with a celebration is felt to be too great.

It is important, though, that the final session is not an anticlimax. Relationships will have developed between all of the collaborative team and they should be recognised and respected sensitively.

Other guides encourage consideration of a closing ceremony or gift. One practitioner, for example, offers each of the couple a plant cutting in a pot. Each plant is actually one half of a whole plant which has been split into two. The imagery is that the two, now separate, came from the same source. The couple, as they take their leave, will go out into their new worlds, post separation, having come from the same source.

These are highly personal issues and practitioners will make their own decisions on whether they are applicable or not.

9.45 CONCLUSION

This book has set out to demonstrate that collaborative law is much more than a mere process of formalised round table meetings. There is a culture within it of curiosity and potent questions.

In the answers to those questions lie the futures of the families that we have the great honour to be invited into for brief periods.

For us, as lawyers or as participants, to be effective within that process, we need to be responsible for ourselves. We need to take responsibility for what we say and do. We need to take responsibility to enable one another to be as effective as they possibly can be so that they too can play their role in the resolution of matters.

We need to take responsibility to speak out, including at those times when it feels difficult to do so, to ensure that our feelings, concerns and even disagreements are known and understood.

For many this will be a new way of communicating and working together.

In the years to come we will see collaborative approaches breaking out from the narrow restraints of family and separation work. In the United States and Canada it is already starting to enter the civil litigation spheres, and so it will here in England and Wales.

We will see leaders in the collaborative spectrum starting to take the collaborative culture of curiosity, respect and questioning into other areas of life and work.

For now though, for ourselves, our clients and their families, we encourage you to only remain curious.

Keep on stretching yourself and your practice. Remember the cycle of experiential development that we explored in Chapter 4. Analyse yourself after each collaborative law meeting. Listen to the feedback. Encourage more challenging feedback.

Ask yourself, or your client or colleague: 'How might we have handled that better?'

Build up your own repertoire of powerful questions to ask the individuals we work with, answers that will transform the collaborative meeting, the client who gave the answer and the future for their family.

Read around communication, negotiation and conflict issues. Invest in yourself and your career through books, training and your own coaching.

Challenge yourself.

How do you really want to work?

Who do you want to be?

How are you going to enable the radical shift in clients to enable them to step out of that victim-villain dynamic and society's other dominant narratives?

How can you truly serve your clients and their families even though they are separating?

I hope this book has served as an inspiration and that it has equipped you with practical skills and a deeper understanding of what happens within the collaborative process.

I hope that it will be a solid rock upon which you can build a truly remarkable practice that will serve not only your clients and their families but also you and your own.

Remember. Hope is not a dirty word.

APPENDIX 1

SAMPLE PARTICIPATION AGREEMENT[1]

The parties to this agreement

Jane Smith

of 45 Compton Way, Villesville

Rachel Lintern

of Mantleys Solicitors, 2 High Street, Villesville

Gary Smith

of 45 Compton Way, Villesville

Roger Singh

of Lewis and Flinksham Solicitors, 13 Queen Square, Villesville

The children

John Gary Smith dob 11/3/05

and Siobhan Jane Smith dob 12/2/03

Our intentions

1 Roger and Rachel are Resolution-trained collaborative lawyers and act in accordance with its principles.

2 Rachel and Roger are independent of each other and each will represent and advise only their own client.

3 We will not go to court to resolve Gary and Jane's differences.

4 We will try to think creatively and constructively to find a fair solution to all issues without court intervention (although we will ask the court to confirm agreed arrangements if this is appropriate).

[1] Courtesy of Resolution, www.resolution.org.uk.

5 Our priority is John and Siobhan's well-being, and, after that, the well-being of Jane and Gary.

Negotiation in good faith

6 We will:

A) make every effort to find amicable solutions that:

a) promote John and Siobhan's best interests;
b) affirm Gary's and Jane's roles as parents; and
c) as far as possible, satisfy Jane and Gary as well as John and Siobhan;

B) be courteous and co-operative;
C) be truthful, open and honest, giving full and frank disclosure and all information that may be relevant to the discussions;
D) jointly instruct all other professionals retained under this agreement, and include an instruction to work in a co-operative way to help resolve problems (this paragraph will apply, for example, to family consultants, mediators, financial advisors, accountants and valuers);
E) use where appropriate Family Consultants, Mediators and such other persons as may help to promote the well-being of Gary and Jane and John and Siobhan.

7 Jane and Gary will:

A) try not to discuss past events;
B) avoid threats, inflammatory language and accusations that do not help us solve the problems;
C) not denigrate or criticise the other parent in front of John and Siobhan;
D) not involve John and Siobhan in the disputes;
E) promote and support a caring, loving and involved relationship between John and Siobhan and the other parent;
F) supply promptly (or within agreed times) all relevant information and documents;
G) disclose without prompting any relevant new information and documents that they receive until the agreement is approved by the court (in accordance with their duty to the court).

8 Roger and Rachel will help their respective clients to identify the information and documents that will help resolve any issues.

Confidentiality

9 The negotiations are confidential and legally privileged. No information about them or learnt from them will be given to the court, friends, colleagues, family or anyone else, unless permitted by paragraphs 10, 11, 12 or 18.

10 If the Legal Services Commission funds Jane or Gary it is entitled to see their file.

11 One of us may have a legal duty to report certain matters to the authorities, and may not be allowed to warn the others. This might arise, for example, if they have reason to believe that a child is suffering (or likely to suffer) significant harm.

12 Gary and Jane agree that details of the case may be used for research, education or training (or any combination of these), but only if information that might identify the family has been removed.

Discussing the financial settlement

13 Whenever Jane and Gary discuss the issues involved in these negotiations outside the meetings with the lawyers, they will inform Rachel and Roger of the discussions. Gary and Jane confirm their intention to limit those discussions pending further minuted agreement to relax this principle.

Legal and other costs

14 Jane and Gary will each pay:

A) their own representative's costs in accordance with the terms of business they have received;
B) half the costs of other professionals instructed under paragraph 6(E), unless otherwise agreed.

15 But:

A) if either Gary or Jane does not have the necessary money to pay these costs, the other will if reasonably possible provide the money, subject to repayment later (unless the repayment is waived under paragraph B) following;
B) it may be agreed during the negotiations that Jane or Gary will pay some or all of the other's costs.

Withdrawal from the process

16 Either Gary or Jane can withdraw from the negotiations under this agreement at any time by telling Roger or Rachel.

17 In order to allow a cooling-off period during which Rachel and Roger will try to rescue the negotiations under this agreement, neither Jane nor Gary may start court proceedings within 21 days of that withdrawal. But this will not apply in an emergency.

18 If a court is involved in the family break-up otherwise than to process the divorce or to make an agreed order:

A) Gary's and Jane's instructions to Roger and Rachel respectively will immediately lapse;
B) Jane and Gary will promptly place themselves on the court record, unless they instruct new lawyers to do so;
C) Rachel and Roger will on request:

a) give their respective former clients the completed document needed to put that client on the court record which that client will file at court immediately;
b) help their respective former clients and any new advisers that they appoint (whether or not they are instructed to act under a collaborative agreement) by providing 1) copies of the papers that represent the disclosure in the case; and 2) a bundle of privileged documents showing the stage reached in the negotiations/discussions;
c) for the avoidance of doubt, any expert's report will form part of the disclosure unless there is agreement to the contrary reached at the time or subsequently to the end of the collaborative process, between the parties or their representative;

and Rachel and Roger will be entitled to charge for so doing; but

D) neither Roger nor Rachel, nor other members of their firms, may be involved (as lawyers or witnesses) in any court proceedings between Gary and Jane;
E) neither Jane nor Gary nor anyone on their behalf will:

a) use in evidence Rachel or Roger's notes except to the extent that they record relevant financial and factual disclosure;
b) call Roger or Rachel to give evidence in any court;
c) refer a judge to the negotiations conducted under this agreement if that judge may impose a final order on any matter that has been the subject of the negotiations; but

F) any financial and/or factual disclosure made within the process may be used in evidence.

19 Rachel and/or Roger will withdraw from this agreement if:

A) their rules of professional conduct require it; or
B) they learn that their client has acted (or is about to act) in breach of this agreement.

But the representative may agree to continue if their client puts the matter right or refrains from the act.

Agreements

20 The agreements reached will only become binding when Gary and Jane have both signed a document recording the terms.

Applicable law

21 The laws of England and Wales govern this agreement.

Gary Smith ....

Jane Smith ....

Roger Singh ....

Rachel Lintern ....

Date

APPENDIX 2

SAMPLE QUESTIONS FOR LAWYERS AND CLIENTS

Over the following pages we suggest a wide range of questions that lawyers might use, and clients might expect to be asked, at various stages of the collaborative law process.

- Screening – is collaborative law appropriate for this client?
  - Emotional readiness
  - TAP issues[1]
  - Approach to dispute resolution
- Social issues relating to separation
- Narrative approaches to separation
- Managing conflict
- Asking for feedback

There is also a collection of questions that clients might ask lawyers that they are thinking of asking to work with them.

These questions are not obligatory. They are merely provided as a stimulus to help us develop ever more curious stances in our collaborative work.

Some of the questions sound strange and contrived when written down on paper. Try reading them out loud, not mumbled or whispered but in your normal speaking voice, or louder. Know what the questions sound like.

Change the words so that the questions become your own instead of lines you have learnt from this book. Remember though to try to avoid 'Why' and closed questions.

---

[1] Trust, Anger or Positional issues; see Chapter 8 at **8.4** et seq.

As you run more collaborative meetings, you can introduce and trial more questions. Make a note of which ones work and which ones do not. If you make up a question that proves to be useful then make a note of that as well.

SCREENING

Emotional readiness

When did all of this happen?

How are you coping?

How did that make you feel?

How do you feel about that now?

What are the prospects of getting back together?

So what does the future hold in store, do you think?

What might it hold it store?

Who else have you been able to speak to about this?

Who else do you think it might be helpful to speak to about this?

Would you like details of a therapist, counsellor or coach to talk to?

TAP issues

Despite everything, how well are you able to trust him or her?

How well are they able to trust you?

What do you think I need to know about him or her?

How well do you think we will be able to work together with him or her?

Where might we expect to have most problems in getting information, if at all?

What issues will be hardest for him or her to talk about?

How do you feel about sitting together with him or her to discuss these things?

Are there any areas he or she will not want to talk about?

What tends to happen when you have disagreements, or used to have disagreements, at home?

How are you able to resolve disagreements?

What happens if either of you doesn't get your way?

Are there any incidents of violence or aggression that you can think of?

What concerns, if any, might he or she have about sitting in a room with us?

Who gets their way more often, usually? How do you feel about that?

How easy will it be for each of you to speak openly when we meet?

How do you think you will feel when disagreements arise within the collaborative process?

* * *

What is the most important thing to you in getting all of this sorted?

Bearing in mind what has happened, how will we be able to ensure that we keep moving forward?

How do you think anger, hurt or blame might get in the way?

How might we be able to manage that?

What things might you get angry or particularly upset about?

How would you like me to respond or what would you like me to do if that happens?

And your partner? What is he or she likely to be most angry or upset about?

How will I know when they are feeling that way? What signs could I look for?

* * *

What is it that made you choose collaborative law? What are the top three reasons?

And your partner?

* * *

How much do you think it might cost to resolve this?

How will you and your partner meet the costs of this?

Approach to dispute resolution

What thoughts have you got about how we might talk these things through?

Have you got an idea already about what it will take for you to agree these matters?

How fixed is that idea? Is that an absolute requirement?

How flexible can you be? And your partner?

* * *

What would a successful outcome look like?

How would you like to be able to relate to your partner once all of this is done?

How would you mark the following issues for importance to you, out of ten? [*Consider using the Wheel model.*[2]]

- Sharing the money
- Independence
- A civil relationship with your partner after the separation
- Your relationship with your children
- Your children's relationship with your partner
- Being able to work together as parents after separation
- Separating with dignity/preserving the good memories you have
- Security

2 See Chapter 4 at **4.12** et seq.

[*Having asked these questions, go deeper. Ask what is important about them, what each entry means to them and the like. The purpose is to move away from fixed positions.*]

* * *

What happens normally when you argue?

How do you normally end arguments?

Are those arguments helpful? Why not?

Would it be more helpful if we could discuss difficult issues in more constructive ways?

What might happen if we could discuss matters without falling into those old arguments?

SOCIAL ISSUES RELATING TO SEPARATION

Who do you speak to about all of this?

What do they think?

What do they say?

What do they think you should do?

What does your own family feel about all of this?

Who else might be involved?

Who will your partner be speaking to?

What might their family be saying?

Have you got friends who have gone through separations?

What were their experiences?

What will they think about your choice to use collaborative law?

Who do you think will be supportive of you in going down this route?

Who might try to talk you out of it? Why? How will you deal with that?

How are your mutual friends dealing with all of this?

Imagine that you reach an agreement. Who else will you need to tell? How will you do that and how might they react?

NARRATIVE APPROACHES TO SEPARATION

How is all of this going to work out, do you think?

How would you like all of this to work out?

What would be the worst ending this whole story could have?

What might we call that bad outcome?

Is there a film, a TV programme or perhaps a friend's own separation that it reminds you of?

How might we avoid that?

What might we call the better outcome?

How do you want your children to perceive your separation?

And in 10 years' time, how would you want your children to look back on this period?

How might we call the kind of relationship that you and your partner could have after this is all over?

* * *

How do you see your partner at the best and the worst of times?

When you are seeing them in that way, how do you feel about them at that time?

Have you got names or adjectives you use to describe them?

Do those names change depending on who you are talking to?

How do they see you and describe you to others, do you think?

When you are seeing them as the worst they could be, how does that make you act?

How do you act differently when you are able to see them in a better light?

Which is more helpful?

How can we spend more time thinking of them in this better light?

Tell me about a time when you felt this way about them? What had happened?

* * *

Why do you think he or she does those things that annoy you?

What else might have been happening?

How would he or she explain what they did?

If we can get to the better outcome, how important is it that we prove what went wrong or who was to blame?

* * *

Between here and getting everything sorted out, what problems might we face?

Once we have got an agreement, what obstacles will you still have to overcome?

What do you think you will do when you meet those obstacles?

How might you best deal with problems in the future?

MANAGING CONFLICT

I can probably guess, but tell me in your own words how you are feeling about that?

What is making you feel that way? Which is the most important thing about it?

What else might be going on for you? And for them?

How would you like to deal with this?

Shall we take a break for a while?

Are you okay?

Tell me again what has happened.

[*And interrupt, asking for clarification. What happened next? This was the same day as the dog got out was it? Was that the blue car? Have I understood that right? And who else was there? Sorry, who is Michelle again? And so on. Asking these details slows down the retelling of the event and creates space to calm down and check understanding.*]

How do we know that he or she said/did/intended that?

Can you think of a time when you have had to deal with this before?

What might we call this problem? [*And make it a specific problem, not a characteristic of the other person.*]

How does this problem lead you to act?

How else might you react to it?

Which would be more helpful?

What was the outcome?

What would he or she say their intentions were in all of this?

What happens if we cannot get past this?

What would help you both to deal with this?

Is this what happens when you try to talk about things just between the two of you?

Can you take me through exactly what is going on, what is happening for each of you?

And what do you think is going on for your partner, right now?

How long have you had these arguments?

Have you ever been able to resolve this?

Who might it be helpful to speak to?

Before we go any further, can you tell me what you have heard your partner just say?

[*To partner*] Is that right? Was there anything else or anything different?

Hypothetically, if your children were here with us now, how would they want to see you working this out?

ASKING FOR FEEDBACK

Has this been helpful?

Has this gone how you had expected?

Has this gone as you had hoped?

What worked well?

What worked best?

What did not work so well?

Was there anything I or my collaborative colleague did that you would rather we did not do again?

What would you like to change for next time we meet?

QUESTIONS CLIENTS MIGHT CONSIDER ASKING POTENTIAL LAWYERS

How much collaborative work have you done?

How many cases have you successfully finished?

Which other lawyers do you work best with? Why?

Do you often work with other professionals within the collaborative process, such as independent financial advisers, therapists, coaches, counsellors, child and parenting experts?

Have you got any articles I can read?

Do you attend local collaborative law meetings with other local professionals?

How frequently?

What will be my role in the collaborative process?

How can you help me in the collaborative process?

Have you got any books and other material you can lend me or recommend?

What additional training have you had other than the compulsory minimum training?

Why do you do collaborative law?

How much of your work is collaborative?

Do you do mediation?

How many of your cases go to court?

Have you worked with my partner's lawyer before? How was it?

APPENDIX 3

THE SEVEN-STEP GUIDE TO PREPARING YOUR ANCHOR STATEMENT

The anchor statement is a short statement that each partner shares with the rest of the collaborative term at an early stage in the first four-way meeting.

One purpose of the anchor statement is to explain why each client has chosen the collaborative process and to share as much or as little as the client wants to about various things that might be important to them.

When anchor statements are used properly they take on several other important roles.

They can set out the hopes and disappointments that each partner holds.

They act as powerful statements about future intentions and dreams.

They tend to be optimistic, albeit tinged with regret.

They set out a vision of the best that could be achieved.

It is worth taking the time to prepare your anchor statement fully. This exercise provides a process to guide you through that. You do not need to follow it to the letter, but hopefully the framework will give you an idea of what you might include.

(1) SET ASIDE TIME TO WORK ON THIS STATEMENT FULLY

The anchor statement's role is to hold you and your partner steady as you work through the collaborative process. For that reason you need to have confidence in it. The preparation work is more than just preparing a few paragraphs of text.

This exercise is designed to help you determine what you and your family want for the future and what you need to do now to achieve that.

Set aside some specified time – try a couple of hours – to consider and reflect upon these questions. Go somewhere where you will not be interrupted and where you feel comfortable. You want to be able to focus on this without being distracted by phone calls, noises, TV, the internet or discomfort.

You might want to prepare the space you are going to be working in as well. Consider having some unobtrusive music, flowers, soft drinks, water, snacks or other treats.

You will need paper and a pen or a device that you can write freely with.

This exercise should not be an ordeal. Do what you can to make it special.

(2) PREPARE YOURSELF

In many ways the anchor statement is a question of imagining the very best outcomes for you, your post-separation relationship and your family and friends. You will need to be in a calm state of mind where you can envision how that might look or feel before you start to commit to writing it down.

Relax.

However you do this is up to you.

You might diarise some exercise, a pampering session, meditation, prayer or a light lunch at a favourite cafe immediately before you settle down for your two hour session.

The point is that you need to be feeling calm and focussed. If you are not relaxed then other pressures will start crowding in on your thoughts and your responses to the questions below may well be influenced by those pressures.

(3) ANSWER THE QUESTIONS SET OUT BELOW

This table sets out several questions designed to stimulate your own thoughts and delve into your very best wishes for the future. It invites you to consider these questions on three levels, namely how you respond with regard to yourself, with regard to the impact on your partner and any children and with regard to the broader family and other social connections, such as friends, colleagues and the like.

Do not try to squeeze your answers into the boxes provided. The table format is used only so you can ensure that you have considered each question against each column.

Question	Regarding yourself	Regarding your partner and children of the family	Regarding the broader family and friends
What are your greatest concerns right now?			
What do you want the outcome to be at the end of this process?			
Where do you want to be in 5, 10 and 20 years' time?			
How do you want to feel about what has happened at the end of this process?			
And again, in 5, 10 and 20 years' time?			
What is, or what are, the most important things to you?			
If you were at a social event in 10 years' time and you overheard people talking about your relationship and separation, what would you like to be hearing them say?			

(4) WHAT NEEDS TO HAPPEN TO ENABLE YOU TO MEET YOUR GOALS FOR YOURSELF, WITH REGARDS TO YOUR POST-SEPARATION RELATIONSHIP, AND YOUR BROADER FAMILY AND FRIENDS?

Here you will reflect upon your answers within the table.

What do you think you need to do, and what do you need to see happen, within the collaborative law process, in order to achieve those outcomes and wishes that you have identified in the table?

(5) DRAFT YOUR INITIAL ANCHOR STATEMENT

There is no fixed format for an anchor statement, but if it helps then consider the following four paragraph structures as a starting point.

Paragraph One (The vision, or the top line)

My dream/hope/vision is that at the end of all of this our family/I/we will be able to …

Paragraph Two (The why, or the benefits)

If we achieve that, then our family/I/we will benefit because …

Paragraph Three (The how, or the commitment)

To achieve that I/we need to be able to [*insert what steps you think you need to take, what kinds of behaviour and approaches to dispute resolution you think you need to adopt*].

Paragraph Four (Why collaborative law?)

I think that the collaborative process will enable us to [*insert the steps and behaviours from paragraph three]* … so that [*insert the vision of top line form paragraph one*] so that … [*insert benefits from paragraph two*].

(6) LEAVE IT FOR A DAY OR TWO AND THEN REVISIT AND POLISH

There is nothing wrong in redrafting this statement. Put it aside for a day or two and then come back to it.

Are you still comfortable with it?

Have you missed anything?

If you have used the four-paragraph framework above then ask yourself if it feels too formulaic. Remember, that framework is designed as a starting point. It will be good enough as a final anchor statement, but it might not feel quite right, almost as if it is too designed. Feel free to rewrite it in your own style.

(7) PRACTICE YOUR ANCHOR STATEMENT

Spend some time reading your anchor statement out loud.

The goal here is not to learn it off by heart, but to be really comfortable with it.

Spending time reading your statement out loud will enable you to hear what it sounds like, in readiness for the first four-way meeting.

Also, as you read and re-read this statement then it will plant itself more firmly into your own consciousness. It will already be beginning to anchor your expectations and hopes before the collaborative meetings have even begun.

APPENDIX 4

RECOMMENDED FURTHER READING

Ahrons, Constance, PhD *We're Still Family* (Harper Collins, 2004)
— *The Good Divorce* (Harper Collins, 1994)
Arbinger Institute, The *The Anatomy of Peace* (Penguin, 2010)
Bishop, Gillian *A Client's Guide to Collaborative Divorce* (Family Law Partnership, 2010)
Coltri, Laurie S *Conflict Diagnosis and Alternative Dispute Resolution* (Pearson Prentice Hall, 2004)
De Bono, Edward *Six Thinking Hats* (Penguin, 2000)
Denny, Neil *Conversational Riffs; Creating Meaning Out of Conflict* (Sunmakers, 2009)
Fisher, Roger and Shapiro, Daniel *Beyond Reason* (Random House, 2005)
Fisher, Roger and Ury, William *Getting to Yes* (Random House, 1999)
Furlong, Gary *The Conflict Resolution Toolbox* (Wiley, 2005)
Goleman, Daniel *Emotional Intelligence* (Bloomsbury, 1995)
Gray, John *Men Are From Mars, Women Are From Venus* (Thorsons, 1993)
Harper, Gary *The Joy of Conflict Resolution* (New Society Publishers, 2004)
Isenhart, Mary Warren and Spangle, Michael *Collaborative Approaches to Resolving Conflict* (Sage Publications, Inc, 2000)
Mosten, Forrest *The Collaborative Law Handbook* (Jossey-Bass, 2009)
Patterson, Kerry, Grenny, Joseph, McMillan, Ron and Switzer, Al *Crucial Conversations* (McGraw Hill, 2002)
Roberts, Geo *Impact! How to Make A Positive Difference* (Xulon Press, 2010)
Ury, William *Getting Past No* (Random House, 1991)
— *The Power of a Positive No* (Hodder and Stoughton, 2007)
Webb, Stuart G and Ousky, Ronald D *The Collaborative Way to Divorce* (Hudson Street Press, 2006)
Whitmore, John *Coaching for Performance* (Nicholas Brealey Publishing, 4th edn, 2009)
Whitworth, Laura, Kimsey-House, Karen, Kimsey-House, Henry and Sandahl, Phillip *Co-Active Coaching* (Davies Black Publishing, 2nd edn, 2007)
Winslade, John and Monk, Gerald *Narrative Mediation* (Jossey-Bass, 2001)
— *Practicing Narrative Mediation* (Jossey-Bass, 2008)

APPENDIX 5

LIST OF TYPICAL MONTHLY OUTGOINGS

Try to fill out the following details as accurately as you can.

Remember that a calendar month is actually 4.33 weeks so make sure that weekly payments are calculated accordingly.

To get an average monthly cost, add together however many different months you can get details for and then divide the total by the number of months you have added.

COST OF HOUSING

	£
Mortgage/rent	
Endowment insurance policy supporting mortgage	
Council tax	
Electricity	
Gas	
Water rates + sewage	
Telephone	
Service charge	
Oil or other fuel cost	
Complete similar details for any other properties owned	

HOUSEHOLD RUNNING COSTS

	£
Food	
Housekeeping goods	
Home insurance – Building	
Home insurance – Contents	
TV licence	
Cleaner	
Gardener	
Any other staff	
Insert others here	

CAR

	£
Insurance	
Tax	
Car maintenance (servicing/MOT/tyres/brakes)	
Petrol	
Car loan payments	
AA or other recovery service	

CHILDREN EXPENSES

	£
Travel to school	
Meals at school	
Uniforms	

Outings and trips	
School fees £ per term	
Other school expenses (please list)	
Clothes	
Optician	
Dentist	
Childminding/nursery fees	
Books	
Activities and clubs	
Christmas and birthdays (incl friends of children)	
Subscriptions/memberships	

PERSONAL EXPENSES

	£
Clothes and shoes	
Hairdressing	
Doctor	
Dentist	
Optician	
Prescription charges	
Dry cleaning	
Entertainment	
Travel to work	
Lunches at work	
Holidays for self or together with family	

Solicitors' fees	
Contributions to pension or other financial planning commitments (please list)	
Subscriptions	

DEBT AND CREDIT CARDS

Money owed to	Reason	Balance owed	Monthly sum

APPENDIX 6

SCHEDULE OF ASSETS AND INCOME

The following table sets out a format that you can use to gather information about relevant assets and income. The table works by calculating the relevant figures and then writing those figures in the joint column for joint assets, or the column for each partner in the case of sole owned assets and income.

Try to ensure that you have supporting documents relating to the details you insert to enable you to prepare, between you, a folder of financial information.

Remember that asking an independent financial adviser to work with the collaborative team can greatly assist in completing this schedule and take a great deal of the legwork away from the lawyers and the couple themselves.

Asset	**Joint names**	**Partner #1**	**Partner #2**	**Notes**
Property				
House value				
Deduct 3% costs of sale				
Deduct mortgage				
Net proceeds of sale				
Any other house				
Deduct 3% costs of sale				
Deduct mortgage				
Net proceeds of sale				

Bank accounts				
Complete one row for each bank or building society account				
Cash				
Money owed to you				

Investments				
Shareholdings in listed companies List number, price and date taken for value of shares				
Unit trusts List number, price and date taken for value of unit trust				
PEPS, ISAs or Tessa. Complete one row for each				

Miscellaneous items				
Car				
Valuable items or collections such as art, jewellery etc				
Owned business				
Value of business				
Share of business				
Other interests				
Consider interests in family trusts or definite future acquisitions				

Debts				
Complete one row for each debt				
Total assets less debts				

Pensions				
Obtain and insert CETV (cash equivalent transfer value) for each pension held				

Income				
Salary				
Child benefit				
Tax credits				
Other benefits				
Annuities				